INSECT MOLECULAR BIOLOGY AND BIOCHEMISTRY

INSECT MOLECULAR BIOLOGY AND BIOCHEMISTRY

EDITED BY

LAWRENCE I. GILBERT

Department of Biology
University of North Carolina
Chapel Hill, NC

ELSEVIER

AMSTERDAM • BOSTON • HEIDELBERG • LONDON • NEW YORK • OXFORD
PARIS • SAN DIEGO • SAN FRANCISCO • SINGAPORE • SYDNEY • TOKYO
Academic Press is an imprint of Elsevier

Academic Press is an imprint of Elsevier
32 Jamestown Road, London NW1 7BY, UK
225 Wyman Street, Waltham, MA 02451, USA
525 B Street, Suite 1800, San Diego, CA 92101-4495, USA

First edition 2012

Notice
No responsibility is assumed by the publisher for any injury and/or damage to persons
or property as a matter of products liability, negligence or otherwise, or from any use or
operation of any methods, products, instructions or ideas contained in the material herein.
Because of rapid advances in the medical sciences, in particular, independent verification of diagnoses
and drug dosages should be made

British Library Cataloguing-in-Publication Data
A catalogue record for this book is available from the British Library

Library of Congress Cataloging-in-Publication Data
A catalog record for this book is available from the Library of Congress

ISBN: 978-0-12-384747-8

For information on all Academic Press publications
visit our website at elsevierdirect.com

Typeset by TNQ Books and Journals Pvt Ltd.
www.tnq.co.in

Printed and bound by CPI Group (UK) Ltd, Croydon, CR0 4YY

CONTENTS

PREFACE

In 2005 the seven-volume series "Comprehensive Molecular Insect Science" appeared and summarized the research in many fields of insect research, including one volume on Biochemistry and Molecular Biology. That volume covered many, but not all, fields, and the newest references were from 2004, with many chapters having 2003 references as the latest in a particular field. The series did very well and chapters were cited quite frequently, although, because of the price and the inability to purchase single volumes, the set was purchased mainly by libraries. In 2010 I was approached by Academic Press to think about bringing two major fields up to date with volumes that could be purchased singly, and would therefore be available to faculty members, scientists in industry and government, postdoctoral researchers, and interested graduate students. I chose *Insect Molecular Biology and Biochemistry* for one volume because of the remarkable advances that have been made in those fields in the past half dozen years.

With the help of outside advisors in these fields, we decided to revise 10 chapters from the series and select five more chapters to bring the volume in line with recent advances. Of these five new chapters, two, by Subba Palli and by Xavier Belles and colleagues, are concerned with techniques and very special molecular mechanisms that influence greatly the ability of the insect to control its development and homeostasis. Another chapter, by Park and Lee, summarizes in a sophisticated but very readable way the immunology of insects, a field that has exploded in the past six years and which was noticeably absent from the Comprehensive series. The other two new chapters are by Yong Zhang and Pat Emery, who deal with circadian rhythms and behavior at the molecular genetic level, and by Philip Jensen, who reviews the role of TGF-β in insect development, again mainly at the molecular genetic level. In most cases the main protagonist is *Drosophila melanogaster*, but where information is available representative insects from other orders are discussed in depth. The 10 updated chapters have been revised with care, and in several cases completely rewritten. The authors are leaders in their research fields, and have worked hard to contribute chapters that they are proud of.

I was mildly surprised that, almost without exception, authors who I invited to contribute to this volume accepted the invitation, and I am as proud of this volume as any of the other 26 volumes I have edited in the past half-century. This volume is splendid, and will be of great help to senior and beginning researchers in the fields covered.

LAWRENCE I. GILBERT
Department of Biology,
University of North Carolina,
Chapel Hill

CONTRIBUTORS

Svend O. Andersen
The Collstrop Foundation, The Royal Danish
Academy of Sciences and Letters, Copenhagen,
Denmark

Yasuyuki Arakane
Division of Plant Biotechnology,
Chonnam National University, Gwangju,
South Korea

Hua Bai
Department of Ecology and Evolutionary Biology,
Brown University, Providence, RI, USA

Xavier Belles
Instituto de Biología Evolutiva (CSIC-UPF),
Barcelona, Spain

Rollie J. Clem
Division of Biology, Kansas State University,
Manhattan, KS, USA

Alexandre S. Cristino
Queensland Brain Institute, The University of
Queensland, Brisbane St Lucia, Queensland,
Australia

Patrick Emery
University of Massachusetts Medical School,
Department of Neurobiology, Worcester, MA, USA

Susan E. Fahrbach
Department of Biology, Wake Forest University,
Winston-Salem, NC, USA

Clélia Ferreira
University of São Paulo, São Paulo, Brazil

René Feyereisen
INRA Sophia Antipolis, France

Stavros J. Hamodrakas
Department of Cell Biology and Biophysics,
Faculty of Biology, University of Athens, Athens,
Greece

Alfred M. Handler
USDA, ARS, Center for Medical, Agricultural,
and Veterinary Entomology, Gainesville,
FL, USA

Vassiliki A. Iconomidou
Department of Cell Biology and Biophysics,
Faculty of Biology, University of Athens, Athens,
Greece

Philip A. Jensen
Department of Biology, Rocky Mountain College,
Billings, MT, USA

Michael R. Kanost
Department of Biochemistry, Kansas State
University, Manhattan, KS, USA

Karl J. Kramer
Department of Biochemistry,
Kansas State University, and USDA-ARS,
Manhattan, KS, USA

Bok Luel Lee
Pusan National University, Busan, Korea

Hans Merzendorfer
University of Osnabrueck, Osnabrueck,
Germany

Subbaratnam Muthukrishnan
Department of Biochemistry,
Kansas State University, Manhattan, KS, USA

John R. Nambu
Department of Biological Sciences, Charles E.
Schmidt College of Science, Florida Atlantic
University, Boca Raton, FL, USA

David A. O'Brochta
University of Maryland, Department of
Entomology and The Institute for Bioscience and
Biotechnology Research, College Park, MD, USA

Subba R. Palli
Department of Entomology, University of
Kentucky, Lexington, KY, USA

Nikos C. Papandreou
Department of Cell Biology and Biophysics,
Faculty of Biology, University of Athens, Athens,
Greece

Ji Won Park
Pusan National University, Busan, Korea

Maria-Dolors Piulachs
Instituto de Biología Evolutiva (CSIC-UPF),
Barcelona, Spain

Mercedes Rubio
Instituto de Biología Evolutiva (CSIC-UPF),
Barcelona, Spain

Robert O. Ryan
Children's Hospital Oakland Research Institute,
Oakland, CA, USA

Lawrence M. Schwartz
Department of Biology, 221 Morrill Science Center,
University of Massachusetts, Amherst, MA, USA

Erica D. Tanaka
Instituto de Biología Evolutiva (CSIC-UPF),
Barcelona, Spain

Walter R. Terra
University of São Paulo, São Paulo, Brazil

Zhijian Tu
Department of Biochemistry, Virginia Tech,
Blacksburg, VA, USA

Dick J. Van der Horst
Utrecht University, Utrecht, The Netherlands

John Wigginton
Department of Entomology,
University of Kentucky, Lexington,
KY, USA

Judith H. Willis
Department of Cellular Biology,
University of Georgia, Athens, GA, USA

Yong Zhang
University of Massachusetts Medical School,
Department of Neurobiology, Worcester, MA, USA

1 Insect Genomics

Subba R Palli
Department of Entomology, University of Kentucky,
Lexington, KY, USA
Hua Bai
Department of Ecology and Evolutionary
Biology, Brown University, Providence, RI, USA
John Wigginton
Department of Entomology, University of Kentucky,
Lexington, KY, USA

Summary

Genomic sequencing has become a routinely used molecular biology tool in many insect science laboratories. In fact, whole-genome sequences for 22 insects have already been completed, and sequencing of genomes of many more insects is in progress. This information explosion on gene sequences has led to the development of bioinformatics and several "omics" disciplines, including proteomics, transcriptomics, metabolomics, and structural genomics. Considerable progress has already been made by utilizing these technologies to address long-standing problems in many areas of molecular entomology. Attempts at integrating these independent approaches into a comprehensive systems biology view or model are just beginning. In this chapter, we provide a brief overview of insect whole-genome sequencing as well as information on 22 insect genomes and recent developments in the fields of insect proteomics, transcriptomics, and structural genomics.

DOI:10.1016/B978-0-12-384747-8.10001-7

1.1. Introduction

Research on insects, especially in the areas of physiology, biochemistry, and molecular biology, has undergone notable transformations during the past two decades. Completion of the sequencing of the first insect genome, the fruit fly *Drosophila melanogaster*, in 2000 was followed by a flurry of activities aimed at sequencing the genomes of several additional insect species. Indeed, genome sequencing has become a routinely used method in molecular biology laboratories. Initial expectations of genome sequencing were that much could be learned by simply looking at the genetic code. In practice, insects are too complex for a complete understanding based on nucleotide sequences alone, and this has led to the realization that insect genome sequences must be complemented with information on mRNA expression as well as the proteins they encode. This has led to the development of a variety of "omics" technologies, including functional genomics, transcriptomics, proteomics, metabolomics, and others. The vast amount of data generated by these technologies has led to a sudden increase in the field of bioinformatics, a field that focuses on the interpretation of biological data. Developments in the World Wide Web have allowed the distribution of this "omics" data, along with analysis, tools to people all over the world. Integrating these data into a holistic view of all the simultaneous processes occurring within an organism allows complex hypotheses to be developed. Instead of breaking down interactions into smaller, more easily understandable units, scientists are moving towards creating models which encompass the totality of an organism's molecular, physical, and chemical phenomena. This movement, known as systems biology, focuses on the integration and analysis of all the available data about an entire biological system, and it aims to paint an authentic and comprehensive portrait of biology.

During the past two decades, research on insects has produced large volumes of information on the genome sequences of several model insects. Genome sequencing allows quantificatation of mRNAs and proteins, as well as predictions on protein structure and function. Attempts to integrate this data into systems biology models are currently just beginning. While it is difficult to cover all the developments in these disciplines, we will try to summarize the latest developments in these existing fields. In the first section of this chapter, insect genome sequencing and the lessons learned from this will be presented. In the next section, analysis of sequenced genomes using "omics" and high-throughput sequencing technologies will be summarized. In the third part of this chapter, an overview of proteomics and structural genomics will be covered. A brief overview of insect systems biology approaches will be presented at the end of this chapter.

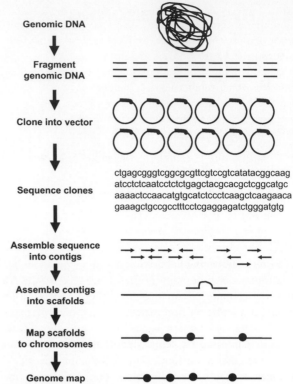

Figure 1 The whole-genome shotgun sequencing (WGS) method begins with isolation of genomic DNA from nuclei isolated from isogenic lines of insects. The DNA is then sheared and size-selected. The size-selected DNA is then ligated to restriction enzyme adaptors and cloned into plasmid vectors. The plasmid DNA is purified and sequenced. The sequences are assembled using bioinformatics tools.

1.2. Genome Sequencing

Almost all insect genomes sequenced to date employed the whole-genome shotgun sequencing (WGS) method (**Figure 1**). Shotgun genome sequencing begins with isolation of high molecular weight genomic DNA from nuclei isolated from isogenic lines of insects. The genomic DNA is then randomly sheared, end-polished with Bal31 nuclease/T4 DNA polymerase primers and, finally, the DNA is size-selected. The size-selected, sheared DNA is then ligated to restriction enzyme adaptors such as the BstX1adaptors. The genomic fragments are then inserted into restriction enzyme-linearized plasmid vectors. The plasmid DNA is purified (generally by the alkaline lysis plasmid purification method), isolated, sequenced, and assembled using bioinformatics tools. Automated Sanger sequencing technology has been the main sequencing method used during the past two decades. Most genomes sequenced to date employed this technology. Sanger sequencing must be distinguished from next generation sequencing technology, which has entered the marketplace during the past four years and is rapidly changing the approaches used to sequence genomes. Genomes sequenced by NGS technologies will be completed more quickly and at a lower price than those from the first few insect genomes.

1.2.1. Genome Assembly

Genomes and transcriptomes are assembled from shorter reads that vary in size, depending on the sequencing technology used. Contigs are created from these short reads by comparing all reads against each other. If sequence identity and overlap length pass a certain threshold value, they are lumped together into a contig by a program called an assembler. Many assembly programs are available, which differ mainly in the details of their implementation and of the algorithms employed. The most commonly used assembler programs are: The Institute for Genomic Research (TIGR) Assembler; the Phrap assembly program developed at the University of Washington; the Celera Assembler; Arachne, the Broad Institute of MIT assembler; Phusion, an assembly program developed by the Sanger Center; and Atlas, an assembly program developed at the Baylor College of Medicine.

The contigs produced by an assembly program are then ordered and oriented along a chromosome using a variety of additional information. The sizes of the fragments generated by the shotgun process are carefully controlled to establish a link between the sequence-reads generated from the ends of the same fragment. In WGS projects, multiple libraries with varying insert sizes are normally sequenced. Additional markers such as ESTs are also used during the assembly of genome sequences. The ultimate goal of any sequencing project is to determine the sequence of every chromosome in a genome at single base-pair resolution. Most often gaps occur within the genome after assembly is completed. These gaps are filled in through directed sequencing experiments using DNA from a variety of sources, including clones isolated from libraries, direct PCR amplification, and other methods.

1.2.2. Homology Detection

After assembly, sequences representing the genome or transcriptome are analyzed for functional interpretation by comparing them with known homologous sequences. Proteins typically carry out the cellular functions encoded in the genome. Protein coding sequences, in the form of open reading frames (ORFs), must first be distinguished from other sequences or those that encode other types of RNA. Transcriptome analysis is simplified by the fact that the sequenced mRNAs have already been processed for intron removal in the cell. Distinguishing the correct ORF where translation occurs, from 5′ and 3′ untranslated regions, is easily accomplished by a blast search against a protein database, or possibly by selecting the longest ORF. Finding genes in eukaryotic genomes is more complex, and presents a unique set of challenges.

1.2.2.1. Genomic ORF detection   Detection of ORFs is more complex in eukaryotes than prokaryotes due to the presence of alternate splicing, poorly understood promoter sequences, and the under-representation of protein coding segments compared to the whole genome. If transcriptome data are available, a number of programs exist to map these sequences back to an organism's genome (Langmead *et al.*, 2009; Clement *et al.*, 2010). This strategy is especially useful when analyzing non-model organisms, or those projects that lack the manpower of worldwide genome sequencing consortiums. In this manner a large number of transcripts can potentially be identified, along with their regulatory and promoter sequences, and information on gene synteny.

De novo gene prediction algorithms often use Hidden Markov Models or other statistical methods to recognize ORFs, which are significantly longer than might be expected by chance. These algorithms also search for sequences containing start and stop codons, polyA tails, promoter sequences, and other characteristics indicative of protein coding segments (Burge and Karlin, 1997). *De novo* gene discovery is partially dependent on the organism used, since compositional differences such as GC content and codon frequency introduce bias, which must be considered for each organism. Artificial intelligence algorithms can be trained to recognize these differences when a sufficient number of protein coding sequences are available. These may originate from transcriptome sequencing, or more traditional approaches such as PCR amplification and Sanger sequencing of mRNAs. Based on a small sample proportion of known genes, artificial intelligence programs can learn the codon bias and splice sites, for example, and extrapolate these findings to the rest of the genome. However, this process is often inaccurate (Korf, 2004).

Comparative genomics is the process of comparing newly sequenced genomes to more well-curated reference genomes. Two highly related species will likely have well conserved protein coding sequences with similar order along a chromosome. The contigs or scaffolds from a newly assembled genome can be mapped to the reference, or the shorter reads can be mapped and assembled in a hybrid approach. Programs that perform this task may often be used to map transcriptome data to a genome, since the two approaches are mechanistically similar.

1.2.2.2. Transcriptome gene annotation   By definition, mRNA represents protein coding sequences, and finding the correct ORF requires only a blast search. However, ribosomal RNA (rRNA) may represent more than 99% of cellular RNA content. The presence of rRNA may be detrimental to the assembly process because stretches of mRNA may overlap, and thus cause erroneously assembled RNA amalgams. Strategies to reduce the amount of sequenced rRNA include mRNA purification and rRNA removal. Oligo (dt) based strategies, such as the Promega PolyATract mRNA isolation kit, use oligo (dt) sequences which bind to the poly A tail of mRNA. The poly T tract is linked to a purification tag, such as biotin, which binds to streptavidin-coated magnetic beads. The beads can be captured, allowing the non-poly adenylated RNA to be washed away. The Invitrogen Ribominus kit uses a similar principle, except oligo sequences complementary to conserved portions of rRNA allow it to be subtracted from total RNA.

During RNA amplification, oligo (dt) primers may be used to increase the proportion of mRNA to total RNA. This process may introduce bias near the 3′ side of mRNA, and thus protocols have been developed to normalize the representation of 5′, 3′, and middle segments of mRNA (Meyer *et al.*, 2009). If the rRNA sequence has already been determined, many assembly programs can be supplied a filter file of rRNA and other detrimental contaminant sequences, such as common vectors, which will be excluded from the assembly process.

1.2.2.3. Homology detection   Annotation is the step of linking sequences with their functional relevance. Since protein homology is the best predictor of function, the NCBI blastx algorithm (Altschul *et al.*, 1990) is a good place to start in predicting homology and thus function. The blastx algorithm translates sequences in all six possible reading frames and compares them against a database of protein sequences.

For less technically inclined users, the blastx algorithm may be most easily implemented in Windows-based programs such as Blast2GO (Conesa *et al.*, 2005; Conesa and Gotz, 2008; http://www.blast2go.org/). Blast2GO offers a comprehensive suite of tools for blasting and advanced functional annotation. However, relying on the NCBI server to perform blast steps often introduces a substantial bottleneck between the server and querying computer. Local blast searches, performed by the end user's computer(s), may significantly reduce annotation time. The blast program suite and associated databases may be downloaded for local blast searches (ftp://ftp.ncbi.nlm.nih.gov/blast/executables/blast+/LATEST/). The NCBI non-redundant protein database is quite large and time consuming to search. Meyer *et al.* (2009) advocate a local approach where sequences are first queried against the smaller, better curated swiss-prot database, and then sequences with no match are blasted against the NR protein database (Meyer *et al.*, 2009). Faster algorithms such as AB-Blast (previously known as WU-Blast) may also speed up the blasting process. After a blastx search, sequences may be compared to other nucleotide sequences (blastn), or translated and compared to a translated sequence to help identify unigenes, or unique sequences. However, blastx is the first choice, since the amino acid sequence is more conserved than the nucleotide sequence. This step will also yield the correct open reading frame of a sequence. In some cases, homologous relationships may be discovered using blastn and tblastn where blastx

did not. The statistically significant expectation value, or the probability that two sequences are related by chance (also called an e value) is an important consideration in blasting, because setting an e value too low may create false relationships, while setting an e value too high may exclude real ones. As sequence length increases, the probability of finding significant blast hits also increases. In practice, blasting at a low e value and small sequence overlap length initially, and then filtering the results based on the distribution of hits obtained, may be beneficial.

1.2.3. Gene Ontology Annotation

Gene Ontology (GO) provides a structured and controlled vocabulary to describe cellular phenomena in terms of biological processes, molecular function, and subcellular localization. These terms do not directly describe the gene or protein; on the contrary they describe phenomena, and if there is sufficient evidence that the product of a gene, a protein, is involved in this phenomenon, then the probability increases that a paralogous protein is involved (Ashburner *et al.*, 2000).

For example, GO analysis for the *Drosophila melanogaster* protein Tango molecular functions indicates that it is a transcription factor which heterodimerizes with other proteins and binds to specific DNA elements and recruits RNA polymerase. The evidence shows what types of experiments or analyses were performed to determine the function. The GO evidence codes can be inferred experimentally from experiments, assays, mutant phenotypes, genetic interactions or expression patterns, as well as computationally from sequence, sequence model, and sequence or structural similarity. The biological processes information shows that Tango is involved in brain, organ, muscle, and neuron development. The cellular components information indicates that Tango's subcellular localization is primarily nuclear. Gene Ontology annotation programs often allow the user to set evidence code weights manually. For example, evidence inferred from direct experiments may provide more confidence than evidence inferred from computational analysis which has been manually curated. Uncurated computational evidence may contain the least confidence level. Tango and its human paralog, the Aryl Hydrocarbon Receptor Nuclear Translocator (ARNT), are both well-studied proteins. However, when using the *Tribolium castaneum* sequence, for example, a good GO mapping algorithm must decide how to report the more relevant information on TANGO without losing pertinent information about the better studied ARNT.

Gene ontology mapping is great when a well-studied paralogous protein is available and the blast e value is low enough to provide statistical confidence in the evolutionary relatedness and conservation of function between two proteins. In our example, the user now has a wealth of information about the *T. castaneum* Tango function, and can design primers for qRTPCR, RNAi, protein expression, or link function to the mRNAs which may have changed between two treatment groups in a transcriptome expression survey such as microarray analysis.

Enzyme codes are a numerical classification for *reactions* that are catalyzed by enzymes, given by the Nomenclature Committee of the International Union of Biochemistry and Molecular Biology (NC-IUBMB) in consultation with the IUPAC-IUBMB Joint Commission on Biochemical Nomenclature (JCBN). Enzyme codes can be inferred from GO relationships.

The Kyoto Encyclopedia of Gene and Genomes (KEGG) is a database of enzymatic, biochemical, and signaling pathways that also maps a variety of other data. KEGG is an integrated database resource consisting of systems, genomic, and chemical information (Kanehisa and Goto, 2000; Kanehisa *et al.*, 2006). The KEGG pathway database consists of hand-drawn maps for cell signaling and communication, ligand receptor interactions, and metabolic pathways gathered from the literature. **Figure 2** shows the pathway for *D. melanogaster* hormone biosynthesis annotated in KEGG. The information in this database could help in interpretation of data from genome analysis employing "omics" methods.

Domain detection algorithms do not require an absolute paralog to predict function, but often use multiple sequence alignments and Hidden Markov Models based on a number of homologous proteins that share common domains. Examples include SMART (Schultz *et al.*, 1998), PFAM (Finn *et al.*, 2010), and the NCBI Conserved Domain Database (CDD) (Marchler-Bauer *et al.*, 2002). Some databases, such as SCOP (Lo Conte *et al.*, 2002), CATH (Martin *et al.*, 1998), and DALI (Holm and Rosenstrom, 2010), focus on structural relationships and evolution. These databases group and classify protein folds based on their structural and evolutionary relatedness. Domain recognition programs have strengths and weaknesses depending on their focus, algorithm implementation, and the database used. Interproscan (Zdobnov and Apweiler, 2001) is a direct or indirect gateway to the majority of these programs and the information they can reveal. Interproscan may be accessed on the web, or through the Blast2GO program suite. Other programs accessed via Interproscan allow the identification of localization signals (i.e., nuclear localization signals), transmembrane spanning domains, sites for post-translational modifications, sequence repeats, intrinsically disordered regions, and many more.

1.2.4. Conserved Domains and Localization Signal Recognition

Conserved domains often act as modular functional units and can be useful in predicting a protein's function.

1.2.5. Fisher's Exact Test

Perturbations in the expression levels between two treatment groups of gene products involved in GO phenomena or KEGG signaling, or which belong to domain/protein

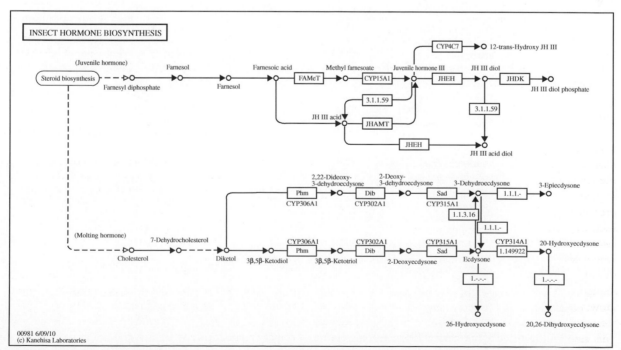

Figure 2 The pathway for *D. melanogaster* hormone biosynthesis annotated in the Kyoto Encyclopedia of Gene and Genomes (KEGG). Reproduced from KEGG database (www.genome.jp/dbget-bin/www_bget?pathway+map00981).

families, can indicate the physiologic effects of the treatment and the mechanisms that are ultimately responsible for changes in phenotypes. mRNA expression changes must be tested for statistical significance to ensure that changes between treatments are not the result of sampling a variable population. Fisher's Exact Test calculates a p-value which corresponds to the probability that functional groups are over-represented by chance. A low p-value might indicate that the over-represented functional groups share some regulatory mechanism which was perturbed by treatment.

1.2.6. Sequenced Genomes

Table 1 lists some sequenced genomes.

Fruit fly, *Drosophila melanogaster*. The *D. melanogaster* sequencing project used several types of sequencing strategies, including sequencing of individual clones, and sequencing of genomic libraries with three insert sizes (Adams *et al.*, 2000). A portion of the *D. melanogaster* genome corresponding to approximately 120 megabases of euchromatin was assembled. This assembled genomic sequence contained 13,600 predicted genes. Some of the proteins coded by these predicted genes showed high similarity with vertebrate homologs involved in processes such as replication, chromosome segregation, and iron metabolism. About 700 transcription factors have been identified based on their sequence similarity with those reported from other organisms. Half of these transcription factors are zinc-finger proteins, and 100 of them contained homoeodomains. Genome sequencing identified 22 additional homeodomain-containing proteins and 4 additional nuclear receptors. Nuclear receptors are sequence-specific ligand-dependent transcription factors that function as both transcriptional activators and repressors, and which regulate many physiological and metabolic processes. The *D. melanogaster* genome encodes 20 nuclear receptor proteins. General translation factors identified in other sequenced genomes are also present in the *D. melanogaster* genome. Interestingly, the *D. melanogaster* genome contained six genes encoding proteins highly similar to the messenger RNA (mRNA) cap-binding protein, eIF4E, suggesting that there may be an added level of complexity to regulation of cap-dependent translation in the fruit fly. The cytochrome P450 monooxygenases (P450s) are a large superfamily of proteins that are involved in synthesis or degradation of hormones and pheromones, as well as the metabolism of natural and synthetic toxins and insecticides (Feyereisen, 2006; see also Chapter 8 in this volume). Eighty-six genes coding for P450 enzymes and four P450 pseudo genes were identified in the *D. melanogaster* genome. About 20% of the proteins encoded by the *D. melanogaster* genome are likely targeted to the cellular membranes, since they contain four or more hydrophobic helices. The largest families of membrane proteins are

Table 1 List of Sequenced Genomes

Common name	Scientific name	Genome size (Mb)	Number of genes predicted	Reference
Beetle, Red flour	*Tribolium castaneum*	160	16404	Richards *et al.*, 2008
Fruit fly	*Drosophila ananassae*	176	15276	Drosophila 12 Genome Consortium, 2007
Fruit fly	*Drosophila erecta*	134	15324	Drosophila 12 Genome Consortium, 2007
Fruit fly	*Drosophila grimshawi*	138	15270	Drosophila 12 Genome Consortium, 2007
Fruit fly	*Drosophila melanogaster*	120	13600	Adams *et al.*, 2000
Fruit fly	*Drosophila mojavensis*	161	14849	Drosophila 12 Genome Consortium, 2007
Fruit fly	*Drosophila persimilis*	138	17325	Drosophila 12 Genome Consortium, 2007
Fruit fly	*Drosophila pseudoobscura*	127	16363	Richards *et al.*, 2005
Fruit fly	*Drosophila sechellia*	115	16884	Drosophila 12 Genome Consortium, 2007
Fruit fly	*Drosophila simulans*	111	15983	Drosophila 12 Genome Consortium, 2007
Fruit fly	*Drosophila virilis*	172	14680	Drosophila 12 Genome Consortium, 2007
Fruit fly	*Drosophila willistoni*	187	15816	Drosophila 12 Genome Consortium, 2007
Fruit fly	*Drosophila yakuba*	127	16423	Drosophila 12 Genome Consortium, 2007
Honey bee	*Apis mellifera*	236	10157	The Honey Bee Genome Consortium, 2006
Louse, body	*Pediculus humanus*	108	10773	Kirkness *et al.*, 2010
Malaria mosquito	*Anopheles gambiae*	278	14000	Holt *et al.*, 2002
Yellow fever mosquito	*Aedes aegypti*	1380	15419	Nene *et al.*, 2007
Southern house mosquito	*Culex quinquefasciatus*	579	18883	Arensburger *et al.*, 2010
Pea aphid	*Acyrothosyphon pisum*	464	10249	The Pea Aphid Genome Consortium, 2010
Wasp, parasitoid	*Nasonia vitripennis* *Nasonia giraulti* *Nasonia longicornis*	240	17279	Werren *et al.*, 2010
Silkworm	*Bombyx mori*	432	14623	The International Silkworm Genome Consortium, 2008

sugar permeases, mitochondrial carrier proteins, and the ATP-binding cassette (ABC) transporters coded by 97, 38, and 48 genes respectively. Among the proteins involved in biosynthetic networks, 31 triacylglycerol lipases that are involved in lipolysis and energy storage and redistribution and 32 uridine diphosphate (UDP) glycosyl transferases (which participate in the production of sterol glycosides and in the biodegradation of hydrophobic compounds) are encoded by the *D. melanogaster* genome. One additional ferritin gene and two additional transferrin genes have been identified by genome sequencing.

In 2005, Richards and colleagues published the genome of a second *Drosophila* species, *Drosophila pseudoobscura* (Richards *et al.*, 2005). In 2007 the Drosophila Genome Consortium completed the sequencing of 10 additional *Drosophila* genomes: *D. sechellia; D. simulans; D. yakuba; D. erecta; D. ananassae; D. persimilis; D. willistoni; D. mojavensis; D. virilis;* and *D. grimshawi* (Drosophila 12 Genome Consortium, 2007). Comparative analysis of sequences from these 10 genomes and the 2 genomes published earlier (*D. melanogaster* and *D. pseudoobscura*) identified many changes in protein-coding genes, noncoding RNA genes, and cis-regulatory regions. Many characteristics of the genomes, such as the overall size, the total number of genes, the distribution of transposable element classes, and the patterns of codon usage, are well conserved among these 12 genomes. Interestingly, a number of genes coding for proteins involved environmental interactions, and reproduction showed rapid change. In these 12 genomes, microRNA genes are more conserved than the protein-coding genes (see Chapter 2 in this volume). Genome-wide alignments of the 12 *Drosophila* species resulted in the prediction and refinement of thousands of protein-coding exons, genes coding for RNAs such as miRNAs, transcriptional regulatory motifs, and functional regulatory regions (Stark *et al.*, 2007). For more information on comparative analysis of 12 *Drosophila* species genomes, the reader is directed to Ashburner's excellent preface article (Ashburner, 2007).

Malaria mosquito, *Anopheles gambiae*. 278 Mb of genome sequence from *An. gambiae* was obtained by the WGS method (Holt *et al.*, 2002). About 10-fold coverage of the genome sequence was achieved. The size of the assembled *An. gambiae* genome is larger than that of *D. melanogaster* (120 Mb). About 14,000 predicted genes were identified in the assembled genome sequence. When compared to the *D. melanogaster* genome, the *An. gambiae* genome contained 100 additional serine proteases, central effectors of innate immunity, and other proteolytic processes (see Chapters 10 and 14 in this volume). The presence of additional serine proteases in *An. gambiae* may be due to differences in feeding behavior, as well as its intimate interactions with both vertebrate hosts and parasites. Also, 36 additional proteins containing fibrinogen domains (carbohydrate-binding lectins that participate in the first

line of defense against pathogens by activating the complement pathway in association with serine proteases) and 24 additional cadherin domain-containing proteins were found in *An. gambiae*. Most of the genes coding for transcription factors, the C2H2 zinc-finger, POZ, Myb-like, basic helix–loop–helix, and homeodomain-containing proteins reported from sequenced genomes are also present in the *An. gambiae* genome. An over-representation of the MYND domain was observed in the *An. gambiae* genome. This domain is predominantly found in chromatin proteins, which are believed to mediate transcriptional repression.

Genes coding for proteins involved in the visual system, structural components of the cell adhesion and contractile machinery, and energy-generating glycolytic enzymes that are required for active food seeking are present in higher numbers in the *An. gambiae* genome when compared with the *D. melanogaster* genome. Genes coding for salivary gland components, as well as anabolic and catabolic enzymes involved in protein and lipid metabolism, are over-represented in the *An. gambiae* genome. Genes coding for proteins involved in insecticide resistance, such as transporters and detoxification enzymes, were also found in higher numbers in the *An. gambiae* genome when compared to their numbers in the *D. melanogaster* genome.

Red flour beetle, *Tribolium castaneum*. The 160-Mb *T. castaneum* genome sequence was obtained by WGS, and contained 16,404 predicted genes (Richards *et al.*, 2008). The *T. castaneum* genome showed expansions in odorant and gustatory receptors, as well as P450s and other detoxification enzyme families (see also Chapter 7 in this volume). In addition, the *T. castaneum* genome contained more ancestral genes involved in cell–cell communication when compared to other insect genomes sequenced to date. RNA interference is systemic in *T. castaneum*, and thus works very well. The SID-1 multi-transmembrane protein involved in double-stranded RNA (dsRNA) uptake in *C. elegans* was not found in *D. melanogaster*. However, three genes that encode proteins similar to SID-1 were found in the *T. castaneum* genome. Expansions of odorant receptors, CYP proteins, proteinases, diuretic hormones, a vasopressin hormone and receptor, and chemoreceptors suggest that these adaptations allowed *T. castaneum* to become a serious pest of stored grain.

Honeybee, *Apis mellifera*. The 236-Mb *A. mellifera* genome was assembled based on 1.8 Gb of sequence obtained by WGS (The Honey Bee Genome Consortium, 2006). About 10,157 potential genes were identified in the assembled genome sequence. Genes coding for most of the highly conserved cell signaling pathways are present in the *A. mellifera* genome. Seventy four genes coding for 96 homeobox domains were identified in the *A. mellifera* genome. When compared to the *D. melanogaster* genome, the *A. mellifera* genome contained more genes coding for odorant receptors and proteins involved in nectar and pollen utilization. This genome also showed fewer genes coding for proteins involved in innate immunity, detoxification enzymes, cuticle-forming proteins, and gustatory receptors.

Parasitoid wasps, *Nasonia vitripennis*, *N. giraulti*, and *N. longicornis*. 240 Mb of *N. vitripennis* genome was assembled from sequences obtained by the Sanger sequencing method (Werren *et al.*, 2010). Sequences from two sibling species, *N. giraulti* and *N. longicornis*, were completed with one-fold Sanger and 12-fold, 45 base-pair (bp) Illumina genome coverage. The assembled genome sequence contained 17,279 predicted genes. About 60% of *Nasonia* genes code for proteins showing high similarity with human proteins, 18% of the genes code for proteins showing similarity with other arthropod homologs, and about 2.4% of *Nasonia* genes code for proteins similar to those in *A. mellifera*, which could therefore be hymenoptera-specific. About 12% of genes code for proteins that showed no similarity with known proteins, and therefore may be *Nasonia*-specific.

Body louse, *Pediculus humanus humanus*. 108 Mb of *P. h. humanus* genome was assembled from 1.3 million pair-end reads from plasmid libraries obtained by WGS (Kirkness *et al.*, 2010). The body louse has the smallest genome size of all the insect genomes sequenced so far. The assembled genome contained 10,773 protein-coding genes and 57 microRNAs. Compared with other insect genomes, the body-louse genome contains significantly fewer genes associated with environmental sensing and response. These proteins include odorant and gustatory receptors and detoxifying enzymes. Only 104 non-sensory G protein-coupled receptors and 3 opsins were identified in *P. h. humanus* genome. This insect has the smallest repertoire of GPCRs identified in any sequenced insect genome to date. Only 10 odorant receptors were detected in *P. h. humanus* genome. Only 37 genes in the *P. h. humanus* genome encode for P450s. Despite its smaller size, the *P. h. humanus* genome contains homologs of all 20 nuclear receptors identified in *D. melanogaster* genome.

Pea aphid, *Acyrthosiphon pisum*. The 464-Mb genome of *A. pisum* was assembled from 4.4 million Sanger sequencing reads (The Pea Aphid Genome Consortium, 2010). Analysis of the *A. pisum* genome showed extensive gene duplication events. As a result, the aphid genome appears to have more genes than any of the previously sequenced insects. Genes coding for proteins involved in chromatin modification, miRNA synthesis, and sugar transport are over-represented in the *A. pisum* genome when compared with other insect genomes sequenced to date. About 20% of the predicted genes in the *A. pisum* genome code for proteins with no significant similarity to other known proteins. Proteins involved in amino acid and purine metabolism are encoded by both host and symbiont genomes at different enzymatic steps. N Selenocysteine biosynthesis is not present in the pea aphid, and selenoproteins are absent. Several genes in the *A. pisum* genome were found to have arisen from bacterial ancestors and some of these genes are highly expressed in bacteriocytes, which may function in the regulation of symbiosis. Interestingly, the genes coding for proteins that function in the IMD pathway of the immune system are absent in the *A. pisum* genome.

Yellow fever Mosquito, *Aedes aegypti*. The 1.38-Gb genome of *Ae. aegypti* was assembled from sequence reads obtained by WGS (Nene *et al.*, 2007). This is the largest insect genome sequenced to date, and is about five times larger than the *An. gambiae* and *D. melanogaster* genomes. Approximately 47% of the *Ae. aegypti* genome consists of transposable elements. The presence of large numbers of transposable elements could have contributed to the larger size of the *Ae. aegypti* genome. About 15,419 predicted genes were identified in the assembled genome. Compared to the genome of *An. gambiae*, an increase in the number of genes encoding odorant binding proteins, cytochrome P450s, and cuticle proteins was observed in the *Ae. aegypti* genome.

Silk moth, *Bombyx mori*. The silkworm genome was sequenced by Japanese and Chinese laboratories simultaneously. The Japanese group used the sequence data derived from WGS to assemble 514 Mbs including gaps, and 387 Mbs without gaps (Mita *et al.*, 2004). Chinese scientists assembled sequences obtained by WGS into a 429-Mb genome (Xia *et al.*, 2004). The two data sets were merged and assembled recently (The International Silkworm Genome, 2008). This resulted in the 8.5-fold sequence coverage of an estimated 432-Mb genome. The repetitive sequence content of this genome was estimated at 43.6%. Gene models numbering 14,623 were predicted using a GLEAN-based algorithm. Among the predicted genes, 3000 of them showed no homologs in insects or vertebrates. The presence of specific tRNA clusters, and several sericin gene clusters, correlates with the main function of this insect: the massive production of silk.

Recently, a consortium of international scientists sequenced the genomic DNA of 40 domesticated and wild silkworm strains to coverage of approximately three-fold. This represents 99.88% of the genome, and led to the development of a single base-pair resolution silkworm genetic variation map (Xia *et al.*, 2009). This effort identified ~16 million single-nucleotide polymorphisms, many indels, and structural variations. These studies showed that domesticated silkworms are genetically different from wild ones; nonetheless, they have managed to maintain large levels of genetic variability. These findings suggest a short domestication event involving a large number of individuals. Candidate genes, numbering 354, that are expressed in the silk gland, midgut, and testes, may have played an important role during domestication.

The southern house mosquito, *Culex quinquefasciatus*. *C. quinquefasciatus* is a vector of important viruses such as the West Nile virus and the St Louis encephalitis virus, and harbors nematodes that cause lymphatic filariasis. Arensburger sequenced and assembled the whole genome of *C. quinquefasciatus* (Arensburger *et al.*, 2010). A larger number of genes, 18,883, reported from the other two mosquito genomes (*Aedes aegypti* and *Anopheles gambiae*), were identified in the assembled *C. quinquefasciatus* genome. An increase in the number of genes coding for olfactory and gustatory receptors, immune proteins, enzymes such as cytosolic glutathione transferases and cytochrome P450s involved in xenobiotic detoxification was observed.

1.3. Genome Analysis

Since its discovery, Sanger sequencing has been largely applied in most genome sequencing projects (Sanger *et al.*, 1977); therefore, a large volume of sequence information from a variety of species has been deposited into various databases. With deciphered full genome sequences for a number of species, scientists could now begin to address biological questions on a genome-wide level. These analyses include the measurement of global gene expression, the identification of functional elements, and the mapping of genome regions associated with quantitative traits. Various new technologies have also been developed to assist with genome analysis. These include DNA microarrays (Schena *et al.*, 1995), serial analysis of gene expression (SAGE) (Schena *et al.*, 1995), chromatin immunoprecipitation microarrays (Ren *et al.*, 2000; Iyer *et al.*, 2001; Lieb *et al.*, 2001), next generation sequencing (NGS) (Margulies *et al.*, 2005; Shendure *et al.*, 2005), genome-wide RNAi screens (Kiger *et al.*, 2003), comparative genomics (Kiger *et al.*, 2003), and metagenomics (Chen and Pachter, 2005). These genomic analysis tools have greatly improved our understanding of how biological and cellular functions are regulated by the RNAs or proteins encoded in an organism's genome. Especially in the agricultural research field, functional genomics studies will enhance our understanding of the biology of insect pests and disease vectors, which in turn will assist the design of future pest control strategies. Here, we will discuss technologies used for functional genomics studies, with an emphasis on forward genetics, DNA microarray, and NGS technologies, and their applications in research on insects.

1.3.1. Forward and Reverse Genetics

The function of genes is often studied using forward genetics approaches. In forward genetic screens, insects are treated with mutagens to induce DNA lesions, followed by a screen to identify mutants with a phenotype of interest. The mutated gene is then identified by employing standard genetic and molecular methods.

Follow-up studies on the mutant phenotype, including molecular analyses of the gene, often lead to determination of its function. Forward genetics approaches have been used for determining the function of many genes. In the fruit fly, *D. melanogaster*, genetic screens have been used for a number of years to discover gene–phenotype associations. With the availability of massive amounts of data derived from whole-genome and omics studies, a systems biology approach needs to be applied to enhance the power of gene function discovery *in vivo*. Mobile elements or chemicals are often used as mutagenesis tools (Ryder and Russell, 2003). The *P* element has been widely used in *D. melanogaster* forward genetics since its development as a tool for transgenesis in 1982 (Rubin and Spradling, 1982). The insertion of *P* elements into the *D. melanogaster* genome allowed subsequent cloning and characterization of a large number of fly genes. *P*-element mediated transgenesis is often used to create mutants by excising the flanking genes based on imprecise mobilization of the *P* elements. *P* elements were also modified to study genes, not only based on a phenotype, but also based on RNA or protein expression patterns, which are often referred to as enhancer trap and gene trap technologies. *P* elements are also being used as mutagenesis agents in a project aimed at generating insertions in every predicted gene in the fruit fly genome.

Recent developments in transgenic techniques focused on the site-specific integration of transgenes at specific genomic sites, which employ recombinases and integrases, have made forward genetics in *D. melanogaster* effective and specific. One of the major drawbacks of *P*-element mediated transgenesis is the non-specific and positional effects caused by inserting exogenous DNA into insect genome. Recently, several methods have been developed to eliminate these unwanted, non-specific effects in transgenic insects. Transgene co-placement was developed by Siegal and Hartl (1996). This method uses two transgenes, a rescue fragment and its mutant version, which are inserted into the same locus by using a *P*-element vector that contains the recognition sites FRT (FLP recombinase recognition site) and loxP (the Cre recombinase recognition site). After integration, FLP can remove one transgene, such as the rescue gene. Cre can remove the other transgene, which may be the mutant version. A method was developed by Golic (Golic *et al.*, 1997) by using FLP recombinase for remobilization of transgene by a donor transposon that contains a transgenic insert together with a marker gene such as *white* flanked by two FRT sites, and an acceptor transposon that contains a second marker and one FRT site. The remobilization of the donor transposon by FLP can be followed by the changes in the expression of *white* gene. The remobilization results in the excision of transgene and its potential integration into the FRT site of the acceptor transposon.

Homologous recombination is the best method for *in vivo* gene targeting, since positional effects can be eliminated completely. Insertional gene targeting (Rong and Golic, 2000) and replacement gene targeting (Gong and Golic, 2003) are two alternative methods that have been developed. Insertional gene targeting results in the insertion of a target gene at a region of homology. Replacement gene targeting results in replacement of endogenous homologous DNA sequences with exogenous DNA through a double reciprocal recombination between two stretches of homologous sequences. Site-specific zinc-finger-nuclease-stimulated gene targeting has been developed to further improve *in vivo* gene targeting (Bibikova *et al.*, 2003; Beumer *et al.*, 2006). The most widely used site-specific integration in *D. melanogaster* employs the bacteriophage Φ C31 integrase. The bacteriophage Φ C31 integrase catalyzes the recombination between the phase attachment site (*attP*), previously integrated into the fly genome, and a bacterial attachment site (*attB*) present in the injected transgenic construct (Groth *et al.*, 2004). A combination of different transgenic methods should aid in *D. melanogaster* functional genomics studies aimed at determining the function of every gene in this insect.

In the reverse genetics approach, studies on the function of the genes start with the gene sequences, rather than a mutant phenotype, which is often used in forward genetics approaches. In this approach, the gene sequence is used to alter the gene function by employing a variety of methods. The effect of the altered gene function on physiological and developmental processes of insects is then determined. Reverse genetics is an excellent complement to forward genetics, and some of the experiments are much easier to perform using reverse genetics rather than forward genetics. For example, RNA interference, a reverse genetics method (covered in Chapter 2 in this volume) is a better method compared to forward genetics to investigate the functions of all the members of a gene family. The availability of whole-genome sequences for a number of insects and the functioning of RNAi in these insects will keep scientists busy studying the functions of all genes in insects during the next few years.

1.3.2. DNA Microarray

In most cases, a group of functionally associated genes share similar expression patterns, which may be temporal, spatial, developmental, or physiological. For example, environmental changes and pathological conditions could alter global gene expression patterns. To understand and characterize the biological roles of an individual gene or a cluster of genes, a high-throughput quantitative method is needed to detect gene expression at the whole-genome level. The DNA microarray technique is one such method that has been developed for monitoring global gene expression patterns. Through robotic printing of thousands of DNA oligonucleotides onto a solid surface, one DNA microarray chip can accommodate more than 50,000 probes (unique DNA sequences). DNA microarrays utilize the principle of Southern blotting (Schena *et al.*, 1995). First, fluorescently labeled probes are synthesized from RNA samples by reverse transcription; the probes are then hybridized to DNA microarrays which contain complementary DNA. After washing away the unbound probes, the intensity of the fluorescent signal for each spot is captured using a microarray scanner. DNA microarrays have been widely used in functional genomics research. In addition to their application on gene expression profiling, DNA microarrays can also be used to identify transcriptional or functional elements in the genome, or identify single nucleotide polymorphisms (SNP) among alleles within or between populations. The applications of DNA microarrays and various other types of arrays are listed in **Table 2**.

1.3.2.1. Global gene expression analysis (transcriptome analysis)

1.3.2.1.1. DNA microarray fabrication. The DNA microarrays used for global gene expression analysis usually contain tens of thousands of probes which cover all the predicted genes in a genome, or sequences representing transcribed regions, also called expressed sequence tags (ESTs). For example, the Affymetrix GeneChip®

Table 2 List of Applications of DNA Microarray

Application	Description	Type of microarray
Gene expression	Measuring global gene expression pattern under various biological conditions	Expression array
ChIP-on-chip	Identifying transcriptional or functional elements at a whole-genome level	Tiling array
DamID	Genome-wide scanning of Adenosine methylation events. Analogously to ChIP-on-chip	DNA methylation array
miRNA profiling	Genome-wide detection of the expression of miRNAs (small non-coding RNAs)	miRNA array
SNP detection	Detecting polymorphisms within a population	SNP array
Pathogen and virus detection	Low-density DNA microarray for the identification of viruses and pathogens	Virus Chip, FluChip

Drosophila Genome 2.0 Array contains over 500,000 data points representing 18,500 transcripts and various SNPs (Affymetrix technical data sheets). DNA microarrays can be prepared by various methods, including photolithography, ink-jet technology, and spotted array technology. Photolithography and ink-jet technologies are used for fabricating so-called oligonucleotide microarrays, which are made by synthesizing or printing short oligonucleotide sequences (25-mer in Affymetrix array or 60-mer in Agilent array) directly onto a solid array surface. The photolithography method is used by Affymetrix and NimbleGen, while the ink-jet print method is used by Agilent. Typically, multiple probes per gene are used in order to achieve precise estimation of gene expression. Long oligonucleotides have better hybridization specificities than short ones, although short oligonucleotides can be printed at a higher density and synthesized at lower cost. In contrast, spotted microarrays are made by synthesizing probes prior to deposition onto the array surface. The probes used for spotted microarrays can be oligonucleotides, cDNA or PCR products. Because of their relatively low cost and flexibility, the spotted microarray technology has been widely used to produce custom arrays in many academic laboratories and facilities. However, spotted microarrays are less uniform and contain low probe density when compared with oligonucleotide arrays. As the cost of custom commercial arrays such as Agilent Custom Gene Expression Microarrays (eArray) has decreased, the use of spotted microarray is decreasing as well.

1.3.2.1.2. Target preparation and hybridization.
Total RNA or mRNA is isolated from experimental samples using commercial TRIzol reagent or RNA isolation and purification kits. Total RNA (1 µg to 15 µg) or mRNA (0.2 µg to 2 µg) is reverse transcribed into first-strand cDNA. For smaller amounts of total starting RNA (10 ng to 100 ng), Affymetrix offers a two-cycle target labeling method to obtain sufficient amounts of labeled targets for DNA hybridization. Then, cDNAs are labeled and hybridized to spotted or oligonucleotide microarrays. In oligonucleotide microarrays, one mRNA sample labeled with one fluorescent dye is analyzed on a single channel. Alternatively, two different fluorescent dyes, such as Cy3 and Cy5, can be used to determine gene expression changes from two different experimental conditions.

1.3.2.1.3. Data analysis.
Although the data analysis methods among commercial microarrays vary, the basic concepts are similar. After hybridization, the fluorescence images are captured by a microarray scanner. The fluorescence intensity data are then corrected and adjusted from the background (noise), which may result from non-specific hybridization or autofluorescence. In two-channel arrays, the fluorescence intensity ratio between two dyes is calculated and adjusted. If the data from a different array or hybridization are to be compared, they need to be normalized before further analysis.

After normalization, various statistical analysis methods can be applied to identify differentially expressed genes between two treatments. Usually, a *t*-test is used for comparing the means of two sample populations, while ANOVA (analysis of variance) is applied for comparing multiple sets of samples or treatments to obtain more accurate variance estimates. Since many genes are tested for statistical differences, multiple test corrections, such as the Bonferroni correction and the Benjamini and Hochberg false discovery rate (FDR) (Benjamini and Hochberg, 1995), are applied to adjust the *P*-value and correct the occurrence of false positives. Bonferroni correction is a very stringent method that uses α/n as the threshold *P*-value for each test where n is the number of tests or the number of genes. In contrast, the Benjamini and Hochberg FDR is less stringent, and the rate of false negative discovery is lower. Various statistical analysis programs are now available from either commercial microarray providers or open source websites. These include GeneSpring from Silicon Genetics (acquired by Agilent in 2004) and Significance Analysis of Microarrays (SAM) (Tusher *et al.*, 2001). Besides differential expression analysis, genes with similar expression patterns can be grouped into one or more clusters using hierarchical clustering methods. Hierarchical clustering analysis helps to visualize gene expression patterns and identify relationships between functionally associated genes (Eisen *et al.*, 1998). On the other hand, programs such as Gene Set Enrichment Analysis (GSEA) are used to determine whether there is a statistically significant, coordinated difference between control and treatment samples for a predefined set of genes that are involved in a similar biological process (Subramanian *et al.*, 2005). Unlike traditional microarray analyses at the single gene level, GSEA has addressed a situation where the fold change between control and treatment samples is small, but there is a concordant difference in the representation of functionally related genes. Several published microarray datasets have been deposited in various online databases, including Gene Expression Omnibus (GEO) at NCBI, ArrayExpress at the European Bioinformatics Institute, and Stanford Genomic Resource at Stanford University. A list of microarray analysis tools and databases is shown in **Table 3**.

1.3.2.1.4. Applications.
The primary goal of developing gene expression microarray technology is to monitor differentially expressed genes at the whole-genome level. Therefore, microarray technology has been used to study the molecular basis of pesticide resistance (Djouaka *et al.*, 2008; Zhu *et al.*, 2010) (**Figure 3**), insect–plant interactions (Held *et al.*, 2004), insect host–parasitoid associations (Lawniczak and Begun, 2004; Barat-Houari *et al.*, 2006; Mahadav *et al.*, 2008; Kankare *et al.*, 2010),

Table 3 List of Microarray Data Analysis Tools and Microarray Databases

Statistical Analysis Programs

GeneSpring	http://www.agilent.com/
SAM	http://www-stat.stanford.edu/~tibs/SAM/
Bioconductor	http://www.bioconductor.org/
Partek	http://www.partek.com/

Cluster and Pathway Analysis Tools

Cluster and TreeView	http://rana.lbl.gov/EisenSoftware.htm
Cluster 3.0	http://bonsai.hgc.jp/~mdehoon/software/cluster/
Java TreeView	http://jtreeview.sourceforge.net/
Gene Set Enrichment Analysis (GSEA)	www.broadinstitute.org/gsea/
Gene Set Analysis (GSA)	http://www-stat.stanford.edu/~tibs/GSA/
Genepattern	http://www.broadinstitute.org/cancer/software/genepattern/
Genecruiser	http://genecruiser.broadinstitute.org/genecruiser3/
Advanced Pathway Painter	http://pathway.painter.gsa-online.de/

Microarray Databases

Gene Expression Omnibus	http://www.ncbi.nlm.nih.gov/geo/
ArrayExpress Archive	http://www.ebi.ac.uk/microarray-as/ae/
Stanford Genomic Resources	http://genome-www.stanford.edu/
Arraytrack	http://www.fda.gov/ScienceResearch/BioinformaticsTools/Arraytrack/
Genevestigator	https://www.genevestigator.com/gv/index.jsp

insect behavior (McDonald and Rosbash, 2001; Etter and Ramaswami, 2002; Dierick and Greenspan, 2006; Adams *et al.*, 2008; Kocher *et al.*, 2008), development and reproduction (White *et al.*, 1999; Kawasaki *et al.*, 2004; Dana *et al.*, 2005; Kijimoto *et al.*, 2009; Bai and Palli, 2010; Parthasarathy *et al.*, 2010a, 2010b), etc. Understanding the mechanisms of pesticide resistance is critical for prolonging the life of existing insecticides, designing novel pest control reagents, and improving control strategies. As a result, several laboratories have begun using microarrays to identify genes responsible for insecticide resistance. For example, using a custom microarray, one cytochrome P450 gene, *CYP6BQ9*, has been identified to be responsible for the majority of deltamethrin resistance in *T. castaneum* (Zhu *et al.*, 2010) (**Figure 3**). Another microarray study discovered that two cytochrome P450 genes, *CYP6P3* and *CYP6M2*, are upregulated in multiple pyrethroid-resistant *Anopheles gambiae* populations collected in Southern Benin and Nigeria (Djouaka *et al.*, 2008). A global view of tissue-specific gene expression profiling has been reported in *Drosophila melanogaster* (Chintapalli *et al.*, 2007). This study identified many genes that are uniquely expressed in specific fly tissues, and provided useful information for understanding the tissue-specific functions of these candidate genes.

Biological processes and cellular functions are rarely regulated by only one or a few genes. Therefore, monitoring the expression changes of a group of genes under different biological conditions could provide useful insights into biological processes and cellular functions. Microarrays have been applied to detect gene expression patterns during insect embryonic development (Furlong *et al.*, 2001; Stathopoulos *et al.*, 2002; Tomancak *et al.*, 2002;

Altenhein *et al.*, 2006; Sandmann *et al.*, 2007) and metamorphosis (White *et al.*, 1999; Butler *et al.*, 2003), under various nutrient conditions (Zinke *et al.*, 2002; Fujikawa *et al.*, 2009), with aging (Weindruch *et al.*, 2001; Pletcher *et al.*, 2002; Terry *et al.*, 2006; Pan *et al.*, 2007), and in many other circumstances.

In combination with newly developed statistical and bioinformatics methods, and gene ontology and signaling pathway databases, microarray technology has also been applied to identify a signaling pathway or a specific cellular function that is altered under various biological conditions (Subramanian *et al.*, 2005). With these approaches, it is possible to discover the interactions between individual pathways and obtain a global network view (Costello *et al.*, 2009; Avet-Rochex *et al.*, 2010).

1.3.2.2. DNA–protein interaction (chromatin immunoprecipitation)

Chromatin immunoprecipitation (ChIP) was developed in the late 1980s (Hebbes *et al.*, 1988) and has been widely applied to the study of protein–DNA interactions *in vivo*. Particularly, transcription factors, histone modifications, and DNA replication-related proteins can be studied using ChIP. By combining ChIP with DNA microarray technology, a process typically called ChIP-on-chip, all the possible DNA-binding sites of a protein of interest throughout the genome can be examined. ChIP-on-chip technology first appeared in 2000 in studies of DNA-binding proteins in the budding yeast, *Saccharomyces cerevisiae* (Ren *et al.*, 2000; Iyer *et al.*, 2001). With the availability of high-density oligonucleotide arrays which contain short sequences representing non-coding regions or entire genomes, ChIP-on-chip has also been applied to the global identification of

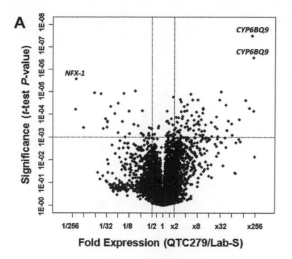

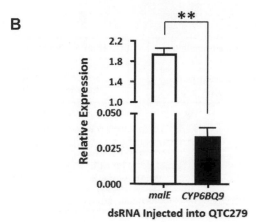

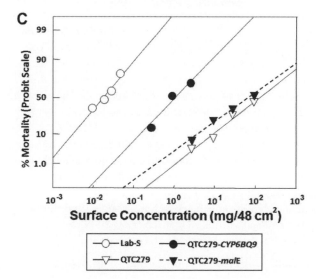

Figure 3 Application of microarray and RNA interference technologies to identify and fight insecticide resistance. Reprinted with permission from Zhu *et al.* (2010).
(A) The V plot of differentially expressed genes identified by microarrays. Fold suppression or overexpression of genes in QTC279 strain when compared with their levels in the Lab-S strain was plotted against the *P* values of the *t*-test. The horizontal bar in the plot shows the nominal significant level 0.001. The vertical bars separate the genes that are a minimum of 2.0-fold difference. Three genes identified by the Bonferroni multiple-testing correction as differentially expressed between resistant and susceptible strains are shown.
(B) Injection of CYP6BBQ9 dsRNA into *Tribolium castaneum* QTC279 beetles reduces CYP6BBQ9 mRNA levels. The mRNA levels of CYP6BQ9 were quantified by qRT-PCR at 5 days after dsRNA injection. The relative mRNA levels were shown as a ratio in comparison with the levels of rp49 mRNA.
(C) Dose–response curves for *T. castaneum* adults exposed to deltamethrin. At 5 days after dsRNA injection, the following were exposed to various doses of deltamethrin: Lab-S (○), a susceptible strain; QTC279 (▽), a deltamethrin-resistant strain; QTC279-CYP6BQ9 RNAi (●), a QTC279 strain injected with CYP6BQ9 dsRNA; and QTC279-malE RNAi (▼), a QTC279 strain injected with malE dsRNA as a control.

other high-throughput technologies. ChIP-on-chip technology will likely contribute to a better understanding of genome organization, including functionally important elements, non-coding RNA, and chromatin markers. This may eventually lead to the comprehensive understanding of gene regulatory networks within an organism's genome.

Many ChIP-on-chip protocols have been published, or are available online. In general, cells or tissues are treated using a reversible cross-linker (e.g., formaldehyde), so that protein and DNA are fixed *in vivo*. Then the protein–DNA complex within the nucleus is extracted and separated from cytoplasm. Purified protein–DNA complexes (referred to as "chromatin" hereafter) are sonicated using a conventional sonicator or Bioruptor® in order to generate DNA fragments that range from 200 to 1000 bp. The sonication conditions need to be pre-adjusted to obtain optimally sized DNA fragments. Before sonication, an aliquot of chromatin needs to be saved as a reference sample (or input samples). Usually a chromatin pre-clean step using protein-A beads is included to remove non-specific binding during the immunoprecipitation step. For the immunoprecipitation step, a certain amount (e.g., 10 μg) of antibody and protein-A beads is added to pre-clean the chromatin. Chromatin bound to protein-A beads is then purified, eluted, and reverse-cross-linked. Since the amount of a single ChIP DNA sample is normally around a few nanograms, and this is not enough for microarray hybridization, an amplification step is required. There are two ways to amplify ChIP DNA: ligation-mediated PCR (LM-PCR) and whole-genome amplification (WGA). The WGA method is considered to have lower background compared to the LM-PCR method (O'Geen

transcriptional regulatory networks in various organisms. These projects include ENCODE (human) (The ENCODE Project Consortium 2004) and modEN-CODE (worm and fly) (Celniker *et al.*, 2009). The goal of these projects is the genome-wide characterization of all possible functional elements using ChIP-on-chip and *et al.*, 2006). Amplified ChIP DNA and Input DNA are then denatured, fluorescently labeled, and hybridized to either a spotted or a oligonucleotide microarray (typically a tiling array). If there is a known target binding site for the protein of interest, the quality of ChIP samples can be assessed using real-time qPCR before submitting the samples for microarray analysis.

The data preprocessing steps of ChIP-on-chip are similar to those used in gene expression microarrays. After microarray scanning and fluorescence intensity recording, the enrichment of each binding site across the genome is obtained by comparing the intensity of each spot between ChIP DNA and Input DNA. Enriched regions can then be further analyzed, including identification of genes associated with each binding region, and conserved motif searching. The enrichment can also be visualized using many free available genome browsers, such as UCSC Genome Browser (http://genome.ucsc.edu/), Integrated Genome Browser (IGB, http://www.bioviz.org/igb/), and Integrative Genomics Viewer (IGV, http://www.broadinstitute.org/igv/). The workflow of a chromatin immunoprecipitation experiment is shown in **Figure 4**.

Antibody quality is a critical factor for successful ChIP-on-chip experiments. Since there are a variety of antibodies for a protein of interest, each with a specific affinity, it is always better to examine all the available antibodies in a small-scale ChIP-PCR experiment. If there are no suitable antibodies for a protein of interest, an epitope-tagged protein can be used (Zhang *et al.*, 2008). In this way, an antibody for the epitope instead of one for the protein of interest can be used in immunoprecipitation. In *Drosophila*, transgenic flies may be generated to express epitope-tagged proteins *in vivo*.

The success of ChIP experiments also depends on the sonication step. It is suggested that 200- to 1000-bp DNA fragments should be obtained after sonication or DNA shearing. Undersonication will result in many large fragments (larger than 1000 bp) and lead to loss of resolution. Oversonication could interfere with the protein–DNA complex formation, and may result in more noise.

As mentioned above, the WGA amplification method is considered better than the LM-PCR method. Due to the bias caused by PCR amplification, the signal-to-noise ratio normally decreases after a PCR reaction; therefore, minimizing the number of PCR cycles is suggested. As reported by O'Geen *et al.* (2006), the WGA amplification method has higher signal-to-noise ratio and more enriched binding sites when compared to the LM-PCR method.

1.3.2.3. DNA–protein interaction (chromatin immuno precipitation)

Due to the availability of whole-genome sequences, the application of ChIP-on-chip technology is mainly used in model insects. ChIP-on-chip has been applied to dissecting the transcriptional regulatory network of

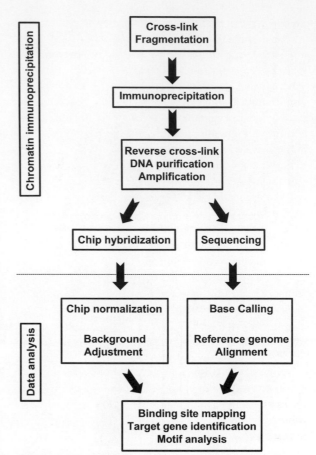

Figure 4 The workflow of a chromatin immunoprecipitation-sequence identification experiment. After cross-linking, the chromatin is precipitated with antibodies; the precipitated chromatin is cross-linked, and the DNA purified and amplified. The amplified DNA is then sequenced and aligned to the reference genome and potential binding sites are identified.

embryogenesis (Sandmann *et al.*, 2007; Zeitlinger *et al.*, 2007; Liu *et al.*, 2009), chromatin modification (Alekseyenko *et al.*, 2008; Smith *et al.*, 2009; Tie *et al.*, 2009), epigenetic silencing (Negre *et al.*, 2006), etc. Interestingly, a high-resolution transcriptional regulatory atlas of mesoderm development was constructed through the analysis of a key set of transcription factors, including Twist, Tinman, Myocyte enhancing factor 2, Bagpipe and Biniou, in the *Drosophila* embryo (Zinzen *et al.*, 2009).

1.3.3. Next Generation Sequencing (NGS)

Although DNA microarray technologies are widely used in many aspects of biological and medical research, there are some limitations. The design of the microarrays is based on our current knowledge of sequenced genomes from computationally predicted raw genome structures. These structures include gene coding regions, introns, enhancers, and non-coding RNAs. Due to a lack of comprehensive knowledge on the chromosome landscape, however, these predictions may or may not be correct. Although some tiling arrays may contain high-density oligonucleotides covering the entire genome, they are normally not cost-effective, particularly in the case of gigantic genomes (e.g., human and many plant genomes). Most importantly, in order to perform a whole-genome analysis, a sequenced genome is an absolute requirement. This becomes a limitation for many non-model organisms that do not have whole-genome sequences.

Fortunately, the breakthrough of revolutionary sequencing technology has overcome this limitation and brought us into a new post-genomics era. Next generation sequencing (NGS), or deep sequencing, was first introduced in 2005 (Margulies *et al.*, 2005; Shendure *et al.*, 2005). When compared to automated Sanger sequencing (or first generation sequencing) (Sanger and Coulson, 1975), NGS technology has dramatically accelerated the sequence speed by increasing the number of sequencing reactions and reducing the reaction volume in one instrument run (Metzker, 2010). Therefore, thousands of sequencing reactions are performed simultaneously, and in some cases NGS is also referred to as massively parallel sequencing. Unlike Sanger sequencing, the incorporation events of fluorescently labeled nucleotides to DNA templates are almost continuously monitored and recorded. More than 100 million short reads (ranging from 35 bp to 300 bp) can be obtained using some NGS technologies. Several NGS platforms, including Roche/454 Life Sciences' GS FLX, Illumina's Solexa GAII, and ABI's SOLiD, are commercially available. Each platform has its own sequencing methods and unique features (see **Table 4**). An overview of NGS technology and various sequencing platforms can be found in a recent review (Metzker, 2010). Here, we will focus on recent applications of NGS technologies in gene expression and ChIP studies.

1.3.3.1. RNA-Seq RNA-sequencing (RNA-Seq) uses NGS technology for transcriptome analysis. In contrast to conventional microarray analysis, RNA-Seq provides much more information, including unpredicted novel transcripts and previously unknown alternatively spliced isoforms. Like other NGS technologies, a cDNA library has to be made from RNA samples by adding adaptor sequences to one or both ends of cDNA. Then, long RNA or cDNA samples need to be fragmented. Small fragments (usually 150–300 bp) are separated by electrophoresis, isolated using the gel extraction method, and then purified for sequencing. After sequencing, which may take from a single day to a week, depending on the platform used, the sequence reads are then aligned to a reference genome, or used for *de novo* assembly if no genome information is available.

Due to the tremendous amount of sequencing data obtained after each sequencing run, there are always challenges in data handling and statistical analysis. Several bioinformatics programs, such as ELAND (by Illumina), SOAP (Li *et al.*, 2008a), and BOWTIE (Langmead *et al.*, 2009), have been developed for mapping the reads to a reference genome. Typically, reads with a single match to the genome sequence will be selected for future analysis. Reads with more than three mismatches, or reads that match to multiple regions of the genome, will be discarded. The mismatches may be due to sequencing errors, polymorphisms, poor sequencing quality, or low expression abundance. The reads can be found within exon regions, exon junctions, and the regions near poly (A)-tails. The expression level for each gene then can be determined by the enrichment of reads across entire ORFs (open reading frames). Like other NGS technologies, RNA-Seq has many advantages over expression microarray analysis. RNA-Seq has very low background, and is cost-effective. It also has better sensitivity to detect genes with very low or high expression levels. Most importantly, RNA-Seq is useful to detect novel and rare transcripts and alternatively spliced transcripts. It also offers great opportunities for the *de novo* transcriptome analysis of non-model organisms.

RNA-Seq technology has been used in a transcriptome analysis of *Aedes aegypti* in response to pollutants and insecticides (David *et al.*, 2010). A *Drosophila melanogaster* 5′-end mRNA transcription database was constructed through RNA-Seq technology, and contains expression profiles of each fly gene at various developmental stages (http://machibase.gi.k.u-tokyo.ac.jp/ [Ahsan *et al.*, 2009]). Roche/454 based pyrosequencing has been widely used to sequence the transcriptome of non-model insects, such as the Glanville fritillary butterfly (Vera *et al.*, 2008).

Table 4　List of Next-Generation Platforms

Platform	Manufactory	Sequencing method	Feature
GS FLX	Roche/454 Life Sciences	Pyrosequencing	Long reads (300–400 bp); fast run time.
Solexa GAII	Illumina	Reversible termination	Short reads (35 or 70 bp); huge reads per run (~20 GB)
SOLiD	ABI	Sequence by ligation	Short reads (~50 bp); huge reads per run (similar to Solexa)
HeliScope	Helicos BioSciences	Reversible termination; single molecule sequencing	No bias introduced from library construction

1.3.3.2. ChIP-Seq Chromatin immunoprecipitation sequencing (ChIP-Seq) is sequencing-based genome-wide mapping of protein–DNA interactions. Similar to the ChIP-on-chip technology mentioned earlier, ChIP-Seq also involves the pull-down of DNA fragments (ChIP DNA) bound by a protein of interest. Instead of hybridizing ChIP DNA to an oligonucleotide microarray, a sequencing library is constructed by adding adaptor sequences to ChIP DNAs, followed by size selection and gel purification. After submitting the library to sequencing, ChIP-Seq raw data are generated, which may contain more than 100 million short reads. These reads will then be aligned to a reference genome, and high quality reads that have a good match to a single genomic region (one to two nucleotide mismatches are allowed) selected. Normally, 60–80% of the total reads can be aligned to a reference genome. The enrichment regions (binding sites) can be obtained by comparing the reads between ChIP DNA and control DNA (e.g., Input or mock DNA samples) in a process called peak calling. Various bioinformatics tools are available for performing peak calling, including PeakSeq (Rozowsky *et al.*, 2009), QuEST (Valouev *et al.*, 2008), CisGenome (Jiang *et al.*, 2010), and Galaxy (Giardine *et al.*, 2005). Finally, the enriched regions (or peaks) can be visualized using genome browsers, as mentioned previously.

ChIP-Seq technology offers many advantages over ChIP-on-chip. The single nucleotide resolution of ChIP-seq data is much higher than that of ChIP-on-chip. Therefore, binding motif analysis is simplified. ChIP-Seq technology also provides more information on protein–DNA interactions, and better genome coverage. Since there is no hybridization step involved, ChIP-Seq normally has less background noise, and can detect a dynamic range of binding events. In contrast, ChIP-on-chip technology has difficulty in distinguishing very low or very high binding events. With technological advancements, ChIP-Seq technology will become less costly for analyzing most genomes. ChIP-Seq has been used in characterizing *MSL*-complex regulatory networks in the X-chromosome of *D. melanogaster* (Alekseyenko *et al.*, 2008), as well as in a genome-wide methylome study of the silkworm, *Bombyx mori* (Xiang *et al.*, 2010). Once the cost of ChIP-Seq declines to prices comparable to ChIP-on-chip, there will be more ChIP-Seq applications in insect research.

1.3.4. Other Methods

In addition to mRNA, there are many non-coding RNAs (ncRNAs) within a genome. These include highly abundant and functionally relevant RNAs such as transfer RNA, ribosomal RNA, microRNAs, and long intergenic non-coding RNAs. Combining functional analysis and high-throughput microarrays or sequencing technologies has allowed the identification and characterization of novel non-coding RNAs (ncRNAs). Many ncRNAs, particularly microRNAs, have been found to be involved in development (Zhang *et al.*, 2009), neurodegeneration (Karres *et al.*, 2007), cell proliferation (Thompson and Cohen, 2006), circadian rhythms (Yang *et al.*, 2008), and host–parasitoid interactions (Gundersen-Rindal and Pedroni, 2010).

High-throughput microarray or sequencing technologies have also been applied to studies on metagenomics, or the study of genetic material recovered from environmental samples (e.g., microflora of the ocean, soil or insect gut). With the help of Roche/454 pyrosequencing technology, the Israeli acute paralysis virus was recently identified, and found to be associated with colony collapse disorder (CCD) in honey bees (Cox-Foster *et al.*, 2007). A large set of bacterial genes with cellulose and xylan hydrolysis functions was identified using pyrosequencing from the hindgut of a wood-feeding higher termite that is closely related to *Nasutitermes ephratae* (Warnecke *et al.*, 2007).

1.4. Proteomics

Proteomics is the study of all proteins present in an organism, and deals with their quantification, identification, and modifications that alter their function. While statistically significant changes in mRNA levels are usually correlated with changes in protein levels, individual proteins can change drastically with little significant correlation at the mRNA level (Bonaldi *et al.*, 2008). Cellular protein abundance is controlled through many different mechanisms. These mechanisms include translational efficiency based in part on regulatory sequences in the 5′ and 3′ untranslated regions of mRNA, and protein degradation through ubiquitination and the 28S proteasome pathway. Post-translational modifications and the presence of interacting partners often alter the function or the functional capacity of a protein.

Modern proteomics relies heavily on mass spectrometry (MS). Mass spectrometry devices measure the mass-to-charge ratio of peptide ions. Mass spectrometry can be used for protein quantitation, identification, and sequencing, and determining the presence of post-translational modifications. Two broad MS strategies, the bottom-up approach and the top-down approach, vary on whether proteolytically digested peptides are analyzed, or the entire protein is sequenced. In the bottom-up approach, peptides of interest are often separated on a two-dimensional (2D) gel, extracted, digested into smaller fragments via trypsin proteolysis, and analyzed by MS. Often, the amino acid sequence and corresponding mass (M) to charge (z) (or M/z) ratio between two trypsin cut sites are sufficient to identify a protein. The mass of the digested peptide is compared against a sequence database containing all genomic open reading frames and their calculated masses. This approach is also known as peptide mass fingerprinting. In the top-down approach, a whole protein can be sequenced using tandem MS, or MS/MS. Tandem MS measures the M/z ratio of a protein ion before fragmentation, and the resulting amino acid or peptide ions after fragmentation. Finally, in shotgun proteomics, a large number of proteins are first digested, then separated by HPLC, and finally analyzed, often by tandem MS.

Proteins need to be separated before MS analysis, and separation is usually accomplished by Liquid Chromatography (LC), High Performance LC (HPLC), or 2D gel electrophoresis. In order to identify proteins with varying abundance between two treatment groups, differential gel electrophoresis (DIGE) can be used, and DIGE can be followed by Matrix Assisted Laser Desorption–Time of Flight (MALDI-TOF) MS analysis (MALDI, matrix-assisted laser desorption/ionization, or TOF, time-of-flight mass spectrometer). In DIGE, proteins from two treatment groups are extracted, mixed with different colored dyes, usually CY3 and CY5, and subsequently run on a 2D polyacrylamide gel which separates proteins based on size and isoelectric focusing point (Gorg *et al.*, 2004) (**Figure 5**). Changes in protein expression can be inferred from changes in the color and intensity of "spots" on the gel, which usually represent one protein. Because the CY3 emission spectrum is in the green range and CY5 fluoresces in the red spectrum, proteins that are equally present in both treatments appear as yellow spots, while those that are up- or downregulated appear as orange spots, and those present in only one treatment group appear red or green. Algorithms have been developed to quantify the spot intensity and protein quantity (Gorg *et al.*, 2000; Herbert *et al.*, 2001; Patton and Beechem, 2002), but the identity of the protein remains unknown

and the spots must therefore be subjected to MS. Similar to mRNA expression measurement, changes in protein levels between two treatment groups must be analyzed statistically for significance.

Differential gel electrophoresis may be followed by peptide mass fingerprinting, or PMF. MALDI-TOF is often coupled to trypsin proteolysis, a bottom-up approach, which is simpler and has greater throughput than MS/MS. After extracting a spot from a 2D gel, the protein must be digested with trypsin, ionized, and finally introduced into the MS device. Introduction can be accomplished by MALDI, or electrospray ionization, and M/z detection may be accomplished by a Time of Flight (TOF) detector. After digestion, the peptide spot is added to a protective matrix. Next, a laser beam converts the protein from a solid molecule into a gas-phase ion with minimal damage to the protein. The matrix protects the protein by absorbing most of the laser energy, and ionizes the protein through a poorly understood mechanism which may involve charge transfer (Knochenmuss, 2006). Mixtures of proteins or digested peptides are further separated by the action of the laser, which only ionizes portions of the matrix, thus reducing the chance of different fragments entering the TOF analyzer at once.

In a typical MALDI-TOF analysis, the laser-based ionization of a peptide fragment accelerates ions into a vacuum where an electrical field is applied perpendicular to the direction of ionization. In this way, all ions have the same potential energy and velocity of zero in the axis towards the mass detector. Potential energy in the form of voltage is equally applied to the ions, which causes them to accelerate towards the TOF detector. Since the voltage applied is uniform, the velocity at which the ions travel is dependent on their mass and charge. The distance traveled from the field to the detector is constant for the

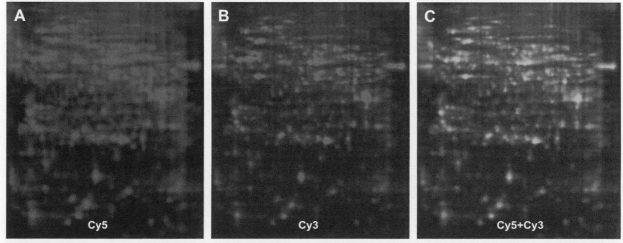

Figure 5 Two-dimensional differential in-gel electrophoresis (2D-DIGE) images of insecticide-susceptible (Cy5-labeled, Panel A) and resistant (Cy3-labeled, Panel B) SF-21 cells treated with insecticide. Panel C is an overlay of the two images. Equal amounts of protein in both cell lines appear yellow (C) and the proteins present in only resistant cells appear green (B), while only susceptible cells appear red (A). Reprinted with permission from Issaq and Veenstra (2008).

same MS instrument. Time is experimentally measured between application of the electric field and arrival at the mass detector. Time is therefore proportional to mass and charge.

The resulting data can often be used to identify proteins. However, the amino acid sequence cannot be determined, since the final peptide masses could result from a number of amino acid combinations. For PMF, a genomic sequence database is required to match the digested peptide mass against known proteins and open reading frames. Tandem mass spectrometry is a popular application for the identification, quantitation, and *de novo* sequencing of proteins. Protein mixtures need not be previously digested enzymatically, and some separation can be achieved by a preliminary mass analyzer inside the MS device. One type of mass analyzer is a quadropole ion trap, which uses DC and AC electrical fields and RF frequencies to trap or capture entering peptide ions. By changing the AC field frequency, peptides of different M/z ratios can be selected, and this is therefore the first M/z analysis, or MS in tandem MS, or MS/MS. In a typical peptide-sequencing experiment, an isolated, selected protein may be fragmented into smaller peptides or even amino acids. Fragmentation may be accomplished by collision-induced dissociation (CID), where the protein is bombarded with neutral ions. Fragmentation can occur at three predictable spots on the protein backbone. The smaller peptides are then caught in a final mass analyzer before detection. The final mass analyzer may be a TOF analyzer or a more sophisticated analyzer. Proteins for tandem MS can be enriched for post-translational modifications, or separated through a number of chromatographic steps. HPLC is often used to separate proteins immediately upstream of MS/MS, and when LC separation is performed on an entire proteome the technique is called shotgun proteomics. Ionization and introduction into MS/MS analyzers from LC separation can often be achieved by electrospray ionization, where the LC solvent evaporates and causes ionization without fragmentation.

1.4.1. Sample Protein Labeling and Separation

Quantification of protein expression changes between two unlabeled treatments is not possible using shotgun proteomics, because the identical proteins have identical M/z ratios. A number of techniques have been developed to uniquely label proteins from a treatment without altering their function. Most of these techniques are applicable to cell culture, while one has been applied to two whole organisms. Stable isotopic labeling in cell lines is a labeling technique that allows protein quantitation between two treatments (Mann, 2006). Cell cultures are supplemented with either natural amino acids (light chain) or stable isotope labeled amino acids which are then incorporated

into proteins (Ong *et al.*, 2002). Deuterium/hydrogen, 12C/13C, and 14N/15N are commonly used non-radioactive isotopes that can be combined to accommodate greater sample numbers. MS is sensitive enough to detect the small mass changes.

Other quantification methods have been developed that label the protein after extraction from the cell. Isotope Coded Affinity Tag (ICAT) makes use of a label that reacts with cysteines, separated by a linker group that contains either deuterium (heavy) or hydrogen (light), and a biotin affinity tag. Proteins are extracted and enzymatically digested, and cysteine containing peptides are purified using streptavadin, and finally subjected to MS (Gygi *et al.*, 1999). Bonaldi *et al.* (2008) used SILAC (stable isotope labeling by amino acids in cell culture) to analyze the *Drosophila* S2 cell line proteome with the use of RNAi, and found that label incorporation did not affect protein expression. Interestingly, overall protein levels changed with little correlation to mRNA changes; however, when statistically significant changes occurred between knockdown and control, the mRNA change was highly correlated with changes in protein concentration. Only two animals have been successfully labeled using SILAC: the mouse and the fruit fly (Gygi *et al.*, 1999; Sury *et al.*, 2010).

1.4.2. Enrichment for PTM

Analyzing an entire proteome from two SILAC treatments and detecting post-translational modifications can be complex due to database searching with an increased number of mass ranges that could uniquely identify a protein. Enrichment may reduce complexity by focusing efforts on a smaller subset of interesting proteins. Antibody-based enrichment is one such method, and antibodies for PTMs can be purchased or custom-designed for specific needs. Examples of PTM antibodies include anti-phosphotyrosine/serine, anti-ubiquitin, etc. Samples may be digested with trypsin before enrichment to further decrease complexity and non-specific interactions. Antibody enrichment can be achieved by either immuno-affinity purification or immunoprecipitation (Zhao and Jensen, 2009). Efficacy may vary between these methods, based on the PTM of interest. Phosphotyrosine proteins fare better when immunoprecipitation is used (Schumacher *et al.*, 2007).

1.4.3. Applications of Proteomics

In parallel to genomics, proteomics provides a global view of protein profiles in an organism. Moreover, newly developed proteomics technologies allow for the deciphering of complicated biological systems, including cellular protein–protein interaction networks and various post-translational modifications. Proteomics technologies have been applied to study protein expression patterns among different insect developmental stages (Zhao *et al.*, 2006; Li *et al.*, 2007; Zhang *et al.*, 2007; Chan and Foster, 2008; Li *et al.*, 2009; Wu *et al.*, 2009) and various insect tissues, such as reproductive tissues (Kelleher *et al.*, 2009; Takemori and Yamamoto, 2009), the nervous system salivary and silk glands (Zhang *et al.*, 2006; Almeras *et al.*, 2009), the cuticle (Holm and Sander, 1997), and hemolymph (Li *et al.*, 2006; Furusawa *et al.*, 2008a). Proteomics has been used to identify novel venom proteins (de Graaf *et al.*, 2010) and salivary gland proteins (Oleaga *et al.*, 2007; Carolan *et al.*, 2009), as well as royal jelly proteins from the honey bee (Furusawa *et al.*, 2008b; Li *et al.*, 2008b; Yu *et al.*, 2010). In addition, proteomics has been applied in studies on insect–plant and host–parasite interactions (Chen *et al.*, 2005; Biron *et al.*, 2005, 2006; Francis *et al.*, 2006; An Nguyen *et al.*, 2007). Interestingly, proteomic-based *de novo* gene discovery has been applied for identifying novel genes that are not predicted by genome annotation (Findlay *et al.*, 2009). The development of powerful phosphoproteomics techniques enables large-scale identification of post-translational modifications, such as phosphorylation (Fu *et al.*, 2009; Rewitz *et al.*, 2009). Insecticide resistance (e.g., Cry toxins produced by the soil bacterium *Bacillus thuringiensis*) has become a serious problem that threatens Bt-based pest control and management. It is important to understand the mode of action of Cry toxins, especially the interaction between Cry toxins and host defense systems. Several studies have applied proteomics technologies to discover Cry binding proteins (McNall and Adang, 2003; Krishnamoorthy *et al.*, 2007; Bayyareddy *et al.*, 2009; Chen *et al.*, 2009) and alterations of larval gut proteins between susceptible and resistant Indian meal moths (Candas *et al.*, 2003).

1.5. Structural Genomics

Structural genomics is the study of the three-dimensional structure of all proteins from a particular organism through a combination of experimental determination and *in silico* modeling. The goal of structural genomics is set by some (Vitkup *et al.*, 2001) as the ability to model 90% of the proteins within a genome through computational techniques using a much smaller number of carefully selected proteins representative of different protein families. Vitkup's survey concluded that, given the structural coverage in the Protein Data Bank (PDB, www.pdb.org; Berman *et al.*, 2002), only about 10% of the amino acids in a genome can be modeled. Based on the rate of 50 structures solved per week (Weissig and Bourne, 1999), and the observation that only 10 of these are non-redundant based on accepted definitions of protein families (Holm and Sander, 1997; Brenner and Levitt, 2000), a realistic application of structural genomics may lie decades in the future.

However, homology modeling is an effective tool for analyzing protein function, especially in the field of entomology. Insects represent a genetically diverse class of organisms, yet comparatively few insect protein structures have been solved to date. The time required to create an accurate homology model can be less than a week – sometimes even a day – and no specialized equipment is required. Models can yield information on ligand and substrate binding, their binding specificity, the evolutionary conservation of residues, the consequences of mutations in regard to pesticide resistance, and potential protein interactions, as well as elucidate targets of interest for further "wet" experiments.

Many of the limitations of modeling correlate with the template–target sequence identity and the subsequent difficulties in obtaining a correct alignment. For example, a protein with 70% sequence identity, or 70 amino acids the same in 100, may yield a target structure that is accurate enough for reasonable positioning of hydrogen atoms given a high-resolution template. Sequences with sequence identity as low as 20% could still be considered useful for many applications, especially when combined with comparative homology data. Docking ligands, pesticides, or drugs into these models is one such task.

Homology modeling has been applied to determine the substrate specificity for two different p450s in *Anopheles gambiae* which shared only 20% identity with their human template (Chiu *et al.*, 2008). P450s are a class of proteins which chemically alter a wide range of substrates, including pesticides, through hydroxylation to facilitate excretion. Mutations in a voltage-gated sodium channel from the house fly *Musca domestica* have been mapped onto homologous structures from mammals, and used to elucidate the role of these mutations in pesticide resistance (O'Reilly *et al.*, 2006). The aryl hydrocarbon receptor, a bHLH PAS transcription factor which controls the expression of proteins related to carcinogen decay, was successfully modeled, and a conserved ligand-binding domain was found. Through structure-based mutagenesis, residues involved in binding the carcinogenic xenobiotic TCDD were successfully elucidated (Pandini *et al.*, 2009). More generally, conservation of residues across evolutionarily diverse organisms, or between highly dissimilar paralogs, may indicate that the residue is important to maintain the three-dimensional fold involved in ligand or substrate binding, or protein–protein interactions.

Homology modeling assumes that the structure of a target protein can be solved based only on its primary amino acid sequence and its structural and evolutionary relatedness to a protein of known structure. Understanding the evolutionary relationships between template and target proteins, and the factors that drove their structural conservation, is extremely useful in homology modeling. Structure is usually more conserved than amino acid sequence, which is more conserved than nucleic acid sequence. One theory suggests that protein folds have evolved the robust ability to retain structure and function in spite of mutations (Taverna and Goldstein, 2002). Fragile folds that collapse in response to a few mutations might be selected against in favor of a robust protein fold which can evolve and adapt.

Proteins with 30% sequence identity, or 30 amino acids the same out of 100, will have similar folds (Sander and Schneider, 1991). Two sequences with greater than 25% sequence ID are considered highly related structures with true evolutionary homology, while those with less than 25% share some structural similarity and arguable homology (Sander and Schneider, 1991). Doolittle (1986) described this zone as the twilight zone, or a range of sequence identities that may be indicative of either divergent or convergent evolution. A sequence and structural analysis of proteins in the PDB found that structurally similar proteins could share sequence ID as little as 7–8%. Random amino acid sequences share about 4% sequence identity, and therefore the percentage of anchor residues, or those strictly required for structural relatedness, is actually only 3–4% (Rost, 1997).

A striking example of this statistic comes from the crystal structures of *E. coli* ribose- and lysine-binding proteins, which share the same fold despite little sequence identity (Kang *et al.*, 1992). Surprisingly, the majority of related homologous structures in the PDB share less than 45% sequence ID (Rost, 1997). Amino acid mutations have been speculated to occur in intrinsically disordered regions, or loops that have little tendency for secondary structure, and have therefore evolved to allow the retention of structure and function. This theory was proven wrong by simulations which showed that, on the contrary, secondary structural elements can be maintained despite mutation accumulations, and in fact mutations in IDPs were much more likely to introduce secondary structure where previously there was none (Schaefer *et al.*, 2010).

Inside of secondary structural elements, genetic drift appears to accumulate mutations in solvent-exposed regions with little functional value. A survey of the mutation rate for all amino acid types found that planar hydrophobic residues are the most conserved, followed by aliphatic residues. Charged residues were the least conserved residue type (Bowie *et al.*, 1990). Some proteins may fold by a mechanism called hydrophobic collapse, where hydrophobic residues nucleate the folding of a protein after or during translation by associating with each other and shielding themselves from water, and thereby shifting charged residues towards the outside (Nolting *et al.*, 1995; Eaton *et al.*, 1996). This process may explain why hydrophobic residues are well conserved.

Solvent-exposed residues are likely less well conserved unless they contribute to functional sites such as interaction interfaces. Sequence and structural conservation at protein–protein interaction interfaces is high. Histone proteins show greater than 98% sequence ID between humans and plants. Histones make ordered contacts with other histones and DNA itself, and thus there is high selection pressure on solvent-exposed residues. Lac repressor has two areas on its solvent-exposed surface that participate in interactions with the lac operator and inducer. These areas are conserved among members of the lac family, with little conservation elsewhere (Kisters-Woike *et al.*, 2000). Conserved patches of solvent-exposed residues can indicate protein interaction domains, and this fact has been exploited by a program called consurf, which can be used to predict interaction interfaces based on a carefully constructed phylogenetic tree, homology models, and a multiple sequence alignment. Interaction domains must evolve reciprocal surfaces in order to continue interacting (Landau *et al.*, 2005). Selection pressure increases as the number of binding partners utilizing the same domain increases more than one (Goh *et al.*, 2000; Kisters Woike 2000).

1.5.1. Analysis of Protein–Ligand Interactions

Small-molecule ligands usually bind in pockets (Kuntz *et al.*, 1982; Lewis, 1991). Ligand functional surfaces are often complementary to their binding space in terms of electrostatics and geometric shape (Altschul *et al.*, 1997). These surfaces are frequently rough in order to fit a large amount of surface area and potential hydrophobic contacts into a defined amount of space (Pettit and Bowie, 1999). Algorithms have been developed to find concave surfaces as potential ligand-binding pockets (Kuntz *et al.*, 1982; Peters *et al.*, 1996). Given the genetic diversity of insects, comparative homology modeling, or comparing the same protein from many different organisms, is a great tool to find ligand-binding pockets.

1.5.2. Cytochrome C: A Case Study

Taxonomists routinely use the protein cytochrome C for DNA bar-coding and species identification because its amino acid sequence tends to be highly conserved among related species, with little variation between members of the same species (Hebert *et al.*, 2003). Why is cytochrome c so conserved? The answer may partially lie in its size, the requirement for a heme-binding pocket, and its role as an interacting partner of proteins involved in both electron transport and apoptosis. As an electron transport protein, it binds a heme group, which can be oxidized or reduced to facilitate electron movement. Despite high sequence conservation, chimpanzee mitochondrial cytochrome oxidase systems suffer a 20% reduction in respiration capacity when introduced into human cell lines (Barrientos *et al.*, 1998). This suggests that the evolution of reciprocal protein interaction interfaces between nuclear and mitochondrial proteins is required. The large number of interacting partners may place conservative selection pressure on these solvent-exposed residues. In the cytochrome c core, 22 of 103 amino acids are implicated in direct heme binding and/or required for the shape and hydrophobicity of the heme pocket and the overall fold. These 22 residues are highly conserved. Two more residues are solvent-exposed charged residues that may participate in partner binding and orientation (Takano and Dickerson, 1981).

1.5.3. Selecting a Template Structure

One easy method for template selection is performing a PSI-BLAST search against the RCSB Protein Data Bank from the NCBI blast homepage. Position Specific Iterative Blast uses a position-specific score matrix derived from the query for sequence comparison against the database of interest. PSI-BLAST can pick up weaker evolutionary relationships, and can give equal weight to the different domains of a protein instead of reporting the stronger more numerous relationships for one domain. PSI-BLAST works by first performing a regular protein blast, and then creating a multiple sequence alignment on the blast data, which are then used to create the position specific score matrix (Altschul *et al.*, 1997; Schaffer *et al.*, 1999). Another convenient feature of PSI-BLAST searches from NCBI is the option to view conserved domains using the conserved domain detection algorithm (CDD). CDD employs Reverse Position Specific Iterated Blast, or Reverse PSI-BLAST or RPS-BLAST (Marchler-Bauer *et al.*, 2002). The two algorithms differ in the derivation of the position-specific score matrix from the database in RPS-BLAST and not from the query in PSI-BLAST (Schaffer, 1999). In the case of large multi-domain proteins it may not be necessary or even possible to model a whole protein due to little sequence conservation in intradomain regions. Some domains are known to fold and function independently of each other, and therefore it may not be necessary to model an entire protein.

1.5.4. Target–Template Sequence Alignment

Correct template–target sequence alignment is a critical factor in model quality. With greater than ~ 50% sequence ID, almost any algorithm will produce a suitable alignment (Rost, 1997) and thereby improve model accuracy. Alignment gaps are detrimental to the modeling process, and placing them in divergent or loop regions can improve model quality. The salign command in Modeller makes use of these two features, as well as placing gaps in solvent-exposed residues (Marti-Renom *et al.*, 2000; Sali, 1995).

1.5.5. Modeling Suite Choice

When choosing a homology modeling software suite, the user should consider the suite's accuracy and ease of use, and the algorithm employed. Target–template pairs with greater than 40% sequence identity produce similar structures regardless of the prediction server used. Modeling suites allow users more precise control over the modeling process, but often require knowledge of scripting languages. Users without sophisticated computer knowledge may want to choose packages with in-depth documentation and user support communities.

As sequence identity approaches the "twilight zone," modeling suite accuracy becomes more important. Servers such as I-Tasser (http://zhanglab.ccmb.med. umich.edu/I-TASSER/ [Zhang, 2008]) and Robetta (http://robetta.bakerlab.org [Kim *et al.*, 2004]), and the Modeller suite (http://www.salilab.org/modeller/ [Sali, 1995]), use an approach to backbone generation that places restraints on values of the model structure. Backbone bond length, and PHI PSI and OMEGA angles, are constricted so that they can fall within a range of values derived from the template structure and a database of sequence structure relationships, also called a probability function. Modeller uses conjugate gradient optimization, beginning with local restraints and extending to global restraints, to optimize Newtonian force. Information on commonly used modeling programs is included in **Table 5**.

1.5.6. Critical Assessment of Protein Structure

Critical Assessment of Protein Structure (CASP) (http://predictioncenter.org/casp8/groups_analysis.cgi) ranks the performance of prediction algorithms for completely automated servers. Some structural biologists choose to submit their experimentally determined structures for assessment in the contest prior to publication. Contestants are given the amino acid sequence of the target protein, and structure predictions are then made by either a human or server. The resulting structure files are compared to the previously determined structure by a number of algorithms, such as Dali (Holm and Rosenstrom, 2010) and Mammoth (Ortiz *et al.*, 2002; Lupyan *et al.*, 2005), which attempt to align the alpha carbon backbone or side chains and then determine the root mean square deviation (RMSD), or a derivative of RMSD, using the three difference dimensional coordinates of each structure file. Comparing two structure files can be somewhat subjective, and thus a number of alignment algorithms are employed. Alignment algorithms, and the databases of protein families that are often created with them, are useful for comparing models against other members to observe evolutionary traits. Protein family structures are also used in the beginning steps of modeling. After finding a suitable template, this structure can be compared to other members of the family.

Table 5 Commonly Used Modeling Programs

Name	Website URL	Citation	Server/user configured	Notes	CASP ranking
Modeller	http://www.salilab.org/modeller/	Sali, 1995	User configured Python scripts	User control is very high; great documentation and user-supported community	N/A
Itasser (Zhang Server)	http://zhanglab.ccmb.med.umich.edu/I-TASSER/	Zhang, 2008	Automated server	Threading approach allows structure predictions when template alignments are weak or non-existent	Highest server ranking in CASP 8; 5th overall
Robetta (BAKER-ROBETTA)	http://robetta.bakerlab.org/	Kim *et al.*, 2004	Automated server	Comparative and *de novo* modeling	2nd highest server rank in CASP 8; 22nd overall
PDFAMS	http://pd-fams.com/	Terashi *et al.*, 2007	User configured scripts, some automation available	Powerful; some software may need to be purchased	4th overall CASP 8
Swiss Model	http://swissmodel.expasy.org/workspace/	Bordoli *et al.*, 2009; Kiefer *et al.*, 2009	Automated server	Accessed via a user-friendly Web Workspace or Deepview (Swiss-PDB-Viewer), a program available in the Microsoft Windows OS	N/A

1.5.7. Structural Determination

X-ray crystallography and nuclear magnetic resonance (NMR) imaging are the two primary methods of structure determination. X-ray crystallography can be used on much larger proteins with much better resolution. Some proteins cannot be expressed in sufficient levels and purified to a level amenable to either crystallography or NMR imaging. Crystallography has the drawbacks that some proteins will not crystallize, and in some cases the structure may actually be modified, or stuck in a single conformation that is not necessarily indicative of the dynamic conformational shifts the protein undergoes.

NMR, on the other hand, can be used to capture many types of motion. Backbone amide shifts have been used to determine ligand binding. Deuterium exchange experiments can reveal the change in solvent accessibility of particular functional groups. Proteins are grown in media containing hydrogen, and NMR recordings are performed in a solution of deuterium-labeled H_2O. Hydrogen–deuterium exchange events can then be monitored. Another advantage of NMR is that structures are not modified by the crystallization process, and are viewed in a more natural aqueous environment. However, not all proteins are easily soluble in solution.

1.6. Metabolomics

Metabolomics involves the high-throughput characterization of all small-molecule metabolites and the products of biochemical pathways. The responses of biological systems to genetic or environmental changes are often reflected in their metabolic profiles. There are three major categories in metabolomics. The first is targeted metabolomics, which documents changes in metabolites in response to environmental conditions the insects encounter. The second, metabolic profiling, qualitatively and quantitatively evaluates metabolic collections. The third, metabolic profiling, collects and analyzes data from crude extracts to classify them based on all metabolites rather than separating them into individual metabolites. Gas chromatography and LC-MS are used for the identification and quantitation of metabolites. Nuclear magnetic resonance methods are employed for *de novo* identification of unknown metabolites. In insects, metabolomics could help in classification, studies on toxicology of insecticides, and safety testing of insecticides, and to monitor effects of genetic and environmental conditions on insect physiological processes.

1.7. Systems Biology

As stated earlier, systems biology takes a holistic view of a system or process by attempting to integrate all the data generated by various independent pathways technologies, and analyzing them together to formulate a hypothesis or model. Researchers working on insects have just begun to apply the systems biology approach to achieve an integrated view on the functioning of insect physiological systems. One such example is the recent study on *D. melanogaster* phagasome. Upon encountering microbes or other antigens, phagocytes internalize these particles into phagosomes to initiate destruction of these immune agents. Stuart and colleagues applied the systems biology approach to address the complex dynamic interactions between proteins present in the phagosomes and their involvement in particle engulfment (Stuart *et al.*, 2007). This analysis identified 617 proteins associated with *D. melanogaster* phagosomes. The 617 phagosome proteins were used to prepare a detailed protein–protein interaction network, and 214 of the 617 phagosome proteins were mapped to a protein–protein interaction network. RNA interference was then employed to determine the contribution of each protein in microbe internalization. RNA interference studies identified gene coding for proteins that are known to function in phagocytosis. In addition, these studies also identified novel regulators of phagocytosis. These pioneering systems biology studies have provided new insights into functional organization of phagosomes. Such holistic approaches applied to various physiological systems in insects may lead to better understanding of the functioning of these systems.

1.8. Conclusions and Future Prospects

The rapid development of next generation sequencing (NGS) technologies during the past four years, following the domination of the automated Sanger sequencing method for almost two decades, could revolutionize the way of thinking about scientific approaches in insect research. The impact of the introduction of NGS technologies into the market is similar to the early days of PCR, with imagination being the only limiting factor for their use. It will be possible to sequence genomes of insects at $1000/genome in the not too distant future. The availability of genome sequences of almost every insect species of interest will help with research in every field of entomology. Advances in omics fields, as well as both forward and reverse genetics and RNA interference (covered in Chapter 2 in this volume) approaches, will also help in advances in research on insects. In the near future, molecular phylogenetics studies will use whole-genome sequences for insect taxonomy. Neurobiologists and physiologists will use systems biology approaches

to understand the complexity of neuronal signaling and other physiological processes.

Acknowledgments

We apologize to those whose work could not be cited owing to space limitations. The research in the Palli laboratory was supported by the National Science Foundation (IBN-0421856), the National Institute of Health (GM070559-06), and the National Research Initiative of the USDA-CSREES (2007-04636). This report is contribution number 11-08-036 from the Kentucky Agricultural Experimental Station.

References

Adams, H. A., Southey, B. R., Robinson, G. E., & Rodriguez-Zas, S. L. (2008). Meta-analysis of genome-wide expression patterns associated with behavioral maturation in honey bees. *BMC Genomics, 9,* 503.

Adams, M. D., Celniker, S. E., Holt, R. A., Evans, C. A., et al. (2000). The genome sequence of *Drosophila melanogaster. Science, 287,* 2185–2195.

Ahsan, B., Saito, T. L., Hashimoto, S., Muramatsu, K., Tsuda, M., et al. (2009). MachiBase: A *Drosophila melanogaster* 5'-end mRNA transcription database. *Nucleic Acids Res., 37,* D49–53.

Alekseyenko, A. A., Peng, S., Larschan, E., Gorchakov, A. A., Lee, O. K., et al. (2008). A sequence motif within chromatin entry sites directs MSL establishment on the *Drosophila* X chromosome. *Cell, 134,* 599–609.

Almeras, L., Fontaine, A., Belghazi, M., Bourdon, S., Boucomont-Chapeaublanc, E., et al. (2009). Salivary gland protein repertoire from *Aedes aegypti* mosquitoes. *Vector Borne Zoonotic Dis., 10,* 391–402.

Altenhein, B., Becker, A., Busold, C., Beckmann, B., Hoheisel, J. D., & Technau, G. M. (2006). Expression profiling of glial genes during *Drosophila* embryogenesis. *Dev. Biol., 296,* 545–560.

Altschul, S. F., Gish, W., Miller, W., Myers, E. W., & Lipman, D. J. (1990). Basic local alignment search tool. *J. Mol. Biol., 215,* 403–410.

Altschul, S. F., Madden, T. L., Schaffer, A. A., Zhang, J., Zhang, Z., et al. (1997). Gapped BLAST and PSI-BLAST: A new generation of protein database search programs. *Nucleic Acids Res., 25,* 3389–3402.

An Nguyen, T. T., Michaud, D., & Cloutier, C. (2007). Proteomic profiling of aphid *Macrosiphum euphorbiae* responses to host-plant-mediated stress induced by defoliation and water deficit. *J. Insect Physiol., 53,* 601–611.

Arensburger, P., Megy, K., Waterhouse, R. M., Abrudan, J., Amedeo, P., Antelo, B., et al. (2010). Sequencing of *Culex quinquefasciatus* establishes a platform for mosquito comparative genomics. *Science, 330,* 86–88.

Ashburner, M. (2007). *Drosophila* genomes by the baker's dozen. *Genetics, 177,* 1263–1268.

Ashburner, M., Ball, C. A., Blake, J. A., Botstein, D., Butler, H., et al. (2000). Gene ontology: Tool for the unification of biology. The Gene Ontology Consortium. *Nat. Genet., 25,* 25–29.

Avet-Rochex, A., Boyer, K., Polesello, C., Gobert, V., Osman, D., et al. (2010). An *in vivo* RNA interference screen identifies gene networks controlling *Drosophila melanogaster* blood cell homeostasis. *BMC Dev. Biol., 10,* 65.

Bai, H., & Palli, S. R. (2010). Functional characterization of bursicon receptor and genome-wide analysis for identification of genes affected by bursicon receptor RNAi. *Dev. Biol., 344,* 248–258.

Barat-Houari, M., Hilliou, F., Jousset, F. X., Sofer, L., Deleury, E., et al. (2006). Gene expression profiling of *Spodoptera frugiperda* hemocytes and fat body using cDNA microarray reveals polydnavirus-associated variations in lepidopteran host genes transcript levels. *BMC Genomics, 7,* 160.

Barrientos, A., Kenyon, L., & Moraes, C. T. (1998). Human xenomitochondrial cybrids. Cellular models of mitochondrial complex I deficiency. *J. Biol. Chem., 273,* 14210–14217.

Bayyareddy, K., Andacht, T. M., Abdullah, M. A., & Adang, M. J. (2009). Proteomic identification of *Bacillus thuringiensis* subsp. *israelensis* toxin Cry4Ba binding proteins in midgut membranes from *Aedes* (*Stegomyia*) *aegypti* Linnaeus (Diptera, Culicidae) larvae. *Insect Biochem. Mol. Biol., 39,* 279–286.

Benjamini, Y., & Hochberg, Y. (1995). Controlling the false discovery rate: A practical and powerful approach to multiple testing. *J.R. Stat. Soc. B, 57,* 289–300.

Berman, H. M., Battistuz, T., Bhat, T. N., Bluhm, W. F., Bourne, P. E., et al. (2002). The Protein Data Bank. *Acta Crystallogr. D Biol. Crystallogr., 58,* 899–907.

Beumer, K., Bhattacharyya, G., Bibikova, M., Trautman, J. K., & Carroll, D. (2006). Efficient gene targeting in *Drosophila* with zinc-finger nucleases. *Genetics, 172,* 2391–2403.

Bibikova, M., Beumer, K., Trautman, J. K., & Carroll, D. (2003). Enhancing gene targeting with designed zinc finger nucleases. *Science, 300,* 764.

Biron, D. G., Marche, L., Ponton, F., Loxdale, H. D., Galeotti, N., et al. (2005). Behavioural manipulation in a grasshopper harbouring hairworm: A proteomics approach. *Proc. Biol. Sci., 272,* 2117–2126.

Biron, D. G., Ponton, F., Marche, L., Galeotti, N., Renault, L., et al. (2006). "Suicide" of crickets harbouring hairworms: A proteomics investigation. *Insect Mol. Biol., 15,* 731–742.

Bonaldi, T., Straub, T., Cox, J., Kumar, C., Becker, P. B., & Mann, M. (2008). Combined use of RNAi and quantitative proteomics to study gene function in *Drosophila. Mol. Cell., 31,* 762–772.

Bordoli, L., Kiefer, F., Arnold, K., Benkert, P., Battey, J., & Schwede, T. (2009). Protein structure homology modeling using SWISS-MODEL workspace. *Nat. Protoc., 4,* 1–13.

Bowie, J. U., Reidhaar-Olson, J. F., Lim, W. A., & Sauer, R. T. (1990). Deciphering the message in protein sequences: Tolerance to amino acid substitutions. *Science, 247,* 1506–1510.

Brenner, S. E., & Levitt, M. (2000). Expectations from structural genomics. *Protein Sci., 9,* 197–200.

Burge, C., & Karlin, S. (1997). Prediction of complete gene structures in human genomic DNA. *J. Mol. Biol., 268,* 78–94.

Butler, M. J., Jacobsen, T. L., Cain, D. M., Jarman, M. G., Hubank, M., et al. (2003). Discovery of genes with highly restricted expression patterns in the *Drosophila* wing disc using DNA oligonucleotide microarrays. *Development, 130,* 659–670.

Candas, M., Loseva, O., Oppert, B., Kosaraju, P., & Bulla, L. A., Jr. (2003). Insect resistance to *Bacillus thuringiensis*: Alterations in the indianmeal moth larval gut proteome. *Mol. Cell Proteomics, 2,* 19–28.

Carolan, J. C., Fitzroy, C. I., Ashton, P. D., Douglas, A. E., & Wilkinson, T. L. (2009). The secreted salivary proteome of the pea aphid *Acyrthosiphon pisum* characterised by mass spectrometry. *Proteomics, 9,* 2457–2467.

Celniker, S. E., Dillon, L. A., Gerstein, M. B., Gunsalus, K. C., Henikoff, S., et al. (2009). Unlocking the secrets of the genome. *Nature, 459,* 927–930.

Chan, Q. W., & Foster, L. J. (2008). Changes in protein expression during honey bee larval development. *Genome Biol., 9,* R156.

Chen, H., Wilkerson, C. G., Kuchar, J. A., Phinney, B. S., & Howe, G. A. (2005). Jasmonate-inducible plant enzymes degrade essential amino acids in the herbivore midgut. *Proc. Natl. Acad. Sci. USA, 102,* 19237–19242.

Chen, K., & Pachter, L. (2005). Bioinformatics for whole-genome shotgun sequencing of microbial communities. *PLoS Comput. Biol., 1,* 106–112.

Chen, L. Z., Liang, G. M., Zhang, J., Wu, K. M., Guo, Y. Y., & Rector, B. G. (2009). Proteomic analysis of novel Cry1Ac binding proteins in *Helicoverpa armigera* (Hubner). *Arch. Insect Biochem. Physiol., 73,* 61–73.

Chintapalli, V. R., Wang, J., & Dow, J. A. (2007). Using Fly-Atlas to identify better *Drosophila melanogaster* models of human disease. *Nat. Genet., 39,* 715–720.

Chiu, T. L., Wen, Z., Rupasinghe, S. G., & Schuler, M. A. (2008). Comparative molecular modeling of *Anopheles gambiae* CYP6Z1, a mosquito P450 capable of metabolizing DDT. *Proc. Natl. Acad. Sci. USA, 105,* 8855–8860.

Clement, N. L., Snell, Q., Clement, M. J., Hollenhorst, P. C., Purwar, J., et al. (2010). The GNUMAP algorithm: Unbiased probabilistic mapping of oligonucleotides from next-generation sequencing. *Bioinformatics, 26,* 38–45.

Conesa, A., & Gotz, S. (2008). Blast2GO: A comprehensive suite for functional analysis in plant genomics. *Intl J. Plant Genomics, 2008,* 619–832.

Conesa, A., Gotz, S., Garcia-Gomez, J. M., Terol, J., Talon, M., & Robles, M. (2005). Blast2GO: A universal tool for annotation, visualization and analysis in functional genomics research. *Bioinformatics, 21,* 3674–3676.

Costello, J. C., Dalkilic, M. M., Beason, S. M., Gehlhausen, J. R., Patwardhan, R., Middha, S., et al. (2009). Gene networks in *Drosophila melanogaster*: Integrating experimental data to predict gene function. *Genome Biol., 10,* R97.

Cox-Foster, D. L., Conlan, S., Holmes, E. C., Palacios, G., Evans, J. D., et al. (2007). A metagenomic survey of microbes in honey bee colony collapse disorder. *Science, 318,* 283–287.

Dana, A. N., Hong, Y. S., Kern, M. K., Hillenmeyer, M. E., Harker, B. W., et al. (2005). Gene expression patterns associated with blood-feeding in the malaria mosquito *Anopheles gambiae. BMC Genomics, 6,* 5.

David, J. P., Coissac, E., Melodelima, C., Poupardin, R., Riaz, M. A., et al. (2010). Transcriptome response to pollutants and insecticides in the dengue vector *Aedes aegypti* using next-generation sequencing technology. *BMC Genomics, 11,* 216.

de Graaf, D. C., Aerts, M., Brunain, M., Desjardins, C. A., Jacobs, F. J., et al. (2010). Insights into the venom composition of the ectoparasitoid wasp *Nasonia vitripennis* from bioinformatic and proteomic studies. *Insect Mol. Biol.*, *19*(Suppl. 1), 11–26.

Dierick, H. A., & Greenspan, R. J. (2006). Molecular analysis of flies selected for aggressive behavior. *Nat. Genet.*, *38*, 1023–1031.

Djouaka, R. F., Bakare, A. A., Coulibaly, O. N., Akogbeto, M. C., Ranson, H., et al. (2008). Expression of the cytochrome P450s, CYP6P3 and CYP6M2 are significantly elevated in multiple pyrethroid resistant populations of *Anopheles gambiae* s.s. from Southern Benin and Nigeria. *BMC Genomics*, *9*, 538.

Doolittle, R. F. (1986). *Of URFs and ORFs: A primer on how to analyze derived amino acid sequences*. Mill Valley, CA: University Science Books.

Drosophila 12 Genome Consortium (2007). Evolution of genes and genomes on the *Drosophila* phylogeny. *Nature*, *450*, 203–218.

Eaton, W. A., Thompson, P. A., Chan, C. K., Hage, S. J., & Hofrichter, J. (1996). Fast events in protein folding. *Structure*, *4*, 1133–1139.

Eisen, M. B., Spellman, P. T., Brown, P. O., & Botstein, D. (1998). Cluster analysis and display of genome-wide expression patterns. *Proc. Natl. Acad. Sci. USA*, *95*, 14863–14868.

Etter, P. D., & Ramaswami, M. (2002). The ups and downs of daily life: Profiling circadian gene expression in *Drosophila*. *Bioessays*, *24*, 494–498.

Feyereisen, R. (2006). Evolution of insect P450. *Biochem. Soc. Trans.*, *34*, 1252–1255.

Findlay, G. D., MacCoss, M. J., & Swanson, W. J. (2009). Proteomic discovery of previously unannotated, rapidly evolving seminal fluid genes in *Drosophila*. *Genome Res.*, *19*, 886–896.

Finn, R. D., Mistry, J., Tate, J., Coggill, P., Heger, A., et al. (2010). The Pfam protein families database. *Nucleic Acids Res.*, *38*, D211–122.

Francis, F., Gerkens, P., Harmel, N., Mazzucchelli, G., De Pauw, E., & Haubruge, E. (2006). Proteomics in *Myzus persicae*: Effect of aphid host plant switch. *Insect Biochem. Mol. Biol.*, *36*, 219–227.

Fu, Q., Liu, P. C., Wang, J. X., Song, Q. S., & Zhao, X. F. (2009). Proteomic identification of differentially expressed and phosphorylated proteins in epidermis involved in larval–pupal metamorphosis of *Helicoverpa armigera*. *BMC Genomics*, *10*, 600.

Fujikawa, K., Takahashi, A., Nishimura, A., Itoh, M., Takano-Shimizu, T., & Ozaki, M. (2009). Characteristics of genes up-regulated and down-regulated after 24 h starvation in the head of *Drosophila*. *Gene*, *446*, 11–17.

Furlong, E. E., Andersen, E. C., Null, B., White, K. P., & Scott, M. P. (2001). Patterns of gene expression during *Drosophila* mesoderm development. *Science*, *293*, 1629–1633.

Furusawa, T., Rakwal, R., Nam, H. W., Hirano, M., Shibato, J., et al. (2008a). Systematic investigation of the hemolymph proteome of *Manduca sexta* at the fifth instar larvae stage using one- and two-dimensional proteomics platforms. *J. Proteome Res.*, *7*, 938–959.

Furusawa, T., Rakwal, R., Nam, H. W., Shibato, J., Agrawal, G. K., et al. (2008b). Comprehensive royal jelly (RJ) proteomics using one- and two-dimensional proteomics platforms reveals novel RJ proteins and potential phospho/glycoproteins. *J. Proteome Res.*, *7*, 3194–3229.

Giardine, B., Riemer, C., Hardison, R. C., Burhans, R., Elnitski, L., et al. (2005). Galaxy: A platform for interactive large-scale genome analysis. *Genome Res.*, *15*, 1451–1455.

Goh, C. S., Bogan, A. A., Joachimiak, M., Walther, D., & Cohen, F. E. (2000). Co-evolution of proteins with their interaction partners. *J. Mol. Biol.*, *299*, 283–293.

Golic, M. M., Rong, Y. S., Petersen, R. B., Lindquist, S. L., & Golic, K. G. (1997). FLP-mediated DNA mobilization to specific target sites in *Drosophila* chromosomes. *Nucleic Acids Res.*, *25*, 3665–3671.

Gong, W. J., & Golic, K. G. (2003). Ends-out, or replacement, gene targeting in *Drosophila*. *Proc. Natl. Acad. Sci. USA*, *100*, 2556–2561.

Gorg, A., Obermaier, C., Boguth, G., Harder, A., Scheibe, B., et al. (2000). The current state of two-dimensional electrophoresis with immobilized pH gradients. *Electrophoresis*, *21*, 1037–1053.

Gorg, A., Weiss, W., & Dunn, M. J. (2004). Current two-dimensional electrophoresis technology for proteomics. *Proteomics*, *4*, 3665–3685.

Groth, A. C., Fish, M., Nusse, R., & Calos, M. P. (2004). Construction of transgenic *Drosophila* by using the site-specific integrase from phage phiC31. *Genetics*, *166*, 1775–1782.

Gundersen-Rindal, D. E., & Pedroni, M. J. (2010). Larval stage *Lymantria dispar* microRNAs differentially expressed in response to parasitization by *Glyptapanteles flavicoxis* parasitoid. *Arch. Virol.*, *155*, 783–787.

Gygi, S. P., Rist, B., Gerber, S. A., Turecek, F., Gelb, M. H., & Aebersold, R. (1999). Quantitative analysis of complex protein mixtures using isotope-coded affinity tags. *Nat. Biotechnol.*, *17*, 994–999.

Hebbes, T. R., Thorne, A. W., & Crane-Robinson, C. (1988). A direct link between core histone acetylation and transcriptionally active chromatin. *EMBO J*, *7*, 1395–1402.

Hebert, P. D., Ratnasingham, S., & deWaard, J. R. (2003). Barcoding animal life: Cytochrome c oxidase subunit 1 divergences among closely related species. *Proc. Biol. Sci.*, *270*(Suppl. 1), S96–S99.

Held, M., Gase, K., & Baldwin, I. T. (2004). Microarrays in ecological research: A case study of a cDNA microarray for plant–herbivore interactions. *BMC Ecol.*, *4*, 13.

Herbert, B. R., Harry, J. L., Packer, N. H., Gooley, A. A., Pedersen, S. K., & Williams, K. L. (2001). What place for polyacrylamide in proteomics? *Trends Biotechnol.*, *19*, S3–9.

Holm, L., & Rosenstrom, P. (2010). Dali server: Conservation mapping in 3D. *Nucleic Acids Res.*, *38*(Suppl.), W545–549.

Holm, L., & Sander, C. (1997). Dali/FSSP classification of three-dimensional protein folds. *Nucleic Acids Res.*, *25*, 231–234.

Holt, R. A., Subramanian, G. M., Halpern, A., Sutton, G. G., Charlab, R., et al. (2002). The genome sequence of the malaria mosquito *Anopheles gambiae*. *Science*, *298*, 129–149.

Issaq, H. J., & Veenstra, T. D. (2008). Two-dimensional poly-acrylamide gel electrophoresis (2D-PAGE): Advances and perspectives. *Biotechniques, 44*, 697–699.

Iyer, V. R., Horak, C. E., Scafe, C. S., Botstein, D., Snyder, M., & Brown, P. O. (2001). Genomic binding sites of the yeast cell-cycle transcription factors SBF and MBF. *Nature, 409*, 533–538.

Jiang, H., Wang, F., Dyer, N. P., & Wong, W. H. (2010). Cis-Genome Browser: A flexible tool for genomic data visualization. *Bioinformatics, 26*, 1781–1782.

Kanehisa, M., & Goto, S. (2000). KEGG: Kyoto Encyclopedia of Genes and Genomes. *Nucleic Acids Res., 28*, 27–30.

Kanehisa, M., Goto, S., Hattori, M., Aoki-Kinoshita, K. F., Itoh, M., et al. (2006). From genomics to chemical genomics: New developments in KEGG. *Nucleic Acids Res., 34*, D354–357.

Kang, C. H., Gokcen, S., & Ames, G. F. (1992). Crystallization and preliminary X-ray studies of the liganded lysine, arginine, ornithine-binding protein from *Salmonella typhimurium. J. Mol. Biol., 225*, 1123–1125.

Kankare, M., Salminen, T., Laiho, A., Vesala, L., & Hoikkala, A. (2010). Changes in gene expression linked with adult reproductive diapause in a northern malt fly species: A candidate gene microarray study. *BMC Ecol., 10*, 3.

Karres, J. S., Hilgers, V., Carrera, I., Treisman, J., & Cohen, S. M. (2007). The conserved microRNA miR-8 tunes atrophin levels to prevent neurodegeneration in *Drosophila. Cell, 131*, 136–145.

Kawasaki, H., Ote, M., Okano, K., Shimada, T., Guo-Xing, Q., & Mita, K. (2004). Change in the expressed gene patterns of the wing disc during the metamorphosis of *Bombyx mori. Gene, 343*, 133–142.

Kelleher, E. S., Watts, T. D., LaFlamme, B. A., Haynes, P. A., & Markow, T. A. (2009). Proteomic analysis of *Drosophila mojavensis* male accessory glands suggests novel classes of seminal fluid proteins. *Insect Biochem. Mol. Biol., 39*, 366–371.

Kiefer, F., Arnold, K., Kunzli, M., Bordoli, L., & Schwede, T. (2009). The SWISS-MODEL Repository and associated resources. *Nucleic Acids Res., 37*, D387–392.

Kiger, A. A., Baum, B., Jones, S., Jones, M. R., Coulson, A., et al. (2003). A functional genomic analysis of cell morphology using RNA interference. *J. Biol., 2*, 27.

Kijimoto, T., Costello, J., Tang, Z., Moczek, A. P., & Andrews, J. (2009). EST and microarray analysis of horn development in Onthophagus beetles. *BMC Genomics, 10*, 504.

Kim, D. E., Chivian, D., & Baker, D. (2004). Protein structure prediction and analysis using the Robetta server. *Nucleic Acids Res., 32*, W526–531.

Kirkness, E. F., Haas, B. J., Sun, W., Braig, H. R., Perotti, M. A., et al. (2010). Genome sequences of the human body louse and its primary endosymbiont provide insights into the permanent parasitic lifestyle. *Proc. Natl. Acad. Sci. USA, 107*, 12168–12173.

Kisters-Woike, B., Vangierdegom, C., & Muller-Hill, B. (2000). On the conservation of protein sequences in evolution. *Trends Biochem. Sci., 25*, 419–421.

Knochenmuss, R. (2006). Ion formation mechanisms in UV-MALDI. *Analyst, 131*, 966–986.

Kocher, S. D., Richard, F. J., Tarpy, D. R., & Grozinger, C. M. (2008). Genomic analysis of post-mating changes in the honey bee queen (*Apis mellifera*). *BMC Genomics, 9*, 232.

Korf, I. (2004). Gene finding in novel genomes. *BMC Bioinformatics, 5*, 59.

Krishnamoorthy, M., Jurat-Fuentes, J. L., McNall, R. J., Andacht, T., & Adang, M. J. (2007). Identification of novel Cry1Ac binding proteins in midgut membranes from *Heliothis virescens* using proteomic analyses. *Insect Biochem. Mol. Biol., 37*, 189–201.

Kuntz, I. D., Blaney, J. M., Oatley, S. J., Langridge, R., & Ferrin, T. E. (1982). A geometric approach to macromolecule–ligand interactions. *J. Mol. Biol., 161*, 269–288.

Landau, M., Mayrose, I., Rosenberg, Y., Glaser, F., Martz, E., et al. (2005). ConSurf 2005: The projection of evolutionary conservation scores of residues on protein structures. *Nucleic Acids Res., 33*, W299–302.

Langmead, B., Trapnell, C., Pop, M., & Salzberg, S. L. (2009). Ultrafast and memory-efficient alignment of short DNA sequences to the human genome. *Genome Biol., 10*, R25.

Lawniczak, M. K., & Begun, D. J. (2004). A genome-wide analysis of courting and mating responses in *Drosophila melanogaster* females. *Genome, 47*, 900–910.

Lewis, R. A. (1991). Clefts and binding sites in protein receptors. *Methods Enzymol., 202*, 126–156.

Li, A. Q., Popova-Butler, A., Dean, D. H., & Denlinger, D. L. (2007). Proteomics of the flesh fly brain reveals an abundance of upregulated heat shock proteins during pupal diapause. *J. Insect Physiol., 53*, 385–391.

Li, J., Zhang, L., Feng, M., Zhang, Z., & Pan, Y. (2009). Identification of the proteome composition occurring during the course of embryonic development of bees (*Apis mellifera*). *Insect Mol. Biol., 18*, 1–9.

Li, R., Li, Y., Kristiansen, K., & Wang, J. (2008a). SOAP: Short oligonucleotide alignment program. *Bioinformatics, 24*, 713–714.

Li, J. K., Feng, M., Zhang, L., Zhang, Z. H., & Pan, Y. H. (2008b). Proteomics analysis of major royal jelly protein changes under different storage conditions. *J. Proteome Res., 7*, 3339–3353.

Li, X. H., Wu, X. F., Yue, W. F., Liu, J. M., Li, G. L., & Miao, Y. G. (2006). Proteomic analysis of the silkworm (*Bombyx mori* L.) hemolymph during developmental stage. *J. Proteome Res., 5*, 2809–2814.

Lieb, J. D., Liu, X., Botstein, D., & Brown, P. O. (2001). Promoter-specific binding of Rap1 revealed by genome-wide maps of protein–DNA association. *Nat. Genet., 28*, 327–334.

Liu, Y. H., Jakobsen, J. S., Valentin, G., Amarantos, I., Gilmour, D. T., & Furlong, E. E. (2009). A systematic analysis of Tinman function reveals Eya and JAK-STAT signaling as essential regulators of muscle development. *Dev. Cell, 16*, 280–291.

Lo Conte, L., Brenner, S. E., Hubbard, T. J.P., Chothia, C., & Murzin, A. G. (2002). SCOP database in 2002: Refinements accommodate structural genomics. *Nucleic Acids Res., 30*, 264–267.

Lupyan, D., Leo-Macias, A., & Ortiz, A. R. (2005). A new progressive-iterative algorithm for multiple structure alignment. *Bioinformatics, 21*, 3255–3263.

Mahadav, A., Gerling, D., Gottlieb, Y., Czosnek, H., & Ghanim, M. (2008). Parasitization by the wasp *Eretmocerus mundus* induces transcription of genes related to immune response and symbiotic bacteria proliferation in the whitefly *Bemisia tabaci. BMC Genomics, 9*, 342.

Mann, M. (2006). Functional and quantitative proteomics using SILAC. *Nat. Rev. Mol. Cell. Biol., 7*, 952–958.

Marchler-Bauer, A., Panchenko, A. R., Shoemaker, B. A., Thiessen, P. A., Geer, L. Y., & Bryant, S. H. (2002). CDD: A database of conserved domain alignments with links to domain three-dimensional structure. *Nucleic Acids Res., 30*, 281–283.

Margulies, M., Egholm, M., Altman, W. E., Attiya, S., Bader, J. S., et al. (2005). Genome sequencing in microfabricated high-density picolitre reactors. *Nature, 437*, 376–380.

Martin, A. C., Orengo, C. A., Hutchinson, E. G., Jones, S., Karmirantzou, M., et al. (1998). Protein folds and functions. *Structure, 6*, 875–884.

Marti-Renom, M. A., Stuart, A. C., Fiser, A., Sanchez, R., Melo, F., & Sali, A. (2000). Comparative protein structure modeling of genes and genomes. *Annu. Rev. Biophys. Biomol. Struct. 29*, 291–325.

McDonald, M. J., & Rosbash, M. (2001). Microarray analysis and organization of circadian gene expression in *Drosophila. Cell, 107*, 567–578.

McNall, R. J., & Adang, M. J. (2003). Identification of novel *Bacillus thuringiensis* Cry1Ac binding proteins in *Manduca sexta* midgut through proteomic analysis. *Insect Biochem. Mol. Biol., 33*, 999–1010.

Metzker, M. L. (2010). Sequencing technologies – the next generation. *Nat. Rev. Genet., 11*, 31–46.

Meyer, E., Aglyamova, G. V., Wang, S., Buchanan-Carter, J., Abrego, D., et al. (2009). Sequencing and *de novo* analysis of a coral larval transcriptome using 454 GSFlx. *BMC Genomics, 10*, 219.

Mita, K., Kasahara, M., Sasaki, S., Nagayasu, Y., Yamada, T., et al. (2004). The genome sequence of silkworm, *Bombyx mori. DNA Res., 11*, 27–35.

Negre, N., Hennetin, J., Sun, L. V., Lavrov, S., Bellis, M., et al. (2006). Chromosomal distribution of PcG proteins during *Drosophila* development. *PLoS Biol., 4*, e170.

Nene, V., Wortman, J. R., Lawson, D., Haas, B., Kodira, C., et al. (2007). Genome sequence of *Aedes aegypti*, a major arbovirus vector. *Science, 316*, 1718–1723.

Nolting, B., Golbik, R., & Fersht, A. R. (1995). Submillisecond events in protein folding. *Proc. Natl. Acad. Sci. USA, 92*, 10668–10672.

O'Geen, H., Nicolet, C. M., Blahnik, K., Green, R., & Farnham, P. J. (2006). Comparison of sample preparation methods for ChIP–chip assays. *Biotechniques, 41*, 577–580.

O'Reilly, A. O., Khambay, B. P., Williamson, M. S., Field, L. M., Wallace, B. A., & Davies, T. G. (2006). Modelling insecticide-binding sites in the voltage-gated sodium channel. *Biochem. J., 396*, 255–263.

Oleaga, A., Escudero-Poblacion, A., Camafeita, E., & Perez-Sanchez, R. (2007). A proteomic approach to the identification of salivary proteins from the argasid ticks *Ornithodoros moubata* and *Ornithodoros erraticus. Insect Biochem. Mol. Biol., 37*, 1149–1159.

Ong, S. E., Blagoev, B., Kratchmarova, I., Kristensen, D. B., Steen, H., et al. (2002). Stable isotope labeling by amino acids in cell culture, SILAC, as a simple and accurate approach to expression proteomics. *Mol. Cell Proteomics, 1*, 376–386.

Ortiz, A. R., Strauss, C. E., & Olmea, O. (2002). MAMMOTH (Matching Molecular Models Obtained from Theory): An automated method for model comparison. *Protein Sci., 11*, 2606–2621.

Pan, F., Chiu, C. H., Pulapura, S., Mehan, M. R., Nunez-Iglesias, J., et al. (2007). Gene Aging Nexus: A web database and data mining platform for microarray data on aging. *Nucleic Acids Res., 35*, D756–759.

Pandini, A., Soshilov, A. A., Song, Y., Zhao, J., Bonati, L., & Denison, M. S. (2009). Detection of the TCDD binding-fingerprint within the Ah receptor ligand binding domain by structurally driven mutagenesis and functional analysis. *Biochemistry, 48*, 5972–5983.

Parthasarathy, R., Sheng, Z., Sun, Z., & Palli, S. R. (2010a). Ecdysteroid regulation of ovarian growth and oocyte maturation in the red flour beetle, *Tribolium castaneum. Insect Biochem. Mol. Biol., 40*, 429–439.

Parthasarathy, R., Sun, Z., Bai, H., & Palli, S. R. (2010b). Juvenile hormone regulation of vitellogenin synthesis in the red flour beetle, *Tribolium castaneum. Insect Biochem. Mol. Biol., 40*, 405–414.

Patton, W. F., & Beechem, J. M. (2002). Rainbow's end: The quest for multiplexed fluorescence quantitative analysis in proteomics. *Curr. Opin. Chem. Biol., 6*, 63–69.

Peters, K. P., Fauck, J., & Frommel, C. (1996). The automatic search for ligand binding sites in proteins of known three-dimensional structure using only geometric criteria. *J. Mol. Biol., 256*, 201–213.

Pettit, F. K., & Bowie, J. U. (1999). Protein surface roughness and small molecular binding sites. *J. Mol. Biol., 285*, 1377–1382.

Pletcher, S. D., Macdonald, S. J., Marguerie, R., Certa, U., Stearns, S. C., et al. (2002). Genome-wide transcript profiles in aging and calorically restricted *Drosophila melanogaster. Curr. Biol., 12*, 712–723.

Ren, B., Robert, F., Wyrick, J. J., Aparicio, O., Jennings, E. G., et al. (2000). Genome-wide location and function of DNA binding proteins. *Science, 290*, 2306–2309.

Rewitz, K. F., Larsen, M. R., Lobner-Olesen, A., Rybczynski, R., O'Connor, M. B., & Gilbert, L. I. (2009). A phosphoproteomics approach to elucidate neuropeptide signal transduction controlling insect metamorphosis. *Insect Biochem. Mol. Biol., 39*, 475–483.

Richards, S., Liu, Y., Bettencourt, B. R., Hradecky, P., Letovsky, S., et al. (2005). Comparative genome sequencing of *Drosophila pseudoobscura*: Chromosomal, gene, and cis-element evolution. *Genome Res., 15*, 1–18.

Richards, S., Gibbs, R. A., Weinstock, G. M., Brown, S. J., Denell, R., et al. (2008). The genome of the model beetle and pest *Tribolium castaneum. Nature, 452*, 949–955.

Rong, Y. S., & Golic, K. G. (2000). Gene targeting by homologous recombination in *Drosophila. Science, 288*, 2013–2018.

Rost, B. (1997). Protein structures sustain evolutionary drift. *Fold Des.*, *2*, S19–24.

Rozowsky, J., Euskirchen, G., Auerbach, R. K., Zhang, Z. D., Gibson, T., et al. (2009). PeakSeq enables systematic scoring of ChIP-seq experiments relative to controls. *Nat. Biotechnol.*, *27*, 66–75.

Rubin, G. M., & Spradling, A. C. (1982). Genetic transformation of *Drosophila* with transposable element vectors. *Science*, *218*, 348–353.

Ryder, E., & Russell, S. (2003). Transposable elements as tools for genomics and genetics in *Drosophila*. *Briefings Funct. Genomics Proteomics*, *2*, 57–71.

Sali, A. (1995). Comparative protein modeling by satisfaction of spatial restraints. *Mol. Med. Today*, *1*, 270–277.

Sander, C., & Schneider, R. (1991). Database of homology-derived protein structures and the structural meaning of sequence alignment. *Proteins*, *9*, 56–68.

Sandmann, T., Girardot, C., Brehme, M., Tongprasit, W., Stolc, V., & Furlong, E. E. (2007). A core transcriptional network for early mesoderm development in *Drosophila melanogaster*. *Genes Dev.*, *21*, 436–449.

Sanger, F., & Coulson, A. R. (1975). A rapid method for determining sequences in DNA by primed synthesis with DNA polymerase. *J. Mol. Biol.*, *94*, 441–448.

Sanger, F., Air, G. M., Barrell, B. G., Brown, N. L., Coulson, A. R., et al. (1977). Nucleotide sequence of bacteriophage phi X174 DNA. *Nature*, *265*, 687–695.

Schaefer, C., Schlessinger, A., & Rost, B. (2010). Protein secondary structure appears to be robust under *in silico* evolution while protein disorder appears not to be. *Bioinformatics*, *26*, 625–631.

Schaffer, A. A., Wolf, Y. I., Ponting, C. P., Koonin, E. V., Aravind, L., & Altschul, S. F. (1999). IMPALA: Matching a protein sequence against a collection of PSI-BLAST-constructed position-specific score matrices. *Bioinformatics*, *15*, 1000–1011.

Schena, M., Shalon, D., Davis, R. W., & Brown, P. O. (1995). Quantitative monitoring of gene expression patterns with a complementary DNA microarray. *Science*, *270*, 467–470.

Schultz, J., Milpetz, F., Bork, P., & Ponting, C. P. (1998). SMART, a simple modular architecture research tool: Identification of signaling domains. *Proc. Natl. Acad. Sci. USA*, *95*, 5857–5864.

Schumacher, J. A., Crockett, D. K., Elenitoba-Johnson, K. S., & Lim, M. S. (2007). Evaluation of enrichment techniques for mass spectrometry: Identification of tyrosine phosphoproteins in cancer cells. *J. Mol. Diagn.*, *9*, 169–177.

Shendure, J., Porreca, G. J., Reppas, N. B., Lin, X., McCutcheon, J. P., et al. (2005). Accurate multiplex polony sequencing of an evolved bacterial genome. *Science*, *309*, 1728–1732.

Siegal, M. L., & Hartl, D. L. (1996). Transgene coplacement and high efficiency site-specific recombination with the Cre/loxP system in *Drosophila*. *Genetics*, *144*, 715–726.

Smith, S. T., Wickramasinghe, P., Olson, A., Loukinov, D., Lin, L., et al. (2009). Genome wide ChIP–chip analyses reveal important roles for CTCF in *Drosophila* genome organization. *Dev. Biol.*, *328*, 518–528.

Stark, A., Lin, M. F., Kheradpour, P., Pedersen, J. S., Parts, L., et al. (2007). Discovery of functional elements in 12 *Drosophila* genomes using evolutionary signatures. *Nature*, *450*, 219–232.

Stathopoulos, A., Van Drenth, M., Erives, A., Markstein, M., & Levine, M. (2002). Whole-genome analysis of dorsal-ventral patterning in the *Drosophila* embryo. *Cell*, *111*, 687–701.

Stuart, L. M., Boulais, J., Charriere, G. M., Hennessy, E. J., Brunet, S., et al. (2007). A systems biology analysis of the *Drosophila* phagosome. *Nature*, *445*, 95–101.

Subramanian, A., Tamayo, P., Mootha, V. K., Mukherjee, S., Ebert, B. L., et al. (2005). Gene set enrichment analysis: A knowledge-based approach for interpreting genome-wide expression profiles. *Proc. Natl. Acad. Sci. USA*, *102*, 15545–15550.

Sury, M. D., Chen, J. X., & Selbach, M. (2010). The SILAC fly allows for accurate protein quantification *in vivo*. *Mol. Cell Proteomics*, *9*, 2173–2183.

Takano, T., & Dickerson, R. E. (1981). Conformation change of cytochrome c. II. Ferricytochrome c refinement at 1.8 A and comparison with the ferrocytochrome structure. *J. Mol. Biol.*, *153*, 95–115.

Takemori, N., & Yamamoto, M. T. (2009). Proteome mapping of the *Drosophila melanogaster* male reproductive system. *Proteomics*, *9*, 2484–2493.

Taverna, D. M., & Goldstein, R. A. (2002). Why are proteins so robust to site mutations? *J. Mol. Biol.*, *315*, 479–484.

Terashi, G., Takeda-Shitaka, M., Kanou, K., Iwadate, M., Takaya, D., et al. (2007). Fams-ace: A combined method to select the best model after remodeling all server models. *Proteins Struct. Funct. Bioinformatics*, *69*, 98–107.

Terry, N. A., Tulina, N., Matunis, E., & DiNardo, S. (2006). Novel regulators revealed by profiling *Drosophila* testis stem cells within their niche. *Dev. Biol.*, *294*, 246–257.

The ENCODE Project Consortium. (2004). The ENCODE (ENCyclopedia Of DNA Elements) Project. *Science*, *306*, 636–640.

The Honey Bee Genome Consortium (2006). Insights into social insects from the genome of the honeybee *Apis mellifera*. *Nature*, *443*, 931–949.

The International Silkworm Genome Consortium (2008). The genome of a lepidopteran model insect, the silkworm *Bombyx mori*. *Insect Biochem. Mol. Biol.*, *38*, 1036–1045.

The Pea Aphid Genome Consortium (2010). Genome sequence of the pea aphid *Acyrthosiphon pisum*. *PLoS Biol.*, *8*, e1000313.

Thompson, B. J., & Cohen, S. M. (2006). The Hippo pathway regulates the bantam microRNA to control cell proliferation and apoptosis in *Drosophila*. *Cell*, *126*, 767–774.

Tie, F., Banerjee, R., Stratton, C. A., Prasad-Sinha, J., Stepanik, V., et al. (2009). CBP-mediated acetylation of histone H3 lysine 27 antagonizes *Drosophila* Polycomb silencing. *Development*, *136*, 3131–3141.

Tomancak, P., Beaton, A., Weiszmann, R., Kwan, E., Shu, S., et al. (2002). Systematic determination of patterns of gene expression during *Drosophila* embryogenesis. *Genome Biol.*, *3*, research0088.

Tusher, V. G., Tibshirani, R., & Chu, G. (2001). Significance analysis of microarrays applied to the ionizing radiation response. *Proc. Natl. Acad. Sci. USA*, *98*, 5116–5121.

Valouev, A., Johnson, D. S., Sundquist, A., Medina, C., Anton, E., et al. (2008). Genome-wide analysis of transcription factor binding sites based on ChIP-Seq data. *Nat. Methods*, *5*, 829–834.

Vera, J. C., Wheat, C. W., Fescemyer, H. W., Frilander, M. J., Crawford, D. L., et al. (2008). Rapid transcriptome characterization for a nonmodel organism using 454 pyrosequencing. *Mol. Ecol.*, *17*, 1636–1647.

Vitkup, D., Melamud, E., Moult, J., & Sander, C. (2001). Completeness in structural genomics. *Nature Struct. Biol.*, *8*, 559–566.

Warnecke, F., Luginbuhl, P., Ivanova, N., Ghassemian, M., Richardson, T. H., et al. (2007). Metagenomic and functional analysis of hindgut microbiota of a wood-feeding higher termite. *Nature*, *450*, 560–565.

Weindruch, R., Kayo, T., Lee, C. K., & Prolla, T. A. (2001). Microarray profiling of gene expression in aging and its alteration by caloric restriction in mice. *J. Nutr.*, *131*, 918S–923S.

Weissig, H., & Bourne, P. E. (1999). An analysis of the Protein Data Bank in search of temporal and global trends. *Bioinformatics*, *15*, 807–831.

Werren, J. H., Richards, S., Desjardins, C. A., Niehuis, O., Gadau, J., et al. (2010). Functional and evolutionary insights from the genomes of three parasitoid *Nasonia* species. *Science*, *327*, 343–348.

White, K. P., Rifkin, S. A., Hurban, P., & Hogness, D. S. (1999). Microarray analysis of *Drosophila* development during metamorphosis. *Science*, *286*, 2179–2184.

Wu, X. F., Li, X. H., Yue, W. F., Roy, B., Li, G. L., et al. (2009). Proteomic identification of the silkworm (*Bombyx mori* L) prothoracic glands during the fifth instar stage. *Biosci. Rep.*, *29*, 121–129.

Xia, Q., Zhou, Z., Lu, C., Cheng, D., Dai, F., et al. (2004). A draft sequence for the genome of the domesticated silkworm (*Bombyx mori*). *Science*, *306*, 1937–1940.

Xia, Q., Guo, Y., Zhang, Z., Li, D., Xuan, Z., et al. (2009). Complete resequencing of 40 genomes reveals domestication events and genes in silkworm (*Bombyx*). *Science*, *326*, 433–436.

Xiang, H., Zhu, J., Chen, Q., Dai, F., Li, X., et al. (2010). Single base-resolution methylome of the silkworm reveals a sparse epigenomic map. *Nat. Biotechnol.*, *28*, 516–520.

Yang, M., Lee, J. E., Padgett, R. W., & Edery, I. (2008). Circadian regulation of a limited set of conserved microRNAs in *Drosophila*. *BMC Genomics*, *9*, 83.

Yu, F., Mao, F., & Jianke, L. (2010). Royal jelly proteome comparison between *A. mellifera ligustica* and *A. cerana cerana*. *J. Proteome Res.*, *9*, 2207–2215.

Zdobnov, E. M., & Apweiler, R. (2001). InterProScan – an integration platform for the signature-recognition methods in InterPro. *Bioinformatics*, *17*, 847–848.

Zeitlinger, J., Zinzen, R. P., Stark, A., Kellis, M., Zhang, H., et al. (2007). Whole-genome ChIP–chip analysis of Dorsal, Twist, and Snail suggests integration of diverse patterning processes in the *Drosophila* embryo. *Genes Dev.*, *21*, 385–390.

Zhang, P., Aso, Y., Yamamoto, K., Banno, Y., Wang, Y., et al. (2006). Proteome analysis of silk gland proteins from the silkworm, *Bombyx mori*. *Proteomics*, *6*, 2586–2599.

Zhang, P., Aso, Y., Jikuya, H., Kusakabe, T., Lee, J. M., et al. (2007). Proteomic profiling of the silkworm skeletal muscle proteins during larval–pupal metamorphosis. *J. Proteome Res.*, *6*, 2295–2303.

Zhang, X., Guo, C., Chen, Y., Shulha, H. P., Schnetz, M. P., et al. (2008). Epitope tagging of endogenous proteins for genome-wide ChIP–chip studies. *Nat. Methods*, *5*, 163–165.

Zhang, Y. (2008). I-TASSER server for protein 3D structure prediction. *BMC Bioinformatics*, *9*, 40.

Zhang, Y., Zhou, X., Ge, X., Jiang, J., Li, M., et al. (2009). Insect-specific microRNA involved in the development of the silkworm *Bombyx mori*. *PLoS One*, *4*, e4677.

Zhao, X. F., He, H. J., Dong, D. J., & Wang, J. X. (2006). Identification of differentially expressed proteins during larval molting of *Helicoverpa armigera*. *J. Proteome Res.*, *5*, 164–169.

Zhao, Y., & Jensen, O. N. (2009). Modification-specific proteomics: Strategies for characterization of post-translational modifications using enrichment techniques. *Proteomics*, *9*, 4632–4641.

Zhu, F., Parthasarathy, R., Bai, H., Woithe, K., Kaussmann, M., et al. (2010). A brain-specific cytochrome P450 responsible for the majority of deltamethrin resistance in the QTC279 strain of *Tribolium castaneum*. *Proc. Natl. Acad. Sci. USA*, *107*, 8557–8562.

Zinke, I., Schutz, C. S., Katzenberger, J. D., Bauer, M., & Pankratz, M. J. (2002). Nutrient control of gene expression in *Drosophila*: Microarray analysis of starvation and sugar-dependent response. *EMBO J.*, *21*, 6162–6173.

Zinzen, R. P., Girardot, C., Gagneur, J., Braun, M., & Furlong, E. E. (2009). Combinatorial binding predicts spatio-temporal cis-regulatory activity. *Nature*, *462*, 65–70.

2 Insect MicroRNAs: From Molecular Mechanisms to Biological Roles

Xavier Belles
Instituto de Biología Evolutiva (CSIC-UPF),
Barcelona, Spain
Alexandre S Cristino
Queensland Brain Institute,
The University of Queensland, Brisbane St Lucia,
Queensland, Australia
Erica D Tanaka
Instituto de Biología Evolutiva (CSIC-UPF),
Barcelona, Spain
Mercedes Rubio
Instituto de Biología Evolutiva (CSIC-UPF),
Barcelona, Spain
Maria-Dolors Piulachs
Instituto de Biología Evolutiva (CSIC-UPF),
Barcelona, Spain

Summary

MicroRNAs (miRNAs) are endogenous, *ca.* 22-nucleotide, single-strand, non-coding RNAs that regulate gene expression by acting post-transcriptionally through base-pairing between the so called "seed" sequence of the miRNA (nucleotides 2–8 at its 5′ end) and its complementary seed match sequence present in the 3′ untranslated region of the target mRNA. Since the discovery of the first miRNAs in the 1990s, a remarkable diversity of miRNAs has been reported in various organisms, including insects, plants, viruses, and vertebrates. Moreover, computational methods have been developed to find new miRNAs as well as mRNA targets. In insects, most miRNAs are involved in modulating a precise dosage of regulatory proteins, thus fine-tuning biological processes like cell proliferation, apoptosis and growth, oogenesis and embryogenesis, nervous system and muscle differentiation, metamorphosis and other morphogenetic processes, and response to biological stress. The miRNA field is still developing, and many questions remain to be solved. Technologies to determine new miRNAs and miRNA targets still need refinement. Further studies are also needed to elucidate the mechanisms regulating miRNA expression, to validate the miRNA targets *in vivo*, and to establish the complex networks that connect miRNAs, mRNAs, and proteins, and that govern the development and function of cells and tissues.

DOI:10.1016/B978-0-12-384747-8.10002-9

2.1. Introduction: The Big World of Small RNAs

Step by step, some of the old paradigms of molecular biology have been falling away. The most significant of these is the central dogma that "one gene equals one protein." It still holds true that most information flows from DNA to proteins through intermediate RNA molecules, but today it is well known that the transcriptome is much more complex and diverse than the genome, thanks to the interplay of a variety of mechanisms. The most thoroughly studied is alternative splicing; that is, the formation of diverse mRNAs through differential splicing of the same RNA precursor, which gives rise to proteins with distinct features. Another factor accounting for transcriptome diversity in quantitative terms and in time and space is the occurrence of transcription factors, sequence-specific DNA-binding factors that usually bind to the promoter region of target genes, thereby activating or repressing their transcription. However, to understand thoroughly the dynamics of the proteome, we have to account for the unknown mechanisms other than simply protein-coding genes and transcription factors. At least, non-coding RNAs (ncRNAs) must also be taken into account in order to have a more complete picture of what is really happening in genomic regulation.

ncRNAs form a heterogeneous group of RNA molecules that are classified into three categories according to their length and function. They range in length from 18 to 25 nucleotides for the group of very small RNAs, which includes short interfering RNAs (siRNAs) and microRNAs (miRNAs); from 20 to 200 nucleotides for the group of small RNAs, which usually play the role of transcriptional

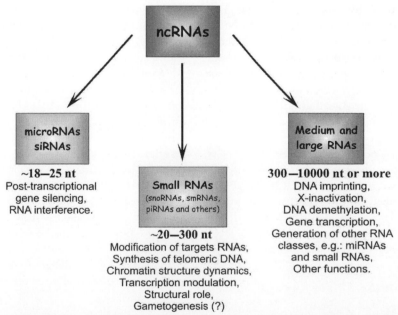

Figure 1 Types of non-coding RNAs (ncRNAs) classified according to their length and functions: very small RNAs – microRNAs and small interfering RNAs (siRNAs); small RNAs; and medium and large RNAs. The corresponding established functions for each type are also indicated. snoRNAs, small nucleolar RNAs; smRNAs, small modulatory RNAs; piRNAs, Piwi-interacting RNAs. Data from Costa (2007).

and translational regulators; and, for the group of medium and large RNAs, up to (and even beyond) 10,000 nucleotides, which are involved in other processes, as detailed in **Figure 1** (Costa, 2007). This chapter deals with the very small RNAs, and, more specifically, with miRNAs.

The history of siRNAs and miRNAs began in the late 1980s, when Jorgensen and colleagues were studying the role of chalcone synthase in the biosynthetic pathway of anthocianin in plants. Anthocianin gives a violet color to petunias, and Jorgensen's team overexpressed chalcone synthase in search of petunias with a deeper violet color. However, they unexpectedly obtained whitish flowers because the expression of chalcone synthase in these transgenic whitish petunias was some 50 times lower than in the wild type, thus suggesting that transgenic chalcone synthase had suppressed the endogenous gene (Jorgensen, 1990). Three years later, but in the field of developmental biology and working on the nematode *Caenorhabditis elegans*, Lee and colleagues (1993) discovered two *lin-4* transcripts, where the smaller, with *ca.* 21 nucleotides, was complementary to seven repeated sequences in the 3′ UTR of the mRNA of the heterochronic gene *lin-14*, which had been identified two years earlier.

These two disparate studies converged in 1998, when Fire and colleagues (1998), also working in *C. elegans*, discovered that the administration of a double-stranded RNA (dsRNA) with a strand complementary to a fragment of an endogenous mRNA can block this mRNA. This phenomenon is now known as RNA interference (RNAi), and its action is mediated by siRNAs of *ca.* 22 nucleotides that derive from dsRNA (Belles, 2010). A year later, while studying post-transcriptional gene silencing as a mechanism of antiviral defense, Hamilton and Baulcombe (1999) noticed the occurrence of antisense viral RNA of *ca.* 25 nucleotides in virus-infected plants. Hamilton and Baulcombe observed that these small RNAs were long enough to convey sequence specificity, and pointed out that they might be key determinants of the gene silencing phenomenon. Further contributions showed that dsRNA-induced mRNA degradation was always mediated by RNAs of 21–23 nucleotides, thus leading researchers to investigate the endogenous source of these small RNAs. Finally, in 2001, three groups working independently (Lagos-Quintana *et al.*, 2001; Lau *et al.*, 2001; Lee and Ambros, 2001) described miRNAs as a novel family of small (*ca.* 22 nucleotides) endogenous RNAs that is diverse in sequence and temporal expression, evolutionarily widespread, and involved in regulating gene expression.

2.1.1. RNAi and siRNAs

The discovery of RNAi in *C. elegans* (Fire *et al.*, 1998) was later extended to other animal groups, namely insects, and the basic mechanisms involved in their action on mRNAs were unveiled step by step in a few years. The biochemical machinery and the effects of RNAi in insects have been the subject of a recent review (Belles, 2010), and we will not deal with them in detail here. In essence, when a long and exogenous dsRNA is delivered to the insect, it is cleaved by the enzyme Dicer-2 into siRNA duplexes of *ca.* 22 nucleotides. These siRNAs then unwind, and single-strand siRNAs bind to Argonaute-2 protein (Ago-2) and assemble into the so-called RNA-induced silencing complex (RISC). The RISC, guided by the siRNA, couples to the target mRNA and degrades it (Belles, 2010). Indeed, the mechanisms of generation and the action of siRNAs are very similar to those of miRNAs (**Figure 2**), with the latter detailed in the following sections.

The advent of RNAi represented a new paradigm in insect functional genomics, because it opened the door for studying non-drosophilid species – that is, species that cannot be genetically transformed, at least not very easily. RNAi experiments are relatively simple, consisting of conveying a dsRNA with a strand complementary to a fragment of the target mRNA to the animal or to cells incubated *in vitro*. After assessing that target mRNA levels have lowered, the study of the phenotype unveils the functions associated with the target mRNA. Experiments can be carried out *in vitro*, where the easiest system involves the incubation of cells with dsRNA added to the medium and then studying the cell behavior; and *in vivo*, where the most straightforward approach consists in delivering the dsRNA to the chosen insect stage (from egg to adult) of the experimental specimen and then examining the resulting phenotype. The approaches *in vivo* have afforded the most spectacular results on insect functional genomics (Belles, 2010).

Kennerdell and Carthew (1998) were the first to use RNAi *in vivo* in insects, studying the genes *frizzled* and *frizzled 2* in the fly *Drosophila melanogaster*. A year later, Brown and colleagues (1999) carried out functional studies of *Hox* genes in the flour beetle *Tribolium castaneum*, and the following year used RNAi for the first time in a hemimetabolan species, the milkweed bug *Oncopeltus fasciatus*, on which Hughes and Kaufman (2000) studied *Hox* gene functions as well. In 2006, RNAi *in vivo* was used for the first time on a phyllogenetically very basal insect, the German cockroach *Blattella germanica* (Ciudad *et al.*, 2006; Martin *et al.*, 2006), which has been shown to be one of the species most sensitive to RNAi. In the past few years an explosion of papers has truly changed the landscape of reverse functional genetics in insects, and has unveiled many gene functions, from development to reproduction, including behavior, coloration, resistance to biological stress, polyphenism, and many others (Belles, 2010).

2.1.2. miRNAs

miRNAs are endogenous, *ca.* 22-nucleotide, single-strand, non-coding RNAs that regulate gene expression on the post-transcriptional level through base-pairing between the seed sequence of the miRNA and its complementary seed match sequence that is present in the 3′ untranslated region (UTR) of the target mRNA.

The first miRNA, lin-4, was discovered in a screen for genes required for post-embryonic development in the nematode *C. elegans* (Lee *et al.*, 1993; Ambros and Horvitz, 1984). The identification of the *lin-4* locus and its regulatory mechanism through the 3′ UTR of *lin-14* mRNA was an interesting finding, although at that time it was almost considered to be a genetic oddity. However, the discovery of another miRNA, let-7, initially in *C. elegans* (Reinhart *et al.*, 2000) and later in various bilaterian species (Pasquinelli *et al.*, 2000), confirmed that, in the case of lin-4 and lin-14, it was not an oddity at all, but rather a new and fundamental layer of the mechanisms regulating gene expression (Lai *et al.*, 2003; Neilson and Sharp, 2008).

The following sections will deal exclusively with miRNAs.

2.2. Biogenesis of miRNAs

miRNAs undergo molecular processing before becoming mature and ready to play their functional role. The pathway of miRNA biogenesis has many commonalities with that of siRNAs, but it is distinct in a number of ways (**Figure 2**). miRNAs are first transcribed as part of a longer primary transcript (pri-miRNA), which folds, forming hairpin structures that correspond to miRNA precursors (pre-miRNAs). pri-miRNAs are then processed in the nucleus and transported to the cytoplasm, where they undergo final maturation (**Figure 2**).

2.2.1. miRNA Processing in the Nucleus

Most miRNA genes are transcribed by RNA polymerase II into pri-miRNA, although in some cases the transcription is mediated by RNA polymerase III (Lee *et al.*, 2004a; Borchert *et al.*, 2006). Usually, pri-miRNAs are several kilobases long, contain local stem-loop structures,

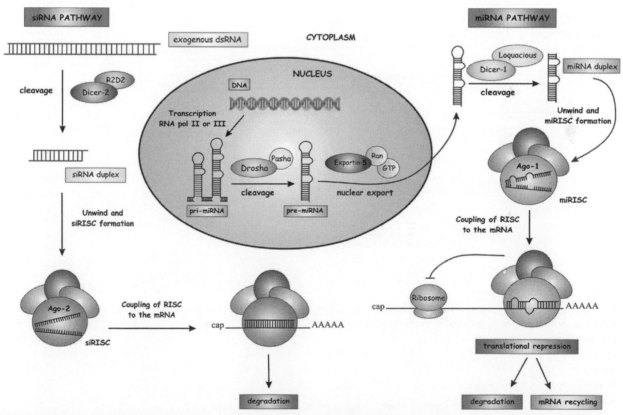

Figure 2 Biogenesis of miRNA and siRNA. miRNA gene is transcribed by RNA Pol II/III into a primary transcript (pri-miRNA) that is processed by Drosha/Pasha and exported to the cytoplasm by Exportin-5. In the cytoplasm, the precursor (pre-miRNA) undergoes the final step of maturation and is cleaved by Dicer-1/Loquacious into an miRNA duplex. After the miRNA duplex unwinds, the mature miRNA is maintained with Argonaute-1 protein (Ago-1) forming a RISC which will be coupled to the target mRNA and will degrade, destabilize, or translationally inhibit it, whereas the miRNA* is released and degraded. On the other hand, siRNA is formed when a long and exogenous double-strand RNA (dsRNA) is cleaved by Dicer-2/R2D2 into an siRNA duplex. Likewise with the miRNA pathway, the siRNA duplex is unwound and single-strand siRNAs are maintain with Argonaute-2 protein (Ago-2) forming a RISC which will recognize the target mRNA and degrade it.

and are polyadenylated and capped, as in current mRNAs (Cai *et al.*, 2004; Lee *et al.*, 2004a), although the cap and the poly(A) tail are removed during miRNA processing. miRNA genes can form clusters in the genome (Behura, 2007), or can be found isolated within an intronic region of protein-coding genes, or in introns and exons of non-coding RNAs (Rodriguez *et al.*, 2004). Moreover, pri-miRNAs can be polycistronic, thus carrying the information of more than one miRNA. In insects, the group of miR-100, let-7, and miR-125 constitutes the best studied example of polycistronic pri-miRNA. The organization of this pri-miRNA is well conserved in many species of insects, and even in vertebrates (**Figure 3**), although the spacer regions between miRNA precursor sequences can vary considerably in structure and length. For example, the distance between the precursor of miR-100 and that of let-7 varies from *ca.* 100 bp in *T. castaneum* to *ca.* 3.9 kb in *Anopheles gambiae*, whereas the distance between the precursor of let-7 and that of miR-125 varies within the range of 250–450 bp (**Figure 3**) (Behura, 2007).

pri-miRNA processing into *ca.* 70- to 80-nucleotide pre-miRNAs takes place exclusively in the nucleus by the action of the microprocessor, a protein complex of *ca.* 500 kDa, which in *D. melanogaster* is composed by the RNase III enzyme Drosha and its partner, Pasha (**Figure 2**) (Denli *et al.*, 2004). In general, insects possess a single *pasha* gene copy, except in the pea aphid *Acyrthosiphon pisum*, where four *pasha*-like genes have been recently reported (Jaubert-Possamai *et al.*, 2010). Pasha protein (also known as DGCR8 in vertebrates) contains two double-stranded RNA-binding domains; it plays an essential role in miRNA processing by recognizing the substrate pri-miRNA and by determining the precise cleavage site, whereas Drosha actually cleaves the pri-miRNA (Denli *et al.*, 2004). The two RNase domains of Drosha cleave the 5′ and 3′ arms of the pri-miRNA 11 base-pairs away from the single-stranded RNA/double-stranded RNA junction at the base of the hairpin stem (**Figure 4**) (Han *et al.*, 2004); of note, a single nucleotide variation in an miRNA precursor stem can block Drosha processing (Duan *et al.*, 2007).

Generally, cleavage of the pri-miRNA by Drosha occurs in an unspliced intronic region before mRNA splicing catalysis (Kim and Kim, 2007). However, some miRNA genes are located within intronic regions which themselves form a hairpin structure. In this special case, the action of Drosha is bypassed during pri-miRNA processing. After splicing of its host mRNA, the miRNA is

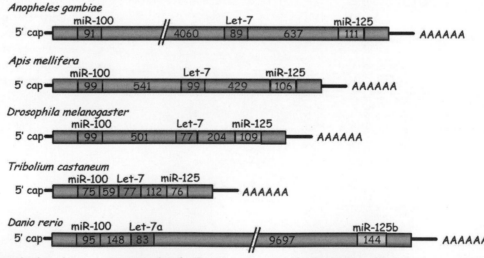

Figure 3 Organization of the primary transcript of miR-100, let-7, and miR-125 cluster in different insect species and the zebrafish. Numbers inside the boxes correspond to the length in base-pairs. The sequences were obtained from the miRBase (http://www.mirbase.org).

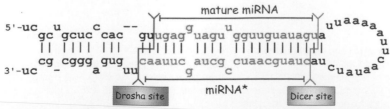

Figure 4 Precursor of miRNA let-7. In one of the arms of the stem-loop of the hairpin the mature miRNA sequence resides (in red); in another is the miRNA* sequence (in blue). The sites of cleavage for both enzymes Drosha and Dicer-1 are shown as purple and green lines, respectively. This *Culex quinquefasciatus* sequence was obtained in miRBase (http://www.mirbase.org/).

released from the intron, then exported from the nucleus to the cytoplasm, and is finally cleaved by Dicer (see below). This class of miRNA, called mirtrons, has been described in flies, nematodes, and mammals (Berezikov *et al.*, 2007; Okamura *et al.*, 2007; Ruby *et al.*, 2007a).

2.2.2. Pre-miRNA Transport from the Nucleus to the Cytoplasm

Once the pri-miRNA is processed in the nucleus, the resulting pre-miRNAs are exported to the cytoplasm by Exportin5 (EXP5) (**Figure 2**), which is a member of the nuclear transport receptor family, in complex with the cofactor Ran-GTP (Yi *et al.*, 2003; Kim, 2004). EXP5 can recognize double-stranded RNA stems longer than 14 base-pairs along with a short 3′ overhang (1–8 nucleotides), which ensures the export of only those pre-miRNAs correctly processed (Lund *et al.*, 2004). Using RNAi, a number of authors have demonstrated the role of EXP5 in the nucleocytoplasmic transport of pre-miRNA. Knockdown of EXP5 mRNA decreases the levels of mature miRNAs, but does not lead to an increase of pre-miRNA levels in the nucleus, which suggests that protecting pre-miRNA from digestion in the nucleus is another important role of EXP5 (Yi *et al.*, 2003; Lund *et al.*, 2004).

2.2.3. miRNA Maturation by Dicer

In the cytoplasm, pre-miRNAs are cleaved by Dicer near the terminal loop (**Figure 4**), thus resulting in the release of a *ca.* 22-nucleotide miRNA duplex with two nucleotides protruding as overhangs at each 3′-end. Dicer is an ATP-dependent multidomain enzyme of the RNase family involved in the cleavage of small double-stranded RNAs. It was identified for the first time in *D. melanogaster* (Bernstein *et al.*, 2001), and two Dicer homologs were later found in this fly, Dicer-1 and Dicer-2, which, in general, are involved in the miRNA and siRNA pathways, respectively (Lee *et al.*, 2004b). RNAi of Dicer-1 in the last instar nymph of the cockroach *B. germanica* inhibits the formation of mature miRNAs and impairs the metamorphic process (Gomez-Orte and Belles, 2009), which confirms not only the role of Dicer-1 in miRNA biogenesis in a phyllogenetically basal insect, but also that of miRNAs in hemimetabolan metamorphosis (see below).

D. melanogaster Dicer-1 interacts with the protein Loquacious (also known as R3D1), which contains three double-stranded RNA-binding domains for pre-miRNA processing. Depletion of Loquacious results in pre-miRNA accumulation in *Drosophila* S2 cells, and immuno-affinity purification experiments revealed that Loquacious locates in a functional pre-miRNA processing complex along with Dicer-1, and stimulates the specific pre-miRNA processing activity (Saito *et al.*, 2005). Of note, both arms of the pre-miRNA stem loop structures are imperfectly paired,

containing G:U wobble pairs and single nucleotide insertions (**Figure 4**). The imperfect base-pairing causes differences in thermodynamic properties and makes one strand of the duplex less stably paired at its 5′ end. Generally, the strand with the lowest thermodynamic stability becomes the mature miRNA (guide strand), whereas the other strand (miRNA* or passenger strand) is degraded.

2.2.4. Regulation of miRNA Biogenesis and Stability

As a general principle, given that most of the miRNA genes are transcribed by RNA Polymerase II, the usual transcription factors associated with Pol II will influence their transcriptional control. A more specific modality of miRNA regulation is the process known as editing, which consists in a post-transcriptional change of RNA sequences caused by deamination of adenosine (A) to inosine (I), thus resulting in alterations in the base-pairing of the transcript. pri-miRNA transcripts modified by ADAR (adenosine deaminase acting on RNAs) have their biogenesis altered in downstream steps. These modifications of the pri-miRNA sequences may block the cleavages by Drosha and Dicer during miRNA maturation; moreover, edited mature miRNAs can recognize other target mRNAs (Kawahara *et al.*, 2007). Therefore, editing is a remarkable regulator of biogenesis, and in addition increases miRNA structural diversity and further extends the diversity of miRNA targets.

Regulation of the miRNA biogenesis pathway also involves feedback mechanisms, like the interplay of Drosha and Pasha, which regulate each other in a circuit of negative feedback. Drosha acts by cleaving a hairpin located in the 5′ UTR of Pasha mRNA. Excess of Drosha decreases Pasha mRNA levels, whereas a reduction of Drosha elicits the reverse effect (Kadener *et al.*, 2009b). Another example of feedback regulation involves human Dicer and the miRNA let-7, wherein Dicer is targeted by let-7 in sites within its coding region (Forman *et al.*, 2008), or the reciprocal regulation showed by let-7 and the RNA-binding protein Lin-28, where let-7 suppresses Lin-28 protein synthesis whereas Lin-28 blocks let-7 maturation. Lin-28 is capable of blocking the cleavages mediated by both Drosha and Dicer; indeed, recombinant Lin-28 can block pri-miRNA processing, whereas knockdown of Lin-28 facilitates the expression of mature let-7 (Viswanathan *et al.*, 2008). Lin-28 acts by inducing uridylation of the let-7 precursor at its 3′ end, which elicits the degradation of the uridylated pre-let-7 because Dicer fails to process hairpin RNA structures with long 3′ extensions (Heo *et al.*, 2008).

Unlike 3′ uridylation, 3′ adenylation may have a stabilizing effect on miRNAs, at least in mammals. For example, a variant of miR-122 possesses a 3′-terminal adenosine that is added by cytoplasmic poly(A) polymerase GLD-2 after unwinding of the miR-122/miR-122* duplex, and this 3′

adenylation appears to prevent shortening, thus stabilizing the miRNA (Katoh et al., 2009). Apparently, adenylation and uridylation are two competing processes; it is interesting, in this sense, that addition of adenine residues in some small RNAs can prevent urydilation (Chen et al., 2000).

2.3. Mechanism of Action of miRNAs

The functional role of a miRNA is ultimately characterized by its effects on the expression of target genes. Currently, the regulatory mechanisms involving miRNAs are related to mRNA cleavage or translational repression by binding to complementary sites usually located on the 3′ UTR region of the mRNA (Carrington and Ambros, 2003; Lai, 2003; Ambros, 2004; Bartel, 2004). In contrast to the inhibitory effects, miRNAs can also stimulate the expression of target genes by upregulation of translation (Vasudevan et al., 2007; Orom et al., 2008). Moreover, miRNAs can also control cell fate by binding to heterogeneous ribonucleoproteins and lifting the translational repression of their target mRNAs; in this way, miRNAs act through a sort of decoy activity that interferes with the function of regulatory proteins (Beitzinger and Meister, 2010; Eiring et al., 2010). The present section, however, will emphasize the more widespread mechanisms, leading to mRNA translational repression, which start when the miRNA binds to Ago-1 protein and with the assembly of the RISC (**Figure 2**).

2.3.1. Argonaute Loading

The Argonaute (Ago) family can be divided into two subfamilies: the Piwi subfamily and the Ago subfamily. Piwi proteins are involved in transposon silencing, and are especially abundant in germ-line cells. Ago-subfamily proteins play key roles in post-transcriptional gene regulation by interacting with siRNAs (see above) and miRNAs, as detailed below.

After Dicer-1-mediated cleavage, the miRNA duplex binds to an Ago-1 protein in the RISC. To form an active RISC, the miRNA duplex has to unwind because only the mature miRNA binds to the Ago-1 protein, whereas the miRNA* is released. In human cells, miRNAs with a high degree of base-pairing in their pre-miRNA hairpin stem are initially processed by Ago-2, which cleaves the 3′ arm of the hairpin (that is, the miRNA* strand) in the middle, thus generating a nicked hairpin (Diederichs and Haber, 2007). In this case, Ago-2 acts before Dicer-1-mediated cleavage and facilitates miRNA duplex dissociation, the removal of nicked strand, and the activation of RISC. These findings elucidated the crucial role of Ago proteins not only during RISC formation, but also in relation to the mechanism that determines which of the two strands will become the survivor mature miRNA.

Identification of the target mRNA by the RISC is based on the complementarity between the mature miRNA and the target mRNA site, and the degree of complementarity determines whether the target mRNA is degraded, destabilized, or translationally inhibited. Binding to Ago-1 greatly enhances miRNA stability, and although little is known about the half-life of individual miRNAs, it is clear that Ago-1 is a limiting factor for endogenous miRNA accumulation due to its protective function.

In D. melanogaster, miRNA* strands may accumulate bound to Ago-2, a protein initially thought to act exclusively in the siRNA pathway. Whether miRNA* binds to Ago-1 or to Ago-2 depends on the miRNA duplex structure, thermodynamic stability, and the identity of first 5′-end nucleotide – i.e., miRNA sequences beginning with cystidine will bind to Ago-2, whereas those beginning with uridine will bind to Ago-1 (Ghildiyal et al., 2010). A number of observations indicate that some miRNA* plays a role in the regulation of gene expression. These observations include that: (1) miRNA* 5′ ends are more defined than their 3′ ends, thus suggesting that there is a seed region involved in regulatory functions (Ruby et al., 2007b; Okamura et al., 2008; Seitz et al., 2008); (2) many miRNA* sequences are evolutionarily conserved (Okamura et al., 2008); (3) in D. melanogaster (Ruby et al., 2007b) and in the basal insect B. germanica (Cristino et al., 2011), tissue concentration of some miRNA* is higher than that of the corresponding miRNA partner.

2.3.2. Repression of Protein Translation

The RISC is the key element that regulates gene expression by repressing protein translation. The first step after RISC formation is the recognition of the target mRNA, mainly through the seed sequence. A number of studies have demonstrated the importance not only of the seed, but also of the whole 5′ region of the miRNA during the interaction with the target mRNA. According to Brennecke and colleagues (2005), there are two categories of miRNA target sites in mRNAs. The first is called the "5′ dominant site," and occurs when there is a near perfect base-pairing in the 5′ end of the miRNA; this category can be subdivided into "canonical" (when both 5′ and 3′ ends have strong base-pairing with the miRNA site) and "seed" (when only the 5′ region presents consistent base-pairing). The second category is called "3′ compensatory," and occurs when base-pairing between the miRNA seed sequence and its corresponding sequence in the target mRNA is weak, and thus a stronger base-pairing in the 3′ region exerts a sort of "compensating" effect.

Initial experiments in C. elegans showed that the miRNAs lin-4 and let-7 repress their respective target mRNAs through interactions with miRNA sites in the 3′ UTR. Subsequently, many other cases of miRNA binding sites in the 3′ UTR of mRNAs were reported, leading to the presumption that this was a general rule. However, recent findings have revealed that miRNAs can repress mRNAs through sites located in the open reading frame (ORF) or in the 5′ UTR (Lee *et al.*, 2009).

The action of RISC on target mRNAs may proceed through different mechanisms. One of them involves post-initiation repression, as shown by experiments carried out in *C. elegans* where lin-4 inhibits the translation of lin-14 mRNA without reducing the mRNA levels and without affecting the shifting of polysomes, thus suggesting that the inhibition of mRNA translation occurs at the elongation step (Wightman *et al.*, 1993; Olsen and Ambros, 1999; Lee *et al.*, 2003). Other details accounting for this mechanism of action have been reported, and a model has been proposed describing the inhibition of ribosome elongation, the induction of ribosome drop-off, and the facilitation of nascent polypeptides proteolysis (Fabian *et al.*, 2009).

The second mechanism of RISC action is the acceleration of target mRNA destabilization, involving: (1) decapping of the m(7)G cap structure in the 5′ end; and/or (2) deadenylation of poly A tail during the initial step of translation (Humphreys *et al.*, 2005). A number of reports using different experimental models have supported this second mechanism; for example, in zebrafish embryos and mammalian cells, miRNAs in the RISC accelerate mRNA deadenylation, which leads to fast mRNA decay (**Figure 5**) (Giraldez *et al.*, 2006; Wu *et al.*, 2006). In *Drosophila* cells both deadenylation and decapping require GW182

protein, CCR4:NOT deadenylase, and the DCP1:DCP2 decapping complexes. Depletion of GW182 in *Drosophila* cells leads to alteration of mRNA expression levels. However, in Ago-1depleted cells, GW182 can still silence the expression of target mRNAs, thus indicating that GW182 acts downstream of Ago-1, and that it is a key component of the miRNA pathway (Behm-Ansmant *et al.*, 2006a).

2.3.3. Processing Bodies and mRNA Storage

In many cases, the last step of miRNA action involves the processing bodies (P-bodies), which are discrete cytoplasmic aggregates that contain enzymes associated to mRNA decay, such as CCR4:NOT complex (deadenylase), DCP1:DCP2 complex (decapping), RCK/p54, and eIF4ET (general translational repressors). The aforementioned GW182 is additionally required for P-body integrity. Apparently, P-bodies are the place where RISC delivers its target mRNA to be degraded or to be stored (**Figure 5**). In human cells, for example, miR-122-repressed mRNAs that are maintained in P-bodies can be released from them under stress conditions, and subsequently be recruited by polysomes (Bhattacharyya *et al.*, 2006). Behm-Ansmant and colleagues (2006b) have proposed a model where RISC binds to target mRNA through interactions with miRNA and Ago-1, and recruits GW182, which labels the

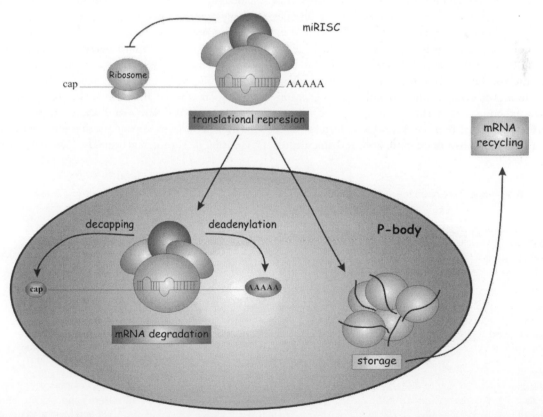

Figure 5 Once the miRISC is formed, the target mRNA can be taken to a special region of cytoplasm known as the P-body, where it will be degraded after decapping and deadenylation, or maintained in the P-body until released from it and recruited to polysomes.

transcript as a target for decay via deadenylation and decapping. Ago-1 and Ago-2 proteins have also been detected in P-bodies (Liu *et al.*, 2005), thus suggesting that both siRNA and miRNA pathways may end in these structures. Nevertheless, this does not mean that P-bodies are crucial for the functioning of these pathways, given that disruption of P-bodies after depletion of Lsm1, which is a key component of them, elicits a dispersion of Ago proteins into the cytoplasm, but does not affect siRNA and miRNA pathways (Chu and Rana, 2006).

2.4. Identification of miRNAs in Insects

Since the discovery of lin-4 and let-7 in the nematode *C. elegans*, a remarkable diversity of miRNAs has been reported in the genomes of various organisms, including insects, plants, viruses, and vertebrates (http://www.mirbase.org). In insects, research on miRNAs was initially limited to *D. melanogaster*, but the availability of sequenced genomes from different species, as well as the development of new bioinformatic tools, has allowed the performance of systematic predictions of miRNAs *in silico*. Accordingly, computational methods based on the evolutionary conservation of genomic sequences and their ability to fold into stable hairpin structures have been applied to species with sequenced genomes, such as a number of nematodes, arthropods, and vertebrates (**Table 1**). Moreover, the development of novel techniques for directional cloning of small RNAs has led to the identification of many other miRNAs (Lagos-Quintana *et al.*, 2001; Lau *et al.*, 2001; Lee and Ambros, 2001).

Nevertheless, the greatest progress came with the advent of high-throughput sequencing technologies and computational methods. Those technologies confirmed most of the miRNA predicted *in silico* in species with the genome reported, made it possible to find new and unexpected

miRNAs, and contributed to the discovery *de novo* of miRNAs in species without the genome sequenced. Therefore, a consistent catalog of miRNAs is now available not only in drosophilids, but also in a selection of species, such as the malaria mosquito (*A. gambiae*), the yellow fever mosquito (*Aedes aegypti*), the pea aphid (*A. pisum*), the vector of West Nile virus (*Culex quinquefasciatus*), the jewel wasp (*Nasonia vitripennis*), the migratory locust (*Locusta migratoria*), the honey bee (*Apis mellifera*), the flour beetle (*T. castaneum*), the silkworm (*Bombyx mori*), and the German cockroach (*B. germanica*) (http://www.mirbase.org; http://www.ncbi.nlm.nih.gov/geo) (Griffiths-Jones, 2006). Both approaches, based on computational methods and high-throughput sequencing, are discussed below.

2.4.1. Computational Methods

The most efficient computational methods for finding miRNA candidates were described in *C. elegans* (MiRscan) (Lim *et al.*, 2003a) and *D. melanogaster* (miRseeker) (Lai *et al.*, 2003). Both methods share conceptual similarities, such as structural and sequence similarity. MiRscan produces an initial set of candidates by sliding a 110-nucleotide window across the *C. elegans* genome and folding those segments that are filtered by the free energy and duplex length. Homologous hairpins are then identified by WU-BLAST in an additional genome which creates a reference set defining the standard features that will finally be used to score and rank all candidate hairpins. Nevertheless, MiRscan was not able to identify more than 50% of the previously known *C. elegans* miRNAs (Lim *et al.*, 2003a). miRseeker was found to be more efficient at identifying genuine miRNAs in two fly species (*D. melanogaster* and *Drosophila pseudoobscura*) by taking into account the conservation across the hairpin (Lai *et al.*, 2003). The method begins by identifying orthologous

Table 1 Algorithms Developed for miRNA Identification

Program	Strategy	Species group	Authors/year
Grad *et al.*	RB	Nematodes	Grad *et al.*, 2003
MiRScan	RB	Nematodes, vertebrates	Lim *et al.*, 2003a, 2003b
miRseeker	RB	Insects (flies)	Lai *et al.*, 2003
Berezikov *et al.*	RB	Human	Berezikov *et al.*, 2005
miPred	RB	Human	Jiang *et al.*, 2007
miRAlign	RB	Metazoan	Wang *et al.*, 2005
ProMIR	HMM	Human	Nam *et al.*, 2005
BayesMiRNAFind	NB	Nematodes, mammals	Yousef *et al.*, 2006
One-ClassMirnaFind	SVM, NB	Human, virus	Yousef *et al.*, 2008
mirCoS-A	SVM	Mammals	Sheng *et al.*, 2007
mir-abela	SVM	Mammals	Sewer *et al.*, 2005
triplet-SVM	SVM	Human	Xue *et al.*, 2005
RNAmicro	SVM	Metazoan	Hertel and Stadler, 2006
miPred	SVM	Human	Ng and Mishra, 2007
MiRFinder	SVM	Human, virus	Huang *et al.*, 2007

HMM, hidden Markov model; NB; Naive Bayes; RB, rule based; SVM, support vector machine.

intergenic and intronic regions of those two fly genomes, and then folding those conserved sequences to identify and score the hairpin structures. The criteria for hairpin evaluation derive from a reference set of known miRNA genes of the two *Drosophila* species. The length of the hairpins and their minimum free energy were first evaluated, and then the distribution of divergent nucleotides was considered to score the candidates. The metrics consist in penalizing divergences depending on where they occur in the pre-miRNA hairpin, as the miRNA arm would tolerate less mutations than the miRNA* arm, which, by itself, would not tolerate more mutations than those observed in the loop region (Lai *et al.*, 2003).

The establishment of guidelines for the experimental validation and annotation of novel miRNA candidates became obviously necessary with the increasing quantity of miRNA genes being identified in various species (Ambros *et al.*, 2003). Thus, an initiative for organizing the information available on miRNA genes was then developed, leading to a database (miRBase, http://www.mirbase.org) where all data regarding miRNA sequences, targets, and gene nomenclature are deposited (Griffiths-Jones *et al.*, 2008).

The large amount of miRNA data available in databases led to the development of a second generation of algorithms based on machine-learning methods. The approach consists in a learning process that identifies the most relevant characteristics and rules from a positive set of miRNA hairpins. Various machine-learning algorithms have been used for miRNA discovery (**Table 1**), the most common being Naïve Bayes (Yousef *et al.*, 2006), support vector machines (Yousef *et al.*, 2008 and references therein), hidden Markov models (HMM) (Nam *et al.*, 2005), genetic programming (Brameier and Wiuf, 2007), and random walks (Jiang *et al.*, 2007).

All these methods contributed somehow to the identification of new miRNAs, despite considerable differences in their trade-off between specificity and sensitivity. The criteria used in all of them were based on actual knowledge of the miRNA biogenesis, and features identified from known miRNAs conserved in at least two species. Indeed, there must be a great number of non-conserved miRNA genes still to be discovered, which may have characteristics and expression profiles substantially different from those of canonical miRNAs. However, the development of a new generation of sequencing technologies is changing the way of thinking about scientific approaches in all fields of biological sciences (Metzker, 2010), including the strategies to find new miRNAs in any species, even those whose genome is not sequenced yet.

2.4.2. High-Throughput Sequencing

Deep-sequencing technologies have created a new paradigm in detecting low-expression or tissue-specific miRNAs, as well as non-canonical and species-specific ones. The most effective algorithms published so far are miRDeep (Yang *et al.*, 2010), MIReNA (Mathelier and Carbone, 2010), and deepBase (Friedlander *et al.*, 2008). Despite varying slightly in their workflow, their general strategy is similar, combining mapping and filtering sequences based on genome annotation, sequence and structure patterns, and properties of miRNA biogenesis.

The identification of miRNAs through deep-sequencing methods is rapidly increasing the catalogs of small RNA sequences for many species from a variety of taxonomic groups. Currently, all deep-sequencing datasets are deposited in the GEO (Gene Expression Omnibus) database at the NCBI (National Center for Biotechnology Information; http://www.ncbi.nlm.nih.gov/geo). At the date of writing (January 2011), there are at least 193 studies of high-throughput sequencing of small RNAs from different eukaryotic species in the GEO database. **Table 2** shows the 14 insect species in the GEO database, and the number of records for each.

Most of the 14 insect species included in **Table 2** have the genome sequenced, or at least have a closely related species with an available genome (e.g., *A. albopictus* and *C. quinquefasciatus*). Two species, *L. migratoria* and *B. germanica*, have no genome sequence available, and identification of miRNAs from deep-sequencing data becomes challenging because none of the methods mentioned above were designed to analyze deep-sequencing data without

Table 2 Insect Species and Number of Records Found in the GEO Database Related to Studies of miRNA Identification

Order	Species	Number of records
Diptera		
	Drosophila melanogaster	21
	Drosophila simulans	1
	Drosophila erecta	1
	Drosophila pseudoobscura	1
	Drosophila virilis	1
	Aedes albopictus	1
	Culex quinquefasciatus	1
Lepidoptera		
	Bombyx mori	2
Hymenoptera		
	Camponotus floridanus	1
	Harpegnathos saltator	1
	Apis mellifera	1
Hemiptera		
	Acyrthosiphon pisum	1
Orthoptera		
	Locusta migratoria	1
Dyctioptera		
	Blattella germanica	1

using a genome sequence as a reference, and the diversity of small RNA types is remarkably high. However, strategies that can identify previously described miRNAs, as well as novel miRNAs on the basis of the number of reads and hairpin features, have recently been proposed (Wei *et al.*, 2009). Genome-independent approaches for miRNA discovery show that we still have a poor understanding of the small RNA world and its regulatory mechanisms in the cell. For example, in the locust *L. migratoria* (Wei *et al.*, 2009) and in the cockroach *B. germanica* (Cristino *et al.*, 2011), sequence read numbers corresponding to miRNA*s were higher than those corresponding to the mature miRNA. Another original finding has been reported in *Drosophila* species (Berezikov *et al.*, 2010), where some miRNA precursors seem not to be processed by RNase III only, given that the usual one- to two-nucleotide 3' overhang does not occur in some sequences represented by a high number of reads.

2.4.3. miRNA Classification

As stated above, identification efforts have led to the description of an impressive number of miRNAs in animals, plants, green algae, fungi, and virus (Griffiths-Jones *et al.*, 2008), and different attempts to classify such a high diversity into families based on structural coincidences have been carried out. As the pattern of nucleotide substitution in miRNA genes is apparently shaped by selective pressures, and considering that the seed is the most important region from a functional point of view (Brennecke *et al.*, 2005; Bartel, 2009), miRNA classification is based on this region. Regarding metazoans, 858 miRNA families are deposited in the miRBase database (v16.0) (Griffiths-Jones *et al.*, 2008), and 254 (30%) of these families are found in at least five species. These records will change with further high-throughput sequencing experiments, but present data indicate that most of the miRNA families

(a total of 562) are found in vertebrates, followed by insects (178 families reported), and then by other metazoan that are phyllogenetically more basal, such as cnidaria, porifera, hemichordata, echinodermata, urochordata, cephalochordata, and nematoda (118 families in all).

The seed region can be more or less conserved in different miRNA families. A good example of a well-conserved seed region is observed in the miRNAs miR-100, miR-125, and let-7 (Behura, 2007). As stated above (see also **Figure 3**), these three miRNAs are often coded by the same polycistronic pri-miRNA that has a conserved organization from invertebrates to vertebrates, which suggest that it is an ancestral pri-miRNA. As expected, the seed region of these miRNAs is highly conserved (**Figure 6**). There are insect-specific miRNA families whose seed region is also very well conserved, as for instance bantam miR-2 and miR-3 (**Figure 6**). The conservation of the seed region occurs not only among paralogous sequences, resulting from intraspecific gene duplication, but also among orthologous sequences arising from speciation events. Of note, the conservation of the seed region is critical for the recognition of mRNA targets, thus the classification of miRNAs into families on the basis of the seed not only contains structural information, but may also reflect functional regularities.

2.5. Target Prediction

In animals, the functional duplexes miRNA:mRNA can occur in a variety of structures where short complementary sequences can be interrupted by gaps and mismatches (Brennecke *et al.*, 2005; Bartel, 2009). Thus, most computational methods have been developed to find target sequences based on the complementarity between the miRNA seed sequence and the mRNA sequence. Several computational approaches estimate the likelihood of miRNA:mRNA duplex formation, mainly based on

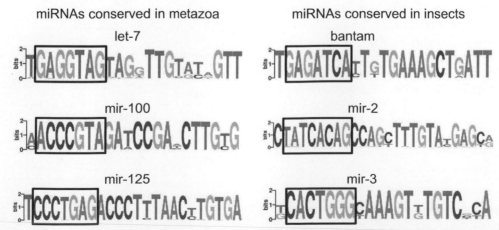

Figure 6 Conservation of miRNA genes on the region corresponding to mature miRNAs in metazoan and insects. The sequence logo is constructed based on the alignment of various miRNA sequences representing the level of nucleotide conservation in each position. The squares indicate the canonical seed regions located at nucleotides 2–8.

sequence complementarity, thermodynamic stability, and evolutionary conservation of the sequence among species (**Table 3**). Machine learning approaches are also used for miRNA target identification. These methods usually combine one or more of the traditional procedures (seed complementarity, thermodynamic stability, and cross-species conservation) with more elaborated probabilistic models (**Table 3**). Also, a new generation of algorithms is integrating high-throughput expression data and computational predictions (Huang *et al.*, 2007; Hammell *et al.*, 2008; van Dongen *et al.*, 2008; Wang and El Naqa, 2008; Bandyopadhyay and Mitra, 2009; H. Liu *et al.*, 2010; Sturm *et al.*, 2010).

To date, miRNA target prediction has been mainly performed by computational approaches, and large numbers of targets have been predicted for most species with the genome sequenced (Bartel, 2009). As a general figure, predictions have suggested that a single miRNA can target 200 mRNAs on average in vertebrates (Krek *et al.*, 2005), whereas in *D. melanogaster* a single miRNA may regulate 54 genes on average (Grun *et al.*, 2005).

2.5.1. microCosm, TargetScan, and PicTar

The miRBase database links miRNAs to targets using microCosm (http://www.ebi.ac.uk/enright-srv/microcosm/), TargetScan (Lewis *et al.*, 2005; Grimson *et al.*, 2007; Friedman *et al.*, 2009) and PicTar (Lewis *et al.*, 2005; Grimson *et al.*, 2007; Friedman *et al.*, 2009) prediction systems. These are therefore the most currently used, and are detailed below.

microCosm, formerly known as miRBase Targets, predicts miRNA targets in the UTR regions of animal genomes from Ensembl database (Hubbard *et al.*, 2007; Flicek *et al.*, 2008). It uses the miRanda algorithm to calculate a score across the miRNA vs UTR alignment (Enright *et al.*, 2003; John *et al.*, 2004; Betel *et al.*, 2008); the energy for the thermodynamic stability of a miRNA:mRNA duplex is calculated by the Vienna RNA folding routines (http://www.tbi.univie.ac.at/RNA/), and the P-values are computed for all targets following the statistical model implemented in RNAhybrid (Rehmsmeier *et al.*, 2004). The Miranda algorithm (Enright *et al.*, 2003; John *et al.*, 2004) is basically divided into three steps. In the first step the miRNAs are aligned against the 3' UTR sequences of the targets, allowing for G:U pairs and short indels. The method does not rely on seed matches, but increases the scaling score for complementarity at the 5' end of the miRNA. The second step computes the thermodynamic stability of the miRNA:mRNA duplex, and the final step reduces the false-positive rate by considering only targets with multiple sites.

TargetScan was the first algorithm that used the concept of seed matches in target prediction (Lewis *et al.*, 2003, 2005). The method only uses miRNAs conserved across different species to scan corresponding 3' UTR sequences.

Table 3 Algorithms Developed for Predicting miRNA Targets

Algorithm	Strategy	Species group	Authors/year
TargetScan	RB	Vertebrates	Lewis *et al.*, 2003
TargetScanS	RB	Vertebrates	Lewis *et al.*, 2005
miRanda	RB	Insects (flies), Human	Enright *et al.*, 2003; John *et al.*, 2004
Diana-microT	RB	Nematodes	Kiriakidou *et al.*, 2004
RNAhybrid	RB	Insects (flies)	Rehmsmeier *et al.*, 2004
MovingTargets	RB	Insects (flies)	Burgler and MacDonald, 2005
MicroInspector	RB	Any species	Rusinov *et al.*, 2005
Nucleus	RB	Insects (flies)	Rajewsky and Socci, 2004
EIMMo	RB	Nematodes, Insects (flies), Vertebrates	Gaidatzis *et al.*, 2007
TargetBoost	BT	Nematodes, Insects (flies)	Saetrom *et al.*, 2005
PicTar	HMM	Nematodes, Insects (flies), Vertebrates	Krek *et al.*, 2005
RNA22	MC	Nematodes, Insects (flies), Vertebrates	Miranda *et al.*, 2006
MicroTar	PD	Any species	Thadani and Tammi, 2006
PITA	PD	Nematodes, Insects (flies), Vertebrates	Kertesz *et al.*, 2007
NBmiRTar	NB	Metazoa	Yousef *et al.*, 2007
miTarget	SVM	Metazoa	Kim *et al.*, 2006
MiRTif	SVM	Metazoa	Yang *et al.*, 2008
mirWIP	E	Nematodes	Hammell *et al.*, 2008
Sylamer	E	Metazoa	van Dongen *et al.*, 2008
GenMiR++	BL, E	Metazoa	Huang *et al.*, 2007
SVMicrO	SVM, E	Mammals	H. Liu *et al.*, 2010
TargetMiner	SVM, E	Human	Bandyopadhyay and Mitra, 2009
MirTarget2	SVM, E	Metazoa	Wang and El Naqa, 2008
TargetSpy	BT, E	Insects (flies), Human	Sturm *et al.*, 2010

BL, Bayesian learning; BT, Boosting technique; E, integration of expression data; HMM, hidden Markov model; MC, Markov chain; PD, pattern discovery; RB, rule based; SVM, support vector machine.

The algorithm defines the seed matches as short segments of seven nucleotides that must have a stringent complementarity to the two to eight nucleotides of the mature miRNA. Then, the remaining miRNA sequence is aligned to the target site, allowing for G : U pairs; the free energy to form a secondary structure in the duplex is predicted by a folding algorithm. A Z-score is calculated on the basis of the number of matches predicted in the same target sequence and respective free energies. Finally, the Z-score is used to rank the candidate targets for each species, and each species is processed in the same way.

PicTar uses a machine learning algorithm to rank target sequences using a HMM maximum likelihood score based on three main steps: (1) the seed matches must expand 7 nucleotides starting at position 1 or 2 in the 5′ end of the miRNA; (2) the minimum free energy of miRNA : mRNA duplexes is used to filter the target sites; and (3) the target sites must locate in overlapping positions across the aligned corresponding 3′ UTR sequences. The target sites that pass the three-step filter are then ranked by the HMM model, which calculates the score considering all segmentations of the target sequence into target sites and background, thus allowing the algorithm to account for multiple binding sites for a single miRNA, as well as several miRNAs targeting the same mRNA.

The current target predictions available in the miRBase by microCosm, TargetScan, and PicTar have some degree of overlap and also of discrepancy that can be due to alignment artifacts, different mRNA UTR and miRNA sequences, and intrinsic differences in the algorithms. In an attempt to provide more updated figures for the distribution of gene targets per miRNA and miRNA per gene target, we analyzed the data from target predictions available in the miRBase (Release 16; Sept 2010), comparing *D. melanogaster* with *Homo sapiens* and *C. elegans*. Results show that the three methods give different average numbers of miRNA-binding sites per mRNA target (19.6, 5.8, and 5.0 for MicroCosm, TargetScan, and PicTar, respectively; **Figure 7**), as well as different numbers of mRNAs targeted by each miRNA (951, 395, and 426 for microCosm, TargetScan, and PicTar, respectively; **Figure 8**). The distribution of the number of miRNA-binding sites per mRNA target (**Figure 7**) is relatively similar among the three methods and the three species studied. Conversely, data on the number of mRNA targeted by an miRNA showed remarkable differences depending on the method, regarding not only the average values, but also and especially their pattern of distribution (**Figure 8**).

2.6. miRNA Functions

Insect model species can be studied through powerful genetic and genomic approaches, the paradigm being the fly *D. melanogaster*. Indeed, the first description of miRNA functions in insects was carried out in this species (Brennecke *et al.*, 2003), by looking at gain-of-function mutants (Lai, 2002; Lai *et al.*, 2005). miRNA functions

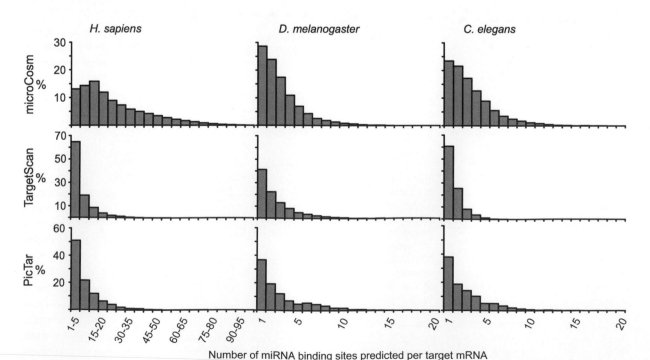

Figure 7 Frequency of the number of miRNA-binding sites in the 3′ UTR of target mRNAs in *Homo sapiens*, *Drosophila melanogaster*, and *Caenorhabditis elegans*, calculated with the three prediction methods available in miRBase: microCosm, TargetScan, and PicTar (Release 16; September 2010).

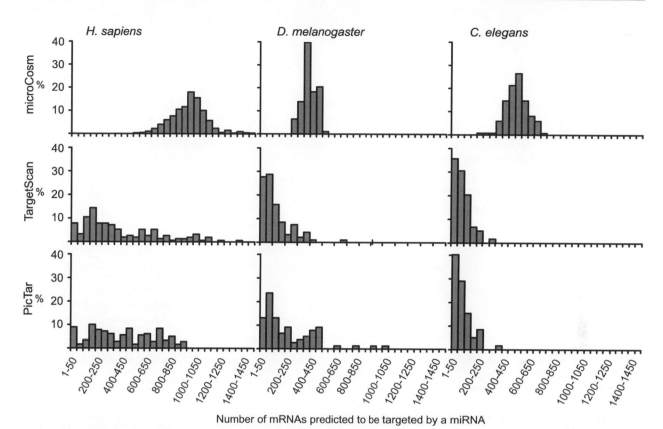

Figure 8 Frequency of the number of mRNAs predicted to be targeted a miRNA in *Homo sapiens*, *Drosophila melanogaster*, and *Caenorhabditis elegans*, calculated with the three prediction methods available in miRBase: microCosm, TargetScan, and PicTar (Release 16; September 2010).

are currently being demonstrated by mutating the genes coding for the miRNAs under study, overexpressing the miRNA of interest, or silencing it using specific anti-miRNAs, and then studying the resulting phenotype. Predicted targets may also be validated by the above methods, including the quantification of the expression of the given target, as well as using *in vitro* systems with luciferase reporter target constructs, where binding of the miRNA to the target sequence is detected by luciferase activity and quantified with colorimetry.

In most cases, functions may be suggested by high-throughput sequencing comparisons in different developing stages, in different organs of the same stage, or in different physiological situations. Studies of this type have been carried out in the silkworm *B. mori* (differences in tissue expression and in different developing stages) (Cao *et al.*, 2008; S. Liu *et al.*, 2010), the pea aphid *A. pisum* (differences in different morphs) (Legeai *et al.*, 2010), the honey bee *A. mellifera* (differences between queens and workers) (Weaver *et al.*, 2007), the migratory locust *L. migratoria* (differences between migratory and solitary phases) (Wei *et al.*, 2009), and the German cockroach *B. germanica* (differences between metamorphic and non-metamorphic instars) (Cristino *et al.*, 2011). Microarray analysis or detailed studies on the developmental expression profiles of particular miRNAs can also suggest their

respective functions (Aravin and Tuschl, 2005; Weaver *et al.*, 2007; He *et al.*, 2008; Yu *et al.*, 2008).

Silencing Dicer-1 expression by RNAi is also a useful approach to studying the influence of the whole set of miRNAs in a given process. This has been achieved in *D. melanogaster*, either *in vivo*, showing, for example, that Dicer-1 plays a general role in ovarian development (Jin and Xie, 2007), or in *Drosophila* cultured cells, where the depletion of Dicer-1 affected the development in both somatic and germ lineages (Lee *et al.*, 2004b). More recently, Dicer-1 depletion by RNAi has been used in the German cockroach, *B. germanica*, to demonstrate the key role of miRNAs in hemimetabolan metamorphosis (see below).

Regarding the functions of particular miRNAs, the data available indicate that most of them appear to be involved in the fine-tuning of biological processes by modulating a precise dosage of regulatory proteins. Probably, they provide robustness to the whole program of gene expression (Hornstein and Shomron, 2006) and resilience to environmental fluctuations, as in the case of miR-7 studied by Li and colleagues (X. Li *et al.*, 2009). However, as revealed by recent general reviews (Bushati and Cohen, 2007; Jaubert *et al.*, 2007), information is still fragmentary, heavily concentrated in the *D. melanogaster* model, and focused on a few biological processes, as detailed in the text below and in **Table 4**, which summarizes cases

Table 4 Functions of miRNA Demonstrated Experimentally*

Function/process	miRNA	Target involved	Authors/year
Cell division of the germinal stem cells	bantam		Hatfield *et al.*, 2005; Shcherbata *et al.*, 2007
Cell division of the germinal stem cells	miR-7, miR-278, miR309	Dacapo	Yu *et al.*, 2009
Germ-line differentiation	miR-7	bam	Pek *et al.*, 2009
Stem cells differentiation	miR-184	Saxophone	Iovino *et al.*, 2009
Axis formation in the egg chamber	miR-184	Gurken	Iovino *et al.*, 2009
Formation of the head and posterior abdominal segments in the embryo	miRs-2/13		Boutla *et al.*, 2003
Embryo segmentation	miR-31, miR-9		Leaman *et al.*, 2005
Embryo growth	miR-6		Leaman *et al.*, 2005
Formation of embryonic cuticle	miR-9		Leaman *et al.*, 2005
Photoreceptor differentiation	miR-7	Yan	Li and Carthew, 2005
Formation of sensory organs	miR-9a	Senseless	Li *et al.*, 2006
Location of CO_2 neurons	miR-279	Nerfin-1	Cayirlioglu *et al.*, 2008
Protection of sense organs from apoptosis	miR-263a/b	Hid	Hilgers *et al.*, 2010
Muscle differentiation	miR-1	Delta	Kwon *et al.*, 2005
Muscle differentiation	miR-133	nPTB	Boutz *et al.*, 2007
Growth	bantam		Hipfner *et al.*, 2002; Edgar, 2006; Thompson and Cohen, 2006
Tissue growth via insulin receptor signaling	miR-278		Teleman *et al.*, 2006
Growth via insulin receptor signaling	miR-8	U-shaped	Hyun *et al.*, 2009
Modulation of ecdysteroid pulses	miR-14	EcR	Varghese and Cohen, 2007
Neuromusculature remodeling during metamorphosis	let-7 (and miR-100, miR-125)		Sokol *et al.*, 2008
Maturation of neuromuscular junctions during metamorphosis	let-7 (and miR-125)	abrupt	Caygill and Johnston, 2008
Wing formation	miR-9a	dLOM	Biryukova *et al.*, 2009
Wing formation	iab-4	*Ultrabithorax*	Ronshaugen *et al.*, 2005
Regulation of circadian rhythms	bantam	clock	Kadener *et al.*, 2009a
Regulation of brain atrophin	miR-8	atrophin	Karres *et al.*, 2007
Anti-apoptotic	Bantam, miR-2	hid	Brennecke *et al.*, 2003, 2005; Stark *et al.*, 2005
Anti-apoptotic in *D. melanogaster*	miR-14	Drice	Xu *et al.*, 2003
Anti-apoptotic in Lepidopteran Sf9 cells	miR-14		Kumarswamy and Chandna, 2010
Anti-apoptotic in the embryo	miR-2, miR-13, miR-11	hid, grim, reaper, sickle	Leaman *et al.*, 2005

*All results refer to *Drosophila melanogaster*, except in the anti-apoptotic action of miR-14, which has been demonstrated also in Sf9 cells of the Lepidopteran *Spodoptera frugiperda*.

where the miRNA function has been demonstrated experimentally.

2.6.1. Germ-Line and Stem Cell Differentiation, Oogenesis

In *D. melanogaster*, cell division of the germinal stem cells (GSC) is under the control of different miRNAs. One of them, bantam, regulates the expression of specific mRNAs in the ovary, being involved in the maintenance of germinal stem cells (Hatfield *et al.*, 2005; Shcherbata *et al.*, 2007). Other miRNAs, like miR-7, miR-278 and miR-309, directly repress Dacapo mRNA through its 3′ UTR, as demonstrated by Yu and colleagues (2009) using luciferase assays. These authors also suggest that bantam and miR-8 regulate Dacapo indirectly, controlling GSC

division; moreover, GSC deficient for miR-278 show a mild, but significant, reduction of cell proliferation. Depletion of miR-7 levels in GSC results in a perturbation of the frequency of Cyclin E-positive GSC, although the kinetics of cell division in miR-7 mutant GSC does not become reduced (Yu *et al.*, 2009).

Another miRNA that plays important roles in stem cell differentiation is miR-184. Depletion of miR-184 in *D. melanogaster* determines that females lay abnormal eggs and become infertile. Stem cell differentiation is impaired due to the increase of Saxophone protein levels. Later, during oogenesis, the absence of mir-184 impairs the axis formation of the egg chamber as a result of altering the expression of Gurken mRNA. In addition, the absence of miR-184 also affects the expression of pair-rule genes required for normal anteroposterior patterning and cellularization of the embryo (Iovino *et al.*, 2009).

Finally, and also in *D. melanogaster*, miR-7 is involved in germ-line differentiation via *maelstrom* and *Bag-of-marbles* (*Bam*) gene products. Maelstrom regulates *Bam* via repression of miR-7, by binding to the miR-7 promoter region (Pek *et al.*, 2009); therefore, *D. melanogaster* mutants for *maelstrom* overexpress *Bam*, which leads to a deficient germ-line differentiation. As expected, a reduction in miR-7 expression rescues this phenotype (Pek *et al.*, 2009)

2.6.2. Embryo Patterning and Morphogenesis

After injecting anti-miDNA-2a and anti-miDNA-13a, *D. melanogaster* embryos exhibited defects in the head and posterior abdominal segments, including cuticle holes and denticle belt malformations. In view of the similarity of the induced phenotypes, Boutla and colleagues (2003) concluded that these related miRNAs, miR-2a and miR-13a, act on the same target genes, together with the also related miR-2b and miR-13b, which form a functional subgroup called miRs-2/13.

Leaman and colleagues (2005) injected antisense 2′O-methyl oligoribonucleotides targeting specific miRNAs into early embryos of *D. melanogaster* in order to screen the function of these miRNAs. Results showed that embryos depleted for miR-31 and miR-9 completed development, but were affected by severe segmentation defects (**Figure 9**). Those injected with miR-9 antisense rarely formed any trace of cuticle, and did not show internal differentiation. Embryos depleted for miR-6 were generally smaller in size than controls and had fewer and abnormally large segments, thus suggesting that apoptotic processes had been enhanced.

2.6.3. Sensory Organs and Functions

In *D. melanogaster*, miR-7 has been localized in early photoreceptors during embryonic eye development. At this developmental stage, miR-7 stimulates photoreceptor differentiation through a reciprocal regulation with *yan*, a gene encoding a transcription factor involved in the differentiation of retinal progenitor cells (Li and Carthew, 2005).

Another miRNA involved in sensory organ development is miR-9. Through both loss-of-function and gain-of-function analyses *in vivo*, Li and colleagues (2006) have reported that miR-9a is responsible for generating precise numbers of sensory organs in *D. melanogaster* embryos and adults. To accomplish this regulatory function, miR-9a represses the translation of *Senseless* mRNA through its 3′ UTR region, thus ensuring a precise differential expression of this gene in sensory organ precursors and in the adjacent epithelial cells (Li *et al.*, 2006).

Neurons that sense CO_2 also provide an interesting case for study. They may have different locations depending on the species. In *D. melanogaster* they are located in the antenna, whereas in mosquitoes they are found in the maxillary palps. Cayirlioglu and colleagues (2008) observed that loss of miR-279 in *D. melanogaster* determines that CO_2 neurons change their location from the antennae to the maxillary palps. The authors suggest that miR-279 downregulates Nerfin-1, a specific target for this miRNA, thus preventing the development of CO_2 neurons in the maxillary palps (Cayirlioglu *et al.*, 2008).

An example of miRNA that ensures developmental robustness during apoptotic tissue pruning is miR-263a/b (Hilgers *et al.*, 2010), which protects sense organs during apoptosis by directly acting upon, and limiting the expression of, the pro-apoptotic gene *hid*. This property

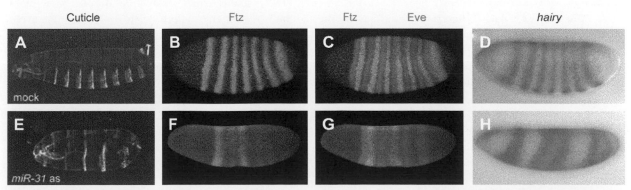

Figure 9 Effects of depletion of miR-31 in *Drosophila melanogaster* embryos. (A) and (E) show Darkfield images of cuticle preparations; (B), (C), (F), and (G) are confocal images of Eve (red) and Ftz (green) stainings; and (D) and (H) *hairy* RNA *in situ* hybridization of blastoderm (2.5-h) embryos. (A)–(D) correspond to controls, and (E)–(H) to *miR-31* antisense-injected embryos. The latter show cuticle defects ranging from partial fusion to complete loss of segments (E). In controls (B) and (C), Eve and Ftz are expressed in seven largely non-overlapping stripes, while *miR-31* antisense-injected embryos (F) and (G) show fewer and weaker stripes that often bleed into each other. The *hairy* transcript pattern also shows fewer stripes (H), indicating that pattern formation is affected upstream of the primary pair rule genes. From Leaman and colleagues (2005), reprinted with permission from *Cell* (Elsevier).

of some miRNAs to buffer fluctuating levels of gene activity makes them well suited to serve a protective function during development (Hilgers *et al.*, 2010).

2.6.4. Muscle Differentiation

In *D. melanogaster*, miR-1, which is one of the best conserved miRNAs in animals, is specifically expressed in the mesoderm during early embryogenesis, and in myogenic precursors and muscle cells in late embryos (Sokol and Ambros, 2005). Depletion of miR-1 using genetic approaches, or by treatment with 2′O-methyl antisense oligonucleotides, resulted in lethality, which implies that miR-1 has essential functions in mesodermally derived tissues (Nguyen and Frasch, 2006).

By analyzing *D. melanogaster* mutants devoid of miR-1, Kwon *et al.* (2005) assessed the essential role of miR-1 for muscle differentiation. They showed that miR-1 regulates the determination of specific cardiac and somatic muscle lineages from pluripotent progenitor cells in early embryogenesis. The Delta protein, a ligand for the Notch signaling pathway, was identified as an miR-1 target in cardiac progenitor cells (Kwon *et al.*, 2005).

Another well-conserved miRNA is miR-133, which is expressed in muscle cells together with miR-1. In *D. melanogaster* embryos, miR-133 plays a key role in controlling alternative splicing during muscle formation, and defining the properties of differentiated muscle cells, through repressing the expression of the splicing factor nPTB during myoblast differentiation into myotubes (Boutz *et al.*, 2007). The results of Boutz and colleagues not only indicate miR-133 directly downregulates a key factor during muscle development, but also establish a role for microRNAs in the control of a developmentally dynamic splicing program.

2.6.5. Growth

In *D. melanogaster*, loss-of-function mutations of the bantam locus are lethal at the early pupal stage, whereas hypomorphic combinations of bantam mutant alleles give rise to adult flies that are smaller than controls (**Figure 10**) and that have deficiencies in fertility (Hipfner *et al.*, 2002). Conversely, overexpression of bantam induces tissue overgrowth due to an increase in cell number. Bantam expression appears to be regulated by the gene *Yorkie*, thus controlling organ growth during development (Edgar, 2006; Thompson and Cohen, 2006).

Related to growth, and also in *D. melanogaster*, miR-278 has been implicated in insulin receptor (InR) signaling, thus contributing to regulation of the energy balance mainly by controling insulin responsiveness. Overexpression of miR-278 promotes tissue growth in the eye and wing imaginal disks, whereas its deficiency leads to a reduction of fat body mass, which is reminiscent of the effect of impaired InR signaling in adipose tissue; the action of miR-278 could be produced through the regulation of *expanded* gene transcripts (Teleman *et al.*, 2006).

More recently, Hyun and colleagues (2009) have reported that miR-8 and its target, U-shaped (USH), regulate body size in *D. melanogaster*. miR-8 null flies are smaller in size and defective in insulin signaling in the fat body. USH inhibits PI3K activity, thus suppressing cell growth. Fat-body-specific expression and clonal analyses showed that miR-8 activates PI3K, thereby promoting fat-cell growth cell-autonomously, and enhancing organismal growth non-cell-autonomously (Hyun *et al.*, 2009).

2.6.6. Metamorphosis: Ecdysteroids and Juvenile Hormone

In insects, molting and metamorphosis are controlled by juvenile hormones and ecdysteroids, usually 20-hydroxyecdysone. Simultaneous expression of miR-125 and let-7 during *D. melanogaster* post-embryonic development is synchronized with the high titer of ecdysteroid pulses that initiate metamorphosis (Bashirullah *et al.*, 2003; Sempere *et al.*, 2003), which suggests that ecdysteroids might regulate the expression of these two miRNAs.

Bashirullah and colleagues (2003), however, showed that miR-125 and let-7 expression is neither dependent on the Ecdysone receptor (EcR) nor inducible by 20-hydroxyecdysone in larval organs incubated *in vitro*. The same authors reported that the expression of both miRNAs can be induced by 20-hydroxyecdysone in *Drosophila* Kc cells, although the induction is considerably delayed with respect to what is observed *in vivo* (Bashirullah *et al.*, 2003). The conclusion of these experiments is that the action of 20-hydroxyecdysone in Kc cells might be indirect, and that miR-125 and let-7 should be directly induced by an unknown temporal signal distinct from the well-known ecdysteroid-EcR cascade.

In a parallel paper, Sempere and colleagues (2003) followed a different approach to study the influence of ecdysteroids on the expression of miR-125, let-7, and miR-100, which are upregulated after the ecdysteroid pulse, as well as of miR-34, which is downregulated. They used the temperature-sensitive *ecd¹* strain that is impaired in ecdysteroid synthesis, and they showed that in *ecd¹* specimens blocked from pupariation by a transfer at 29°C, miR-125, let-7, and miR-100 were detected at much lower levels, whereas miR-34 was detected at much higher levels, compared with the wild type. Sempere and colleagues (2003) also studied the possible role of *Broad complex*, an early inducible gene in the ecdysteroid cascade, using *npr⁶* specimens, which lack all Broad complex factors. Results showed that miR-125, let-7, and miR-100 were detected at much lower levels (and miR-34 at much higher levels) in homozygous *npr⁶* specimens than in *npr⁶/+* or wild type specimens. With the same experimental approach, Sempere and colleagues (2003)

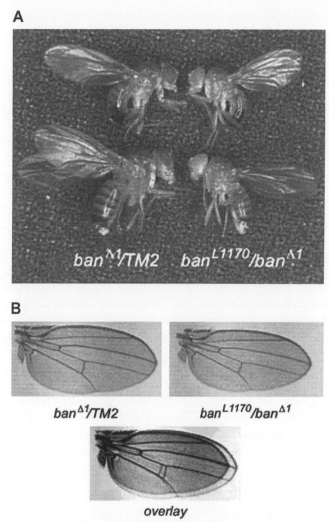

Figure 10 In *Drosophila melanogaster*, loss-of-function mutations of the bantam locus are lethal at the early pupal stage, whereas hypomorphic combinations of bantam mutant alleles give rise to adult flies that are smaller than controls. Panel (A) shows the reduced body size of male and female flies of *ban^L1170^/ban^Δ1^* (defective for bantam), with respect to *ban^Δ1^/TM2* siblings (control). Panel (B) compares wing sizes in both combinations. In the overlay, the *ban^L1170^/ban^Δ1^* wing is shown in red and the *ban^Δ1^/TM2* wing in green. From Hipfner and colleagues (2002), reprinted with permission from the Genetics Society of America.

concluded that ecdysteroids and Broad complex activity are required for temporal upregulation of miR-125, let-7, and miR-100, and downregulation of miR-34. Additional experiments carried out by these authors with *Drosophila* S2 cells showed that incubation times longer than 30 h with 20-hydroxyecdysone correlated with increased levels of miR-125, let-7, and miR-100, whereas miR-34 was detected at very low levels at all times studied. Moreover, the addition of methoprene, a juvenile hormone analog, enhanced the expression of miR-34 and reduced the ecdysteroid-stimulatory effect on the expression of miR-125, let-7, and miR-100 (Sempere *et al.*, 2003). In these experiments, Broad complex was shown to be necessary for enhancing the activity of 20-hydroxyecdysone.

Discrepancies between the two papers (Bashirullah *et al.*, 2003; Sempere *et al.*, 2003) look more apparent than real, given that the respective sets of results, which emerge from quite different experimental approaches, are not incompatible. Indeed, those of Sempere and colleagues (2003) do not discard an indirect action of ecdysteroids, which is the hypothesis postulated by Bashirullah and colleagues (2003).

A miRNA clearly associated to ecdysteroid pulses in *D. melanogaster* is miR-14. In this fly, ecdysteroid signaling through the EcR seems to act via a positive autoregulatory loop that increases EcR levels, thus optimizing the effect of ecdysteroid pulses. In this context, miR-14 modulates this loop by limiting the expression of EcR, whose mRNA contains three miR-14 sites in the 3′ UTR. In turn, ecdysteroid signaling, through EcR, downregulates miR-14. This modulatory action of miR-14 may be crucial due to the intrinsic lability of the positive autorregulatory loop that controls ecdysteroid signaling (Varghese and Cohen, 2007).

2.6.7. Metamorphosis: Morphogenesis

Work by Sokol and colleagues (2008) showed that the *D. melanogaster* let-7-Complex locus (let-7-C, comprising let-7, miR-100, and miR-125; see **Figure 3**) is mainly expressed in the pupal and adult neuromusculature. let-7-C knockout flies look morphologically normal, but display defects in different adult behaviors (like flight and motility) and in fertility. Importantly, their neuromusculature clearly shows juvenile features, which suggests that an important function of let-7-C is to ensure the appropriate remodeling of the abdominal neuromusculature during the larval-to-adult transition. The study also showed that this function is carried out predominantly by let-7 alone (Sokol *et al.*, 2008).

In a related work, Caygill and Johnston (2008) obtained a *D. melanogaster* mutant that lacks let-7 and miR-125 activities and shows a pleiotropic phenotype that arises during metamorphosis. These authors showed that the loss of let-7 and miR-125 results in temporal delays in the terminal cell-cycle exit in the wing, and in the maturation of neuromuscular junctions of imaginal abdominal muscles. The authors focused on the latter process by identifying the *abrupt* (*ab*) gene (which encodes a nuclear protein) as a let-7 target, and by providing evidence showing that let-7 regulates the maturation rate of abdominal neuromuscular junctions during metamorphosis by regulating *ab* expression (Caygill and Johnston, 2008).

Wing morphogenesis has been studied by Biryukova and colleagues (2009), who described that miR-9a regulates *D. melanogaster* wing development through a functional target site in the 3′ UTR of the LIM only (dLOM) mRNA. dLMO is a transcription cofactor that directly inhibits the activity of Apterous, the factor required for the proper wing dorsal identity. Deletions of the 3′ UTR that remove the miR-9a site generate gain-of-function dLMO mutants associated with high levels of dLMO mRNA and protein. These mutants lack wing margins, a phenotype that is characteristic of null miR-9a mutants. Of note, miR-9a and dLMO are co-expressed in wing disks and interact genetically for controlling wing development; thus, the absence of miR-9a results in overexpression of dLMO, while gain-of-function miR-9a mutant suppresses dLMO expression. The data suggest that miR-9a ensures a precise dosage of dLMO during *D. melanogaster* wing development (Biryukova *et al.*, 2009).

Another miRNA involved in wing morphogenesis of *D. melanogaster* is iab-4. Sequence analysis suggested that iab-4 could regulate *Ultrabithorax* (*Ubx*), and expression pattern studies of iab-4 and *Ubx* showed that they are complementary in critical developmental moments. Direct evidence for an interaction between iab-4 and *Ubx*

was obtained with luciferase assays. Finally, ectopic expression of iab-4 miRNA in haltere disks caused a homeotic transformation of halteres to wings, which occurs when *Ubx* expression is reduced (Ronshaugen *et al.*, 2005).

As stated above, RNAi experiments that reduced Dicer-1 expression in the last instar nymph of *B. germanica* depleted miRNA levels, and the next molt, instead of giving the adult stage, gave supernumerary nymphs. These were morphologically similar to the supernumerary nymphs obtained after treating the last instar nymph with juvenile hormone (**Figure 11**). The RNAi experiments with Dicer-1 indicate that miRNAs are crucial for hemimetabolan metamorphosis (Gomez-Orte and Belles, 2009).

2.6.8. Behavior

A recent paper by Kadener and colleagues (2009a) addresses the contribution of miRNAs to the regulation of circadian rhythms. The authors first knocked down the miRNA biogenesis pathway in *D. melanogaster* circadian tissues, which severely affected behavioral rhythms, thus indicating that miRNAs function in circadian timekeeping. To identify miRNA–mRNA pairs that might be important for this regulation, immunoprecipitation of Ago-1, followed by microarray analysis, led to identification of a number of mRNAs presumably under miRNA control. These included three core clock mRNAs: clock; vrille; and clockworkorange. To identify miRNAs involved in circadian timekeeping, the authors inhibited miRNA biogenesis in circadian tissues and then carried out a tiling array analysis. Behavioral and molecular experiments showed that bantam has a role in the core circadian pacemaker, and S2 cell biochemical assays indicated that bantam regulates the translation of clock by targeting three sites in the clock 3′ UTR (Kadener *et al.*, 2009a).

In a work addressed to study of the role of miR-8 in *D. melanogaster*, Karres and colleagues (2007) identified atrophin (also known as grunge) as a direct target of miR-8. miR-8 mutant phenotypes show high levels of apoptosis in the brain, and behavioral defects, like impaired capability for climbing, which are attributable to elevated atrophin activity. Decrease of atrophin levels in miR-8-expressing cells to below the level generated by miR-8 regulation is detrimental, which points to a sort of "tuning target" relationship between them (Karres *et al.*, 2007).

2.6.9. Polyphenism, Caste Differentiation, and Sexual Differences

Legeai and colleagues (2010) suggested that miRNA might participate in the regulation of aphid polyphenism, and studied the expression of miRNA in different female morphs of *A. pisum* using microarray approaches. Most (95%) of the miRNA tested (*n* = 149) had similar expression in different morphs, but some of them, including

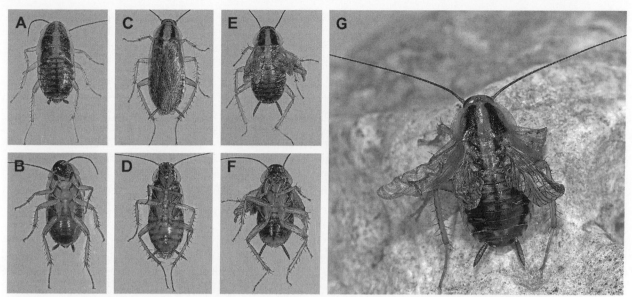

Figure 11 Inhibition of *Blattella germanica* metamorphosis after impairing miRNA maturation by depleting Dicer-1 expression with RNAi approaches in sixth (last) instar nymph. Dorsal and ventral view of normal sixth instar nymph (A, B), normal adult (C, D) and seventh instar supernumerary nymphoid (E, F) resulting from metamorphosis inhibition. The nymphoids resemble those obtained after treating the last instar nymph with juvenile hormone (G). Photos from Albert Masó; data from Gomez-Orte and Belles (2009).

miR*s, were differentially expressed – like let-7 and miR-100, which were upregulated in oviparae specimens, and miR2a-1, which was downregulated. The comparison between two parthenogenetic morphs gave three miR-NAs (miR-34, miR-X47, and miR-X103) and two miR*s (miR307* and miRX52*) that showed differential expression. While miR307* was upregulated in virginoparae, the others were downregulated with respect to the sexuparae morph.

Using total reads from Solexa deep sequencing, Wei and colleagues (2009) compared miRNA expression in the gregarious and solitary phases of the migratory locust, *L. migratoria*. In the gregarious phase, canonical miRNAs were expressed at levels between 1.5- and 2-fold higher than in the solitary phase; the most prominent differences were found in miR-276, miR-125, miR-1, let-7, and miR-315. Interestingly, miR-1 is a muscle-specific miRNA and miR-315 is a potent Wingless signaling activator, at least in *D. melanogaster*; therefore, the differences in flying power between gregarious and solitary locusts may be related by the action of these two miRNAs. However, most of the differences concern new unannotated small RNAs, which are much more abundant in the solitary phase, although the functions of these miRNA candidates remain unknown.

In the honey bee *A. mellifera*, expression profiles of miRNAs in workers and queens have been compared using quantitative RT-PCR, in adult body parts (head, abdomen, thorax) as well as in the whole body of the pupal stage (Weaver *et al.*, 2007). Results highlighted the differential expression, between queens and workers, in the abdomen,

which is probably related to the location of the ovaries and the differential fecundity of the two castes. Regarding particular miRNAs, miR-71 shows strong expression in worker pupae; comparing adult body parts, miR-71 has a higher expression in the head and thorax of adult workers, whereas expression in the abdomen is higher in the queen caste. Conversely, miR-9a is highly expressed in the thorax and abdomen of workers, while their expression levels are the highest in the thorax of workers.

In the silkworm *B. mori*, Liu and colleagues studied the expression of miRNA adult males and females using microarray approaches, and found that the expression of some 20 miRNAs was significantly higher in the body wall of males (S. Liu *et al.*, 2010). This differential expression was assessed for a selection of 10 miRNAs (including bantam, miR-1, miR-13a, and miR-2a) using Northern blot. Microarray analysis also revealed that the expression of 13 miRNAs was significantly higher in ovaries, whereas only 4 were differentially expressed in testes. Differences between sexes were also found in other tissues, including Malpighian tubules, head, midgut, fat body, or silk gland. However, the authors pointed out that differences in miRNA expression might be due to individual differences in the metabolic state, because the expression of some of these miRNAs is influenced by nutritional status (Cheung *et al.*, 2009).

2.6.10. Response to Biological Stress

Larvae of the moth *Lymantria dispar* show differentiated miRNA expression after wasp parasitization

(Gundersen-Rindal and Pedroni, 2010). Microarray studies revealed that miR-1, miR-184, and miR-277 are highly upregulated in larval hemocytes, whereas miR-279 and let-7 are highly downregulated. Expression changes were assessed in hemolymph, fat body, brain, and midgut from infected larva, with respect to controls, using qRT-PCR. Of all the tissues analyzed from parasitized specimens, the midgut was the one that showed least miRNA activity. miR-1 was upregulated in all tissues from parasitized specimens, whereas miR-277 was the most strongly upregulated in the fat body. Expression of miR-279 was variable in the different tissues; it was remarkably upregulated in fat body but clearly downregulated in hemolymph, and it had a negligible expression in brain and midgut. Two human herpes virus-associated miRNAs (hcmv-miR-UL70 3p and kshv-miR-K12-3) were upregulated in hemocytes of parasitized *L. dispar* (Gundersen-Rindal and Pedroni, 2010). These are among many miRNAs that have been hypothesized to act as suppressors of the immediate and early genes that respond to a viral infection (Murphy *et al.*, 2008).

The response of miRNAs to an infection has also been studied in mosquitoes. In *A. gambiae*, expression of miR-34, miR-1174, and miR-1175 decreases after *Plasmodium* infection, while that of miR-989 increases (Winter *et al.*, 2007). Minor changes in miRNA expression have been observed in *C. quinquefasciatus* after West Nile virus infection (Skalsky *et al.*, 2010), although miR-989 showed a 2.8-fold downregulation and miR-92 appeared somewhat upregulated. The expression of these two miRNAs has been studied in different mosquito species, and results have shown that miR-989 expression is restricted to females, and predominantly to the ovary (Winter *et al.*, 2007; Mead and Tu, 2008), although it was later detected in the midgut of *Ae. aegypti* (S. Li *et al.*, 2009). miR-92 has been related to embryonic development in *Ae. aegypti* (S. Li *et al.*, 2009) and *B. mori* (Liu *et al.*, 2009). Results of deregulation of miR-989 and miR-92 suggested to Skalsky and colleagues (2010) that their targets participate in mediating flavivirus infection of the mosquito host.

2.6.11. Apoptosis

A group of *D. melanogaster* miRNAs including, miR-278, miR-14, bantam, and miR-2, regulate cell proliferation and apoptosis, targeting a number of pro-apoptotic genes, like *hid*, which is repressed by bantam and miR-2 miRNAs (Brennecke *et al.*, 2003, 2005; Stark *et al.*, 2005).

One of the functions of mir-14 in *D. melanogaster* is suppressing cell death; therefore, loss of miR-14 is associated with a reduced lifespan, stress sensitivity, and increased levels of the apoptotic effector caspase Drice (Xu *et al.*, 2003). The same anti-apoptotic function has been found in Lepidopteran (*Spodoptera frugiperda*) Sf9 cells, where miR-14 is required for constitutive cell survival

(Kumarswamy and Chandna, 2010). However, the results do not exclude that additional miRNAs might also contribute to regulating Lepidopteran cell survival and death.

Finally, experiments depleting miRNA functions by injection of miRNA antisense nucleic acids in early embryos, which permits systematic loss-of-function analysis *in vivo*, have identified the miR-2/13 family and miR-6 as controlling apoptosis during *D. melanogaster* embryonic development through post-transcriptional repression of the proapoptotic proteins hid, grim, reaper, and sickle (Leaman *et al.*, 2005).

2.7. Conclusions and Perspectives

There are reasons to believe that the still-scarce data available on miRNAs are just the tip of an iceberg. Nevertheless, rapidly expanding information is making it increasingly obvious that miRNAs' contribution to the genomic output is not a sort of genetic oddity or "transcriptional background noise," but a class of key post-transcriptional regulators of gene expression. Indeed, genomic regulation cannot be completely understood without incorporating the role of miRNAs, which constitute a regulatory layer that works in concert with the mRNA and protein network. However, the field of miRNA study is still in the development phase, and there are many aspects – the bulk of the iceberg – that require further research.

The miRNA machinery appears to be more complex than previously thought, and rapid progress is unveiling many unexpected details. An example concerning the mechanisms responsible for stabilized or reduced miRNA expression is the discovery of specific cis-acting modifications and trans-acting proteins that affect miRNA half-life, which are revealing new elements that contribute to their homeostasis (Kai and Pasquinelli, 2010), and the identification of Dicer-independent miRNA biogenesis pathways, such as those using the catalytic activity of Ago-2 (Cheloufi *et al.*, 2010; Cifuentes *et al.*, 2010). Therefore, recent data suggest that such mechanistic aspects will have more surprises revealed as we continue to expand our understanding of them.

Moreover, further studies are required to elucidate how miRNA genes are regulated. There are contributions studying the influence of transcription factors acting on the promoter region of miRNA genes, like that of Pek and colleagues (2009) on the aforementioned repressor action of Maelstrom on the miR-7 promoter region. However, more studies in this line are needed if we wish to better understand the regulatory mechanisms mediated by miRNAs. Faster progress seems predictable in the field of cataloging miRNAs. The challenge is to find the unexpected, non-conserved miRNAs, and the new generation of algorithms will have to combine not only high-throughput approaches and powerful computational methods, but also expression data, genomic location, and structural and sequence features, as approached in the studies of Brennecke and Cohen (2003), Friedlander and colleagues (2008), and Mathelier and Carbone (2010).

miRNA target prediction started more than two decades ago with the serendipitous findings that emerged from miRNA target recognition (Wightman *et al.*, 1991, 1993; Lee *et al.*, 1993). The key principles were then applied to computational methods for miRNA target prediction (Bartel, 2009), and these methods soon allowed the prediction of hundreds of miRNA targets. However, computational prediction of miRNA targets still relies on the few principles defined more than 20 years ago, and, arguably, this will not help to unveil novel aspects of miRNA target mechanisms. Thus, unbiased approaches to studying the interaction of miRNA and target would be valuable in order to identify new principles of miRNA-target recognition, and to improve the systems for target prediction, as in the creative approach of Orom and Lund (2007).

Finally, the phase of predicting putative targets *in silico* following computational methods must be followed by experimental work to validate the predictions and to identify targets *in vivo*. In this line, specific miRNA silencing will be one of the most useful approaches, and entomologists and other non-biomedical researchers will benefit from the miRNA antagonists that are being designed in the context of biomedical studies in search of therapeutic agents against human diseases (Gao and Huang, 2009). In this sense, *D. melanogaster* and other insect species will continue to be the favorite models, given the advantages they offer, especially straightforward manipulation. Validation of targets will contribute to elucidating the place and role of miRNAs in the molecular network that regulates the development and homeostasis of biological processes. The networks describing the interaction of miRNAs, mRNAs, and proteins are presumably highly organized and complex, and their their study therefore represents a formidable challenge. However, it will be a worthwhile effort, because these networks are possibly the best approximation to the living world that is available with present means.

There is a fairly widespread opinion that proteins are what really matter in the functional landscape that shapes fitness, so transcript abundance is only useful as a mere proxy for the activity of the corresponding proteins (Feder and Walser, 2005). However, the expanding universe of small silencing RNAs (Ghildiyal and Zamore, 2009) and their widespread functional roles in genomic regulation show this to be yet one more old paradigm that is about to fall.

Acknowledgments

Financial support from the Ministry of Science and Innovation, Spain (Projects BFU2008-00484 to MDP and CGL2008-03517/BOS to XB) is gratefully acknowledged.

EDT has a post-doctoral contract from the CSIC (JAE-doc program) and MR a pre-doctoral research grant from the CSIC (JAE-pre program). Thanks are due to Stephen Cohen for providing **Figure 10**, and to Albert Masó for the photographs in **Figure 11**.

References

Ambros, V. (2004). The functions of animal microRNAs. *Nature, 431*, 350–355.

Ambros, V., & Horvitz, H. R. (1984). Heterochronic mutants of the nematode *Caenorhabditis elegans. Science, 226*, 409–416.

Ambros, V., Bartel, B., Bartel, D. P., Burge, C. B., Carrington, J. C., et al. (2003). A uniform system for microRNA annotation. *RNA, 9*, 277–279.

Aravin, A., & Tuschl, T. (2005). Identification and characterization of small RNAs involved in RNA silencing. *FEBS Lett., 579*, 5830–5840.

Bandyopadhyay, S., & Mitra, R. (2009). TargetMiner: microRNA target prediction with systematic identification of tissue-specific negative examples. *Bioinformatics, 25*, 2625–2631.

Bartel, D. P. (2004). MicroRNAs: Genomics, biogenesis, mechanism, and function. *Cell, 116*, 281–297.

Bartel, D. P. (2009). MicroRNAs: Target recognition and regulatory functions. *Cell, 136*, 215–233.

Bashirullah, A., Pasquinelli, A. E., Kiger, A. A., Perrimon, N., Ruvkun, G., et al. (2003). Coordinate regulation of small temporal RNAs at the onset of *Drosophila* metamorphosis. *Dev. Biol., 259*, 1–8.

Behm-Ansmant, I., Rehwinkel, J., Doerks, T., Stark, A., Bork, P., et al. (2006a). mRNA degradation by miRNAs and GW182 requires both CCR4:NOT deadenylase and DCP1:DCP2 decapping complexes. *Genes Dev., 20*, 1885–1898.

Behm-Ansmant, I., Rehwinkel, J., & Izaurralde, E. (2006b). MicroRNAs silence gene expression by repressing protein expression and/or by promoting mRNA decay. *Cold Spring Harb. Symp. Quant Biol., 71*, 523–530.

Behura, S. K. (2007). Insect microRNAs: Structure, function and evolution. *Insect Biochem. Mol. Biol., 37*, 3–9.

Beitzinger, M., & Meister, G. (2010). Preview. MicroRNAs: From decay to decoy. *Cell, 140*, 612–614.

Belles, X. (2010). Beyond *Drosophila*: RNAi *in vivo* and functional genomics in insects. *Annu. Rev. Entomol., 55*, 111–128.

Berezikov, E., Guryev, V., van de Belt, J., Wienholds, E., Plasterk, R. H., & Cuppen, E. (2005). Phylogenetic shadowing and computational identification of human microRNA genes. *Cell, 120*, 21–24.

Berezikov, E., Chung, W. J., Willis, J., Cuppen, E., & Lai, E. C. (2007). Mammalian mirtron genes. *Mol Cell, 28*, 328–336.

Berezikov, E., Liu, N., Flynt, A. S., Hodges, E., Rooks, M., et al. (2010). Evolutionary flux of canonical microRNAs and mirtrons in *Drosophila. Nat. Genet., 42*, 6–9, author reply 9–10.

Bernstein, E., Caudy, A. A., Hammond, S. M., & Hannon, G. J. (2001). Role for a bidentate ribonuclease in the initiation step of RNA interference. *Nature, 409*, 363–366.

Betel, D., Wilson, M., Gabow, A., Marks, D. S., & Sander, C. (2008). The microRNA.org resource: Targets and expression. *Nucleic Acids Res.*, *36*, D149–153.

Bhattacharyya, S. N., Habermacher, R., Martine, U., Closs, E. I., & Filipowicz, W. (2006). Relief of microRNA-mediated translational repression in human cells subjected to stress. *Cell*, *125*, 1111–1124.

Biryukova, I., Asmar, J., Abdesselem, H., & Heitzler, P. (2009). *Drosophila* mir-9a regulates wing development via fine-tuning expression of the LIM only factor, dLMO. *Dev. Biol.*, *327*, 487–496.

Borchert, G. M., Lanier, W., & Davidson, B. L. (2006). RNA polymerase III transcribes human microRNAs. *Nat. Struct. Mol. Biol.*, *13*, 1097–1101.

Boutla, A., Delidakis, C., & Tabler, M. (2003). Developmental defects by antisense-mediated inactivation of micro-RNAs 2 and 13 in *Drosophila* and the identification of putative target genes. *Nucleic Acids Res.*, *31*, 4973–4980.

Boutz, P. L., Chawla, G., Stoilov, P., & Black, D. L. (2007). MicroRNAs regulate the expression of the alternative splicing factor nPTB during muscle development. *Genes Dev.*, *21*, 71–84.

Brameier, M., & Wiuf, C. (2007). *Ab initio* identification of human microRNAs based on structure motifs. *BMC Bioinformatics*, *8*, 478.

Brennecke, J., & Cohen, S. M. (2003). Towards a complete description of the microRNA complement of animal genomes. *Genome Biol.*, *4*, 228.

Brennecke, J., Hipfner, D. R., Stark, A., Russell, R. B., & Cohen, S. M. (2003). bantam encodes a developmentally regulated microRNA that controls cell proliferation and regulates the proapoptotic gene hid in *Drosophila*. *Cell*, *113*, 25–36.

Brennecke, J., Stark, A., Russell, R. B., & Cohen, S. M. (2005). Principles of microRNA-target recognition. *PLoS Biol.*, *3*, e85.

Brown, S., Holtzman, S., Kaufman, T., & Denell, R. (1999). Characterization of the *Tribolium Deformed* ortholog and its ability to directly regulate *Deformed* target genes in the rescue of a *Drosophila Deformed* null mutant. *Dev. Genes Evol.*, *209*, 389–398.

Burgler, C., & Macdonald, P. M. (2005). Prediction and verification of microRNA targets by MovingTargets, a highly adaptable prediction method. *BMC Genomics*, *6*, 88.

Bushati, N., & Cohen, S. M. (2007). microRNA functions. *Annu. Rev. Cell. Dev. Biol.*, *23*, 175–205.

Cai, X., Hagedorn, C. H., & Cullen, B. R. (2004). Human microRNAs are processed from capped, polyadenylated transcripts that can also function as mRNAs. *RNA*, *10*, 1957–1966.

Cao, J., Tong, C., Wu, X., Lv, J., Yang, Z., et al. (2008). Identification of conserved microRNAs in *Bombyx mori* (silkworm) and regulation of fibroin L chain production by microRNAs in heterologous system. *Insect Biochem. Mol. Biol.*, *38*, 1066–1071.

Carrington, J. C., & Ambros, V. (2003). Role of microRNAs in plant and animal development. *Science*, *301*, 336–338.

Caygill, E. E., & Johnston, L. A. (2008). Temporal regulation of metamorphic processes in *Drosophila* by the let-7 and miR-125 heterochronic microRNAs. *Curr. Biol.*, *18*, 943–950.

Cayirlioglu, P., Kadow, I. G., Zhan, X., Okamura, K., Suh, G. S., et al. (2008). Hybrid neurons in a microRNA mutant are putative evolutionary intermediates in insect CO_2 sensory systems. *Science*, *319*, 1256–1260.

Cheloufi, S., Dos Santos, C. O., Chong, M. M., & Hannon, G. J. (2010). A Dicer-independent miRNA biogenesis pathway that requires Ago catalysis. *Nature*, *465*, 584–589.

Chen, Y., Sinha, K., Perumal, K., & Reddy, R. (2000). Effect of 3′ terminal adenylic acid residue on the uridylation of human small RNAs *in vitro* and in frog oocytes. *RNA*, *6*, 1277–1288.

Cheung, L., Gustavsson, C., Norstedt, G., & Tollet-Egnell, P. (2009). Sex-different and growth hormone-regulated expression of microRNA in rat liver. *BMC Mol. Biol.*, *10*, 13.

Chu, C. Y., & Rana, T. M. (2006). Translation repression in human cells by microRNA-induced gene silencing requires RCK/p54. *PLoS Biol.*, *4*, e210.

Cifuentes, D., Xue, H., Taylor, D. W., Patnode, H., Mishima, Y., et al. (2010). A novel miRNA processing pathway independent of Dicer requires Argonaute2 catalytic activity. *Science*, *328*, 1694–1698.

Ciudad, L., Piulachs, M. D., & Belles, X. (2006). Systemic RNAi of the cockroach vitellogenin receptor results in a phenotype similar to that of the *Drosophila* yolkless mutant. *FEBS J.*, *273*, 325–335.

Costa, F. F. (2007). Non-coding RNAs: Lost in translation? *Gene*, *386*, 1–10.

Cristno, A. S., Tanaka, E. D., Rubio, M., Piulachs, M.-D., & Belles, X. (2011). Deep sequencing of organ- and stage-specific microRNAs in the evolutionarily basal insect *Blattella germanica* (L.). PLOS One, 6(4), e19350.

Denli, A. M., Tops, B. B., Plasterk, R. H., Ketting, R. F., & Hannon, G. J. (2004). Processing of primary microRNAs by the Microprocessor complex. *Nature*, *432*, 231–235.

Diederichs, S., & Haber, D. A. (2007). Dual role for argonautes in microRNA processing and posttranscriptional regulation of microRNA expression. *Cell*, *131*, 1097–1108.

Duan, R., Pak, C., & Jin, P. (2007). Single nucleotide polymorphism associated with mature miR-125a alters the processing of pri-miRNA. *Hum. Mol. Genet.*, *16*, 1124–1131.

Edgar, B. A. (2006). From cell structure to transcription: Hippo forges a new path. *Cell*, *124*, 267–273.

Eiring, A. M., Harb, J. G., Neviani, P., Garton, C., Oaks, J. J., et al. (2010). miR-328 functions as an RNA decoy to modulate hnRNP E2 regulation of mRNA translation in leukemic blasts. *Cell*, *140*, 652–665.

Enright, A. J., John, B., Gaul, U., Tuschl, T., Sander, C., et al. (2003). MicroRNA targets in *Drosophila*. *Genome Biol.*, *5*, R1.

Fabian, M. R., Mathonnet, G., Sundermeier, T., Mathys, H., Zipprich, J. T., et al. (2009). Mammalian miRNA RISC recruits CAF1 and PABP to affect PABP-dependent deadenylation. *Mol. Cell*, *35*, 868–880.

Feder, M. E., & Walser, J. C. (2005). The biological limitations of transcriptomics in elucidating stress and stress responses. *J. Evol. Biol.*, *18*, 901–910.

Fire, A., Xu, S., Montgomery, M. K., Kostas, S. A., Driver, S. E., et al. (1998). Potent and specific genetic interference by double-stranded RNA in *Caenorhabditis elegans*. *Nature*, *391*, 806–811.

Flicek, P., Aken, B. L., Beal, K., Ballester, B., Caccamo, M., et al. (2008). Ensembl 2008. *Nucleic Acids Res.*, *36*, D707–714.

Forman, J. J., Legesse-Miller, A., & Coller, H. A. (2008). A search for conserved sequences in coding regions reveals that the let-7 microRNA targets Dicer within its coding sequence. *Proc. Natl. Acad. Sci. USA*, *105*, 14879–14884.

Friedlander, M. R., Chen, W., Adamidi, C., Maaskola, J., Einspanier, R., et al. (2008). Discovering microRNAs from deep sequencing data using miRDeep. *Nat. Biotechnol.*, *26*, 407–415.

Friedman, R. C., Farh, K. K., Burge, C. B., & Bartel, D. P. (2009). Most mammalian mRNAs are conserved targets of microRNAs. *Genome Res.*, *19*, 92–105.

Gaidatzis, D., van Nimwegen, E., Hausser, J., & Zavolan, M. (2007). Inference of miRNA targets using evolutionary conservation and pathway analysis. *BMC Bioinformatics*, *8*, 69.

Gao, K., & Huang, L. (2009). Nonviral methods for siRNA delivery. *Mol. Pharm.*, *6*, 651–658.

Ghildiyal, M., & Zamore, P. D. (2009). Small silencing RNAs: An expanding universe. *Nat. Rev. Genet.*, *10*, 94–108.

Ghildiyal, M., Xu, J., Seitz, H., Weng, Z., & Zamore, P. D. (2010). Sorting of *Drosophila* small silencing RNAs partitions microRNA* strands into the RNA interference pathway. *RNA*, *16*, 43–56.

Giraldez, A. J., Mishima, Y., Rihel, J., Grocock, R. J., Van Dongen, S., et al. (2006). Zebrafish MiR-430 promotes deadenylation and clearance of maternal mRNAs. *Science*, *312*, 75–79.

Gomez-Orte, E., & Belles, X. (2009). MicroRNA-dependent metamorphosis in hemimetabolan insects. *Proc. Natl. Acad. Sci. USA*, *106*, 21678–21682.

Grad, Y., Aach, J., Hayes, G. D., Reinhart, B. J., Church, G. M., Ruvkun, G., & Kim, J. (2003). Computational and experimental identification of *C. elegans* microRNAs. *Mol. Cell.*, *11*, 1253–1263.

Griffiths-Jones, S. (2006). miRBase: The microRNA sequence database. *Methods Mol. Biol.*, *342*, 129–138.

Griffiths-Jones, S., Saini, H. K., van Dongen, S., & Enright, A. J. (2008). miRBase: Tools for microRNA genomics. *Nucleic Acids Res.*, *36*, D154–158.

Grimson, A., Farh, K. K., Johnston, W. K., Garrett-Engele, P., Lim, L. P., et al. (2007). MicroRNA targeting specificity in mammals: Determinants beyond seed pairing. *Mol. Cell*, *27*, 91–105.

Grun, D., Wang, Y. L., Langenberger, D., Gunsalus, K. C., & Rajewsky, N. (2005). microRNA target predictions across seven *Drosophila* species and comparison to mammalian targets. *PLoS Comput. Biol.*, *1*, e13.

Gundersen-Rindal, D. E., & Pedroni, M. J. (2010). Larval stage *Lymantria dispar* microRNAs differentially expressed in response to parasitization by *Glyptapanteles flavicoxis* parasitoid. *Arch. Virol.*, *155*, 787–787.

Hamilton, A. J., & Baulcombe, D. C. (1999). A species of small antisense RNA in posttranscriptional gene silencing in plants. *Science*, *286*, 950–952.

Hammell, M., Long, D., Zhang, L., Lee, A., Carmack, C. S., et al. (2008). mirWIP: microRNA target prediction based on microRNA-containing ribonucleoprotein-enriched transcripts. *Nat. Methods*, *5*, 813–819.

Han, J., Lee, Y., Yeom, K. H., Kim, Y. K., Jin, H., et al. (2004). The Drosha–DGCR8 complex in primary microRNA processing. *Genes Dev.*, *18*, 3016–3027.

Hatfield, S. D., Shcherbata, H. R., Fischer, K. A., Nakahara, K., Carthew, R. W., & Ruohola-Baker, H. (2005). Stem cell division is regulated by the microRNA pathway. *Nature*, *435*, 974–978.

He, P. A., Nie, Z., Chen, J., Lv, Z., Sheng, Q., et al. (2008). Identification and characteristics of microRNAs from *Bombyx mori*. *BMC Genomics*, *9*, 248.

Heo, I., Joo, C., Cho, J., Ha, M., Han, J., et al. (2008). Lin28 mediates the terminal uridylation of let-7 precursor microRNA. *Mol. Cell*, *32*, 276–284.

Hertel, J., & Stadler, P. F. (2006). Hairpins in a haystack, recognizing microRNA precursors in comparative genomics data. *Bioinformatics*, *22*, e197–202.

Hilgers, V., Bushati, N., & Cohen, S. M. (2010). *Drosophila* microRNAs 263a/b confer robustness during development by protecting nascent sense organs from apoptosis. *PLoS Biol.*, *8*, e1000396.

Hipfner, D. R., Weigmann, K., & Cohen, S. M. (2002). The bantam gene regulates *Drosophila* growth. *Genetics*, *161*, 1527–1537.

Hornstein, E., & Shomron, N. (2006). Canalization of development by microRNAs. *Nat. Genet.*, *38*(Suppl), S20–24.

Huang, J. C., Babak, T., Corson, T. W., Chua, G., Khan, S., et al. (2007). Using expression profiling data to identify human microRNA targets. *Nat. Methods*, *4*, 1045–1049.

Hubbard, T. J., Aken, B. L., Beal, K., Ballester, B., Caccamo, M., et al. (2007). Ensembl 2007. *Nucleic Acids Res.*, *35*, D610–617.

Hughes, C. L., & Kaufman, T. C. (2000). RNAi analysis of *Deformed*, *proboscipedia* and *Sex combs reduced* in the milkweed bug *Oncopeltus fasciatus*: Novel roles for Hox genes in the hemipteran head. *Development*, *127*, 3683–3694.

Humphreys, D. T., Westman, B. J., Martin, D. I., & Preiss, T. (2005). MicroRNAs control translation initiation by inhibiting eukaryotic initiation factor 4E/cap and poly(A) tail function. *Proc. Natl. Acad. Sci. USA*, *102*, 16961–16966.

Hyun, S., Lee, J. H., Jin, H., Nam, J., Namkoong, B., et al. (2009). Conserved MicroRNA miR-8/miR-200 and its target USH/FOG2 control growth by regulating PI3K. *Cell*, *139*, 1096–1108.

Iovino, N., Pane, A., & Gaul, U. (2009). miR-184 has multiple roles in *Drosophila* female germline development. *Dev. Cell*, *17*, 123–133.

Jaubert, S., Mereau, A., Antoniewski, C., & Tagu, D. (2007). MicroRNAs in *Drosophila*: The magic wand to enter the Chamber of Secrets? *Biochimie*, *89*, 1211–1220.

Jaubert-Possamai, S., Rispe, C., Tanguy, S., Gordon, K., Walsh, T., et al. (2010). Expansion of the miRNA pathway in the hemipteran insect *Acyrthosiphon pisum*. *Mol. Biol. Evol.*, *27*, 979–987.

Jiang, P., Wu, H., Wang, W., Ma, W., Sun, X., et al. (2007). MiPred: Classification of real and pseudo microRNA precursors using random forest prediction model with combined features. *Nucleic Acids Res.*, *35*, W339–344.

Jin, Z., & Xie, T. (2007). Dcr-1 maintains *Drosophila* ovarian stem cells. *Curr. Biol.*, *17*, 539–544.

John, B., Enright, A. J., Aravin, A., Tuschl, T., Sander, C., et al. (2004). Human microRNA targets. *PLoS Biol.*, *2*, e363.

Jorgensen, R. (1990). Altered gene expression in plants due to trans interactions between homologous genes. *Trends Biotechnol.*, *8*, 340–344.

Kadener, S., Menet, J. S., Sugino, K., Horwich, M. D., Weissbein, U., et al. (2009a). A role for microRNAs in the *Drosophila* circadian clock. *Genes Dev.*, *23*, 2179–2191.

Kadener, S., Rodriguez, J., Abruzzi, K. C., Khodor, Y. L., Sugino, K., et al. (2009b). Genome-wide identification of targets of the drosha-pasha/DGCR8 complex. *RNA*, *15*, 537–545.

Kai, Z. S., & Pasquinelli, A. E. (2010). MicroRNA assassins: Factors that regulate the disappearance of miRNAs. *Nat. Struct. Mol. Biol.*, *17*, 5–10.

Karres, J. S., Hilgers, V., Carrera, I., Treisman, J., & Cohen, S. M. (2007). The conserved microRNA miR-8 tunes atrophin levels to prevent neurodegeneration in *Drosophila*. *Cell*, *131*, 136–145.

Katoh, T., Sakaguchi, Y., Miyauchi, K., Suzuki, T., Kashiwabara, S., et al. (2009). Selective stabilization of mammalian microRNAs by 3′ adenylation mediated by the cytoplasmic poly(A) polymerase GLD-2. *Genes Dev.*, *23*, 433–438.

Kawahara, Y., Zinshteyn, B., Sethupathy, P., Iizasa, H., Hatzigeorgiou, A. G., et al. (2007). Redirection of silencing targets by adenosine-to-inosine editing of miRNAs. *Science*, *315*, 1137–1140.

Kennerdell, J. R., & Carthew, R. W. (1998). Use of dsRNA-mediated genetic interference to demonstrate that *frizzled* and *frizzled 2* act in the *wingless* pathway. *Cell*, *95*, 1017–1026.

Kertesz, M., Iovino, N., Unnerstall, U., Gaul, U., & Segal, E. (2007). The role of site accessibility in microRNA target recognition. *Nat. Genet.*, *39*, 1278–1284.

Kim, S. K., Nam, J. W., Rhee, J. K., Lee, W. J., & Zhang, B. T. (2006). miTarget, microRNA target gene prediction using a support vector machine. *BMC Bioinformatics*, *7*, 411.

Kim, V. N. (2004). MicroRNA precursors in motion: exportin-5 mediates their nuclear export. *Trends Cell. Biol.*, *14*, 156–159.

Kim, Y. K., & Kim, V. N. (2007). Processing of intronic microRNAs. *EMBO J.*, *26*, 775–783.

Kiriakidou, M., Nelson, P. T., Kouranov, A., Fitziev, P., Bouyioukos, C., et al. (2004). A combined computational-experimental approach predicts human microRNA targets. *Genes Dev.*, *18*, 1165–1178.

Krek, A., Grun, D., Poy, M. N., Wolf, R., Rosenberg, L., et al. (2005). Combinatorial microRNA target predictions. *Nat. Genet.*, *37*, 495–500.

Kumarswamy, R., & Chandna, S. (2010). Inhibition of microRNA-14 contributes to actinomycin-D-induced apoptosis in the Sf9 insect cell line. *Cell Biol. Intl.*, *34*, 851–857.

Kwon, C., Han, Z., Olson, E. N., & Srivastava, D. (2005). MicroRNA1 influences cardiac differentiation in *Drosophila* and regulates Notch signaling. *Proc. Natl. Acad. Sci. USA*, *102*, 18986–18991.

Lagos-Quintana, M., Rauhut, R., Lendeckel, W., & Tuschl, T. (2001). Identification of novel genes coding for small expressed RNAs. *Science*, *294*, 853–858.

Lai, E. C. (2002). MicroRNAs are complementary to 3′ UTR sequence motifs that mediate negative post-transcriptional regulation. *Nat. Genet.*, *30*, 363–364.

Lai, E. C. (2003). microRNAs: Runts of the genome assert themselves. *Curr. Biol.*, *13*, R925–936.

Lai, E. C., Tomancak, P., Williams, R. W., & Rubin, G. M. (2003). Computational identification of *Drosophila* microRNA genes. *Genome Biol.*, *4*, R42.

Lai, E. C., Tam, B., & Rubin, G. M. (2005). Pervasive regulation of *Drosophila* Notch target genes by GY-box-, Brd-box-, and K-box-class microRNAs. *Genes Dev.*, *19*, 1067–1080.

Lau, N. C., Lim, L. P., Weinstein, E. G., & Bartel, D. P. (2001). An abundant class of tiny RNAs with probable regulatory roles in *Caenorhabditis elegans*. *Science*, *294*, 858–862.

Leaman, D., Chen, P. Y., Fak, J., Yalcin, A., Pearce, M., et al. (2005). Antisense-mediated depletion reveals essential and specific functions of microRNAs in *Drosophila* development. *Cell*, *121*, 1097–1108.

Lee, I., Ajay, S. S., Yook, J. I., Kim, H. S., Hong, S. H., et al. (2009). New class of microRNA targets containing simultaneous 5′-UTR and 3′-UTR interaction sites. *Genome Res.*, *19*, 1175–1183.

Lee, R. C., & Ambros, V. (2001). An extensive class of small RNAs in *Caenorhabditis elegans*. *Science*, *294*, 862–864.

Lee, R. C., Feinbaum, R. L., & Ambros, V. (1993). The *C. elegans* heterochronic gene lin-4 encodes small RNAs with antisense complementarity to lin-14. *Cell*, *75*, 843–854.

Lee, Y., Ahn, C., Han, J., Choi, H., Kim, J., et al. (2003). The nuclear RNase III Drosha initiates microRNA processing. *Nature*, *425*, 415–419.

Lee, Y., Kim, M., Han, J., Yeom, K. H., Lee, S., et al. (2004a). MicroRNA genes are transcribed by RNA polymerase II *EMBO J.*, *23*, 4051–4060.

Lee, Y. S., Nakahara, K., Pham, J. W., Kim, K., He, Z., et al. (2004b). Distinct roles for *Drosophila* Dicer-1 and Dicer-2 in the siRNA/miRNA silencing pathways. *Cell*, *117*, 69–81.

Legeai, F., Rizk, G., Walsh, T., Edwards, O., Gordon, K., et al. (2010). Bioinformatic prediction, deep sequencing of microRNAs and expression analysis during phenotypic plasticity in the pea aphid, *Acyrthosiphon pisum*. *BMC Genomics*, *11*, 281.

Lewis, B. P., Shih, I. H., Jones-Rhoades, M. W., Bartel, D. P., & Burge, C. B. (2003). Prediction of mammalian microRNA targets. *Cell*, *115*, 787–798.

Lewis, B. P., Burge, C. B., & Bartel, D. P. (2005). Conserved seed pairing, often flanked by adenosines, indicates that thousands of human genes are microRNA targets. *Cell*, *120*, 15–20.

Li, S., Mead, E. A., Liang, S., & Tu, Z. (2009). Direct sequencing and expression analysis of a large number of miRNAs in *Aedes aegypti* and a multi-species survey of novel mosquito miRNAs. *BMC Genomics*, *10*, 581.

Li, X., & Carthew, R. W. (2005). A microRNA mediates EGF receptor signaling and promotes photoreceptor differentiation in the *Drosophila* eye. *Cell*, *123*, 1267–1277.

Li, X., Cassidy, J. J., Reinke, C. A., Fischboeck, S., & Carthew, R. W. (2009). A microRNA imparts robustness against environmental fluctuation during development. *Cell*, *137*, 273–282.

Li, Y., Wang, F., Lee, J. A., & Gao, F. B. (2006). MicroRNA-9a ensures the precise specification of sensory organ precursors in *Drosophila*. *Genes Dev.*, *20*, 2793–2805.

Lim, L. P., Lau, N. C., Weinstein, E. G., Abdelhakim, A., Yekta, S., et al. (2003a). The microRNAs of *Caenorhabditis elegans*. *Genes Dev.*, *17*, 991–1008.

Lim, L. P., Glasner, M. E., Yekta, S., Burge, C. B., & Bartel, D. P. (2003b). Vertebrate microRNA genes. *Science, 299,* 1540.

Liu, H., Yue, D., Chen, Y., Gao, S. J., & Huang, Y. (2010). Improving performance of mammalian microRNA target prediction. *BMC Bioinformatics, 11,* 476.

Liu, J., Valencia-Sanchez, M. A., Hannon, G. J., & Parker, R. (2005). MicroRNA-dependent localization of targeted mRNAs to mammalian P-bodies. *Nat. Cell. Biol., 7,* 719–723.

Liu, S., Zhang, L., Li, Q., Zhao, P., Duan, J., et al. (2009). MicroRNA expression profiling during the life cycle of the silkworm (*Bombyx mori*). *BMC Genomics, 10,* 455.

Liu, S., Gao, S., Zhang, D., Yin, J., Xiang, Z., et al. (2010). MicroRNAs show diverse and dynamic expression patterns in multiple tissues of *Bombyx mori*. *BMC Genomics, 11,* 85.

Lund, E., Guttinger, S., Calado, A., Dahlberg, J. E., & Kutay, U. (2004). Nuclear export of microRNA precursors. *Science, 303,* 95–98.

Martin, D., Maestro, O., Cruz, J., Mane-Padros, D., & Belles, X. (2006). RNAi studies reveal a conserved role for RXR in molting in the cockroach *Blattella germanica*. *J. Insect Physiol., 52,* 410–416.

Mathelier, A., & Carbone, A. (2010). MIReNA: Finding microRNAs with high accuracy and no learning at genome scale and from deep sequencing data. *Bioinformatics, 26,* 2226–2234.

Mead, E. A., & Tu, Z. (2008). Cloning, characterization, and expression of microRNAs from the Asian malaria mosquito *Anopheles stephensi*. *BMC Genomics, 9,* 244.

Metzker, M. L. (2010). Sequencing technologies – the next generation. *Nat. Rev. Genet., 11,* 31–46.

Miranda, K. C., Huynh, T., Tay, Y., Ang, Y. S., Tam, W. L., et al. (2006). A pattern-based method for the identification of microRNA binding sites and their corresponding heteroduplexes. *Cell, 126,* 1203–1217.

Murphy, E., Vanicek, J., Robins, H., Shenk, T., & Levine, A. J. (2008). Suppression of immediate-early viral gene expression by herpesvirus-coded microRNAs: Implications for latency. *Proc. Natl. Acad. Sci. USA, 105,* 5453–5458.

Nam, J. W., Shin, K. R., Han, J., Lee, Y., Kim, V. N., et al. (2005). Human microRNA prediction through a probabilistic co-learning model of sequence and structure. *Nucleic Acids Res., 33,* 3570–3581.

Neilson, J. R., & Sharp, P. A. (2008). Small RNA regulators of gene expression. *Cell, 134,* 899–902.

Ng, K. L., & Mishra, S. K. (2007). *De novo* SVM classification of precursor microRNAs from genomic pseudo hairpins using global and intrinsic folding measures. *Bioinformatics, 23,* 1321–1330.

Nguyen, H. T., & Frasch, M. (2006). MicroRNAs in muscle differentiation: Lessons from *Drosophila* and beyond. *Curr. Opin. Genet. Dev., 16,* 533–539.

Okamura, K., Hagen, J. W., Duan, H., Tyler, D. M., & Lai, E. C. (2007). The mirtron pathway generates microRNA-class regulatory RNAs in *Drosophila*. *Cell, 130,* 89–100.

Okamura, K., Phillips, M. D., Tyler, D. M., Duan, H., Chou, Y. T., et al. (2008). The regulatory activity of microRNA* species has substantial influence on microRNA and 3′ UTR evolution. *Nat. Struct. Mol. Biol., 15,* 354–363.

Olsen, P. H., & Ambros, V. (1999). The lin-4 regulatory RNA controls developmental timing in *Caenorhabditis elegans* by blocking LIN-14 protein synthesis after the initiation of translation. *Dev. Biol., 216,* 671–680.

Orom, U. A., & Lund, A. H. (2007). Isolation of microRNA targets using biotinylated synthetic microRNAs. *Methods, 43,* 162–165.

Orom, U. A., Nielsen, F. C., & Lund, A. H. (2008). MicroRNA-10a binds the 5′ UTR of ribosomal protein mRNAs and enhances their translation. *Mol. Cell, 30,* 460–471.

Pasquinelli, A. E., Reinhart, B. J., Slack, F., Martindale, M. Q., Kuroda, M. I., et al. (2000). Conservation of the sequence and temporal expression of let-7 heterochronic regulatory RNA. *Nature, 408,* 86–89.

Pek, J. W., Lim, A. K., & Kai, T. (2009). *Drosophila* maelstrom ensures proper germline stem cell lineage differentiation by repressing microRNA-7. *Dev. Cell, 17,* 417–424.

Rajewsky, N., & Socci, N. D. (2004). Computational identification of microRNA targets. *Dev. Biol., 267,* 529–535.

Rehmsmeier, M., Steffen, P., Hochsmann, M., & Giegerich, R. (2004). Fast and effective prediction of microRNA/target duplexes. *RNA, 10,* 1507–1517.

Reinhart, B. J., Slack, F. J., Basson, M., Pasquinelli, A. E., Bettinger, J. C., et al. (2000). The 21-nucleotide let-7 RNA regulates developmental timing in *Caenorhabditis elegans*. *Nature, 403,* 901–906.

Rodriguez, A., Griffiths-Jones, S., Ashurst, J. L., & Bradley, A. (2004). Identification of mammalian microRNA host genes and transcription units. *Genome Res., 14,* 1902–1910.

Ronshaugen, M., Biemar, F., Piel, J., Levine, M., & Lai, E. C. (2005). The *Drosophila* microRNA iab-4 causes a dominant homeotic transformation of halteres to wings. *Genes Dev., 19,* 2947–2952.

Ruby, J. G., Jan, C. H., & Bartel, D. P. (2007a). Intronic microRNA precursors that bypass Drosha processing. *Nature, 448,* 83–86.

Ruby, J. G., Stark, A., Johnston, W. K., Kellis, M., Bartel, D. P., et al. (2007b). Evolution, biogenesis, expression, and target predictions of a substantially expanded set of *Drosophila* microRNAs. *Genome Res., 17,* 1850–1864.

Rusinov, V., Baev, V., Minkov, I. N., & Tabler, M. (2005). MicroInspector, a web tool for detection of miRNA binding sites in an RNA sequence. *Nucleic Acids Res., 33,* 696–700.

Saetrom, O., Snove, O., Jr., & Saetrom, P. (2005). Weighted sequence motifs as an improved seeding step in microRNA target prediction algorithms. *RNA, 11,* 995–1003.

Saito, K., Ishizuka, A., Siomi, H., & Siomi, M. C. (2005). Processing of pre-microRNAs by the Dicer-1-Loquacious complex in *Drosophila* cells. *PLoS Biol., 3,* e235.

Seitz, H., Ghildiyal, M., & Zamore, P. D. (2008). Argonaute loading improves the 5′ precision of both MicroRNAs and their miRNA* strands in flies. *Curr. Biol., 18,* 147–151.

Sempere, L. F., Sokol, N. S., Dubrovsky, E. B., Berger, E. M., & Ambros, V. (2003). Temporal regulation of microRNA expression in *Drosophila melanogaster* mediated by hormonal signals and broad-Complex gene activity. *Dev. Biol., 259,* 9–18.

Sewer, A., Paul, N., Landgraf, P., Aravin, A., Pfeffer, S., et al. (2005). Identification of clustered microRNAs using an *ab initio* prediction method. *BMC Bioinformatics, 6,* 267.

Shcherbata, H. R., Ward, E. J., Fischer, K. A., Yu, J. Y., Reynolds, S. H., et al. (2007). Stage-specific differences in the requirements for germline stem cell maintenance in the *Drosophila* ovary. *Cell Stem Cell, 1,* 698–709.

Sheng, Y., Engstrom, P. G., & Lenhard, B. (2007). Mammalian microRNA prediction through a support vector machine model of sequence and structure. *PLoS One, 2,* e946.

Skalsky, R. L., Vanlandingham, D. L., Scholle, F., Higgs, S., & Cullen, B. R. (2010). Identification of microRNAs expressed in two mosquito vectors, *Aedes albopictus* and *Culex quinquefasciatus. BMC Genomics, 11,* 119.

Sokol, N. S., & Ambros, V. (2005). Mesodermally expressed *Drosophila* microRNA-1 is regulated by Twist and is required in muscles during larval growth. *Genes Dev., 19,* 2343–2354.

Sokol, N. S., Xu, P., Jan, Y. N., & Ambros, V. (2008). *Drosophila* let-7 microRNA is required for remodeling of the neuromusculature during metamorphosis. *Genes Dev., 22,* 1591–1596.

Stark, A., Brennecke, J., Bushati, N., Russell, R. B., & Cohen, S. M. (2005). Animal MicroRNAs confer robustness to gene expression and have a significant impact on 3′ UTR evolution. *Cell, 123,* 1133–1146.

Sturm, M., Hackenberg, M., Langenberger, D., & Frishman, D. (2010). TargetSpy: A supervised machine learning approach for microRNA target prediction. *BMC Bioinformatics, 11,* 292.

Teleman, A. A., Maitra, S., & Cohen, S. M. (2006). *Drosophila* lacking microRNA miR-278 are defective in energy homeostasis. *Genes Dev., 20,* 417–422.

Thadani, R., & Tammi, M. T. (2006). MicroTar, predicting microRNA targets from RNA duplexes. *BMC Bioinformatics, 7*(Suppl. 5), S20.

Thompson, B. J., & Cohen, S. M. (2006). The Hippo pathway regulates the bantam microRNA to control cell proliferation and apoptosis in *Drosophila. Cell, 126,* 767–774.

van Dongen, S., Abreu-Goodger, C., & Enright, A. J. (2008). Detecting microRNA binding and siRNA off-target effects from expression data. *Nat. Methods, 5,* 1023–1025.

Varghese, J., & Cohen, S. M. (2007). microRNA miR-14 acts to modulate a positive autoregulatory loop controlling steroid hormone signaling in *Drosophila. Genes Dev., 21,* 2277–2282.

Vasudevan, S., Tong, Y., & Steitz, J. A. (2007). Switching from repression to activation: microRNAs can up-regulate translation. *Science, 318,* 1931–1934.

Viswanathan, S. R., Daley, G. Q., & Gregory, R. I. (2008). Selective blockade of microRNA processing by Lin28. *Science, 320,* 97–100.

Wang, X., & El Naqa, I. M. (2008). Prediction of both conserved and nonconserved microRNA targets in animals. *Bioinformatics, 24,* 325–332.

Wang, X., Zhang, J., Li, F., Gu, J., He, T., et al. (2005). MicroRNA identification based on sequence and structure alignment. *Bioinformatics, 21,* 3610–3614.

Weaver, D. B., Anzola, J. M., Evans, J. D., Reid, J. G., Reese, J. T., et al. (2007). Computational and transcriptional evidence for microRNAs in the honey bee genome. *Genome Biol., 8,* R97.

Wei, Y., Chen, S., Yang, P., Ma, Z., & Kang, L. (2009). Characterization and comparative profiling of the small RNA transcriptomes in two phases of locust. *Genome Biol., 10,* R6.

Wightman, B., Burglin, T. R., Gatto, J., Arasu, P., & Ruvkun, G. (1991). Negative regulatory sequences in the lin-14 3′-untranslated region are necessary to generate a temporal switch during *Caenorhabditis elegans* development. *Genes Dev., 5,* 1813–1824.

Wightman, B., Ha, I., & Ruvkun, G. (1993). Posttranscriptional regulation of the heterochronic gene lin-14 by lin-4 mediates temporal pattern formation in *C. elegans. Cell, 75,* 855–862.

Winter, F., Edaye, S., Huttenhofer, A., & Brunel, C. (2007). *Anopheles gambiae* miRNAs as actors of defence reaction against *Plasmodium* invasion. *Nucleic Acids Res., 35,* 6953–6962.

Wu, L., Fan, J., & Belasco, J. G. (2006). MicroRNAs direct rapid deadenylation of mRNA. *Proc. Natl. Acad. Sci. USA, 103,* 4034–4039.

Xu, P., Vernooy, S. Y., Guo, M., & Hay, B. A. (2003). The *Drosophila* microRNA Mir-14 suppresses cell death and is required for normal fat metabolism. *Curr. Biol., 13,* 790–795.

Xue, C., Li, F., He, T., Liu, G. P., Li, Y., & Zhang, X. (2005). Classification of real and pseudo microRNA precursors using local structure-sequence features and support vector machine. *BMC Bioinformatics, 6,* 310.

Yang, J. H., Shao, P., Zhou, H., Chen, Y. Q., & Qu, L. H. (2010). deepBase: A database for deeply annotating and mining deep sequencing data. *Nucleic Acids Res., 38,* D123–130.

Yang, Y., Wang, Y. P., & Li, K. B. (2008). MiRTif, a support vector machine-based microRNA target interaction filter. *BMC Bioinformatics, 9*(Suppl. 12), S4.

Yi, R., Qin, Y., Macara, I. G., & Cullen, B. R. (2003). Exportin-5 mediates the nuclear export of pre-microRNAs and short hairpin RNAs. *Genes Dev., 17,* 3011–3016.

Yousef, M., Nebozhyn, M., Shatkay, H., Kanterakis, S., Showe, L. C., et al. (2006). Combining multi-species genomic data for microRNA identification using a Naive Bayes classifier. *Bioinformatics, 22,* 1325–1334.

Yousef, M., Jung, S., Kossenkov, A. V., Showe, L. C., & Showe, M. K. (2007). Naive Bayes for microRNA target predictions-machine learning for microRNA targets. *Bioinformatics, 23,* 2987–2992.

Yousef, M., Jung, S., Showe, L. C., & Showe, M. K. (2008). Learning from positive examples when the negative class is undetermined-microRNA gene identification. *Algorithms Mol. Biol., 3,* 2.

Yu, J. Y., Reynolds, S. H., Hatfield, S. D., Shcherbata, H. R., Fischer, K. A., et al. (2009). Dicer-1-dependent Dacapo suppression acts downstream of Insulin receptor in regulating cell division of *Drosophila* germline stem cells. *Development, 136,* 1497–1507.

Yu, X., Zhou, Q., Li, S. C., Luo, Q., Cai, Y., et al. (2008). The silkworm (*Bombyx mori*) microRNAs and their expressions in multiple developmental stages. *PLoS One, 3,* e2997.

3 Insect Transposable Elements

Zhijian Tu
Department of Biochemistry,
Virginia Tech, Blacksburg, VA, USA

Abbreviations

env, envelope protein
endo-siRNA, endogenous small interfering RNA
EST, expressed sequence tag
gag, group-associated antigene, or group-specific antigen
IN, integrase
LINEs, long interspersed repetitive (or nuclear) elements
LTR, long terminal repeat
MITEs, miniature inverted-repeat transposable elements
ORF, open reading frame

piRNA, Piwi-interacting RNA
PR, protease
RH, RNase H
RT, reverse transcriptase
SINEs, short interspersed repetitive (or nuclear) elements
siRNA, small interfering RNA
TE, transposable element
TIR, terminal inverted repeats
TSD, target site duplication

DOI:10.1016/B978-0-12-384747-8.10003-0

3.1. Introduction

More than half a century ago, Barbara McClintock's observation of unstable mutations in maize led to the discovery of two mobile genetic elements, *Activator* (*Ac*) and *Dissociator* (*Ds*) (McClintock, 1948, 1950). Her discovery of these mobile segments of DNA, later named transposable elements (TEs), set forth the revolutionary concept of a fluid and dynamic genome. Six decades later, as biology is entering the post-genomic era, there is renewed and rapidly growing appreciation for the tremendous diversity of TEs and their evolutionary impact.

Being mobile, TEs have the ability to replicate and spread in the genome as primarily "selfish" genetic units (Doolittle and Sapienza, 1980; Orgel and Crick, 1980). They tend to occupy significant portions of the eukaryotic genome. For example, at least 46% of the human genome (Lander *et al.*, 2001) and 47% of the yellow fever mosquito genome (Nene *et al.*, 2007) are TE-derived sequences. The relative abundance and diversity of TEs have contributed to the differences in the structure and size of eukaryotic genomes (Kidwell, 2002; Feschotte and Pritham, 2007). TE insertion and recombination are major sources of potentially detrimental mutations, and the host genomes have evolved sophisticated mechanisms to control TE activity (see, for example, Hartl *et al.*, 1997; Malone and Hannon, 2009). The same insertional or recombinatory activities by TEs generate a great deal of genetic and genomic plasticity, and provide the raw material for adaptive evolution (Kidwell and Lisch, 2000; Brookfield, 2005; Feschotte and Pritham, 2007; Lin *et al.*, 2007; Cordaux and Batzer, 2009; Gonzalez and Petrov, 2009). For example, TEs have reshaped the human genome by ectopic rearrangements, by creating new genes, and by modifying and shuffling existing genes (Lander *et al.*, 2001; Muotri *et al.*, 2007; Cordaux and Batzer, 2009). In some cases, TEs had been co-opted to perform critical functions in the biology of their host. One well-documented example is the generation of the extensive array of immunoglobulins and T-cell receptors by V(D)J recombination, which is evolved from an ancient transposition system (Gellert, 2002; Fugmann, 2010). Another example is the maintenance of telomeric structures in *Drosophila melanogaster* by site-specific insertions of two TEs (Pardue and DeBaryshe, 2002; Mason *et al.*, 2008). Therefore, the "selfish" TEs could evolve a wide spectrum of relationships with their hosts, ranging from "junk parasites" to "molecular symbionts" (Brookfield, 1995; Kidwell and Lisch, 2000). Moreover, the apparent arms race between TEs and the host genomes has driven the evolution of the recently discovered piRNA (Piwi-interacting RNA) and endogenous small interfering RNA (endo-siRNA) pathways, which may have had profound impacts on gene regulation and epigenetic silencing (Aravin *et al.*, 2007; Nishida *et al.*, 2007; Pelisson *et al.*, 2007; Yin and Lin, 2007; Brennecke *et al.*, 2008; Chung *et al.*, 2008; Ghildiyal *et al.*, 2008; Klattenhoff *et al.*, 2009; Lau *et al.*, 2009; Lisch, 2009; Malone and Hannon, 2009; Zeh *et al.*, 2009).

The intricate dynamic between TEs and their host genomes is further complicated by the fact that some TEs are capable of crossing species barriers to spread in a new genome. Such a process is referred to as horizontal (or lateral) transfer, which is distinct from the vertical transmission of genetic material from ancestral species/organisms to their descendants. Horizontal transfer may be an important part of the life cycle of some TEs, and it may contribute to their continued success during evolution (Silva *et al.*, 2004). The recent explosion of genome sequencing projects has provided convenient resources and revealed additional examples and novel insights into horizontal transfer (e.g., Biedler *et al.*, 2007; Loreto *et al.*, 2008; Bartolome *et al.*, 2009; Gilbert *et al.*, 2010; Schaak *et al.*, 2010; Thomas *et al.*, 2010a).

From an applied perspective, TEs have been used as tools to genetically manipulate cells/organisms, taking advantage of their ability to integrate cognate DNA in the genome. A well-known example is the transformation system derived from the *D. melanogaster P* transposable element, which has been instrumental to our understanding of this model organism by providing transformation and mutagenesis tools (see Chapter 4). In addition, some TEs have been used as genetic markers for mapping and population studies, taking advantage of their dimorphic insertion states (presence and absence of an insertion) and their interspersed distribution in the genome. For example, the human *Alu* elements have been shown to be useful population genetic markers (Batzer *et al.*, 1994; Batzer and Deininger, 2002; Salem *et al.*, 2003; Ray, 2007). Similar types of markers have been used to trace the explosive speciation of the Cichlid fishes and other vertebrates (Shedlock and Okada, 2000; Terai *et al.*, 2003). TEs have also been used as markers in insects to study incipient speciation, and to map resistance genes (Barnes *et al.*, 2005; Bonin *et al.*, 2008, 2009; Santolamazza *et al.*, 2008).

In this chapter, I provide an update to the previous review (Tu, 2005) and focus on recent advances in the study of insect TEs. A brief introduction on TE classification and transposition mechanisms will be followed by sections that descibe the current approaches to studying insect TEs, and sections that highlight the impact and the evolutionary dynamics of TEs in insect genomes. Applications of TEs in genetic and molecular analysis of insects will be discussed towards the end of the chapter. Readers may consult recent reviews (see Chapter 4 in this volume; also Tu and Coates, 2004; Feschotte and Pritham, 2007; Atkinson, 2008; Loreto *et al.*, 2008; Malone and Hannon, 2009; Zeh *et al.*, 2009; Schaack *et al.*, 2010), and the second edition of a book on Mobile DNA (Craig *et al.*, 2002) for details on related topics.

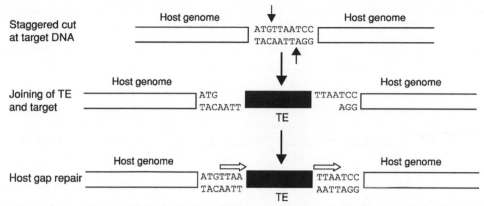

Figure 1 Mechanism of generating target site duplication (TSD). Both sides of the TSD are not part of the TE sequence; they are target sequences duplicated upon a TE insertion. Most TEs create TSDs, although the *Helitron* DNA transposons and some non-LTR retrotransposons do not. See recent reviews (Kapitonov and Jurka, 2001; Craig, 2002; Eickbush and Malik, 2002) for illustrations of the transposition mechanisms of different types of TEs.

3.2. Classification and Transposition Mechanisms of Eukaryotic Transposable Elements

TEs can be categorized as Class I RNA-mediated or Class II DNA-mediated elements, according to their transposition mechanisms (Finnegan, 1992). The transposition of RNA-mediated TEs involves a reverse transcription step, which generates cDNA from RNA molecules (Eickbush and Malik, 2002). The cDNA molecules are then integrated in the genome, allowing replicative amplification. The transposition of DNA-mediated elements is directly from DNA to DNA, which does not involve an RNA intermediate (Craig, 2002). In most cases, both classes of TEs will create target site duplication (TSD) upon their insertion in the genome (**Figure 1**). Both DNA-mediated and RNA-mediated elements can be further categorized into different groups. There have been several reviews on different classes of TEs (Deininger and Roy-Engel, 2002; Eickbush and Malik, 2002; Feschotte *et al.*, 2002; Robertson, 2002). All groups of TEs discussed here have been found in various species of insects.

3.2.1. Class I RNA-Mediated TEs

RNA-mediated TEs include long terminal repeat (LTR) retrotransposons, non-LTR retrotransposons, and short interspersed repetitive/nuclear elements (SINEs). Non-LTR retrotransposons are also referred to as retroposons or long interspersed repetitive/nuclear elements (LINEs). The structural features of the three groups of RNA-mediated TEs are illustrated in **Figure 2**, using representatives from different insects. All RNA-mediated TEs produce RNA transcripts that are reverse transcribed into cDNA to be integrated in the genome (Eickbush and Malik, 2002). Detailed mechanisms used by LTR

and non-LTR retrotransposons are elegantly described in recent reviews (Eickbush, 2002; Voytas and Boeke, 2002; Eickbush and Jamburuthugoda, 2008).

3.2.1.1. LTR retrotransposons   LTR retrotransposons transpose through a mechanism much like that used by retroviruses. The LTRs in the LTR retrotransposons are generally 200–500 bp long, and are involved in all aspects of their life cycle that include providing promoter sequences and transcription termination signals (Eickbush and Malik, 2002). As shown in **Figure 2**, LTR retrotransposons encode a *pol* (polymerase)-like protein that contains reverse transcriptase (RT), ribonuclease H (RNase H), protease (PR), and integrase (IN) domains that are important for their retrotransposition. The RT domain performs the key function of reverse transcription, and its sequence has been used for phylogenetic classification of LTR retrotransposons into four clades: Ty1/copia; Ty3/*gypsy*; BEL; and DIRS (Eickbush and Malik, 2002). The IN domain is responsible for inserting the cDNA copy into the host genome. In addition to the *pol*-like protein, LTR retrotransposons encode an additional protein related to the retroviral *gag* (group-associated antigene, or group-specific antigen) protein that binds nucleic acids or forms the nucleocapsid shell. Some LTR retrotransposons also have an *env* (envelope)-like fragment that encodes a transmembrane receptor-binding protein that allows the transmission of retroviruses. Some of the LTR retrotransposons that encode an *env* protein are in fact retroviruses (Eickbush and Malik, 2002). Some LTR retrotransposons use a tyrosine recombinase instead of the integrase to integrate into the host genome (Eickbush and Jamburuthugoda, 2008).

3.2.1.2. Non-LTR retrotransposons   Non-LTR retrotransposons, or LINEs, or retroposons, are generally

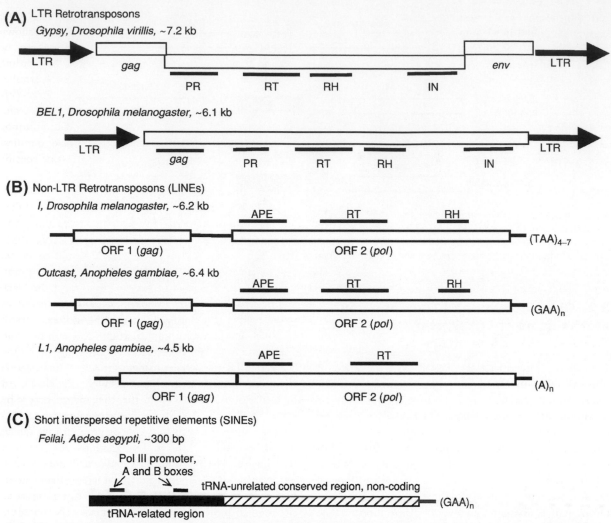

Figure 2 Structural characteristics of representative Class I RNA-mediated transposable elements in insects. Representatives are shown from three major groups, long terminal repeat (LTR) retrotransposons (A), non-LTR retrotransposons (B), and short interspersed repetitive elements (SINEs, (C)). The name of each representative element, its host species, and its approximate length are shown as the heading. Open reading frames (ORFs) are shown as open boxes. *Env*, envelope protein; *gag*, group-associated antigene, or group-specific antigen; IN, integrase; LINEs, long interspersed repetitive elements; LTR, long terminal repeat; PR, protease; RH, RNase H; RT, reverse transcriptase. The elements are not drawn to scale. References for information on these RNA-mediated elements are as following: *BEL1* (Davis and Judd, 1995); *Feilai* (Tu, 1999); *Gypsy* (Mizrokhi and Mazo, 1991); *I* (Fawcett *et al.*, 1986); *L1* (Biedler and Tu, 2003); *Outcast* (Biedler and Tu, 2003).

3–8 kilobases long, and have been found in virtually all eukaryotes studied. Like the LTR retrotransposons, most non-LTR retrotransposons also have a *pol*-like protein that includes an RT domain which is essential for their retrotransposition. The RT domain has been used for phylogenetic classification of non-LTR retrotransposons into 17 clades, most of which probably date back to the Precambrian era, approximately 600 million years ago (Eickbush and Malik, 2002; Biedler and Tu, 2003). Some elements also have an RNase H and/or AP endonuclease (APE) domain encoded in the *pol*-like open reading frame. In addition to the *pol*-like protein, many non-LTR retrotransposons encode an additional protein related to the retroviral *gag* protein. Studies of a *gag*-like protein

from L1 retrotransposon in mice show that it acts as a nucleic acid chaperone (Martin and Bushman, 2001). Other typical structural characteristics found in various non-LTR families are internal pol II promoters and 3′ ends containing AATAAA polyadenylation signals, poly (A) tails, or simple tandem repeats. Target Primed Reverse Transcription has been proposed as the mechanism of retrotransposition for R2 of *Bombyx mori*, and this may be generally true for all non-LTR elements (Luan *et al.*, 1993; Eickbush, 2002; Eickbush and Jamburuthugoda, 2008). Because they transpose by Target Primed Reverse Transcription, some non-LTR retrotransposons could rely rather heavily on host DNA repair mechanisms. This relationship with the host may give non-LTR retrotransposons some flexibility with regard to the domains required in an autonomous element (Eickbush and Malik, 2002). Some non-LTR retrotransposons, such as R2, are site-specific, because their endonucleases make precise cleavage at specific targets (Eickbush, 2002).

3.2.1.3. SINEs

SINEs are generally between 100 and 500 bp long. Unlike LTR and non-LTR retrotransposons, SINEs do not have any coding potential. SINEs may have been borrowing the retrotransposition machinery from autonomous non-LTR retrotransposons, which may be facilitated by similar sequences or structures at the 3′ ends of a SINE and its "partner" non-LTR retrotransposon (Ohshima *et al.*, 1996; Okada and Hamada, 1997; Dewannieux *et al.*, 2003; Dewannieux and Heidmann, 2005). Unlike non-LTR retrotransposons that use internal Pol II promoters, SINE transcription is directed from their own Pol III promoters which are similar to those found in small RNA genes. SINEs can be further divided into three groups based on similarities of their 5′ sequences to different types of small RNA genes. Elements such as the primate *Alu* family share sequence similarities with 7SL RNA (Jurka, 1995), while most other SINEs belong to a different group that share sequence similarities to tRNA molecules (Adams *et al.*, 1986; Okada, 1991; Tu, 1999; Tu *et al.*, 2004; Luchetti and Mantovani, 2009; Xu *et al.*, 2010). Recently, a new group of SINEs, named *SINE3*, have been discovered in the zebrafish genome, which share similarities to 5S rRNA (Kapitonov and Jurka, 2003a). Some non-LTR retrotransposons tend to generate truncated copies due to incomplete reverse transcription during cDNA synthesis. Although these short copies of RNA-mediated TEs are also called SINEs (Malik and Eickbush, 1998), they should not be confused with the true SINEs that use Pol III promoters.

3.2.2. Class II DNA-Mediated TEs

DNA-mediated TEs include cut-and-paste DNA transposons (**Figure 3**), miniature inverted-repeat TEs (MITEs; e.g., Tu, 2001a; Coates *et al.*, 2010a), *Helitrons* (Kapitonov and Jurka, 2001, 2003a; Thomas *et al.*, 2010a), and a recently discovered group called *Mavericks* or *Polintons* (Feschotte and Pritham, 2005; Kapitonov and Jurka, 2006; Pritham *et al.*, 2007). All Class II TEs transpose directly from DNA to DNA, and no RNA intermediate is involved (Craig, 2002).

3.2.2.1. Cut-and-paste DNA transposons

DNA transposons such as *P*, *hobo*, and *mariner* are usually characterized by 10- to 200-bp terminal inverted repeats (TIRs) flanking one or more open reading frames that encode a transposase. They usually transpose by a cut-and-paste mechanism, and their copy number can be increased through a repair mechanism (Finnegan, 1992; Craig, 2002; Zhou *et al.*, 2004; Hickman *et al.*, 2005; Mitra *et al.*, 2008; Richardson *et al.*, 2009). As shown in **Figure 3**, cut-and-paste DNA transposons can be subdivided into several families or superfamiles according to their transposase sequences. The families/superfamilies that have been found in insects include *IS630-Tc1-mariner*, *hAT*, *Merlin*, *piggyBac*, *PIF/Harbinger*, *P*, and *Transib* (Shao and Tu, 2001; Robertson, 2002; Kapitonov and Jurka, 2003a; Feschotte, 2004). These families/superfamilies are also characterized by TSDs of specific sequence or length.

3.2.2.2. MITEs

Miniature inverted-repeat TEs (MITEs) are widely distributed in plants, vertebrates, and invertebrates (Oosumi *et al.*, 1995; Wessler *et al.*, 1995; Smit and Riggs, 1996; Tu, 1997, 2001a; Yang *et al.*, 2009; Coates *et al.*, 2010b). Most MITEs share common structural characteristics, such as TIRs, small size, lack of coding potential, AT richness, and the potential to form stable secondary structures (Wessler *et al.*, 1995). MITEs may have been "borrowing" the transposition machinery of autonomous DNA transposons by taking advantage of shared TIRs (MacRae and Clegg, 1992; Feschotte and Mouches, 2000a; Zhang *et al.*, 2001). An alternative hypothesis suggests that they may transpose by a hairpin DNA intermediate produced from the folding back of single-stranded DNA during replication, which may better explain how MITEs could achieve immensely high copy numbers in some genomes (Izsvak *et al.*, 1999). However, more recent evidence clearly favors cross-mobilization by autonomous transposons, and suggests that internal features of MITEs may help them achieve high transposition activity (Yang *et al.*, 2009). One obvious source of MITEs is internal deleted autonomous DNA transposons (Feschotte and Mouches, 2000a). In this case, MITEs are basically non-autonomous deletion derivatives of DNA transposons. Recent studies show that the similarities between many MITEs and their putative autonomous partners are restricted to the TIRs (Feschotte *et al.*, 2003). Although subsequent loss of autonomous partners in the genome remains a possible explanation for the lack of internal sequence similarity between MITEs and their putative autonomous partners, two other explanations are perhaps more plausible. First, MITEs could originate *de novo* from chance mutation or recombination events resulting in the association of TIRs flanking unrelated segments of DNA (MacRae and Clegg, 1992; Tu, 2000; Feschotte *et al.*, 2003). Alternatively, these MITEs could originate from abortive gap-repair following the transposition of DNA transposons, which has been shown to occasionally introduce transposon-unrelated sequences (Rubin and Levy, 1997).

3.2.2.3. *Helitrons*: the rolling circle transposons

Helitrons and related transposons have recently been

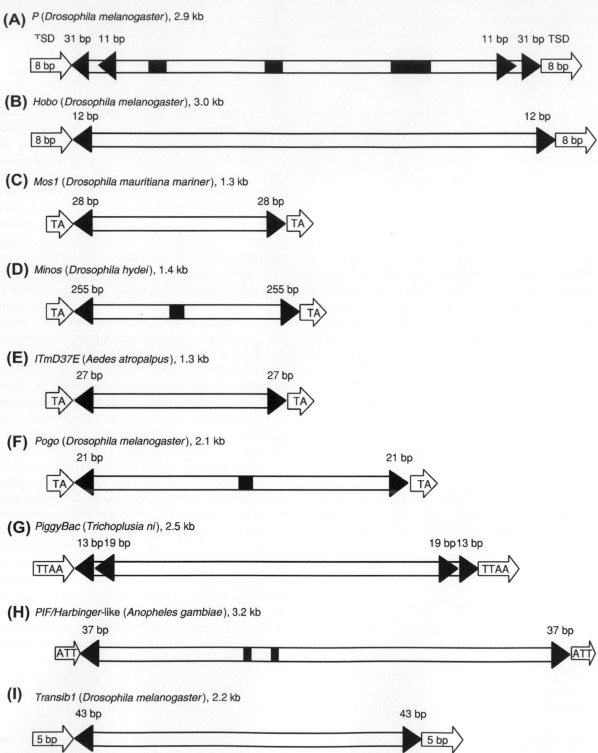

Figure 3 Structural characteristics of representative cut-and-paste DNA transposons in insects. The name of each representative transposon, its host species, and its approximate length are shown as the heading of each panel. Open arrows indicate target site duplications (TSDs); filled triangles indicate terminal and subterminal inverted repeats. The lengths of these inverted repeats are marked. Exons are shown as open boxes, and introns are shown as filled black boxes. 5′ and 3′ untranslated regions are not shown. The elements are not drawn to scale. References for information on these transposons are as following: *P* (Engels, 1989); *hobo* (Streck *et al.*, 1986); *mariner Mos1* (Jacobson *et al.*, 1986); *minos* (Franz and Savakis, 1991); *ITmD37E* (Shao and Tu, 2001); *pogo* (Tudor *et al.*, 1992; Feschotte and Mouches, 2000a); *piggyBac* (Cary *et al.*, 1989); *PIF/Harbinger*-like (Biedler *et al.*, unpublished); *Transib1* (Kapitonov and Jurka, 2003a). Two recently discovered families, *Gamol* (Coy and Tu, 2005) and *Merlin* (Feschotte, 2004), are not shown.

discovered in insects and plants, which appear to use a rolling-circle mechanism of transposition (Le *et al.*, 2000; Kapitonov and Jurka, 2001, 2003b; Coates *et al.*, 2010b). Instead of cut-and-paste transposase, *Helitrons* encode proteins similar to helicase, ssDNA-binding protein, and replication initiation protein. These proteins facilitate the rolling-circle replication of *Helitrons*, a mechanism previously described for the bacterial *IS91* transposons (Garcillan-Barcia *et al.*, 2002).

3.2.2.4. *Polintons/Mavericks*: giant DNA transposons that encode integrase as well as DNA polymerase

A group of large DNA transposons that were first discovered in *Tetrahymena* (Wuitschick *et al.*, 2002) have recently been shown to be broadly distributed in metazoan, fungi, and various single-cell eukaryotes (Feschotte and Pritham, 2005; Kapitonov and Jurka, 2006; Pritham *et al.*, 2007). These elements are either called *Politons* or *Mavericks*, and share several features, including 6-bp TSD, long TIRs, and coding sequences for integrase, DNA polymerase, and a few other proteins. They are sometimes 20 kb in length, and appear to be related to adenoviruses, bacteriophages, and eukaryotic linear plasmids. It is proposed that an excised *Polinton/Maverick* can self-replicate with its own polymerase and integrate into the genome using its integrase.

3.2.3. Related Topics

3.2.3.1. *Foldback* elements

Drosophila Foldback elements are characterized by very long inverted repeats (Truett *et al.*, 1981). It is not known how *Foldback* elements transpose, although the presence of long inverted repeats indicates a possible DNA-mediated mechanism. Some researchers group *Foldback* elements as a distinct class, namely Class III (Kaminker *et al.*, 2002).

3.2.3.2. What is a family?

Before moving ahead with discussions on the discovery and diversity of insect TEs in the next two sections, it may be helpful to clarify the use of the term "family" in the context of TEs. The term "family" is often used to refer to a group of related TEs in diverse organisms that usually share conserved amino acid sequences in their transposase or reverse transcriptase. The *mariner* family is such an example. A TE also consists of many copies generated by transposition events in a genome. Therefore, these related copies are sometimes also referred to as a family. Some families consist of multiple distinct groups that are subdivided into subfamilies. Obviously, relatedness is a relative concept in evolution. A working definition is needed in each case until a universal family/subfamily definition is developed.

3.3. Methods to Uncover and Characterize Insect TEs

3.3.1. Early Discoveries and General Criteria

Before the availability of the large amount of genomic sequence data, TEs were often discovered by serendipitous observations during genetic experiments. As described in the introduction, McClintock's observation of the unstable mutations in maize led to her discovery of two mobile genetic elements, *Ac* and *Ds*, although the molecular characterization of these elements came many years later (reviewed in Fedoroff, 1989). Similarly, the observation of an unstable *white-peach* eye-color mutation in *Drosophila mauritiana* led to the discovery of the *mariner* transposon (Hartl, 1989). The *piggyBac* transposon was discovered as an insertion in a baculovirus after passage through a cell line of the cabbage looper *Trichoplusia ni* (reviewed by Fraser, 2000). In a slightly different vein, the *D. melanogaster P* and *I* elements were discovered because of their association with a genetic phenomena called hybrid dysgenesis, which refers to a group of abnormal traits, including high mutation rates and sterility, in crosses of certain strains (Kidwell, 1977; Finnegan, 1989). The genetic mutations described above, albeit rare, tend to identify active transposition events that resulted from active TEs in the genome.

The repetitive nature of TEs can also be used for their discovery and isolation, although not all repetitive elements are TEs. When DNA sequences are available, TEs can be identified on the basis of either similarity to known TEs, or common structural characteristics. In some cases, evidence for past TE insertion events could be identified on the basis of sequence analysis, which further supports the mobile history of a particular element. The criteria and methods described in this section are not unique to insects. However, it may be necessary to visit this topic here because of the lack of a systematic review on these issues, and because of the growing interest in TE analysis in the current genomic environment.

3.3.2. Experimental Methods to Isolate and Characterize Repetitive Elements

Several experimental approaches have been used to discover TEs on the basis of their repetitive nature. Although relatively straightforward, these methods may not clearly distinguish between TEs and other repetitive sequences in the genome. In other words, the repeats discovered using these methods are not always TEs. One way to discover repeats in the genome is to isolate visible bands in an agarose gel running a sample of restriction enzyme digested genomic DNA. This is based on the assumption that only highly reiterated sequences containing two or more conserved recognition sites for the restriction enzyme will produce a visible band amongst the smear of digested genomic DNA. The bands can be cut out from the gel and purified for cloning and sequencing. Another approach to search for repeats is to screen a genomic library using labeled genomic DNA as probe. This approach can be effectively used to identify abundant or highly repetitive sequences in the genome, which is based on the principle that only the repetitive fraction of the genome will produce a sufficient amount of labeled fragments that will generate hybridization signals during the screening (Gale, 1987; Cockburn and Mitchell, 1989). A third method is to use *Cot* analysis to help identify repetitive sequences in the genome, which is based on DNA reassociation kinetics (Adams *et al.*, 1986; Peterson *et al.*, 2002). For example, *Cot* analysis of genomic DNA can be performed to isolate moderately a repetitive portion of the genome that tends to contain TEs, and a subgenomic library can be constructed using this fraction of genomic DNA to search for potential TEs. Several methods can be used to identify and isolate TEs on the basis of information derived from related TEs. For example, homologous TE probes may be used in Southern blotting and genomic library screening experiments to identify related TEs. PCR using primers that are conserved between related TEs can also be used to isolate different members of a TE family.

3.3.3. Computational Approaches to Discover and Analyze TEs

The completion of a number of insect genome projects and the ongoing genome revolution fueled by the rapidly improving "next-generation" sequencing technologies provide an ever-expanding sea of data that can be explored to identify interspersed TEs. As described below, a number of new tools have been developed which represent a shift from merely masking TEs (RepeatMasker; reviewed by Jurka, 2000) to the discovery, annotation, and genomic analysis of TEs. The use of bioinformatics tools provides great advantages by allowing analysis of TEs in the entire genome, and by allowing quick surveys of a large number of TE families to identify the most promising candidates for discovering active TEs (see section 3.5) and for population analysis (section 3.9). It should be noted that these approaches are not limited to fully sequenced genomes. Because of the repetitive nature of TEs, sequences from a small fraction of a genome tend to contain a large number of TEs that may be discovered using bioinformatic approaches described here. Of course, a greater number of sequences and longer assembly would be beneficial in analyzing low-copy number or long TE sequences.

3.3.3.1. Homology-dependent approaches   Searching for TEs in a genome on the basis of similarities to known

elements discovered in different species is relatively straightforward. However, given their diversity and abundance, systematic computational approaches are necessary for efficient and comprehensive analysis. One such program was reported (Berezikov *et al.*, 2000) that uses profile hidden Markov models to find all sequences matching the full-length reverse transcriptase with the conserved FYXDD motif common to all reverse transcriptases. We previously developed a BLAST-based systematic approach to simultaneously identify and classify TEs (Biedler and Tu, 2003). This approach incorporates multi-query BLAST (Altschul *et al.*, 1997) and a few computer program modules (available at tefam.biochem.vt.edu) that organize BLAST output, retrieve sequence fragment, and mask database for identified TEs. The method was successfully used to discover and characterize non-LTR retrotransposons in the *An. gambiae* genome assembly. More recently, other programs that automate the discovery and annotation of TEs within a genome assembly have also been reported (e.g., Rho *et al.*, 2007; Rho and Tang, 2009).

3.3.3.2. Homology-independent methods   There are a few computer programs that uncover certain groups of TEs based on their structural characteristics, rather than specific sequence homologies. For example, FINDMITE1 searches the database for inverted repeats flanked by user-defined direct repeats within a specified distance (Tu, 2001a). There is also a program, named MAK, that uncovers MITEs as well as reporting the associations of MITEs with neighboring genes and related autonomous DNA transposons (Yang and Hall, 2003). LTR_STRUC is a program that identifies LTR retrotransposons on the basis of the presence of long terminal repeats (most LTRs contain TG…CA termini), target site duplications, and additional information such as primer binding site and polypurine tract (McCarthy and McDonald, 2003). Although it is not designed to uncover solo LTRs or truncated non-LTR retrotransposons, the program offers a rapid and efficient approach to systematically identify and characterize LTR retrotransposons in a given genome. It can be used as a discovery tool for new families of LTR retrotransposons. Recent programs such as LTR_FINDER, LTRharvest, and LTRdigest offer further improvements with regard to efficiency and sensitivity in the discovery and annotation of LTR retrotransposons (Xu and Wang, 2007; Ellinghaus *et al.*, 2008; Steinbiss *et al.*, 2009).

There are a small number of programs that identify TE sequences on the basis of their repetitive nature in the genome. The most commonly used programs include Recon, ReAS, RepeatGluer, RepeatScout RepeatFinder, and PILER (Volfovsky *et al.*, 2001; Bao and Eddy, 2002; Pevzner *et al.*, 2004; Edgar and Myers, 2005; Li *et al.*, 2005; Price *et al.*, 2005). For example, RepeatFinder uses a clustering method to analyze repetitive sequences (Volfovsky *et al.*, 2001), and RECON (Bao and Eddy, 2002) uses a multiple sequence alignment algorithm to identify all repetitive sequences. RepeatScout is a user-friendly and rapid repeat finding program that uncovers repeats by extending consensus seed sequences (Price *et al.*, 2005). ReAS can be used for TE discovery from whole-genome shotgun sequences. Saha and colleagues compared all six *de novo* repeat discovery programs, and found RepeatScout to be most efficient for analysis of genome assemblies and ReAS to be most efficient for analysis of shotgun sequences (Saha *et al.*, 2008). In addition to the repeat finding programs discussed above, there is also a novel approach that is based on comparison of whole-genome alignments of closely related *Drosophila* species to identify TE insertions as revealed by disruptions of conservation (Caspi and Pachter, 2006). This approach can potentially identify the boundaries of TE insertions and allow the inference of the age of insertion. It may become widely used as more genome assemblies of closely related species are made available. It is often a daunting task to classify or annotate a large number of TEs uncovered using the *ab initio* or *de novo* approaches mentioned above. Programs such as TEpipe (Biedler and Tu, 2003), REPCLASS (Feschotte *et al.*, 2009), and MGESCAN-non-LTR (Rho *et al.*, 2009) are designed to automate the TE classification process.

3.4. Diversity and Characteristics of Insect TEs

3.4.1. Overview

Virtually all classes and types of eukaryotic TEs have been found in insects. Insect TEs such as *copia*, *gypsy*, *I*, *R1*, *P*, *mariner*, *hobo*, *piggyBac*, and *transib* are the founding members of several diverse families/superfamilies that were later shown to have broad distributions in eukaryotes. In addition, recent studies have revealed a few novel and intriguing TEs in insects, which are described in detail below. The previous review (Tu, 2005) provided a relatively extensive compilation of the two classes of TEs in insects. It is not possible to discuss here all the new insect TEs discovered since then; instead, I will focus on recent advances and interesting features of some novel insect TEs, and describe new insights obtained from comparative genomic analysis.

3.4.2. Recent Advances

3.4.2.1. Class I TEs Recent discovery of the use of tyrosine recombinase instead of integrase in some LTR retrotransposons further highlighted the flexibility in domain acquisition by LTR elements (Eickbush and Jamburuthugoda, 2008). The acquisition of the *env*-like protein by some LTR retrotransposons such as *gypsy* confers the ability to leave the cell and become infectious retroviruses (Eickbush and Malik, 2002). There are a few recent surveys that revealed great diversity of LTR retrotransposons in sequenced insect genomes (Tubio *et al.*, 2004, 2005; Xu *et al.*, 2005; Nene *et al.*, 2007; Tribolium Genome Sequencing Consortium, 2008; Minervini *et al.*, 2009; Arensburger *et al.*, 2010). Of the 17 clades of non-LTR retrotransposons, 12 have been found in insects (Eickbush and Malik, 2002; Biedler and Tu, 2003). In fact, the founding members of many of these clades were discovered in insects. Two new clades, named *Loner* and *Outcast*, were discovered in *An. gambiae* (Biedler and Tu, 2003). Recent surveys of the genomes of the flour beetle, the silkworm, and the Culicinae mosquito also revealed highly diverse non-LTR retrotransposons (Nene *et al.*, 2007; Osanai-Futahashi *et al.*, 2008; Tribolium Genome Sequencing Consortium, 2008; Arensburger *et al.*, 2010). Insect SINEs characterized so far all belong to the tRNA-related group – for example, a SINE discovered in *Ae. aegypti*, named *Feilai*, consists of a tRNA-related promoter region, a tRNA-unrelated conserved region, and a triplet tandem repeat at its 3′ end (**Figure 2**). The *Twin* SINE family, which was discovered in *Culex pipiens* (Feschotte *et al.*, 2001), consists of two tRNA-related regions separated by a 39-bp spacer. *SINE200* from *An. gambiae* contains only one of the two conserved boxes found in tRNA-related Pol III promoters (Santolamazza *et al.*, 2008). A recently discovered SINE in the silkworm is frequently found in the untranslated exons (UTRs) of genes (Xu *et al.*, 2010).

3.4.2.2. Class II TEs Several cut-and-paste DNA transposons from insects are the founding members of their respective families/superfamilies that have broad distributions. The families/superfamilies that have been found in insects include *IS630-Tc1-mariner*, *hAT*, *Merlin*, *piggyBac*, *PIF/Harbinger*, *P*, and *Transib* (Shao and Tu, 2001; Robertson, 2002; Kapitonov and Jurka, 2003a; Feschotte, 2004). Conserved transposase sequences and TSDs of specific sequence or length are the hallmarks of each family/superfamily. (Some of these families are also described in Chapter 4 in this volume.) The structural characteristics of representative elements from each family are shown in **Figure 3**. Recent surveys showed broad distribution of *piggyBac* transposons in insects (Handler *et al.*, 2008; Wang *et al.*, 2008, 2010). Recent expansion and reclassification of the *IS630-Tc1-mariner* superfamily will be discussed below as an example of the diversity of transposon superfamilies (Shao and Tu, 2001; Coy and Tu, 2005). Structural analysis of the *Hermes* (*hAT*) transposase and the DNA-transposase complex of *Mos1* (*mariner*) further illustrated the molecular mechanisms of the cut-and-paste process (Hickman *et al.*, 2005; Richardson *et al.*, 2009). Episomal *hAT* elements, recently recovered in insects, may be maternally transmitted and influence transposition (O'Brochta *et al.*, 2009). MITEs that share similar TSDs and TIRs with cut-and-paste DNA transposons in the *IS630-Tc1-mariner*, *hAT*, *piggyBac*, and *PIF/Harbinger* families have been found in mosquitoes (Tu, 1997, 2001a; Holt *et al.*, 2002; Nene *et al.*, 2007; Arensburger *et al.*, 2010). A large number of MITEs have been discovered in silkworm (Coates *et al.*, 2010a; Han *et al.*, 2010). A MITE that generates TA-specific TSDs has also been reported in a Coleopteran insect (Braquart *et al.*, 1999). Two hAT-like MITEs have been found in *D. willistoni* (Holyoake and Kidwell, 2003), and a deletion-derivative of the pogo transposon has been found in *D. melanogaster* (Feschotte *et al.*, 2002). As in mosquitoes (Tu, 1997), MITEs were found to be associated with genes in *Helicoverpa zea* (Chen and Li, 2007). Advances in the complex DNA transposons *Helitrons* and *Polintons/Mavericks* are discussed in section 3.4.3.

3.4.3. Intriguing Insect TEs

3.4.3.1. Two intriguing families of Class I TEs: *Maque* and *Penelope*
A family of very short interspersed repetitive elements named *Maque* has recently been found in *An. gambiae*. There are approximately 220 copies of *Maque*. Only approximately 60 bp long, *Maque* has the appearance of a distinct transposition unit. The majority of *Maque* elements were flanked by 9- to 14-bp TSDs. *Maque* has several characteristics of non-LTR retrotransposons, such as TSDs of variable length, imprecise 5′ terminus, and CAA simple repeats at the 3′ end. The evolutionary origin of *Maque* and the differences between *Maque* and other known retro-elements including SINEs is not yet known. We suggest that the 5′ end of *Maque* represents a strong stop position that causes frequent premature termination of reverse transcription (Tu, 2001b). Although no autonomous non-LTR retrotransposons have been found that share similar 3′ sequences with *Maque*, there is a family of non-LTR retrotransposons, *Ag-I-2* (Biedler and Tu, 2003), that have the same CAA tandem repeats at their 3′ termini. It is possible that short sequences such as *Maque* which contain just the reverse transcriptase recognition signal could potentially contribute to the genesis of some primordial SINEs (Tu, 2001b). Insertion polymorphism of this element and the SINE200 was used to study the incipient speciation between the M and S molecular forms of *An. gambiae* (Barnes *et al.*, 2005; Santolamazza *et al.*, 2008).

Penelope, another intriguing family, was discovered as a TE involved in the hybrid dysgenesis of crosses between field-collected and laboratory strains of *D. virilis* (Evgen'ev *et al.*, 1997). It has a reverse transcriptase that is grouped with the reverse transcriptase from telomerase

(Arkhipova *et al.*, 2003). More strikingly, *Penelope* and related elements in bdelloid rotifers are able to retain their introns, which is inconsistent with a transposition mechanism involving an RNA intermediate. It was proposed that the Uri endonuclease domain found in all *Penelope*-like elements may allow them, at least in part, to use a DNA-mediated mechanism similar to that used by group I introns (Arkhipova *et al.*, 2003). On the basis of these unique features and phylogenetic analysis of *Penelope*-like elements in diverse eukaryotes, *Penelope* was classified as a unique group that is distinct from LTR and non-LTR retrotransposons (Evgen'ev and Arkhipova, 2005).

3.4.3.2. Classification of the *IS630-Tc1-mariner* (*ITm*) superfamily
It was previously shown that some prokaryotic *IS* elements, eukaryotic *Tc1* and *mariner* transposons, and eukaryotic retrotransposons and retroviruses form a megafamily which share similar signature sequences or motifs in the catalytic domain of their respective transposase and integrase (Capy *et al.*, 1996, 1997). The common motif for this transposase–integrase megafamily is a conserved D(Asp)DE(Glu) or DDD catalytic triad. The distance between the first two Ds is variable while the distance between the last two residues in the catalytic triad is mostly invariable for a given transposon family in eukaryotes, indicating functional importance. Within this megafamily, the eukaryotic DNA transposon families *Tc1* and *mariner* and the bacterial *IS630* element and its relatives in prokaryotes and ciliates comprise a superfamily, the *IS630-Tc1-mariner* superfamily, which is based on overall transposase similarities and a common TA dinucleotide insertion target (Henikoff, 1992; Doak *et al.*, 1994; Robertson and Lampe, 1995; Capy *et al.*, 1996; Shao and Tu, 2001). *Tc1*-like elements identified in fungi, invertebrates, and vertebrates all contain a DD34E motif, while most *mariner* elements identified in flatworm, insects, and vertebrates contain a DD34D motif. A few TEs that contain DD37D and DD39D motifs were previously regarded as basal subfamilies – the max subfamily and mori subfamily, respectively – of the *mariner* family (Robertson, 2002). We have reported a novel transposon named *ITmD37E* in a wide range of mosquito species (Shao and Tu, 2001). The *ITmD37E* transposases contain a conserved DD37E catalytic motif. Sequence comparisons and phylogenetic analyses suggest that *ITmD37E* is a new family, and that the mori subfamily (DD37D) and max subfamily (DD39D) of *mariner* may also be classified as two distinct families, namely the *ITmD37D* and *ITmD39D* families (**Figure 4**). The recognition of the three new families, *ITmD37E*, *ITmD37D*, *and ITmD37D*, is consistent with the fact that they share family-specific catalytic motifs and similar TIRs. Claudianos and colleagues also noticed the need for reclassification of the DD37D transposons, and named them the *maT* family (Claudianos *et al.*, 2002). A group

(A)

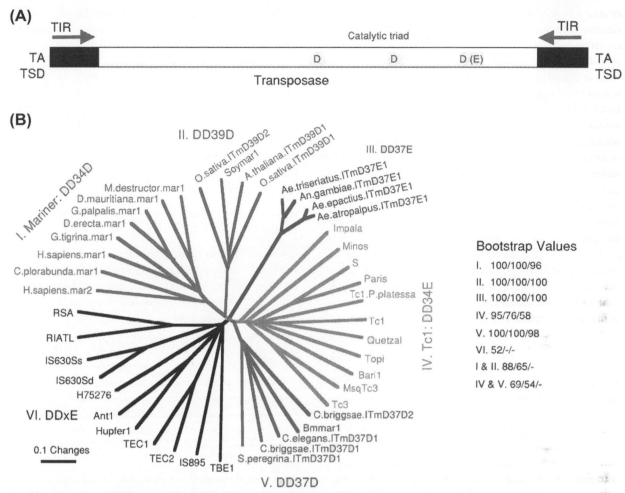

(B)

Figure 4 Structural features and classification of the *IS630-Tc1-mariner* superfamily. (A) Structural features. The catalytic triad in the transposase is highlighted. The characteristic TA target site duplications (TSDs) flanking an *IS630-Tc1-mariner* are shown. The terminal inverted repeats (TIRs) specify the boundaries of the element. Possible introns are not shown. (B). Phylogenetic relationship between members of the *IS630-Tc1-mariner* superfamily on the basis of the catalytic domain. The alignment used here was previously described (Shao and Tu, 2001). The tree shown is an unrooted phylogram constructed using a minimum evolution algorithm. Two additional methods, neighbor-joining and maximum parsimony, were also used. Confidence of the groupings was estimated using 500 bootstrap replications. The bootstrap value represents the percent of times that branches were grouped together at a particular node. The first, second, and third numbers represent the bootstrap value derived from minimum evolution, neighbor-joining, and maximum parsimony analysis, respectively. Only the values for major groupings are shown. Various colors indicate different clades. All phylogenetic analyses were conducted using PAUP 4.0 b8 (Swofford, 2001). Note that a recently described group of transposons that contain a DD41D catalytic triad is not included here; they are a distinct group related to the DD37D transoposons (Gomulski *et al.*, 2001). Also not shown is the *Gambol* family, which has the same DD34E triad as the *Tc1* family but belongs to its own distinct family (Coy and Tu, 2005).

of transposons that contain a DD41D catalytic motif have been found in the medfly *Ceratitis rosa*, establishing yet another family (Gomulski *et al.*, 2001; Robertson and Walden, 2003); namely, the *ITmD41D* family. In summary, according to recent analyses, the *IS630-Tc1-mariner* superfamily can be organized into seven families, *ITmD37E*, *ITmD37D*, *ITmD39D*, *ITmD41D*, *Tc1*, *mariner*, and *pogo*, and an unresolved clade which includes bacterial *IS630*-like elements and some fungal and ciliate transposons (**Figure 4**). *Pogo* is an interesting case, as it has a unique N-terminal DNA-binding domain and a long C-terminal domain rich in acidic residues,

although it contains a DDxD catalytic domain related to *IS630-Tc1-mariner* transposons (Smit and Riggs, 1996). Recently discovered *Gambol* elements are distinct from the *Tc1* elements, according to phylogenetic analysis, although both families contain the DD34E catalytic triad (Coy and Tu, 2005).

3.4.3.3. *Microuli*, a miniature subterminal inverted-repeat TE   *Microuli* is a family of small (~200 bp) and highly AT rich (68.8–72.6%) TEs found in *Ae. aegypti* that do not have any coding capacity (Tu and Orphanidis, 2001). There is a 61- to 62-bp internal subterminal inverted-repeat as well as a 7-bp subterminal inverted-repeat 11 bp from the two termini. In addition, there are three imperfect subterminal direct repeats near the 5′ end. All of the above characteristics clearly resemble the structural features of MITEs. The only feature that separates *Microuli* from MITES is that *Microuli* elements lack TIRs. Therefore, we use the phrase "miniature subterminal inverted-repeat transposable elements," or MSITEs, to refer to the structural characteristics of the *Microuli* elements. Short insertion sequences that contain subterminal inverted-repeats but lack TIRs have been identified in the genomes of rice and a *Culex* mosquito (Song *et al.*, 1998; Feschotte and Mouches, 2000b). Of the 19 nucleotides at the 5′ (and only 5′) terminus of *Microuli*, 14 are identical to the TIR of *Wuneng*, a previously characterized MITE in *Ae. aegypti* (Tu, 1997). Both *Microuli* and *Wuneng* insert specifically into the TTAA target. It has been suggested that MITES and the autonomous DNA transposons share the same transposition machinery based on common TIRs (Feschotte *et al.*, 2002). Then, how did *Microuli* transpose without the TIRs? The three subterminal direct repeats could potentially be the binding sites for transposases, because subterminal inverted repeats and subterminal direct repeats have been shown to bind transposases in several autonomous DNA transposons (Morgan and Middleton, 1990; Beall and Rio, 1997; Becker and Kunze, 1997). It remains unclear how the termini of *Microuli* are determined at the strand cleavage step without the TIR. The TTAA target duplication plus a 3-bp TIR are essential for the excision of the autonomous transposon *piggyBac* (Bauser *et al.*, 1999). Therefore, it is possible that *Microuli* may also be able to use the TTAA target sequence as part of the signal for recombination. It is tempting to hypothesize that some MITEs could evolve from MSITEs through mutation and/or recombination events at the termini which would result in TIRs. Similar elements with subterminal inverted repeats have been recently found in *Helicoverpa zea* (Coates *et al.*, 2010a).

3.4.3.4. Complex DNA transposons: *Helitrons* and *Polintons/Mavericks* *Helitrons* use a rolling-circle mechanism for transposition, and they have been found in *D. melanogaster*, *An. gambiae*, and Lepidopteran species (Kapitonov and Jurka, 2001, 2003a; Coates *et al.*, 2010b). Insect *Helitrons* have several characteristics, including short specific terminal sequences (5′ TC and 3′ CTAG), a 3′ hairpin, and the lack of TSDs. Instead of a cut-and-paste transposase, *Helitron1* in *An. gambiae* encodes an intronless protein including domains similar to helicase and replication initiation protein. There are approximately 100 copies of *Helitron* elements in *A. gambiae*, which form 10 distinct families (Kapitonov and Jurka, 2003a).

Politons/Mavericks are broadly distributed in metazoa, fungi, and various single-cell eukaryotes (Feschotte and

Pritham, 2005; Kapitonov and Jurka, 2006; Pritham *et al.*, 2007). These elements can be 20 kb in length and share several features, including 6-bp TSD, long TIRs, and coding sequences for integrase, DNA polymerase, and a few other proteins. They appear to be related to adenoviruses, bacteriophages, and eukaryotic linear plasmids. It is proposed that an excised *Polinton/Maverick* can self-replicate with its own polymerase and integrate into the genome using its integrase. This group represents the most complex DNA transposons to date, and they are found in *Drosophila* and *Tribolium* (Pritham *et al.*, 2007)

3.4.4. Insights from Comparative Genomic Analysis

There are 31 insect genome assemblies available at the National Center for Biotechnology Information (NCBI) Genome Project Database (ncbi.nlm.nih.gov/entrez): 19 genome assemblies are available from Dipteran species, including 1 Hessian fly (http://www.ncbi.nlm.nih.gov/g enomeprj/45867) and 6 mosquitoes (Holt *et al.*, 2002; Nene *et al.*, 2007; Arensburger *et al.*, 2010; Lawniczak *et al.*, 2010; http://www.ncbi.nlm.nih.gov/genome prj/46227); and 12 *Drosophila* (Drosophila 12 Genomes Consortium, 2007). There are assemblies from seven Hymenopteran species, including the honeybee (Honeybee Genome Sequencing Consortium, 2006), three ants (Bonasio *et al.*, 2010; http://www.ncbi.nlm.nih.gov/gen omeprj/48091), and three wasps (Werren *et al.*, 2010). There are also assemblies from one Lepidopteran species, *Bombyx mori* (Xia *et al.*, 2004; Mita *et al.*, 2004), and one Coleopteran species, *Tribolium castaneum* (Tribolium Genome Sequencing Consortium, 2008). Assemblies from three hemimetabolous insects are available, including one Phthiraptern insect body louse (Kirkness *et al.*, 2010) and two Hemipteran (The International Aphid Genomics Consortium, 2010; http://www.ncbi.nlm.nih. gov/genomeprj/13648). In addition, a number of insect genomes are being sequenced using "next-generation" approaches, and it is anticipated that rapid expansion of sequenced genomes will bring tremendous opportunities to the investigation of TE diversity and evolution. Whole-genome comparative analysis of insect TEs is still in its early stages, and a few interesting observations are highlighted below. Systematic analysis of the 12 *Drosophila* genomes revealed that while the TE content varies from 2.7% to ~25% of the host genomes, the relative abundance of different groups of TEs is conserved across most of the species (Drosophila 12 Genomes Consortium, 2007). Comprehensive analysis identified over 100 potential horizontal transfer events by more than 20 TEs among the 12 *Drosophila* species, most of which involved DNA transposons and LTR retrotransposons (Loreto *et al.*, 2008; Bartolome *et al.*, 2009). Systematic comparison of multiple aligned genomes revealed TE insertion sites across the entire genomes, and supported a hypothesis that most TEs in *D. melanogaster* are recently active (Caspi and Pachter, 2006). The published genomes of *Anopheles, Culex,* and *Aedes* mosquitoes vary by five-fold in size, ranging from ~270 Mbp for *An. gambiae* (Holt *et al.,* 2002) to ~500 Mbp for *C. quinquefasciatus* (Arensburger *et al.,* 2010), and ~1300 Mbp for *Ae. aegypti* (Nene *et al.,* 2007). TE contents in these three species are 11–16%, 29%, and 47% of the assembled genomes, respectively, indicating that TEs contributed significantly to the genome size variations among mosquito species. While 16% of the *Ae. aegypti* genome is occupied by MITE-like elements, cut-and-paste DNA transposons represent only 3% of the genome, suggesting that a small number of DNA transposons may be responsible for cross-mobilizing a large number of non-autonomous MITE-like sequences (Nene *et al.,* 2007). Systematic comparisons also revealed an apparent horizontal transfer event between *Aedes* and *Anopheles* mosquitoes involving an *ITmD37E* DNA transposon (Biedler and Tu, 2007). Among the sequenced Hymenopteran species, the honeybee genome contains only ~7% repetitive sequences while repeat contents range from 15 to 27% in the ants and wasps (Honeybee Genome Sequencing Consortium, 2006; Bonasio *et al.,* 2010; Werren *et al.,* 2010). The parasitic body louse harbors only a very small number of TEs, which occupy 1% of its 110-Mbp genome (Kirkness *et al.,* 2010).

3.5. Search for Active TEs in Insect Genomes

Active TEs may be used as tools for the genetic manipulation of insects for basic and applied research (see section 3.9). In addition, the behavior of TEs in host genomes and their spread in natural populations may be studied by monitoring active TE families. It is therefore highly desirable to isolate active copies of TEs. As described in section 3.3, TEs discovered from observations of genetic mutations tend to result from active transposition events. Although several active TEs were discovered in this manner, this discovery process relies heavily on fortuitous events. Several methods that may facilitate the search for active TEs in insect genomes are described below.

3.5.1. Identification of Potentially Active TEs on the Basis of Bioinformatic Analysis

As discussed above, the ongoing genome revolution has produced an immense quantity of sequence data from which diverse TEs can be identified in various insect genomes. The computational programs described in section 3.3 can greatly facilitate the discovery and characterization of a large number of TE families. Unfortunately, the vast majority of TEs have accumulated inactivating mutations during evolution, rendering the discovery of active TEs a task similar to finding needles in a haystack. Bioinformatic analysis can provide leads to potentially active candidates that can be studied further. For example, using a semi-automated reiterative search strategy, we identified many potentially active families of non-LTR retrotransposons in the *An. gambiae* genome (Biedler and Tu, 2003). Here, candidate families were identified based on sequence characteristics, which include the presence of full-length elements, intact open reading frames, multiple copies with high nucleotide identity, and the presence of TSDs. High nucleotide identity indicates recent amplification from a source element, without enough time for divergence caused by nucleotide substitution and other mutations. It should be emphasized that sequence analysis can only provide leads for further analysis. For example, high sequence identity between copies of a TE family may not always indicate recent transposition activity because it can also result from gene conversion events. Using the bioinformatics principles described above, a mosquito *hAT* element named *Herves* and an ant *mariner* element named *Mboumar* were identified and subsequently shown to support transposition (Arensburger *et al.,* 2005; Munoz-Lopez *et al.,* 2008).

3.5.2. Detection of TE Transcription

Transcription is a required step during transposition of the RNA-mediated TEs. Although DNA-mediated TEs do not use RNA as an intermediate, transcription is required for production of transposase proteins. Therefore, the detection of transcription may offer further support for an active family in both classes of TEs. Transcription can be inferred if a match is found in an expressed sequence tag (EST) database to a TE sequence from the same organism. For example, 21 families of non-LTR retrotransposons had significant hits when BLAST searches were carried out against over 94,000 *An. gambiae* ESTs downloaded from NCBI (Biedler and Tu, 2003). Comparisons of TEs against high-throughput illumina sequencing databases may also reveal TE transcription. Transcription of TEs may also be detected experimentally by RT-PCR and Northern blot. The source of mRNA may affect the outcome of these experiments, because the activity of some TEs may be temporally and spatially controlled. Recent analysis showed that transcription of the *hobo* transposon may be developmentally regulated in *D. melanogaster* (Depra *et al.,* 2009). It has been shown that TE activity can be elevated during the culturing of mammalian and plant cells (Wessler, 1996; Grandbastien, 1998; Liu and Wendel, 2000; Kazazian and Goodier, 2002). Different cell lines are available for a number of insect species. One caveat of the above approach is that transcripts shown by either experimental detection or EST analysis could arise from spurious transcription. These transcripts could originate by transcription from a nearby host promoter.

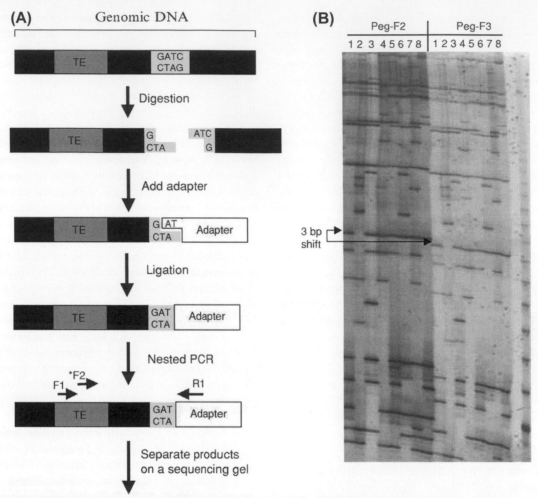

Figure 5 TE display, a method to scan multiple insertion sites of a TE in the genome. (A) Principle of TE display, which is a modified form of Amplified Fragment Length Polymorphism (AFLP). The difference is that TE-specific primers (F1 and F2) are used in addition to the adapter primer (R1). F2 is labeled as shown by the asterisk. (B). Partial image of a TE display using primers for the *Pegasus* element with eight female individuals from an *Anopheles gambiae* colony (GAMCAM) originally collected from Cameroon (Biedler *et al.*, 2003). The eight samples on the left are amplified with a *Pegasus*-specific primer, Peg-F2. The eight samples on the right are the same as those on the left except they were amplified with primer Peg-F3, which is designed to amplify a product smaller by three bases. The three-base shift is clearly observable. A size marker is shown on the right. Bands from a TE display gel were re-amplified and sequenced, showing that they contained *Pegasus* sequences as well as flanking genomic and adapter sequences in the expected order (not shown). Co-migrating bands among different individuals had the same flanking genomic sequence, indicating that they were from the same genomic locus. Note that *Pegasus* is a MITE.

3.5.3. Detection of *in vivo* Transposition Events by TE Display

TE display (Van den Broeck *et al.*, 1998; Casa *et al.*, 2000; Biedler *et al.*, 2003) is a sensitive and reproducible experimental method to detect TE insertions (**Figure 5**). TE display is a powerful tool for genome-wide analysis of TE insertions, and for detection of new insertions due to transposition (De Keukeleire *et al.*, 2001). It offers a higher degree of sensitivity and resolution than genomic Southern analysis. TE display has been used to detect somatic transposition (De Keukeleire *et al.*, 2001; Sethuraman *et al.*, 2007). This is done simply by looking for the presence of new bands that represent newly transposed copies

of a TE, although a caveat of this approach is that change of restriction site may also result in new TE display bands. The same method may also be used to identify germ-line transposition by comparing TE display patterns of parent insects with the patterns of a large number of offspring. Alternatively, one could take advantage of the possibility that some TEs are activated in cell culture. Using TE display, it may be possible to identify active families by comparing the relative abundance of a TE in cultured cells with that in individuals from different strains of the same species (Jiang *et al.*, 2003). This approach is based on the assumption that some TE families may be more active in cultured cells than in live organisms.

3.5.4. Detection of *in vivo* Transposition Events by Inverse PCR

Actively transposing DNA-mediated TEs can be identified as extrachromsomal DNA in the form of linear or circular intermediates or byproducts (Arca *et al.*, 1997; Gorbunova and Levy, 1997; O'Brochta *et al.*, 2009). Using a set of outward-orienting primers within the TE, the circular extrachromosomal copies may be amplified, which may serve as evidence of active excision or transposition. However, head-to-head copies of the same TE in the genome could also produce PCR products when outward-orienting primers are used, which need to be ruled out by sequencing and further analysis.

3.5.5. Transposition Assay, Reconstruction, and Genetic Screen

Transposition assays can be used to directly assess the functionalities of both the *cis*- (TIRs) and the *trans*- (transposase) components of a DNA transposon, allowing the demonstration of autonomous transposition events. This topic is reviewed in Chapter 4. In addition, transposition assays have also been established for the detection of retrotransposition of non-LTR retrotransposons (Jensen *et al.*, 1994; Ostertag *et al.*, 2000). A molecular reconstruction approach has been developed to restore inactivated copies of vertebrate transposons named *Sleeping Beauty* (Ivics *et al.*, 1997) and *Harbinger* (Sinzelle *et al.*, 2008). This approach may become increasingly feasible as the cost of gene synthesis continues to drop. As an alternative, a genetic screen based on a bacterial system has been developed to identify hyperactive copies of an insect *mariner* transposon among randomly mutated copies (Lampe *et al.*, 1999). This approach can potentially be used to screen for active copies of transposons that do not require specific host factors. In summary, the progress in insect genome projects and the development and application of the methods described in this section will greatly facilitate searches for active TEs. The task of finding needles in the haystack could potentially be replaced by targeted and efficient investigations.

3.6. Evolution of Insect TEs

The evolutionary dynamics of TEs are complex, in part because of their replication and their interactions with the host genome. The intricate dynamics between TEs and their host genomes are further complicated by the ability of some TEs to cross species barriers and spread to the genome of a new species by horizontal transfer. Horizontal transfer may be an important part of the life cycle of some TEs, and contribute to their continued success during evolution (Silva *et al.*, 2004; Schaack *et al.*, 2010). While the broad distribution of both RNA-mediated TEs and DNA-mediated TEs in all eukaryotic groups is evidence of the long-term evolutionary success of TEs, different TEs may have adopted different strategies, for which several insect TEs in both classes provide good examples.

3.6.1. Genomic Considerations of TE Evolution

It has been hypothesized that TE insertions may present three types of potential deleterious effects, including: (1) insertional mutagenesis, which may disrupt gene function and/or regulation; (2) transcriptional/translational cost of the production of TE transcripts and proteins; and (3) ectopic recombination between homologous copies of TEs in different chromosomal locations, which may result in duplication, deletion, and new linkage relationships between genes (Nuzhdin, 1999; Kidwell and Lisch, 2001; Bartolome *et al.*, 2002; Petrov *et al.*, 2003; Feschotte and Pritham, 2007). The costs of having TEs may also include the costs associated with DNA replication when TEs occupy a large fraction of the genome. Obviously, these hypotheses are not mutually exclusive. This section discusses the intra-genomic dynamics of TE–host interaction. The population dynamics affecting the spread of TEs in insects, which is also important for TE evolution, will be discussed in section 3.7.

3.6.1.1. Self-regulation of insect TEs   Self-regulation has been shown for *Drosophila mariner* and *P* elements (Hartl *et al.*, 1997; Kidwell and Lisch, 2001). In the case of *Drosophila P* element, self-regulation is achieved through the activities of at least two types of element-encoded repressors. In the case of *mariner*, several mechanisms may be involved, including overproduction inhibition (an increase in the amount of transposase results in a decrease in net transposase activity), missense mutation effects (defective transposase encoded by missense copies interfering with functional transposase), and titration effects by inactive copies. In this regard, it is interesting to note that several hyperactive mutants of an active *mariner*, originally discovered from the horn fly, have been isolated (Lampe *et al.*, 1999). This suggests that the horn fly *mariner* has not evolved for maximal activity.

3.6.1.2. Host-control of insect TEs and small RNA pathways   RNA interference (RNAi), a mechanism that confers post-transcriptional suppression on the basis of homology to small fragments of double-stranded RNA, has long been implicated as a host defense mechanism against a broad spectrum of TEs in the nemotode *Caenorhabditis elegans* (Ketting *et al.*, 1999; Tabara *et al.*, 1999). Recent studies in *Drosphila* uncovered diverse and complex small RNA pathways that control TEs in germ-line and somatic tissues (reviewed in Malone and Hannon, 2009; Forstemann, 2010). The invoement of small RNAs in the control of TE activity in *Drosophila*

was initially proposed on the basis of cosuppression of the *I* element by an increasing number of *I*-related transgenes (Jensen *et al.*, 1999; Labrador and Corces, 2002). Repeat-associated small interfering RNAs (rasiRNA) were discovered in the *Drosphila* germ-line; these RNAs bind Piwi-type proteins (Piwi and Aubergine), and were later called Piwi-interacting RNAs, or piRNAs (Vagin *et al.*, 2006; Brennecke *et al.*, 2007; Gunawardane *et al.*, 2007). These piRNAs correspond to TEs in different groups, including *roo*, *I*, and *gypsy*. Piwi and aubergine mutant flies were shown to de-repress *gypsy*, *TART*, and *P* elements (reviewed in Malone and Hannon, 2009). piRNAs mostly originate from piRNA clusters, which consist of many truncated, nested, and inactivated TE remnants in the *Drosophila* genome. In the germ-line, piRNAs are amplified through a ping-pong mechanism involving Piwi, Aubergine, and Ago3. piRNAs are also found in the somatic support cells of the ovary, where no ping-pong amplification is detected because of the lack of Aubergine and Ago3 (Malone *et al.*, 2009; Li *et al.*, 2009). In the somatic support cells, these piRNAs correspond to sequences that reside in the *flamenco/COM* locus, which suppress *gypsy*, *ZAM*, and *idefix* elements (Pelisson *et al.*, 2007; Desset *et al.*, 2008). These observations link one of the best early examples of host TE control (*flamenco* over *gypsy*, Bucheton, 1995) with the small RNA pathway. piRNAs may be inherited maternally, conferring TE suppression to the offspring epigenetically. This epigenetic control was shown in the *P*- and *I*-element-mediated hybrid dysgenesis (Brennecke *et al.*, 2008). In essence, when males that contain active *P* or *I* elements mate with females that have no piRNAs against these TEs, reduced fertility will result as a consequence of uncontrolled *P*- or *I*-element transposition in the offspring.

piRNAs differ from the originally discovered siRNAs because piRNAs are longer in their length (24–29 nt versus 21–22 nt) and bind Piwi-type proteins. Endogenous siRNAs (endo-siRNAs) that correspond to TEs and other repetitive sequences have also been found in the *Drosophila* germ-line and somatic cells (Czech *et al.*, 2008; Ghildiyal *et al.*, 2008; Kawamura *et al.*, 2008; Okamura *et al.*, 2008). These endo-siRNAs may be derived from sense and antisense transcription of repeats as well as host genes, which may regulate gene expression and repress TE activity, depending on the specific target. There is also an indication that an RNA-dependent RNA polymerase in *Drosophila* may produce dsRNAs that will be processed to make endo-siRNAs (Lipardi and Paterson, 2009).

Two other points related to host control of TE activities are worth noting. The first is the link between TE silencing by piRNAs and heterochromatin. Klattenhoff and colleagues (2009) recently reported that an HP1 (heterochromatin protein 1) family protein is required for transposon silencing, piRNA production and amplification. The second significant discovery is that the alterations of the

Hsp90 chaperone machinery affect the piRNA pathway and lead to transposon activation and mutation (Specchia *et al.*, 2010). The authors hypothesize that Hsp90 may act as a genetic buffering system for TE activity, thus supporting the concept of canalization during development.

3.6.1.3. Non-random distribution of insect TEs
Patterns of non-random TE distribution have been shown in both *D. melanogaster* and *An. gambiae* (Bartolome *et al.*, 2002; Holt *et al.*, 2002; Kapitonov and Jurka, 2003a). TEs tend to accumulate in heterochromatin. Such a distribution bias could result either from preferential TE insertion, or from selection against insertions in euchromatic regions, or from both. Bartolome and colleagues suggest that the abundance of TEs is more strongly associated with local recombination rates (Bartolome *et al.*, 2002), which are low in heterochromatic regions, rather than with gene density. They argue that this association is consistent with the hypothesis that selection against harmful effects of ectopic recombination is a major force opposing TE spread. However, selection against insertional mutagenesis is also at work, as shown by the absence of insertions in coding regions. The insertional bias of *P* elements has been recently demonstrated during genome-scale *P* mutagenesis analysis (Spradling *et al.*, 1995, 1999). Therefore, insertion bias may contribute to the biased pattern of TE distribution in insects. A related topic here is the suggestion that concentrations of TE insertions in the *Drosophila* Y-chromosome may have contributed to the evolutionary process leading to its inactivation (Labrador and Corces, 2002). It should be noted that not all TEs have a bias towards heterochromatic or recombination-deprived regions. On the basis of analysis of limited gene sequences, it was shown that *Ae. aegypti* MITEs tend to be associated with the non-coding regions within or near genes (Tu, 1997), which is similar to what has been observed for plant MITEs (Zhang *et al.*, 2000).

3.6.1.4. Autonomous and non-autonomous TEs
In addition to the genomic interactions described above, most TEs have to contend with the fact that defective copies are often generated during or after transposition. This process could contribute to self-regulation, as discussed above. It can also lead to total inactivation and ultimate extinction of a TE, as the inactive TE population eventually overwhelms the active copies (Eickbush and Malik, 2002; Feschotte and Pritham, 2007). Therefore, the replicative ability that is responsible for the success of the TE may also lead to its inactivation in a genome. Interestingly, some non-autonomous TEs have been very successful with regard to amplification, although the mechanisms that contribute to their success are not entirely clear. For example, SINEs found in insect genomes, including *Ae. aegypti* and *B. mori*, all contain thousands

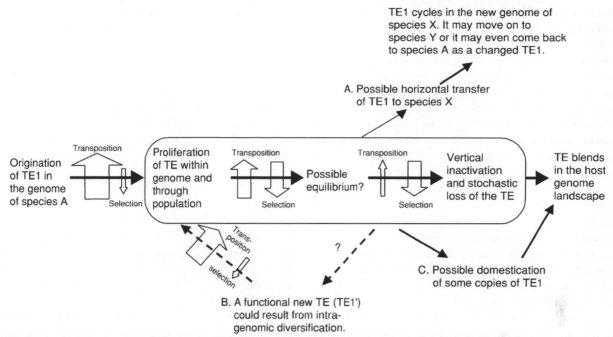

Figure 6 A model of the evolutionary dynamics of TEs in eukaryotic genomes. This hypothetical model incorporates recent work by several groups (Hartl *et al.*, 1997; Lampe *et al.*, 2001; Silva *et al.*, 2004). Some aspects of the model are better suited for DNA transposons that generally have a high propensity for horizontal transfer.

of copies (Adams *et al.*, 1986; Tu, 1999). Similarly, most of the MITEs found in insects are also highly reiterated, although low copy-number families are also found (Tu, 1997, 2000, 2001a; Nene *et al.*, 2007; Han *et al.*, 2010). The small size of MITEs and SINEs may confer less deleterious effects on the host, either because they are less efficient substrates for homologous recombination (Petrov *et al.*, 2003) or because their impact on neighboring genes may be less severe. Therefore, reduced selection pressure as well as other properties inherent to MITEs and SINEs could contribute to their apparent success. It is a fascinating question as to how SINEs and MITEs affect the evolution of the autonomous TEs that mobilize them.

3.6.2. Vertical Transmission and Horizontal Transfer of Insect TEs

As described above, the replicative ability that is responsible for the success of a TE may in some cases lead to its inactivation by generating defective copies or by activating host control mechanisms. Therefore, the ability to escape vertical inactivation by invading a new genome would greatly enhance the evolutionary success of DNA transposons. However, not all TEs have adopted this life cycle of invasion, amplification, senescence, and new invasion (**Figure 6**; see also Hartl *et al.*, 1997; Eickbush and Malik, 2002; Robertson, 2002; Schaack *et al.*, 2010).

3.6.2.1. Detection of horizontal transfer The occurrence of horizontal transfer can be supported in

various degrees by three types of evidence (Silva *et al.*, 2004; Schaack *et al.*, 2010). First, detection of elements with a high level of sequence similarity in divergent taxa will offer strong support for horizontal transfer, although variable rates of sequence change should be considered. Second, detection of phylogenetic incongruence between TEs and their hosts will also provide relatively strong support for horizontal transfer. However, this alone will not be convincing, especially in light of the high levels of intra-genomic diversity of TE families observed in insects. In other words, the existence of multiple TE lineages could confound the phylogenetic analysis, as paralogous lineages may be treated as orthologous ones. Finally, horizontal transfer may be inferred when "patchy" distribution of a TE among closely related taxa is observed. This type of support is weak, as loss of TEs from sister taxa may result from a phenomenon similar to assortment of an ancestral polymorphism (Silva *et al.*, 2004).

3.6.2.2. Horizontal transfer and vertical transmission in insects: differences between different groups of TEs The first case of eukaryotic horizontal transfer was reported in *Drosophila*, where the *P* element was shown to have invaded the *D. melanogaster* genome during the last century from a species in the *D. willistoni* group (reviewed in Kidwell, 1992). Evidence for this lateral event includes all three types of support described above, and is therefore widely accepted (Silva *et al.*, 2004). Further analyses of a large number of *P* element sequences from many *Drosophila* species showed that many horizontal transfer

events must have occurred to account for the current distribution pattern of *P* in *Drosophila* (Silva and Kidwell, 2000; Silva *et al.*, 2004). Another spectacular case of horizontal transfer involves the *mariner* transposons, which was also first discovered in insects. The *mariner* transposon family has been implicated in hundreds or more horizontal transfer events among a wide range of animal species, including a large number of insects across different orders (Robertson, 1993, 2002). Other examples of horizontal transfer of DNA transposons, which involve insects, include *ITmD37E*, *Harrow*, *Helitrons*, *hobo*, *Hosimary*, and *piggyBac* (Bonnivard *et al.*, 2000; Handler and McCombs, 2000; Biedler *et al.*, 2007; Gilbert *et al.*, 2010; Mota *et al.*, 2010; Thomas *et al.*, 2010b). Horizontal transfer has also been shown for LTR retrotransposons in insects. An earlier example involves the *Drosophila copia* element (Jordan *et al.*, 1999). *Copia* elements in *D. melanogaster* and *D. willistoni*, two divergent species that separated more than 40 million years ago, showed less than 1% nucleotide difference. It appears that *copia* jumped from *D. melanogaster* to *D. willistoni*. Horizontal transfer of the *Drosophila gypsy* element has also been reported (Terzian *et al.*, 2000; Vazquez-Manrique *et al.*, 2000). Evidence for horizontal transfer of LTR retrotransposons is accumulating (e.g., Kotnova *et al.*, 2007; Cordeiro *et al.*, 2008). As described in section 3.4.4, comprehensive analysis identified over 100 potential horizontal transfer events by more than 20 TEs among the 12 *Drosophila* species (Loreto *et al.*, 2008; Bartolome *et al.*, 2009). Schaack and collegues (2010) counted 67 horizontal transfer events involving LTR retrotransposons, 58 horizontal transfer events involving DNA transposons, and only 3 events involving non-LTR elements in *Drosophila*.

3.6.2.3. Possible reasons for differing propensities for horizontal transfer

Two reasons have been proposed to explain the apparent differences in the prevalence of horizontal transfer events between DNA transposons and non-LTR retrotransposons (Eickbush and Malik, 2002, and herein). The first is that DNA transposons need horizontal transfer for their long-term survival, but non-LTR retrotransposons appear not to be dependent on such rare events in evolution. Defective copies of DNA transposons retain the ability to be transposed as long as they have the *cis*-acting signals such as the TIRs. This indiscrimination leads to the inevitable fate of inactivation of the entire transposon family. Therefore, horizontal transfer offers a much-needed escape from the above vertical inactivation, which greatly enhances the evolutionary success of DNA transposons. On the other hand, it has been shown that the reverse transcriptase of non-LTR retrotransposons tends to associate with the mRNA molecules from which they were translated (Wei *et al.*, 2001). This *cis*-preference would bias

transposition events in favor of the active elements, thus providing a mechanism to longer sustain the non-LTR retrotransposons. However, the *cis*-preference is not enough to prevent the highly successful retrotransposition of SINEs in insects (Adams *et al.*, 1986; Tu, 1999) and other organisms (Lander *et al.*, 2001), which presumably borrow the retrotransposition machinery from non-LTR retrotransposons. It will be interesting to see how SINEs affect the evolution of their non-LTR retrotransposon "partners." The second explanation is that the transposition process of non-LTR retrotransposons does not involve an extrachromosomal DNA intermediate, which may be important in horizontal transfer (Eickbush and Malik, 2002; see also below). In addition, DNA transposons use their transposase for integration, while non-LTR retrotransposons require more extensive involvement of host repair machinery. If the repair machinery involved in retrotransposition is species-specific, the activity of a non-LTR retrotransposon may be more restricted to its original host. Therefore, DNA transposons may be more predisposed to horizontal transfer than non-LTR retrotransposons. LTR retrotransposons form an extrachromosomal DNA intermediate, and use transposase-like integrase for integration. Therefore, LTR retrotransposons have access to the same horizontal transfer mechanisms as the DNA transposons, although their life cycle may not require horizontal transfer because defective copies are thought not to be a major factor (Eickbush and Malik, 2002). In addition, some LTR elements may acquire *env* protein and convert to infectious viruses that could be transmitted horizontally. It should be noted that the above are general statements, and the propensity to horizontal transfer may vary among individual families within the three groups discussed here.

3.6.2.4. Mechanisms of horizontal transfer

Mechanisms of horizontal transfer are poorly understood, although direct transfer of the extrachromosomal DNA intermediate and indirect transfer through a viral vector have been proposed as possible mechanisms (Eickbush and Malik, 2002; Silva *et al.*, 2004). Geographical and temporal overlap between the donor and recipient host species may be essential. An intriguing case of horizontal transfer of a *mariner* element between a parasitoid wasp and its lepidopteran host offers a good example of such overlap (Yoshiyama *et al.*, 2001). Similarly, four TEs discovered in the kissing bug *Rodnius prolixus* were shown to be nearly identical to those found in the bug's vertebrate hosts (Gilbert *et al.*, 2010). The intracellular symbiont *Wolbachia* has also been suggested as one possible vector for horizontal transfer (reviewed in Schaack *et al.*, 2010).

As the genome sequences rapidly accumulate, the numbers of examples of horizontal transfer are likely to increase. With the rapid progress in high-throughput sequencing technology and the continuing reduction of sequencing cost, it is perhaps time to move beyond "accidental" discovery of horizontal transfer and perform broad low-coverage genome sequencing surveys of many species with ecological overlap. Such surveys will identify repetitive sequences that show unexpectedly high identity between species, which will lead to candidates of very recent horizontal transfer events. This in turn will likely offer opportunities to investigate the mechanisms and circumstances of horizontal transfer, because the factors required for such lateral transfer may still be accessible for examination.

3.6.3. Other Possible Evolutionary Strategies

In addition to horizontal transfer and vertical extinction, recent studies suggest that there might be a third way, or an alternative strategy, which may be adopted by some TEs (Lampe *et al.*, 2001). On the basis of the loss of interaction between *mariner* transposons of slightly changed TIRs, it was proposed that intra-specific or intra-genomic diversification of *mariner* transposons may allow the newly diverged *mariner* to start a new lineage. Although this requires the coevolutionary events to occur in both the transposase and the TIRs, this scenario would provide the transposon the opportunity to escape vertical inactivation, because it is now virtually a brand new element in a virgin genome owing to the loss of interaction between itself and its relatives in the genome. Genome sequencing has provided increasing opportunities to survey the diversity of different families of TEs. Our recent analysis showed a large number of lineages of non-LTR retrotransposons of the CR1 and Jockey clades in *An. gambiae* (Biedler and Tu, 2003). Given the presence of multiple recently active lineages within the CR1 and Jockey clades, it is tempting to speculate that the observed diversity may be driven by competition among different non-LTR families, or by attempts to escape suppressive mechanisms imposed by the host.

On the other hand, some TEs are recruited for host functions, and thus become "domesticated" (Lander *et al.*, 2001). This type of molecular domestication is the ultimate case of trading "freedom" for "security." It allows TEs to sustain and positively impact the host, examples of which will be discussed in section 3.8. Strictly speaking, these domesticated TEs are no longer TEs. However, it is theoretically possible that these "domesticated" TEs could revert back to their "old ways" on rare occasions.

3.6.4. Understanding the Intra-Genomic Diversity of Insect TEs

High levels of TE diversity have been reported in the many insect genomes (see, for example, Holt *et al.*, 2002; Kaminker *et al.*, 2002; Kapitonov and Jurka,

2003a; Nene *et al.*, 2007; Arensburger *et al.*, 2010). The evolutionary process that generated this diversity may also be quite variable. It is possible that the evolution of some TEs may be a complex mix of both vertical transmission and horizontal transfer events. Parsing out the results of intra-genomic diversification from those of horizontal transfer events may require additional data from related species. Understanding the process responsible for the intra-genomic diversity of insect TEs and the potential interactions between different TE families in insect genomes will be both challenging and rewarding. A summary of the current hypothesis on TE evolution is illustrated in **Figure 6**. Some aspects of this model are better suited for DNA transposons that have a high propensity for horizontal transfer.

3.7. TEs in Insect Populations

3.7.1. Fundamental Questions and Practical Relevance

In general, the increase of TE copy number through transposition is balanced by selective forces against the potential genetic load of TEs on host fitness (Nuzhdin, 1999). The previous section discussed the control of TE transposition rate and other mechanisms to minimize their deleterious effects. This section attempts to describe the population dynamics affecting the spread of TEs in insect genomes. Earlier work on TEs in *Drosophila* populations suggest that the copy numbers in euchromatic regions are low, and most euchromatic copies exist at very low frequency (< 5%) in the population (summarized in Petrov *et al.*, 2003). This was interpreted as evidence of selection against individual TE copies in nature (Charlesworth and Langley, 1989). It has been hypothesized that this selection is against three types of potentially overlapping deleterious effects by TEs, including insertional mutagenesis, transcriptional or translational cost, and ectopic recombinations between similar copies of TEs in different chromosomal regions (Charlesworth and Langley, 1989; Nuzhdin, 1999; Kidwell and Lisch, 2001; Petrov *et al.*, 2003). For quite some time, one of the major questions in this field has been parsing out the main factors containing the spread of TEs in natural populations. However, recent analysis showed that many *Drosophila* TEs have high population frequencies and provide adaptive roles during evolution (reviewed in Gonzalez and Petrov, 2009). From an applied perspective, TEs have been proposed as tools to genetically drive the spread of beneficial genes through insect populations to control infectious diseases (e.g., Ashburner *et al.*, 1998; Alphey *et al.*, 2002). For such a sophisticated approach to work, it is important to understand the population dynamics of TEs in their insect hosts.

3.7.2. Experimental Approaches

In situ hybridization of the *Drosophila* polytene chromosomes was the main workhorse in early studies of the population dynamic of different TE families (Charlesworth and Langley, 1989). Although extremely useful, this method only works in species with accessible polytene chromosomes, and it is not efficient in detecting short regions of sequence similarities to the probe (Petrov *et al.*, 2003). An alternative method, genomic Southern blotting, was also used to study TE insertions and excisions in *Drosophila* (Maside *et al.*, 2001). Although Southern blotting is a good method to estimate TE copy numbers, it is not reliable when the numbers are high. In addition, it may not be able to detect low-frequency sites in cases where multiple small insects have to be pooled to obtain enough high-quality genomic DNA. However, a single *Drosophila* can provide enough DNA for a Southern experiment.

TE display, a genome-scale detection method for TE insertions, has been used recently in *Drosophila* and mosquitoes (Biedler *et al.*, 2003; Yang and Nuzhdin, 2003; Bonin *et al.*, 2008; Subramanian *et al.*, 2008). Because it is a PCR-based method, TE display allows investigation of multiple TE families using genomic DNA isolated from individual insects. Although extremely powerful, this method cannot reliably distinguish homozygous from heterozygous insertions. However, TE display will allow recovery of a specific insertion site by sequencing the corresponding band. The sequence flanking the TE copy can be used to locate the specific site by searching the genome database, if available, or by inverse PCR. Therefore, sequences flanking a TE at a specific locus can be used as primers to amplify genomic DNA isolated from an individual sample (**Figure 5**). When the PCR products are run on an agarose gel, individuals with insertions at both alleles will show a single high molecular mass band, while individuals with no insertions at either allele will give a single low molecular mass band (**Figure 5**). Individuals that have heterozygous alleles will give both bands. Thus, this locus-specific approach is co-dominant. When the genome sequence or a bulk of genomic sequences are available for an insect species, it is not absolutely necessary to couple locus-specific PCR with TE display, because a number of TE insertion sites will already have been available for analysis (Petrov *et al.*, 2003). However, TE display can facilitate the investigation by providing a rapid scan of a large number of loci, and by providing initial assessment of the level of polymorphism.

3.7.3. Recent Advances

Population studies mainly of *Drosophila* TEs suggest that selection against ectopic recombination may be one of the major factors in containing TE copy number in nature (Bartolome *et al.*, 2002; Kidwell and Lisch, 2001; Rizzon *et al.*, 2002; Petrov *et al.*, 2003). The previously mentioned small RNA pathways may also be a critical factor controlling TE copy numbers (Lee and Langley, 2010). Recent surveys in *D. melanogaster* revealed an unexpected number of sites that are either fixed or of high frequency (Petrov *et al.*, 2003; Yang and Nuzhdin, 2003), and a number of examples of TE insertions providing adaptive roles have been discovered (e.g., Maside *et al.*, 2001; Petrov *et al.*, 2003; Lerman and Feder, 2005; Gonzalez *et al.*, 2008, 2010; Gonzalez and Petrov, 2009). Much more will be learned as genome sequences from individual insects within natural populations become available.

It may reasonable to assume that TE copy number and insertion frequency in a population are highly dynamic parameters that can be influenced by TE-specific factors such as its intrinsic ability for transposition and self-regulation, by species- or genome-specific factors such as deletion and recombination rate and genomic control of transposition, and of course by effective population size and other ecological factors. As most population surveys only reflect a cross-section during evolution, the relative stage of a TE in its life cycle (**Figure 6**) is also an important consideration (Vieira *et al.*, 1999).

3.8. Impact of TEs in Insects

3.8.1. TEs and Genome Size and Organization

As described in the introduction, TEs are integral and significant components of eukaryotic genomes. For example, at least 46% of the human genome (Lander *et al.*, 2001) is TE-derived sequences. It has been proposed that the differing TE abundance may account for the "C-value paradox," which reflects the discrepancy between genome size or DNA content of an organism (as indicated by the C-value) and its biological complexity (Kidwell, 2002). In other words, organisms with similar genetic/biological complexity may have huge variations in genome size due to differences in TE content. Here, some preliminary information gleaned from comparative analysis between different species of mosquitoes is described briefly. The published genomes of *Anopheles*, *Culex*, and *Aedes* mosquitoes vary by five-fold in size, ranging from ~270 Mbp in *An. gambiae* (Holt *et al.*, 2002) to ~500 Mbp in *C. quinquefasciatus* (Arensburger *et al.*, 2010) and ~1300 Mbp in *Ae. aegypti* (Nene *et al.*, 2007). TE contents in these three species are 11–16%, 29%, and 47% of the assembled genomes, respectively, indicating that TEs contributed significantly to the genome size variations among mosquito species. Varied amounts of repetitive elements have also been shown to be a major factor for the nearly three-fold intra-specific differences of genome size in populations of the Asian tiger mosquito, *Ae. albopictus* (Rai and Black, 1999). Such an intra-specific variation provides a rare opportunity to study the contribution of TEs to genome evolution at the initial stage of the evolutionary process.

As noted before, TEs can induce chromosomal rearrangements through ectopic recombination and other mechanisms (Gray, 2000). Such rearrangements could result in gross reorganization of the chromosomes, which may have a significant evolutionary impact, as suggested by McClintock (McClintock, 1984). A DNA TE named *Odysseus* was found adjacent to the distal breakpoint of a naturally occurring paracentric chromosomal inversion that is characteristic of *An. arabiensis*, one of the cryptic species in the *An. gambiae* complex (Mathiopoulos *et al.*, 1998). Similar evidence of TE involvement in chromosomal rearrangement has been reported in *Drosophila* (Lim and Simmons, 1994; Caceres *et al.*, 2001; Delprat *et al.*, 2009). It is proposed that TEs may indeed be important in generating chromosomal rearrangements in nature, which may result in reproductive isolation and genetic changes that allow *An. gambiae* to exploit a range of ecological niches (Mathiopoulos *et al.*, 1999).

3.8.2. Evolutionary Impact

The tremendous potential for TEs to generate genetic diversity has long been recognized (McClintock, 1956). They can cause spontaneous mutation, recombination, chromosomal rearrangement, and hybrid dysgenesis (Bregliano *et al.*, 1980; Engels, 1989; Shiroishi *et al.*, 1993; Mathiopoulos *et al.*, 1998). TEs have been generally regarded as "selfish" DNA since the early 1980s (Doolittle and Sapienza, 1980; Orgel and Crick, 1980), as opposed to the original "controlling elements" hypothesis stating that TEs provide the physical basis controlling gene action and mutation (McClintock, 1956). This change of attitude was mainly due to the realization that TEs are somewhat "independent" genetic units, and that their replicative ability allows them to spread even when they are not beneficial to the host organism. The question of whether the "selfish" TEs are just "junk DNA" to the host, or can play important and even adaptive roles in organismal evolution, is at the heart of the debate. Thanks in part to recent studies of insect TEs, it is increasingly clear that there may be a middle ground between the "junk DNA" and the "controlling elements" hypotheses. The host genome can be viewed as an ecological community with complex host–TE and TE–TE interactions (Brookfield, 1995; Kidwell and Lisch, 2000). Given the opportunistic nature of the evolutionary process, "selfish" elements could develop a wide spectrum of relationships with their host. They

could be a "junk parasite," a "molecular symbiont," or something in between.

Examples of TEs being co-opted to contribute to organismal biology continue to accumulate. The introduction mentions the early examples of telomere maintenance in *Drosophila* site-specific insertion retrotransposons (Biessmann *et al.*, 1992a, 1992b; Levis *et al.*, 1993). A fascinating genome-wide analysis showed that retrotransposition was responsible for the creation of a significant number of new functional genes in *Drosophila* (Betran *et al.*, 2002). Some TE copies can be "domesticated" and take on a host function (also see section 3.6.3) – for example, one of the *P* element repeats appears to have evolved the function of transcription factors (Miller *et al.*, 1995). Other examples of "domesticated" TEs found in non-insect species include the RAG1 and RAG2 genes involved in V(D)J recombination in vertebrate lymphocytes, and the more than 40 new genes derived from TEs in the human genome (Kidwell and Lisch, 2000; Lander *et al.*, 2001; Gellert, 2002). More than 20 adaptive and fixed TE insertions have been found in *Drosophila* (reviewed in Gonzalez and Petrov, 2009), and a recent genome-wide survey identified 10 TEs associated with adaptations to temperate environments in *Drosophila* (Gonzalez *et al.*, 2010). TE insertions have been shown to be associated with insecticide resistance in the moth *Helicoverpa zea* (Chen and Li, 2007), and *Culex quinquefasciatus* mosquitoes (Itokawa *et al.*, 2010).

In addition to possible insertional mutagenesis associated with TE transposition and imprecise excision, TEs can also serve as substrates for homologus recombination that can result in chromosomal deletion, duplication, and inversion. Polymorphic chromosomal inversions are common in *Drosophila* and *An. gambiae*. TEs are often implicated in the generation of these inversions. It is proposed that these natural chromosomal inversions may result in reproductive isolation and perhaps genetic changes that allow *An. gambiae* to exploit a range of ecological niches (Mathiopoulos *et al.*, 1999).

The mere presence of TEs may represent a powerful genetic force with which the genome has been evolving. The arms race between TEs and the host genomes may drive the evolution of genetic and epigenetic control mechanisms that may be important to the host organisms. For example, recently discovered piRNA and endosiRNA pathways may have had a profound impact on gene regulation and epigenetic silencing (Aravin *et al.*, 2007; Nishida *et al.*, 2007; Pelisson *et al.*, 2007; Yin and Lin, 2007; Brennecke *et al.*, 2008; Chung *et al.*, 2008; Ghildiyal *et al.*, 2008; Klattenhoff *et al.*, 2009; Lau *et al.*, 2009; Lisch, 2009; Malone and Hannon, 2009; Zeh *et al.*, 2009). On the basis of their broad distribution in bacteria, archea, and eukaryotes (Craig *et al.*, 2002), it is safe to assume that TEs have long been evolving together with the immensely diverse life forms on this planet, and will continue to do so.

3.9. Applications of Insect TEs

3.9.1. Endogenous TEs and Genetic Manipulation of Insects

P element-based transgenic and mutagenesis tools in *D. melanogaster* have played a major role in the tremendous success of this tiny fly as a model organism for genetic analysis. A limited number of DNA transposons, such as *hobo*, *mariner*, *minos*, and *piggyBac*, which have a broader host range than *P*, have been developed as tools in insects (see Chapter 4). Further analysis of TEs in insect genomes may expand the pool of active DNA transposons, which may be used to generate a set of tools with diverse features that can be used collectively for a variety of genetic analyses in different insects. In addition to simply transforming an insect, active TEs mentioned above are used to construct specific vectors to be used in gene trapping, enhancer trapping, and genome-wide insertional mutagenesis studies (Spradling *et al.*, 1999; Klinakis *et al.*, 2000; Horn *et al.*, 2003; Bonin and Mann, 2004). These analyses are powerful ways to investigate gene function and regulation on a genome-scale.

In addition to providing possible new active TEs to be used as tools for genetic manipulation of insects, a better understanding of endogenous TEs will allow better-informed usage of current transposon-based genetic tools. Interactions between exogenous and endogenous transposons that share similar TIRs have been shown to be a potential problem (Sundararajan *et al.*, 1999; Jasinskiene *et al.*, 2000). Such interactions could be significant in light of the discovery of a diverse range of DNA TEs in a few insects in which genetic manipulation is being actively pursued (see, for example, Nene *et al.*, 2007). Analyses of endogenous insect TEs will lead to better-informed design of transposon-based transformation tools that reduce instability resulting from interactions with endogenous TEs (Ashburner *et al.*, 1998; Atkinson *et al.*, 2001). It is also hoped that the non-Mendelian inheritance of TEs could help beneficial transgenes sweep through insect populations, as the *P* element did in *Drosophila* (Ribeiro and Kidwell, 1994; Engels, 1997). Such a strategy is being investigated in the context of driving refractory genes into mosquito populations to control mosquito-borne infectious diseases (Ashburner *et al.*, 1998; Alphey *et al.*, 2002), although concerns, including disassociation between transposon and the beneficial gene, will need to be addressed (Marshall, 2008). A better understanding of endogenous TEs may be important to help achieve sustained success of such sophisticated genetic approaches.

3.9.2. SINE Insertion Polymorphism as Polymorphic Genetic Markers

The genetic differences and the pattern of gene flow between insect populations are of fundamental importance to a number of entomological questions, ranging from evolution to practical applications. Single nucleotide polymorphisms (SNPs) are powerful markers for population genetic analysis, especially for insects with a large amount of sequence data available. Polymorphic insertion sites of interspersed TEs are potentially rich sources of a different type of genetic markers for population and genetic mapping studies. The discussion here is focused on SINEs, which are especially useful for the reasons described below. In population studies, sequences flanking a SINE at a specific locus are used as primers to amplify genomic DNA isolated from an individual sample. When the PCR products are run on an agarose gel, the genotype of an individual will be revealed on the basis of the number and size of bands. Thus, this locus-specific PCR assay may be used for co-dominant markers that reveal the dimorphism (insertion vs non-insertion) at a specific site. SINEs including the human *Alu* elements have been shown to be powerful genetic markers (e.g., Batzer *et al.*, 1994; Roy-Engel *et al.*, 2001; Batzer and Deininger, 2002; Salem *et al.*, 2003). The ability of SINE insertion polymorphic markers to differentiate recently separated human populations is a good indication of their power (e.g., de Pancorbo *et al.*, 2001; Nasidze *et al.*, 2001; Watkins *et al.*, 2001). The locus-specific PCR assay of SINE insertions has a few potential advantages over the popular microsatellite markers. The same SINE insertions are identical by descent. The probability that different SINEs of the same size independently insert into the same chromosomal location may be negligible (Stoneking *et al.*, 1997; York *et al.*, 1999; de Pancorbo *et al.*, 2001; Watkins *et al.*, 2001). Moreover, the ancestral state of the SINE insertion polymorphism is likely to be the absence of a SINE because of general lack of excision, although exceptions do exist (Medstrand *et al.*, 2002). The ability to distinguish the ancestral versus derived states provides additional resolving power to address population genetic questions (York *et al.*, 1999). One potential limitation of this approach could be the removal of a SINE insertion by rare recombination or gene conversion events, which may be confused with the non-insertion state. Such events may be revealed during the PCR analysis. Sequence analysis at the insertion site will also allow the investigation of this possibility. *Maque* and *SINE200* have been successfully used to study the incipient speciation between the M and S forms of *An. gambiae* (Barnes *et al.*, 2005; Santolamazza *et al.*, 2008).

The development of a TE-anchored PCR approach, or TE display, has made it possible to directly screen for TE insertion polymorphism in a few species, including insects

(**Figure 5**; see also Biedler *et al.*, 2003; Yang and Nuzhdin, 2003; Arensburger *et al.*, 2005; Bonin *et al.*, 2008). TE display efficiently scans a large number of loci in the genome, which makes it a very good tool for genotyping. However, as a population genetic tool it has a major limitation common for dominant markers such as AFLP and RAPD; namely, the inability to distinguish between heterozygous and homozygous insertions, rendering the detection of population genetic structure difficult. However, TE display can be used as a direct screen to identify potential polymorphic insertion sites. Therefore, TE display in conjunction with the development of locus-specific PCR markers that are co-dominant will help TE insertion polymorphism markers reach their full potential as population genomic tools for insects.

3.9.3. SINE Insertions as Phylogenetic Markers

TE insertions, more specifically SINE insertions, have been used as molecular systematic tools to trace the evolutionary relationship between whales and Artiodactyla (Shimamura *et al.*, 1997; Nikaido *et al.*, 1999), and between Salmonid fishes (Murata *et al.*, 1993). Perhaps the most impressive use of SINE insertions is the resolution of the evolutionary relationship of one of the major tribes of the African cichlid fishes that have evolved through an explosive adaptive radiation (Takahashi *et al.*, 1998; Terai *et al.*, 2003). To obtain TE insertion information for molecular systematics, locus-specific PCR, described above, is used. Here, an RNA-mediated TE such as a SINE is again better suited because its transposition does not involve excision. Therefore, the ancestral state is known to be the absence of a SINE (Shedlock and Okada, 2000; Nishihara and Okada, 2008). The basic concept of this approach is illustrated in **Figure 7**. The insertion state can be easily determined using an agarose gel or melt-curve analysis. When a fixed insertion is found at a particular site in species 1 and

2 but not in species 3, it can be inferred that 1 and 2 are sister taxa. One of the requirements of the above approach is that the TE insertion site should be fixed within a species. TE display, a fingerprinting method described earlier (**Figure 5**), may be used to search for such sites. Potential problems such as non-specific deletions, gene conversions, and sorting of ancestral polymorphisms can either be detected during the PCR and gel electrophoresis analysis, or mitigated by surveying multiple loci (Shedlock and Okada, 2000; Nishihara and Okada, 2008). Sequences of the SINEs themselves in the loci used for the systematic analysis may provide further phylogenetic information. The usefulness of a particular SINE family in molecular systematics studies is dependent on its distribution and lifespan in the taxonomic group of interest. The tremendous diversity of insect species offers interesting challenges to evolutionary biologists. For example, a number of medically and economically important insect organisms exist as cryptic species complexes (Munstermann and Conn, 1997; Krzywinski and Besansky, 2003). New phylogenetic tools that can be integrated with methods using conventional characters such as morphology and DNA sequences will undoubtedly be of significance. Although SINEs have not been extensively studied in insects, highly repetitive SINEs have already been characterized in different orders, including many medically and economically important species. Therefore, SINEs may provide useful markers for molecular systematic analysis of insects, one of the most diverse groups of life forms on this planet.

In summary, TEs have been successfully used as vectors for genetic manipulation of insects and other organisms. A better understanding of insect TEs will allow better-informed usage of the currently available TE-based tools. The application of TEs as population and phylogenetic markers is at an early stage. Although these markers are promising tools, their scope of application, their resolving power and reliability depend on a better understanding of the population and evolutionary dynamics of the TEs. Therefore, fundamental studies discussed in previous sections also have significant implications for the applications of TE-based molecular tools. As in so many areas of biology, the availability of genome sequences from related species as well as individuals within populations will greatly facilitate the investigation and application of insect TEs.

3.10. Summary

The past few years have witnessed an explosive growth both in the development of TE-based genetic and molecular tools and in our fundamental understanding of the diversity, regulation, and impact of TEs in insects. Studies of TEs in insects, especially in *D. melanogaster*, have more than once led to discoveries that broadly impacted the field of TE research. In addition, studies in *D. melanogaster* also significantly contributed to the

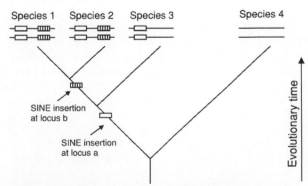

Figure 7 A schematic illustration of the principle of using SINE insertions as molecular systematic markers. Monophyletic relationships between species may be inferred on the basis of shared SINE insertions (see section 3.9.3 for detailed explanation).

discovery of novel small RNA pathways and the functions of small RNAs in TE suppression, gene regulation, and epigenetic silencing. The availability of new bioinformatic and experimental tools and the rapidly expanding genome revolution provide an exciting opportunity for the discovery of novel TEs in a wide range of insects, for identification of evidence of TE horizontal transfer, and for in-depth molecular and genomic analysis of these mobile genetic elements. Only through comparative genomic approaches can we come close to a full appreciation of the complex and intricate dynamics governing the evolution of diverse TEs in insect genomes. From an applied perspective, systematic analysis of TEs in many insect species has significant economic and health implications because of the importance of these insects in disease transmission and agriculture.

Acknowledgment

Work in the author's laboratory is supported by NIH grants AI42121 and AI53203, the Jeffress Foundation, and the Virginia Agricultural Experimental Station.

References

Adams, D. S., Eickbush, T. H., Herrera, R. J., & Lizardi, P. M. (1986). A highly reiterated family of transcribed oligo(A)-terminated, interspersed DNA elements in the genome of *Bombyx mori. J. Mol. Biol., 187*, 465–478.

Alphey, L., Beard, C. B., Billingsley, P., Coetzee, M., Crisanti, A., et al. (2002). Malaria control with genetically manipulated insect vectors. *Science, 298*, 119–121.

Altschul, S. F., Madden, T. L., Schaffer, A. A., Zhang, J., Zhang, Z., et al. (1997). Gapped BLAST and PSI-BLAST: A new generation of protein database search programs. *Nucleic Acids Res., 25*, 3389–3402.

Aravin, A. A., Hannon, G. J., & Brennecke, J. (2007). The Piwi–piRNA pathway provides an adaptive defense in the transposon arms race. *Science, 318*, 761–644.

Arca, B., Zabalou, S., Loukeris, T. G., & Savakis, C. (1997). Mobilization of a Minos transposon in *Drosophila melanogaster* chromosomes and chromatid repair by heteroduplex formation. *Genetics, 145*, 267–279.

Arensburger, P., Kim, Y. J., Orsetti, J., Aluvihare, C., O'Brochta, D. A., & Atkinson, P. W. (2005). An active transposable element, Herves, from the African malaria mosquito *Anopheles gambiae. Genetics, 169*, 697–708.

Arensburger, P., Megy, K., Waterhouse, R. M., Abrudan, J., Amedeo, P., et al. (2010). Sequencing of *Culex quinquefasciatus* establishes a platform for mosquito comparative genomics. *Science, 330*, 86–88.

Arkhipova, I. R., Pyatkov, K. I., Meselson, M., & Evgen'ev, M. B. (2003). Retroelements containing introns in diverse invertebrate taxa. *Nat. Genet., 33*, 123–124.

Ashburner, M., Hoy, M. A., & Peloquin, J. J. (1998). Prospects for the genetic transformation of arthropods. *Insect Mol. Biol., 7*, 201–213.

Atkinson, P. W. (2008). Proposed uses of transposons in insect and medical biotechnology. *Adv. Exp. Med. Biol., 627*, 60–70.

Atkinson, P. W., Pinkerton, A. C., & O'Brochta, D. A. (2001). Genetic transformation systems in insects. *Annu. Rev. Entomol., 46*, 317–346.

Bao, Z., & Eddy, S. R. (2002). Automated *de novo* identification of repeat sequence families in sequenced genomes. *Genome Res., 12*, 1269–1276.

Barnes, M. J., Lobo, N. F., Coulibaly, M. B., Sagnon, N. F., Costantini, C., & Besansky, N. J. (2005). SINE insertion polymorphism on the X chromosome differentiates *Anopheles gambiae* molecular forms. *Insect Mol. Biol., 14*, 353–363.

Bartolome, C., Maside, X., & Charlesworth, B. (2002). On the abundance and distribution of transposable elements in the genome of *Drosophila melanogaster. Mol. Biol. Evol., 19*, 926–937.

Bartolome, C., Bello, X., & Maside, X. (2009). Widespread evidence for horizontal transfer of transposable elements across *Drosophila* genomes. *Genome Biol., 10*, R22.

Batzer, M. A., & Deininger, P. L. (2002). Alu repeats and human genomic diversity. *Nat. Rev. Genet., 3*, 370–379.

Batzer, M. A., Stoneking, M., Alegria-Hartman, M., Bazan, H., Kass, D. H., et al. (1994). African origin of human-specific polymorphic Alu insertions. *Proc. Natl. Acad. Sci. USA, 91*, 12288–12292.

Bauser, C. A., Elick, T. A., & Fraser, M. J. (1999). Proteins from nuclear extracts of two lepidopteran cell lines recognize the ends of TTAA-specific transposons piggyBac and tagalong. *Insect Mol. Biol., 8*, 223–230.

Beall, E. L., & Rio, D. C. (1997). *Drosophila* P-element transposase is a novel site-specific. *Genes Dev., 11*, 2137–2151.

Becker, H. A., & Kunze, R. (1997). Maize Activator transposase has a bipartite DNA binding domain that recognizes subterminal sequences and the terminal inverted repeats. *Mol. Gen. Genet., 254*, 219–230.

Berezikov, E., Bucheton, A., & Busseau, I. (2000). A search for reverse transcriptase-coding sequences reveals new non-LTR retrotransposons in the genome of *Drosophila melanogaster. Genome Biol., 1*, RESEARCH0012.

Betran, E., Thornton, K., & Long, M. (2002). Retroposed new genes out of the X in *Drosophila. Genome Res., 12*, 1854–1859.

Biedler, J., & Tu, Z. (2003). Non-LTR retrotransposons in the African malaria mosquito, *Anopheles gambiae*: Unprecedented diversity and evidence of recent activity. *Mol. Biol. Evol., 20*, 1811–1825.

Biedler, J., & Tu, Z. (2007). The Juan non-LTR retrotransposon in mosquitoes: Genomic impact, vertical transmission and indications of recent and widespread activity. *BMC Evol. Biol., 7*, 112.

Biedler, J., Qi, Y., Holligan, D., Della Torre, A., Wessler, S., & Tu, Z. (2003). Transposable element (TE) display and rapid detection of TE insertion polymorphism in the *Anopheles gambiae* species complex. *Insect Mol. Biol., 12*, 211–216.

Biedler, J. K., Shao, H., & Tu, Z. (2007). Evolution and horizontal transfer of a DD37E DNA transposon in mosquitoes. *Genetics, 177*, 2553–2558.

Biessmann, H., Champion, L. E., O'Hair, M., Ikenaga, K., Kasravi, B., & Mason, J. M. (1992a). Frequent transpositions of *Drosophila melanogaster* HeT-A transposable elements to receding chromosome ends. *EMBO J., 11*, 4459–4469.

Biessmann, H., Valgeirsdottir, K., Lofsky, A., Chin, C., Ginther, B., et al. (1992b). HeT-A, a transposable element specifically involved in "healing" broken chromosome ends in *Drosophila melanogaster. Mol. Cell. Biol., 12,* 3910–3918.

Bonasio, R., Zhang, G., Ye, C., Mutti, N. S., Fang, X., et al. (2010). Genomic comparison of the ants *Camponotus floridanus* and *Harpegnathos saltator. Science, 329,* 1068–1071.

Bonin, A., Paris, M., Despres, L., Tetreau, G., David, J. P., & Kilian, A. (2008). A MITE-based genotyping method to reveal hundreds of DNA polymorphisms in an animal genome after a few generations of artificial selection. *BMC Genomics, 9,* 459.

Bonin, A., Paris, M., Tetreau, G., David, J. P., & Despres, L. (2009). Candidate genes revealed by a genome scan for mosquito resistance to a bacterial insecticide: sequence and gene expression variations. *BMC Genomics, 10,* 551.

Bonin, C. P., & Mann, R. S. (2004). A piggyBac transposon gene trap for the analysis of gene expression and function in *Drosophila. Genetics, 167,* 1801–1811.

Bonnivard, E., Bazin, C., Denis, B., & Higuet, D. (2000). A scenario for the hobo transposable element invasion, deduced from the structure of natural populations of *Drosophila melanogaster* using tandem TPE repeats. *Genet. Res., 75,* 13–23.

Braquart, C., Royer, V., & Bouhin, H. (1999). DEC: A new miniature inverted-repeat transposable element from the genome of the beetle *Tenebrio molitor. Insect Mol. Biol., 8,* 571–574.

Bregliano, J. C., Picard, G., Bucheton, A., Pelisson, A., Lavige, J. M., & L'Heritier, P. (1980). Hybrid dysgenesis in *Drosophila melanogaster. Science, 207,* 606–611.

Brennecke, J., Aravin, A. A., Stark, A., Dus, M., Kellis, M., et al. (2007). Discrete small RNA-generating loci as master regulators of transposon activity in *Drosophila. Cell, 128,* 1089–1103.

Brennecke, J., Malone, C. D., Aravin, A. A., Sachidanandam, R., Stark, A., & Hannon, G. J. (2008). An epigenetic role for maternally inherited piRNAs in transposon silencing. *Science, 322,* 1387–1392.

Brookfield, J. F. (1995). Transposable elements as selfish DNA. In D. J. Sherratt (Ed.). *Mobile Genetic Elements* (pp. 130–153). Oxford, UK: Oxford University Press.

Brookfield, J. F. (2005). Evolutionary forces generating sequence homogeneity and heterogeneity within retrotransposon families. *Cytogenet. Genome Res., 110,* 383–391.

Bucheton, A. (1995). The relationship between the flamenco gene and gypsy in *Drosophila*: How to tame a retrovirus. *Trends Genet., 11,* 349–353.

Caceres, M., Puig, M., & Ruiz, A. (2001). Molecular characterization of two natural hotspots in the buzzatii genome induced by transposon insertions. *Genome Res., 11,* 1353–1364.

Capy, P., Vitalis, R., Langin, T., Higuet, D., & Bazin, C. (1996). Relationships between transposable elements based upon the integrase-transposase domains: Is there a common ancestor? *J. Mol. Evol., 42,* 359–368.

Capy, P., Langin, T., Higuet, D., Maurer, P., & Bazin, C. (1997). Do the integrases of LTR-retrotransposons and class II element transposases have a common ancestor? *Genetica, 100,* 63–72.

Cary, L. C., Goebel, M., Corsaro, B. G., Wang, H. G., Rosen, E., & Fraser, M. J. (1989). Transposon mutagenesis of baculoviruses: Analysis of *Trichoplusia ni* transposon IFP2 insertions within the FP-locus of nuclear polyhedrosis viruses. *Virology, 172,* 156–169.

Casa, A. M., Brouwer, C., Nagel, A., Wang, L., Zhang, Q., et al. (2000). Inaugural article: The MITE family heartbreaker (Hbr): Molecular markers in maize. *Proc. Natl. Acad. Sci. USA, 97,* 10083–10089.

Caspi, A., & Pachter, L. (2006). Identification of transposable elements using multiple alignments of related genomes. *Genome Res., 16,* 260–270.

Charlesworth, B., & Langley, C. H. (1989). The population genetics of *Drosophila* transposable elements. *Annu. Rev. Genet., 23,* 251–287.

Chen, S., & Li, X. (2007). Transposable elements are enriched within or in close proximity to xenobiotic-metabolizing cytochrome P450 genes. *BMC Evol. Biol., 7,* 46.

Chung, W. J., Okamura, K., Martin, R., & Lai, E. C. (2008). Endogenous RNA interference provides a somatic defense against *Drosophila* transposons. *Curr. Biol., 18,* 795–802.

Claudianos, C., Brownlie, J., Russell, R., Oakeshott, J., & Whyard, S. (2002). maT – a clade of transposons intermediate between mariner and Tc1. *Mol. Biol. Evol., 19,* 2101–2109.

Coates, B. S., Kroemer, J. A., Sumerford, D. V., & Hellmich, R. L. (2010a). A novel class of miniature inverted repeat transposable elements (MITEs) that contain hitchhiking (GTCY)(n) microsatellites. *Insect Mol. Biol., 20,* 15–27.

Coates, B. S., Sumerford, D. V., Hellmich, R. L., & Lewis, L. C. (2010b). A helitron-like transposon superfamily from lepidoptera disrupts (GAAA)(n) microsatellites and is responsible for flanking sequence similarity within a microsatellite family. *J. Mol. Evol., 70,* 275–288.

Cockburn, A. F., & Mitchell, S. E. (1989). Repetitive DNA interspersion patterns in Diptera. *Arch. Insect Biochem. Physiol., 10,* 105–113.

Cordaux, R., & Batzer, M. A. (2009). The impact of retrotransposons on human genome evolution. *Nat. Rev. Genet., 10,* 691–703.

Cordeiro, J., Robe, L. J., Loreto, E. L., & Valente, V. L. (2008). The LTR retrotransposon micropia in the cardini group of *Drosophila* (Diptera: Drosophilidae): A possible case of horizontal transfer. *Genetica, 134,* 335–344.

Coy, M. R., & Tu, Z. (2005). Gambol and Tc1 are two distinct families of DD34E transposons: Analysis of the *Anopheles gambiae* genome expands the diversity of the IS630-Tc1-mariner superfamily. *Insect Mol. Biol., 14,* 537–546.

Craig, N. (2002). Mobile DNA: An introduction. In N. Craig, R. Craigie, M. Gellert, & A. Lambowitz (Eds.), *Mobile DNA II* (pp. 3–11). Herndon, VA: American Society for Microbiology Press.

Craig, N., Craigie, R., Gellert, M., & Lambowitz, A. (2002). *Mobile DNA II.* Herndon, VA: American Society for Microbiology Press.

Czech, B., Malone, C. D., Zhou, R., Stark, A., Schlingeheyde, C., et al. (2008). An endogenous small interfering RNA pathway in *Drosophila. Nature, 453,* 798–802.

Davis, P. S., & Judd, B. H. (1995). Molecular characterization of the 3C region between white and roughest loci of *Drosophila melanogaster*. *Drosoph. Inf. Serv.*, *76*, 130–134.

Deininger, P., & Roy-Engel, A. (2002). Mobile elements in plant and animal genomes. In N. Craig, R. Craigie, M. Gellert, & A. Lambowitz (Eds.). *Mobile DNA II* (pp. 1074–1092). Herndon, VA: American Society for Microbiology Press.

De Keukeleire, P., Maes, T., Sauer, M., Zethof, J., Van Montagu, M., & Gerats, T. (2001). Analysis by Transposon Display of the behavior of the dTph1 element family during ontogeny and inbreeding of *Petunia hybrida*. *Mol. Genet. Genomics*, *265*, 72–81.

Delprat, A., Negre, B., Puig, M., & Ruiz, A. (2009). The transposon Galileo generates natural chromosomal inversions in *Drosophila* by ectopic recombination. *PLoS One*, *4*, e7883.

de Pancorbo, M. M., Lopez-Martinez, M., Martinez-Bouzas, C., Castro, A., Fernandez-Fernandez, I., et al. (2001). The Basques according to polymorphic Alu insertions. *Hum. Genet.*, *109*, 224–233.

Depra, M., Valente, V. L., Margis, R., & Loreto, E. L. (2009). The hobo transposon and hobo-related elements are expressed as developmental genes in *Drosophila*. *Gene*, *448*, 57–63.

Desset, S., Buchon, N., Meignin, C., Coiffet, M., & Vaury, C. (2008). *Drosophila melanogaster* the COM locus directs the somatic silencing of two retrotransposons through both Piwi-dependent and -independent pathways. *PLoS One*, *3*, e1526.

Dewannieux, M., & Heidmann, T. (2005). L1-mediated retrotransposition of murine B1 and B2 SINEs recapitulated in cultured cells. *J. Mol. Biol.*, *349*, 241–247.

Dewannieux, M., Esnault, C., & Heidmann, T. (2003). LINE-mediated retrotransposition of marked Alu sequences. *Nat. Genet.*, *35*, 41–48.

Doak, T. G., Doerder, F. P., Jahn, C. L., & Herrick, G. (1994). A proposed superfamily of transposase genes: Transposon-like ciliated protozoa and a common "D35E" motif. *Proc. Natl. Acad. Sci. USA*, *91*, 942–946.

Doolittle, W. F., & Sapienza, C. (1980). Selfish genes, the phenotype paradigm and genome evolution. *Nature*, *284*, 601–603.

Drosophila 12 Genome Consortium (2007). Evolution of genes and genomes on the *Drosophila* phylogeny. *Nature*, *450*, 203–218.

Edgar, R. C., & Myers, E. W. (2005). PILER: Identification and classification of genomic repeats. *Bioinformatics*, *21*(Suppl. 1), i152–8.

Eickbush, T. (2002). R2 and related site-specific non-long terminal repeat retrotransposons. In N. Craig, R. Craigie, M. Gellert, & A. Lambowitz (Eds.). *Mobile DNA II* (pp. 813–835). Herndon, VA: American Society for Microbiology Press.

Eickbush, T., & Malik, H. (2002). Origins and evolution of retrotransposons. In N. Craig, R. Craigie, M. Gellert, & A. Lambowitz (Eds.). *Mobile DNA II* (pp. 1111–1144). Herndon, VA: American Society for Microbiology Press.

Eickbush, T. H., & Jamburuthugoda, V. K. (2008). The diversity of retrotransposons and the properties of their reverse transcriptases. *Virus Res.*, *134*, 221–234.

Ellinghaus, D., Kurtz, S., & Willhoeft, U. (2008). LTRharvest, an efficient and flexible software for *de novo* detection of LTR retrotransposons. *BMC Bioinformatics*, *9*, 18.

Engels, W. R. (1989). *P* elements in *Drosophila melanogaster*. In D. E. Berg, & M. J. Howe (Eds.). *Mobile DNA* (pp. 437–484). Herndon, VA: American Society for Microbiology Press.

Engels, W. R. (1997). Invasions of *P* elements. *Genetics*, *145*, 11–15.

Evgen'ev, M. B., & Arkhipova, I. R. (2005). Penelope-like elements – a new class of retroelements: Distribution, function and possible evolutionary significance. *Cytogenet. Genome Res.*, *110*, 510–521.

Evgen'ev, M. B., Zelentsova, H., Shostak, N., Kozitsina, M., Barskyi, V., et al. (1997). Penelope, a new family of transposable elements and its possible role in hybrid dysgenesis in *Drosophila virilis*. *Proc. Natl. Acad. Sci. USA*, *94*, 196–201.

Fawcett, D. H., Lister, C. K., Kellett, E., & Finnegan, D. J. (1986). Transposable elements controlling I-R hybrid dysgenesis in *D. melanogaster* are similar to mammalian LINEs. *Cell*, *47*, 1007–1015.

Fedoroff, N. (1989). Maize transposable elements. In D. E. Berg, & M. J. Howe (Eds.). *Mobile DNA* (pp. 375–411). Herndon, VA: American Society for Microbiology Press.

Feschotte, C. (2004). Merlin, a new superfamily of DNA transposons identified in diverse animal genomes and related to bacterial IS1016 insertion sequences. *Mol. Biol. Evol.*, *21*, 1769–1780.

Feschotte, C., & Mouches, C. (2000a). Evidence that a family of miniature inverted-repeat transposable elements (MITEs) from the *Arabidopsis thaliana* genome has arisen from a pogo-like DNA transposon. *Mol. Biol. Evol.*, *17*, 730–737.

Feschotte, C., & Mouches, C. (2000b). Recent amplification of miniature inverted-repeat transposable elements in the vector mosquito *Culex pipiens*: Characterization of the Mimo family. *Gene*, *250*, 109–116.

Feschotte, C., & Pritham, E. J. (2005). Non-mammalian c-integrases are encoded by giant transposable elements. *Trends Genet.*, *21*, 551–552.

Feschotte, C., & Pritham, E. J. (2007). DNA transposons and the evolution of eukaryotic genomes. *Annu. Rev. Genet.*, *41*, 331–368.

Feschotte, C., Fourrier, N., Desmons, I., & Mouches, C. (2001). Birth of a retroposon: The Twin SINE family from the vector mosquito *Culex pipiens* may have originated from a dimeric tRNA precursor. *Mol. Biol. Evol.*, *18*, 74–84.

Feschotte, C., Zhang, X., & Wessler, S. (2002). Miniature inverted-repeat transposable elements and their relationship to established DNA transposons. In N. Craig, R. Craigie, M. Gellert, & A. Lambowitz (Eds.). *Mobile DNA II* (pp. 1147–1158). Herndon, VA: American Society for Microbiology Press.

Feschotte, C., Swamy, L., & Wessler, S. R. (2003). Genome-wide analysis of mariner-like transposable elements in rice reveals complex relationships with stowaway Miniature Inverted Repeat Transposable Elements (MITEs). *Genetics*, *163*, 747–758.

Feschotte, C., Keswani, U., Ranganathan, N., Guibotsy, M. L., & Levine, D. (2009). Exploring repetitive DNA landscapes using REPCLASS, a tool that automates the classification of transposable elements in eukaryotic genomes. *Genome Biol. Evol.*, *1*, 205–220.

Finnegan, D. J. (1989). The I factor and I-R hybrid dysgenesis in *Drosophila melanogaster*. In D. E. Berg, & M. J. Howe (Eds.). *Mobile DNA* (pp. 503–517). Herndon, VA: American Society for Microbiology Press.

Finnegan, D. J. (1992). Transposable elements. *Curr. Opin. Genet. Dev.*, *2*, 861–867.

Forstemann, K. (2010). Transposon defense in *Drosophila* somatic cells: A model for distinction of self and non-self in the genome. *RNA Biol.*, *7*, 158–161.

Franz, G., & Savakis, C. (1991). Minos, a new transposable element from *Drosophila hydei*, is a member of the Tc1-like family of transposons. *Nucleic Acids Res.*, *19*, 6646.

Fraser, M. J. (2000). The TTAA-specific family of transposable elements: Identification, functional characterization, and utility for transformation of insects. In A. M. Handler, & A. A. James (Eds.). *Insect Transgenesis: Methods and Applications* (pp. 249). New York, NY: CRC Press.

Fugmann, S. D. (2010). The origins of the Rag genes – from transposition to V(D)J recombination. *Semin. Immunol.*, *22*, 10–16.

Gale, K. R. (1987). *Characterization of the Mosquito, Aedes aegypti, at the Molecular Level of Genetics*. Liverpool, UK: University of Liverpool.

Garcillan-Barcia, M. P., Bernales, I., Mendiola, M. V., & De La Cruz, F. (2002). IS91 rolling-circle transposition. In N. Craig, R. Craigie, M. Gellert, & A. Lambowitz (Eds.). *Mobile DNA II* (pp. 891–904). Herndon, VA: American Society for Microbiology Press.

Gellert, M. (2002). V(D)J recombination. In N. Craig, R. Craigie, M. Gellert, & A. Lambowitz (Eds.). *Mobile DNA II* (pp. 705–729). Herndon, VA: American Society for Microbiology Press.

Ghildiyal, M., Seitz, H., Horwich, M. D., Li, C., Du, T., et al. (2008). Endogenous siRNAs derived from transposons and mRNAs in *Drosophila* somatic cells. *Science*, *320*, 1077–1081.

Gilbert, C., Schaack, S., Pace, J. K., II, Brindley, P. J., & Feschotte, C. (2010). A role for host–parasite interactions in the horizontal transfer of transposons across phyla. *Nature*, *464*, 1347–1350.

Gomulski, L. M., Torti, C., Bonizzoni, M., Moralli, D., Raimondi, E., et al. (2001). A new basal subfamily of mariner elements in *Ceratitis rosa* and other tephritid flies. *J. Mol. Evol.*, *53*, 597–606.

Gonzalez, J., & Petrov, D. A. (2009). The adaptive role of transposable elements in the *Drosophila* genome. *Gene*, *448*, 124–133.

Gonzalez, J., Lenkov, K., Lipatov, M., Macpherson, J. M., & Petrov, D. A. (2008). High rate of recent transposable element-induced adaptation in *Drosophila melanogaster*. *PLoS Biol.*, *6*, e251.

Gonzalez, J., Macpherson, J. M., & Petrov, D. A. (2009). A recent adaptive transposable element insertion near highly conserved developmental loci in *Drosophila melanogaster*. *Mol. Biol. Evol.*, *26*, 1949–1961.

Gonzalez, J., Karasov, T. L., Messer, P. W., & Petrov, D. A. (2010). Genome-wide patterns of adaptation to temperate environments associated with transposable elements in *Drosophila*. *PLoS Genet*, *6*, e1000905.

Gorbunova, V., & Levy, A. A. (1997). Circularized Ac/Ds transposons: Formation, structure and fate. *Genetics*, *145*, 1161–1169.

Grandbastien, M. A. (1998). Activation of plant retrotransposons under stress condiitons. *Trends Plant Sci.*, *3*, 181–187.

Gray, Y. H. (2000). It takes two transposons to tango: Transposable-element-mediated chromosomal rearrangements. *Trends Genet.*, *16*, 461–468.

Gunawardane, L. S., Saito, K., Nishida, K. M., Miyoshi, K., Kawamura, Y., et al. (2007). A slicer-mediated mechanism for repeat-associated siRNA 5 end formation in *Drosophila*. *Science*, *315*, 1587–1590.

Han, M. J., Shen, Y. H., Gao, Y. H., Chen, L. Y., Xiang, Z. H., & Zhang, Z. (2010). Burst expansion, distribution and diversification of MITEs in the silkworm genome. *BMC Genomics*, *11*, 520.

Handler, A. M., & McCombs, S. D. (2000). The piggyBac transposon mediates germ-line transformation in the Oriental fruit fly and closely related elements exist in its genome. *Insect Mol. Biol.*, *9*, 605–612.

Handler, A. M., Zimowska, G. J., & Armstrong, K. F. (2008). Highly similar piggyBac elements in Bactrocera that share a common lineage with elements in noctuid moths. *Insect Mol. Biol.*, *17*, 387–393.

Hartl, D. (1989). Transposable element *mariner* in *Drosophila* species. In D. E. Berg, & M. J. Howe (Eds.). *Mobile DNA* (pp. 531–536). Herndon, VA: American Society for Microbiology Press.

Hartl, D. L., Lozovskaya, E. R., Nurminsky, D. I., & Lohe, A. R. (1997). What restricts the activity of mariner-like transposable elements? *Trends Genet.*, *13*, 197–201.

Henikoff, S. (1992). Detection of *Caenorhabditis* transposon homologs in diverse organisms. *New Biol.*, *4*, 382–388.

Hickman, A. B., Perez, Z. N., Zhou, L., Musingarimi, P., Ghirlando, R., et al. (2005). Molecular architecture of a eukaryotic DNA transposase. *Nat. Struct. Mol. Biol.*, *12*, 715–721.

Holt, R. A., Subramanian, G. M., Halpern, A., Sutton, G. G., Charlab, R., et al. (2002). The genome sequence of the malaria mosquito *Anopheles gambiae*. *Science*, *298*, 129–149.

Holyoake, A. J., & Kidwell, M. G. (2003). Vege and Mar: Two Novel hAT MITE families from *Drosophila willistoni*. *Mol. Biol. Evol.*, *20*, 163–167.

Honeybee Genome Sequencing Consortium (2006). Insights into social insects from the genome of the honeybee *Apis mellifera*. *Nature*, *443*, 931–949.

Horn, C., Offen, N., Nystedt, S., Hacker, U., & Wimmer, E. A. (2003). piggyBac-based insertional mutagenesis and enhancer detection as a tool for functional insect genomics. *Genetics*, *163*, 647–661.

Itokawa, K., Komagata, O., Kasai, S., Okamura, Y., Masada, M., & Tomita, T. (2010). Genomic structures of Cyp9m10 in pyrethroid resistant and susceptible strains of *Culex quinquefasciatus*. *Insect Biochem. Mol. Biol.*, *40*, 631–640.

Ivics, Z., Hackett, P. B., Plasterk, R. H., & Izsvak, Z. (1997). Molecular reconstruction of Sleeping Beauty, a Tc1-like transposon from fish, and its transposition in human cells. *Cell, 91,* 501–510.

Izsvak, Z., Ivics, Z., Shimoda, N., Mohn, D., Okamoto, H., & Hackett, P. B. (1999). Short inverted-repeat transposable elements in teleost fish and implications for a mechanism of their amplification. *J. Mol. Evol., 48,* 13–21.

Jacobson, J. W., Medhora, M. M., & Hartl, D. L. (1986). Molecular structure of a somatically unstable transposable element in *Drosophila. Proc. Natl. Acad. Sci. USA, 83,* 8684–8688.

Jasinskiene, N., Coates, C. J., & James, A. A. (2000). Structure of hermes integrations in the germline of the yellow fever mosquito, *Aedes aegypti. Insect Mol. Biol., 9,* 11–18.

Jensen, S., Cavarec, L., Dhellin, O., & Heidmann, T. (1994). Retrotransposition of a marked *Drosophila* line-like I element in cells in culture. *Nucleic Acids Res., 22,* 1484–1488.

Jensen, S., Gassama, M. P., & Heidmann, T. (1999). Cosuppression of I transposon activity in *Drosophila* by I-containing sense and antisense transgenes. *Genetics, 153,* 1767–1774.

Jiang, N., Bao, Z., Zhang, X., Hirochika, H., Eddy, S. R., et al. (2003). An active DNA transposon family in rice. *Nature, 421,* 163–167.

Jordan, I. K., Matyunina, L. V., & McDonald, J. F. (1999). Evidence for the recent horizontal transfer of long terminal repeat retrotransposon. *Proc. Natl. Acad. Sci. USA, 96,* 12621–12625.

Jurka, J. (1995). Origin and evolution of *Alu* repetitive elements. In R. J. Maraia (Ed.). *The Impact of Short Interspersed Elements (SINEs) on the Host Genome.* Austin, TX: R.G. Landes Company.

Jurka, J. (2000). Repbase update: A database and an electronic journal of elements. *Trends Genet., 16,* 418–420.

Kaminker, J. S., Bergman, C. M., Kronmiller, B., Carlson, J., Svirskas, R., et al. (2002). The transposable elements of the *Drosophila melanogaster* genomics perspective. *Genome Biol. 3,* RESEARCH0084–0084.

Kapitonov, V. V., & Jurka, J. (2001). Rolling-circle transposons in eukaryotes. *Proc. Natl. Acad. Sci. USA, 98,* 8714–8719.

Kapitonov, V. V., & Jurka, J. (2003a). A novel class of SINE elements derived from 5S rRNA. *Mol. Biol. Evol., 20,* 694–702.

Kapitonov, V. V., & Jurka, J. (2003b). Molecular paleontology of transposable elements in the *Drosophila melanogaster* genome. *Proc. Natl. Acad. Sci. USA, 100,* 6569–6574.

Kapitonov, V. V., & Jurka, J. (2006). Self-synthesizing DNA transposons in eukaryotes. *Proc. Natl. Acad. Sci. USA, 103,* 4540–4545.

Kawamura, Y., Saito, K., Kin, T., Ono, Y., Asai, K., et al. (2008). *Drosophila* endogenous small RNAs bind to Argonaute 2 in somatic cells. *Nature, 453,* 793–797.

Kazazian, H. H., Jr., & Goodier, J. L. (2002). LINE drive. retrotransposition and genome instability. *Cell, 110,* 277–280.

Ketting, R. F., Haverkamp, T. H., van Luenen, H. G., & Plasterk, R. H. (1999). Mut-7 of *C. elegans,* required for transposon silencing and RNA interference, is a homolog of Werner syndrome helicase and RNaseD. *Cell, 99,* 133–141.

Kidwell, M. G. (1977). Reciprocal differences in female recombination associated with dysgenesis in *Drosophila melanogaster. Genet. Res., 30,* 77–88.

Kidwell, M. G. (1992). Horizontal transfer of *P* elements and other short inverted repeat transposons. *Genetica, 86,* 275–286.

Kidwell, M. G. (2002). Transposable elements and the evolution of genome size in eukaryotes. *Genetica, 115,* 49–63.

Kidwell, M. G., & Lisch, D. R. (2000). Transposable elements and host genome evolution. *Trends Ecol. Evol., 15,* 95–99.

Kidwell, M. G., & Lisch, D. R. (2001). Perspective: Transposable elements, parasitic DNA, and genome evolution. *Evolution Intl. J. Org. Evolution, 55,* 1–24.

Kidwell, M. G., & Lisch, D. R. (2002). Transposable elements as sources of genomic variation. In N. Craig, R. Craigie, M. Gellert, & A. Lambowitz (Eds.). *Mobile DNA II* (pp. 59–90). Herndon, VA: American Society for Microbiology Press.

Kirkness, E. F., Haas, B. J., Sun, W., Braig, H. R., Perotti, M. A., et al. (2010). Genome sequences of the human body louse and its primary endosymbiont provide insights into the permanent parasitic lifestyle. *Proc. Natl. Acad. Sci. USA, 107,* 12168–12173.

Klattenhoff, C., Xi, H., Li, C., Lee, S., Xu, J., et al. (2009). The *Drosophila* HP1 homolog Rhino is required for transposon silencing and piRNA production by dual-strand clusters. *Cell, 138,* 1137–1149.

Klinakis, A. G., Loukeris, T. G., Pavlopoulos, A., & Savakis, C. (2000). Mobility assays confirm the broad host-range activity of the transposable element and validate new transformation tools. *Insect Mol. Biol., 9,* 269–275.

Kotnova, A. P., Glukhov, I. A., Karpova, N. N., Salenko, V. B., Lyubomirskaya, N. V., & Ilyin, Y. V. (2007). Evidence for recent horizontal transfer of gypsy-homologous LTR-retrotransposon gtwin into *Drosophila erecta* followed by its amplification with multiple aberrations. *Gene, 396,* 39–45.

Krzywinski, J., & Besansky, N. J. (2003). Molecular systematics of Anopheles: From subgenera to subpopulations. *Annu. Rev. Entomol. 48,* 111–139.

Labrador, M., & Corces, V. (2002). Interactions between transposable elements and the host genome. In N. Craig, R. Craigie, M. Gellert, & A. Lambowitz (Eds.). *Mobile DNA II* (pp. 1008–1023). Herndon, VA: American Society for Microbiology Press.

Lampe, D. J., Akerley, B. J., Rubin, E. J., Mekalanos, J. J., & Robertson, H. M. (1999). Hyperactive transposase mutants of the Himar1 mariner transposon. *Proc. Natl. Acad. Sci. USA, 96,* 11428–11433.

Lampe, D. J., Walden, K. K., & Robertson, H. M. (2001). Loss of transposase-DNA interaction may underlie the divergence of mariner family transposable elements and the ability of more than one mariner to occupy the same genome. *Mol. Biol. Evol., 18,* 954–961.

Lander, E. S., Linton, L. M., Birren, B., Nusbaum, C., Zody, M. C., et al. (2001). Initial sequencing and analysis of the human genome. *Nature, 409,* 860–921.

Lau, N. C., Ohsumi, T., Borowsky, M., Kingston, R. E., & Blower, M. D. (2009). Systematic and single cell analysis of *Xenopus* Piwi-interacting RNAs and Xiwi. *EMBO J, 28,* 2945–2958.

Lawniczak, M. K.N., Emrich, S. J., Holloway, A. K., Regier, A. P., Olson, M., et al. (2010). Widespread divergence between incipient *Anopheles gambiae* species revealed by whole genome sequences. *Science, 330,* 512–514.

Le, Q. H., Wright, S., Yu, Z., & Bureau, T. (2000). Transposon diversity in *Arabidopsis thaliana. Proc. Natl. Acad. Sci. USA, 97,* 7376–7381.

Lee, Y. C., & Langley, C. H. (2010). Transposable elements in natural populations of *Drosophila melanogaster. Philos. Trans. R. Soc. Lond. B. Biol. Sci., 365,* 1219–1228.

Lerman, D. N., & Feder, M. E. (2005). Naturally occurring transposable elements disrupt hsp70 promoter function in *Drosophila melanogaster. Mol. Biol. Evol., 22,* 776–783.

Levis, R. W., Ganesan, R., Houtchens, K., Tolar, L. A., & Sheen, F. M. (1993). Transposons in place of telomeric repeats at a *Drosophila* telomere. *Cell, 75,* 1083–1093.

Li, C., Vagin, V. V., Lee, S., Xu, J., Ma, S., et al. (2009). Collapse of germline piRNAs in the absence of Argonaute3 reveals somatic piRNAs in flies. *Cell, 137,* 509–521.

Li, R., Ye, J., Li, S., Wang, J., Han, Y., et al. (2005). ReAS: Recovery of ancestral sequences for transposable elements from the unassembled reads of a whole genome shotgun. *PLoS Comput. Biol., 1,* e43.

Lim, J. K., & Simmons, M. J. (1994). Gross chromosome rearrangements mediated by transposable elements in *Drosophila melanogaster. Bioessays, 16,* 269–275.

Lin, R., Ding, L., Casola, C., Ripoll, D. R., Feschotte, C., & Wang, H. (2007). Transposase-derived transcription factors regulate light signaling in *Arabidopsis. Science, 318,* 1302–1305.

Lipardi, C., & Paterson, B. M. (2009). Identification of an RNA-dependent RNA polymerase in *Drosophila* involved in RNAi and transposon suppression. *Proc. Natl. Acad. Sci. USA, 106,* 15645–15650.

Lisch, D. (2009). Epigenetic regulation of transposable elements in plants. *Annu. Rev. Plant Biol., 60,* 43–66.

Liu, B., & Wendel, J. F. (2000). Retrotransposon activation followed by rapid repression in introgressed rice plants. *Genome, 43,* 874–880.

Loreto, E. L., Carareto, C. M., & Capy, P. (2008). Revisiting horizontal transfer of transposable elements in *Drosophila. Heredity, 100,* 545–554.

Luan, D. D., Korman, M. H., Jakubczak, J. L., & Eickbush, T. H. (1993). Reverse transcription of R2Bm RNA is primed by a nick at the chromosomal target site: A mechanism for non-LTR retrotransposition. *Cell, 72,* 595–605.

Luchetti, A., & Mantovani, B. (2009). Talua SINE biology in the genome of the Reticulitermes subterranean termites (Isoptera, Rhinotermitidae). *J. Mol. Evol., 69,* 589–600.

MacRae, A. F., & Clegg, M. T. (1992). Evolution of Ac and Ds1 elements in select grasses (Poaceae). *Genetica, 86,* 55–66.

Malik, H. S., & Eickbush, T. H. (1998). The RTE class of non-LTR retrotransposons is widely distributed in animals and is the origin of many SINEs. *Mol. Biol. Evol., 15,* 1123–1134.

Malone, C. D., & Hannon, G. J. (2009). Small RNAs as guardians of the genome. *Cell, 136,* 656–668.

Malone, C. D., Brennecke, J., Dus, M., Stark, A., McCombie, W. R., et al. (2009). Specialized piRNA pathways act in germline and somatic tissues of the *Drosophila* ovary. *Cell, 137,* 522–535.

Marshall, J. M. (2008). The impact of dissociation on transposon-mediated disease control strategies. *Genetics, 178,* 1673–1682.

Martin, S. L., & Bushman, F. D. (2001). Nucleic acid chaperone activity of the ORF1 protein from the mouse LINE-1 retrotransposon. *Mol. Cell. Biol., 21,* 467–475.

Maside, X., Bartolome, C., Assimacopoulos, S., & Charlesworth, B. (2001). Rates of movement and distribution of transposable elements in *Drosophila melanogaster: In situ* hybridization vs Southern blotting data. *Genet. Res., 78,* 121–136.

Mason, J. M., Frydrychova, R. C., & Biessmann, H. (2008). *Drosophila* telomeres: An exception providing new insights. *Bioessays, 30,* 25–37.

Mathiopoulos, K. D., della Torre, A., Predazzi, V., Petrarca, V., & Coluzzi, M. (1998). Cloning of inversion breakpoints in the *Anopheles gambiae* complex traces a transposable element at the inversion junction. *Proc. Natl. Acad. Sci. USA, 95,* 12444–12449.

Mathiopoulos, K. D., della Torre, A., Santolamazza, F., Predazzi, V., Petrarca, V., & Coluzzi, M. (1999). Are chromosomal inversions induced by transposable elements? A paradigm from the malaria mosquito *Anopheles gambiae. Parassitologia, 41,* 119–123.

McCarthy, E. M., & McDonald, J. F. (2003). LTR_STRUC: A novel search and identification program for LTR retrotransposons. *Bioinformatics, 19,* 362–367.

McClintock, B. (1948). Mutable loci in maize. Carnegie Inst. Wash. *Year Book, 47,* 155–169.

McClintock, B. (1950). The origin and behavior of mutable loci in maize. *Proc. Natl. Acad. Sci. USA, 36,* 344–355.

McClintock, B. (1956). Controlling elements and the gene. Cold Spring Harbor Symp. *Quant. Biol., 8,* 58–74.

McClintock, B. (1984). The significance of responses of the genome to challenge. *Science, 226,* 792–801.

Medstrand, P., van de Lagemaat, L. N., & Mager, D. L. (2002). Retroelement distributions in the human genome: Variations associated with age and proximity to genes. *Genome Res., 12,* 1483–1495.

Miller, W. J., Paricio, N., Hagemann, S., Martinez-Sebastian, M. J., Pinsker, W., & de Frutos, R. (1995). Structure and expression of clustered *P* element homologues in *Drosophila subobscura* and *Drosophila guanche. Gene, 156,* 167–174.

Minervini, C. F., Viggiano, L., Caizzi, R., & Marsano, R. M. (2009). Identification of novel LTR retrotransposons in the genome of *Aedes aegypti. Gene, 440,* 42–49.

Mita, K., Kasahara, M., Sasaki, S., Nagayasu, Y., Yamada, T., et al. (2004). The genome sequence of silkworm, *Bombyx mori. DNA Res., 11,* 27–35.

Mitra, R., Fain-Thornton, J., & Craig, N. L. (2008). piggyBac can bypass DNA synthesis during cut and paste transposition. *EMBO J., 27,* 1097–1099.

Mizrokhi, L. J., & Mazo, A. M. (1991). Cloning and analysis of the mobile element gypsy from *D. virilis. Nucleic Acids Res., 19,* 913–916.

Morgan, G. T., & Middleton, K. M. (1990). Short interspersed repeats from Xenopus that contain multiple motifs are related to known transposable elements. *Nucleic Acids Res., 18,* 5781–5786.

Mota, N. R., Ludwig, A., Valente, V. L., & Loreto, E. L. (2010). Harrow: New *Drosophila* hAT transposons involved in horizontal transfer. *Insect Mol. Biol., 19,* 217–228.

Munoz-Lopez, M., Siddique, A., Bischerour, J., Lorite, P., Chalmers, R., & Palomeque, T. (2008). Transposition of Mboumar-9: Identification of a new naturally active mariner-family transposon. *J. Mol. Biol.*, *382*, 567–572.

Munstermann, L. E., & Conn, J. E. (1997). Systematics of mosquito disease vectors (Diptera, Culicidae): Impact of molecular biology and cladistic analysis. *Annu. Rev. Entomol*, *42*, 351–369.

Muotri, A. R., Marchetto, M. C., Coufal, N. G., & Gage, F. H. (2007). The necessary junk: New functions for transposable elements. *Hum. Mol. Genet.*, 16 Spec. No. 2, R159–167.

Murata, S., Takasaki, N., Saitoh, M., & Okada, N. (1993). Determination of the phylogenetic relationships among Pacific salmonids by using short interspersed elements (SINEs) as temporal landmarks of evolution. *Proc. Natl. Acad. Sci. USA*, *90*, 6995–6999.

Nasidze, I., Risch, G. M., Robichaux, M., Sherry, S. T., Batzer, M. A., & Stoneking, M. (2001). Alu insertion polymorphisms and the genetic structure of humans from the *Caucasus. Eur. J. Hum. Genet.*, *9*, 267–272.

Nene, V., Wortman, J. R., Lawson, D., Haas, B., Kodira, C., et al. (2007). Genome sequence of *Aedes aegypti*, a major arbovirus vector. *Science*, *316*, 1718–1723.

Nikaido, M., Rooney, A. P., & Okada, N. (1999). Phylogenetic relationships among cetartiodactyls based on insertions of short and long interpersed elements: Hippopotamuses are the closest extant relatives of whales. *Proc. Natl. Acad. Sci. USA*, *96*, 10261–10266.

Nishida, K. M., Saito, K., Mori, T., Kawamura, Y., Nagami-Okada, T., et al. (2007). Gene silencing mechanisms mediated by Aubergine piRNA complexes in *Drosophila* male gonad. *RNA*, *13*, 1911–1922.

Nishihara, H., & Okada, N. (2008). Retroposons: Genetic footprints on the evolutionary paths of life. *Methods Mol. Biol.*, *422*, 201–225.

Nuzhdin, S. V. (1999). Sure facts, speculations, and open questions about the evolution of transposable element copy number. *Genetica*, *107*, 129–137.

O'Brochta, D. A., Stosic, C. D., Pilitt, K., Subramanian, R. A., Hice, R. H., & Atkinson, P. W. (2009). Transpositionally active episomal hAT elements. *BMC Mol. Biol.*, *10*, 108.

Ohshima, K., Hamada, M., Terai, Y., & Okada, N. (1996). The 3′ ends of tRNA-derived short interspersed repetitive elements are derived from the 3′ ends of long interspersed repetitive elements. *Mol. Cell. Biol.*, *16*, 3756–3764.

Okada, N. (1991). SINEs. *Curr. Opin. Genet. Dev.*, *1*, 498–504.

Okada, N., & Hamada, M. (1997). The 3′ ends of tRNA-derived SINEs originated from the 3′ ends of LINEs: A new example from the bovine genome. *J. Mol. Evol.*, *44*(Suppl. 1), S52–S56.

Okamura, K., Chung, W. J., Ruby, J. G., Guo, H., Bartel, D. P., & Lai, E. C. (2008). The *Drosophila* hairpin RNA pathway generates endogenous short interfering RNAs. *Nature*, *453*, 803–806.

Oosumi, T., Garlick, B., & Belknap, W. R. (1995). Identification and characterization of putative transposable DNA elements in solanaceous plants and *Caenorhabditis elegans. Proc. Natl. Acad. Sci. USA*, *92*, 8886–8890.

Orgel, L., & Crick, F. (1980). Selfish DNA: The ultimate parasite. *Nature*, *284*, 604–607.

Osanai-Futahashi, M., Suetsugu, Y., Mita, K., & Fujiwara, H. (2008). Genome-wide screening and characterization of transposable elements and their distribution analysis in the silkworm, *Bombyx mori. Insect Biochem. Mol. Biol.*, *38*, 1046–1057.

Ostertag, E. M., Prak, E. T., DeBerardinis, R. J., Moran, J. V., & Kazazian, H. H., Jr. (2000). Determination of L1 retrotransposition kinetics in cultured cells. *Nucleic Acids Res.*, *28*, 1418–1423.

Pardue, M. -L., & DeBaryshe, P. (2002). Telomeres and transposable elements. In N. Craig, R. Craigie, M. Gellert, & A. Lambowitz (Eds.). *Mobile DNA II* (pp. 870–887). Herndon, VA: American Society for Microbiology Press.

Pelisson, A., Sarot, E., Payen-Groschene, G., & Bucheton, A. (2007). A novel repeat-associated small interfering RNA-mediated silencing pathway downregulates complementary sense gypsy transcripts in somatic cells of the *Drosophila* ovary. *J. Virol.*, *81*, 1951–1960.

Peterson, D. G., Schulze, S. R., Sciara, E. B., Lee, S. A., Bowers, J. E., et al. (2002). Integration of Cot analysis, DNA cloning, and high-throughput facilitates genome characterization and gene discovery. *Genome Res.*, *12*, 795–807.

Petrov, D. A., Aminetzach, Y. T., Davis, J. C., Bensasson, D., & Hirsh, A. E. (2003). Size matters: Non-LTR retrotransposable elements and ectopic recombination in *Drosophila. Mol. Biol. Evol.*, *20*, 880–892.

Pevzner, P. A., Tang, H., & Tesler, G. (2004). De novo repeat classification and fragment assembly. *Genome Res.*, *14*, 1786–1796.

Price, A. L., Jones, N. C., & Pevzner, P. A. (2005). *De novo* identification of repeat families in large genomes. *Bioinformatics*, *21*(Suppl. 1), i351–358.

Pritham, E. J., Putliwala, T., & Feschotte, C. (2007). Mavericks, a novel class of giant transposable elements widespread in eukaryotes and related to DNA viruses. *Gene*, *390*, 3–17.

Rai, K. S., & Black, W. C.T. (1999). Mosquito genomes: Structure, organization, and evolution. *Adv. Genet.*, *41*, 1–33.

Ray, D. A. (2007). SINEs of progress: Mobile element applications to molecular ecology. *Mol. Ecol.*, *16*, 19–33.

Rho, M., & Tang, H. (2009). MGEScan-non-LTR: Computational identification and classification of autonomous non-LTR retrotransposons in eukaryotic genomes. *Nucleic Acids Res.*, *37*, e143.

Rho, M., Choi, J. H., Kim, S., Lynch, M., & Tang, H. (2007). *De novo* identification of LTR retrotransposons in eukaryotic genomes. *BMC Genomics*, *8*, 90.

Ribeiro, J. M., & Kidwell, M. G. (1994). Transposable elements as population drive mechanisms: Specification of critical parameter values. *J. Med. Entomol.*, *31*, 10–16.

Richardson, J. M., Colloms, S. D., Finnegan, D. J., & Walkinshaw, M. D. (2009). Molecular architecture of the Mos1 paired-end complex: The structural basis of DNA transposition in a eukaryote. *Cell*, *138*, 1096–1108.

Rizzon, C., Marais, G., Gouy, M., & Biemont, C. (2002). Recombination rate and the distribution of transposable elements *Drosophila melanogaster* genome. *Genome Res.*, *12*, 400–407.

Robertson, H. (2002). Evolution of DNA transposons in eukaryotes. In N. Craig, R. Craigie, M. Gellert, & A. Lambowitz (Eds.). *Mobile DNA II* (pp. 1093–1110). Herndon, VA: American Society for Microbiology Press.

Robertson, H. M. (1993). The mariner transposable element is widespread in insects. *Nature, 362,* 241–245.

Robertson, H. M., & Lampe, D. J. (1995). Distribution of transposable elements in arthropods. *Annu. Rev. Entomol, 40,* 333–357.

Robertson, H. M., & Walden, K. K. (2003). Bmmar6, a second mori subfamily mariner transposon from the *Bombyx mori. Insect Mol. Biol., 12,* 167–171.

Roy-Engel, A. M., Carroll, M. L., Vogel, E., Garber, R. K., Nguyen, S. V., et al. (2001). Alu insertion polymorphisms for the study of human genomic diversity. *Genetics, 159,* 279–290.

Rubin, E., & Levy, A. A. (1997). Abortive gap repair: Underlying mechanism for Ds element formation. *Mol. Cell. Biol., 17,* 6294–6302.

Saha, S., Bridges, S., Magbanua, Z. V., & Peterson, D. G. (2008). Empirical comparison of *ab initio* repeat finding programs. *Nucleic Acids Res., 36,* 2284–2294.

Salem, A. H., Kilroy, G. E., Watkins, W. S., Jorde, L. B., & Batzer, M. A. (2003). Recently integrated Alu elements and human genomic diversity. *Mol. Biol. Evol., 20,* 1349–1361.

Santolamazza, F., Mancini, E., Simard, F., Qi, Y., Tu, Z., & della Torre, A. (2008). Insertion polymorphisms of SINE200 retrotransposons within speciation islands of *Anopheles gambiae* molecular forms. *Malaria J, 7,* 163.

Schaack, S., Gilbert, C., & Feschotte, C. (2010). Promiscuous DNA: Horizontal transfer of transposable elements and why it matters for eukaryotic evolution. *Trends Ecol. Evol., 25,* 537–546.

Sethuraman, N., Fraser, M. J., Jr., Eggleston, P., & O'Brochta, D. A. (2007). Post-integration stability of piggyBac in *Aedes aegypti. Insect Biochem. Mol. Biol., 37,* 941–951.

Shao, H., & Tu, Z. (2001). Expanding the diversity of the IS630-Tc1-mariner superfamily: Discovery of a unique DD37E transposon and reclassification of the DD37D and DD39D transposons. *Genetics, 159,* 1103–1115.

Shedlock, A. M., & Okada, N. (2000). SINE insertions: Powerful tools for molecular systematics. *Bioessays, 22,* 148–160.

Shimamura, M., Yasue, H., Ohshima, K., Abe, H., Kato, H., et al. (1997). Molecular evidence from retroposons that whales form a clade within even-toed ungulates. *Nature, 388,* 666–670.

Shiroishi, T., Sagai, T., & Moriwaki, K. (1993). Hotspots of meiotic recombination in the mouse major histocompatibility complex. *Genetica, 88,* 187–196.

Silva, J. C., & Kidwell, M. G. (2000). Horizontal transfer and selection in the evolution of *P* elements. *Mol. Biol. Evol., 17,* 1542–1557.

Silva, J. C., & Kidwell, M. G. (2004). Evolution of *P* elements in natural populations of *Drosophila willistoni* and *D. sturtevanti. Genetics, 168,* 1323–1335.

Silva, J. C., Loreto, E. L., & Clark, J. B. (2004). Factors that affect the horizontal transfer of transposable elements. *Curr. Issues Mol. Biol., 6,* 57–71.

Sinzelle, L., Kapitonov, V. V., Grzela, D. P., Jursch, T., Jurka, J., et al. (2008). Transposition of a reconstructed Harbinger element in human cells and functional homology with two transposon-derived cellular genes. *Proc. Natl. Acad. Sci. USA, 105,* 4715–4720.

Smit, A. F., & Riggs, A. D. (1996). Tiggers and DNA transposon fossils in the human genome. *Proc. Natl. Acad. Sci. USA, 93,* 1443–1448.

Song, W. Y., Pi, L. Y., Bureau, T. E., & Ronald, P. C. (1998). Identification and characterization of 14 transposon-like noncoding regions of members of the Xa21 family of disease genes in rice. *Mol. Gen. Genet., 258,* 449–456.

Specchia, V., Piacentini, L., Tritto, P., Fanti, L., D'Alessandro, R., et al. (2010). Hsp90 prevents phenotypic variation by suppressing the mutagenic activity of transposons. *Nature, 463,* 662–665.

Spradling, A. C., Stern, D. M., Kiss, I., Roote, J., Laverty, T., & Rubin, G. M. (1995). Gene disruptions using P transposable elements: An integral component of the *Drosophila* genome project. *Proc. Natl. Acad. Sci. USA, 92,* 10824–10830.

Spradling, A. C., Stern, D., Beaton, A., Rhem, E. J., Laverty, T., et al. (1999). The Berkeley *Drosophila* Genome Project gene disruption project: Single P-element insertions mutating 25% of vital *Drosophila* genes. *Genetics, 153,* 135–177.

Steinbiss, S., Willhoeft, U., Gremme, G., & Kurtz, S. (2009). Fine-grained annotation and classification of *de novo* predicted LTR retrotransposons. *Nucleic Acids Res., 37,* 7002–7013.

Stoneking, M., Fontius, J. J., Clifford, S. L., Soodyall, H., Arcot, S. S., et al. (1997). Alu insertion polymorphisms and human evolution: Evidence for a larger population size in Africa. *Genome Res., 7,* 1061–1071.

Streck, R. D., MacGaffey, J. E., & Beckendorf, S. K. (1986). The structure of hobo transposable elements and their insertion. *EMBO J, 5,* 3615–3623.

Subramanian, R. A., Akala, O. O., Adejinmi, J. O., & O'Brochta, D. A. (2008). Topi, an IS630/Tc1/mariner-type transposable element in the African malaria mosquito, *Anopheles gambiae. Gene., 423,* 63–71.

Sundararajan, P., Atkinson, P. W., & O'Brochta, D. A. (1999). Transposable element interactions in insects: Crossmobilization of hobo and Hermes. *Insect Mol. Biol., 8,* 359–368.

Swofford, D. L. (2001). *Phylogenetic Analysis Using Parsimony (*and Other Methods).* Sunderland, MA: Sinauer Associates.

Tabara, H., Sarkissian, M., Kelly, W. G., Fleenor, J., Grishok, A., et al. (1999). The rde-1 gene, RNA interference, and transposon silencing in *C. elegans. Cell, 99,* 123–132.

Takahashi, K., Terai, Y., Nishida, M., & Okada, N. (1998). A novel family of short interspersed repetitive elements (SINEs) from cichlids: The patterns of insertion of SINEs at orthologous loci support the proposed monophyly of four major groups of cichlid fishes in Lake Tanganyika. *Mol. Biol. Evol., 15,* 391–407.

Terai, Y., Takahashi, K., Nishida, M., Sato, T., & Okada, N. (2003). Using SINEs to probe ancient explosive speciation: "Hidden" radiation of African cichlids? *Mol. Biol. Evol., 20,* 924–930.

Terzian, C., Ferraz, C., Demaille, J., & Bucheton, A. (2000). Evolution of the Gypsy endogenous retrovirus in the *Drosophila melanogaster* subgroup. *Mol. Biol. Evol., 17,* 908–914.

The International Aphid Genomics Consortium (2010). Genome sequence of the pea aphid *Acyrthosiphon pisum*. *PLoS Biol*, *8*, e1000313.

Thomas, J., Schaack, S., & Pritham, E. J. (2010a). Pervasive horizontal transfer of rolling-circle transposons among animals. *Genome Biol Evol.*, *2*, 656–664.

Thomas, X., Hedhili, S., Beuf, L., Demattei, M. V., Laparra, H., et al. (2010b). The mariner Mos1 transposase produced in tobacco is active *in vitro*. *Genetica*, *138*, 519–530.

Tribolium Genome Sequencing Consortium (2008). The genome of the model beetle and pest *Tribolium castaneum*. *Nature*, *452*, 949–955.

Truett, M. A., Jones, R. S., & Potter, S. S. (1981). Unusual structure of the FB family of transposable elements in *Drosophila*. *Cell*, *24*, 753–763.

Tu, Z. (1997). Three novel families of miniature inverted-repeat transposable elements are associated with genes of the yellow fever mosquito, *Aedes aegypti*. *Proc. Natl. Acad. Sci. USA*, *94*, 7475–7480.

Tu, Z. (1999). Genomic and evolutionary analysis of Feilai, a diverse family of highly reiterated SINEs in the yellow fever mosquito, *Aedes aegypti*. *Mol. Biol. Evol.*, *16*, 760–772.

Tu, Z. (2000). Molecular and evolutionary analysis of two divergent subfamilies of a novel miniature inverted repeat transposable element in the yellow fever mosquito, *Aedes aegypti*. *Mol. Biol. Evol.*, *17*, 1313–1325.

Tu, Z. (2001a). Eight novel families of miniature inverted repeat transposable elements in the African malaria mosquito, *Anopheles gambiae*. *Proc. Natl. Acad. Sci. USA*, *98*, 1699–1704.

Tu, Z. (2001b). Maque, a family of extremely short interspersed repetitive elements: Characterization, possible mechanism of transposition, and evolutionary implications. *Gene*, *263*, 247–253.

Tu, Z. (2005). Insect transposable elements. In L. I. Gilbert, K. Iatrou, & S. Gill (Eds.). *Insect Transposable Elements in Comprehensive Molecular Insect Science* (pp. 395–436). Oxford, UK: Elsevier.

Tu, Z., & Coates, C. (2004). Mosquito transposable elements. *Insect Biochem. Mol. Biol.*, *34*, 631–644.

Tu, Z., & Orphanidis, S. P. (2001). Microuli, a family of miniature subterminal inverted-repeat transposable elements (MSITEs): Transposition without terminal inverted repeats. *Mol. Biol. Evol.*, *18*, 893–895.

Tu, Z., Li, S., & Mao, C. (2004). The changing tails of a novel short interspersed element in *Aedes aegypti*: Genomic evidence for slippage retrotransposition and the relationship between 3′ tandem repeats and the poly(dA) tail. *Genetics*, *168*, 2037–2047.

Tubio, J. M., Costas, J. C., & Naveira, H. F. (2004). Evolution of the mdg1 lineage of the Ty3/gypsy group of LTR retrotransposons in *Anopheles gambiae*. *Gene*, *330*, 123–131.

Tubio, J. M., Naveira, H., & Costas, J. (2005). Structural and evolutionary analyses of the Ty3/gypsy group of LTR retrotransposons in the genome of *Anopheles gambiae*. *Mol. Biol. Evol.*, *22*, 29–39.

Tudor, M., Lobocka, M., Goodell, M., Pettitt, J., & O'Hare, K. (1992). The pogo transposable element family of *Drosophila melanogaster*. *Mol. Gen. Genet.*, *232*, 126–134.

Vagin, V. V., Sigova, A., Li, C., Seitz, H., Gvozdev, V., & Zamore, P. D. (2006). A distinct small RNA pathway silences selfish genetic elements in the germline. *Science*, *313*, 320–324.

Van den Broeck, D., Maes, T., Sauer, M., Zethof, J., De Keukeleire, P., et al. (1998). Transposon Display identifies individual transposable elements in high copy number lines. *Plant J.*, *13*, 121–129.

Vazquez-Manrique, R. P., Hernandez, M., Martinez-Sebastian, M. J., & de Frutos, R. (2000). Evolution of gypsy endogenous retrovirus in the *Drosophila obscura* species group. *Mol. Biol. Evol.*, *17*, 1185–1193.

Vieira, C., Lepetit, D., Dumont, S., & Biemont, C. (1999). Wake up of transposable elements following *Drosophila simulans* worldwide colonization. *Mol. Biol. Evol.*, *16*, 1251–1255.

Volfovsky, N., Haas, B. J., & Salzberg, S. L. (2001). A clustering method for repeat analysis in DNA sequences. *Genome Bio.*, *2*, RESEARCH0027.

Voytas, D., & Boeke, J. (2002). Ty1 and Ty5 of *Saccharomyces cerevisiae*. In N. Craig, R. Craigie, M. Gellert, & A. Lambowitz (Eds.). *Mobile DNA II* (pp. 631–662). Herndon, VA: American Society for Microbiology Press.

Wang, J., Du, Y., Wang, S., Brown, S. J., & Park, Y. (2008). Large diversity of the piggyBac-like elements in the genome of *Tribolium castaneum*. *Insect Biochem. Mol. Biol.*, *38*, 490–498.

Wang, J., Miller, E. D., Simmons, G. S., Miller, T. A., Tabashnik, B. E., & Park, Y. (2010). piggyBac-like elements in the pink bollworm, *Pectinophora gossypiella*. *Insect Mol. Biol.*, *19*, 177–184.

Watkins, W. S., Ricker, C. E., Bamshad, M. J., Carroll, M. L., Nguyen, S. V., et al. (2001). Patterns of ancestral human diversity: An analysis of Alu-insertion and restriction-site polymorphisms. *Am. J. Hum. Genet.*, *68*, 738–752.

Wei, W., Gilbert, N., Ooi, S. L., Lawler, J. F., Ostertag, E. M., et al. (2001). Human L1 retrotransposition: cis preference versus trans complementation. *Mol. Cell. Biol.*, *21*, 1429–1439.

Werren, J. H., Richards, S., Desjardins, C. A., Niehuis, O., Gadau, J., et al. (2010). Functional and evolutionary insights from the genomes of three parasitoid *Nasonia* species. *Science*, *327*, 343–248.

Wessler, S. R. (1996). Turned on by stress. Plant retrotransposons. *Curr. Biol.*, *6*, 959–961.

Wessler, S. R., Bureau, T. E., & White, S. E. (1995). LTR-retrotransposons and MITEs: Important players in the evolution of plant genomes. *Curr. Opin. Genet. Dev.*, *5*, 814–821.

Wuitschick, J. D., Gershan, J. A., Lochowicz, A. J., Li, S., & Karrer, K. M. (2002). A novel family of mobile genetic elements is limited to the germline genome in *Tetrahymena thermophila*. *Nucleic Acids Res.*, *30*, 2524–2537.

Xia, Q., Zhou, Z., Lu, C., Cheng, D., Dai, F., et al. (2004). A draft sequence for the genome of the domesticated silkworm (*Bombyx mori*). *Science*, *306*, 1937–1940.

Xu, J., Xia, Q., Li, J., Zhou, Z., & Pan, G. (2005). Survey of long terminal repeat retrotransposons of domesticated silkworm (*Bombyx mori*). *Insect Biochem. Mol. Biol.*, *35*, 921–929.

Xu, J., Liu, T., Li, D., Zhang, Z., Xia, Q., & Zhou, Z. (2010). BmSE, a SINE family with 3′ ends of (ATTT) repeats in domesticated silkworm (*Bombyx mori*). *J. Genet. Genomics, 37*, 125–135.

Xu, Z., & Wang, H. (2007). LTR_FINDER: An efficient tool for the prediction of full-length LTR retrotransposons. *Nucleic Acids Res., 35*, W265–W268.

Yang, G., & Hall, T. C. (2003). MAK, a computational tool kit for automated MITE analysis. *Nucleic Acids Res., 31*, 3659–3665.

Yang, G., Nagel, D. H., Feschotte, C., Hancock, C. N., & Wessler, S. R. (2009). Tuned for transposition: Molecular determinants underlying the hyperactivity of a Stowaway MITE. *Science, 325*, 1391–1394.

Yang, H. P., & Nuzhdin, S. V. (2003). Fitness costs of Doc expression are insufficient to stabilize its copy number in *Drosophila melanogaster. Mol. Biol. Evol., 20*, 800–804.

Yin, H., & Lin, H. (2007). An epigenetic activation role of Piwi and a Piwi-associated piRNA in *Drosophila melanogaster. Nature, 450*, 304–308.

York, D. S., Blum, V. M., Low, J. A., Rowold, D. J., Puzyrev, V., et al. (1999). Phylogenetic signals from point mutations and polymorphic Alu insertions. *Genetica, 107*, 163–170.

Yoshiyama, M., Tu, Z., Kainoh, Y., Honda, H., Shono, T., & Kimura, K. (2001). Possible horizontal transfer of a transposable element from host to parasitoid. *Mol. Biol. Evol., 18*, 1952–1958.

Zeh, D. W., Zeh, J. A., & Ishida, Y. (2009). Transposable elements and an epigenetic basis for punctuated equilibria. *Bioessays, 31*, 715–726.

Zhang, Q., Arbuckle, J., & Wessler, S. R. (2000). Recent, extensive, and preferential insertion of members of the miniature inverted-repeat transposable element family Heartbreaker into genic regions of maize. *Proc. Natl. Acad. Sci. USA, 97*, 1160–1165.

Zhang, X., Feschotte, C., Zhang, Q., Jiang, N., Eggleston, W. B., & Wessler, S. R. (2001). P instability factor: An active maize transposon system associated with the amplification of Tourist-like MITEs and a new superfamily of transposases. *Proc. Natl. Acad. Sci. USA, 98*, 12572–12577.

Zhou, L., Mitra, R., Atkinson, P. W., Hickman, A. B., Dyda, F., & Craig, N. L. (2004). Transposition of hAT elements links transposable elements and V(D)J recombination. *Nature, 432*, 995–1001.

4 Transposable Elements for Insect Transformation

Alfred M Handler
USDA, ARS, Center for Medical, Agricultural,
and Veterinary Entomology, Gainesville, FL, USA
David A O'Brochta
University of Maryland, Department of Entomology
and The Institute for Bioscience and Biotechnology
Research, College Park, MD, USA

Summary

The germ-lines of more than 35 species from five orders of insects have been genetically transformed, using vectors derived from Class II transposable elements. Initially the *P* and *hobo* vector systems developed for *D. melanogaster* were not applicable to other species, but four transposons found in other species, *Hermes, Minos, Mos1*, and *piggy-Bac*, were found to be widely functional in most insects. Genetic marker discovery and development have been equally important to vector development. Originally, cloned eye-color genes from *Drosophila* that complemented existing mutations in other insects were used, but now more widely applicable dominant-acting fluorescent protein genes are effective transformation markers and reporters for gene expression. Transformation technology is advancing at a fortuitous time when genomics is providing resources necessary for transgenic strain development in pest species to control their population size and behavior. Transposon-based transformation methods are also advancing insertional mutagenesis techniques, such as enhancer traps and transposon tagging, to facilitate the gene discovery and functional analysis that provides these resources. Together, efficient and routine methods for transposon-mediated germ-line transformation and genomics analysis should provide tools critical to the advancement of our understanding and control of insect species.

DOI:10.1016/B978-0-12-384747-8.10004-2

4.1. Introduction

The ability to create genetically transformed organisms has played a central role in the history of modern genetics; in particular, to our understanding of gene expression and development. Indeed, the pioneering transformation experiments of *Pneumoccus* by Griffith (1928), and subsequent systematic analyses by Avery, MacLeod, and McCarty (1944), which showed transformation from a "rough" to a "smooth" bacterial cell wall phenotype, were instrumental in defining DNA as the inherited genetic material. The importance of these initial transformation experiments to prokaryotic genetic analysis was widely appreciated, and continued studies by many other laboratories laid the foundation for modern molecular biology.

The importance of "transformation" technology to eukaryotic genetic studies was apparent, and several of the initial attempts to create transgenic animals were performed in insects, though the means of achieving and assessing insect transformation were not straightforward. The primary reasons why these early attempts to create transgenic insects were largely unsuccessful were the inability to isolate and reproduce individual genetic elements that could be used as transformation vectors and markers, and the lack of efficient means of introducing DNA into germ cells. Most of the initial studies of insect transformation relied on soaking embryos or larvae, with visible mutant phenotypes, in solutions of genomic DNA from wild type individuals in hopes of reverting the mutant phenotype. The first experiments performed in *Bombyx* and *Ephestia* met with some success, where mutant wing color pattern phenotypes were reverted in some organisms, though inheritance was inconsistent and transformation events could not be unequivocally confirmed (Caspari and Nawa, 1965; Nawa and Yamada, 1968; Nawa *et al.*, 1971). Similar results were obtained in studies of *Drosophila* (Fox and Yoon, 1966, 1970), and for all of these initial experiment it is most likely the observed phenotypic changes resulted from extrachromosomal maintenance of introduced DNA in the somatic tissue by an unknown mechanism. A different approach involving the microinjection of wild type genomic DNA into embryos homozygous for a recessive eye-color mutation (*vermilion*) resulted in transformants with a reversion to the normal red eye-color phenotype. While the reversion event was genetically mapped away from the mutant locus, a thorough molecular analysis to verify a transformation event, before the lines were lost, was not achieved and so the nature of the phenomenon observed in this experiment remains unexplained (Germeraad *et al.*, 1976).

In the mid-1970s a turning point in insect science occurred with the extension of molecular genetic analysis to *Drosophila melanogaster*. These early studies and subsequent studies not only provided many of the tools and reagents necessary for developing and critically assessing genetic transformation in insects, but also emphasized the need for a technology that would facilitate a more complete understanding of the genes being isolated using recombinant DNA methods. One technology that was clearly needed was a means to stably integrate DNA molecules into the chromosomes of germ cells, resulting in heritable germ-line transformation. The simple introduction of raw linearized DNA into pre-blastoderm embryos in the hope of fortuitous recombination into host chromosomes was clearly not reliable. Interest was growing, however, in the use of mobile genetic elements as vectors for DNA integration, including retrotransposons and transposons that were being isolated in *Drosophila* for the first time. Foremost among these was the *P* transposable element, isolated from certain mutant alleles of the *white* gene. The subsequent testing and success of transformation mediated by the *P* element in the *Drosophila* germ-line proved to be a dramatic turning point in the genetic analysis of an insect species (see **Figure 1** for the general germ-line transformation scheme). The eventual impact of this technology on understanding genetic mechanisms in all eukaryotic systems cannot be understated. The success with *P* in *Drosophila* gave hope that this system could be straightforwardly extended to genetic manipulation of other insect species, and especially those highly important to agriculture and human health. While there was reason for optimism, we now realize that this was a naïve expectation, given what we now understand about the natural history of *P* elements relative to other Class II transposable elements – in particular, its extremely limited distribution and its dependence on species-specific host factors. The inability of *P* to function in non-drosophilids, however, was a motivating force to more completely understand transposon regulation, and the identification and testing of new vector systems. These included other transposable elements, as well as viral and bacterial vectors.

The development of routine methods for insect gene-transfer was probably delayed by a decade due to attention being focused exclusively on the *P* element. Yet this delay has resulted in a more varied toolbox of vectors and markers that now allow nearly routine transformation for many important species, and the potential for transformation of most insects (see Handler, 2001). Indeed, some of the tools developed for testing the *P* element, in particular embryonic mobility assays, are now routinely used for initial tests for function of other vectors in an insect species before more laborious and time-consuming transformation experiments are attempted.

The creation of this varied toolbox first related to the potential need for different vector and marker systems for different insect species. We now realize that the future of genetic analysis will depend on multiple vector and marker systems for each of these species, since genomics and functional genomics studies will require multiple systems

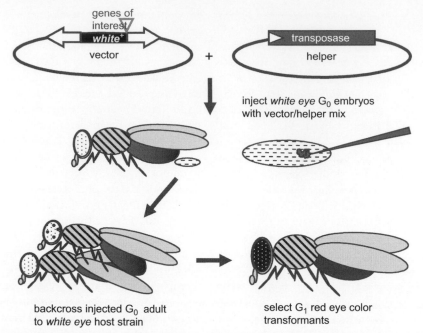

Figure 1 Diagram of the germ-line transformation method using a transposon-based vector marked with a wild type *white eye+* gene, and a transposase helper, in a mutant *white eye−* host strain. Both vector and helper are non-autonomous in that the vector has a non-functional transposase, and the helper has one or both terminal sequences deleted. Vector and helper plasmids are mixed in 3:1 to 5:1 proportions at a total concentration of <1 µg/ml in an injection buffer, and microinjected into preblastoderm *we−* host G_0 embryos. G_0 adults (which may show mosaic ommatidial pigmentation) are backcrossed to *we−* host strain flies with G_1 progeny screened for complete eye pigmentation. Transgenic mutant-rescue eye pigmentation may be weaker than wild type eye due to position effect suppression. See text for more complete methods.

for DNA integration and reporters for gene expression. Indeed, germ-line transformation is essential for the insertional mutagenesis and functional genomics studies that are critical underpinnings for both assessing genomic architecture and relating sequences to gene expression. Notably, the continuing functional analysis of the *Drosophila* genome now relies on the vectors and markers, described in this chapter, that were first developed for non-drosophilid insect species.

4.2. *P* Element Transformation

4.2.1. *P* Element

The use of transposable element-based vectors for *Drosophila* transformation followed the discovery of short inverted terminal repeat-type elements similar to the *Activator* (*Ac*) element discovered in maize by McClinotck (see Federoff, 1989). The first such element to be discovered in insects was the *P* element, the factor responsible for hybrid dysgenesis that occurred in crosses of males from a *P* strain (containing *p*aternal [*P*] factors associated with hybrid dysgenesis) with females from an *M* strain (devoid of *P* factor) (Kidwell *et al.*, 1977). The identification of *P* sequences resulted from the molecular analysis of *P*-induced *white* mutations that occurred in dysgenic hybrids (Rubin *et al.*, 1982). While the initial

P elements isolated as insertion sequences were incomplete, non-autonomous elements, complete functional elements were later isolated and characterized by O'Hare and Rubin (1983).

P is 2907 bp in length with 31-bp terminal inverted repeats (ITR) and 11-bp subterminal inverted repeats that occur approximately 125 bp from each terminus (see **Figure 2**). Other repeat sequences exist within *P*, but their functional significance, if any, remains unknown. A defining signature for *P*, as with other transposable elements, is the nature of its insertion site, which consists of an 8-bp direct repeat duplication. The extensive use of *P* for transformation and transposon mutagenesis has shown the element to have a distinctly non-random pattern of integration. It is now clear that *P* elements are blind to a significant fraction of the genome, and new gene vectors are being employed in *Drosophila* to complement these limitations. *P* elements and all transposable elements currently used as insect gene vectors belong to a general group of transposable elements known as Class II short inverted terminal repeat transposons (see Finnegan, 1989). These elements transpose via a DNA intermediate, and generally utilize a cut-and-paste mechanism that creates a duplication of the insertion site. These are distinguished from Class I elements, or retrotransposons, that transpose via reverse transcription an RNA intermediate, and may or may not have long terminal inverted repeats.

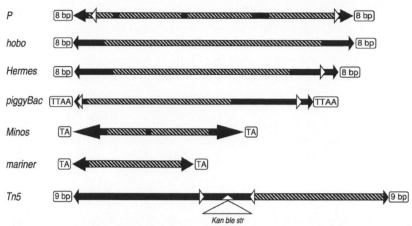

Figure 2 Diagram of transposable elements currently in use for the germ-line transformation of insect species. The left arms represent the 5′ termini, and right arms represent the 3′ termini. Transposon sizes and specific internal elements are shown in relative positions but are not at precise scale. Major structural elements include duplicated insertion sites (open boxes); inverted terminal repeat sequences (black arrowheads); internal subterminal repeat sequences (white arrowheads); transposase coding region (boxed diagonals); and intron sequences (black boxes). The *Tn5* element is a composite transposable element consisting of two functional elements flanking three antibiotic resistance genes. Refer to text for specific details on nucleotide lengths and relative positions.

The original use of *P* for germ-line transformation was accomplished by inserting a marker gene into the element so that it did not disrupt activity of the terminal sequences or the transposase gene. The *rosy*⁺ gene was inserted at the 3′ end of the transposase-coding region, but upstream of the 3′ subterminal inverted repeat sequence. Plasmids containing this vector were injected into preblastoderm (syncitial) embryos homozygous for *ry*⁻ so the *P* vector could transpose into germ cell nuclei. Germ-line transformation events were identified in the following generation (G₁) by virtue of reversion of the mutant *ry*⁻ eye-color phenotype to wild type. These experiments not only proved the feasibility of transposon-mediated transformation, but also permitted structure–function relationships within the *P* element to be determined (Karess and Rubin, 1984). The *P* transcriptional unit was found to be composed of four exons separated by three introns. Further analysis determined the cause of *P*'s germ-line only activity to be due to the absence of splicing of the third intron in somatic cells. The absence of third-intron splicing in the soma results in production of non-functional truncated transposase polypeptides in these tissues (Rio *et al.*, 1986).

While the original *P* vector allowed efficient transformation, the presence of a functional transposase gene within the vector made the system self-mobilizable (autonomous) and inherently unstable, allowing potential excision or transposition of the original insertion event. Subsequent vector development resulted in a binary system in which the transposase coding region was either deleted from the vector or made defective by insertion of a marker gene. The ability of the transposase to act *in trans* allowed transposase to be provided on a separate plasmid (helper) that could facilitate vector integrations when co-introduced with the vector-containing plasmid into the

same nucleus (Rubin and Spradling, 1982). Integrations would remain stable if the helper did not integrate, but the original helpers, such as pπ25.1, were autonomous *P* elements themselves that could integrate along with the vector. While helper integration was diminished by injecting several-fold higher concentrations of vector plasmid, this possibility was only eliminated with the creation of defective helpers having one or both of their terminal sequences deleted (known as "wings-clipped" helpers). The first of these was pπ25.7wc, which was immobilized by deleting 3′ terminal sequences (Karess and Rubin, 1984). This prototype vector system served as a model for binary systems of non-autonomous vector:helper elements used for all the transposon-based transformation systems currently in use.

A notable characteristic of *P* elements was not only their discontinuous intraspecific distribution (*P* and *M* strains), but also their discontinuous interspecific distribution. Based on distribution patterns, it has become apparent that *P* elements were recently introduced into *D. melanogaster* from *D. willistoni* by an unknown mechanism (Daniels and Strasbaugh, 1986). Regardless of the mechanism, since the 1950s *P* elements have thoroughly invaded wild populations of *D. melanogaster* (Anxolabéhère *et al.*, 1988), and without the existence of *M* strain laboratory stocks that were removed from nature before this time, the development of *P* vectors may never have been realized. This is due to the repression of *P* mobility in *P*-containing strains that was first observed in hybrid dysgenesis studies, which also showed that movement was not repressed in *M* strains devoid of *P*. The basis for *P*-strain repression appears to be due to a number of factors, including repressor protein synthesis, transposase titration by resident defective elements, and transcriptional control of transposase gene transcription (Handler *et al.*, 1993a; Simmons *et al.*, 2002; Castro and Carareto, 2004; Jensen *et al.*, 2008). As will be discussed, other vector systems in use have thus far been shown to be widely functional in several orders of insects, and the presence of the same or a related transposon in a host insect does not necessarily repress vector transposition. In this and several other aspects, the *P* vector system appears to be the exception rather than the rule for transposon-mediated gene transfer in insects.

4.2.2. *P* Vectors and Markers

Regardless of regulatory differences between *P* and other transposon vector systems currently in use, methods developed for *P* transformation of *Drosophila* serve as a paradigm for all other insect vector systems (see **Figure 1**). Those familiar with *Drosophila* transformation will be in a good position to attempt these methods in other insects. Current techniques developed for other insect species are variations on a theme, though, as we describe, considerable modifications have been made. Several comprehensive reviews are available for more specific details on the structure, function, and use of *P* for transformation in *Drosophila*, which are highly relevant to the understanding and use of other vector systems (see Karess, 1985; Spradling, 1986; Engels 1989; Handler and O'Brochta, 1991; Venken and Bellen, 2005, 2007). Particularly useful are the books and methods manual by Ashburner (1989a, 1989b) that review the various vectors, markers and methodologies used for *Drosophila* transformation, as well as early techniques used to manipulate *Drosophila* embryos. This information is especially applicable to other insect systems.

The first consideration for transformation is the design of vector and helper plasmids, and the marker system used for transformant selection. The first *P* vectors and helpers were actually autonomous vectors, which was probably a useful starting point, since the actual sequence requirements for vector mobility and transposase function were unknown. As noted, the first non-autonomous helper had a 3′ terminal deletion that prevented its transposition, providing greater control over vector stability. However, this source of transposase was inefficient until it was placed under *hsp70* regulation that allowed transposase induction by heat shock (Steller and Pirrotta, 1986). All other vector system helper constructs have similarly taken advantage of heat shock promoters, mostly from the *D. melanogaster hsp70* gene, but other *hsp* promoters have been tested, including those from the host species being transformed. Other constitutive promoters such as those from the genes for *actin* and *α1-tubulin* have proven successful for helper transposase regulation, and will be discussed further on.

While sufficient transposase production is critical for transposition, the structure of the vector is equally important, and, for some, very subtle changes from the autonomous vector can dramatically decrease or eliminate mobility. These variations include critical sequences (typically in the termini and subtermini), and placement and amount of exogenous DNA inserted within the termini. For some vectors, the amount of plasmid DNA external to the vector can affect transposition rates. Subsequent to the initial test of several *P* vectors, the terminal sequence requirements for *P* mobility were determined to include 138 bp of the 5′ end and 216 bp of the 3′ end. While the inverted repeat sequences within these terminal regions are identical, the adjacent sequences were found not to be interchangeable in terms of vector mobility (Mullins *et al.*, 1989). Of interest was the discovery that the strongest binding affinity for the *P* transposase was at sequences approximately 50 bp internal to the terminal repeats (Rio and Rubin, 1988; Kaufman *et al.*, 1989). While the minimal sequences required for mobility may be used in vectors, typically the rate of mobility decreases with the decreased length of terminal sequence. Specific sequences may be required for binding of transposase or other nuclear factors, and conformational changes needed for recombination may be dependent upon sequence length and position.

P-vector mobility was also found to be influenced by the amount of exogenous DNA inserted between the termini, with transformation frequency diminishing with increasing size. Initial tests with 8-kb vectors marked with *rosy* yielded transformation frequencies of approximately 50% per fertile G_0, while use of 15-kb vectors resulted in 20% frequencies (see Spradling, 1986). Larger vectors could transpose, but frequencies approached 1% or less.

Of equal importance to creating an efficient vector system is having marker genes and appropriate host strains that will allow efficient and unambiguous identification or selection of transgenic individuals. Indeed, the genetic resources available for *Drosophila* also provided cloned wild type DNA and appropriate mutant hosts for use in visible mutant-rescue marker systems that made testing *P* transformation possible. As noted, the first of these used the *ry+* eye-color gene, but this required a relatively large genomic fragment of nearly 8 kb. The *white* (*w*) eye-color gene was then tested, but this required a genomic sequence that was longer than *ry*, and resultant transformation frequencies were relatively low (Hazelrigg *et al.*, 1984; Pirrotta *et al.*, 1985). New *w* markers, known as mini-*white*, which had the large first intron deleted, decreased the marker insert to 4 kb, resulting in much more efficient transformation, and placing the mini-*white* marker under *hsp70* regulation increased efficiencies further (Klemenz *et al.*, 1987). Use of *white* markers, especially in CaSpeR vectors (Pirrotta, 1988), has been a mainstay of *Drosophila* transformation, yet expression of the *w* gene in particular is subject to position effect variegation/suppression (PEV) that typically diminishes eye pigmentation. PEV, indeed, was originally discovered as a result of translocating w^+ proximal to heterochromatin (Green, 1996), and it routinely manifests itself in w^- flies transformed with w^+. This effect has been observed with use of eye-pigmentation markers in several other insect species as well.

Other markers based upon chemical selections or enzymatic activity were also developed for *Drosophila*, though none have found routine use. These included alcohol dehydrogenase (*Adh*) (Goldberg *et al.*, 1983) and dopa decarboxylase (*Ddc*) (Scholnick *et al.*, 1983), which complemented existing mutations; and neomycin phosphotransferase (NPT or *neo*) (Steller and Pirrotta, 1985), β-galactosidase (Lis *et al.*, 1983), organophosphorus dehydrogenase (*opd*) (Benedict *et al.*, 1995), and dieldrin-resistance (*Rdl*) (ffrench-Constant *et al.*, 1991), which are dominant selections not requiring pre-existing mutations (see ffrench-Constant and Benedict, 2000)

4.2.3. *P* Transformation of Non-Drosophilids

Given the straightforward procedures for transforming *Drosophila* with *P* elements, there were high expectations that the system would function in other insects. The ability to test this was facilitated by the development of the neomycin (G418)-resistance marker system (Steller and Pirrotta, 1985), and neomycin resistance-containing *P* vectors were widely tested in tephritid flies and mosquitoes (see Walker, 1990; Handler and O'Brochta, 1991). Unfortunately, the neo-resistance system was generally unreliable, and recovery of resistant individuals that were not transgenic was common. In three mosquito species, however, neomycin-resistant transgenic insects were recovered, but they arose from rare transposition-independent recombination events (Miller *et al.*, 1987; McGrane *et al.*, 1988; Morris *et al.*, 1989). Other dominant chemical resistance markers, including *opd* and *Rdl*, which had some success in *Drosophila*, were also tested, but no transformation events could be verified in other insects. A major limitation of these experiments was that, given the numerous variables involved, it was impossible to determine which components in the system were failing. This limitation led to efforts to systematically determine whether the transposon vector system was indeed functional in host embryos, which resulted in the development of rapid transposon mobility assays as described below. The first of these assays tested *P* excision in drosophilid and non-drosophilid embryos, revealing that *P* function decreased in drosophilids as a function of relatedness to *D. melanogaster*, with no function evident in non-drosophilids (O'Brochta and Handler, 1988; Handler *et al.*, 1993b). These results were the first indication that for transposon-mediated germ-line transformation to succeed in non-drosophilids, new vector systems would have to be created from existing and newly discovered transposon systems.

4.3. Excision and Transposition Assays for Vector Mobility

Assessing the ability of an insect gene vector to function in a particular species can be challenging. The procedures required to create a transgenic insect using transposable element-based gene vectors require a great deal of technical skill, and the ability to perform basic genetic manipulations. Depending on the insect, its generation time, and its amenability to being reared in the laboratory, the process of genetic transformation can be quite lengthy. At the early stages of developing non-drosophilid transformation technology, there was little experience in manipulating and injecting the embryos of the various non-drosophilid species of insects. In addition, the genetic markers available to select for or recognize transgenic insects were limited, and none could confidently be expected to function optimally in the species being tested at that time. Consequently, early efforts to test the functionality of potential gene vectors by attempting to create transgenic insects required simultaneous success in dealing with a number of daunting challenges. The failure of these efforts to yield a transgenic insect could not, unfortunately, be ascribed to the failure of any one particular step in the process (see Handler and O'Brochta, 1991). These efforts, therefore, did not represent an isolated test of the gene vector, since failure to obtain a transgenic insect may have been due to a failure in DNA delivery, expression of the genetic marker, or the failure of the transposable element vector system. Technology development under these conditions was very difficult. What was needed was an experimental system that permitted the activity of the transposable element system to be assessed in the species of interest, independent of any prospective genetic marker system and DNA delivery system. Such a system was developed for investigating the mobility properties of the *D. melanogaster P* element, and was very adaptable to other transposable element and insect systems (see **Figure 3**).

The system developed for *P* elements involved transfecting *Drosophila* cells with a mixture of two plasmids; one containing a *P* element inserted into the coding region of the lacZα peptide of a common cloning vector, and a second containing the *P* element transposase gene under the regulatory control of a strong promoter (Rio *et al.*, 1986). Transient expression of the transposase gene resulted in the production of transposase, catalyzing the excision of *P* elements from the "excision indicator plasmids." Subsequent recovery of the injected plasmids from the cells, followed by their introduction into an appropriate strain of *E. coli*, permitted plasmids that had lost the *P* element through excision to be recognized by virtue of their restored lacZα peptide coding capacity. This transient *P* element excision assay was readily adaptable to use in *Drosophila* embryos through the process of direct microinjection of preblastoderm embryos, and it played a

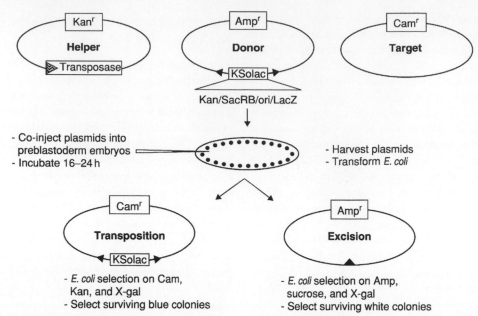

Figure 3 Plasmid-based transposable element mobility assays. A mixture of three plasmids is co-injected into preblastoderm embryos to ensure incorporation into nuclei. After approximately 24 hours, the plasmids are extracted from the embryos and introduced into *E. coli*. Transient expression of the transposase gene on the helper plasmid in the developing embryos results in the production of functional transposase. If the transposase catalyzes excision and transposition of the element, excision will result in the loss of element-specific markers on the donor plasmid. In the example shown, sucrose sensitivity, β-galactosidase activity, and kanamycin resistance are lost, and others could be used. Transposition results in the target plasmid acquiring all of the element-specific markers. In this example, the target plasmid is from a Gram-positive bacteria and is incapable of replicating in *E. coli* unless it acquires the origin of replication present on the element. Assays can be completed in 3 days, and rates of movement of 0.001% or greater are routinely detectable.

critical role in assessing the functionality of the *P* element system in a variety of drosophilid and non-drosophilid insect systems (O'Brochta and Handler, 1988).

As originally configured, the excision assay only permitted the identification and recovery of excision events that resulted in the restoration of the open reading frame of the lacZα peptide reporter gene. Various modifications in this basic assay were adopted that permitted precise and imprecise excisions to be identified and recovered (O'Brochta *et al.*, 1991). For example, marker genes, such as the *E. coli* lacZα peptide coding region, *E. coli supF*, sucrase (*SacRB*) from *Bacillus subtilis*, and streptomycin sensitivity, were incorporated into the transposable element (O'Brochta *et al.*, 1991; Coates *et al.*, 1997; Sundararajan *et al.*, 1999). Plasmids recovered that lacked marker gene expression were usually excision events. Further refinements of the excision assay involved the use of transposable element-specific restriction endonuclease sites as a means for selecting for excision events. Digesting plasmids recovered from embryos with restriction endonucleases with sites only in the transposable element was a very powerful method of physically removing plasmids that had not undergone excision from the pool of plasmids recovered from embryos and used to transform *E. coli*. Each restriction site was essentially a single dominant genetic marker, and therefore transposable elements

with multiple restriction sites provided a very powerful system for selecting against plasmids that had not undergone excision (D. O'Brochta, unpublished).

Continued development of element mobility assays led to assays in which inter-plasmid transposition could be measured. These assays involved the co-injection of a transposase-encoding helper plasmid, an element "donor" plasmid, and a "target" plasmid. Typically, the target plasmid contains a gene whose inactivation results in a selectable phenotype. For example, the *SacRB* gene has been used because its inactivation eliminates sucrose sensitivity. If the donor element also contains unique genetic markers, then transposition events would lead to a recombinant plasmid with a new combination of a variety of markers (Saville *et al.*, 1999). Perhaps the most powerful transposition assay developed for assessing transposable elements in insect embryos involved the use of a genetic marker cassette containing a plasmid origin of replication, an antibiotic resistance marker, and the lacZα peptide coding region, in combination with a target consisting of a Gram-positive plasmid (pGDV1) (Sarkar *et al.*, 1997a). pGDV1 contains an origin of replication that cannot function in *E. coli*, although it does have a chloramphenicol resistance gene that is functional in this species. Transposition of the marked transposable element into pGDV1 converts it into a functional replicon in *E. coli*. Because of the absolute *cis*-dependence of origins of replication, and the complete inability of pGDV1 to replicate in *E. coli*, transposition events can be readily detected even at low frequencies.

Transient mobility assays are now a standard for defining vector competence in insect embryos, and in particular when assessing a vector in a species for the first time. For this application, transposition assays provide the most information relevant to the potential for successful germ-line transformation, and can be used as a system to test helper construct function. As noted below, however, there may be differing constraints on plasmid and chromosomal transpositions for particular transposons. The use of these assays for analyzing transposon function is discussed in more detail in the relevant sections below.

Embryonic assays also provide an essential test system for assessing potential transgene instability by mobilizing or cross-mobilizing systems within a host genome, which is critical information for the risk analysis of transgenic insects being considered for release. The importance of excision assays for this purpose became evident by the *hobo* excision assays in *M. domestica* that revealed the existence of the *Hermes* element (Atkinson *et al.*, 1993), and the subsequent assays that defined the interaction between the two transposons (Sundararajan *et al.*, 1999). Since cross-mobilizing systems do not always promote precise excisions, assays that reveal imprecise as well as precise excisions are most sensitive for this purpose. Since successful transposition may depend on precise excision, transposition assays may only reveal the existence of mobilizing systems that have a high level of functional relatedness.

4.4. Transformation Marker Systems

The availability and development of selectable marker systems has played a large part in recent advancements in insect transformation, which have been equal in importance to vector development. The rapid implementation and expansion of *P* transformation in *Drosophila* was possible, in large part, due to the availability of several eye-color mutant-rescue systems. These systems depend on the transgenic expression of the dominant-acting wild type gene for an eye-color mutation present in the host strain (see Sarkar and Collins, 2000). Successful transformation of non-drosophilid species was similarly dependent upon the development of analogous systems, with the first transformations in *Ceratitis capitata* and *Aedes aegypti* relying on *white* and *cinnabar* mutant-rescue systems, respectively. While chemical resistance markers were used initially for non-drosophilid transformation, and can be highly useful for specific applications, their inefficiency and inconsistency when used alone provided ambiguous results for several species (see ffrench-Constant and Benedict, 2000). Eye-color marker systems are generally efficient and reliable, and cloned wild type genes

from *Drosophila* often complement orthologous mutant alleles in other insects; however, only a handful of species have stable mutant strains that can serve as suitable hosts for mutant-rescue strategies. The most significant advancement in marker gene development for the wide use of insect transformation has been the development of fluorescent protein markers (see Higgs and Sinkins, 2000; Horn *et al.*, 2002). As dominant-acting neomorphs that do not depend on pre-existing mutations, they are directly useful in almost all host strains. When compared to the *white* eye-color marker in *Drosophila*, the enhanced green fluorescent protein (EGFP) gene seemed to be less affected by position effect suppression, and thus has the additional advantage of more reliable detection (Handler and Harrell, 1999). Certainly for the forseeable future, fluorescent protein markers will continue to be the markers of choice for most insect transformation strategies.

4.4.1. Eye-Color Markers

The first insect transformations used mutant-rescue systems to identify transformant individuals, but in these experiments total genomic DNA was used, rather unreliably, to complement mutations in the respective host strains. The most reliable of these, however, was reversion of the *vermilion* eye-color mutation in *D. melanogaster* (Germeraad *et al.*, 1976). The success of the initial *P* element transformations in *Drosophila* also depended on reversion of eye-color mutant strains, but the use of cloned *rosy* and *white* genomic DNA within the vector plasmid allowed for much greater efficiency and reliability. The first non-drosophilid transformations in medfly (Loukeris *et al.*, 1995a; Handler *et al.*, 1998, Michel *et al.*, 2001) similarly relied on use of the wild type medfly *white* gene cDNA that was placed under *Drosophila hsp70* regulation (Zwiebel *et al.*, 1995). This gene complemented a mutant allele in a *white eye* medfly host strain that was isolated more than 20 years earlier. The medfly *white* gene also complemented the orthologous gene mutation in the oriental fruit fly, yielding, in one line, a nearly complete reversion (Handler and McCombs, 2000). The first transformations of *Ae. aegypti* used a kynurenine hydroxylase⁻, *white* mutant host strain, but for these tests the complementing marker was the wild type form of the *D. melanogaster cinnabar* gene (Cornel *et al.*, 1997). The *D. melanogaster vermilion* gene, which encodes tryptophan oxygensase (*to*), has also been used to complement the orthologous *green* eye-color mutation in *Musca domestica* (White *et al.*, 1996), and the *Anopheles gambiae* tryptophan oxygensase gene complements *vermilion* in *Drosophila* (Besansky *et al.*, 1997). The *vermilion* and *cinnabar* orthologs have also been cloned from *Tribolium*, and while the *white* mutation in this species is complemented by tryptophan oxygensase, no pre-existing eye-color mutation is complemented by kynurenine hydroxylase (Lorenzen *et al.*, 2002). The use of eye-color mutant-rescue systems has certainly been critical to initial advances in insect transformation, and these markers should have continued utility for those species that have been successfully tested. The use of these markers, however, for development of insect transformation in other species will be limited by the availability of suitable mutant host strains.

4.4.2. Chemical Selections

Previous to the development of mutant-rescue marker systems, transformant selections in non-drosophilid insects focused on genes that could confer resistance to particular chemicals or drugs. Importantly, these types of selections could be used for screening transformants *en masse* by providing the selectable chemical or drug in culture media. Ideally, only transformed individuals would survive the selection, allowing rapid screening of large numbers of G_1 insects. For vectors that are inefficient and insects that are difficult to rear, the efficient screening of populations can be essential to identifying transformant individuals. The first drug resistance selection tested used the bacterial neomycin phosphotransferase gene (NPT II or *neomycin^r*), which conferred resistance by inactivation of the neomycin analog G418 (or Geneticin) (Steller and Pirrotta, 1985). This seemed straightforward, since the selection and *hsneo* marker system (putting NPT II under heat shock regulation) had already been developed and tested in *Drosophila* for mass transformant screens, and the bacterial resistance gene was thought to be functional in most eukaryotes. The initial *P* transformations in *Drosophila* using the pUChsneo vector were generally reliable; however, the marker was not easily transferable to other species. G418 resistance was highly variable, most likely due to species differences in diet, physiology, and symbiotic bacteria, and indeed, variation in resistance in transformed *Drosophila* has been attributed to strains of yeast used in culture media (see Ashburner, 1989a). Other chemical resistance markers, including *organophosphorus dehydrogenase* (*opd*), conferring resistance to paraoxan (Phillips *et al.*, 1990; Benedict *et al.*, 1995), and the gene for dieldrin resistance (*Rdl*) (ffrench-Constant *et al.*, 1991), which were initially tested in *Drosophila* were also problematic when tested in other species. These failures were due in large part to ineffective vector systems, but a common attribute in these studies was the selection of individuals having non-vector related or natural resistance to the respective chemical. While naturally resistant insects could be distinguished from transformed insects by molecular genotype tests, the recurrence of resistant insects in subsequent generations would make use of the transgenic strains highly impractical.

While the problems cited made chemical resistance selections frustrating for several species, and they have not been used for any recent transformation experiments,

some successes were reported and the need for mass screening still exists. The initial tests for *P* transformation in several mosquito species used the pUChsneo vector with G418-resistant transformants being selected, though transformation frequencies were low and all of them resulted from fortuitous recombination events and not *P*-mediated transposition (Miller *et al.*, 1987; McGrane *et al.*, 1988; Morris *et al.*, 1989). Nonetheless, chemical selections can be very powerful, and, if reliable, they would dramatically improve the efficiency of transformation screens for most insects. It is quite possible that many species will not be amenable to current transformation techniques without markers that allow selection *en masse*. A potential means of increasing the reliability of chemical resistance screens would be to link a resistance marker to a visible marker within the vector. Initial G_1 transformants could be screened *en masse* for chemical resistance, with surviving individuals verified as transformants and maintained in culture using the visible marker. We have begun to test this type of marking by linking the *hsneo* construct with a red fluorescent protein marker in the *piggyBac* vector. Thus far, initial results in *Drosophila* are highly encouraging (A. Handler and R. Harrell, unpublished).

Of the enzyme systems tested for chemical selection in *Drosophila* that might be extended to other insects, the alcohol dehydrogenase (*Adh*) system might have the most promise (Goldberg *et al.*, 1983). An *Adh* marker gene can complement the *adh* mutation in *Drosophila*, eliminating lethal sensitivity to ethanol treatment in mutant hosts. An *adh* gene has been cloned from the medfly, and a strategy has been developed to use it for genetic sexing by male-specific overexpression (Christophides *et al.*, 2001). Conceivably, a similar strategy could be extended to transformant selections, though its use would be limited to medfly and possibly other tephritid species.

4.4.3. Fluorescent Protein Markers

The dramatic advancement of insect transformation in recent years has been due primarily to the development of fluorescent protein markers which are dominant-acting neomorphs that do not depend on pre-existing mutations. The first of these to be tested was the green fluorescent protein (GFP) gene that was isolated from the jellyfish *Aequorea victoria* (Prasher *et al.*, 1992), and exhibited heterologous functionality in the nematode *Caenorhabditis elegans* (Chalfie *et al.*, 1994). GFP expression was tested in transformed *Drosophila*, where it was successfully used as a reporter of gene expression (Plautz *et al.*, 1996; Hazelrigg *et al.*, 1998). GFP was first tested in non-drosophilid insects when GFP-marker Sindbis viruses were successfully used to infect the mosquito *Ae. aegypti* (Higgs *et al.*, 1996). The dramatic somatic expression of GFP in adults was highly encouraging for the further use of GFP for germ-line transformants.

The use of GFP to detect germ-line transformation events was first tested in *Drosophila* using a construct that placed a modified form of GFP, enhanced GFP (EGFP), with a nuclear localizing sequence, under the regulatory control of the promoter from *polyubiquitin* (Lee *et al.*, 1988; Handler and Harrell, 1999; see also Davis *et al.*, 1995). The creation and use of the *piggyBac* vector pB[Dm*w*, PUbnlsEGFP] in *D. melanogaster* allowed for a direct comparison of EGFP expression as a transformation marker to that from the visible mini-*white* marker. The results from this experiment indicated that not only was the PUbnlsEGFP marker efficient and easily detectable under epifluorescense optics, but also many of the G_1 transformants that expressed GFP did not express a detectable level of *white*[+] (A. Handler and R. Harrell, unpublished). Although the biological basis of this observation is not known, this result provided encouraging evidence for the use of GFP as a marker in non-drosophilids. Several subsequent transformation experiments using EGFP regulated by a variety of promoters in *piggyBac*, *Hermes*, and *Minos* vectors confirmed these expectations. Notably, fluorescent protein marker genes allowed germ-line transformation to be tested in several species that otherwise had no visible marker systems, such as the Caribbean fruit fly *Anastrepha suspensa*, which was transformed with pB[PUbnlsEGFP] (Handler and Harrell, 2000). This vector was subsequently tested in *Lucilia cuprina* (Heinrich *et al.*, 2002) and *An. albimanus* (Perera *et al.*, 2002). Similarly, a *Hermes* vector marked with EGFP regulated by the *Drosophila actin5C* promoter was first tested in *Drosophila* (Pinkerton *et al.*, 2000), and was

then used to efficiently select transformants in *Ae. aegypti* (Pinkerton *et al.*, 2000), *Stomoxys calcitrans* (O'Brochta *et al.*, 2000), and *Culex quinquefasciatus* (Allen *et al.*, 2001). A *Minos* vector marked with *actin5C*-EGFP was used to select *An. stephensi* transformants (Catteruccia *et al.*, 2000), and a *piggyBac* vector marked with EGFP under *Bombyx actin 3A* promoter regulation was used to transform the lepidopteran species *Bombyx mori* (Tamura *et al.*, 2000) and *Pectinophora gossypiella* (Peloquin *et al.*, 2000).

Both the *polyubiquitin* and *actin* promoters have activity in all tissues throughout development, making insects marked in this fashion particularly useful for some applications, such as the marking of insects used in biocontrol release programs (see Handler, 2002a). However, the detection of these markers can occasionally be difficult due to quenching or masking of fluorescence by melanized cuticle or scales. Fluorescent protein expression regulated by strong tissue-specific promoters has proven particularly valuable. Foremost among this type of marker is a series of fluorescent protein open reading frames under the regulatory control of the artificial promoter 3xP3, derived from the *Drosophila eyeless* gene (Sheng *et al.*, 1997; Horn *et al.*, 2000). Fluorescent protein expressed using 3xP3 is found primarily in the larval nervous system, and the eyes and ocelli of adults. *piggyBac*, *Hermes*, and *Mos1* vectors containing 3xP3-EGFP were first used to transform *D. melanogaster* and *T. castaneum* (*piggyBac* and *Hermes*) (Berghammer *et al.*, 1999), and have been widely used in the creation of many species of transgenic insects (see **Table 1**). The particular strengths and weaknesses for a

Table 1   Transposon Vectors and Markers Currently used for the Germ-Line Transformation of Various Insect Species

Transposon	Host species	Marker	Reference(s)
Hermes	Aedes aegypti	Dm-cinnabar[+]	Jasinskiene et al., 1998
		actin5C-EGFP	Pinkerton et al., 2000
	Anopheles stephensi	Act5CEGFP	R. Harrell and D. O'Brochta, unpublished
	Bicyclus anynana	3xP3-EGFP	Marcus et al., 2004
	Ceratitis capitata	Cc-white[+]	Michel et al., 2001
	Culex quinquefasciatus	actin5C-EGFP	Allen et al., 2001
	Drosophila melanogaster	Dm-white[+]	O'Brochta et al., 1996
		actin5C-EGFP	Pinkerton et al., 2000
		3xP3-EGFP	Horn et al., 2000
	Stomoxys calcitrans	actin5C-EGFP	O'Brochta et al., 2000
	Tribolium castaneum	3xP3-EGFP	Berghammer et al., 1999
Herves	Drosophila melanogaster	3xP3-EGFP	Arensburger et al., 2005
hobo	Drosophila. melanogaster	Dm-mini-white[+]	Blackman et al., 1989
	Drosophila virilis	Dm-mini-white[+]	Lozovskaya et al., 1996; Gomez and Handler, 1997
hopper	Anastrepha suspensa	PUb-DsRed	A. Handler and R. Harrell, unpublished
	Drosophila melanogaster	PUb-DsRed	A. Handler and R. Harrell, unpublished
mariner (Mos1)	Aedes aegypti	Dm-cinnabar[+]	Coates et al., 1998
	Drosophila melanogaster	Dm-white[+]	Garza et al., 1991; Lidholm et al., 1993
		3xP3-EGFP	Horn et al., 2000
	Drosophila virilis	Dm-white[+]	Lohe and Hartl, 1996a
	Musca domestica	pMos1 (unmarked)	Yoshiyama et al., 2000

(Continued)

Table 1   Transposon Vectors and Markers Currently used for the Germ-Line Transformation of Various Insect Species—cont'd

Transposon	Host species	Marker	Reference(s)
Minos	Anopheles stephensi	actin5C-EGFP	Catteruccia et al., 2000
	Bombyx mori	actin3(A3)-EGFP	Uchino et al., 2007
	Ceratitis capitata	Cc-white+	Loukeris et al., 1995b
	Drosophila melanogaster	Dm-white+	Loukeris et al., 1995a
	Tribolium castaneum	3xP3-EGFP	Pavlopoulos et al., 2004
P-element	Drosophila melanogaster	Dm-rosy+	Rubin and Spradling, 1982
		Dm-white+	Hazelrigg et al., 1984; Pirrotta et al., 1985
		Dm-hsp70-mini-white+	Klemenz et al., 1987
		pUChsneo	Steller and Pirrotta, 1985
	Drosophila simulans	Dm-rosy+	Scavarda and Hartl, 1984
piggyBac	Aedes aegypti	Dm-cinnabar+	Lobo et al., 2002
		3xP3-EGFP	Kokoza et al., 2001
	Aedes albopictus	3xP3-ECFP	Labbé et al., 2010
	Aedes fluviatilis	3xP3-EGFP	Rodrigues et al., 2006
	Anastrepha ludens	ubiquitin-CopGreen/ PhiYFP/J-Red	Condon et al., 2007
		PUb-nls-EGFP/DsRed	Meza et al., 2010
	Anastrepha suspensa	PUb-nls-EGFP	Handler and Harrell, 2000
	Anopheles albimanus	PUb-nls-EGFP	Perera et al., 2002
	Anopheles gambiae	hr5-ie1:EGFP	Grossman et al., 2001
	Anopheles stephensi	actin5C-DsRed	Nolan et al., 2002
	Athalia rosae	BmA3-EGFP, hsp70-GFP	Sumitani et al., 2003
	Bactrocera dorsalis	Cc-white+	Handler and McCombs, 2000
		PUb-nls-EGFP	Handler and McCombs, unpublished
	Bactrocera oleae	tTA/EGFP	Koukidou et al., 2006
	Bicyclus anynana	3xP3-EGFP	Marcus et al., 2004
	Bombyx mori	BmA3-EGFP	Tamura et al., 2000
		3xP3-EGFP	Thomas et al., 2002; Uhlirova et al., 2002
	Ceratitis capitata	Cc-white+	Handler et al., 1998
		PUb-nls-EGFP	A. Handler and R. Krasteva, unpublished
		PUb-DsRed1	Schetelig et al., 2009
	Cochliomyia hominivorax	PUb-nls-EGFP	Allen et al., 2004
	Cydia pomonella	3xP3-EGFP	Ferguson et al., 2010
	Drosophila ananassae		
	Drosophila erecta	3xP3-EC/GFP	Holtzman et al., 2010
	Drosophila melanogaster	Dm-white+, PUb-nls-EGFP	Handler and Harrell, 1999
		PUb-DsRed1	Handler and Harrell, 2001
		3xP3-EGFP	Horn et al., 2000
		3xP3-EYFP	Horn and Wimmer, 2000
		3xP3-ECFP	Horn and Wimmer, 2000
		3xP3-DsRed	Horn et al., 2002
	Drosophila mojaviensis	3xP3-EC/GFP	Holtzman et al., 2010
	Drosophila pseudoobscura	3xP3-EC/GFP	Holtzman et al., 2010
	Drosophila sechellia	3xP3-EC/GFP	Holtzman et al., 2010
	Drosophila simulans	3xP3-EC/GFP	Holtzman et al., 2010
	Drosophila virilis	3xP3-EC/GFP	Holtzman et al., 2010
	Drosophila willistoni	3xP3-EC/GFP	Holtzman et al., 2010
	Drosophila yakuba	3xP3-EC/GFP	Holtzman et al., 2010
	Harmonia axyridis	3xP3-EGFP	Kuwayama et al., 2006
	Lucilia cuprina	PUb-nls-EGFP	Heinrich et al., 2002
	Lucilia sericata	Lchsp83-ZsGreen	Concha et al., 2010
	Musca domestica	3xP3-EGFP	Hediger et al., 2000
	Pectinophora gossypiella	BmA3-EGFP	Peloquin et al., 2000
	Plutella xylostella	Hrie1DsRed/Opei2Zs Green	Martins et al., 2010
	Tribolium castaneum	3xP3-EGFP	Berghammer et al., 1999; Lorenzen et al., 2003
Tn5	Aedes aegypti	3xP3-DsRed	Rowan et al., 2004

marker construct such as 3xP3-EGFP are evident from experiments, where it enabled the selection of transgenic *Bombyx* embryos prior to larval hatching (Thomas *et al.*, 2002), while it is almost undetectable in *Ae. aegypti* adults having normal eye pigmentation (Kokoza *et al.*, 2001). It must therefore be recognized that the utility of fluorescent protein markers must be considered in the context of the host insect's structure and physiology during development.

Fluorescent protein genetic markers tend to be more sensitive indicators of genetic transformation than eye-color markers; however, they are subject to qualitative and quantitative variation in their expression. Tissue-specific variation in transgene expression is likely due to local chromatin structure impacting access of promoters to essential transcription factors, while expression of transgenes in unexpected cells and tissue is likely due to the influence of local enhancer. For example, *polyubiquitin*-regulated EGFP expression is most intense in the thoracic flight muscles of adult *D. melanogaster* and tephritid fruit flies. In adult transgenic Caribbean fruit flies containing *PUb*-EGFP, EGFP was only observed in the thorax, and spectrofluorometric assays revealed as much as five-fold differences in fluorescence among lines with equal copy numbers of transgenes (Handler and Harrell, 2000). In contrast to typical thoracic expression in tephritid flies, *PUb*-EGFP expression in adult transgenic *L. cuprina* was limited to female ovaries (Heinrich *et al.*, 2002). PUb-DsRed expression in one transgenic medfly line was most intense in tarsi, while in another it was most intense at the tracheal apertures at the dorsal/ventral midline of the abdomen (A. Handler and R. Krasteva, unpublished). In *T. castaneum*, 3xP3-EGFP expression is typically in the optic lobes and brain, though several lines have shown atypical muscle-specific expression throughout development (Lorenzen *et al*, 2003). In various transgenic lines of *An. stephensi*, the 3xP3-EGFP marker has shown atypical expression in the pylorus and epidermal cells, and in a subset of cells in the rectum (D. O'Brochta, W. Kim, and H. Koo, unpublished).

The use of GFP will certainly continue to be a useful and popular insect transformation marker, but there is also a need for a variety of distinguishable fluorescent protein markers to permit the detection of multiple independent transgenes, and, when used in concert, for conditional gene expression systems and gene discovery methods, such as enhancer traps (Bellen *et al.*, 1989; Wilson *et al.*, 1989; Brand *et al.*, 1994). After testing 3xP3-EGFP, the 3xP3 promoter was linked to the GFP red-shifted variants that emit blue (BFP), cyan (CFP), and yellow (YFP) fluorescence, which were tested in *Drosophila*, and have proven useful individually as reporters and for identifying transformants (Horn and Wimmer, 2000). BFP and GFP have distinct enough emission spectra to be used together, though BFP photobleaches quickly and is not useful for

many applications. While use of EGFP with ECFP is also problematic, ECFP and EYFP can be distinguished when using appropriate filter sets. For details on appropriate filter sets for particular applications, see Horn *et al.* (2002), and the website for Chroma Technology Corp. (Brattleboro, VT; www.chroma.com), which manufactures filters for most of the stereozoom fluorescence microscopes used for insect studies.

The most spectrally distinct fluorescent protein from GFP and its variants is a red fluorescent protein (RFP), known as DsRed, isolated from the Indo-Pacific sea coral *Discosoma striata* (Matz *et al.* 1999). It was first tested in insects by linking it to the *polyubiquitin* promoter in a *piggyBac* vector (pB[PUb-DsRed1]) and tested in *Drosophila*, where it exhibited highly intense expression (Handler and Harrell, 2001). Importantly, DsRed expression was completely distinguishable from EGFP when the two transgenic lines were interbred, and when co-expressed as an hsp70-Gal4/UAS-DsRed reporter in lines having vectors marked with EGFP. DsRed and its variants have since been incorporated into several mosquito and fruit fly species (Nolan *et al.*, 2002; A. Handler, unpublished). Some of the original RFP variants include those found in a mutagenesis screen for rapid maturation and increased solubility (DsRed.T1/T3/T4), though their relative brightness is less intense than the wild type form (Bevis and Glick, 2002). The DsRed.T4 variant is available as DsRed-Express (Clontech), along with further variant forms which include monomeric RFPs that are preferred for fusion protein labeling (Strack *et al.*, 2008). Both EGFP and DsRed are highly stable and generally resistant to photobleaching, and could be detected in tephritid flies several weeks after death, though DsRed and its variants are the relatively more stable of the two. Notably, PUb-DsRed.T3-marked transformant Caribbean fruit flies were unambiguously distinguished from unmarked wild type flies after being kept in liquid traps (torula yeast borax and propylene glycol) in field conditions for up to 3 weeks (Nirmala *et al.*, 2010). This is highly advantageous for the use of these genes as markers for released insects that might only be retrieved several weeks after death in traps. A drawback for fluorescent proteins, and DsRed in particular, is that they require oligomerization and slow maturation that can take up to 48 hours, resulting in low intensity in early development. However, variants of DsRed with shorter maturation times (Campbell *et al.*, 2002), and new fluorescent proteins with enhanced properties for specific applications, are becoming available on a consistent basis (see Matz *et al.*, 2002; Chudakov *et al.*, 2010).

In addition to providing new markers that are more easily identifiable, additional distinguishable markers will be invaluable to new methods of vector manipulation (see section 4.5.6). These currently include vectors for post-integration stabilization of transposon vectors, requiring either two or three markers (Handler *et al.*, 2004; Dafa'alla *et al.*, 2006), and repeatable targeting of genomic insertion sites requiring independent markers for each transgene insertion (Horn and Handler, 2005; Nimmo *et al.*, 2006; Schetelig *et al.*, 2010). These new and variant fluorescent proteins, many of which are available from the Clontech Living Colors® collection, include proteins isolated from various reef corals and sea anenomes, and rapidly maturing monomeric forms of the previously discovered fluorescent proteins. The AmCyan and ZsGreen FP markers were used for the first time in a study that showed that the *D. melanogaster* scs/scs' and gypsy insulators, and the chicken *β-globin* HS4 insulator, are effective means to minimize genomic position effect suppression of transgene expression in *piggyBac* vectors (Sarkar *et al.*, 2006). ZsGreen, isolated from an *Anthozoa* reef coral, has since been placed under *L. cuprina hsp83* promoter regulation to more efficiently select *L. cuprina* transformants compared to use of PUb-nls-EGFP (Concha *et al.*, 2010). This argues for enhanced transformant selection with new fluorescent proteins under conspecific promoter regulation, which has been further supported by use of the same marker to identify the first *L. sericata* transformants (Concha *et al.*, 2010).

4.4.3.1 Detection methods for fluorescent proteins
Once heterologous expression of GFP in nematodes was discovered, it was realized that use of the marker for whole body analysis of gene expression would require an optical system allowing a large depth of field and a stage with working space for culture plates. Up to this time, most epifluorescence systems were linked to compound or inverted microscopes that had limited field depth and capability to manipulate organisms under observation. This led to the development of an epifluorescence module using a mercury lamp that could be attached to a Leica stereozoom microscope system. Most major microscope manufacturers now market integrated epifluorescent stereozoom microscopes with capabilities for several filter systems.

A lower-cost alternative for GFP screening is use of a lamp module using ultra-bright blue-light emitting diodes (LEDs) with barrier filters, which attaches to the objective lens of most stereozoom microscopes (BLS Ltd, Budapest, Hungary). It costs considerably less than a mercury lamp system, but at present only has capabilities for detecting GFP and YFP.

The use of fluorescent protein markers, and especially multiple markers, will be greatly aided by the use of fluorescence-activated embryo sorters. A device that was first developed to sort *Drosophila* embryos expressing GFP (Furlong *et al.*, 2001) has been modified and commercially marketed for *Drosophila* and other organisms as the COPAS system by Union Biometrica (Somerville, MA). The latest sorting machines are highly sensitive, having

the ability not only to distinguish different fluorescent proteins, but also to discriminate between levels of fluorescence from the same protein. Thus, these systems may have enormous importance to the straightforward screening for transgenics, and more sophisticated assays such as those for enhancer traps. Practical applications could include the screening of released transgenic insects caught in traps (in systems adapted for adults), or for genetic-sexing of embryos having a Y-linked or male-specific fluorescent marker.

4.5. Transposon Vectors

4.5.1. *Hermes*

4.5.1.1. Discovery, description, and characteristics
Hermes is a member of the *hAT* family of transposable elements, and is related to the *hobo* element of *D. melanogaster*, the *Ac* element from maize, and the *Tam3* element from *Antirrhinum majus* (Warren *et al.*, 1994). The initial interest in this family of elements by those concerned with creating new insect gene vectors stemmed from two observations. First, during the middle and late 1980s, the mobility characteristics of the *Ac/Ds* element system were being extensively studied because the element was recognized as having great potential to serve as a gene-analysis and gene-finding tool in maize and other plants. In addition, the mobility properties of *Ac/Ds* were being extensively tested in species of plants other than maize, and in almost every case evidence for *Ac/Ds* mobility was obtained (Fedoroff, 1989). *Ac/Ds* appeared to be a transposable element with a very broad host range – unlike, for example, the *P* element from *D. melanogaster*, which only functions in closely related species (O'Brochta and Handler, 1988). Because transposable elements with broad host ranges were of interest to those attempting to develop insect transformation technology, *Ac*-like elements warranted attention. The second significant observation at this time was that the *hobo* element from *D. melanogaster* had notable DNA sequence similarity to *Ac/Ds*, suggesting that it was a distant relative of this broadly active element (Calvi *et al.*, 1991). Investigation into the host range of *hobo* using plasmid-based mobility assays (as described above) ensued (O'Brochta *et al.*, 1994). It was during the investigation of *hobo* that *Hermes* was discovered (Atkinson *et al.*, 1993). Atkinson and colleagues performed plasmid-based *hobo* excision assays in embryos of the housefly *M. domestica* as part of an initial attempt to assess the host range of *hobo*. Assays were performed in the presence of *hobo*-encoded transposase and *hobo* excision events were recovered, suggesting that *hobo*, like *Ac/Ds*, would have a broad host range. However, when the assays were performed without providing *hobo*-encoded transposase, *hobo* excision events were still recovered in *M. domestica*. The movement of *hobo* in the

absence of *hobo*-transposase was completely dependent upon the inverted terminal repeats of *hobo*, and the resulting excision events had all of the characteristics of a transposase-mediated process. It was proposed that *M. domestica* embryos contained a *hobo* transposase activity, and that this activity arose from the transposase gene of an endogenous *hobo*-like transposable element (Atkinson *et al.*, 1993). These investigators were eventually able to confirm their hypothesis, and the element they discovered was called *Hermes* (Warren *et al.*, 1994).

Hermes is 2749 bp in length, and is organized like other Class II transposable elements in that it contains inverted terminal repeats and a transposase-coding region (see **Figure 2**). It contains 17-bp ITRs, with 10 of the distal 12 nucleotides being identical to the 12-bp ITRs of *hobo*. *Hermes* encodes for a transposase with a predicted size of 72 kDa, and, based on the amino acid sequence, is 55% identical and 71% similar to *hobo* transposase (Warren *et al.*, 1994). The cross-mobilization of *hobo* by *Hermes* transposase that was proposed by Atkinson *et al.* (1993) was tested directly by Sundararajan *et al.* (1999). These investigators used plasmid-based excision assays in *D. melanogaster* embryos to show that *hobo* transposase could mobilize *Hermes* elements, and that *Hermes* transposase could mobilize *hobo* elements (Sundararajan *et al.*, 1999). The phenomenon of cross-mobilization has important implications for the future use of transposable element-based gene vectors in non-drosophilid insects, and will be discussed further on. As is typical of transposable elements, *Hermes* is present as a middle repetitive sequence within the genomes of multiple strains of *M. domestica*, and in all populations of *M. domestica* examined there appeared to be full-length copies of the element (Subramanian *et al.*, 2009). *Hermes* appears to be active in some strains of *M. domestica*, since excised *Hermes* elements in the form of covalently closed circles (episomes) were readily detected (O'Brochta *et al.*, 2009).

4.5.1.2. Patterns of integration

The integration behavior of *Hermes* has been examined in a variety of contexts. Sarkar *et al.* (1997a, 1997b) tested the ability of *Hermes* to transpose, using a plasmid-based assay, in five species of Diptera. They recovered transpositions of *Hermes* in the target plasmid at a frequency of approximately 10^{-3} in all species tested. In addition, they examined the distribution of 127 independent transposition events into the 2.8-kb plasmid used as a target in their assay, and observed a distinctly non-random pattern of integrations. Most notable was the existence of 3 sites that were targets for *Hermes* integration 10 or more times each. In an experiment in which any site used twice or more was considered a hotspot for integration, the 3 sites used 10 or more times constitute sites with unusual characteristics. The precise nature of those characteristics, however, could not be defined. The sites shared four of eight nucleotides

of the target site in common (GTNNNNAC); however, other sites with this nucleotide composition were not equally attractive as integration sites, indicating that other factors must be influencing target choice. Saville *et al.* (1999) demonstrated that sequences flanking *hobo* integration hotspots were critical for determining the targeting characteristics of a site. These investigators were able to move an 8-bp *hobo* target site from plasmid to plasmid without losing its target characteristics, as long as they included 20 bp of flanking sequence on each side of the target. It was suggested that proximity to a preferred integration site increased the likelihood of a site being used as a target (Sarkar *et al.*, 1997a). They found that sites 80- and 160-bp flanking the integration hotspot were also preferred integration sites. The authors suggested that nucleosomal organization of the target contributes significantly to the target site selection process, and contributes to the local juxtaposition of hotspots and flanking DNA.

4.5.1.3. Structure–function relationships

Many Class II transposable elements contain a distinct amino acid motif within their catalytic domains, consisting of two aspartate residues and a glutamate. This DD35E motif can be found in many, but not all, Class II transposable elements. The presence of this motif in *Hermes* transposase was initially unclear. Bigot *et al.* (1996) proposed the existence of a DDE motif among members of the *hAT* family; however, they proposed that the second aspartate was replaced by a serine in *Ac*, *hobo*, and *Hermes*. Capy *et al.* (1996) concluded that *hAT* elements, like *P* elements from *Drosophila*, do not contain the DDE motif, based on sequence alignments; Lerat *et al.* (1999) supported this conclusion based on the lack of similarity in predicted secondary structure of the transposase of members of the *mariner*/*Tc* superfamily and *hobo* transposase. Michel *et al.* (2003) examined experimentally the importance of D402, S535, and E572 to the proper functioning of *Hermes* transposase. They found that mutations D402N and E572Q abolished transposase activity, while the mutations S535A and S535D had no effect on transposase activity. The work of Michel *et al.* (2003) provided the first experimental data to support the hypothesis that the positive charges of residues D402 and E572 are required for transposition. The authors concluded, based on these data, that D402, S535, and E572 do not constitute the catalytic center of *Hermes* transposase, because one of the residues was not essential for activity.

Zhou *et al.* (2004), using purified *Hermes* transposase protein and *in vitro* transposition reactions, found that, as expected, the element underwent cut-and-paste transposition, but this led to the creation of hairpin structures at the ends of donor DNA following excision (Zhou *et al.*, 2004). Based on the structure of the reaction products and an analysis of the amino acid sequence, Zhou *et al.*

(2004) concluded that there were significant similarities among *Hermes* transposase, the V(D)J recombinase RAG, and retroviral integrases (DDE transposases). The successful crystalization of *Hermes* transposase and the determination of the protein's structure confirmed the presence of a retroviral integrase fold, and clearly links this element, albeit distantly, to other transposable elements containing that protein fold (Hickman *et al.*, 2005).

Because *Hermes* transposase acts within the nucleus, it is expected to contain a nuclear localization signal to direct the mature transposase from the ribosome to the nucleus. Deletion and site-directed mutagenesis analysis were performed and demonstrated that the *Hermes* nuclear localization signal is located at the amino acid end of the protein and divided among three domains (Michel and Atkinson, 2003).

The inverted terminal repeats of transposable elements play an essential role in their mobility. Altering the sequence of ITRs can, depending on the element, lead to loss-of-function, hyperactivity of the element, or switching of the mode of transposition from a cut-and-paste mechanism to a replicative mechanism. *Hermes* contains imperfect ITRs, with a two base-pair mismatch within the ITR (Warren *et al.*, 1994). In addition, a naturally occurring polymorphism in the terminal nucleotide of the right 3′ ITR exists. Elements with a cytidine in the terminal position of the right ITR have no activity within *D. melanogaster*, but are capable of undergoing an aberrant form of transposition in mosquitoes. Small pentanucleotide motifs in the subterminal regions of both *Hermes* and *hobo* have been found to be important for the mobilization of *Hermes* and *hobo*. The sequences GTGGC and GTGAC are interspersed throughout the subterminal region of the element, and similar repeats are present in the subterminal regions of *Ac* and are known to be transposase-binding sites. In *Hermes*, altering a single repeat can eliminate transpositional activity (Atkinson *et al.*, 2001).

Hermes transposase is capable of dimerizing and one region of the protein critical for dimerization is located in the C-terminus of the protein, including amino acids 551–569. This region is not only essential for dimerization, but is also required for transposition activity. A second region that affects dimerization is located in the N-terminus of the protein, within the first 252 amino acids of the transposase. However, this region apparently plays a non-specific role in dimerization (P. W. Atkinson and K. Michel, personal communication). More recently, Hickman *et al.* (2005) found that transposition of *Hermes* was only observed when the protein formed hexamers.

4.5.1.4. Host range of *Hermes* *Hermes* has a wide insect host range, and has been found to function (as measured by either plasmid-based mobility assays or germ-line transformation) in at least 13 species of insects, including 11 flies, 1 beetle and 1 moth (Atkinson

et al., 2001). *Hermes* functions rather efficiently in *D. melanogaster*, and transforms this species at rates of 20–40% (O'Brochta *et al.*, 1996). In all other species tested the efficiency of transformation was considerably lower, and tended to be less than 10%. For example, *Tribolium casteneaum* was transformed at a rate of 1%, *Ae. aegypti* at 5%, *C. quinquefasciatus* at 11%, *C. capitata* at 3%, *S. calcitrans* at 4%, and *Bicyclus anynana* at 10.2% (Atkinson *et al.*, 2001; Marcus *et al.*, 2004). In all insects except mosquitoes, *Hermes* appeared to use a standard cut-and-paste type mechanism, as is typical of most Class II transposable elements. Such integrations are characterized by the movement of only those sequences delimited by the inverted terminal repeats, and the integrated elements are flanked by direct duplications of 8 bp. Integration of *Hermes* into the germ-line of *Ae. aegypti* and *C. quinquefasciatus* appears to occur by a non-canonical mechanism resulting in the integration of DNA sequences originally flanking the element on the donor plasmid. The amount of flanking DNA that accompanies the integration of *Hermes* in these mosquito species varies. In some cases, two tandem copies of the *Hermes* element were transferred to the chromosome and each copy was separated by plasmid DNA sequences (Jasinskiene *et al.*, 2000). Although these transposition reactions are unusual they are dependent upon *Hermes* transposase, since the introduction of *Hermes*-containing plasmid DNA in the absence of *Hermes* transposase failed to yield transformation events. The germ-line integration behavior of *Hermes* in mosquitoes is not unique, however; other elements being used as gene vectors, such as *mariner* and *piggyBac*, have occasionally shown similar behavior in *Ae. aegypti* (D. O'Brochta, unpublished). Transposition assays performed with plasmids in developing mosquito embryos and in mosquito cell lines showed that *Hermes* could transpose via a canonical cut-and-paste type mechanism under these conditions (Sarkar *et al.*, 1997b). The basis for the difference in types of integration events between plasmid-based transposition assays and chromosomal integrations is unknown, but may reflect differences in somatic and germ cells. In *Aedes*, canonical cut-and-paste transposition has been readily detected in the somatic tissues of insects containing an autonomous element. Germ-line transposition in these same insects has not been detected. It has been suggested that mosquitoes might contain endogenous *hAT* elements that affect the ability of *Hermes* elements to be integrated precisely. An alternative suggestion is that *Hermes* may have a second mode of transposition, as do the transposable elements *Tn7*, *IS903*, and *Mu*, which utilize a replicative mechanism of integration. Such a mechanism would result in integration products that resemble those observed in the germ-line of *Ae. aegypti* and *C. quinquefasciutus*. Replicative transposition of *Hermes* has not been demonstrated experimentally, and direct tests of the "alternate mechanism" hypothesis have not been reported.

Hermes' activity is not limited to insects; the element has be shown to be active in yeast (*Schizosaccharomyces pombe*) and planaria (*Girardia tigrina*), suggesting that it will have broad utility, and not just as an insect gene vector (González-Estévez *et al.*, 2003; Evertts *et al.*, 2007; Park *et al.*, 2009).

4.5.1.5. Post-integration behavior

Once integrated into the genome of *D. melanogaster*, *Hermes* maintains its ability to be remobilized and has shown mobility characteristics similar to those of other transposable elements. Following the introduction of an autonomous *Hermes* element in which the transposase gene was under *hsp70* promoter regulation, and also contained an EGFP marker gene under constitutive regulatory control of the *actin5C* promoter, Guimond *et al.* (2003) found the element continued to transpose in the germ-line at a rate of 0.03 jumps per element per generation. The element used in this study was also active in the somatic tissue, and the authors used this as a means of collecting approximately 250 independent transposition events. Analysis of somatic integration events revealed a number of interesting patterns. First, it was found that transpositions were clustered around the original integration event. On average, 39% of the *Hermes* transpositions recovered were intrachromosomal and 17% were within the same numbered polytene chromosome division. Of the new insertions, 10% were at sites within 2 kb of the donor element, indicating that *Hermes*, like other transposable elements, shows the characteristic of local hopping. Local hopping refers to the tendency of some elements to preferentially integrate into closely linked sites. Local hopping has been described for a number of elements and is likely to be a general characteristic of Class II transposable elements, although the mechanistic basis for this behavior is unknown. Certain regions of the *D. melanogaster* genome, as defined by numbered divisions of the polytene chromosomes, are preferred as integration sites, with these regions being repeatedly targeted by *Hermes*. The observed clustering of independent transposition events in regions of the chromosome seems to reflect undefined aspects of the transposition process that might be influenced by the chromatin landscape. With one exception, the clustering observed by Guimond *et al.* (2003) was not correlated with any common feature of the chromosomes or the genes within a region. This type of non-random pattern of integration with regional differences has also been reported for other elements. Interestingly, there does not seem to be any strong correlation between the preferred insertion-site regions of the elements *P*, *hobo*, and *Hermes*, at least with respect to chromosome 3 of *D. melanogaster* (see Figure 7 in Guimond *et al.*, 2003). Guimond *et al.* (2003) also observed a notable clustering

of integrations in polytene chromosome division 5. Of the 11 integration events recovered from division 5 (3.2% of all the transposition events examined), 8 were within the 2.7-kb segment of DNA upstream of the cytoplasmic actin gene, *actin5C*. This same 2.7-kb segment of the 5′ regulatory region of *actin5C* was also present within the autonomous *Hermes* element, as a promoter for the EGFP marker that the investigators tracked as it jumped within the genome.

The strong clustering of transpositions in a target sequence that is homologous to a sequence contained within the vector has been referred to as "homing." This type of target-site selection bias was first described for *P* elements, and has been reported on a number of occasions. It was initially reported as a strong bias in the integration site distribution of a number of primary germ-line integration events in which a *P* element containing the *engrailed* gene preferentially integrated into the *engrailed* region of the host genome (Hama *et al.*, 1990; Kassis *et al.*, 1992). A similar biasing of integration site selection was also observed with *P* elements containing *Antennapedia* and *Bithorax* regulatory sequences (Engstrom *et al.*, 1992; Bender and Hudson, 2000). More recently, Taillebourg and Dura (1999) reported a remarkable example of homing of a remobilized *P* element in *D. melanogaster*. This element contained either an 11-kb or a 1.6-kb fragment of the 5′ region of the *linotte* gene, and it was found that 20% of the remobilized elements integrated into the 5′ region of the *linotte* gene. Insertions in this case were highly localized, and most occurred within a 36-bp fragment of the *linotte* regulatory region. *Hermes* homing indicates that the phenomenon is not element-specific, but may be a general characteristic of Class II elements. Guimond *et al.* (2003) suggested that homing was a special case of local hopping, and the physical proximity between donor elements and target sites seems to underlie the phenomenon of local hopping. The presence of transgene regulatory sequences (e.g., *actin5C* 5′ region) may promote tethering of the donor elements to similar regulatory regions via proteins with common DNA-binding sites. Deliberate tethering of transposable elements to selected sequences may be a means to regulate target-site selection and to minimize the detrimental mutagenic effects of transposable element integration (Bushman, 1994; Kaminski *et al.*, 2002).

The post-integration behavior of the same autonomous *Hermes* element described above in *Ae. aegypti* had quite different characteristics. In this case germ-line transposition of the autonomous *Hermes* element was never detected, and it should be noted that the primary integration events in the germ-line involved the integration of DNA sequences flanking the element (Jasinskiene *et al.*, 1998, 2000). Despite the fact that the element was intact and functional transposase was expressed, the element was immobile in the germ-line. This was not the case, however, in the soma of *Ae. aegypti*, where *Hermes* excision and cut-and-paste transpositions were readily detected. Transposition events in the soma had all the hallmarks of Class II cut-and-paste integration. Only those sequences precisely delimited by the ITRs moved, and integration resulted in the creation of 8-bp direct duplications at the target site. Excision of *Hermes* was imprecise, and led, in some cases, to the creation of small deletions. The basis for the difference in behavior of the *Hermes* element in the germ-line versus the somatic tissue of *Ae. aegypti* is unknown. Clearly the post-integration behavior of *Hermes* in this species will influence how this element will be employed, and in situations where germ-line stability is essential *Hermes* will be particularly useful. It will not be useful, in its present form, for constructing gene-finding tools such as enhancer and promoter traps that rely heavily on transposable element vector remobilization to be effective.

4.5.1.6. Extrachromosomal forms of *Hermes*   Excision of *Hermes* in *M. domestica*, and autonomous *Hermes* elements in *D. melanogaster* and *Ae. aegypti*, leads to the formation of circularized *Hermes* elements in which the terminal inverted repeats are covalently jointed end-to-end in various ways following the excision reaction (O'Brochta *et al.*, 2009). The most common configuration results in the ends being joined end-to-end with a short spacer sequence between them. The spacer sequence was most often 1, 3, or 4 bp, but could also be as much as 200 base pairs. The extrachromosomal *Hermes* elements found in *M. domestica* are particularly interesting, because they have been found in all populations tested and in great abundance in somatic tissue. These data provide evidence for the somatic activity of *Hermes* in the insects from which it was originally isolated. Circularized forms of excised transposable elements of a number of types have been reported in the past (Sundraresan and Freeling, 1987). For example, circularized forms of *Ac/Ds* have been described, as well as *Minos* (Arca *et al.*, 1997; Gorbunova and Levy, 1997), yet the significance of extrachromosomal forms of transposable elements has remained unclear. In some cases the circularized elements do not contain intact terminal inverted repeats, and consequently the elements are not expected to be integration-competent. Based on rather limited data, it has generally been concluded that such forms represent byproducts of aborted or interrupted transposition reactions. A recent study of the extrachromosomal forms of *Hermes* suggests that these elements may have some biological significance. Some circularized forms of *Hermes* elements with intact inverted terminal repeats were found to be capable of integrating into the genome of *D. melanogaster*, indicating that they could contribute to forward transposition (O'Brochta *et al.*, 2009). The ability of circularized forms of excised *Hermes* elements to reintegrate may impact the potential

of this element to be transmitted both vertically and horizontally. Circular, extrachromosomal forms of *Hermes* were readily detected in unfertilized eggs of *M. domestica* that contain native genomic copies of the *Hermes* element, and in *D. melanogaster* that contain active autonomous *Hermes* elements, strongly suggesting that they are capable of being transmitted maternally. Clearly, maternal transmission of active, integration-competent extrachromosomal forms of *Hermes* has the potential to facilitate an increase in frequency of the transposable element within populations; however, to date there are no data regarding the significance of extrachromosomal forms on element dynamics in populations.

4.5.1.7. *hAT* elements in other insects   *hAT* elements have been widely detected in insect genomes. The Queensland fruit fly *Bactrocera tryoni* contains members of at least two distinct *hAT*-like transposable elements (Pinkerton *et al.*, 1999). *Homer* is a 3789-kb element whose sequence is 53% identical to *Hermes* and 54% identical to *hobo*. The transposase coding region is approximately 53% identical and 71% similar to the transposases of *Hermes* and *hobo*, respectively. Similarly, the ITRs of *Homer*, which are 12 bp in length, are identical to those of the *hobo* and *Hermes* elements at 10 of 12 positions. There are also *Homer*-like elements within *B. tryoni*. There are fewer than 10 copies per genome, and, while these elements have not been fully characterized, a conceptual translation of the transposase of this *Homer*-like element reveals 48% identity and 66% similarity to the transposase of *Homer*. These *Homer*-like elements are as similar to *hobo* as they are to *Homer*. Although *Homer* appears to be weakly functional in *D. melanogaster*, based on plasmid-based excision assays, all *Homer*-like elements contain inactivating frameshift mutations.

The Australian sheep blowfly *L. cuprina* contains a non-functional *hAT* element called *hermit*. *Hermit* was initially found by low stringency hybridization screening of an *L. cuprina* genomic library using a DNA probe homologous to *hobo* (Coates *et al.*, 1996). *Hermit* is 2716 bp in length and contains perfect 15-bp ITR, the distal 12 of which are identical to the *hobo* ITRs at 10 of 12 positions. Although inactive because of frameshift mutations within the transposase coding region, its amino acid sequence is 42% identical and 64% similar to *hobo* transposase. *Hermit* is unusual in that it is present as a unique sequence within *L. cuprina*, in contrast to multiple copies that exist for most transposons. Although present only once within this species, it does appear to have arisen within the genome as a result of transposition, since the existing copy of the element is flanked by an 8-bp direct duplication of a sequence that is similar to the consensus target site duplication derived from other *hAT* elements. *Hermit* appears to have become inactivated soon after integrating into the *L. cuprina* genome.

Several *hAT* elements have been discovered in tephritid fruit flies, using a PCR approach similar to that used to discover *Hermes* (Handler and Gomez, 1996). Of these elements, a complete *hAT* transposon (*hopper*) was isolated from a genomic library of the wild Kahuku strain of the Oriental fruit fly *B. dorsalis*, using the Bd-HRE PCR product as a hybridization probe (Handler and Gomez, 1997). A complete 3120-kb element was isolated, having 19-bp ITRs; however, the putative transposase-coding region is frameshifted and does not have a duplicated 8-bp insertion site, suggesting that it had accumulated mutations and was non-functional. The Kahuku sequence was used to isolate additional *hopper* elements using an inverse and direct PCR approach, and a new 3131-bp *hopper* was isolated from the *B. dorsalis white eye* strain (Handler, 2003). This element has an uninterrupted coding region and an 8-bp duplicated insertion site consistent with possible function. Preliminary experiments in which transformants have been generated in *D. melanogaster* and *A. suspensa* using a *hopper*[we] vector marked with DsRed, and an *hsp-hopper*[we] helper, support autonomous function for the *hopper*[we] element (A. Handler and R. Harrell, unpublished).

Notably, *hopper* is highly diverged from all other known insect *hAT* elements, and its transposase is distantly, yet equally, related to the coding regions of *hobo* and *Ac*. Of the terminal 12 nucleotides only 5 are identical to those of *hobo*, while 6 are identical to the ITRs of *Homer*. Handler and Gomez (1996) inferred the presence of an active *hobo*-like transposable element system in the Mediterranean fruit fly *Ceratitis capitata*, because non-autonomous *hobo* elements from *D. melanogaster* were active in excision assays performed in medfly embryos in the absence of any experimentally provided *hobo* transposase. The element *Cchobo* was subsequently isolated, and its transposase coding region was found to be 99.6% and 73.3% identical to *hobo* and another *C. capitata hAT* element, *CcHRE*, respectively (Gomulski *et al.*, 2004). No subsequent tests of *Cchobo's* functionality as an insect gene vector have been reported.

hAT elements have been also reported in the human malaria vector, *An. gambiae*. Approximately 25 sequences resembling *hAT* transposases were discovered, although none appeared to be part of an intact transposable element. More recently, however, genomic DNA database-search criteria were used based on unique aspects of *hAT* transposable elements such as length and spacing of inverted terminal repeats, and the characteristics of *hAT* element target sites. This search revealed a *hAT* element *An. gambiae* that contained perfect 12-bp ITRs flanked by 8-bp direct duplications and a 603 amino acid transposase open reading frame that appeared to contain no internal stop codons. This element (*Herves*) is most closely related to *hopper* and is transpositionally active in *D. melanogaster* (Arensburger *et al.*, 2005). Subramanian *et al.* (2007) examined copy-numbers, integration-site polymorphisms, and nucleotide diversity of *Herves* in individual *An. gambiae* collected largely in East Africa, and concluded that *Herves* appears to have been introduced into this mosquito lineage prior to the recent diversification of species that now form the *An. gambiae* species complex. Integration-site polymorphism data are consistent with the element having been active in the recent past, although the authors did not test whether the element is active in contemporary populations of *An. gambiae*.

hAT transposable elements are well-represented in the genus *Drosophila*; Oritz and colleagues reported finding multiple new *hAT* elements in the genomes of 10 species of *Drosophila* for which whole-genome DNA sequence was available (Ortiz and Loreto, 2009; Ortiz *et al.*, 2010). Two of the elements discovered by these investigators were related to the *Herves* element from *An. gambiae*, and, based on the apparent structural integrity of these elements, they were thought to be active (Depra *et al.*, 2010). More extensive surveys for the presence of these *Herves*-like sequences (called *hosimary* by Depra *et al.*, 2010) in 52 species in the family Drosophilidae revealed the presence of *hosimary* in members of the *melanogaster* species group and in distantly related *Zaprionus indianus*. The high degree of sequence similarity among *hosimary* elements in *Drosophila* and *Zaprionus* suggests that horizontal transfer may have played a role in the history of these elements. Other examples of *hAT* elements within drosophilids having high sequence similarity and discontinuous interspecific distributions, suggesting horizontal transfer, have also been reported. For example, Mota *et al.* (2010) studied the evolution of *hAT* elements related to *Homo3* from *Drosophila mojavensis*, and *Howilli3* from *Drosophila willistoni*, in 65 species of drosophilids, and found a high degree of DNA sequence similarity among elements isolated from different subgenera. These authors suggested that horizontal transfer was the best explanation for their observations. Although much of the study of insect *hAT* elements has been performed in Diptera, other orders of insects also harbor *hAT* elements, but none have been used as a gene vector (Borsatti *et al.*, 2003).

4.5.2. *piggyBac*

4.5.2.1. Discovery of *piggyBac* and other TTAA-specific elements
Similar to several other insect transposable element systems, the *piggyBac* element was discovered fortuitously in association with a mutant phenotype. However, unlike all the other transposons used for insect transformation, the mutant phenotype was the result of a functional element that had transposed into an infectious organism. Fraser and colleagues (see Fraser, 2000) discovered several FP (Few Polyhedra) mutations in the baculoviruses *Autographa californica* nucleopolyhedrovirus (AcNPV) and *Galleria mellonella*

nucleopolyhedrovirus (GmNPV) after passage through the *Trichoplusia ni* cell line, TN-368 (Fraser *et al.*, 1983, 1985). Among these elements that inserted specifically into tetranucleotide TTAA sites was *piggyBac* (then named IFP2), which transposed into AcNPV. Although it might be assumed that IFP2 was an autonomous functional element based on its mobility, another TTAA insertion-site element, *tagalong* (then called TFP3), discovered in AcNPV and GmNPV, was later found not to have an uninterrupted transposase coding region, and thus had to be mobilized by another functional TFP3, or related element. Autonomous functional elements have not yet been found for *tagalong*, though the original IFP2 *piggyBac* element was indeed functional (Wang *et al.*, 1989; Wang and Fraser, 1993). All the *piggyBac* elements discovered in TN-368 were found to be identical to IFP2, having a length of 2472 kb with 13-bp perfect ITRs and 19-bp subterminal repeats located 31 bp from the 5′ ITR and 3 bp from the 3′ ITR (Cary *et al.*, 1989; see **Figure 2**). Notably, five *piggyBac*-like elements have been isolated from *T. ni* larval genomic DNA, but thus far none have been found to be identical to IFP2 (Zimowska and Handler, 2006).

The IFP2 transposase coding region exists as a single reading frame of 2.1 kb that encodes a protein with a predicted molecular mass of 64 kDa. The functionality of *piggyBac* and the precise nature of its transposition was further verified by a series of viral and plasmid transposition and excision assays. A *piggyBac* indicator plasmid marked with *polh/lacZ* was used in *Spodoptera frugiperda* SF21AE cell line assays which showed that the original *piggyBac* element, within the p3E1.2 plasmid, could mobilize the marked element. These assays proved that the 3E1 *piggyBac* element encoded a functional transposase, and defined the element's TTAA insertion-site specificity and the precise nature of its transposition. Importantly, these assays also showed directly that *piggyBac* could be mobilized in other lepidopteran species (Fraser *et al.*, 1995), indicating that it might function similarly as a vector for germ-line transformation. This was a critical realization given the failure of *P* to be mobilized in non-drosophilids, which was consistent with its failure as a vector in these species.

The highly precise nature of *piggyBac* transposition is unique among known transposons, many of which excise in a fashion that leaves staggered ends at the donor site. The necessary filling-in of these ends for target joining and gap repair by DNA synthesis often results in mutations, and this can be a desired effect for transposon-induced mutagenesis. Recent *in vitro* tests in yeast, using purified transposase protein, indicate that *piggyBac* excision results in complementary TTAA overhangs at the donor site, which are restored precisely by ligation without the need for DNA synthesis (Mitra *et al.*, 2008). The mechanism for this is similar to that used by the widespread

DDD/DDE recombinase transposons, which was surprising, given their lack of sequence similarity to *piggyBac*. Mutations affecting the *piggyBac* D268 and D346 residues, however, suggest that they have Mg^{2+}-dependent catalytic function. The actual mechanistic relationship between *piggyBac* and the DDD/DDE transposon family, as well as other aspects of its transpositional activity, await further study. Given the growing importance of *piggyBac* to genetic transformation of a wide array of organisms, and especially its potential for gene therapy in human stem cells (Feschotte, 2006), this knowledge should be rapidly forthcoming.

4.5.2.2. *piggyBac* transformation The failure of *P* vectors to transform non-drosophilid species made the testing of other available transposon systems a high priority. The other systems found to be functional in non-drosophilids, however, were first successfully tested for gene-transfer vector function in *Drosophila*. For *piggyBac*, germ-line transformation was first attempted in the Mediterranean fruit fly *C. capitata*. This was possible due to the availability of a marker system that had been previously tested by medfly transformation with the *Minos* transposon vector. The medfly *white*⁺ gene cDNA was linked to the *Drosophila hsp70* promoter, and used as a mutant-rescue system in a *white eye* host strain (Loukeris *et al.*, 1995a; Zwiebel *et al.*, 1995). In the absence of data for the minimal sequence requirements for *piggyBac* mobility, the first *piggyBac* vector was constructed by insertion of the 3.6-kb hsp-*white*⁺ cDNA marker into the unique *Hpa*I site within *piggyBac* in the p3E1.2 plasmid. None of the *piggyBac* sequences was deleted, although the insertion interrupted the coding region and eliminated the expression of functional transposase. Construction of the first *piggyBac* helper plasmid involved a simple deletion of the 5′ terminal inverted repeat resulting from a *Sac*I digestion and religation of p3E1.2. There is some uncertainty as to whether the upstream *Sac*I site cuts within the *piggyBac* promoter (Cary *et al.*, 1989), yet transposase expression was indeed sufficient to support germ-line transpositions from the vector plasmid. The first experiment with this helper in medfly resulted in one transgenic line at a transformation frequency of 5% per fertile G_0; however, sibling sublines exhibited two and three independent integrations (Handler *et al.*, 1998). This experiment with a *piggyBac*-regulated helper was repeated, and five additional G_1 lines were isolated at approximately the same frequency (5% per fertile G_0). These attempts at *piggyBac* transformation yielded relatively low transformation frequencies, but it was notable that a lepidopteran transposon vector system had autonomous function in a dipteran species.

Subsequent to the medfly transformation, *piggyBac* transformation was tested in *Drosophila* using the mini-*white* marker from that species (Handler and Harrell,

1999). Using the self-regulated helper, transformants were isolated at a similar frequency of 1–3%, but tests with a *hsp70*-regulated transposase increased the frequency to above 25%, consistent with *P* and *hobo* transformations using heat shock promoted transposase.

4.5.2.2.1. Dipteran transformations

With a more highly effective helper, the *white*$^+$-marked *piggyBac* vector tested in medfly was subsequently tested in a *white eye* mutant strain of another tephrtid pest, the oriental fruit fly *Bactrocera dorsalis* (Handler and McCombs, 1998). Although the transformation frequency of *B. dorsalis* was somewhat lower than that observed in *Drosophila* using the same phspBac helper (26%), it was discovered that *B. dorsalis* genome contained multiple *piggyBac*-like elements that might have had a repressive effect on *piggyBac* transposition. Several other tephritid pest species have since been transformed with *piggyBac* vectors containing fluorescent protein expressing transgenes, including the Caribbean fruit fly *Anastrepha suspensa* and the Mexican fruit fly *A. ludens*. The transformation of *A. suspensa* was the first non-drosophilid transformation to use a fluorescent protein expression transgene as a marker (PUb-nls-EGFP; Handler and Harrell, 2000). The transformation of *A. ludens* was done with vectors allowing post-integration stabilization and sperm-specific marking using a *beta2-tubulin*-regulated DsRed marker (Condon *et al.*, 2007; Zimowska *et al.*, 2009; Meza *et al.*, 2010). Other tephritids that have been transformed include the Queensland fruit fly *Bactrocera tryoni* (Raphael *et al.*, 2010), and recent transformations of medfly and caribfly created insects with sperm-specific expression of DsRed (Scolari *et al.*, 2008; Zimowska *et al.*, 2008) or site-specific recombination sites, *attP* (Schetelig *et al.*, 2009, 2010).

Other dipteran species transformed with *piggyBac* include several of medical and agricultural importance, such as the mosquitoes *Ae. aegypti* (Kokoza *et al.*, 2001; Lobo *et al.*, 2002), *An. gambiae* (Grossman *et al.*, 2001), *An. albimnaus* (Perera *et al.*, 2002), *An. stephensi* (Nolan *et al.*, 2002), and *Aedes fluviatilis* (Rodrigues *et al.*, 2006). *An. gambiae* was of particular interest, given its medical importance as a major malaria vector, and its relatively low transformation frequency of ~1% in initial experiments was discouraging (with many anecdotal reports of failure). A more recent transformation yielded several founder transformant lines at a frequency range of 4–18%, representing a considerable improvement (Lombardo *et al.*, 2009). This experiment took advantage of improvements in mosquito transformation methodology discussed in detail by Lobo *et al.* (2006). Other successful *piggyBac*-mediated transformations of dipterans include *M. domestica* (Hediger *et al.*, 2001), *Lucilia cuprina* (Heinrich *et al.*, 2002), *L. sericata* (Concha *et al.*, 2010), the New World screwworm *Cochliomyia hominivorax* (Allen *et al.*, 2004; Handler *et al.*, 2009), and a number of *Drosophila* species (Holtzman *et al.*, 2010).

4.5.2.2.2. Lepidopteran transformations

Given that *piggyBac* was first isolated from a lepidopteran species, there was some optimism that it would be functional as a vector in other moth species that had not yet been transformed using the other transposon vectors originally discovered in dipteran species. Function was first tested by transposition assays in the pink bollworm *P. gossypiella* (Thibault *et al.*, 1999), which then led to successful germ-line transformation of this species using the phspBac helper and a vector marked with EGFP regulated by the *Bombyx actinA3* promoter (Peloquin *et al.*, 2000). Concurrent experiments were also performed in the silkmoth *B. mori* using a similar *actinA3*-regulated EGFP marker, but for this species transformation was achieved with an *actinA3*-regulated transposase helper (Tamura *et al.*, 2000). *Bombyx mori* has since been transformed routinely for a variety of studies, making it the most widely transformed species with *piggyBac*-based vectors. Some of the transformed lines include a UAS-Gal4 gene expression system (Imamura *et al.*, 2003), an enhancer-trap system (Uchino *et al.*, 2008), an inheritable heat shock inducible RNAi system (Dai *et al.*, 2007), and the production of a recombinant Spider dragline silk (Wen *et al.*, 2010). More recent lepidopteran *piggyBac* transformations include the codling moth *Cydia pomonella* (Ferguson *et al.*, 2010), *Plutella xylostella* (Martins *et al.*, 2010), and the first transformation of a butterfly, *Bicyclus anynana* (Marcus *et al.*, 2004).

4.5.2.2.3. Coleopteran, Hymenopteran and Orthopteran transformations

Other insect species transformed with *piggyBac* vectors include the coleopterans *T. castaneum* (red flour beetle; Berghammer *et al.*, 1999; Lorenzen *et al.*, 2003), and *Harmonia axyridis* (ladybird beetle; Kuwayama *et al.*, 2006). The only hymenopteran transformed thus far is the sawfly *Athalia rosae* (Sumitani *et al.*, 2003). For *Tribolium*, *piggyBac* vectors have been used for large-scale enhancer-trap screens (Lorenzen *et al.*, 2007) as well as large-scale insertional mutagenesis (Trauner *et al.*, 2009), which should be invaluable to the functional genomic analysis of this species. In *Harmonia*, effective RNAi activity against a GFP marker was demonstrated.

4.5.2.2.4. General considerations for piggyBac transformations

Notably, many of the species transformed with a *piggyBac* vector used a helper regulated by the *Drosophila hsp70* promoter, and with vectors marked with EGFP, though other fluorescent proteins have since been used for some as well. Although most of these transformations occurred at frequencies of between 3% and 5% per fertile G_0, dramatic differences between species have been observed, and in some of the same species performed by different laboratories. As noted, a single transformant line was reported for *An. gambiae*, at a frequency of approximately 1% (Grossman *et al.*, 2001), while transformation in *An. albimanus* occurred at frequencies ranging from 20% to 40% (Perera *et al.*, 2002). The first transformations of *Tribolium* occurred at the unusually high frequency of 60% (Berghammer *et al.*, 1999).

Many of the transformations were preceded by testing *piggyBac* function by embryonic transposition assays, which were first developed for *piggyBac* mobility in the pink bollworm (Thibault *et al.*, 1999). As discussed previously, these assays can rapidly assess the relative mobility of *piggyBac* in a specific host species in a few days. Positive results from these assays provided some assurance that more tedious and time-consuming transformation experiments had some likelihood of success. For some studies the assays were also used to test promoter function in helper plasmids, or provide insights into insertion site specificity, or determine the likelihood of a particular vector construct retaining function in the absence of specific sequences (Lobo *et al.*, 2001). For example, *piggyBac* helper promoters were tested by transposition assays and germ-line transformation in *D. melanogaster* and *L. cuprina* (Li *et al.*, 2001a; Heinrich *et al.*, 2002). It was found that in *Drosophila* an *hsp70*-regulated helper yielded the highest transposition frequency, while a constitutive *α1-tubulin*-regulated helper was more effective for germ-line transformation. By comparison, in *Lucilia* the *hsp70* helper was most effective for both plasmid and germ-line transpositions, while the *Drosophila α1-tub*-helper failed to support transformation. More recent comparisons of promoter function in *Lucilia* have been based on germ-line transformation, showing that the *L. cuprina hsp83* promoter is more effective relative to the *Drosophila hsp70* for both transposase and ZsGreen marker activity (Concha *et al.*, 2010). Transposition assays have also shown target site preferences among the TTAA sites within the pGDV1 target plasmid, and assays in *Drosophila* showed a bias for sites having A or T nucleotides at positions −3, −1, +1, and +3 relative to TTAA (Li *et al.*, 2001a). However, a sequence analysis of 45 genomic integration sites in *Tribolium*, after *piggyBac* vector remobilization, failed to show this bias (Lorenzen *et al.*, 2003), which may be an indication of variances between plasmid and chromosomal transpositions, or species specificity for insertion site preference.

Mobility assays also provide a rapid means of testing sequence requirements for vector mobility, and allow modifications for more efficient vector function. Since vector mobility is known to be negatively affected by increasing size, this information should allow minimal vectors to be created that retain optimal function. However, minimal sequence requirements for plasmid transpositions may differ from those for chromosomal transposition. For example, excision and transposition assays performed in *T. ni* embryos showed that the *piggyBac* inverted terminal repeat and subterminal repeat sequences were sufficient for transposition (35 bp from the 5′ terminus and 63 bp from the 3′ terminus), but that an outside spacer region between the ITRs of greater than 40 bp is necessary for optimal transposition from a plasmid (Li *et al.*, 2001b). Use of similar vectors in *Drosophila*, however, did not result in germ-line transformants (A. Handler, unpublished). The minimal sequence requirements for *piggyBac* transformation verified thus far for *Drosophila* are 300 bp from the 5′ terminus and 250 bp from the 3′ terminus (Li *et al.*, 2005).

4.5.2.3. Post-integration behavior of *piggyBac*

The post-integration behavior of *piggyBac* vectors has been investigated in a number of species, including *Drosophila melanogaster*, *Ceratitis capitata*, *Tribolium castaneum*, *Bombyx mori*, and *Aedes aegypti* (Thibault *et al.*, 2004; Lorenzen *et al.*, 2007; Sethuraman *et al.*, 2007; Uchino *et al.*, 2008; Schetelig *et al.*, 2009; Trauner *et al.*, 2009). In *Drosophila*, *piggyBac* has been used extensively to generate insertions in a large number of genes throughout the genome (Thibault *et al.*, 2004). These studies have shown that *piggyBac* can remobilize in this species and that it has integration site preferences that are complementary to the widely used *P* element, increasing its value as a functional genomics tool. Notably, *piggyBac* was a more effective gene-disruption tool than *P* elements because, unlike *P* elements, they did not preferentially insert into the 5′ region of genes. Remobilization in the germ-line of *D. melanogaster* was efficient when transposase was provided by a transposase open reading frame located on a chromosome and regulated by a promoter active in germ cells. New transposition events were recovered from 60–80% of the germ-lines tested. Consequently, these investigators were able to generate over 18,000 *piggyBac* insertions. Although *piggyBac* can be remobilized, excision always results in the perfect restoration of the chromosome to its pre-integration state. This is a unique aspect of the *piggyBac* system and quite unlike all other insect gene vectors, in which excision often results in small, and sometimes large, perturbations of the genomic sequences around the site of element excision. Although excision-mediated addition or deletion of sequences can be a useful way of creating allelic series, this is not an option with *piggyBac* because of its tendency to excise precisely. Although Thibault *et al.* (2004) reported efficient remobilization, they used a limited number of initial elements to generate the 18,000 transpositions recovered during their experiment. If a somewhat larger sample of integrated *piggyBac* elements is examined, one finds that the rates of *piggyBac* remobilization in *D. melanogaster* vary widely and depend greatly on where in the genome the element is located. Esnault *et al.* (2010) measured the remobilization activity of 20 identical *piggyBac* elements on the X-chromosome and found that excision/transposition activities varied over two orders of magnitude, though almost all of the variance observed was due to chromosomal position effects. The effects of the vector's position in the genome also affected the levels of gene expression from genes within the vector, but these effects were not correlated with levels of vector remobilization. Esnault *et al.* (2011) also showed that no more than approximately 500 bp of flanking chromosomal DNA are responsible for the observed position-dependent variance in element activity. An element could be transplanted to other genomic positions and would retain its original remobilization activity as long as 500–1000 bp of the original flanking chromosomal DNA accompanied the transplanted element. Thus, these authors found that *piggyBac* was sensitive to its local DNA sequence context, and that this context-effect was portable within the genome. In addition, they found that the context effect was also portable to plasmids. *piggyBac* elements in high mobility contexts were more efficient gene vectors in *D. melanogaster* than identical elements in low mobility contexts (Esnault *et al.* in press).

Lorenzen *et al.* (2007) recovered transposition events from 97% of the germ-lines of transgenic *T. castaneum* containing a single *piggyBac* element and a chromosomal source of *piggyBac* transposase. Using a similar strategy, Trauner *et al.* (2009) produced over 6500 new *piggyBac* insertions in *T. castaneum* as part of a large-scale effort to identify genes involved in development. Although the phenomenon of local hopping was not reported for *piggyBac* transposition in *D. melanogaster*, it was observed in *T. castaneum* (Thibault *et al.*, 2004; Trauner *et al.*, 2009). While active and efficient gene vectors are essential for such large-scale gene finding efforts, equally important is the ability to rear and maintain large numbers of unique genetic lines. This is not possible for many insect species.

With similar interests in using *piggyBac* remobilization as a tool for identifying genes through enhancer trapping and insertional mutagenesis, Uchino *et al.* (2008) were able to create *B. mori* lines ubiquitously expressing *piggyBac* transposase, and lines containing enhancer-trap constructs. They found that the average maximum frequency of transposition was approximately 42%. Although they only generated a small number of lines (105) relative to similar studies with *D. melanogaster* and *T. castaneum*, it appeared that *piggyBac* did not prefer to transpose locally, and that it did appear to prefer intergenic regions and repetitive DNA over coding and genic sequences. Nonetheless, *piggyBac* proved to be an effective tool for identifying genes, based on enhancer trapping in this lepidopteran species.

In the Mediterranean fruit fly *C. capitata*, integrated *piggyBac* elements can also be remobilized when supplied with functional *piggyBac* transposase (Schetelig *et al.*, 2009). These investigators were not remobilizing *piggyBac* for the purposes of gene-finding, but as part of a strategy for stabilizing integrated transgenes in which excision of the element left a previously integrated transgene that was no longer flanked by functional terminal inverted repeats of the *piggyBac* element (Schetelig *et al.*, 2009). Remobilization was stimulated by injecting transposase expressing plasmids into presumptive germ cells and screening for element mobility in the next generation.

Although *piggyBac* elements appear to have great potential in insects for being used in applications requiring the remobilization of integrated elements (enhancer/promoter trapping, mutagenesis, transgene stabilization, gene drive), their behavior in *Aedes aegypti* is notably different. Sethuraman *et al.* (2007) attempted to remobilize the *piggyBac* elements in five transgenic lines of *A. aegypti* by introducing, through genetic crosses, chromosomally-located *piggyBac* transposase genes. Testing multiple combinations of *piggyBac* reporter elements and transposase-expressing transgenes, these investigators failed to detect any evidence of *piggyBac* transposition. This unexpected stability of *piggyBac* following its integration into the genome of *Ae. aegypti* was confirmed in transgenic *Ae. aegypti* cell lines which contained integrated *piggyBac* elements and were transfected with plasmids containing the same *piggyBac* transposase gene that had been integrated into the genome and used by Sethuraman *et al.* (2007) (D. O'Brochta and Palavasam, unpublished data). Although there was no remobilization of chromosomally located *piggyBac* elements in *Ae. aegypti* cell lines in the presence of *piggyBac* transposase, plasmid-borne *piggyBac* elements could remobilize (excise) under these conditions, confirming the presence of functional transposase. These data suggest that the chromosomal context of integrated elements is playing an important role in determining their potential to remobilize in *Ae. aegypti*, which is consistent with the results of Esnault and colleagues (in press). Interestingly when *piggyBac* elements and 1 kb of flanking chromosomal DNA were transplanted from *Ae. aegypti* into the genome of *D. melanogaster*, they were now capable of high levels of remobilization activity (excision and transposition) (A. Palavasam, C. Esnault and D. O'Brochta, unpublished), confirming the functionality of the integrated elements that were formerly in *Ae. aegypti*, and suggesting that the local context effect described by Esnault *et al.* (in press) is species-specific.

4.5.2.4. Phylogenetic distribution of *piggyBac* and implications for transgene stability

Similar to other transposons used for transformation, *piggyBac* is a member of a larger family (or superfamily) of related elements, such as the *mariner/Tc1* or *hAT* families. The *piggyBac* superfamily includes *piggyBac*-like elements that are highly similar to *piggyBac*, as well as more diverged *piggyBac*-related elements (though use of this terminology has not been consistent). *piggyBac*-related elements were first discovered in *T. ni* genomic DNA, where five different, though nearly identical, *piggyBac* elements were discovered and sequenced (Zimowska and Handler,

2006). One of these elements had a single amino acid change in the transposase open reading frame that did not affect the functionality of the protein (G. Zimowska and A. Handler, unpublished). Other elements nearly identical to *piggyBac* were originally identified in the tephritid species *B. dorsalis sensu strictu*, where Southern analysis of transgenic lines and the host strain revealed at least 8–10 *piggyBac*-like elements in the genome (Handler and McCombs, 2000). PCR analysis of these elements from *B. dorsalis s. s.* and more recent sequencing of *piggyBac*-like elements from 14 species throughout the *B. dorsalis* species complex (consisting in total of 70 or more species) have led to the finding that all have 94% or greater nucleotide sequence identity to the original *T. ni piggyBac* (Bonizzoni *et al.*, 2007; Handler *et al.*, 2008). Yet none were found to be identical to IFP2, and only one, from *B. minuta*, was found to have an intact transposase open reading frame that has yet to be proven functional. The isolation of some complete *B. dorsalis* s.s. *piggyBac*-like elements as genomic clones and by inverse PCR indicates that these are complete elements with conserved terminal and subterminal sequences that are integrated into duplicated TTAA insertion sites.

The extensive evolutionary distance between *T. ni* and *Bactrocera* strongly suggests that the transposon moved recently between these species by horizontal transmission, and the separation of their geographical habitats raises the possibility that this movement may have been mediated by intermediary organisms. The likelihod that *piggyBac* elements exist in a wide range of insects, if not other animals, was supported by a database search for related sequences. A Southern blot survey (using IFP2 sequences as probe) for closely related *piggyBac*-like elements in more than 50 species showed the most clear evidence for multiple *piggyBac* elements in the fall armyworm *Spodoptera frugiperda*, but hybridization patterns suggested that most of the elements are defective and non-functional (A. Handler, unpublished; see Handler, 2002b). This was supported by isolation of highly similar *piggyBac*-like sequences in *S. frugiperda*, *Helicoverpa zea*, *H. armigera*, and *Macdunnoughia crassisigna* (Zimowska and Handler, 2006; Wu et al., 2008), with a potentially functional element in *M. crassisigna*. Interestingly, these sequences share more similarities with the *piggyBac*-like elements in Bactrocera than with IFP2 (and the other *T. ni piggyBacs*), suggesting that these elements arose from a distinct lineage (Handler *et al.*, 2008). More highly diverged *piggyBac*-related elements have also been found in the moths Heliothis virescens (Wang *et al.*, 2006), Helicoverpa armigera (Sun *et al.*, 2008), and Pectinophora gossypiella (Wang *et al.*, 2010), with some having elements with uninterrupted transposase open reading frames and intact terminal sequences leaving open the possibility that they are competent to transpose. Indeed, we have found that at least one *piggyBac*-like element discovered in a

larval *T. ni* genome, having a single amino acid residue change relative to IFP2 (Zimowska and Handler, 2006), is functional, based on transformation helper function (G. Zimowska and A. Handler, unpublished). The discovery of functional transposable elements *in vivo* is uncommon, likely due to their creating a genetic load resulting in organismal lethality, and thus mechanisms may exist in *T. ni* to repress *piggyBac* mobility. However, if IFP2 or a predecessor does exist *in vivo*, its presence in a derivative cell line is not surprising.

It is also intriguing to consider how horizontal transmission of *piggyBac* may have occurred, considering that the element was originally discovered by virtue of its transposition into an infectious baculovirus. This could potentially explain a distribution among lepidopterans, but the movement between moths and flies remains a mystery, although baculoviruses are capable of infecting (although not replicating) a wide range of organisms (Laakkonen *et al.*, 2008). Understanding the interspecies movement of *piggyBac*, as well as all other vectors used for practical application, will be critical to understanding and eliminating risk associated with the release of transgenic insects.

4.5.3. *mariner*

4.5.3.1. Discovery, description, characteristics The *mariner* element was first discovered as an insertion element responsible for the *white*-peach (w^{pch}) mutant allele of *D. mauritiana* (Haymer and Marsh, 1986; Jacobson *et al.*, 1986). This particular allele was interesting when discovered because it was highly unstable, with reversions to wild type occurring at a frequency of approximately 10^{-3} per gene per generation. *white*-peach individuals also had a high frequency of mosaic eyes, at an approximate frequency of 10^{-3}, suggesting somatic instability. Molecular analysis of the w^{pch} allele indicated that it was the result of a 1286-bp transposable element insertion into the 5′ untranslated leader region of the *white* gene (Jacobson *et al.*, 1986; see **Figure 2**). The *mariner* element is a Class II type transposable element with 28-bp imperfect inverted repeats with four mismatches. The element recovered from w^{pch} contained a single open reading frame capable of encoding a 346 amino acid polypeptide (Jacobson *et al.*, 1986). While the original w^{pch} was highly unstable, another strain of *D. mauritiana* was discovered in which mosaicism of the eyes occurred in every fly (Bryan *et al.*, 1987). This mosaicism factor was found to be hereditable, and was referred to as *Mos1* (Mosaic eyes). *Mos1* was a dominant autosomal factor on chromosome 3, and was subsequently found to be identical to *mariner* except for six amino acid differences in the putative transposase coding region (Medhora *et al.*, 1988). *Mos1* encodes for a functional transposase, while the 346 amino acid polypeptide of the w^{pch} *mariner* element was not a functional transposase.

One of the most notable characteristics of *mariner* and *mariner*-like elements (*MLE*s) is their widespread distribution. *MLE*s are found not only in insects and invertebrates, but also in vertebrates and plants (Robertson, 2000). Not long after the *D. mauritiana mariner* elements were described, a related element was discovered in the *cecropin* gene of the moth *Hyalophora cecropia* (Lidholm *et al.*, 1991). Based on the sequence comparison between the *mariner* elements from *D. mauritiana* and *H. cecropia*, Robertson (1993) designed degenerate PCR primers and surveyed 404 species of insects for the presence of related sequences (Robertson, 1993). He found that 64 of the genomes examined contained *MLE*s, and within this group are five subgroups referred to as the *mauritiana, cecropia, mellifera, irritans*, and *capitata* subgroups (Robertson and MacLeod, 1993). Since that original analysis insect *MLE*s have continued to be discovered, and currently there are two additional subgroups recognized, known as *mori* and *briggsae* (Lampe *et al.*, 2000). Additional subgroups are likely to be recognized in the future as additional representatives of this family of elements are found. As genome sequence data have accumulated, *MLE*s continue to be discovered (Liu *et al.*, 2004; Zakharkin *et al.*, 2004; Coy and Tu, 2005; Rouleux-Bonnin *et al.*, 2005; Wang *et al.*, 2005; Mittapalli *et al.*, 2006; Ren *et al.*, 2006; Haine *et al.*, 2007; Carr, 2008; Rezende-Teixeira *et al.*, 2008, 2010; Subramanian *et al.*, 2008; Rivera-Vega and Mittapalli, 2010), and a recent analysis resulted in the identification of 15 subgroups within the *mariner* family of transposable elements (Rouault *et al.*, 2009). Elements from different subgroups are typically about 50% identical at the nucleotide sequence level, while the transposases encoded by elements from different subgroups are usually between 25 and 45% identical at the amino acid level. A notable feature of the phylogenetic relationships of the *MLE*s is their incongruence with the phylogenetic relationships of the insects from which they were isolated. The implication is that many of these elements were introduced into their host genome via a horizontal gene transfer event (Robertson and Lampe, 1995a). The abundant examples of horizontal transfer of *mariner* elements have led to the conclusion that such transfers occur relatively frequently. Hartl *et al.* (1997) estimated that the rate of horizontal transmission of *MLE*s is about the same as the rate of speciation, at least within the *D. melanogaster* species subgroup. The widespread occurrence of horizontal transmission of *MLE*s has been proposed to be critical for the long-term survival of these elements. Horizontal transmission provides a means for invading naïve genomes, where element proliferation can occur before inactivating influences of mutation and host regulation can occur (Hartl *et al.*, 1997). This model continues to gain support from data describing the distribution and evolution of *MLE*s in insects (Lampe *et al.*, 2003).

Although hundreds of *MLE*s have been reported, only three (*Mos1* from *D. mauritiana, Himar1* from *Haematobia irritans*, and *Famar1* from *Forficula auricularia*) have been demonstrated to be functional or active. *Haemotobia irritans* contains approximately 17,000 copies of *Himar1*, although all the copies examined were highly defective. Functional elements could be reconstructed based on the consensus sequence of *Himar1*, and then constructed by modifying the closely related *Cpmar1* element from the green lacewing, *Chrysoperla plorabunda*, to match the *Himar* consensus sequence (Robertson and Lampe, 1995b; Lampe *et al.*, 1998). Purification of the transposase from a bacterial expression system and its use in an *in vitro* mobility assay demonstrated the functionality of the *Himar1* protein and the inverted terminal repeats of the element, but the elements were not active in insect cells (Lampe *et al.*, 1996). Taking advantage of the ability of *MLE*s from insects to excise and transpose in bacteria, Barry *et al.* (2004) were able to screen approximately 2000 MLE open reading frames in the *Famar1* group of elements from the earwig, *Forficula auricularia. Famar1* is an abundant *MLE* in *F. auricularia*, with over 40,000 copies per genome. A total of 45 functional transposase open reading frames were discovered, and determining the sequence of 20 revealed unexpected diversity. As many as nine amino acid changes separated the ancestral element and the most diverged functional transposase (Barry *et al.*, 2004). There were also differences in the relative activity of the elements in *E. coli*, ranging over 10-fold. The most active *Famar1* element was twice as active as *Himar1* in *E.coli* (Barry *et al.*, 2004). While *F. auricularia* contains a diverse set of functional *Famar1* transposase coding regions, those proteins appear to be evolving neutrally, consistent with current models of transposable element evolution (Eickbush and Malik, 2002; Barry *et al.*, 2004). The ability of *Famar1* elements to function in insects remains untested.

4.5.3.2. Structure–function relationships The transposases of *MLE*s belong to a large group of integrases and transposases that share a significant feature of their catalytic domains. Specifically, *MLE*s contain the highly conserved D, D, 35E amino acid motif within the active site of the protein, which is part of a conserved protein structure referred to as the integrase fold (Robertson, 2000; Richardson, 2006, 2009). This part of the active site interacts with divalent cations that are essential for catalysis. Transposase binds to the ITRs of the element, and gel retardation assays were used to assess the binding activity of eight mutant transposases with deletions at the N- or C-termini. Mutational and structural studies have led to a detailed understanding of the functional organization of the transposase protein, and a number of important details concerning the mechanism of *Mos1* transposition (Auge-Gouillou *et al.*, 2001a, 2005; Dawson and Finnegan, 2003; Lipkow *et al.*, 2004; Butler *et al.*, 2006; Richardson *et al.*, 2009). They were able to show that amino acids 1–141 were sufficient for binding to the ITRs. The ITR-binding domain of *Mos1* transposase differs somewhat from that of *Tc1* elements in that it is composed of two different structural motifs, a helix–turn–helix motif and an α-helical region (Auge-Gouillou *et al.*, 2001a).

The structural organization of *Mos1* appears to be important in determining the level of activity of the element. For example, the ITRs of *Mos1* are not identical and differ in sequence at four positions, which affects the activity of the element *in vitro*. Auge-Gouillou *et al.* (2001b) reported a 10-fold higher affinity of *Mos1* transposase for the 3′ ITR compared to the 5′ ITR. In addition, modified 5′ ITRs that were made to resemble 3′ ITRs at one of the four variable positions resulted in an increase in transposase binding. These investigators also showed that a *Mos1* element with two 3′ ITRs had 104 times the transposition activity of the native ITRs (Auge-Gouillou *et al.*, 2001b; Sinzelle *et al.*, 2008; Casteret *et al.*, 2009). This hyperactive double-ended configuration has not been tested *in vivo*, and did not result in increased element activity when tested in insects (Pledger *et al.*, 2004).

The detailed molecular understanding of *MLEs* and their movement is beginning to permit rationally designed variants to be created and tested (Germon *et al.*, 2009). Hyperactive transposase mutants of the *Himar1* transposase were reported (Lampe *et al.*, 1999), and one of the mutants contained two amino acid changes (at positions 131 and 137) in the ITR-binding domain of the protein. Although not tested directly, it is possible that these hyperactive mutants result in increased binding of the transposase, and, consequently, higher rates of movement. Paradoxically, neither *Himar1* nor any of the hyperactive mutants showed any transpositional activity in insects (Lampe *et al.*, 2000). The organization of *MLEs* and the transgenes they carry impact the activity of the elements (Casteret *et al.*, 2009). Like most other transposable elements that have been tested, the *MLEs* have preferred integration sites, resulting in their non-random distribution in DNA target molecules (Crenes *et al.*, 2009, 2010).

4.5.3.3. Host range of *mariner*

The widespread distribution of *MLEs* in nature and the frequent examples of their horizontal transfer between species would seem to indicate that these elements have a broad host range. Empirical studies in which *Mos1* has been employed as a gene vector in a wide variety of organisms support this conclusion. *Mos1* has been used successfully to create transgenic *D. melanogaster*, *D. virilis*, *Ae. Aegypti*, and *M. domestica* (Lidholm *et al.*, 1993; Lohe and Hartl, 1996a; Coates *et al.*, 1998; Yoshiyama *et al.*, 2000). In each of these species, the frequency of transformation was approximately 5%. This element has also been

used to create transgenic *B. mori* cells in culture (Wang *et al.*, 2000). In addition to transgenic insects, *Mos1* has been used to create transgenic *Leishmania*, *Plasmodium*, zebrafish, and chickens (Gueiros-Filhos and Beverley, 1997; Fadool *et al.*, 1998; Sherman *et al.*, 1998; Mamoun *et al.*, 2000). Similarly, the *Himar1* element has been shown to function in *E. coli*, Archaebacteria, and human cells (Zhang *et al.*, 1998, 2000; Rubin *et al.*, 1999). *Himar1*, however, has not been shown to be active in *D. melanogaster* or any other insect species, for reasons that are not at all clear (Lampe *et al.*, 2000).

4.5.3.4. Post-integration behavior

The post-integration behavior of *Mos1* has been investigated in *D. melanogaster* and *Ae. aegypti*. *Mariner* gene vectors used to create transgenic *D. melanogaster* have been found to be uncommonly stable, even in the presence of functional transposase. Lidholm *et al.* (1993) created two lines of transgenic *D. melanogaster* with a *mariner* vector derived from *Mos1* and containing the mini-*white* gene as a genetic marker. When these lines were crossed to *Mos1* transposase-expressing lines, eye mosaicism was found in only 1% of the progeny, while these same *Mos1* expressing lines resulted in 100% mosaicism of the w^{pch} element. Similarly, germ-line transposition occurred at rates of less than 1% (Lidholm *et al.*, 1993), and Lohe *et al.* (1995) reported similar evidence for post-integration stability of *mariner* vectors. Lozovsky *et al.* (2002) suggested, after investigating the post-integration mobility of a number of *mariner* vectors containing different genetic markers in different locations within the element, that *mariner* mobility is highly dependent upon critical spacing of subterminal sequences and the inverted repeats. They found that vectors with simple insertions of exogenous DNA of varying lengths and in varying positions showed levels of somatic and germ-line excision that were at least 100-fold lower than that observed with uninterrupted *mariner* elements. Only vectors consisting of two (almost complete) elements flanking the marker gene showed detectable levels of both somatic and germ-line mobility. Approximately 10% of the insects with these composite vectors had mosaic eyes when transposase was present. Germ-line excision rates of approximately 0.04% were observed in these same insects. Again, these values are considerably less than those reported for uninterrupted elements. In addition to the potential importance of subterminal sequence spacing (Lozovsky *et al.*, 2002), Lohe and Hartl (2002) suggested that efficient mobilization of *mariner in vivo* also depends on the presence of critical sequences located quite distant from the inverted repeats. Based on the mobility characteristics of about 20 *mariner* elements with a wide range of internal deletions, they concluded that there are three regions within the element that play an important role in *cis*. Region I is approximately 350 bp in length, and is located 200 bp from the left 5′ inverted terminal repeat. Region II is approximately 50 bp in length, and located approximately 500 bp from the right 3′ ITR. Region III is about 125 nucleotides in length, and located approximately 200 bp from the right ITR (Lohe and Hartl, 2002). While the presence of subterminal sequences that play a critical role in the movement of many Class II transposable elements is not unusual, what is uncommon in the case of *mariner* is the location of these *cis*-critical sequences. Their dispersed distribution within the element is unique, and, consequently, manipulating the element for the purposes of creating gene vectors and associated tools without disrupting these important relationships may be difficult.

The post-integration mobility of *Mos1* can also be regulated by non-structural aspects of the system, including "overproduction inhibition" and "dominant-negative complementation." Increasing the copy number of *Mos1* in the genome resulted in a 25% decrease in the rate of germ-line excision. Copy number increases in *Mos1* presumably lead to increased transposase levels, and, by an unknown mechanism, to the inhibition of excision (Lohe and Hartl, 1996b). High concentrations of transposase may lead to non-specific associations of the protein resulting in inactive oligomers of transposase. In addition, the presence of mutated forms of *Mos1* transposase can repress the activity of functional transposase. Because the transposases of other transposable elements act as dimers or multimers, it is thought that mutated *Mos1* transposases may become incorporated into multimers with functional transposases, thereby inactivating the entire complex (Lohe and Hartl, 1996b).

The possibility that transposase overproduction may negatively affect its own activity is a highly important concept in terms of vector system development. Most systems have the helper transposase under strong promoter regulation to optimize transpositional activity, though this may, indeed, be counterproductive. For *mariner* vectors, and potentially other systems, optimal transformation may require testing various helper promoters and a range of plasmid concentrations.

The post-integration mobility properties of *mariner* were also examined in the yellow fever mosquito *Ae. aegypti* (Wilson *et al.*, 2003). As part of an effort to create an enhancer trapping and gene discovery technology for *Ae. aegypti*, Wilson and colleagues created non-autonomous *mariner*-containing lines, and lines expressing *Mos1* transposase. By creating heterozygotes between these two lines they attempted to detect and recover germ-line transposition events, but only a single germ-line transposition event was recovered after screening 14,000 progeny. Somatic transpositions were detected, and while precise estimates of rates of somatic transposition were not possible because of the detection method, the authors observed fewer than one event per individual, which they estimated to be an indication of a very low rate

of movement. The vectors used by Wilson and colleagues resembled the simple vectors reported by Lozovsky *et al.* (2002) that had apparently disrupted spacing of the ITRs, and partial deletions of *cis*-critical sequences described by Lohe and Hartl (2002).

While the post-integration stability of *mariner* has been described in two species, and appears to be a general mobility characteristic of this element and not a reflection of a species-specific host effect, paradoxical observations remain to be explained. First, the use of *mariner* as a primary germ-line transformation vector in non-drosophilid insects and in non-insect systems is an effective means for creating transgenic organisms. Indeed, the host range of *mariner* as a gene transformation vector is unrivaled by any of the other gene vectors currently employed for insect transformation. *Mariner* has been used as a gene vector in microbes, protozoans, insects, and vertebrates. The rate of germ-line transformation using *mariner*-based vectors in insects is approximately 10% or less, and is comparable to the efficiency of *Hermes*, *Minos*, and *piggyBac* gene vectors. This raises the question of whether *mariner* vectors present on plasmids behave the same as *mariner* vectors integrated into insect chromosomes. Given the rates of germ-line integration from plasmids, it appears that the *mariner* vectors being used are not suffering from "critical spacing/critical sequence" defects. In addition, the behavior of *mariner in vitro* also differs from the behavior of chromosomally integrated elements. Tosi and Beverly (2000) demonstrated that only 64 nucleotides from the left end and 33 nucleotides from the right end of *mariner* were essential for transposition of a 1.1-kb vector *in vitro*. The rate of transposition of a minimal *mariner* vector *in vitro* was only two-fold less than that of a vector containing essentially a complete *mariner* element. These results suggest that *mariner* mobility has relatively simple sequence requirements, and that the role of subterminal sequences is minimal *in vitro*. These apparently conflicting data suggest that host factors may play an important role in the transposition process *in vivo*, and may influence the relative importance of *cis* sequences in the *mariner* transposition process. The broad distribution of *MLEs* and the host range of *mariner/Mos1* suggest, however, that host factors play little role in the movement of these elements.

The post-integration behavior of *mariner/Mos1* seems to indicate that this element will not be a good candidate for developing gene-finding tools such as promoter/enhancer trapping and transposon tagging systems in *Ae. aegypti* and perhaps other insects. On the other hand, if a high level of post-integration stability is desired, then *mariner* is an appropriate element to consider in insects. The potential of this element to be lost through excision or transposition is low, even in the presence of functional *mariner* transposase. As currently configured and used, *mariner* vectors may be considered suicide vectors in insects, since they essentially become dysfunctional upon integration.

4.5.3.5. *MLEs* in other insects   While hundreds of *MLEs* have been described, few have been shown to be functional. The original *mariner* element from the *white*-peach allele was transpositionally competent, although it did not produce a functional transposase. *Mos1* is a functional autonomous element, and has been the basis for constructing all *mariner* gene vectors that function in insects. *Himar1* is a functional element from the *irritans* subgroup that was reconstructed based on multiple sequence comparisons of elements within this group. It has not been shown to be functional in insects, despite significant efforts to do so. Lampe *et al.* (2000) report that at least eight other elements from the other subgroups are likely to be active, or to be made active by minor modifications.

4.5.4. Minos

The first germ-line transformation of a non-drosophilid insect mediated by a transposon-based vector system was achieved with the *Minos* element. *Minos* was originally isolated as a fortuitous discovery in *D. hydei* during the sequencing of the non-coding region of a ribosomal gene (Franz and Savakis, 1991). *Minos* was found to be a 1.4-kb element having, unlike the other Class II transposons used as vectors, relatively long inverted terminal repeats of 255 bp, with its transcriptional unit consisting of two exons (see **Figure 2**). Additional *Minos* elements were isolated from *D. hydei*, having small variations of one or two nucleotides, though the new elements had a transition change that restored the normal reading frame, allowing translation of a functional transposase. The sequence homology, general structure, and TA insertion-site specificity placed *Minos* within the *Tc* transposon family (Franz *et al.*, 1994). *Minos* was first used to transform *D. melanogaster*, with *Minos*-mediated events demonstrated by sequencing insertion sites and remobilization of integrations (Loukeris *et al.*, 1995b). The first non-drosophilid transformation with *Minos* was achieved in a medfly *white eye* host strain using a cDNA clone for the medfly *white* gene as a marker (Zwiebel *et al.*, 1995), at an approximate frequency of 1–3% per fertile G_0 (Loukeris *et al.*, 1995a). *Minos* has since been used with fluorescent protein markers to transform another tephritid pest species, the olive fruit fly *Bactrocera oleae* (Koukidou *et al.*, 2006); the silkworm moth *Bombyx mori* (Uchino *et al.*, 2007); *An. stephensi* (Catteruccia *et al.*, 2000); and *T. castaneum* (Pavlopoulos *et al.*, 2004). Recently, transformation frequencies have been substantially increased in *Drosophila* and medfly by use of *in vitro* synthesized transposase mRNA as a helper (Kapetanaki *et al.*, 2002).

Although *Minos* has been only occasionally used for insect transformation, embryonic and cell-line mobility assays in several insect species in the Diptera, Lepidoptera, and Orthoptera have indicated a broad range of functions.

Notably, *Minos* transposition in the cricket *Gryllus bimaculatus* was driven by transposase regulated by a *Gryllus* actin gene promoter, and not by the *Drosophila hsp70* promoter that has been widely used in dipterans (Zhang *et al.*, 2002). The broad function of the *Minos* vector is further supported by its ability to transpose in a mouse germ-line and in invertebrates (Drabek *et al.*, 2003; Pavlopoulos *et al.*, 2007).

Minos structure places it within the *mariner/Tc* transposon superfamily, though knowledge of the distribution of *Minos* is thus far limited to the genus *Drosophila* (Arca and Savakis, 2000). In *Drosophila*, *Minos* is clearly widely distributed in the *Drosophila* and *Sophophora* subgenus, though discontinuously in the *Sophophora*. As noted for the *hAT*, *mariner*, and *piggyBac* elements, *Minos* may have also undergone horizontal transfer between *Drosophila* species.

4.5.5. Tn5

Tn5 is one of a number of very well-characterized transposable elements from prokaryotes. Recently, hyperactive forms of this element have been created in the laboratory that have proven to be the basis for the development of a number of commercially useful genomics tools (Epicentre, Madison, WI; http://www.epicentre.com) (Goryshin and Reznikoff, 1998). *Tn5*-based genomics tools can be used in a wide variety of bacterial species, and, given the system's independence from host encoded factors, might be applicable to eukaryotic systems as well (Goryshin *et al.*, 2000). Efforts to use *Tn5* as an insect gene vector have been successful.

Tn5 is a prokaryotic transposon, 5.8 kb in length, that is often referred to as a composite transposon because it consists of five independently functional units (for review, see Reznikoff, 2000; see also **Figure 2**). It contains three antibiotic resistance genes that are flanked by 1.5-kb inverted repeat sequences. Each inverted repeat is actually a copy of an *IS50* insertion sequence, which are themselves functional transposons. Each *IS50* element contains 19-bp terminal sequences known as OE (outside end) and IE (inside end), and while OE and IE are very similar, they are not identical. *IS50* also encodes for two proteins: transposase (Tnp) is 476 amino acids long and catalyzes transposition, while the second protein is an inhibitor of transposition (Inh). The *IS50* elements present at each end of *Tn5* are not identical, and only *IS50R* is fully functional. *IS50L* contains an ochre codon that prematurely terminates the Tnp and Inh proteins, resulting in a loss of function of both proteins.

The transposition reaction and all of the components involved in the reaction have been studied in great detail (Reznikoff *et al.*, 1999). Transposition proceeds by a cut-and-paste process involving binding of Tnp to the end sequences followed by dimerization of the bound Tnp to form a synaptic complex. Cleavage at the ends of the element results in an excised transposon with bound transposase that interacts with a target DNA molecule. Strand transfer results in the integration of the element into the target, and, *in vitro*, this reaction requires only a donor element, a target DNA molecule, transposase, and Mg^{2+} (Goryshin and Reznikoff, 1998). Modifications of both the transposase and the terminal 19-bp sequences have led to the creation of *Tn5* elements consisting of little more than two copies of end sequences that can be mobilized 1000-fold more efficiently than an unmodified *Tn5* element. This hyperactive *Tn5* system has been developed into a powerful tool for genetic analysis of a variety of organisms. *Tn5* has been attractive as a broad host-range genomics tool because its pattern of integration is random and its biochemical requirements very simple. *Tn5* has been shown to function in a variety of bacterial and non-bacterial systems.

Current insect transformation protocols consist of microinjecting a mixture of two plasmids into preblastoderm embryos (see section 4.6.3). One plasmid contains a non-autonomous transposable element with the transgenes and genetic markers of interest, while the second plasmid contains a copy of the transposase gene. Transient expression of the transposase gene is required post-injection, and is followed by element excision and integration. Previous experiments examining the frequency of element excision in *Hermes*, *mariner*, *Minos*, and *piggyBac* from plasmids injected into insect embryos along with "helper" plasmids showed that only one plasmid per 1000 injected underwent an excision event. Therefore, 99.9% of the donor plasmids introduced into insect embryos contributed nothing to the transformation efforts. The introduction of pre-excised elements configured as active intermediates, such as synaptic complexes, was considered a means to permit higher integration rates and overall efficiency of transformation.

Transgenic *Ae. aegypti* were created using a *Tn5* vector containing DsRed under 3xP3 regulatory control (Rowan *et al.*, 2004). Pre-excised vectors in the form of synaptic complexes were injected into preblastoderm embryos; 900 adults were obtained from the injected embryos, and families consisting of approximately 10 G_0 individuals were established. Two families of G_0 individuals produced transgenic progeny for an estimated transformation frequency of 0.22% (2/900). Analysis of the transgenic progeny showed that multiple integrations of *Tn5* occurred in each line. The patterns of integrations were complex, with evidence of the *Tn5* vector integrating into *Tn5* vector sequences. The integration of the vector into copies of itself, followed by the integration of the resulting concatamers, was very unusual, and in no case was a simple cut-and-paste integration of the *Tn5* vector found with characteristic 9-bp direct duplications flanking the element. The complex pattern of *Tn5* integration was thought

to be a direct consequence of injecting pre-assembled intermediates that were inactive in the absence of Mg^{2+}. Therefore, as soon as the synaptic complexes were injected they became activated, and the first target sequences the elements were likely to encounter were other *Tn5* synaptic complexes. At the time of injection, *Ae. aegypti* embryos only contained approximately four to eight nuclei, making genomic target DNA relatively rare. Furthermore, the synaptic complexes injected were expected to have a very short half-life. Therefore, although active intermediates were being introduced, a number of factors contributed to the inefficiency observed with this system, including a short half-life of the active intermediate and low numbers of genomic target sequences. Injecting binary plasmid systems (as is done with *Hermes*, *mariner*, *Minos*, and *piggyBac*), while relatively inefficient in producing active transposition intermediates, achieves persistence over an extended period of time. Consequently, more target genomes are exposed to active vectors over a longer period of time, resulting in higher transformation rates. The limitation of injecting synaptic complexes is unlikely to be specific to the *Tn5* system, and similar approaches with other insect gene vectors are likely to encounter similar problems. It should be noted, however, that the results of Rowan and O'Brochta demonstrate that *Tn5* is functional in insects, and, while injecting active intermediates is not recommended, using *Tn5* in a more conventional binary plasmid system consisting of a donor and helper plasmids is likely to be a viable option for creating transgenic insects.

4.5.6. Improved Transposon Vectors for Basic and Applied Transformation

4.5.6.1. Stabilization of transposon vectors Unlike transformation systems, in which genomic integrations result from the recombination of introduced DNA, transposon-based vectors are subject to remobilization by the intended or unintended presence of functional transposase. Vector remobilization is a desired result for insertional mutagenesis protocols that require the repetitive insertion, or "hopping," of the vector after injection of helper plasmid or mating to a jumpstarter strain having a genomic source of transposase (Brand *et al.*, 1994; Horn *et al.*, 2003). However, for the development of transgenic strains for most applied purposes, and especially for programs requiring their field release, stable vector insertions are required, if not essential, for maintaining stable lines and ensuring that transgenes do not move between species by horizontal interspecies transfer. While these may be rare events, horizontal transmission is thought to be a natural phenomenon responsible for the movement and spread of transposable elements among species (Robertson and Lampe, 1995b; Lampe *et al.*, 2003). For non-autonomous vectors whose transposase coding region is deleted or interrupted, such mobilization could be catalyzed *in trans* by an unintended source. This could be from the undetected presence of a functional transposon in the genome of the host insect, but, more dauntingly, by a similar transposon that is less likely to be detected, whose transposase can cross-mobilize the vector. Such cross-mobilization has indeed been experimentally demonstrated for the *hobo* and *Hermes* elements within the *hAT* transposon family (Sundararajan *et al.*, 1999).

While cross-mobilization of a defective non-autonomous vector may be rare in nature, the large population scale and pressures of mass-reared transgenics for biocontrol release programs could be favorable for the selection of such events (Robinson *et al.*, 2004), raising serious ecological concerns for transgenic insect release. Thus, the post-integration stabilization of transposon-based vectors was thought to be a critical need for the applied use of transgenic strains (Handler, 2004). Two similar approaches have taken advantage of mobilization requiring the transposon 5′ (or left arm) and 3′ (or right arm) inverted terminal repeat sequences. If either one or both of the termini are deleted, transposon mobility – or remobilization – is eliminated. This was first achieved in *Drosophila* with a *piggyBac* vector having an internal tandem duplication of the left (L) 5′ terminus, and a single right (R) 3′ terminus in an L1-L2-R1 orientation, with distinguishable marker genes in between L1-L2 and L2-R1. This resulted in a vector, pBac{L1-PUbDsRed1-L2-3xP3-ECFP-R1}, that could have the L1-PUbDsRed1 transgene sequence stabilized by remobilization, or deletion, of the L2-3xP3-ECFP-R1 sequences (Handler *et al.*, 2004). This was possible by first integrating the entire vector into a host genome, with subsequent remobilization of the L2-3xP3-ECFP-R1 subvector by mating to a jumpstarter strain (having phspBac helper integrated in a *Minos* vector). This deleted the ECFP marker along with L2, and the only 3′ (R1) terminus, thus stabilizing the remaining L1-PUbDsRed1 sequences with respect to any unintended source of transposase. Extensive efforts to remobilize the L1-PUbDsRed1 sequences proved their stability, and the same stabilization method was also achieved in the tephritid pest species *A. ludens* (Meza *et al.*, 2010).

The single terminus deletion method was subsequently modified so that none of the original vector sequences remained, by essentially creating a dual vector with intervening DNA sequences in an L1-marker1-R1-marker2-L2-marker3-R2 orientation (Dafa'alla *et al.*, 2006). For this vector, remobilization of both the L1-marker1-R1 and L2-marker3-R2 subvectors resulted in only the marker3 DNA left integrated in the genome, with no transposon sequences remaining to facilitate mobilization. While theoretically an improvement over the single-arm deletion method, the dual-arm deletion method is more cumbersome, and likely offers only a minor advantage in terms of transgene stability.

4.5.6.2. Vectors for gene targeting   Transposon-based vector insertions in a host genome are typically random, though different transposons have varying insertion site sequence biases. This lack of specificity is highly useful for insertional mutagenesis screens such as transposon-tagging and enhancer traps, making this type of transformation a valuable tool for functional genomic analysis. However, there are negative effects of random integrations for both basic and applied studies. Gene expression from different insertion sites often varies due to genomic position and enhancer effects that result in abnormal, often suppressed, gene expression or misexpression with respect to tissue or development. This makes comparative gene expression and functional analyses problematic, if not unreliable, and makes creation of optimal strains for applied use difficult. Furthermore, insertional mutations often have deleterious, if not lethal, effects on transgenic strain viability and fitness, compromising their use for biological control programs (Catteruccia *et al.*, 2003; Irvin *et al.*, 2004).

The disadvantages of random vector insertions can be minimized by targeting transgene integrations to predefined genomic sites devoid of known coding or regulatory function that are minimally affected by position and enhancer effects. Targeting transgene insertions also allows true allelic comparisons of gene expression when studying gene structure–function relationships. The need for targeting motivated the development of site-specific targeting systems in insects based upon the phiC31 integrase (Groth *et al.*, 2004), and the *FRT*/FLP (Horn and Handler, 2005) and *loxP*/Cre (Oberstein *et al.*, 2005) recombinase systems. All of these systems have been tested, initially in *Drosophila*, by introducing recombination sites into the genome with transposon vectors. Secondary insertions were then made, with plasmids having the appropriate recombination sites co-injected with their respective integrase or recombinase helpers. For the applied use of transgenic strains, the strategy would be to create several strains with differing target sites that would be characterized in terms of transgene expression, in addition to strain viability and fitness. Optimal target site strains would then be used for subsequent integrations and genetic manipulations.

Initially, target sites were created by using the same dual heterospecific *FRT* (Senecoff *et al.*, 1985) or *loxP* (Hoess *et al.*, 1985) recombination sites in the target site and donor plasmid. Both systems have recombination sites that consist of two 13-bp inverted repeats separated by an 8-bp spacer that specifically recombine with an identical site in the presence of recombinase. Heterospecific sites have sequence variations in the spacer regions, and since only identical sites can recombine, use of the dual sites allows only reliable double-recombination. This results in the exchange of DNA between the sites in the donor plasmid and the sites in genomic target, also known as recombinase-mediated cassette exchange (RMCE). This was assessed in the *FRT* system by interconvertible fluorescent protein markers (Horn and Handler, 2005) that could also be used in non-drosophilids, while the *loxP* system used *Drosophila* mutant-rescue markers (Oberstein *et al.*, 2005). DNA introduced by *FRT* RMCE contained a terminal *piggyBac* sequence that allowed subsequent stabilization of the target site *piggyBac* vector. While RMCE is highly effective in *Drosophila*, thus far it has not been successfully tested in other insect species, though there is no theoretical reason for it not to be equally effective.

The phiC31 integrase-based system has also been tested in *Drosophila* (Groth *et al.*, 2004), where genomic *attP* sites, introduced by *P* element transformaton, were targeted in a unidirectional fashion in the presence integrase by plasmid vectors having an *attB* site. Targeting occurred at relatively high frequencies, though drawbacks of this system are that the entire *attB*-containing plasmid is inserted into the *attP*-target site, additional heterospecific attachment sites do not exist that might allow re-targeting of the same site, and occasionally genomic "pseudo" *attP* sites may be targeted. Still, using dual *attB* and *attP* sites in the target and donor sequences, an RMCE system has been created (Bateman *et al.*, 2006). However, the sequence exchange is irreversible, though further targeting may be possible using the *FRT* or *loxP* systems. Nevertheless, the fC31 single recombination system has been successfully tested in mosquito species (Nimmo *et al.*, 2006; Labbé *et al.*, 2010) and a tephritid species (Schetelig *et al.*, 2010). In addition, medfly targeting has been used to introduce a *piggyBac* terminus, allowing target site stabilization.

4.6. Transformation Methodology

The technical methodology for insect transformation has largely remained the same or has only been slightly modified from the techniques originally used to transform *Drosophila*. The references cited for *P* transformation are relevant to this, as well as several articles that focus on methods for non-drosophilid transformation (Handler and O'Brochta, 1991; Morris *et al.*, 1997; Handler, 2000). The most variable aspect of this method is the preparation of embryos for DNA microinjection, though, arguably, the lack of new techniques for DNA introduction has been the primary limitation in the more widespread use of the technology. While all successful insect transformations have utilized microinjection, variations on this method have been necessary for different types of embryos, and most of the procedures must be tested empirically and modified for particular insect species. This may be extended to different strains and for a variety of local ambient conditions, including temperature and humidity. The apparatus for microinjection is usually the same for all species, though a wide variety of variations and modifications are possible and sometimes required.

The basic equipment includes an inverted microscope or a stereozoom microscope with a mechanical stage having a magnification up to 60–80×, a micromanipulator that is adjustable in three axes with an appropriate needle holder, and a means to transmit the DNA into the egg. For dechorionated eggs, transmitted light allows precise positioning of the needle within the egg posterior, while direct illumination is needed for non-dechorionated eggs, which typically include those of mosquitoes and moths.

The standard for gene transfer methodology in general, and embryo microinjection in particular, was originally developed for *Drosophila*. The standard method involves collecting preblastoderm embryos within 30 minutes of oviposition, and dechorionating them either manually or chemically. The timing of egg collection and DNA injection is related to the need to inject into pre-blastoderm embryos during a phase of nuclear divisions previous to cellularization. This allows the injected DNA to be taken up into the nuclei, and specifically into the primordial germ cell nuclei that are the gamete progenitors. For *Drosophila*, cellularization of the pole cells begins at approximately 90 minutes after fertilization at 25°C, with blastoderm formation occurring about 30 minutes afterwards. The timing of these events and the location of the pole cells varies among insects, and thus some knowledge of early embryogenesis in the desired host insect is highly advantageous. In the absence of this information for a particular species, the most prudent time of injection would be the earliest time after oviposition that does not compromise viability.

4.6.1. Embryo Preparation

Manual dechorionation of *Drosophila* eggs is achieved by gently rolling the eggs on double-stick tape with a forcep until they pop out. While gentle on the eggs and requiring little desiccation time, manual dechorionation is tedious, and has not been applicable to any other insect. Chemical dechorionation is typically achieved by soaking eggs in a 50% bleach solution (2.5% hypochlorite) for 2–4 minutes and washing them at least three times in 0.02% Triton-X 100. Tephritid fruit fly eggs usually have thinner chorions that can be dechorionated in 30% bleach (1.25% hypochlorite) in 2–3 minutes, but this must be determined empirically, since they are easily over-bleached, resulting in death, either directly or after injection. Some species, such as *M. domestica*, can be only partially dechorionated, but bleached eggs can be released from the chorion by agitation. We have found the simplest and most precise method for bleach dechorionation with rapid washes is by using a 42-mm Buchner funnel with a filter flask attached to a water vacuum. Eggs can be washed into the funnel on filter paper and swirled within the funnel, with the solution gently sucked out by regulating the water flow or the seal between the funnel and flask. The last wash is done on black filter paper that allows the eggs to be easily detected, which facilitates their mounting for injection (see below).

Many insect eggs cannot be dechorionated without a high level of lethality, and must be injected without dechorionation. These include those of most moth and mosquito species. *Drosophila* and tephritid flies can, similarly, be injected without dechorionation, and while embryo viability after injection is often lower than for dechorionated eggs, the frequency of transformation in surviving embryos is often higher. It is more difficult to determine a precise site for injection in non-dechorionated eggs, though this can be aided by adding food coloring to the DNA injection mix.

After dechorionation, fruit fly embryos are typically placed on a thin strip (~1 mm) of double-stick tape placed on a microscope slide or 22 × 30 mm cover slip, though use of a cover slip is more versatile for subsequent operations. A thin strip of tape is suggested due to anecdotal reports of toxic solvents from the tape affecting survival, though some particular tapes are considered to be non-toxic (3M Double Coated Tape 415; 3M, St Paul, MN) and some are useful for particular applications, such as the aqueous conditions needed for mosquito eggs. Adhesives resistant to moisture include Toupee tape (TopStick™, Vapon Inc.) and Tegaderm (3M). When eggs are injected under oil, the tape strip is placed within a thick rectangle, created with a wax pencil, that can retain the oil. It is important that the wax fence is not breached by oil when overlaying the eggs, since the loss of oil will result in embryo death.

Where possible, eggs are placed on the tape in an orientation with their posterior ends facing outwards towards the needle, but at a slight angle. All fruit fly eggs must be desiccated to some extent before injection. The interior of the egg is normally under positive pressure, and yolk and injected DNA will invariably flow out after injection without desiccation. This will result in lethality, sterility (from loss of pole plasm), or the lack of transformation if the plasmid DNA is lost. The time and type of desiccation, however, must be evaluated empirically, and sometimes varied during the course of an injection period. A major factor for dechorionated eggs is the length of time they are kept on moist filter paper before being placed on the tape. Typically, we desiccate embryos on one strip of tape (15–20 embryos) for 8–10 minutes. Depending on the ease of injection, the time can be varied by 1–2 minutes. In ambient conditions that are humid, it may be necessary to desiccate in a closed chamber with a drying agent (e.g., drierite), with or without a gentle vacuum. An important consideration is that a very short variation in the time for desiccation can be the difference between perfect desiccation and over-desiccation resulting in death, and that the optimal desiccation time will vary for different eggs on the tape. Thus, it is unlikely that all the eggs will respond well to the set conditions, which must be modified so that the majority of eggs can be injected

with DNA at a high level of survival and fertility. After the determined time for desiccation, the eggs must be immediately placed under Halocarbon 700 oil, or oil of similar density, to stop the desiccation process. Desiccation of most non-dechorionated eggs is more challenging, and one approach is to soak eggs in 4-M NaCl for several minutes. In contrast, for non-dechorionated mosquito eggs desiccation can occur within 1–2 minutes after removal from water, which is evidenced by slight dimpling of the egg surface, and this must be observed to avoid over-desiccation. Due to the rapidity of desiccation, mosquito eggs are typically arranged on moist filter paper and blotted together onto a taped cover slip from above, and after desiccation the eggs are submerged in Halocarbon oil. Non-dechorionated eggs from many species do not require oil, and it may be lethal for some insects such as moths, yet we find oil submersion helpful for survival of *Drosophila* and tephritid flies.

4.6.2. Needles

The type of needle and its preparation is possibly the most important component of successful embryo injections. Most dechorionated fruit fly eggs can be injected easily with borosilicate needles that are drawn out to a fine tip and broken off to a 1- to 2-μm opening. Opening the tip is typically achieved by scraping the needle against the edge of the slide carrying the eggs to be injected. Opening the needle by beveling, however, creates consistently sharp tips that are much more important for non-dechorionated eggs, and stronger alumina-silicate and quartz needles also provide an improvement to easily pierce chorions or tough vitelline membranes. Beveled needles are also critical when a large tip-opening is required for large plasmids that are susceptible to shearing. Preparation of borosilicate needles, pulled from 25-μl capillary stock that has been silanized, can be achieved with several types of vertical or horizontal needle pullers, and we find the Sutter Model P-30 (Sutter Instruments, Novato, CA) vertical micropipette puller to be highly effective. Alumina-silicate needles, and certainly quartz needles, require more sophisticated pullers that allow for fine programmable adjustment of high filament temperatures and pulling force, and the Sutter Models P-97 and P-2000 fulfill this need. Several needle bevelers are available, with the Sutter BV-10 used by many labs.

4.6.3. DNA Preparation and Injection

A mixture of highly purified vector and helper plasmid DNA is essential to embryo survival. This is achieved most optimally by purifying plasmid twice through cesium chloride gradients or a solid-phase anion exchange chromatography column. These have the advantage of high yields of DNA, but the disadvantage of specialized equipment and long preparation times. Successful transformation has been achieved with plasmids prepared with silica-gel membrane kits from Qiagen Corp. (Valencia, CA), but their successful use has been inconsistent, with failures possibly related to the type of host bacteria and its growth conditions. The Qiagen Endotoxin-free plasmid preparation systems allow additional purity, and we routinely use this system for successful plasmid injection.

Purified plasmid concentration must be accurately titered and verified by gel electrophoresis previous to injection-mix preparation. Appropriate amounts of vector and helper plasmid are ethanol precipitated, washed several times in 70% ethanol, and resuspended in injection buffer. Injection buffer has typically been the same as that originally used for *Drosophila* (5 mM KCl, 0.1 mM sodium phosphate, pH 6.8), though this may not be optimal for other insects, and embryo survival should be assessed by control injections. Total DNA concentration for injection should not exceed 1 mg/ml, using two- to four-fold higher concentration of vector to helper (e.g., 600 ng/μl vector to 200 ng/μl helper). Higher DNA concentrations are inadvisable, since they are subject to shearing during injection and may clog the needle, and the nucleic acids and/or contaminants can be toxic to the embryo. High transposase levels may also have a negative effect on transposition, as with the overproduction-inhibition phenomenon observed with *mariner* (Lohe and Hartl, 1996b).

Previous to injection the DNA mixture should be filtered through a 0.45-μm membrane, or centrifuged, before loading into the injection needle. Typically, DNA is back-filled into the injection needle using a drawn-out silanized 100-μl microcapillary, and a microliter of DNA should be sufficient for injecting hundreds of eggs.

4.6.3.1. DNA injection

The microinjection of DNA into embryos requires a system that forces a minute amount of DNA through the needle in a highly controllable fashion. Remarkably, many *Drosophila* labs simply use a mounted syringe and tubing filled with oil connected to a needle holder, with manual pressure applied. This system is successful due to accumulated expertise and the efficiency of transformation in the species, but would probably be less useful for injecting more sensitive embryos that transform less easily. Regulated air-pressure systems are available that are economical and allow highly controlled and rapid DNA injection. We use the PicoPump from WPI, which is most versatile in allowing positive and negative (with vacuum) pressure, and a hold capability that prevents backflow into the needle resulting in clogging (especially by yolk). A less expensive system can be constructed from Clippard components (Clippard Instrument Laboratory, Inc., Cincinnati, OH) that uses a simple air-pressure regulator and electronic valve and switch (see Handler, 2000). Needle holders from WPI can be used with both systems (MPH-3 and MPH-1, respectively).

All embryo injections are performed on a microscope with a mechanical stage, with the injection needle mounted on a micromanipulator. Microscopes first used for *Drosophila* transformation were inverted or compound microscopes, but the availability of a useful mechanical stage and stage adaptor for the Olympus SZ stereozoom microscope makes this the most versatile choice (the Olympus stage can be mounted on most stereomicroscopes). The micromanipulator can be free-standing next to the stage, or mounted on the microscope base. It allows the precise positioning of the needle at the desired point of entry into the egg, while the actual injection occurs by using the mechanical stage to push the egg into the needle. Piezo Translators, which were developed for rapid and automatic intracellular injection, may be more efficient for some embryos, and will obviate the need for a mechanical stage (Peloquin *et al.*, 1997). The WPI MPM20 translator used with the PV820 PicoPump allows a fully automated system for egg penetration, DNA injection, and needle withdrawal.

4.6.4. Post-Injection Treatment

After injection, the cover slip can be placed in a covered petri dish (but not sealed) with moist filter paper. We find use of square dishes with black filter paper to the easiest for up to six cover slips and simple observation of the embryos and hatched larvae. For injected embryos submerged in oil, oxygen concentration may be a limiting factor for development, if not viability. This can be ameliorated by reducing crowding of eggs on the cover slip, or by incubation in a portable hat-box tissue culture chamber that is humidified and under slight positive pressure with oxygen. For eggs without oil, oxygen saturation without pressure is advisable.

Most helper constructs have the transposase gene under heat shock regulation. The *Drosophila hsp70* promoter is a constitutive promoter that is active in the absence of heat shock (but also responds to anoxia, which may occur in embryos under oil), and transformation is possible with most vectors with or without heat shock treatment. If heat shock is desirable, it should be noted that the optimal temperature varies for different species. For example, *hsp70* responds optimally at 37°C in *Drosophila*, but at 39°C in medfly (Papadimitriou *et al.*, 1998). Injected embryos should be incubated for at least 4–6 hours after injection before heat shock, or after overnight incubation. Optimal temperatures for insect development vary, but the lowest temperatures possible can be beneficial to survival, and the injection process can slow development by 50% or more. Thus, larval hatching may be delayed considerably, and hatching should be monitored for several days after the expected time before discarding embryos.

Hatched larvae can be placed on normal culture media, though they may be weak and require careful handling and soft diet. Rearing of putative transgenic lines is typically achieved by back-crossing to the parental line in small group matings, or individual mating if a determination of transformation frequency is required. Inbreeding of G_0s can minimize rearing efforts, but this may be complicated by a high rate of infertility, which is typically close to 50% after fruit fly injections.

4.6.5. Improvements for Transformation Methodology

Dramatic progress has been made in transformation technology for non-drosophilid insects, and it appears that the vectors and markers in use should be widely applicable. Nevertheless, transformation of many other insect species will be highly challenging, primarily due to limitations in delivery of DNA into pre-blastoderm embryos. As noted, to date all successful non-drosophilid transformations have resulted from embryonic microinjection of DNA, but for many species current injection techniques are likely to result in high levels of lethality or sterility. Experimentation with alternative methods has been reported, though, arguably, none has been tested exhaustively for germ-line transformation, or vector systems were used that are now known to be ineffective. The most promising method is biolistics, where eggs are bombarded with micropellets encapsulated by DNA, which was first developed as a ballistics method to transform plant cells (Klein *et al.*, 1987). Ballistics is based upon a "shotgun" technique for bombardment, and it is the only non-injection method successfully used to transform an insect. This was a *P* transformation of *Drosophila*, though only a single transformant line (at an unknown frequency) was created, and the technique never gained wide applicability (Baldarelli and Lengyel, 1990). This was most likely due to the high efficiency of *P* transformation of *Drosophila* by microinjection, eliminating the need for an alternative technique.

Mosquito eggs are considerably more difficult to inject, and a significant effort was made to modify a biolistics approach to DNA delivery in *An. gambiae*, using a burst of pressurized helium for bombardments (Miahle and Miller, 1994). This technique was effective in introducing plasmid DNA into mosquito eggs, yielding high levels of transient expression of a reporter gene. Biolistics was subsequently used for transient expression in specific tissues, allowing the testing of fibroin gene promoters in the *B. mori* silk gland (Horard *et al.*, 1994; Kravariti *et al.*, 2001). Recent advances have included the use of a rigid macro-carrier in the Bio-Rad PDS/1000-Helium biolistics apparatus that minimizes the blast effect in soft tissue (Thomas *et al.*, 2001). This allows greater micropellet penetration into insect tissues, with improved survival. Despite these advances in delivering DNA into eggs and tissue, biolistics has yet to yield a germ-line transformant. A more recent modification of biolistics using the PDS-1000/He system with the Hepta adapter (BioRad) has, for the first time, yielded repeatable 3–4% transformation frequencies in *Drosophila* (Yuen *et al.*, 2008). This is highly encouraging, though use of this system in non-drosophilid species has not been reported, and conditions may require modification for other types of embryos.

The only other method reported for DNA delivery is electroporation, which, similar to biolistics, has resulted in high levels of transient expression of plasmid-encoded genes in *Drosophila* (Kamdar *et al.*, 1992), as well as in *Helicoverpa zea* and *M. domestica* (Leopold *et al.*, 1996). Though transformation has not been reported, as with biolistics, it is not apparent that this was seriously tested, or if functional vector systems were used (certainly for non-drosophilids). Electroporation techniques have also advanced in recent years, with DNA transferred into many different tissue types from a variety of organisms using new electroporation chamber designs and electric field parameters.

These recent advances with both biolistics and electroporation are highly encouraging that new efforts will have greater chances for success, and they deserve high priority for testing. Both methods also have the advantage, if successful, of the ability to deliver DNA simultaneously to multiple embryos, ranging from hundreds to thousands, depending on the species. This would be highly beneficial to all transformation experiments, but especially so for species that transform at low frequencies. These methods could also be used in cellularized embryos after blastoderm formation in insects having embryos that cannot be easily handled or collected in the preblastoderm stage.

Other approaches to DNA delivery can include the incorporation of vector/helper DNA into bacterial or viral carriers, which may be delivered by maternal injection or feeding. Variations on microinjection that might be required for ovoviviparous insects include maternal injection into ovaries or abdominal hemocoel (Presnail and Hoy, 1994), and use of liposomes might allow injection into cellularized embryos (Felgner *et al.*, 1987). All of these techniques should be re-evaluated with the use of vectors and markers now known to be highly efficient in non-drosophilid systems.

4.7. Summary

After concerted efforts for more than 30 years to achieve gene transfer in non-drosophilid insects, only in the past 15 years or so have these efforts been fruitful. Since 1995 the germ-lines of nearly 20 species in 4 orders of insects have been transformed, and this number may be only limited by the insects of current experimental and applied interest. Unlike plant and vertebrate animal systems that allow relatively efficient genomic integration of introduced DNA, insect systems have generally relied on vector-mediated integrations, and the only vectors found reliable for germ-line transformation are those based on transposable elements. Curiously, the two main vector systems developed for routine use in *D. melanogaster*, and originally discovered in that species, *P* and *hobo*, have not been applicable as vectors to any other species. Yet four other transposons found in non-melanogaster or non-drosophilid species are widely functional in insects and, for some, other organisms. Their discovery has been of enormous importance to the wider use of transformation technology, since little progress would have been made if most vector systems were specific to a particular host. Equal in importance to the advancements in vector development has been concurrent progress in genetic marker discovery and development. This began with the finding that cloned eye-color genes from *Drosophila* could complement existing mutations in other insects, and has continued with the more recent use of several fluorescent protein genes that are widely applicable as markers for transformation and reporters for gene expression.

The advancement of these techniques comes at a fortuitous time when genomics is providing a wealth of genetic infomation and resources that might be used to create transgenic strains of pest and beneficial insects to control their population size and behavior. As part of these efforts, genetic transformation is also critical to functional genomics studies that will provide information essential to understanding the biological function of genetic material, and relating specific genomic elements to those functions. Techniques such as enhancer traps and transposon tagging, which rely on remobilizable insertional mutagenesis, are only possible with transposon-based vector systems, and other techniques such as RNAi are greatly facilitated by these systems. Together, routine methods for transposon-mediated germ-line transformation and genomics analysis should provide the tools for dramatic progress in our understanding and control of insect species.

Acknowledgments

We gratefully acknowledge support from the USDA-NIFA-AFRI Competitive Grants Program and National Institutes of Health.

References

Allen, M. L., O'Brochta, D. A., Atkinson, P. W., & Levesque, C. S. (2001). Stable, germ-line transformation of *Culex quinquefasciatus* (Diptera: Culicidae). *J. Med. Entomol.*, *38*, 701–710.

Allen, M. L., Handler, A. M., Berkebile, D. R., & Skoda, S. R. (2004). *piggyBac* transformation of the New World screwworm, *Cochliomyia hominivorax*, produces multiple distinct mutant strains. *Med. Vet. Entomol.*, *18*, 1–9.

Anxolabéhère, D., Kidwell, M., & Periquet, G. (1988). Molecular characteristics of diverse populations are consistent with the hypothesis of a recent invasion of *Drosophila melanogaster* by mobile *P* elements. *Mol. Biol. Evol.*, *5*, 252–269.

Arca, B., & Savakis, C. (2000). Distribution of the transposable element *Minos* in the genus *Drosophila*. *Genetica*, *108*, 263–267.

Arca, B., Zabolou, S., Loukeris, T. G., & Savakis, C. (1997). Mobilization of a *Minos* transposon in *Drosophila melanogaster* chromosomes and chromatid repair by heteroduplex formation. *Genetics*, *145*, 267–279.

Arensburger, P., Kim, Y. J., Orsetti, J., Aluvihare, C., O'Brochta, D. A., & Atkinson, P. W. (2005). An active transposable element, *Herves*, from the African malaria mosquito *Anopheles gambiae*. *Genetics*, *169*, 697–708.

Ashburner, M. (1989a). *Drosophila: A Laboratory Handbook*. Cold Spring Harbor, NY: Cold Spring Harbor Laboratory Press.

Ashburner, M. (1989b). *Drosophila: A Laboratory Manual*. Cold Spring Harbor, NY: Cold Spring Harbor Laboratory Press.

Atkinson, P. W., Warren, W. D., & O'Brochta, D. A. (1993). The *hobo* transposable element of *Drosophila* can be cross-mobilized in houseflies and excises like the *Ac* element of maize. *Proc. Natl. Acad. Sci. USA*, *90*, 9693–9697.

Atkinson, P. W., Pinkerton, A. C., & O'Brochta, D. A. (2001). Genetic transformation systems in insects. *Ann. Rev. Entomol.*, *46*, 317–346.

Auge-Gouillou, C., Hamelin, M. -H., Demattei, M. -V., Periquet, G., & Bigot, Y. (2001a). The ITR binding domain of the *mariner* Mos-1 transposase. *Mol. Gen. Genomics*, *265*, 58–65.

Auge-Gouillou, C., Hamelin, M. -H., Demattei, M. -V., Periquet, G., & Bigot, Y. (2001b). The wild-type conformation of the Mos-1 inverted terminal repeats is suboptimal for transposition in bacteria. *Mol. Gen. Genomics*, *265*, 51–57.

Auge-Gouillou, C., Brillet, B., Hamelin, M. H., & Bigot, Y. (2005). Assembly of the *mariner Mos1* synaptic complex. *Mol. Cell. Biol.*, *25*, 2861–2870.

Avery, T., Macleod, C. M., & McCarty, M. (1944). Studies on the chemical nature of the substance inducing transformation of pneumococcal types. I. Induction of transformation by a desoxyribonucleic acid fraction isolated from pneumococcus type III. *J. Exper. Med.*, *79*, 137–158.

Baldarelli, R. M., & Lengyel, J. A. (1990). Transient expression of DNA after ballistic introduction into *Drosophila* embryos. *Nucleic Acids Res.*, *18*, 5903–5904.

Barry, E. G., Witherspoon, D. J., & Lampe, D. J. (2004). A bacterial genetic screen identifies functional coding sequences of the insect *mariner* transposable element Famar1 amplified from the genome of the earwig, *Forficula auricularia*. *Genetics*, *166*, 823–833.

Bateman, J. R., Lee, A. M., & Wu, C-t (2006). Site-specific transformation of *Drosophila* via phiC31 integrase-mediated cassette exchange. *Genetics*, *173*, 769–777.

Bellen, H. J., O'Kane, C. J., Wilson, C., Grossniklaus, U., Pearson, R. K., & Gehring, W. J. (1989). P-element-mediated enhancer detection: A versatile method to study development in *Drosophila*. *Develop.*, *3*, 1288–1300.

Bender, W., & Hudson, A. (2000). *P* element homing in the *Drosophila* bithorax complex. *Develop.*, *127*, 3981–3992.

Benedict, M. Q., Salazar, C. E., & Collins, F. H. (1995). A new dominant selectable marker for genetic transformation: *Hsp70-opd*. *Insect Biochem. Mol. Biol.*, *25*, 1061–1065.

Berghammer, A. J., Klingler, M., & Wimmer, E. A. (1999). A universal marker for transgenic insects. *Nature*, *402*, 370–371.

Besansky, N. J., Mukabayire, O., Benedict, M. Q., Rafferty, C. S., Hamm, D. M., & McNitt, L. (1997). The Anopheles gambiae tryptophan oxygenase gene expressed from a baculovirus promoter complements. *Drosophila melanogaster vermilion*. *Insect Biochem. Mol. Biol.*, *27*, 803–805.

Bevis, B. J., & Glick, B. S. (2002). Rapidly maturing variants of the Discosoma red fluorescent protein (DsRed). *Nat. Biotechnol.*, *20*, 83–87.

Bigot, Y., Auge-Gouillou, C., & Periquet, G. (1996). Computer analyses reveal a *hobo*-like element in the nematode *Caenorhabditis elegans*, which presents a conserved transposase domain common with the *Tc1-Mariner* transposon family. *Gene*, *174*, 265–271.

Blackman, R. K., Macy, M., Koehler, D., Grimaila, R., & Gelbart, W. M. (1989). Identification of a fully functional *hobo* transposable element and its use for germ-line transformation of *Drosophila*. *EMBO J.*, *8*, 211–217.

Bonizzoni, M., Gomulski, L. M., Malacrida, A. R., Capy, P., & Gasperi, G. (2007). Highly similar *piggyBac* transposase-like sequences from various Bactrocera (Diptera, Tephritidae) species. *Insect Mol. Biol.*, *16*, 645–650.

Borsatti, F., Azzoni, P., & Mandrioli, M. (2003). Identification of a new *hobo* element in the cabbage moth, *Mamestra brassicae* (Lepidoptera). *Hereditas.*, *139*, 151–155.

Brand, A. H., Manoukian, A. S., & Perrimon, N. (1994). Ectopic expression in *Drosophila*. *Meth. Cell Biol.*, *44*, 635–654.

Bryan, G. J., Jacobson, J. W., & Hartl, D. L. (1987). Heritable somatic excision of a *Drosophila* transposon. *Science*, *235*, 1636–1638.

Bushman, F. D. (1994). Tethering human immunodeficiency virus 1 integrase to a DNA site directs integration to nearby sequences. *Proc. Natl. Acad. Sci.*, *91*, 9233–9237.

Butler, M. G., Chakraborty, S. A., & Lampe, D. J. (2006). The N-terminus of *Himar1 mariner* transposase mediates multiple activities during transposition. *Genetica*, *127*, 351–366.

Calvi, B. R., Hong, T. J., Findley, S. D., & Gelbart, W. M. (1991). Evidence for a common evolutionary origin of inverted repeat transposons in *Drosophila* and plants: *hobo*, *Activator*, and *Tam3*. *Cell*, *66*, 465–471.

Campbell, R. E., Tour, O., Palmer, A. E., Steinbach, P. A., Baird, G. S., et al. (2002). A monomeric red fluorescent protein. *Proc. Natl. Acad. Sci. USA*, *99*, 7877–7882.

Capy, P., Vitalis, R., Langin, T., Higuet, D., & Bazin, C. (1996). Relationships between transposable elements based upon the integrase-transposase domains: Is there a common ancestor? *J. Mol. Evol.*, *42*, 359–368.

Carr, M. (2008). Multiple subfamilies of *mariner* transposable elements are present in stalk-eyed flies (Diptera: Diopsidae). *Genetica*, *132*, 113–122.

Cary, L. C., Goebel, M., Corsaro, H. H., Wang, H. H., Rosen, E., & Fraser, M. J. (1989). Transposon mutagenesis of baculoviruses: Analysis of *Trichoplusia ni* transposon IFP2 insertions within the FP-Locus of nuclear polyhedrosis viruses. *Virology*, *161*, 8–17.

Caspari, E., & Nawa, S. (1965). A method to demonstrate transformation in Ephestia. *Z. Naturforsch*, *206*, 281–284.

Casteret, S., Chbab, N., Cambefort, J., Auge-Gouillou, C., Bigot, Y., & Rouleux-Bonnin, F. (2009). Physical properties of DNA components affecting the transposition efficiency of the *mariner Mos1* element. *Mole. Gen. Genom.*, *282*, 531–546.

Castro, J. P., & Carareto, C. M.A. (2004). *Drosophila melanogaster P* transposable elements: Mechanisms of transposition and regulation. *Genetica*, *12*, 107–118.

Catteruccia, F., Nolan, T., Loukeris, T. G., Blass, C., Savakis, C., et al. (2000). Stable germline transformation of the malaria mosquito *Anopheles stephensi*. *Nature*, *405*, 959–962.

Catteruccia, F., Godfray, H. C., & Crisanti, A. (2003). Impact of genetic manipulation on the fitness of *Anopheles stephensi* mosquitoes. *Science*, *299*, 1225–1227.

Chalfie, M., Tu, Y., Euskirchen, G., Ward, W., & Prasher, D. C. (1994). Green fluorescent protein as a marker for gene expression. *Science*, *263*, 802–805.

Christophides, G. K., Savakis, C., Mintzas, A. C., & Komitopoulou, K. (2001). Expression and function of the *Drosophila melanogaster* ADH in male *Ceratitis capitata* adults: A potential strategy for medfly genetic sexing based on gene-transfer technology. *Insect Mol. Biol.*, *10*, 249–254.

Chudakov, D. M., Matz, M. V., Lukyanov, S., & Lukyanov, K. A. (2010). Fluorescent proteins and their applications in imaging living cells and tissues. *Physiol. Rev.*, *90*, 1103–1163.

Coates, C. J., Johnson, K. N., Perkins, H. D., Howells, A. J., O'Brochta, D. A., & Atkinson, P. W. (1996). The *hermit* transposable element of the Australian sheep blowfly, *Lucilia cuprina*, belongs to the *hAT* family of transposable elements. *Genetica*, *97*, 23–31.

Coates, C. J., Turney, C. L., Frommer, M., O'Brochta, D. A., & Atkinson, P. W. (1997). Interplasmid transposition of the *mariner* transposable element in non-drosophilid insects. *Mol. Gen. Genet.*, *253*, 728–733.

Coates, C. J., Jasinskiene, N., Miyashiro, L., & James, A. A. (1998). *Mariner* transposition and transformation of the yellow fever mosquito, *Aedes aegypti*. *Proc. Natl. Acad. Sci. USA*, *95*, 3742–3751.

Concha, C., Belikoff, E. J., Carey, B. -L., Li, F., Schiemann, A. H., & Scott, M. J. (2010). Efficient germ-line transformation of the economically important pest species *Lucilia cuprina* and *Lucilia sericata* (Diptera, Calliphoridae). *Insect Biochem. Mol. Biol.*, in press.

Condon, K. C., Condon, G. C., Dafa'alla, T. H., Forrester, O. T., Phillips, C. E., et al. (2007). Germ-line transformation of the Mexican fruit fly. *Insect Mol. Biol.*, *16*, 573–580.

Cornel, A. J., Benedict, M. Q., Rafferty, C. S., Howells, A. J., & Collins, F. H. (1997). Transient expression of the *Drosophila melanogaster cinnabar* gene rescues eye color in the white eye (WE) strain of *Aedes aegypti*. *Insect Biochem. Mol. Biol.*, *27*, 993–997.

Coy, M. R., & Tu, Z. (2005). Gambol and Tc1 are two distinct families of DD34E transposons: Analysis of the *Anopheles gambiae* genome expands the diversity of the *IS630-Tc1-mariner* superfamily. *Insect Mol. Biol.*, *14*, 537–546.

Crenes, G., Ivo, D., Herisson, J., Dion, S., Renault, S., et al. (2009). The bacterial Tn9 chloramphenicol resistance gene: An attractive DNA segment for *Mos1 mariner* insertions. *Mol. Genet. Genom.*, 281, 315–328.

Crenes, G., Moundras, C., Demattei, M. V., Bigot, Y., Petit, A., & Renault, S. (2010). Target site selection by the *mariner*-like element, *Mos1*. *Genetica*, 138, 509–517.

Dafa'alla, T. H., Condon, G. C., Condon, K. C., Phillips, C. E., Morrison, N. I., et al. (2006). Transposon-free insertions for insect genetic engineering. *Nat. Biotechnol.*, 24, 820–821.

Dai, H., Jiang, R., Wang, J., Xu, G., Cao, M., et al. (2007). Development of a heat shock inducible and inheritable RNAi system in silkworm. *Biomol. Eng.*, 24, 625–630.

Daniels, S. B., & Strausbaugh, L. D. (1986). The distribution of P-element sequences in *Drosophila*: The *willistoni* and *saltans* species groups. *J. Mol. Evol.*, 23, 138–148.

Davis, I., Girdham, C. H., & O'Farrell, P. H. (1995). A nuclear GFP that marks nuclei in living *Drosophila* embryos; maternal supply overcomes a delay in the appearance of zygotic fluorescence. *Devel. Biol.*, 170, 726–729.

Dawson, A., & Finnegan, D. J. (2003). Excision of the *Drosophila mariner* transposon *Mos1*: Comparison with bacterial transposition and V(D)J recombination. *Mol. Cell*, 11, 225–235.

Depra, M., Panzera, Y., Ludwig, A., Valente, V. L. S., & Loreto, E. L. S. (2010). *hosimary*: A new *hAT* transposon group involved in horizontal transfer. *Molec. Genet. Genom.*, 283, 451–459.

Drabek, D., Zagoraiou, L., deWit, T., Langeveld, A., Roumpaki, C., et al. (2003). Transposition of the *Drosophila hydei Minos* transposon in the mouse germ line. *Genomics*, 81, 108–111.

Eickbush, T. H., & Malik, H. S. (2002). Origins and evolution of retrotransposons. In N. L. Craig, R. Craige, M. Gellert, & A. M. Lambowitz (Eds.). *Mobile DNA II* (pp. 1111–1146). Washington, DC: ASM Press.

Engels, W. R. (1989). *P* elements in *Drosophila melanogaster*. In D. E. Berg, & M. M. Howe (Eds.). *Mobile DNA* (p. 439). Washington, DC: American Society of Microbiology.

Engstrom, Y., Schneuwly, S., & Gehring, W. (1992). Spatial and temporal expression of an Antennapedia LacZ gene construct integrated into the endogenous Antennapedia gene of *Drosophila melanogaster*. *Roux's Arch. Dev. Biol.*, 201, 65–80.

Esnault, C., Palavesam, A., Pilitt, K., & O'Brochta, D. A. (2011). Intrinsic characteristics of neighboring DNA modulate transposable element activity in *Drosophila melanogaster*. *Genetics*, 187, 319–331.

Evertts, A. G., Plymire, C., Craig, N. L., & Levin, H. L. (2007). The *Hermes* transposon of *Musca domestica* is an efficient tool for the mutagenesis of *Schizosaccharomyces pombe*. *Genetics*, 177, 2519–2523.

Fadool, J. M., Hartl, D. L., & Dowling, J. E. (1998). Transposition of the *mariner* element from *Drosophila mauritiana* in zebrafish. *Proc. Natl. Acad. Sci. USA*, 95, 5182–5186.

Federoff, N. (1989). Maize transposable elements. In D. E. Berg, & M. M. Howe (Eds.). *Mobile DNA* (pp. 375). Washington, DC: American Society for Microbiology.

Felgner, P. L., Gadek, T. R., Holm, M., Roman, R., Chan, H. W., et al. (1987). Lipofection: A highly efficient, lipid-mediated DNA-transfection procedure. *Proc. Natl. Acad. Sci. USA*, 84, 7413–7417.

Ferguson, H. J., Neven, L. G., Thibault, S. T., Mohammed, A., & Fraser, M. (2010). Genetic transformation of the codling moth, *Cydia pomonella* L., with *piggyBac* EGFP. *Transgenic Res.*, 20, 1, 201–214.

Feschotte, C. (2006). The *piggyBac* transposon holds promise for human gene therapy. *Proc. Natl. Acad. Sci. USA*, 103, 14981–14982.

ffrench-Constant, R. H., & Benedict, M. Q. (2000). Resistance genes as candidates for insect transgenesis. In A. M. Handler, & A. A. James (Eds.). *Insect Transgenesis: Methods and Applications* (pp. 109). Boca Raton, FL: CRC Press.

ffrench-Constant, R. H., Mortlock, D. P., Shaffer, C. D., MacIntyre, R. J., & Roush, R. T. (1991). Molecular cloning and transformation of cyclodiene resistance in *Drosophila*: An invertebrate GABA$_A$ receptor locus. *Proc. Natl. Acad. Sci. USA*, 88, 7209–7213.

Finnegan, D. J. (1989). Eucaryotic transposable elements and genome evolution. *Trends Genet.*, 5, 103–107.

Fox, A. S., & Yoon, S. B. (1966). Specific genetic effects of DNA in *Drosophila melanogaster*. *Genetics*, 53, 897–911.

Fox, A. S., & Yoon, S. B. (1970). DNA-induced transformation in *Drosophila*: Locus specificity and the establishment of transformed stocks. *Proc. Natl. Acad. Sci. USA*, 67, 1608–1615.

Franz, G., & Savakis, C. (1991). *Minos*, a new transposable element from *Drosophila hydei*, is a member of the Tc-1-like family of transposons. *Nucl. Acids Res.*, 19, 6646.

Franz, G., Loukeris, T. G., Dialektaki, G., Thompson, C. R.L., & Savakis, C. (1994). Mobile *Minos* elements from *Drosophila hydei* encode a two-exon transposase with similarity to the paired DNA-binding domain. *Proc. Natl. Acad. Sci. USA*, 91, 4746–4750.

Fraser, M. J. (2000). The TTAA-specific family of transposable elemnts: Identification, functional characterization, and utility for transformation of insects. In A. M. Handler, & A. A. James (Eds.). *Insect Transgenesis: Methods and Applications* (p. 249). Boca Raton, FL: CRC Press.

Fraser, M. J., Smith, G. E., & Summers, M. D. (1983). Acquisition of host-cell DNA sequences by baculoviruses – relationship between host DNA insertions and FP mutants of *Autographa-californica* and *Galleria-mellonella* nuclear polyhedrosis viruses. *J. Virol.*, 47, 287–300.

Fraser, M. J., Brusca, J. S., Smith, G. E., & Summers, M. D. (1985). Transposon-mediated mutagenesis of a baculovirus. *Virology*, 145, 356–361.

Fraser, M. J., Cary, L., Boonvisudhi, K., & Wang, H. G. (1995). Assay for movement of Lepidopteran transposon IFP2 in insect cells using a baculovirus genome as a target DNA. *Virology*, 211, 397–407.

Furlong, E. E., Profitt, D., & Scott, M. P. (2001). Automated sorting of live transgenic embryos. *Nat. Biotechnol.*, 19, 153–156.

Garza, D., Medhora, M., Koga, A., & Hartl, D. L. (1991). Introduction of the transposable element *mariner* into the germline of *Drosophila melanogaster*. *Genetics*, 128, 303–310.

Germeraad, S. (1976). Genetic transformation in *Drosophila* by microinjection of DNA. *Nature*, 262, 229–231.

Germon, S., Bouchet, N., Casteret, S., Carpentier, G., Adet, J., et al. (2009). *Mariner Mos1* transposase optimization by rational mutagenesis. *Genetica, 137*, 265–276.

Goldberg, D. A., Posakony, J. W., & Maniatis, T. (1983). Correct developmental expression of a cloned alcohol dehydrogenase gene transduced into the *Drosophila* germ line. *Cell, 34*, 59–73.

Gomez, S. P., & Handler, A. M. (1997). A *Drosophila melanogaster hobo-white*⁺ vector mediates low frequency gene transfer in *D. virilis* with full interspecific *white*⁺ complementation. *Insect Mol. Biol., 6*, 1–8.

Gomulski, L. M., Torti, C., Murelli, V., Bonizzoni, M., Gasperi, G., & Malacrida, A. R. (2004). Medfly transposable elements: Diversity, evolution, genomic impact and possible applications. *Insect Biochem. Mol. Biol., 34*, 139–148.

González-Estévez, C., Momose, T., Gehring, W. J., & Saló, E. (20003). Transgenic planarian lines obtained by electroporation using transposon-derived vectors and an eye-specific GFP marker. *Proc. Natl. Acad. Sci. USA, 100*, 14046–14051.

Gorbunova, V., & Levy, A. A. (1997). Circularized *Ac/Ds* transposons: Formation, structure and fate. *Genetics, 145*, 1161–1169.

Goryshin, I. Y., & Reznikoff, W. S. (1998). Tn5 *in vitro* transposition. *J. Biol. Chem., 273*, 7367–7374.

Goryshin, I. Y., Jendrisak, J., Hoffman, L. M., Meis, R., & Reznikoff, W. S. (2000). Insertional transposon mutagenesis by electroporation of released *Tn5* transposition complexes. *Nature Biotech., 18*, 97–100.

Green, M. M. (1996). The "Genesis of the White-Eyed Mutant" in *Drosophila melanogaster*: A reappraisal. *Genetics, 142*, 329–331.

Griffith, F. (1928). Significance of pneumococcal types. *J. Hyg. Camb., 27*, 113–159.

Grossman, G. L., Rafferty, C. S., Clayton, J. R., Stevens, T. K., Mukabayire, O., & Benedict, M. Q. (2001). Germline transformation of the malaria vector, *Anopheles gambiae*, with the *piggyBac* transposable element. *Insect Mol. Biol., 10*, 597–604.

Groth, A., Fish, M., Nusse, R., & Calos, M. P. (2004). Construction of transgenic *Drosophila* by using the site-specific integrase from phage PhiC31. *Genetics, 166*, 1775.

Gueiros-Filho, F. J., & Beverley, S. M. (1997). Trans-kingdom transposition of the *Drosophila* element *mariner* within the protozoan Leishmania. *Science, 276*, 1716–1719.

Guimond, N., Bideshi, D. K., Pinkerton, A. C., Atkinson, P. W., & O'Brochta, D. A. (2003). Patterns of *Hermes* transposition in *Drosophila melanogaster*. *Mol. Gen. Genet., 268*, 779–790.

Haine, E. R., Kabat, P., & Cook, J. M. (2007). Diverse *Mariner*-like elements in fig wasps. *Insect Mol. Biol., 16*, 743–752.

Hama, C., Ali, Z., & Kornberg, T. B. (1990). Region-specific recombination and expression are directed by portions of the Drosophila *engrailed* promoter. *Genes. Dev., 4*, 1079–1093.

Handler, A. M. (2000). An introduction to the history and methodology of insect gene transfer. In A. M. Handler, & A. A. James (Eds.). *Insect Transgenesis: Methods and Applications* (p. 3). Boca Raton, FL: CRC Press.

Handler, A. M. (2001). A current perspective on insect gene transfer. *Insect Biochem. Mol. Biol., 31*, 111–128.

Handler, A. M. (2002a). Prospects for using genetic transformation for improved SIT and new biocontrol methods. *Genetica, 116*, 137–149.

Handler, A. M. (2002b). Use of the *piggyBac* transposon for germ-line transformation of insects. *Insect Biochem. Mol. Biol., 32*, 1211–1220.

Handler, A. M. (2003). Isolation and analysis of a new *hopper hAT* transposon from the *Bactrocera dorsalis* white eye strain. *Genetica, 118*, 17–24.

Handler, A. M. (2004). Understanding and improving transgene stability and expression in insects for SIT and conditional lethal release programs. *Insect Biochem. Mol. Biol., 34*, 121–130.

Handler, A. M., & Gomez, S. P. (1996). The *hobo* transposable element excises and has related elements in tephritid species. *Genetics, 143*, 1339–1347.

Handler, A. M., & Gomez, S. P. (1997). A new *hobo, Activator, Tam*3 transposable element, *hopper*, from *Bactrocera dorsalis* is distantly related to *hobo* and Ac. *Gene. 185*, 133–135.

Handler, A. M., & Harrell, R. A. (1999). Germline transformation of *Drosophila melanogaster* with the *piggyBac* transposon vector. *Insect Mol. Biol., 8*, 449–458.

Handler, A. M., & Harrell, R. A. (2000). Transformation of the Caribbean fruit fly with a *piggyBac* transposon vector marked with polyubiquitin-regulated GFP. *Insect Biochem. Mol. Biol., 31*, 199–205.

Handler, A. M., & Harrell, R. A. (2001). Polyubiquitin-regulated DsRed marker for transgenic insects. *Biotechniques, 31*, 820–828.

Handler, A. M., & McCombs, S. D. (2000). The *piggyBac* transposon mediates germ-line transformation in the Oriental fruit fly and closely related elements exist in its genome. *Insect Mol. Biol., 9*, 605–612.

Handler, A. M., & O'Brochta, D. A. (1991). Prospects for gene transformation in insects. *Annu. Rev. Entomol., 36*, 159–183.

Handler, A. M., Gomez, S. P., & O'Brochta, D. A. (1993a). Negative regulation of *P* element excision by the somatic product and terminal sequences of *P* in *Drosophila melanogaster*. *Mol. Gen. Genet., 237*, 145–151.

Handler, A. M., Gomez, S. P., & O'Brochta, D. A. (1993b). A functional analysis of the P-element gene-transfer vector in insects. *Arch. Insect Biochem. Physiol., 22*, 373–384.

Handler, A. M., McCombs, S. D., Fraser, M. J., & Saul, S. H. (1998). The lepidopteran transposon vector, *piggyBac*, mediates germ-line transformation in the Mediterranean fruitfly. *Proc. Natl. Acad. Sci. USA, 95*, 7520–7525.

Handler, A. M., Zimowska, G. J., & Horn, C. (2004). Post-integration stabilization of a transposon vector by terminal sequence deletion in *Drosophila melanogaster*. *Nat. Biotechnol., 22*, 1150–1154.

Handler, A. M., Zimowska, G. J., & Armstrong, K. F. (2008). Highly similar *piggyBac* elements in Bactrocera that share a common lineage with elements in noctuid moths. *Insect Mol. Biol., 17*, 387–393.

Handler, A. M., Allen, M. L., & Skoda, S. R. (2009). Development and utilization of transgenic New World screwworm, *Cochliomyia hominivorax*. *Med. Vet. Entomol. Suppl., 1*, 98–105.

Hartl, D. L., Lohe, A. R., & Lozovskaya, E. R. (1997). Modern thoughts on an ancyent marinere: Function, evolution, regulation. *Annu. Rev. Genet., 31*, 337–358.

Haymer, D. S., & Marsh, J. L. (1986). Germ line and somatic instability of a *white* mutation in *Drosophila mauritiana* due to a transposable element. *Dev. Genet., 6,* 281–291.

Hazelrigg, T., Levis, R., & Rubin, G. M. (1984). Transformation of *white* locus DNA in *Drosophila*: Dosage compensation, zeste interaction, and position effects. *Cell, 64,* 1083–1092.

Hazelrigg, T., Liu, N., Hong, Y., & Wang, S. (1998). GFP expression in *Drosophila* tissues: Time requirements for formation of a fluorescent product. *Dev. Biol., 199,* 245–249.

Hediger, M., Niessen, M., Wimmer, E. A., Dübendorfer, A., & Bopp, D. (2000). Genetic transformation of the housefly *Musca domestica* with the lepidopteran derived transposon *piggyBac*. *Insect Mol. Biol., 10,* 113–119.

Heinrich, J. C., Li, X., Henry, R. A., Haack, N., Stringfellow, L., et al. (2002). Germ-line transformation of the Australian sheep blowfly *Lucilia cuprina*. *Insect Mol. Biol., 11,* 1–10.

Hickman, A. B., Perez, Z. N., Zhou, L. Q., Musingarimi, P., Ghirlando, R., et al. (2005). Molecular architecture of a eukaryotic DNA transposase. *Nat. Struct. Mol. Biol., 12,* 715–721.

Higgs, S., & Sinkins, D. L. (2000). Green fluorescent protein (GFP) as a marker for transgenic insects. In A. M. Handler, & A. A. James (Eds.). *Insect Transgenesis: Methods and Applications* (p. 93). Boca Raton, FL: CRC Press.

Higgs, S., Traul, D., Davis, B. S., Kamrud, K. I., Wilcox, C. L., & Beaty, B. J. (1996). Green fluorescent protein expressed in living mosquitoes – without the requirement of transformation. *Biotechniques, 21,* 660–664.

Hoess, R., Wierzbicki, A., & Abremski, K. (1985). Formation of small circular DNA molecules via an *in vitro* site-specific recombination system. *Gene, 40,* 325–329.

Holtzman, S., Miller, D., Eisman, R., Kuwayama, H., Niimi, T., & Kaufman, T. C. (2010). Transgenic tools for members of the genus *Drosophila* with sequenced genomes. *Fly, 4,* 1–14.

Horard, B., Mange, A., Pelissier, B., & Couble, P. (1994). *Bombyx* gene promoter analysis in transplanted silkgland transformed by particle delivery system. *Insect Mol. Biol., 3,* 261–265.

Horn, C., & Handler, A. M. (2005). Site-specific genomic targeting in *Drosophila*. *Proc. Natl. Acad. Sci. USA, 102,* 12483–12488.

Horn, C., & Wimmer, E. A. (2000). A versatile vector set for animal transgenesis. *Dev. Genes Evol., 210,* 630–637.

Horn, C., Jaunich, B., & Wimmer, E. A. (2000). Highly sensitive, fluorescent transformation marker for *Drosophila* transgenesis. *Dev. Genes Evol., 210,* 623–629.

Horn, C., Schmid, B., Pogoda, F. S., & Wimmer, E. A. (2002). Fluorescent transformation markers for insect transgenesis. *Insect Biochem. Mol. Biol., 32,* 1221–1235.

Horn, C., Offen, N., Nystedt, S., Häcker, U., & Wimmer, E. A. (2003). *piggyBac*-based insertional mutagenesis and enhancer detection as a tool for functional insect genomics. *Genetics, 163,* 647–661.

Imamura, M., Nakai, J., Inoue, S., Quan, G. X., Kanda, T., & Tamura, T. (2003). Targeted gene expression using the GAL4/UAS system in the silkworm *Bombyx mori*. *Genetics, 165,* 1329–1340.

Irvin, N., Hoddle, M. S., O'Brochta, D. A., Carey, B., & Atkinson, P. W. (2004). Assessing fitness costs for transgenic *Aedes aegypti* expressing the GFP marker and transposase genes. *Proc. Natl. Acad. Sci. USA, 101,* 891–896.

Jacobson, J. W., Medhora, M. M., & Hartl, D. L. (1986). Molecular structure of somatically unstable transposable element in *Drosophila*. *Proc. Natl. Acad. Sci. USA, 83,* 8684–8688.

Jasinskiene, N., Coates, C. J., Benedict, M. Q., Cornel, A. J., Rafferty, C. S., et al. (1998). Stable transposon mediated transformation of the yellow fever mosquito, *Aedes aegypti*, using the *Hermes* element from the housefly. *Proc. Natl. Acad. Sci. USA, 95,* 3743–3747.

Jasinskiene, N., Coates, C. J., & James, A. A. (2000). Structure of *Hermes* integrations in the germline of the yellow fever mosquito, *Aedes aegypti*. *Insect Mol. Biol., 9,* 11–18.

Jensen, P. A., Stuart, J. R., Goodpaster, M. P., Goodman, J. W., & Simmons, M. J. (2008). Cytotype regulation of *P* transposable elements in *Drosophila melanogaster*: Repressor polypeptides or piRNAs? *Genetics, 179,* 1785–1793.

Kamdar, P., Von Allmen, G., & Finnerty, V. (1992). Transient expression of DNA in *Drosophila* via electroporation. *Nucleic Acids Res., 11,* 3526.

Kaminski, J. M., Huber, M. R., Summers, J. B., & Ward, M. B. (2002). Design of a nonviral vector for site-selective, efficient integration into the human genome. *FASEB J., 16,* 1242–1247.

Kapetanaki, M. G., Loukeris, T. G., Livadaras, I., & Savakis, C. (2002). High frequencies of *Minos* transposon mobilization are obtained in insects by using *in vitro* synthesized mRNA as a source of transposase. *Nucleic Acids Res., 30,* 3333–3340.

Karess, R. E. (1985). *P* element mediated germ line transformation of *Drosophila*. In D. M. Glover (Ed.). *DNA Cloning Volume II: A Practical Approach* (p. 121). Oxford, UK: IRL Press.

Karess, R. E., & Rubin, G. R. (1984). Analysis of *P* transposable element functions in *Drosophila*. *Cell, 38,* 135–146.

Kassis, J. A., Noll, E., VanSickle, E. P., Odenwald, W. F., & Perrimon, N. (1992). Altering the insertional specificity of a *Drosophila* transposable element. *Proc. Natl. Acad. Sci. USA, 89,* 1919–1923.

Kaufman, P. K., Doll, R. F., & Rio, D. C. (1989). *Drosophila P* element transposase recognizes internal *P* element DNA sequences. *Cell, 59,* 359–371.

Kidwell, M. G., Kidwell, J. F., & Sved, J. A. (1977). Hybrid dysgenesis in *Drosophila melanogaster*: A syndrome of aberrant traits including mutation, sterility, and male recombination. *Genetics, 86,* 813–833.

Klein, T. M., Wolf, E. D., Wu, R., & Sanford, J. C. (1987). High-velocity microprojectiles for delivering nucleic acids into living cells. *Nature, 327,* 70–73.

Klemenz, R., Weber, U., & Gehring, W. J. (1987). The white gene as a marker in a new P-element vector for gene transfer in *Drosophila*. *Nucleic Acids Res., 15,* 3947–3959.

Kokoza, V., Ahmed, A., Wimmer, E. A., & Raikhel, A. S. (2001). Efficient transformation of the yellow fever mosquito *Aedes aegypti* using the *piggyBac* transposable element vector pBac[3xP3-EGFP afm]. *Insect Biochem. Mol. Biol., 31,* 1137–1143.

Koukidou, M., Klinakis, A., Reboulakis, C., Zagoraiou, L., Tavernarakis, N., et al. (2006). Germ line transformation of the olive fly *Bactrocera oleae* using a versatile transgenesis marker. *Insect Mol. Biol.*, *15*, 95–103.

Kravariti, L., Thomas, J., Sourmeli, S., Rodakis, G. C., Mauchamp, B., et al. (2001). The biolistic method as a tool for testing the differential activity of putative silkmoth chorion gene promoters. *Insect Biochem. Mol. Biol.*, *31*, 473–479.

Kuwayama, H., Yaginuma, T., Yamashita, O., & Niimi, T. (2006). Germ-line transformation and RNAi of the ladybird beetle, *Harmonia axyridis*. *Insect Mol. Biol.*, *15*, 507–512.

Laakkonen, J. P., Kaikkonen, M. U., Ronkainen, P. H. A., Ihalainen, T. O., Niskanen, E. A., et al. (2008). Baculovirus-mediated immediate-early gene expression and nuclear reorganization in human cells. *Cell Microbiol.*, *10*, 667–681.

Labbé, G. M., Nimmo, D. D., & Alphey, L. (2010). *piggybac*- and *PhiC31*-mediated genetic transformation of the Asian tiger mosquito, *Aedes albopictus* (Skuse). *PLoS Negl. Trop. Dis.*, *4*, e788.

Lampe, D. J., Churchill, M. E., & Robertson, H. M. (1996). A purified *mariner* transposase is sufficient to mediate transposition *in vitro*. *EMBO J.*, *15*, 5470–5479.

Lampe, D. J., Grant, T. E., & Robertson, H. M. (1998). Factors affecting transposition of the *Himar1 mariner* transposon *in vitro*. *Genetics*, *149*, 179–187.

Lampe, D. J., Akerley, B. J., Rubin, E. J., Mekalanos, J. J., & Robertson, H. M. (1999). Hyperactive transposase mutants of the *Himar1 mariner* transposon. *Proc. Natl. Acad. Sci. USA*, *96*, 1142–1433.

Lampe, D. J., Walden, K. K. O., Sherwood, J. M., & Robertson, H. M. (2000). Genetic engineering of insects with *mariner* transposons. In A. M. Handler, & A. A. James (Eds.). *Insect Transgenesis: Methods and Applications* (p. 237). Boca Raton, FL: CRC Press.

Lampe, D. J., Witherspoon, D. J., Soto-Adames, F. N., & Robertson, H. M. (2003). Recent horizontal transfer of mellifera subfamily *mariner* transposons into insect lineages representing four different orders shows that selection acts only during horizontal transfer. *Mol. Biol. Evol.*, *20*, 554–562.

Lee, H., Simon, J. A., & Lis, J. T. (1988). Structure and expression of ubiquitin genes of *Drosophila melanogaster*. *Mol. Cell. Biol.*, *8*, 4727–4735.

Leopold, R. A., Hughes, K. J., & DeVault, J. D. (1996). Using electroporation and a slot cuvette to deliver plasmid DNA to insect embryos. *Genet. Anal.*, *12*, 197–200.

Lerat, E., Brunet, F., Bazin, C., & Capy, P. (1999). Is the evolution of transposable elements modular? *Genetica*, *107*, 15–25.

Li, X., Heinrich, J. C., & Scott, M. J. (2001a). *piggyBac*-mediated transposition in *Drosophila melanogaster*: An evaluation of the use of constitutive promoters to control transposase gene expression. *Insect Mol. Biol.*, *10*, 447–456.

Li, X., Lobo, N., Bauser, C. A., & Fraser, M. J., Jr. (2001b). The minimum internal and external sequence requirements for transposition of the eukaryotic transformation vector *piggyBac*. *Mol. Genet. Genomics*, *266*, 190–198.

Li, X., Harrell, R. A., Handler, A. M., Beam, T., Hennessy, K., & Fraser, M. J., Jr. (2005). *piggyBac* internal sequences are necessary for efficient transformation of target genomes. *Insect Mol. Biol.*, *14*, 17–30.

Lidholm, D. A., Gudmundsson, G. H., & Boman, H. G. (1991). A highly repetitive, *mariner*-like element in the genome of *Hyalophora cecropia*. *J. Biol. Chem.*, *266*, 11518–11521.

Lidholm, D. -A., Lohe, A. R., & Hartl, D. L. (1993). The transposable element *mariner* mediates germline transformation in *Drosophila melanogaster*. *Genetics*, *134*, 859–868.

Lipkow, K., Buisine, N., Lampe, D. J., & Chalmers, R. (2004). Early intermediates of *mariner* transposition: Catalysis without synapsis of the transposon ends suggests a novel architecture of the synaptic complex. *Mol. Cell. Biol.*, *24*, 8301–8311.

Lis, J. T., Simon, J. A., & Sutton, C. A. (1983). New heat shock puffs and β-galactosidase activity resulting from transformation of *Drosophila* with an hsp70-LacZ hybrid gene. *Cell*, *35*, 403–410.

Liu, N., Pridgeon, J. W., Wang, H., Liu, Z., & Zhang, L. (2004). Identification of *mariner* elements from house flies (*Musca domestica*) and German cockroaches (*Blattella germanica*). *Insect Mol. Biol.*, *13*, 443–447.

Lobo, N., Li, X., Hua-Van, A., & Fraser, M. J., Jr. (2001). Mobility of the *piggyBac* transposon in embryos of the vectors of Dengue fever (*Aedes albopictus*) and La Crosse encephalitis (*Ae. triseriatus*). *Mol. Genet. Genomics*, *265*, 66–71.

Lobo, N. F., Hua-Van, A., Li, X., Nolen, B. M., & Fraser, M. J., Jr. (2002). Germ line transformation of the yellow fever mosquito, *Aedes aegypti*, mediated by transpositional insertion of a *piggyBac* vector. *Insect Mol. Biol.*, *11*, 133–913.

Lobo, N. F., Clayton, J. R., Fraser, M. J., Kafatos, F. C., & Collins, F. H. (2006). High efficiency germ-line transformation of mosquitoes. *Nat. Protoc.*, *1*, 1312–1317.

Lohe, A. R., & Hartl, D. L. (1996a). Germline transformation of *Drosophila virilis* with the transposable element *mariner*. *Genetics*, *143*, 365–374.

Lohe, A. R., & Hartl, D. L. (1996b). Autoregulation of *mariner* transposase activity by overproduction and dominant-negative complementation. *Mol. Biol. Evol.*, *13*, 549–555.

Lohe, A. R., & Hartl, D. L. (2002). Efficient mobilization of *mariner in vivo* requires multiple internal sequences. *Genetics*, *160*, 519–526.

Lohe, A. R., Lidholm, D. A., & Hartl, D. L. (1995). Genotypic effects, maternal effects and grandmaternal effects of immobilized derivatives of the transposable element *mariner*. *Genetics*, *140*, 183–192.

Lombardo, F., Lycett, G. J., Lanfrancotti, A., Coluzzi, M., & Arcà, B. (2009). Analysis of apyrase 5′ upstream region validates improved *Anopheles gambiae* transformation technique. *BMC Res. Notes*, *2*, 24.

Lorenzen, M. D., Brown, S. J., Denell, R. E., & Beeman, R. W. (2002). Cloning and characterization of the *Tribolium castaneum* eye-color genes encoding tryptophan oxygenase and kynurenine 3-monooxygenase. *Genetics*, *160*, 225–234.

Lorenzen, M. D., Berghammer, A. J., Brown, S. J., Denell, R. E., Klingler, M., & Beeman, R. W. (2003). *piggyBac*-mediated germ-line transformation in the beetle *Tribolium castaneum*. *Insect Mol. Biol.*, *32*, 1211–1220.

Lorenzen, M. D., Kimzey, T., Shippy, T. D., Brown, S. J., Denell, R. E., & Beeman, R. W. (2007). *piggyBac*-based insertional mutagenesis in *Tribolium castaneum* using donor/helper hybrids. *Insect Mol. Biol.*, *16*, 265–275.

Loukeris, T. G., Livadaras, I., Arca, B., Zabalou, S., & Savakis, C. (1995a). Gene transfer into the medfly, *Ceratitis capitata*, with a *Drosophila hydei* transposable element. *Science, 270*, 2002–2005.

Loukeris, T. G., Arca, B., Livadras, I., Dialektaki, G., & Savakis, C. (1995b). Introduction of the transposable element *Minos* into the germ line of *Drosophila melanogaster*. *Proc. Natl. Acad. Sci. USA, 92*, 9485–9489.

Lozovskaya, E. R., Nurminsky, D. I., Hartl, D. L., & Sullivan, D. T. (1996). Germline transformation of *Drosophila virilis* mediated by the transposable element *hobo*. *Genetics, 142*, 173–177.

Lozovsky, E. R., Nurminsky, D., Wimmer, E. A., & Hartl, D. L. (2002). Unexpected stability of *mariner* transgenes in *Drosophila*. *Genetics, 160*, 527–535.

Mamoun, C. B., Guzman, I. Y., Beverly, S. M., & Goldberg, D. E. (2000). Transposition of the Drosophila element *mariner* within the human malaria parasite *Plasmodium falciparum*. *Mol. Biochem. Parasitol., 110*, 405–407.

Marcus, J. M., Ramos, D. M., & Monteiro, A. (2004). Germline transformation of the butterfly *Bicyclus anynana*. *Proc. R. Soc. Lond. Series B Biol. Sci., 271*, S263–265.

Martins, S., Naish, N., Morrison, N., & Alphey, L. (2010). Transgenesis of the diamondback moth using the *piggyBac* transposable element. *Antenna, 34*, 143.

Matz, M. V., Fradkov, A. F., Labas, Y. A., Savitsky, A. P., Zaraisky, A. G., et al. (1999). Fluorescent proteins from nonbioluminescent Anthozoa species. *Nat. Biotechnol., 17*, 969–973.

Matz, M. V., Lukyanov, K. A., & Lukyanov, S. A. (2002). Family of the green fluorescent protein: Journey to the end of the rainbow. *Bioessays, 24*, 953–959.

McGrane, V., Carlson, J. O., Miller, B. R., & Beaty, B. J. (1988). Microinjection of DNA into *Aedes triseriatus* ova and detection of integration. *Am. J. Trop. Med. Hyg., 39*, 502–510.

Medhora, M. M., MacPeek, A. H., & Hartl, D. L. (1988). Excision of the *Drosophila* transposable element *mariner*: identification and characterization of the *Mos* factor. *EMBO J., 7*, 2185–2189.

Meza, J. S., Nirmala, X., Zimowska, G. J., Zepeda-Cisneros, C. S., & Handler, A. M. (2010). Development of transgenic strains for the biological control of the Mexican fruit fly, *Anastrepha ludens. Genetica, 139, 1*, 53–62 .

Miahle, E., & Miller, L. H. (1994). Biolistic techniques for transfection of mosquito embryos (*Anopheles gambiae*). *Biotechniques, 16*, 924–931.

Michel, K., & Atkinson, P. W. (2003). Nuclear localization of the *Hermes* transposase depends on basic amino acid residues at the N-terminus of the protein. *J. Cell Biochem., 89*, 778–790.

Michel, K., Stamenova, A., Pinkerton, A. C., Franz, G., Robinson, A. S., et al. (2001). *Hermes*-mediated germ-line transformation of the Mediterranean fruit fly *Ceratitis capitata*. *Insect Mol. Biol., 10*, 155–162.

Michel, K., O'Brochta, D. A., & Atkinson, P. W. (2003). The C-terminus of the *Hermes* transposase contains a protein multimerization domain. *Insect Biochem. Mol. Biol., 33*, 959–970.

Miller, L. H., Sakai, R. K., Romans, P., Gwadz, R. W., Kantoff, P., & Coon, H. G. (1987). Stable integration and expression of a bacterial gene in the mosquito *Anopheles gambiae*. *Science, 237*, 779–781.

Mitra, R., Fain-Thornton, J., & Craig, N. L. (2008). *piggyBac* can bypass DNA synthesis during cut and paste transposition. *EMBO J., 27*, 1097–1109.

Mittapalli, O., Shukle, R. H., & Wise, I. L. (2006). Identification of *mariner*-like elements in *Sitodiplosis mosellana* (Diptera: Cecidomyiidae). *Can. Entomol., 138*, 138–146.

Morris, A. C. (1997). Microinjection of mosquito embryos. In J. M. Crampton, C. B. Beard, & C. Louis (Eds.). *Molecular Biology of Insect Disease Vectors: A Methods Manual*. New York, NY: Chapman and Hall.

Morris, A. C., Eggleston, P., & Crampton, J. M. (1989). Genetic transformation of the mosquito *Aedes aegypti* by micro-injection of DNA. *Med. Vet. Entomol., 3*, 1–7.

Mota, N. R., Ludwig, A., Valente, V. L. D. S., & Loreto, E. L. S. (2010). *harrow*: New Drosophila *hAT* transposons involved in horizontal transfer. *Insect Mol. Biol., 19*, 217–228.

Mullins, M. C., Rio, D. C., & Rubin, G. M. (1989). *Cis*-acting DNA sequence requirements for P-element transposition. *Cell, 3*, 729–738.

Nawa, S., & Yamada, S. (1968). Hereditary change in *Ephestia* after treatment with DNA. *Genetics, 58*, 573–584.

Nawa, S., Sakaguchi, B., Yamada, M. A., & Tsujita, M. (1971). Hereditary change in *Bombyx* after treatment with DNA. *Genetics, 67*, 221–234.

Nimmo, D. D., Alphey, L., Meredith, J. M., & Eggleston, P. (2006). High efficiency site-specific genetic engineering of the mosquito genome. *Insect Mol. Biol., 15*, 129–136.

Nirmala, X., Olson, S., Holler, T. C., Cho, K. H., & Handler, A. M. (2010). A DsRed fluorescent protein marker under *polyubiquitin* promoter regulation allows visual and amplified gene detection of transgenic Caribbean fruit flies in liquid traps. *Biocontrol.*, in press.

Nolan, T., Bower, T. M., Brown, A. E., Crisanti, A., & Catteruccia, F. (2002). *piggyBac*-mediated germline transformation of the malaria mosquito *Anopheles stephensi* using the red fluorescent protein dsRED as a selectable marker. *J. Biol. Chem., 277*, 8759–8762.

Oberstein, A., Pare, A., Kaplan, L., & Small, S. (2005). Site-specific transgenesis by Cre-mediated recombination in *Drosophila*. *Nat. Methods, 2*, 583–585.

O'Brochta, D. A., & Handler, A. M. (1988). Mobility of *P* elements in drosophilids and nondrosophilids. *Proc. Natl. Acad. Sci. USA, 85*, 6052–6056.

O'Brochta, D. A., Gomez, S. P., & Handler, A. M. (1991). *P* element excision in *Drosophila melanogaster* and related drosophilids. *Mol. Gen. Genet., 225*, 387–394.

O'Brochta, D. A., Warren, W. D., Saville, K. J., & Atkinson, P. W. (1994). Interplasmid transposition of *Drosophila hobo* elements in non-drosophilid insects. *Mol. Gen. Genet., 244*, 9–14.

O'Brochta, D. A., Warren, W. D., Saville, K. J., & Atkinson, P. W. (1996). *Hermes*, a functional non-drosophilid insect gene vector. *Genetics, 142*, 907–914.

O'Brochta, D. A., Atkinson, P. W., & Lehane, M. J. (2000). Transformation of *Stomoxys calcitrans* with a *Hermes* gene vector. *Insect Mol. Biol., 9*, 531–538.

O'Brochta, D. A., Stosic, C. D., Pilitt, K., Subramanian, R. A., Hice, R. H., & Atkinson, P. W. (2009). Transpositionally active episomal *hAT* elements. *BMC Mol. Biol., 10, 108*.

O'Hare, K., & Rubin, G. M. (1983). Structures of *P* transposable elements and their sites of insertion and excision in the *Drosophila melanogaster* genome. *Cell*, *34*, 25–35.

Ortiz, M. D., & Loreto, E. L. S. (2009). Characterization of new hAT transposable elements in 12 *Drosophila* genomes. *Genetica*, *135*, 67–75.

Ortiz, M. D., Lorenzatto, K. R., Correa, B. R.S., & Loreto, E. L. S. (2010). *hAT* transposable elements and their derivatives: An analysis in the 12 *Drosophila* genomes. *Genetica*, *138*, 649–655.

Papadimitriou, E., Kritikou, D., Mavroidis, M., Zacharopoulou, A., & Mintzas, A. C. (1998). The heat shock 70 gene family in the Mediterranean fruit fly *Ceratitis capitata*. *Insect Mol. Biol.*, *7*, 279–290.

Park, J. M., Evertts, A. G., & Levin, H. L. (2009). The *Hermes* transposon of *Musca domestica* and its use as a mutagen of *Schizosaccharomyces pombe*. *Methods*, *49*, 243–247.

Pavlopoulos, A., Berghammer, A. J., Averof, M., & Klingler, M. (2004). Efficient transformation of the beetle *Tribolium castaneum* using the *Minos* transposable element: Quantitative and qualitative analysis of genomic integration events. *Genetics*, *167*, 737–746.

Pavlopoulos, A., Oehler, S., Kapetanaki, M. G., & Savakis, C. (2007). The DNA transposon *Minos* as a tool for transgenesis and functional genomic analysis in vertebrates and invertebrates. *Genome Biol.*, (Suppl. 1), S2.

Peloquin, J. J., Thibault, S. T., Schouest, L. P., Jr., & Miller, T. A. (1997). Electromechanical microinjection of pink bollworm *Pectinophora gossypiella* embryos increases survival. *Biotechniques*, *22*, 496–499.

Peloquin, J. J., Thibault, S. T., Staten, R., & Miller, T. A. (2000). Germ-line transformation of pink bollworm (Lepidoptera: Gelechiidae) mediated by the *piggyBac* transposable element. *Insect Mol. Biol.*, *9*, 323–333.

Perera, O. P., Harrell, R. A., & Handler, A. M. (2002). Germline transformation of the South American malaria vector, *Anopheles albimanus*, with a *piggyBac/EGFP* transposon vector is routine and highly efficient. *Insect Mol. Biol.*, *11*, 291–297.

Phillips, J. P., Xin, J. H., Kirby, K., Milne, C. P., Krell, P., & Wild, J. R. (1990). Transfer and expression of an organophosphate insecticide degrading gene from *Pseudomonas* in *Drosophila melanogaster*. *Proc. Natl. Acad. Sci. USA*, *87*, 8155–8159.

Pinkerton, A. C., Whyard, S., Mende, H. M., Coates, C. J., O'Brochta, D. A., & Atkinson, P. W. (1999). The Queensland fruit fly, *Bactrocera tryoni*, contains multiple members of the *hAT* family of transposable elements. *Insect Mol. Biol.*, *8*, 423–434.

Pinkerton, A. C., Michel, K., O'Brochta, D. A., & Atkinson, P. W. (2000). Green fluorescent protein as a genetic marker in transgenic *Aedes aegypti*. *Insect Mol. Biol.*, *9*, 1–10.

Pirrotta, V. (1988). Vectors for P-mediated transformation in *Drosophila*. In R. L. Rodriguez, & D. T. Denhardt (Eds.), *Vectors – A Survey of Molecular Cloning Vectors and their Uses* (p. 437). Oxford, UK: Butterworths.

Pirrotta, V., Steller, H., & Bozzetti, M. P. (1985). Multiple upstream regulatory elements control the expression of the *Drosophila white* gene. *EMBO J.*, *4*, 3501–3508.

Plautz, J. D., Day, R. N., Dailey, G. M., Welsh, S. B., Hall, J. C., et al. (1996). Green fluorescent protein and its derivatives as versatile markers for gene expression in living *Drosophila melanogaster*, plant and mammalian cells. *Gene.*, *173*, 83–87.

Pledger, D. W., Fu, Y. Q., & Coates, C. J. (2004). Analyses of cis-acting elements that affect the transposition of *Mos1 mariner* transposons *in vivo*. *Mol. Genet. Genom.*, *272*, 67–75.

Prasher, D. C., Eckenrode, V. K., Ward, W. W., Prendergast, F. G., & Cormier, M. J. (1992). Primary structure of the *Aequorea victoria* green fluorescent protein. *Gene.*, *111*, 229–233.

Presnail, J. K., & Hoy, M. A. (1994). Transmission of injected DNA sequences to multiple eggs of *Metaseiulus occidentalis* and *Amblyseius finlandicus* (Acari: Phytoseiidae) following maternal microinjection. *Exp. Appl. Acarol.*, *18*, 319–330.

Raphael, K. A., Shearman, D. C., Streamer, K., Morrow, J. L., Handler, A. M., & Frommer, M. (2010). Germ-line transformation of the Queensland fruit fly, *Bactrocera tryoni*, using a *piggyBac* vector in the presence of endogenous *piggyBac* elements. *Genetica*, *139*, 91–97.

Ren, X., Park, Y., & Miller, T. A. (2006). Intact *mariner*-like element in tobacco budworm, *Heliothis virescens* (Lepidoptera: Noctuidae). *Insect Mol. Biol.*, *15*, 743–748.

Rezende-Teixeira, P., Siviero, F., Andrade, A., Santelli, R. V., & Machado-Santelli, G. M. (2008). *Mariner*-like elements in *Rhynchosciara americana* (Sciaridae) genome: Molecular and cytological aspects. *Genetica*, *133*, 137–145.

Rezende-Teixeira, P., Lauand, C., Siviero, F., & Machado-Santelli, G. M. (2010). Normal and defective *mariner*-like elements in *Rhynchosciara* species (Sciaridae, Diptera). *Genet. Mol. Res.*, *9*, 849–857.

Reznikoff, W. S. (2000). Tn5 transposition. In N. L. Craig, R. Craige, M. Gellert, & A. M. Lambowitz (Eds.), *Mobile DNA II* (pp. 403–422). Washington, DC: ASM Press.

Reznikoff, W. S., Bhasin, A., Davies, D. R., Goryshin, I. Y., Mahnke, L. A., et al. (1999). Tn5: A molecular window on transposition. *Biochem. Biophys. Res. Comm.*, *266*, 729–734.

Richardson, J. M., Dawson, A., O'Hagan, N., Taylor, P., Finnegan, D. J., & Walkinshaw, M. D. (2006). Mechanism of *Mos1* transposition: Insights from structural analysis. *EMBO J.*, *25*, 1324–1334.

Richardson, J. M., Colloms, S. D., Finnegan, D. J., & Walkinshaw, M. D. (2009). Molecular architecture of the *Mos1* paired-end complex: The structural basis of DNA transposition in a eukaryote. *Cell*, *138*, 1096–1108.

Rio, D. C., & Rubin, G. M. (1988). Identification and purification of a *Drosophila* protein that binds to the terminal 31-base-pair inverted repeats of the transposable element *Proc. Natl. Acad. Sci. USA*, *85*, 8929–8933.

Rio, D. C., Laski, F. A., & Rubin, G. M. (1986). Identification and immunochemical analysis of biologically active *Drosophila P* element transposase. *Cell*, *44*, 21–32.

Rivera-Vega, L., & Mittapalli, O. (2010). Molecular characterization of *mariner*-like elements in emerald ash borer, *Agrilus planipennis* (Coleoptera, Polyphaga). *Arch. Insect Biochem. Physiol.*, *74*, 205–216.

Robertson, H. M. (1993). The *mariner* transposable element is widespread in insects. *Nature*, *362*, 241–245.

Robertson, H. M. (2000). Evolution of DNA transposons in eukaryotes. In N. L. Craig, R. Craige, M. Gellert, & A. M. Lambowitz (Eds.), *Mobile DNA II* (p. 1093). Washington, DC: ASM Press.

Robertson, H. M., & Lampe, D. J. (1995a). Distribution of transposable elements in arthropods. *Annu. Rev. Entomol.*, *40*, 333–357.

Robertson, H. M., & Lampe, D. J. (1995b). Recent horizontal transfer of a *mariner* element between Diptera and Neuroptera. *Mol. Biol. Evol.*, *12*, 850–862.

Robertson, H. M., & MacLeod, E. G. (1993). Five major subfamilies of *mariner* transposable elements in insects, including the Mediterranean fruit fly, and related arthropods. *Insect Mol. Biol.*, *2*, 125–139.

Robinson, A. S., Franz, G., & Atkinson, P. W. (2004). Insect transgenesis and its potential role in agriculture and human health. *Insect Biochem. Mol. Biol.*, *34*, 113–120.

Rodrigues, F. G., Oliveira, S. B., Rocha, B. C., & Moreira, L. A. (2006). Germline transformation of *Aedes fluviatilis* (Diptera: Culicidae) with the *piggyBac* transposable element. *Mem. Inst. Oswaldo Cruz*, *101*, 755–757.

Rouault, J. D., Casse, N., Chenais, B., Hua-Van, A., Filee, J., & Capy, P. (2009). Automatic classification within families of transposable elements: Application to the *mariner* family. *Gene*, *448*, 227–232.

Rouleux-Bonnin, F., Petit, A., Demattei, M. V., & Bigot, Y. (2005). Evolution of full-length and deleted forms of the *Mariner*-like element, Botmar1, in the genome of the bumble bee, *Bombus terrestris* (Hymenoptera: Apidae). *J. Mol. Evol.*, *60*, 736–747.

Rowan, K., Orsetti, J., Atkinson, P. W., & O'Brochta, D. A. (2004). Tn5 as an insect gene vector. *Insect Biochem. Mol. Biol.*, *34*, 695–705.

Rubin, E. J., Akerley, B. J., Novik, V. N., Lampe, D. J., Husson, R. N., & Mekalanos, J. J. (1999). *In vivo* transposition of *mariner*-based elements in enteric bacteria and mycobacteria. *Proc. Natl. Acad. Sci. USA*, *96*, 1645–1650.

Rubin, G. M., & Spradling, A. C. (1982). Genetic transformation of *Drosophila* with transposable element vectors. *Science*, *218*, 348–353.

Rubin, G. M., Kidwell, M. G., & Bingham, P. M. (1982). The molecular basis of P-M hybrid dysgenesis: The nature of induced mutations. *Cell*, *29*, 987–994.

Sarkar, A., & Collins, F. C. (2000). Eye color genes for selection of transgenic insects. In A. M. Handler, & A. A. James (Eds.), *Insect Transgenesis: Methods and Applications* (p. 79). Boca Raton, FL: CRC Press.

Sarkar, A., Coates, C. J., Whyard, S., Willhoeft, U., Atkinson, P. W., & O'Brochta, D. A. (1997a). The *Hermes* element from *Musca domestica* can transpose in four families of cyclorrhaphan flies. *Genetica*, *99*, 15–29.

Sarkar, A., Yardley, K., Atkinson, P. W., James, A. A., & O'Brochta, D. A. (1997b). Transposition of the *Hermes* element in embryos of the vector mosquito, *Aedes aegypti. Insect Biochem. Mol. Biol.*, *27*, 359–363.

Sarkar, A., Atapattu, A., Belikoff, E. J., Heinrich, J. C., Li, X., et al. (2006). Insulated *piggyBac* vectors for insect transgenesis. *BMC Biotechnol.*, *6*, 27.

Saville, K. J., Warren, W. D., Atkinson, P. W., & O'Brochta, D. A. (1999). Integration specificity of the *hobo* element of *Drosophila melanogaster* is dependent on sequences flanking the target site. *Genetica*, *105*, 133–147.

Scavarda, N. J., & Hartl, D. L. (1984). Interspecific DNA transformation in *Drosophila. Proc. Natl. Acad. Sci. USA*, *81*, 7515–7519.

Schetelig, M. F., Scolari, F., Handler, A. M., Kittelmann, S., Gasperi, G., & Wimmer, E. A. (2009). Site-specific recombination for the modification of transgenic strains of the Mediterranean fruit fly *Ceratitis capitata. Proc. Natl. Acad. Sci. USA*, *106*, 18171–18176.

Schetelig, M. F., Götschel, F., Viktorinová, I., Handler, A. M., & Wimmer, E. A. (2010). Recombination technologies for enhanced transgene stability in bioengineered insects. *Genetica*, *139*, 71–78.

Scholnick, S. B., Morgan, B. A., & Hirsh, J. (1983). The cloned Dopa decarboxylase gene is developmentally regulated when reintegrated into the *Drosophila* genome. *Cell*, *34*, 37–45.

Scolari, F., Schetelig, M. F., Bertin, S., Malacrida, A. R., Gasperi, G., & Wimmer, E. A. (2008). Fluorescent sperm marking to improve the fight against the pest insect *Ceratitis capitata* (Wiedemann; Diptera: Tephritidae). *N. Biotechnol.*, *25*, 76–84.

Senecoff, J. F., Bruckner, R. C., & Cox, M. M. (1985). The FLP recombinase of the yeast 2-micron plasmid: Characterization of its recombination site. *Proc. Natl. Acad. Sci. USA*, *82*, 7270–7274.

Sethuraman, N., Fraser, M. J., Jr., Eggleston, P., & O'Brochta, D. A. (2007). Post-integration stability of *piggyBac* in *Aedes aegypti. Insect Biochem. Mol. Biol.*, *37*, 941–951.

Sheng, G., Thouvenot, E., Schmucker, D., Wilson, D. S., & Desplan, C. (1997). Direct regulation of rhodopsin 1 by Pax-6/eyeless in *Drosophila*: Evidence for a conserved function in photoreceptors. *Genes. Dev.*, *11*, 1122–1131.

Sherman, A., Dawson, A., Mather, C., Gilhooley, H., Li, Y., et al. (1998). Transposition of the *Drosophila* element *mariner* into the chicken germ line. *Nature Biotech.*, *16*, 1050–1053.

Simmons, M. J., Haley, K. J., Grimes, C. D., Raymond, J. D., & Niemi, J. B. (2002). A *hobo* transgene that encodes the P-element transposase in *Drosophila melanogaster*: Autoregulation and cytotype control of transposase activity. *Genetics*, *161*, 195–204.

Sinzelle, L., Jegot, G., Brillet, B., Rouleux-Bonnin, F., Bigot, Y., & Auge-Gouillou, C. (2008). Factors acting on *Mos1* transposition efficiency. *BMC Mol. Biol.*, *9*, 106.

Spradling, A. C. (1986). *P* element-mediated transformation. In D. B. Roberts (Ed.), *Drosophila: A Practical Approach* (p. 175). Oxford, UK: IRL Press.

Steller, H., & Pirrotta, V. (1985). A transposable *P* vector that confers selectable G418 resistance to *Drosophila* larvae. *EMBO J.*, *4*, 167–171.

Steller, H., & Pirrotta, V. (1986). *P* transposons controlled by the heat shock promoter. *Mol. Cell Biol.*, *6*, 1640–1649.

Strack, R. L., Strongin, D. E., Bhattacharyya, D., Tao, W., Berman, A., et al. (2008). A noncytotoxic DsRed variant for whole-cell labeling. *Nat. Methods*, *5*, 955–957.

Subramanian, R. A., Arensburger, P., Atkinson, P. W., & O'Brochta, D. A. (2007). Transposable element dynamics of the hAT element Herves in the human malaria vector *Anopheles gambiae* s.s. *Genetics, 176*, 2477–2487.

Subramanian, R. A., Akala, O. O., Adejinmi, J. O., & O'Brochta, D. A. (2008). *Topi*, an *IS630/Tc1/mariner*-type transposable element in the African malaria mosquito, *Anopheles gambiae*. *Gene, 15*, 63–71.

Subramanian, R. A., Cathcart, L. A., Krafsur, E. S., Atkinson, P. W., & O'Brochta, D. A. (2009). *Hermes* transposon distribution and structure in *Musca domestica*. *J. Heredity, 100*, 473–480.

Sumitani, M., Yamamoto, D. S., Oishi, K., Lee, J. M., & Hatakeyama, M. (2003). Germline transformation of the sawfly, *Athalia rosae* (Hymenoptera: Symphyta), mediated by a *piggyBac*-derived vector. *Insect Biochem. Mol. Biol., 33*, 449–458.

Sun, Z. C., Wu, M., Miller, T. A., & Han, Z. J. (2008). *piggyBac*-like elements in cotton bollworm, *Helicoverpa armigera* (Hübner). *Insect Mol. Biol., 17*, 9–18.

Sundararajan, P., Atkinson, P. W., & O'Brochta, D. A. (1999). Transposable element interactions in insects: Crossmobilization of *hobo* and *Hermes*. *Insect Mol. Biol., 8*, 359–368.

Sundaresan, V., & Freeling, M. (1987). An extrachromosomal form of the Mu transposon of maize. *Proc. Natl. Acad. Sci. USA, 84*, 4924–4928.

Taillebourg, E., & Dura, J. M. (1999). A novel mechanism for *P* element homing in *Drosophila*. *Proc. Natl. Acad. Sci. USA, 96*, 6856–6861.

Tamura, T., Thibert, T., Royer, C., Kanda, T., Eappen, A., et al. (2000). A *piggybac* element-derived vector efficiently promotes germ-line transformation in the silkworm *Bombyx mori* L. *Nature Biotech., 18*, 81–84.

Thibault, S. T., Luu, H. T., Vann, N., & Miller, T. A. (1999). Precise excision and transposition of *piggyBac* in pink bollworm embryos. *Insect Mol. Biol., 8*, 119–123.

Thibault, S. T., Singer, M. A., Miyazaki, W. Y., et al. (2004). A complementary transposon tool kit for *Drosophila melanogaster* using *P* and *piggyBac*. *Nat Genet., 36*, 283–287.

Thomas, J. L., Bardou, J., L'hoste, S., Mauchamp, B., & Chavancy, G. (2001). A helium burst biolistic device adapted to penetrate fragile insect tissues. *J. Insect Sci., 1*, 9.

Thomas, J. L., Da Rocha, M., Besse, A., Mauchamp, B., & Chavancy, G. (2002). 3xP3-EGFP marker facilitates screening for transgenic silkworm *Bombyx mori* L. from the embryonic stage onwards. *Insect Biochem. Mol. Biol., 32*, 247–253.

Tosi, L. R.O., & Beverly, S. M. (2000). *Cis* and *trans* factors affecting *Mos1 mariner* evolution and transposition *in vitro*, and its potential for functional genomics. *Nucleic Acid Res., 28*, 784–790.

Trauner, J., Schinko, J., Lorenzen, M. D., Shippy, T. D., Wimmer, E. A., et al. (2009). Large-scale insertional mutagenesis of a coleopteran stored grain pest, the red flour beetle *Tribolium castaneum*, identifies embryonic lethal mutations and enhancer traps. *BMC Biol., 7*, 73.

Uchino, K., Imamura, M., Shimizu, K., Kanda, T., & Tamura, T. (2007). Germ line transformation of the silkworm, *Bombyx mori*, using the transposable element *Minos*. *Mol. Genet. Genomics, 277*, 213–220.

Uchino, K., Sezutsu, H., Imamura, M., Kobayashi, I., Tatematsu, K., et al. (2008). Construction of a *piggyBac*-based enhancer trap system for the analysis of gene function in silkworm *Bombyx mori*. *Insect Biochem. Mol. Biol., 38*, 1165–1173.

Uhlirova, M., Asahina, M., Riddiford, L. M., & Jindra, M. (2002). Heat-inducible transgenic expression in the silkmoth *Bombyx mori*. *Dev. Genes. Evol., 212*, 145–151.

Venken, K. J.T., & Bellen, H. J. (2005). Emerging technologies for gene manipulation in *Drosophila melanogaster*. *Nature Rev. Gen., 6*, 167–178.

Venken, K. J.T., & Bellen, H. J. (2007). Transgenesis upgrades for *Drosophila melanogaster*. *Develop, 134*, 3571–3584.

Walker, V. K. (1990). Gene transfer in insects. In K. Maramorosch (Ed.), *Advances in Cell Culture* (*Vol. 7*, p. 87). New York, NY: Academic Press.

Wang, H. H., & Fraser, M. J. (1993). TTAA serves as the target site for TFP3 lepidopteran transposon insertions in both nuclear polyhedrosis virus and *Trichoplusia ni* genomes. *Insect Mol. Biol., 1*, 1–7.

Wang, H. H., Fraser, M. J., & Cary, L. C. (1989). Transposon mutagenesis of baculoviruses: Analysis of TFP3 lepidopteran insertions at the FP locus of nuclear polyhedrosis viruses. *Gene, 81*, 97–108.

Wang, J., Staten, R., Miller, T. A., & Park, Y. (2005). Inactivated *mariner*-like elements (MLE) in pink bollworm, *Pectinophora gossypiella*. *Insect Mol. Biol., 14*, 547–553.

Wang, J., Ren, X., Miller, T. A., & Park, Y. (2006). *piggyBac*-like elements in the tobacco budworm, *Heliothis virescens* (Fabricius). *Insect Mol. Biol., 15*, 435–443.

Wang, J., Miller, E. D., Simmons, G. S., Miller, T. A., Tabashnik, B. E., & Park, Y. (2010). *piggyBac*-like elements in the pink bollworm, *Pectinophora gossypiella*. *Insect Mol. Biol., 19*, 177–184.

Wang, W., Swevers, L., & Iatrou, K. (2000). *Mariner* (*Mos1*) transposase and genomic integration of foreign gene sequences in *Bombyx mori* cells. *Insect Mol. Biol., 9*, 145–155.

Warren, W. D., Atkinson, P. W., & O'Brochta, D. A. (1994). The *Hermes* transposable element from the house fly, *Musca domestica*, is a short inverted repeat-type element of the *hobo*, *Ac*, and *Tam3* (*hAT*) element family. *Genet. Res. Camb., 64*, 87–97.

Wen, H., Lan, X., Zhang, Y., Zhao, T., Wang, Y., et al. (2010). Transgenic silkworms (*Bombyx mori*) produce recombinant spider dragline silk in cocoons. *Mol. Biol. Rep., 37*, 1815–1821.

White, L. D., Coates, C. J., Atkinson, P. W., & O'Brochta, D. A. (1996). An eye color gene for the detection of transgenic non-drosophilid insects. *Insect Biochem. Mol. Biol., 26*, 641–644.

Wilson, C., Pearson, R. K., Bellen, H. J., O'Kane, C. J., Grossniklaus, U., & Gehring, W. J. (1989). P-element-mediated enhancer detection: An efficient method for isolating and characterizing developmentally regulated genes in *Drosophila*. *Genes Develop., 3*, 1301–1313.

Wilson, R., Orsetti, J., Klocko, A. D., Aluvihare, C., Peckham, E., et al. (2003). Post-integration behavior of a *Mos1* gene vector in *Aedes aegypti*. *Insect Biochem. Mol. Biol., 33*, 853–863.

Wu, M., Sun, Z. C., Hu, C. L., Zhang, G. F., & Han, Z. J. (2008). An active *piggyBac*-like element in Macdunnoughia rassisigna. *Insect Sci.*, *15*, 521–528.

Yoshiyama, M., Honda, H., & Kimura, K. (2000). Successful transformation of the housefly, *Musca domestica* (Diptera: Muscidae) with the transposable element, *mariner*. *Appl. Entomol. Zool.*, *35*, 321–325.

Yuen, J. L., Read, S. A., Brubacher, J. L., Singh, A. D., & Whyard, S. (2008). Biolistics for high-throughput transformation and RNA interference in *Drosophila melanogaster*. *Fly*, *2*, 247–254.

Zakharkin, S. O., Willis, R. L., Litvinova, O. V., Jinwal, U. K., Headley, V. V., & Benes, H. (2004). Identification of two *mariner*-like elements in the genome of the mosquito *Ochlerotatus atropalpus*. *Insect Biochem. Mol. Biol.*, *34*, 377–386.

Zhang, J. K., Pritchett, M. S., Lampe, D. J., Robertson, H. M., & Metcalf, W. W. (2000). *In vivo* transposon mutagenesis of the methanogenic archeon *Methanosarcina acetivorans C2A* using a modified version of the insect *mariner*-family transposable element *Himar1*. *Proc. Natl. Acad. Sci. USA*, *97*, 9665–9670.

Zhang, H., Shinmyo, Y., Hirose, A., Mito, T., Inoue, Y., et al. (2002). Extrachromosomal transposition of the transposable element *Minos* in embryos of the cricket *Gryllus bimaculatus*. *Dev. Growth Diff.*, *44*, 409–417.

Zhang, L., Sankar, U., Lampe, D. J., Robertson, H. M., & Graham, F. L. (1998). The *Himar1 mariner* transposase cloned in a recombinant adenovirus vector is functional in mammalian cells. *Nucleic Acids Res.*, *226*, 3687–3693.

Zhou, L. Q., Mitra, R., Atkinson, P. W., Hickman, A. B., Dyda, F., & Craig, N. L. (2004). Transposition of *hAT* elements links transposable elements and V(D)J recombination. *Nature*, *432*, 995–1001.

Zimowska, G. J., & Handler, A. M. (2006). Highly conserved *piggyBac* elements in noctuid species of Lepidoptera. *Insect Biochem. Mol. Biol.*, *36*, 421–428.

Zimowska, G. J., Nirmala, X., & Handler, A. M. (2009). The beta2-tubulin gene from three tephritid fruit fly species and use of its promoter for sperm marking. *Insect Biochem. Mol. Biol.*, *39*, 508–515.

Zwiebel, L. J., Saccone, G., Zacharapoulou, A., Besansky, N. J., Favia, G., et al. (1995). The *white* gene of *Ceratitis capitata*: A phenotypic marker for germline transformation. *Science*, *270*, 2005–2008.

5 Cuticular Proteins

Judith H Willis
Department of Cellular Biology,
University of Georgia, Athens, GA, USA
Nikos C Papandreou
Department of Cell Biology and Biophysics,
Faculty of Biology, University of Athens,
Athens, Greece
Vassiliki A Iconomidou
Department of Cell Biology and Biophysics,
Faculty of Biology, University of Athens,
Athens, Greece
Stavros J Hamodrakas
Department of Cell Biology and Biophysics,
Faculty of Biology, University of Athens,
Athens, Greece

5.1. Introduction

In the first edition of this series, Silvert (1985) outlined several major areas of uncertainty regarding cuticular proteins. The questions raised were: Were proteins extracted from cuticle authentic cuticular proteins, or might some be contaminants of adhering cells and hemolymph? Was the epidermis the sole site of synthesis of cuticular proteins, or were some synthesized in other tissues and transported to the cuticle? What was the relation among cuticular proteins of various developmental stages? Did cuticular proteins share common structural features?

That review presented all the cuticular protein sequence data then available – four complete and three partial sequences from *Drosophila melanogaster*, and one partial sequence from *Sarcophaga bullata*. The considerable sequence similarity seen with those limited data indicated that cuticular protein genes belonged to multi-gene families, and the even more limited genomic information revealed that similar genes were adjacent on a chromosome.

Progress over the next two decades was impressive, but not surprising, given the advances in relevant techniques. Elegant immunolocalization analyses solved the

DOI:10.1016/B978-0-12-384747-8.10005-4

problem of the sources of cuticular proteins. The 2005 version of this article reported that over 300 cuticular protein sequences had been recognized, from 6 orders and over 20 species of insects. Progress since 2005 has been predominantly in identifying new cuticular protein sequences. Since then, several whole genomes have been annotated and hundreds of new sequences have been posted, based on EST (expressed sequence tag) analyses. It is no longer possible to list the available sequences. Rather, this chapter will summarize what was learned from these data. Some areas of analysis have not progressed far since the 2005 version, and will be repeated here without revision.

5.2. Cuticle Structure and Synthesis

5.2.1. Cuticle Morphology

5.2.1.1. Terminology The descriptive terms used here to describe the regions of cuticle have been simplified according to Locke's (2001) cogent suggestions for new nomenclature. He proposes the use of the term "envelope" to describe the outermost layer of cuticle, rather than the previous term "cuticulin." At the start of each molt cycle, the smooth apical plasma membrane forms microvilli with plaques at their tips where the new envelope assembles. This discrete layer of 10–30 nm serves not only to protect the underlying epidermis from molting fluid enzymes that begin to digest the old cuticle, but also, as Locke points out, affects "resistance to abrasion and infection, penetration of insecticides, permeability, surface reflectivity, and physical colors." The sequences and properties of its constituent proteins remain unknown.

Next formed is the epicuticle, about 1 μm in thickness. This chitin-free layer (but see Section 5.2.1.3) is stabilized by quinones. It was formerly referred to as the "inner epicuticle," with cuticulin being the outer.

Former arguments about the precise distinction between exo- and endo-cuticle are eliminated by Locke's lumping together of the inner regions of the cuticle under the term "procuticle," encompassing both pre-ecdysial and post-ecdysial secretions. The procuticle, then, is the region that combines chitin and cuticular proteins in various combinations, and becomes sclerotized (Andersen, 2010a; see also Chapter 6 in this volume) and pigmented to varying degrees. This is the region depicted in electron micrographs showing stacks of precisely oriented lamellae. According to Locke (2001), it is the secretion of chitin fibers by apical microvilli that apparently bend in concert across the epithelial sheet to orient the laminae that gather into lamellae (Neville, 1975; Locke, 1998). Alternative mechanisms have been proposed by Moussian (2010), involving movement of the "chitin synthesis complex" across the cell surface or merely self-assembly. While knowledge of the process of secreting and assembling such a highly ordered structure is limited, details about the proteins associated with the lamellae are now voluminous.

5.2.1.2. Growth of the cuticle within an instar Central to the issue of cuticle structure is the important fact that considerable cuticle growth can occur during an intermolt period (Williams, 1980), some of it by a smoothing out of macro- and microscopic folds and pleats (Carter and Locke, 1993). During intra-instar growth, new cuticular proteins are interspersed among the old, necessitating a model of chitin–protein and protein–protein interactions that will permit such intussusception (Condoulis and Locke, 1966; Wolfgang and Riddiford, 1986).

5.2.1.3. Localization of cuticular proteins within the cuticle Precise localization of cuticular proteins within the cuticle, and even within cellular organelles, has been made possible with immunogold labeling of electron-microscopic sections. Here, a specific primary antibody is bound to the sections and visualized with a secondary antibody conjugated to colloidal gold particles.

Antibodies have been raised against extracts of whole cuticle or isolated electrophoretic bands, and the specificity of each antibody ascertained with Western blots. While each polyclonal antibody raised against a single band was specific for the immunizing protein, monoclonals raised against cuticular extracts frequently reacted with more than one electrophoretic band.

One concern with immunolocalization is that as cuticular proteins become modified in the cuticle by binding to chitin or by becoming sclerotized, the immunizing epitopes might become masked – a problem that should be more serious with monoclonal than with polyclonal antibodies. All groups recognized that while the presence of an antigen is significant, its absence may reflect no more than such masking.

This concern is significant when one considers results of immunolocalization in the assembly zone, the region of cuticle directly above the microvilli. It is here that chitin secreted from the tips of the microvilli interacts with cuticular proteins secreted into the perimicrovillar space. Immunolocalization studies revealed only a few of the cuticular proteins within the perimicrovillar space, but the same ones and others were abundant in the assembly zone directly above it (Locke *et al.*, 1994; Locke, 1998). The authors' conclusion was that the assembly zone "is where we should expect proteins to unravel and expose most epitopes in preparation for assuming a new configuration as they stabilize in the maturing cuticle." Wolfgang *et al.* (1986, 1987) found two *Drosophila melanogaster* cuticular proteins exclusively in this zone, and suggested they might function in cuticle assembly. Locke *et al.* (1994) point out that it was common for antibodies raised against *Calpodes ethlius* proteins to react more strongly with the assembly zone than with more mature regions of cuticle, where sclerotization and chitin binding might mask epitopes. Thus, further substantial evidence than the failure to detect a protein in more mature regions is needed to confirm that it belonged exclusively to the assembly zone.

It was known from earlier work on protein and mRNA distribution that cuticles from different metamorphic stages and different anatomical regions had different cuticular proteins, and that there may be a change in cuticular proteins synthesized by a single cell within a molt cycle (for review, see Willis, 1996). Such a transition in proteins synthesized is especially apparent at the time of ecdysis, and, in some insects, late in the instar. Consistent with this, immunolocalization revealed different proteins in morphologically distinct early and late lamellae in *D. melanogaster* pupae, and *Tenebrio molitor* and *Manduca sexta* larvae (Doctor *et al.*, 1985; Fristrom *et al.*, 1986; Wolfgang and Riddiford, 1986; Wolfgang *et al.*, 1986; Lemoine *et al.*, 1989, 1993; Bouhin *et al.*, 1992a, 1992b; Rondot *et al.*, 1998). Only two proteins with known sequence are among this group: TMACP22 (P26968.1) and TMLPCP22 (P80686.2).

Csikos and colleagues (1999) have used immunohistochemistry to follow some of *Manduca's* cuticular proteins throughout the molt cycle. These proteins are obviously in a dynamic state as they move from epidermis to cuticle to molting fluid to fat body, and then apparently back to cuticle via the hemolymph. More detailed studies are needed to learn if the same molecules make the return trip, and whether their initial passage from molting fluid into the hemolymph is solely via uptake and then basal secretion by the epidermis, or whether the midgut plays a role, since lepidopteran larvae drink their molting fluid (Cornell and Pan, 1983).

The findings with epicuticle, the first region to be secreted beneath the envelope, were complex. None of the monoclonal antibodies that recognized *Tenebrio* cuticular proteins reacted with epicuticle (Lemoine *et al.*, 1990). On the other hand, arylphorin from *Calpodes* has been localized to epicuticle and no other cuticular region (Leung *et al.*, 1989), and several proteins, of unknown sequence, were found both in the epicuticle and in the lamellar regions of the procuticle in *D. melanogaster* (Fristrom *et al.*, 1986) and *Calpodes* (Locke *et al.*, 1994). This finding of cuticular proteins in both epicuticle and lamellar regions was surprising, since the epicuticle had always been described as lacking chitin (*cf.* Fristrom *et al.*, 1986, Fristrom and Fristrom, 1993) and thus was expected to have unique proteins. A study of moth olfactory sensilla detected chitin in the procuticle with gold-conjugated wheat germ agglutinin, but it was not found in the epicuticle (Steinbrecht and Stankiewicz, 1999).

In addition to temporal differences in the secretion of cuticular proteins by single cells, there may be regional differences in the cuticle secreted by single cells. Individual epidermal cells of the articulating membranes

(intersegmental membranes) in *Tenebrio* secrete a cuticle with sclerotized cones embedded in softer cuticle. Two of the classes of monoclonal antibodies raised against *Tenebrio's* larval and pupal cuticular proteins recognized proteins in these cones. The same antibodies recognized proteins in cuticles in other regions that were destined to be sclerotized. Different antibodies recognized the proteins in the softer cuticle (Lemoine *et al.*, 1990, 1993).

Locke *et al.* (1994) were able, using carefully reconstructed sections of *Calpodes* larval cuticle, to distinguish one protein (C36) that was found with the same distribution as the chitin microfibrils that had been visualized with wheat germ agglutinin (WGA), a lectin that recognizes N-acetylglucosamine, while other antigens failed to show this distribution. Notably, only C36 isolated from cuticle reacted with WGA on lectin blots. Based on this evidence, Locke *et al.* (1994) suggest that the isolated protein may have obtained its N-acetylglucosamine from chitin.

5.2.1.4. Cuticles formed following disruption of normal metamorphosis

Treatment of many insects with juvenile hormone (JH) causes them to resynthesize a cuticle with a morphology characteristic of the current metamorphic stage, rather than the next. Thus, in *Tenebrio*, treatment of pupae with JH prior to pupal–adult apolysis causes the formation of a second pupa rather than an adult. Earlier work revealed that these second pupae had proteins with the same electrophoretic mobility as those extracted from normal pupae (Roberts and Willis, 1980; Lemoine *et al.*, 1989). A combination of Northern analysis and *in situ* hybridization demonstrated that second pupae have the same cuticular protein mRNAs and protein localization as normal pupae (Lemoine *et al.*, 1993; Rondot *et al.*, 1998). Adult cuticular proteins are not deposited in these cuticles, and the adult mRNAs do not appear (Lemoine *et al.*, 1989, 1993; Bouhin *et al.*, 1992a, 1992b; Charles *et al.*, 1992). Some JH-treated *Tenebrio* pupae form two cuticles, the first pupal-like in morphology and the second with adult features. The adult-like cuticle was shown with immunolocalization to have TMACP22 (P26968.1) (Bouhin *et al.*, 1992a). If JH is applied too late to form a perfect second pupa, the next cuticle formed will be a composite with morphological features of two metamorphic stages (Willis *et al.*, 1982). Bouhin *et al.* (1992b) found that all the epidermal cells laying down such a composite cuticle had mRNAs for TMACP22.

Zhou and Riddiford (2002) used Northern analysis to characterize the somewhat nondescript cuticles made by *D. melanogaster* that had been manipulated by misexpressing the gene *br* (*broad*), which codes for a transcription factor that first appears before the larval/pupal molt in flies and moths. By following mRNAs for the adult cuticular protein ACP65A (CG10297) or the pupal cuticular protein Edg78E (CG7673), they were able to demonstrate the essential role of *br* in directing pupal development, and thereby clarified the perplexing action of juvenoids in the higher Diptera.

5.2.2. The Site of Synthesis of Cuticular Proteins

One of the unresolved issues addressed in Silvert's (1985) review was the site of synthesis of cuticular proteins. This might appear to be a trivial issue, for one would expect that the epidermis that underlies the cuticle would synthesize the cuticular proteins. There are, however, reports in the literature that proteins found in the hemolymph were present in cuticle, and even that labeled proteins injected into the hemolymph would appear in cuticle. Silvert discussed the possibility that the injected protein had been broken down and resynthesized so that the cuticular protein was labeled solely because its constituent amino acids had come from a labeled pool.

Five methods have now provided data that address the site of synthesis of cuticular proteins. The most common method is to learn in what tissues and at which stages mRNA is present for a particular cuticular protein by detecting its presence via Northern analysis, RT-PCR, or qRT-PCR. This method is so common that specific examples will not be given. The second method is to incubate epidermis or integument *in vitro* with radioactive amino acids, separate the proteins, and compare the electrophoretic mobility of the labeled proteins to proteins isolated from cleaned cuticles. A third method is to isolate mRNAs from tissues and translate these *in vitro* with commercially available wheat germ extracts or rabbit reticulocytes, and compare the translation products to known cuticular proteins. The fourth method is *in situ* hybridization, and the fifth is immunolocalization to visualize proteins within the endoplasmic reticulum and Golgi apparatus.

The first three methods suffer from the possibility that tracheae and adhering tissues, fat body, muscles, and hemocytes contribute to the mRNA pool. Both labeling methods suffer from the problem that cuticular proteins are notoriously sensitive to solubilizing buffer and gel conditions (pH, urea concentration) (Cox and Willis, 1987a), and, unless cuticular protein standards and labeled translation products are mixed prior to electrophoresis, they may not show identical electrophoretic mobility even in adjacent lanes. Some workers have precipitated labeled translation products with antibodies raised against extracts of cuticle or individual cuticular proteins, then solublized the precipitate, run it on a gel, and detected the labeled product with fluorography. Csikos *et al.* (1999) used Western blots of translation products to identify cuticular proteins. Since cuticular proteins are destined for secretion from cells, they have a signal peptide that is cleaved before the protein is secreted into the cuticle. Hence, translation products made *in vitro* will be larger than the protein extracted from cuticle. There are two methods to circumvent this problem. The translation products can have their signal peptides cleaved by adding a preparation of canine microsomes, or antibodies against cuticular proteins (specific or against an extract) can be used to precipitate the translation products before they are solubilized and run on a gel. Either method allows some certainty in the comparison of these *in vitro* translation products with authentic cuticular proteins. It was also found that some commercial preparations of wheat germ extract have endogenous signal peptide processing activity (Binger and Willis, 1990).

Frequently, ^{35}S-methionine was used for metabolic labeling of integument and for *in vitro* translation. This is an unfortunate choice, as most mature cuticular proteins lack methionine residues (see section 5.3.2.1). The initiator methionine will be lost, along with the entire signal peptide. Clear differences in labeling patterns with ^{35}S-methionine and ^{3}H-leucine have been found, with none of the major proteins from pharate adult cuticle of *D. melanogaster* or from larval cuticles of *H. cecropia* showing methionine labeling (Roter *et al.*,1985; Willis, 1999). Why, then, did several studies find all of the known cuticular proteins labeled with methionine? Perhaps the finding that ^{35}S-methionine can donate its label to a variety of amino acids in preformed proteins (Browder *et al.*,1992; Kalinich and McClain, 1992) explains its appearance, and suggests that it needs to be used with caution for such studies with cuticular proteins.

The fourth method is *in situ* hybridization, where specific mRNAs can be identified in the epidermis. *In situ* hybridization allows one to be somewhat more discerning about the site of synthesis of a cuticular protein, because it is possible to monitor the presence or absence of a particular mRNA at the level of an individual cell. With this technique, integument is fixed and sectioned, and then probed with a labeled cDNA or cRNA, allowing the identification of particular regions of the epidermis by examining the morphology of the overlying cuticle. With most detection methods, contaminating tissues and precise regions of the epidermis can be identified, and the presence of the particular mRNA in them can be assessed. Thus, this technique identifies the location of the mRNAs recognized by the specific probe used. It was this technique that revealed the precision with which mRNAs are produced, for abrupt boundaries of expression occur between sclerites and intersegmental membranes (Rebers *et al.*, 1997), or at muscle insertion zones (Horodyski and Riddiford, 1989), or next to specialized epidermal cells (Horodyski and Riddiford, 1989; Rebers *et al.*, 1997). This technique even revealed the presence of mRNA for cuticular proteins in epithelia of imaginal discs from young larvae (Gu and Willis, 2003). A limitation of the technique is that some cRNA probes bind to the cuticle itself, possibly obscuring detection of mRNA in the underlying epidermis (Fechtel *et al.*, 1989, Gu and

Table 1   Evidence for the Association of Location or Type of Cuticle and Sequence Class of Some Cuticular Proteins

Species	Protein	Sequence Class	Localization[a]	Nature of Evidence[b]	When Deposited	Reference
Bombyx mori	BMLCP18	RR-1	Imaginal discs	EST		Gu and Willis (2003)
Drosophila melanogaster	EDG-78	RR-1	Larval and imaginal cells of prepupa	ISH		Fechtel et al. (1989)
Drosophila melanogaster	EDG-84	RR-2	Imaginal disc cells	ISH		Fechtel et al. (1989)
Drosophila melanogaster	PCP	RR-1	Prepupal thorax and abdomen	ISH		Henikoff et al. (1986)
Hyalophora cecropia	HCCP12	RR-1	Soft cuticle; imaginal discs	CD and ISH		Cox and Willis (1985), Gu and Willis (2003)
Hyalophora cecropia	HCCP66	RR-2	Hard cuticle	CD and ISH		Cox and Willis (1985), Gu and Willis (2003)
Locusta migratoria	LM-ACP7	RR-2	Hard cuticle	CD		Andersen et al. (1995)
Locusta migratoria	LM-ACP8	RR-2	Hard cuticle	CD		Andersen et al. (1995)
Locusta migratoria	LM-ACP19	RR-2	Hard cuticle	CD		Andersen et al. (1995)
Manduca sexta	CP14.6	RR-1	Soft cuticle	ISH		Rebers et al. (1997)
Manduca sexta	LCP 16/17	RR-1	Soft cuticle	ISH		Horodyski and Riddiford (1989)
Tenebrio molitor	ACP17	Glycine-rich	Hard cuticle	ISH	Strongest post-ecdysis	Mathelin et al. (1995, 1998)
Tenebrio molitor	ACP20	RR-2	Hard cuticle	ISH	Primarily pre-ecdysis	Charles et al. (1992)
Tenebrio molitor	ACP-22	RR-2	Hard cuticle	ISH, mAB	Pre-ecdysis	Bouhin et al. (1992a,1992b)
Tenebrio molitor	TMLPCP22	51 aa motif	Hard and soft cuticle pre-ecdysis, then only soft cuticle	ISH, mAB	Primarily pre-ecdysis	Rondot et al. (1998)
Tenebrio molitor	TMLPCP23	51 aa motif	Hard and soft cuticle	ISH	Only pre-ecdysis	Rondot et al. (1998)
Tenebrio molitor	TMLPCP29	RR-3 and 18-residue motif	Hard and soft cuticle, except not posterior borders of sclerites	ISH	Post-ecdysis	Mathelin et al. (1998)

[a]For in situ hybridization, cuticle type was determined by nature of cuticle overlying the epidermis.
[b]CD, careful dissection prior to extraction of proteins; ISH, in situ hybridization used to to localize mRNA; mAB, monoceonal antibody immunolocalization; EST, from Bombyx EST project (Mita et al. 2002).

Willis, 2003). Fechtel *et al.* (1989) found this artifact to be cuticle-type- as well as strand- and probe-specific. Results from several species are summarized in **Table 1**.

The fifth method, immunolocalization, was described earlier in conjunction with localization of specific proteins within the cuticle, but it can also be used to identify the site of synthesis by looking for a particular protein within the endoplasmic reticulum or Golgi apparatus (Sass *et al.*, 1994a, 1994b).

The results from mRNA detection, metabolic tissue labeling, and *in vitro* translations reveal that all cuticular proteins with known sequences or for which specific probes are available are synthesized by the integumental preparations. Different proteins are synthesized at different times in a molt cycle, and in different anatomical regions, and there are some cuticular proteins whose synthesis is stage-specific. Differences in the presence of mRNA parallel the appearance of labeled proteins, indicating that much of the temporal and spatial control of cuticular protein synthesis is at the level of transcription. As mentioned above, however, all three of these methods are limited by the possible contamination of tissues by non-epidermal cells, and by their inability to address heterogeneity of cell types within the epidermis.

A microarray analysis of isolated hemocytes from *An. gambiae* revealed the presence of mRNA for nine cuticular proteins (Baton *et al.*, 2009). Transcripts for *AgamCPR*26 and *AgamCPR*90 were significantly higher in adults challenged with heat-killed *Micrococcus luteus* than in naïve individuals. A massive study on hemocytes in *D. melanogaster* found significant levels of transcript for *DmelLCP*1-4 in hemocytes from both naïve and bacteria-challenged larvae (Irving *et al.*, 2005).

Studies that have combined tissue labeling or *in vitro* translations with immunolocalization have at last clarified the relationship between hemolymph and cuticular proteins with identical electrophoretic and immunological properties. The most comprehensive studies of protein trafficking were carried out in *Calpodes*, and revealed four classes of exported proteins that are handled by the epidermis.

These findings are so important that the experimental methodology is worth discussing. The first approach used was to seal sheets of final instar integument into a bathing chamber so there could be no leakage from the cut edges of the tissue, and then find what proteins were made in a 2-hour exposure to ^{35}S-methionine. Three classes of proteins were identified with this procedure; one was secreted exclusively into the cuticle (C class), a second appeared in the bathing fluid and hence had been secreted basally (B class), while the third was secreted in both directions (BD class) (Palli and Locke, 1987). Immunolocalization of numerous other *Calpodes* proteins (of unknown sequence) confirmed the existence of these three routing classes of epidermal proteins. A fourth class, the T class, was identified for proteins transported into cuticle but not synthesized by the epidermis. Its presence eliminated any concerns that the classes might be artifacts from labeling with ^{35}S-methionine (Sass *et al.*, 1993).

One member of the T class (T66) was studied in more detail. It was localized by immunogold throughout the cuticle, and, although found in epidermal cells, was not found in association with the Golgi apparatus, confirming its transcellular transport, rather than synthesis by the epidermis. A subsequent study identified the exclusive site of its synthesis as spherulocytes (Sass *et al.*, 1994a).

Whether the BD proteins are secreted from both apical and basal borders of epidermal cells is still not clear. Locke (1998, 2003) now favors the possibility that all secretion is apical, where the Golgi are concentrated, and that the secreted proteins are subsequently taken back into the cell from perimicrovillar space and transported in vesicles to the basal surface, where the contents are released into the hemolymph.

In conclusion, it is now clear that the epidermis can synthesize both cuticular and hemolymph proteins. It can also transport proteins made in tissues other than epidermis from hemolymph to cuticle.

5.2.3. Tracheal Cuticular Proteins

An often-neglected source of cuticle in insects is the tracheal system. Since tracheae are associated with all insect tissues, caution is needed in interpreting the significance of the presence of mRNAs or cuticular proteins from non-integumental tissues. Cox and Willis (1985) recognized that some of the proteins from tracheae had the same isoelectric points as proteins isolated from integumentary cuticle. A further study was carried out a decade later by Sass *et al.* (1994b), combining electrophoretic

analysis with immunogold labeling. Chitin was localized with wheat germ agglutinin, and found in all regions of tracheae and tracheoles except the taenidial cushion. Antibodies that had been raised against individual electrophoretic bands from integumentary extracts represented proteins from all four classes of integumentary peptides. Some C proteins, those from the surface cuticle, were found associated with chitin, but only in taenidia; other C proteins were in the general matrix, with and without chitin. The B and BD peptides were only found in the taenidial cushion, the region lacking chitin. It appears that hemolymph peptides that are synthesized by the epidermis may be tracheal cuticle precursors. The one T protein studied (T66, made in spherulocytes) was also found in the general matrix. An important insight from this study was the conclusion that: "The extremely thin tracheal epithelium suggests that transepithelial transport might supply proteins to the tracheal cuticle more evenly than Golgi complex secretions" (Sass *et al.*, 1994b). Analysis of tracheal morphogenesis is an active field that has recently been reviewed (Centanin *et al.*, 2010; Moussian, 2010; see also Ghabrial *et al.*, 2003). Little information is available about the proteins that contribute to tracheae. Gasp (a member of the CPAP-3 family) was found to be restricted to tracheae in *D. melanogaster* (Barry *et al.*, 1999), but transcripts from its clear ortholog in the lepidopteran *Choristoneura fumierana* were associated primarily with the body surface epidermis (Nisole *et al.*, 2010).

5.3. Classes of Proteins Found in Cuticles

5.3.1. Non-Structural Proteins

Some representative non-structural proteins that have been identified in cuticle are listed in **Table 2**.

5.3.1.1. Pigments Proteins from three classes of pigments used in cuticle – insecticyanins and two different yellow proteins – have been sequenced. The insecticyanins are blue pigments made by the epidermis and secreted into both hemolymph and cuticle. They are easily extracted from cuticle with aqueous buffers. Members of the lipocalin family, they are present as tetramers with the gamma isomer of biliverdin IX situated in a hydrophobic pocket. In the cuticle, in cooperation with carotenes, they confer green coloration. Their structure has been determined to 2.6 Å by X-ray diffraction (Holden *et al.*, 1987), making them structurally the best characterized cuticular proteins. Two genes code for insecticyanins in *Manduca* (Li and Riddiford, 1992).

The yellow protein in *D. melanogaster*, coded by the *y* gene (CG3757), has been localized with immunocytochemistry in cuticles destined to become melanized (Kornezos and Chia, 1992). Thus, it was found in

Table 2   Characteristics of Some Non-Structural Proteins that have been found in Cuticle

Species	Protein Name	Number of Amino Acids[a]	Function	Sequence Method[b]	Identifier[c]
Schistocerca gregaria	Putative carotene binding protein	250	Transfers carotene into cuticle	DS	13959427
Caliphora vicinia	ARYLPHORIN A4	743	Found in cuticle	CT	114232
	ARYLPHORIN C223	743		CT	114236
Drosophila melanogaster	YELLOW	520	Positions melanin pigment in cuticle	CT	140623
Bombyx mori	CECROPIN A	41	Defense protein	CT	2493573
	CECROPIN B	41	Defense protein	CT	1705754
	PROPHENOLOXIDASE	675	Metanization enzyme	CT	13591614
Calpodes ethilus	CECP 22	169	Cuticle digestion	CT	4104409
Manduca sexta	ARYLPHORINα	684		CT	114240
	ARYLPHORINβ	687		CT	1168527
	INSECTICYANIN A	189	Blue pigment	CT	124151
	INSECTICYANIN B	189	Blue pigment	CT	124527
	SCOLEXIN A	279	Serine protease immune protein	CT	4262357
	SCOLEXIN B	279	Serine protease immune protein	CT	4262359

[a]Sequence length of mature peptide; signal peptides were deleted using data from authors or SignalP V2.0(http://www.cbs.dtu.dk/services/SignalP 2.0/).
[b]DS, direct sequencing of protein; CT, conceptual translation of a cDNA, genomic region, or EST product.
[c]Protein sequences and additional annotation can be found at: http://www.ncbi.nlm.nih.gov/protein

association with larval mouth hooks, denticle belts, and Keilin's organs. Mutants of *y* lack black pigment in the affected cuticular region. Mutant analysis revealed two classes of mutants; those that affect all types of cuticle at all stages, and those affecting only particular areas of specific stages. At least 40 different adult cuticular structures could express their color independently (Nash, 1976), and the regulatory regions responsible for some of the stage and regional specificity have been identified (Geyer and Corces, 1987). The yellow protein has been described as a structural component of the cuticle that interacts with products from the gene, *ebony*, a beta-alanyl-dopamine synthase, to allow melanin to be deposited. Flybase (http://flybase.bio.indiana.edu/) reports that 1005 different alleles of *y* have been described, in 775 references, beginning in 1916. The complete sequence of *y* has been determined for 13 species of *Drosophila* in addition to *D. melanogaster*. An examination of *y* expression revealed that both cis- and trans-regulation are responsible for differences in pigmentation patterns among different species (Wittkopp *et al.*, 2002, 2009). There is no evidence for a known chitin-binding domain in the yellow protein; the only domain recognized is pfam03022 (major royal jelly protein). Although the sequence for yellow is 37% identical and 56% similar to a dopachrome conversion enzyme from *Aedes aegypti* that is involved in the melanotic encapsulation immune response, yellow itself evidently is devoid of enzyme activity (Han *et al.*, 2002). As a further complication, there are 13 other genes in *D. melanogaster* related to *y*; most of their products do not seem to affect

pigmentation. *T. castaneum* also has 14 *y* homologs, and a comprehensive study using RNAi and mass spectrophotometric analyses revealed diverse activities, many involving the cuticle, with only the ortholog of *Dmely* having a role in cuticle melanization (Arakane *et al.*, 2010). Orthologs of *Dmely* also play a role in cuticle pigmentation in *B. mori* and *Papilio xuthus* (Futahashi and Fujiwara, 2005; Futahashi *et al.*, 2008).

Another distinct cuticular protein (P82886.1) implicated in pigmentation, putatively beta-carotene binding, has been isolated from extracts of cuticle from mature adult *Schistocerca gregaria* using column chromatography to isolate a protein that was yellow in color. It bears significant sequence similarity to various insect juvenile-hormone-binding proteins, as well as odorant-binding proteins. Wybrandt and Andersen (2001) suggest that it is involved in transport of carotenes into epidermis and then the cuticle.

5.3.1.2. Enzymes   Some of the enzymes involved in sclerotization have been identified in cuticle. Since they are discussed by Andersen in Chapter 6, and in a recent review (Andersen, 2010a), they will not be considered here.

Some enzymes that belong to the molting fluid become evident as the electrophoretic banding pattern of proteins isolated from cuticle changes as *Calpodes* initiates molting at the end of the fifth instar, with the most conspicuous change being the appearance of a band of 19 kDa. Antibodies raised against this protein were used to isolate a cDNA from a library cloned in an expression vector.

The conceptual translation revealed a "cuticular molt protein" (AAD02029.1, also called CEPP22). Its sequence suggested it might have amidase activity. Further analysis revealed that the protein was present in the cuticle before each molt, and was also found in molting fluid. Marcu and Locke (1998, 1999) present evidence that this protein may be activated by proteolysis, and speculate that it may function to cleave an amidic bond between N-acetylglucosamine from chitin and amino acids in cuticular proteins.

Enzymes involved in digesting the old cuticle are temporary residents in cuticle. These include proteases and chitinases. Their interaction is discussed by Marcu and Locke (1998).

5.3.1.3. Defense proteins

Also found in the cuticle are components of the insect defense system. In one study, cuticle was removed from *Bombyx* larvae 24 hours after they had been abraded with emery paper and exposed to bacteria. The antibacterial peptide, cecropin, was purified from the cuticles (Lee and Brey, 1994). Both pro-phenoloxidase and a zymogen form of a serine protease capable of activating it have been extracted from *Bombyx* larval cuticle. Colloidal gold secondary antibodies revealed that the pro-phenoloxidase was localized throughout the epi- and procuticle, and in a conspicuous orderly array on the basal side of the helicoidal chitin lamellae. An extra-epidermal source is likely for this enzyme since no labeling was found in the epidermis, and neither was mRNA detected in the epidermal cells. It is assumed to function in the melanization that occurs in response to injury (Ashida and Brey, 1995).

Molnar *et al.* (2001) presented immunological evidence for a protein related to the defense protein scolexin in the cuticle of *Manduca*. This protein exists in two forms in *Manduca*, but the antibody used did not distinguish between them.

The cuticle also appears to be the repository for a peptide (HCP; GI:240104242; 2RPS_A) that stimulates aggregation and movement of hemocytes in the moth *Pseudaletia separate* (*Mythimna separate*) (Nakatogawa *et al.*, 2009).

5.3.1.4. Arylphorins

The final class of non-structural proteins is the arylphorins, proteins with high content of aromatic amino acids and some lipid. These proteins, first identified from hemolymph, have been of special interest since the discovery by Scheller *et al.* (1980) that although calliphorin (the arylophorin from *Calliphora*) was found in cuticle it seemed to come from the hemolymph, because labeled calliphorin injected into the hemolymph appeared in cuticle. But there is also evidence that the epidermis is capable of synthesizing arylphorins, for Riddiford and Hice (1985) had detected arylphorin mRNA in the epidermis of *Manduca*.

Palli and Locke (1987) used an anti-arylphorin antibody to identify an 82-kDa protein made in *Calpodes* integumental sheets *in vitro* that appeared in both cuticle and media; thus, arylphorin appeared to be a bi-directionally secreted integumentary protein. Next, colloidal gold secondary antibodies were used to visualize the location of anti-arylphorin in ultrathin sections of various tissues (Leung *et al.*, 1989). The resolution afforded by this method made it possible to recognize arylphorin in epicuticle (but not lamellar cuticle) in the Golgi complexes of the fat body, and to show by quantitating gold particles that it was also found in Golgi complexes of epidermis, midgut, pericardial cells, and hemocytes, as well as the meshwork of fibrous cuticle in tracheae. Thus, while the possibility remains that some arylphorin is transported from hemolymph to cuticle, it need not be, for the epidermis itself is capable of synthesizing and secreting this protein. These studies further demonstrated that a given protein can be synthesized by multiple tissues. Whether it is the same gene that functions in all tissues remains to be determined.

The role of arylphorin remains unknown. It is generally assumed to be participating in sclerotization because of its high tyrosine content. Is it degraded in the cuticle so that its constituent amino acids are released, or does it remain an integral part of the cuticle? The latter is favored by the available evidence because calliphorin has been shown to bind strongly to chitin *in vitro* (Agrawal and Scheller, 1986), and no breakdown products were detected after injection of labeled calliphorin (Konig *et al.*, 1986).

5.3.2. Structural Proteins

5.3.2.1. Overview and families of cuticular proteins

More than a decade ago, a comprehensive and insightful review of cuticular proteins presented the complete sequence and full citation for all 40 cuticular proteins known at that time, and identified features that remain their hallmarks (Andersen *et al.*, 1995). Most of the structural cuticular proteins whose sequences were known in 1995 came from the efforts of Svend Andersen and his group, and were based on direct sequencing of purified cuticular proteins. These data provided the starting point for subsequent analyses, for features identified in those early studies led to the assignment of predicted protein sequences as corresponding to putative structural cuticular proteins. The 2005 version of this review provided information about 139 cuticular proteins, and many were based on sequences from cDNAs or short stretches of genomic DNA. Some of the sequences had been confirmed, indeed had their isolation guided, by N-terminal sequences from proteins isolated from cuticle. Now, annotation of several insect genomes is complete. There are EST projects for multiple insect species. Fortunately, proteomic analyses on cuticle preparations have confirmed that many of the sequences designated as coding for cuticular proteins are indeed coding for authentic rather than putative cuticular proteins. Proteomic studies also identified new families of cuticular proteins. A few analyses of mutant forms or animals with RNAi depleted transcripts have added to the confirmation of specific roles for specific cuticular proteins.

Thus, the number of structural cuticular proteins sequences has increased from fewer than 200 to several thousand, which recently have been organized into 13 fairly well-defined families, with several more as yet not classified (Willis, 2010). Some general comments on cuticular protein nomenclature will be followed by a definition of, and comments about, each family.

While nomenclature of cuticular proteins is not standardized it is improving, and the recognition and definition of distinct families (Willis, 2010) should aid in establishing relationships of the cuticular proteins within and among species. Now that we know that multiple genes may code for proteins with very similar or even identical sequences, it is recommended that the early practice of naming proteins with numbers that correspond to presumed orthologs in other species be abandoned until annotation of whole genomes is complete. Also unwise is calling them LCP or ACP because they were first identified in a larva or adult, because in many cases stage-specificity vanished as further studies were carried out. If one has a whole-genome sequence, the proteins in each family can be named in the order that the genes are located on chromosomes, but so far that practice has only been followed for *Bombyx*. At the very least, the prefix for the family followed by a number provides a useful and informative name. A four-letter abbreviation for the genus and species should precede that name when a paper deals with more than a single species.

A final complication is whether two almost identical proteins are allelic variants, or products of two distinct genes. In some cases an "isoform" has been described. Genomic sequences, however, revealed that stretches coding for proteins of almost identical sequence may be linked on a chromosome (Charles *et al.*, 1997; Dotson *et al.*, 1998; Cornman *et al.*, 2008; Cornman and Willis, 2008, 2009; Futahashi *et al.*, 2008). Only when one has a well-annotated genome is it possible to learn if two similar sequences represent distinct genes or alleles of a single gene. As one goes from ESTs to genomes, expansion and contraction of names will probably occur. A more extensive discussion of cuticle protein nomenclature can be found in a recent review (Willis, 2010).

The 2005 version of this review included a table that listed all known cuticular protein sequences except for those from whole-genome analyses that were just becoming available. Such a table would now exceed the length of this version, for each of the sequenced insect genomes has well over 100 structural cuticular proteins, and the EST data for dozens more insects also have numerous proteins that are their homologs. **Table 3** gives a numerical summary of the cuticular proteins in some of the annotated genomes. Details on the characteristics of the families are discussed below. One interesting feature on numbers was unearthed by Cornman (2009), who compared numbers of CP genes in seven *Drosophila* species and compared them to numbers in the other two Diptera whose genomes are well annotated. Numbers of the CPR family in the *Drosophila* species ranged from 100 to 104 genes, while *Ae. aegypti* had about 50% more than the 156 in *An. gambiae*. The divergence time between the two lower Diptera is estimated to be 95 my, while members of the genus *Drosophila* are believed to have shared a common ancestor about 40 my ago.

Most of the proteins now described as cuticular were classified by their "discoverers" or computer-driven annotation because their sequences (or a part thereof) were similar to a cuticular protein already in the databases; obviously, such proteins should only be described as putative cuticular proteins until additional evidence is available. Over 90% of the *An. gambiae* cuticular proteins have been confirmed as authentic because peptides corresponding to them were found in extracts of cuticles by tandem mass spectrometry (He *et al.*, 2007). A smaller number of the *Bombyx* proteins have also been confirmed using chitin-binding proteins as starting material (Tang *et al.*, 2010), and many more are known to be authentic based on pre-genomic analyses (Futahashi *et al.*, 2008). Proteomics analyses are being carried out for *Tribolium* (Dittmer, personal communication). The presence of a signal peptide is essential for a cuticular protein, and coupled with compelling sequence similarity is strong evidence that the proteins have been correctly classified as putative cuticle proteins.

One feature of structural cuticular proteins frequently mentioned is that they lack cysteine and methionine residues in the mature protein; Andersen (2005) suggested that the reactivity of cystine and cysteine with ortho-quinones could interfere with sclerotization. Thus, the recent finding of the CPAP1 and CPAP3 families with one or three easily recognizable domains each with six cysteines revealed an unappreciated type of cuticular protein. Moreover, the CPCFC family first recognized with BcNCP1 has two or three similar motifs each with two conservatively spaced cysteine residues. While many cuticular proteins are quite short (< 200 amino acids), that early generalization too needs revision. There is an enormous CP, dumpy (CG33196), with 22,971 amino acids, that anchors muscle to cuticle in *D. melanogaster*. Among the more conventional cuticular proteins, even the CPR family in *An. gambiae* has 16% of its mature proteins with between 200 and 300 amino acids, while 10% have over 300, with the largest (AgamCPR140) having 837 (Cornman *et al.*, 2008). This large cuticular protein

Table 3 Approximate Number of Genes in Different Cuticular Protein Families in Species with Manual Annotation of Cuticular Proteins in Whole-Genome Data

Section of chapter	CPR 3.2.2	CPF + CPFL 3.2.3	TWDL 3.2.4	CPLCG 3.2.5	CPLCW 3.2.6	CPLCA 3.2.7	CPLCP 3.2.8	CPG 3.2.9	APIDERMIN 3.2.10	CPAP1 3.2.11	CPAP3 (OBSTRUCTOR) 3.2.11	CPCFC 3.2.12	OTHER 3.2.13	TOTAL
An. gambiae	156	11	12	27	9	3	4 +23?	0	0	0	7	1	10	240+
B. mori	148	5	4	0	0	0	7	18*	0	0	1	1	33	217
D. melanogaster	101	3	27	3	0	11	5	0	0	2	6	1	?	159
A. mellifera	32	3	2	0	0	0	2	0	3	0	5	0	?	47
N. vitripennis	62	4	2	0	0	0	3	0	3	0	6	0	?	80
T. castaneum	101	8	3	2	0	0	4	0	0	10	7	2	?	137

*Gly-Rich family from *Bombyx* is really a composite of possibly three families (see text). The 6 that have been identified as CPLCPs were deleted from this number, and only the 18 restricted to lepidoptera that have several GGY repeats were included. Absence of additional defining features prevented searches in other groups.

Sources: Togawa et al., 2007; Futahashi et al., 2008; Cornman et al., 2008, Cornman and Willis, 2009, Jasrapuria et al., 2010; *T. castaneum* based on personal communication from N. Dittmer and M. Kanost; Willis (unpublished).

Table 4 Presence of Cuticular Protein Families and Features in Different Groups of Insects

	Diptera		Lepido-	Coleo-	Hymeno-	Hemi-	Ortho-	Dictyo-	Phthira-	Collembola
	Brachycera	Nematocera								
CPR	+	+	+	+	+	+	+	+	+	+
CPF/CPFL	+	+	+	+	+	+	+	+	+	id
TWDL	+	+	+	+	+	+	+	+	+	id
CPLCA	+	+	no	no	no	no	no	no	no	id
CPLCG	+	+	no	+	no	no	no	+	no	id
CPLCW	no	+	no	no	no	no	no	no	no	id
CPLCP	+	+	+	+	+				+	
GPG			+							
apidermin	no	no	no	no	+	no	id	id	no	id
CPAP 1	+	+	+	+	+	+	+	+	+	+
CPAP3	+	+	+	+	+	+	+	+	+	+
CPCFC	+	+	+	+	no	+	+	+	+	+
18 aa motif	+	+	+	+	+	+	+	+	+	id
CP with >3 AAP[AVL]	+	+	+	+	+	+	+	id	+	id

This table is revised from Willis, 2010.

Final syllable ptera was removed from names of most orders.

Data were obtained from Blast searches in addition to analyses in: Togawa et al., 2007; Futahashi et al., 2008; Cornman et al. 2008, Cornman and Willis, 2009; Carmon et al., 2007, Jasrapuria et al., 2010.

id = insufficient data available to record absence; empty boxes indicate that motifs were insufficiently well defined to allow a search.

has an ortholog in *Pediculus humanus* (XP_002432942.1) of the same length. The situation in *Bombyx* is somewhat similar to *Anopheles*; of 148 CPR family members, 22% have 200–300 amino acids and 13% have over 300, with the largest (BmorCPR146) having 1618 (Futahashi *et al.*, 2008).

Thirteen families of cuticular proteins have now been recognized, and the characteristics and history of each will be described below. Two – CPR and CPF – were recognized early. The proteomics study of He *et al.* (2007) revealed peptides from several dozen more possible cuticular proteins. These were annotated and their temporal expression patterns determined, and they have been separated into five distinct families, described in detail in Cornman and Willis (2009). Most of these proteins have extensive regions of Low sequence Complexity, and have been named CPLC followed by a final initial to designate one of four distinct families. The fifth low complexity family retained the original name TWDL. Most of the families described in that paper have turned up in other insect orders. As mentioned above, three families of cuticular proteins with conserved cysteine residues have been identified: CPAP1 and CPAP3 (Jasrapuria *et al.*, 2010); and CPCFC. There are glycine-rich cuticular proteins that do not belong to any of these families, a small family (apidermin) so far restricted to Hymenoptera (Kucharski *et al.*, 2007), and then a few other cuticular protein sequences that have not yet been assigned to families. It is intriguing that members of most of these families are restricted to arthropods, some in only one or two insect orders, while others are fairly widely distributed (**Table 4**).

Many cuticular protein sequences are available at the website CuticleDB (http://bioinformatics2.biol.uoa.gr/cuticle DB/index.jsp), which allows a variety of different search strategies (Magkrioti *et al.*, 2004).

A convenient way to illustrate diagnostic features is with WebLogos (Schneider and Stephens, 1990; Crooks *et al.*, 2004), and these will be presented in **Figures 1–3**. A summary of these features that can be used for an initial BLAST search to learn if a database has cuticular protein sequences is available in Supplementary Information File 1 in Willis (2010).

5.3.2.2. CPR family: Proteins with the R&R consensus By far the most common family of cuticular proteins is that containing the R&R Consensus. The name comes from a 28-aa motif, first recognized by Rebers and Riddiford (1988) in six cuticular proteins. The original R&R Consensus is part of a longer conserved sequence, pfam00379. A valuable website, Pfam (http://pfam.janelia.org/family/), has used hidden Markov modeling to define motifs characteristic of particular classes of proteins (Bateman *et al.*, 2002). In accordance with recent nomenclature, this extended consensus region of about 63 amino acids will be referred to hereafter as the R&R Consensus. When a protein sequence is searched against non-redundant protein sequences using blastp at the BLAST server (http://www.ncbi.nlm.nih.gov/BLAST/), the first information that is presented is an indication of matches to pfam entries. The pfam sequence that allows annotators to classify a protein as a cuticular protein in the CPR family is pfam00379, a 68-aa sequence that includes the extended R&R Consensus. It also goes under the name "chitin_bind_4," for reasons that will become apparent in section 5.5.4. Pfam00379 was obviously based on proteins of both RR-1 and RR-2 classes, for it matches neither particularly well. This makes it particularly useful for a preliminary classification of a putative cuticular protein sequence.

An indication of the importance of the R&R Consensus comes from the Pfam website. It reports 2456 distinct proteins with the Consensus from 67 different species of arthropods (http://pfam.janelia.org/family?acc=PF00379#tabview=tab6). This is an underestimate, because close to 100 sequences from Hymenoptera are absent. The CPR family is restricted to arthropods. The one exception, *Xenopus* NP_001090156.1, is due to a contaminating sequence from *Drosophila erecta* (Willis, 2010).

While 98% of the entries have only a single occurrence of the R&R Consensus, the exceptions are interesting. The most notable exception is a protein from the tailfin of the prawn *Penaeus japonicus*. The entire sequence of this protein is made up of 14 consecutive pfam00379 motifs (Ikeya *et al.*, 2001). A protein from the horseshoe crab *Tachypleus tridentatus* (BAE44187.1) has five Consensus regions (Iijima *et al.*, 2005), and the current annotation of the *Ixodes scapularis* genome reports several instances in a single predicted protein. Manual annotation revealed that most insect genes predicted to code for a protein with more than a single Consensus region actually coded for multiple proteins, easily recognized by standard markers of gene and transcript boundaries. There remains a small number of insect proteins that genuinely appear to have two Consensus regions, and the only one with three has orthologs in several species (Willis, 2010). When only a single Consensus region is present, it can be found near the N- or C-terminus, or within the protein. Three distinct forms of the Consensus have been recognized and named by Andersen (1998, 2000): RR-1, RR-2, and RR-3. RR-1-bearing proteins have been isolated from flexible cuticles, while RR-2 proteins have been associated with hard cuticle. This generalization was based on relatively few cases, and it has also been suggested that RR-2 proteins will contribute to exocuticle while RR-1 will be found predominantly in endocuticle (Andersen, 2000). This issue has not been resolved, even with the extensive expression data that are now available (Togawa *et al.*, 2007). Hopefully, immunolocalization data (see section 5.2.1.3) will prove helpful. The RR-2 Consensus region is far more conserved in length and sequence than the one from RR-1 proteins, as can be seen in the WebLogos in **Figure 1**. The website CuticleDB provides a tool using Hidden Markov Modeling to learn if a protein is RR-1 or RR-2 (Karouzou *et al.*, 2007).

Within the CPR family, numerous proteins can be identified that have orthologs in several species, some with distinct Consensus regions and other features (Cornman and Willis, 2008; Zhang and Pelletier, 2010).

The wealth of information on cuticular protein sequences and the unraveling of how the structure of some contributes to the interaction of chitin and protein (see section 5.5) is only a beginning. Essential properties of cuticle remain to be explained, and important questions raised in the older literatures about various means of achieving cuticle plasticity and the importance of hydration in cuticle stabilization must not be forgotten (Vincent, 2002, and references therein).

An especially interesting member of the CPR family is the resilin gene. The name "resilin" has been given to the rubber-like proteins responsible for the elasticity of jumping fleas and vibrating wings. Analysis of resilin-bearing cuticles in froghoppers (*Aphrophora alni* and *Philaenus spumarius*) concludes that resilin can function in two quite different ways. It is used:

> as an energy buffer in rhythmically active, fast mechanical movements, such as those of the wings during flight or the tymbals in cicadas . . . The almost perfect elastic recovery of resilin and its extreme resistance to mechanical fatigue mean that it can return nearly all of the power put into it for the next cycle of movement. The second role . . . is in providing a flexible material that is combined with the stiffer chitinous cuticle in a composite structure.
>
> **(Burrows *et al.*, 2008)**

The first identification of a complete sequence for resilin was carried out by Ardell and Andersen (2001), who used peptides from locust (*Schistocerca gregaria*) and cockroach (*Periplaneta americana*) resilin to identify a likely homolog in *D. melanogaster*. The peptides came from the R&R Consensus region. The protein they identified was CG15920. Its 18 N-terminal copies of a 15-residue repeat and 13 C-terminal copies of a 13-residue repeat were predicted to contribute to a beta-spiral, a common form for proteins with elastic properties (Ardell and Andersen, 2001). The corresponding gene produces two transcripts; one lacks over two-thirds of the start of the Consensus region. Two groups have studied the physical properties of CG15920 and its repeat regions, and showed that they have the properties one would expect of highly elastic proteins (Elvin *et al.*, 2005; Qin *et al.*, 2009).

The identification of resilin in other species is complicated. Two recent analyses, one brief (Willis, 2010) and the other detailed (Andersen, 2010b), emphasize the difficulties and reach different conclusions about some possible homologs. Andersen emphasizes the need for repeat regions that would underlie the elastic properties, while Willis focused on the R&R Consensus region that showed such conservation between *Schistocerca*, *Periplaneta*, and *D. melanogaster*. Both conclude that an authentic resilin gene should code for the Consensus, although alternative splicing may eliminate it in some of its transcripts. A major complication is that Lyons *et al.* (2007)

(A) RR-1 *Bombyx mori*

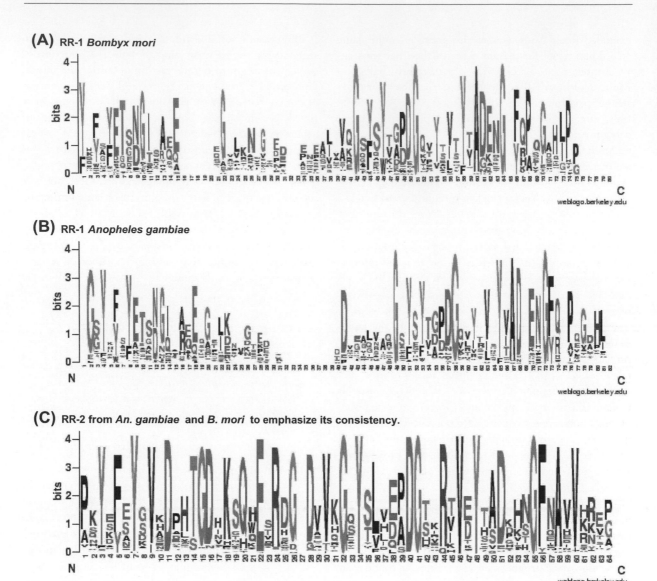

(B) RR-1 *Anopheles gambiae*

(C) RR-2 from *An. gambiae* and *B. mori* to emphasize its consistency.

Figure 1 Comparison of the highly conserved RR-2 Consensus with the more variable one from RR-1 proteins. WebLogos were constructed at <http://www.weblogo.berkeleky.edu/logo.cgi.> (Schneider and Stephens, 1990; Crooks *et al.*, 2004). (A) RR-1 Consensus regions from 52 sequences from *B mori*. (B) RR-1 Consensus regions from 51 sequences from *An. gambiae*. (C) WebLogo constructed from 87 *B. mori* and 101 *An. gambi*ae sequences. Panels reprinted from Willis (2010), with permission.

found highly elastic physical properties of an *An. gambiae* protein coded for by an EST BX61916.1, but the corresponding gene, AGAP002367, lacks the Consensus region and is most closely related to a *D. melanogaster* protein (CG7709) that has been characterized as a mucin. The *Anopheles* protein with the closest similarity (74%) to the Consensus region of Dmelresilin is AgamCPR152, but this protein lacks any repeats. Andersen identified AgamCPR140, which has many repeats, as a possible resilin, but the Consensus region is only 34% similar to Dmelresilin.

One consistent property of resilin is its ability to fluoresce due to di- and tri-tyrosine cross-links (Andersen and Weis-Fogh, 1964), so a combination of studies that

establish anatomical location of candidate proteins along with the physical properties should make it possible to sort out what sequences are truly resilin, and identify possible differences when it is serving its different roles.

5.3.2.3. CPF and CPFL families   A motif corresponding to a 51-aa repeat first recognized by Andersen *et al.* (1997) has been identified in a modified form in at least 9 orders of insects. However, the common repeat is somewhat shorter, at 42–44 amino acids, so the original name for the CPF family has been retained, with the F now referring to forty rather than to fifty. A detailed discussion of this family can be found in Togawa *et al.* (2007), where it is pointed out that in addition to the conserved motif of about 42 amino acids, the proteins are also similar in the amino acids near their carboxyl-termini.

The C-terminal region characteristic of the CPF family has also been found in other cuticular proteins that lack the defining consensus. Togawa *et al.* (2007) named these CPFL, for CPF-like. All four CPF proteins and six of the seven CPFL proteins in *An. gambiae* have been verified as authentic cuticular proteins, based on shared peptides identified in a tandem mass spectrometry analysis of cast cuticles (He *et al.*, 2007).

CPF and/or CPFL proteins have been identified throughout the hexapoda, including collembola and diplura (**Table 4**), but not yet in Crustacea or Chelicerata.

5.3.2.4. TWDL family

One of the families of low-complexity proteins previously had been identified in *D. melanogaster* and named TWDL after the tubby phenotype in one of its mutants that reminded the authors of Tweedle Dee (Guan *et al.*, 2006). There are 27 members of this family in *D. melanogaster*, 12 in *An. gambiae*, and fewer in other insects (**Table 3**). Their relationships are discussed in detail in Cornman and Willis (2009). Four conserved regions were defined by Guan and colleagues, and they remain diagnostic of the proteins across the Insecta (**Figure 2**). One member of the TWDL family (BmorCPT1) has been identified in a proteomics analysis of larval chitin-binding proteins, and a recombinant version binds chitin in an *in vitro* assay (Tang *et al.*, 2010).

5.3.2.5. CPLCG family

The largest of the new CP families is CPLCG, recognized by a conserved G-x(2)-H-(x2)-P (Cornman and Willis, 2009). The x residues are restricted to just a few amino acids, and a sequence logo, encompassing a longer stretch of conserved amino acids, is shown in **Figure 2**. Two members of this family had been reported in *D. melanogaster* (Qiu and Hardin, 1995), 3 are now recognized, along with 27 in *An. gambiae*. The *D. melanogaster* sequences had been named Dacp-1 and -2, but members of the family are not restricted to adults, and the CPLCG name is more accurate. Furthermore, the family is not restricted to the Diptera, but was identified in other orders of insects and the crustacean *Daphnia* (**Tables 3, 4**).

5.3.2.6. CPLCW family

Another small family, CPLCW, appears to be restricted to mosquitoes (**Table 4**). The WebLogo (**Figure 2**) shows the invariant W after which it was named, but several other amino acids in a 29-aa region are also almost invariant. Its nine genes are clustered in *An. gambiae* interspersed among some members of the CPLCG family, but the protein sequences of CPLCG and CPLCW families are distinct, having an average similarity of only 20% (Cornman and Willis, 2009).

5.3.2.7. CPLCA family

The CPLCA family has from 13% to 26% alanine residues, but this number is not higher than in some members of other families; rather, the family is best identified by the presence of the retinin domain (pfam04527), although the *D. melanogaster* protein retinin is an outlier in the phylogeny of the group (Cornman and Willis, 2009). A WebLogo more typical of the group has been created (**Figure 3**). While the first published account of this family (Cornman and Willis, 2009) stated that it is restricted to Diptera, there is clearly an EST in *Daphnia* that has a sequence corresponding to the WebLogo (FE341353.1).

5.3.2.8. CPLCP family

This is the most problematic of the cuticular protein families. Peptides corresponding to four genes turned up in the proteomics analysis of proteins from larval head capsules and cast pupal cuticles of *An. gambiae* (He *et al.* 2007). An additional 23 genes coding for related proteins are also present in *An. gambiae*, but none have yet been confirmed by proteomics, although their expression profiles resemble those of authentic low complexity cuticular proteins (Cornman and Willis, 2009). Members of the family have a high density of PV and PY pairs, but additional features described by Cornman and Willis appear to be restricted to mosquitoes where both *Aedes* and *Culex* have been found to have larger families (Cornman and Willis, 2009; Willis, 2010).

5.3.2.9. CPG, the glycine-rich protein family

A group of 28 genes enriched in GGGG or GGxGG repeats was described in *B. mori* (Futahashi *et al.*, 2008), but the group appears heterogeneous because six proteins with only zero to three repeats appear to belong to the CPLCP family; these were identified after that paper was published (Willis, 2010). Another subset of 18 appears to be lepidopteran-specific, and these can appropriately be designated as CPGs (see Willis, 2010, Supplementary Material 2, for details).

5.3.2.10. Apidermin family

Three apidermins, small (6.1–9.2 kDa), highly hydrophobic, and with at least 30% alanine content were described in *Apis mellifera* (Kucharski *et al.*, 2007), and now three have been found in *Nasonia*, but as presently annotated, they are much larger (23–39 kDa). Members of the family do not have an obvious structure; rather, they were recognized by chromosomal linkage, and their role in the cuticle was confirmed with RT-PCR on cuticle-forming tissue. At present they have only been identified in Hymenoptera (**Table 4**). Their designation as a family thus is based on the initial publication, not the normal criterion of shared sequence similarity, and so it is not possible to evaluate the significance of numerous EST sequences from the beetle, *Diaprepes abbreviatus* that are somewhat similar to *A. mellifera* apidermin 1 (e.g., CN474619.1).

(A) TWDL FAMILY (sequence is continuous for 103 residues)

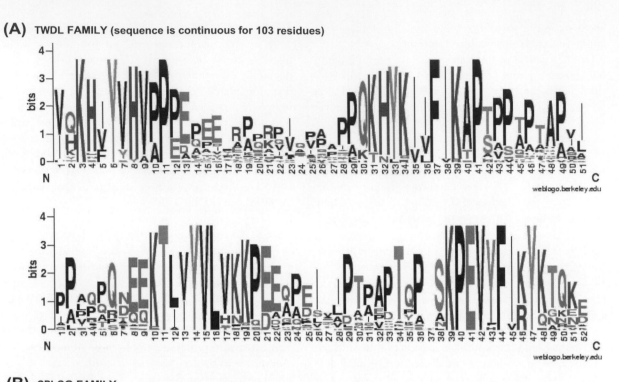

(B) CPLCG FAMILY

(C) CPLCW FAMILY

Figure 2 WebLogos (see **Figure 1**) for three cuticular protein families. (A) TWDL family. Twenty-four sequences from eight species in six orders of insects were used. The continuous sequence was split to facilitate recognition of the four conserved regions. (B) CPLCG family. Note the highly conserved GHPG at residues 5, 8, 11, 14. Eighty-six sequences from dipterans were used. (C) CPLCW family. The 26 CPLCW sequences of this mosquito-restricted family were used. Unlike other WebLogos, the alignment for this one required gaps of five or eight amino acids between positions 16 and 25 to accommodate the longer *Ae. aegypti* sequences. Panels (A) and (B) reprinted from Willis (2010), with permission; some modified from Cornman and Willis (2009). More details are in those references.

(A) CPLCA

(B) CPCFC REPEAT REGIONS

(C) 18 AMINO ACID REPEAT

Figure 3 WebLogos (see **Figure 1**) for two cuticular protein families and one motif. (A) CPLCA family. The WebLogo is based on three sequences from each of four species, *An. gambiae*, *Ae. aegypti*, *C. pipiens*, and *D. melanogaster*, that had the closest match to AgamCPLCA1. This region corresponds to the retinin domain. (B) WebLogo for CPCFC family. Data from the single occurrence of this protein in individual genera of eight insect orders, plus the two occurrences in *T. castaneum* and *Heliconius melpomene*. All three (two in Coleoptera and Lepidoptera) repeat regions from each protein were used. (C) The 18 amino acid repeat from 40 sequences from 26 proteins from 5 insect orders and 2 crustaceans. Panels (A) and (C) reprinted from Willis (2010), with permission.

5.3.2.11. CPAP1 and CPAP3 families A recent publication has identified two more families of cuticular proteins: CPAP1 and CPAP3 (Jasrapuria *et al.*, 2010). They are unusual in that they have multiple cysteine residues, an amino acid rarely found in cuticular proteins from the other families. The families were named because they resemble some peritrophins, hence are peritrophin-like, but the spacing of the cysteines is distinct. The names come from Cuticular Proteins Analagous to Peritrophins. Comparable groups of six cysteines have been demonstrated to form a chitin-binding domain called the "peritrophin A domain," or "type 2 chitin-binding domain"

(ChtBD2), with the six cysteines assumed to form three disulfide bridges. An exhaustive search for proteins with this domain was carried out in *Tribolium* accompanied by RT-PCR analysis of their temporal and spatial distributions. It yielded, in addition to members of the two new families of cuticular proteins, several genuine peritrophins, as well as chitinases and chitin deacetylases. It is assumed that the ChtBD2 domains in all these proteins bind chitin, but this has only been demonstrated experimentally for a chitinase (Arakane *et al.*, 2003) and a CPAP3 protein from another species (Nisole *et al.*, 2010; see also section 5.5.4). So far, the CPAP1 family, with only one ChtBD2 domain, has only been identified in beetles, but the CPAP3 family, with three ChtBD2 domains, is more widespread (**Table 4**). Indeed, its motifs are found outside the arthropods (Jasrapuria *et al.*, 2010). The founding members of the CPAP3 family were a group of proteins, named obstructers, in *D. melanogaster* (Barry *et al.*, 1999; Behr and Hoch, 2005), among them a protein, Gasp, found in tracheae.

5.3.2.12. CPCFC family There is a third cuticular protein family with well-conserved cysteine residues. The founding member is BcNCP1, first identified in *Blaberus craniifer* (Jensen *et al.*, 1997). It has three repeat regions, each with a pair of cysteines separated by five other amino acids; the first and fourth amino acids in each repeat are proline. In a recent publication (Willis, 2010) family status was not recognized, because at that time there were only single occurrences of BcNCP1 orthologs in any species, and a family must have paralogs within a species. Now that criterion has been met in *Heliconius melpomene* and *Tribolium*, and likely in another beetle, *Diaprepes abbreviates*, each with two related genes. This chapter recognizes family status for these paralogs. Several other species of beetles and moths have good orthologs, and in every case the middle cys-bearing region is missing. We are naming this family CPCFC in recognition of the two or three pairs of cysteines that are separated by five amino acids. A WebLogo is shown in **Figure 3**.

5.3.2.13. Cuticular proteins not assigned to families There remain some cuticular proteins that have not reached the criterion for belonging to families. Among them are three proteins identified with proteomics in *An. gambiae* (described in Cornman and Willis, 2009), and a group called CPH (cuticular protein hypothetical) in *Bombyx mori* (Futahashi *et al.*, 2008). Some of the CPH can now be assigned to families; others remain unclassified.

5.3.3. Motifs Found in Cuticular Proteins that do not Define Families

The review by Andersen *et al.* (1995) was the first to assemble a variety of motifs found in cuticular proteins. It is now possible to distinguish among two classes of motifs. The first defines a family such as the CPR and CPF families, while the second includes motifs that occur commonly in cuticular proteins but are found in more than one family; many of these are very short. It is this second class that will be discussed in this section.

The most common short motif described by Andersen *et al.* (1995) was A-A-P-(A/V). Once cuticular proteins of *An. gambiae* had been annotated, it was necessary to expand that motif to A-A-P-(A/V/L). While one or two instances of that motif are found in many proteins,

especially chorion proteins, the occurrence of three or more in a single protein appears to be restricted to cuticular proteins (Willis, 2010). The function of this motif was discussed by Andersen *et al.* (1995) who concluded:

> *A relevant feature of the Ala-Ala-Pro-Ala motif appears to be a strong tendency to form turns; several conformations can be present in equilibrium, indicating low energy barriers between the conformations. When the sequence occurs regularly in a protein, as it does in many of the CPs as well as in other structural proteins, it can be suggested that the result will be proteins folded in a more or less regular helix, which is easily and reversibly deformed by external forces, thereby resembling elastin.*

Andersen *et al.* (1995) recognized several sequences with stretches of glycine, leucine, and tyrosine, beginning G-Y-G-L- or G-L-L-G. Other cuticular proteins are also high in glycine, but with less regular motifs; these are designated by the number of consecutive Gs. Proteins enriched in glycine residues are found in a variety of structures, such as plant cell walls, cockroach ootheca, and silk (see Bouhin *et al.*, 1992a, for discussion). Subsequent to their 1995 review, Andersen and his colleagues recognized two additional motifs.

Three copies of an 18-residue motif were found in a *B. mori* protein (PCP, now named BmorCPH31) by Nakato *et al.* (1992). Subsequently, Andersen (2000) recognized the repeat in a small number of cuticular proteins from four orders of insects and two crustaceans. A sequence logo based on its occurrence in 27 proteins from 5 orders of insects and 2 crustaceans is shown in **Figure 3**. These proteins include some with the R&R Consensus, especially those assigned as RR-3, as well as others, like BmorCPH31 and four other *B. mori* cuticular proteins, that do not have this Consensus (Futahashi *et al.*, 2008).

A recent analysis (Cornman, 2010) analyzed the short motifs GYR and YLP in several *Drosophila* species in relation to cuticular proteins and other classes of proteins.

5.3.4. Glycosylation of Cuticular Proteins

Glycosylation of cuticular proteins was first reported by Trim in 1941, and then in limited subsequent reports (see Cox and Willis, 1987b, for review). In recent years, post-translational modifications of cuticular proteins have been determined by staining gels with periodic acid Schiff (PAS), by using labeled lectins to probe blots of electrophoretically separated proteins, or by discovering discrepancies in masses of peptide fragments experimentally determined by MALDI-MS analysis and calculated from Edman sequencing.

Most of the major cuticular proteins seen on gels stained with Coomassie Blue are not recognized by PAS or lectins, while some minor ones are glycosylated. This was true for *H. cecropia*, where PAS staining revealed glycosylated proteins in extracts of flexible cuticles of *H. cecropia* and a screen with eight lectins revealed the presence of mannose and N-acetylgalactosamine, with more limited binding to N-acetylglucosamine, galactose, and fucose, in a few of the proteins from all stages (Cox and Willis, 1987b). A comparable study in *Tenebrio* revealed one major band of water-soluble larval and pupal cuticular proteins that had N-acetylglucosamine, and a few other bands were weakly visualized with lectins; none of the proteins from adult cuticle reacted with the lectins (Lemoine *et al.*, 1990). In another Coleopteran, *Anthonomus grandis*, glycosylation was found in cuticular proteins extracted from all three metamorphic stages (Stiles, 1991). In yet another coleopteran, *T. castaneum*, the BioRad Immun-Blot kit for glycoprotein detection revealed multiple bands on a blotted 1D SDS gel; none of the abundant bands below 30 kDa were stained (Missios *et al.*, 2000). In *Calpodes*, all the BD peptides but very few of the C class proteins (see section 5.2.2) extracted from the cuticle were associated with α-D-glucose and α-D-mannose, just like most of the hemolymph proteins. Some of each class appeared to be modified with N-acetylglucosamine. T66, a protein synthesized in spherulocytes, transported to epidermis, and then secreted into the cuticle, however, was not glycosylated. In none of these species is the amino acid sequence of a glycosylated protein known.

Sequence-related information about glycosylation is available for cuticular proteins isolated from locusts and *Manduca* where the direct analysis of residues had been used. In *Locusta migratoria*, one to three threonine residues were modified in the protein LM-ACP-abd4. In each case, the modification was with a moiety with a mass of 203, identified as N-acetylglucosamine (Talbo *et al.*, 1991). Each of the three threonine residues occurred in association with proline (FPTPPP, LATLPPTPE). All eight of the cuticular proteins that have been sequenced from *S. gregaria* nymphs had evidence for glycosylation with a moiety with a mass of 203, all at a threonine residue found in a cluster of prolines (Andersen, 1998). Three proteins recently isolated from *Manduca* were similarly shown to be glycosylated on threonines also in proline-rich regions. Surprisingly, in these cases masses of the adducts were varied (184, 188, and 189) and their nature was not determined (Suderman *et al.*, 2003). In all of these cases, the available evidence indicates that the threonine residues had been *O*-glycosylated. The significance of such glycosylation awaits further elucidation.

5.4. Genomic Information

5.4.1. Introduction

The first four cuticular proteins whose complete sequences were determined were also the first to have their genes described (Snyder *et al.*, 1982). The wealth of experimental detail and thoughtful discussion in that paper make it a classic in the cuticular protein literature. These four genes for *D. melanogaster* cuticular proteins LCP-1, -2, -3, and -4 were found to occupy 7.9-kb of DNA, along with what appeared to be a pseudogene. Each gene had a single intron, and that intron interrupted the protein-coding region between the third and fourth amino acids. *LCP-1* and *-2* were in the opposite orientation of *LCP-3* and *-4*. The nucleic acid sequences in the protein coding regions for LCP-1 and -2 were 91% identical, and for LCP-3 and -4 were 85% identical, with similarity between the two groups about 60%. For the non-coding regions of the mRNAs, the 5′ upstream regions had more sequence similarity than the 3′ downstream. A consensus poly(A) addition site, AATAAA, was found for two of the genes, 110 bp from the stop codon, while similar but not identical sequences (AATACA, AGTAAA) were found for the other two. The four genes were all expressed in the third instar, and several short, shared elements were found in their 5′ regions upstream from the transcription start site. Snyder *et al.* (1982) also speculated on the origin of the cluster through gene duplication and inversion. These features of those four genes (coding for RR-1 proteins) have turned out to be the common elements of most of the cuticular protein genes that are known – hence linkage, shared and divergent orientation, an intron that interrupts the signal peptide, presence of a pseudogene in the cluster, atypical poly(A) addition sites, and divergence of 3′-untranslated regions have been found for cuticular protein genes in Diptera, Lepidoptera, and Coleoptera.

5.4.2. Chromosomal Linkage of Cuticular Protein Genes

In addition to the four *D. melanogaster* genes discussed in the previous section, several more instances of linked cuticular proteins genes were described prior to sequencing entire genomes. In some cases the evidence for these genes was restricted to cross-hybridization of the genomic fragment, and complete sequences were not known for all the members.

A detailed analysis of the cluster of genes at 65A allowed Charles *et al.* (1997, 1998) to describe important features that most likely contributed to the multiplication and diversification of cuticular protein genes. Twelve genes were identified in a stretch of 22 kb, with the direction of transcription, or more accurately the strand used, being: > < < < < < < > > > > >. The third gene in the cluster appeared to be a pseudogene. Several important features were found: the number of *Lcp-b* genes within the cluster was variable among different strains of *D. melanogaster*; and some genes lacked introns, had tracks of As at the 3′ end and short flanking direct repeats. These features are consistent with their having arisen by retrotransposition.

Now that there are complete sequence data for the entire 65A region the situation has been shown to be even more complex, and comparison with six other *Drosophila* species has provided new insights (Cornman, 2009). Eighteen CPR genes are present in the 65A region of *D. melanogaster*; seven of these are present in most or all seven species as one-to-one orthologs, with their chromosomal order conserved. Others have orthologs only within one of the two species groups analyzed. Others are found scattered among the array, with paralogs only within one or two species, and this analysis, of course, could not deal with the variation in copy number within a species. Cornman confirmed the findings of Charles *et al.* (1997) that some of the genes lacked introns, and assessed the possibility the latter raised that retroposition played a role in the formation of this array, but concluded that "retrogenes do not appear to contribute substantially to the distinctive pattern of evolution within these arrays."

The consequences of gene duplication in terms of gene expression are an important issue. It could be that duplicated genes were preserved to boost the amount of product made in the short period that the single-layer epidermis is secreting cuticle. Alternatively, duplication may allow for precise regulation of expression of genes both spatially and temporally. Subtle differences in protein sequence may be advantageous for particular structures. A detailed analysis of mRNA levels with Northern blot analysis demonstrated that some members of the *65A* cuticular protein cluster have quite different patterns of expression. *Acp* was expressed only in adults. Expression was not detected for *Lcp-a*; all other *Lcp* genes were expressed in all larval stages, and all but *Lcp-b* and *-f* also contributed to pupal cuticle (Charles *et al.*, 1998).

One of the major findings to come out of whole-genome sequencing was two different forms of chromosomal linkage of genes for cuticular proteins. Data from *An. gambiae*, *D. melanogaster*, and *B. mori* revealed that many CP genes are found adjacent to one another. Such genes have been described as being in tandem arrays, and both RR-1 and RR-2 genes are clustered in this manner, always in separate arrays (Cornman *et al.*, 2008).

In mosquitoes, there are numerous instances of sequence clusters – groups of genes that are very similar in sequence. Members are generally, but not always, found adjacent within a tandem array. Eight clusters (with 4–16 members) of RR-2 genes were identified in *An. gambiae* (Cornman *et al.*, 2008), and comparable clusters were also present in *Ae. aegypti* and *Culex pipiens*. There are no clusters in *D. melanogaster* coding for more than three proteins with almost identical sequences, but three small sequence clusters with a total of 15 RR-2 genes are present in the *B. mori* genome (Futahashi *et al.*, 2008). The suggestion was made that the *Anopheles* sequence clusters serve to facilitate accumulation of mRNA in a brief period of time, while the *Bombyx* workers speculated that different members of the clusters might be used to build specific structures (Futahashi and Fujiwara, 2008; Futahashi *et al.*, 2008). A detailed analysis of sequence clusters in *An. gambiae* can be found in Cornman and Willis (2008).

It is not only CPR genes that are found in tandem arrays. There is a large tandem array on chromosome 3R in *An. gambiae* that has all 27 CPLCG genes and all 9 CPLCW genes. Members of the two families are interspersed, and in the array are an additional 10 unrelated genes. Twelve of the CPLCG genes belong to a sequence cluster, and, despite their considerable similarity (86% identity at the nucleotide level), they are dispersed throughout the tandem array and interspersed with the CPLCW genes that form another sequence cluster with at least 92% sequence identity at the protein level (Cornman and Willis, 2009).

5.4.3. Intron Structure of Cuticular Protein Genes

Genomic sequence data are now available for hundreds of cuticular proteins. Intron position has only been analyzed in detail for *An. gambiae* (Cornman and Willis, 2008) and *B. mori* (Futahashi *et al.*, 2008). An early prediction that genes for cuticular proteins would have no more than 2 introns was incorrect, for several have been identified with 5 or more, and one, *BmorCPR146*, has 13. Nonetheless, the number is usually low, averaging 2.3 for *An. gambiae* CPRs and 2.4 for that family in *B. mori*. Cuticular proteins in other *An. gambiae* cuticular protein families generally have only two exons. The most common position for the first intron is interrupting the signal peptide. Whether this conserved position represents something important awaits further exploration, but there are several ways the intron might be important (Charles, 2010). One possibility is that it contains information needed for transcription. Direct evidence that this is the case for one gene comes from an analysis of the *DmelACP65A* gene. Expression is suppressed in the absence of the intron that occurs after coding for the first four amino acids of the signal peptide, and is restored if the intron is added upstream of the transcription start site (Bruey-Sedano *et al.*, 2005).

Another common position is at or near the start of the aromatic triad. Some genes that lack the intron interrupting the signal peptide have one in this region. An early analysis of intron position led Charles *et al.* (1997) to postulate that the primitive condition for introns in insect cuticular proteins would be two; over time, some genes lost one, some the other, and some lost both or arrived in the genome by retrotransposition.

There is also a *D. melanogaster* cuticular protein whose gene is located within the region corresponding to the first intron of *Gart* (now named *ade3*, CG31628), a gene that encodes proteins involved in the purine pathway. The gene for this RR-1 protein (*Pcp*, CG3440) is read off the opposite strand and has its own intron, conventionally placed interrupting the signal peptide (Henikoff *et al.*, 1986). A comparably placed gene with 70% amino acid sequence identity is found in *D. pseudoobscura* (Henikoff and Eghtedarzadeh, 1987).

5.4.4. Regulatory Elements

One of the attractions of studying cuticular proteins is that they are secreted at precise times in the molt cycle, and are thus candidates for genes under hormonal control (Riddiford, 1994; Togawa *et al.*, 2008). It would be expected, therefore, that some might have hormone response elements. Imperfect matches to ecdysteroid response elements (EcRE) from *D. melanogaster* were found on two of its cuticular protein genes – *EDG78* and *EDG84* (Apple and Fristrom, 1991). These genes are activated in imaginal discs exposed to a pulse of ecdysteroids, but if exposed to continuous hormone, no message appears. The two cuticular protein genes that have been studied in *H. cecropia* have regions close to their transcription start sites that resemble EcREs (Binger and Willis, 1994; Lampe and Willis, 1994), and upstream from *MSCP14.6* are also two regions that match (Rebers *et al.*, 1997).

It is now apparent that the regulatory regions controlling response to ecdysteroids encompass more than just an EcRE. Indeed, the EcRE itself can also recognize β-FTZ-F1, a protein induced in response to ecdysteroid stimulation that has been shown to be a major regulator of cuticular protein synthesis (reviewed in Charles, 2010). Charles (2010) discusses the evidence for this and other transcription factors (BR, DHR38, OCT, SVP) that bind upstream of cuticular protein genes.

Both *Bombyx PCP* (now *BmorCPH31*) and *H. cecropia HCCP66* have response elements for members of the POU family of receptors (Nakato *et al.*, 1992; Lampe and Willis, 1994). POU proteins are transcription factors used for tissue-specific regulation in mammals (Scholer, 1991). Gel mobility shift experiments established that there was a protein in epidermal cells that could bind to this element (Lampe and Willis, 1994).

Numerous additional genes, some of them coding for transcription factors, have been implicated in cuticle formation in *D. melanogaster* (Moussian, 2010).

Now that genomic sequence information is available, identification of regulatory elements and verification of their action is underway.

5.5. Interactions of Cuticular Proteins with Components of Cuticle

One of the most challenging aspects in the study of the cuticle is the elucidation of interactions among cuticular proteins and cuticle's non-proteinaceous components. The most abundant is the CPR family, which is characterized by the presence of the R&R Consensus (see section 5.3.2.2). The abundance of sequences bearing the R&R Consensus in cuticles formed by every species of arthropod examined led several workers to suggest that the role of the R&R Consensus might be to bind to chitin, and this has now been confirmed with recombinant proteins (see section 5.5.4).

Of particular interest among the other families of cuticular proteins that lack the R&R Consensus is the CPF gene family, now recognized by a 44-aa sequence motif (Togawa *et al.*, 2007) (see section 5.3.2.3). As discussed below, CPFs may interact with other components of cuticle, such as sex pheromones (Hall, 1994; Greenspan and Ferveur, 2000) or cuticular lipids, acting as possible repositories (Papandreou *et al.*, 2010).

Various approaches have been followed to gather information about the interactions of cuticular proteins with other components of cuticle. The first was to analyze cuticular protein sequences with appropriate software to predict their secondary structure. The second approach was to use spectroscopic techniques on cuticular components to gain information about the conformation of their protein constituents *in situ*, and compare experimental information with predictions. Third, the tertiary structures of cuticular proteins have been modeled, and the fourth route was a direct experimental approach to test whether proteins exhibiting the extended Consensus could bind to chitin. Such analyses are restricted primarily to the CPR and CPF families, the only ones to be discussed here.

5.5.1. Secondary Structure Predictions

Prediction of secondary structure was carried out on the extended R&R Consensus region of the cuticular proteins now classified in the CPR family (see Iconomidou *et al.*, 1999, for details of proteins analyzed, programs used, and pictorial representation of results.)

The results indicated that the extended R&R domain of cuticular proteins has a considerable proportion of β-pleated sheet structure and a total absence of α-helix. Other features revealed include the presence of glycines and histidines at the predicted β-turn/loop regions. Glycines are considered good turn/loop formers (Chou and Fasman, 1974a, 1974b), while histidines, which in this case are "exposed," are certainly involved in cuticular sclerotization and in the variations of the water-binding capacity of cuticle and the interactions of its constituent proteins (Andersen, 2005). Also, the β-sheets exhibit an amphipathic character – i.e., one face is polar, the other non-polar. Alternating residues along a strand point in the opposite direction on the two faces of a β-sheet. With these proteins, it is the aromatic or hydrophobic residues that alternate with other, sometimes hydrophilic, residues. The aromatic rings are thus positioned to stack against faces of the saccharide rings of chitin. This type of interaction is fairly common in protein–saccharide complexes (Vyas, 1991; Hamodrakas *et al.*, 1997; Tews *et al.*, 1997).

The suggestion that cuticular proteins adopt a β-sheet conformation is not new. Fraenkel and Rudall (1947) provided evidence from X-ray diffraction that the protein associated with chitin on intact cuticle has a β-type of structure.

5.5.2. Experimental Studies of Cuticular Protein Secondary Structure

The next step in probing the structure of cuticular proteins involved direct measurements on intact cuticles, on proteins extracted from them with a strong denaturing buffer with 8M guanidine hydrochloride, and on the extracted cuticle. The cuticles came from the flexible abdominal cuticle of larvae of *H. cecropia*, and extracts have HCCP12, a RR-1 protein, as a major constituent (Cox and Willis, 1985; Binger and Willis, 1994). The same prediction programs described above were used on the sequence for HCCP12, and it indicated that the entire protein had a considerable proportion of β-pleated sheet and total absence of α-helix. Fourier-transform Raman spectroscopy (FT-Raman), attenuated total reflectance infrared spectroscopy (ATR FT-IR), and circular dichroism spectroscopy (CD) were carried out on these preparations (Iconomidou *et al.*, 2001). These techniques eliminated problems that had been found previously with more conventional laser-Raman spectra due to the high fluorescent background associated with cuticle.

The FT-Raman spectra of both the intact and extracted cuticle were dominated by the contribution of bands due to chitin. Certain features of the Raman spectrum of the intact cuticle signified the presence of proteins. The protein contribution to the spectrum of intact cuticle was revealed by subtracting the spectrum of the extracted cuticle, after scaling the discrete chitin bands of both preparations. The comparison of this difference spectrum to that from the isolated proteins revealed striking similarities, suggesting that the former gave a reliable physical picture of the cuticle protein vibrations in the native state. While Iconomidou *et al.* (2001) presented a detailed analysis of the spectra and the basis for each assignment, only a few features will be reviewed here. Several of the spectral bands could be attributed to side-chain vibrations of amino acids with aromatic rings, tyrosine, phenylalanine, and tryptophan; others were typical of β-sheet structure and others could be assigned to β-turns or coil. The absence of bands at characteristic positions indicates that α-helical structures are not favored.

Results from ATR-FT-IR spectra from the extracted proteins were in good agreement with their FT-Raman spectra. These spectra had been obtained on lyophilized samples. The CD spectrum, on the other hand, was obtained with proteins solubilized in water. Detailed analysis of the CD spectrum indicated a high percentage (54%) of β-sheet conformation with a small contribution of α-helix (~13%). The contributions of β-turns/loops and random coil were estimated as 24% and 9%, respectively (Iconomidou *et al.*, 2001). These results demonstrated that the main structural element of cuticle proteins is the antiparallel β-pleated sheet. Comparable results were obtained from lyophilized proteins and intact cuticles, and from proteins in solutions, thus negating the concern that lyophilization might increase the β-sheet content of proteins as discussed by Griebenow *et al.* (1999). These direct measurements confirm the results from secondary structure prediction discussed above in section 5.5.1.

These findings are in accord with the prediction of Atkins (1985) that the antiparallel β-pleated sheet part of cuticular proteins would bind to α-chitin. His proposal was based mainly on a 2D lattice matching between the surface of α-chitin and the antiparallel β-pleated sheet structure of cuticular proteins.

There seem to have been several independent solutions in nature whereby chitin binds to protein; in all, surface aromatic residues appear to be significant (Shen and Jacobs-Lorena, 1999). In several cases, β-sheets have been implicated. The chitin-binding motifs of two lectins studied at atomic resolution contain a two-stranded β-sheet (Suetake *et al.*, 2000). In bacterial chitinases, an antiparallel β-sheet barrel has also been postulated to play an important role in "holding" the chitin chain in place to facilitate catalysis. Four conserved tryptophans on the surface of the β-sheet are assumed to interact firmly with chitin, "guiding" the long chitin chains towards the catalytic "groove" (Perrakis *et al.*, 1997; Uchiyama *et al.*, 2001).

5.5.3. Modeling of Cuticular Proteins

5.5.3.1. CPR protein models Secondary structure prediction and experimental data summarized above (see sections 5.5.1 and 5.5.2) indicated that β-pleated sheet is most probably the underlying molecular conformation of the members of the CPR family, and that this conformation is most probably involved in β-sheet/chitin-chain interactions of the cuticular proteins with the chitin filaments (Iconomidou *et al.*, 1999, 2001). Can this information be translated into a three-dimensional model?

Unexpectedly, distant sequence similarities of the extended R&R Consensus from several CPR proteins were found with a lipocalin, bovine plasma retinol-binding protein (RBP) (Hamodrakas *et al.*, 2002). Lipocalins are members of a family of extracellular proteins, typically small (160–200 residues), with low sequence similarity among family members (frequently <20%). They exhibit several common molecular recognition properties, and, while they were classified mainly as transport proteins, it is now clear that they have various functions (Flower,

1996). The lipocalin fold is a highly symmetrical all-β structure dominated by a single eight-stranded antiparallel up-and-down β-sheet barrel (Flower *et al.*, 2000). Fairly recently, it was found that lipocalins are characterized by two hydrophobic "clusters" of residues, the "inner" and the "outer" clusters (Adam *et al.*, 2008).

The first attempt utilized HCCP12, an RR-1 protein leading to a construction of a structural model that corresponds to the "extended R&R Consensus" (Hamodrakas *et al.*, 2002). The original model (**Figure 4A**) comprises the C-terminal 66 residues (out of 89 in total) of HCCP12, and has many advantages since it corresponds to the full sequence of the "extended R&R Consensus" (see section 5.3.2.2). This work was extended to RR-2 proteins, leading to comparable results, as shown for AGCP2b (Agam-CPR97) in **Figure 4D** (Iconomidou *et al.*, 2005).

Low-resolution docking experiments of an extended N-acetylglucosamine tetramer to the model of HCCP12, utilizing the docking program GRAMM (Vakser, 1996), revealed that the proposed model for cuticle proteins accommodates, perpendicularly to the half-barrel β-strands, at least one extended chitin chain (**Figure 4A**) (Hamodrakas *et al.*, 2002).

Homology modeling results indicate that the basic structural motif of the CPR family is an antiparallel β-sheet structure with a "cleft" full of conserved aromatic residues that form "flat" hydrophobic surfaces on one "face," perfectly positioned to stack against faces of the saccharide rings of chitin. One unpredicted feature in the model is a short two-turn α-helix at the C-terminus of the extended R&R Consensus. This C-terminal part of the model is reminiscent in some respects of the chitin-binding domain of an invertebrate chitin-binding lectin, a two stranded β-sheet followed by a helical turn (Suetake *et al.*, 2000). More detailed docking experiments (Iconomidou *et al.*, 2005), utilizing GRAMM (Vakser, 1996), showed that chitin protein chains may run parallel to the β-strands of the half-β-barrel (**Figures 4B, 4C**). Thus, β-barrels of cuticle proteins may intervene between the long chitin chains in cuticle without disrupting continuity. This parallel arrangement of cuticle protein β-strands with the chitin chains agrees with observations made by Atkins, over 20 years ago (Atkins, 1985), from X-ray diffraction patterns.

The inherent twist of the half-barrel β-sheet of the cuticle proteins and its observed packing arrangement at an angle with the chitin chains may provide a molecular basis for the morphological observation of a helicoidal twist in cuticle. These models were also subjected to analysis (Iconomidou *et al.*, 2005) of the positions of histidine residues, since they might play a role in cuticle sclerotization (Neville, 1975; Andersen *et al.*, 1995; Andersen, 2005) and appear to be very conserved in RR-2 sequences.

The general remarks that arise from the analysis are that histidines are positioned "exposed" either in turns or at the edges of the half-barrel or its periphery, permitting interactions with chitin involved this way in cuticle sclerotization (**Figure 4D**). Alternatively, they could be involved in the variations of the water-binding capacity of cuticle and the interactions of its constituent proteins, because small changes of pH can affect the ionization of their imidazole groups (Andersen *et al.*, 1995).

These observations are in excellent agreement with the predictions made several years ago for the role of histidines from secondary structure predictions (Iconomidou *et al.*, 1999), and strengthen further the value of the models previously proposed for CPR proteins (Hamodrakas *et al.*, 2002).

5.5.3.2. CPF protein models

The next obvious step involved attempts at elucidating the structural motifs and possible functions of cuticular proteins that belong to families where the "extended R&R Consensus" is absent. An appropriate choice was the CPF family of cuticular proteins (see section 5.3.2.3) (Togawa *et al.*, 2007). This family of cuticular proteins is of particular interest because they are expressed just before pupal or adult ecdysis, suggesting that these families are most probably components of the outer layer of pupal and adult cuticles – that is, they are likely located in the epi- or exo-cuticle. Actually, the epicuticle is one cuticular region that lacks chitin, suggesting that the CPF family of proteins may interact with components of the cuticle other than chitin.

Similarly to CPR proteins, members of the CPF family share significant sequence similarity to the crystallographically solved structure of bovine retinol-binding protein (RBP), which belongs to the class of lipocalins. The models of two proteins, AgamCPF3 from *Anopheles gambiae* and a CPF homolog, CG8541, from *Drosophila melanogaster*, were constructed based on this similarity (Papandreou *et al.*, 2010). The derived models (**Figures 5A** and **5B**) indicate that the basic folding motif of CPFs is most probably an antiparallel, up-and-down, β-sheet full-barrel structure, unlike the proposed half-barrel for the CPR family.

The next step involved a high-resolution experiment, utilizing GRAMM (Vakser, 1996), of the proposed model with a NAG tetramer (Papandreou *et al.*, 2010). The results (**Figure 6B**) indicated that the tetramer does not fit into the binding pocket of the CPFs; rather, the CPFs might interact loosely with chitin chains, with their β-strands lying parallel to the chitin chains, in agreement with experimental observations (Atkins, 1985). Further evidence against a role of the CPFs in direct binding to chitin comes from failure of recombinant CPF proteins to bind to chitin (Togawa *et al.*, 2007). Comparative structural information in the paper by Papandreou *et al.* (2010) also indicates that carbohydrates should not bind in the pocket. Protein–carbohydrate interactions involve aromatic residues, and in the cleft of the

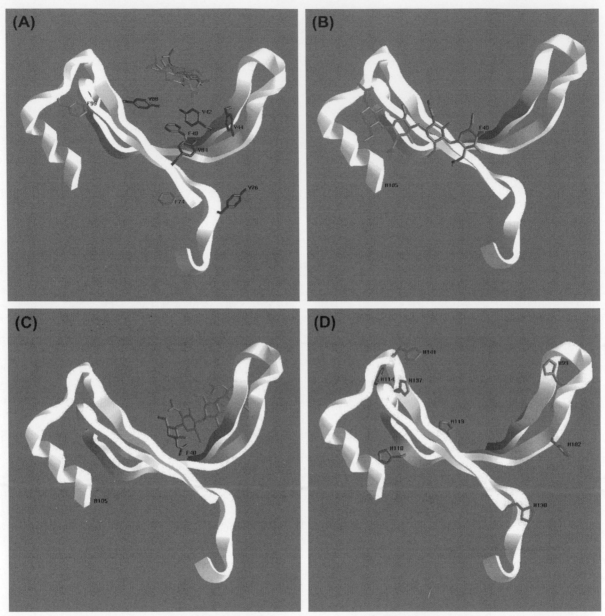

Figure 4 Ribbon models of cuticular proteins derived from homology modeling. (A) A ribbon model of cuticle protein structure, displayed using GRASP (Nicholls *et al.*, 1991). The structure of the representative RR-1 cuticle protein HCCP12 was modeled on that of bovine retinol-binding protein (RBP; PDB code 1FEN) (Zanotti *et al.*, 1994), utilizing the program WHAT IF (Vriend, 1990). Further details are in Hamodrakas *et al.* (2002). The side chains of several aromatic residues are shown and numbered, following the numbering scheme of the unprocessed HCCP12 sequence. The model structure has a "cleft" full of aromatic residues, which form "flat" surfaces of aromatic rings (upper side), ideally suited for cuticle protein–chitin chain interactions, and an outer surface (lower side), which should be important for protein–protein interactions in cuticle. The model is a complex of HCCP12 with an N-acetyl glucosamine (NAG) tetramer in an extended conformation. The complex was derived from a "low-resolution" docking experiment of a NAG tetramer, in an extended conformation, with the model of HCCP12, utilizing the docking program GRAMM (Vakser, 1996) and the default parameters of the program. (B) and (C) Two more possible complexes of HCCP12 with a NAG tetramer in an extended conformation derived from a "high-resolution" docking experiment, utilizing the program GRAMM (Vakser, 1996) and the default parameters of the program for "high resolution." The two models presented in (B) and (C) are the two "top on the list," most favorable complexes, whereas third on the list is a structure similar to that of (A). The one in (B) has the NAG tetramer more or less parallel to the last β-strand of the HCCP12 half β-barrel model, whereas that in (C) has the NAG tetramer more or less parallel to the first β-strand of the HCCP12 half β-barrel model. Note that in both (B) and (C) the chitin chain runs parallel to the β-strands, whereas in (A) the chain is arranged perpendicular to the β-strands. (D) A display of a model of the RR-2 protein AGCP2b. The numbering is that of the unprocessed protein. Histidine (H) side chains are shown as "ball and sticks," in red, with their corresponding numbering (from Willis *et al.*, 2005).

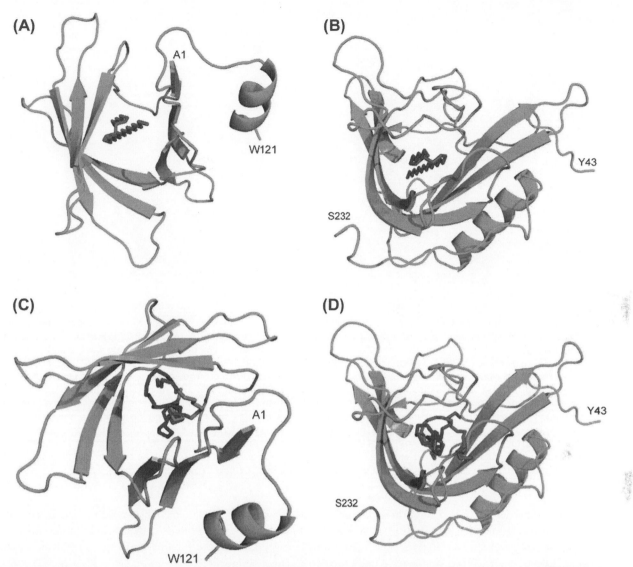

Figure 5 (A) A ribbon model of the cuticular protein AgamCPF3 structure (green), displayed using the software PyMOL (Delano, 2005). The model was modeled on that of bovine retinol-binding protein (RBP; PDB code 1FEN (Zanotti *et al.*, 1994) utilizing the software Modeller v9.2 (Sali and Blundell, 1993). The entire secreted protein, from A1 to W121, is shown in the model. It is complexed with 7(Z),11(Z)-heptacosadiene (7,11-HD), shown in red. The complex was derived from a docking experiment of 7,11-HD, with the model of AgamCPF3, utilizing the docking software Autodock4.2 (Morris *et al.*, 2009). The ligand is inside the "pocket" of the β-barrel of AgamCPF3. The ligand was considered as rigid, in its minimum energy conformation. The ligand represents a cluster of 4 out of 10 best solutions (runs). (B) A ribbon model of the CPF protein DmelCG8541 structure (green), constructed and displayed as in **Figure 5(A)**. The model comprises 190 of 257 residues of the secreted protein, from Y43 to S232. It is complexed with 7,11-HD, shown in blue. Details of the docking experiment that produced this complex are as in **Figure 5(A)**. The ligand represents a cluster of 7 out of 10 best solutions (runs). (C) A ribbon model of the cuticular protein AgamCPF3 structure (green), constructed and displayed as in **Figure 5(A)**. The entire secreted protein, from A1 to W121, is shown in the model. The complex was derived from a docking experiment of 7,11-HD, with the model of AgamCPF3, utilizing the docking software Autodock4.2 (Morris *et al.*, 2009). Two out of 10 best solutions (runs) for the ligand are shown in red and blue, respectively, inside the "pocket" of the β-barrel of AgamCPF3. The remaining eight solutions also show the ligand to reside inside the "pocket." The 7,11-HD ligand was considered as flexible (all rotable bonds were set free). (D) A ribbon model of the cuticular protein DmelCG8541 structure (green), constructed and displayed as in **Figure 5(B)**. The model comprises 190 of 257 residues of the secreted protein, from Y43 to S232. It is complexed with 7,11-HD. All other details of the docking experiment that produced the complex are as in **Figure 5(A)**. Two out of 10 best solutions (runs) for the ligand are shown in magenta and blue, respectively, inside the "pocket" of the β-barrel of DmelCG8541. The remaining eight solutions also show the ligand to reside inside the "pocket." Reproduced from Papandreou *et al.* (2010), with permission.

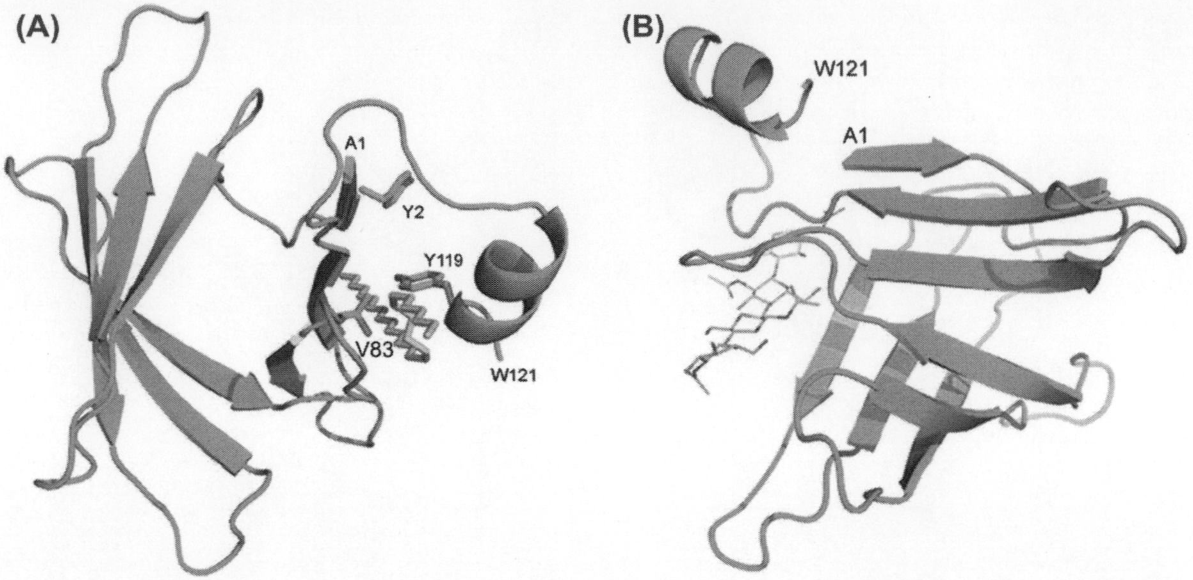

Figure 6 (A) A ribbon model of the cuticular protein AgamCPF3 structure (green), constructed and displayed as in **Figure 5(A)**. The entire secreted protein, from A1 to W121, is shown in the model. The complex was derived from a docking experiment of 7,11-HD (shown in cyan) with the model of AgamCPF3, utilizing the docking software Autodock4.2 (Morris *et al.*, 2009). It shows the ligand, outside the β-barrel of AgamCPF3, in contact with the "hydrophobic outer cluster" (see **Table 1** in Papandreou *et al.*, 2010). The side chains of three hydrophophic residues of the conserved "hydrophobic outer cluster," Y2, V83, and Y119, are shown. The ligand was considered as rigid, in its minimum energy conformation. The ligand represents the cluster of the remaining 6 out of 10 best solutions (see **Figure 5(A)**). (B) A complex of AgamCPF3 (ribbon model shown in green) with a NAG tetramer (ball and stick model) in an extended conformation (taken as a chitin analog). The complex was derived from a "high resolution" docking experiment, utilizing the docking software GRAMM (Vakser, 1996) and the default parameters of the program, displayed using PyMol (Delano, 2005). The model presented is the "top of the list," most favorable complex. Note that the "chitin chain" runs parallel to the β-strands, of at least half of the β-barrel, in agreement with experimentally derived data (Atkins, 1985). No solution was obtained with the "chitin chain" into the pocket of the β-barrel. The entire secreted protein, from A1 to W121, is shown in the model. Reproduced from Papandreou *et al.* (2010), with permission.

half-barrel model of HCCP12 are three critical aromatic residues, (Hamodrakas *et al.*, 2002); the model of HCCP66 has four aromatic residues in its cleft (Iconomidou *et al.*, 2005), whereas comparable residues in the two CPFs, AgamCPF3 and DmelCG8541, are hydrophobic but not aromatic.

Therefore, the questions that arise are, what is the functional role of the CPF proteins, and what fits within the cavity of the barrel? One possible function is that they intercalate among the chitin crystallites and chitin-binding proteins of the procuticle. However, this does not explain why they should form a binding pocket. Alternatively, if CPFs are components of the epicuticle, they could perhaps bind, as lipocalins do, to the lipoidal molecules, which are known to act as female contact sex pheromones in certain insect species (Antony and Jallon, 1982; Antony *et al.*, 1985) and are primarily located in the epicuticle (Andersen, 1979). We attempted to dock 7(Z), 11(Z)-heptacosadiene (7,11-HD), the predominant female-specific sex pheromone of *D. melanogaster* (Antony *et al.*, 1985), to the derived models of the *D. melanogaster* CPF protein, CG8541, utilizing GRAMM (Vakser, 1996) and Autodock4.2 (Morris *et al.*, 2009). The pheromone was considered both as rigid and flexible. Docking results

showed that this interaction is possible, indeed energetically favorable, and that 7,11-HD could fit into the binding pocket of the β-barrel or in the outer hydrophobic cluster (**Figures 5B, D**).

Complex formation between AgamCPF3 and 7,11-HD is also favored, although the molecular nature of sex pheromones in *An. gambiae*, if they exist, remains unknown, suggesting that a similar structure could easily bind to AgamCPF3, either inside the pocket (**Figures 5A, C**) or outside (**Figure 6A**). Microarray analyses have found significantly different levels of CPF3 transcript in adults of the incipient species M and S, and within the same form following a blood meal or in response to mating (Cassone *et al.*, 2008; Marinotti *et al.*, 2006; Rogers *et al.*, 2008). On the other hand, it is surprising that an epicuticular component would continue to be made and secreted into outer regions of the cuticle days after adult eclosion. An alternative occupant of the CPF binding pocket might just be intracuticular lipids that are present throughout the cuticle. Several of these cuticular lipids have chemical structures very similar to 7,11-HD (Hadley, 1981). Therefore, they would fit easily into the pocket of the β-barrel of the CPFs, or bind to their "outer hydrophobic cluster" (**Figures 5, 6A**).

Why do the proposed models correspond to a half-barrel model for CPRs and a full barrel for CPFs? The CPR Consensus region alone (<70 aa) was used, for that is the region of the protein that matches closely to retinol-binding protein; it is far too short to form a full barrel. By contrast, the CPF match is far longer, and compatible with a full barrel.

5.5.4. Fusion Proteins Establish a Role for the Extended R&R Consensus

Predictions of secondary and tertiary structure and experimental evidence supporting them (discussed above in sections 5.1–5.3) established that the extended R&R Consensus has the properties to serve as a chitin-binding motif. In particular, the planar surfaces of the predicted β-sheets will expose aromatic residues positioned for protein–chitin interaction. The ultimate test of these predictions would be to show that the extended consensus region is sufficient to confer chitin binding on a protein.

Rebers and Willis (2001) investigated this possibility by creating fusion proteins using the extended R&R Consensus from the *An. gambiae* putative cuticular protein, AGCP2b (Dotson *et al.*, 1998; now annotated as AgamCPR97). First, they expressed this protein in *E. coli* and isolated it from cell lysates. The construct used coded for the complete protein minus the predicted signal peptide, and had a histidine tag added to the N-terminus to facilitate purification. AGCP2b is a protein of 222 amino acids, with an RR-2 type of consensus. The purified protein bound to chitin beads, and could be eluted from these beads with 8M urea or boiling SDS. This established unequivocally that AGCP2b was a chitin-binding protein. Chitin binding previously had been obtained with mixtures of protein extracted from cuticles of two beetles and *D. melanogaster* (Hackman, 1955; Fristrom *et al.*, 1978; Hackman and Goldberg, 1978).

The next, and essential, step was to create a fusion protein uniting a protein that did not bind to chitin with the extended R&R Consensus region. Such a fusion was created between glutathione-S-transferase (GST) and 65 amino acids for AGCP2b – covering the region of pfam00379, the extended R&R Consensus:

APANYEFSYSVHDEH**TG**DIKSQHETR*RH*-
GDEVH G Q Y S L L D S D G H Q R I V D -
Y̲HADHHTGF̲NAVVRREP

GST and the fusion protein were each affinity purified using a glutathione-sepharose column. GST alone did not bind to chitin but the fusion protein did, requiring denaturing agents for release.

Other experiments defined in more detail the requirements for converting GST into a chitin-binding protein. A shorter fragment of AGCP2b, 40 amino acids (underlined above), with the strict R&R Consensus (shown in italics) did not bind chitin. Nor did the full construct when either the Y and F (highlighted) of the strict R&R Consensus or the T and D (bolded) of the extended consensus were "mutated" to alanine (Rebers and Willis, 2001).

In addition to establishing a function of the extended R&R Consensus, the experiments with "mutant" forms also provided confirmation of key elements in the models discussed in section 5.5.3. Substitution of the two conserved aromatic residues, postulated to be contact points with chitin, abolished chitin binding. With the TD "mutations," alanines were substituted for two other conserved residues. These flank a glycine that is conserved in position in the "extended consensus" of all hard and many soft cuticles (Iconomidou *et al.*, 1999). According to the proposed model (**Figure 4A**), these two polar residues would point away from the hydrophobic "cleft" and thus should not participate in chitin binding. It should be noted, however, that this glycine is located at a sharp turn, at the end of the second β-strand (in the vicinity of H102 of **Figure 4D**). The substitution of two polar residues by two alanines may result in destruction of this turn and to improper folding, thus leading to a structure not capable of binding chitin.

These experiments established, at last, that the extended R&R Consensus is sufficient to confer chitin-binding properties on a protein, and thereby resolved years of speculation on the importance of this region. Since then, comparable experiments have been done with other proteins in the CPR family, both RR-1 and RR-2 forms; all confirm that the extended R&R Consensus can bind chitin (Togawa *et al.*, 2004, 2007; Qin *et al.*, 2009).

5.5.5. Members of Other Cuticular Protein Families Analyzed for Chitin Binding

Data are now available that identify members of other CP families as capable of binding chitin. Most notable was the finding that a recombinant BmorCPT1 bound to chitin (Tang *et al.*, 2010). Given that both CPAP1 and CPAP3 families have ChtBD2 domains (see section 5.3.2.11), it is expected that their members will also bind chitin, but this has only been demonstrated experimentally for a recombinant form of the gasp homolog, a member of the CPAP3 family, from *Choristoneura fumiferana* (Nisole *et al.*, 2010).

In contrast, using the same methodology, Togawa *et al.* (2007) failed to demonstrate that either AgamCPF1 or AgamCPF3 could bind to chitin. While the CPR proteins are easily purified after expression in *E. coli*, the CPF proteins required use of the Pierce Refolding Kit® for proper solubilization. Therefore, their failure to bind could be due to improper refolding, although the information from homology modeling is consistent with a lack of chitin binding (see section 5.5.3.2).

Chitinase, some lectins and proteins from peritrophic membranes all bind chitin (for review, see Shen and Jacobs-Lorena, 1999). What is unique about the extended R&R Consensus and members of the TWDL family is that they lack cysteine residues. These residues serve essential roles in the other types of chitin-binding proteins, forming disulfide bonds that hold the protein in the proper configuration for binding. While these other chitin-binding proteins have weak sequence similarities to one another, they do not approach the sequence conservation seen in the R&R Consensus throughout the arthropods, or the TWDL consensus in the groups in which it is found. Rebers and Willis (2001) suggested that the conservation of the R&R Consensus (shown in **Figure 1**) could well be due to the need to preserve a precise conformation of the chitin-binding domain in the absence of stabilizing disulfide bonds, and the same reasoning could now be applied to the TWDL sequences where consensus regions are evident (**Figure 2A**).

5.5.6. Summary of Interaction Studies

Four different types of data have been presented in section 5.5 analyzing the extended R&R Consensus: secondary structure predictions of anti-parallel β-sheets (section 5.5.1), experimental spectroscopic evidence from cuticles and cuticle extracts for the predominance of such β-sheets in cuticular protein conformation (section 5.5.2), models showing organization of the consensus into a half β-barrel with a groove that can accommodate chitin (section 5.5.3.1), and direct demonstration that the extended consensus is sufficient to confer chitin binding on a protein (section 5.5.4). These four types of data are all in agreement that the highly conserved amino acid sequence of the extended R&R Consensus forms a novel chitin-binding domain, albeit one that displays an essential feature of other proteins that interact with chitin – namely, the presentation of aromatic residues in a planar surface. Crystal structures of the cuticular protein–chitin complex are needed to assure that these inferences are correct.

5.6. Summary and Future Challenges

This chapter has summarized the wealth of information about cuticular proteins amassed since Silvert's review in 1985. Most striking is that the several hundred-fold increase in sequences for structural cuticular proteins has revealed that the majority have a conserved domain (pfam00379) that is an extended version of the R&R Consensus. We now know that proteins with the R&R Consensus interact with chitin, and we can predict in some detail the features of their sequence that confer this property. We have not yet begun to analyze how the regions outside the Consensus contribute to cuticular properties. There is also direct experimental evidence that

a member of the TWDL family also binds chitin. However, we have yet to learn how proteins from other families contribute to cuticle structure, or how members of the different families interact with one another and other constituents of the cuticle.

Cuticular protein transcripts are turning up as major indicators of differential gene expression in analyses of insecticide resistance (Vontas *et al.*, 2007; Zhang *et al.*, 2008; Awolola *et al.*, 2009), desiccation resistance (Zhang *et al.* 2008), resistance to heavy metals (Shaw *et al.*, 2007; Roelofs *et al.*, 2009), response to changing photoperiod (Gallot *et al.* 2010), and even strain differences and mating (Cassone *et al.*, 2008; Rogers *et al.*, 2008). Obviously, we need to understand how the individual cuticular proteins are contributing in such major ways to such important events.

Cuticular proteins with pfam00379 are one of the largest multigene families found in *D. melanogaster* (Lespinet *et al.*, 2002), and their numbers are far larger in mosquitoes and *Bombyx*. We need more information about whether this multiplicity serves to allow rapid synthesis of cuticle, or whether different genes are used to construct cuticles in different regions. If the latter, the question becomes whether subtle differences in sequence are important for different cuticular properties, or if gene multiplication has been exploited to allow precise temporal and spatial control. We also need to learn how two hymenopterans, *Apis mellifera* and *Nasonia vitripennis*, manage with far fewer genes for cuticular proteins; is it their protected larval and pupal stages, or something else? The elegant immunolocalization studies that have been carried out were done with antibodies against proteins whose sequences for the most part are unknown. Now that we recognize that several genes may have almost identical sequences, we have to be very careful in designing specific probes for use in Northern analyses, for *in situ* hybridization, for qRT-PCR, and for immunolocalization, if our goal is to learn the use to which each individual gene is put.

Cuticular protein sequences are certain to be described in ever-increasing numbers as more insect genomes are analyzed. Describers need to be careful to submit to databases an indication of whether assignment as a cuticular protein is based on sequence alone, or on some type of corroborating evidence. It would be helpful if there were a more consistent system for naming cuticular proteins. At the very least, each protein should have a designation of genus and species, and a unique number ideally preceded by the gene family name – e.g., AgamCPR52, DmelTWDL12.

A wealth of sequence information is available already for cuticular proteins, but many challenges lie ahead for those who wish to continue to further our understanding of how the diverse forms and properties of cuticle are constructed extracellularly as these proteins self-assemble in proximity to chitin, and make specific contributions to the properties of the exoskeleton.

References

Adam, B., Charloteaux, B., Beaufays, J., Vanhamme, L., Godfroid, E., et al. (2008). Distantly related lipocalins share two conserved clusters of hydrophobic residues: Use in homology modeling. *BMC Struct. Biol.*, 8, 1.

Agrawal, O. P., & Scheller, K. (1986). The formation of the chitin-arylphorin complex *in vitro*. In R. Muzzarelli, C. Jeuniaux, & C. W. Gooday (Eds.). *Chitin in Nature and Technology* (pp. 316–320). New York, NY: Plenum Press.

Andersen, S. O. (1979). Biochemistry of insect cuticle. *Annu. Rev. Entomol.*, 24, 29–61.

Andersen, S. O. (1998). Amino acid sequence studies on endocuticular proteins form the desert locust *Schistocerca gregaria. Insect Biochem. Mol. Biol.*, 28, 421–434.

Andersen, S. O. (2000). Studies on proteins in post-ecdysial nymphal cuticle of locust, *Locusta migratoria*, and cockroach *Blaberus cranifer. Insect Biochem. Mol. Biol.*, 30, 569–577.

Andersen, S. O. (2005). Cuticular sclerotization and tanning. In L. I. Gilbert, K. Iatrou, & S. S. Gill (Eds.). *Comprehensive Molecular Insect Science* (Vol. 4, pp. 145–170). Amsterdam, The Netherlands: Elsevier.

Andersen, S. O. (2010a). Insect cuticular sclerotization: A review. *Insect Biochem. Mol. Biol.*, 40, 166–178.

Andersen, S. O. (2010b). Studies on resilin-like gene products in insects. *Insect Biochem. Mol. Biol.*, 40, 541–551.

Andersen, S. O., & Weis-Fogh, T. (1964). Resilin. A rubberlike protein in arthropod cuticle. *Adv. Insect Physiol.*, 2, 1–65.

Andersen, S. O., Hojrup, P., & Roepstorff, P. (1995). Insect cuticular proteins. *Insect Biochem. Mol. Biol.*, 25, 153–176.

Andersen, S. O., Rafn, K., & Roepstorff, P. (1997). Sequence studies of proteins from larval and pupal cuticle of the yellow meal worm, *Tenebrio molitor. Insect Biochem. Mol. Biol.*, 27, 121–131.

Antony, C., & Jallon, J. M. (1982). The chemical basis for sex recognition in *Drosophila melanogaster. J. Insect Physiol.*, 28, 873–880.

Antony, C., Davis, T. L., Carlson, D. A., Pechine, J. M., & Jallon, J. M. (1985). Compared behavioral responses of male *Drosophila melanogaster* (Canton S) to natural and synthetic aphrodisiacs. *J. Chem. Ecol.*, 11, 1617–1629.

Apple, R. T., & Fristrom, J. W. (1991). 20-hydroxyecdysone is required for, and negatively regulates, transcription of *Drosophila* pupal cuticle protein genes. *Develop. Biol.*, 146, 569–582.

Arakane, Y., Zhu, Q., Matsumiya, M., Muthukrishnan, S., & Kramer, K. J. (2003). Properties of catalytic, linker and chitin-binding domains of insect chitinase. *Insect Biochem. Mol. Biol.*, 33, 631–648.

Arakane, Y., Dittmer, N. T., Tomoyasu, Y., Kramer, K. J., Muthukrishnan, S., et al. (2010). Identification, mRNA expression and functional analysis of several yellow family genes in *Tribolium castaneum. Insect Biochem. Mol. Biol.*, 40, 259–266.

Ardell, D. H., & Andersen, S. O. (2001). Tentative identification of a resilin gene in *Drosophila melanogaster. Insect Biochem. Mol. Biol.*, 31, 965–970.

Ashida, M., & Brey, P. T. (1995). Role of the integument in insect defense: Pro-phenol oxidase cascade in the cuticular matrix. *Proc. Natl. Acad. Sci. USA*, 92, 10698–10702.

Atkins, E. D. T. (1985). Conformations in polysaccharides and complex carbohydrates. *Proc. Intl. Symp. Biomol. Struct. Interactions, Suppl. J. Biosci.*, 8, 375–387.

Awolola, T. S., Oduola, O. A., Strode, C., Koekemoer, L. L., Brooke, B., & Ranson, H. (2009). Evidence of multiple pyrethroid resistance mechanisms in the malaria vector *Anopheles gambiae* sensu stricto from Nigeria. *Trans. R. Soc. Trop. Med. Hyg.*, 103, 1139–1145.

Barry, M. K., Triplett, A. A., & Christensen, A. C. (1999). A peritrophin-like protein expressed in the embryonic tracheae of *Drosophila melanogaster. Insect Biochem. Mol. Biol.*, 29, 319–327.

Bateman, A., Birney, E., Cerruti, L., Durbin, R., Etwiller, L., et al. (2002). The Pfam protein families database. *Nucleic Acids Res.*, 30, 276–280.

Baton, L. A., Robertson, A., Warr, E., Strand, M. R., & Dimopoulos, G. (2009). Genome-wide transcriptomic profiling of *Anopheles gambiae* hemocytes reveals pathogen-specific signatures upon bacterial challenge and *Plasmodium berghei* infection. *BMC Genomics*, 10, 257.

Behr, M., & Hoch, M. (2005). Identification of the novel evolutionary conserved *obstructor* multigene family in invertebrates. *FEBS Lett.*, 579, 6827–6833.

Binger, L. C., & Willis, J. H. (1990). *In vitro* translation of epidermal RNAs from different anatomical regions and metamorphic stages of *Hyalophora cecropia. Insect Biochem.*, 20, 573–583.

Binger, L. C., & Willis, J. H. (1994). Identification of the cDNA, gene and promoter for a major protein from flexible cuticles of the giant silkmoth *Hyalophora cecropia. Insect Biochem. Mol. Biol.*, 24, 989–1000.

Bouhin, H., Charles, J. -P., Quennedey, B., Courrent, A., & Delachambre, J. (1992a). Characterization of a cDNA clone encoding a glycine-rich cuticular protein of *Tenebrio molitor*: Developmental expression and effect of a juvenile hormone analogue. *Insect Molec. Biol.*, 1, 53–62.

Bouhin, H., Charles, J. -P., Quennedey, B., & Delachambre, J. (1992b). Developmental profiles of epidermal mRNAs during the pupal–adult molt of *Tenebrio molitor* and isolation of a cDNA encoding an adult cuticular protein: Effects of a juvenile hormone analogue. *Develop. Biol.*, 149, 112–122.

Browder, L. W., Wilkes, J., & Rodenhiser, D. I. (1992). Preparative labeling of proteins with [^{35}S]methionine. *Anal. Biochem.*, 204, 85–89.

Bruey-Sedano, N., Alabouvette, J., Lestradet, M., Hong, L., Girard, A., et al. (2005). The *Drosophila ACP*65 *A* cuticle gene: Deletion scanning analysis of cis-regulatory sequences and regulation by DHR38. *Genesis*, 43, 17–27.

Burrows, M., Shaw, S. R., & Sutton, G. P. (2008). Resilin and chitinous cuticle form a composite structure for energy storage in jumping by froghopper insects. *BMC Biol.*, 6, 41.

Carter, D., & Locke, M. (1993). Why caterpillars do not grow short and fat. *Intl. J. Insect Morphol. Embryol.*, 22, 81–102.

Cassone, B. J., Mouline, K., Hahn, M. W., White, B. J., Pombi, M., et al. (2008). Differential gene expression in incipient species of *Anopheles gambiae. Mol. Ecol.*, 17, 2491–2504.

Centanin, L., Gorr, T. A., & Wappner, P. (2010). Tracheal remodelling in response to hypoxia. *J. Insect Physiol.*, 56, 447–454.

Charles, J. -P. (2010). The regulation of expression of insect cuticle protein genes. *Insect Biochem. Mol. Biol.*, *40*, 205–213.

Charles, J. -P., Bouhin, H., Quennedey, B., Courrent, A., & Delachambre, J. (1992). cDNA cloning and deduced amino acid sequence of a major, glycine-rich cuticular protein from the coleopteran *Tenebrio molitor*. Temporal and spatial distribution of the transcript during metamorphosis. *Eur. J. Biochem.*, *206*, 813–819.

Charles, J. -P., Chihara, C., Nejad, S., & Riddiford, L. M. (1997). A cluster of cuticle protein genes of *Drosophila melanogaster* at 65A: Sequence, structure and evolution. *Genetics*, *147*, 1213–1226.

Charles, J. -P., Chihara, C., Nejad, S., & Riddiford, L. M. (1998). Identification of proteins and developmental expression of RNAs encoded by the 65A cuticle protein gene cluster in *Drosophila melanogaster*. *Insect Biochem. Mol. Biol.*, *28*, 131–138.

Chou, P., & Fasman, G. D. (1974a). Conformational parameters for amino acids in helical, β-sheet and random coil regions calculated from proteins. *Biochemistry*, *13*, 211–221.

Chou, P., & Fasman, G. D. (1974b). Prediction of protein conformation. *Biochemistry*, *13*, 222–245.

Condoulis, W. V., & Locke, M. (1966). The deposition of endocuticle in an insect, *Calpodes ethlius* Stoll (Lepidoptera: Hesperiidae). *J. Insect Physiol.*, *12*, 311–323.

Cornell, J. C., & Pan, M. L. (1983). The disappearance of moulting fluid in the tobacco hornworm, *Manduca sexta*. *J. Exp. Biol.*, *107*, 501–504.

Cornman, R. S. (2009). Molecular evolution of *Drosophila* cuticular protein genes. *Plos ONE*, *4*, e8345.

Cornman, R. S. (2010). The distribution of GYR- and YLP-like motifs in *Drosophila* suggests a general role in cuticle assembly and other protein–protein interactions. *PLoS ONE*, *5*, e12536.

Cornman, R. S., & Willis, J. H. (2008). Extensive gene amplification and concerted evolution within the CPR family of cuticular proteins in mosquitoes. *Insect Biochem. Mol. Biol.*, *38*, 661–676.

Cornman, R. S., & Willis, J. H. (2009). Annotation and analysis of low-complexity protein families of *Anopheles gambiae* that are associated with cuticle. *Insect Mol. Biol.*, *18*, 607–622.

Cornman, R. S., Togawa, T., Dunn, W. A., He, N., Emmons, A. C., & Willis, J. H. (2008). Annotation and analysis of a large cuticular protein family with the R&R Consensus in *Anopheles gambiae*. *BMC Genomics*, *9*, 22.

Cox, D. L., & Willis, J. H. (1985). The cuticular proteins of *Hyalophora cecropia* from different anatomical regions and metamorphic stages. *Insect Biochem.*, *15*, 349–362.

Cox, D. L., & Willis, J. H. (1987a). Analysis of the cuticular proteins of *Hyalophora cecropia* with two dimensional electrophoresis. *Insect Biochem.*, *17*, 457–468.

Cox, D. L., & Willis, J. H. (1987b). Post-translational modifications of the cuticular proteins of *Hyalophora cecropia* from different anatomical regions and metamorphic stages. *Insect Biochem.*, *17*, 469–484.

Crooks, G. E., Hon, G., Chandonia, J. M., & Brenner, S. E. (2004). WebLogo: A sequence logo generator. *Genome Res.*, *14*, 1188–1190.

Csikos, G., Molnar, K., Borhegyi, N. H., Talian, G. C., & Sass, M. (1999). Insect cuticle, an *in vivo* model of protein trafficking. *J. Cell Sci.*, *112*, 2113–2124.

Delano, W. L., 2005. *The PyMOL Molecular Graphics System*, DeLano Scientific LLC, South San Francisco, CA 94080–1918, USA.

Doctor, J., Fristrom, D., & Fristrom, J. W. (1985). The pupal cuticle of *Drosophila*: Biphasic synthesis of pupal cuticle proteins *in vivo* and *in vitro* in response to 20-hydroxyecdysone. *J. Cell Biol.*, *101*, 189–200.

Dotson, E. M., Cornel, A. J., Willis, J. H., & Collins, F. H. (1998). A family of pupal-specific cuticular protein genes in the mosquito *Anopheles gambiae*. *Insect Biochem. Mol. Biol.*, *28*, 459–472.

Elvin, C. M., Carr, A. G., Huson, M. G., Maxwell, J. M., Pearson, R. D., et al. (2005). Synthesis and properties of cross-linked recombinant pro-resilin. *Nature*, *437*, 999–1002.

Fechtel, K., Fristrom, D. K., & Fristrom, J. W. (1989). Prepupal differentiation in *Drosophila*: Distinct cell types elaborate a shared structure, the pupal cuticle, but accumulate transcripts in unique patterns. *Development*, *106*, 649–656.

Flower, D. R. (1996). The lipocalin protein family: Structure and function. *Biochem. J.*, *318*, 1–14.

Flower, D. R., North, A. C., & Sansom, C. E. (2000). The lipocalin protein family: Structural and sequence overview. *Biochim. Biophys. Acta.*, *1482*, 9–24.

Fraenkel, G., & Rudall, K. M. (1947). The structure of insect cuticles. *Proc. R. Soc. B*, *134*, 111–143.

Fristrom, D., & Fristrom, J. W. (1993). The metamorphic development of the adult epidermis. In M. Bate, & A. Martinez Arias (Eds.). *The Development of Drosophila melanogaster* (pp. 843–897). Cold Spring Harbor, NY: Cold Spring Harbor Laboratory Press.

Fristrom, D., Doctor, J., & Fristrom, J. W. (1986). Procuticle proteins and chitin-like material in the inner epicuticle of the *Drosophila* pupal cuticle. *Tissue and Cell*, *18*, 531–543.

Fristrom, J. W., Hill, R. J., & Watt, F. (1978). The procuticle of *Drosophila*: Heterogeneity of urea-soluble proteins. *Biochemistry*, *17*, 3917–3924.

Futahashi, R., & Fujiwara, H. (2005). Melanin-synthesis enzymes coregulate stage-specific larval cuticular markings in the swallowtail butterfly, *Papilio xuthus*. *Dev. Genes Evol.*, *215*, 519–529.

Futahashi, R., & Fujiwara, H. (2008). Identification of stage-specific larval camouflage associated genes in the swallowtail butterfly *Papilio xuthus*. *Dev. Genes Evol.*, *218*, 491–504.

Futahashi, R., Okamoto, S., Kawasaki, H., Zhong, Y. S., Iwanaga, M., et al. (2008). Genome-wide identification of cuticular protein genes in the silkworm, *Bombyx mori*. *Insect Biochem. Mol. Biol.*, *38*, 1138–1146.

Gallot, A., Rispe, C., Leterme, N., Gauthier, J. P., Jaubert-Possamai, S., & Tagu, D. (2010). Cuticular proteins and seasonal photoperiodism in aphids. *Insect Biochem. Mol. Biol.*, *40*, 235–240.

Geyer, P. G., & Corces, V. G. (1987). Separate regulatory elements are responsible for the complex pattern of tissue-specific and developmental transcription of the *yellow* locus in *Drosophila melanogaster*. *Genes and Develop.*, *1*, 996–1004.

Ghabrial, A., Luschnig, S., Metzstein, M. M., & Krasnow, M. A. (2003). Branching morphogenesis of the *Drosophila* tracheal system. *Annu. Rev. Cell Dev. Biol., 19*, 623–647.

Greenspan, R. J., & Ferveur, J. F. (2000). Courtship in *Drosophila*. *Annu. Rev. Genetics, 34*, 205–232.

Griebenow, K., Santos, A. M., & Carrasquillo, K. G. (1999). Secondary structure of proteins in the amorphous dehydrated state probed by FTIR spectroscopy. Dehydration-induced structural changes and their prevention. *Intl. J. Vibr. Spectrosc., 3*, 1–34.

Gu, S., & Willis, J. H. (2003). Distribution of cuticular protein mRNAs in silk moth integument and imaginal discs. *Insect Biochem. Mol. Biol., 33*, 1177–1188.

Guan, X., Middlebrooks, B. W., Alexander, S., & Wasserman, S. A. (2006). Mutation of TweedleD, a member of an unconventional cuticle protein family, alters body shape in *Drosophila*. *Proc. Natl. Acad. Sci. USA, 103*, 16794–16799.

Hackman, R. H. (1955). Studies on chitin. III. Absorption of proteins to chitin. *Austral. J. Biol. Sci., 8*, 530–536.

Hackman, R. H., & Goldberg, M. (1978). The non-covalent binding of two insect cuticular proteins by a chitin. *Insect Biochem., 8*, 353–357.

Hadley, N. F. (1981). Cuticular lipids of terrestrial plants and arthropods: A comparison of their structure, composition and waterproofing function. *Biol. Rev., 56*, 23–47.

Hall, J. C. (1994). The mating of a fly. *Science, 264*, 1702–1714.

Hamodrakas, S. J., Kanellopoulos, P. N., Pavlou, K., & Tucker, P. A. (1997). The crystal structure of the complex of Concanavalin A with 4-methylumbelliferyl-α-D-glucopyranoside. *J. Struct. Biol., 118*, 23–30.

Hamodrakas, S. J., Willis, J. H., & Iconomidou, V. A. (2002). A structural model of the chitin-binding domain of cuticle proteins. *Insect Biochem. Mol. Biol., 32*, 1577–1583.

Han, Q., Fang, J., Ding, H., Johnson, J. K., Christensen, B. M., & Li, J. (2002). Identification of *Drosophila melanogaster* yellow-f and yellow-f2 proteins as dopachrome-conversion enzymes. *Biochem. J., 368*, 333–340.

He, N., Botelho, J. M., McNall, R. J., Belozerov, V., Dunn, W. A., et al. (2007). Proteomic analysis of cast cuticles from *Anopheles gambiae* by tandem mass spectrometry. *Insect Biochem. Mol. Biol., 37*, 135–146.

Henikoff, S., & Eghtedarzadeh, M. K. (1987). Conserved arrangement of nested genes at the *Drosophila Gart* locus. *Genetics, 117*, 711–725.

Henikoff, S., Keene, M. A., Fechtel, K., & Fristrom, J. W. (1986). Gene within a gene: Nested *Drosophila* genes encode unrelated proteins on opposite DNA strands. *Cell, 44*, 33–42.

Holden, H. M., Rypniewski, W. R., Law, J. H., & Rayment, I. (1987). The molecular structure of insecticyanin from the tobacco hornworm *Manduca sexta* L. at 2.6 A resolution. *EMBO J., 6*, 1565–1570.

Horodyski, F. M., & Riddiford, L. M. (1989). Expression and hormonal control of a new larval cuticular multigene family at the onset of metamorphosis of the tobacco hornworm. *Develop. Biol., 132*, 292–303.

Iconomidou, V. A., Willis, J. H., & Hamodrakas, S. J. (1999). Is β-pleated sheet the molecular conformation which dictates formation of helicoidal cuticle? *Insect Biochem. Mol. Biol., 29*, 285–292.

Iconomidou, V. A., Chryssikos, G. D., Gionis, V., Willis, J. H., & Hamodrakas, S. J. (2001). "Soft"-cuticle protein secondary structure as revealed by FT-Raman, ATR-FT-IR and CD spectroscopy. *Insect Biochem. Mol. Biol., 31*, 877–885.

Iconomidou, V. A., Willis, J. H., & Hamodrakas, S. J. (2005). Unique features of the structural model of "hard" cuticle proteins: Implications for chitin–protein interactions and cross-linking in cuticle. *Insect Biochem. Mol. Biol., 35*, 553–560.

Iijima, M., Hashimoto, T., Matsuda, Y., Nagai, T., Yamano, Y., et al. (2005). Comprehensive sequence analysis of horseshoe crab cuticular proteins and their involvement in transglutaminase-dependent cross-linking. *FEBS J., 272*, 4774–4786.

Ikeya, T., Persson, P., Kono, M., & Watanabe, T. (2001). The *DD5* gene of the decapod crustacean *Penaeus japonicus* encodes a putative exoskeletal protein with a novel tandem repeat structure. *Comp. Biochem. Physiol. B, 128*, 379–388.

Irving, P., Ubeda, J. M., Doucet, D., Troxler, L., Lagueux, M., et al. (2005). New insights into *Drosophila* larval haemocyte functions through genome-wide analysis. *Cell Microbiol., 7*, 335–350.

Jasrapuria, S., Arakane, Y., Osman, G., Kramer, K. J., Beeman, R. W., & Muthukrishnan, S. (2010). Genes encoding proteins with peritrophin A-type chitin-binding domains in *Tribolium castaneum* are grouped into three distinct families based on phylogeny, expression and function. *Insect Biochem. Mol. Biol., 40*, 214–227.

Jensen, U. G., Rothmann, A., Skou, L., Andersen, S. O., Roepstorff, P., & Hojrup, P. (1997). Cuticular proteins from the giant cockroach, *Blaberus craniifer. Insect Biochem. Mol. Biol., 27*, 109–120.

Kalinich, J. F., & McClain, D. E. (1992). An *in vitro* method for radiolabeling proteins with ^{35}S. *Anal. Biochem., 205*, 208–212.

Karouzou, M. V., Spyropoulos, Y., Iconomidou, V. A., Cornman, R. S., Hamodrakas, S. J., & Willis, J. H. (2007). *Drosophila* cuticular proteins with the R&R Consensus: Annotation and classification with a new tool for discriminating RR-1 and RR-2 sequences. *Insect Biochem. Mol. Biol., 37*, 754–760.

Konig, M., Agrawal, O. P., Schenkel, H., & Scheller, K. (1986). Incorporation of calliphorin into the cuticle of the developing blowfly, *Calliphora vicinia. Roux's Arch. Develop. Biol., 195*, 296–301.

Kornezos, A., & Chia, W. (1992). Apical secretion and association of the *Drosophila yellow* gene product with developing larval cuticle structures during embryogenesis. *Mol. Gen. Genet., 235*, 397–405.

Kucharski, R., Maleszka, J., & Maleszka, R. (2007). Novel cuticular proteins revealed by the honey bee genome. *Insect Biochem. Mol. Biol., 37*, 128–134.

Lampe, D. J., & Willis, J. H. (1994). Characterization of a cDNA and gene encoding a cuticular protein from rigid cuticles of the giant silkmoth, *Hyalophora cecropia. Insect Biochem. Mol. Biol., 24*, 419–435.

Lee, W. -J., & Brey, P. T. (1994). Isolation and identification of cecropin antibacterial peptides from the extracellular matrix of insect integument. *Anal. Biochem., 217*, 231–235.

Lemoine, A., Millot, C., Curie, G., & Delachambre, J. (1989). A monoclonal antibody against an adult-specific cuticular protein of *Tenebrio molitor* (Insecta: Coleoptera). *Develop. Biol., 136*, 546–554.

Lemoine, A., Millot, C., Curie, G., & Delachambre, J. (1990). Spatial and temporal variations in cuticle proteins as revealed by monoclonal antibodies, immunoblotting analysis and ultrastructural immunolocalization in a beetle, *Tenebrio molitor*. *Tissue and Cell*, *22*, 177–189.

Lemoine, A., Millot, C., Curie, G., Massonneau, V., & Delachambre, J. (1993). Monoclonal antibodies recognizing larval- and pupal-specific cuticular proteins of *Tenebrio molitor* (Insecta: Coleoptera). *Roux's Arch. Dev. Biol.*, *203*, 92–99.

Lespinet, O., Wolf, Y. I., Koonin, E. V., & Aravind, L. (2002). The role of lineage-specific gene family expansion in the evolution of eukaryotes. *Genome Res.*, *12*, 1048–1059.

Leung, H., Palli, S. R., & Locke, M. (1989). The localization of arylphorin in an insect, *Calpodes ethlius*. *J. Insect Physiol.*, *35*, 223–231.

Li, W., & Riddiford, L. M. (1992). Two distinct genes encode two major isoelectric forms of insecticyanin in the tobacco hornworm, *Manduca sexta*. *Eur. J. Biochem.*, *205*, 491–499.

Locke, M. (1998). Epidermis. In F. W. Harrison, & M. Locke (Eds.). *Microscopic Anatomy of Invertebrates* (Vol. 11A, pp. 75–138). New York, NY: Wiley-Liss.

Locke, M. (2001). The Wigglesworth Lecture: Insects for studying fundamental problems in biology. *J. Insect Physiol.*, *47*, 495–507.

Locke, M. (2003). Surface membranes, Golgi complexes and vacuolar systems. *Annu. Rev. Entomol.*, *48*, 1–27.

Locke, M., Kiss, A., & Sass, M. (1994). The cuticular localization of integument peptides from particular routing categories. *Tissue and Cell*, *26*, 707–734.

Lyons, R. E., Lesieur, E., Kim, M., Wong, D. C., Huson, M. G., et al. (2007). Design and facile production of recombinant resilin-like polypeptides: Gene construction and a rapid protein purification method. *Protein Eng. Des. Sel.*, *20*, 25–32.

Magkrioti, C. K., Spyropoulos, I. C., Iconomidou, V. A., Willis, J. H., & Hamodrakas, S. J. (2004). cuticleDB: A relational database of Arthropod cuticular proteins. *BMC Bioinformatics*, *5*, 138.

Marcu, O., & Locke, M. (1998). A cuticular protein from the moulting stages of an insect. *Insect Biochem. Mol. Biol.*, *28*, 659–669.

Marcu, O., & Locke, M. (1999). The origin, transport and cleavage of the molt-associated cuticular protein CECP22 from *Calpodes ethlius* (Lepidoptera: Hesperridae). *J. Insect Physiol.*, *45*, 861–870.

Marinotti, O., Calvo, E., Nguyen, Q. K., Dissanayake, S., Ribeiro, J. M., & James, A. A. (2006). Genome-wide analysis of gene expression in adult *Anopheles gambiae*. *Insect Mol. Biol.*, *15*, 1–12.

Mathelin, J., Bouhin, H., Quennedey, B., Courrent, A., & Delachambre, J. (1995). Identification, sequence and mRNA expression pattern during metamorphosis of a cDNA encoding a glycine-rich cuticular protein in *Tenebrio molitor*. Gene, 156, 259–264.

Mathelin, J., Quennedey, B., Bouhin, H., & Delachambre, J. (1998). Characterization of two new cuticular genes specifically expressed during the post-ecdysial molting period in *Tenebrio molitor*. Gene, 211, 351–359.

Mita, K., Morimyo, M., Okano, K., Koika, Y., Nohata, J., Suzuki, M. G., & Shimade, T. (2002). Construction of an EST database for *Bombyx Mori* and its applications. *Current Sci.* 83, 426–431.

Missios, S., Davidson, H. C., Linder, D., Mortimer, L., Okobi, A. O., & Doctor, J. S. (2000). Characterization of cuticular proteins in the red flour beetle, *Tribolium castaneum*. *Insect Biochem. Mol. Biol.*, *30*, 47–56.

Molnar, K., Borhegyi, N. H., Csikos, G., & Sass, M. (2001). The immunoprotein scolexin and its synthesizing sites – the midgut epithelium and the epidermis. *Acta Biologica Hungarica*, *52*, 473–484.

Morris, G. M., Huey, R., Lindstrom, W., Sanner, M. F., Belew, R. K., et al. (2009). AutoDock4 and AutoDockTools4: Automated docking with selective receptor flexibility. *J. Comput. Chem.*, *30*, 2785–2791.

Moussian, B. (2010). Recent advances in understanding mechanisms of insect cuticle differentiation. *Insect Biochem. Mol. Biol.*, *40*, 363–375.

Nakato, H., Izumi, S., & Tomino, S. (1992). Structure and expression of gene coding for a pupal cuticle protein of *Bombyx mori*. *Biochim. Biophys. Acta.*, *1132*, 161–167.

Nakatogawa, S., Oda, Y., Kamiya, M., Kamijima, T., Aizawa, T., et al. (2009). A novel peptide mediates aggregation and migration of hemocytes from an insect. *Curr. Biol.*, *19*, 779–785.

Nash, W. G. (1976). Patterns of pigmentation color states regulated by the *y* locus in *Drosophila melanogaster*. *Develop. Biol.*, *48*, 336–343.

Neville, A. C. (1975). *Biology of the Arthropod Cuticle*. New York, NY: Springer-Verlag.

Nicholls, A., Sharp, K. A., & Honig, B. (1991). Protein folding and association: Insights from the interfacial and thermodynamic properties of hydrocarbons. *Proteins: Struct. Funct. Genet.*, *11*, 281–296.

Nisole, A., Stewart, D., Bowman, S., Zhang, D., Krell, P. J., et al. (2010). Cloning and characterization of a Gasp homolog from the spruce budworm, *Choristoneura fumiferana*, and its putative role in cuticle formation. *J. Insect Physiol.*, *56*, 1427–1435.

Palli, S. R., & Locke, M. (1987). The synthesis of hemolymph proteins by the larval epidermis of an insect *Calpodes ethlius* (Lepidoptera: Hesperiidae). *Insect Biochem.*, *17*, 711–722.

Papandreou, N. C., Iconomidou, V. A., Willis, J. H., & Hamodrakas, S. J. (2010). A possible structural model of members of the CPF family of cuticular proteins implicating binding to components other than chitin. *J. Insect Physiol.*, *56*, 1420–1426.

Perrakis, A., Ouzounis, C., & Wilson, K. S. (1997). Evolution of immunoglobulin-like modules in chitinases: Their structural flexibility and functional implications. *Folding and Design*, *2*, 291–294.

Qin, G., Lapidot, S., Numata, K., Hu, X., Meirovitch, S., et al. (2009). Expression, cross-linking, and characterization of recombinant chitin binding resilin. *Biomacromolecules*, *10*, 3227–3234.

Qiu, J., & Hardin, P. E. (1995). Temporal and spatial expression of an adult cuticle protein gene from *Drosophila* suggests that its protein product may impart some specialized cuticle function. *Develop. Biol.*, *167*, 416–425.

Rebers, J. F., & Riddiford, L. M. (1988). Structure and expression of a *Manduca sexta* larval cuticle gene homologous to *Drosophila* cuticle genes. *J. Mol. Biol.*, *203*, 411–423.

Rebers, J. E., & Willis, J. H. (2001). A conserved domain in arthropod cuticular proteins binds chitin. *Insect Biochem. Mol. Biol.*, *31*, 1083–1093.

Rebers, J. E., Niu, J., & Riddiford, L. M. (1997). Structure and spatial expression of the *Manduca sexta* MSCP14.6 cuticle gene. *Insect Biochem. Mol. Biol.*, *27*, 229–240.

Riddiford, L. M. (1994). Cellular and molecular actions of juvenile hormone I. General considerations and premetamorphic actions. *Adv. Insect Physiol.*, *24*, 213–274.

Riddiford, L. M., & Hice, R. H. (1985). Developmental profiles of the mRNAs for *Manduca* arylphorin and two other storage proteins during the final larval instar of *Manduca sexta*. *Insect Biochem.*, *15*, 489–502.

Roberts, P. E., & Willis, J. H. (1980). Effects of juvenile hormone, ecdysterone, actinomycin D, and *mitomycin* C on the cuticular proteins of *Tenebrio molitor*. *J. Embryol. Exp. Morph.*, *56*, 107–123.

Roelofs, D., Janssens, T. K., Timmermans, M. J., Nota, B., Marien, J., et al. (2009). Adaptive differences in gene expression associated with heavy metal tolerance in the soil arthropod *Orchesella cincta*. *Mol. Ecol.*, *18*, 3227–3239.

Rogers, D. W., Whitten, M. M., Thailayil, J., Soichot, J., Levashina, E. A., & Catteruccia, F. (2008). Molecular and cellular components of the mating machinery in *Anopheles gambiae* females. *Proc. Natl. Acad. Sci. USA*, *105*, 19390–19395.

Rondot, I., Quennedey, B., & Delachambre, J. (1998). Structure, organization and expression of two clustered cuticle protein genes during the metamorphosis of an insect *Tenebrio molitor*. *Eur. J. Biochem.*, *254*, 304–312.

Roter, A. H., Spofford, J. B., & Swift, H. (1985). Synthesis of the major adult cuticle proteins of *Drosophila melanogaster* during hypoderm differentiation. *Develop. Biol.*, *107*, 420–431.

Sali, A., & Blundell, T. L. (1993). Comparative protein modelling by satisfaction of spatial restraints. *J. Molec. Biol.*, *234*, 779–815.

Sass, M., Kiss, A., & Locke, M. (1993). Classes of integument peptides. *Insect Biochem. Mol. Biol.*, *23*, 845–857.

Sass, M., Kiss, A., & Locke, M. (1994a). Integument and hemocyte peptides. *J. Insect Physiol.*, *40*, 407–421.

Sass, M., Kiss, A., & Locke, M. (1994b). The localization of surface integument peptides in tracheae and tracheoles. *J. Insect Physiol.*, *40*, 561–575.

Scheller, K., Zimmermann, H. -P., & Sekeris, C. E. (1980). Calliphorin, a protein involved in the cuticle formation of the blowfly, *Calliphora vicinia*. *Z. Naturforsch.*, *35 c*, 387–389.

Schneider, T. D., & Stephens, R. M. (1990). Sequence logos: A new way to display consensus sequences. *Nucl. Acids Res.*, *18*, 6097–6110.

Scholer, H. R. (1991). Octamania: The POU factors in murine development. *Trends Genet.*, *7*, 323–328.

Shaw, J. R., Colbourne, J. K., Davey, J. C., Glaholt, S. P., Hampton, T. H., et al. (2007). Gene response profiles for *Daphnia pulex* exposed to the environmental stressor cadmium reveals novel crustacean metallothioneins. *BMC Genomics*, *8*, 477.

Shen, Z., & Jacobs-Lorena, M. (1999). Evolution of chitin-binding proteins in invertebrates. *J. Mol. Evol.*, *48*, 341–347.

Silvert, D. J. (1985). Cuticular proteins during postembryonic development. In G. A. Kerkut, & L. I. Gilbert (Eds.). *Comprehensive Insect Physiology Biochemistry and Pharmacology* (Vol. 2, pp. 239–254). Oxford, UK: Pergamon Press.

Snyder, M., Hunkapiller, M., Yuen, D., Silvert, D., Fristrom, J., & Davidson, N. (1982). Cuticle protein genes of *Drosophila*: Structure, organization and evolution of four clustered genes. *Cell*, *29*, 1027–1040.

Steinbrecht, R. A., & Stankiewicz, B. A. (1999). Molecular composition of the wall of insect olfactory sensilla – the chitin question. *J. Insect Physiol.*, *45*, 785–790.

Stiles, B. (1991). Cuticle proteins of the boll weevil, *Anthonomus grandis*, abdomen: Structural similarities and glycosylation. *Insect Biochem.*, *21*, 249–258.

Suderman, R. J., Andersen, S. O., Hopkins, T. L., Kanost, M. R., & Kramer, K. J. (2003). Characterization and cDNA cloning of three major proteins from pharate pupal cuticle of *Manduca sexta*. *Insect Biochem. Mol. Biol.*, *33*, 331–343.

Suetake, T., Tsuda, S., Kawabata, S., Miura, K., Iwanaga, S., et al. (2000). Chitin-binding proteins in invertebrates and plants comprise a common chitin-binding structural motif. *J. Biol. Chem.*, *275*, 17929–17932.

Talbo, G., Hojrup, P., Rahbek-Nielsen, H., Andersen, S. O., & Roepstorff, P. (1991). Determination of the covalent structure of an N- and C- terminally blocked glycoprotein from endocuticle of *Locusta migratoria*. Combined use of plasma desorption mass spectrometry and Edman degradation to study post-translationally modified proteins. *Eur. J. Biochem.*, *195*, 495–504.

Tang, L., Liang, J., Zhan, Z., Xiang, Z., & He, N. (2010). Identification of the chitin-binding proteins from the larval proteins of silkworm, *Bombyx mori*. *Insect Biochem. Mol. Biol.*, *40*, 228–234.

Tews, I., Scheltiga, T., Perrakis, A., Wilson, K. S., & Dijkstra, B. W. (1997). Substrate-assisted catalysis unifies 2 families of chitinolytic enzymes. *J. Am. Chem. Soc.*, *119*, 7954–7959.

Togawa, T., Nakato, H., & Izumi, S. (2004). Analysis of the chitin recognition mechanism of cuticle proteins from the soft cuticle of the silkworm, *Bombyx mori*. *Insect Biochem. Mol. Biol.*, *34*, 1059–1067.

Togawa, T., Dunn, W. A., Emmons, A. C., & Willis, J. H. (2007). CPF and CPFL, two related gene families encoding cuticular proteins of *Anopheles gambiae* and other insects. *Insect Biochem. Mol. Biol.*, *37*, 675–688.

Togawa, T., Dunn, W. A., Emmons, A. C., Nagao, J., & Willis, J. H. (2008). Developmental expression patterns of cuticular protein genes with the R&R Consensus from *Anopheles gambiae*. *Insect Biochem. Mol. Biol.*, *38*, 508–519.

Trim, A. R. H. (1941). Studies in the chemistry of insect cuticle. I. Some general observations on certain arthropod cuticles with special reference to the characterization of the proteins. *Biochem. J.*, *35*, 1088–1098.

Uchiyama, T., Katouno, F., Nikaidou, N., Nonaka, T., Sugiyama, J., & Watanabe, T. (2001). Roles of the exposed aromatic residues in crystalline chitin hydrolysis by chitinase A from *Serratia marcescens* 2170. *J. Biol. Chem.*, *276*, 41343–41349.

Vakser, I. A. (1996). Low-resolution docking: Prediction of complexes for undetermined structures. *Biopolymers*, *39*, 455–464.

Vincent, J. F. V. (2002). Arthropod cuticle: A natural composite shell system. *Composites: Part A, 33,* 1311–1315.

Vontas, J., David, J. P., Nikou, D., Hemingway, J., Christophides, G. K., et al. (2007). Transcriptional analysis of insecticide resistance in *Anopheles stephensi* using cross-species microarray hybridization. *Insect Mol. Biol., 16,* 315–324.

Vriend, G. (1990). WHAT IF: A molecular modeling and drug design package. *J. Mol. Graph., 8,* 52–56.

Vyas, N. K. (1991). Atomic features of protein–carbohydrate interactions. *Curr. Opin. Struct. Biol., 1,* 723–740.

Williams, C. M. (1980). Growth in insects. In M. Locke, & D. S. Smith (Eds.). *Insect Biology in the Future* (pp. 369–383). London, UK: Academic Press.

Willis, J. H. (1996). Metamorphosis of the cuticle, its proteins, and their genes. In L. I. Gilbert, B. G. Atkinson, & J. Tata (Eds.). *Metamorphosis/Post-Embryonic Reprogramming of Gene Expression in Amphibian and Insect Cell* (pp. 253–282). London, UK: Academic Press.

Willis, J. H. (1999). Cuticular proteins in insects and crustaceans. *Amer. Zool., 39,* 600–609.

Willis, J. H. (2010). Structural cuticular proteins from arthropods: Annotation, nomenclature, and sequence characteristics in the genomics era. *Insect Biochem. Mol. Biol., 40,* 189–204.

Willis, J. H., Rezaur, R., & Sehnal, F. (1982). Juvenoids cause some insects to form composite cuticles. *J. Embryol. Exp. Morph., 71,* 25–40.

Willis, J. H., Iconomidou, V. A., Smith, R. F., & Hamodrakas, S. J. (2005). Cuticular proteins. In L. I. Gilbert, K. Iatrou, & S. S. Gill (Eds.). *Comprehensive Molecular Insect Science* (Vol. 4, pp. 79–109). Amsterdam, The Netherlands: Elsevier.

Wittkopp, P. J., Vaccaro, K., & Carroll, S. B. (2002). Evolution of *yellow* gene regulation and pigmentation in *Drosophila*. *Current Biol., 12,* 1547–1556.

Wittkopp, P. J., Stewart, E. E., Arnold, L. L., Neidert, A. Haerum, B. K., et al. (2009). Intraspecific polymorphism to interspecific divergence: genetics of pigmentation in *Drosophila*. *Science, 326,* 540–544.

Wolfgang, W. J., & Riddiford, L. M. (1986). Larval cuticular morphogenesis in the tobacco hornworm, *Manduca sexta*, and its hormonal regulation. *Develop. Biol., 113,* 305–316.

Wolfgang, W. J., Fristrom, D., & Fristrom, J. W. (1986). The pupal cuticle of *Drosophila*: Differential ultrastructural immunolocalization of cuticle proteins. *J. Cell Biol., 102,* 306–311.

Wolfgang, W. J., Fristrom, D., & Fristrom, J. W. (1987). An assembly zone antigen of the insect cuticle. *Tissue and Cell, 19,* 827–838.

Wybrandt, G. B., & Andersen, S. O. (2001). Purification and sequence determination of a yellow protein from sexually mature males of the desert locust, *Schistocerca gregaria*. *Insect Biochem. Mol. Biol., 31,* 1183–1189.

Zanotti, G., Marcello, M., Malpeli, G., Folli, C., Sartori, G., & Berni, R. (1994). Crystallographic studies on complexes between retinoids and plasma retinol-binding protein. *J. Biol. Chem., 269,* 29613–29620.

Zhang, J., Goyer, C., & Pelletier, Y. (2008). Environmental stresses induce the expression of putative glycine-rich insect cuticular protein genes in adult *Leptinotarsa decemlineata* (Say). *Insect Mol. Biol., 17,* 209–216.

Zhang, J., & Pelletier, Y. (2010). Characterization of cuticular chitin-binding proteins of *Leptinotarsa decemlineta* (Say) and post-ecdysial transcript levels at different developmental stages. *Insect Mol. Biol., 19,* 517–525.

Zhou, X., & Riddiford, L. M. (2002). Broad specifies pupal development and mediates the "status quo" action of juvenile hormone on the pupal–adult transformation in *Drosophila* and *Manduca*. *Development, 129,* 2259–2269.

6   Cuticular Sclerotization and Tanning

Svend O Andersen
The Collstrop Foundation,
The Royal Danish Academy of Sciences and Letters,
Copenhagen, Denmark

Summary

The physical properties of insect cuticles are to a large extent determined by the degree of sclerotization (stabilization). The process of sclerotization often takes place shortly after eclosion, but may also occur before ecdysis or in connection with puparium formation. During sclerotization the two dopamine derivatives, *N*-acetyldopamine (NADA) and *N*-β-alanyldopamine (NBAD), are incorporated into the cuticular matrix. Incorporation of NBAD involves oxidation to its *o*-quinone followed by isomerization to a *p*-quinone methide, and both quinone derivatives can react with nucleophilic amino acids in the cuticular proteins. Incorporation of NADA involves oxidation to its *o*-quinone, isomerization to a *p*-quinone methide followed by isomerization to α,β-dehydro-*N*-acetyldopamine (dehydro-NADA) and oxidation of the latter to the unsaturated quinones, dehydro-NADA *o*-quinone and dehydro-NADA *p*-quinone methide. The dehydro-NADA *p*-quinone methide reacts readily with various nucleophilic groups, resulting in formation of inter-protein cross-links and polymers.

The pronounced diversity of insect cuticles indicates that several variations of the above-mentioned scheme are used for stabilizing cuticles, and a comparison of the details of the sclerotization processes occurring in different types of cuticle will probably be rewarding.

DOI:10.1016/B978-0-12-384747-8.10006-6

6.1. Introduction

The cuticle covers the insect body as an effective barrier between the animal and its surroundings; it provides protection against desiccation, microorganisms, and predators, and as an exoskeleton it provides attachment sites for muscles. Cuticle can occur as relatively hard and stiff regions, the sclerites, separated by more flexible and pliable cuticular regions, the arthrodial membranes, which make the various forms of locomotion possible. Marked differences in mechanical properties can be present on the microscopic level; two neighboring epidermal cells can produce cuticle with contrasting properties, indicating that cuticular composition is precisely controlled on the cellular level. The mechanical properties of the various cuticular regions are presumably optimal with respect to the forces to which they are exposed during the normal life of the animal; proper flight can only be sustained when the various wing regions have a near-optimal balance between stiffness and flexibility. If the wing material is locally too soft or too stiff, the varying air pressure during the wing strokes will not cause the wings to bend to the shapes needed for generating optimal lift.

The mechanical properties of cuticle are determined by the interplay of many factors, such as cuticular thickness, relative amounts of chitin and proteins, chitin architecture, protein composition, water content, intracuticular pH, degree of sclerotization, and other secondary modifications. Sclerotization of insect cuticle has been reviewed several times (Sugumaran, 1988, 1998; Andersen, 1990, 2005, 2010; Hopkins and Kramer, 1992; Andersen *et al.*, 1996), but many aspects of the process are still poorly understood. The present review will attempt to give an up-to-date presentation of the sclerotization problems, and to draw attention to some of the problems that need to be investigated in more detail.

Cuticular sclerotization is a chemical process by which certain regions of the insect cuticle are transformed irreversibly from a pliant material into a stiffer and harder structure, characterized by decreased deformability, decreased extractability of the matrix proteins, and increased resistance to enzymatic degradation. During sclerotization the color of the cuticle may change; some cuticles remain nearly colorless, and some become lighter or darker shades of brown or black. The term "tanning" is often used synonymously with sclerotization, but sometimes it is specifically used for the processes whereby brownish (tan) cuticles are formed. Sclerotization often takes place in connection with molting, starting just after the new, as yet unsclerotized, cuticle has been expanded to its final size and shape, but some specialized cuticular regions are sclerotized while the insect is still in its pharate state inside the old cuticle. Such pre-ecdysially sclerotized regions cannot be expanded post-ecdysially, but help the insect to escape from the exuvium. The dipteran puparium

is an example of a soft larval cuticle which is sclerotized at the end of the last larval instar to form a hard protective case, inside which metamorphosis to pupa and adult can take place. Sclerotization of structural materials in insects is not restricted to cuticle; other materials, such as egg cases, chorions, and silks, may be stabilized by chemical processes closely related to cuticular sclerotization.

6.2. A Model for Cuticular Sclerotization

During the past 70 years several models have been proposed for the chemical reactions occurring in the insect cuticle during the sclerotization process, and although many details of the individual steps in the reactions still are controversial or unexplored, there is general agreement concerning the main features of the process. The currently accepted sclerotization model is shown in **Figures 1** and **2**, and the main features are as follows: the amino acid tyrosine (**1**) is hydroxylated to 3,4-dihydroxyphenylalanine (DOPA, **2**), which by decarboxylation is transformed to dopamine (**3**), a compound of central importance for both sclerotization and melanin formation. Dopamine can be N-acylated to either N-acetyldopamine (NADA, **4**) or N-β-alanyldopamine (NBAD, **5**), and both can serve as precursors in the sclerotization process. They are enzymatically oxidized to the corresponding o-quinones (**6**), which can react with available nucleophilic groups, whereby the catecholic structure is regained and the nucleophile is linked to the aromatic ring (**11**). The o-quinones of NADA and NBAD may also be enzymatically isomerized to the corresponding p-quinone methides (**7**), after which the β-position of the side chain can react with nucleophiles (**12**). The p-quinone methide of NADA can also be enzymatically isomerized to a side-chain-unsaturated catechol derivative, α,β-dehydro-N-acetyldopamine (dehydro-NADA, **8**), but it is doubtful whether the p-quinone methide of NBAD is isomerized to α,β-dehydro-N-β-alanyldopamine (dehydro-NBAD) to any significant

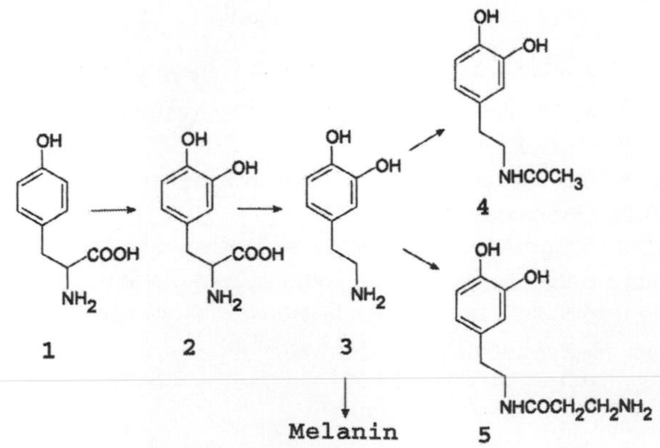

Figure 1 Formation of sclerotization precursors NADA (**4**) and NBAD (**5**) from tyrosine (**1**).

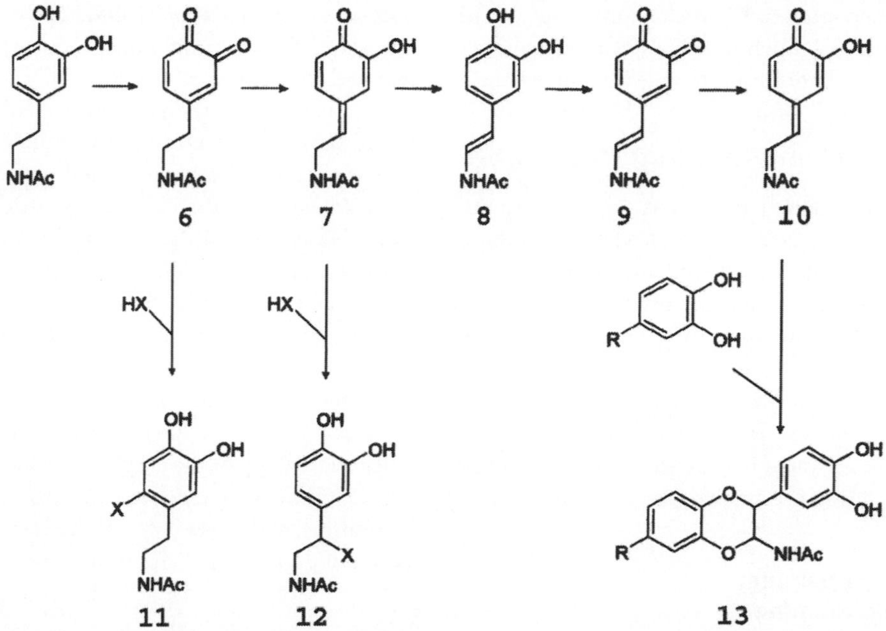

Figure 2 Suggested scheme for formation of acyldopamine derivatives and concomitant formation of adducts during cuticular sclerotization. **(6):** o-quinone; **(7):** p-quinone methide; **(8):** dehydro-NADA; **(9):** dehydro-NADA o-quinone; **(10):** dehydro-NADA p-quinone methide; **(11):** C-6 substituted acyldopamine adduct; **(12):** β-substituted acyldopamine adduct. **(13):** dihydroxyphenyl-dihydrobenzodioxine derivative.

extent. Dehydro-NADA can be oxidized to the corresponding o-quinone (**9**) and p-quinone methide (**10**). The latter can readily react with available catechols to give dihydroxyphenyl-dihydrobenzodioxine derivatives (**13**), and it will probably also react with nucleophilic amino acids to give various substitution products.

The cuticular nucleophilic groups which can react with the quinones formed from NADA and NBAD include the histidine imidazole group and free amino groups, such as terminal amino groups in proteins, ε-amino groups in lysine residues, and the amino group in β-alanine. The phenolic group of tyrosine may furthermore react with the p-quinone methide of dehydro-NADA. The quinones may react with water and possibly also with hydroxyl groups and free amino groups in chitin, although little evidence exists for reactions with chitin. *In vitro* incubations have shown that the quinones may also react with available catechols. The various reactions between quinones and nucleophilic residues in the cuticular proteins result in the proteins being more or less covered by aromatic residues depending upon the degree of sclerotization; some of the quinones may be involved in crosslinking the cuticular proteins and perhaps also in forming links between proteins and chitin, and some will only be linked to a single protein molecule, thereby increasing its hydrophobicity without being part of a covalent cross-link. During sclerotization, most of the water-filled spaces between the matrix proteins in the presclerotized cuticle will be filled with polymerized catecholic material. As a result of these processes the interactions between

the cuticular components become stronger, the peptide chains become more difficult to deform, and the proteins can no longer move relative to each other or to the chitin system. Together, all these changes contribute to making the material stiffer and more resistant to degradation.

The various reactions involved in this model will be discussed in more detail in the following sections, with the main emphasis on aspects where the evidence is insufficient or missing, or where some observations disagree with the scheme, to indicate the areas where more research is needed. The appearance and properties of cuticle from different body regions of the same animal vary widely, and a considerable part of this variation, such as differences in coloration and mechanical properties, is probably due to quantitative and qualitative differences in the sclerotization process. There is no compelling reason to expect that exactly the same detailed process is used for stabilization in all types of solid cuticle, and generalizations based upon results obtained with a single or a few insect species can easily be misleading. Many cuticular types will have to be analyzed to help us understand how the individual steps involved in sclerotization can be modified to give the local variations between cuticular regions in a given insect, and between cuticles from different insect species. It will also be important to study how the reactions are controlled to give the optimal degree of sclerotization. Most of the results and ideas presented in this chapter have been obtained by studies involving material from relatively few insect species, such as cuticle from blowfly larvae and puparia, cuticle from pupae of the moth *Manduca sexta*, and locust femur cuticle, and detailed studies of cuticle from other species will probably result in a much more varied and fascinating picture of cuticular sclerotization than is presented here.

6.3. Sclerotization (Tanning) Precursors

The terms "sclerotization agents" and "tanning agents" were originally used for the compounds which are secreted from the epidermal cells into the cuticle, where they are enzymatically oxidized to products sufficiently reactive to form covalent links to proteins and chitin. There is now a tendency to restrict the term "sclerotization agents" to the reactive species directly involved in forming links to the cuticular components. The compounds secreted from the epidermis to be activated in the cuticle shall accordingly be called "sclerotization precursors" (Sugumaran, 1998).

6.3.1. N-Acetyldopamine and N-β-Alanyldopamine

The first discovered and most common precursor for cuticular sclerotization is NADA (4), which is synthesized by N-acetylation of dopamine. The central role of NADA in sclerotization was demonstrated by Karlson's research group (for review, see Karlson and Sekeris, 1976). They showed that NADA is incorporated in the puparial cuticle of the blowfly *Calliphora vicina* during its sclerotization, and that radioactively labeled tyrosine is metabolized to NADA when injected into last-instar larvae shortly before puparium formation, and degraded when injected into younger larvae. The rate-limiting step was shown to be the enzymatically catalyzed decarboxylation of DOPA controlled by the steroid hormone ecdysone (Karlson and Sekeris, 1962; Fragoulis and Sekeris, 1975). NADA was shown to be involved in cuticular sclerotization in several other insect species, such as the desert locust, *Schistocerca gregaria* (Karlson and Schlossberger-Raecke, 1962; Schlossberger-Raecke and Karlson, 1964). Incorporation of NADA into cuticle can be a very efficient process; after injection of radioactive NADA into young adult locusts, about 80% of the total radioactivity could later be recovered from the sclerotized cuticle (Andersen, 1971). NADA appears to be involved in cuticular sclerotization in all insect species investigated.

The amino acid β-alanine was reported as a constituent of several types of sclerotized cuticle (Karlson *et al.*, 1969; Bodnaryk, 1971; Hackman and Goldberg, 1971; Srivastava, 1971), and it was suspected to participate in the sclerotization process (Andersen, 1979a). Hopkins *et al.* (1982) showed that the β-alanyl derivative of dopamine, NBAD (5), is a sclerotizing precursor in the cuticle of *M. sexta* pupae, thus accounting for the presence of β-alanine in hydrolysates of the fully sclerotized cuticle. NBAD is also a sclerotization precursor in the other cuticles from which β-alanine can be released by acid hydrolysis, such as the cuticle of the red flour beetle, *Tribolium*

castaneum (Kramer *et al.*, 1984). The synthesis and utilization of NBAD during pupation of *M. sexta* have been reported (Krueger *et al.*, 1989).

The first step in the synthesis of NADA and NBAD is hydroxylation of tyrosine to DOPA (2); this process can be catalyzed both by the o-diphenoloxidase, tyrosinase, which can also catalyze the oxidation of DOPA and other o-diphenols to o-quinones, and by the enzyme tyrosine hydroxylase. Tyrosinases play an important role in the defense systems of insects, and tyrosine hydroxylase appears to be the enzyme responsible for synthesis of the sclerotization precursors. The enzyme is present in epidermal cells, and both hardness and coloration of *T. castaneum* adult cuticle are significantly diminished when the activity of the enzyme is reduced by RNA interference (Gorman and Arakane, 2010).

Tyrosine hydroxylase is also important for sclerotization and darkening of the adult cuticle of *D. melanogaster*; it has been demonstrated that production of tyrosine hydroxylase in the epidermal cells in the pharate adult fruit fly is induced by the neurohormone CCAP (crustacean cardioactive peptide), and soon after ecdysis the sclerotization process is initiated by the neurohormone bursicon, which induces activation of the tyrosine hydroxylase by phosphorylation of a serine residue (Davis *et al.*, 2007). It is interesting that no such phosphorylation of tyrosine hydroxylase is needed for initiation of sclerotization of the puparium of *D. melanogaster*, a process that appears to be controlled by release of tyrosine from O-phoshotyrosine (Davis *et al.*, 2007). The hormone bursicon has functions other than induction of sclerotization, such as plasticization and stretching of wings after ecdysis, formation of melanin, and deposition of endocuticle (Dai *et al.*, 2008; Honegger *et al.*, 2008).

The DOPA residues formed in the epidermal cells are not directly involved in sclerotization, but are transformed to dopamine (3) by a DOPA-decarboxylase, and at least in *T. castaneum* this decarboxylase is essential for production of both NADA and NBAD (Arakane *et al.*, 2009). Another important decarboxylase in the epidermal cells is an aspartate-1-decarboxylase, which transforms aspartic acid to β-alanine. Reducing its activity by means of RNA interference resulted in adult beetles with a completely black cuticle instead of the rust-red cuticle of control animals. The black beetles contained less NBAD and more dopamine than untreated animals, and formation of the black color (melanins) could be prevented by injection of β-alanine. Mechanical tests of the elytra of black animals indicated that the cuticle was less cross-linked than in the controls (Arakane *et al.*, 2009).

The N-acetyldopamine synthetase, which acetylates dopamine to NADA, and the N-β-alanyldopamine synthetase, which β-alanylates dopamine to NBAD, are present in epidermal cells beneath sclerotizing cuticle (Krueger *et al.*, 1989; Wappner *et al.*, 1996; Pérez *et al.*, 2002), and it appears that insect epidermis is equipped with all the enzymes necessary for producing the sclerotization precursors. The acylation reactions can also occur in tissues other than epidermis; for instance, both NADA and NBAD can be produced within the nervous system (Krueger *et al.*, 1990).

The two sclerotizing compounds, NADA and NBAD, are used together in many cuticular types, but the cuticle of some insects, such as the locusts *Schistocerca gregaria* and *Locusta migratoria*, appears to be exclusively sclerotized by NADA, and no β-alanine has been obtained from their acid hydrolysates. No cuticle has yet been reported to be sclerotized exclusively by NBAD. A correlation has been reported between the intensity of brown color of the fully sclerotized cuticle and the amounts of NBAD taking part in the sclerotization process: cuticles that are sclerotized exclusively by NADA are often colorless or very lightly straw-colored, and when NBAD dominates in the process dark brown cuticles are formed (Brunet, 1980; Hopkins *et al.*, 1984). Czapla *et al.* (1990) reported that cuticular strength in five differently colored strains of the cockroach *Blatella germanica* correlated well with the concentrations of β-alanine and NBANE, and melanization correlated with dopamine concentration. During sclerotization, cuticular strength, as well as cuticular concentrations of β-alanine and NBAD, increased more rapidly in the rust-red wild type of *T. castaneum* than in the black mutant strain, whereas cuticular dopamine increased more rapidly in the black mutant than in the wild type (Roseland *et al.*, 1987).

Significant amounts of sclerotization precursors are often present as conjugates before the onset of sclerotization. The conjugates can be glucosides, phosphates, or sulfates (Brunet, 1980; Kramer and Hopkins, 1987), they are not easily oxidized and have to be hydrolyzed to free catechols before they can take part in sclerotization. It is assumed that the catechol conjugates serve as a storage reservoir of catecholamines ready to be used when the need for sclerotization arises (Brunet, 1980). A dopamine conjugate, identified as the 3-*O*-sulfate ester, is present in the hemolymph of newly ecdysed cockroaches, and its concentration decreases rapidly during sclerotization of the cockroach cuticle. The sulfate moiety is not transferred into the cuticle; removal of sulfate and acylation of the liberated dopamine to NADA and/or NBAD will most likely take place in the epidermal cells (Bodnaryk and Brunet, 1974; Czapla *et al.*, 1988, 1989).

Hopkins *et al.* (1984) reported that a large fraction of the various catecholamines in *M. sexta* hemolymph and cuticle is present as acid labile conjugates. In larval and pupal hemolymph these conjugates are mainly 3-*O*-glucosides together with small amounts of 4-*O*-glucosides, whereas adult hemolymph contains more of the 4-*O*-glucoside than of the 3-*O*-glucoside (Hopkins *et al.*, 1995). Both conjugated and unconjugated forms of NADA and NBAD are present in *M. sexta* hemolymph, but are only present in low amounts in the cuticle

(Hopkins *et al.*, 1984), indicating that the epidermal cells contain a β-glucosidase able to hydrolyze the conjugates to unconjugated catecholamines and glucose.

6.3.2. Putative Sclerotization Precursors

So far, convincing evidence that they function as cuticular sclerotization precursors has only been obtained for NADA and NBAD, but other compounds have been described as likely sclerotization precursor candidates, such as *N*-acetyl-norepinephrine (NANE) (**14**), *N*-β-alanyl-norepinephrine (NBANE) (**15**), and 3,4-dihydroxyphenylethanol (DOPET) (**16**) (**Figure 3**). Probably, they all have some role in sclerotization, but their formation and metabolism needs to be studied in more detail.

6.3.2.1. *N*-acetylnorepinephrine (NANE) and *N*-β-alanylnorepinephrine (NBANE)
NANE and NBANE are special cases among cuticular catechols, because they can be considered as both by-products of the sclerotization process and precursors for sclerotization. They have been reported to occur both free and as *O*-glucosides in hemolymph and integument in several insects (Hopkins *et al.*, 1984, 1995; Morgan *et al.*, 1987; Czapla *et al.*, 1989). NANE and NBANE can be generated within the cuticle, when the enzymatically produced *p*-quinone methides of NADA and NBAD react with water instead of either reacting with cuticular proteins or isomerising to dehydro-derivatives, and they can also be produced by hydrolysis of unidentified labile products of the sclerotization process. Mild acid treatment of sclerotized cuticle can release NANE and NBANE from the cuticular structure, probably due to hydrolysis of a bond between the β-position of the catechols and some cuticular constituent. The nature of the bond is uncertain, but it could be an ether linkage connecting the acyldopamine side chain to chitin. Formation of an ether, β-methoxy-NADA, occurs when isolated pieces of cuticle or extracted cuticular enzymes act upon NADA in the presence of methanol; the compound is acid labile, and is readily hydrolyzed to free NANE (Andersen, 1989a; Sugumaran *et al.*, 1989a).

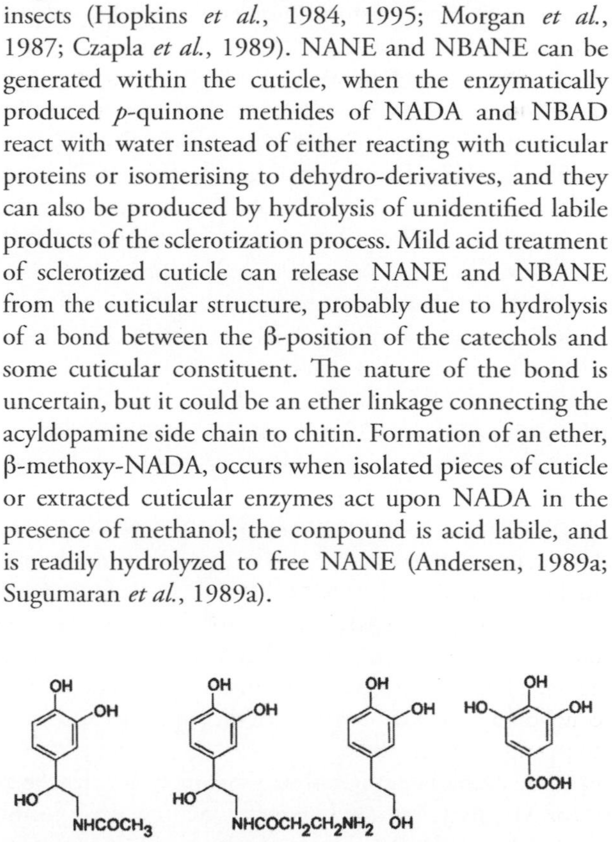

Figure 3 Hypothetical sclerotization precursors. (**14**): *N*-acetyl-norepinephrine (NANE); (**15**): *N*-β-alanyl-norepinephrine (NBANE); (**16**): 3,4-dihydroxyphenylethanol (DOPET); (**17**): gallic acid.

NANE can be covalently incorporated into the cuticular matrix during sclerotization, indicating that the compound can serve as a sclerotization precursor (Andersen, 1971), and this is probably also the case for NBANE. When radioactively labeled NANE was injected into newly ecdysed locusts a significant fraction of the radioactivity (about 15%) was incorporated into the cuticle, and hydrolysis was needed to release the activity. Acid hydrolysis of cuticle from locusts injected with labeled norepinephrine resulted in the release of both labeled norepinephrine and arterenone, whereas little radioactivity was present in the neutral ketocatechol fraction. This is in contrast to parallel experiments where labeled dopamine was injected into locusts and nearly all the radioactivity was recovered as neutral ketocatechols, indicating that the cuticular enzymes can catalyze the incorporation of norepinephrine and NANE into the cuticular matrix, but not as efficiently and not by the same route as incorporation of dopamine and NADA.

6.3.2.2. Dihydroxyphenylethanol (DOPET) The third putative sclerotization precursor, 3,4-dihydroxyphenylethanol (DOPET, **16**), is present in the hemolymph and integument of some insects; it has been obtained from cuticle of the cockroach *Periplaneta americana* (Atkinson *et al.*, 1973a; Czapla *et al.*, 1988) and the beetle *Pachynoda sinuata* (Andersen and Roepstorff, 1978), and can function as substrate for the cuticular phenoloxidases. Extraction and acid hydrolysis of sclerotized cuticles have yielded various DOPET derivatives, suggesting that DOPET can be transported into the cuticle and incorporated into the cuticular matrix during sclerotization. Adducts of DOPET and histidine have been obtained by acid hydrolysis of sclerotized *Manduca* pupal cuticle and identified by means of mass spectrometry (Kerwin *et al.*, 1999), and a dihydroxyphenyl-dihydrobenzodioxine-type adduct of DOPET and NADA was extracted from sclerotized beetle (*P. sinuata*) cuticle (Andersen and Roepstorff, 1981), but the metabolism of DOPET in insects needs to be studied in much more detail.

The relative roles of dopamine, NANE, NBANE, and DOPET compared to the two major sclerotization precursors, NADA and NBAD, have never been properly established. The compounds are probably only of minor importance for the mechanical properties of sclerotized cuticles, but their involvement in the sclerotization process could play a role in fine-tuning of the cuticular properties.

6.3.2.3. Other sclerotization precursors It has been reported that improved growth of the tree locust *Anacridium melanorhodon* can be obtained by addition of gallic acid (**17**) (**Figure 3**) and other plant phenols to its food, and that the ingested plant phenols are incorporated into the cuticle and may contribute to its stabilization (Bernays *et al.*, 1980; Bernays and Woodhead, 1982).

6.4. Transport of Sclerotization Precursors to the Cuticle

The cuticle is constantly exposed to external forces varying in intensity and direction during insect movements; the mechanical properties of the cuticular regions must correspond to the forces they are exposed to, and to guarantee this a local regulation of the degree of sclerotization must be present. The sclerotization can be regulated via availability of sclerotization precursors, for instance by control of local synthesis of precursors and their uptake from hemolymph into epidermis, and by control of the transport of precursors from epidermal cells to cuticular matrix, where they will be exposed to the enzymes converting them to sclerotization agents. Precursors for sclerotization can be synthesized by the epidermal cells, but they may also be produced by other cell types, such as hemocytes and/or fat body cells.

The *white pupa* mutant of the Mediterranean fruit fly, *Ceratitis capitata*, does not sclerotize its puparium, but normal larval and adult cuticles are produced. The concentrations of the various catecholamines are very low in the mutant puparial cuticle compared to the wild type strain, whereas the concentrations in the hemolymph of NADA, NBAD, and dopamine are about 10 times higher in the mutant than in the wild type, indicating that the mutant is defective in a system transporting catecholamines from hemolymph to puparial cuticle (Wappner *et al.*, 1995). Precursors, such as dopamine, NADA, and NBAD, injected into the hemocoel of locusts shortly before or during cuticular sclerotization, are rapidly taken up by the epidermal cells and transported into the cuticle, whereas precursors injected several days before the start of sclerotization are mainly degraded or modified by glycosylation or phosphorylation. Such precursor conjugates may either remain in the animal to serve as a reserve pool of sclerotizing material, or be excreted via the Malphigian tubules by the standard detoxification mechanisms. The different fates of NADA injected before or during ecdysis could be explained by different activities of transport systems located in the apical or basolateral membranes of the epidermal cells.

It is thus likely that the epidermal cells possess specific transport systems controlling transfer of the right amounts of sclerotizing precursors into the cuticle at the right time, and presumably such transport systems are present in the basolateral cell membrane to facilitate uptake from the hemolymph and in the apical cell membrane to control the transport of sclerotizing precursors from the cells into the cuticle. Active diphenoloxidases (laccases) are, in some insects, present in the unsclerotized pharate cuticle (Andersen, 1972a, 1979b, and unpublished data), and presumably the precursors are only transported into these cuticles when sclerotization is initiated following ecdysis. Is activation of post-ecdysial transport of sclerotization precursors into the cuticle governed by the presence of the hormone bursicon?

In cuticles sclerotized before ecdysis, the sclerotization precursors might be transported without delay from the epidermal cells into the pharate cuticle to be oxidized and incorporated. It is not known whether the pre-ecdysial transport occurs via specific transporters in the apical membrane or whether it happens by passive diffusion.

It has been reported that catechol derivatives attached to proteins can be transported intact from hemolymph to the cuticular structure to serve as combined matrix components and sclerotization precursors (Koeppe and Mills, 1972; Koeppe and Gilbert, 1974; Bailey *et al.*, 1999). Such transport indicates that receptors able to recognize the catechol–protein complexes are present in the membranes of the epidermal cells.

6.5. Cuticular Enzymes and Sclerotization

The study of cuticular enzymes has to a large extent been concerned with characterization of enzymes assumed to play a role in cuticular sclerotization, mainly those involved in oxidation of catechols, and there has been a tendency to neglect the possibility that enzymes not directly involved in sclerotization may play a role in cuticular metabolism. Enzymes such as glucose oxidase, catalase, and superoxide dismutase have been reported from locust cuticle (Candy, 1979), but have not been studied in much detail. Catechol oxidation is an important step in sclerotization, as well as in wound healing and immune responses in insects, and it is often difficult to decide whether a given cuticular phenoloxidase activity is involved in sclerotization or whether its main role is to take part in defense reactions. It is therefore necessary to be careful when interpreting the observations reported for cuticular enzymes.

6.5.1. *ortho*-Diphenoloxidases

After being transported from the epidermal cells into the cuticle the catecholic sclerotization precursors may encounter different oxidative enzymes, such as *o*-diphenoloxidases, laccases, and peroxidases, capable of oxidizing them to quinones, but the relative roles of the enzymes are still uncertain. Insects contain inactive pro-enzymes for *o*-diphenoloxidases both in hemolymph and in cuticle, which can be activated by a cascade of processes involving limited proteolysis and initiated by wounding or by the presence of small amounts of microbial cell-wall components (Ashida and Dohke, 1980; Ashida and Brey, 1995). The *o*-diphenoloxidases are able to oxidize a wide range of *o*-diphenols, but not *p*-diphenols; they can hydroxylate monophenols, such as tyrosine and tyramine, to *o*-diphenols, and they are readily inhibited by thioureas

and sodium diethyldithiocarbamate. They have been isolated and characterized from soft, non-sclerotizing cuticles, such as larval cuticle of *Bombyx mori* (Ashida and Brey, 1995; Asano and Ashida, 2001a), sclerotizing pupal cuticle of *Manduca sexta* (Aso *et al.*, 1984; Morgan *et al.*, 1990), and blowfly and fleshfly larval cuticles (Barrett, 1987a, 1987b, 1991). The amino acid sequences of *o*-diphenoloxidases from various insect species have been deduced from the corresponding DNA sequences (Fujimoto *et al.*, 1995; Hall *et al.*, 1995; Kawabata *et al.*, 1995). The insect *o*-diphenoloxidases resemble diphenoloxidases (tyrosinases) from other organisms, but differ with regard to substrate specificity and amino acid sequence (Sugumaran, 1998; Chase *et al.*, 2000).

The established sequences of insect prophenoloxidase genes indicate that the gene products do not possess an *N*-terminal signal peptide sequence (Sugumaran, 1998), in contrast to what is observed for most proteins destined for export from cells. It appears that the prophenoloxidases are released by cell rupture before they are transformed into active enzymes (Kanost and Gorman, 2008). The silkmoth *B. mori* has genes for two *o*-diphenoloxidase proenzymes, and the products of both genes are present in both hemolymph and cuticle of *B. mori* larvae; the proenzymes appear to be synthesized in hemocytes, and can be transported into the cuticle via the epidermal cells. A difference between the cuticular and hemolymphal proenzymes is that some methionine residues, which in the hemolymphal proenzymes are unmodified, are oxidized to methionine sulfoxides in the cuticular proenzymes. When activated, the cuticular enzymes have nearly the same substrate specificity as the hemolymphal enzymes; a difference between the two groups of proenzymes is that the oxidized cuticular form cannot be transported back across the epidermal cell layer, indicating that the epidermal transport of the proenzymes is a one-way traffic, from hemolymph to cuticle (Asano and Ashida, 2001a, 2001b).

The enzymatic properties of the diphenoloxidases purified from hemolymph and from pharate pupal cuticle of *M. sexta* are very similar, suggesting a close relationship between the enzymes (Aso *et al.*, 1985; Morgan *et al.*, 1990), and it seems probable that they, like the *B. mori* enzymes, are derived from the same gene(s).

Both cuticular and hemolymphal *o*-diphenoloxidases become very sticky when activated; they tend to aggregate and to stick to any available surface and macromolecule, thereby hindering diffusion of the active enzymes from the site where they were activated. The *o*-diphenoloxidase activity in the larval cuticle of the blowfly *Lucilia cuprina* was localized to epicuticular filaments (Binnington and Barrett, 1988). Some activity was also observed in the procuticle, but only when the cuticle had been damaged beforehand, and the activity was limited to the close neighborhood of the wound, indicating that wounding is needed to activate the enzyme, and that the active enzyme remains in the vicinity of the wound.

Using immunocytochemical methods, the prophenoloxidase in larval cuticle of *B. mori* could be demonstrated both in the non-lamellate endocuticle, where it was randomly distributed, and in an orderly arrayed pattern in the lamellate endocuticle; it appeared to be absent from the cuticulin layer and the epidermal cells (Ashida and Brey, 1995).

The role of the cuticular *o*-diphenoloxidases in the sclerotization process is problematic; their presence as inactive proenzymes, which have to be activated, their close relationship to the hemolymphal phenoloxidases, and their abundance in non-sclerotizing cuticle suggest that their role is to take part in defense against wounding and microorganisms, and not to be involved in sclerotization.

6.5.2. Laccases

Laccase-type phenoloxidases have been reported to be present in various dipteran larval cuticles shortly before and during puparium formation, such as larval cuticles of *Drosophila virilis* (Yamazaki, 1969), *D. melanogaster* (Sugumaran *et al.*, 1992), *Calliphora vicina* (Barrett and Andersen, 1981), *Sarcophaga bullata* (Barrett, 1987a), and *L. cuprina* (Barrett, 1987b), and laccases have also been described from pupal cuticles of *B. mori* (Yamazaki, 1972) and *M. sexta* (Thomas *et al.*, 1989), as well as from adult cuticle of the locust *S. gregaria* (Andersen, 1978). Insect laccases are structurally related to laccases of plant and fungal origin, and, in contrast to the insect *o*-diphenoloxidases, the laccase gene products contain a typical signal peptide sequence, indicating that the enzymes are secreted into the extracellular space.

Laccase activity appears in larval cuticles of *D. virilis* (Yamazaki, 1969) and *L. cuprina* (Binnington and Barrett, 1988) shortly before pupariation, and in both species the enzyme activity decreases gradually as puparial sclerotization progresses. Laccase activity was demonstrated in pharate cuticle of adult locusts, *S. gregaria,* a few days before ecdysis, and it remained at high levels for at least 2 weeks after ecdysis. Activity has also been demonstrated in nymphal exuviae, indicating that the locust laccase is not inactivated by sclerotization (S.O. Andersen, unpublished data).

Insect laccases are active towards a broad spectrum of *o*- and *p*-diphenols: NBAD and NADA are among the best *o*-diphenolic substrates tested, and methyl-hydroquinone is the best *p*-diphenolic substrate. Insect laccases are not inhibited by such compounds as thiourea, phenylthiourea, and sodium diethyldithiocarbamate, which are effective inhibitors of *o*-diphenoloxidases, but they are inhibited by carbon monoxide and low concentrations of fluorides, cyanides, and azides (Yamazaki, 1972; Andersen, 1978; Barrett, 1987a; Barrett and Andersen,

1981). The laccases are not affected by treatments which will inactivate many other enzymes; the *S. gregaria* laccase remains active after blocking available amino and phenolic groups by dinitrophenylation or dansylation, and it survives temperatures up to about 70°C, but it is inactivated by treatment with tetranitromethane, which nitrates tyrosine residues (Andersen, 1979b). The laccases appear to be firmly linked to the cuticular structure; typically they cannot be extracted by conventional protein extractants, but are readily extracted after limited tryptic digestion of the yet-unhardened cuticle (Yamazaki, 1972; Andersen, 1978). The enzyme was obtained from *C. vicina* larval cuticle by prolonged extraction at pH 8 without addition of any protease, but as latent protease activity is present in the cuticle, the release of laccase from the cuticular residue may be due to proteolysis (Barrett and Andersen, 1981). The enzyme is not released by tryptic digestion of already sclerotized cuticle.

The ultrastructural localization of laccase activity has been studied in the *L. cuprina* larval cuticle (Binnington and Barrett, 1988), where enzyme activity was observed in the inner epicuticle of late third instar larvae (about to pupariate), but not in epicuticle of younger larvae. The laccase activity in *L. cuprina* larval cuticle could be demonstrated without prior activation, in contrast to the cuticular *o*-diphenoloxidases, indicating that the laccase is not deposited as an inactive proenzyme in this insect, and neither is an inactive proenzyme likely to be present in pharate locust cuticle since enzyme activity was demonstrated without any activating treatment. Both full-length and amino-terminally truncated recombinant forms of *M. sexta* cuticular laccase were expressed and purified, and both forms had activities towards NADA and NBAD similar to that of the enzyme purified from pharate pupal cuticle, indicating that the *M. sexta* enzyme is not produced as an inactive proenzyme (Dittmer *et al.*, 2009). A pro-laccase has been purified and partially characterized from cuticle of newly pupated pupae of *B. mori* (Ashida and Yamazaki, 1990; Yatsu and Asano, 2009). The inactive pro-laccase could be activated by treatment with various proteolytic enzymes, and the substrate specificities of the laccase variants obtained depended upon the protease used for activation.

Definitive proof that a cuticular laccase is responsible for cuticular sclerotization was obtained by demonstrating that inhibition of the *laccase 2* gene product by means of RNA interference could prevent cuticular sclerotization in the beetle *T. castaneum* (Arakane *et al.*, 2005). RNA interference inhibition of the *laccase 1* gene product as well as the two *o*-diphenoloxidases in *T. castaneum* had no influence on cuticular sclerotization. Involvement of the *laccase 2* gene in sclerotization has also been demonstrated in the beetle *Monochamus alternatus* (Niu *et al.*, 2008), in the mosquito *Culex pipiens* (Pan *et al.*, 2009), in *M. sexta* (Dittmer *et al.*, 2009) and in *B. mori* (Yatsu and Asano,

2009), confirming that the diphenoloxidase responsible for cuticular sclerotization is laccase 2.

6.5.3. Cuticular Peroxidases

Several routes for the oxidation of catechols to *o*-quinones can be advantageous for an insect, especially when the different routes are regulated independently, and sclerotization by means of peroxidase activity could be such an alternative route. Peroxidase activity has been demonstrated by histochemical methods in proleg spines of *Calpodes ethlius* larvae (Locke, 1969), and in larval and pupal cuticle of *Galleria mellonella* and *Protoformia terraenovae* (Grossmüller and Messner, 1978; Messner and Janda, 1991; Messner and Kerstan, 1991), peroxidase activity can also be observed intracellularly in different cell types in insects. It is not known whether the cuticular peroxidase activities are identical to the intracellular enzymes, as the cuticular activities have never been properly characterized. Proteins can be cross-linked by means of the peroxidase system, and it has been suggested that the enzyme could be involved in cuticular sclerotization (Hasson and Sugumaran, 1987). A peroxidase is likely to be involved in the cross-linking of the rubber-like elastic cuticular protein resilin (Andersen, 1966; Coles, 1966). This cuticular protein is cross-linked by oxidative coupling of tyrosine residues during its extracellular deposition, and the tyrosine radicals needed for the coupling are likely to be formed by a peroxidase-catalyzed oxidation process. Peroxidases can also oxidize catechols to semiquinone radicals, two of which may readily dismutate to form an *o*-quinone and a catechol. The enzyme needs hydrogen peroxide as one of its substrates, and Candy (1979) reported that locust cuticle contains glucose oxidase activity, which oxidizes glucose to D-gluconate with concomitant production of hydrogen peroxide. It was suggested that the hydrogen peroxide produced may participate in sclerotization reactions. Candy (1979) also reported that other enzymes involved in hydrogen peroxide metabolism, such as peroxidase, catalase, and superoxide dismutase, are present in locust cuticle.

Peroxidase activity in solid cuticle may be involved in the production of dityrosine cross-links, and in oxidizing catechols to quinones for sclerotization. Small amounts of dityrosine have been obtained from sclerotized locust cuticle (Andersen, 2004a), and dityrosine as well as brominated dityrosines have been obtained from the hardened exocuticle of the crab *Cancer pagurus* (Welinder et al., 1976). When a protein from *M. sexta* pupal cuticle was treated with a fungal laccase, a product was formed that reacted with a monoclonal antibody against dityrosine (Suderman et al., 2010), suggesting that laccases can oxidize tyrosine to dityrosine without involvement of hydrogen peroxide. Laccase-catalyzed formation of dityrosine residues was also described by Mattinen et al. (2005). It is

not known whether dityrosine formation plays any important roles in sclerotization of solid cuticles, but it appears to be important for stabilization of other structural materials in insects. The eggshells of *D. melanogaster* are stabilized by formation of dityrosine cross-links between the protein chains (Petri et al., 1976; Mindrinos et al., 1980), and the hardening of *Aedes aegypti* egg chorion includes both peroxidase-mediated protein cross-linking through dityrosine formation, and diphenoloxidase-catalyzed chorion melanization (Li et al., 1996). The hydrogen peroxide necessary for dityrosine formation in *A. aegypti* chorion is produced in an enzymatic process by which NADH is oxidized with concomitant reduction of molecular oxygen to hydrogen peroxide. The necessary supply of NADH for this process is provided by enzyme-catalyzed oxidation of malate coupled to reduction of NAD$^+$ (Han et al., 2000a, 2000b). It is unknown whether a similar system for providing hydrogen peroxide is involved in sclerotization of some insect cuticles.

6.5.4. *ortho*-Quinones and *para*-Quinone Methides

The three types of oxidases, *o*-diphenoloxidases, laccases, and peroxidases, can all oxidize NADA and NBAD to their respective *o*-quinones; when the oxidation process is catalyzed by laccases and peroxidases free radicals are produced, but this is not the case when oxidations are catalyzed by *o*-diphenoloxidases. The *o*-quinones are reactive compounds, which react spontaneously with nucleophilic groups, and they can be enzymatically isomerized to *p*-quinone methides, which will also react with nucleophilic compounds. It has been reported that the laccase in *M. sexta* pupal cuticle oxidizes *o*-diphenols to a mixture of *o*-quinones and *p*-quinone methides, indicating that a specific isomerase is not an absolute requirement for *p*-quinone methide formation (Thomas et al., 1989). An enzyme catalyzing *o*-quinone isomerization has been partially characterized from larval cuticle of *H. cecropia* (Andersen, 1989a) and *S. bullata* (Sugumaran, 1987; Saul and Sugumaran, 1988), both cuticles being of the soft, pliant type. The enzyme is also present in the hemolymph of *S. bullata* (Saul and Sugumaran, 1989a, 1990), where it participates in defense reactions. So far the enzyme has not been demonstrated in cuticles which are sclerotized in connection with ecdysis, and the available evidence is insufficient to decide whether the isomerase is involved in cuticular sclerotization or whether its function is restricted to defense purposes.

The various adducts obtained by incubating samples of insect cuticle with catechols and nucleophiles will be discussed below, but it seems reasonable to ask whether the two β-hydroxylated compounds, NANE and NBANE, present in various sclerotized cuticles and in hemolymph, are formed in the cuticle by reactions between water and the *p*-quinone methides of NADA and NBAD, respectively, or whether they are synthesized outside the cuticle – for instance in the epidermal cells – and transported to the cuticle. In the former case they will simply be by-products of the sclerotization process, and in the latter case they could be alternative sclerotization compounds.

6.5.5. Dehydro-NADA and Dehydro-NBAD

The *p*-quinone methide formed by isomerization of NADA-*o*-quinone can be further isomerized to α,β-dehydro-NADA (**8**), a NADA derivative carrying a double bond between the α- and β-carbon atoms of the side chain. The enzyme responsible for this isomerization has been called *N*-acetyldopamine quinone methide/1,2-dehydro-*N*-acetyldopamine tautomerase (Saul and Sugumaran, 1989b). The activity has been reported to be present in larval cuticle of *S. bullata* (Saul and Sugumaran, 1989b, 1989c) and *D. melanogaster* (Sugumaran et al., 1992). Small amounts of an enzyme activity catalyzing the isomerization of NBAD *p*-quinone methide to dehydro-NBAD has been demonstrated in extracts of *C. vicina* larval cuticle (Ricketts and Sugumaran, 1994). Several cuticles, which are sclerotized by mixtures of NADA and NBAD, will readily convert NADA to the dehydro-derivative, while conversion of NBAD only occurs to a minor extent, if at all (Andersen, 1989b; Andersen et al., 1996).

The dehydro-NADA formed during cuticular sclerotization can be oxidized by *o*-diphenoloxidases as well as laccases, and the resulting side-chain-unsaturated quinones may react spontaneously with other catechols to give substituted dihydroxyphenyl-dihydrobenzodioxines (**13**) (Andersen and Roepstorff, 1982; Sugumaran et al, 1988). The presence of various dihydroxyphenyl-dihydrobenzodioxine derivatives in naturally sclerotized cuticle indicates that oxidation products of dehydro-NADA tend to react with available catechols during sclerotization (Andersen and Roepstorff, 1981, 1982; Andersen, 1985), and that presence of dihydroxyphenyl-benzodioxine derivatives in sclerotized cuticles can be used as an indication for the presence of a dehydro-NADA forming activity.

The sclerotization of some types of cuticle is dominated by reactions involving the dehydro-NADA quinones, while sclerotization of other types of cuticle is dominated by reactions of NADA and NBAD *o*- and *p*-quinones with matrix proteins. The ability to form benzodioxine derivatives from NADA is much more pronounced in such cuticular types as locust femur, *H. cecropia* larval head capsule, and *D. melanogaster* puparia, than in locust mandibles, *D. melanogaster* larval cuticle, and *T. molitor* pupal cuticle (Andersen, 1989c, 1989d). The observed differences may be due to different amounts of the enzymes involved in the process, but it has been suggested that the cuticular diphenoloxidases, isomerases, and tautomerases occur together in large enzyme complexes, enabling one

enzyme to deliver its products directly to the next enzyme for further processing (Andersen et al., 1996; Sugumaran, 1998). The evidence for such complexes is insufficient, but should not be disregarded.

6.5.6. Catechol Adducts Formed *in vitro*

In the original sclerotization scheme, suggested by Pryor (1940a, 1940b), *o*-quinones react preferably with ε-amino groups from lysine residues, but both *in vitro* and *in vivo* studies have shown that the imidazole group in histidine residues is the preferred group for reactions with both *o*-quinones and *p*-quinone methides, and that other groups can also take part.

Incubation of NADA and *N*-acetylcysteine with blowfly (*Sarcophaga bullata*) larval cuticle resulted in formation of an adduct (**18**) where the sulfur atom in *N*-acetylcysteine is linked to the 5-position of the NADA moiety (Sugumaran et al., 1989b), indicating that SH-groups are good acceptors for the *o*-quinone of NADA. Electrochemical oxidation of dopamine in the presence of *N*-acetylcysteine gave a mixture of C-5 and C-2 (**19**) monoadducts together with a disubstituted product, 2,5-S,S′-di(*N*-acetylcysteinyl)dopamine (**20**) (**Figure 4** Xu et al., 1996a; Huang et al., 1998;). Since the monoadducts are more readily oxidized to quinones than the parent catechol, a monoadduct formed between an oxidized catechol and a protein-linked cysteine will be more prone to be reoxidized to quinone than a free catechol, and after reoxidation it can react with a nucleophilic group in a neighboring protein chain to form a covalent cross-link between the proteins. Thus cysteine–catechol based cross-links are possible, but they are not likely to play an important role in cross-linking cuticular proteins since cysteine residues are rare in cuticular proteins. Perhaps the scarcity of cysteines in cuticular proteins is related to the

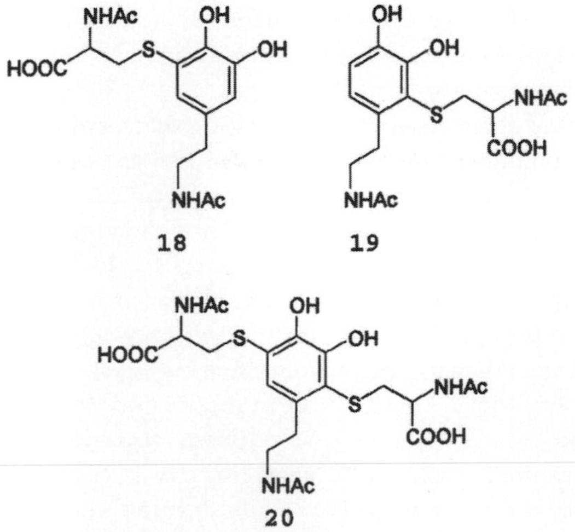

Figure 4 Structure of adducts formed between *N*-acetylcysteine and NADA.

readiness with which they react with *o*-quinones. When locust cuticle is incubated with NADA and benzenesulfinic acid, the oxidized NADA is trapped by adduct formation with the sulfinic acid, and only when all sulfinic acid has been consumed will *o*-quinones be available for reaction with cuticular proteins, and for isomerization and further metabolism to polymeric compounds (S.O. Andersen, unpublished data). If significant amounts of free SH-groups were present in the cuticular matrix proteins, that could in a similar way delay further metabolism of the *o*-quinones, and cause sub-optimal sclerotization.

Methionine residues can also react with *o*-quinones to form adducts, and such adducts have been suggested to take part in sclerotization (Gupta and Vithayathil, 1982; Sugumaran and Nelson, 1998), but so far no methionine-containing adducts have been reported from sclerotized cuticles.

Electrochemical oxidation of NADA in the presence of *N*-acetylhistidine gave mono-adducts where a nitrogen atom in the imidazole ring is linked to either the C-6 (**21**) or the C-2 ring position (**22**) in NADA (**Figure 5**), the C-6 position being the preferred position (Xu *et al.*, 1996b). Electrochemical characterization of the C-6 and C-2 *N*-acetylhistidine-NADA adducts showed that both adducts are more reluctant than NADA itself to be oxidized (Xu *et al.*, 1996b), indicating that formation of adducts involving two *N*-acetylhistidine residues linked to the same NADA residue is rather unlikely.

Using *H. cecropia* larval cuticle to oxidize a mixture of NADA and *N*-acetylhistidine resulted in a mixture of products, and both the C-6 ring position and the β-position of the side chain (**23**) (**Figure 5**) were involved in adduct formation (Andersen *et al.*, 1991, 1992a). The formation of a side-chain adduct indicates that the β-position was activated, probably due to formation of NADA *p*-quinone methide, demonstrating that cuticular oxidation of catechols is more complex than electrochemical oxidation.

Incubation of larval cuticle of *H. cecropia* with NADA and compounds containing a free amino group, such as α-*N*-acetyllysine and β-alanine, resulted in formation of several products, and adducts were identified with the amino groups linked to either the 6-position of the ring

in NADA (**24**) or to the β-position of the side chain (**25**) (**Figure 6**). The former adducts have a quinoid structure, indicating that the initially formed catecholic adducts are spontaneously oxidized by exposure to air. The adducts containing an amino acid linked to the β-position of the side chain were stable and not readily oxidized (Andersen *et al.*, 1992b). A product obtained by prolonged incubation of cuticle with NADA and α-*N*-acetyllysine was identified as a 4-phenylphenoxasin-2-one (**26**) (**Figure 6**), a compound composed of three NADA residues joined to one α-*N*-acetyllysine residue (Peter *et al.*, 1992).

The structure of one of the products formed during incubation of blowfly larval cuticle with NADA indicated that the *o*-quinone of NADA can react with water to form *N*-acetyl-3,4,6-trihydroxyphenylethylamine (6-hydroxy-NADA), which will couple oxidatively with another NADA residue to give the dimeric compound (**27**) (**Figure 7** Andersen *et al.*, 1992c;). Products indicating formation of 6-hydroxy-NADA have not yet been obtained from *in vivo* sclerotized cuticles, and *in vitro* formation is probably due to the large excess of water in the incubation medium, contrasting with the relatively low water content in sclerotizing cuticle (30–40% of the cuticular wet weight (Andersen, 1981)). The incubation of NADA with blowfly larval cuticle also resulted in formation of small amounts of a product (**28**) (**Figure 7**) consisting of two NADA-residues linked together via their 6-positions (Andersen *et al.*, 1992c). A corresponding 2,6′-linked dimer of NADA was suggested as intermediate in formation of the above-mentioned 4-phenylphenoxasin-2-one (**26**) (Peter *et al.*, 1992).

6.5.7. Various Catechol Derivatives Obtained from Sclerotized Cuticles

During the process of sclerotization most of the catecholic material will be firmly linked to the cuticular matrix and can only be solubilized by degrading the cuticle, for instance by hydrolysis. A large variety of catecholic derivatives are released by acid hydrolysis of sclerotized cuticle; some are simple catechols and some catecholic adducts containing amino acids. The structures of the derivatives can indicate whether or not they are likely to

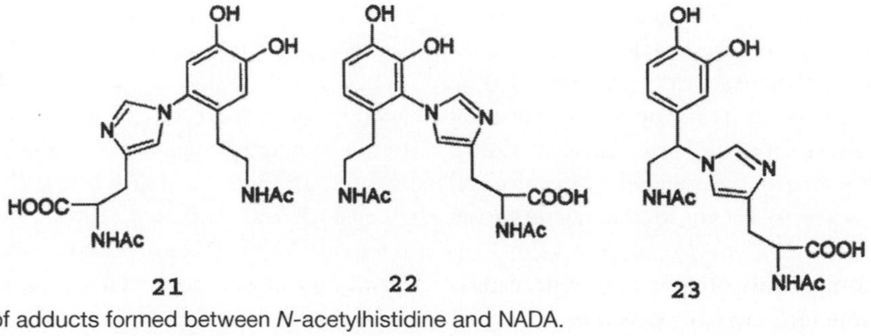

Figure 5 Structure of adducts formed between *N*-acetylhistidine and NADA.

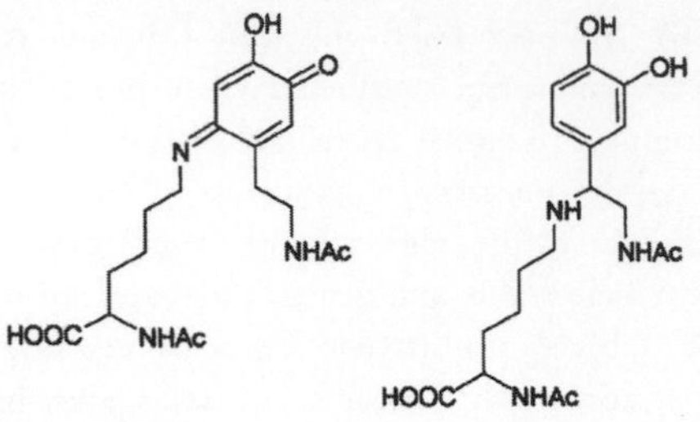

Figure 6 Structure of products formed from α-*N*-acetyllysine and oxidized NADA.

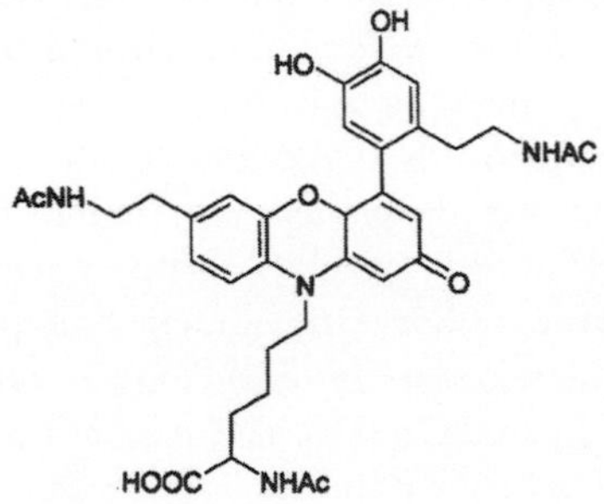

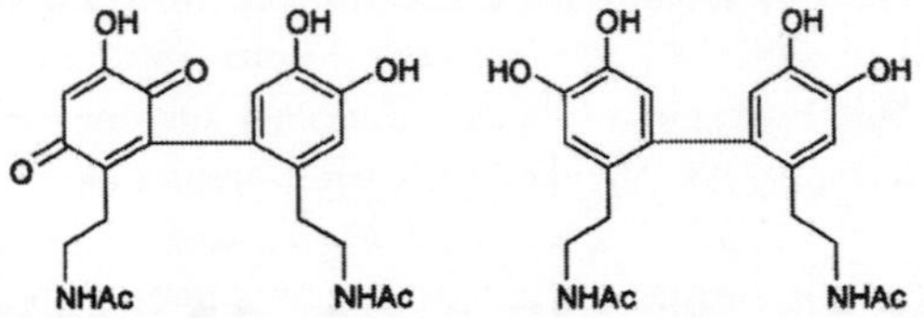

Figure 7 Products formed by incubation of NADA together with cuticle.

be degradation products of materials derived from the sclerotizing precursors, and may also indicate which reactions are responsible for their formation. Among the compounds extracted from sclerotized cuticle we find unused sclerotization precursors, intermediates of the sclerotization process, by-products from the process, and degradation products of protein-bound cross-links and polymers. It is likely that some of the extracted compounds have functions unrelated to sclerotization; catechols can be precursors for pigments, such as papiliochromes (Umebachi, 1993), or they can function as antioxidants protecting the epicuticular lipids from autoxidation (Atkinson *et al.*, 1973b). 3,4-Dihydroxyphenyl acetic acid, which is present in the solid cuticle of beetle species (Andersen, 1975), may serve the latter purpose, as it apparently does not take part in the sclerotization process (Barrett, 1984a, 1990). It can be problematic to decide whether catechols obtained from cuticles are related to the sclerotization process or not.

The mixture of compounds obtained by mild extraction of sclerotized cuticles (such as extraction in boiling water or neutral salt solutions) consists of compounds that are less modified than those obtained by acidic extraction at elevated temperatures, but a critical and careful interpretation of their structures will be necessary in all cases. Extraction with dilute acids tends to give higher yields of catechols than extraction with water, but the compounds identified in the extracts are often the same (Atkinson *et al.*, 1973a). The higher yields obtained by acidic solvents may be due partly to swelling of the cuticular material at low pH values, resulting in easier liberation of trapped compounds, and partly to hydrolysis of acid-labile bonds. The sclerotization precursors NADA and NBAD, and their hydroxylated derivatives NANE and NBANE, are typical extraction products, and 3,4-dihydroxybenzoic acid and 3,4-dihydroxybenzaldehyde have been extracted from several types of sclerotized cuticle; they are probably formed during the sclerotization process by extensive oxidative degradation of the side chain of the sclerotizing precursors. The precise reaction pathway for their formation is not known, but their occurrence in sclerotized cuticles indicates that the intracuticular environment is highly oxidative during the sclerotization process. The presence in cuticular extracts of dopamine and norepinephrine may be due to deacetylation of the acylated forms, or they may have been transferred directly from epidermal cells to the cuticle, either by active transport across the apical cell membrane or by passive leakage through the cell membrane. The black cuticles of the *Drosophila* mutants *black* and *ebony* contain elevated levels of dopamine (Wright, 1987), which must have been transferred to the cuticle in the non-acylated state.

Quite large amounts of ketocatechols, such as arterenone (**29**), DOPKET (**30**), *N*-acetylarterenone (**33**), 3,4-dihydroxyphenylglyoxal (**31**), 3,4-dihydroxyphenylglyoxylic acid (**32**), and 3,4-dihydroxyphenylketoethylacetate (**34**) (**Figure 8**), can be obtained from sclerotized cuticle by hydrolysis with dilute hydrochloric acid (Andersen, 1970, 1971; Andersen and Barrett, 1971; Andersen and Roepstorff, 1978). The yields of ketocatechols can amount to several percent of the cuticular dry weight (Andersen, 1975; Barrett, 1977), and the type of ketocatechols obtained depends upon the exact conditions of hydrolysis. It has been argued that they are degradation products of a common precursor in the cuticle (Andersen, 1971).

More complex catechol derivatives can be obtained by using milder conditions to extract sclerotized cuticle, such as cold concentrated formic acid or boiling dilute acetic acid. From such extracts a number of dimeric compounds of the dihydroxyphenyl-dihydrobenzodioxine type were isolated and identified (Andersen and Roepstorff, 1981; Roepstorff and Andersen, 1981), and a related trimeric compound (**35**) (**Figure 9**) was obtained by formic acid extraction of sclerotized locust cuticle (Andersen *et al.*, 1992a). Ketocatechols are readily formed when such

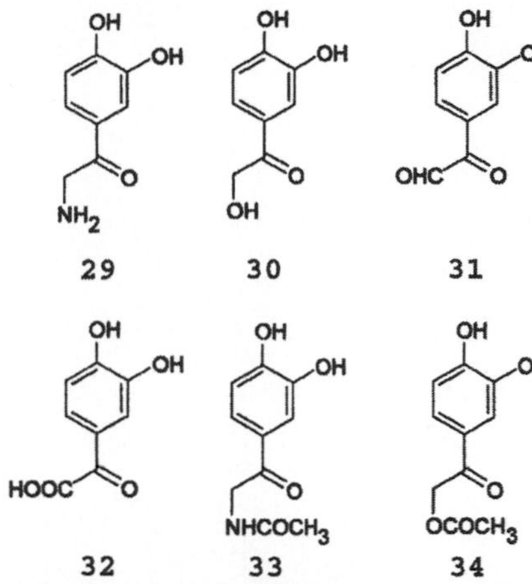

Figure 8 Ketocatechols obtained by acid hydrolysis of sclerotized cuticle. (**29**): arterenone; (**30**): 3,4-dihydroxyketoethanol (DOPKET); (**31**): 3,4-dihydroxyphenylglyoxal; (**32**): 3,4-dihydroxyphenylglyoxylic acid; (**33**): N-acetylarteremone; (**34**): O-acetyl-dihydroxyphenylketoethanol.

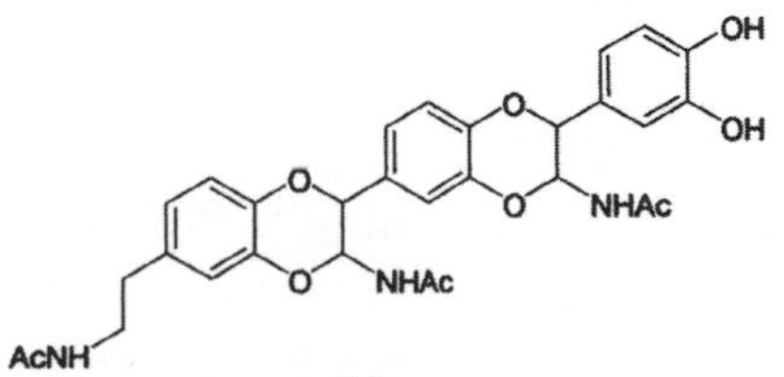

Figure 9 Structure of NADA-trimer.

dimers and oligomers are hydrolyzed with acid (Andersen and Roepstorff, 1981), but the amount of ketocatechols obtained by hydrolysis of the extracted benzodioxine derivatives is only a small fraction of the amount produced by acid hydrolysis of intact sclerotized cuticle, indicating that the major part of ketocatechols obtainable from cuticle is derived from catecholic material covalently linked to the cuticular proteins and chitin. Since the various benzodioxine dimers can be formed *in vitro* by reacting oxidized dehydro-NADA with catechols, it is likely that oxidized dehydro-NADA will react with protein-linked catechols to form protein-linked benzodioxine dimers and higher oligomers.

Acid hydrolysis of *M. sexta* pupal cuticle has yielded adducts containing histidine linked to either the C-6 ring position (**36**) or the β-position (**37**) of dopamine, or to the corresponding positions in DOPET (**38** and **39**) (**Figure 10**), demonstrating that adduct formation to histidine residues occurs during *in vivo* sclerotization (Xu *et al.*, 1997; Kerwin *et al.*, 1999), and indicating that DOPET may have a role as a sclerotization precursor. Direct evidence that covalent bonds are formed between acyldopamines and histidine residues during sclerotization had previously been obtained by solid state NMR studies, utilizing incorporation of isotopically

labeled dopamine, histidine, and β-alanine into sclerotizing pupal cuticle of *M. sexta* (Schaefer *et al.*, 1987; Christensen *et al.*, 1991). The NMR spectra demonstrated the presence of bonds between nitrogen atoms in the imidazole ring of histidine and ring-positions or the β-position of the dopamine side chain. Formation of covalent bonds involving the amino group of β-alanine and the ε-amino group of lysine was also indicated. Furthermore, catecholamine-containing proteins from sclerotizing *M. sexta* pupal cuticle have been purified and partially characterized, and NBANE was released from these proteins by mild acid hydrolysis, indicating the presence of a bond between the β-position of the side chain of NBAD and some amino acid residue in the proteins (Okot-Kotber *et al.*, 1996).

Several amino acid-containing adducts, some of which have a ketocatecholic structure, were obtained by acid hydrolysis of locust and *Tenebrio* sclerotized cuticle. Some of the adducts contained histidine residues linked via their imidazole ring to the β-position of various catechols, such as dopamine (**37**), DOPET (**39**), and 3,4-dihydroxyphenyl-acetaldehyde (DOPALD, **40**) (Andersen and Roepstorff, 2007; **Figure 10**). Other adducts were derivatives of 3,4-dihydroxyacetophenone with an amino acid residue linked to the α-carbon atom, and the amino acid could be glycine and β-alanine (**41** and **42**) linked via their amino groups, lysine linked via its ε-amino group (**43**), or tyrosine linked via its phenolic group (**44**) (Andersen, 2007; Andersen and Roepstorff, 2007; **Figure 11**). Several more amino acid-containing adducts were released during the hydrolysis, but they have not been identified. No adducts with an amino acid residue linked to the aromatic ring have yet been obtained from hydrolysates of *S. gregaria* and *T. molitor* sclerotized cuticles, and neither have adducts containing histidine linked to the α-carbon atom of 3,4-dihydroxyacetophenone been observed, whereas histidine is the only amino acid that so far has been found linked to the β-position of the side chain.

The adducts containing an amino acid residue linked to 3,4-dihydroxyacetophenone are a sort of ketocatechol, and presumably the keto-group is formed during hydrolysis of the cuticles, as keto-groups are not normally present in sclerotized cuticle. The adducts have been suggested to be hydrolytic degradation products of cross-linking materials, consisting of NADA residues linked to cuticular proteins via both α- and β-positions (Andersen, 2007; Andersen and Roepstorff, 2007). Such cross-links will be formed if oxidized dehydro-NADA residues react with two nucleophilic amino acids in the cuticular proteins, and a scheme was proposed for how such cross-links are degraded to ketocatecholic adducts during acid hydrolysis.

The histidine-dopamine β-adduct is presumably a hydrolytic degradation product of NADA and NBAD adducts formed when the corresponding *p*-quinone

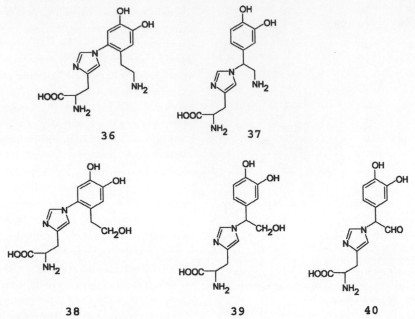

Figure 10 Structure of histidine-containing catecholic adducts obtained by hydrolysis of sclerotized cuticles.

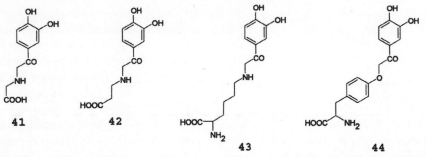

Figure 11 Structure of amino acid-containing derivatives of 3, 4-dihydroxyacetophenone obtained by hydrolysis of sclerotized cuticles.

methides reacted with histidine residues in the cuticular proteins. Such protein-linked catechols may possibly be reoxidized to quinones and react with another nucleophilic group, thereby forming a cross-link, but no evidence has yet been presented for the existence of such cross-links.

Quantitative determination of the various amino acid-containing adducts obtained by hydrolysis of 16 different types of sclerotized cuticle showed that cuticles from adult locusts and cockroaches yielded large amounts of ketocatecholic adducts and only little β-histidine-dopamine, indicating that utilization of dehydro-NADA is the main pathway for their sclerotization. Cuticles from *M. sexta* and *H. cecropia* pupae, *T. molitor* larvae and pupae, and *C. vicina* puparia gave mainly β-histidine-dopamine and only small amounts of the ketocatecholic adducts, indicating that sclerotization of these cuticles mainly involves *p*-quinone methides. Adult beetles (*T. molitor* and *Pachynoda sinuata*) gave significant amounts of both types of adducts, indicating that both *p*-quinone methides and dehydro-NADA are to a significant extent involved in the sclerotization (Andersen, 2008).

The number of insect cuticles that have been analyzed is too small to allow any firm conclusions, but the results obtained so far indicate that the *p*-quinone methide of dehydro-NADA is the main sclerotizing agent of cuticles where NADA is the only or the dominating sclerotization precursor, and that the *o*-quinones and *p*-quinone methides formed from NADA in these cuticles mainly play roles as precursors for dehydro-NADA. The *o*-quinones and *p*-quinone methides of NBAD appear to be the dominating sclerotization agents in those cuticles where NBAD is the important precursor for sclerotization. In cuticles where both NADA and NBAD are involved to a significant extent in sclerotization, NBAD *o*-quinones and *p*-quinone methides, as well as dehydro-NADA *p*-quinone methide, will contribute more or less equally to the process.

6.6. Control of Sclerotization

The mechanical properties of the various cuticular regions tend to differ significantly, varying from the very hard and resistant mandibles to the soft and flexible arthrodial membranes, and the cuticular properties are likely to be optimized with respect to the functions of the regions. The mechanical properties of cuticles are to a major extent determined by the degree of sclerotization, indicating that initiation as well as duration and degree of sclerotization are precisely controlled. In some cases initiation of sclerotization is controlled by hormones: ecdysone in the case of puparium sclerotization; and bursicon in the case of post-ecdysial sclerotization. The duration of sclerotization may also be hormonally controlled, perhaps by a decrease in hormone titer. The amounts of sclerotizing material incorporated per milligram of cuticle can be used as a measure of the degree of sclerotization, and is probably determined by local factors and not by hormones. Factors such as the amounts of sclerotizing precursors available and the capacity of the epidermis to transport sclerotizing precursors into the cuticle are likely to be important. The degree of sclerotization is apparently not determined by the amount cross-linking enzymes present in the cuticle, since the enzyme activity in various cuticular regions of adult locust does not correlate with the amounts of keto-catechols obtainable (Andersen, 1974b).

6.6.1. Post-Ecdysial Sclerotization

Many cuticular regions are expanded to a new and larger size when the insect has emerged from the old cuticle (Cottrell, 1964), and most insects start cuticular expansion as soon as they have escaped from the exuvium; however, in some insects the period from end of ecdysis to fully expanded cuticle can be prolonged – for instance in flies, where the newly emerged adult has to dig its way through the substratum in which it pupariated before it can expand and harden its wings. The signal for initiating sclerotization of the expanded cuticle after ecdysis is release of the neurohormone bursicon from the central nervous system. Bursicon has a pronounced influence on the activities of the epidermal cells; it has been reported that absence of bursicon results in the failure of endocuticle deposition, as well as lack of melanin production and of sclerotization of the cuticle (Fraenkel and Hsiao, 1965; Fogal and Fraenkel, 1969), and it was suggested that bursicon is involved in the control of tyrosine hydroxylation to DOPA (Seligman et al., 1969). The molecular structure of bursicon was later established to be a heterodimer of two cystine knot protein subunits (Luo et al., 2005; Mendive et al., 2005), and it was shown that the hormone is responsible for initiating a number of processes occurring in insects immediately or soon after ecdysis (Dai et al., 2008; Honegger et al., 2008), one of them being

activation of the epidermal tyrosine hydroxylase by phosphorylation of a serine residue (Davis et al., 2007).

In some insects sclerotization stops when the pre-ecdysially deposited cuticle has been sclerotized to form exocuticle, although deposition of endocuticle may continue for several days, resulting in a sclerotized exocuticle and a non-sclerotized endocuticle. In other insects cuticular sclerotization continues during endocuticle deposition, with the result that both exo- and endocuticle become sclerotized, although not to the same extent. Sclerotization of femur cuticle in adult locusts (S. gregaria) continues for at least 10–12 days after ecdysis, and both exo- and endocuticle are sclerotized (Andersen and Barrett, 1971), in contrast to sclerotization of femur cuticle of fifth instar nymphs of the same species, which lasts for only a single day, and deposition of unsclerotized endocuticle continues for about 4–5 days (Andersen, 1973). Accordingly, the endocuticular proteins are readily extracted from femurs of mature nymphs, whereas little protein can be extracted from femurs of mature adults. This difference is probably related to the different fate of these two types of cuticle. The nymphal cuticle will to a large extent be degraded in preparation for the next ecdysis, and sclerotized cuticle is more resistant to enzymatic degradation than non-sclerotized cuticle. The adult cuticle has to last for the remaining life of the animal, and there is no apparent advantage in having an easily degradable endocuticle. The leg cuticle of adult locusts is also exposed to stronger mechanical forces than the cuticle of nymphal legs, and so it may be an advantage for adult locusts to have both layers of the leg cuticle sclerotized, although not to the same extent. A similar difference in sclerotization in adults and younger instars is probably present in other insect species. It has not yet been established how the duration of the post-ecdysial sclerotization period is controlled, but it could well be by regulation of the bursicon titer.

6.6.2. Pre-Ecdysial Sclerotization

The sclerotization of some cuticular regions may start during the pharate stage before the insect has ecdysed, and after ecdysis such regions will have retained the size and shape they obtained during the pre-ecdysial deposition of cuticular materials (Cottrell, 1964). Pre-ecdysial sclerotization may be limited to small local regions, such as mandibles and spines, or to larger cuticular regions, covering thorax, head, and legs. Further sclerotization of such already sclerotized regions can continue after ecdysis, so that the already stiffened material is further strengthened. Pre-ecdysial sclerotization has been studied in several cuticular regions in adult honeybees (Apis mellifera adansonii) (Andersen et al., 1981); the matrix proteins in these regions resist extraction with solvents that do not degrade the cuticle, indicating that the proteins have become cross-linked. Acid hydrolysis of pre-ecdysially sclerotized pharate cuticles sampled during ecdysis yielded significant amounts of ketocatechols, and still more was obtained by hydrolysis of the corresponding cuticular regions from mature worker bees, where endocuticle deposition was complete, indicating that similar types of sclerotization are involved in both pre- and post-ecdysial stabilization. No pre-ecdysial sclerotization was observed in wings; their sclerotization occurred rapidly after ecdysis and was nearly complete when the bees left the cell in which they had pupated, indicating that insect wings are not stabilized before they have been expanded to their proper size after emergence. In contrast to the other cuticular regions in bees, the sclerotization of the wing cuticle is probably controlled by release of the neurohormone bursicon, as described for the wings in other insect species (Honegger *et al.*, 2008).

6.6.3. Puparial Sclerotization

The soft, pliable cuticle of the last larval instar of higher Diptera is modified to a hard and non-deformable material during puparium formation. Pupal and adult development occurs inside the puparium, the function of which is mainly to protect the animal during the developmental processes. No pronounced regional differentiation in cuticular sclerotization appears to occur during puparium formation. The biochemical processes of puparium sclerotization are very similar to those involved in the sclerotization of adult cuticle, but the two systems differ in how they are controlled (Sekeris, 1991). Puparium sclerotization in blowflies begins with an increase in ecdysteroid titer, which induces the expression of the enzyme DOPA decarboxylase, catalyzing the decarboxylation of DOPA to dopamine. The latter is then acylated to the sclerotization precursors NADA and NBAD, which are utilized for sclerotization by processes identical to those used for sclerotization of cuticle from adult insects. The flies *Musca autumnalis* and *M. fergusoni* harden their puparia by the deposition of calcium and magnesium phosphates; phenolic sclerotization seems not to be involved (Gilby and McKellar, 1976; Darlington *et al.*, 1983).

6.6.4. Balance Between Cuticular Enzymes

The sclerotization occurring in the local cuticular regions depends partly upon the available amounts of the two precursor compounds NADA and NBAD, and partly upon the balance between the various enzyme activities in the cuticle. An attempt to study some of these questions was made more than 30 years ago (Andersen, 1974a, 1974b), but the methods available at that time were insufficient to give conclusive answers. To study the extent to which the ring or the side chain were involved in adduct formation, NADA labeled with tritium either on the aromatic ring or on the β-position of the side chain was used, as it was

assumed that sclerotization via an *o*-quinone will result in tritium release from the ring, whereas release of tritium from the NADA side chain was expected to occur without quinone formation. Different types of cuticle were found to differ in their ability to release tritium from the two positions; cuticles that were lightly colored after sclerotization mainly released tritium from the side chain, and dark-brown cuticles released tritium mainly from the ring system (Andersen, 1974a). It is now evident that tritium release from the β-position of the side chain occurs during formation of the *p*-quinone methide of NADA and will be dependent on the amounts of *o*-quinone isomerase activity present, and that tritium release from the aromatic ring will occur during adduct formation between an *o*-quinone and a nucleophilic compound. Tritium release from the aromatic ring will be a measure of the fraction of NADA which after oxidation by cuticular diphenoloxidases is not isomerized to a *p*-quinone methide, and tritium release from the β-position of the side chain will be a combined measure of *p*-quinone methide formation and dehydro-NADA formation. Tritium release from the α-position will be a measure of dehydro-NADA formation.

Locust (*S. gregaria*) femur cuticle was found to release little tritium from the ring system of NADA, and significant and nearly equal amounts of tritium from the α- and β-positions of the NADA side chain (Andersen, 1974a and unpublished data), indicating that most of the NADA used for sclerotization by this insect is converted to 1,2-dehydro-NADA. Since NADA and NBAD appear to be treated differently during cuticular sclerotization, it will be necessary to use α- as well as β-tritiated forms of both NADA and NBAD to determine the quantitative importance of *p*-quinone methide sclerotization relative to sclerotization involving the dehydro-derivatives. Such determinations have, to my knowledge, not yet been reported.

6.6.5. Intensity of Cuticular Sclerotization

As the mechanical properties of cuticles are important for proper function, it would be an advantage to be able to determine the degree of sclerotization of well-defined cuticular regions. One way of doing this would be to determine the amounts of NADA and NBAD which are being incorporated, but it is difficult to obtain a precise measure of how much is really incorporated, how much of it forms cross-links, and how much forms mono-adducts. Solid state NMR studies may be used for such measurements (Schaefer *et al.*, 1987; Christensen *et al.*, 1991), but the techniques are not generally available. Determinations of degree of sclerotization of various cuticular samples were attempted by measuring the amounts of neutral ketocatechols released by acid hydrolysis from cuticular samples (Andersen, 1974b, 1975; Barrett, 1977, 1980), but this method only determines the part of total sclerotization that results in formation of polymeric dehydro-NADA oxidation products. An indication of the quantitative role of NBAD might be obtained by determining the amounts of β-alanine released during acid hydrolysis, but some β-alanine will probably not be released by hydrolysis since it may take part in formation of acid-stable adducts (Christensen *et al.*, 1991). Although analyses for ketocatechols and β-alanine cannot give the absolute values, they can indicate the relative roles of the two sclerotization types they represent.

Ketocatechols were not obtained from pharate adult femur cuticle of the locust *S. gregaria*, but later, during maturation of the locusts, a steady increase in yield of ketocatechols was observed, reaching a constant level after about 1 week, when the ketocatechol yield was about 0.2 μmol/mg dry cuticle (Andersen and Barrett, 1971), indicating a close relationship between ketocatechol formation and sclerotization in locust cuticle.

All cuticular regions of 10- to 12-day-old locusts yielded some ketocatechols upon hydrolysis: the lowest yields were obtained from abdominal intersegmental membranes (0.02% of the dry weight); abdominal sclerites gave 0.38% of the dry weight; and the highest values were obtained from the dorsal mesothorax (5.25% of the dry weight) and the mandibles (3.36% of the dry weight) (Andersen, 1974b). The results are in agreement with the expectation that regions where strength and hardness are essential for proper function are also the regions giving the highest yields of ketocatechols.

S. gregaria mature nymphal cuticle gave much lower amounts of ketocatechols than corresponding samples from mature adults, probably because nymphal cuticle is not exposed to deforming forces comparable to those experienced by adult cuticle during flight. The only exception was the nymphal mandibles; 3.7% of their dry weight was recovered as ketocatechols, similar to what was obtained from adult mandibles (Andersen, 1974b).

Quantitative ketocatechol determinations have been performed on several other insect species, such as the beetles *Tenebrio molitor* and *Pachynoda epphipiata* (Andersen, 1975). Most regions of the exuviae of a cicada, *Tibicen pruinosa*, gave ketocatechol values between 6% and 7% of the exuvial dry weight (Barrett, 1977), but exuvial cuticle from the cicada compound eyes gave a ketocatechol yield (11.8% of the dry weight). The relatively high values obtained from exuviae agree with the notion that sclerotization is generally more pronounced in exocuticle than in endocuticle. Barrett (1984b) reported that hydrolysis of the head capsule from larvae of the red-humped oakworm, *Symmerista cannicosta*, gave 4.6% of the most abundant ketocatechol (2-hydroxy-3′,4′-dihydroxyacetophenone, DOPKET), and that exuviae from the larval body gave only 0.3% of this compound. The ketocatechols obtained from various soft cuticles are most likely derived from the epicuticle.

An attempt has been made to obtain a measure of the relative rate of sclerotization by determining the amounts of radioactivity incorporated into the various cuticular regions after injection of a single dose of labeled dopamine or NADA (Andersen, 1974b). This can give a picture of the competition of the various regions for the available sclerotization precursor at the time of injection, but it cannot measure the total sclerotization occurring in the regions. Soon after ecdysis, labeled dopamine was injected into nymphal and adult *S. gregaria*; 24 hours later the amounts of radioactivity incorporated into the various cuticular regions were determined. Good agreement was observed between the amounts of radioactivity incorporated and the amounts of ketocatechols that could be recovered from corresponding regions of non-injected animals of the same age. The only exception was that less ketocatechol was obtained from adult mandibles than expected from their ability to incorporate radioactive dopamine. The mandibles are dark brown, and their sclerotization is presumably different from that of the uncolored body cuticle. The results indicate that the yield of ketocatechols can be a useful measure of degree of sclerotization in some types of cuticle.

The regional pattern of incorporation of labeled NADA into locust cuticle was similar to that obtained for labeled dopamine, both patterns changed similarly during maturation of the animals (Andersen, 1974b), and the incorporation of NADA decreased with time, but with markedly different velocities in the different regions.

No convincing correlation was observed between the rate at which the cuticular regions released tritium from β-labeled NADA *in vitro* and either the yield of ketocatechols obtained by hydrolysis of these regions or the *in vivo* uptake of labeled dopamine and NADA, suggesting that it is not the amount of sclerotizing enzymes that is the main determining factor for the degree of sclerotization, but the local availability of sclerotizing precursors.

6.7. Cuticular Darkening

Cuticular sclerotization is often accompanied by darkening of the cuticle to various shades of brown and black. Completely black colors are likely to be due to deposition of melanins formed by oxidation of dopamine, and formation of brownish colors is closely connected to the process of sclerotization. Melanin-like materials are of common occurrence in cuticular structures, and they may be present as microscopic granules or be distributed homogeneously within the cuticular matrix (Kayser-Wegmann, 1976; Kayser-Wegmann and Kayser, 1983; Hiruma and Riddiford, 1988). The granules are produced within epidermal cells and transported to the subepicuticular space via long cellular projections (Curtis *et al.*, 1984; Kayser, 1985), and the granular melanin appears to be linked to granular proteins and not to proteins in the cuticular matrix. The evenly distributed melanins seem to be formed *in situ* within the cuticular matrix, and covalent links are probably present between the polymeric melanin and matrix proteins, thus rendering the proteins more stable and insoluble, and contributing to both darkening and increased mechanical stiffness of the cuticle. In some cases, it can be difficult to discern between the process of melanization and the process of sclerotization.

No melanins are deposited in the cuticle of an albino mutant of the locust *S. gregaria* (Malek, 1957), and cuticular sclerotization is not affected; cuticular incorporation of radioactive tyrosine is nearly the same in wild type and albino mutant of *S. gregaria* (Karlson and Schlossberger-Raecke, 1962), showing that melanization does not play a major role in sclerotization. Injection of a tyrosinase inhibitor, phenylthiourea, into larvae of *Protoformia terraenovae* prevents melanin deposition during puparium formation, but does not affect hardening or the appearance of brown color (Dennell, 1958). However, handling of small samples of wild type and albino cuticle from locust nymphs indicates that melanization may somehow influence the mechanical properties, as the melanized samples appear to be more brittle than the unmelanized samples (S.O. Andersen, unpublished data). A quantitative comparison of the mechanical properties may reveal to what extent melanization contributes to the physical properties of the material.

Melanins can be formed by polymerization of oxidation products of either DOPA or dopamine; oxidation of the two precursors leads to formation of dopachrome and dopamine chrome, respectively, which can be transformed into 5,6-dihydroxyindole by the enzyme, dopachrome conversion factor. Dihydroxyindole can be polymerized via formation of an *o*-quinone to darkly colored melanins, and intermediates formed in the process may react with nucleophilic compounds; the presence of sulfur-containing amino acids may result in formation of reddish-brown melanins, and it is likely that other nucleophilic compounds may also be involved.

Dopamine appears to be the main precursor for cuticular melanins, whereas DOPA is the precursor for the melanins formed during wound healing and hemolymphal defense reactions. In *Drosophila*, two members of the *yellow* gene family, *yellow-f* and *yellow-f2*, code for dopachrome-conversion enzymes (Han *et al.*, 2002), and mutations of these genes block formation of cuticular melanin. The *yellow-y* gene product is also necessary for cuticular melanin formation, but appears to be without dopachrome-conversion activity.

The black body color of the *D. melanogaster* mutants *black* and *ebony* is due to the inability of these mutants to produce sufficient NBAD for cuticular sclerotization: *black* is defective in the synthesis of β-alanine, and can be rescued by injection of β-alanine; while *ebony* is defective in the enzyme NBAD-synthetase, and cannot be rescued

by injection of β-alanine (Wright, 1987). The result is that in both mutants some of the dopamine which cannot be used for NBAD synthesis is used for cuticular melanin production. Both stiffness and puncture-resistance of the cuticle are decreased in the mutants due to inefficient sclerotization, and electron microscope studies show that the cuticular chitin lamellae are abnormally wide and diffuse (Jacobs, 1978, 1980, 1985), indicating that even if dopamine-derived melanin can take part in cuticular stabilization, the result is inferior to the material obtained by NBAD sclerotization.

The *tan* mutant of *D. melanogaster*, which is characterized by absence of the wild type cuticular melanin pattern, has low activity of the enzyme *N*-β-alanyldopamine-hydrolase, which catalyzes the hydrolysis of NBAD to β-alanine and dopamine (Wright, 1987). The enzyme systems responsible for NBAD synthesis and NBAD hydrolysis, respectively, are probably located in different compartments, and the presence of melanin in some but not all cuticular regions of wild type fruit flies can be explained by the presence of the NBAD-hydrolase within the matrix of the melanizing regions, where it will hydrolyze some of the NBAD secreted from the epidermal cells, creating a dopamine concentration sufficient to stimulate a localized melanin production.

The formation of brown colors appears to be directly linked to cuticular sclerotization, and several suggestions have been put forward to explain why some cuticles become brown during sclerotization while other cuticles remain nearly colorless. It has been suggested that a correlation exists between the use of NBAD as a sclerotization precursor and the intensity of brown color of the sclerotized cuticle (Brunet, 1980; Morgan *et al.*, 1987; Hopkins and Kramer, 1992), and it has also been suggested that a dark-brown cuticular color indicates that the aromatic rings of the sclerotization precursors are linked directly to cuticular proteins via quinone formation, whereas formation of links between the NADA side chain and the proteins results in a colorless cuticle (Andersen, 1974a). The suggestion was based upon the observation that colorless or lightly colored cuticles preferentially release tritium from the side chain of NADA, and that dark-brown cuticles preferentially release tritium from the ring positions.

The colorless benzodioxine-type compounds, which on acid hydrolysis yield ketocatechols, are preferentially formed from NADA and only to a minor extent, if at all, from NBAD (Andersen, 1989b). Most of the NADA residues will rapidly be processed to dehydro-NADA, resulting in formation of colorless protein-linked derivatives. During sclerotization NBAD can be expected to form links to cuticular proteins via ring positions and the β-position of the side chain, and some NBAD-derived quinones may also react with the amino group from another NBAD residue, or maybe with its own amino group. The formation of brown color due to reactions involving the aromatic rings will thus depend both on the amounts of available NBAD, and on the balance between quinone-forming enzymes and the enzyme activity catalyzing formation of dehydro-NADA (Andersen, 1989c, 1989d).

6.8. Cuticular Sclerotization in Insects Compared to That in Other Arthropods

The release of ketocatechols during acid hydrolysis can be used as an indication of the involvement of dehydro-NADA in sclerotization, and has been applied to determine how widespread the occurrence is of this variant of cuticular sclerotization. Ketocatechols have been obtained in varying amounts from cuticular samples of all pterygote insects studied so far, and the wings were found to be especially good sources of ketocatechols. None of the apterygote insects analyzed, representing Thysanura, Collembola, and Diplura, gave any measurable amounts of ketocatechols, and neither did the sclerotized cuticle of non-insectan arthropods, such as Decapoda, Isopoda, Araneae, Xiphosura, and Acarina (Andersen, 1985). The presence of ketocatechol-yielding material in only pterygote cuticles indicates that development of the ability to fly occurred together with the development of the use of dehydro-NADA in sclerotization, resulting in a form of sclerotin which combines strength, toughness, and lightness to an optimal degree for flight purposes. The suggestion needs to be investigated in more detail, and a more detailed characterization of the sclerotization process(es) in cuticle of non-insectan arthropods is also needed. Little is known of sclerotization in crustaceans, and even less is known for other arthropod groups.

6.9. Unsolved Problems

The schemes for sclerotization of insect cuticle shown in **Figures 1** and **2** can account for most of the observations and experimental results that have been reported. It is likely that we have now obtained a reasonable understanding of the main features of the chemical processes occurring during sclerotization, but some observations are difficult to reconcile with the suggested scheme. This may be due to faulty observations or errors in interpretation, but it is also possible that different types of sclerotization occur in some cuticles serving specialized functions, or there may be essential, but unrecognized, elements in the common sclerotization scheme. In any case, the unexplained observations deserve a critical study before our understanding of the chemistry of sclerotization is satisfactory. It is a weakness in the suggested sclerotization scheme that it builds upon a combination of individual reactions, which were characterized in different types of cuticles from relatively few insect species, and we have not yet obtained sufficient evidence that all the reported reactions, and only these reactions, are of general occurrence in cuticular sclerotization.

6.9.1. Alternative Pathway for Dehydro-NADA Formation

The sclerotization process is depicted in **Figure 2** as a linear reaction chain, where reactive intermediates formed at different steps in the chain can react with and modify cuticular proteins, and it is proposed that the different sclerotization patterns observed in various types of cuticles depend mainly upon the absolute and relative amounts of the two sclerotization precursors, NADA and NBAD, and the enzyme activities involved. When the activity of o-quinone isomerase is larger than the p-quinone methide tautomerase activity, p-quinone methides are expected to be produced more rapidly than they are isomerized to dehydro-NADA, and this will favor reactions between the p-quinone methide and proteins. If the tautomerase is the most active enzyme, only a small fraction of the NADA p-quinone methides will have an opportunity to react with proteins before they are isomerized to dehydro-NADA.

A decrease in the consumption of NADA and an accumulation of dehydro-NADA is observed when the laccase in locust cuticle is inhibited by sodium azide (Andersen, 1989c), indicating that sodium azide inhibits the oxidation of dehydro-NADA more than the oxidation of NADA. This was an unexpected observation, since dehydro-NADA is a better substrate for the cuticular laccase than is NADA (Andersen and Roepstorff, 1982). To explain the observation it was suggested that dehydro-NADA is formed not by isomerization of the p-quinone methide, but directly from NADA by means of a special enzyme, a desaturase (Andersen and Roepstorff, 1982; Andersen *et al.*, 1996). The observation can also be explained by assuming that oxidation of NADA and dehydro-NADA is catalyzed by two different enzymes (probably both being laccases), and that the dehydro-NADA oxidizing enzyme is much more sensitive to azide inhibition than the NADA oxidizing enzyme. This could result in accumulation of dehydro-NADA, but will demand a strict compartmentalization for the two enzyme activities, as the surplus of dehydro-NADA will have to be prevented from gaining access to the NADA oxidizing laccase (Andersen, 2010).

6.9.2. Extracuticular Synthesis of Catechol–Protein Conjugates for Sclerotization

Another problem to be studied in more detail is whether and how protein-bound catecholic derivatives are transferred from hemolymph to cuticle to participate in sclerotization. It has been reported that the epidermal cells can transfer proteins, arylphorins, from hemolymph to the cuticle (Scheller *et al.*, 1980; Schenkel *et al.*, 1983; König *et al.*, 1986; Peter and Scheller, 1991), and apparently such transfer can also occur for proteins to which catechols are linked (Koeppe and Mills, 1972; Koeppe and Gilbert,

1974; Bailey *et al.*, 1999). If the protein-linked catechols are oxidized to quinones after arrival in the cuticle, it can be assumed that they react with nucleophilic residues in the other proteins in the cuticle and thus participate in sclerotin formation, but so far it is not known for certain how, where, and whether such protein–catechol conjugates are formed, and whether, after transfer to the cuticle, they take part in sclerotization. Diffusion problems and steric hindrance may make it difficult for the protein-bound catecholic residues to get access to the cuticular diphenoloxidases, but they may be oxidized by encountering small, easily diffusable quinones formed by enzyme catalyzed oxidation of free low-molecular weight catechols.

6.9.3. Importance of Cuticular Dehydration

Fraenkel and Rudall (1940) reported a significant decrease in cuticular water content in connection with puparium sclerotization in blowflies, and later it was argued that formation of chemical cross-links cannot fully explain the changes in mechanical properties occurring during sclerotization, and that controlled dehydration of the cuticular matrix could be the most important factor in the stabilization of cuticle (Hillerton and Vincent, 1979; Vincent and Hillerton, 1979; Vincent, 1980). Dehydration may be caused by increased hydrophobicity of cuticular proteins due to reaction with the enzymatically formed quinones, by filling the initially water-filled interstices between protein molecules with polymerized catechols, and by water being actively transported out of the cuticle by some transport system residing in the apical cell membrane of the epidermis. All three mechanisms are likely to be involved in cuticular dehydration. Water transport coupled to active transport of ions appears to be the process which can be most precisely controlled, and may be the most important dehydration mechanism. Precise control of the content of water in various cuticles is certainly important for their mechanical properties, but it still seems unlikely that dehydration of cuticles can explain all the major changes occurring during sclerotization.

6.9.4. Lipids and Sclerotization

Lipids may play an essential role in cuticular sclerotization (Wigglesworth, 1985, 1988) – a possibility that tends to be overlooked, and should be studied in more detail. The epicuticle mainly consists of proteins and lipids connected to each other to form a thin, extremely resistant and inextractable layer. It is not known how the lipids and proteins are linked together, but the resistance towards hydrolytic degradation indicates that stable covalent bonds between lipids and proteins play an important role. Semiquinones and other free radicals are likely to be formed during laccase catalyzed oxidation of catechols to quinones, and

free radicals may react with unsaturated lipids, resulting in stable lipophenolic complexes. Such reactions could be part of the stabilization of the epicuticle, and they could contribute to making the connections between epicuticle and the underlying procuticle more stable and secure.

6.9.5. Chlorotyrosines in Sclerotized Cuticles

Halogen-containing derivatives of tyrosine are common in structural proteins from marine animals (Hunt, 1984), but have also been reported from insect sclerotized cuticles (Andersen, 1972b, 2004b). Analysis of various cuticular regions from mature *S. gregaria* showed that they all contained significant amounts of both 3-monochlorotyrosine and 3,5-dichlorotyrosine; the highest amounts were obtained from femur and tibia cuticle, and the lowest amounts were obtained from the mandibles. Moderate amounts of chlorotyrosines were obtained from arthrodial membranes and from tracheae sampled from nymphal exuviae. The formation of chlorotyrosines starts in the pharate stage of adult locusts, and the amounts obtained from a single femur increase rapidly during the first day after ecdysis; thereafter, the rate of increase continues at a lower rate for at least 16 days. Mono- and dichlorotyrosine have also been obtained from cuticle from larvae, pupae, and adults of *T. molitor* and *H. cecropia*, from cockroach nymphs (*Blaberus craniifer*), and from exuviae of *Rhodnius prolixus* (Andersen, 2004b).

Chlorination of tyrosine to mono- and dichlorotyrosine is an oxidative process and it occurs during the period of sclerotization, but no convincing correlation was observed between rate of sclerotization and rate of chlorination, indicating that formation of chlorotyrosines is not just a by-product of the sclerotization process. Incubation of pieces of locust cuticle together with NADA and sodium chloride did not increase their content of chlorotyrosines, although they readily incorporated NADA. The route for formation of cuticular chlorotyrosines is still unknown, and neither is it known what function they have in the cuticle, if any at all.

6.9.6. Metals in Cuticle

Some cuticular regions in insects can be stabilized by incorporation of metals, especially zinc and manganese (Hillerton and Vincent, 1982; Hillerton *et al.*, 1984; Quicke *et al.*, 1998; Schofield, 2001). The metals are mainly found in regions where hardness and wear resistance are important, such as the cutting edge of the mandibles in plant-feeding insects (Schofield *et al.*, 2002; Cribb *et al*, 2008) and the ovipositors of some hymenopterous insects (Quicke *et al.*, 1998). In adult leaf-cutter ants, the cuticular deposition of zinc is reported to be post-ecdysial and occurring after pigmentation of the mandibular teeth, suggesting that zinc deposition and sclerotization are separate hardening processes (Schofield *et al.*, 2002). It has been reported that the presence of metals increases the cuticular hardness (Schofield *et al.*, 2002; Cribb *et al*, 2008), but Cribb *et al.* (2010) report that the cutting edge of the larval mandibles of the beetle *Pseudotaenia frenchi*, which do not contain transition metals, is harder than that of mandibles of the adult beetle, which do contain manganese. It is not known how the larval mandibles obtain their extreme hardness. Little is known of the links between the metals and the cuticular matrix; maybe the metals form complexes with histidine residues in the cuticular proteins, and maybe they form complexes with catecholic groups introduced during cuticular sclerotization. So far, transition metals have only been reported from some specialized and sclerotized regions, and incorporation of metals can probably be considered as a way of stabilizing cuticles supplementing sclerotization.

References

Andersen, S. O. (1966). Covalent cross-links in a structural protein, resilin. *Acta Physiol.*, *66*(Suppl. 263), 1–81, Scand.

Andersen, S. O. (1970). Isolation of arterenone (2-amino-3′,4′-dihydroxyacetophenone) from hydrolysates of sclerotized insect cuticle. *J. Insect Physiol.*, *16*, 1951–1959.

Andersen, S. O. (1971). Phenolic compounds isolated from insect hard cuticle and their relationship to the sclerotization process. *Insect Biochem.*, *1*, 157–170.

Andersen, S. O. (1972a). An enzyme from locust cuticle involved in the formation of cross-links from *N*-acetyldopamine. *J. Insect Physiol.*, *18*, 527–540.

Andersen, S. O. (1972b). 3-Chlorotyrosine in insect cuticular proteins. *Acta Chem. Scand.*, *26*, 3097–3100.

Andersen, S. O. (1973). Comparison between the sclerotization of adult and larval cuticle in *Schistocerca gregaria*. *J. Insect Physiol.*, *19*, 1603–1614.

Andersen, S. O. (1974a). Evidence for two mechanisms of sclerotization in insect cuticle. *Nature*, *251*, 507–508.

Andersen, S. O. (1974b). Cuticular sclerotization in larval and adult locusts, *Schistocerca gregaria*. *J.Insect Physiol.*, *20*, 1537–1552.

Andersen, S. O. (1975). Cuticular sclerotization in the beetles, *Pachynoda epphipiata* and *Tenebrio molitor*. *J.Insect Physiol.*, *21*, 1225–1232.

Andersen, S. O. (1978). Characterization of trypsin-solubilized phenoloxidase from locust cuticle. *Insect Biochem.*, *8*, 143–148.

Andersen, S. O. (1979a). Biochemistry of insect cuticle. *Annu. Rev. Entomol.*, *24*, 29–61.

Andersen, S. O. (1979b). Characterization of the sclerotization enzyme(s) in locust cuticle. *Insect Biochem.*, *9*, 233–239.

Andersen, S. O. (1981). The stabilization of locust cuticle. *J. Insect Physiol.*, *27*, 393–396.

Andersen, S. O. (1985). Sclerotization and tanning of the cuticle. In G. A. Kerkut, & L. I. Gilbert (Eds.). *Comprehensive Insect Physiology, Biochemistry, and Pharmacology* (Vol. 3, pp. 59–74). Oxford, UK: Pergamon Press.

Andersen, S. O. (1989a). Investigation of an ortho-quinone isomerase from larval cuticle of the American silkmoth, *Hyalophora cecropia*. *Insect Biochem.*, *19*, 803–808.

Andersen, S. O. (1989b). Oxidation of *N*-β-alanyldopamine by insect cuticles and its role in cuticular sclerotization. *Insect Biochem.*, *19*, 581–586.

Andersen, S. O. (1989c). Enzymatic activities in locust cuticle involved in sclerotization. *Insect Biochem.*, *19*, 59–67.

Andersen, S. O. (1989d). Enzymatic activities involved in incorporation of *N*-acetyldopamine into insect cuticle during sclerotization. *Insect Biochem.*, *19*, 375–382.

Andersen, S. O. (1990). Sclerotization of insect cuticle. In E. Ohnishi, & H. Ishizaki (Eds.). *Molting and Metamorphosis* (pp. 133–155). Tokyo, Japan: Japan Scientific Societies Press.

Andersen, S. O. (2004a). Regional differences in degree of resilin cross-linking in the desert locust, *Schistocerca gregaria*. *Insect Biochem. Mol. Biol.*, *34*, 459–466.

Andersen, S. O. (2004b). Chlorinated tyrosine derivatives in insect cuticle. *Insect Biochem. Mol. Biol.*, *34*, 1079–1087.

Andersen, S. O. (2005). Cuticular sclerotization and tanning. In L. I. Gilbert, K. Iatrou, & S. S. Gill (Eds.). *Comprehensive Molecular Insect Science* (Vol. 4, pp. 145–170). Oxford, UK: Elsevier Pergamon Press.

Andersen, S. O. (2007). Involvement of tyrosine residues, *N*-terminal amino acids, and β-alanine in insect cuticular sclerotization. *Insect Biochem. Mol. Biol.*, *37*, 969–974.

Andersen, S. O. (2008). Quantitative determination of catecholic degradation products from insect sclerotized cuticles. *Insect Biochem. Mol. Biol.*, *38*, 877–882.

Andersen, S. O. (2010). Insect cuticular sclerotization: A review. *Insect Biochem. Mol. Biol.*, *40*, 166–178.

Andersen, S. O., & Barrett, F. M. (1971). The isolation of keto-catechols from insect cuticle and their possible role in sclerotization. *J. Insect Physiol.*, *17*, 69–83.

Andersen, S. O., & Roepstorff, P. (1978). Phenolic compounds released by mild acid hydrolysis from sclerotized cuticle: Purification, structure, and possible origin from cross-links. *Insect Biochem.*, *8*, 99–104.

Andersen, S. O., & Roepstorff, P. (1981). Sclerotization of insect cuticle. 2. Isolation and identification of phenolic dimers from sclerotized insect cuticle. *Insect Biochem.*, *11*, 25–31.

Andersen, S. O., & Roepstorff, P. (1982). Sclerotization of insect cuticle. 3. An unsaturated derivative of *N*-acetyldopamine and its role in sclerotization. *Insect Biochem.*, *12*, 269–276.

Andersen, S. O., & Roepstorff, P. (2007). Aspects of cuticular sclerotization in the locust, *Schistocerca gregaria*, and the beetle, *Tenebrio molitor*. *Insect Biochem. Mol. Biol.*, *37*, 3, 223–234.

Andersen, S. O., Thompson, P. R., & Hepburn, H. R. (1981). Cuticular sclerotization in the honey bee (*Apis mellifera adansonii*). *J. Comp. Physiol.*, *145*, 17–20.

Andersen, S. O., Jacobsen, J. P., Roepstorff, P., & Peter, M. G. (1991). Catecholamine–protein conjugates: Isolation of an adduct of *N*-acetylhistidine to the side chain of *N*-acetyldopamine from an insect-enzyme catalyzed reaction. *Tetrahedron Lett.*, *32*, 4287–4290.

Andersen, S. O., Peter, M. G., & Roepstorff, P. (1992a). Cuticle-catalyzed coupling between *N*-acetylhistidine and *N*-acetyldopamine. *Insect Biochem. Mol. Biol.*, *22*, 459–469.

Andersen, S. O., Jacobsen, J. P., & Roepstorff, P. (1992b). Coupling reactions between amino compounds and N-acetyldopamine catalyzed by cuticular enzymes. *Insect Biochem. Mol. Biol.*, *22*, 517–527.

Andersen, S. O., Jacobsen, J. P., Bojesen, G., & Roepstorff, P. (1992c). Phenoloxidase catalyzed coupling of catechols: Identification of novel coupling products. *Biochim. Biophys. Acta*, *1118*, 134–138.

Andersen, S. O., Peter, M. G., & Roepstorff, P. (1996). Cuticular sclerotization in insects. *Comp. Biochem. Physiol. B*, *113*, 689–705.

Arakane, Y., Muthukrishnan, S., Beeman, R. W., Kanost, M. R., & Kramer, K. J. (2005). *Laccase 2* is the phenoloxidase gene required for beetle cuticle tanning. *Proc. Natl. Acad. Sci. USA*, *102*, 11337–11342.

Arakane, Y., Lomakin, J., Beeman, R. W., Muthukrishnan, S., Gehrke, S. H., et al. (2009). Molecular and functional analyses of amino acid decarboxylases involved in cuticle tanning in *Tribolium castaneum*. *J. Biol. Chem.*, *284*, 16584–16594.

Asano, T., & Ashida, M. (2001a). Cuticular pro-phenoloxidase of the silkworm, *Bombyx mori*: Purification and demonstration of its transport from hemolymph. *J. Biol. Chem.*, *276*, 11100–11112.

Asano, T., & Ashida, M. (2001b). Transepithelially transported pro-phenoloxidase in the cuticle of the silkworm, *Bombyx mori*: Identification of its methionyl residues oxidized to methionine sulfoxides. *J. Biol. Chem.*, *276*, 11113–11125.

Ashida, M., & Brey, P. T. (1995). Role of the integument in insect defense: Pro-phenol oxidase cascade in the cuticular matrix. *Proc. Natl. Acad. Sci. USA*, *92*, 10698–10702.

Ashida, M., & Dohke, K. (1980). Activation of pro-phenoloxidase by the activating enzyme of the silk worm, *Bombyx mori*. *Insect Biochem.*, *10*, 37–47.

Ashida, M., & Yamazaki, H. I. (1990). Biochemistry of the phenoloxidase system in insects: With special reference to its activation. In E. Ohnishi, & H. Ishizaki (Eds.). *Molting and Metamorphosis* (pp. 239–265). Tokyo, Japan: Japan Scientific Societies Press.

Aso, Y., Kramer, K. J., Hopkins, T. L., & Whetzel, S. Z. (1984). Properties of tyrosinase and DOPA quinone imine conversion factor from pharate pupal cuticle of *Manduca sexta* L. *Insect Biochem.*, *14*, 463–472.

Aso, Y., Kramer, K. J., Hopkins, T. L., & Lookhart, G. L. (1985). Characterization of haemolymph protyrosinase and a cuticular activator from *Manduca sexta* (L.). *Insect Biochem.*, *15*, 9–17.

Atkinson, P. W., Brown, W. V., & Gilby, A. R. (1973a). Phenolic compounds from insect cuticle: Identification of some lipid antioxidants. *Insect Biochem.*, *3*, 309–315.

Atkinson, P. W., Brown, W. V., & Gilby, A. R. (1973b). Autoxidation of insect cuticular lipids: Stabilization of alkyl dienes by 3,4-dihydricphenols. *Insect Biochem.*, *3*, 103–112.

Bailey, W. D., Kimbrough, T. D., & Mills, R. R. (1999). Catechol conjugation with hemolymph proteins and their incorporation into the cuticle of the American cockroach, *Periplaneta americana*. *Comp. Biochem. Physiol. C*, *122*, 139–145.

Barrett, F. M. (1977). Recovery of ketocatechols from exuviae of last instar larvae of the cicada, *Tibicen pruinosa*. *Insect Biochem.*, *7*, 209–214.

Barrett, F. M. (1980). Recovery of phenolic compounds from exuviae of the spruce budworm, *Choristoneura fumiferana* (Lepidoptera: Tortricidae). *Can. Entomol.*, *112*, 151–157.

Barrett, F. M. (1984a). Metabolic origin of 3,4-dihydroxyphenylacetic acid in *Tenebrio molitor* and differences in its content in larval, pupal, and adult cuticle. *Can. J. Zool.*, *62*, 1005–1010.

Barrett, F. M. (1984b). Purification of phenolic compounds and a phenoloxidase from larval cuticle of the red-humped oakworm, *Symmerista cannicosta* Francl. *Arch. Insect Biochem. Physiol.*, *1*, 213–223.

Barrett, F. M. (1987a). Characterization of phenoloxidases from larval cuticle of *Sarcophaga bullata* and a comparison with cuticular enzymes from other species. *Can J. Zool.*, *65*, 1158–1166.

Barrett, F. M. (1987b). Phenoloxidases from larval cuticle of the sheep blowfly, *Lucilia cuprina*: Characterization, developmental changes, and inhibition by antiphenoloxidase antibodies. *Arch. Insect Biochem. Physiol.*, *5*, 99–118.

Barrett, F. M. (1990). Incorporation of various putative precursors into cuticular 3,4-dihydroxyphenylacetic acid of adult *Tenebrio molitor*. *Insect Biochem.*, *20*, 645–652.

Barrett, F. M. (1991). Phenoloxidases and the integument. In K. Binnington, & A. Retnakaran (Eds.). *Physiology of the Insect Epidermis* (pp. 195–212). Melbourne, Australia: CSIRO.

Barrett, F. M., & Andersen, S. O. (1981). Phenoloxidases in larval cuticle of the blowfly, *Calliphora vicina*. *Insect Biochem.*, *11*, 17–23.

Bernays, E. A., & Woodhead, S. (1982). Incorporation of dietary phenols into the cuticle in the tree locust *Anacridium melanorhodon*. *J. Insect Physiol.*, *28*, 601–606.

Bernays, E. A., Chamberlain, D. J., & McCarthy, P. (1980). The differential effects of ingested tannic acid on different species of Acridoidea. *Entomol. Experiment. Applic*, *28*, 158–166.

Binnington, K. C., & Barrett, F. M. (1988). Ultrastructural localization of phenoloxidases in cuticle and haemopoietic tissue of the blowfly *Lucilia cuprina*. *Tissue Cell*, *20*, 405–419.

Bodnaryk, R. P. (1971). N-terminal β-alanine in the puparium of the fly *Sarcophaga bullata*: Evidence from kinetic studies of its release by partial acid hydrolysis. *Insect Biochem.*, *1*, 228–236.

Bodnaryk, R. P., & Brunet, P. C. J. (1974). 3-O-Hydrosulphato-4-hydroxyphenylethylamine (dopamine 3-O-sulphate); a metabolite involved in the sclerotization of insect cuticle. *Biochem. J.*, *138*, 463–469.

Brunet, P. C. J. (1980). The metabolism of the aromatic amino acids concerned in the cross-linking of insect cuticle. *Insect Biochem.*, *10*, 467–500.

Candy, D. J. (1979). Glucose oxidase and other enzymes of hydrogen peroxide metabolism from cuticle of *Schistocerca gregaria*. *Insect Biochem.*, *9*, 661–665.

Chase, M. R., Raina, K., Bruno, J., & Sugumaran, M. (2000). Purification, characterization and molecular cloning of prophenoloxidases from *Sarcophaga bullata*. *Insect Biochem. Mol. Biol.*, *30*, 953–967.

Christensen, A. M., Schaefer, J., Kramer, K. J., Morgan, T. D., & Hopkins, T. L. (1991). Detection of cross-links in insect cuticle by REDOR NMR spectroscopy. *J. Am. Chem. Soc.*, *113*, 6799–6802.

Coles, G. C. (1966). Studies on resilin biosynthesis. *J. Insect Physiol.*, *12*, 679–691.

Cottrell, C. B. (1964). Insect ecdysis with particular emphasis on cuticular hardening and darkening. *Adv. Insect Physiol.*, *2*, 175–218.

Cribb, B. W., Stewart, A., Huang, H., Truss, R., Noller, B., et al. (2008). Insect mandibles – comparative mechanical properties and links with metal incorporation. *Naturwissenschaften*, *95*, 17–23.

Cribb, B. W., Lin, C. -L., Rintoul, L., Rasch, R., Hasenpusch, J., & Huang, H. (2010). Hardness in arthropod exoskeletons in the absence of transition metals. *Acta Biomaterialia*, *6*, 3152–3156.

Curtis, A. T., Hori, M., Green, J. M., Wolfgang, W. J., Hiruma, K., & Riddiford, L. M. (1984). Ecdysteroid regulation of the onset of cuticular melanization in allatectomized and black mutant *Manduca sexta* larvae. *J. Insect Physiol.*, *30*, 597–606.

Czapla, T. H., Hopkins, T. L., Kramer, K. J., & Morgan, T. D. (1988). Diphenols in hemolymph and cuticle during development and cuticle tanning of *Periplaneta americana* (L.) and other cockroach species. *Arch. Insect Biochem. Physiol.*, *7*, 13–28.

Czapla, T. H., Hopkins, T. L., & Kramer, K. J. (1989). Catecholamines and related o-diphenols in the hemolymph and cuticle of the cockroach *Leucophaea maderae* (F.) during sclerotization and pigmentation. *Insect Biochem.*, *19*, 509–515.

Czapla, T. H., Hopkins, T. L., & Kramer, K. J. (1990). Cuticular strength and pigmentation of five strains of adult *Blatella germanica* (L.) during sclerotization: Correlations with catecholamines, beta-alanine and food deprivation. *J. Insect Physiol.*, *36*, 647–654.

Dai, L., Dewey, E. M., Zitnan, D., Luo, C. -W., Honegger, H. -W., & Adams, M. E. (2008). Identification, developmental expression, and functions of bursicon in the tobacco hawkmoth, *Manduca sexta*. *J. Comp. Neurol.*, *506*, 759–774.

Darlington, M. V., Meyer, H. J., Graf, G., & Freeman, T. P. (1983). The calcified puparium of the face fly, *Musca autumnalis* (Diptera: Muscidae). *J. Insect Physiol.*, *29*, 157–162.

Davis, M. M., O'Keefe, S. L., Primrose, D. A., & Hodgetts, R. B. (2007). A neuropeptide hormone cascade controls the precise onset of post-eclosion cuticular tanning in *Drosophila melanogaster*. *Development*, *134*, 4395–4404.

Dennell, R. (1958). The amino acid metabolism of a developing insect cuticle: The larval cuticle and puparium of *Calliphora vomitoria*. 3. The formation of the puparium. *Proc. R Soc. B*, *148*, 176–183.

Dittmer, N. T., Gorman, M. J., & Kanost, M. R. (2009). Characterization of endogenous and recombinant forms of laccase-2, a multicopper oxidase from the tobacco hornworm, *Manduca sexta*. *Insect Biochem. Mol. Biol.*, *39*, 596–606.

Fogal, W., & Fraenkel, G. (1969). The role of bursicon in melanization and endocuticle formation in the adult fleshfly, *Sarcophaga bullata*. *J. Insect Physiol.*, *15*, 1235–1247.

Fraenkel, G., & Hsiao, C. (1965). Bursicon, a hormone which mediates tanning of the cuticle in the adult fly and other insects. *J. Insect Physiol.*, *11*, 513–556.

Fraenkel, G., & Rudall, K. M. (1940). A study of the physical and chemical properties of insect cuticle. *Proc. R. Soc. B*, *129*, 1–35.

Fragoulis, E. G., & Sekeris, C. E. (1975). Induction of dopa (3,4-dihydroxyphenylalanine) decarboxylase in blowfly integument by ecdysone. *Biochem. J.*, *146*, 121–126.

Fujimoto, K., Okino, N., Kawabata, S., & Ohnishi, E. (1995). Nucleotide sequence of the cDNA encoding the proenzyme of phenoloxidase A_1 of *Drosophila melanogaster*. *Proc. Natl. Acad. Sci. USA*, *92*, 7769–7773.

Gilby, A. R., & McKellar, J. W. (1976). The calcified puparium of a fly. *J. Insect Physiol.*, *22*, 1465–1468.

Gorman, M. J., & Arakane, Y. (2010). Tyrosine hydroxylase is required for cuticle sclerotization and pigmentation in *Tribolium castaneum*. *Insect Biochem. Mol. Biol.*, *40*, 267–273.

Grossmüller, M., & Messner, B. (1978). Zur quantitativen Verteilung der Peroxidase- und Phenoloxidase-Aktivität in der Kutikula von *Galleria mellonella*-Larven und -Puppen: Ein Beitrag zur Sklerotisierung der Insektenkutikula. *Zool. Jb. Physiol.*, *82*, 16–22.

Gupta, M. N., & Vithayathil, P. J. (1982). Isolation and characterization of a methionine adduct of DOPA o-quinone. *Bioorg. Chem.*, *11*, 101–107.

Hackman, R. H., & Goldberg, M. (1971). Studies on the hardening and darkening of insect cuticles. *J. Insect Physiol.*, *17*, 335–347.

Hall, M., Scott, T., Sugumaran, M., Söderhall, K., & Law, J. H. (1995). Proenzyme of *Manduca sexta* phenoloxidase: Purification, activation, substrate specificity of the active enzyme and molecular cloning. *Proc. Natl. Acad. Sci. USA*, *92*, 7764–7768.

Han, Q., Li, G., & Li, J. (2000a). Chorion peroxidase-mediated $NADH/O_2$ oxidoreduction cooperated by chorion malate dehydrogenase-catalyzed NADH production: A feasible pathway leading to H_2O_2 formation during chorion hardening in *Aedes aegypti* mosquitoes. *Biochim. Biophys. Acta*, *1523*, 246–253.

Han, Q., Li, G., & Li, J. (2000b). Purification and characterization of chorion peroxidase from *Aedes aegypti* eggs. *Arch. Biochem. Biophys.*, *378*, 107–115.

Han, Q., Fang, J., Ding, H., Johnson, J. K., & Christensen, B. M. (2002). Identification of *Drosophila melanogaster* yellow-f and yellow-f2 proteins as dopachrome-conversion enzymes. *Biochem. J.*, *368*, 333–340.

Hasson, C., & Sugumaran, M. (1987). Protein cross-linking by peroxidase: Possible mechanism for sclerotization of insect cuticle. *Arch. Insect Biochem. Physiol.*, *5*, 13–28.

Hillerton, J. E., & Vincent, J. F.V. (1979). The stabilization of insect cuticles. *J. Insect Physiol.*, *25*, 957–963.

Hillerton, J. E., & Vincent, J. F.V. (1982). The specific location of zinc in insect mandibles. *J. Exp. Biol.*, *101*, 333–336.

Hillerton, J. E., Robertson, B., & Vincent, J. F.V. (1984). The presence of zinc or manganese as the predominant metal in the mandibles of adult, stored-product beetles. *J. Stored Prod. Res.*, *20*, 133–137.

Hiruma, K., & Riddiford, L. M. (1988). Granular phenoloxidase involved in cuticular melanization in the tobacco hornworm: Regulation of its synthesis in the epidermis by juvenile hormone. *Devel. Biol.*, *130*, 87–97.

Honegger, H. -W., Dewey, E. M., & Ewer, J. (2008). Bursicon, the tanning hormone of insects: Recent advances following the discovery of its molecular identity. *J. Comp. Physiol. A*, *194*, 989–1005.

Hopkins, T. L., & Kramer, K. J. (1992). Insect cuticle sclerotization. *Annu. Rev. Entomol., 37*, 273–302.

Hopkins, T. L., Morgan, T. D., Aso, Y., & Kramer, K. J. (1982). *N*-β-alanyldopamine: Major role in insect cuticle tanning. *Science, 217*, 364–366.

Hopkins, T. L., Morgan, T. D., & Kramer, K. J. (1984). Catecholamines in haemolymph and cuticle during larval, pupal, and adult development of *Manduca sexta* (L.). *Insect Biochem., 14*, 533–540.

Hopkins, T. L., Morgan, T. D., Mueller, D. D., Tomer, K. B., & Kramer, K. J. (1995). Identification of catecholamine β-glucosides in the hemolymph of the tobacco hornworm, *Manduca sexta* (L.) during development. *Insect Biochem. Mol. Biol., 25*, 29–37.

Huang, X., Xu, R., Hawley, M. D., Hopkins, T. L., & Kramer, K. J. (1998). Electrochemical oxidation of *N*-acyldopamines and regioselective reactions of their quinones with *N*-acetylcysteine and thiourea. *Arch. Biochem. Biophys., 352*, 19–30.

Hunt, S. (1984). Halogenated tyrosine derivatives in invertebrate scleroproteins: Isolation and identification. *Methods Enzymol., 107*, 413–438.

Jacobs, M. E. (1978). β-Alanine tanning of *Drosophila* cuticles and chitin. *Insect Biochem., 8*, 37–41.

Jacobs, M. E. (1980). Influence of β-alanine on ultrastructure, tanning, and melanization of *Drosophila melanogaster* cuticles. *Biochem. Genet., 18*, 56–76.

Jacobs, M. E. (1985). Role of β-alanine in cuticular tanning, sclerotization, and temperature regulation in *Drosophila melanogaster. J. Insect Physiol., 31*, 509–515.

Kanost, M. R., & Gorman, M. J. (2008). Phenoloxidases in insect immunity. In N. E. Beckage (Ed.). *Insect Immunology* (pp. 69–96). Oxford, UK: Elsevier.

Karlson, P., & Schlossberger-Raecke, I. (1962). Zum Tyrosin-stoffwechsel der Insekten. 8. Die Sklerotisierung der Cuticula bei der Wildform und der Albinomutante von *Schistocerca gregaria* Forsk. *J. Insect Physiol., 8*, 441–452.

Karlson, P., & Sekeris, C. E. (1962). *N*-acetyl-dopamine as sclerotizing agent of the insect cuticle. *Nature, 195*, 183–184.

Karlson, P., & Sekeris, C. E. (1976). Control of tyrosine metabolism and cuticle sclerotization by ecdysone. In H. P. Hepburn (Ed.). *The Insect Integument* (pp. 145–156). London: Elsevier.

Karlson, P., Sekeri, K. E., & Marmaras, V. I. (1969). Die Aminosäurezusammensetzung verschiedener Proteinfraktionen aus der Cuticula von *Calliphora erythrocephala* in verschiedenen Entwicklungsstadien. *J. Insect Physiol., 15*, 319–323.

Kawabata, T., Yasuhara, Y., Ochiai, M., Matsuura, S., & Ashida, M. (1995). Molecular cloning of insect pro-phenoloxidase: A copper-containing protein homologous to arthropod hemocyanin. *Proc. Natl. Acad. Sci. USA, 92*, 7774–7778.

Kayser, H. (1985). Pigments. In G. A. Kerkut, & L. I. Gilbert (Eds.). *Comprehensive Insect Physiology, Biochemistry, and Pharmacology* (Vol. 10, pp. 367–415). Oxford, UK: Pergamon Press.

Kayser-Wegmann, I. (1976). Differences in black pigmentation in lepidopteran cuticles as revealed by light and electron microscopy. *Cell Tissue Res., 171*, 513–521.

Kayser-Wegmann, I., & Kayser, H. (1983). Black pigmentation of insect cuticles: A view based on microscopic and autoradiographic studies. In K. Scheller (Ed.). *The Larval Serum Proteins of Insects: Function, Biosynthesis, Genetics* (pp. 151–167). Stuttgart, Germany: Georg Thieme Verlag.

Kerwin, J. L., Turecek, F., Xu, R., Kramer, K. J., Hopkins, T. L., et al. (1999). Mass spectrometric analysis of catechol-histidine adducts from insect cuticle. *Anal. Biochem., 268*, 229–237.

Koeppe, J. K., & Gilbert, L. I. (1974). Metabolism and protein transport of a possible pupal tanning agent in *Manduca sexta. J. Insect Physiol., 20*, 981–992.

Koeppe, J. K., & Mills, R. R. (1972). Hormonal control of tanning by the American cockroach: Probable bursicon mediated translocation of protein bound phenols. *J. Insect Physiol., 18*, 465–469.

König, M., Agrawal, O. P., Schenkel, H., & Scheller, K. (1986). Incorporation of calliphorin into the cuticle of the developing blowfly, *Calliphora vicina. Roux's Arch. Devel. Biol., 195*, 296–301.

Kramer, K. J., & Hopkins, T. L. (1987). Tyrosine metabolism for insect cuticle tanning. *Arch. Insect Biochem. Physiol., 6*, 279–301.

Kramer, K. J., Morgan, T. D., Hopkins, T. L., Roseland, C. R., Aso, Y., et al. (1984). Catecholamines and β-alanine in the red flour beetle: Roles in cuticle sclerotization and melanization. *Insect Biochem., 14*, 293–298.

Krueger, R. A., Kramer, K. J., Hopkins, T. L., & Speirs, R. D. (1989). *N*-β-alanyldopamine levels and synthesis in integument and other tissues of *Manduca sexta* (L.) during the larval-pupal transformation. *Insect Biochem., 19*, 169–175.

Krueger, R. R., Kramer, K. J., Hopkins, T. L., & Speirs, R. D. (1990). *N*-β-alanyldopamine and *N*-acetyldopamine occurrence and synthesis in the central nervous system of *Manduca sexta* (L.) *Insect Biochem., 20*, 605–610.

Li, J., Hodgeman, B. A., & Christensen, B. M. (1996). Involvement of peroxidase in chorion hardening in *Aedes aegypti. Insect Biochem. Mol. Biol., 26*, 309–317.

Locke, M. (1969). The localization of a peroxidase associated with hard cuticle formation in an insect, *Calpodes ethlius* Stoll, Lepidoptera, Hesperiidae. *Tissue and Cell, 1*, 555–574.

Luo, C.-W., Dewey, E. M., Sudo, S., Ewer, J., Hu, X., et al. (2005). Bursicon, the insect cuticle-hardening hormone, is a heterodimeric cystine knot protein that activates G protein-coupled receptor LGR2. *Proc. Natl. Acad. Sci. USA, 102*, 2820–2825.

Malek, S. R.A. (1957). Sclerotization and melanization: Two independent processes in the cuticle of the desert locust. *Nature, 180*, 237.

Mattinen, M.-L., Kruus, K., Buchert, J., Nielsen, J. H., Andersen, H. J., & Steffensen, C. L. (2005). Laccase-catalyzed polymerization of tyrosine-containing peptides. *FEBS J., 272*, 3640–3650.

Mendive, F. M., Loy, T. V., Claeysen, S., Poels, J., Williamson, M., et al. (2005). *Drosophila* molting neurohormone bursicon is a heterodimer and the natural agonist of the orphan receptor DLGR2. *FEBS Lett., 579*, 2171–2176.

Messner, B., & Janda, V. (1991). Der Peroxidase-Nachweis in der Kutikula von *Galleria mellonella* (Insecta: Lepidoptera) und *Protophormia terraenovae* (Insecta: Diptera): Ein Beitrag zur Sklerotisierung der Insekten-kutikula. *Zool. Jb. Physiol., 95*, 31–37.

Messner, B., & Kerstan, U. (1991). Die Peroxidase im Sklerotisierungsprozess von Hartteilen verschiedener wirbelloser Tiere. *Zool. Jb. Physiol.*, *95*, 23–29.

Mindrinos, M. N., Petri, W. H., Galanopoulos, V. K., Lombard, M. F., & Margaritis, L. H. (1980). Crosslinking of the *Drosophila* chorion involves a peroxidase. *Roux's Arch. Devel. Biol.*, *189*, 187–196.

Morgan, T. D., Hopkins, T. L., Kramer, K. J., Roseland, C. R., Czapla, T. H., et al. (1987). *N*-β-Alanylnorepinephrine biosynthesis in insect cuticle and possible role in sclerotization. *Insect Biochem.*, *17*, 255–263.

Morgan, T. D., Thomas, B. R., Yonekura, M., Czapla, T. H., Kramer, K. J., & Hopkins, T. L. (1990). Soluble tyrosinases from pharate pupal integument of the tobacco hornworm, *Manduca sexta* (L.) *Insect Biochem.*, *20*, 251–260.

Niu, B. -L., Shen, W. -F., Liu, Y., Weng, H. -B., He, L. -H., et al. (2008). Cloning and RNAi-mediated functional characterization of *MaLac2* of the pine sawyer, *Monochamus alternatus*. *Insect Mol. Biol.*, *17*, 303–312.

Okot-Kotber, B. M., Morgan, T. D., Hopkins, T. L., & Kramer, K. J. (1996). Catecholamine-containing proteins from the pharate pupal cuticle of the tobacco hornworm, *Manduca sexta*. *Insect Biochem. Mol. Biol.*, *26*, 475–484.

Pan, C., Zhou, Y., & Mo, J. (2009). The clone of laccase gene and its potential function in cuticular penetration resistance of *Culex pipiens pallens* to fenvalerate. *Pestic. Biochem. Physiol.*, *93*, 105–111.

Pérez, M., Wappner, P., & Quesada-Allué, L. A. (2002). Catecholamine β-alanyl ligase in the medfly *Ceratitis capitata*. *Insect Biochem. Mol. Biol.*, *32*, 617–625.

Peter, M. G., Andersen, S. O., Hartmann, R., Miessner, M., & Roepstorff, P. (1992). Catechol–protein conjugates: Isolation of 4-phenylphenoxazin-2-ones from oxidative coupling of *N*-acetyldopamine with aliphatic amino acids. *Tetrahedron*, *48*, 8927–8934.

Peter, M. G., & Scheller, K. (1991). Arylphorins and the integument. In K. Binnington, & A. Retnakaran (Eds.). *Physiology of the Insect Epidermis* (pp. 113–122). Melbourne, Australia: CSIRO.

Petri, W. H., Wyman, A. R., & Kafatos, F. C. (1976). Specific protein synthesis in cellular differentiation. 3. The egg shell proteins in *Drosophila melanogaster* and their program of synthesis. *Devel. Biol.*, *49*, 185–199.

Pryor, M. G.M. (1940a). On the hardening of the ootheca of *Blatta orientalis. Proc. R. Soc. B*, *128*, 378–392.

Pryor, M. G.M. (1940b). On the hardening of cuticle of insects. *Proc. R. Soc. B*, *128*, 393–407.

Quicke, D. L.J., Wyeth, P., Fawke, J. D., Basibuyuk, H. H., & Vincent, J. F.V. (1998). Manganese and zinc in the ovipositors and mandibles of hymenopterous insects. *Zool. J. Linnean Soc.*, *124*, 387–396.

Ricketts, D., & Sugumaran, M. (1994). 1,2-Dehydro-*N*-β-alanyldopamine as a new intermediate in insect cuticular sclerotization. *J. Biol. Chem.*, *269*, 22217–22221.

Roepstorff, P., & Andersen, S. O. (1981). Electron impact, chemical ionization and field desorption mass spectrometry of substituted dihydroxyphenyl-benzodioxins isolated from insect cuticle: Occurrence of thermal decomposition and oligomerization reactions in the mass spectrometer. *Biomed. Mass Spectrom.*, *8*, 174–178.

Roseland, C. R., Kramer, K. J., & Hopkins, T. L. (1987). Cuticular strength and pigmentation of rust-red and black strains of *Tribolium castaneum*: Correlation with catecholamine and β-alanine content. *Insect Biochem.*, *17*, 21–28.

Saul, S. J., & Sugumaran, M. (1988). A novel quinone: Quinone methide isomerase generates quinone methides in insect cuticle. *FEBS Lett.*, *237*, 155–158.

Saul, S. J., & Sugumaran, M. (1989a). *o*-Quinone/quinone methide isomerase: A novel enzyme preventing the destruction of self-matter by phenoloxidase-generated quinones during immune response in insects. *FEBS Lett.*, *249*, 155–158.

Saul, S. J., & Sugumaran, M. (1989b). *N*-acetyldopamine quinone methide/1,2-dehydro-N-acetyldopamine tautomerase: A new enzyme involved in sclerotization of insect cuticle. *FEBS Lett.*, *255*, 340–344.

Saul, S. J., & Sugumaran, M. (1989c). Characterization of a new enzyme system that desaturates the side chain of *N*-acetyldopamine. *FEBS Lett.*, *251*, 69–73.

Saul, S. J., & Sugumaran, M. (1990). 4-Alkyl- *o*-quinone/2-hydroxy-*p*-quinone methide isomerase from the larval hemolymph of *Sarcophaga bullata*. 1. Purification and characterization of enzyme-catalyzed reaction. *J. Biol. Chem.*, *265*, 16992–16999.

Schaefer, J., Kramer, K. J., Garbow, J. R., Jacob, G. S., Stejskal, E. O., et al. (1987). Aromatic links in insect cuticle: Detection by solid state ^{13}C and ^{15}N NMR. *Science*, *235*, 1200–1204.

Scheller, K., Zimmermann, H. -P., & Sekeris, C. E. (1980). Calliphorin, a protein involved in the cuticle formation of the blowfly, *Calliphora vicina. Z. Naturforsch. C*, *35*, 387–389.

Schenkel, H., Myllek, C., König, M., Hausberg, P., & Scheller, K. (1983). Calliphorin: Studies on its biosynthesis and function. In K. Scheller (Ed.). *The Larval Serum Proteins of Insects: Functions, Biosynthesis, Genetics* (pp. 18–39). Stuttgart, Germany: Georg Thieme Verlag.

Schlossberger-Raecke, I., & Karlson, P. (1964). Zum Tyrosin-stoffwechsel der Insekten. 13. Radioautographische Lokalisation von Tyrosinmetaboliten in der Cuticula von *Schistocerca gregaria* Forsk. *J. Insect Physiol.*, *10*, 261–266.

Schofield, R. M.S. (2001). Metals in cuticular structures. In P. Brownell, & G. Polis (Eds.). *Scorpion Biology and Research* (pp. 234–256). Oxford, UK: Oxford University Press.

Schofield, R. M.S., Nesson, M. H., & Richardson, K. A. (2002). Tooth hardness increases with zinc-content in mandibles of young adult leaf-cutter ants. *Naturwissenschaften*, *89*, 579–583.

Sekeris, C. E. (1991). The role of molting hormone in sclerotization in insects. In A. P. Gupta (Ed.). *Morphogenetic Hormones of Arthropods: Roles in Histogenesis, Organogenesis and Morphogenesis* (Vol 1, pp. 150–212). New Brunswick, NJ: Rutgers University Press, Part 3.

Seligman, M., Friedman, S., & Fraenkel, G. (1969). Bursicon mediation of tyrosine hydroxylation during tanning of the adult cuticle of the fly, *Sarcophaga bullata. J. Insect Physiol.*, *15*, 553–561.

Srivastava, R. P. (1971). The amino acid composition of cuticular proteins of different developmental stages of *Galleria mellonella. J. Insect Physiol.*, *17*, 189–196.

Suderman, R. J., Dittmer, N. T., Kramer, K. J., & Kanost, M. R. (2010). Model reactions for insect cuticle sclerotization: Participation of amino groups in the cross-linking of *Manduca sexta* cuticle protein MsCP36. *Insect Biochem. Mol. Biol.*, *40*, 252–258.

Sugumaran, M. (1987). Quinone methide sclerotization: A revised mechanism for β-sclerotization of insect cuticle. *Bioorgan. Chem.*, *15*, 194–211.

Sugumaran, M. (1988). Molecular mechanisms for cuticular sclerotization. *Adv. Insect Physiol*, *21*, 179–231.

Sugumaran, M. (1998). Unified mechanism for sclerotization of insect cuticle. *Adv. Insect Physiol.*, *27*, 229–334.

Sugumaran, M., & Nelson, E. (1998). Model sclerotization studies. 4. Generation of *N*-acetylmethionyl catechol adducts during tyrosinase-catalyzed oxidation of catechols in the presence of *N*-acetylmethionine. *Arch. Insect Biochem. Physiol.*, *38*, 44–52.

Sugumaran, M., Hennigan, B., Semensi, V., & Dali, H. (1988). On the nature of nonenzymatic and enzymatic oxidation of the putative sclerotization precursor, 1,2-dehydro-*N*-acetyldopamine. *Arch. Insect Biochem. Biophys.*, *8*, 89–100.

Sugumaran, M., Saul, S. P., & Semensi, V. (1989a). Trapping of transiently formed quinone methide during enzymatic conversion of *N*-acetyldopamine to *N*-acetylnorepinephrine. *FEBS Lett.*, *252*, 135–138.

Sugumaran, M., Dali, H., & Semensi, V. (1989b). Chemical and cuticular phenoloxidase-mediated synthesis of cysteinylcatechol adducts. *Arch. Insect Biochem. Physiol.*, *11*, 127–137.

Sugumaran, M., Giglio, L., Kundzicz, H., Saul, S. P., & Semensi, V. (1992). Studies on the enzymes involved in puparial cuticle sclerotization in *Drosophila melanogaster*. *Arch. Insect Biochem. Physiol.*, *19*, 271–283.

Thomas, B. R., Yonekura, M., Morgan, T. D., Czapla, T. H., Hopkins, T. L., & Kramer, K. J. (1989). A trypsin-solubilized laccase from pharate pupal integument of the tobacco hornworm, *Manduca sexta*. *Insect Biochem.*, *19*, 611–622.

Umebachi, Y. (1993). The third way of dopamine. *Trends Comp. Biochem. Physiol.*, *1*, 709–720.

Vincent, J. F.V. (1980). Insect cuticle: A paradigm for natural composites. *Symp. Soc. Exp. Biol.*, *34*, 183–209.

Vincent, J. F.V., & Hillerton, J. E. (1979). The tanning of insect cuticle: A critical review and a revised mechanism. *J. Insect Physiol.*, *25*, 653–658.

Wappner, P., Kramer, K. J., Hopkins, T. L., Merritt, M., Schaefer, J., & Quesada-Allué, L. A. (1995). *White Pupa*: A *Ceratitis capitata* mutant lacking catecholamines for tanning the puparium. *Insect Biochem. Mol. Biol.*, *25*, 365–373.

Wappner, P., Kramer, K. J., Manso, F., Hopkins, T. L., & Quesada-Allué, L. A. (1996). *N*-β-alanyldopamine metabolism for puparial tanning in wild-type and mutant *Niger* strains of the Mediterranean fruit fly, *Ceratitis capitata*. *Insect Biochem. Mol. Biol.*, *26*, 585–592.

Welinder, B. S., Roepstorff, P., & Andersen, S. O. (1976). The crustacean cuticle. 4. Isolation and identification of crosslinks from *Cancer pagurus* cuticle. *Comp. Biochem. Physiol. B*, *53*, 529–533.

Wigglesworth, V. B. (1985). Sclerotin and lipid in the waterproofing of the insect cuticle. *Tissue and Cell*, *17*, 227–248.

Wigglesworth, V. B. (1988). The source of lipids and polyphenols for the insect cuticle: The role of fat body, oenocytes and oenocytoids. *Tissue and Cell*, *20*, 919–932.

Wright, T. R.F. (1987). The genetics of biogenic amine metabolism, sclerotization and melanization in *Drosophila melanogaster*. *Adv. Genet.*, *24*, 127–222.

Xu, R., Huang, X., Kramer, K. J., & Hawley, M. D. (1996a). Characterization of products from the reactions of dopamine quinone with *N*-acetylcysteine. *Bioorgan. Chem.*, *24*, 110–126.

Xu, R., Huang, X., Morgan, T. D., Prakash, O., Kramer, K. J., & Hawley, M. D. (1996b). Characterization of products from the reactions of *N*-acetyldopamine quinone with *N*-acetylhistidine. *Arch. Biochem. Biophys.*, *329*, 56–64.

Xu, R., Huang, X., Hopkins, T. L., & Kramer, K. J. (1997). Catecholamine and histidyl protein cross-linked structures in sclerotized insect cuticle. *Insect Biochem. Mol. Biol.*, *27*, 101–108.

Yamazaki, H. I. (1969). The cuticular phenoloxidase in *Drosophila virilis*. *J. Insect Physiol.*, *15*, 2203–2211.

Yamazaki, H. I. (1972). Cuticular phenoloxidase from the silkworm *Bombyx mori*: Properties, solubilization, and purification. *Insect Biochem.*, *2*, 431–444.

Yatsu, J., & Asano, T. (2009). Cuticle laccase of the silkworm, *Bombyx mori*: Purification, gene identification and presence of its inactive precursor in the cuticle. *Insect Biochem. Mol. Biol.*, *39*, 254–262.

7   Chitin Metabolism in Insects

Subbaratnam Muthukrishnan
Kansas State University, Manhattan, KS, USA
Hans Merzendorfer
University of Osnabrueck, Osnabrueck, Germany
Yasuyuki Arakane
Chonnam National University, Gwangju, South Korea
Karl J Kramer
Kansas State University, and USDA-ARS,
Manhattan, KS, USA

7.1. Introduction

"Chitin Metabolism in Insects" was the title of chapters in both the original edition of the *Comprehensive Insect Physiology, Biochemistry and Pharmacology* series published in 1985 and the follow-up *Comprehensive Molecular Insect Science* series in 2005 (Kramer *et al.*, 1985; Kramer and Muthukrishnan, 2005). Since 2005 substantial progress in gaining additional understanding of this topic has continued to take place, primarily through the application of the techniques of molecular genetics, functional genomics, proteomics, transcriptomics, metabolomics, and biotechnology to an assortment of studies focused on insect chitin metabolism. Several other reviews have also been published that have reported on some of the advances that have taken place (Dahiya *et al.*, 2006; Merzendorfer, 2006, 2009; Arakane and Muthukrishnan, 2010). Most interestingly, the list of genes and gene products found to be involved in insect chitin metabolism has been lengthened

significantly. In this chapter we will highlight some of the more recent and important findings, with emphasis on results obtained from studies conducted on the synthesis, structure, physical state, modification, organization, and degradation of chitin in insect tissues, as well as the interplay of chitin with chitin-binding proteins, the regulation of genes responsible for chitin metabolism, and, finally, the targeting of chitin metabolism for insect-control purposes.

7.2. Chitin Structure and Occurrence

Chitin is the major polysaccharide present in insects and many other invertebrates as well as in several microbes, including fungi. Structurally, it is the simplest of the glycosaminoglycans, being a $\beta(1\rightarrow4)$ linked linear homopolymer of N-acetylglucosamine (GlcNAc, $[C_8H_{13}O_5N]_n$, where $n \gg 1$). It serves as the skeletal polysaccharide of several animal phyla, such as the Arthropoda, Annelida, Molluska,

DOI:10.1016/B978-0-12-384747-8.10007-8

and Coelenterata. In several groups of fungi, chitin replaces cellulose as the structural polysaccharide. In insects, it is found in the body wall or cuticle, gut lining or peritrophic matrix (PM), salivary gland, trachea, eggshells, and muscle attachment points. In the course of evolution, insects have made excellent use of the rigidity and chemical stability of the polymeric chitin to assemble both hard and soft extracellular structures such as the cuticle (exoskeleton) and PM respectively, both of which enable insects to be protected from the environment while allowing for growth, mobility, respiration, and communication. All of these structures are primarily composites of chitin fibers and proteins with varying degrees of hydration and trace materials distributed along the structures. The insolubility and structural complexity of the cuticle has limited its study. However, sclerotized cuticle can be modeled as an interpenetrating network of chitin fibers with embedded cross-linked protein and pigments. Both synthesis and degradation of chitin take place at multiple developmental stages in the cuticle and the PM. It is usually synthesized as portions of the old endocuticle and PM and trachaea are resorbed, and the digested materials are recycled. Although primarily composed of poly-GlcNAc, chitin also can contain a small percentage of unsubstituted (or N-deacetylated) glucosamine (GlcNAc) residues, making it a GlcNAc-GlcN heteropolymer (Muzzarelli, 1973; Fukamizo et al., 1986). When the epidermal and gut cells synthesize and secrete a particular form of chitin consisting of antiparallel chains or alpha-chitin, the chains are assembled into microfibrils and then into sheets. As layers of chitin are added, the sheets are cross-oriented relative to one another at a constant angle to form a helicoidal bundle (known as the Bouligand structure), which can contribute to the formation of an extremely strong, plywood-like material.

Although there is no doubt that there are strong noncovalent interactions between chitin and chitin-binding proteins, there is only weak indirect evidence that there are covalent interactions between them. The evidence so far for direct involvement of chitin in cross-links to proteins has been inconclusive. Results of solid state NMR and chemical analyses have indicated the presence of trace levels of aromatic amino acids in chitin preparations, suggesting that those amino acids were there because they were involved in protein cross-links with chitin (Schaefer et al., 1987). Additional spectroscopic evidence for glucosamine–catecholamine adducts derived from chitin–protein cross-links in cuticle was obtained using electrospray mass spectrometry and tandem mass spectrometry (Kerwin et al., 1999). However, those observations have not been investigated further. More direct evidence for chitin–protein cross-links from studies of intact cuticle instead of degraded or digested samples is needed before the precise nature of the covalent interactions of cuticular proteins with chitin fibers can be resolved (Demolliens et al., 2008).

Alpha-chitin fibers, because of their hydrophilic nature, are generally highly hydrated. Chitin dehydration via impregnation of hydrophobic proteins probably contributes to tissue stiffening and deplasticization (Vincent, 2009). In addition, the formation of a cross-linked and interpenetrating protein network in the dehydrated composite leads to additional hardening (Andersen, 2010); thus, chemical bonds surely play a crucial role in cuticle mechanics by increasing the load carried by the proteins and by providing a hydrophobic "coating" around the chitin nanofibers, thus preventing softening of the latter by water adsorption. Chitin nanofibrils probably form the initial template, similar to glass or carbon fiber mats in composite processing. Filler proteins and catechols are then secreted through the chitinous procuticle. Once oxidation of catechols to quinones and quinone methides has occurred, cross-linking and hardening of the extracellular matrix ensues. As sclerotization proceeds, water is progressively expelled. The precise role of water removal on the structural properties of the cuticle is not fully understood, in part because the effect of water on individual components of the composite is poorly understood, but some progress is starting to take place. Also, the individual contributions of chitin and protein to the mechanical properties are unknown. In the hydrated state, there is considerable variation in moduli reported for chitosan/chitin scaffolds (Wu et al., 2006). There is a difference of several orders of magnitude in the stiffness of chitin/chitosan between the fully hydrated state, where it is present as a porous, water-saturated scaffold, and the dry state. To mimic the action of catechols to stiffen chitosan scaffolds, Wu et al. (2005) achieved a two-fold increase in stiffness after treatment of chitosan films with oxidized catechols. Although there was a significant increase in stiffness, it was less than the increase observed from insect cuticle tanning. Recently, dynamic mechanical analysis of insect cuticle during maturation revealed that while the water content has an important role in determining cuticle mechanical properties, the tanning reactions themselves contribute substantially to these properties beyond simply inducing dehydration (Lomakin et al., 2011). Cuticle, whether tanned or untanned, increases in hardness while drying, but the increase is generally less than that observed from tanning alone.

7.3. Chitin Synthesis

Although extensive knowledge on the precise molecular mechanism of chitin synthesis is lacking, substantial progress has been made regarding the function and regulation of several genes involved in the chitin biosynthetic pathway. In the past 10 years, many genes coding for key enzymes of this pathway have been isolated and sequenced from various insect species. Analyses of their expression in different tissues during development have provided the first clues about their function. The availability of *Drosophila melanogaster* (fruit fly) mutants defective in some of these genes, together with the ability to specifically silence their expression by RNAi in the fly and other species, has boosted our understanding of this process. Most progress has been made on chitin synthases (CHSs), which have been identified in a variety of organisms, including fungi, nematodes, mollusks, and insects. Amino acid sequence similarities have been the principal tools used for identifying CHSs, which form a subfamily within a larger group of the glycosyltransferases (family GT2) that catalyze the transfer of a sugar moiety from an activated sugar donor onto saccharide or non-saccharide acceptors (Coutinho *et al.*, 2003; Cantarel *et al.*, 2009). CHS has not been an easy enzyme to assay, which has made its study rather difficult. Traditionally, CHS activity is measured by a radioactive assay using [^{14}C]- or [^{3}H]-labeled UDP-GlcNAc as the precursor followed by quantification of insoluble radiolabeled chitin after acid precipitation. Alternately, a high-throughput non-radioactive assay is available, which involves binding of synthesized chitin to a wheat germ agglutinin (WGA)-coated surface, followed by detection of the polymer with a horseradish peroxidase–WGA conjugate (Lucero *et al.*, 2002). Also, the direct incorporation of fluorescently labeled substrates, such as certain dansyl-UDP-GlcNAc analogs, may prove to be useful for developing fluorescence-based enzyme assays (Yeager and Finney, 2005). The paucity of information concerning the enzyme's biochemical and kinetic properties was mainly due to the inability to obtain active soluble CHS preparations. Recently, however, a purification and solubilization protocol has been developed, which allowed purifying CHS-B from the midgut of *Manduca sexta* (tobacco hornworm) as an active, oligomeric complex (Maue *et al.*, 2009). In addition, first attempts to heterologously express CHSs from protists and fungi in yeast systems turned out to be successful (Van Dellen *et al.*, 2006; Martinez-Rucobo *et al.*, 2009; Barreto *et al.*, 2010). These purification and expression protocols should facilitate greater progress in insect CHS studies in the future.

7.3.1. Sites of Chitin Biosynthesis

The epidermis and the midgut are two major tissues where chitin synthesis occurs in insects. Epidermal cells are responsible for the deposition of new cuticle during each molt, and the midgut cells are generally associated with the formation of the PM during feeding. Both the cuticle and the PM contain chitin microfibrils, which function as a matrix that binds numerous cuticle and PM proteins. However, chitin is associated with other tissues as well, including the head skeleton, foregut, hindgut, trachea, wing hinges, salivary glands, and mouthparts of adults and/or larvae. In early development, chitin is additionally found in the cuticle of the developing larva within the embryo, as well as in the extra-embryonic serosal cuticle and the eggshells (Wilson and Cryan, 1997; Moreira *et al.*, 2007;

Rezende *et al.* 2008). In general, it is assumed that the cells closest to the site where chitin is found are responsible for its biosynthesis. However, this interpretation is somewhat complicated by the fact that assembly of chitin microfibrils occurs in the extracellular space and is influenced by proteins that organize their deposition (see section 7.6).

7.3.1.1. Chitin synthesis in the epidermis and tracheal system   Chitin is a major constituent of the cuticle, the outermost layer of insects, which serves as an exoskeleton and protects against various harming agents. Within the cuticle, chitin is mainly found in the procuticle, with higher amounts in the endocuticle than in the exocuticle, but is absent from the epicuticle (Sass *et al.*, 1994). Chitin deposition in the cuticle was recently reinvestigated in *Drosophila* embryos in an ultrastructural study using electron microscopy and gold-conjugated wheat germ agglutinin (gold-WGA), which binds to GlcNAc residues in chitin and glycoproteins (Schwarz and Moussian, 2007). In agreement with previous findings, gold particles could only be detected in the procuticle but not in the epicuticle. The gross architecture of the procuticle is established mainly by consecutive layers of chitin bundles of microfibrils embedded in a matrix of cuticle proteins. The orientation of a single lamina of chitin microfibrils can be twisted in relation to the neighboring layers above and below it by different angles in different insect species, giving rise to helicoidal or pseudo-orthogonal textures. Much of what we know on cuticle differentiation derives from ultrastructural studies of cuticle renewal during insect molting (Locke, 2001; Moussian, 2010). The classical concept of cuticle formation is based on three sequential phases. First, the envelope is laid down at the plasma membrane surface, usually above electron-dense plaques at the tips of microvilli, which were postulated to carry the chitin-synthesizing machinery (Locke, 1991). Then, the epicuticle is assembled beneath the envelope. Finally, the procuticle, which is considerably thicker than the other two layers, is assembled and oriented at the cell surface. However, a recent ultrastructural study of cuticle differentiation in *Drosophila* embryos revealed a slightly different picture, as envelope, epicuticle, and procuticle are partially formed in parallel in the first phase, then the cuticle thickens in the second phase, and in a third phase the chitin laminae acquire their final orientation (reviewed in Moussian *et al.*, 2006). Interestingly, the apical membrane of the embryonic epidermis does not form microvilli-like protrusions. Instead, it exhibits longitudinal microtubule-stabilized furrows, which were called apical undulae and are oriented perpendicular to the first layers of chitin microfibrils (Schwarz and Moussian, 2007). These apical undulae may have a crucial role in determining the orientation of chitin microfibrils, at least in the embryonic cuticle. Factors that affect the shape of the apical membrane, such as syntaxin 1A, indirectly affect chitin orientation, presumably by

interfering with the transport of proteins involved in cuticle or chitin assembly (Moussian *et al.*, 2007).

During embryogenesis, chitin synthesis also plays a role for tracheal morphogenesis. Chitin is also found in the tracheal cuticle, which has been thought to have a composition similar to that of the epidermal cuticle. This point needs clarification by direct chemical analysis of tracheal cuticle. However, it came as a surprise when two research groups reported independently that chitin forms a transient lumenal matrix during tracheal development in *Drosophila* embryos (Devine *et al.*, 2005; Tonning *et al.*, 2005). The lumenal chitin appears to be necessary to control tube size, diameter, and shape by orchestrating the function of surrounding tracheal cells. *Drosophila* genetics, in combination with different microscopic techniques, have proven most valuable in dissecting cuticle differentiation, and yielded a number of factors that are involved in controlling this process. Some of these factors will be discussed in more detail later in this chapter (see section 7.6.).

In addition to the histochemical detection of chitin with colored or fluorescent compounds that bind to chitin with different specificities, the expression of CHS genes has been used to identify chitin-synthesizing tissues. *CHS* gene expression was analyzed in various insects by RT-PCR, Northern blots, and *in situ* hybridization. These studies clearly demonstrated that epidermal and tracheal cells express *CHS* genes, and hence confirmed that these epithelia are sites of chitin biosynthesis. The first cDNA encoding an insect chitin synthase was identified by Tellam and colleagues (2000) in *Lucilia cuprina* (sheep blow fly), and termed *LcCHS1*. RT-PCR using total RNA preparations from the carcass and trachea indicated expression of *LcCHS1* in these tissues. *In situ* hybridization revealed a strong signal for the *LcCHS1* mRNA in a single layer of epidermal cells immediately underneath the procuticle. Similar results were obtained for the expression of homologous *CHS* (also referred to as *CHS-A*) genes from other insect sources, including *D. melanogaster*, *M. sexta*, *Spodoptera frugiperda* (fall armyworm), and *T. castaneum* (red flour beetle) (Ibrahim *et al.*, 2000; Gagou *et al.*, 2002; Zhu *et al.*, 2002; Arakane *et al.*, 2004; Bolognesi *et al.*, 2005; Hogenkamp *et al.*, 2005; Zimoch *et al.*, 2005). In agreement with the detection of chitin in eggs, *CHS* gene expression was reported during embryogenesis by RT-PCR using RNA from *Lucilia sericata* and *Aedes aegypti* eggs (Moreira *et al.*, 2007; Tarone *et al.*, 2007; Rezende *et al.*, 2008).

7.3.1.2. Chitin synthesis in the midgut
Chitin is a component of the insect PM, and accounts for about 3–13% (w/w) of its dry weight. There are two patterns of PM production in insects. Type I PMs are synthesized and delaminated throughout the entire midgut epithelium. Type II PMs are formed as a continuous lining of the gut, which is produced by a specialized region of the anterior midgut called the cardia (Lehane, 1997). The most detailed picture of chitin

synthesis and its association with PM proteins has emerged from observations using transmission, scanning electron, light, and fluorescence microscopy (TEM, SEM, LM, and FM, respectively) in three lepidopteran species; namely, *Ostrinia nubilalis* (European corn borer), *Trichoplusia ni* (cabbage looper), and *M. sexta* (Harper and Hopkins, 1997; Harper *et al.*, 1998; Harper and Granados, 1999; Wang and Granados, 2000; Hopkins and Harper, 2001; Zimoch and Merzendorfer, 2002). TEM in combination with gold-WGA staining demonstrated that the PM of *O. nubilalis* contains a fibrous, chitin-containing matrix that appears first at the tips of the microvilli of the midgut epithelial cells just past the stomadeal valves, and is rapidly assimilated into a thin PM surrounding the food bolus (Harper and Hopkins, 1997). The PM becomes thicker and multilayered in the middle and posterior regions of the midgut. The orthogonal lattice of chitin meshwork is slightly larger than the diameter of the microvilli. SEM and LM studies revealed that the PM delaminates from the tips of the microvilli. This observation suggests that microvilli serve as sites (and possibly as templates) for the organization of the PM by laying down a matrix of chitin microfibrils, which associate with PM proteins. A similar pattern of delamination of the PM containing both chitin and intestinal mucins was demonstrated in larvae of *T. ni* (Harper and Granados, 1999; Wang and Granados, 2000). Incorporating WGA into the diet can interrupt formation of the PM. WGA-fed *O. nubilalis* larvae exhibited an unorganized PM, which was multilayered and thicker than the normal PM (Hopkins and Harper, 2001). WGA was actually associated with the PM as well as with the microvillar surface, as revealed by immunostaining with antibodies specific for WGA. Because there was very little WGA within the epithelial cells, the interaction of WGA appears to be extracellular. Presumably, WGA interferes with the formation of the organized chitin network and/or the association of PM proteins with the chitin network, leading to a reduced protein association with the PM (Harper *et al.*, 1998). There was also extensive disintegration of the microvilli, and the appearance of dark inclusion bodies, as well as apparent microvillar fragments within the thickened multilayered PM. Species such as *M. sexta*, which secrete multiple and thickened PMs that are somewhat randomly organized, tolerated WGA better, and sequestered larger amounts of WGA within the multilayered PM (Hopkins and Harper, 2001).

As in the case of epidermal chitin synthesis, RT-PCR, Northern blots and *in situ* hybridization demonstrated the expression of a gene encoding a midgut specific CHS form. This gene was originally identified in *D. melanogaster*, but its expression and function were characterized in *Aedes aegypti*, *M. sexta*, and *T. castaneum* (Ibrahim *et al.*, 2000; Zimoch and Merzendorfer, 2002; Arakane *et al.*, 2004), as well as more recently in *S. exigua*, *S. frugiperda*, and *O. nubilalis* (Bolognesi *et al.*, 2005; Kumar *et al.*, 2008; Khajuria *et al.*, 2010).

The first evidence that midgut cells express a *CHS* gene was provided by Ibrahim *et al.* (2000) for female *Ae. aegypti* mosquitoes dissected several hours after a blood meal. *In situ* hybridizations with an antisense RNA probe for *AeCHS2* (*CHS-B*) in blood-fed mosquitoes localized the mRNA at the apical site of midgut epithelial cells. Likewise, *in situ* hybridization with an antisense RNA probe to *MsCHS2* (*CHS-B*) from *M. sexta* revealed that high levels of transcripts for this gene are present in apical regions of the columnar cells of the anterior midgut but completely absent in the epidermis or tracheal system of *M. sexta* larvae (Zimoch and Merzendorfer, 2002). An antibody to the catalytic domain of the *M. sexta*, CHS was used to detect the enzyme in midgut brush border membranes at the extreme apical ends of microvilli, a result suggestive of some special compartment or possibly apical membrane-associated vesicles. In line with its assumed role in PM formation during feeding stages, *MsCHS2* mRNA was detected in the midgut of feeding but not of starving or molting larvae (Zimoch *et al.*, 2005). Similar expression patterns were reported for *S. exigua*, *S. frugiperda*, and *O. nubilalis* by RT-PCR (Bolognesi *et al.*, 2005; Kumar *et al.*, 2008; Khajuria *et al.*, 2010). In *S. frugiperda*, chitin could be stained in the PM only when *SfCHS2* (*CHS-B*) expression was detectable (Bolognesi *et al.*, 2005). From the finding that *TcCHS2* (encoding *CHS-B*) expression was observed in *T. castaneum* only in late larvae and adults, but not in pupal stages, where chitin is synthesized during cuticle formation, it was concluded that *TcCHS-B* functions in the course of PM formation in the midgut (Arakane *et al.*, 2004), a hypothesis further substantiated by RNAi experiments (Arakane *et al.*, 2005; see also section 7.3.4.3).

7.3.2. Chitin Biosynthetic Pathway

It has been assumed that most parts of the chitin biosynthetic pathway of insects would be similar or identical to the Leloir pathway, which has been worked out extensively in fungi and other microbes (**Figure 1**). This appears to be the case except for some minor details (Palli and Retnakaran, 1999). The source of the sugar residues for chitin synthesis can be traced to fat body glycogen, which is acted upon by glycogen phosphorylase. Glucose-1-P produced by this reaction is converted to trehalose, which is released into the hemolymph. Trehalose, the extracellular source of sugar in many insect species, is acted upon by a trehalase, which is widely distributed in insect tissues, including the epidermis and gut, to yield intracellular glucose (Becker *et al.*, 1996). This view was recently substantiated by Chen *et al.* (2010), who showed that the RNAi-induced knockdown of the expression of two trehalase-encoding genes, *SeTre1* and *SeTre2*, caused downregulation of the CHS-encoding genes *SeCHS1* and *SeCHS2*, respectively, and led to reduced chitin levels in the cuticle and the PM. The conversion of glucose to fructose-6-P

needed for chitin synthesis involves two glycolytic enzymes present in the cytosol. These enzymes are hexokinase and glucose-6-P isomerase, which convert glucose to fructose-6-P. From the latter, the chitin biosynthetic pathway branches off, with the first enzyme catalyzing this branch being glutamine fructose-6-phosphate amidotransferase (GFAT, E. C. 2.6.1.16), which might be thought of as the first committed step in amino sugar biosynthesis. The conversion of fructose-6-P to GlcNAc phosphate involves amination, acetyl transfer, and an isomerization step, which moves the phosphate from C-6 to C-1 (catalyzed by a phospho-N-acetylglucosamine mutase). The conversion of this compound to the nucleotide sugar derivative follows the standard pathway and leads to the formation of an UDP-derivative of GlcNAc, which serves as the substrate for CHS. The entire chitin biosynthetic pathway is outlined in **Figure 1**. The involvement of dolichol-linked GlcNAc as a precursor for chitin was proposed quite some time ago (Horst, 1983), but that hypothesis has received very limited experimental support (Quesada-Allue, 1982). At this point, this possibility remains unproven. Similarly, the requirement for a primer to which the GlcNAc residues can be transferred also remains speculative. Based on the model for glycogen biosynthesis, which requires glycogenin as the primer (Gibbons *et al.*, 2002), CHS or an associated protein may fulfill this priming function. Because each sugar residue in chitin is rotated 180° relative to the preceding sugar, which requires CHS to accommodate a alternating "up/down" configuration, another precursor, UDP-chitobiose, has been proposed to be a disaccharide donor during biosynthesis (Chang *et al.*, 2003). Evaluation of radiolabeled UDP-chitobiose as a CHS substrate in yeast, however, revealed that it was not incorporated into chitin. Nevertheless, by testing monomeric and dimeric uridine-derived nucleoside inhibitors as mechanistic probes Yeager and Finney (2004) found a 10-fold greater inhibition for the dimeric inhibitor than the corresponding monomeric inhibitor. However, both inhibitors bound with low affinities in the millimolar range. The stereochemical problem in chitin synthesis of adding GlcNAc to the growing chain in two opposite orientations resembles the situation with hyaluronan synthases (HAS), which produce the hyaluronan polymer from two different monosaccharides, UDP-GlcNAc and UDP-glucuronic acid. HASs are "dual action" glycosyltransferases that accomplish hyaluronan biosynthesis by two substrate-binding and active sites (Weigel and DeAngelis, 2007). As class I HASs are related to chitin synthase, two binding sites for alternating GlcNAc orientations may also occur in CHSs.

7.3.2.1. Key enzymes The biosynthetic pathway of chitin can be thought of as consisting of three subreactions. The first set leads to the formation of the amino sugar, GlcNAc, the second to its activated form UDP-GlcNAc, and the last yields the polymeric chitin

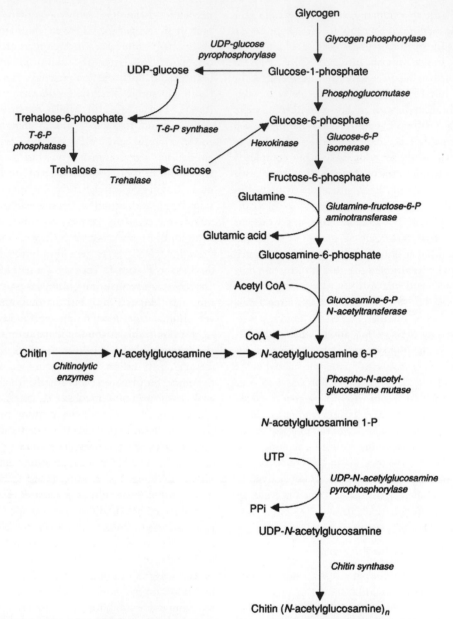

Figure 1 Biosynthetic pathway for chitin in insects starting from glycogen, trehalose, and recycled chitin.

from the amino sugar. The rate-limiting enzyme in the first subreaction appears to be glutamine-fructose-6-phosphate aminotransferase (GFAT, EC 2.6.1.16), which is found in the cytosol. The critical enzyme in the second subreaction is UDP-N-acetylglucosamine pyrophosphorylase (UAP, EC 2.7.7.23), which is also found in the cytosol, and that in the last subreaction is CHS (EC 2.4.1.16), which is localized in its active form at the plasma membrane. Not surprisingly, these three enzymes appear to be major sites of regulation of chitin synthesis.

7.3.2.2. Function and regulation of GFATs In *Drosophila*, two genes encoding GFAT (*Gfat1* and *Gfat2*) have been identified (Adams *et al.*, 2000; Graack *et al.*,

2001). Both of these genes are on chromosome 3, but they are present at different locations. Their intron–exon organizations are different, as are the amino acid sequences of the encoded proteins. GFAT consists of two separate domains: an N-terminal domain that has both glutamine-binding and aminotransferase motifs identified in GFATs from other sources; and a C-terminal domain with both fructose-6-phosphate binding and isomerase motifs. *Gfat1* is expressed in embryos in the developing tracheal system, cuticle-forming tissues, and corpora cells of larval salivary glands (Graack *et al.*, 2001). The major regulation of GFAT1 appears to be post-translational. When *Gfat1* was expressed in yeast cells, the resulting enzyme was feedback-inhibited by UDP-GlcNAc, and was stimulated by protein kinase A (PKA). Even though it has not been demonstrated that there is a phosphorylated form of GFAT1 which is susceptible to feedback inhibition by UDP-GlcNAc, this possibility remains viable. However, the situation may be complicated by overlapping kinase activities, as recently a novel, highly conserved phosphorylation site was identified, which accounts for *in vivo* phosphorylation of human GFAT1 overexpressed in insect cells by protein kinases other than PKA (Li *et al.*, 2007). Examination of a mutant that mimics phosphorylation at this site demonstrated that the modification stimulates glucosamine-6-phosphate-synthesizing activity, but has no effect on UDP-GlcNAc inhibition.

Another insect species in which *Gfat1* function and regulation has been analyzed in more detail is *Aedes aegypti*. The mosquito gene has no introns, and the promoter appears to contain sequence elements related to ecdysteroid response elements (EcRE) as well as E74 and Broad complex Z4 elements. E74 and Broad complex Z4 proteins are transcription factors known to be upregulated by ecdysone (Thummel, 1996). Two Gfat1 transcripts with different sizes were observed in Northern blot analyses of RNA from adult females, and their levels increased further after blood feeding (Kato *et al.*, 2002). Since ecdysteroid titers increase following blood feeding, it is possible that this gene is under the control of ecdysteroids, either directly or indirectly. Silencing of gene expression by dsRNA injection additionally revealed that GFAT1 is necessary for chitin synthesis in the course of PM formation in the midgut, which occurs in female mosquitoes in response to a blood meal (Kato *et al.*, 2006). Feedback inhibition of chitin synthesis by UDP-GlcNAc has also been reported in this study, indicating that the mosquito enzyme is likely to be regulated in a manner similar to the *Drosophila* enzyme.

7.3.2.3. Function and regulation of UAPs Insect genomes usually possess only one gene encoding UDP-GlcNAc pyrophosphorylase (UAP). The known exception is *T. castaneum*, which has two *UAP* genes. The first phenotypes for defects in the *UAP* gene were described in *Drosophila*, where this gene was alternately termed *mummy*, *cabrio*, or *cystic*, according to three phenotypes that were identified in independent genetic screens for genes involved in tracheal, epidermal, and CNS development (Nüsslein-Vollhard *et al.*, 1984; Hummel *et al.*, 1999; Beitel and Krasnow, 2000). While *cystic* was originally recognized to be important for tracheal morphogenesis and tube size control, *mummy* and *cabrio* mutants were reported to exhibit severe defects in cuticle formation and CNS development of the embryo. Eventually, all of these genes were shown to be allelic by Araújo *et al.* (2005) and Schimmepfeng *et al.* (2006), and the gene encoding UAP is now consistently named *mummy* (*mmy*). Interestingly, the *mmy* mutant phenotype is similar to that of the so-called "halloween"

mutants, which fail to produce the morphogenetic hormone 20-hydroxyecdysone (Gilbert, 2004). UAP functions in apical extracellular matrix formation by producing UDP-GlcNAc needed for chitin synthesis and for protein glycosylation. Consequently, deletion or defects in *mmy* can lead to the complete absence of chitin in the cuticle and tracheal lumen, as evidenced by a lack of WGA staining in mutant embryos or larvae carrying a single nucleotide substitution leading to the exchange of glycine to valine at position 261 (Tonning *et al.*, 2006). Moreover, the epithelial organization is affected in *mmy* mutants, as adherens junctions between epidermal cells appear wider than in wild type embryos, and the characteristic ladder-like structure of the septate junctions is missing. Additionally, a membrane-integral septate junction component (Fas3) is delocalized in the mutant, indicating that *mmy* may have an additional function in proper localization of membrane-bound septate junction components (Tonning *et al.*, 2006). Expression of *mmy* is hormonally regulated in apical extracellular matrix-differentiating tissues, and selectively upregulated when chitinous material is deposited during development. It is possible that the enzyme is also regulated at the post-translational level by uridine, as this nucleic acid base was shown to be an effective inhibitor for the yeast enzyme (Yamamoto *et al.*, 1980). In *Ae. aegypti* the gene encoding UAP is constitutively expressed throughout all life stages, and blood feeding does not significantly alter mRNA levels (Kato *et al.*, 2005). The cDNA was cloned and the enzyme expressed as a recombinant enzyme, allowing determination of substrate specificity. The enzyme uses GlcNAc-1-P as a substrate, but it also exhibited low activity when incubated with Glc-1-P. In *T. castaneum* two UAP isoforms were identified, which share 60% identical amino acids but differ significantly in their developmental and tissue-specific expression patterns, as well as in function, as revealed by RNAi studies (Arakane *et al.*, 2010). While the knockdown of *TcUAP1* transcripts caused arrested development at the larval–larval, larval–pupal, and pupal–adult molts, knockdown of *TcUAP2* retarded larval growth or resulted in pupal paralysis. Results of chitin-staining experiments in cuticle and PM indicated that chitin deposition is prevented only when *TcUAP1*, but not when *TcUAP2*, expression was blocked. However, both genes are essential for beetle development and survival. TcUAP1 obviously is required for chitin synthesis in the course of cuticle and PM formation, whereas TcUAP2 appears to have other critical roles, presumably in glycosylation of proteins.

7.3.3. Chitin Synthases: Organization of Genes and Biochemical Properties

7.3.3.1. Number and organization of *CHS*-encoding genes *CHS* genes from numerous unicellular and filamentous species of fungi have been isolated and characterized (reviewed in Roncero, 2002; Horiuchi, 2009). Genome sequencing revealed three to nine *CHS* genes per individual fungal species, which were categorized into seven gene classes. In contrast, nematode, mollusk, crustacean, and insect genomes contain only one or two *CHS* genes per species (**Figure 2A**). Since Tellam *et al.* (2000) published the first cDNA sequence for a CHS from *Lucilia cuprina* (sheep blowfly), cDNA sequences for CHSs have been reported from numerous invertebrates, and the availability of an increasing number of genome sequences has provided additional information on *CHS* genes. Nematode *CHS*s were from two filarial pathogens, *Brugia malayi*, and *Dirofilaria immitis*, the plant parasite *Meloidogyne artiellia* and *Caenorhabditis elegans* (Harris *et al.*, 2000; Harris and Fuhrman, 2002; Veronico *et al.*, 2001). In both *D. immitis* and *M. artiella*, there is currently only evidence for a single gene, but in *B. malyai* and *C. elegans*, two genes were identified. *CeCHS1* is required for eggshell formation, whereas *CeCHS2* is needed to form the grinder in the ectodermal pharynx (Zhang *et al.*, 2005). CHS sequences from crustaceans and chelicerates were deduced from the *Daphnia pulex* and *Ixodes scapularis* genome projects, both of which have two *CHS* genes. Likewise, all insect genomes available so far harbor two *CHS* genes, which have been divided into class A and class B genes, with the latter appearing to be the more ancient form (**Figure 2A**).

The insect species from which complete cDNAs for CHSs have been isolated are *L. cuprina* (Tellam *et al.*, 2000), *D. melanogaster* (Gagou *et al.*, 2002), *Ae. aegypti* (Ibrahim *et al.*, 2000), *Anopheles quadrimaculatus* (Zhang and Zhu, 2006), *M. sexta* (Zhu *et al.*, 2002), *S. frugiperda* (Bolognesi *et al.*, 2005), *Spodoptera exigua* (Chen at al., 2007; Kumar *et al.*, 2008) and *T. castaneum* (Arakane *et al.*, 2004). Genomic sequences from *Anopheles gambiae*, *T. castaneum*, *D. melanogaster* and *M. sexta*, which were deduced from available genome projects or obtained by individual nucleotide sequencing, were used to determine the organization of *CHS* genes in these species (**Figure 3**).

The overall structure of *CHS* genes varies among different insect species and gene classes. The numbers of exons range from 8 to 24, with lengths from 46 bp to more than 3000 bp. While most genes contain at least some exons that contribute longer ORFs, the lepidopteran *CHS* genes appear more fragmentized, because they contain a higher number of shorter exons (Zhu *et al.*, 2002; Kumar *et al.*, 2008). Insect *CHS-A* genes have two mutually exclusive exons, resulting in two mRNA splice variants. Both exons code for 59 amino acids comprising extracellular, transmembrane and intracellular domains, the latter being located near the carboxyl terminus of the protein. One major difference between the two exons that are alternately spliced is that all of the *b* forms code for segments that have a site for *N*-linked glycosylation just before the transmembrane helix,

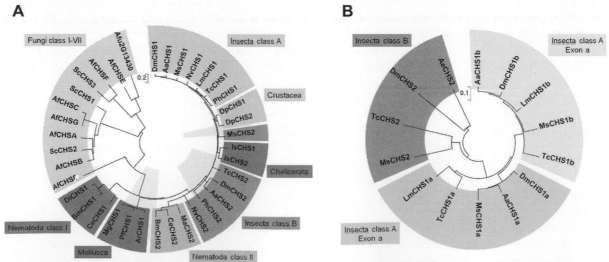

Figure 2 Phylogenetic trees of CHS proteins and conserved exons. The trees are based on ClustalW alignments and were performed with the neighbor joining method. Bootstrap tests of phylogeny were performed with 10,000 replications., (A) Bootstrap consensus tree of CHS proteins from fungi, nematodes, mollusks and arthropods., (B) Bootstrap consensus tree of exons a and b found in class A CHS genes, and the corresponding region of class B CHS genes. *Aa, Aedes aegypti* (XP_001662200.1, XP_001651163.1); *Af, Aspergillus fumigatus* (XP_749322.1, XP_746604.1, XP_748263.1, XP_752630.1, CAA70736.1, XP_747364.1, XP_754184.1, XP_755676.1); *Ar, Atrina rigida* (AAY86556.1); *Bm, Brugia malayi* (XP_001898491.1, AAS77206.1); *Ce, Caenorhabditis elegans* (NP_492113.2, NP_493682.2); *Dm, Drosophila melanogaster* (AAG22215.3, AAF51798.2); *Di, Dirofilaria immitis* (AAG39382.1); *Dp, Daphnia pulex* (NCBI_GNO_134384, NCBI_GNO_326244); *Is, Ixodes scapularis* (XP_002405234.1; XP_002405231.1); *Lm, Locusta migratoria* (ACY38589.1); *Ms, Manduca sexta* (AAL38051.2, AAX20091.1); *Ma, Meloidogyne artiellia* (AAG40111.1); *Mg, Mytilus galloprovincialis* (ABQ08059.1); *Nv, Nasonia vitripennis* (XP_001602290.1, XP_001602181.1); *Pf, Pinctada fucata* (BAF73720.1); *Ph, Pediculus humanus corporis* (XP_002423597.1), XP_002423604.1); *Sc, Saccharomyces cerevisiae* (NP_014207.1, NP_009594.1, NP_009579.1); *Tc, Tribolium castaneum* (AAQ55059.1, AAQ55061.1).

whereas none of the *a* forms do. The precise physiological significance of alternate exon usage and potential glycosylation in CHS expression and function is still unknown, even though it is clear that there is developmental regulation of alternate exon usage (see section 7.3.4.2.).

7.3.3.2. Modular structure of chitin synthases

CHSs are members of family GT2 of the glycosyltransferases (Coutinho *et al.*, 2003), which generally utilize a mechanism where inversion of the anomeric configuration of the sugar donor occurs. The protein fold (termed GT-A) for this family is considered to be two associated β/α/β domains that form a continuous central sheet of at least eight β-strands. The GT-A enzymes share a common ribose/metal ion-coordinating motif (termed DxD motif) as well as another carboxylate residue that acts as a catalytic base. The general organization of CHSs has been deduced from a comparison of amino acid sequences of these enzymes from several species of insects, nematodes, and yeasts (Merzendorfer, 2006). These enzymes have three distinguishable domains: an N-terminal domain with moderate sequence conservation among different species and containing several transmembrane segments; a central catalytic domain that is believed to be orientated toward the cytoplasm; and a C-terminal domain with multiple transmembrane segments (**Figure 4**). The catalytic domain contains several highly conserved stretches including GT2 consensus sequences, which have been suggested to be involved in binding of UDP, the donor and acceptor saccharides, and the product. They include sequences similar to the Walker A and B motifs for binding of the nucleotide moiety (Walker *et al.*, 1982), sequences similar to the DXD and G(X)4(Y/F) R motifs likely involved in substrate binding, the GEDRxx(T/S) motif at the acceptor binding site, and the (Q/R)XXRW motif involved in product binding. The latter motif is present only in processive GTs. While the transmembrane segments in the N-terminal domain show different patterns among different insect species, the transmembrane segments in the C-terminal domain are remarkably conserved both with respect to their location and the spacing between adjacent transmembrane segments. Particularly striking is the fact that five such transmembrane segments are found in a cluster immediately following the catalytic domain, and two more segments are located closer to the C-terminus. The cluster of five transmembrane helices spanning the membrane, known as 5-TMS (5-transmembrane spans), has been suggested to be involved in the extrusion of the polymerized chitin chains across the plasma membrane to the exterior of the cell, as has been proposed for the extrusion of cellulose (Richmond, 2000). Following the last transmembrane helix of the 5-TMS, a sequence similar to the (S/T)WGT(R/K) motif found in fungal chitin synthases is located at the extracellular site. The CHSs derived from class A genes were predicted to have a coiled-coil region following the 5-TMS region (Zhu *et al.*, 2002; Arakane *et al.*, 2004). Also, all of the genes encoding the class A CHSs have two alternate exons (corresponding to alternate exon 7 of *D. melanogaster*, exon 8 of *T. castaneum*, exon 6 of *A. gambiae*, and the exon 20 homolog of *M. sexta*). The alternate exons are located on the C-terminal side of the 5-TMS region, and encode the next transmembrane segment and flanking

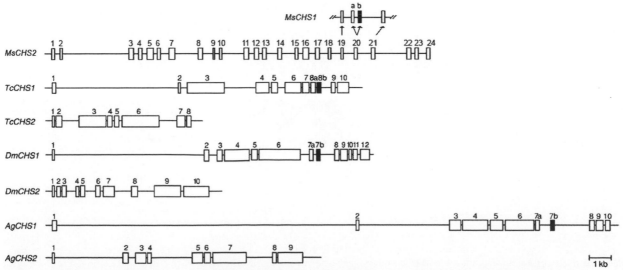

Figure 3 Schematic diagram of the organization of insect *CHS-A* and *CHS-B* genes. The exon–intron organization was deduced from comparisons of available cDNA and genomic sequences. Boxes indicate exons; lines indicate introns. The second of the two alternative exons (8b) of *TcCHS1*, *DmCHS1* (*7b*), *AgCHS2* (*7b*), and *MsCHS1* (homolog of exon 20 from *MsCHS2*, *20b*) are indicated as closed boxes (modified according to Arakane *et al.*, 2004, and Hogenkamp *et al.*, 2005). *Ag, Anopheles gambiae; Dm, Drosophila melanogaster; Ms, Manduca sexta; Tc, Tribolium castaneum.*

sequences (**Figure 4**). The alternate exon-encoded regions of the CHS proteins differ in sequence by as much as 30%, and most of these differences are in the regions flanking the transmembrane segment. This finding suggests that the proteins may differ in their ability to interact with cytosolic or extracellular proteins, which might regulate chitin synthesis, transport, and/or organization. An attractive hypothesis is that these flanking sequences may influence the plasma membrane location of a CHS by interacting with cytoskeletal elements, or perhaps by generation of extracellular vesicles involved in chitin assembly.

7.3.3.3. Zymogenic properties of chitin synthase

In numerous fungal and insect systems, chitin synthesis is activated by trypsin and other serine proteases, suggesting that CHS is produced as a zymogen (reviewed in Merzendorfer, 2006). However, there is very little knowledge on the significance of this phenomenon in arthropods. In yeast, which has three *CHS* genes, proteolytic activation by trypsin has been reported for Chs1 and Chs2 (Cabib and Farkas, 1971; Sburlati and Cabib, 1986). With Chs3, the situation is more complicated, as the zymogenic properties appear to depend on UDP-GlcNAc and additional proteins, such as the regulatory subunit Chs4 (Choi *et al.*, 1994; Ono *et al.*, 2000). However, no endogenous proteinase has been identified that would cleave the CHS zymogen. The

zymogenic properties of yeast Chs2 and Chs3 have been reinvestigated recently. For Chs2 it was demonstrated that trypsin acts on a soluble protease that, once activated, stimulates Chs2 activity (Martínez-Rucobo *et al.*, 2009). Another study reports a role of the CaaX proteinase Ste24 in chitin synthesis (Meissner *et al.*, 2010). Ste24 is a membrane-integral protease of the endoplasmic reticulum, which is known to be involved in proteolytic maturation of the yeast mating factor **a**. Yeast two-hybrid studies have indicated, however, that Ste24 interacts with Chs3. The interacting domain was mapped to a cytosolic region that immediately precedes the catalytic domain of Chs3. Deletion of *ste24* led to Calcofluor white (CFW) resistance and decreased chitin levels, whereas overexpression led to CWF hypersensitivity and increased chitin levels. The CFW phenotype of wild type cells could be rescued by expressing the homologous gene from *T. castaneum* in *ste24*Δ cells, indicating orthologous functions. Although Ste24 directly binds to Chs3, it appears not to be a substrate of the protease. Instead, genetic experiments indicate that Chs4 is cleaved by Ste24 in a prenylation-dependent manner at its C-terminal CaaX motif, and that this processing is required for intracellular transport of Chs3 to the plasma membrane (Meissner *et al.*, 2010). Addition of trypsin to cell-free extracts obtained from different insect species such as *Diaprepes abbreviatus*, *M. sexta*, *T. castaneum*, and *Stomoxys calcitrans* leads to the stimulation of chitin synthesis by

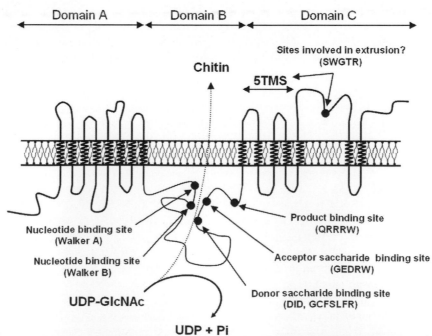

Figure 4 Structural model of the tripartite domain organization of *Drosophila* DmCHS1. The N-terminal domain A of *Drosophila* contains 8 transmembrane helices (TMHs), but this number varies, between different insect species, from 7 to 10. The central domain B is facing the cytoplasm, and forms the catalytic site. The ensuing domain C contains 5 + 2 TMHs, and the C-terminus is located at the extraplasmatic site. Generally, all TMHs are highly conserved in insects. Putative motifs involved in nucleotide, donor, acceptor, and product binding are indicated. The polymer is synthesized in the cytosol and the chitin chain needs to be translocated across the membrane, a process that might require the 5TMS cluster and the extrusion motif SWGTR.

30–50% (Cohen and Casida 1980b; Mayer *et al.*, 1980; Ward *et al.*, 1991; Zimoch *et al.*, 2005). In *Manduca*, trypsin-dependent stimulation of chitin synthesis was observed in crude midgut extracts, but not in membrane fractions of the midgut. However, it could be restored by re-adding a soluble fraction, suggesting that trypsin does not directly act on CHS but on a soluble protein that in turn stimulates chitin synthesis, which is similar to the recent finding of a soluble factor that activates yeast Chs2 (Zimoch *et al.* 2005; Martínez-Rucobo *et al.*, 2009). Attempts to directly purify and identify the soluble factor from *M. sexta* have failed. However, a chympotrypsin-like peptidase (MsCTLP-1) was identified in the midgut, which binds to the extracellular C-terminal domain of MsCHS2 (Broehan *et al.*, 2007). MsCTLP-1 is secreted into the gut lumen when the larvae start to feed, and it stimulates chitin synthesis after being proteolytically activated by trypsin (Broehan *et al.*, 2008). In line with the assumption that CHS enzymes are produced as zymogens, denaturing gel electrophoresis and immunoblotting of oligomeric CHS complexes purified from the midgut of *M. sexta* yielded a distinct pattern of CHS fragments, which is consistent with the assumption that the CHS monomer is cleaved twice during maturation, and that the resulting three fragments are part of the active enzyme (Maue *et al.*, 2009).

7.3.4. Chitin Synthases: Regulation and Function

7.3.4.1. Regulation of chitin synthase gene expression
Insect class *CHS1* and *CHS2* genes encoding CHS-A and CHS-B enzymes, respectively, are expressed in different tissues and exhibit different patterns of expression during development. Although technical difficulties associated with the isolation of specific tissues free of other contaminating tissues (mainly trachea) initially hampered their unambiguous assignment, some general conclusions can be drawn from studies investigating *CHS* gene expression in different species and stages of development. *CHS* genes are expressed at all stages of growth, including embryonic, larval, pupal, and adult stages. *CHS1* genes are expressed over a wider range of developmental stages (Tellam *et al.*, 2000; Gagou *et al.*, 2002; Zhu *et al.*, 2002). *CHS2* genes are not expressed in the embryonic or pupal stages but are expressed in the larval stages, especially during feeding in the last instar and in adults, including blood-fed mosquitoes (Ibrahim *et al.*, 2000; Zimoch and Merzendorfer, 2002; Arakane *et al.*, 2004). These developmental differences in *CHS1* and *CHS2* expression prompted the assumption that insect *CHS* genes have specialized functions in different tissues or at different developmental stages. Accordingly, *LcCHS1* is expressed only in the carcass (larva minus internal tissues) and trachea of *L. cuprina*, but not in salivary gland, crop, cardia, midgut, or hindgut (Tellam *et al.*,

2000). In blood-fed female mosquitoes, a gene encoding a CHS-B enzyme is expressed in epithelial cells of the midgut (Ibrahim *et al.*, 2000). In *T. castaneum*, *TcCHS1* is expressed in embryos, larvae, pupae, and young adults, but not in mature adults (more than a month old), while *TcCHS2* is expressed at early and late larval stages as well as in adult stages, but not in embryos and pupae (Arakane *et al.*, 2004). Similar expression profiles were reported in two lepidopteran insects for the *CHS* genes *SeCHS1* in *S. exigua* and *SfCHS2* in *S. frugiperda* (Bolognesi *et al.*, 2005; Chen *et al.*, 2007). Tissue-specific expression was also investigated systematically in *M. sexta* (Zhu *et al.*, 2002; Hogenkamp *et al.*, 2005; Zimoch *et al.*, 2005). These studies demonstrated that *MsCHS1* is expressed in epidermal cells and in the tracheal system of larvae and pupae, whereas *MsCHS2* is expressed only in midgut tissue. Transcriptional regulation of *CHS* expression has been suggested to be mediated by ecdysone-responsive elements in the upstream regions of both *Drosophila* genes (Merzendorfer and Zimoch, 2003; also see section 7.7). For *krotzkopf verkehrt* (*kkv*), the gene encoding DmCHS1, another mode of transcriptional control appears to exist, as it is strongly upregulated in epidermal cells surrounding wounds caused by microinjection needles (Pearson *et al.*, 2009). What is remarkable, however, is that *kkv* uses a fundamentally different signaling pathway for wound activation than other genes involved in wound healing, such as *ddc* and *ple* coding for dopa decarboxylase and tyrosine hydroxylase, respectively. While the latter two genes require the JUN/FOS and grainy head (GRH) transcription factors to induce the wound response, transcriptional activities of the identified wound enhancer in the *kkv* upstream region was not affected by these transcription factors (Pearson *et al.*, 2009).

A more recent finding is that a chitinous serosal cuticle containing chitin is produced very early in development of *Aedes aegypti* (Rezende *et al.*, 2008). The serosal cuticle was shown to contain chitin and to be responsible for the development of desiccation tolerance of mosquito eggs. The serosal chitin is apparently the product of a class A CHS derived from the *CHS1* gene. This burst of chitin synthesis occurs long before organogenesis and before formation of the larval cuticle. Chitin has also been detected in eggs, eggshells, and ovaries of *Aedes aegypti* (Moreira *et al.*, 2007). In ovaries and eggs of *T. castaneum*, we have detected transcripts of *TcCHS-A* (our unpublished data).

To summarize, the analysis of expression patterns of the two *CHS* genes in different tissues and periods of development of several insects suggests that class A CHS enzymes are synthesized by epidermal cells when cuticle deposition occurs in embryos, larvae, pupae, and young adults, whereas class B enzymes are produced by the midgut epithelial cells in the course of PM formation in the larval and adult stages and is probably limited to these feeding stages.

7.3.4.2. Tissue-specific expression of alternate exons
The genes encoding class A CHSs from *D. melanogaster*, *A. gambiae*, *Ae. aegypti*, *T. castaneum*, and *M. sexta*, but not the genes encoding class B CHSs, exhibit two alternately spliced exons, which are highly conserved between different insect species (**Figure 2B**). Each exon encodes a 59-amino acid segment following the 5-TMS region. This segment contains a 20-aa transmembrane region and flanking sequences. In addition, the presence of a predicted coiled–coil region immediately following the 5-TMS region in the CHSs encoded by those genes that have the alternate exons suggests a link between these two structural features, and the possibility of regulation of alternate exon usage. In agreement with this idea, transcripts containing either one of these exons have been detected in *T. castaneum*, *M. sexta*, and, more recently, in *Ae. aegypti* (Arakane *et al.*, 2004; Hogenkamp *et al.*, 2005; Zimoch *et al.*, 2005; Chen *et al.*, 2007; Rezende *et al.*, 2008). In *T. castaneum* embryos, transcripts with either exon 8a or 8b were detected, whereas in last instar larvae and prepupae, only exon 8a transcripts were present. In the pupal stage, however, transcripts with exon 8a or exon 8b were abundant, along with trace amounts of a transcript with both exons. In mature adults none of these transcripts were detected, whereas *TcCHS2* transcripts were easily detected (Arakane *et al.*, 2004). Injection of dsRNA specific to either one of both alternately spliced mRNAs revealed that splice variant 8a of *TcCHS1* is required for both the larval–pupal and pupal–adult molts, whereas splice variant 8b is required only for the latter. This finding, together with the relative amounts of these mRNAs, suggested that the splice variant with exon 8a contributes mostly to pupal cuticular chitin synthesis. Nevertheless, the variant with exon 8b appears to have a vital role in the emergence of the adult from the pupal cuticle, which obviously cannot be fulfilled by the exon 8a isoform alone. With regard to relative amounts of both splice variants, similar results were observed in fifth instar larvae of *M. sexta* (Hogenkamp *et al.*, 2005; Zimoch *et al.*, 2005). RT-PCR based detection of the alternately spliced transcripts at different developmental stages in the epidermis revealed that the ratio of mRNA levels for both splice variants varies during development, with *MsCHS1* exon 20a being more predominant generally than that with exon 20b (Hogenkamp *et al.*, 2005). Tracheal cells also express both variants of *MsCHS1*, but, in this tissue, *MsCHS1* with exon 20b is more abundant (Zimoch *et al.*, 2005). The latter finding was confirmed also in *L. migratoria* (Zhang *et al.*, 2010a). When *LmCHS1* expression was silenced by dsRNA injection into second instar nymphs, the locusts developed three distinct phenotypes exhibiting severe molting defects and eventually died. While the knockdown of *LmCHS1a* expression revealed phenotypes similar to those for *LmCHS1*, the knockdown of *LmCHS1* transcripts with

alternate exon b, which is more abundant than the one with alternate exon a in tracheal tissue, led to crimped cuticles. The major finding of that study, however, was that the function of insect CHSs and their alternate exons are conserved in both holo- and hemimetabolous insects.

As discussed above, AaCHS1 accounts for chitin synthesis in the course of serosal cuticle formation in *Ae. aegypti* embryos, and two splice variants containing either exon 6a or exon 6b have been identified. Quantitative PCR showed that at the moment of serosal cuticle formation, splice variant 6a is predominantly expressed. The biochemical basis for a specific function, however, remains unknown.

7.3.4.3. Knockout mutants and RNAi reveal differential functions of *CHS* genes *Drosophila* mutants and RNAi experiments were extremely helpful in analyzing the differential functions of the two *CHS* genes. EMS mutagenesis and screening of the resultant mutant embryos for defects in epidermal differentiation and cuticular patterning helped to identify genes involved in controlling cuticle morphology (Jüergens *et al.*, 1984; Nüsslein-Volhard *et al.*, 1984; Wieschaus *et al.*, 1984; Ostrowski *et al.*, 2002). These genes include *kkv*, *knickkopf* (*knk*), *grainy head* (*grh*), retroactive (*rtv*), and zepellin (*zep*), some of which will be discussed later, in section 7.6. Mutations in these genes resulted in poor cuticle integrity and reversal of embryonic orientation in the egg to varying degrees. Generally, homozygous mutant embryos failed to hatch. When these mutant embryos were mechanically devitellinized, the cuticles became grossly enlarged, yielding the "blimp" phenotype. Interestingly, embryos derived from wild type females treated with diflubenzuron or lufenuron displayed a similar "blimp-like" phenotype when devitellinized, indicating that either genetic or chemical disruption of chitin deposition leads to this phenotype (Ostrowski *et al.*, 2002; Gangishetti *et al.*, 2009). Also, inhibition of chitin synthesis in *D. melanogaster* embryos induced by the CHS-specific inhibitor nikkomycine Z leads to cuticle defects, as they are similar to those observed in *Drosophila kkv* mutants (Tonning *et al.*, 2006; Gangishetti *et al.*, 2009). Ostrowski *et al.* (2002) characterized the *kkv* gene and identified it as a *CHS*-like gene, and Moussian *et al.* (2005a) finally showed that this class A CHS is essential for chitin synthesis in epidermal and tracheal cuticles. Careful analysis of the ultrastructure of the embryonic cuticle of *kkv* mutants confirmed that chitin synthesis by a class A CHS is essential for procuticle formation. Another interesting finding was that in *kkv* mutants the cuticle frequently detaches from underlying epidermal or tracheal cells, suggesting that chitin is also required for anchoring the cuticle (Moussian *et al.*, 2005a). In addition, the head skeleton of *kkv* mutant embryos is undersized and deformed, and sclerotization and pigmentation are impaired. An unexpected finding was that *kkv* is required for tracheal tube expansion, which starts before chitin is actually deposited in the tracheal cuticle during embryogenesis (Devine *et al.*, 2005; Tonning *et al.*, 2005). This finding suggested that chitin has an additional function in early tracheal morphogenesis. Histological stainings with Congo red, WGA, or a fluorescence-labeled chitin-binding domain revealed that the tracheal lumen contains chitin cables before the tracheal cuticle is formed. The loss of lumenal chitin evidently affects subapical cytoskeletal organization of tracheal cells. Therefore, it was hypothesized that the chitinous lumenal matrix is sensed by tracheal cells to coordinate cytoskeletal organization, which controls the diameter size of the tracheae.

RNAi experiments to interfere with *CHS* expression and investigate CHS function have been performed in *T. castaneum*, *A. aegypti*, *A. gambiae*, *S. exigua*, *O. nubilalis*, and *Locusta migratoria*. In *T. castaneum*, injection of dsRNA for either *TcCHS1* or *TcCHS2* into young larvae, penultimate instar larvae, and prepupae resulted in a substantial knockdown of *TcCHS1* and *TcCHS2* mRNA levels. *TcCHS1*-specific RNAi disrupted larval–larval, larval–pupal, and pupal–adult molts, and caused a significant reduction in total chitin content (Arakane *et al.*, 2005). Interestingly, the phenotypes differed significantly depending on whether the insects were injected in the penultimate larval, last larval, or prepupal instar. The first of these groups failed to pupate and died without any splitting of the old larval cuticle, while the second group initiated the larval–pupal molt, but the pupae died without shedding their larval exuviae, although splitting of the old cuticle had occurred. The third group failed to carry out the pupal–adult molt, and died as pharate adults trapped in their pupal exuviae. In contrast, *TcCHS2* dsRNA injection into last instar larvae or prepupae had no effect on pupal or adult development, but when injected into penultimate instar, the larvae shrank in size and died without molting to the last instar. As the knockdown affected only immature or penultimate larvae, it was suggested that *TcCHS2* knockdown impairs chitin synthesis necessary for PM formation. Indeed, when midguts prepared from last instar larvae treated with dsRNA to *TcCHS1* and *TcCHS2* were stained with a fluorescein-conjugated chitin-binding domain, a fluorescent PM was detected in larvae treated with dsRNA for *TcCHS1*, but not following RNAi for *TcCHS2* (Arakane *et al.*, 2005). These experiments provided strong evidence that the *Tribolium* chitin synthase genes, *TcCHS1* and *TcCHS2*, have different functions, as they are involved in the synthesis of chitin in epidermal/tracheal cuticles and midgut PM, respectively. In a succeeding study, dsRNA for either one of the two *CHS* genes was injected into young and old female adults to investigate effects on egg-laying and embryogenesis (Arakane *et al.*, 2008). When dsRNA for *TcCHS1* was injected into young female adults (less than

10 days old), the beetles died without laying any eggs. When older female adults were injected, the beetles developed normally and laid eggs that were drastically reduced in chitin content and failed to hatch. The embryos had a twisted and enlarged blimp-like phenotype (**Figure 5**). Hence, *TcCHS1* appears to have roles in the development of embryos and adults, in addition to its role in cuticle formation. Interestingly, injection of dsRNA for *TcCHS2* into adults led to a significant reduction in chitin content of the PM, and caused death after 2 weeks. The female beetles treated with dsRNA for *TcCHS2* also failed to lay eggs, presumably due to starvation, because the fat body was significantly depleted due to autophagy (Arakane *et al.*, 2008). Similar to the situation in the beetle, RNAi experiments to knock down the *AaCHS2* transcripts of *Ae. aegypti* showed that it is required in female mosquitoes for the *de novo* synthesis of the PM after a blood meal (Kato *et al.*, 2006).

As the function of CHSs is vital for insect development and survival, RNAi-mediated knockdown of *CHS* genes could be a powerful approach in pest control. Based on the observation that chitin synthesis can be blocked by dsRNA injection in mosquitoes, Zhang *et al.* (2010b) developed a method to generate a systemic knockdown of *CHS* gene expression in *A. gambiae* larvae by feeding nanoparticles consisting of chitosan and dsRNA specific for the target gene. In line with the presumed function of both CHSs in cuticle and PM syntheses, the larvae became more susceptible to diflubenzuron and to Calcofluor White (CFW), when *AgCHS1* or *AgCHS2* expression, respectively, was inhibited. Another promising approach would be to feed bacteria expressing dsRNA to target genes, as was originally performed with *C. elegans* (Timmons and Fire, 1998). Indeed, when *E. coli* bacteria expressing dsRNA to *SeCHS1* were fed to larvae of the lepidopteran pest *S. exigua*, the survival rate was decreased as they advanced in development (Tian *et al.*, 2009).

7.4. Chitin Degradation and Modification

Insects must periodically replace their old cuticle with a new one because it is too rigid to allow for growth. Key to this process is the elaboration of the molting fluid with an assortment of chtitinases and proteases. Chitinases are among a group of proteins that insects use to digest the structural polysaccharide in their exoskeletons and gut linings during the molting process (Kramer *et al.*, 1985; Kramer and Koga, 1986; Kramer and Muthukrishnan, 1997; Fukamizo, 2000). Precise regulation of chitin metabolism is a complex and intricate process that is critical for insect growth, metamorphosis, organogenesis, and survival (Arakane and Muthukrishnan, 2010). Chitin content, which fluctuates throughout the life cycle of the insect, is directly influenced not only by chitin synthases

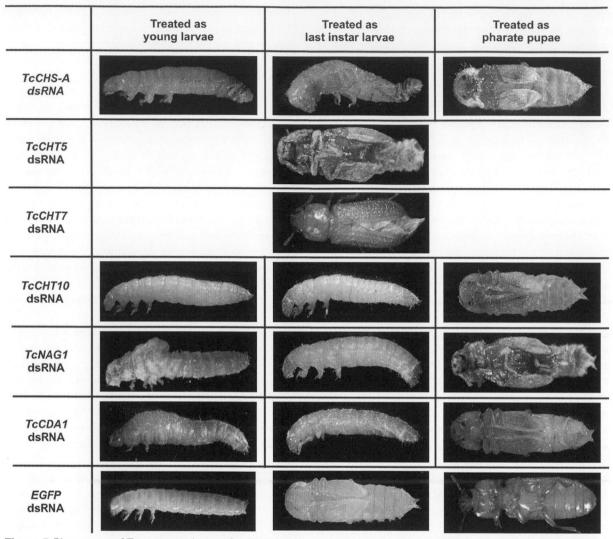

	Treated as young larvae	Treated as last instar larvae	Treated as pharate pupae
TcCHS-A dsRNA			
TcCHT5 dsRNA			
TcCHT7 dsRNA			
TcCHT10 dsRNA			
TcNAG1 dsRNA			
TcCDA1 dsRNA			
EGFP dsRNA			

Figure 5 Phenotypes of *T. castaneum* larvae after RNAi for genes of chitin metabolism. dsRNAs for the indicated genes (200 ng per insect, $n = 20$) were injected into penultimate instar larvae (young larvae), last instar larvae, pharate pupae as indicated above each panel. All animals injected with dsRNA for *CHS-A*, *TcCHT10*, *TcNAG1*, and *TcCDA1* died at the ensuing molt. Unlike RNAi of *TcCHT10*, injection of dsRNA (200 ng per insect) for *TcCHT5* into penultimate instar and last instar larvae as well as pharate pupae prevented only adult molt. When dsRNA for *TcCHT7* (200 ng per insect) was injected into pharate pupae, normal phenotypes were observed in the pupal stage. However, unlike buffer-injected controls, *TcCHT7* dsRNA-treated insects failed to expand their adult elytra and their wings did not fold properly (modified from Zhu *et al.*, 2008c). Animals injected with control dsRNA for *EGFP* developed in a normal fashion, and had no mortality or abnormal phenotype.

(CHSs), but also by chitinases (CHTs, EC 3.2.1.14) and β-*N*-acetylglucosaminidases (NAGs, EC 3.2.1.52). Chitin is digested in the cuticle and PM to GlcNAc by a binary enzyme system composed of CHT and NAG (Fukamizo and Kramer, 1985a, 1985b; Filho *et al.*, 2002). The former enzyme from molting fluid hydrolyzes chitin into oligosaccharides, whereas the latter, which is also found in the molting fluid, further degrades the oligomers to the monomer from the non-reducing end. In some cases, additional unrelated proteins that possess one or more chitin-binding domains (CBD), but are devoid of chitinolytic activity, enhance degradation of chitin (Vaaje-Kolstad *et al.*, 2005). This system also probably operates in the gut during degradation of PM, and increases the porosity of the PM. It may also help in the digestion of chitin-containing prey (Bolognesi *et al.*, 2005; Khajuria *et al.*, 2010).

The precise control of chitin content is critical not only for the survival of the insect, but also for optimal function of individual anatomical structures such as wings and other appendages. In addition, modulation of the physical properties of chitin-containing structures of insects is accomplished, in part, by the deacetylation of the polysaccharide by chitin deacetylases (CDAs, EC 3.5.1.41). Partially deacetylated chitin may have different protein-binding and physical properties than those of chitin. The process of partially deacetylating chitin and the importance of this modification for insect growth and development have emerged as new areas of research in insect molecular science (Luschnig *et al.*, 2006; Wang *et al.*, 2006; Arakane *et al.*, 2009).

7.4.1. Insect Chitinases

7.4.1.1. Cloning of genes encoding insect chitinases and chitinase-like proteins Since the first report of an insect chitinase, its cDNA and its corresponding gene from *M. sexta* (*MsCHT5*) (Koga *et al.*, 1987; Kramer *et al.*, 1993; Choi *et al.*, 1997; Kramer and Muthukrishnan, 1997), numerous insect *CHT* genes and cDNAs have been cloned and characterized from several insect species belonging to different orders, including dipterans, lepidopterans, coleopterans, hemipterans, and hymenopterans (Kramer and Muthukrishnan, 2005). The organization of most of these genes is very similar to that of *MsCHT5*, and most of the proteins display a domain architecture consisting of catalytic, linker, and/or chitin-binding domains (CBD) similar to MsCHT5. These genes/enzymes include epidermal chitinases from the silkworm *Bombyx mori* (Kim *et al.*, 1998; Abdel-Banat and Koga, 2001), the fall webworm *Hyphantria cunea* (Kim *et al.*, 1998), wasp venom from *Chelonus* sp. (Krishnan *et al.*, 1994), the common cutworm *Spodoptera litura* (Shinoda *et al.*, 2001), the fall armyworm *Spodoptera frugiperda* (Bolognesi *et al.*, 2005), a molt-associated chitinase from the spruce budworm *Choristoneura fumiferana* (Zheng *et al.*, 2002), and midgut-associated chitinases from the malaria mosquito *A. gambiae* (Shen and Jacobs-Lorena, 1997), yellow fever mosquito *Ae. aegypti* (de la Vega *et al.*, 1998; Khajuria *et al.*, 2010), the beetle *Phaedon cochleariae* (Girard and Jouanin, 1999), and the sand fly *Lutzomyia longipalpis* (Ramalho-Ortigao and Traub-Cseko, 2003), as well as several deduced from *Drosophila* genome data. A smaller linkerless fat body-specific chitinase from the

tsetse fly *Glossina morsitans* (Yan *et al.*, 2002), and a very large epidermal chitinase with five copies of the catalytic domain and multiple chitin-binding domain from the yellow mealworm *Tenebrio molitor* (Royer *et al.*, 2002), have also been described.

Daimon *et al.* (2003) described a gene encoding another type of chitinase from the silkworm, BmCHT-h. The encoded chitinase shared extensive similarities with microbial and baculoviral chitinases (73% amino acid sequence identity to *Serratia marcescens* chitinase, and 63% identity to *Autographa californica* nuclear polyhedrosis virus chitinase). Even though this enzyme had the signature sequence characteristic of a family 18 chitinase, it had a rather low percentage of sequence identity with the family of insect chitinases. It was suggested that an ancestral species of *B. mori* acquired this chitinase gene via horizontal gene transfer from *Serratia* or a baculovirus. A gene encoding a CHT-like protein that is highly related to BmCHT-h was also found in the pea aphid, *Acyrthosiphon pisum* (Nakabachi *et al.*, 2010).

Only after the complete genome sequences became available was it recognized that insect genomes contain a large number of genes encoding CHT-like proteins widely divergent not only in their DNA and amino acid sequences, but also in the organization of their domains (Zhu *et al.*, 2004, 2008a; Arakane and Muthukrishnan, 2010). The number of *CHT* genes per insect genome is in the range of 7 to 24 for *D. melanogaster*, *A. gambiae*, *Ae. aegypti*, *B. mori*, *A. pisum*, and *T. castaneum*. This range excludes genes encoding CHT-like proteins whose consensus sequences are poorly conserved (see section 7.1.2.2; Khajuria *et al.*, 2010; Nakabachi *et al.*, 2010; Zhu *et al.*, 2004, 2008a). The 22 genes that encode CHTs or chitinase-like proteins (CHLPs) in *T. castaneum* have been divided into eight subgroups, based on sequence similarity and domain organization (**Figure 6**) (Arakane

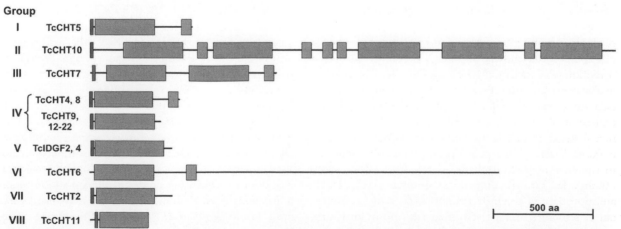

Figure 6 Domain organization of *T. castaneum* chitinase gene family. The program SMART was used to analyze the identified domains. TcCHT7 and TcCHT11 have a single transmembrane span at the N-terminal region. Blue boxes, signal peptide; pink boxes, catalytic domain; green boxes, chitin binding domain; red boxes, transmembrane span; lines, linker regions.

and Muthukrishnan, 2010). The chitinases in all insect species can be similarly classified into multiple groups (**Figure 7**). There is only one copy of the gene encoding a group I chitinase (CHT5) in all species except for *A. gambiae*, *Ae. aegypti*, and the human body louse *Pediculus humanus corporis*, in which obvious gene duplications have occurred, resulting in one to four additional copies (Khajuria *et al.*, 2010). To date, only one gene representing each of the groups II, III, VI, VII, and VIII (CHT10, 7, 6, and 11, respectively) has been found in various insect species. Interestingly, in addition to the group III *CHT* genes (CHT7s with two catalytic domains) identified in fully sequenced insect genomes such as *T. castaneum*, *D. melanogaster*, *A. gambiae*, *Ae. aegypti*, *C. pipiens*, *A. mellifera*, *N. vitripennis*, *A. pisum*, and *P. corporis*, orthologs have also been found in non-insect arthropod genomes, including those of the crustacean water flea *Daphnia pulex*, and the arachnid deer tick *Ixodes scapularis*, indicating an ancient origin of CHT7 that predates separation of the class Chelicerata more than five million years ago. Group IV appears to be the largest group in all insect species studied, containing 5, 8, 10, and 14 genes in *D. melanogaster*, *A. gambiae*, *Ae. aegypti*, and *T. castaneum*, respectively. The sole exception so far is *A. pisum*. No chitinase gene encoding a protein that belongs to this group was identified in *A. pisum* (Nakabachi *et al.*, 2010). In *T. castaneum*, most group IV *CHT* genes form a large cluster within a small region of the genome, suggesting the occurrence of a recent gene duplication event. Group V is composed of the genes encoding CHLPs such as imaginal disc growth factors (IDGFs). The number of genes for this group ranges from one in *B. mori* and *A. pisum* to as many as six in *D. melanogaster* (Arakane and Muthukrishnan, 2010; Nakabachi *et al.*, 2010).

7.4.1.2. Domain organization of insect chitinases

Insect CHTs belong to family-18 glycosylhydrolases (the GH-18 super family) and function in hydrolysis of chitin in the exoskeleton and PM-associated chitin in the midgut, utilizing an endo-type cleavage mechanism during the molting process (Kramer and Muthukrishnan, 1997, 2005). Members of the CHT family contain a multidomain structural organization that includes a leader peptide and/or a transmembrane span, one to five catalytic domains (GH-18), multiple Ser/Thr-rich linker regions that are usually heavily glycosylated, and zero to seven six-cysteine-containing chitin-binding domains (CBDs) related to the peritrophin A domain (**Figure 6**; Royer *et al.*, 2002; Arakane *et al.*, 2003; Zhu *et al.*, 2008b). The catalytic domains of all insect CHTs, which are comprised of about 370 amino acids, assume a $\beta 8 \alpha 8$-barrel structure and possess signature motifs of family 18 glycosylhydrolases (Kramer *et al.*, 1993; Perrakis *et al.*, 1994, Terwissscha van Scheltinga *et al.*, 1994; de la Vega *et al.*, 1998; Fusetti *et al.*, 2002; Varela *et al.*,

2002; Tsai *et al.*, 2004; Arakane and Muthukrishnan, 2010). The consensus sequence for conserved motif I is KXX(V/L/I)A(V/L)GGW in the $\beta 3$-strand, where X is a non-conserved amino acid. The conserved motif II is FDG(L/F)DLDWE(Y/F)P, which is known to be located in or near the catalytic site ($\beta 4$-strand) of the enzyme, with a glutamate residue (E) being the most critical residue in this motif as the putative proton donor in the catalytic mechanism (Watanabe *et al.*, 1993; Lu *et al.*, 2002; Zhang *et al.*, 2002). Conserved motifs III and IV are MXYDL(R/H)G in the $\beta 6$-strand and GAM(T/V) WA(I/L)DMDD in the $\beta 8$-strand.

CBDs found in insect CHTs all belong to carbohydrate-binding module 14 (CBM-14, pfam 01607; ChtBD2 family = SMART family 00494, Boraston *et al.*, 2004). Insect CBDs are only about 60 amino acids long and have less conserved amino acid sequences, with the exception of the six cysteines and several aromatic residues whose relative locations are highly conserved (Jasrapuria *et al.*, 2010). The proposed function(s) of the CBD is to help anchor the enzyme onto the insoluble chitin to enhance chitin degradation efficiency (Linder *et al.*, 1996; Arakane *et al.*, 2003). As described in section 7.4.1.1, based on the amino acid sequence similarity and domain architecture, insect CHTs can be classified into eight groups (**Figures 6** and **7**). Group I CHTs (CHT5s) represent the prototypical and enzymatically characterized CHTs purified from molting fluid and/or integument of *M. sexta* and *B. mori* (Koga *et al.*, 1983, 1997). All of these group members contain a signal peptide, one catalytic domain, a Ser/Thr-rich linker region, and one CBD. Group II CHTs (CHT10s) are rather diverse in their domain architecture, and have four or five catalytic domains, together with four to seven CBDs. Dipterans and *A. pisum* (hemiptera) appear to be unique in having only four catalytic domains and four CBDs. The domain corresponding to the most N-terminal catalytic domain and one CBD found in group II chitinases from other species appear to be missing in the dipteran CHT10s (Zhu *et al.*, 2008b; Arakane and Muthukrishnan, 2010; Nakabachi *et al.*, 2010). The second catalytic unit of all CHT10s (the first catalytic unit in the case of the dipteran and *A. pisum* proteins) is predicted to lack chitinolytic activity due to a substitution of the most critical amino acid residue glutamate (E) with asparagine (N) in conserved motif II. Group III CHTs (CHT7s) possess two catalytic domains and one C-terminal CBD. The first catalytic domains of the group III proteins from all insect species studied share greater sequence similarity with each other than they do to the second catalytic domain, suggesting a unique function and/or evolutionary origin for each of the catalytic domains. Unlike most insect CHTs, CHT7s are predicted to have an N-terminal transmembrane segment, and are likely to be membrane-bound proteins. Indeed, recombinant *T. castaneum* CHT7 (TcCHT7) that was expressed in Hi-5 insect

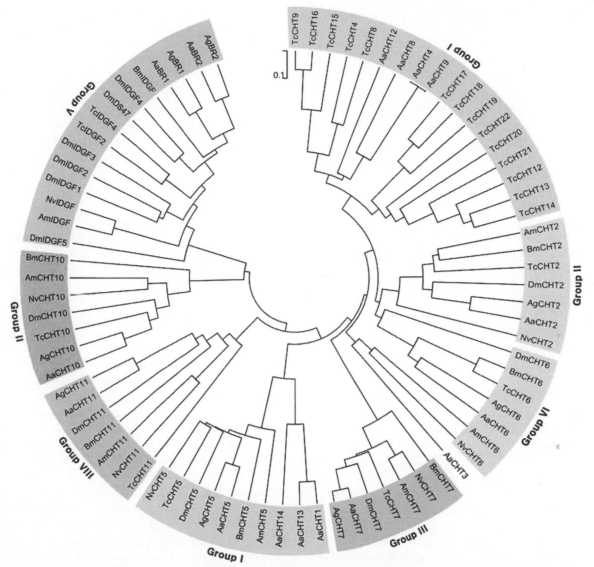

Figure 7 Phylogenetic analysis of putative chitinases and chitinase-like proteins (IDGFs) in insects. ClustalW software was used to perform multiple sequence alignments prior to phylogenetic analysis. The phylogenetic tree was constructed by MEGA 4.0 software using UPGMA (Tamura *et al.*, 2007). Protein sequences obtained from GenBank as follows: *Tribolium castaneum*, TcCHT2 (AY873913); TcCHT4 (EF125543); TcCHT5 (AY675073); TcCHT6 (AY873916); TcCHT7 (DQ659247); TcCHT8 (DQ659248); TcCHT9 (DQ659249); TcCHT10 (DQ659250); TcCHT11 (DQ659251);TcCHT12 (XM_967709); TcCHT13 (DQ659252); TcCHT14 (XM_967912); TcCHT15 (XM_967984); TcCHT16 (AY873915); TcCHT17 (XP_972719); TcCHT18 (XP_973161); TcCHT19 (XP_973119); TcCHT20 (NP_001034516); TcCHT21 (NP_001034517); TcCHT22 (NP_001038095); TcIDGF2 (DQ659253); TcIDGF4 (DQ659254); *Aedes aegypti*, AaCHT1 (XP_001656232); AaCHT2 (XP_001662520); AaCHT3 (XP_001663568); AaCHT4 (XP_001663099); AaCHT5 (XP_001656234); AaCHT6 (XP_001662588); AaCHT7 (XP_001650020); AaCHT8 (XP_001663098); AaCHT9 (XP_001663099); AaCHT10 (XP_001655973); AaCHT11 (XP_001654045); AaCHT12 (XP_001658836); AaCHT13 (XP_001656231); AaCHT14 (XP_001656233); AaBR1 (XP_001660745); AaBR2 (XP_001660748); *Apis mellifera*, AmCHT2 (XP_623744); AmCHT5 (XP_623995); AmCHT6 (XP_393252); AmCHT7 (XP_396925); AmCHT10 (XP_395734); AmCHT11 (XP_395707); AmIDGF (XP_396769); *Drosophila melanogaster*, DmCHT2 (NP_477298); DmCHT5 (NP_650314); DmCHT6 (NP_572598); DmCHT7 (NP_647768); DmCHT10 (NP_001036422); DmCHT11 (NP_572361); DmIDGF1 (NP_477258); DmIDGF2 (NP_477257); DmIDGF3 (NP_723967); DmIDGF4 (NP_727374); DmIDGF5 (NP_611321); DmDS47 (NM_057733); *Bombyx mori*, BmCHT2 (BGIBMGA009695); BmCHT5 (BGIBMGA010240); BmCHT6 (BGIBMGA009890); BmCHT7 (BGIBMGA005539); BmCHT10 (BGIBMGA006874); BmCHT11 (BGIBMGA005859); BmIDGF (BGIBMGA000648); *Anopheles gambiae*, AgCHT2 (XP_315650); AgCHT5 (XP_001237469); AgCHT6 (AGAP000198); AgCHT7 (XP_308858); AgCHT10 (XP_001238192); AgCHT11 (XP_310662); AgBR1 (AAS80137); AgBR2 (AY496421); *Nasonia vitripennis*, NvCHT2 (XP_001601416); NvCHT5 (NP_001155084); NvCHT6 (); NvCHT7 (XP_001604515); NvCHT10 (XR_036825); NvCHT11 (XP_001604954); NvIDGF (XP_001599305).

cells using the baculovirus protein expression system was found to be in the cell pellet rather than in the medium, as expected for secreted proteins. The washed cell pellet containing recombinant TcCHT7 could hydrolyze chitin added to the culture medium, suggesting that the catalytic domains of this putative membrane-bound protein face the extracellular space (Arakane, unpublished data). Group IV CHTs comprise the largest and most divergent group of proteins. CHTs in this group have a signal peptide and one catalytic domain. Most (but not all) of the members lack a CBD (**Figure 6**). Group V chitinase-like proteins (CHLPs) include the imaginal disc growth factors (IDGFs) and the hemocyte aggregation inhibitor protein (HAIP, Kanost *et al.*, 1994; Pan *et al.*, 2010). CHLPs have a signal peptide, one catalytic domain, and no CBDs. Like other family-18 proteins, the crystal structure of *D. melanogaster* IDGF2 and homology modeling of all proteins in this group revealed the $\beta_8\alpha_8$-TIM barrel structure (Varela *et al.*, 2002). However, members of this group have an additional loop sequence located between the β4-strand and the α4-helix immediately after conserved region II. Although these proteins possess all four of the family-18 conserved motifs, the glutamate residue in conserved motif II is substituted by a glutamine in all members of the group, with the exception of two *T. castaneum* IDGFs (TcIDGF2 and TcIDGF4; Zhu *et al.*, 2008b). TcIDGF2 and TcIDGF4 retain the glutamate residue in conserved region II but lack chitinase activity, either due to a D to A substitution in the conserved motif II, or to an extra loop stretching between the β4-strand and the α4-helix that possibly interferes with a productive substrate–enzyme interaction (Zhu *et al.*, 2008a), or both. Group VI CHTs (CHT6s) exhibit a domain architecture similar to that of group I (a signal peptide, one catalytic domain, and one CBD), but they have a very long C-terminal stretch (e.g., 1819 amino acids in length after the CBD in TcCHT6) that has no predicted conserved domain (**Figure 6**) except for the *A. pisum* enzyme, which possesses an additional CBD at the C-terminal region (Nakabachi *et al.*, 2010). Group VII CHTs (CHT2s) possess a domain architecture similar to that of group IV CHTs, which have a signal peptide, one catalytic domain, and no CBDs. They are classified as a separate group because phylogenetic analysis clearly indicates that these CHTs form a different clade near group II CHT10s. Group VIII CHTs (CHT11s) have one catalytic domain and no CBD. Interestingly, they have a predicted transmembrane segment instead of a signal peptide at the N-terminus, and they fall into a branch next to group III (CHT7s), all of which are predicted to be membrane-bound proteins.

7.4.1.3. Gene expression and functions of insect chitinases
The redundancy of genes for CHTs raises important questions about their functions. Several insect CHT cDNAs have been obtained from epidermis, gut, and fat body, and extensively characterized (Kramer and Muthukrishnan, 2005). The epidermal endochitinases presumably function in turnover of the old cuticle, as these enzymes are found in the molting fluid along with *N*-acetylglucosaminidases, whereas the gut CHTs are thought to participate in the breakdown of chitin in the PM. In *T. castaneum*, tissue specificity and developmental patterns of expression of the 22 *TcCHT* and *TcCHLP* genes were analyzed by RT-PCR using cDNAs prepared from RNAs isolated at different developmental stages, such as embryo, larva, pharate pupa, pupa, and adult (Zhu *et al.*, 2008c; Arakane and Muthukrishnan, 2010). The group I gene *TcCHT5*, group II gene *TcCHT10*, group III gene *TcCHT7*, group V genes *TcIDGF2* and *TcIDGF4*, group VI gene *TcCHT6*, group VII gene *TcCHT2*, and group VIII gene *TcCHT11* are expressed at all stages analyzed, with some variation, whereas all group IV genes (*TcCHTs 2, 4, 8, 9*, and *12 to 22*) were predominantly expressed in the feeding stages (larva and adult). In addition, all chitinase genes belonging to group IV were expressed in larval gut tissue but not in the carcass (whole body minus gut), suggesting a possible function of these TcCHTs in PM-associated chitin turnover or digestion of dietary chitin (Zhu *et al.*, 2008c). Khajuria *et al.* (2010) recently reported that orally feeding dsRNA for a midgut-specific chitinase gene (encoding a group IV CHT) from larvae of *O. nubilalis* (*OnCHT*) significantly reduced the transcript levels of this gene and led to a significant increase of chitin content in the PM. The body weight of dsRNA *OnCHT*-fed larvae was decreased by 54% as compared with that of control dsRNA *GFP*-fed larvae, suggesting that some group IV CHTs are critical for regulating PM-chitin content, insect growth, and development. Interestingly, *A. pisum* appears to have no group IV *CHT* genes (Nakabachi *et al.*, 2010). *A. pisum* (hemipteran) possesses a perimicrovillar membrane (PMM) that is devoid of chitin, suggesting that group IV CHTs may not play a role in the PM turnover. Instead, one *CHT* gene, *ApCHT6* (encoding a group VIII CHT), was highly expressed in the midgut of *A. pisum*. Similarly, *TcCHT11* (encoding a group VIII CHT) was expressed in larval midgut, but not in the carcass (Arakane and Muthukrishnan, 2010). Group VIII CHTs, as well as group VI CHTs, may play critical roles in PM/PMM chitin degradation and turnover.

RNAi for group IV chitinases in *T. castaneum* for individual chitinases (and some combinations of chitinases) failed to produce any visible phenotypes, perhaps reflecting the redundant functions of this large group of chitinolytic enzymes. In contrast, injection of dsRNA for all chitinases belonging to groups I, II, III, and V resulted in unique lethal phenotypes. The most severe molting defect was observed after injection of dsRNA for *TcCHT10* (encoding a group II CHT). Injections of dsRNA for *TcCHT10* prevented the embryo from hatching and also averted all types of molts, including larval–larval, larval–pupal, and pupal–adult, depending on the timing of administration of the dsRNA (**Figure 5**; Zhu *et al.*, 2008c). These results suggest a critical role for group II CHTs at every molt and developmental stage. Other CHTs (e.g., CHT5, also expressed in the epidermis) could not compensate for the loss of function of a group II CHT.

Unlike RNAi for *TcCHT10*, injection of ds*TcCHT5* (encoding a group I CHT) prevented only the pupal–adult molt (**Figure 5**). Although the gene encoding this prototypical CHT was expressed throughout all developmental stages, and the corresponding enzymes from several other insect species have been found in larval molting fluid, the failure to obtain a larval–larval or larval–pupal molting arrest probably indicates that one or more of the other CHTs (e.g., group II CHT, TcCHT10) could compensate for TcCHT5 at all molts except during adult eclosion. Group III CHTs, which appear to encode membrane-bound enzymes with two catalytic domains and one CBD at the C-terminus, appear to be critical for tissue differentiation, rather than chitin degradation associated with molting. Indeed, in *D. melanogaster*, expression of the *DmCHT7* (CG1869) gene increased more than 40-fold in the wing during the 32- to 40-h pupal wing differentiation period (Ren *et al.*, 2005). In *T. castaneum*, injection of dsRNA for *CHT7* resulted in a defective elytral and hindwing expansion without affecting molting (**Figure 5**; Zhu *et al.*, 2008c). Group V is composed of IDGFs that are known to be involved in cell proliferation and differentiation (Kawamura *et al.*, 1999; Zhang *et al.*, 2006). It is worthy of note that although group V CHTs have no chitinolytic activity (Zhu *et al.*, 2008b), they appear to be important for the adult molt. Injection of dsRNA for one of these *CHLP*s in *T. castaneum*, *TcIDGF4*, prevented adult eclosion (Zhu *et al.*, 2008c). It is possible that TcIDGF4 may be required for tracheal proliferation during adult metamorphosis. Two *A. gambiae* proteins, AgBR1 and AgBR2, which belong to this group, were induced specifically in the hemolymph by bacterial challenge (Shi and Paskewitz, 2004), suggesting that some members of the CHLP group (and/or members of other CHT groups) may have a role in the immune response.

7.4.2. Insect *N*-Acetylglucosaminidases

7.4.2.1. Phylogenetic analysis of insect *N*-acetylglucosaminidases
Beta-*N*-acetylglucosaminidases (NAGs; EC 3.2.1.30) have been defined as enzymes that release – acetylglucosamine residues from the non-reducing end of chitooligosaccharides and from glycoproteins with terminal *N*-acetylglucosamines. Insect NAGs are members of the family-20 hexosaminidase super-family of the glycosylhydrolases of the Carbohydrate Active

Enzymes database, CAZY (Coutinho and Henrissat, 1999; Cantarel., *et al.*, 2009). These enzymes have been detected in the molting fluid, hemolymph, integument, and gut tissues of several species of insects (Kramer and Koga, 1986; Hogenkamp *et al.*, 2008), and cooperate with CHTs to hydrolyze chitin to generate monomers of N-acetylglucosamine (Fukamizo and Kramer, 1985a, 1985b). Insect CHTs are unable to convert the chitin substrate completely to GlcNAc monomers. Therefore, NAG is the enzyme primarily responsible for the production of the monomer from chitooligosaccharides for recycling. Kinetic studies with *M. sexta* CHT (MsCHT5, group I CHT) have revealed that this enzyme is subject to substrate and/or product inhibition when chitooligosaccharides and/or colloidal chitin are utilized as substrates (Koga *et al.*, 1982, 1983; Arakane *et al.*, 2003). Therefore, one of the potential functions of NAGs may be to prevent the accumulation of chitooligosaccharides at concentrations that are high enough to interfere with efficient degradation of chitin by CHT (Kramer and Muthukrishnan, 2005).

cDNAs for epidermal *β-N*-acetylglucosaminidases of *B. mori*, *B. mandarina*, *T. ni*, and *M. sexta* have been isolated and characterized (Nagamatsu *et al.*, 1995; Zen *et al.*, 1996; Goo *et al.*, 1999; Hogenkamp *et al.*, 2008). A NAG also has been detected in the gut of *Ae. aegypti*, where its activity increased dramatically upon blood feeding (Filho *et al.*, 2002). A search of the *D. melanogaster*, *A. gambiae*, *Ae. aegypti*, *Culex pipiens*, *A. mellifera*, *N. vitripennis*, *B. mori*, and *T. castaneum* genome databases revealed the presence of multiple *NAG* genes, as well as the genes encoding *β-N*-acetylhexosaminidases (*HEX*s) in these species (Hogenkamp *et al.*, 2008). Phylogenetic analysis of NAGs from these insects indicates that NAGs can be classified into four distinct groups – NAG group I (NAG1), NAG group II (NAG2), *N*-glycan processing NAGs (FDL) (group III, Leonard *et al.*, 2006), and HEX group IV – according to their amino acid sequences (**Figure 8**). To date, only a single gene representing each of the groups I, II, and III has been found in the various insect species, with the exception of *C. pipiens*, which appears to have three genes encoding NAG-like proteins closely related to group I NAGs. Group I is composed of the enzymatically well-characterized NAGs, including NAGs from *M. sexta* (MsNAG1) and *B. mori* (BmNAG1). DmHEXO2, which has been shown to have NAG activity (Mark *et al.*, 2003; Leonard *et al.*, 2006), was placed in group II. Group III is composed of the *D. melanogaster* fused lobes protein (DmFDL), along with the fused lobes (fdl) homologs of other insect species (Leonard *et al.*, 2006). All of the proteins belonging to this group possess a predicted transmembrane anchor and a signal anchor, except for a signal peptide that can be found in NAGs belonging to groups I, II, and IV. In *T. castaneum*, TcNAG3 could not be unambiguously assigned to any of

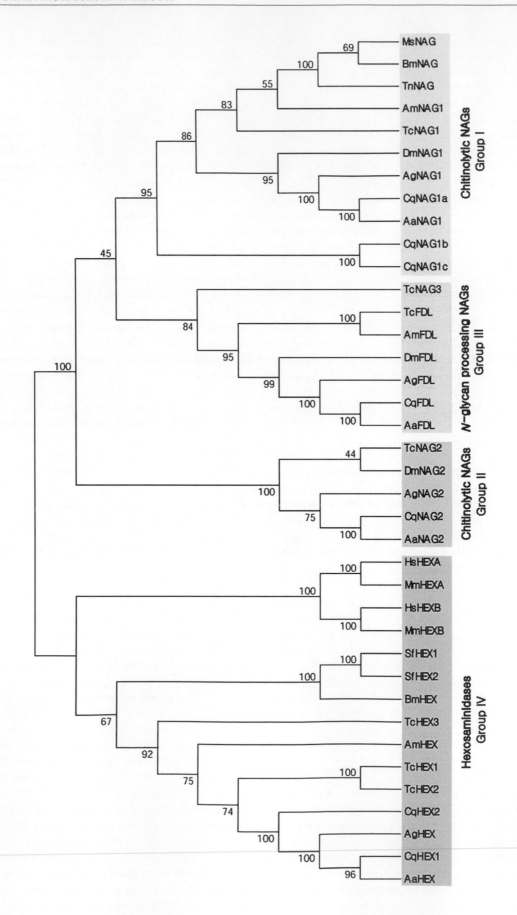

Figure 8 Phylogenetic analysis of NAGs and hexoaminidases in *Tribolium*, other insects and metazoans. MEGA4.0 (Tamura *et al.*, 2007) was used to construct the consensus phylogenetic tree using UPGMA. Bootstrap analyses of 1000 replications are shown. Protein sequences extracted from GenBank include: MsNAG, *Manduca sexta* (AY368703); BmNAG, *Bombyx mori* (genbank: AF326597); TnNAG, *Trichoplusia ni* (AY078172); AmNAG1, *Apis mellifera* (XM_624790); TcNAG1, *Tribolium castaneum* (EF592536); DmNAG1 (DmHEXO1), *Drosophila melanogaster* (NM_079200); AgNAG1, *Anopheles gambiae* (XP_315391); CqNAG1a, *Culex quinquefasciatus* (XP_001864406); AaNAG1, *Aedes aegypti* (EAT43909); CqNAG1b, *Culex quinquefasciatus* (XP_001864407); CqNAG1c, *Culex quinquefasciatus* (XP_001866097); TcNAG3, *Tribolium castaneum* (EF592538); TcFDL, *Tribolium castaneum* (EF592539); AmFDL, *Apis mellifera* (XP_394963); DmFDL, *Drosophila melanogaster* (NP_725178); AgFDL, *Anopheles gambiae* XP_308677); CqFDL, *Culex quinquefasciatus* (XP_001850423); AaFDL, *Aedes aegypti* (EAT36388); TcNAG2, *Tribolium castaneum* (EF592537); DmNAG2 (DmHEXO2), *Drosophila melanogaster* (NM_080342); AgNAG2, *Anopheles gambiae* (XM_307483); CqNAG2, *Culex quinquefasciatus* (XP_001842710); AaNAG2, *Aedes aegypti* (EAT40440); HsHEXA, *Homo sapiens* (NM_000520); HsHEXB, *Homo sapiens* (NM_000521); MsHEXA, *Mus musculus* (NM_010421); MsHEXB, *Mus musculus* (NM_010422); SfHEX1, *Spodoptera frugiperda* (DQ183187); SfHEX2, *Spodoptera frugiperda* (DQ249307); BmHEX, *Bombyx mori* (AY601817); TcHEX3, *Tribolium castaneum* (XM_970565); AmHEX, *Apis mellifera* (XM_001122538); TcHEX1, *Tribolium castaneum* (XM_970563); TcHEX2, *Tribolium castaneum* (XM_970567); CqHEX2, *Culex quinquefasciatus* (XP_001867058); AgHEX, *Anopheles gambiae* (XM_319210); CqFEX1, *Culex quinquefasciatus* (XP_001867057); and AaHEX, *Aedes aegypti* (EAT43655).

the three subgroups. TcNAG3 is more closely related to TcFDL than to TcNAG1 and TcNAG2, but the TcFDL and TcNAG3 genes are present on different linkage groups (**Figure 8**, Hogenkamp *et al.*, 2008).

7.4.2.2. Expression and functional analysis of insect *N*-acetylglucosaminidases

Hogenkamp and colleagues (2008) performed dsRNA-mediated post-transcriptional downregulation (RNAi) of transcripts for all four *NAG* genes from a single insect species (*T. castaneum*) to study the functions of insect NAGs. Injection of a dsRNA corresponding to any one *TcNAG* gene resulted in substantial downregulation of the target transcript without significantly affecting the levels of the other *TcNAG* transcripts. Depletion of transcripts for any one of the targeted genes produced lethal molting arrest phenotypes. However, some of the injected insects did succeed in completing each type of molt (larval–larval, larval–pupal, and pupal–adult). TcNAG1 appeared to be most critical in chitin catabolism during molting. Administration of dsRNA for *TcNAG1* resulted in developmental arrest, and more than 80% of the insects died at the time of the next molt (**Figure 5**). During each type of molt, larval–larval, larval–pupal, and pupal–adult, the insects were unable to completely shed their exoskeleton. The pupa–adult molting phenotype produced by injection of dsRNA for *TcNAG1* is strikingly similar to that obtained in RNAi studies with ds*TcCHT5* (**Figure 5**; see section 7.4.1.3). Insects injected with dsRNA for *TcCHT5* also failed to shed their old cuticle, and the new cuticle was visible underneath the old cuticle (Zhu *et al.*, 2008c; Arakane and Muthukrishnan, 2010). It has been shown that in *M. sexta*, CHT is susceptible to oligosaccharide inhibition (Koga *et al.*, 1982, 1983; Arakane *et al.*, 2003). Injection of dsRNA for *TcNAG1* may result in the accumulation of chitiooligosaccharides in the molting fluid, and therefore it may cause inhibition of TcCHT5

activity, resulting in a phenotype similar to that observed in dsRNA for *TcCHT5*-treated insects. The high level of expression of *TcNAG1*, its phylogenetic relationship to other well-characterized molting-associated insect NAGs (**Figure 8**), and the phenotypic effect of knocking down *TcNAG1* transcripts suggest that, among all of the TcNAGs, TcNAG1 (group I NAG) is the enzyme primarily responsible for the efficient degradation of cuticular chitin, in concert with TcCHT5 (group I CHT), in *T. castaneum*, and that this may be the case in other insect species as well.

Although TcNAG1 is most likely to be the principal NAG for catabolism of cuticle-associated chitin, the other three NAGs identified in *T. castaneum* also appear to play important and perhaps indispensable roles in cuticle turnover and development. Injection of dsRNA for *TcNAG2* (encoding a group II NAG orthologous to DmHEXO2) prevents all types of molts, especially the pupal–adult molt. Like the phenotype produced by injection of dsRNA for *TcNAG1* (**Figure 5**), more than 75% of the animals treated with dsRNA for *TcNAG2* were unable to fully shed the old pupal cuticle. Since injection of dsRNA for *TcNAG2* did not change the level of *TcNAG1* transcripts, TcNAG1 could not compensate for the lack of TcNAG2 in adult eclosion in *T. castaneum*. In addition, *TcNAG2* transcript level in the midgut is relatively higher than that in the carcass (whole body minus midgut), suggesting *TcNAG2* as well as *TcNAG1*, which are highly expressed in both tissues, also play critical roles in the PM-associated chitin turnover.

Group III consists of the insect orthologs of the *D. melanogaster* fused lobes gene, *DmFDL*. The FDL proteins are predicted to be membrane-bound, with a single transmembrane helix located near the N-terminus. Furthermore, ultracentrifugation experiments on a lepidopteran protein from the culture media of *Sf*9 and *Sf*21 cells indicated that a major portion of the NAG activity resided in the membrane fraction (Altmann *et al.*, 1995; Tomiya *et al.*, 2006). This lepidopteran NAG was capable of effectively hydrolyzing chitotriose-PA (pyridylamino), while the recombinant DmFDL was unable to digest chitotriose (Leonard *et al.*, 2006). The latter hydrolyzed only the GlcNAc residue attached to the α-1,3-linked mannose of the core pentasaccharide of *N*-glycans. No cleavage activity of any other GlcNAc residues was observed, including the GlcNAc residue attached to the α-1,6-linked mannose of the core pentasaccharide. Furthermore, DmFDL did not catalyze the endo-type hydrolysis of the N,N′-diacetylchitobiosyl unit in the high-mannose pentasaccharide core. A similar *N*-glycan substrate specificity for the terminal GlcNAc attached to the α-1,3-linked mannose was observed in membrane-bound β-*N*-acetylhexosaminidases from several lepidopteran insect cell lines, including Sf21, Bm-N, and Mb-0503 (Altmann *et al.*, 1995; Tomiya *et al.*, 2006). Taken together, FDLs may play a critical role in *N*-glycan processing.

Unlike RNAi for *TcNAG1* (group I NAG), injection of dsRNA for *TcFDL* exhibits a small percentage (10–20%) of lethal molting defect phenotypes at the larval–larval and larval–pupal molts (Hogenkamp *et al.*, 2008). Much higher mortality (80%), however, was observed at the pupal–adult molting stage, indicating that TcFDL plays an essential role for adult eclosion. The transcript level of *TcFDL* in the midgut was relatively low compared to that of the carcass. Therefore, the observed lethal phenotype at the pharate adult stage may be a direct result of the knockdown of this transcript in the cuticular epidermal cells, rather than in the gut lining cells. If TcFDL does in fact play a role in chitin turnover in the cuticle, then this protein may be secreted and not membrane-bound. Indeed, Leonard and colleagues (2006) have observed that DmFDL is, to a large extent, secreted into the extracellular space. Whether there is another point of regulation at the level of release of membrane-bound FDLs is an interesting possibility.

Another *T. castaneum* NAG, *TcNAG3*, has not been unambiguously assigned to any of the three NAG groups (**Figure 8**). Similar to *TcNAG2* (group II NAG), *TcNAG3* is also expressed at a significantly higher level in the larval midgut than in the carcass (Hogenkamp *et al.*, 2008). Furthermore, an analysis of the developmental pattern of expression of *TcNAG3* indicated that it is primarily expressed during the larval stages. Unlike RNAi for the other three *TcNAGs*, injection of ds*TcNAG3* did not consistently result in lethal phenotypes, and the majority of dsRNA-injected insects survived to adults with no visible phenotypic changes. However, a small number of individuals (approximately 20%) did exhibit a lethal larval phenotype similar to that of *TcNAG1* RNAi (**Figure 5**). In addition, a few insects (approximately 10%) exhibited a lethal pharate adult molting phenotype after dsRNA

TcNAG3 injection. These insects were unable to fully shed their old pupal cuticle, similar to the phenotypes observed after dsRNA *TcNAG1* and dsRNA *TcNAG2* injections. The *TcNAG3* gene is expressed predominantly in the larval stages, with only trace levels of expression in the pupal and adult stages (Hogenkamp *et al.*, 2008). In other insect species analyzed, only genes that can be classified into groups *NAG1*, *NAG2*, and *FDL* have been identified (**Figure 8**). Therefore, *TcNAG3* appears to be unique, and its relatively high expression in the midgut compared to the carcass suggests that it may be specialized for the turnover of PM-associated chitin rather than cuticular chitin during larval stages.

7.4.3. Insect Chitin Deacetylases

7.4.3.1. Phylogenetic analysis and domain organization of chitin deacetylases
The extracellular matrix (ECM) of the insect exoskeleton is modified in different ways to give the cuticle its proper physiological and mechanical properties – namely, rigidity and thickness, or flexibility and thinness (Kramer and Muthukrishnan, 2005). Chitin deacetylases (CDAs, EC 3.5.1.41) are secreted metalloproteins that belong to a family of extracellular chitin-modifying enzymes that catalyze the *N*-deacetylation of chitin to form chitosan, a polymer of β-1,4-linked D-glucosamine residues with electrostatic properties very different from chitin. This modification might contribute to the affinity of chitosan for a variety of cuticular proteins distinct from those that bind specifically to chitin. CDAs have been well characterized in various fungi and bacteria (Caufrier *et al.*, 2003), and belong to the carbohydrate esterase family 4 (CE4) of the CAZY database (www.cazy.org; Cantarel *et al.*, 2009). CE4 esterases catalyze deacetylation of different carbohydrate substrates, such as chitin, acetylxylan, and bacterial peptidoglycan. Chitooligosaccharide deacetylases and NodB, a nodulation protein from *Rhizobium*, belong to this family, and possess a similar catalytic domain (John *et al.*, 1993).

The first cDNA encoding an insect CDA-like protein (TnPM-P42, also referred to as TnCDA9) was characterized from the PM in the cabbage looper, *Trichoplusia ni*, only 5 years ago (Guo *et al.*, 2005). Since then, several genes/cDNAs encoding insect CDAs have been identified from different species (Luschnig *et al.*, 2006; Wang *et al.*, 2006; Campbell *et al.*, 2008; Dixit *et al.*, 2008; Toprak *et al.*, 2008; Jakubowska *et al.*, 2010). A comparative analysis of *CDA* gene families in several insect species with fully sequenced genomes, including Diptera, Coleoptera, Hymenoptera, and Lepidoptera, revealed that the number of *CDA* genes varies with species. Based on amino acid sequence similarity, insect CDAs are classified into five groups, I to V (**Figure 9**; Dixit *et al.*, 2008; Jakubowska *et al.*, 2010).

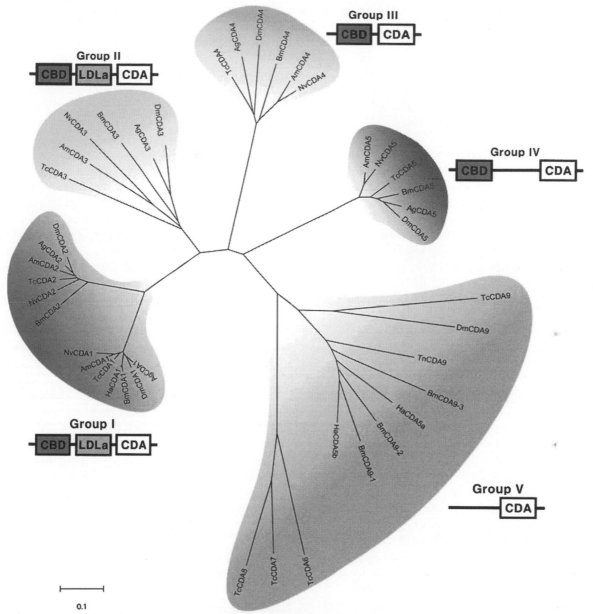

Figure 9 A phylogenetic tree of putative CDAs from different insects. A consensus phylogenetic tree was constructed using neighbor-joining method in the software MEGA 4.0 (Tamura *et al.*, 2007). Protein sequences obtained from GenBank as follows; NvCDA1, *Nasonia vitripennis* (XP_001604765); AmCDA1, *Apis mellifera* (XP_391915); TcCDA1, *Tribolium castaneum* (ABU2522); HaCDA1, *Helicoverpa armigera* (ADB43610); BmCDA1, *Bombyx mori* (BGIBMGA006213); DmCDA1, *Drosophila melanogaster* (NP_730444); AgCDA1, *Anopheles gambiae* (XP_320597); DmCDA2, *Drosophila melanogaster* (NP_001163469); AgCDA2, *Anopheles gambiae* (XP_320596); AmCDA2, *Apis mellifera* (XP_623723); TcCDA2, *Tribolium castaneum* (ABU25224); NvCDA2, *Nasonia vitripennis* (XP_001604838); BmCDA2, *Bombyx mori* (BGIBMGA006214); DmCDA3, *Drosophila melanogaster* (NP_609806); AgCDA3, *Anopheles gambiae* (XP_317336); BmCDA3, *Bombyx mori* (BGIBMGA008988); NvCDA3, *Nasonia vitripennis* (XP_001606617); AmCDA3, *Apis mellifera* (XP_001121246); TcCDA3, *Tribolium castaneum* (ABW74145); TcCDA4, *Tribolium castaneum* (ABW74146); AgCDA4, *Anopheles gambiae* (XP_310753); DmCDA4, *Drosophila melanogaster* (NP_728468); BmCDA4, *Bombyx mori* (BGIBMGA010573); AmCDA4, *Apis mellifera* (XP_001120478); NvCDA4, *Nasonia vitripennis* (XP_001607989); AmCDA5, *Apis mellifera* (XP_624655); NvCDA5, *Nasonia vitripennis* (XP_001603918); TcCDA5, *Tribolium castaneum* (ABW74147); BmCDA5, *Bombyx mori* (BGIBMGA002696); AgCDA5, *Anopheles gambiae* (XP_316929); DmCDA5, *Drosophila melanogaster* (NP_001097044); TcCDA6, *Tribolium castaneum* (ABW74149); TcCDA7, *Tribolium castaneum* (ABW74150); TcCDA8, *Tribolium castaneum* (ABW74151); TcCDA9, *Tribolium castaneum* (ABW74152); DmCDA9, *Drosophila melanogaster* (NP_611192); TnCDA9, *Trichoplusia ni* (AAY46199); BmCDA9-3, *Bombyx mori* (BGIBMGA013758); HaCDA5a, *Helicoverpa armigera* (ADB43611); HaCDA5b, *Helicoverpa armigera* (ADB43612); BmCDA9-1, *Bombyx mori* (BGIBMGA013756); BmCDA9-2, *Bombyx mori* (BGIBMGA013757).

Group I CDAs (CDA1s and CDA2s) consist of *D. melanogaster Serpentine* (*DmSerp*) and *Vermiform* (*DmVerm*) (referred to as *DmCDA1* and *DmCDA2*, respectively) and their orthologs (*CDAs 1* and *2*) from each species. All group I CDAs have a chitin-binding peritrophin-A domain (CBD), a low-density lipoprotein receptor class A domain (LDLa), and a CDA catalytic domain. There are two to four transcript variants produced by alternative splicing and/or exon skipping from the *CDA2* pre-mRNAs (Dixit *et al.*, 2008). Group II, III, and IV families are represented by only one CDA in each species, namely CDA3, CDA4, and CDA5, respectively. Although, like group I CDAs, CDA3s also possess a single copy of each of the three domains, the overall amino acid sequence identity is only about 38% with CDA1s and CDA2s (amino acid sequence identity between CDA1s and CDA2s is about 60%). Group III enzymes (CDA4s) have a single copy of the CBD and the CDA catalytic domain, but lack an LDLa domain. Group IV CDAs (CDA5s), like CDA4s, each possess a single CBD and a single CDA catalytic domain. These two domains, however, are connected by a long Ser/Thr/Pro/Gln-rich linker (e.g., about 2400 amino acids in AgCDA5), which results in CDA5s being the largest CDA proteins. At least three insect species, *D. melanogaster*, *A. mellifera*, and *T. castaneum*, have more than one isoform of CDA5 due to alternative splicing and/or exon skipping during the processing of pre-mRNA for these genes. Group V consists of two subgroups. One subgroup includes the CDA9s. Two CDAs (HaCDA5a and HaCDA5b), identified recently by proteomic analysis and EST sequence analysis of the PM of the cotton bollworm *Helicoverpa armigera* (Campbell *et al.*, 2008; Jakubowska *et al.*, 2010), also belong to this CDA9 subgroup of group V (**Figure 9**). Interestingly, two lepidopterans, *B. mori* and *H. armigera*, appear to have multiple genes related to CDA9. The other subgroup of group V consists of paralogs from *T. castaneum* only (TcCDAs 6, 7, and 8), and not from other insect species. All the proteins belonging to this group have only a CDA catalytic domain, and no CBD or LDLa domains.

7.4.3.2. Functional analysis of insect chitin deacetylases

Developmental patterns and tissue-specific expression of different *CDA* genes in the same species suggest that the chitin deacetylases may have specific functions. In *D. melanogaster*, the two group I genes, *DmSerp* (*DmCDA1*) and *DmVerm* (*DmCDA2*), are required for normal tracheal tube development and morphology (Lusching *et al.*, 2006; Wang *et al.*, 2006). *D. melanogaster* mutants lacking either *serp* or *verm* exhibited excessively long and tortuous embryonic tracheal tubes. In *T. castaneum*, injection of dsRNA for *TcCDA1* or *TcCDA2*, which are predominantly expressed in epidermis and tracheae, prevented all types of molts, including larval–larval, larval–pupal, and pupal–adult (**Figure 5**; Arakane *et al.*,

2009). Furthermore, alternative exon-specific RNAi for *TcCDA2* (*TcCDA2a* and *TcCDA2b*) revealed functional specialization of the isoforms for this CDA. Unlike exon non-specific RNAi for *TcCDA2*, injection of dsRNAs specific for either one of alternative exons did not prevent any molts, suggesting that the proteins TcCDA2a and TcCDA2b could compensate for each other. However, the resulting adults exhibited different abnormal phenotypes. RNAi for *TcCDA2a* affected only femoral–tibial joint movement, while dsRNA for *TcCDA2b* resulted in elytra with crinkled and rough dorsal surfaces (Arakane *et al.*, 2009). These results suggest that group I CDAs play critical roles in maintaining the structural integrity of the cuticular chitin laminae and chitin fibers of the tracheal tube. It is possible that there are unique cuticular proteins that preferentially bind to deacetylated portions of chitin, whereas others preferentially bind to fully acetylated chitin. These proteins may help to organize the chitinous cuticular layers and provide the proper rigidity and/or flexibility in different regions of the cuticle.

Injection of a mixture of dsRNAs for *T. castaneum* group V CDAs, *TcCDAs 6, 7, 8*, and *9*, which are all predominantly expressed in the gut, significantly reduced the transcript levels of individual *CDAs*. However, no adverse effects on the appearance, behavior, or survival of these dsRNA-treated insects were observed (Arakane *et al.*, 2009). Interestingly, Jakubowska *et al.* (2010) observed that one of the group V (CDA9 subgroup) *CDA* genes from *H. armigera* (*HaCDA5a*) was downregulated by baculovirus infection in larvae. Like TnCDA9, HaCDA5a had a strong binding affinity for chitin, although it lacks any predicted chitin-binding domain. Incubation of the PM from *S. frugiperda* with recombinant HaCDA5a increased PM permeability in a concentration-dependent manner. Infection of insects with a recombinant baculovirus carrying this gene significantly increased the speed of kill for *S. frugiperda* and *S. exigua*. Together, these observations indicate that the group V CDA, HaCDA5a, may have a role in determining PM structure/morphology or permeability. For instance, downregulation of transcripts for this gene after pathogen attack resulted in reduced PM permeability, presumably to avoid pathogen infection. Additional studies in the future may reveal the physiological functions of the many CDAs belonging to groups II, III, and IV.

7.5. Chitin-Binding Proteins

Chitin is almost always found in association with numerous proteins that influence the overall mechanical and physicochemical properties of the chitin–protein matrix, which can range from very rigid (e.g., head capsule and mouth parts) to fully flexible (e.g., larval body and wing cuticle). Since chitin is an extracellular matrix polysaccharide, the proteins that have an affinity for chitin are expected to be extracellularly secreted proteins. This is generally true, with the constraint that some CBPs can be in vesicles or storage granules between the time they are synthesized and when they are secreted or released into the extracellular space by exocytosis.

There are three broad groups of insect proteins containing sequence motifs that have been associated with chitin-binding ability. The first group consists of a very large number of insect cuticular proteins, belonging to the CPR family, containing a consensus sequence(s) known as the extended Rebers & Riddiford Consensus (R&R Consensus) of a stretch of about 70 amino acids that defines pfam 00379 (Willis, 2010; see also Chapter 5). The second group of proteins contains an amino acid sequence motif known as the "peritrophin A" motif (Tellam *et al.*, 1999). To avoid confusion about its biological role(s), this motif will be referred to as the ChtBD2 domain in this chapter, because it is found not only in the group of proteins extracted from the peritrophic matrix, but also in proteins extracted from (or expressed in) cuticle-forming tissues. Proteins with the ChtBD2 motif are further subdivided intro three groups: peritrophic matrix proteins (with 1–19 ChtBD2 domains, determined to date); cuticular proteins analogous to peritrophins-3 (with 3 ChtBD2 domains); and cuticular proteins analogous to peritrophins-1 (with 1 ChtBD2 domain) (Jasrapuria *et al.*, 2010). This domain consists of a linear sequence of about 60 amino acids with 6 cysteines and conserved spacings between successive cysteine residues. The ChtBD2 domain defines family 14 of carbohydrate-binding proteins with chitin-binding ability (CBM14; pfam01607; SMART 00494). The second group also includes enzymes of chitin metabolism (chitinases, chitin deacetylases, and a protease) that have one or more ChtBD2 domains in addition to their catalytic domains. The third group of chitin-binding proteins consists of the family of antimicrobial peptides related to tachystatins from horseshoe crab (denoted as A1, A2, B1, B2, and C subfamilies), as well as the calcium channel antagonists, agatoxins from spider venom. Tachystatins are expressed in hemocytes, where they are stored in the form of small granules and are released into the hemolymph upon an immune stimulus. This group of proteins with six cysteines and a high affinity for chitin has a triple-stranded β-sheet structure with an inhibitory cysteine knot motif (Fujitani *et al.*, 2007). This structure is quite different from the peritrophin A motif and tachystatin (see below), and belongs to pfam 11478. They are not associated with cuticle or the PM, but they do play a major role in immune defense against bacteria, fungi, and other pathogens.

Representative members of each of the three groups of chitin-binding proteins have been extracted from the cuticle or the PM, or isolated from hemocytes. They have also been expressed in bacterial or other hosts, and some of the purified proteins have been shown to have chitin-binding ability. Several proteins belonging to the first and second groups of chitin-binding proteins are only predicted from known cDNA or genomic sequences and have not been biochemically characterized, largely as a result of difficulties associated with extracting them from highly sclerotized cuticular preparations or exuviae. The following sections will focus on the proteins of the second group of proteins with ChtBD2 motifs, and also include a limited discussion of group 3 chitin-binding proteins. A discussion on the first group of cuticular proteins with the R&R or other consensus motifs is kept to a minimum, because it is the subject of Chapter 5 in this book (Willis, 2010).

7.5.1. Chitin-Binding Proteins with the R&R Consensus

The CPR family of cuticular proteins is generally rich in histidines and devoid of cysteines. The absence of cysteines has been regarded as a defining characteristic of this group of proteins, with rare exceptions. The number of cuticular proteins belonging to the CPR subfamily in different insects varies widely, ranging from 32 in *A. mellifera* to >150 in *A. gambiae* (see Chapter 5), indicating a genus-specific expansion of specific families of cuticular proteins. Among the many families of cuticular proteins in insects, only some members of the CPR family with the R&R Consensus have been unequivocally shown to bind to chitin (summarized in Chapter 5). A member each of the Tweedle family from *B. mori* (Tang *et al.*, 2010) and one protein of the CPAP family (see below) have also been shown to possess chitin-binding ability. Modeling studies using the 65-aa long R&R Consensus have led to the notion that this region assumes a half-barrel structure into which a liner chain of *N*-acetylglucosamines can be fitted using van der Waals interactions between the sugar oligomer and the hydrophobic rings of conserved aromatic amino acids in this consensus (Iconomidou *et al.*, 2005). In an interesting study, Rebers and Willis (2001) demonstrated that the addition of this consensus sequence alone to glutathione-S-transferase resulted in acquisition of an affinity for chitin by this chimeric protein.

7.5.2. Peritrophic Matrix Proteins

The second group of proteins with the ChtBD2 motif is the family of proteins known as "peritrophins" that can be extracted from the PM using strong denaturing/chaotropic reagents, such as 6-M urea or 6-M guanidine hydrochloride (Tellam *et al.*, 1999). The extracted PMPs or recombinantly expressed PMPs have chitin-binding activity (Elvin *et al.*, 1996; Wijffels *et al.*, 2001; Wang *et al.*, 2004). This motif was shown to be responsible for binding to chitin by expressing a single ChtBD2 domain of *Trichoplusia ni* peritrophin, CBP1, in an insect cell line, and demonstrating its chitin-binding ability (Wang *et al.*, 2004). Proteins with multiple ChtBD2 domains are commonly found strongly associated with the PM. Not all of them are actually extractable, even with strong chaotropic agents. Some require extraction with strong organic solvents, such as anhydrous trifluoromethanesulfonic acid, which also deglycosylates O-linked glycoproteins (Campbell *et al.*, 2008).

The number of ChtBD2 domains in insect PMPs varies from 1 to as many as 19 in the bertha armyworm *Mamestra configurata* (Shi *et al.*, 2004; Dinglasan *et al.*, 2009; Venancio *et al.*, 2009; Jasrapuria *et al.*, 2010; Toprak *et al.*, 2010). Some PMPs have multiple ChtBD2 repeats in a tandem arrangement with short spacers rich in P, S, and T residues. Some of these linkers are potential sites of O-glycosylation. Other PMPs have mucin domains interspersed between ChtBD2 domains in various patterns of alternating ChtBD2 and mucin domains (Wang *et al.*, 2004; Venancio *et al.*, 2009). PMPs with only one or two ChtBD2 domains have also been reported (Jasrapuria *et al.*, 2010; Toprak *et al.*, 2010). The number of PMPs in different species is variable. Both *Ae. aegypti* and *D. melanogaster* have been predicted to have about 65 PMPs, though many of these may not be components of the PM (Venancio *et al.*, 2009). Detailed expression studies of all proteins with ChtBD2 domains in *T. castnaeum* have demonstrated that there are only 11 *bona fide* PMPs in this beetle (Jasrapuria *et al.*, 2010). Direct proteomic analysis of >200 proteins extracted from PMs dissected from adult *A. gambiae* females fed a protein-free diet has revealed the presence of only 12 PMPs, with the number of ChtBD2 repeats ranging from 1 to 4. It is likely that the total number of PMPs in insects is in the range of 10–20, although it can't be ruled out that additional *PMP* genes are expressed in the gut. However, their conceptual protein products were not detected in proteomic analyses because they were still in the insoluble pellet after extraction with detergents used in an extensive study (Dinglasan *et al.*, 2009). Interestingly, different *PMP* genes of *T. castaneum* were not expressed uniformly through the length of the midgut, with some *PMP*s being expressed in the anterior midgut, whereas others coding for proteins with multiple ChtBD2 domains were expressed in the posterior midgut (Jasrapuria *et al.*, 2010). Whether this differential spatial expression results in altered permeability of the PM along the length of the midgut remains to be investigated.

7.5.3. Cuticular Proteins Analogous to Peritrophins (CPAPs)

In addition to the *PMP* genes, which are expressed exclusively in the midgut lining cells, there are other genes encoding proteins with ChtBD2 domains, which are expressed in tissues other than the midgut. All of these proteins are predicted to have a cleavable signal peptide, and are expected to be capable of interacting with extracellular chitin. These genes are expressed predominantly in epidermal tissue as well as in other cuticle-forming tissues, including tracheae, elytra, hindwings, and hindgut. These genes have been subdivided into two groups, *CPAP1* and *CPAP3*, to reflect the fact that they encode proteins with one or three ChtBD2 domains, respectively (Jasrapuria *et al.*, 2010). *CPAP3* is the new name given to the orthologs of the previously characterized *D. melanogaster* "obstructor" or "gasp" gene family.

Mutants of the *D. melanogaster CPAP3-C* gene are embryo-lethal, and have been reported to exhibit cuticular defects (Barry *et al.*, 1999; Behr and Hoch, 2005). In *D. melanogaster* there are 10 genes encoding CPAP3 proteins, which can be further subdivided into two groups of 5 genes each. Only orthologs for the first group (*CPAP3-A*, *CPAP3-B*, *CPAP3-C*, *CPAP3-D*, and *CPAP-E*) are present in insects other than *Drosophila* species. There are significant variations in the expression profiles of these genes in different cuticle-forming tissues and/or developmental stages, suggesting functional differences among the CPAP3 proteins. RNA interference studies carried out in *T. castaneum* are consistent with such specialized functions of individual CPAP3 proteins (Jasrapuria *et al.*, unpublished data).

While it is expected that the CPAP3 proteins with three ChtBD2s will bind to chitin strongly, this has been demonstrated for only one recombinant protein from the spruce budworm *Choristoneura fumiferana*, which was expressed in *E. coli* (Nisole *et al.*, 2010). However, only a minor percentage of the His-tagged protein bound to the chitin, with the major portion appearing in the flow-through fraction, perhaps indicating that not all molecules of this recombinant protein had folded properly to exhibit strong chitin-binding activity. So far, there is no report of expression of this class of proteins in an insect cell system that may overcome the problem of misfolding as demonstrated for two PMP proteins with 10 and 12 repeats of ChtBD2 domains (Wang *et al.*, 2004).

A second group of genes encoding proteins with one ChtBD2 domain, referred to as the *CPAP1* family proteins, has been characterized extensively using a bioinformatics analysis of the *T. castaneum* genome (Jasrapuria *et al.*, 2010). These proteins vary extensively in size, and in the location of the ChtBD2 domain. Like CPAP3, they are also expressed in cuticle-forming tissues and have putative cleavable signal sequences consistent with a role involving interactions with chitin. So far, there are no reports on the chitin-binding ability of these proteins. Only some of these proteins have orthologs in *D. melanogaster*, casting doubt on whether these proteins are ubiquitous in insects. However, RNAi studies have produced lethal phenotypes when transcripts for 3 of the 10 genes encoding CPAP1 proteins were depleted in *T. castaneum* (Jasrapuria, unpublished data).

7.5.4. Enzymes of Chitin Metabolism

Enzymes of chitin metabolism, including some members of the chitinase and chitin deacetylase families, have the ChtBD2 motif (Kramer *et al.*, 1993; Campbell *et al.*, 2008; Dixit *et al.*, 2008; Zhu *et al.*, 2008a). The presence of one or more copies of this ChtBD2 motif has been suggested to increase the affinity of enzymes of chitin metabolism for the insoluble substrate, chitin, and to increase the processivity of these extracellular enzymes. Support for this idea comes from the drastic loss of ability to bind to insoluble chitin upon removal of the region containing the ChtBD2 motif from the C-terminal region of an *M. sexta* chitinase, which follows the catalytic domain. A C-terminal fragment of only 58 amino acids with this domain did bind to colloidal chitin, and addition of one or two copies of this domain to the chitinase catalytic domain progressively increased the affinity of the chitinase to colloidal chitin (Arakane *et al.*, 2003). While most of these enzymes have only one ChtBD2 motif, one class of insect chitinases (group II) has four or five ChtBD2 motifs dispersed among multiple catalytic domains (Royer *et al.*, 2002; Zhu *et al.* 2008a). A role for these multiple ChtBD2s in facilitating the depolymerizing chitin crystallites has been suggested (Arakane and Muthukrishnan, 2010).

7.5.5. Role of Secondary Structure of ChtBD2 Motif in Binding to Chitin

Tertiary structures based on 2D-NMR studies in solution are available for only two insect proteins with ChtBD2 domains with high affinity for chitin; namely, tachycitin and scarabacin. The antimicrobial peptide tachycitin from the horseshoe crab, which has a structure different from the tachystatins, is 76aa long, and has a higher K_m for chitin binding than tachystatin – $19.5\,\mu M$ versus $4.3\,\mu M$, respectively (Kawabata *et al.*, 2003). The second chitin-binding antimicrobial peptide for which an NMR-deduced structure is available is scarabacin from the coconut rhinocerous beetle *Oryctes rhinoceros*, which is 36aa long and has a $K_d = 1.3\,\mu M$ (Hemmi *et al.*, 2003). A comparison of these two structures with that of another chitin-binding minimal fragment called hevein-32, from the rubber latex protein hevein, has provided some interesting insights about the role of a part of the ChtBD2 motif in chitin binding.

Tachycitin has 10 cysteines in the form of 5 disulfide bonds, and has significant similarity to several peritrophins from a wide spectrum of insects, including PMP3 of *T. castaneum*, with which it shares 51% amino acid sequence identity. Of these 10 cysteines, 5 are in perfect register with the linear arrangement of the cysteines in PMP3 ChtBD2 domains, without introducing gaps in either sequence, except for the first cysteine in the motif.

More importantly, the amino acid sequence from positions 40 to 60 of tachycitin, which includes one disulfide bond, shares significant similarities to those of several other peritrophins from a wide range of insect species (Suetake *et al.*, 2000). Furthermore, the three-dimensional structure of this stretch of 21 amino acids is nearly identical to that of a hevein-32 from positions 20 to 32 (Aboitiz *et al.*, 2004). Both proteins have two anti-parallel β-sheets followed by a short α-helix in this region, which also includes a disulfide bond. Several aromatic amino acids that have been shown to contact the oligosaccharide ligands $(GlcNAc)_{3-6}$ are also conserved in the two sequences.

The 3D structure of scarabacin reveals the presence of only one disulfide bond between cys18 and cys29. The C-terminal half of this peptide from cys18 to ser 36 also has a secondary structure consisting of two anti-parallel β-sheets and a short α-helical turn super-imposable on hevein-32 or tachycitin (Hemmi *et al.*, 2003). These data suggest that only the C-terminal half of the ChtBD2 domain may be critical for chitin binding. Consistent with this interpretation is the finding that the N-terminal half of tachycitin has a completely different 3D structure, consisting of a three-stranded β-sheet while retaining the hevein/scarabacin-like chitin-binding motif on the C-terminal domain (Suetake *et al.*, 2000; Hemmi *et al.*, 2003). These data suggest that all three chitin-binding proteins (hevein, scarabacin, and tachycitin) share a common chitin-binding secondary/tertiary structure, even though they do not have extensive amino acid sequence identity. By extrapolation, we expect that all of the proteins with ChtBD2 domains will also have this structural motif consisting of two anti-parallel β-sheets and a short α-helical turn.

A protein from the vestimentiferan *Riftia pachyptila* has been shown to bind specifically to β-chitin, but not to α-chitin or cellulose (Chamoy *et al.*, 2001). The sequence of this protein includes a cysteine-rich region that resembles the C-terminal region of many mammalian chitinases, and is likely to be a chitin-biding motif. However, it does not have the consensus sequence or the characteristic spacing between adjacent cysteines of the ChtBD2 motif, and may represent yet another type of chitin-binding domain.

7.6. Chitin-Organizing Proteins

In addition to the CPR proteins with the R&R Consensus and the CPAP proteins with the ChtBD2 motif, which are expected to interact with chitin, some additional proteins may be associated with chitin, and help to organize it into bundles and the laminae that are characteristic of a mature procuticle. Two proteins encoded by *Knickkopf* (*Knk*) and *Retroactive* (*Rtv*) genes are known to be involved in this process in *D. melanogaster* (Ostrowski *et al.*, 2002; Moussian *et al.*, 2005b; Tonning *et al.*, 2005). Mutations in these two genes result in a dilated cuticle and loss of the fibrillar organization of tracheal chitin, and death of the developing embryo. Transmission electron microscopic analyses of the developing embryonic cuticle in these mutants revealed loss of the laminar architecture of chitin and the accumulation of electron-dense material in the procuticle.

How do these two proteins function to organize the cuticle-associated chitin? The domain organization of these proteins and their predicted properties offer some hints. The 75-kDa KNK is a GPI-anchored membrane protein with a multidomain architecture consisting of two DM13 domains, a dopamine monooxygenase N-terminal domain (DOMON domain), and a unique C-terminal region that has not been associated with any well-characterized domain. However, this sequence has some similarities to plastocyanin (Moussian, 2010). Interestingly, this region also has some sequence similarity to several plant proteins that possess DM13 and DOMON domains as well as a cytochrome b561 domain. It is possible that Knickkopf and its orthologs are extracellular proteins that may have a role in oxidation–reduction reactions perhaps involving dopamine. *T. castaneum* KNK expressed in a baculovirus-insect cell expression system does bind to colloidal chitin (Chaudhari, unpublished data). RNAi of this *KNK* gene results in loss of chitin, and this loss appears to be due to the protective effect of KNK on chitin against degradation by chitinolytic enzymes. The distribution of this protein between the procuticle and plasma membrane is consistent with such a chitin-protective role.

RTV is also a membrane-bound protein with a single C-terminal transmembrane domain, which localizes this protein to the apical surface of the plasma membrane (Schwarz and Moussian, 2007). RTV mutants have a spindle-shaped body, and often the cuticle separates from the epidermal layer underneath. This protein, which is about 150 amino acids long, has 10 cysteines, belongs to the neurotoxin-like SCOP superfamily of proteins, and has a β-sandwich structure with 2 and 3 β-strands in the 2 β–sheets. Its ability to bind to chitin has not been demonstrated, but the six aromatic amino acids present in the loops indicate such a possibility.

7.7. Hormonal Regulation of Chitin Metabolism

Chitinolytic activity in the molting fluid rises just prior to each molt and falls shortly thereafter. These changes parallel the increasing and falling ecdysteroid titers prior to ecdysis, as observed initially by Kimura (1976). A direct role for ecdysteroids in inducing chitinase expression was demonstrated using *M. sexta* larval abdomens that were precluded from receiving hormonal signals from the brain by a ligature below the second thoracic segment. Injection

of 20-hydroxyecdysone (20HE) into these ligated abdomens resulted in a sharp and rapid increase in transcripts for chitinase. This increase was abolished by a simultaneous injection of a juvenile hormone mimic (Fukamizo & Kramer, 1987). Koga *et al.* (1992) reported a similar induction of chitinase by ecdysteroid, utilizing isolated *Bombyx* abdomens. Zheng *et al.* (2003) observed that injection of an ecdysteroid agonist resulted in induction of expression of a chitinase gene in epidermal tissue of *C. afumiferana*, and demonstrated the accumulation of chitinase in molting fluid. It appears that hormonal regulation of chitinase genes occurs in a broad range of insect species. However, the presence of multiple genes encoding chitinases was not appreciated when these early studies were done, and it was not apparent which class of chitinases was induced by the ecdysteroid treatment. Based on our present knowledge about the tissue specificity of expression of different groups of chitinases, it is likely that these early studies were only focused on the expression of group I chitinases.

A group II chitinase gene with five catalytic domains from the beetle *Tenebrio molitor* has also been shown to be hormonally regulated (Royer *et al.*, 2002). During pupal–adult metamorphosis, the abundance of transcripts for this gene paralleled the changes in ecdysteroid (20HE) titers during metamorphosis. Interestingly, even topical application of the JH analog, methoprene, induced transcripts for this chitinase within 8 hours after treatment. These results are somewhat contradictory to the studies on *M. sexta* chitinase, in which JH had no inductive effect on chitinase transcript levels (Kramer *et al.*, 1993). In *B. mori*, another chitinase gene, *BmChiR1*, required 20HE for induction, and was suppressed by the simultaneous application of a JH analog (Takahashi *et al.*, 2002). Even though this chitinase was reported to have only two inactive catalytic domains and one CBD, our bioinformatics analysis (Merzendorfer, unpublished data) indicated that this gene actually encodes a protein with five catalytic domains and seven CBDs, and appears to be a group II chitinase. A recent study on the regulation of chitinase gene expression in a shrimp species demonstrated that it is induced by ecdysteroids. Hence, ecdysteroids may be required for induction of chitinases in most arthropods (Priya *et al.*, 2009). While it is clear that the expression of more than one chitinase gene is controlled by ecdysteroid and possibly by JH, it is likely that these effects are mediated through one or more transcription factors induced by ecdysteroids (Riddiford *et al.*, 2003). However, there are no published reports on the identification of hormone response elements in the promoters of any of the insect chitinase genes.

There is little evidence to support the idea that hormones play a direct role in the control of chitin synthesis. Instead, chitin synthesis is initiated at about the time of (or prior to) apolysis, when new cuticle is being deposited.

In general, chitin synthesis reaches peak levels in between molts when new cuticle is being synthesized at the maximal rate. In larval stages, this is also the period when PM-associated chitin is synthesized. Thus, both CHS-A and CHS-B levels are high during feeding periods in larval stages. In the pupal stage *CHS-B* transcripts are undetectable, whereas levels of transcripts for *CHS-A* have multiple peaks roughly corresponding to periods of synthesis of pupal cuticle, adult epidermal cuticle, and tracheal chitin (Hogenkamp *et al.*, 2005; Arakane *et al.*, 2008).

7.8. Chitin Metabolism and Insect Control

7.8.1. Inhibition of Chitin Synthesis

The absence of chitin in animals and plants has led to the development of insect control strategies that target enzymes involved in the synthesis, modification, and degradation of chitin. Several membrane proteins that are likely to be involved in the assembly of chitin in the procuticle, or regulation of chitin metabolism, may also be attractive targets. Compounds that directly or indirectly interfere with chitin biosynthesis include peptidyl nucleosides, acylureas, thiadiazines, and different kinds of chitin-binding molecules. The peptidyl nucleosides were isolated originally from different *Streptomyces* species, and include polyoxins and nikkomycins (Hori *et al.*, 1971; Dahn *et al.*, 1976). They are substrate analogs resembling the structure of UDP-GlcNAc, and competitively inhibit chitin synthases of fungal and insect sources, with nikkomycin being the most potent inhibitor (Cohen, 2001). As peptidyl nucleosides that exhibit low permeability across the hydrophobic epicuticle are easily degraded in the intestine and show toxic side effects in vertebrates, they have not been developed further to control insect pests, but some of them are in use as fungicides in agriculture (Zhang and Miller, 1999; Cohen, 2001; Ruiz-Herrera and San-Blas, 2003). In contrast, since the discovery of the high insecticidal potential of diflubenzuron in the early 1970s by Dutch scientists, various acylurea derivatives, such as lufenuron, novaluron, and hexaflumuron, have been developed commercially for controlling agricultural pests (Palli and Retnakaran, 1999). They have been shown to inhibit chitin synthesis and to disturb cuticle formation, causing abortive molting. Ultrastructural analysis revealed defects in chitin synthesis, abnormal deposition of endocuticular layers, and impaired PM formation. Studies with these "chitin synthesis inhibitors" have provided some insights concerning the role of chitin in development, and its biological function. In particular, the use of the acylurea compound lufenuron has provided substantial information on chitin synthesis during *Drosophila* development (Wilson and Cryan, 1997). The

effects of this insect growth regulator were complex and variable, depending on the developmental stage and dose at which the insects were exposed to this agent. When newly hatched larvae were reared on a diet containing very low concentrations of lufenuron, the larvae did not die until the second or third instar, and some pupariated even though the pupae were abnormally compressed. Pharate adults either failed to eclose or died shortly after emergence, and had deformed legs. The flight ability of the emerged adults was also affected when the larvae were exposed to very low concentrations of lufenuron. First and second instar larvae fed higher concentrations of lufenuron had normal growth and physical activity for several hours, but the insects died at about the time of the next ecdysis. Third instar larvae fed high concentrations of lufenuron underwent pupariation, but the puparia had an abnormal appearance, and the anterior spiracles failed to evert. Strikingly, adults showed no mortality and had no flight disability even when fed high levels of lufenuron, indicating that once all chitin-containing structures had been formed, this "chitin inhibitor" had very little effect on morphology and function. Thus, insect development is affected by lufenuron at all stages when chitin synthesis occurs. Another phase of insect development affected by this compound was egg hatching, even though oviposition was normal. The embryos completed development, but failed to rupture the vitelline membrane. In an ultrastructural study of acylurea effects on *Drosophila* embryogenesis, Gangishetti and colleagues have shown recently that egg hatching is completely abolished after treating female flies with a high dose of lufenuron and mating them with untreated males (Gangishetti *et al.*, 2009). In line with its lower insecticidal activity, the same treatment performed with diflubenzuron resulted in a constant rate of larval survival. Overall, the hatching rates depended on the dosage of the insecticides. The embryonic phenotypes were grouped into five classes: (1) hatching wild type larvae; (2) non-hatching larvae that appeared slightly bloated after being released manually from the eggshells; (3) non-hatching larvae with a strongly melanized head skeleton and a cuticle detached from the epidermis, which is similar to *knk* and *rtv* phenotypes (see section 7.6); (4) non-hatching larvae with a crumbled head skeleton and detached cuticle, which is similar to the *kkv* phenotype (see section 7.3.4.3. and **Figure 5**); and (5) non-hatching larvae with strong segmentation and morphological defects. The latter phenotypes were indistinguishable from the effects of the nucleoside peptide antibiotic nikkomycin, which is a competitive inhibitor of chitin synthase. Electron microscopy revealed that the treatment with lower doses of the insecticides affected cuticle thickness and orientation of microfibrils, while higher doses disrupted chitin synthesis completely, as evidenced by the lack of Calcofluor white fluorescence in the cuticle (Gangishetti *et al.*, 2009). Interestingly, no changes in *kkv* and *mummy* gene expression were observed, but the expression of certain genes encoding cytochrome P450 enzymes was substantially upregulated, indicating that the respective enzymes are involved in diflubenzuron and lufenuron detoxification. Similar results were also observed in *Tribolium*, where diflubenzuron fed to larvae did not significantly influence *TcCHS1* or *TcCHS2* expression, but did affect mRNA levels for certain cytochrome P450 enzymes (merzendorfer, unpublished data). In contrast to *Drosophila* and *Tribolium*, RT-PCR and Northern blot analyses carried out with *A. quadrimaculatus* revealed a two-fold upregulation of *AqCHS1* mRNA levels in response to a high dose of diflubenzuron, while the chitin content in surviving larvae decreased in a dose-dependent manner (Zhang and Zhu, 2006). The observed increase in *AqCHS1* mRNA levels associated with a decrease in chitin content corroborates the common view that acylurea insecticides affect chitin synthesis at a post-transcriptional level. Hence, diflubenzuron-induced *AqCHS1* expression may serve as a mechanism to compensate for chitin deficiency.

Several studies have aimed to elucidate the underlying mechanism of the insecticidal activity of diflubenzuron. Diflubenzuron efficiently blocks chitin synthesis, as the incorporation of radiolabeled sugars into the growing chitin chain is inhibited (Post and Vincent, 1973; Hajjar and Casida, 1978; Mayer *et al.*, 1980; Clarke and Jewess, 1990). However, in contrast to peptidyl nucleosides that block chitin polymerization, diflubenzuron obviously does not affect the catalytic step, because chitin synthesis is not impaired in cell-free systems (Cohen and Casida 1980a; Mayer *et al.*, 1980; Kitahara *et al.*, 1983; Zimoch *et al.*, 2005). It also does not interfere with any of the metabolic reactions yielding UDP-N-acetylglucosamine, and neither does it affect chitin synthesis in fungi (Verloop and Ferrel, 1977; Cohen, 1987). Based on these and other findings, it was suggested that diflubenzuron acts at a post-catalytic step of chitin synthesis (Cohen, 2001). Many other mechanisms for the action of diflubenzuron have been suggested, including effects on glycolytic enzymes, chitinases, phenoloxidases, hormonal sites, and microsomal oxidases (Ishaaya and Cohen, 1974; Ishaaya and Ascher, 1977; Mitlin *et al.*, 1977; DeLoach *et al.*, 1981; Soltani, 1984). Studies using imaginal discs and cell-free systems indicated that benzoylphenylureas inhibit ecdysteroid-dependent GlcNAc incorporation into chitin (Mikolajczyk *et al.*, 1994; Oberlander and Silhacek, 1998). These results indicated that acylurea compounds target ecdysone-dependent sites, which eventually leads to inhibition of chitin formation. However, direct proof for this hypothesis is lacking. On the basis of competitive binding assays performed with glibenclamide, a more recent study suggested that a sulfonylurea receptor might be the target for diflubenzuron (Abo-Elghar *et al.*, 2004). As the sulfonylurea receptors (SURs) may also

act as regulatory subunits of inward rectifying potassium channels in insects (Akasaka *et al.*, 2006), inhibition of a SUR could alter the membrane potential in such a way that Ca^{2+} homeostasis and eventually protein secretion required for cuticle and PM formation is affected. In line with this assumption, glibenclamide as well as diflubenzuron were found to affect Ca^{2+} uptake by isolated cuticular vesicles from the German cockroach *Blatella germanica* (Abo-Elghar *et al.*, 2004). Although the significance of this finding remains uncertain, future research following up on this hypothesis may elucidate the target site of acylureas.

Another chemical group of "chitin synthesis inhibitors" comprises thiadiazine derivatives, such as buprofezin (Applaud), which is used as an insecticide that specifically acts on sucking insects such as homopterans and hemipterans (Kanno, 1981). Although quite different in chemical structure, the effect of buprofezin resembles that of acylureas, as it blocks incorporation of radiolabeled chitin precursors and interferes with insect development. However, buprofezin may have a different target site in insects, as it also blocks acetylcholinesterase (AChE) activity. The activity of AChE in crude homogenates from the whitefly *Bemisia tabaci* was significantly inhibited by buprofezin at a concentration of $0.5\,\mu M$ (Cottage and Gunning, 2006). Strikingly, inhibition was not observed in buprofezin-resistant flies.

Chitin-binding molecules interfere with the microfibril assembly, and hence block chitin deposition at its final step. There are polysaccharide-binding dyes, such as Calcofluor White (CFW), Congo red or primuline, which interfere with chitin crystallization by disrupting hydrogen bond formation and hence perturbing microfibril assembly (Vermeulen and Wessels, 1986). Accordingly, these dyes were reported to impair fungal cell wall morphogenesis (Selitrennikoff, 1984; Roncero and Duran, 1985). In insects, the process of PM formation appears to be particularly susceptible to CFW, and its effects were studied in flies, mosquitoes and caterpillars. Injection of as little as $0.05\,\mu g$ CFW into *Calliphora erythrocephala* flies led to perturbations of PM formation and increased permeabilities for FITC-labeled dextrans with molecular masses ranging between 17 and 32 kDa (Zimmermann and Peters, 1987). However, in contrast to other PM-disrupting agents such as dithiothreitol or chitinase, changes in PM permeabilities for FITC-labeled dextrans with a molecular mass of 2 MDa were not observed when mosquito larvae were treated with CFW or Congo red (Edwards and Jacobs-Lorena, 2000). In *L. cuprina* the PM structure was not affected, although the larvae showed growth retardation and a reduction in lifespan (Tellam and Eisenmann, 2000). In the mite *Acarus siro*, combinations of diflubenzuron and CFW were most effective in reducing chitin content of the PM (Sobotnik *et al.*, 2008). Hence, combinations of CFW with other insecticidal compounds affecting chitin synthesis may prove to be a useful strategy for insect control. Disruption of the PM structure was consistently reported in various lepidopteran species (Wang and Granados, 2000; Bolognesi *et al.*, 2001; Zhu *et al.*, 2007). When larvae of *T. ni* and *S. exigua* were fed with a CFW-containing diet, an increase in PM permeability was observed and the larvae became more susceptible to baculoviral infections. Interestingly, a significant amount of proteins was released upon CFW treatment, which may explain altered permeabilities (Wang and Granados, 2000; Zhu *et al.*, 2007). Next to chitin-binding dyes, numerous sugar-binding proteins (lectins) from animals and plants such as galectins, WGA, and chitinase-like lectins bind chitin or chitosan because of their high preference for GlcNAc. Like CFW, they disrupt PM formation in numerous cases, and therefore have been investigated for their insecticidal potential (Cohen, 2010). The effects of WGA on PM formation are summarized in section 7.3.1.2. However, these types of proteins also bind to glycoproteins and proteoglycans present in the PM, and hence their particular mode of action is difficult to asseess *in vivo*.

7.8.2. Exploiting Chitinases for Insect Control

Chitinases have been used in a variety of ways for insect control and other purposes (Kramer and Muthukrishnan, 1997; Gooday, 1999). Several chitinase inhibitors with biological activity have been identified based on natural products chemistry (Spindler and Spindler-Barth, 1999), such as allosamidin, which mimics the carbohydrate substrate (Rao *et al.*, 2003), and cyclic peptides (Houston *et al.*, 2002). Although useful for biochemical studies, none of these chitin catabolic inhibitors have been developed for commercial use, primarily because of the high cost of production and potential side effects. As we learn more details about chitinase catalysis, it might become more economically feasible to develop and optimize chitinase inhibitors for insect pest management.

Fungi and plants use chitinases for establishing infection and as a defense against invading pathogens, respectively. Entomopathogens secrete a plethora of extracellular proteins with potential activity in insect hosts. One of these proteins is chitinase, which is used by fungi such as *Metarhizium anisopliae* to help penetrate the host cuticle and render host tissues suitable for consumption (St Leger *et al.*, 1996; Krieger de Moraes *et al.*, 2003). Among the 10 most frequent transcripts in a strain of *M. anisopliae* are 3 encoding chitinases and a chitosanase (Freimoser *et al.*, 2003a). However, when *M. anisopliae* was transformed to overexpress its native chitinase, the pathogenicity towards the tobacco hornworm was unaltered, suggesting that wild type levels of chitinase are not limiting for cuticle penetration (Screen *et al.*, 2001). Another fungal species, *Conidiobolus coronatus*, also produces both endo- and exo-acting chitinolytic enzymes during growth

on insect cuticle (Freimoser *et al.*, 2003b). Apparently, both *M. anisopliae* and *C. coronatus* produce a chitinolytic enzyme system to degrade cuticular components.

Both microbial and insect chitinases have been shown to enhance the toxicity of the entomopathogenic bacterium *Bacillus thuringiensis* (Bt) (Regev *et al.*, 1996; Tantimavanich *et al.*, 1997; Ding *et al.*, 1998; Sampson and Gooday, 1998; Wiwat *et al.*, 2000). For example, when the chitinolytic activities of several strains of *B. thuringiensis* were compared with their insecticidal activity, it was determined that the enzyme could enhance the toxicity of Bt to *S. exigua* larvae by more than two-fold (Liu *et al.*, 2002). Microbial chitinases have been used in mixing experiments to increase the potency of entomopathogenic microorganisms (Kramer and Muthukrishnan 1997). Synergistic effects between chitinolytic enzymes and microbial insecticides were reported as early as the 1970s. Bacterial chitinolytic enzymes were first used to enhance the activity of Bt and a baculovirus. Larvae of *C. fumiferana* died more rapidly when exposed to chitinase–Bt mixtures than when exposed to the enzyme or bacterium alone (Smirnoff and Valero, 1972; Lysenko, 1976; Morris, 1976). Mortality of gypsy moth, *Lymantria dispar*, larvae was enhanced when chitinase was mixed with Bt, relative to treatment with Bt alone, in laboratory experiments (Dubois, 1977). The toxic effect was correlated positively with enzyme levels (Gunner *et al.*, 1985). The larvicidal activity of a nuclear polyhedrosis virus toward *L. dispar* larvae was increased about five-fold when it was administered with a bacterial chitinase (Shapiro *et al.*, 1987).

Inducible chitinolytic enzymes from bacteria cause insect mortality under certain conditions. These enzymes may compromise the structural integrity of the PM barrier and improve the effectiveness of a Bt toxin by enhancing contact of the toxin molecules with their epithelial membrane receptors. For example, five chitinolytic bacterial strains isolated from midguts of *Spodoptera littoralis* induced a synergistic increase in larval mortality when combined with Bt spore-crystal suspensions relative to either an individual bacterial strain or a Bt suspension alone (Sneh *et al.*, 1983). An enhanced toxic effect toward *S. littoralis* also resulted when a combination of low levels of a truncated recombinant Bt toxin and a bacterial endochitinase was incorporated into a semisynthetic insect diet (Regev *et al.*, 1996). Crude chitinase preparations from *B. circulans* enhanced the toxicity of Bt *kurstaki* toward diamondback moth larvae (Wiwat *et al.*, 1996). Liu *et al.* (2002) reported that several strains of Bt produced their own chitinases, which had synergistic larvicidal activity with the endotoxins.

A family-18 insect chitinase has been used as an enhancer of baculovirus toxicity and as a host plant resistance factor in transgenic plants. Introduction of an *M. sexta* chitinase cDNA into *Autographa californica* multiple nuclear polyhedrosis viral (AcMNPV) DNA accelerated the rate of killing of fall armyworm compared to the wild type virus (Gopalakrishnan *et al.*, 1995). Baculoviral chitinases themselves play a role in liquefaction of insect hosts (Hawtin *et al.*, 1997; Thomas *et al.*, 2000). A constitutively expressed exochitinase from *B. thuringiensis* potentiated the insecticidal effect of the vegetative insecticidal protein Vip when they were fed to neonate larvae of *S. litura* (Arora *et al.*, 2003). Mutagenesis of the AcMNPV chitinase gene resulted in cessation of liquefaction of infected *T. ni* larvae, supporting a role of chitinase in viral spreading (Thomas *et al.*, 2000). When diet containing AcMNPV chitinase expressed in *E. coli* was fed to *B. mori* larvae, a dose-dependent increase in loss of integrity of the PM was observed. Even at a dose of 1 mg/g of larvae, there was 100% mortality (Rao *et al.*, 2004).

Tobacco budworms were killed when reared on transgenic tobacco expressing a truncated, enzymatically active form of *M. sexta* class I chitinase (Ding *et al.*, 1998). A synergistic interaction between insect chitinase expressed in transgenic tobacco plants and Bt applied as a spray at sublethal levels occurred when using the tobacco hornworm as the test insect. In contrast to results obtained with the tobacco budworm, studies with the hornworm revealed no consistent differences in larval growth or foliar damage when the insects were reared on first-generation transgenic chitinase-positive tobacco plants as compared to chitinase-negative control plants. When Bt toxin was applied at levels where no growth inhibition was observed on control plants, chitinase-positive plants had significantly less foliar damage and lower larval biomass production. These results indicated that the insect chitinase transgene did potentiate the effect of sublethal doses of Bt toxin, and *vice versa* (Ding *et al.*, 1998), but chitinase was not very effective on its own as a biocontrol agent. Tomato plants have been transformed with fungal chitinase genes with concomitant enhancements in resistance to insect pests (Gongora *et al.*, 2001). Effects observed include reduced growth rates and increased mortality, as well as a decrease in plant height and flowering time, with an increase in the number of flowers and fruits (Gongora and Broadway, 2002). Chitinase-secreting bacteria have been used to suppress herbivorous insect pests. A strain of *Enterobacter cloacae* transformed with a chitinase gene digested the chitinous membranes of phytophagous ladybird beetles, *Epilachna vigintioctopunctata*, and also suppressed leaf feeding and oviposition when the beetles ingested transformed bacteria entrapped in alginate microbeads sprayed on tomato seedlings (Otsu *et al.*, 2003). When pure chitinase from tomato moth larvae was injected into larvae, decreased cuticle thickness and 100% mortality was observed even at a low dose (2.5 µg/g). Insects fed this protein exhibited reductions in growth and food consumption (Fitches *et al.*, 2004). Acaricidal activity of a purified chitinase from a hard tick, *Haemophysalis longicornis*, has also been demonstrated (Assenga *et al.*,

2006). Immunization with this chitinase as the antigen protected mice from tick infections (You *et al.*, 2009).

Several GlcNAc-specific lectins from plants have been evaluated for insect toxicity (Harper *et al.*, 1998; Macedo *et al.*, 2003). These proteins appear to disrupt the integrity of the PM by binding to chitin or glycan receptors on the surface of cells lining the insect gut. Moreover, they may bind to glycosylated digestive enzymes and inhibit their activity. Another type of plant chitin-binding protein is the seed storage protein vicilin, which is actually a family of oligomeric proteins with variable degrees of glycosylation (Macedo *et al.*, 1993; Shutov *et al.*, 1995). Some vicilins are insecticidal to bruchid beetles and stalk borers (Sales *et al.*, 2001; Mota *et al.*, 2003). Apparently, these proteins bind to the PM, causing developmental abnormalities and reduced survival rates. To date, no non-enzymatic carbohydrate-binding protein derived from an insect has been evaluated for biocidal activity. A novel approach has been proposed to develop strategies for insect control by utilizing chitin-binding molecules to specifically target formation of the PM. CFW, a chemical whitener with chitin-binding properties, was used as a model compound in the diet to inhibit PM formation in *T. ni*, and to increase larval susceptibility to baculovirus infection (Wang and Granados, 2000). It was also effective in suppressing PM formation in *S. frugiperda*, and at the same time in preventing the establishment of a decreasing gradient of proteinases along the midgut tissue (Bolognesi *et al.*, 2001).

A protease from *A. gambiae* with a chitin-binding domain has been described, which may be involved in insect defense (Danielli *et al.*, 2000). This 147-kDa protein, sp22D, is expressed in a variety of tissues, most strongly in hemocytes, and is secreted into the hemolymph. Upon bacterial infection, the transcripts for this protein increase by about two-fold, suggesting a role in insect defense. This protein has a multidomain organization that includes two copies of an N-terminal ChtBD2 domain, a C-terminal protease domain, and several receptor domains. It binds strongly to chitin, and undergoes complex proteolytic processing during pupal to adult metamorphosis. It has been proposed that exposure of this protease to chitin may regulate its activity during tissue remodeling or wounding.

Two synthetic peptides were found to inhibit *A. gambiae* midgut chitinase, and also to block sporogonic development of the human malaria parasite *Plasmodium falciparum* and avian malaria parasite *P. gallinaceum*, when the peptides were fed to infected mosquitoes (Bhatnagar *et al.*, 2003). The design of these peptides was based on the putative proregion sequence of mosquito midgut chitinase. The results indicated that expression of chitinase inhibitory peptides in transgenic mosquitoes might alter the vectorial capacity of mosquitoes to transmit malaria.

7.9. Future Studies and Concluding Remarks

Although substantial progress in studies of insect chitin metabolism has occurred since the initial edition of *Comprehensive Insect Physiology, Biochemistry and Pharmacology* was published in 1985, we still do not know much about how chitin is produced and transported across the cell membrane so that it can interact perfectly with other components for assembly of supramolecular extracellular matrices such as the exoskeleton and PM. These structures are still very much biochemical puzzles in which we do not understand well how the various components come together during morphogenesis, or are digested during the molting process. Hopefully, this chapter will stimulate more effort to gain an understanding of how insects utilize chitin metabolism for growth and development, and also to facilitate development of materials that may perturb insect chitin metabolism for pest management purposes.

Since 2005, many questions have been answered about the biosynthesis of insect chitin, including: why do insects have two genes for CHS, and at what developmental stages are the various CHSs produced? However, we know little about the unique properties and functions of each CHS. Of particular interest is the role of alternate splicing in generating different isoforms of CHSs from the same gene. The developmental cues that control alternate splicing and how they affect chitin synthesis and/or deposition will be the subject of future studies. Attempts to express full-length *CHS* genes in heterologous systems for the production of active recombinant enzymes or subdomains has met with very limited success, probably because CHSs are membrane-bound proteins. The recent finding that proteolytic processing may be necessary for CHS activation may also have contributed to this lack of success (Broehan *et al.*, 2007, 2008). In the future, the availability of pure proteins and molecular probes for specific CHSs would facilitate a better understanding of chitin biosynthesis and its regulation.

Two other significant questions about the regulation of insect chitin biosynthesis are: what is the mechanism of the initiation phase, and is there an autocatalytic initiator molecule? Like glycogen synthesis, chitin synthesis may involve both initiation and elongation phases. As the initiator of glycogen synthesis, glycogenin transfers glucose from UDP-glucose to itself to generate an oligosaccharide–protein primer for elongation (Gibbons *et al.*, 2002). Like chitin synthase, glycogenin is a glycosyltransferase, which raises the question of whether chitin synthase has an autocatalytic function similar to glycogenin, and whether there is a separate "chitinogenin"-like protein. Another possibility is the participation of a lipid primer for chitin synthesis. Cellulose synthesis in plants involves the transfer of lipid-linked cellulodextrins to a growing glucan chain (Read and Bacic, 2002). The lipid in this case is sitosterol-β-glucoside. No lipid primer has been identified to date for insect chitin synthesis.

Little is known about the catalytic mechanism of any insect CHS. Once active insect CHS-related recombinant proteins can be produced in a cell line, site-directed mutagenesis can be used to probe for essential residues in the catalytic and regulatory domains. It is likely that acidic amino acids play critical roles in CHS catalysis in a manner comparable to those identified in other glycosyltransferases (Hefner *et al.*, 2002) and in yeast chitin synthases (Nagahashi *et al.*, 1995).

One of the major unanswered questions about insect chitinolytic enzymes and chitin deaceylases is: why are there so many of these enzymes? Some species have more than 20 chitinase or chitinase-like genes, and we only know the function of a few of them. Chitinolytic enzymes are gaining importance for their biotechnological applications in agriculture and healthcare (Dahiya *et al.*, 2006). Additional success in using chitinases from both insects and other organisms for different applications depends on a better understanding of their biochemistry and regulation so that their useful properties can be optimized through genetic and biochemical engineering. Reasons for the rather high number of chitinolytic and chitin deacetylase enzymes with various domain structures are not fully understood.

So far there has been little success in using chitinase in pest-control applications, but it may prove more useful as an enhancer protein in a cocktail with other biopesticides targeted at the cuticle or gut (Fiandra *et al.*, 2010; Di Maro *et al.*, 2010). Also, only a few catalytic domains or chitin-binding domains, or various combinations thereof (domain shuffling and/or swapping), have been evaluated for biocidal activity, and thus further toxicological experimentation after recombinations is warranted (Zakariassen *et al.*, 2009; Li and Greene, 2010; Neeraja *et al.*, 2010). With good progress occurring in regard to functional analysis from RNAi studies, the ability to choose an appropriate target gene or protein associated with insect chitin metabolism that can be exploited to achieve targeted and selective control of pest insects has improved.

A hypothetical model for chitin-containing extracellular matrices in insects is the following: a fiber-reinforced composite structure whereby chitin fibers form the initial scaffold that is subsequently impregnated with a blend of proteins into which some components of lower abundance, such as water, catechols, lipids, pigments, and minerals, are interspersed. For a soft hydrated material such as the PM and trachea, chitin/chitosan and protein are the major components that associate primarily non-covalently via hydrogen bonding, as well as through hydrophobic and electrostatic interactions with relatively little protein cross-linking. The chitin-organizing proteins may have a role in the precise arrangement of the individual laminar layers of chitin, as well as their relative orientation with the layers above and below it. For matrices that become sclerotized, such as tanned cuticle, catechols are incorporated and oxidized to quinones and quinone methides, which subsequently cross-link the proteins, and perhaps chitin/chitosan as well. Future studies are needed to characterize more fully the covalent and non-covalent interactions and reactions of chitin, protein, lipid, mineral salts, and oxidized catechols (chitin–water, chitin–protein, chitin–catechol, chitin–lipid, chitin–pigment, chitin–mineral interactions) from appropriate secretory tissues. Results from such studies will provide critical insights into the anabolic and catabolic pathways by which the chitin–protein composite is formed and recycled, as well as into the bioinspired fabrication of environmentally sustainable load-bearing materials whose formulation is based, at least in part, on insect chitin chemistry and metabolism.

Acknowledgments

We thank Dr. Mi Young Nho for helping with the phylogenetic tree construction and The National Science Foundation for financial support.

References

Abdel-Banat, B. M., & Koga, D. (2001). A genomic clone for a chitinase gene from the silkworm, *Bombyx mori*: Structural organization identifies functional motifs. *Insect Biochem. Mol. Biol., 31*, 497–508.

Abo-Elghar, G. E., Fujiyoshi, P., & Matsumura, F. (2004). Significance of the sulfonylurea receptor (SUR) as the target of diflubenzuron in chitin synthesis inhibition in *Drosophila melanogaster* and *Blattella germanica. Insect Biochem. Mol. Biol., 34*, 743–752.

Aboitiz, N., Vila-Perello, M., Groves, P., Asensio, J. L., Andreu, D., et al. (2004). NMR and modeling studies of protein–carbohydrate interactions: Synthesis, three-dimensional structure, and recognition properties of a minimum hevein domain with binding affinity for chitooligosaccharides. *Chembiochem., 5*, 1245–1255.

Adams, M. D., Celniker, S. E., Holt, R. A., Evans, C. A., Gocayne, J. D., et al. (2000). The genome sequence of *Drosophila melanogaster. Science, 287*, 2185–2195.

Akasaka, T., Klinedinst, S., Ocorr, K., Bustamante, E. L., Kim, S. K., & Bodmer, R. (2006). The ATP-sensitive potassium (KATP) channel-encoded dSUR gene is required for *Drosophila* heart function and is regulated by tinman. *Proc. Natl. Acad. Sci. USA, 103*, 11999–12004.

Altmann, F., Schwihla, H., Staudacher, E., Glossl, J., & Marz, L. (1995). Insect cells contain an unusual, membrane-bound beta-*N*-acetylglucosaminidase probably involved in the processing of protein *N*-glycans. *J. Biol. Chem., 270*, 17344–17349.

Andersen, S. O. (2010). Insect cuticular sclerotization: A review. *Insect Biochem. Mol. Biol., 40*, 166–178.

Arakane, Y., & Muthukrishnan, S. (2010). Insect chitinase and chitinase-like proteins. *Cell. Mol. Life Sci., 67*, 201–216.

Arakane, Y., Zhu, Q., Matsumiya, M., Muthukrishnan, S., & Kramer, K. J. (2003). Properties of catalytic, linker and chitin-binding domains of insect chitinase. *Insect Biochem. Mol. Biol., 33*, 631–648.

Arakane, Y., Hogenkamp, D. G., Zhu, Y. C., Kramer, K. J., Specht, C. A., et al. (2004). Characterization of two chitin synthase genes of the red flour beetle, *Tribolium castaneum*, and alternate exon usage in one of the genes during development. *Insect Biochem. Mol. Biol., 34*, 291–304.

Arakane, Y., Muthukrishnan, S., Kramer, K. J., Specht, C. A., Tomoyasu, Y., et al. (2005). The *Tribolium* chitin synthase genes *TcCHS1* and *TcCHS2* are specialized for synthesis of epidermal cuticle and midgut peritrophic matrix. *Insect Mol. Biol., 14*, 453–463.

Arakane, Y., Specht, C. A., Kramer, K. J., Muthukrishnan, S., & Beeman, R. W. (2008). Chitin synthases are required for survival, fecundity and egg hatch in the red flour beetle, *Tribolium castaneum. Insect Biochem. Mol. Biol., 38*, 959–962.

Arakane, Y., Dixit, R., Begum, K., Park, Y., Specht, C. A., et al. (2009). Analysis of functions of the chitin deacetylase gene family in *Tribolium castaneum. Insect Biochem. Mol. Biol., 39*, 355–365.

Arakane, Y., Baguinon, M., Jasrapuria, S., Chaudhari, S., Doyungan, A., et al. (2011). Both UDP N-acetylglucosamine pyrophosphorylases of *Tribolium castaneum* are critical for molting, survival and fecundity. *Insect Biochem. Mol. Biol., 41*, 42–50.

Araújo, S. J., Aslam, H., Tear, G., & Casanova, J. (2005). *mummy/cystic* encodes an enzyme required for chitin and glycan synthesis, involved in trachea, embryonic cuticle and CNS development – analysis of its role in *Drosophila* tracheal morphogenesis. *Dev. Biol., 288*, 179–193.

Arora, N., Ahmad, T., Rajagopal, R., & Bhatnagar, R. K. (2003). A constitutively expressed 36 kDa exochitinase from *Bacillus thuringiensis* HD-1. *Biochem. Biophys. Res. Commun., 307*, 620–625.

Assenga, S. P., You, M., Shy, C. H., Yamagishi, J., Sakaguchi, T., et al. (2006). The use of a recombinant baculovirus expressing a chitinase from the hard tick *Haemaphysalis longicornis* and its potential application as a bioacaricide for tick control. *Parasitol. Res., 98*, 111–118.

Barreto, L., Sorais, F., Salazar, V., San-Blas, G., & Nino-Vega, G. A. (2010). Expression of *Paracoccidioides brasiliensis* CHS3 in a *Saccharomyces cerevisiae* chs3 null mutant demonstrates its functionality as a chitin synthase gene. *Yeast, 27*, 293–300.

Barry, M. K., Triplett, A. A., & Christensen, A. C. (1999). A peritrophin-like protein expressed in the embryonic tracheae of *Drosophila melanogaster. Insect Biochem. Mol. Biol., 29*, 319–327.

Becker, A., Schloder, P., Steele, J. E., & Wegener, G. (1996). The regulation of trehalose metabolism in insects. *Exs, 52*, 433–439.

Behr, M., & Hoch, M. (2005). Identification of the novel evolutionary conserved *obstructor* multigene family in invertebrates. *FEBS Lett., 579*, 6827–6833.

Beitel, G. J., & Krasnow, M. A. (2000). Genetic control of epithelial tube size in the *Drosophila* tracheal system. *Development, 127*, 3271–3282.

Bhatnagar, R. K., Arora, N., Sachidanand, S., Shahabuddin, M., Keister, D., & Chauhan, V. S. (2003). Synthetic propeptide inhibits mosquito midgut chitinase and blocks sporogonic development of malaria parasite. *Biochem. Biophys. Res. Commun., 304*, 783–787.

Bolognesi, R., Ribeiro, A. F., Terra, W. R., & Ferreira, C. (2001). The peritrophic membrane of *Spodoptera frugiperda*: Secretion of peritrophins and role in immobilization and recycling digestive enzymes. *Arch. Insect Biochem. Physiol., 47*, 62–75.

Bolognesi, R., Arakane, Y., Muthukrishnan, S., Kramer, K. J., Terra, W. R., & Ferreira, C. (2005). Sequences of cDNAs and expression of genes encoding chitin synthase and chitinase in the midgut of *Spodoptera frugiperda*. *Insect Biochem. Mol. Biol., 35*, 1249–1259.

Boraston, A. B., Bolam, D. N., Gilbert, H. J., & Davies, G. J. (2004). Carbohydrate-binding modules: Fine-tuning polysaccharide recognition. *Biochem. J., 382*, 769–781.

Broehan, G., Zimoch, L., Wessels, A., Ertas, B., & Merzendorfer, H. (2007). A chymotrypsin-like serine protease interacts with the chitin synthase from the midgut of the tobacco hornworm. *J. Exp. Biol., 210*, 3636–3643.

Broehan, G., Kemper, M., Driemeier, D., Vogelpohl, I., & Merzendorfer, H. (2008). Cloning and expression analysis of midgut chymotrypsin-like proteinases in the tobacco hornworm. *J. Insect Physiol., 54*, 1243–1252.

Cabib, E., & Farkas, V. (1971). The control of morphogenesis: An enzymatic mechanism for the initiation of septum formation in yeast. *Proc. Natl. Acad. Sci. USA, 68*, 2052–2056.

Campbell, P. M., Cao, A. T., Hines, E. R., East, P. D., & Gordon, K. H. (2008). Proteomic analysis of the peritrophic matrix from the gut of the caterpillar, *Helicoverpa armigera*. *Insect Biochem. Mol. Biol., 38*, 950–958.

Cantarel, B. L., Coutinho, P. M., Rancurel, C., Bernard, T., Lombard, V., & Henrissat, B. (2009). The Carbohydrate-Active EnZymes database (CAZy): An expert resource for glycogenomics. *Nucleic Acids Res., 37*, D233–238.

Caufrier, F., Martinou, A., Dupont, C., & Bouriotis, V. (2003). Carbohydrate esterase family 4 enzymes: Substrate specificity. *Carbohydr. Res., 338*, 687–692.

Chamoy, L., Nicolai, M., Ravaux, J., Quennedey, B., Gaill, F., & Delachambre, J. (2001). A novel chitin-binding protein from the vestimentiferan *Riftia pachyptila* interacts specifically with beta-chitin. Cloning, expression, characterization. *J. Biol. Chem., 276*, 8051–8058.

Chang, R., Yeager, A. R., & Finney, N. S. (2003). Probing the mechanism of a fungal glycosyltransferase essential for cell wall biosynthesis. UDP-chitobiose is not a substrate for chitin synthase. *Org. Biomol. Chem., 1*, 39–41.

Chen, J., Tang, B., Chen, H., Yao, Q., Huang, X., et al. (2010). Different functions of the insect soluble and membrane-bound trehalase genes in chitin biosynthesis revealed by RNA interference. *PLoS One, 5*, e10133.

Chen, X., Yang, X., Senthil Kumar, N., Tang, B., Sun, X., et al. (2007). The class A chitin synthase gene of *Spodoptera exigua*: Molecular cloning and expression patterns. *Insect Biochem. Mol. Biol., 37*, 409–417.

Choi, H. K., Choi, K. H., Kramer, K. J., & Muthukrishnan, S. (1997). Isolation and characterization of a genomic clone for the gene of an insect molting enzyme, chitinase. *Insect Biochem. Mol. Biol., 27*, 37–47.

Choi, W. J., Sburlati, A., & Cabib, E. (1994). Chitin synthase 3 from yeast has zymogenic properties that depend on both the *CAL1* and the *CAL3* genes. *Proc. Natl. Acad. Sci. USA, 91*, 4727–4730.

Clarke, B. S., & Jewess, P. J. (1990). The inhibition of chitin synthesis in *Spodoptera littoralis* larvae by flufenoxuron, teflubenzuron and diflubenzuron. *Pestic. Sci., 28*, 377–388.

Cohen, E. (1987). Chitin biochemistry: Synthesis and inhibition. *Annu. Rev. Entomol., 32*, 71–93.

Cohen, E. (2001). Chitin synthesis and inhibition: A revisit. *Pest Manag. Sci., 57*, 946–950.

Cohen, E. (2010). Chitin biochemistry: Synthesis, hydrolysis and inhibition. In C. Jérôme, & J. S. Stephen (Eds.), *Advances in Insect Physiology* (Vol. 38, pp. 5–74). New York, NY: Academic Press.

Cohen, E., & Casida, J. E. (1980a). Inhibition of *Tribolium* gut chitin synthetase. *Pestic. Biochem. Physiol., 13*, 129–136.

Cohen, E., & Casida, J. E. (1980b). Properties of *Tribolium* gut chitin synthetase. *Pestic. Biochem. Physiol., 13*, 121–128.

Cottage, E. L., & Gunning, R. V. (2006). Buprofezin inhibits acetylcholinesterase activity in B-biotype *Bemisia tabaci*. *J. Mol. Neurosci., 30*, 39–40.

Coutinho, P. M., & Henrissat, B. (1999). Carbohydrate-active enzymes: An integrated database approach. In H. H. Gilbert, G. J. Davies, H. Henrissat, & B. Svensson (Eds.), *Recent Advances in Carbohydrate Bioengineering* (pp. 3–12). Cambridge, UK: Royal Society of Chemistry.

Coutinho, P. M., Deleury, E., Davies, G. J., & Henrissat, B. (2003). An evolving hierarchical family classification for glycosyltransferases. *J. Mol. Biol., 328*, 307–317.

Dahiya, N., Tewari, R., & Hoondal, G. S. (2006). Biotechnological aspects of chitinolytic enzymes: A review. *Appl. Microbiol. Biotechnol., 71*, 773–782.

Dahn, U., Hagenmaier, H., Hohne, H., Konig, W. A., Wolf, G., & Zahner, H. (1976). Stoffwechselprodukte von Mikroorganismen. 154. Mitteilung. Nikkomycin, ein neuer Hemmstoff der Chitinsynthese bei Pilzen. *Arch. Microbiol., 107*, 143–160.

Daimon, T., Hamada, K., Mita, K., Okano, K., Suzuki, M. G., et al. (2003). A *Bombyx mori* gene, BmChi-h, encodes a protein homologous to bacterial and baculovirus chitinases. *Insect Biochem. Mol. Biol., 33*, 749–759.

Danielli, A., Loukeris, T. G., Lagueux, M., Müeller, H. M., Richman, A., & Kafatos, F. C. (2000). A modular chitin-binding protease associated with hemocytes and hemolymph in the mosquito, *Anopheles gambiae*. *Proc. Natl. Acad. Sci. USA, 97*, 7136–7141.

de la Vega, H., Specht, C. A., Liu, Y., & Robbins, P. W. (1998). Chitinases are a multi-gene family in *Aedes*, *Anopheles* and *Drosophila*. *Insect Mol. Biol., 7*, 233–239.

DeLoach, J. R., Meola, S. M., Mayer, R. T., & Thompson, J. M. (1981). Inhibition of DNA synthesis by diflubenzuron in pupae of the stable fly *Stomoxys calcitrans* (L.). *Pestic. Biochem. Physiol., 15*, 172–180.

Demolliens, A., Boucher, C., Durocher, Y., Jolicoeur, M., Buschmann, M. D., & De Crescenzo, G. (2008). Tyrosinase-catalyzed synthesis of a universal coil-chitosan bioconjugate for protein immobilization. *Bioconjug. Chem., 19*, 1849–1854.

Devine, W. P., Lubarsky, B., Shaw, K., Luschnig, S., Messina, L., & Krasnow, M. A. (2005). Requirement for chitin biosynthesis in epithelial tube morphogenesis. *Proc. Natl. Acad. Sci. USA*, *102*, 17014–17019.

Di Maro, A., Terracciano, I., Sticco, L., Fiandra, L., Ruocco, M., et al. (2010). Purification and characterization of a viral chitinase active against plant pathogens and herbivores from transgenic tobacco. *J. Biotechnol.*, *147*, 1–6.

Ding, X., Gopalakrishnan, B., Johnson, L. B., White, F. F., Wang, X., et al. (1998). Insect resistance of transgenic tobacco expressing an insect chitinase gene. *Transgenic Res.*, *7*, 77–84.

Dinglasan, R. R., Devenport, M., Florens, L., Johnson, J. R., McHugh, C. A., et al. (2009). The *Anopheles gambiae* adult midgut peritrophic matrix proteome. *Insect Biochem. Mol. Biol.*, *39*, 125–134.

Dixit, R., Arakane, Y., Specht, C. A., Richard, C., Kramer, K. J., et al. (2008). Domain organization and phylogenetic analysis of proteins from the chitin deacetylase gene family of *Tribolium castaneum* and three other species of insects. *Insect Biochem. Mol. Biol.*, *38*, 440–451.

Dubois, N. R. (1977). *Pathogenicity of selected resident microorganisms of Lymantria dispar after induction for chitinase, PhD thesis*. Amherst, MA: University of Massachusetts.

Edwards, M. J., & Jacobs-Lorena, M. (2000). Permeability and disruption of the peritrophic matrix and caecal membrane from *Aedes aegypti* and *Anopheles gambiae* mosquito larvae. *J. Insect Physiol.*, *46*, 1313–1320.

Elvin, C. M., Vuocolo, T., Pearson, R. D., East, I. J., Riding, G. A., et al. (1996). Characterization of a major peritrophic membrane protein, peritrophin-44, from the larvae of *Lucilia cuprina*. cDNA and deduced amino acid sequences. *J. Biol. Chem.*, *271*, 8925–8935.

Fiandra, L., Terracciano, I., Fanti, P., Garonna, A., Ferracane, L., et al. (2010). A viral chitinase enhances oral activity of TMOF. *Insect Biochem. Mol. Biol.*, *40*, 533–540.

Filho, B. P., Lemos, F. J., Secundino, N. F., Pascoa, V., Pereira, S. T., & Pimenta, P. F. (2002). Presence of chitinase and beta-*N*-acetylglucosaminidase in the *Aedes aegypti*. A chitinolytic system involving peritrophic matrix formation and degradation. *Insect Biochem. Mol. Biol.*, *32*, 1723–1729.

Fitches, E., Wilkinson, H., Bell, H., Bown, D. P., Gatehouse, J. A., & Edwards, J. P. (2004). Cloning, expression and functional characterisation of chitinase from larvae of tomato moth (*Lacanobia oleracea*): A demonstration of the insecticidal activity of insect chitinase. *Insect Biochem. Mol. Biol.*, *34*, 1037–1050.

Freimoser, F. M., Screen, S., Bagga, S., Hu, G., & St Leger, R. J. (2003a). Expressed sequence tag (EST) analysis of two subspecies of *Metarhizium anisopliae* reveals a plethora of secreted proteins with potential activity in insect hosts. *Microbiology*, *149*, 239–247.

Freimoser, F. M., Screen, S., Hu, G., & St Leger, R. (2003b). EST analysis of genes expressed by the zygomycete pathogen *Conidiobolus coronatus* during growth on insect cuticle. *Microbiology*, *149*, 1893–1900.

Fujitani, N., Kouno, T., Nakahara, T., Takaya, K., Osaki, T., et al. (2007). The solution structure of horseshoe crab antimicrobial peptide tachystatin B with an inhibitory cystine-knot motif. *J. Pept. Sci.*, *13*, 269–279.

Fukamizo, T. (2000). Chitinolytic enzymes: Catalysis, substrate binding, and their application. *Curr. Protein Pept. Sci.*, *1*, 105–124.

Fukamizo, T., & Kramer, K. J. (1985a). Mechanism of chitin hydrolysis by the binary chitinase system in insect moulting fluid. *Insect Biochem.*, *15*, 141–145.

Fukamizo, T., & Kramer, K. J. (1985b). Mechanism of chitin oligosaccharide hydrolysis by the binary enzyme chitinase system in insect moulting fluid. *Insect Biochem.*, *15*, 1–7.

Fukamizo, T., & Kramer, K. J. (1987). Effect of 20-hydroxyecdysone on chitinase and β-N-acetylglucosaminidase during the larval–pupal transformation of *Manduca sexta* (L.). *Insect Biochem.*, *17*, 547–550.

Fukamizo, T., Kramer, K. J., Mueller, D. D., Schaefer, J., Garbow, J., & Jacob, G. S. (1986). Analysis of chitin structure by nuclear magnetic resonance spectroscopy and chitinolytic enzyme digestion. *Arch. Biochem. Biophys.*, *249*, 15–26.

Fusetti, F., von Moeller, H., Houston, D., Rozeboom, H. J., Dijkstra, B. W., et al. (2002). Structure of human chitotriosidase. Implications for specific inhibitor design and function of mammalian chitinase-like lectins. *J. Biol. Chem.*, *277*, 25537–25544.

Gagou, M. E., Kapsetaki, M., Turberg, A., & Kafetzopoulos, D. (2002). Stage-specific expression of the chitin synthase *DmeChSA* and *DmeChSB* genes during the onset of *Drosophila* metamorphosis. *Insect Biochem. Mol. Biol.*, *32*, 141–146.

Gangishetti, U., Breitenbach, S., Zander, M., Saheb, S. K., Müeller, U., et al. (2009). Effects of benzoylphenylurea on chitin synthesis and orientation in the cuticle of the *Drosophila* larva. *Eur. J. Cell Biol.*, *88*, 167–180.

Gibbons, B. J., Roach, P. J., & Hurley, T. D. (2002). Crystal structure of the autocatalytic initiator of glycogen biosynthesis, glycogenin. *J. Mol. Biol.*, *319*, 463–477.

Gilbert, L. I. (2004). Halloween genes encode P450 enzymes that mediate steroid hormone biosynthesis in *Drosophila melanogaster*. *Mol. Cell Endocrinol.*, *215*, 1–10.

Girard, C., & Jouanin, L. (1999). Molecular cloning of a gut-specific chitinase cDNA from the beetle *Phaedon cochleariae*. *Insect Biochem. Mol. Biol.*, *29*, 549–556.

Gongora, C. E., & Broadway, R. M. (2002). Plant growth and development influenced by transgenic insertion of bacterial chitinolytic enzymes. *Mol. Breeding*, *9*, 123–135.

Gongora, C. E., Wang, S., Barbehenn, R. V., & Broadway, R. M. (2001). Chitinolytic enzymes from *Streptomyces albidoflavus* expressed in tomato plants: Effects on *Trichoplusia ni*. *Entomol. Experimentia Appl.*, *99*, 193–204.

Goo, T. -W., Hwang, J. -S., Sung, G. -B., Yun, E. -Y., Bang, H. -S., et al. (1999). Molecular cloning and characterization of a gene encoding a beta-*N*-acetylglucosaminidase homolog from *Bombyx mandarina*. *Korean J. Seri-cult. Sci.*, *41*, 147–153.

Gooday, G. W. (1999). Aggressive and defensive roles for chitinases. In R. A. A. Muzzarelli, & P. Jolles (Eds.), *Chitin and Chitinases* (pp. 157–169). Basel, Switzerland: Birkhäuserverlag.

Gopalakrishnan, B., Muthukrishnan, S., & Kramer, K. J. (1995). Baculovirus-mediated expression of a *Manduca sexta* chitinase gene: Properties of the recombinant protein. *Insect Biochem. Mol. Biol.*, *25*, 255–265.

Graack, H. R., Cinque, U., & Kress, H. (2001). Functional regulation of glutamine:fructose-6-phosphate aminotransferase 1 (GFAT1) of *Drosophila melanogaster* in a UDP-N-acetylglucosamine and cAMP-dependent manner. *Biochem. J., 360*, 401–412.

Gunner, H. P., Met, M. Z., & Berger, S. (1985). In D. G. Gamble, & F. B. Lewis (Eds.), *Microbial Control of Spruce Budworms. US Forest Service GTR-NE-100* (p. 102). Broomall, PA: Northeastern Forest Experiment Station.

Guo, W., Li, G., Pang, Y., & Wang, P. (2005). A novel chitin-binding protein identified from the peritrophic membrane of the cabbage looper, *Trichoplusia ni. Insect Biochem. Mol. Biol., 35*, 1224–1234.

Hajjar, N. P., & Casida, J. E. (1978). Insecticidal benzoylphenyl ureas: Structure–activity relationships of chitin synthesis inhibitors. *Science, 200*, 1499–1500.

Harper, M. S., & Granados, R. R. (1999). Peritrophic membrane structure and formation of larval *Trichoplusia ni* with an investigation on the secretion patterns of a PM mucin. *Tissue Cell, 31*, 202–211.

Harper, M. S., & Hopkins, T. L. (1997). Peritrophic membrane structure and secretion in European corn borer larvae (*Ostrinia nubilalis*). *Tissue Cell, 29*, 463–475.

Harper, M. S., Hopkins, T. L., & Czapla, T. H. (1998). Effect of wheat germ agglutinin on formation and structure of the peritrophic membrane in European corn borer (*Ostrinia nubilalis*) larvae. *Tissue Cell, 30*, 166–176.

Harris, M. T., & Fuhrman, J. A. (2002). Structure and expression of chitin synthase in the parasitic nematode *Dirofilaria immitis. Mol. Biochem. Parasitol., 122*, 231–234.

Harris, M. T., Lai, K., Arnold, K., Martinez, H. F., Specht, C. A., & Fuhrman, J. A. (2000). Chitin synthase in the filarial parasite, *Brugia malayi. Mol. Biochem. Parasitol., 111*, 351–362.

Hawtin, R. E., Zarkowska, T., Arnold, K., Thomas, C. J., Gooday, G. W., et al. (1997). Liquefaction of *Autographa californica* nucleopolyhedrovirus-infected insects is dependent on the integrity of virus-encoded chitinase and cathepsin genes. *Virology, 238*, 243–253.

Hefner, T., Arend, J., Warzecha, H., Siems, K., & Stockigt, J. (2002). Arbutin synthase, a novel member of the NRD-1beta glycosyltransferase family, is a unique multifunctional enzyme converting various natural products and xenobiotics. *Bioorg. Med. Chem., 10*, 1731–1741.

Hemmi, H., Ishibashi, J., Tomie, T., & Yamakawa, M. (2003). Structural basis for new pattern of conserved amino acid residues related to chitin-binding in the antifungal peptide from the coconut rhinoceros beetle *Oryctes rhinoceros. J. Biol. Chem., 278*, 22820–22827.

Hogenkamp, D. G., Arakane, Y., Zimoch, L., Merzendorfer, H., Kramer, K. J., et al. (2005). Chitin synthase genes in *Manduca sexta*: Characterization of a gut-specific transcript and differential tissue expression of alternately spliced mRNAs during development. *Insect Biochem. Mol. Biol., 35*, 529–540.

Hogenkamp, D. G., Arakane, Y., Kramer, K. J., Muthukrishnan, S., & Beeman, R. W. (2008). Characterization and expression of the β-N-acetylhexosaminidase gene family of *Tribolium castaneum. Insect Biochem. Mol. Biol., 38*, 478–489.

Hopkins, T. L., & Harper, M. S. (2001). Lepidopteran peritrophic membranes and effects of dietary wheat germ agglutinin on their formation and structure. *Arch. Insect Biochem. Physiol., 47*, 100–109.

Hori, M., Kakiki, K., Suzuki, S., & Misato, T. (1971). Studies on the mode of action of polyoxins. Part III. Relation of polyoxin structure to chitin synthatase inhibition. *Agric. Biol. Chem., 35*, 1280–1291.

Horiuchi, H. (2009). Functional diversity of chitin synthases of *Aspergillus nidulans* in hyphal growth, conidiophore development and septum formation. *Med. Mycol. 47(Suppl, 1)*, S47–S52.

Horst, M. N. (1983). The biosynthesis of crustacean chitin. Isolation and characterization of polyprenol-linked intermediates from brine shrimp microsomes. *Arch. Biochem. Biophys., 223*, 254–263.

Houston, D. R., Eggleston, I., Synstad, B., Eijsink, V. G., & van Aalten, D. M. (2002). The cyclic dipeptide CI-4 [cyclo-(l-Arg-d-Pro)] inhibits family 18 chitinases by structural mimicry of a reaction intermediate. *Biochem. J., 368*, 23–27.

Hummel, T., Schimmelpfeng, K., & Klämbt, C. (1999). Commissure formation in the embryonic CNS of *Drosophila. Dev. Biol., 209*, 381–398.

Ibrahim, G. H., Smartt, C. T., Kiley, L. M., & Christensen, B. M. (2000). Cloning and characterization of a chitin synthase cDNA from the mosquito *Aedes aegypti. Insect Biochem. Mol. Biol., 30*, 1213–1222.

Iconomidou, V. A., Willis, J. H., & Hamodrakas, S. J. (2005). Unique features of the structural model of "hard" cuticle proteins: Implications for chitin-protein interactions and crosslinking in cuticle. *Insect Biochem. Mol. Biol., 35*, 553–360.

Ishaaya, I., & Ascher, K. (1977). Effect of diflubenzuron on growth and carbohydrate hydrolases of *Tribolium castaneum. Phytoparasitica, 5*, 149–158.

Ishaaya, I., & Cohen, E. (1974). Dietary TH-6040 alters cuticle composition and enzyme activity of housefly larval cuticle. *Pestic. Biochem. Physiol., 4*, 484–490.

Jakubowska, A. K., Caccia, S., Gordon, K. H., Ferre, J., & Herrero, S. (2010). Downregulation of a chitin deacetylase-like protein in response to baculovirus infection and its application for improving baculovirus infectivity. *J. Virol., 84*, 2547–2555.

Jasrapuria, S., Arakane, Y., Osman, G., Kramer, K. J., Beeman, R. W., & Muthukrishnan, S. (2010). Genes encoding proteins with peritrophin A-type chitin-binding domains in *Tribolium castaneum* are grouped into three distinct families based on phylogeny, expression and function. *Insect Biochem. Mol. Biol., 40*, 214–227.

John, M., Rohrig, H., Schmidt, J., Wieneke, U., & Schell, J. (1993). *Rhizobium* NodB protein involved in nodulation signal synthesis is a chitooligosaccharide deacetylase. *Proc. Natl. Acad. Sci. USA, 90*, 625–629.

Jüergens, G., Wieschaus, E., Nuesslein-Volhard, C., & Kluding, H. (1984). Mutations affecting the pattern of the larval cuticle in *Drosophila melanogaster*: II. Zygotic loci on the third chromosome. *Roux's Arch. Dev. Biol., 193*, 283–295.

Kanno, H., Asai, T., Maekawa, S., 1981. 2-tert-butylimino-3-isopropyl-perhydro-1,3,5-thaidiazin-4-one (NNI 750), a new insecticide. In *Brighton Crop Protection Conference, Pests and Diseases*, Vol. 1, (pp. 56–59). Brighton, UK.

Kanost, M. R., Zepp, M. K., Ladendorff, N. E., & Andersson, L. A. (1994). Isolation and characterization of a hemocyte aggregation inhibitor from hemolymph of *Manduca sexta* larvae. *Arch. Insect Biochem. Physiol.*, *27*, 123–136.

Kato, N., Dasgupta, R., Smartt, C. T., & Christensen, B. M. (2002). Glucosamine:fructose-6-phosphate aminotransferase: Gene characterization, chitin biosynthesis and peritrophic matrix formation in *Aedes aegypti*. *Insect Mol. Biol.*, *11*, 207–216.

Kato, N., Mueller, C. R., Wessely, V., Lan, Q., & Christensen, B. M. (2005). *Aedes aegypti* phosphohexomutases and uridine diphosphate-hexose pyrophosphorylases: Comparison of primary sequences, substrate specificities and temporal transcription. *Insect Mol. Biol.*, *14*, 615–624.

Kato, N., Mueller, C. R., Fuchs, J. F., Wessely, V., Lan, Q., & Christensen, B. M. (2006). Regulatory mechanisms of chitin biosynthesis and roles of chitin in peritrophic matrix formation in the midgut of adult *Aedes aegypti*. *Insect Biochem. Mol. Biol.*, *36*, 1–9.

Kawabata, S., Osaki, T., & Iwanaga, S. (2003). Innate immunity. In R. A.B. Ezekowitz, & J. Hoffman (Eds.). *Innate Immunity in the Horseshoe Crab* (pp. 109–125). Totowa, NJ: Humana Press.

Kawamura, K., Shibata, T., Saget, O., Peel, D., & Bryant, P. J. (1999). A new family of growth factors produced by the fat body and active on *Drosophila* imaginal disc cells. *Development*, *126*, 211–219.

Kerwin, J. L., Whitney, D. L., & Sheikh, A. (1999). Mass spectrometric profiling of glucosamine, glucosamine polymers and their catecholamine adducts. Model reactions and cuticular hydrolysates of *Toxorhynchites amboinensis* (Culicidae) pupae. *Insect Biochem Mol. Biol.*, *29*, 599–607.

Khajuria, C., Buschman, L. L., Chen, M. S., Muthukrishnan, S., & Zhu, K. Y. (2010). A gut-specific chitinase gene essential for regulation of chitin content of peritrophic matrix and growth of *Ostrinia nubilalis* larvae. *Insect Biochem. Mol. Biol.*, *40*, 621–629.

Kim, M. G., Shin, S. W., Bae, K. S., Kim, S. C., & Park, H. Y. (1998). Molecular cloning of chitinase cDNAs from the silkworm, *Bombyx mori* and the fall webworm, *Hyphantria cunea*. *Insect Biochem. Mol. Biol.*, *28*, 163–171.

Kimura, S. (1976). Insect haemolymph exo-beta-*N*-acetylglucosaminidase from *Bombyx mori*. Purification and properties. *Biochim. Biophys. Acta*, *446*, 399–406.

Kitahara, K., Nakagawa, Y., Nishioka, T., & Fujita, T. (1983). Cultured integument of *Chilo suppressalis* as a bioassay system of insect growth regulators. *Agric. Biol Chem.*, *47*, 1583–1589.

Koga, D., Mai, M. S., Dziadik-Turner, C., & Kramer, K. J. (1982). Kinetics and mechanism of exochitinase and β-N-acetylhexosaminidase from the tobacco hornworm, *Manduca sexta* L., Lepidoptera: Sphingidae. *Insect Biochem.*, *12*, 493–499.

Koga, D., Jilka, J., & Kramer, K. J. (1983). Insect endochitinases: glycoproteins from moulting fluid, integument and pupal haemolymph of *Manduca sexta* L. *Insect Biochem.*, *13*, 295–305.

Koga, D., Isogai, A., Sakuda, S., Matsumoto, S., & Suzuki, A. (1987). Specific inhibition of *Bombyx mori* chitinase by allosamidin. *Agric. Biol. Chem.*, *51*, 471–476.

Koga, D., Funakoshi, T., Mizuki, K., Ide, A., Kramer, K. J., et al. (1992). Immunoblot analysis of chitinolytic enzymes in integument and molting fluid of the silkworm, *Bombyx mori*, and the tobacco hornworm, *Manduca sexta*. *Insect Biochem. Mol. Biol.*, *22*, 305–311.

Koga, D., Sasaki, Y., Uchiumi, Y., Hirai, N., Arakane, Y., & Nagamatsu, Y. (1997). Purification and characterization of *Bombyx mori* chitinases. *Insect Biochem. Mol. Biol.*, *27*, 757–767.

Kramer, K. J., & Koga, D. (1986). Insect chitin: Physical state, synthesis, degradation and metabolic regulation. *Insect Biochem.*, *16*, 851–877.

Kramer, K. J., & Muthukrishnan, S. (1997). Insect chitinases: Molecular biology and potential use as biopesticides. *Insect Biochem. Mol. Biol.*, *27*, 887–900.

Kramer, K. J., & Muthukrishnan, S. (2005). Chitin metabolism in insects. In L. I. Gilbert, K. Iatrou, & S. Gill (Eds.). *Comprehensive Molecular Insect Science* (Vol. 4, pp. 111–144). Oxford, UK: Elsevier.

Kramer, K. J., Dziadik-Turner, C., & Koga, D. (1985). Chitin metabolism in insects. In G. A. Kerkut, & L. I. Gilbert (Eds.), *Comprehensive Insect Physiology, Biochemistry, and Pharmacology* (Vol. 3, pp. 75–115). Oxford, UK: Pergamon Press.

Kramer, K. J., Corpuz, L., Choi, H. K., & Muthukrishnan, S. (1993). Sequence of a cDNA and expression of the gene encoding epidermal and gut chitinases of *Manduca sexta*. *Insect Biochem. Mol. Biol.*, *23*, 691–701.

Krieger de Moraes, C., Schrank, A., & Vainstein, M. H. (2003). Regulation of extracellular chitinases and proteases in the entomopathogen and acaricide *Metarhizium anisopliae*. *Curr. Microbiol.*, *46*, 205–210.

Krishnan, A., Nair, P. N., & Jones, D. (1994). Isolation, cloning, and characterization of new chitinase stored in active form in chitin-lined venom reservoir. *J. Biol. Chem.*, *269*, 20971–20976.

Kumar, N. S., Tang, B., Chen, X., Tian, H., & Zhang, W. (2008). Molecular cloning, expression pattern and comparative analysis of chitin synthase gene B in *Spodoptera exigua*. *Comp. Biochem. Physiol. B*, *149*, 447–453.

Lehane, M. J. (1997). Peritrophic matrix structure and function. *Annu. Rev. Entomol.*, *42*, 525–550.

Leonard, R., Rendic, D., Rabouille, C., Wilson, I. B., Preat, T., & Altmann, F. (2006). The *Drosophila* fused lobes gene encodes an *N*-acetylglucosaminidase involved in *N*-glycan processing. *J. Biol. Chem.*, *281*, 4867–4875.

Li, H., & Greene, L. H. (2010). Sequence and structural analysis of the chitinase insertion domain reveals two conserved motifs involved in chitin-binding. *PLoS One*, *5*, e8654.

Li, Y., Roux, C., Lazereg, S., LeCaer, J. P., Laprevote, O., et al. (2007). Identification of a novel serine phosphorylation site in human glutamine:fructose-6-phosphate amidotransferase isoform 1. *Biochemistry*, *46*, 13163–13169.

Linder, M., Salovuori, I., Ruohonen, L., & Teeri, T. T. (1996). Characterization of a double cellulose-binding domain. Synergistic high affinity binding to crystalline cellulose. *J. Biol. Chem.*, *271*, 21268–21272.

Liu, M., Cai, Q. X., Liu, H. Z., Zhang, B. H., Yan, J. P., & Yuan, Z. M. (2002). Chitinolytic activities in *Bacillus thuringiensis* and their synergistic effects on larvicidal activity. *J. Appl. Microbiol.*, *93*, 374–379.

Locke, M. (1991). Insect epidermal cells. In K. Binnington, & A. Retnakaran (Eds.). *Physiology of the Insect Epidermis* (pp. 1–22). Melbourne, Australia: CRISCO Publications.

Locke, M. (2001). The Wigglesworth Lecture: Insects for studying fundamental problems in biology. *J. Insect Physiol.*, *47*, 495–507.

Lomakin, J., Huber, P. A., Eicher, C., Arakane, Y., Krames, K. J., Beeman, R. W., Kanost, M. R., & Gehrke, S. H. (2011). Mechanical properties of the beetle elytron, a biological composite material. *Biomacromolecules*, *12*, 321–335.

Lu, Y. M., Zen, K. C., Muthukrishnan, S., & Kramer, K. J. (2002). Site-directed mutagenesis and functional analysis of active site acidic amino acid residues D142, D144 and E146 in *Manduca sexta* (tobacco hornworm) chitinase. *Insect Biochem. Mol. Biol.*, *32*, 1369–1382.

Lucero, H. A., Kuranda, M. J., & Bulik, D. A. (2002). A non-radioactive, high throughput assay for chitin synthase activity. *Anal. Biochem.*, *305*, 97–105.

Luschnig, S., Batz, T., Armbruster, K., & Krasnow, M. A. (2006). *serpentine* and *vermiform* encode matrix proteins with chitin binding and deacetylation domains that limit tracheal tube length in *Drosophila*. *Curr. Biol.*, *16*, 186–194.

Lysenko, O. (1976). Chitinase of *Serratia marcescens* and its toxicity to insects. *J. Invert. Pathol.*, *27*, 385–386.

Macedo, M. L.R., Andrade, L. B.S., Moraes, R. A., & Xavier-Filho, J. (1993). Vicilin variants and the resistance of cowpea (*Vigna unguiculata*) seeds to the cowpea weevil (*Callosobruchus maculatus*). *Comp. Biochem. Physiol. C*, *105*, 89–94.

Macedo, M. L.R., Damico, D. C.S., Freire, M. D.G.M., Toyama, M. H., Marangoni, S., et al. (2003). Purification and characterization of an *N*-acetylglucosamine-binding lectin from *Koelreuteria paniculata* seeds and its effect on the larval development of *Callosobruchus maculatus* (Coleoptera: Bruchidae) and *Anagasta kuehniella* (Lepidoptera: Pyralidae). *J. Agric. Food Chem.*, *51*, 2980–2986.

Mark, B. L., Mahuran, D. J., Cherney, M. M., Zhao, D., Knapp, S., & James, M. N. (2003). Crystal structure of human beta-hexosaminidase B: Understanding the molecular basis of Sandhoff and Tay-Sachs disease. *J. Mol. Biol.*, *327*, 1093–1109.

Martínez-Rucobo, F. W., Eckhardt-Strelau, L., & Terwissccha van Scheltinga, A. C. (2009). Yeast chitin synthase 2 activity is modulated by proteolysis and phosphorylation. *Biochem. J.*, *417*, 547–554.

Maue, L., Meissner, D., & Merzendorfer, H. (2009). Purification of an active, oligomeric chitin synthase complex from the midgut of the tobacco hornworm. *Insect Biochem. Mol. Biol.*, *39*, 654–659.

Mayer, R. T., Chen, A. C., & DeLoach, J. R. (1980). Characterization of a chitin synthase from the stable fly, *Stomoxys calcitrans* (L.). *Insect Biochem.*, *10*, 549–556.

Meissner, D., Odman-Naresh, J., Vogelpohl, I., & Merzendorfer, H. (2010). A novel role of the yeast CaaX protease Ste24 in chitin synthesis. *Mol. Biol. Cell*, *21*, 2425–2433.

Merzendorfer, H. (2006). Insect chitin synthases: A review. *J. Comp. Physiol. B*, *176*, 1–15.

Merzendorfer, H. (2009). Chitin. In H. -J. Gabius (Ed.). *The Sugar Code: Fundamentals of Glycosciences* (pp. 217–229). Weinheim, Germany: Wiley-VCH.

Merzendorfer, H., & Zimoch, L. (2003). Chitin metabolism in insects: Structure, function and regulation of chitin synthases and chitinases. *J. Exp. Biol.*, *206*, 4393–4412.

Mikolajczyk, P., Oberlander, H., Silhacek, D. L., Ishaaya, I., & Shaaya, E. (1994). Chitin synthesis in *Spodoptera frugiperda* wing imaginal discs. I. Chlorfluazuron, diflubenzuron and teflubenzuron inhibit incorporation but not uptake of [14C]-N-acetyl-D-glucosamine. *Arch. Insect Biochem. Physiol.*, *25*, 245–258.

Mitlin, N., Wiygul, G., & Haynes, J. W. (1977). Inhibition of DNA synthesis in boll weevils (*Anthonomus grandis boheman*) sterilized by dimilin. *Pestic. Biochem. Physiol.*, *7*, 559–563.

Moreira, M. F., Dos Santos, A. S., Marotta, H. R., Mansur, J. F., Ramos, I. B., et al. (2007). A chitin-like component in *Aedes aegypti* eggshells, eggs and ovaries. *Insect Biochem. Mol. Biol.*, *37*, 1249–1261.

Morris, O. N. (1976). A 2-year study of the efficacy of *Bacillus thuringiensis*-chitinase combinations in spruce budworm (*Choristoneura fumiferana*) control. *Can. Entomol.*, *108*, 225.

Mota, A. C., Damatta, R. A., Lima Filho, M., Silva, C. P., & Xavier-Filho, J. (2003). Cowpea (*Vigna unguiculata*) vicilins bind to the peritrophic membrane of larval sugarcane stalk borer (*Diatraea saccharalis*). *J. Insect Physiol.*, *49*, 873–880.

Moussian, B. (2010). Recent advances in understanding mechanisms of insect cuticle differentiation. *Insect Biochem. Mol. Biol.*, *40*, 363–375.

Moussian, B., Schwarz, H., Bartoszewski, S., & Nusslein-Volhard, C. (2005a). Involvement of chitin in exoskeleton morphogenesis in *Drosophila melanogaster*. *J. Morphol.*, *264*, 117–130.

Moussian, B., Tang, E., Tonning, A., Helms, S., Schwarz, H., et al. (2005b). *Drosophila Knickkopf* and *Retroactive* are needed for epithelial tube growth and cuticle differentiation through their specific requirement for chitin filament organization. *Development*, *133*, 163–171.

Moussian, B., Seifarth, C., Mueller, U., Berger, J., & Schwarz, H. (2006). Cuticle differentiation during *Drosophila* embryogenesis. *Arthropod. Struct. Dev.*, *35*, 137–152.

Moussian, B., Veerkamp, J., Mueller, U., & Schwarz, H. (2007). Assembly of the *Drosophila* larval exoskeleton requires controlled secretion and shaping of the apical plasma membrane. *Matrix Biol.*, *26*, 337–347.

Muzzarelli, R. A.A. (1973). Chitin. In R. A. A. Muzzarelli (Ed.), *Natural Chelating Polymers: Alginic Acid, Chitin, and Chitosan* (pp. 83–252). New York, NY: Pergamon Press.

Nagahashi, S., Sudoh, M., Ono, N., Sawada, R., Yamaguchi, E., et al. (1995). Characterization of chitin synthase 2 of *Saccharomyces cerevisiae*. Implication of two highly conserved domains as possible catalytic sites. *J. Biol. Chem.*, *270*, 13961–13967.

Nagamatsu, Y., Yanagisawa, I., Kimoto, M., Okamoto, E., & Koga, D. (1995). Purification of a chitooligosaccharidolytic beta-*N*-acetylglucosaminidase from *Bombyx mori* larvae during metamorphosis and the nucleotide sequence of its cDNA. *Biosci. Biotechnol. Biochem.*, *59*, 219–225.

Nakabachi, A., Shigenobu, S., & Miyagishima, S. (2010). Chitinase-like proteins encoded in the genome of the pea aphid, *Acyrthosiphon pisum*. *Insect Mol. Biol. 19(Suppl. 2)*, 175–185.

Neeraja, C., Subramanyam, R., Moerschbacher, B. M., & Podile, A. R. (2010). Swapping the chitin-binding domain in *Bacillus* chitinases improves the substrate binding affinity and conformational stability. *Mol. Biosyst., 6*, 1492–1502.

Nisole, A., Stewart, D., Bowman, S., Zhang, D., Krell, P. J., et al. (2010). Cloning and characterization of a Gasp homolog from the spruce budworm, *Choristoneura fumiferana*, and its putative role in cuticle formation. *J. Insect Physiol., 56*, 1427–1435.

Nüesslein-Volhard, C., Wieschaus, E., & Kluding, H. (1984). Mutations affecting the pattern of the larval cuticle in *Drosophila melanogaster*. 1. Zygotic loci on the second chromosome. *Roux's Arch. Dev. Biol., 193*, 267–282.

Oberlander, H., & Silhacek, D. L. (1998). New perspectives on the mode of action of benzoylphenylurea insecticides. In I. Ishaaya, & D. Degheele (Eds.). *Applied Agriculture: Insecticides with Novel Modes of Action* (pp. 92–105). Berlin, Germany: Springer.

Ono, N., Yabe, T., Sudoh, M., Nakajima, T., Yamada-Okabe, T., et al. (2000). The yeast Chs4 protein stimulates the trypsin-sensitive activity of chitin synthase 3 through an apparent protein–protein interaction. *Microbiology, 146*, 385–391.

Ostrowski, S., Dierick, H. A., & Bejsovec, A. (2002). Genetic control of cuticle formation during embryonic development of *Drosophila melanogaster. Genetics, 161*, 171–182.

Otsu, Y., Mori, H., Komuta, K., Shimizu, H., Nogawa, S., et al. (2003). Suppression of leaf feeding and oviposition of phytophagous ladybird beetles (Coleoptera: Coccinellidae) by chitinase gene-transformed phylloplane bacteria and their specific bacteriophages entrapped in alginate gel beads. *J. Econ. Entomol., 96*, 555–563.

Palli, S. R., & Retnakaran, A. (1999). Molecular and biochemical aspects of chitin synthesis inhibition. *Exs, 87*, 85–98.

Pan, Y., Chen, K., Xia, H., Yao, Q., Gao, L., et al. (2010). Molecular cloning, expression and characterization of BmIDGF gene from *Bombyx mori. Z. Naturforsch. C, 65*, 277–283.

Pearson, J. C., Juarez, M. T., Kim, M., Drivenes, O., & McGinnis, W. (2009). Multiple transcription factor codes activate epidermal wound-response genes in *Drosophila. Proc. Natl. Acad. Sci. USA, 106*, 2224–2229.

Perrakis, A., Tews, I., Dauter, Z., Oppenheim, A. B., Chet, I., et al. (1994). Crystal structure of a bacterial chitinase at 2.3 A resolution. *Structure, 2*, 1169–1180.

Post, L. C., & Vincent, W. R. (1973). A new insecticide inhibits chitin synthesis. *Naturwissenschaften, 60*, 431–432.

Priya, T. A., Li, F., Zhang, J., Wang, B., Zhao, C., & Xiang, J. (2009). Molecular characterization and effect of RNA interference of retinoid X receptor (RXR) on E75 and chitinase gene expression in Chinese shrimp *Fenneropenaeus chinensis. Comp. Biochem. Physiol. B, 153*, 121–129.

Quesada-Allue, L. A. (1982). The inhibition of insect chitin synthesis by tunicamycin. *Biochem. Biophys. Res. Commun., 105*, 312–319.

Ramalho-Ortigao, J. M., & Traub-Cseko, Y. M. (2003). Molecular characterization of Llchit1, a midgut chitinase cDNA from the leishmaniasis vector *Lutzomyia longipalpis. Insect Biochem. Mol. Biol., 33*, 279–287.

Rao, F. V., Houston, D. R., Boot, R. G., Aerts, J. M., Sakuda, S., & Van Aalten, D. M. (2003). Crystal structures of allosamidin derivatives in complex with human macrophage chitinase. *J. Biol. Chem., 14*, 20110–20116.

Rao, R., Fiandra, L., Giordana, B., de Eguileor, M., Congiu, T., et al. (2004). AcMNPV ChiA protein disrupts the peritrophic membrane and alters midgut physiology of *Bombyx mori* larvae. *Insect Biochem. Mol. Biol., 34*, 1205–1213.

Read, S. M., & Bacic, T. (2002). Plant biology. Prime time for cellulose. *Science, 295*, 59–60.

Rebers, J. E., & Willis, J. H. (2001). A conserved domain in arthropod cuticular proteins binds chitin. *Insect Biochem. Mol. Biol., 31*, 1083–1093.

Regev, A., Keller, M., Strizhov, N., Sneh, B., Prudovsky, E., et al. (1996). Synergistic activity of a *Bacillus thuringiensis* delta-endotoxin and a bacterial endochitinase against *Spodoptera littoralis* larvae. *Appl. Environ. Microbiol., 62*, 3581–3586.

Ren, N., Zhu, C., Lee, H., & Adler, P. N. (2005). Gene expression during *Drosophila* wing morphogenesis and differentiation. *Genetics, 171*, 625–638.

Rezende, G. L., Martins, A. J., Gentile, C., Farnesi, L. C., Pelajo-Machado, M., et al. (2008). Embryonic desiccation resistance in *Aedes aegypti*: Presumptive role of the chitinized serosal cuticle. *BMC Dev. Biol., 8*, 82.

Richmond, T. (2000). Higher plant cellulose synthases. *Genome Biol., 1*, reviews 3001.1–3001.6.

Riddiford, L. M., Hiruma, K., Zhou, X., & Nelson, C. A. (2003). Insights into the molecular basis of the hormonal control of molting and metamorphosis from *Manduca sexta* and *Drosophila melanogaster. Insect Biochem. Mol. Biol., 33*, 1327–1338.

Roncero, C. (2002). The genetic complexity of chitin synthesis in fungi. *Curr. Genet., 41*, 367–378.

Roncero, C., & Duran, A. (1985). Effect of Calcofluor white and Congo red on fungal cell wall morphogenesis: *In vivo* activation of chitin polymerization. *J. Bacteriol., 163*, 1180–1185.

Royer, V., Fraichard, S., & Bouhin, H. (2002). A novel putative insect chitinase with multiple catalytic domains: Hormonal regulation during metamorphosis. *Biochem. J., 366*, 921–928.

Ruiz-Herrera, J., & San-Blas, G. (2003). Chitin synthesis as target for antifungal drugs. *Curr. Drug Targets Infect. Disord., 3*, 77–91.

Sales, M. P., Pimenta, P. P., Paes, N. S., Grossi-de-Sa, M. F., & Xavier-Filho, J. (2001). Vicilins (7S storage globulins) of cowpea (*Vigna unguiculata*) seeds bind to chitinous structures of the midgut of *Callosobruchus maculatus* (Coleoptera: Bruchidae) larvae. *Braz. J. Med. Biol. Res., 34*, 27–34.

Sampson, M. N., & Gooday, G. W. (1998). Involvement of chitinases of *Bacillus thuringiensis* during pathogenesis in insects. *Microbiology, 144*(Pt 8), 2189–2194.

Sass, M., Kiss, A., & Locke, M. (1994). The localization of surface integument peptides in tracheae and tracheoles. *J. Insect Physiol., 40*, 5621–5675.

Sburlati, A., & Cabib, E. (1986). Chitin synthetase 2, a presumptive participant in septum formation in *Saccharomyces cerevisiae. J. Biol. Chem., 261*, 15147–15152.

Schaefer, J., Kramer, K. J., Garbow, J. R., Jacob, G. S., Stejskal, E. O., et al. (1987). Aromatic cross-links in insect cuticle: Detection by solid-state ^{13}C and ^{15}N NMR. *Science, 235,* 1200–1204.

Schimmelpfeng, K., Strunk, M., Stork, T., & Klämbt, C. (2006). *Mummy* encodes an UDP-N-acetylglucosamine-dipohosphorylase and is required during *Drosophila* dorsal closure and nervous system development. *Mech. Dev., 123,* 487–499.

Schwarz, H., & Moussian, B. (2007). Electron-microscopic and genetic dissection of arthropod cuticle differentiation. In A. Méndez-Vilas, & J. Díaz (Eds.). *Modern Research and Educational Topics in Microscopy* (Vol. 3). Badajoz, Spain: Formatex Research Center.

Screen, S. E., Hu, G., & St Leger, R. J. (2001). Transformants of *Metarhizium anisopliae sf. anisopliae* overexpressing chitinase from *Metarhizium anisopliae sf. acridum* show early induction of native chitinase but are not altered in pathogenicity to Manduca sexta. *J. Invertebr. Pathol., 78,* 260–266.

Selitrennikoff, C. P. (1984). Calcofluor white inhibits *Neurospora* chitin synthetase activity. *Exp. Mycol., 8,* 269–272.

Shapiro, M., Preisler, H. K., & Robertson, J. L. (1987). Enhancement of baculovirus activity on gypsy moth (Lepidoptera: Limantidae) by chitinase. *J. Econ. Entomol., 80,* 1113–1115.

Shen, Z., & Jacobs-Lorena, M. (1997). Characterization of a novel gut-specific chitinase gene from the human malaria vector *Anopheles gambiae. J. Biol. Chem., 272,* 28895–28900.

Shi, L., & Paskewitz, S. M. (2004). Identification and molecular characterization of two immune-responsive chitinase-like proteins from *Anopheles gambiae. Insect Mol. Biol., 13,* 387–398.

Shi, X., Chamankhah, M., Visal-Shah, S., Hemmingsen, S. M., Erlandson, M., et al. (2004). Modeling the structure of the type I peritrophic matrix: Characterization of a *Mamestra configurata* intestinal mucin and a novel peritrophin containing 19 chitin binding domains. *Insect Biochem. Mol. Biol., 34,* 1101–1115.

Shinoda, T., Kobayashi, J., Matsui, M., & Chinzei, Y. (2001). Cloning and functional expression of a chitinase cDNA from the common cutworm, *Spodoptera litura*, using a recombinant baculovirus lacking the virus-encoded chitinase gene. *Insect Biochem. Mol. Biol., 31,* 521–532.

Shutov, A. D., Kakhovskaya, I. A., Braun, H., Baumlein, H., & Muntz, K. (1995). Legumin-like and vicilin-like seed storage proteins: Evidence for a common single-domain ancestral gene. *J. Mol. Evol., 41,* 1057–1069.

Smirnoff, W. A., & Valero, J. R. (1972). Metabolic disturbances in *Choristoneura fumiferana* Clemens infected by *Bacillus thuringiensis* alone or with added chitinase. *Rev. Can. Biol., 31,* 163–169.

Sneh, B., Schuster, S., & Gross, S. (1983). Biological control of *Spodoptera littoralis* (Boisd.) (Lepidoptera: Noctuidae) by *Bacillus thuringiensis subsp. entomocidus* and *Bracon hebetor* Say (Hymenoptera: Braconidae). *Z. Ang. Entomol., 96,* 77–83.

Sobotnik, J., Kudlikova-Krizkova, I., Vancova, M., Munzbergova, Z., & Hubert, J. (2008). Chitin in the peritrophic membrane of *Acarus siro* (Acari: Acaridae) as a target for novel acaricides. *J. Econ. Entomol., 101,* 1028–1033.

Soltani, N. (1984). Effects of ingested diflubenzuron on the longevity and the peritrophic membrane of adult mealworms (*Tenebrio molitor* L.). *Pestic. Sci., 15,* 221–225.

Spindler, K. D., & Spindler-Barth, M. (1999). Inhibitors of chitinases. *Exs, 87,* 201–209.

St Leger, R. J., Joshi, L., Bidochka, M. J., Rizzo, N. W., & Roberts, D. W. (1996). Characterization and ultrastructural localization of chitinases from *Metarhizium anisopliae, M. flavoviride*, and *Beauveria bassiana* during fungal invasion of host (*Manduca sexta*) Cuticle. *Appl. Environ. Microbiol., 62,* 907–912.

Suetake, T., Tsuda, S., Kawabata, S., Miura, K., Iwanaga, S., et al. (2000). Chitin-binding proteins in invertebrates and plants comprise a common chitin-binding structural motif. *J. Biol. Chem., 275,* 17929–17932.

Takahashi, M., Kiuchi, M., & Kamimura, M. (2002). A new chitinase-related gene, BmChiR1, is induced in the *Bombyx mori* anterior silk gland at molt and metamorphosis by ecdysteroid. *Insect Biochem. Mol. Biol., 32,* 147–151.

Tamura, K., Dudley, J., Nei, M., & Kumar, S. (2007). MEGA4: Molecular Evolutionary Genetics Analysis (MEGA) software version 4.0. *Mol. Biol. Evol., 24,* 1596–1599.

Tang, L., Liang, J., Zhan, Z., Xiang, Z., & He, N. (2010). Identification of the chitin-binding proteins from the larval proteins of silkworm, *Bombyx mori. Insect Biochem. Mol. Biol., 40,* 228–234.

Tantimavanich, S., Pantuwatana, S., Bhumiratana, A., & Panbangred, W. (1997). Cloning of a chitinase gene into *Bacillus thuringiensis subsp. aizawai* for enhanced insecticidal activity. *J. Gen. Appl. Microbiol., 43,* 341–347.

Tarone, A. M., Jennings, K. C., & Foran, D. R. (2007). Aging blow fly eggs using gene expression: A feasibility study. *J. Forensic Sci., 52,* 1350–1354.

Tellam, R. L., & Eisemann, C. (2000). Chitin is only a minor component of the peritrophic matrix from larvae of *Lucilia cuprina. Insect Biochem. Mol. Biol., 30,* 1189–1201.

Tellam, R. L., Wijffels, G., & Willadsen, P. (1999). Peritrophic matrix proteins. *Insect Biochem. Mol. Biol., 29,* 87–101.

Tellam, R. L., Vuocolo, T., Johnson, S. E., Jarmey, J., & Pearson, R. D. (2000). Insect chitin synthase cDNA sequence, gene organization and expression. *Eur. J. Biochem., 267,* 6025–6043.

Terwissscha van Scheltinga, A. C., Kalk, K. H., Beintema, J. J., & Dijkstra, B. W. (1994). Crystal structures of hevamine, a plant defence protein with chitinase and lysozyme activity, and its complex with an inhibitor. *Structure, 2,* 1181–1189.

Thomas, C. J., Gooday, G. W., King, L. A., & Possee, R. D. (2000). Mutagenesis of the active site coding region of the *Autographa californica* nucleopolyhedrovirus *chiA* gene. *J. Gen. Virol., 81,* 1403–1411.

Thummel, C. S. (1996). Files on steroids – *Drosophila* metamorphosis and the mechanisms of steroid hormone action. *Trends Genet., 12,* 306–310.

Tian, H., Peng, H., Yao, Q., Chen, H., Xie, Q., et al. (2009). Developmental control of a lepidopteran pest *Spodoptera exigua* by ingestion of bacteria expressing dsRNA of a nonmidgut gene. *PLoS One, 4,* e6225.

Timmons, L., & Fire, A. (1998). Specific interference by ingested dsRNA. *Nature, 395,* 854.

Tomiya, N., Narang, S., Park, J., Abdul-Rahman, B., Choi, O., et al. (2006). Purification, characterization, and cloning of a *Spodoptera frugiperda* Sf9 β-*N*-acetylhexosaminidase that hydrolyzes terminal N-acetylglucosamine on the N-glycan core. *J. Biol. Chem., 281,* 19545–19560.

Tonning, A., Hemphala, J., Tang, E., Nannmark, U., Samakovlis, C., & Uv, A. (2005). A transient luminal chitinous matrix is required to model epithelial tube diameter in the *Drosophila* trachea. *Dev. Cell, 9,* 423–430.

Tonning, A., Helms, S., Schwarz, H., Uv, A. E., & Moussian, B. (2006). Hormonal regulation of *mummy* is needed for apical extracellular matrix formation and epithelial morphogenesis in *Drosophila. Development, 133,* 331–341.

Toprak, U., Baldwin, D., Erlandson, M., Gillott, C., Hou, X., et al. (2008). A chitin deacetylase and putative insect intestinal lipases are components of the *Mamestra configurata* (Lepidoptera: Noctuidae) peritrophic matrix. *Insect Mol. Biol., 17,* 573–585.

Toprak, U., Baldwin, D., Erlandson, M., Gillott, C., Harris, S., & Hegedus, D. D. (2010). Expression patterns of genes encoding proteins with peritrophin A domains and protein localization in *Mamestra configurata. J. Insect Physiol., 56,* 1427–1435.

Tsai, M. L., Liaw, S. H., & Chang, N. C. (2004). The crystal structure of Ym1 at 1.31 A resolution. *J. Struct. Biol., 148,* 290–296.

Vaaje-Kolstad, G., Horn, S. J., van Aalten, D. M., Synstad, B., & Eijsink, V. G. (2005). The non-catalytic chitin-binding protein CBP21 from *Serratia marcescens* is essential for chitin degradation. *J. Biol. Chem., 280,* 28492–28497.

Van Dellen, K. L., Bulik, D. A., Specht, C. A., Robbins, P. W., & Samuelson, J. C. (2006). Heterologous expression of an *Entamoeba histolytica* chitin synthase in *Saccharomyces cerevisiae. Eukaryot. Cell, 5,* 203–206.

Varela, P. F., Llera, A. S., Mariuzza, R. A., & Tormo, J. (2002). Crystal structure of imaginal disc growth factor-2. A member of a new family of growth-promoting glycoproteins from *Drosophila melanogaster. J. Biol. Chem., 277,* 13229–13236.

Venancio, T. M., Cristofoletti, P. T., Ferreira, C., Verjovski-Almeida, S., & Terra, W. R. (2009). The *Aedes aegypti* larval transcriptome: A comparative perspective with emphasis on trypsins and the domain structure of peritrophins. *Insect Mol. Biol., 18,* 33–44.

Verloop, A., & Ferrell, C. D. (1977). Benzoylphenyl ureas – A new group of larvicides interfering with chitin deposition. In J. R. Plimmer (Ed.), *Pesticide Chemistry in the 20th Century,* Vol. 37, (pp. 237–270). Washington, DC: ACS Symposium Series American Chemical Society.

Vermeulen, C. A., & Wessels, J. G. (1986). Chitin biosynthesis by a fungal membrane preparation. Evidence for a transient non-crystalline state of chitin. *Eur. J. Biochem., 158,* 411–415.

Veronico, P., Gray, L. J., Jones, J. T., Bazzicalupo, P., Arbucci, S., et al. (2001). Nematode chitin synthases: Gene structure, expression and function in *Caenorhabditis elegans* and the plant parasitic nematode *Meloidogyne artiellia. Mol. Genet. Genom., 266,* 28–34.

Vincent, J. (2009). If it's tanned it must be dry: A critique. *J. Adhesion, 85,* 755–769.

Walker, J. E., Saraste, M., Runswick, M. J., & Gay, N. J. (1982). Distantly related sequences in the alpha- and beta-subunits of ATP synthase, myosin, kinases and other ATP-requiring enzymes and a common nucleotide binding fold. *EMBO J., 1,* 945–951.

Wang, P., & Granados, R. R. (2000). Calcofluor disrupts the midgut defense system in insects. *Insect Biochem. Mol. Biol., 30,* 135–143.

Wang, P., Li, G., & Granados, R. R. (2004). Identification of two new peritrophic membrane proteins from larval *Trichoplusia ni*: Structural characteristics and their functions in the protease rich insect gut. *Insect Biochem. Mol. Biol., 34,* 215–227.

Wang, S., Jayaram, S. A., Hemphala, J., Senti, K. A., Tsarouhas, V., et al. (2006). Septate-junction-dependent luminal deposition of chitin deacetylases restricts tube elongation in the *Drosophila* trachea. *Curr. Biol., 16,* 180–185.

Ward, G. B., Mayer, R. T., Feldlaufer, M. F., & Svoboda, J. A. (1991). Gut chitin synthase and sterols from larvae of *Diaprepes abbreviatus* (Coleoptera: Curculionidae). *Arch. Insect Biochem. Physiol., 18,* 105–117.

Watanabe, T., Kobori, K., Miyashita, K., Fujii, T., Sakai, H., et al. (1993). Identification of glutamic acid 204 and aspartic acid 200 in chitinase A1 of Bacillus circulans WL-12 as essential residues for chitinase activity. *J. Biol. Chem., 268,* 18567–18572.

Weigel, P. H., & DeAngelis, P. L. (2007). Hyaluronan synthases: A decade-plus of novel glycosyltransferases. *J. Biol. Chem., 282,* 36777–36781.

Wieschaus, E., Nüesslein-Volhard, C., & Jüergens, G. (1984). Mutations affecting the pattern of the larval cuticle in *Drosophila melanogaster*: Zygotic loci on the X-chromosome and the fourth chromosome. *Roux's Arch. Dev. Biol., 193,* 296–307.

Wijffels, G., Eisemann, C., Riding, G., Pearson, R., Jones, A., et al. (2001). A novel family of chitin-binding proteins from insect type 2 peritrophic matrix. cDNA sequences, chitin binding activity, and cellular localization. *J. Biol. Chem., 276,* 15527–15536.

Willis, J. H. (2010). Structural cuticular proteins from arthropods: Annotation, nomenclature, and sequence characteristics in the genomics era. *Insect Biochem. Mol. Biol., 40,* 189–204.

Wilson, T. G., & Cryan, J. R. (1997). Lufenuron, a chitin-synthesis inhibitor, interrupts development of *Drosophila melanogaster. J. Exp. Zool., 278,* 37–44.

Wiwat, C., Thaithanun, S., Pantuwatana, S., & Bhumiratana, A. (2000). Toxicity of chitinase-producing *Bacillus thuringiensis* ssp. *kurstaki* HD-1 (G) toward *Plutella xylostella. J. Invertebr. Pathol., 76,* 270–277.

Wu, L. Q., Ghodssi, R., Elabd, Y. A., & Payne, G. F. (2005). Biomimetic pattern transfer. *Adv. Func. Mater., 15,* 189–195.

Wu, L. Q., McDermott, M. K., Zhu, C., Ghodssi, R., & Payne, G. F. (2006). Mimicking biological phenol reaction cascades to confer mechanical function. *Adv. Func. Mater., 16,* 1967–1974.

Yamamoto, K., Moriguchi, M., Kawai, H., & Tochikura, T. (1980). Inhibition of UDP-N-acetylglucosamine pyrophosphorylase by uridine. *Biochim. Biophys. Acta, 614,* 367–372.

Yan, J., Cheng, Q., Narashimhan, S., Li, C. B., & Aksoy, S. (2002). Cloning and functional expression of a fat body-specific chitinase cDNA from the tsetse fly, *Glossina morsitans morsitans. Insect Biochem. Mol. Biol., 32*, 979–989.

Yeager, A. R., & Finney, N. S. (2004). The first direct evaluation of the two-active site mechanism for chitin synthase. *J. Org. Chem., 69*, 613–618.

Yeager, A. R., & Finney, N. S. (2005). Synthesis of fluorescently labeled UDP-GlcNAc analogues and their evaluation as chitin synthase substrates. *J. Org. Chem., 70*, 1269–1275.

You, M., & Fujisaki, K. (2009). Vaccination effects of recombinant chitinase protein from the hard tick *Haemaphysalis longicornis* (Acari: Ixodidae). *J. Vet. Med. Sci., 71*, 709–712.

Zakariassen, H., Aam, B. B., Horn, S. J., Varum, K. M., Sorlie, M., & Eijsink, V. G. (2009). Aromatic residues in the catalytic center of chitinase A from *Serratia marcescens* affect processivity, enzyme activity, and biomass converting efficiency. *J. Biol. Chem., 284*, 10610–10617.

Zen, K. C., Choi, H. K., Krishnamachary, N., Muthukrishnan, S., & Kramer, K. J. (1996). Cloning, expression, and hormonal regulation of an insect beta-*N*-acetylglucosaminidase gene. *Insect Biochem. Mol. Biol., 26*, 435–444.

Zhang, D., & Miller, M. J. (1999). Polyoxins and nikkomycins: Progress in synthetic and biological studies. *Curr. Pharm. Des., 5*, 73–99.

Zhang, H., Huang, X., Fukamizo, T., Muthukrishnan, S., & Kramer, K. J. (2002). Site-directed mutagenesis and functional analysis of an active site tryptophan of insect chitinase. *Insect Biochem. Mol. Biol., 32*, 1477–1488.

Zhang, J., & Zhu, K. Y. (2006). Characterization of a chitin synthase cDNA and its increased mRNA level associated with decreased chitin synthesis in *Anopheles quadrimaculatus* exposed to diflubenzuron. *Insect Biochem. Mol. Biol., 36*, 712–725.

Zhang, J., Iwai, S., Tsugehara, T., & Takeda, M. (2006). MbIDGF, a novel member of the imaginal disc growth factor family in *Mamestra brassicae*, stimulates cell proliferation in two lepidopteran cell lines without insulin. *Insect Biochem. Mol. Biol., 36*, 536–546.

Zhang, J., Liu, X., Li, D., Guo, Y., Ma, E., & Zhu, K. Y. (2010a). Silencing of two alternative splicing-derived mRNA variants of chitin synthase 1 gene by RNAi is lethal to the oriental migratory locust, *Locusta migratoria manilensis* (Meyen). *Insect Biochem. Mol. Biol., 40*, 824–833.

Zhang, X., Zhang, J., & Zhu, K. Y. (2010b). Chitosan/double-stranded RNA nanoparticle-mediated RNA interference to silence chitin synthase genes through larval feeding in the African malaria mosquito (*Anopheles gambiae*). *Insect Mol. Biol., 19*, 683–693.

Zhang, Y., Foster, J. M., Nelson, L. S., Ma, D., & Carlow, C. K. (2005). The chitin synthase genes *chs-1* and *chs-2* are essential for *C. elegans* development and responsible for chitin deposition in the eggshell and pharynx, respectively. *Dev. Biol., 285*, 330–339.

Zheng, Y., Zheng, S., Cheng, X., Ladd, T., Lingohr, E. J., et al. (2002). A molt-associated chitinase cDNA from the spruce budworm, *Choristoneura fumiferana. Insect Biochem. Mol. Biol., 32*, 1813–1823.

Zheng, Y. P., Retnakaran, A., Krell, P. J., Arif, B. M., Primavera, M., & Feng, Q. L. (2003). Temporal, spatial and induced expression of chitinase in the spruce budworm, *Choristoneura fumiferana. J. Insect Physiol., 49*, 241–247.

Zhu, Q., Deng, Y., Vanka, P., Brown, S. J., Muthukrishnan, S., & Kramer, K. J. (2004). Computational identification of novel chitinase-like proteins in the *Drosophila melanogaster* genome. *Bioinformatics, 20*, 161–169.

Zhu, Q., Arakane, Y., Banerjee, D., Beeman, R. W., Kramer, K. J., & Muthukrishnan, S. (2008a). Domain organization and phylogenetic analysis of the chitinase-like family of proteins in three species of insects. *Insect Biochem. Mol. Biol., 38*, 452–466.

Zhu, Q., Arakane, Y., Beeman, R. W., Kramer, K. J., & Muthukrishnan, S. (2008b). Characterization of recombinant chitinase-like proteins of *Drosophila melanogaster* and *Tribolium castaneum. Insect Biochem. Mol. Biol., 38*, 467–477.

Zhu, Q., Arakane, Y., Beeman, R. W., Kramer, K. J., & Muthukrishnan, S. (2008c). Functional specialization among insect chitinase family genes revealed by RNA interference. *Proc. Natl. Acad. Sci. USA, 105*, 6650–6655.

Zhu, R., Liu, K., Peng, J., Yang, H., & Hong, H. (2007). Optical brightener M2R destroys the peritrophic membrane of *Spodoptera exigua* (Lepidoptera: Noctuidae) larvae. *Pest Manag. Sci., 63*, 296–300.

Zhu, Y. C., Specht, C. A., Dittmer, N. T., Muthukrishnan, S., Kanost, M. R., & Kramer, K. J. (2002). Sequence of a cDNA and expression of the gene encoding a putative epidermal chitin synthase of *Manduca sexta. Insect Biochem. Mol. Biol., 32*, 1497–1506.

Zimmermann, D., & Peters, W. (1987). Fine structure and permeability of peritrophic membranes of *Calliphora erythrocephala* (Meigen) (Insecta: Diptera) after inhibition of chitin and protein synthesis. *Comp. Biochem. Physiol. B, 86B*, 353–360.

Zimoch, L., & Merzendorfer, H. (2002). Immunolocalization of chitin synthase in the tobacco hornworm. *Cell Tissue Res., 308*, 287–297.

Zimoch, L., Hogenkamp, D. G., Kramer, K. J., Muthukrishnan, S., & Merzendorfer, H. (2005). Regulation of chitin synthesis in the larval midgut of *Manduca sexta. Insect Biochem. Mol. Biol., 35*, 515–527.

8 Insect CYP Genes and P450 Enzymes

René Feyereisen
INRA Sophia Antipolis, France

8.1. Introduction

8.1.1. Overview

Cytochrome P450, or *CYP* genes, constitute one of the largest family of genes, with representatives in virtually all living organisms, from bacteria to protists, plants, fungi, and animals (Werck-Reichhart and Feyereisen, 2000). An ever-growing number of P450 sequences is available, and the role of P450 enzymes is being documented in an increasing number of physiological processes. The human genome carries about 57 CYP genes, and insect genomes can carry from 36 CYP genes in the body louse *Pediculus humanus* (Lee *et al.*, 2010) to 170 in a mosquito (Arensburger *et al.*, 2010). Each P450 protein is the product of a distinct CYP gene, and P450 diversity is the result of successive gene (or genome) duplications followed by sequence divergence. The typically 45- to 55-kDa P450 proteins are heme-thiolate enzymes. Their essential common feature is the absorbance peak near 450 nm of their Fe^{II}-CO complex for which they are named (Omura and Sato, 1964). P450 enzymes are best known for their monooxygenase role, catalyzing the transfer of one atom of molecular oxygen to a substrate and reducing the other to water. The simple stoichiometry [1] commonly describes the monooxygenase or mixed-function oxidase reaction of P450.

$$RH + O_2 + NADPH + H^+ \rightarrow ROH + H_2O + NADP^+ \quad (1)$$

However, oxygen atom transfer is not the only catalytic function of P450 enzymes. They also show activity as oxidases, reductases, desaturases, isomerases, etc., and collectively are known to catalyze at least 60 chemically distinct reactions (Ortiz de Montellano, 1995; Mansuy, 1998; Guengerich, 2001). Because of their complex catalytic mechanism, P450 enzymes often – perhaps always – generate superoxide or hydrogen peroxide from "unsuccessful," or uncoupled, reactions, leading to oxidative stress in cells. There are soluble forms of P450 (in bacteria), and membrane-bound forms (in microsomes and mitochondria of eukaryotes). Most animal P450s are dependent on

DOI:10.1016/B978-0-12-384747-8.10008-X

redox partners for their supply of reducing equivalents (NADPH cyt P450 reductase and cyt b_5 in microsomes; a ferredoxin and a ferredoxin reductase in mitochondria). Some bacterial and fungal P450 enzymes are fusion proteins with a variety of redox partners, whereas some P450 enzymes act directly on their substrates without the need for dioxygen or reducing equivalents (Hannemann *et al.*, 2007; Munro *et al.*, 2007).

Many P450 proteins are specialized in the metabolism of endogenous substrates (steroid hormones, lipids, etc.), but much of their notoriety has been associated with the metabolism or detoxification of xenobiotics (natural products, drugs, pesticides, etc.). In insects, they are involved in very many cases of insecticide resistance. In well-studied cases, P450 enzymes generate *more* toxic compounds – for example, organophosphate inhibitors of cholinesterase from less toxic phosphorothioate insecticides.

Foreign compounds, as well as endogenous metabolites, can induce the transcription of P450 genes through complex interactions with members of the nuclear receptor family and with βHLH-PAS proteins, such as the PXR (pregnane X receptor) and the Ah (aryl hydrocarbon) receptors of vertebrates, respectively. The regulation of P450 gene expression by chemicals allows animals to respond to changing environments directly by mounting a detoxification defense, and indirectly by adapting their basal metabolism, hormone balance, and, hence, rate of development and reproduction. Foreign compounds can also act as inhibitors of P450 enzymes, and many plant natural products are known to inhibit P450 activities more or less specifically, so that virtually any diet is a mixture of P450 substrates, inhibitors, and inducers.

There are many excellent reviews and books clearly presenting the state of knowledge on P450, generally with an emphasis on mammalian and bacterial enzymes (e.g., Ortiz de Montellano, 2005). Reliance on the advances made in non-insect systems is both necessary and risky – necessary because the structural and functional homology of P450 enzymes allows general principles to emerge, and risky because such principles may not always apply to peculiar aspects of insect P450 evolution or function. The chapter will certainly fail to be comprehensive, in view of the enormous literature pertaining to P450, but will try to convey the state of our current understanding as a basis for future research. Many aspects of P450 biology are intertwined and complex (structure, activity, induction), so the linear and compartmentalized treatment of the subject as done here will probably fail to pass muster with P450 novice and specialist alike. Indeed, the interpretations and choices of examples cited are mine, and the reader is invited to browse through text and references repeatedly. To help fill the enormous gaps in our knowledge on insect P450, the reader is also invited to join the field of P450 research.

This is an update of the *Comprehensive Molecular Insect Science* (Vol. 4) chapter on insect cytochrome P450 written in 2003 (Feyereisen, 2005). In that chapter, the development of P450 research on insects in the previous 20 years was outlined, and is not repeated here. References to the 2005 chapter indicate sections that have been trimmed to increase clarity. Another review on insect detoxification mechanisms is that of Li *et al.* (2007).

The first cloning and sequencing of a P450 cDNA (rat CYP2B1) was in 1982 (Fuji-Kuriyama *et al.*, 1982), and that of an insect P450 (house fly CYP6A1) followed in 1989 (Feyereisen *et al.*, 1989). The first crystal structure of a soluble bacterial P450, P450cam, was published in 1985 (Poulos *et al.*, 1985), and the first structure of a mammalian microsomal P450 engineered for solubility followed in 2000 (Williams *et al.*, 2000a). Heterologous expression of cloned insect P450 cDNAs in 1994 (Andersen *et al.*, 1994; Ma *et al.*, 1994) then allowed the rational study of individual P450 enzymes. Insect genome sequencing projects, starting with that of *Drosophila melanogaster* in 2000, finally revealed the cast of P450 characters in insects (Tijet *et al.*, 2001). Ten years later, the gap in our knowledge between sequences and functions has dramatically increased.

8.1.2. Vocabulary

Although P450 enzymes and CYP genes are displacing mixed-function oxidases (MFOs), microsomal oxidases, polysubstrate monooxygenases (PSMO), or simply monooxygenases in the vocabulary, words of caution are needed at the outset. The word "cytochrome" was initially given to the liver P-450 pigment (Omura, 1993; Estabrook, 1996), and it is still associated with P450 enzymes. However, P450s are formally not cytochromes, but rather heme-thiolate proteins (Mansuy, 1998). Not all P450-dependent reactions are monooxygenations, and not all monooxygenases are P450 enzymes (Mansuy and Renaud, 1995). In particular, flavin monooxygenases (FMOs) are NADPH-dependent enzymes that catalyze some reactions similar or identical to those catalyzed by P450 enzymes. In mammalian species, FMOs are microsomal enzymes best known for their N- and S-oxidation activities (Ziegler, 2002). Little is known about FMOs in insects. In the cinnabar moth *Tyria jacobaea*, a soluble FMO specifically N-oxidizes pyrolizidine alkaloids such as senecionine, seneciphylline, monocrotaline, and axillarine (Lindigkeit *et al.*, 1997; Naumann *et al.*, 2002). Only two *FMO* genes have been recognized in the *Drosophila* genome, and recombinant DmFMO-2 produced in *E. coli* is active in methimazole sulfoxidation assay (Scharf *et al.*, 2004). The *Bombyx mori* genome has three *FMO* genes (Sehlmeyer *et al.*, 2010).

A word of caution, too, about the "microsomal" oxidases. Insect P450s can be found in microsomal membranes as well as in mitochondria, and subcellular localization does not provide any information on the

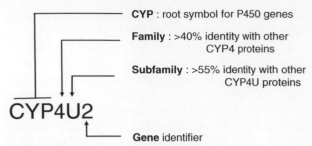

CYP : root symbol for P450 genes

Family : >40% identity with other CYP4 proteins

Subfamily : >55% identity with other CYP4U proteins

CYP4U2

Gene identifier

Figure 1 Scheme of the P450 nomenclature.

particular physiological role or catalytic competence of a P450 enzyme.

8.1.3. Nomenclature

A nomenclature of P450 genes and proteins was introduced when only 65 sequences were known (Nebert *et al.*, 1987). Gene families were initially designated by Roman numerals, but the proliferation of diverse sequences rapidly became discouraging, even to those versed in classics. The rules of nomenclature were then revised to their current form (Nebert *et al.*, 1991; Nelson *et al.*, 1993, 1996) where a CYP prefix, followed by an arabic numeral, designates the family (all members nominally > 40% identical), a capital letter designates the subfamily (all members nominally > 55% identical), and an arabic numeral designates the individual *gene* (*all italics*) or message and protein (no italics) (**Figure 1**). The confusing designation "CYP450" should be avoided. Different P450 enzymes are generally products of different genes; they are not isozymes or isoforms. The identity (%) rules for family and subfamily designations are not strictly adhered to, but names, once adopted, are rarely changed. Initially, many insect P450s were arbitrarily lumped into the CYP6 and the CYP4 families, even though they had less than 40% amino acid identity with CYP6A1 or with vertebrate CYP4 proteins. Naming genes in the lumper mode made the CYP6 and CYP4 families the largest ones in insects by a cascade effect. CYP6B1 is only 32% identical to CYP6A1 (Cohen *et al.*, 1992), so placing it in the CYP6 family "forced" many subsequent sequences into that family even if they did not meet the 40% criterion. The splitter mode prevailed at the completion of genome projects, which led to a new proliferation of CYP families in insects: the CYP300 series. A termite P450 claimed the welcoming designation of CYP4U2 (GenBank AF046011). Gotoh (1993) introduced a useful nomenclature of higher order than CYP families, the E (for eukaryotic type) and B (for bacterial type) "classes" and subclasses (I, II, III, etc.), that regroup CYP families on the basis of phylogeny. Nelson (1998) has similarly introduced the notion of phylogenetically related CYP families as "clans," but the precise criteria for naming Gotoh's "classes" and Nelson's "clans" have not been defined. The proliferation of CYP families

is now so great that the "clan" designation becomes a useful heuristic guide, and will be used here.

Alleles of a gene are named as subscripts v1, v2 (e.g., *CYP6B1v2*; Cohen *et al.*, 1992). In a practical shortcut, *CYP6CM1vQ* and *vB* designate the alleles of this gene found in the Q and B biotypes, respectively, of *Bemisia tabaci* (Karunker *et al.*, 2008). Pseudogenes are noted by the suffix P. This suffix is added to the closest paralog that is an active gene, e.g., *CYP9E2* and *CYP9E2P1* in *Blattella germanica* (Wen *et al.*, 2001). However, this is not always done, as the closest paralog is sometimes not easily recognized. Solo exons, detritus exons, or internal duplicated or partial exons have a specific nomenclature as well (Nelson *et al.*, 2004), and this has not yet been applied to insect P450 gene fragments. In following the tradition that predates the CYP nomenclature, P450 enzymes can be named with a small suffix, such as P450cam, the camphor hydroxylase of *Pseudomonas putida* later named CYP101A1; P450BM3, the fatty acid hydroxylase of *Bacillus megaterium* (CYP102A1); or P450scc, the cholesterol side-chain cleavage enzyme (CYP11A1). In insects, few P450 enzymes have been named in this way. P450Lpr is the predominant P450 in the pyrethroid-resistant strain Learn-Pyr of the house fly, later identified as CYP6D1 (Tomita and Scott, 1995). P450hyd (Reed *et al.*, 1994) is the P450 forming hydrocarbons in the house fly. In the *Drosophila* gene nomenclature (Lindsley and Zimm, 1992; FlyBase) the initial letter is capitalized; hence *CYP6A1* in the house fly and *Cyp6a2* in *Drosophila*. The *Drosophila* nomenclature is precise in the importance it gives to priority. Therefore, genes described before being characterized as P450 and before obtaining a CYP designation can be called by their original name – for instance, the Halloween genes (see section 8.4.1.1). For those genes, the use of the CYP synonym that is favored by the community and rationalizes the nomenclature is easier in comparative and biochemical studies. For instance, *spook* (*spo*) is the original Halloween designation of the *Drosophila* gene later identified as *Cyp307a1*. However, *spookier* and *spookiest* have no priority over the CYP name in the literature, and are only confusing designations for *Cyp307a2* and *CYP307B1*; the latter is not even a *Drosophila* gene. Similarly, it is wrong to misrepresent *Cyp4d21* as a novel gene and name it "sex-specific enzyme 1 (*sxe1*)" (Fujii and Amrein, 2002) when it is not sex-specific. It is then adding to confusion to designate this gene as a fatty acid ω-hydroxylase (Fujii *et al.*, 2008) when there is not a shred of evidence that CYP4D21 has this activity.

The CYP nomenclature of Nebert *et al.* (1987) was clearly designed to reflect the evolutionary relationships between the genes as evidenced by the degree of sequence identity of the proteins they encoded. As such, the use of CYP families and clans is a Darwinian nomenclature. P450 proteins have been categorized by some authors and by InterPro into classes and groups (B, E, I, II, III, etc.)

that reflect the types of electron delivery to the active site. There are many types and combinations of redox partners for P450 enzymes (Hannemann *et al.*, 2007), and these are not related to the phylogeny of the P450s, so that the InterPro designations (e.g., IPR002403 "Cyt_P450_E_grp-IV") are mostly wrong and/or misleading. IPR001128 Cyt_P450 is the only correct "parent" designation that contains the conserved site of P450s, IPR017972 (Prosite PS00086). The eight other InterPro numbers are best ignored. Similarly, the GeneOntology term GO:0004497 "monooxygenase activity" is probably the most accurate, if not exclusive, molecular function designation for CYP genes. The 50 or so child terms for GO:0004497 are also best ignored in the analysis of insect genes. Researchers who discover CYP genes or P450 proteins in GenBank through BLAST searches can easily be confused by the many misleading annotations and characteristics given to these entries by their similarity to other genes. Only the presence of the cl12078 (P450 superfamily) or pfam00067 (Cytochrome P450) domains are precise and unambiguous.

8.2. Diversity and Evolution of Insect CYP Genes

8.2.1. Diversity of CYP Genes

8.2.1.1. P450 sequences from classical cloning techniques and PCR approaches
The first insect P450s cloned and sequenced were CYP6A1 from *Musca domestica* in 1989 (Feyereisen *et al.*, 1989); CYP4C1 from *Blaberus discoidalis* in 1991 (Bradfield *et al.*, 1991); and CYP4D1 and CYP6A2 from *Drosophila melanogaster* (Gandhi *et al.*, 1992; Waters *et al.*, 1992), as well as CYP6B1 from *Papilio polyxenes*, in 1992 (Cohen *et al.*, 1992). These sequences and many that followed in the literature were obtained, sometimes serendipitously, by classical cloning techniques (Feyereisen, 2005).

By 1994, sufficient information about vertebrate and insect P450s allowed the isolation of fragments of P450 cDNAs and genes by PCR methods, with degenerate oligonucleotides corresponding to consensus sequences. This has often been used as a first step in the isolation of full-length P450 clones in insects and ticks (Feyereisen, 2005). A variety of approaches were taken, two of which were particularly successful. In the first, sequences in the I helix and surrounding the conserved cysteine were used to isolate PCR products of about 450–500 bp that coded for about 130 amino acids, or almost a third of the full-length P450. In the second, the sequence around the conserved cysteine was used to design a first primer, with oligo(dT) serving as anchor on the poly(A)$^+$ message. Fragments of varying length were obtained by this 3′-RACE strategy; the C-terminal 30- to 50-amino acid sequence of the P450, and a variable 3′ UTR sequence

(see Snyder *et al.*, 1996, for review). The PCR approaches that first led to the description of 17 new *CYP4* genes in *Anopheles albimanus* (Scott *et al.*, 1994) were then applied in many species (Feyereisen, 2005). A large number of partial P450 sequences were thus generated, published, and deposited in GenBank. This has led to two related problems. The first is that small sequence differences can be found between very closely related PCR products. Are these artefactual, or do they represent allelic variation? In the study on *A. albimanus*, for instance, 64 clones were sequenced, of which 47 encoded P450 fragments, describing 17 genes (Scott *et al.*, 1994). In some clones, up to seven nucleotide differences from the closest sequence were seen. A few nucleotides may not just differentiate two alleles of the same gene, but may be sufficient to differentiate two genes, as the complete sequence of the *Drosophila* and *Anopheles gambiae* genomes subsequently showed. This causes a nomenclature problem, when two distinct genes are prematurely described as allelic variants of a single gene. The second problem is that the P450 fragments obtained by PCR, one-third of the full sequence or less, make the calculation of percentage identity (the base of the nomenclature rules) difficult. In the *A. albimanus* study (Scott *et al.*, 1994) this problem was resolved by establishing a function that derives identity over the full-length P450 from the percentage identity of the PCR product. Further validations of this approach have been presented (Danielson *et al.*, 1999; Fogleman and Danielson, 2000). The practice of giving an official CYP designation to short, partial P450 sequences has now been abandoned.

8.2.1.2. Genome sequences: CYPome diversity revealed
The *D. melanogaster* complete genome sequence was published in 2000 (Adams *et al.*, 2000). Annotation of the P450 sequences (Tijet *et al.*, 2001) gave 90 sequences, of which 83 appeared to code for potentially functional P450s. Seven sequences were either partial sequences or obvious pseudogenes; 40 genomic sequences were not represented by ESTs. New releases of the genome have slightly modified this count (see section 8.2.1.3). A similar annotation of the P450 genes from the complete *A. gambiae* genome published in 2002 (Holt *et al.*, 2002) gave 111 genes (Ranson *et al.*, 2002a), of which 5 are thought to represent pseudogenes. An increasing number of insect genomes have become available, at an increasing pace. This, and the continuous addition of P450 sequences from EST projects from various species, is redefining the approaches to P450 research, starting with annotation.

The need for "manual" annotations (e.g., Claudianos *et al.*, 2006; Strode *et al.*, 2006; Lee *et al.*, 2010; Oakeshott *et al.*, 2010) remains critical because of the error-prone gene- and transcript-calling programs, as noted for P450s by Gotoh (1998). Despite marked improvements in these programs (and the accumulation of correct sequences on which these programs can rely), some of the sources of errors remain. They typically include the fusion of two neighboring P450 genes into one, or the truncation of a gene. Potentially confusing is the proliferation of GenBank accessions for transcripts and proteins identified *in silico*, with only scant notice that they represent little more than a software-digested genomic sequence and are highly redundant. Initial errors are then compounded by their "transmission" to new genomes. Collectively, these difficulties remain a major challenge to the complete description of the P450 gene complement (CYPome) of a species. Comparison of the intron/exon structure of closely related genes, and alignments with EST sequences when these are available, can facilitate P450 gene annotation in a majority of cases. Independent evidence, such as the cloning of full-length cDNAs, or functional expression, is rarely available to resolve annotation problems, so that the completion of a CYPome annotation remains an ongoing task. A detailed transcriptomic analysis of *Drosophila* that uses several complementary techniques (RNAseq, tiling microarrays, and cDNA sequencing) can significantly increase our understanding of the *Drosophila* CYPome (Graveley *et al.*, 2011), and will require careful gene-by-gene reannotation. New sequencing technologies have also yielded many new P450 sequences from organisms for which the complete genome is not yet available. For instance, Zagrobelny *et al.* (2009) obtained a large number of P450 transcript sequences from larvae of *Zygaena filipendulae* by 454 sequencing, and Pauchet *et al.* (2010) similarly obtained 36 new P450 sequences from *Manduca sexta*. The assembly of the usually short reads

from new sequencing techniques risks assembling chimeric P450s from closely related genes or alleles, so that PCR confirmation of the full-length sequences may be needed.

8.2.1.3. CYPome size The CYPome size of an insect genome is not a definite number (Feyereisen, 2011). Insects can survive with small CYPomes even in toxic environments. The human body louse *Pediculus humanus*, with 36 CYP genes, is known to become highly resistant to many classes of insecticides (Lee *et al.*, 2010), and the honey bee, with just 46 CYP genes (Claudianos *et al.*, 2006), is not more sensitive than other species in a comparison of the toxicity of 62 insecticides (Hardstone and Scott, 2010).

Table 1 gives the approximate number of P450 genes in insect genomes. The numbers are approximations for two reasons – one trivial, one less so (Feyereisen, 2011). The trivial reason is that coverage and quality of assembly of newly sequenced genomes is quite variable, so that numbers of P450 genes can change with new releases of the genomes. In some cases this means that a new sequence brings new genes, but more often it means that a gene fragment or putative pseudogene is "promoted" to a *bona fide* gene. An example is *Cyp307a2* in *Drosophila melanogaster*, first thought to be a pseudogene, then recognized as a full gene when more DNA sequence from heterochromatin (not available in the initial release of the *Drosophila* genome) was annotated. Similarly, a sequence gap adjacent to *Cyp12d1*, when filled, revealed a second *Cyp12d* gene with just three (non-silent) nucleotide substitutions. Presumably the very high identity of these two

Table 1 Approximate Numbers of CYP Genes in Various Species, and Number of Genes per CYP Clan

	CYPome Size	CYP2 Clan	mitochondrial CYP Clan	CYP3 Clan	CYP4 Clan	Reference*
Insecta						
Drosophila melanogaster	88	7	11	36	32	Tijet *et al.* (2001)
Anopheles gambiae	105	10	9	40	46	Ranson *et al.* (2002a)
Aedes aegypti	160	12	9	82	57	Strode et al. (2008)
Culex quinquefasciatus	170					Arensburger *et al.* (2010)
Bombyx mori	85	7	12	30	36	
Apis mellifera	46	8	6	28	4	Claudianos *et al.* (2006)
Nasonia vitripennis	92	7	7	48	30	Oakeshott *et al.* (2010)
Camponotus floridanus	132					Bonasio *et al.* (2010)
Harpegnathos saltator	93					Bonasio *et al.* (2010)
Tribolium castaneum	134	8	9	72	45	Tribolium Genome Sequencing Consortium (2008)
Acyrthosiphon pisum	64	10	8	23	23	
Pediculus humanus	36	8	8	11	9	Lee *et al.* (2010)
Crustacea						
Daphnia pulex	75	20	6	12	37	Baldwin *et al.* (2009)
Acari						
Tetranychus urticae	86	48	5	10	23	

*Data collected from genome sequence papers, annotation papers and further analysis

adjacent genes caused a problem in the initial assembly, resulting in a gap. Such problems are found for all genes and all genomes, and only for very few organisms can the CYPome annotation be considered "finished." New generation sequencing technologies may increase this problem, because few genomes of higher eukaryotes will deserve the close attention and depth of coverage that major model species such as *Drosophila* are receiving. The less trivial reason (Feyereisen, 2011) is that there is polymorphism in gene copy number (see section 8.2.1.4), so that the number of genes in the CYPome of an organism depends on one's viewpoint. The "reference" genomes may have more or less CYP genes than are found in a natural population. Several cases of CYP copy number variation (or even amplification) are now known. A more difficult type of polymorphism is copy number polymorphism involving large (> 100-kb) segments of genome, as in the initial sequence of *A. gambiae*.

8.2.1.4. Genomic variation

8.2.1.4.1. Allelic variants Apparent allelic variants of cloned P450 cDNAs and genes are very frequent, and were already being described in the earliest studies of insect P450 (Cohen *et al.*, 1992, 1994; Cohen and Feyereisen, 1995). Examples of this variation were most striking in *A. gambiae*, because the sequence of the PEST strain included several "scaffolds" that were not included in the genome assembly. These scaffolds represent large tracts of heterozygosity ("dual haplotype regions;" Holt *et al.*, 2002), probably resulting from the mosaic nature (contributions from different *A. gambiae* cytotypes) of the PEST strain that was sequenced. Interestingly, some of these scaffolds harbor P450 genes, even P450 gene clusters, and a comparison of the various haplotypes revealed not only considerable differences, but also variations in the total number of P450 genes. In other words, P450 gene duplication events have occured independently in different mosquito cytotypes. A gene recently converted to a pseudogene in one population or strain as a result of a debilitating mutation or transposable element insertion may still be active in another population. The *A. gambiae* genome is highly polymorphic, with an SNP frequency of 1 every 26 bp in CYP genes (Wilding *et al.*, 2009).

8.2.1.4.2. Copy number variation The CYPome size, or total number of P450 genes in a species, is therefore a relative number that is genotype-dependent. Techniques such as high density tiling arrays have revealed high numbers of copy number polymorphisms. In *Drosophila*, 2658 such polymorphisms, half covering genes, were detected in a survey of 15 lines (Emerson *et al.*, 2008). This study revealed the duplication of *Cyp28d2* and the deletion of *Cyp6a17* in some lines, and the duplication of the *Cyp6g1* and *Cyp6a2* pair of genes. A detailed account of

the *Cyp6g1* duplication (Schmidt *et al.*, 2010) is discussed below (see section 8.4.5.3). CYP gene duplications have been reported in mosquitoes (Wondji *et al.*, 2009; Itokawa *et al.*, 2010), and amplifications in an aphid (Puinean *et al.*, 2010).

8.2.1.4.3. Alternative splicing There is little evidence for alternative splicing of insect P450 transcripts as additional means of generating diversity. One example is the *Cyp4d1* gene that utilizes two alternate first exons, and ESTs for each transcript type have been found. The first cDNA cloned (Gandhi *et al.*, 1992) uses a proximal first exon and corresponds to transcript Cyp4d1-PA, whereas several ESTs of transcript Cyp4d1-PB use a more distal first exon instead. Cyp4d1-PA is expressed predominantly in the midgut of feeding third instar larvae, and Cyp4d1-PB is expressed more in the fat body (Chung *et al.*, 2009). The two predicted proteins differ only from the N-terminal to the beginning of SRS 1 (see section 8.3.1). The consequences of this alternative splicing in terms of catalytic competence remain to be examined. The recent detailed analysis of the *Drosophila* transcriptome (Graveley *et al.*, 2011) indicates that 35 CYP genes are found as multiple isoforms, so that *Cyp4d1* may just be one of many future examples.

8.2.1.4.4. Pseudogenes In *B. germanica*, three pseudogenes of *CYP4C21* and two pseudogenes of *CYP9E2* have been described (Wen *et al.*, 2001). These pseudogenes are characterized by deletions and/or the presence of several stop codons in the open reading frame. *CYP9E2P2*, for instance, has just 10 nucleotide differences from *CYP9E2*, but 2 base deletions render this gene non-functional. The processed pseudogene *CYP4W1P* in the cattle tick has a 191-bp deletion, but only three other nt changes in the open reading frame (Crampton *et al.*, 1999). Another tick pseudogene, *CYP319A1P*, contains two DNA insertions in the open reading frame and also appears of recent origin (He *et al.*, 2002). CYP genes missing key residues essential for activity are also probable pseudogenes; for instance, *Pediculus humanus CYP358A1* misses the cysteine ligand to the heme, and therefore should be considered a pseudogene. *Drosophila* is peculiar in its ability to delete genomic DNA at a high rate, and the proportion of pseudogenes in the genome is low – 1 pseudogene for 130 proteins, as compared to 1 for 19 in *C. elegans* (Harrison *et al.*, 2003). For P450s alone, *Drosophila* has a far lower proportion of pseudogenes than humans and mice that have as many CYP genes as pseudogenes. However, the proportion of pseudogenes in the P450 family is high (four to six pseudogenes) in *Drosophila* when compared to other gene families – an indication of rapid turnover of P450 genes. Precise annotation of CYP pseudogenes in insect genomes is lagging even further than the precise annotation of genes, although pseudogenes give important information on the dynamics of CYP evolution.

8.2.2. CYPome Evolution

The main driver of CYPome evolution is of course gene duplication, followed by divergence (by neofunctionalization or subfunctionalization) or death (pseudogenization or deletion) (review in Feyereisen, 2011); many mechanisms can generate paralogous genes in the CYPome. These include tandem duplications, including unequal crossing over of various lengths, transposable element-induced non-allelic recombinations, and chromosome duplications, as well as retropositions. Lynch and Conery (2000), who notably excluded large gene families from their estimates, reported rates of 0.0023 duplications per gene per MY (million years) for *Drosophila*, with a half-life of a gene duplicate of about 2.9 MY. Hahn *et al.* (2007), using data from gene families in the 12 *Drosophila* genomes, calculated a rate of duplication of 0.0012/gene per MY. Similarly, Osada and Innan (2008), in a detailed study of the *Drosophila* genus taking into account possible gene conversions, found a duplication rate of 0.001/gene per MY. This rate of duplication thus appears quite robust and is comparable to the spontaneous mutation rate of nucleotide level – an observation that was certainly unexpected

a decade ago. As a result, using *Drosophila* as an example, we can estimate the duplication rate in the P450 family (about 85 genes) to generate a new P450 every 5–12 MY. The dataset of the 12 species in the *Drosophila* genus allows a direct verification of this estimate, with multiple gains and losses over 60 MY. In a detailed study of *Drosophila* P450s, Chung *et al.* (2009) found that only one-third of the 53 genes they studied had a 1:1 ortholog in the 11 other *Drosophila* species. This figure is consistent with the rates of gene duplication estimated for the whole genome.

8.2.2.1. Four clans of insect CYP genes We now have access to *Drosophila melanogaster* and 11 related *Drosophila* species, 3 species of mosquitoes, 4 Hymenoptera (the honey bee, a wasp, and two ants), and representatives of other major orders of insects, Lepidoptera, Coleoptera, Hemiptera, and Phthiraptera. In addition, genomes from one crustacean species (*Daphnia pulex*) and representative Acari (the cattle tick *Ixodes scapularis* and the two-spotted spider mite, *Tetranychus urticae*) have been sequenced. These genomes are not all complete or published, and several others are nearing completion. **Figure 2** shows a maximum likelihood tree of 37 insect P450 protein

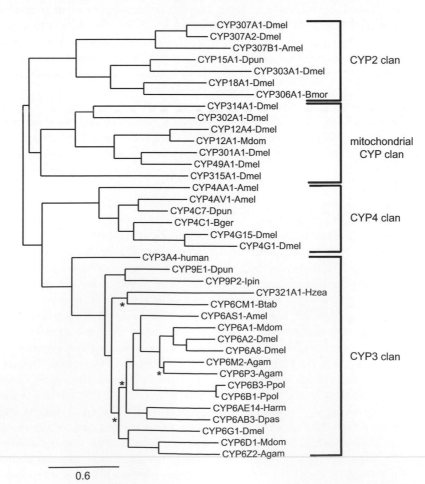

0.6

Figure 2 Four clans of CYP genes in insects. The tree shows a maximum likelihood analysis of selected insect P450s discussed in this chapter. Muscle alignment and PhyML analysis. Branches with less than 75% bootstrap support are indicated by *.

sequences (a sampling of P450s that are discussed here). It typically reveals that insect P450 sequences are distributed in four major clades (Feyereisen, 2006) that are strongly supported by bootstrap analysis. These correspond to four CYP "clans." These four clans are named after the founding family in vertebrates (CYP3, CYP4, CYP2 clans) or their subcellular location (mitochondrial CYP clan), and are briefly described below.

8.2.2.1.1. CYP3 clan

A large group of insect P450s comprises the CYP6, CYP9, and CYP28 families, as well as the CYP308-310 families, and now over 30 families into the CYP400s (**Table 2**). It is most closely related to vertebrate CYP3 and CYP5 families. Because of the initial "lumping" of CYP6B1 into the CYP6 family and later "splitting" of the CYP300s, this group is rather heterogeneous with regard to CYP families and subfamilies. Few of the CYP6 enzymes have been characterized (see **Table 5**, below), and in many (but not all) studies they are shown to metabolize xenobiotics and plant natural compounds. Genes from this group appear to share the characteristics of "environmental response genes" as defined by Berenbaum (2002); specifically, (1) very high diversity, 2) proliferation by duplication events, (3) rapid rates of evolution, (4) occurrence in gene clusters, and (5) tissue- or temporal-specific expression. Of course, these characteristics are not independent of each other, and they are difficult to measure objectively. Nonetheless, the multiple paralogs from these families can be dated to no more than 150–200 MY ago, and this gives a good idea of the dynamic nature of this clade's evolution.

Table 2 Four Clans of CYP Genes in Insects, with CYP Family Numbers

CYP2 Clan	Mitochondrial CYP Clan	CYP3 Clan	CYP4 Clan
15	12	6	4
18	49	9	311-313
303-307	301-302	28	316
343	314-315	308-310	318
359	333-334	317	325
369	339	321	340-341
	353	324	349-352
	366	329	380
		332	367
		336-338	405
		345-348	411-412
		354	
		357-358	
		365	
		395-400	
		408	
		413	

Numbers as of 2010. Data analyzed from http://drnelson.uthsc.edu/cytochrome P450.html

8.2.2.1.2. CYP4 clan

The CYP 4 clan includes the large CYP4 family that has members from the vertebrates and insects, but also several P450 families from *C. elegans*, as well as the insect CYP311, 312, 313, 316, etc. – now over 20 families (**Table 2**). This group of sequences is highly diversified, perhaps even more so than the CYP3 clan, and some *CYP4* genes are clearly inducible by xenobiotics. However, specialized physiological functions are recognized (Sutherland *et al.*, 1998) or proposed (Bradfield *et al.*, 1991) for some enzymes of this group. This remains the least well-studied group of insect P450 enzymes.

8.2.2.1.3. Mitochondrial CYP clan

The mitochondrial P450s of vertebrates and insects (as well as of other Metazoa; see Feyereisen, 2011) are monophyletic. Within the mitochondrial clade, the CYP12 family appears to behave like families in the CYP3 and CYP4 clans, as a rapidly evolving group of paralogous genes. Other families of mitochondrial P450s (CYP302, CYP314, CYP315) are now clearly linked to the ecdysteroid metabolism pathway (see section 8.4.1.1). Mitochondrial P450s of insects thus evolved differently from the vertebrate mitochondrial P450s, and subcellular localization in mitochondria can no longer be considered as evidence for a role in endocrine physiology. These sequences are all derived from an ancestral, probably CYP2 clan microsomal P450, and not from a soluble bacterial P450. The origins of mitochondrial CYP clan P450s are discussed in Feyereisen (2011).

8.2.2.1.4. CYP2 clan

This clan includes vertebrate microsomal CYP1, CYP2, CYP17, and CYP21, as well as insect CYP15, CYP18, and several families in the CYP300 series. Some of these insect sequences represent enzymes involved in essential physiological functions (e.g., CYP15, CYP18, CYP306, CYP307), as are CYP17 and CYP21 of vertebrates. The *CYP2* genes, however, are more diverse, and widely considered as "environmental response genes" in mammalian species. They are also more diverse in *Daphnia* and Acari than in insects (**Table 1**).

As shown in **Table 1**, the numbers of genes in the CYP2 clan and in the mitochondrial CYP clan are relatively stable in insects, whereas the CYP3 and CYP4 clans show large variations in gene numbers. **Table 2** shows the current (end 2010) list of CYP family names in each clan. Despite the lumping of genes into the CYP6 and CYP4 families that occured in the early 1990s, it is still clear that the number of gene families (sequences with >40% sequence identity) is greater in the CYP3 and CYP4 clans than in the other two clans. Although more genomes from hemimetabolous insects, bristletails and silverfish, need to be sequenced, it appears that the "sequence space" is not yet saturated in the CYP3 and CYP4 clans, but fewer CYP families remain to be discovered in the CYP2 and mitochondrial CYP clans. Surprises are possible, given the low ratio of representative genomes versus insect diversity, but it is likely that the distribution of CYP genes into just four clans will remain a distinctive characteristic of the insect CYPomes. In the crustacean *Daphnia pulex* (Baldwin *et al.*, 2009) and the two spotted spidermite *Tetranychus urticae* a similar distribution in just four clans is observed, although the relative numbers of genes in each clan differ significantly from insects.

8.2.2.2. Blooms of CYP subfamilies Conspicuous in all CYPome annotations is the presence of one or more CYP subfamilies that appear both abundant and lineage-specific. Even the honey bee CYPome, despite a comparatively low (46) number of CYP genes, has 15 members of the CYP6AS subfamily. This was called a CYP "bloom," as a more evocative term than "recent, phylogenetically independent proliferation of close paralogs" or "lineage specific gene family expansion" (Feyereisen, 2011). Other examples of several such blooms in a diversity of species are the 12 *CYP6A* genes in the fruit fly, 13 *CYP6BQ* genes in the red flour beetle *Tribolium castaneum*, and 19 *CYP4AB* genes in the jewel wasp *Nasonia vitripennis*. These CYP blooms are one of the most striking features of P450 evolution, and their origin, by mostly stochastic processes, has been discussed (Feyereisen, 2011). On average, a duplicated gene has twice the probability of being duplicated again than a singleton, and, as the size of the family grows, the probability of further growth by duplication is increased. Therefore, blooms, as seen for CYP families or subfamilies, can be the result of an initial trigger, and do not require the long-term maintenance of this trigger, as the bloom has been unleashed. These blooms result in a power-law distribution of CYP family sizes in the CYPomes of virtually all organisms.

8.2.2.3. Genomic organization: clusters The processes of sequential tandem gene duplication events can lead to large clusters of CYP genes on chromosomes, and these are often striking landmarks of the CYPomes. The presence of P450 genes in clusters was revealed in early studies with *Drosophila* and the house fly (Frolov and Alatortsev, 1994; Cohen and Feyereisen, 1995). Further evidence was obtained by *in situ* hybridization to polytene chromosomes of *Drosophila* and *A. gambiae* (Dunkov *et al.*, 1996; Ranson *et al.*, 2002b), and ultimately by the analysis of P450 genes in the complete genome sequences of *Drosophila* and subsequently sequenced genomes. The largest *Drosophila* P450 cluster carries nine genes, eight *CYP6A* genes and *CYP317A1*, at 51D on the right arm of chromosome 2. Six of these genes (*Cyp6a17* to *Cyp6a21*) are coordinately regulated during the circadian rhythm (Ueda *et al.*, 2002). In *A. gambiae*, the largest cluster carries 14 P450 genes of the CYP6 family and another contains 12 *CYP325* genes and 2 pseudogenes. Only

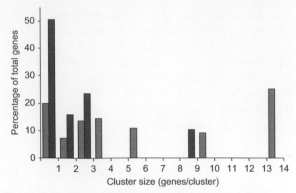

Figure 3 Clusters of P450 genes and singletons (cluster size = 1) genes in the *Drosophila melanogaster* (blue) and *Anopheles gambiae* (red) genomes. Adapted from Ranson *et al.*, 2002a.

22 genes are present as singletons, with 16 clusters of 4 or more genes in the mosquito (Ranson *et al.*, 2002a) **(Figure 3)**. In *Papilio glaucus*, *CYP6B4v2* and *CYP6B5v1* are clustered within 10 kb of each other (Hung *et al.*, 1996). They are 99.3% identical at the nucleotide level, with 98% identity of their single 732-bp intron and 95% identity over 616 bp of the promoter region. The proteins they encode differ by just one amino acid. *CYP6B4* and *CYP6B5* are thus recently duplicated genes that have not yet diverged substantially in sequence (Hung *et al.*, 1996). Similar close pairs are found in *Drosophila* – for example, *Cyp12d1* and *Cyp12d2*, which are 2 kb apart and differ by only three nucleotides, leading to three changes at the amino acid level. In *A. gambiae*, *CYP6AF1* and *CYP6AF2* are 99.8% identical at the nucleotide level, and differ by just one amino acid (Ranson *et al.*, 2002a).

The comparison of syntenic regions of the genome in closely related species reveals how rapidly these clusters can evolve. The CYP9A cluster in Lepidoptera is an example where three species, one distant (*Bombyx mori*) and two more closely related (*Helicoverpa armigera* and *Spodoptera frugiperda*), have been carefully analyzed (d'Alencon *et al.*, 2010). At least eight gene duplications have taken place in about 80 MY, to yield four, five, and nine genes in the three species, respectively, with only three ortholog pairs still recognizable in the two more closely (± 20 MY) related species. Some clusters of duplicated genes can be maintained over very long times without being spread out by recombination. The head-to-head pair of close paralogs *CYP306A1* and *CYP18A1* (Niwa *et al.*, 2004) is conserved as a cluster in all insects studied so far (except in *A. gambiae* which has lost the *CYP18A1* gene), and is even found in the crustacean *Daphnia pulex* (Rewitz and Gilbert, 2008), thus dating this cluster to well over 500 MY. These genes encode an ecdysteroid C-25 hydroxylase and C-26 hydroxylase/oxidase respectively, so that a slight change in substrate regioselectivity sealed the neofunctionalization of the duplicated gene, with the C-25 hydroxylase being the presumed ancestor.

Two recently duplicated genes may undergo conversion, if they do not diverge quickly enough (Walsh, 1987). Gene conversion at the 5′ end of the *CYP6B8* and *CYP6B28* genes of *Helicoverpa zea* (X. Li *et al.*, 2002a) has been suggested by the lower level of nt differences in the first half of the first exon, as compared to the rest of the sequence. Exon-specific deficit in variation among pairs of P450 is indicative of a gene conversion event (Matsunaga *et al.*, 1990). Changes in regulatory patterns (induction, tissue-specific expression) can be observed for recently duplicated genes (W. Li *et al.*, 2002), and may lead to their independent evolution by subfunctionalization.

Although events such as unequal crossing-over can lead to gene duplications, there are other mechanisms, such as retrotransposition. Capture of a spliced mRNA by a retrotransposon can reintroduce an intronless sequence into the genome, where it may evolve further, or die a pseudogene. This process generally occurs in germ cells, so it should be limited to genes that are expressed in those cells. The *Drosophila Cyp4g1* gene, at the tip of the X-chromosome (1B3), may be a case in point. It lacks introns, and its closest paralog, *Cyp4g15*, has five introns and is located at 10C2–3.

8.2.2.4. P450 gene orthologs

With so much gene duplication, are there still any P450 orthologs in related species? (Orthology = the "same" gene in different species that has only diverged as a result of speciation.) The availability of two completely sequenced genomes made it possible to identify those members of the P450 family that were truly orthologous. This is not as easy as it seems, because, in some borderline cases, two formally orthologous genes may be sufficiently related to their closest formal paralog in the same species as to make the evolutionary connection between the two pairs unclear. Nonetheless, it came as a surprise to find a very small number of 1:1 pairs of orthologous P450 genes between *D. melanogaster* and *A. gambiae* (Ranson *et al.*, 2002a). Zdobnov *et al.* (2002) indicate a genome-wide level of 44–47% orthologous genes, but for the P450 genes only nine orthologs were found, of which five are mitochondrial P450s (Ranson *et al.*, 2002a). When two P450 genes are recognized as orthologs even though the two species diverged about 250 million years ago, the most likely explanation is that there is a strong evolutionary constraint that has maintained this orthologous relationship. A similar or identical physiological function for the orthologous pair of P450 enzymes may represent such a constraint. Three of the five mitochondrial P450s indeed have a recognized function that is predicted to be identical in all insects: CYP302A1; CYP314A1; and CYP315A1 are hydroxylases of the ecdysteroid biosynthetic pathway (see section 8.4.1). There are several cases where one gene in one species has two "orthologs" in the other; this is the case when a gene duplication or a gene loss occured in

just one of the two species after speciation and millions of years of separate evolutionary history. These cases merit special attention, because knowledge of the function of one of the P450s may quickly lead to understanding the function of its alter egos. They also merit attention for the possible confusion they create in the nomenclature. Two genes with the same name can be paralogs when their evolutionary history is better understood. A case in point is *D. melanogaster* CYP307A1, the ortholog of *Drosophila sechellia* CYP307A1, but paralog of *A. gambiae* CYP307A1. The latter is an ortholog of *Drosophila melanogaster* CYP307A2.

8.2.2.5. Intron/exon organization of CYP genes

The intron–exon organization of P450 genes is a useful tool in the analysis of P450 phylogeny, as shown by the systematic studies in *C. elegans* (Gotoh, 1998), *Arabidopsis thaliana* (Paquette *et al.*, 2000), and *Drosophila* (Tijet *et al.*, 2001), and by a comparison of the Halloween family of P450s (Rewitz *et al.*, 2007). Intron sequence comparisons, as well as sequence comparisons of 5′ flanking sequences have also helped clarify the evolutionary relationships of very closely related *CYP6B* genes of *Papilio* species (W. Li *et al.*, 2002). Multiple events of intron loss and gain can be deduced from a comparison of the intron–exon organization of orthologous pairs of P450 genes for *Drosophila* and *A. gambiae* (Ranson *et al.*, 2002a). The intron phase is non-biased. CYP introns follow the GT/AG rule, except in rare cases like the first intron in the *Drosophila Cyp9c1* gene (Tijet *et al.*, 2001) and the first intron in the *Cyp6a8* gene (Maitra *et al.*, 2002). The latter was recognized by comparison with the full length cDNA, it does not conform to usual sequence patterns, and has an AT/TC splice junction. This intron is very short (36 bp), and potentially represents 12 additional in-frame codons (Maitra *et al.*, 2002).

8.3. P450 Enzymes

8.3.1. P450 Protein Structure

The sequence identity of distantly-related P450 proteins can be as low as that predicted from the random assortment of two sets of 500 or so amino acids. This is because there are very few absolutely conserved amino acids; in fact, only one is absolutely conserved, the Cys ligand to the heme (Rupasinghe *et al.*, 2006). In insect sequences available to date, these are found in five conserved motifs of the protein (**Figure 4**): the **WxxxR** motif, the GxE/DTT/S motif, the **ExLR** motif, the PxxFxP**E**/D**R**F motif, and the PFxxGxRx**C**xG/A motif. Despite this tremendous overall sequence diversity, the increasing number of crystal structures for P450 proteins, mostly soluble forms from bacteria and P450s from mammalian species, reveals a quite high conservation of the three-dimensional

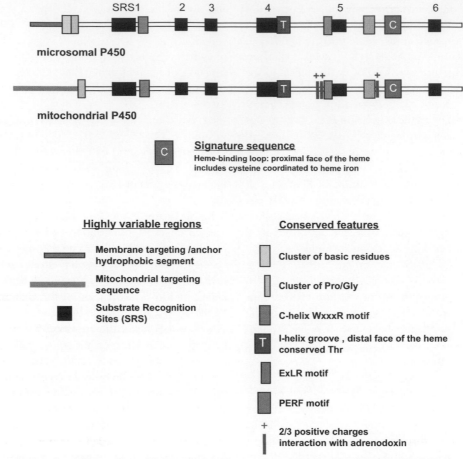

Figure 4 Conserved and variable regions of P450 proteins illustrated over their primary structure (sequence). Adapted from Feyereisen, 2005.

structure. **Figure 5** shows the structure of P450-BM3 (CYP102A1), a well-studied P450.

The description of the structure essentially follows the nomenclature of the P450cam protein, the camphor hydroxylase of *Pseudomonas putida* (Poulos *et al.*, 1985). The first motif, **W**xxx**R**, is located in the C-helix, and the Arg is thought to form a charge pair with the propionate of the heme. This motif is easily discernible in multiple alignments. The second conserved motif, GxE/D**TT**/S, surrounds a conserved threonine in the middle of the long helix I that runs on top of the plane of the heme, over pyrrole ring B. The third conserved motif, **ExLR**, is located in helix K. It is thought to stabilize the overall structure through a set of salt bridge interactions (**E–R–R**) with the fourth conserved motif, PxxFx**PE/DRF** (often PERF, but R is sometimes replaced by H or N), which is located after the K′ helix in the "meander" facing the **ExLR** motif (Hasemann *et al.*, 1995). The fifth conserved motif, PFxxGxRx**C**xG/A, precedes helix L and carries the cysteine (thiolate) ligand to the heme iron on the opposite side of helix I. The cysteine ligand is responsible for the typical 450-nm (hence P450) absorption of the Fe^{II}–CO

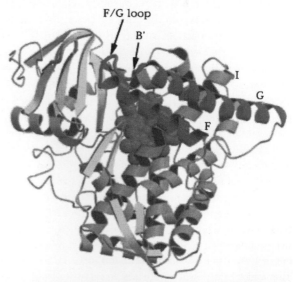

Figure 5 Schematic diagram of P450BM3 (CYP102A1) in the partially open conformation. The B′, F, G, and I helices are labeled. The F and G helices and the F′G loop are highlighted in magenta. These helices slide over the surface of the I helix, which leads to an open–close motion of the access channel leading to the active site. Reproduced from Poulos, 2003.

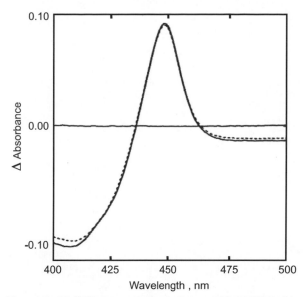

Figure 6 CO-difference spectrum of recombinant CYP12A1. The P450 was reduced with either sodium dithionite (full line) or with bovine adrenodoxin, adrenodoxin reductase and NADPH (dashed line).

Microsomal P450 *e.g.*, CYP6A1

MDFGSFLLYALGVLASLALYFVRWNFGYWKRRGIPHEEPH

NH₂ — HYDROPHOBIC + +++G P P

20 amino acids

Mitochondrial P450 *e.g.*, CYP12A1

MIKYKQYSRAIVALRQRGAQQYSTNVTNASQPDVKAT......

NH₂ + + + + + ?

20 amino acids

Processing site

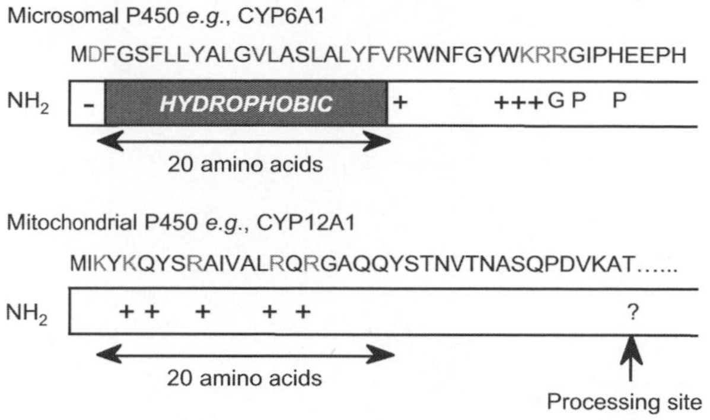

Figure 7 Scheme of the N-terminal sequence of microsomal and mitochondrial P450 proteins.

complex of P450 (**Figure 6**). This heme binding loop is the most conserved portion of the protein, often considered as "signature" for P450 proteins. The term P420 is given to P450 proteins in which the heme is not liganded, the Cys is reduced, and the enzyme is thus inactivated (often irreversibly). Deviations from the consensus sequences of these five motifs deserve special attention. For instance, the conserved CYP301A1 of several species have an unusual Y instead of F in the canonical PFxxGxRxCxG/A motif before the Cys axial ligand to the heme. Several insect P450 sequences, notably CYP307 and CYP321, and some CYP9 sequences, lack the conserved threonine in helix I. Examples of P450 that lack this threonine are CYP74 (allene oxide synthase) and CYP5 (thromboxane synthase), which do not depend on molecular oxygen for activity, and CYP107A1 and CYP158A1, which do, but substitute a substrate hydroxyl group for that of the missing threonine.

P450 proteins are also characterized by their N-terminal sequence (**Figure 7**). Those targeted to the endoplasmic reticulum have a stretch of about 20 hydrophobic amino acids preceding one or two charged residues, which serve as a halt-transfer signal, followed by a short motif of prolines and glycines. The latter serves as a "hinge" that slaps the globular domain of the protein onto the surface of the membrane while the N-terminus is anchored through it. The presence of the PGPP hinge is necessary for proper heme incorporation and assembly of functional P450 in the cell (Yamazaki *et al.*, 1993; Chen *et al.*, 1998). A hydrophobic region between helices F and G is thought to penetrate the lipid bilayer, thus increasing the contact of the P450 with the hydrophobic environment from

which many substrates can enter the active site (Williams *et al.*, 2000a).

The N-terminal sequence of P450 proteins targeted to mitochondria is usually somewhat longer, and shows several charged residues (**Figure 7**). The mature mitochondrial protein is proteolytically cleaved at a position that has not been formally recognized for insect mitochondrial P450s to date, but is known for several mitochondrial P450s of vertebrate species. Mitochondrial P450 proteins are also characterized by a pair of charged amino acids in the K helix: R391 and K395 in CYP12A1. Homologous amino acids in mammalian P450scc (K377 and K381) are responsible for the high affinity to the ferrodoxin-type (adrenodoxin) electron donor (Wada and Waterman, 1992; Pikuleva *et al.*, 1999). An additional positively charged residue (R454 of CYP12A1) is homologous to R418 of CYP27A1, shown to increase affinity to adrenodoxin even further (Pikuleva *et al.*, 1999). The insect mitochondrial clan CYPs (**Table 2**) is most closely related to the mammalian mitochondrial P450 (CYP11, CYP24, and CYP27 families). Subcellular localization of CYP12A1 by immunogold histochemistry with antibodies raised against the CYP12A1 protein produced in bacteria established the mitochondrial nature of CYP12A1 (Guzov *et al.*, 1998).

There is evidence that some vertebrate microsomal P450s can be both microsomal *and* mitochondrial, as suggested by observations from Avadhani's group (Anandatheerthavarada *et al.*, 1997). Indeed, a number of mammalian P450s of the CYP1, CYP2, and CYP3 families can be found in mitochondria, where they have a longer half-life than on the ER. Several mechanisms for this dual localization have been proposed, all involving posttranslational modifications. These include endoprotease cleavage of an N-terminal sequence that reveals a cryptic mitochondrial targeting sequence (CYP1A1/2), or internal PKA phosphorylation that unmasks such a sequence (CYP2A5, 2B1, 2C11). Other microsomal P450s are also directed to mitochondria by unclear mechanisms (Genter *et al.*, 2006; Neve and Ingelman-Sundberg, 2007; Seliskar and Rozman, 2007).

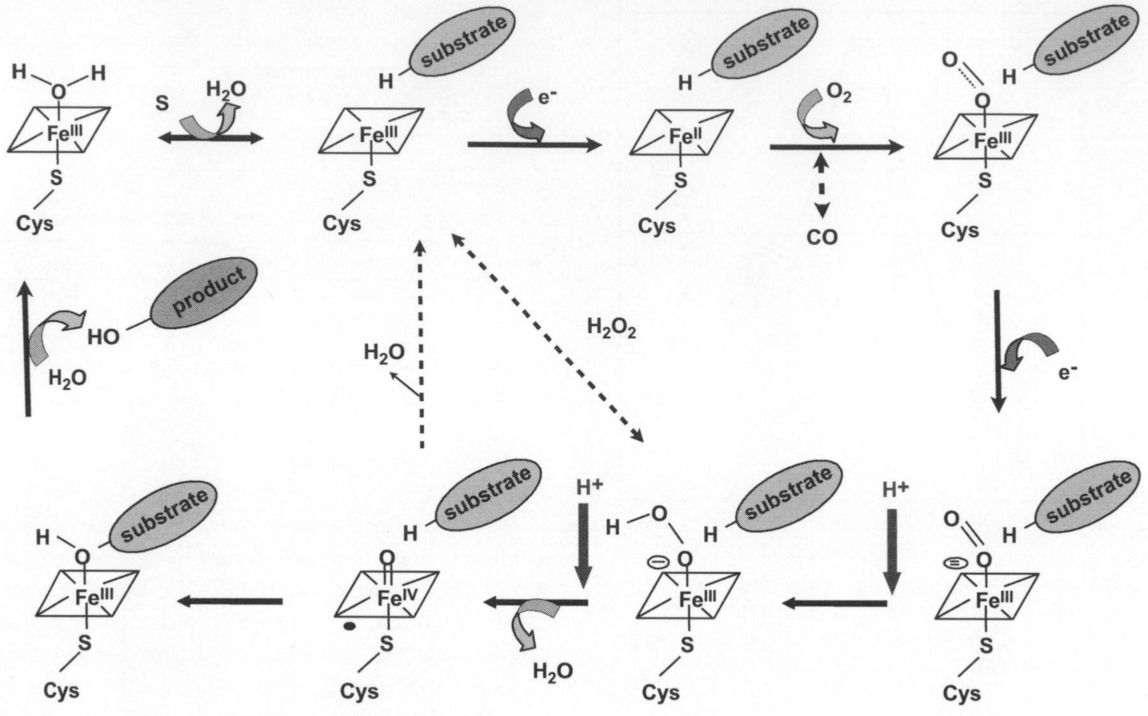

Figure 8 Catalytic cycle of P450 enzymes in monooxygenation reactions (see text for details). Adapted from Feyereisen, 2005.

8.3.2. Catalytic Mechanisms of P450 Enzymes

Little work on insect P450 has focused on the catalytic cycle, and the mechanism derived from our understanding of the bacterial and mammalian P450 enzymes (Ortiz de Montellano, 2005) will be briefly summarized (**Figure 8**). The oxidized P450 is a mixture of two forms: a low spin (Fe^{III}) form with water as the sixth coordinated ligand on the opposide side of the Cys thiolate ligand; and a high spin (Fe^{II}) pentacoordinated form. Substrate binding displaces water from the sixth liganding position, leading to a shift to high spin. This shift can be observed (Type I spectrum) and is accompanied by a decrease in the redox potential of P450. The P450–substrate complex receives a first electron from a redox partner (P450 reductase or adrenodoxin), and ferrous P450 (Fe^{II}) then binds O_2. At this step CO can compete with O_2 for binding to P450;

its binding leads to a stable complex, with an absorption maximum at 450 nm (**Figure 6**), which is catalytically inactive. CO can be displaced by light irradiation at 450 nm. The P450–O_2–substrate complex then accepts a second electron (from P450 reductase or in some cases cytochrome b_5, or from adrenodoxin) to form a ferric peroxide anion. After protonation, a ferric hydroperoxo complex then leads to the activated oxygen form(s) of the enzyme. Although the formal reaction is the insertion of an atom of oxygen into the substrate, the other atom being reduced to water (hence the term "mixed-function oxidase"), the precise nature of the oxidizing species has long remained elusive. Recently, a key intermediate named compound I (by analogy to compound I of peroxidases) was characterized. This intermediate is an iron (IV) oxo species with a delocalized oxidizing equivalent (Rittle and Green, 2010). Hydroxylation of an unactivated C–H bond therefore follows a "rebound" mechanism, where hydrogen is abstracted from the substrate by compound I, forming an iron (IV) hydroxide that then recombines quickly with the substrate radical. The types of reactions currently known to be catalyzed by insect P450 enzymes are listed in **Table 3**.

8.3.3. P450 Reaction Stoichiometry

P450 enzymes, whether microsomal or mitochondrial, need to interact with redox partners for their supply of reducing equivalents from NADPH. **Figure 9**

Interspersed throughout the globular domain of the P450 proteins are six regions with a low degree of sequence similarity, covering about 16% of the total length of the protein (**Figure 4**). Initially recognized in CYP2 proteins by Gotoh (1992), these are called Substrate Recognition Sites (SRSs), and this designation has been generically extended to other P450s. However, they do not have precise sizes and boundaries across CYP families, and their usefulness in studying P450 evolution is no longer compelling (Kirischian et al., 2011).

Table 3   Enzymatic Reactions Catalyzed by Insect P450 Enzymes

Reaction Catalyzed	P450
oxidase activity	
O_2 to H_2O, H_2O_2, $O_2^{\cdot-}$	CYP6A1 (and probably most P450 enzymes)
aliphatic hydroxylation	
C–H hydroxylation	CYP4C7, CYP6A1, CYP6A2, CYP6A8, CYP6G1, CYP6M2, CYP6CM1vQ, CYP9T2, CYP12A1, CYP18A1, CYP302A1, CYP306A1, CYP312A1, CYP314A1, CYP315A1
O-dealkylation	CYP6A1, CYP6D1, CYP6A5, CYP6B4, CYP6B17, CYP6B21, CYP6G1, CYP6Z2, CYP6CM1vQ, CYP9A12, CYP9A14, CYP12A1,CYP321A1
dehalogenation	CYP6G1
epoxidation	CYP6A1, CYP6A2, CYP6B8, CYP6B27,CYP6AB3, CYP6AB11, CYP9E1,CYP12A1, CYP15A1, CYP321A1
aromatic hydroxylation	CYP6D1, CYP6G1, CYP6M2
heteroatom oxidation and dealkylation	
phosphorothioate ester oxidation	CYP6A1, CYP6A2, CYP6D1, CYP12A1,
N-dealkylation	CYP6A5, CYP12A1
N-oxidation	+ (nicotine)
S-oxidation	+ (phorate)
aldehyde oxidation	CYP18A1
complex and atypical reactions	
cyanogenic glucoside biosynthesis:	
(1) Val/Ile to oximes	CYP405A2
(2) oximes to cyanohydrin	CYP332A3
aryl ether cleavage	CYP6M2
carbon–carbon cleavage	+? (sterols, ecdysteroid)
decarbonylation with C–C cleavage	CYP4G1
aromatization	+ (defensive steroids)
reduction	-
endoperoxide isomerization	-

+ indicates metabolism by microsomal P450, but specific enzyme not identified (substrate indicated), - indicates no evidence to date

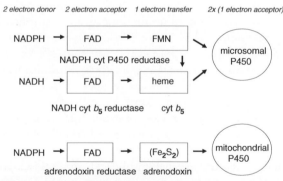

Figure 9 Mitochondrial and microsomal P450 redox partners.

schematically illustrates the two types of electron-transfer complexes thus formed, and a description of the redox partners is provided in section 8.3.5. The obligatory role of P450 reductase in catalysis of the microsomal P450 has been proven in reconstitution experiments, but the role of phospholipids is less clear, and has not been specifically studied for insect P450s. Activity of the mitochondrial CYP12A1 also showed absolute dependence on reconstitution in the presence of (bovine) mitochondrial redox partners (Guzov *et al.*, 1998).

The stoichiometry [1] that commonly describes the monooxygenase (*sensu* Hayaishi) or mixed-function oxidation (*sensu* Mason) reaction of P450 has not been confirmed experimentally for any insect P450. A more complex stoichiometry would take into account the "leakage" of activated oxygen species as superoxide, hydrogen peroxide, and water at the expense of NADPH during catalysis. In a "well coupled" reaction, as in [1], these by-products would not be formed. In "uncoupled" reactions, the catalytic cycle is short-circuited. The reduced P450–O_2–substrate complex can release superoxide, and this aborts the cycle. The ferric hydroperoxo complex can release hydrogen peroxide in what is called the peroxide shunt (Ortiz de Montellano, 2005). Under some experimental conditions, NADPH and molecular oxygen can be substituted by organic hydroperoxides (e.g., cumene hydroperoxide) by taking the peroxide shunt directly from the P450–substrate complex, but this does not work for all P450s and all reactions (Ortiz de Montellano, 2005). Finally, compound I can also be reduced and turn P450 into a terminal oxidase, releasing one more water molecule. It is assumed, but not generally proven, that a specialized P450 metabolizing its favorite substrate would follow stoichiometry [1]. An approximate balance (the parameters measured are underlined) for CYP6A1 epoxidation of heptachlor gives the following results (Murataliev *et al.*, 2008):

$$1 \text{ heptachlor} + 13 \text{ O}_2 + 11.2 \text{ NADPH} \Rightarrow$$
$$1 \text{ heptachlor epoxide} + 1.5 \text{ H}_2\text{O} +$$
$$11.2 \text{ NADP}^+ + 9.8 \text{ H}_2\text{O}_2 \quad (2)$$

and for testosterone hydroxylation under the same experimental conditions:

$$1 \text{ } testosterone + 6 \text{ } \underline{O_2} + 4.7 \text{ } \underline{NADPH} \Rightarrow$$
$$1 \text{ } hydroxytestosterone + 0.5 \text{ H}_2\text{O} +$$
$$4.7 \text{ NADP}^+ + 4.2 \text{ } \underline{H_2O_2} \quad (3)$$

These stoichiometries show that a P450 such as CYP6A1 can be simultaneously an oxidase and a monooxygenase. These coupling stoichiometries (Murataliev *et al.*, 2008) are dependent on the ratio of P450 and P450 reductase, as well as on the presence or absence of cytochrome b_5 (section 8.3.5.2). Note that these stoichiometries are not balanced, reflecting experimental error in the measurements. The uncoupling of monooxygenation is highly likely to be a common feature of insect P450 enzymes that metabolize xenobiotics of synthetic or plant origin. The generation of reactive oxygen species is a corollary of active P450 metabolism.

8.3.4. Substrate Selectivity and Homology Models of P450 Enzymes

Metabolism of a substrate by a P450 enzyme produced in a heterologous system is first a qualitative description of substrate selectivity. A negative result, when it is reported, should be interpreted with more caution. It can take considerable effort to achieve optimal reaction conditions, and this can sometimes turn a negative result into a positive one. The two first and best-characterized insect P450 enzymes to be expressed are informative in this regard; CYP6A1 did not metabolize diazinon with partially purified P450 and P450 reductase (Andersen *et al.*, 1994), but improvements in purification and reconstitution revealed that it could (Sabourault *et al.*, 2001), and also allowed a more quantitative description of CYP6A1 catalysis (Murataliev *et al.*, 2008). CYP6B1 appeared to be selective in the metabolism of linear furanocoumarins when expressed with the endogenous levels of P450 reductase in the baculovirus system (Ma *et al.*, 1994), but optimal levels of added P450 reductase increased metabolism of linear furanocoumarins 32-fold, and revealed very substantial metabolism of angular furanocoumarin metabolism as well (Wen *et al.*, 2003).

Consequently, a quantitative description of substrate selectivity and of kinetic parameters of insect P450 enzymes is not easily achieved. Homology modeling of P450 structure is therefore increasingly useful to describe and predict P450 function. However, homology modeling is only as good as the predicted three-dimensional

similarity between an unknown P450 and its known model chosen from among the well-characterized X-ray structures of bacterial or mammalian P450. When one amino acid change can change substrate specificity, and when sequence identity with known structures is 30% or lower, then reliance on three-dimensional structure conservation is high. Furthermore, increasing evidence for significant conformational changes in P450 structure upon substrate binding and catalysis is difficult to integrate, even in molecular dynamics simulations.

The lepidopteran CYP6B subfamily that comprises many furanocoumarin-metabolizing P450s has been the subject of the most comprehensive analysis by homology modeling (Chen *et al.*, 2002; Baudry *et al.*, 2003; W. Li *et al.*, 2004, Wen *et al.*, 2006; Mao *et al.*, 2007a, 2007b). These studies collectively show that the predicted structure of the more specific CYP6B1 is better suited to metabolize linear furanocoumarins, and that the substrate binding pocket of the more generalist CYP6B4 and CYP6B8 are predicted to be larger and more flexible (reviewed in Li *et al.*, 2007). These studies have been supported by site-directed mutagenesis of residues in the SRS1 region (Chen *et al.*, 2002; Pan *et al.*, 2004; Wen *et al.*, 2005). An enzyme from another subfamily, CYP6AB3, also capable of metabolizing furanocoumarins, showed a similar active site architecture and polarity (W. Li *et al.*, 2004; Mao *et al.*, 2006a, 2007b). A comparison of homology models for CYP6B8 and CYP321A1 describes substrate binding cavities predicted to be more spacious in these two enzymes that metabolize a wider range of compounds (Rupasinghe *et al.*, 2007). Molecular homology models have also corroborated the different substrate specificities of mosquito CYP6Z1 and CYP6Z2 (Chiu *et al.*, 2008; McLaughlin *et al.*, 2008) and of honey bee CYP6AS3 that metabolizes quercetin (Mao *et al.*, 2009). Modeling based on the human CYP3A4 structure, docking, and molecular dynamics simulations correctly predicted the regioselectivity of imidacloprid hydroxylation by CYP6CM1vQ (Karunker *et al.*, 2009). A model of CYP6G1 showed a V-shaped active site, smaller than that of CYP3A4 but rationalizing the metabolism of DDT and imidacloprid by this enzyme (Jones *et al.*, 2010).

Information on substrate access to the active site and active site topology can also be inferred from spectral studies (section 8.6.1.2) and biochemical studies (**Figure 10**). The active site topology of CYP6A1 was studied by a technique developed in Ortiz de Montellano's group (Ortiz de Montellano and Graham-Lorence, 1993). The enzyme is first incubated with phenyldiazene to form a phenyl–iron complex. Ferricyanide-induced *in situ* migration of the phenyl group to the porphyrin nitrogens causes the formation of covalent adducts than can be separated by HPLC (**Figure 10**). The *N*-phenyl protoporphyrin IX adducts of CYP6A1 were formed in a $17:25:33:24$ ratio of the $N_B:N_A:N_C:N_D$ isomers (Andersen *et al.*,

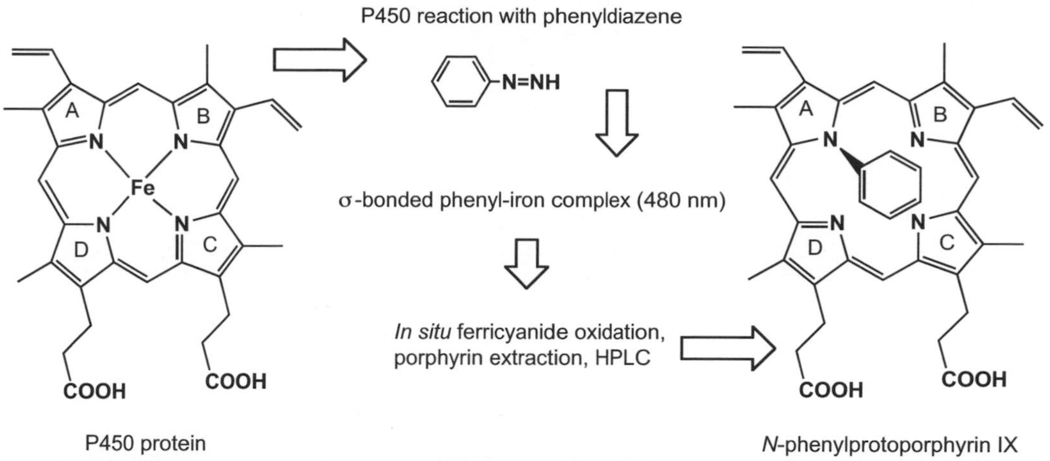

Figure 10 Active site topology of a P450 enzyme tested by formation of phenyl-porphyrin adducts. In this example, an adduct with pyrrole ring A is formed. The ratio of the four possible adducts indicates the degree of encumbrance in the native P450 protein.

1997). Thus, in the native protein, all four pyrrole groups are somewhat exposed, whereas in several other P450s, labeling is more specific. Specific labeling indicates that the protein encumbers more space on top of the heme prosthetic group – for example, leaving only one pyrrole ring exposed, as in P450scc (Pikuleva *et al.*, 1995). The type of labeling seen with CYP6A1 indicates less encumbrance by the protein on top of the heme, as in CYP3A4 (Schrag and Wienkers, 2000). These experiments suggest that the active site of CYP6A1 is relatively accessible, and not severely constrained. Indeed, CYP6A1 metabolizes flat steroids, and bulky cyclodiene insecticides, as well as a variety of sesquiterpenoids (see **Table 5**, below) (Andersen *et al.*, 1994, 1997; Jacobsen *et al.*, 2006; Murataliev *et al.*, 2008). High uncoupling of electron transfer relative to monooxygenation may be a result of this broad substrate specificity.

8.3.5. P450 Redox Partners

8.3.5.1. NADPH cytochrome P450 reductase (P450 reductase) P450 reductase (EC 1.6.2.4) belongs to a family of flavoproteins utilizing both FAD and FMN as cofactors. These diflavin reductases emerged from the ancestral fusion of a gene coding for a ferredoxin reductase with its NADP(H) and FAD binding domains, with a gene coding for a flavodoxin with its FMN domain. This origin of the enzyme was first proposed by Porter and Kasper (1986), based on their analysis of the rat P450 reductase sequence. The fusion is dramatically illustrated by the three-dimensional structure of P450 reductase (Wang *et al.*, 1997), where the domains are clearly distinguished (**Figure 11**). The architecture of this domain fusion has been found in a handful of other enzymes (Murataliev *et al.*, 2004), and in some cases further fusion with a P450 gene has led to self-sufficient P450 proteins (Munro *et al.*, 2007).

The insect P450 reductases sequenced to date are clearly orthologous to the mammalian P450 reductases, with an overall amino acid sequence identity of 54% for the house fly P450 reductase, first cloned and sequenced in 1993 (Koener *et al.*, 1993). There is a single P450 reductase gene in insect genomes, although there are other diflavin reductases. The house fly P450 reductase gene codes for a protein of 671 amino acids, and was mapped to chromosome III. The P450 reductases of other insects are very similar to the house fly enzyme – 82% identity for the *D. melanogaster* enzyme (Hovemann *et al.*, 1997), 57% identity for the *B. mori* enzyme (Horike *et al.*, 2000), and 75% identity for the *A. gambiae* P450 reductase (Nikou *et al.*, 2003). The insect, mammalian, and yeast enzymes are functionally interchangeable in reconstituted systems of the purified proteins or in heterologous expression systems. However, no detailed study has documented how *well* a mammalian or yeast P450 reductase can support the activity of an insect P450 when compared to the cognate insect P450 reductase.

Early attempts to purify and characterize the enzyme from microsomes of house fly abdomens were hampered by the facile proteolytic cleavage of the N-terminal portion of the protein. This hydrophobic peptide anchors the reductase in the membrane, and its removal abolishes the ability of the remainder of the protein ("soluble" or "tryptic" reductase) to reduce P450s. The proteolytically-cleaved reductase nonetheless retains the ability to reduce artificial electron acceptors such as cytochrome c, DCPIP, or ferricyanide (Hodgson, 1985). Heterologous expression of the cloned P450 reductase has been achieved in *E. coli* as an N-terminal fusion with bacterial pelB signal sequence (Andersen *et al.*, 1994; McLaughlin *et al.*, 2008) or as a C-terminal 6xHis-tagged protein (Kaewpa *et al.*, 2007), and as the native enzyme in the baculovirus expression system (Wen *et al.*, 2003). The *Anopheles*

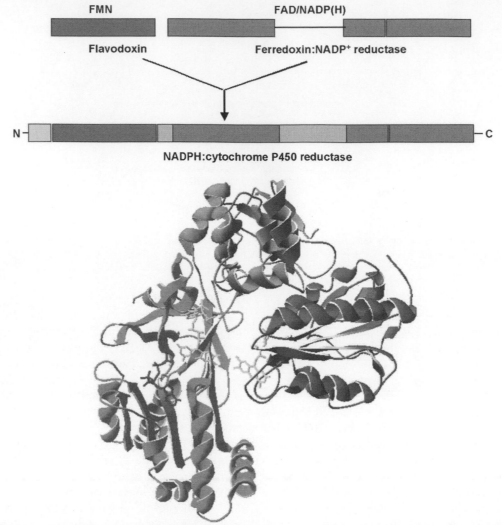

Figure 11 Structure of NADPH cytochrome P450 reductase. Top: evolutionary origin of the FMN (blue), FAD and NADP(H) (green) binding domains of the protein (yellow: membrane anchor; grey: connecting domain). Bottom: three-dimensional structure of the enzyme with the domains identified by color. The bound substrate NADPH (red) and cofators FAD and FMN (yellow) are indicated. Reproduced from Murataliev *et al.* (2004), with permission.

minimus enzyme is relatively unstable and prone to lose FMN, but mutagenesis (L86F and L219F) to the residues found in other insect reductases increases FMN retention (Sarapusit *et al.*, 2008). A purification scheme (Murataliev *et al.*, 1999) has produced quantities of enzyme sufficient for a detailed catalytic characterization of the enzyme's functioning and of its reconstitution with redox partners (**Figure 12**).

P450 reductase is an obligatory partner of microsomal P450 enzymes. Antisera to *Spodoptera eridania* or house fly P450 reductase inhibit all P450-dependent activities tested (Crankshaw *et al.*, 1981; Feyereisen and Vincent, 1984). RNAi knockdown of P450 reductase in adult *A. gambiae* increases permethrin susceptibility (Lycett *et al.*, 2006), suggesting P450 activity as a major determinant of its toxicity. Immunoinhibition with P450 reductase antibodies serves as a strong indication of microsomal

P450 involvement in NADPH-dependent activities, such as ecdysone 20-hydroxylation in the cockroach and in *B. mori* eggs (Halliday *et al.*, 1986; Horike and Sonobe, 1999; Horike *et al.*, 2000), and (*Z*)-9-tricosene biosynthesis in the house fly (Reed *et al.*, 1994). P450 reductase immunoinhibition could serve as a tool to distinguish P450-dependent activities from flavin monooxygenase (FMO) activities in microsomes from insect sources. P450 reductase can also transfer electrons to cytochrome b_5 (section 8.3.5.2) and to other microsomal enzymes, such as heme oxygenase.

8.3.5.1.1. P450 interaction with P450 reductase
In insect microsomes, the ratio of P450 enzymes to P450 reductase is about 6–18 to 1 (Feyereisen *et al.*, 2005). In this ratio, all P450 enzymes are summed, so that the actual ratio of one specific P450 enzyme to P450 reductase is

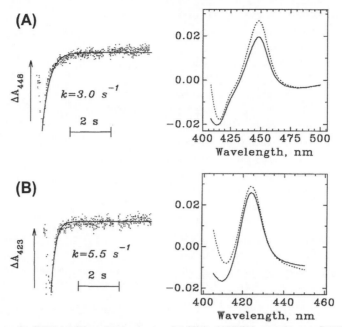

Figure 12 Reduction of house fly CYP6A1 (A) and cytochrome *b*5 (B) by NADPH cytochrome P450 reductase. The recombinant proteins expressed in *E.coli* were reconstituted *in vitro*. On the left, the kinetics of reduction measured by stopped-flow spectrophotometry are shown with the calculated first-order rate constants. On the right, the end point difference spectra of CYP6A1 and cytochrome *b*5 after reduction by P450 reductase (solid line) and sodium dithionite (dotted line). Adapted from Guzov *et al.*, 1996.

probably smaller. The rate of the overall microsomal P450 reaction (two transfers of one electron) is relatively slow so that dissociation of the P450–P450 reductase complex is possible between the first and the second electron transfer. Indeed, cytochrome b_5 can replace P450 reductase for the supply of the second electron in some cases. The effect of varying the P450:P450 reductase ratio on catalytic rates was measured in a reconstituted system for heptachlor epoxidation by CYP6A1. The rate of epoxidation was determined by the concentration of the binary complex of P450 and P450 reductase, with the same high rate being observed in the presence of an excess of either protein (**Figure 13**). The half-saturating concentration of either protein was about 0.1 μM in the presence of cytochrome b_5 (Murataliev *et al.*, 2008). This is in good agreement with the K_m of 0.14 and 0.5 μM for P450 reductase in the presence and absence of cytochrome b_5 measured previously (Guzov *et al.*, 1996). Co-infection of Sf9 cells with baculoviruses carrying lepidopteran CYP6B1 and house fly P450 reductase has revealed that highest catalytic activity was achieved at equivalent, moderate, multiplicities of infection for the two viruses (Wen *et al.*, 2003; Mao *et al.*, 2006a). Higher enzymatic activities of cell lysates towards furanocoumarins were not achieved when either protein was produced in excess. This result can be explained in part by documented limitations of the cell's ability to host, fold, and provide cofactors for both P450 and reductase, but it mostly supports the idea that highest activity is achieved for the highest concentration of the binary complex of the two partners.

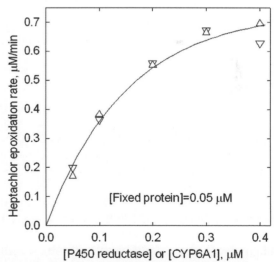

Figure 13 Heptachlor epoxidation by *E. coli*-expressed recombinant house fly CYP6A1 and NADPH cytochrome P450 reductase. A reconstituted system containing variable concentration of CYP6A1 (Δ or P450 reductase ∇) and a fixed concentration (0.05 μM) of the reciprocal partner, and a cytochrome *b*5 concentration of 1.0 μM was incubated in the presence of NADPH. Reproduced from Murataliev *et al.* (2008), with permission.

8.3.5.1.2. P450 reductase functioning   P450 reductase accepts two electrons from NADPH – more precisely, it accepts a hydride ion (one proton and two electrons) – and donates the two electrons, one at a time, to P450 enzymes. P450 reductase is therefore an enzyme with two substrates, NADPH and the electron acceptor (P450 or artificial acceptor such as cyt c); and two products, NADP⁺ and the reduced electron acceptor. With two bound flavins and a pathway of electron transfer NADPH > FAD > FMN > P450 (or cyt c), its reduction state during catalysis can theoretically vary between the fully oxidized state (0 el.) and the fully reduced state (4 el.). Studies with the purified recombinant house fly P450 reductase (Murataliev and Feyereisen, 1999, 2000; Murataliev *et al.*, 1999; reviewed in Murataliev *et al.*, 2004) have shed light on two questions posed by this electron transfer function: what is the kinetic mechanism of this two-substrate enzyme (Ping-Pong or sequential Bi-Bi), and what are the respective reduction states of the two flavins during catalysis?

In the Ping-Pong mechanism, the first product of the reaction must be released before the second substrate binds to the enzyme, and no ternary complex is formed. In sequential Bi-Bi mechanisms, both substrates bind to the enzyme to form a ternary complex. Although several kinetic mechanisms have been proposed (Hodgson, 1985), a careful study of the recombinant house fly P450 reductase clearly established a sequential random Bi-Bi mechanism (Murataliev *et al.*, 1999). The great sensitivity of the enzyme to ionic strength hampers the comparison of different studies (Murataliev *et al.*, 2004). Moreover, inhibition of the enzyme by phosphate above physiological (mM) concentrations (Murataliev and Feyereisen, 2000) makes studies at high phosphate concentrations (e.g., 0.3 M; Sarapusit *et al.*, 2008, 2010) physiologically irrelevant, and results in both excessive K_m values for NADPH, and incorrect kinetic mechanisms.

The formation of a ternary complex of NADPH, P450 reductase, and the electron acceptor suggested a role for reduced nucleotide binding in the catalysis of fast electron transfer. The rate of cytochrome c reduction was shown to equal the rate of hydride ion transfer from the nucleotide donor to FAD (Murataliev *et al.*, 1999). A faster electron transfer rate was observed with NADPH as compared to NADH (Murataliev *et al.*, 1999) and the 2′-phosphate was shown to contribute more than half of the free energy of binding (Murataliev and Feyereisen, 2000). The affinity of the oxidized P450 reductase was 10× higher for NADPH than for NADP⁺ (Murataliev *et al.*, 1999), and a conformational change induced by NADPH binding and important for fast catalysis was suggested by these studies. Recent work on active and inactive P450 reductases from various sources has indeed provided evidence for large conformational changes in P450 reductase. One would be the originally described "closed" conformation optimized for interflavin exchange (see **Figure 11**), and another would be a more "open" conformation, better suited for electron transfer to P450 (Aigrain *et al.*, 2009; Ellis *et al.*, 2009; Hamdane *et al.*, 2009). In addition to the open/closed conformational changes of the FAD and FMN domains relative to each other, there is also a postulated

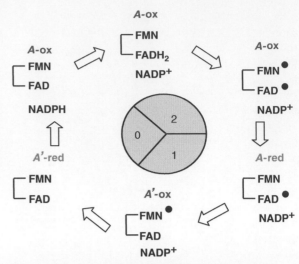

Figure 14 Reduction state of NADPH cytochrome P450 reductase during catalysis, the "0–2–1–0" cycle. The enzyme cycles between a fully oxidized state and a two-electron reduced state. The electron acceptor (A and A′ = P450 or cyt c) receives one electron at a time from an FMN semiquinone form (FMN•) of the enzyme. The release of NADP⁺ is shown here to occur at the last step, but may occur earlier. See text for details, and Murataliev *et al.* (2004) for review.

rotation of the FMN cofactor around conserved Asp187 of yeast, (Asp205 of housefly) between the buried position and a more exposed position that has been described as a second FMN binding site (Lamb *et al.*, 2006; Ivanov *et al.*, 2010). Whether the reductase works by flapping domains or flipping FMN (or both?) remains conjectural.

The state of reduction of the flavins of P450 reductase during catalysis was deduced from kinetic experiments, rates of NADPH oxidation, and EPR measurements of flavin semiquinone (free radical) levels. These revealed the existence of a catalytically competent FMN semiquinone, different from the "blue" neutral FMN semiquinone (known as the air-stable semiquinone), which is not a catalytically relevant form of the house fly enzyme (Murataliev and Feyereisen, 1999). Furthermore, the detailed studies of house fly P450 reductase led to a proposed catalytic cycle where the reduction state of the enzyme does not exceed 2 el., and where an FMN semiquinone, and not an FMN hydroquinone, serves as electron donor to the acceptor P450 or cytochrome c. This "0–2–1–0 cycle" (**Figure 14**) likely represents the general mechanism of P450 reductases, with strong evidence that it operates in P450BM3 and in other P450 reductases as well (Murataliev *et al.*, 2004).

8.3.5.2. Cytochrome *b*5 In contrast to P450 reductase, the role of cytochrome b_5 as a partner in P450-dependent reactions is considerably more complex. There is a single cytochrome b_5 gene in insect genomes, although there are several genes with cyt b_5-like domains. The house fly cytochrome b_5 is a 134-aa protein (Guzov *et al.*, 1996) with

48% sequence identity with the orthologous rat cytochrome b_5. Its N-terminal domain of about 100 residues is the heme-binding domain, which is about 60% identical to that of the vertebrate cytochrome b_5. Its C-terminal portion is a hydrophobic membrane anchor. A probable fatty acid desaturase-cyt b_5 fusion protein has been misidentified as the *Drosophila* cytochrome b_5 (Kula *et al.*, 1995; Kula and Rozek, 2000), but the correct ortholog (CG2140) is 76% identical to house fly cytochrome b_5. The *H. armigera* cytochrome b_5 (Ranasinghe and Hobbs, 1999a) is 127 amino acids in length, and 51% identical to the house fly cytochrome b_5, and the *A. gambiae* cytochrome b_5 is 54% identical (Nikou *et al.*, 2003). The known insect cytochrome b_5 sequences differ at their C-terminal from both the vertebrate microsomal and outer mitochondrial membrane cytochrome b_5 sequences, so that inferences about the subcellular targeting of the insect protein (Wang *et al.*, 2003) would seem premature.

The house fly cytochrome b_5 protein was produced in *E. coli*, purified, and fully characterized (Guzov *et al.*, 1996). Absorption spectroscopy and EPR revealed properties very similar to b_5 cytochromes from vertebrates. NMR spectra indicated that the orientation of the heme in the protein relative to its α,γ *meso* axis is about $1:1$. This means that the protein is present in two forms of approximately equal abundance, which result from two modes of insertion of the non-covalently bound heme in the protein between the two coordinating histidines (face-up and face-down). Expression of the heme-binding domain in *E. coli* revealed that the heme is kinetically trapped in a $1.2:1$ ratio of the two isomers, and that this orientation results from the selective binding of heme by the apoprotein (Wang *et al.*, 2003). The X-ray crystal structure of house fly cytochrome b_5 was solved at 1.55 Å (PDB 2IBJ), revealing a small hydrophobic patch around Met 71 that contributes to the high stability of the holoprotein (Wang *et al.*, 2007).

A redox potential of −26 mV was measured by cyclic voltammetry on a treated gold electrode in the presence of hexamminechromium(III) chloride, and was verified by classical electrochemical titration. Stopped flow spectrophotometry showed that the cytochrome b_5 is reduced by house fly P450 reductase at a high rate (5.5/s) (Guzov *et al.*, 1996) (**Figure 12**).

Cytochrome b_5 can also be reduced by its own reductase, an NADH-dependent FAD flavoprotein, and can therefore provide either NADH- or NADPH-derived electrons to P450 enzymes. NADH-cytochrome b_5 reductase (EC 1.6.2.2) has been studied in *Ceratitis capitata* and *M. domestica* (Megias *et al.*, 1984; Zhang and Scott, 1996a). The N-terminus sequence of the purified house fly enzyme aligns to an internal sequence of the *Drosophila* enzyme (CG5946), indicating that it represents a proteolytically processed form. NADH-cytochrome b_5 reductase and cytochrome b_5 are also known to provide

electrons to other acceptors, such as fatty acid desaturases and elongases.

8.3.5.2.1. Role of cytochrome b5 Depending on the P450 enzyme and on the reaction catalyzed, cytochrome b_5 may be inhibitory, or without effect, or its presence may be obligatory. Cytochrome b_5 can have a quantitative effect on overall reaction rates, and/or a qualitative role on the type of reaction catalyzed and the ratio of the reaction products. The role of cytochrome b_5 may or may not depend on its redox (electron transfer) properties. It can also influence the overall stoichiometry of the P450 reaction, in particular the "coupling rate," − i.e., the utilization and fate of electrons from NADPH relative to monooxygenation. Cytochrome b_5 should therefore be regarded as an important *modulator* of microsomal P450 systems. General reviews of the role of cytochrome b_5 in P450 reactions are available (Porter, 2002; Schenkman and Jansson, 2003), and known examples of this modulator role in insect systems follow.

The relative contribution of NADH in P450 reactions, but more importantly the NADH synergism of NADPH-dependent reactions that is occasionally observed, is probably attributable to cytochrome b_5 as redox partner. Indeed, the K_m of the P450 reductase for NADH is 1000-fold higher than for NADPH, and the V_{max} is 10-fold lower (Murataliev *et al.*, 1999), so the contribution of NADH under normal conditions is probably channeled by NADH-cytochrome b_5 reductase and cytochrome b_5.

An anti-cytochrome b_5 antiserum severely inhibited (up to 90%) methoxycoumarin and ethoxycoumarin O-dealkylation and benzo[a]pyrene hydroxylation, but not methoxyresorufin and ethoxyresorufin O-dealkylation, when assayed in microsomes of the house fly LPR strain (Zhang and Scott, 1994). This antiserum also inhibits cypermethrin 4′-hydroxylation by these CYP6D1-enriched microsomes (Zhang and Scott, 1996b).

House fly cytochrome b_5 stimulates heptachlor epoxidation and steroid hydroxylation when reconstituted with cytochrome P450 reductase, house fly CYP6A1, and phospholipids (Guzov *et al.*, 1996; Murataliev *et al.*, 2008). Stimulation of cyclodiene epoxidation and diazinon metabolism were also observed with *Drosophila* CYP6A2 expressed with the baculovirus system (Dunkov *et al.*, 1997). Cytochrome b_5 is efficiently reduced by P450 reductase (**Figure 12**), but it does not increase the rate of P450 reduction by P450 reductase. Because of its small redox potential, cytochrome b_5 is unlikely to play an important role in delivering the first electron to P450 catalysis, and its stimulatory role probably involves an increased rate of transfer of the second electron. Cytochrome b_5 decreases the apparent K_m for P450 reductase and increases the V_{max} for epoxidation at constant CYP6A1 concentrations (Guzov *et al.*, 1996). The results suggest a role for cytochrome b_5 in the P450 reductase–P450 interactions.

Whereas heptachlor epoxidation by CYP6A1 was increased 2- to 3-fold by the addition of cytochrome b_5, the hydroxylation of testosterone, androstenedione, and progesterone was stimulated 7- to 10-fold. The addition of cytochrome b_5 increased the ratio of 2β-hydroxylation over 15β-hydroxylation of testosterone. This suggests that cytochrome b_5 can have an effect on CYP6A1 conformation, probably altering the interaction of the binding site with either the C-17 hydroxyl (decreased) or the C-3 carbonyl (increased) group. Interestingly, the effect of cytochrome b_5 on hydroxylation regioselectivity was also obtained with apo-b_5 (cytochrome b_5 depleted of heme, and therefore redox incompetent), whereas the effect on turnover number was only much smaller with apo-b_5 (Murataliev et al., 2008). The effect of apo-b_5 is not due to heme transfer from P450 to apo-b_5, and in fact both apo-b_5 and (holo) cytochrome b_5 were shown to stabilize the ferrous–CO complex of CYP6A1, decreasing the rate its conversion to P420 (Murataliev et al., 2008).

Cytochrome b_5 increases the coupling stoichiometry of CYP6A1 catalysis. In the heptachlor epoxidation assay, coupling (NADPH or O_2 used/heptachlor epoxide formed) increased from < 8% to over 25%. This effect is even more pronounced for testosterone, where coupling efficiency in the presence of cytochrome b_5 can reach 84% (Murataliev et al., 2008). The effect of cytochrome b_5 results in a decrease in H_2O_2 production in both assays. The exact site of H_2O_2 production (P450 reductase or CYP6A1) is not known. Cytochrome b_5 also stimulates the CYP6CM1vQ-catalyzed dealkylation of several alkoxycoumarins and resorufin (Karunker et al., 2009), and the metabolism of permethrin and deltamethrin by CYP6M2 (Stevenson et al., 2011).

Coordinated induction and/or overexpression of cytochrome b_5 and P450 genes has been reported (Liu and Scott, 1996; Kasai et al., 1998; Ranasinghe and Hobbs, 1999a, 1999b; Nikou et al., 2003), and this indicates that the effects of cytochrome b_5 seen in vitro are probably significant in vivo as well.

8.3.5.3. Redox partners of mitochondrial P450   The redox partners of mitochondrial P450s are adrenodoxin reductase, an NADPH-dependent FAD flavoprotein, and adrenodoxin, a [2Fe-2S] ferredoxin-type iron sulfur protein. These are named for the two redox partners of mammalian adrenal mitochondrial P450s, and this designation has been liberally bestowed on proteins from animals that do not have adrenals. Insect adrenodoxin reductase and adrenodoxin have not been functionally characterized, but their bovine orthologs are capable of supporting the activity of an insect mitochondrial P450, house fly CYP12A1 (Guzov et al., 1998). The reduction of CYP12A1 is rapid and efficient with bovine adrenodoxin

reductase/adrenodoxin, while under the same conditions house fly microsomal P450 reductase is only marginally effective.

There is one adrenodoxin ortholog in insect genomes. The Drosophila adrenodoxin ortholog (CG1319) was revealed by the complete genome sequence at 64B1 as a 152-aa protein, 61% identical to the bovine protein. EPR spectroscopic evidence for the presence of an adrenodoxin-like protein in fat body mitochondria of Spodoptera littoralis has been presented (Shergill et al., 1995).

Insect genomes also carry a single adrenodoxin reductase ortholog. The Drosophila P-element induced mutant dare1 (for defective in the avoidance of repellents) was found to encode Drosophila adrenodoxin reductase (Freeman et al., 1999). Strong dare mutants undergo developmental arrest, and this phenotype is largely rescued by feeding 20-hydroxyecdysone. Decreasing by half the wild type expression of dare blocks the olfactory response. The gene is expressed at low levels in all tissues of the adult fly, including the brain and the antennae. Highest expression is found in the prothoracic gland portion of the ring gland of third instar larvae, as well as in the nurse cells of adult ovaries. These tissues are known to require mitochondrial P450s (CYP302A1, CYP315A1) for ecdysteroid production. The 55-kDa protein encoded by dare is 42% identical to the human enzyme.

8.4. P450 Functions

Ever since the pioneering work of Agosin and Terriere in the early 1960s on microsomal enzymes that hydroxylate DDT and naphthalene (review in Agosin, 1985), the enzymes now recognized as P450 have been best known for their role in xenobiotic metabolism in insects. They are often denoted as "detoxification enzymes." This designation not only neglects the importance of P450 enzymes in basic physiological processes, as pointed out for drug-metabolizing enzymes in general by Nebert (1991); it also neglects the conceptual difference between metabolism (what the enzyme does to the chemical) and toxicity (what the chemical does to the biological system). As noted before (Feyereisen, 1999) "there are many cases where P450 enzymes act as anything but 'detoxification enzymes'. The easy dichotomy between biosynthetic and detoxification functions of P450s reflects more the teleological tendencies of the observer than the phylogeny or biochemistry of the enzymes".

In most studies, the P450 substrates tested are either endogenous compounds and their analogs, or xenobiotics. Rarely are both types of substrates tested on the same P450, and indeed, the thermodynamic reality of enzyme–substrate interaction makes the quest for a formally complete description of the chemical diversity of substrates of any enzyme illusory. A future understanding of P450 functions perhaps will be based more on their evolutionary trajectories, with some P450s being presently constrained by their proven physiological role (but how long have they played this role?), and some P450s being positively selected for their present adaptive role in detoxification (of plant chemicals in evolutionary time, or of insecticides in historical time). Except for a handful of genes, often represented by 1:1 orthologs between species, our knowledge of insect P450 function is not sufficient to classify individual P450s or CYP families rationally along such evolutionary lines. Therefore, the following description may be found arbitrary.

8.4.1. Metabolism of Signal Molecules

Physiological functions are best known for a series of CYP genes involved in biosynthesis, activation, and catabolism of the molting hormones or ecdysteroids. These genes are among the most conserved CYP genes in insects, although their evolution and the biochemistry of the enzymes they encode are not fully understood. These genes belong to the CYP2 clan (CYP307, CYP306, CYP18) and to the mitochondrial CYP clan (CYP302, CYP314, CYP315) (**Figure 15**).

8.4.1.1. Ecdysteroid biosynthesis

8.4.1.1.1. Ecdysone biosynthesis: biochemistry The hydroxylation of 2-deoxyecdysone to ecdysone in *Locusta migratoria* was observed in the prothoracic glands (PG) as expected, and in the Malpighian tubules, midgut, fat body, and epidermal tissues as well (Kappler *et al.*, 1986). The C-2 hydroxylation was characterized as a mitochondrial P450 in larval Malpighian tubules, and in ovarian

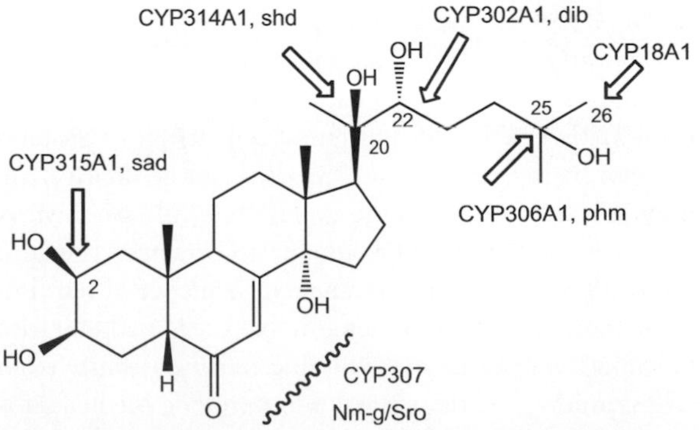

Figure 15 The insect molting hormone 20-hydroxyecdysone with the sites of hydroxylation of its precursors by CYP306A1, CYP315A1, and CYP302A1, activation from ecdysone by CYP314A1 and inactivation by CYP18A1 (by hydroxylation and further oxidation to ecdysonoic acids). The reactions catalyzed by the CYP307 enzymes and by the short chain dehydrogenase/reductase nm-g/sro are not known. See text for details.

follicle cells of vitellogenic females (Kappler *et al.*, 1986), as well as in the prothoracic glands (Kappler *et al.*, 1988). This P450 activity is peculiar in its very low sensitivity to CO inhibition, but virtually stoichiometric incorporation of one atom of molecular oxygen was demonstrated (Kabbouh *et al.*, 1987). The biochemical characterization of this P450 activity is supportive of the idea that the same gene product is responsible for C-2 hydroxylation in the Malphighian tubules and in the PG. Two further hydroxylase activities have been characterized in *L. migratoria* prothoracic glands as typical P450 enzymes (Kappler *et al.*, 1988). With 2,22,25-trideoxyecdysone as substrate for C-25 hydroxylation and 2,22,-dideoxyecdysone as substrate for C-22 hydroxylation, these two activities were traced to the microsomal and mitochondrial fractions respectively (Kappler *et al.*, 1988). The specificity of the three enzymes is suggested by their low K_m (0.5–2.5 µM), but significant competitive inhibition of the C-2- and C-22-hydroxylations by several ecdysteroids indicates that their place and substrate(s) in a grid or in a linear pathway is still conjectural, as noted by Rees (1995). Evidence for the P450 nature of the C-25, C-22, and C-2 hydroxylases and their subcellular localization in *Manduca sexta* PG has been presented (Grieneisen *et al.*, 1993). That study also suggested that the 7,8-dehydrogenation of cholesterol was a microsomal, NADPH-dependent P450 enzyme. However, the protein encoded by the *neverland* gene (*nvd*) first characterized in *B. mori* and *Drosophila* (Yoshiyama *et al.*, 2006) is the best candidate to catalyze this reaction. Neverland is a protein containing both a Rieske [2Fe-2S] domain and a non-heme iron domain. There is, to date, no *biochemical* evidence for the involvement of P450 enzymes in the biosynthetic steps that occur in Dennis Horn's famed "black box" between the 7,8-desaturation and the ultimate hydroxylations.

8.4.1.1.2. Ecdysone biosynthesis: Molecular genetics Namiki *et al.* (2005) identified a gene selectively expressed in *B. mori* prothoracic glands by differential display. This gene, *CYP307A1*, belonging to the CYP2 clan, was most closely related to *Drosophila Cyp307a1*, which corresponds to the Halloween gene *spook* (*spo*). Both *B. mori* and *Drosophila* CYP307A1 cDNAs rescued the *spo* lethality when expressed in transgenic flies under a ubiquitous promoter, indicating that they catalyzed the same reaction. Namiki and colleagues suggested that CYP307A1 might be implicated in the early portion of the ecdysteroid pathway or "black box" (**Figure 15**). However, *Drosophila Cyp307a1* is not expressed in the PG cells of the ring gland, only in the embryonic amnioserosa and in the adult ovary. The conundrum this posed was solved by Ono *et al.* (2006), who assembled a complete sequence for the *Drosophila Cyp307a2* gene from newly released heterochromatic DNA sequence and from the pseudogene *Cyp307a2p* annotated in the initial *Drosophila* genome (Tijet *et al.*,

2001). *Cyp307a2* is expressed in the PG cells of the ring gland, where it therefore "replaces" the missing *Cyp307a1* expression (Ono *et al.*, 2006). The evolutionary history of the CYP307 family is quite complex, and still unfolding with each new genome release (Rewitz *et al.*, 2007; Sztal *et al.*, 2007; Rewitz and Gilbert, 2008). A *CYP307* gene with three exons was duplicated in the common ancestor of insects. This gene, *CYP307B1*, is now found in *Pediculus*, *Tribolium*, honey bee, and mosquitoes, but was lost in the *Nasonia* and *Bombyx* lineages. Its ortholog is also found in *Daphnia pulex*. The duplicated gene, with just two exons, is found in *Pediculus*, *Tribolium*, and mosquitoes, but was lost in the *Apis* lineage. In higher Diptera the *CYP307B1* gene is also lost, and its two-exon paralog (*Cyp307a2*) was duplicated by retrotranposition, giving the intronless *spo* (*Cyp307a1*). The Sophophora subgenus to which *D. melanogaster* belongs therefore carries two genes, *spo* (*Cyp307a1*) and *Cyp307a2*. In the *Drosophila* subgenus, however, a second retrotransposition of *Cyp307a2* occurred, giving *Cyp307a3*, while *spo* was lost. The various *CYP307* genes are probably all redundant in function, but each has a specific temporal and tissue expression pattern, and they exemplify the subfunctionalization model of gene duplication. The microsomal CYP307 enzymes work in concert with the short-chain dehydrogenase/reductase (SDR) enzyme encoded by *non-molting glossy* (in *B. mori*) or *shroud* (a Halloween gene in *Drosophila*) (Niwa *et al.*, 2010). The precise biochemistry catalyzed by this P450 and SDR is still unknown, but is thought to constitute the "black box" between 7-dehydrocholesterol and 2,22,25-trideoxyecdysone (or "ketodiol"). The CYP307 proteins are unique in having a WxxxQ instead of WxxxR motif in the C helix, and lacking the highly conserved threonine in the I helix, suggesting that these P450s have an unusual catalytic site and reaction.

CYP307A1 is rapidly phosphorylated in *M. sexta* prothoracic glands after PTTH treatment (Rewitz *et al.*, 2009), whereas other Halloween gene products (CYP306A1, CYP315A1, CYP302A1) are not. Two phosphorylation sites, T165 or S168, and S438, can be identified on CYP307A1, but it is not clear whether phosphorylation is a specific response to PTTH, or whether the effect observed simply reflects the higher amounts of CYP307A1 protein after peptide treatment. Furthermore, the phosphorylation sites are not conserved in all CYP307 proteins.

The microsomal C-25 hydroxylase was identified as CYP306A1, a CYP2 clan P450 in *B. mori* and in *Drosophila* (**Figure 15**). It is encoded in *Drosophila* by a member of the Halloween group of genes, the *phantom* (*phm*) gene (Niwa *et al.*, 2004; Warren *et al.*, 2004). In *B. mori*, microarray analysis and differential display of the PG transcriptome showed that *CYP306A1* was expressed predominantly (but not solely) in the PG, and at times of high ecdysteroid production. The *Drosophila* ortholog

Cyp306a1 was shown to be truncated by an early stop codon in the *phm* mutant, and the *phm* mutation was rescued in transgenic flies by transformation with either *B. mori* or *Drosophila* wild type CYP306A1. *Drosophila* S2 cells transfected with the cDNA from either species were able to hydroxylate 2,22,25-trideoxyecdysone (ketodiol) to 2,22-dideoxyecdysone (Niwa *et al.*, 2004; Warren *et al.*, 2004). The C-25 hydroxylase has a relatively loose substrate specificity, as 3β-hydroxy-5β-cholest-7-en-6-one was also hydroxylated at the C-25 position (Niwa *et al.*, 2004). The 3α,5α- and 3β,5α-isomers of the ketodiol were also metabolized (Warren *et al.*, 2004).

CYP302A1, identified as the C-22 hydroxylase, is encoded by the Halloween gene called *disembodied* (*dib*) (Chavez *et al.*, 2000; Warren *et al.*, 2002). Similarly, the *shadow* (*sad*) gene encodes CYP315A1, the C-2 hydroxylase (Warren *et al.*, 2002). Both genes are mitochondrial CYP clan P450s. The *dib* gene was identified by classical molecular genetics approaches starting from the genetic locus, ultimately leading to a mitochondrial P450 sequence (Chavez *et al.*, 2000). The identification of the *sad* gene exploited the match between the genetic map and the sequence from the *Drosophila* genome project (Warren *et al.*, 2002). Mutations leading to stop codons in the sequence confirmed the identification of the genes. Both *dib* and *sad* genes are expressed in embryos, as well as in the larval ring gland and follicle cells of adult ovaries. Transient expression of *dib* and *sad* in *Drosophila* S2 cells, coupled with the use of radiolabeled substrates 2,22-dideoxyecdysone and 2-deoxyecdysone, respectively, was used to identify the reactions catalyzed (Warren *et al.*, 2002). The *B. mori* ortholog CYP302A1, produced in S2 cells, was also shown to convert 2,22-dideoxyecdysone to 2-deoxyedysone. The *CYP302A1* gene was rapidly induced by prothoracicotropic hormone (PTTH) *in vitro* (Niwa *et al.*, 2005). As opposed to *CYP307*, the *CYP302A1*, *306A1*, and *315A1* genes have a single ortholog in all species studied to date.

A common feature of the initial identification of these enzymes is the relative paucity of biochemical characterization. Their reported substrate specificity and regioselectivity of hydroxylations is convincing but qualitative, and their relative position in the overall biosynthesis of ecdysteroids is often deduced from partial evidence. The commonly depicted linear, unbranched, sequence of reactions from cholesterol should therefore be considered tentative. Fortunately, modern metabolomic techniques applied on ecdysteroidogenic tissues may soon provide further information on this point.

8.4.1.2. Ecdysteroid metabolism

8.4.1.2.1. Ecdysone 20-monooxygenase activity (E20MO)
The conversion of ecdysone to 20-hydroxyecdysone does not occur in the prothoracic glands, but

does occur in many peripheral tissues, such as the fat body, midgut, and Malpighian tubules. The P450 nature of the enzyme catalyzing the 20-hydroxylation of ecdysone was well established by 1985 (Smith, 1985), following the initial reports of 1977 (Bollenbacher *et al.*, 1977; Feyereisen, 1977; Johnson and Rees, 1977). An NADPH- and O$_2$-dependent enzyme system, inhibited by typical pharmacological P450 inhibitors, was studied in several insect species. The evidence for P450 involvement was strengthened by the light-sensitive carbon monoxide inhibition observed in some of the most thorough studies (Feyereisen and Durst, 1978; Smith *et al.*, 1979; Greenwood and Rees, 1984). The agreement on the P450 nature of the reaction was accompanied by a lack of consensus on the subcellular localization of E20MO. Some studies showed a microsomal activity, other studies showed a mitochondrial activity, and yet other studies indicated the presence of both microsomal and mitochondrial activities in the same tissue (Smith, 1985; Lafont *et al.* 2005). For instance, it has been reported to be mostly microsomal in imaginal discs of *Pieris brassicae* (Blais and Lafont, 1986), and in *Gryllus bimaculatus* midgut (Liebrich and Hoffmann, 1991). In the midgut of *Diploptera punctata* (Halliday *et al.*, 1986) and in embryos of *B. mori* (Horike and Sonobe, 1999) the activity is essentially microsomal, and can be inhibited by antibodies to insect P450 reductases. This is clear evidence that the enzyme derives its reducing equivalents from the usual microsomal redox partner in those cases.

In *Spodoptera littoralis* fat body, E20MO activity is predominantly mitochondrial, with a small amount of microsomal activity (Hoggard and Rees, 1988). The mitochondrial E20MO activity was reportedly inhibited by antibodies to vertebrate P450scc (CYP11A), P45011β (CYP11B), adrenodoxin, and adrenodoxin reductase (Chen *et al.*, 1994), despite the considerable sequence divergence predicted between the vertebrate and insect proteins. The immunodetection of polypeptides significantly larger than predicted (e.g., P450 at 82 kDa and adrenodoxin at 73 kDa) was also a surprising feature of that study. In whole-body homogenates of third-instar *Drosophila*, and in the midgut of larval *S. frugiperda*, the E20MO activity is distributed 1:3 between mitochondrial and microsomal fractions (Smith, 1985; Yu, 1995), and it is also distributed in both fractions in larvae of the house fly and of the flesh fly *Neobellieria bullata* (Darvas *et al.*, 1993). Weirich *et al.* (1996) compared the apparent K_m, and specific activities of the mitochondrial and microsomal E20MO activities of *M. sexta* larval midgut. They concluded that at physiological ecdysone titers, despite a higher specific activity, the mitochondrial E20MO would contribute less than one-eighth the activity of the microsomal form.

The existence of several genes encoding P450s with E20MO activity could account for the dual localization

and complex regulation of the enzyme's activity. On the other hand, alternative splicing or post-translational modifications of a single gene product could also lead to the microsomal and mitochondrial forms, and this seems more likely, given the unusual N-terminal sequences of CYP314A1 proteins (see below). Interestingly, both microsomal and mitochondrial activities of the *Spodoptera littoralis* fat body E20MO were reported to be reversibly activated by phosphorylation (Hoggard and Rees, 1988; Hoggard *et al.*, 1989). Cases of post-translational modifications of P450 enzymes by reversible phosphorylation are not very common, and this observation would suggest that the *S. littoralis* fat body E20MOs are products of the same gene.

Petryk and colleagues (2003) identified *Drosophila* CYP314A1 as the product of the *shade* (*shd*) gene, a member of the Halloween group of developmental mutants (**Figure 15**). Expression of CYP314A1 in *Drosophila* S2 cells enabled the NADPH-dependent hydroxylation of ecdysone to 20-hydroxyecdysone by cell homogenates. RNA *in situ* hybridization shows that the *shd* gene is not expressed in the ring glands, but expression is seen in the gut, fat body, and Malpighian tubules. In embryos, *shd* is expressed primarily in epidermal cells by the time of germband extension. Embryonic lethality of *shd* mutants indicates that *Cyp314a1* encodes the only significant E20MO activity in *Drosophila* at that stage. A CYP314A1 protein modified at the C-terminus by the addition of three copies of the hemaglutinin epitope was targeted to mitochondria of S2 cells (Petryk *et al.*, 2003). *Cyp314a1* has single clear orthologs in all insect genomes sequenced to date, with the possible exception of the pea aphid, which may carry a *CYP314A1* duplicate. The *M. sexta* CYP314A1 ortholog was expressed functionally in S2 cells, and its tissue and developmental expression suggests it is the only E20MO encoding gene (Rewitz *et al.*, 2006). The CYP314A1 of *B. mori* was also functionally expressed in the baculovirus system, and shown to hydroxylate ecdysone; the gene is not expressed in diapause-arrested embryos (Maeda *et al.*, 2008).

The sequences of the predicted CYP314A1 proteins are unusual. They are clearly members of the mitochondrial CYP clan, and have several intron positions in common with other mitochondrial P450 genes (Ranson *et al.*, 2002a; Rewitz *et al.*, 2007). However, they have usually just one or two of the three positively charged residues thought to confer high affinity to adrenodoxin (section 8.3.1). Their exact N-terminal sequence is also variable and somewhat unclear in the absence of EST sequences or proteomic data to confirm the annotation prediction of the N-terminus.

Regulation of E20MO activity, developmental changes, induction, and inhibition are discussed by Lafont *et al.* (2005) and Rewitz *et al.* (2006).

8.4.1.2.2. Ecdysteroid catabolism The C-26 hydroxylation of 20-hydroxyecdysone has been characterized as a typical microsomal P450 activity in an epithelial cell line of *Chironomus tentans* (Kayser *et al.*, 1997). However, a dual localization (microsomal and mitochondrial) was reported for ecdysteroid 26-hydroxylase in *M. sexta* midgut (Williams *et al.*, 1997, 2000b). The reaction has a low K_m for 20-hydroxyecdysone ($\sim 1\,\mu M$). The *C. tentans* cell line metabolizes 20,26-dihydroxyecdysone further, to two less polar metabolites produced in a constant $3:1$ ratio. This metabolism is NADPH-dependent, and inhibited by azole compounds. The two metabolites are diastereomers of a cyclic hemiacetal formed (non-enzymatically) by the reaction of the C-22 hydroxyl group with a C-26 aldehyde (Kayser *et al.*, 2002). The C-26 aldehyde is a presumed intermediate in the conversion of C-26 ecdysteroids to C-26 ecdysonoic acids, a common inactivation product of ecdysteroids (Rees, 1995; Lafont *et al.*, 2005). It is now clearly established that the P450 conversion of the C-26 hydroxyl to the C-26 carboxylic acid is carried out by the same enzyme that initially hydroxylates 20-hydroxyecdysone. This enzyme was shown to be encoded by the *Cyp18a1* gene in *Drosophila* (Guittard *et al.*, 2011), a gene initially designated *Eig17-1* (Hurban and Thummel, 1993) because it is an *e*cdysone-*i*nducible *g*ene. The expression of *Cyp18a1* clearly pulses closely after each ecdysteroid peak (Bassett *et al.*, 1997) (**Figure 16**). *Cyp18a1* has a single ortholog in most species. It is absent in *A. gambiae* (Claudianos *et al.*, 2006), but has a duplicated paralog in *B. mori*. Guittard *et al.* (2011) showed conversion of 20-hydroxyecdysone and 20,26-dihydroxyecdysone to 20-hydroxyecdysonoic acid by S2 cells transfected by CYP18A1. *Cyp18a1* null alleles or RNAi knockdown of CYP18A1, as well as ectopic overexpression of *Cyp18a1*, causes severe phenotypes and lethality, thus highlighting the important regulatory role of the enzyme in ecdysteroid inactivation (Rewitz *et al.*, 2010; Guittard *et al.*, 2011). *CYP18A1* is expressed in the PG of Lepidoptera (Davies *et al.*, 2006) and of *Drosophila* (Guittard *et al.*, 2011), but the function of this ecdysteroid-inactivating P450 in ecdysteroidogenic cells is currently unknown.

Side-chain cleavage of ecdysteroids has been reported as an inactivation route, but despite the analogy to the reaction catalyzed by vertebrate P450scc (CYP11A) – i.e., the C–C bond cleavage at a vicinal C-20,C-22 diol – there is currently no information on the enzymology of this reaction. P450scc is a unique enzyme in that it catalyzes not just the C–C bond cleavage, but also the two preceding hydroxylations of cholesterol at C-20 and C-22 (in that order).

8.4.1.3. P450 and juvenile hormone metabolism

The landmark chemical feature of juvenile hormones is the presence of the epoxide group, and it is well established that epoxidation of the JH precursors in the corpora allata (CA) is catalyzed by a P450 enzyme. The biochemical evidence was presented in two studies on the allatal enzyme, one with *Blaberus giganteus* CA homogenates (Hammock, 1975) and the other with *L. migratoria* CA (Feyereisen *et al.*, 1981), which established its microsomal nature. A number of compounds that took advantage of the P450 nature of the epoxidase have been tested as inhibitors of juvenile hormone biosynthesis (Hammock and Mumby, 1978; Brooks *et al.*, 1985; Pratt *et al.*, 1990; Unnithan *et al.*, 1995). The allatal epoxidase was viewed as an attractive new target for "biorational" insecticides, because it is an insect-specific target, and because its inhibition should lead to desirable effects – precocious metamorphosis and adult sterility akin to the effects of precocenes (Bowers *et al.*, 1976). A serious pursuit of this goal would only be possible with a molecular characterization of the epoxidase and its heterologous expression that would permit screening at a scale not possible by the laborious dissection of the glands. Photoaffinity labeling was first developed as a technique that could facilitate purification and subsequent cloning of the epoxidase. Bifunctional compounds, with a substituted imidazole to coordinate the heme iron and a diazirine or a benzophenone group to label the substrate binding site, were synthesized and tested (Andersen *et al.*, 1995). These compounds were potent inhibitors of methyl farnesoate epoxidation to JH III in the CA of *Diploptera punctata*, and selectively labeled a 55-kDa protein.

8.4.1.3.1. CYP15: The epoxidase

Eventual cloning of the epoxidase was, however, achieved by a less subtle route: the sequencing of ESTs from *D. punctata* CA. Sequencing of > 900 ESTs from the CA of vitellogenic females yielded 3 ESTs matching a P450 sequence by BLAST analysis. A full-length cDNA of this P450 was expressed in *E.coli* (Helvig *et al.*, 2004a). This P450, named CYP15A1, has

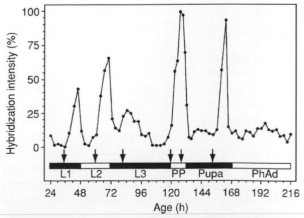

Figure 16 Developmental pattern of *Cyp18a1* expression in *Drosophila*. Peaks of *Cyp18a1* expression follow each surge in endogenous molting hormone level. Reproduced from Bassett *et al.* (1997), with permission.

a high affinity for methyl farnesoate, showing a type I spectrum with a K_s of 6 μM. The enzyme, when reconstituted with fly P450 reductase, catalyzed the NADPH-dependent epoxidation of 2E,6E-methyl farnesoate to JH III. The epoxidation is highly stereoselective (98:2) to the natural 10R enantiomer over its diastereomer. The enzyme also has a high substrate specificity, epoxidizing the 2E,6E isomer preferentially over the 2Z,6E isomer, and accepting no other substrate tested, including farnesoic acid. The rank order of inhibition of CYP15A1 activity by substituted imidazoles is identical to that of JH biosynthesis by isolated CA (Helvig et al., 2004a). CYP15A1 is expressed selectively in the CA of adult D. punctata, Schistocerca gregaria, and Blattella germanica (Helvig et al., 2004a; Maestro et al., 2010; Marchal et al., 2011), as well as in embryos of the latter species. RNAi silencing of the desert locust S. gregaria ortholog of CYP15A1 causes a decrease of spontaneous and farnesoic acid-stimulated JH III synthesis, as well as an accumulation of methyl farnesoate accumulation in the glands (Marchal et al., 2011).

A single CYP15 ortholog is found in most available insect genomes, with the notable exception of the Drosophila genus. When compared to other insects, higher Diptera (Cyclorrhapha) characteristically muddle the whole picture of JH endocrinology, in terms of mode of action as well as biosynthesis. A JH III-bisepoxide has been identified as a product of the ring glands, which include the CA (Jones and Jones, 2007). Epoxidation of a farnesoid molecule can be achieved by other enzymes (see below), so it is conceivable that CYP15 was lost in higher Diptera and this loss compensated by the utilization of non-epoxidized farnesoids, or the recruitment within or outside the CA of other P450 enzymes, or both. In their survey of P450 expression by in situ hybridization in Drosophila, Chung et al. (2009) discovered that Cyp6g2 is specifically expressed in the CA portion of the ring glands. The catalytic activity of CYP6G2 is currently unknown, but one hypothesis would be that it has replaced CYP15 as epoxidase. Its suppression by RNAi causes larval/pupal lethality (Chung et al., 2009). CYP6G2 does not appear to be a highly specific enzyme, however, because its transgenic, ectopic overexpression results in nitenpyram resistance (Daborn et al., 2007). B. mori CYP15C1 catalyzes farnesoic acid epoxidation (T. Shinoda, personal communication) as expected from Lepidoptera, where epoxidation takes place before esterification by JHAMT (juvenile hormone acid methyl transferase). The CYP15 enzymes have therefore evolved a lineage-specific substrate selectivity. The case of Hemiptera may be revealing as well. A "skipped bisepoxide" of methyl farnesoate was identified as a product of the CA of the pentatomid bug Plautia stali (Kotaki et al., 2009). The enzyme(s) catalyzing the stereoselective 10R and 2R,3S epoxidations are still unknown.

8.4.1.3.2. CYP4C7: A cockroach terpenoid ω-hydroxylase
Expression of the CYP15A1 gene is high when JH synthetic levels are high, but another P450 gene, CYP4C7, also selectively expressed in the CA of D. punctata (Sutherland et al., 1998), has an expression pattern that mirrors the pattern of JH synthesis. The recombinant CYP4C7 enzyme produced in E. coli metabolized a variety of sesquiterpenoids, but not mono- or diterpenes. In addition to metabolizing JH precursors, farnesol, farnesal, farnesoic acid, and methyl farnesoate, it also metabolized JH III to a major metabolite identified as (10E)-12-hydroxy-JH III (Sutherland et al., 1998). Although this ω-hydroxylated JH III has not been identified as a product of the CA of D. punctata, L. migratoria CA are known to produce several hydroxylated JHs in the radiochemical assay in vitro (Darrouzet et al., 1997; Mauchamp et al., 1999). These hydroxy-JHs, 8'-OH-, 12'-OH- (Darrouzet et al., 1997) and 4'-OH-JH III (Mauchamp et al., 1999), may be major products of the CA after JH III itself, and their role is unknown. Their presence suggests that locust CA have a P450 homologous to cockroach CYP4C7 that has a lower regioselectivity. Indeed, the hydroxy-JHs can be synthesized by locust CA from JH III, and this hydroxylation is inhibited by CO and piperonyl butoxide (Couillaud et al., 1996).

The tight physiological regulation of CYP4C7 expression in adult female D. punctata (Sutherland et al., 2000) indicates that the terpenoid ω-hydroxylase has an important function to play at the end of vitellogenesis, and at the time of impending chorionation and ovulation. It was hypothesized (Sutherland et al., 1998, 2000) that this hydroxylation was a first step in the inactivation of the very large amounts of JH and JH precursors present in the CA of this species after the peak of JH synthesis.

8.4.1.3.3. Metabolism of farnesoids
The study of JH metabolism has long been dominated by esterases and epoxide hydrolases, but early work on insecticide-resistant strains of the house fly revealed oxidative metabolism as well (for review, see Hammock, 1985). Evidence that house fly CYP6A1 efficiently metabolizes sesquiterpenoids, including JH to its 6,7-epoxide, confirms these early studies (Andersen et al., 1997). Hydroxylation and epoxidation of JHs are thus confirmed P450-mediated metabolic pathways for these hormones.

A comparison of P450 enzymes that metabolize methyl farnesoate (**Table 4**) shows their catalytic versatility. One is extremely substrate-specific (CYP15A1), another is not (CYP6A1). Three of them are epoxidases, one is a hydroxylase on this substrate. CYP6A1 also has activity as a hydroxylase – but not on this substrate. One is a stereoselective epoxidase (CYP15A1), another lacks product enantioselectivity (CYP9E1). A chimeric CYP102A1/4C7 mutant makes both 12-hydroxy-MF and JH III from methyl farnesoate (Chen et al., 2010).

Table 4 Specificity of P450 Enzymes Towards Methyl Farnesoate (MF)

P450	Substrate	Reaction	Product Formed
CYP15A1	2E, 6E-MF	Epoxidation	10R-epoxide
CYP4C7	Sesquiterpenoids	ω-hydroxylation	12E-OH
CYP9E1	2Z>2E-MF (1.5:1)	Epoxidation	10-epoxide 10R:10S (1:1)
CYP6A1	MF or JH all 4 isomers	Epoxidation	6 or 10- epoxide 10S:10R(3:1)
CYP102A1/4C7 chimera	MF	Epoxidation and ω-hydroxylation	10,11-epoxide (30%)+12E-OH (70%)

Data from Helvig *et al.* (2004a) and Chen *et al.* (2010)

8.4.1.3.4. CYP303A1: Morphogen metabolism?

CYP303A1, a member of the CYP2 clan to which CYP15 belongs, was shown to be encoded by the *Drosophila nompH* gene (Willingham and Keil, 2004). One of several *no mechanoreceptor potentials* (*nomp*) mutants, *nompH* was mapped to a region (35F6-12) encompassing the *Cyp303a1* gene. The mutation that impairs the development and structure of external sensory organs was rescued by the wild type gene, and shown in two alleles to be caused by non-conservative substitutions G454R and G311E. Both mutations are expected to severely impair or abolish P450 activity. *Cyp303a1* is expressed selectively at the apical region of the socket cells of sensory bristles, and its role is predicted to be developmental. Interestingly, CYP303A1 is one of few highly conserved P450s found in all insect species sequenced to date, but not in *Daphnia pulex*. It is tempting to hypothesize that it metabolizes a small signal molecule.

8.4.1.4. Biosynthesis of long-chain hydrocarbons

The cuticle of insects can be characterized by the blend of hydrocarbons that are deposited on the epicuticle to provide waterproofing. This blend can serve as a subtle tool in chemical taxonomy (Lockey, 1991). In some insects, cuticular or exocrine alkanes and alkenes can serve as allomones, or even pheromones. To cite but one example, the Dufour gland secretions of the leaf-cutting ant *Atta laevigata* are deposited on foraging trails. This trail pheromone comprises n-heptadecane, (*Z*)-9-nonadecene, 8,11-nonadecadiene, and (*Z*)-9-tricosene (Salzemann *et al.*, 1992). Despite their apparently simple structure, the biosynthesis of these hydrocarbons is complex, and the enzymes involved are still under intense study. Schematically (**Figure 17**), long chain (18–28 carbons) fatty acid CoA esters are first reduced to their aldehyde by an acyl-CoA reductase, and they are subsequently shortened to form C-1 hydrocarbons (Reed *et al.*, 1994). The conversion of (*Z*)-tetracosenoyl-CoA (from a C24 : 1 fatty acid) to (*Z*)-9-tricosene (a C23 : 1 alkene) has been characterized in house fly epidermal microsomes. The role of the aldehyde tetracosenal as an intermediate is evidenced by trapping experiments with hydroxylamine. The next step is NADPH- and O_2-dependent, and is truly an oxidative decarbonylation reaction, which releases the terminal carbon as CO_2 and not as CO (Reed *et al.*, 1994).

Inhibition by carbon monoxide and anti-P450 reductase antibodies are strongly indicative of a P450 reaction, and this most peculiar enzyme was called P450hyd (Reed *et al.*, 1994). The reaction does not involve a terminal desaturation, because both C-2 protons of the aldehyde are retained in the hydrocarbon product, and the aldehydic proton is transfered to the product (Reed *et al.*, 1995). This type of oxidative decarbonylation reaction is certainly universal in insects, and a mechanistically identical conversion of octadecanal to heptadecane has been documented in flies, cockroaches, crickets, and termites (Mpuru *et al.*, 1996).

The oenocytes are a recognized site of hydrocarbon biosynthesis (Wigglesworth, 1970; Fan *et al.*, 2003), and are known as a site of very high expression of *Cyp4g1* (Gutierrez *et al.*, 2007; Chung *et al.*, 2009) and P450 reductase (Lycett *et al.*, 2006). Selective RNAi suppression of *Cyp4g1* or P450 reductase expression in the oenocytes of *Drosophila* causes massive lethality at eclosion by dessication, and causes a massive decrease in hydrocarbon content of the epicuticle. The oxidative decarbonylation reaction is catalyzed by recombinant CYP4G1, identifying CYP4G1 and its orthologs, present in all insect species but not in *Daphnia*, as P450hyd (unpublished results).

8.4.1.5. Pheromone metabolism

In female house flies the (*Z*)-9-tricosene produced by P450hyd is a major component of the sex pheromone, being responsible for inducing the courtship ritual and the males' striking activity. Additional components of the pheromone are the sex recognition factors (*Z*)-9,10-epoxytricosane and (*Z*)-14-tricosen-10-one. These compounds are obviously derived from (*Z*)-9-tricosene by oxidative metabolism (Blomquist *et al.*, 1984). A microsomal P450 was shown to oxidize the alkene from either side at a distance of 9/10 carbons in-chain (Ahmad *et al.*, 1987), but the structural requirements of the enzyme for its substrate are otherwise strict (Latli and Prestwich, 1991). This P450 activity is found in various tissues of the male and female, including the epidermis and fat body, but the highest specific activity is found in male antennae (Ahmad *et al.*, 1987). The C23 epoxide and ketone are absent internally in females, but accumulate on the surface of the cuticle (Mpuru *et al.*, 2001), suggesting the localization of this P450 activity in epidermal cells of female flies.

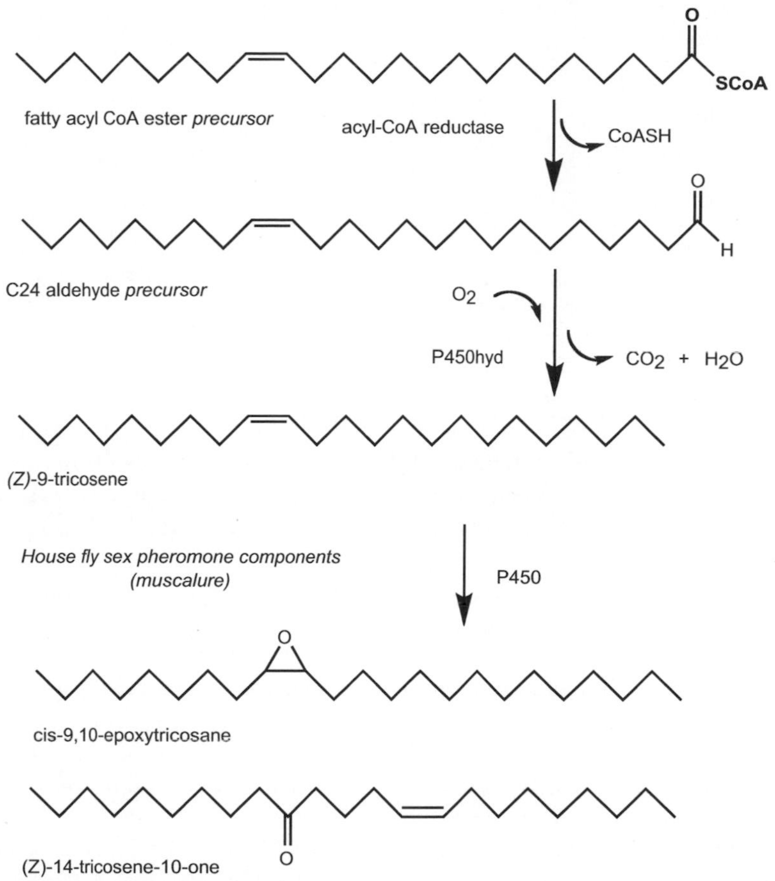

fatty acyl CoA ester *precursor*

acyl-CoA reductase

CoASH

C24 aldehyde *precursor*

O_2

P450hyd

CO_2 + H_2O

(*Z*)-9-tricosene

House fly sex pheromone components (muscalure)

P450

cis-9,10-epoxytricosane

(*Z*)-14-tricosene-10-one

Figure 17 Biosynthesis of (*Z*)-9-tricosene and components of the house fly sex pheromone. P450hyd acts as a decarbonylase to produce (*Z*)-9-tricosene from an aldehyde precursor. Another P450 attacks (*Z*)-9-tricosene from either side to give the two components of muscalure.

More generally, P450 enzymes are probably involved in the biosynthesis of many insect pheromones and allomones – for example, epoxides of polyunsaturated hydrocarbons in arctiid moths, disparlure of the gypsy moth (Brattsten, 1979a; Jurenka *et al.*, 2003), or monoterpenes in bark beetles (Brattsten, 1979a, White *et al.*, 1979). The colonization of pine trees by bark beetles involves the detoxification of the tree resin terpenoids and the biosynthesis of aggregation pheromones that allow a mass attack to overcome the tree's defenses. P450 enzymes participate in both processes, and, because the aggregation pheromone components are monoterpenoids, it is possible that pheromone biosynthetic P450s in bark beetles evolved from resin detoxifying P450s. Elements to evaluate this intriguing hypothesis are now accumulating. In the pine engraver beetle, *Ips pini*, a P450 induced in males by feeding on host phloem and expressed in the (pheromone producing) midgut was shown to hydroxylate myrcene to ipsdienol, a major component of the aggregation pheromone (Sandstrom *et al.*, 2006). This P450, CYP9T2, is highly stereoselective, producing 81% 4*R*-(–)-ipsdienol (**Figure 18**). A closely related P450, CYP9T1, from the pinyon ips, *Ips confusus*, has a similar pattern of expression,

CYP9T1 85%
Ips confusus

CYP9T2 81%
Ips pini

myrcene

(4R)-(–)-ipsdienol

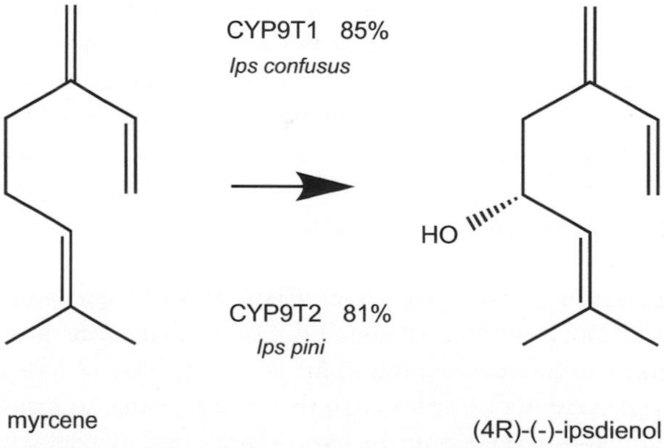

Figure 18 Hydroxylation of myrcene to ipsdienol by bark beetle P450 enzymes. The percentage enantioselectivity for the major product is given for each P450.

and was also shown to convert myrcene to ipsdienol with an 85% product enantioselectivity for 4*R*-(–)-ipsdienol (Sandstrom *et al.*, 2008). Several P450s are induced by phloem feeding in male *Ips paraconfusus*, most dramatically CYP9T1, but also CYP4AY2 and CYP4BG1 (Huber *et al.*, 2007), suggesting a coordinated detoxification and pheromone synthesis response. However, most pheromone

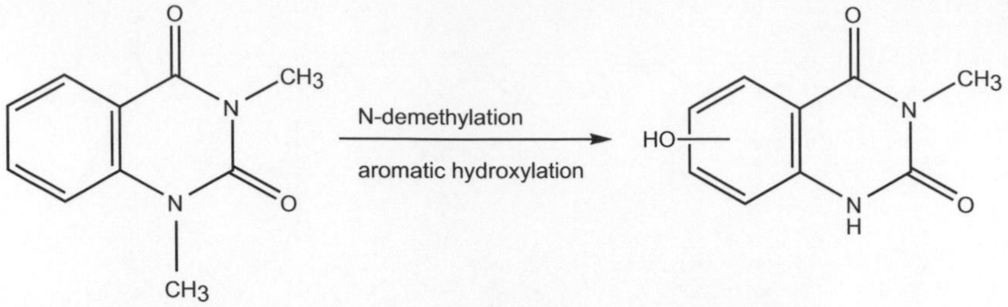

Figure 19 P450-dependent N-demethylation and ring hydroxylation of the *Phyllopertha diversa* sex pheromone by antennal microsomes of male scarab beetles. See text and Wojtasek and Leal (1999) for details.

components are biosynthesized *de novo* rather than from the tree's monoterpenes, and the enantiomeric composition of the aggregation pheromones is not determined by the CYP9T1 or T2 enzymes (reviewed in Blomquist *et al.*, 2010). In the mountain pine beetle *Dendroctonus ponderosae*, epoxidation of 6(*Z*)-nonen-2-one to the precursor of the pheromone component *exo*-brevicomin is also a probable P450 reaction, and *CYP6CR1* is a candidate gene for this epoxidase (Aw *et al.*, 2010).

In the honey bee, caste-specific ω and ω-1 hydroxylations of fatty acids to mandibular pheromones are chemically typical of P450 reactions (Plettner *et al.*, 1998). These enzymes involved in pheromone biosynthesis are probably exquisitely specific, but they have not been identified yet, and are unlikely to be the candidates proposed (Malka *et al.*, 2009).

Evidence for pheromone catabolism by P450 enzymes is also accumulating. As shown for the metabolism of (*Z*)-9-tricosene in the house fly, a distinction between biosynthesis and catabolism can be purely semantic in the case of a biogenetic succession of chemicals that have different signaling functions. Compound B, which is a metabolite of compound A, may have less or none of compound A's activity, but may have its own specific biological activity. Nonetheless, pheromones, as signal molecules, need to be metabolically inactivated, and this catabolism may need to occur in the antennae themselves.

Several P450 mRNAs have been identified in insect antennae. In *Drosophila*, a partial P450 cDNA, along with a UDP-glycosyltransferase and a short chain dehydrogenase/reductase, were found by Northern blot analysis to be preferentially expressed in the third antennal segments, with lower expression in legs (Wang *et al.*, 1999). This P450 cDNA corresponds to *Cyp6w1*, for which many ESTs have been identified in a head cDNA library. In *Mamestra brassicae*, four P450 cDNAs and P450 reductase were cloned from antennae: CYP4G20, CYP4L4, CYP4S4, and CYP9A13 (Maibeche-Coisne *et al.*, 2002, 2005). These genes are strongly expressed in the sensilla trichodea, as shown by *in situ* hybridization. Whereas *CYP4S4* expression is restricted to the antennae, the expression of the other genes is more widespread. Five P450 ESTs representing three P450 genes were found

in a small EST project from male *M. sexta* antennae (Robertson *et al.*, 1999), further evidence that antennae may harbor several P450 enzymes. *CYP341A2* was preferentially expressed in the tarsal chemosensilla of *Papilio xuthus* (Ono *et al.*, 2005). High levels of P450 reductase expression have been noted in *Drosophila* antennae (Hovemann *et al.*, 1997), and adrenodoxin reductase is also expressed in antennae (Freeman *et al.*, 1999). Chemosensory P450s are therefore an active enzyme system, complete with their redox partners. The physiological role of these P450s is still formally unknown, but the presence of other enzymes generally associated with detoxification processes in chemosensory organs (Robertson *et al.*, 1999; Wang *et al.*, 1999; Ono *et al.*, 2005) indeed suggests that P450 enzymes may serve as *odorant-degrading enzymes* (ODE) in insects, along with esterases, aldehyde oxidases, glucosyl-transferases, etc.

Functional evidence for P450 enzymes as ODEs is still scant, apart from one pioneering study. Antennal microsomes of the pale brown chafer *Phyllopertha diversa* metabolize the alkaloid sex pheromone by a P450 enzyme (Wojtasek and Leal, 1999). This enzyme is specifically produced in male antennae, and its activity is not detected in other scarab beetles or lepidopteran species (**Figure 19**). Three P450 cDNAs were cloned from male antennae, CYP4AW1, CYP4AW2, and CYP6AT1, with CYP4AW1 showing antennal-specific expression (Maibeche-Coisne *et al.*, 2004). Electrophysiological recordings of the olfactory receptor neurons for the pheromone revealed that treatment by infusion of the P450 inhibitor metyrapone caused anosmia of the neurons, strongly suggesting that metabolic clearance of the pheromone had been impaired (Maibeche-Coisne *et al.*, 2004). Although a model for the extracellular (sensillar lymph side) localization of the pheromone-degrading P450 was presented, there is no kinetic or cytological evidence for this unusual but not impossible topology in the beetle antenna.

8.4.1.6. Metabolism of fatty acids and lipids

P450-catalyzed fatty acid ω- and ω-1 hydroxylations have been reported in insects (Feyereisen, 2005). Recombinant CYP4C7, CYP6A1, and CYP12A1 lack lauric acid hydroxylase activity (Andersen *et al.*, 1997; Guzov *et al.*,

1998; Sutherland *et al.*, 1998), but CYP6A8 of *Drosophila* was shown to catalyze laurate ω-1 hydroxylation (Helvig *et al.*, 2004b), indicating that fatty acid hydroxylase activity is neither restricted to CYP4 clan P450s, nor easily predictable. Unfortunately, fatty acid hydroxylase activity has been assigned without evidence to several insect CYP4 family P450s in databases and the literature.

Although the role of vertebrate P450 enzymes in the metabolism of arachidonic acid and eicosanoids is well established (Capdevila *et al.*, 2002), there is currently no indication for a similar function of P450 in insects. Plant-derived tocopherols are selectively metabolized in *Drosophila* by ω-hydroxylation, resulting in the accumulation of α-tocopherol (vitamin E) (Parker *et al.*, 2005).

There is little biochemical information on the enzymes responsible for the dealkylation of phytosterols to cholesterol. The reactions involved (desaturation, epoxidation, C–C bond cleavage) have been elegantly described in terms of their chemistry (Ikezawa *et al.*, 1993). They are well within the catalytic competence of P450 enzymes, and, whether these dealkylation reactions are catalyzed by P450 enzymes or not, this remains a challenging area of research in view of their central role in the nutritional physiology of phytophagous species.

Fatty acid–amino acid conjugates (FACs) are elicitors of plant volatile production found in the oral secretions of caterpillars, and may play a role in nitrogen assimilation in the gut. Some FACs in macrolepidoptera are hydroxylated (Yoshinaga *et al.*, 2010). The biosynthesis of volicitin from the plant-derived fatty acid linolenic acid involves C-17 hydroxylation and glutamine conjugation (Pare *et al.*, 1998). This hydroxylation may be catalyzed by a P450 enzyme (Ishikawa *et al.*, 2009).

8.4.1.7. Defensive compounds The tremendous variety of chemicals used for defense of insects against predation is well documented, but the enzymes involved in their *de novo* synthesis or in their transformation from ingested precursors are less studied. It is very likely that P450 enzymes may have found there a fertile ground for their chemical prowess. Just two examples are described.

The cyanogenic glycosides found as defensive compounds in many insect species (Nahrstedt, 1988) are derived from either sequestration of the plant compounds, or *de novo* biosynthesis. For instance, the cyanogenic glycosides linamarin and lotaustralin, found in *Heliconius* butterflies and *Zygaena* moths, are clearly derived from the amino acids valine and isoleucine. Their biosynthetic pathway in *Zygaena filipendula* was recently shown to be an extraordinary example of convergent evolution with the pathway found in plants. In plants, just two multifunctional P450 enzymes are sufficient to convert the amino acid by N-hydroxylations and further metabolism to the aldoxime, and then by metabolism to the nitrile, and final C-hydroxylation to the hydroxynitrile substrate

of a conjugating UDP-glucosyl transferase (**Figure 20**). For dhurrin biosynthesis from tyrosine in sorghum, the two multifunctional P450s are CYP79A1 and CYP71E1 (Kahn *et al.*, 1997). The plant pathway thus "channels" the substrates through the two P450s with little escape of the aldoxime intermediate. In insects, the aldoximes (17–22%) and nitriles (25–28%) are very efficiently incorporated *in vivo* (Davis and Nahrstedt, 1987; Holzkamp and Nahrstedt, 1994). In *Z. filipendula*, all these reactions are catalyzed by two P450 enzymes and a UDP glucosyl transferase that partition the same reactions as in plants among themselves (Jensen *et al.*, 2011). The first multifunctional P450, CYP405A2, is a CYP4 clan P450 unrelated to CYP79A1, and the second multifunctional P450, CYP332A3, is a CYP3 clan P450 unrelated to CYP71E1 (**Figure 20**). Remarkably, plants and insects have thus evolved, separately, highly specialized P450

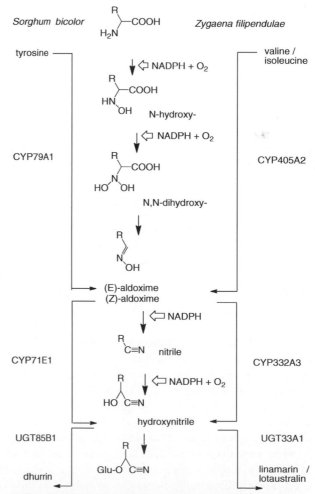

Figure 20 Biosynthesis of linamarin and lotaustralin from valine and isoleucine in *Zygaena filipendulae* (right), compared to the biosynthesis of dhurrin from tyrosine in *Sorghum bicolor* (left). R = side chain of the amino acid. Convergent evolution led to two multifunctional P450 enzymes in insects and plants.

enzymes catalyzing the same reactions, and assembled them into this chemically complex pathway.

Another comparison of interest will be that of enzymes involved in the biosynthesis of insect defensive steroids with the well-characterized steroid metabolizing enzymes of vertebrates. A number of aquatic Coleoptera (Dysticidae) and Hemiptera (Belastomatidae) synthesize a variety of steroids, mostly pregnanes (Scrimshaw and Kerfoot, 1987). With cholesterol as the presumed precursor in these carnivorous insects, one may envisage the evolution of an insect side-chain cleavage enzyme, of a C-21 hydroxylase, and of a C17-C21 lyase that would catalyze reactions identical to those of CYP11A, CYP21, and CYP17. Some defensive steroids also have an aromatic A-ring, 7α or 15α hydroxyl groups. A pregnene-3β,20β-diol glucoside is synthesized from cholesterol in female pupae of *M. sexta* (Thompson *et al.*, 1985). The role of this compound is unknown, but its synthesis strongly suggests the existence of a C20-C22 side-chain cleavage enzyme in Lepidoptera as well.

8.4.2. Xenobiotic Metabolism: Activation and Inactivation

8.4.2.1. Natural products The metabolism of plant toxins by insects has been reviewed extensively (see, for example, Brattsten, 1979b; Dowd *et al.*, 1983; Ahmad, 1986; Ahmad *et al.*, 1986; Mullin, 1986; Yu, 1986; Li *et al.*, 2007). Relatively few studies have directly assessed the role of P450 enzymes in the metabolism of natural compounds by more than one or two criteria, such as microsomal localization, NADPH- and O$_2$-dependence, and inhibition by piperonyl butoxide. In most cases, metabolism is associated with detoxification – for example, the metabolism of xanthotoxin in *Papilio polyxenes*, *Papilio multicaudatus*, *S. frugiperda*, and *Depressaria pastinacella* (Bull *et al.*, 1986; Nitao, 1990; Mao *et al.*, 2006b), the metabolism of α-terthienyl in larvae of three lepidopteran species (Iyengar *et al.*, 1990), and the metabolism of nicotine in *M. sexta* larvae (Snyder *et al.*, 1993). α- and β-thujones are detoxified by P450, as evidenced by synergism of their toxicity by three P450 inhibitors in *Drosophila* and by the lower toxicity of six of their metabolites (Hold *et al.*, 2001). Studies with flavone (Wheeler *et al.*, 1993), monoterpenes (Harwood *et al.*, 1990), and the alkaloid carnegine (Danielson *et al.*, 1995) clearly show, however, that evidence for metabolism by P450 *in vitro* may not be sufficient to define an *in vivo* toxicological outcome.

Natural products in the diet can act as inducers of P450 as well as of other enzymes (e.g., glutathione S-tansferases), and as a result of this induction, the metabolism of the inducing compound or of co-ingested plant compounds can change dramatically over time (Brattsten *et al.*, 1977). Differences between acute and chronic toxicity are thus often the result of altered expression patterns (quantitative and qualitative) of P450 genes. This was demonstrated, for instance, in studies on *Spodoptera* larvae (Brattsten, 1983; Gunderson *et al.*, 1986). The monoterpene pulegone and its metabolite menthofuran are more acutely toxic to *S. eridania* than to *S. frugiperda*, but the reverse is true for chronic toxicity. The toxicity of aflatoxin B1 (AFB1) to *H. zea* larvae is reduced to various degrees by co-ingestion of xanthotoxin or coumarin that induce CYP genes, including *CYP321A1* and *CYP6B8* (Zeng *et al.*, 2007; Wen *et al.*, 2009), and by piperonyl butoxide (Zeng *et al.*, 2007). This is an instance where the protective effect of ingested piperonyl butoxide may be through induction of P450, rather than inhibition of bioactivation. AFB1 is metabolized by CYP321A1 (but not CYP6B8) to the O-demethylated AFP1 and another minor metabolite. There is currently no evidence for bioactivation of AFB1 to the potentially toxic epoxide (AFBO) in *H. zea* (Niu *et al.*, 2008). Paradoxically, salicylate and jasmonate induce both *CYP321A1* and *CYP6B8*, but fail to protect the larvae from AFB1 toxicity (Zeng *et al.*, 2009). These two genes therefore do not determine the toxicodynamics of AFB1 in *H. zea*.

A study by Yu (1987) compared the metabolism of a large number of plant chemicals of different chemical classes by *S. frugiperda* and *Anticarsia gemmatalis* (velvetbean caterpillar) microsomes. Two indirect methods were used: on one hand, the NADPH-dependent decrease in substrate; on the other, the substrate-induced NADPH oxidation. This metabolism is inhibited by piperonyl butoxide and by carbon monoxide, and induced by a number of chemicals, particularly indole-3-carbinol, strongly suggesting P450 involvement. Such indirect methods are very useful as screening tools, as a first step towards a more thorough characterization of metabolism. However, they give no qualitative indication regarding the chemical fate of the substrate, nor quantitative indication of the levels of metabolism, as NADPH consumption is correlated to the coupling rate of the reaction, rather than to the rate of product formation (section 8.3.3). Clearly, insect P450 enzymes as a whole are capable of metabolizing a tremendous variety of naturally occuring chemicals, but the role of individual enzymes and their catalytic competence is only starting to be understood. Several P450 enzymes have now been shown to metabolize natural compounds (**Table 5**), although in too many cases the metabolites have not been identified and activity is measured only as substrate consumption. Furanocoumarin metabolism, in particular, is discussed below.

Natural products are not just an endless catalog of P450 substrates and inducers; they also comprise a varied and complex set of inhibitors of P450 enzymes. These inhibitors range from "classical" reversible inhibitors to substrates that are activated to chemically reactive, cytotoxic forms (see, for example, Neal and Wu, 1994).

Table 5 Heterologous Expression of Insect P450

Expression System	P450 Produced	From	Substrate Metabolized	Reference	Redox Partner(s)
E.coli	CYP4C7	*Diploptera punctata*	sesquiterpenoids	Sutherland *et al.* (1998)	MdR, Mdb5
	CYP6A1	*Musca domestica*	aldrin, heptachlor, sesquiterpenoids, diazinon, testosterone, progesterone, androstenedione, 7-propoxycoumarin, 1-bromochlordene, chlordene, 1-hydroxychlordene, isodrin, pisatin, chlorfenapyr	Andersen *et al.* 1994); Andersen *et al.* (1997); Sabourault *et al.* (2001); Feyereisen (2005); Jacobsen *et al.* (2006); Murataliev *et al.* (2008)	MdR (Mdb5)
	CYP6A5	*Musca domestica*	benzphetamine, *p*-chloro-*N*-methylaniline, methoxyresorufin	Feyereisen (2005)	MdR
	CYP9E1	*Diploptera punctata*	sesquiterpenoids	Feyereisen (2005)	MdR
	CYP12A1	*Musca domestica*	aldrin, heptachlor, diazinon, azinphosmethyl, amitraz, progesterone, testosterone, 7-alkoxycoumarins	Guzov *et al.* (1998)	bovine AdR/Adx
E.coli membranes	CYP15A1	*Diploptera punctata*	*t t*, methyl farnesoate	Helvig *et al.* (2004a)	MdR
	CYP6M2	*Anopheles gambiae*	pyrethroids	Stevenson *et al.* (2011)	AgR Agb5
	CYP6P3	*Anopheles gambiae*	pyrethroids	Muller *et al.* (2008b)	
	CYP6Z2	*Anopheles gambiae*	alkoxyresorufins	McLaughlin *et al.* (2008)	
	CYP6CM1vQ	*Bemisia tabaci*	imidacloprid, alkoxycoumarins and resorufins	Karunker *et al.* (2009)	AgR Agb5
baculovirus	CYP6A2	*Drosophila melanogaster*	aldrin, heptachlor, diazinon	Dunkov *et al.* (1997)	MdR
	CYP6B1	*Papilio polyxenes*	furanocoumarins, anaphthoflavone, flavone, coumarin, diazinon	Ma *et al.* (1994); Hung *et al.* (1997); Wen *et al.* (2003); Li *et al.* (2004); Wen *et al.* (2005)	(MdR)
	CYP6B3	*Papilio polyxenes*	furanocoumarins, anaphthoflavone	Wen *et al.* (2006)	MdR
	CYP6B4	*Papilio glaucus*	furanocoumarins	Hung *et al.*, (1997)	MdR
	CYP6B8	*Helicoverpa zea*	furanocoumarins, ethoxycoumarin chlorogenic acid, quercetin, rutin, xanthotoxin, flavone, indole-3-carbinol, acypermethrin, diazinon, aldrin	Li *et al.*(2003) Li *et al.* (2004)	MdR
	CYP6B17	*Papilio glaucus*	furanocoumarins, ethoxycoumarin	Li *et al.* (2003)	MdR
	CYP6B21	*Papilio glaucus*	furanocoumarins, ethoxycoumarin	Li *et al.* (2003)	MdR
	CYP6B25	*Papilio canadensis*	furanocoumarins	Li *et al.* (2003)	MdR
	CYP6B27	*Helicoverpa zea*	aldrin, diazinon, carbaryl, acypermethrin	Wen *et al.*, (2009)	MdR
	CYP6B33	*Papilio multicaudatus*	furanocoumarins	Mao *et al.* (2008a)	MdR
	CYP6Z1	*Anopheles gambiae*	DDT, carbaryl, xanthotoxin	Chiu *et al.* (2008)	MdR Dmb5
	CYP6Z2	*Anopheles gambiae*	carbaryl	Chiu *et al.* (2008)	MdR Dmb5
	CYP6AA3	*Anopheles minimus*	deltamethrin	Boonsuepsakul *et al.* (2008)	AmR
	CYP6AB3	*Depressaria pastinacella*	imperatorin	Mao *et al.*, (2006b)	MdR
	CYP6AB3v2	*Depressaria pastinicella*	imperatorin, myristicin	Mao *et al.*, (2007a, 2008b)	MdR

(Continued)

Table 5 Heterologous Expression of Insect P450—cont'd

Expression System	P450 Produced	From	Substrate Metabolized	Reference	Redox Partner(s)
	CYP6AB11	Amyelois transitella	imperatorin, piperonyl butoxide	Niu et al., (2011)	MdR
	CYP6AS1,3,4,10	Apis mellifera	quercetin	Mao et al.(2009)	MdR
	CYP6BQ9	Tribolium castaneum	deltamethrin, benzyloxyresorufin	Zhu et al. (2010)	MdR
	CYP9T1	Ips confusus	myrcene	Sandstrom et al. (2008)	MdR
	CYP9T2	Ips pini	myrcene	Sandstrom et al. (2006)	MdR
	CYP314A1	Bombyx mori	ecdysone	Maeda et al. (2008)	endogenous
	CYP321A1	Helicoverpa zea	furanocoumarins, anaphthoflavone, acypermethrin, aflatoxin B1, aldrin, diazinon	Sasabe et al. (2004); Rupasinghe et al.(2007); Niu et al. (2008)	MdR
yeast	CYP6A2	Drosophila melanogaster	aflatoxinB1, 7,12-dimethylbenz[a]anthracene, 3-amino-1- methyl-5Hpyrido[4,3-b]-indole	Saner et al., (1996)	ScR
	CYP6A8	Drosophila melanogaster	lauric acid, aldrin	Helvig et al.(2004b)	DmR
	CYP6D1	Musca domestica	methoxyresorufin	Smith and Scott (1997)	ScR
	CYP9A12	Helicoverpa armigera	p-nitroanisole, methoxyresorufin, esfenvaler ate	Yang et al.(2008)	ScR
	CYP9A14	Helicoverpa armigera	p-nitroanisole, methoxyresorufin, esfenvalerate	Yang et al.(2008)	ScR
transfected S2 cells	CYP302A1	Drosophila melanogaster, Bombyx mori, Anopheles gambiae	2,22-dideoxyecdysone	Warren et al., (2002); Niwa et al. (2005); Pondeville et al. (2008)	endogenous
	CYP306A1	Drosophila melanogaster, Bombyx mori	2,22,25-trideoxyecdysone	Niwa et al. (2004); Warren et al. (2004)	endogenous
	CYP314A1	Drosophila melanogaster, Anopheles gambiae	ecdysone	Petryk et al., (2003); Pondeville et al. (2008)	endogenous
	CYP315A1	Drosophila melanogaster, Anopheles gambiae	2-deoxyecdysone / 2,22-dideoxyecdysone	Warren et al. (2002); Pondeville et al., (2008)	endogenous
transfected tobacco cells	CYP6G1	Drosophila melanogaster	imidacloprid, DDT, methoxychlor	Joussen et al. (2008, 2010)	endogenous

See section 6.2 for description of the heterologous expression systems. Redox partners: R =P450 reductase, b5 = cytochrome b5 : Md; M. domestica; Ag: A. gambiae; Am: A. minimus; Sc: Saccharomyces cerevisiae; Dm: Drosophila melanogaster. Endogenous = unidentified redox partner(s) from the host system. AdR/Adx = adrenodoxin reductase and adrenodoxin.

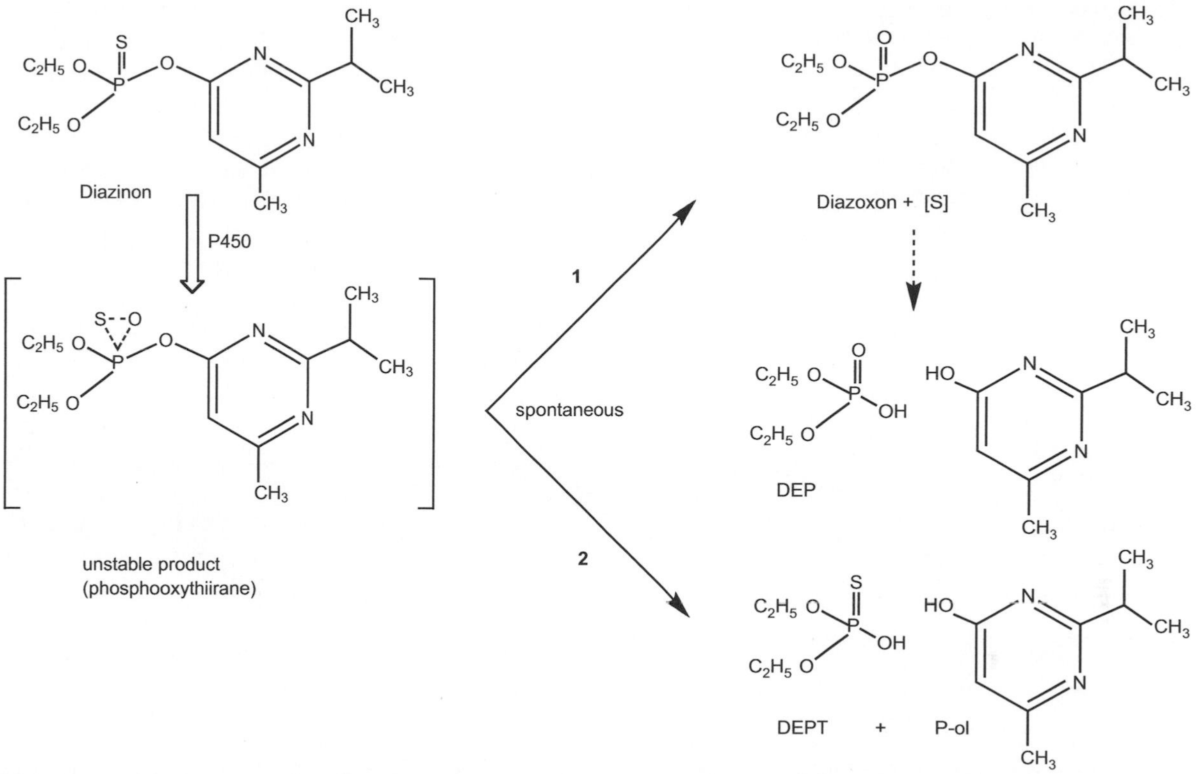

Figure 21 Metabolism of diazinon by cytochrome P450. Following an insertion of oxygen into the substrate, a reactive intermediate collapses (1) by desulfuration, or (2) by cleavage of the ester linkage. DEP, diethylphosphate; DEPT, diethylphosphorothioate; P-ol, 2-isopropoxy-4-methyl-6-hydroxypyrimidine; [S], reactive form of sulfur released during the reaction. Diazoxon can be further converted to DEP and P-ol by the same or another P450 in a subsequent reaction, or by a phosphotriester hydrolase. DEP may also be formed by spontaneous degradation of the initial product of diazinon monooxygenation. The ratio of outcomes (1) and (2) and the fate of the reactive sulfur depends on the P450 enzyme and on the type of OP substrate (see **Table 6**). Alternative routes of P450 metabolism of diazinon that do not involve the P=S bond are not shown.

8.4.2.2. Insecticides and other xenobiotics

The metabolism of insecticides by P450 enzymes is very often a key factor in determining toxicity to insects and to non-target species, but it can also represent a key step in the chain of events between contact, penetration, and interaction at the target site. The classical example is probably the metabolism of phosphorothioate insecticides. In many cases, the active ingredients of organophosphorus insecticides are phosphorothioate (P=S) compounds (also known as phosphorothionates), whereas the molecule active at the acetylcholinesterase target site is the corresponding phosphate (P=O). It has long been recognized that the P=S to P=O conversion is a P450-dependent reaction. In the case of diazinon, this desulfuration has been studied for three heterologously expressed insect P450 enzymes (Dunkov *et al.*, 1997; Guzov *et al.*, 1998; Sabourault *et al.*, 2001). All three P450s metabolized diazinon not just to diazoxon, the metabolite resulting from desulfuration, but also to a second metabolite resulting from oxidative ester cleavage. Similarly, antibodies to house fly CYP6D1 inhibit the microsomal desulfuration of chlorpyriphos, as well as its oxidative ester cleavage (Hatano and Scott, 1993). The

mechanism of P450-dependent desulfuration is believed to involve the initial attack of the P=S bond by an activated oxygen species of P450, leading to an unstable and therefore hypothetical phosphooxythiirane product (**Figure 21**). The collapse of this product can lead to two possible outcomes: (1) the replacement of sulfur by oxygen in the organophosphate product with the release of a reactive form of sulfur; and, (2) the cleavage of the phosphate ester (or thioester) link with the substituent of highest electron-withdrawing properties, the "leaving group."

Outcome (1) can be viewed as "activation," because the P=S to P=O desulfuration produces an inhibitor of acetylcholinesterase often several orders of magnitude more potent than the P=S parent compound. However, the fate of this product of "activation" depends on the histological proximity to the target, sequestration, excretion, and further metabolism of the phosphate (by oxidative or hydrolytic enzymes). Kinetic evidence with the heterologously expressed CYP6A1, CYP6A2, and CYP12A1, as well as immunological evidence with CYP6D1, indicates that these P450 enzymes do not metabolize the P=O product of the parent P=S compound any further.

Table 6 Desulfuration and Oxidative Ester Cleavage of Organophosphorus Insecticides by P450 Enzymes and Microsomes

P450 Enzyme	Ratio of OP Desulfuration/ Oxidative Ester Cleavage	Reference
human CYP2C19	8.5 d, 1.30 p, 0.14 c	Kappers et al., (2001); Tang et al. (2001); Mutch et al. (2003)
human CYP3A4	3.0 d, 0.50 p, 0.66 c	ibid.
human CYP2B6	0.7 d, 0.01 p, 3.38 c	ibid.
Drosophila CYP6A2	0.92 d	Dunkov et al. (1997)
house fly CYP6A1	0.37 d	Sabourault et al. (2001)
house fly CYP12A1	0.69 d	Guzov et al. (1998)
house fly CYP6D1	2.0 c	Hatano and Scott, (1993)
house fly CSMA* microsomes	0.95 f	Ugaki et al. (1985)
house fly Akita-f** microsomes	0.59 f	ibid.
Heliothis virescens* microsomes	1.90 mp	Konno and Dauterman (1989)
Heliothis virescens NC-86** microsomes	1.32 mp	ibid.

d: diazinon; p: ethyl parathion; c: chlorpyriphos;
f: fenitrothion; mp: methyl parathion
*susceptible strain; **resistant strain

Outcome (2) is, without question, a detoxification, because the oxidative cleavage of the "leaving group" produces compounds unable to inhibit acetylcholinesterase, the dialkylphosphorothioate and/or dialkylphosphate. The ratio of outcomes (1) and (2) appear to be P450-specific *and* substrate-specific (**Table 6**), suggesting that the collapse of the unstable initial product of P450 attack is influenced by the active site environment. Theoretically, some P450 enzymes may very strongly favor outcome (1) or (2), or *vice versa*, thus qualifying as relatively "clean" activators or detoxifiers, but there is, to date, little direct evidence from the insect toxicological literature for such P450 enzymes (see, however, Oi *et al.*, 1990).

Furthermore, the sulfur released by the reaction can covalently bind either to neighboring proteins, thus leading to cellular damage, or to the P450 protein itself (at least in vertebrate liver where this specific aspect has been studied; Kamataki and Neal, 1976). Parathion causes NADPH-dependent inhibition of methoxyresorufin O-demethylation activity (a P450Lpr-selective activity) in the house fly, whereas chlorpyriphos does not (Scott *et al.*, 2000). Therefore, it is not just the fate of the initial P450 metabolite of the P=S compound that depends on the P450 and the OP, but also the fate of the sulfur released, that can vary.

Changes in the level of expression of P450 genes may be sufficient to change this delicate balance between activation and inactivation *in vivo*. For instance, fenitrothion resistance in the Akita-f strain of the house fly is related to an increase in oxidative ester cleavage over desulfuration measured in abdominal microsomes (Ugaki *et al.*, 1985). In *Heliothis virescens*, methyl parathion resistance in the NC-86 strain, which has an unchanged level of total P450, is related to a replacement of a set of P450 enzymes with high desulfuration activity by a set of P450 enzymes that metabolize less parathion, and do so with a lower desulfuration/oxidative ester cleavage ratio (Konno *et al.*, 1989) (**Table 6**).

P450 enzymes that metabolize OPs can metabolize other insecticides as well, and this sometimes leads to potentially useful interactions. Thus enhanced detoxification of dicofol in spider mites can lead to enhanced chlorpyriphos activation, and hence negative cross-resistance (Hatano *et al.*, 1992). Similarly, permethrin resistance in horn flies is suppressible by piperonyl butoxide, and negatively related to diazinon toxicity (Cilek *et al.*, 1995). In *H. armigera* populations from West Africa, triazophos shows negative cross-resistance with pyrethroids, and in this case the synergism shown by the OP towards the pyrethroids appears due to an enhanced activation to the oxon form (Martin *et al.*, 2003). These interactions were observed *in vivo* or with microsomes, but it is likely that they do reflect the properties of single P450 enzymes with broad substrate specificity, rather than the fortuitous coordinate regulation of different P450 enzymes with distinct specificities.

Organophosphorus compounds such as disulfoton and fenthion can also be activated by thioether oxidation (formation of sulfoxide and sulfone), but it is not clear whether these reactions are catalyzed in insects by a P450 or by a flavin monooxygenase (FMO). Further examples of oxidative bioactivation of organophosphorus compounds have been discussed (Drabek and Neumann, 1985).

Pyrethroid metabolism by P450 enzymes is a well known pathway of degradation in insects, and several have been identified (**Table 5**). Hydroxylations and further metabolism make pyrethroid metabolism (often of isomer mixtures) an analytical challenge, and this is shown for the single enantiomer of deltamethrin (**Figure 22**).

The toxicity of fipronil to house flies is increased six-fold by the synergist piperonyl butoxide, whereas the desulfinyl photodegradation product is not substantially detoxified by P450 (Hainzl and Casida, 1996; Hainzl *et al.*, 1998). Conversion of fipronil to its sulfone appears to be catalyzed by a P450 enzyme in *Ostrinia nubilalis* (Durham *et al.*, 2002) and in *Diabrotica virgifera* (Scharf *et al.*, 2001). In the latter, the toxicity of fipronil sulfone is about the same as that of the parent compound, and piperonyl butoxide has only a marginal effect as synergist. In contrast,

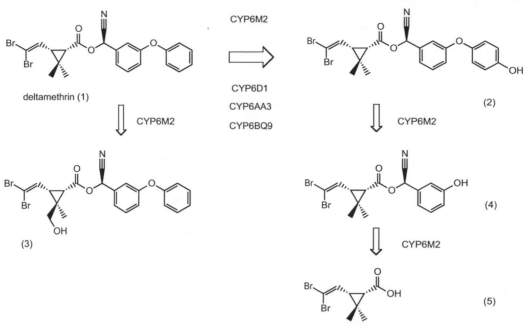

Figure 22 Metabolism of deltamethrin by insect P450 enzymes: (1) deltamethrin; (2) 4′-hydroxydeltamethrin; (3) *trans*-hydroxymethyl-deltamethrin; (4) cyano(3-hydroxyphenyl)methyl deltamethrate; (5) deltamethric acid.

synergists antagonize the toxicity of fipronil in *B. germanica*, suggesting that oxidation to the sulfone represents an activation step in this species (Valles *et al.*, 1997).

The now banned cyclodiene insecticides aldrin, heptachlor, and isodrin are epoxidized by P450 enzymes to the environmentally stable, toxic epoxides, dieldrin, heptachlor epoxide, and endrin, respectively (Brooks, 1979; Drabek and Neumann, 1985). Recombinant CYP6A1, -A2, -A8, -B8, and -B27, CYP12A1, and CYP321A1 can catalyze these epoxidations (**Table 5**). Examples of proinsecticide metabolism include the activation of chlorfenapyr by N-dealkylation (Black *et al.*, 1994), and of diafenthiuron by S-oxidation (Kayser and Eilinger, 2001). In each case, the insect P450-dependent activation is a key in the selective toxicity of these proinsecticides that target mitochondrial respiration. Recombinant house fly CYP6A1 catalyzes the activation of chlorfenapyr. In *H. virescens*, toxicity of chlorfenapyr is negatively correlated with cypermethrin toxicity (Pimprale *et al.*, 1997). Genetic analysis indicates that a single factor is involved, so the same P450 that activates chlorfenapyr may also detoxify cypermethrin in this species (**Figure 23**). A similar case of negative cross-resistance of chlorfenapyr in a pyrethroid-resistant strain has been reported in the horn-fly *Haematobia irritans* (Sheppard and Joyce, 1998).

The metabolism of imidacloprid is also of interest, particularly in relation to resistance. Piperonyl butoxide can synergize the toxicity of imidacloprid, and two P450 enzymes, CYP6G1 of *D. melanogaster* and CYP-6CM1vQ of *Bemisia tabaci*, have been shown to metabolize this neonicotinoid (Joussen *et al.*, 2008; Karunker *et al.*, 2009) (**Figure 24**). Hydroxylations at the 4 and

5 positions can lead to the olefinic metabolite or to the dihydroxylated metabolite. In the whitefly the 5-hydroxy metabolite is not toxic, but the 4-hydroxy metabolite is as toxic as the parent compound, so regioselectivity may be of importance.

In vivo synergism by piperonyl butoxide, a typical inhibitor of P450 enzymes (section 8.6.3), is often used to implicate a P450-mediated detoxification, and there are innumerable examples in the literature. The inference is much stronger when two unrelated synergists are used *in vitro*, and when metabolites of the pesticide are identified. For instance, pyriproxifen is hydroxylated by fat body and midgut microsomes of larval house flies to 4′-OH-pyriproxyfen and 5″OH-pyriproxyfen, and these activities are inhibited by PB and TCPPE (Zhang *et al.*, 1998).

The study of xenobiotic metabolism by individual P450 enzymes expressed in heterologous systems is summarized in **Table 5**. Most of these enzymes are from the CYP3 clan, family CYP6, with the notable exception of the mitochondrial CYP12. Whereas the CYP6 enzymes clearly comprise some enzymes with "broad and overlapping" substrate specificity, even closely related enzymes of this family can differ substantially in their catalytic competence. The task of predicting which xenobiotic or natural product will be metabolized by which type of P450 is currently not possible.

8.4.3. P450 and Host Plant Specialization

8.4.3.1. The Krieger hypothesis and beyond The interactions of plants and insects, and, more specifically, the role of plant chemistry in the specialization of

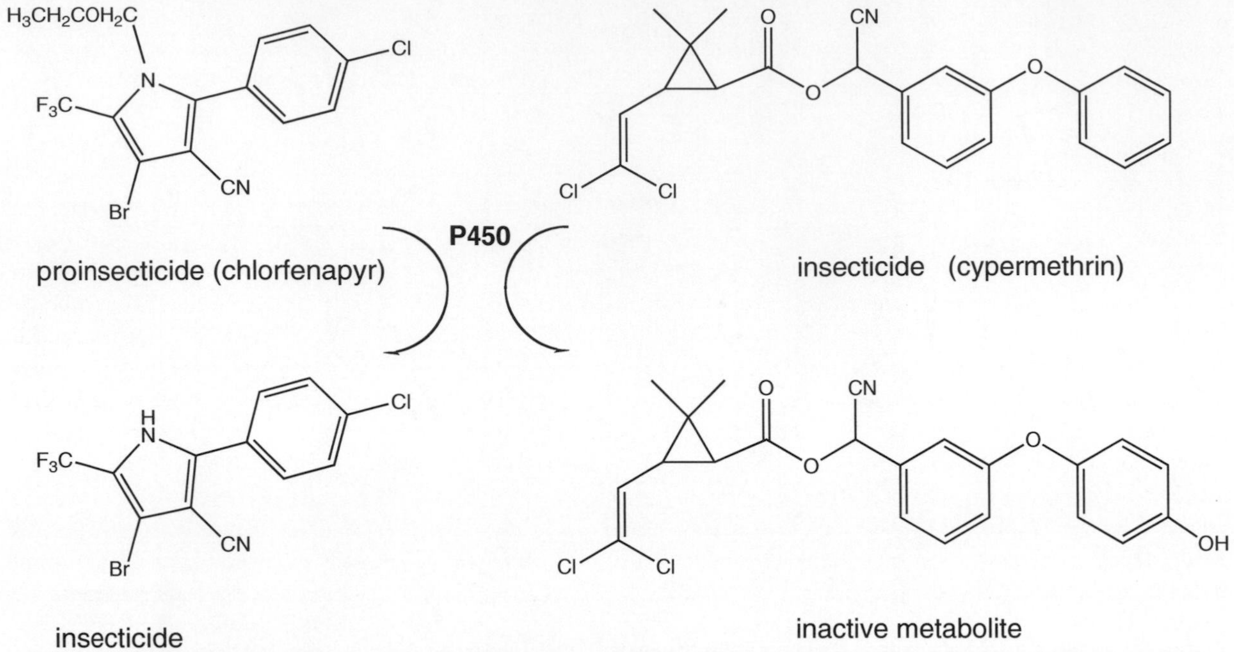

Figure 23 Chlorfenapyr and cypermethrin metabolism. The same P450 in *Heliothis virescens* probably activates the pyrrole and inactivates the pyrethroid, resulting in negative cross-resistance.

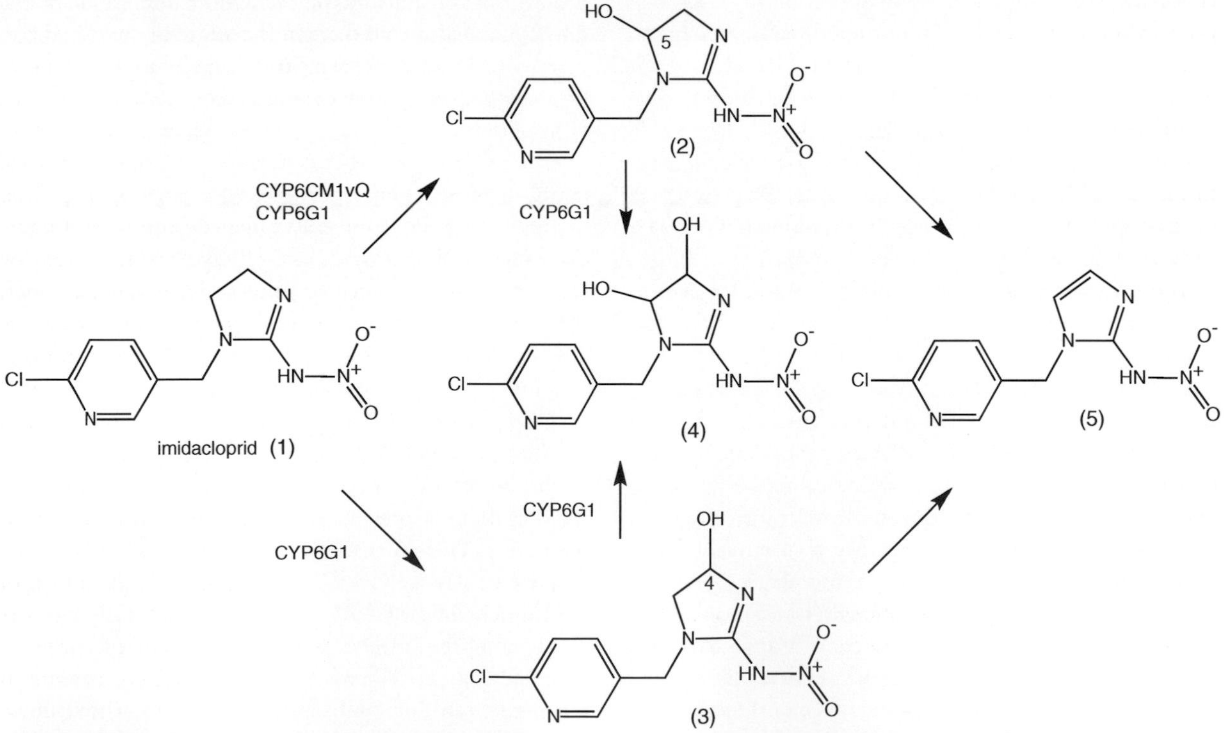

Figure 24 Metabolism of imidacloprid by insect P450 enzymes: (1) imidacloprid; (2) 5-hydroxyimidacloprid; (3) 4-hydroxyimidacloprid; (4) dihydroxyimidacloprid; (5) non-enzymatically derived dehydroimidacloprid.

phytophagous insects have generated a vast literature. "Secondary" plant substances are variously seen to regulate insect behavior and/or to serve as weapons in a coevolutionary "arms race" (Dethier, 1954; Fraenkel, 1959; Ehrlich and Raven, 1964; Jermy, 1984; Bernays and Graham, 1988). In chemical ecology alone, "no other area

is quite so rife with theory" (Berenbaum, 1995). Many of the theories and some of the experiments implicitly or explicitly deal with the insect's ability to metabolize plant secondary substances by P450 and other enzymes. In the case of behavioral cues, we are far from understanding the true importance of P450 enzymes in the integration of chemosensory information – for example, as "odorant degrading enzymes." In the case of detoxification, however, the landmark paper of Krieger *et al.* (1971) can be seen as echoing the Fraenkel (1959) paper, by exposing the *raison d'être* of P450 enzymes. They stated that "higher activities of midgut microsomal oxidase enzymes in polyphagous than in monophagous species indicates that the natural function of these enzymes is to detoxify natural insecticides present in the larval food plants." In that 1971 study, aldrin epoxidation was measured in gut homogenates of last instar larvae from 35 species of Lepidoptera. Polyphagous species had on average a 15-times higher activity than monophagous species. This trend was seen in sucking insects as well, with a 20-fold lower aldrin epoxidase activity in the oleander aphid *Aphis nerii* when compared to the potato aphid *Myzus euphorbiae* or to the green peach aphid *Myzus persicae* (Mullin, 1986). The former is a specialist feeder on two plant families, Asclepiadaceae and Apocyanaceae, whereas the latter two are generalists found on 30–72 plant families. The concept extended to other detoxification enzymes, and was broadened to cover prey/predator – for example, in mites, where the predatory mite has a five times lower aldrin epoxidase activity than its herbivorous prey (Mullin *et al.*, 1982). The toxicity of the natural phototoxin α-terthienyl is inversely proportional to the level of its metabolism in Lepidoptera, and is related to diet breadth. Metabolism is highest in *O. nubilalis*, which feeds on numerous phototoxic Asteraceae; lower in *H. virescens*, which has a broad diet, including some Asteraceae that are non-phototoxic; and lowest in *M. sexta*, a specialist of Solanaceae (Iyengar *et al.*, 1990).

The conceptual framework of Krieger and colleagues has been challenged (Gould, 1984) and defended (Ahmad, 1986). An alternative view (Berenbaum *et al.*, 1992) proposes that aldrin epoxidation represents "P450s with broad substrate specificity [that] are most abundant in insects that encounter a wide range of host plant metabolites." A careful repetition of the Krieger experiments on lepidopterous larvae from 58 species of New South Wales failed to show significant differences in aldrin epoxidation between monophagous and polyphagous species (Rose, 1985). High activity in both monophagous and polyphagous species was invariably linked to the presence of monoterpenes in the host diet. The evidence presented in the sections below indicates that polyphagous and oligophagous species alike rely on the ability to draw on a great diversity of P450 genes, encoding a great diversity of specific and less specific enzymes, and regulated by a great diversity of environmental sensing mechanisms – i.e., induction. The ability to induce P450 enzymes and deal with a wide range of toxic chemicals in the diet has been thought to present a "metabolic load" for polyphagous species, with specialists restricting their "detoxification energy" to one or a few harmful substrates (see, for example, Whittaker

and Feeny, 1971). However, careful studies in both oligophagous and polyphagous species have refuted the concept of induction as imposing a metabolic load (e.g., Neal, 1987; Appel and Martin, 1992).

Global approaches to the comparison of generalist and specialist herbivores feeding on plants with higher or lower levels of chemical defenses are now available, as shown by the elegant studies of Govind *et al.* (2010). The transcriptional responses of neonates of the specialist *M. sexta* and generalist *H. zea* fed for 24 hours on *Nicotiana attenuata* plants, wild type or progressively suppressed in their jasmonate response, were compared by microarrays. This type of approach will allow a finer analysis of CYP gene responses as they are integrated in the overall physiological response of the organism.

8.4.3.2. Host plant chemistry and herbivore P450
Cactophilic *Drosophila* species from the Sonoran Desert are specialized to specific columnar cactus hosts by their dependency on unusual sterols (*D. pachea*) or by their unique ability to detoxify their hosts' allelochemicals, notably isoquinoline alkaloids and triterpene glycosides (Frank and Fogleman, 1992; Fogleman *et al.*, 1998). P450-mediated detoxification was shown by the loss of larval viability in media that contained both allelochemicals and piperonyl butoxide, and by the induction of total P450 or alkaloid metabolism by the cactus allelochemicals or by phenobarbital (Frank and Fogleman, 1992; Fogleman *et al.*, 1998). Several P450s of the *CYP4*, *CYP6*, *CYP9*, and *CYP28* families are induced by cactus-derived isoquinoline alkaloids and by phenobarbital, but not by triterpene glycosides; only a *CYP9* gene was induced by alkaloids and not by phenobarbital (Danielson *et al.*, 1997, 1998; Fogleman *et al.*, 1998). The capacity to detoxify isoquinoline alkaloids was not related to DDT or propoxur tolerance, and while phenobarbital induced P450s capable of metabolizing the alkaloid carnegine in *D. melanogaster*, this was not sufficient to produce *in vivo* tolerance (Danielson *et al.*, 1995). Selection of *D. melanogaster* with saguaro cactus alkaloids over 16 generations, however, led to P450-mediated resistance to the cactus alkaloids (Fogleman, 2000). These studies suggest the evolution of specific responses in the cactophilic species involving the recruitment of a phylogenetically unrelated subset of P450 genes in each instance of specialization of a fly species on its host cactus. For instance, *CYP28A1* is induced in *D. mettleri* by rotting senita cactus, and *CYP4D10* is induced by decaying saguaro cactus tissue. In a population genetic survey, evidence for positive selection was obtained for *CYP28A1*, but not for *CYP4D10* (Bono *et al.*, 2008).

The oligophagous tobacco hornworm (*M. sexta*) feeds essentially on Solanaceae, and its adaptation to the high levels of insecticidal nicotine found in tobacco depends largely on metabolic detoxification, although other tolerance mechanisms may be contributing as well (Snyder and Glendinning, 1996). Hornworm larvae fed a nicotine-free artificial diet (naïve insects) are rapidly poisoned by the ingestion of a nicotine-supplemented diet, but this diet is not deterrent. Poisoning is evidenced by convulsions and inhibition of feeding. The small amount of ingested nicotine induces its own metabolism, so that approximately 36 hours later the larvae resume feeding normally, without further signs of poisoning. The inhibition of feeding and its resumption after nicotine exposure is directly related to P450 induction. Indeed, treatment with piperonyl butoxide, which itself has no effect on feeding, inhibits the increase in nicotine-diet consumption that occurs once nicotine metabolism has been induced (Snyder and Glendinning, 1996). Naïve insects metabolize nicotine to nicotine 1-N-oxide at a low level, whereas nicotine-fed insects metabolize it further to cotinine-N-oxide at a higher level (Snyder et al., 1994). These reactions are catalyzed by one or more P450 enzymes (Snyder et al., 1993). The effects of nicotine on marker P450 activities are complex: the metabolism of three substrates is induced at low nicotine levels, seven are only induced at higher levels, and three are unaffected (Snyder et al., 1993).

P450 induction has also been inferred in the polyphagous spider mite *Tetranychus urticae*, where the performance of a bean-adapted population on tomato was severely compromised by piperonyl butoxide (Agrawal et al., 2002). The P450 inhibitor did not reduce acceptance of tomato as a host, nor did it reduce the performance of the bean-adapted population on bean, strongly suggesting a post-ingestive induction of P450 as a mechanism of acclimatization to the novel host. In the polyphagous noctuid *S. frugiperda*, ingestion of indole 3-carbinol increases once the continuous exposure to this toxic compound has induced P450 enzymes (Glendinning and Slansky, 1995). Larvae of the major cotton pest *H. armigera* can grow in the presence of gossypol only by inducing P450 activities, and *CYP6AE14* expression in the midgut correlates with growth on gossypol medium. RNAi suppression of *CYP6AE14* expression achieved by feeding on tobacco leaves or *Arabidopsis* expressing a CYP6AE14 dsRNA construct causes a delay in larval growth (Y. B. Mao et al. 2007). These experiments were verified in CYP6AE14 dsRNA transgenic cotton, with impaired growth of *H. armigera* larvae and reduced plant damage being reported (Mao et al., 2010).

8.4.3.3. *Papilio* species and furanocoumarins The adaptation of herbivores to toxic components of their host plants is best documented in the genus *Papilio*. The black swallowtail, *P. polyxenes*, feeds on host plants from just two families, the Apiaceae (Umbelliferae) and the Rutaceae. These plants are phytochemically similar, particularly in their ability to synthesize furanocoumarins. Biogenetically derived from umbelliferone (7-hydroxycoumarin), the

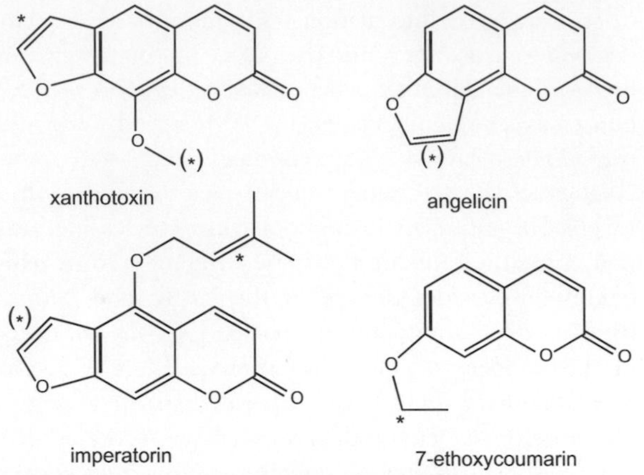

Figure 25 Structures of the furanocoumarins xanthothotoxin, angelicin and imperatorin, and of the model P450 substrate 7-ethoxycoumarin. Sites of known (and probable) P450 oxidation are shown by an asterisk.

linear furanocoumarins (related to psoralen) and angular furanocoumarins (related to angelicin) are toxic to non-adapted herbivores (Berenbaum, 1990) (**Figure 25**). This toxicity can be enhanced by light, as furanocoumarins are best known for their UV photoreactivity leading to adduct formation with macromolecules, particularly DNA. Xanthotoxin, a linear furanocoumarin, induces its own metabolism in a dose-dependent fashion when added to the diet of *P. polyxenes* larvae (Cohen et al., 1989). This P450-dependent metabolism proceeds probably by an initial epoxidation of the furan ring, followed by further oxidative attack and opening of the ring, leading to non-toxic hydroxylated carboxylic acids (Ivie et al., 1983; Bull et al., 1986). Inducible xanthotoxin metabolism is observed in all leaf-feeding stages of *P. polyxenes*, and is higher in early instars (Harrison et al., 2001). Xanthotoxin induces its own metabolism in the midgut, but also in the fat body and integument (Petersen et al., 2001). The metabolism of bergapten and sphondin is also induced by dietary xanthotoxin. Levels of total midgut microsomal P450 are unaffected by xanthotoxin, and photoactivation is not required for induction (Cohen et al., 1989). The metabolism of xanthotoxin is 10 times faster in *P. polyxenes* than in the non-adapted *S. frugiperda* (Bull et al., 1986). *P. polyxenes* microsomal P450 are also less sensitive to the inhibitory effects of xanthotoxin. NADPH-dependent metabolism of xanthotoxin leads to an uncharacterized reactive metabolite that can covalently bind P450 or neighboring macromolecules – i.e., xanthotoxin can act as "suicide substrate" (Neal and Wu, 1994). This NADPH-dependent covalent labeling of microsomal proteins is seven times higher in *M. sexta* than in *P. polyxenes*. Inhibition of aldrin epoxidation and *p*-nitroanisole O-demethylation by xanthotoxin is also 6- and 300-fold higher, respectively, in *M. sexta* than in *P. polyxenes* (Zumwalt and Neal, 1993).

P. polyxenes is thus a specialist feeder well adapted to feed on xanthotoxin-containing plants. Allelic variants of two P450 genes were cloned from *P. polyxenes* midgut: *CYP6B1* (Cohen *et al.*, 1992), and *CYP6B3*, which is 88% identical to *CYP6B1* (Hung *et al.*, 1995a). Both CYP genes are inducible by furanocoumarins (Hung *et al.*, 1995b; Harrison *et al.*, 2001; Petersen *et al.*, 2001), with *CYP6B3* induction restricted to the fat body (Petersen *et al.*, 2001). CYP6B1 metabolizes the linear furanocoumarins bergapten, xanthotoxin, isopimpinellin, psoralen, trioxsalen, visnagin, and khellin. CYP6B1 metabolizes the angular furanocoumarins angelicin and sphondin as well, but 5–10 times less efficiently than xanthotoxin (Ma *et al.*, 1994; Wen *et al.*, 2003). Angelicin is also less efficiently metabolized *in vivo* (Li *et al.*, 2003), and *P. polyxenes* is less adapted to it (Berenbaum and Feeny, 1981). Although generally less active than CYP6B1, CYP6B3 metabolizes both linear and angular furanocoumarins to almost the same extent (Wen *et al.*, 2006).

Compared to the specialist *P. polyxenes*, *Papilio glaucus* is a generalist that encounters furanocoumarin-containing plants (e.g., hoptree, *Ptelea trifoliata*) only occasionally. Xanthotoxin nevertheless induces its own metabolism in *P. glaucus* (Cohen *et al.*, 1992). *P. glaucus* is highly polyphagous, reported to feed on over 34 plant families, and therefore offers a interesting contrast to *P. polyxenes*. Esterase, glutathione S-transferase, and P450 activities are highly variable and dependent on the species of deciduous tree foliage that this species feeds on (Lindroth, 1989). *P. glaucus* has significant levels of linear and angular furanocoumarin metabolism, which are highly inducible by xanthotoxin (Hung *et al.*, 1997). A series of nine *CYP6B* genes and some presumed allelic variants were cloned from *P. glaucus* (Hung *et al.*, 1996, 1997; Li *et al.*, 2001; W. Li *et al.*, 2002). The first two genes, *CYP6B4* and *CYP6B5*, are products of a recent gene duplication event, and their promoter region is very similar (Hung *et al.*, 1996). Six additional and closely related members of the *CYP6B* subfamily were cloned from *Papilio canadensis*, another generalist very closely related to *P. glaucus* but not known to feed on plants containing furanocoumarins (Li *et al.*, 2001; W. Li *et al.*, 2002). Xanthotoxin induced *CYP6B4*- and *CYP6B17*-like genes in both species, but the level of furanocoumarin metabolism was lower in *P. canadensis* (Li *et al.*, 2001). This wide spectrum of CYP6B enzymes represents a broad range of activities towards furanocoumarin substrates. Whereas CYP6B4 of *P. glaucus* efficiently metabolizes both linear and angular furanocoumarins (Hung *et al.*, 1997), CYP6B17 of *P. glaucus*, and CYP6B21 and -25 from *P. canadiensis*, have a more modest catalytic capacity (Li *et al.*, 2003). In contrast, *Papilio multicaudatus*, an oligophagous species that can feed on hoptree as host and is thought to be basal in the

P. glaucus clade, actively metabolizes furanocoumarins (Mao *et al.*, 2006b). Among the six CYP6 cDNAs cloned from xanthotoxin-induced midgut, CYP6B33 actively metabolizes linear and angular furanocoumarins (Mao *et al.*, 2007b). *Papilio troilus*, a more distant relative of the former three species that specializes on Lauraceae that lack furanocoumarins has undetectable basal or induced xanthotoxin metabolism (Cohen *et al.*, 1992).

The P450 enzymes of the CYP6B subfamily from *Papilio* species fit the description of enzymes with "broad and overlapping specificity," at least towards furanocoumarins. CYP6B1 metabolizes other chemicals as well, notably flavones and diazinon, at a high rate (X. Li *et al.*, 2004). Their range of catalytic competence is quite variable, however. CYP6B21 and CYPB25 metabolize angelicin at a similar rate (0.4–0.5 nmol/min per nmol P450), but whereas CYP6B21 also metabolizes 7-ethoxycoumarin at a similar rate (0.5 nmol/min per nmol P450), CYP6B25 docs not have appreciable 7-dealkylation activity (Li *et al.*, 2003) (**Figure 25**).

8.4.3.4. Furanocoumarins and other insect species The metabolism of furanocoumarins or the inducibility of P450 by these compounds is not restricted to Papilionidae. Xanthotoxin induces its own metabolism in the parsnip webworm *Depressaria pastinacella* (Nitao, 1989). This species belongs to the Oecophoridae, and is a specialist feeder on three genera of furanocoumarin-containing Apiaceae. It is highly tolerant to these compounds, and metabolizes them not just by opening the furan ring; in the case of sphondin, it is also capable of O-demethylation (Nitao *et al.*, 2003). Although furanocoumarin metabolism is inducible, the basal (uninduced) activity is high (Nitao, 1989), and the response is a general one, with little discrimination of the type of furanocoumarin inducer or the type of furanocoumarin metabolized (Cianfrogna *et al.*, 2002). The P450 enzymes involved in furanocoumarin metabolism by *D. pastinacella* include CYP6AB3 (Mao *et al.*, 2006a, 2007b), which metabolizes imperatorin (**Figure 25**) on its isopentenyl side chain rather than on its furan ring. They must therefore include other P450s as well to metabolize most furanocoumarins that lack such a side chain. CYP6AB3 also metabolizes myristicin (not a furanocoumarin) on its propenyl side chain, at an even higher rate than imperatorin (Mao *et al.*, 2008b).

A species that does not encounter furanocoumarins, the solanaceous oligophage *M. sexta*, responds to xanthotoxin by inducing *CYP9A4* and *CYP9A5* (Stevens *et al.*, 2000). The generalist *S. frugiperda* also induces P450 as well as glutathione S-transferases in response to xanthotoxin (Yu, 1984; Kirby and Ottea, 1995). It has low basal P450-mediated xanthotoxin metabolism, but this metabolism is inducible by a variety of compounds, including terpenes and flavone (Yu, 1987). In the highly polyphagous *H. zea*, a similar situation is encountered. Xanthotoxin metabolism is low, but inducible by itself as well as by phenobarbital and α-cypermethrin (Li *et al.*, 2000b). A number of CYP genes have been cloned from *H. zea*. Among them, *CYP6B8* and *CYP321A1* are inducible as by furanocoumarins and other chemicals (Li *et al.*, 2000a; Wen *et al.*, 2009). Both P450s metabolize linear and angular furanocoumarins, as well as a number of other natural compounds and pesticides (Li *et al.*, 2004; Sasabe *et al.*, 2004; Rupasinghe *et al.*, 2007). Metabolism of linear and angular furanocoumarins is therefore shared by many insects, and probably many P450 enzymes.

8.4.4. Host Plant, Induction, and Pesticides

The adaptive plasticity conferred by the inducibility of P450 enzymes on different diets can have important consequences for insect control and the bionomics of pest insects. It is far from being just an ecological oddity, or an interesting set of tales of insect natural history. It is well recognized that the same insect species fed different (host) plants will show differences in its response to pesticides (Ahmad, 1986; Yu, 1986; Lindroth, 1991), and that these differences often reflect the induction of P450 enzymes, as well as of other enzymes, glutathione S-transferases, epoxide hydrolases, etc. The complexity of plant chemistry makes it difficult to account for the contribution of each individual chemical to this response, and key components are often analyzed first (see, for example, Moldenke *et al.*, 1992). Similarly, the multiplicity of P450 genes and the range of P450 enzyme specificity make it difficult to predict the outcome of exposure to a plant chemical. The toxicological importance of the plant diet on the herbivore's P450 status (induction, inhibition) is well recognized in pharmacology, where the joint use of chemical therapy and traditional herbs can have unpredicted outcomes (Zhou *et al.*, 2003).

Larvae of the European corn borer *Ostrinia nubilalis* fed leaves from corn varieties with increasing DIMBOA content, and thus increasing levels of resistance to leaf damage, had correspondingly increased levels of total midgut P450 and *p*-nitroanisole O-demethylation activity (Feng *et al.*, 1992). These studies suggest that constitutive host plant resistance may affect the insect response to xenobiotics. In addition, the induction of host plant defense by insect damage may itself be a signal for induction of herbivore P450 enzymes, as shown in *H. zea*. The plant defense signal molecules jasmonate and salicylate induce *CYP6B8*, *-B9*, *-B27*, and *-B28* (X. Li *et al.*, 2002b) in both fat body and midgut. The response to salicylate is relatively specific, as *p*-hydroxybenzoate, but not methylparaben, also acts as inducer.

Treatment with 2-tridecanone, a toxic allelochemical from trichomes of wild tomato, protects *H. zea* larvae against carbaryl toxicity (Kennedy, 1984), and *H. virescens*

larvae against diazinon toxicity (Riskallah *et al.*, 1986a). In *H. virescens* larvae, the compound caused both qualitative and quantitative changes in P450 spectral properties (Riskallah *et al.*, 1986b), an induction confirmed by its effect on specific P450 genes in the gut of *M. sexta* larvae (Snyder *et al.*, 1995; Stevens *et al.*, 2000). Larvae of *H. virescens* with a genetic resistance to 2-tridecanone have increased P450 levels and P450 marker activities (benzphetamine demethylation, benzo[a]pyrene hydroxylation, phorate sulfoxidation), and these can be further increased by feeding 2-tridecanone (Rose *et al.*, 1991). Larvae of *H. zea* exposed to various natural products in the diet show induced *CYP6B8* and *CYP321A1*, as well as reduced diazinon and carbaryl toxicity (Zeng *et al.*, 2007; Wen *et al.*, 2009). A laboratory population of *H. zea* can rapidly display increased tolerance to α-cypermethrin by selection of an increased P450 detoxification ability with a high dose of dietary xanthotoxin (Li *et al.*, 2000b).

Beyond the host plant of herbivores, it is the whole biotic and chemical environment that determines the response of an insect to pesticide exposure. Herbicides and insecticide solvents can serve as inducers (Brattsten and Wilkinson, 1977; Kao *et al.*, 1995; Miota *et al.*, 2000). Aquatic larvae are exposed to natural or anthropogenic compounds that alter their P450 detoxification profile (David *et al.*, 2006; Poupardin *et al.*, 2008, 2010; Riaz *et al.*, 2009). Virus infection affects P450 levels (Brattsten, 1987), and the expression of several P450 genes is affected during the immune response (see section 8.5.1).

8.4.5. Insecticide Resistance

8.4.5.1. Phenotype, genotype, and causal relationships Insecticide resistance is achieved in a selected strain or population: (1) by an alteration of the target site; (2) by an alteration of the effective dose of insecticide that reaches the target; or (3) by a combination of the two. The resistance phenotypes have long been analyzed according to these useful biochemical and physiological criteria. At the molecular genetic level, several classes of mutations can account for these phenotypes (Taylor and Feyereisen, 1996), and a causal relationship between a discrete mutation and resistance has been clearly established for several cases of target site resistance (ffrench-Constant *et al.*, 1999). The precise molecular mutations responsible for P450-mediated insecticide resistance are only beginning to be explored. The *cis*-mutation (*Accord* insertion) in the 5' UTR of the *Cyp6g1* gene in *Drosophila* causing upregulation of expression (Daborn *et al.*, 2002) is currently the most detailed account, and is discussed below. Although all classes of molecular mutations (structural, up- or downregulation; Taylor and Feyereisen, 1996) can be theoretically involved in P450-mediated resistance, it has been difficult finding, and establishing the role of resistance mutations for P450 genes.

Traditionally, the first line of evidence for a role of a P450 enzyme in resistance has been the use of an insecticide synergist (e.g., piperonyl butoxide), with suppression or decrease in the level of resistance by treatment with the synergist being diagnostic. In cases too many to list here, this initial and indirect evidence is probably correct; however, there are cases where piperonyl butoxide synergism has not been explained by increased detoxification (Kennaugh *et al.*, 1993). Piperonyl butoxide may also be a poor inhibitor of the P450(s) responsible for resistance, or may inhibit some esterase activity (see section 8.6.3 below), so the use of a second, unrelated synergist may be warranted (Brown *et al.*, 1996; Zhang *et al.*, 1997). In addition, the synergist as P450 inhibitor can decrease the activation of a proinsecticide, so that lack of resistance suppression can be misleading. Chlorpyriphos resistance in *D. melanogaster* from Israeli vineyards maps to the right arm of chromosome 2, and is enhanced rather than supressed by piperonyl butoxide (Ringo *et al.*, 1995). An independent and additional line of evidence is the measurement of total P450 levels or metabolism of selected model substrates, with an increase in either or both being viewed as diagnostic. Again, such evidence is tantalizing but indirect, and the absence of change uninformative.

An increase in the metabolism of the insecticide itself in the resistant strain is more conclusive, and has been demonstrated in many studies. For instance, permethrin metabolism to 4′-hydroxypermethrin was higher in microsomes from *Culex quinquefasciatus* larvae that are highly resistant to permethrin (Kasai *et al.*, 1998) than in their susceptible counterparts. Total P450 and cytochrome b_5 levels were 2.5 times higher in the resistant strain. Both permethrin toxicity and metabolism were inhibited by two unrelated synergists, TCPPE and piperonyl butoxide. A similarly convincing approach was taken to show P450 involvement in the resistance of house fly larvae of the YPPF strain to pyriproxifen. Gut and fat body microsomes were shown to metabolize the IGR to 4′-OH-pyriproxyfen and 5″-OH-pyriproxyfen at higher rates than microsomes of the susceptible strains, and this metabolism was synergist-suppressible (Zhang *et al.*, 1998). The major dominant resistance factor was linked to chromosome 2 in that strain (Zhang *et al.*, 1997).

Increased levels of transcripts for one or more P450 genes in insecticide resistant strains have now been reported in many cases (**Table 7**). DNA microarrays covering most genes encoding detoxification enzymes, large EST collections, or the whole genome have also been used in several species to accelerate the discovery of differentially expressed (mostly overexpressed) CYP genes in several species. These include *Drosophila*, mosquitoes, *T. castaneum*, and *M. persicae*. These studies suggest that

Table 7 CYP Genes Overexpressed in Insecticide-Resistant Strains

	Species	P450 Overexpressed	Strain	Resistance Pattern	Reference
Diptera, flies	*Musca domestica*	CYP6A1	Rutgers and other strains	OP, carbamates, IGR	Feyereisen *et al.* (1989); Cariño *et al.*, (1992); Sabourault *et al.* (2001)
		CYP6A5v2, CYP6A36	ALHF	pyrethroids	Zhu and Liu (2008); Zhu *et al.*, (2008a)
		CYP6D1, CYP6D3	LPR	pyrethroids	Liu and Scott, (1996); Kasai and Scott (2001b)
		CYP6D1,	NG98, Georgia	pyrethroids	Kasai and Scott (2000)
		CYP6D1, CYP6D3v2, CYP6A24	YS, YPER, Hachinohe	pyrethroids	Kamiya *et al.* (2001) Shono *et al.* (2002)
		CYP12A1	Rutgers		Guzov *et al.* (1998)
	Drosophila melanogaster	Cyp6a2	91R	DDT	Waters *et al.* (1992)
			MHIII-D23	malathion	Maitra *et al.* (2000)
		Cyp6a8	91R		Maitra *et al.* (1996)
			MHIII-D23		Maitra *et al.* (2000)
			Wis-1lab	DDT	Le Goff *et al.* (2003)
		Cyp6g1	Hikone R and many lines	DDT	Daborn *et al.* (2001, 2002); Catania *et al.* (2004)
			WC2	lufenuron, propoxur	Daborn *et al.* (2002)
			EMS1, and many lines	imidacloprid	Daborn *et al.* (2002); Catania *et al.* (2004)

Table 7   CYP Genes Overexpressed in Insecticide-Resistant Strains—cont'd

Species	P450 Overexpressed	Strain	Resistance Pattern	Reference
Diptera, mosquitoes		Wisconsin-1, 91-R	DDT	Brandt et al. (2002)
		Innisfail	diazinon	Pyke et al. (2004)
	Cyp12d1/2	Wisconsin-1, 91-R	DDT	Brandt et al. (2002); Le Goff et al. (2003)
	Cyp12a4	NB16	lufenuron	Bogwitz et al. (2005)
Drosophila simulans	CYP6G1	OV1	DDT, imidacloprid, malathion	Le Goff et al. (2003)
Anopheles gambiae	CYP6Z1, CYP325A3	RSP	pyrethroids	Nikou et al. (2003); David et al. (2005)
	CYP6M2, CYP6P3	various	pyrethroids	Djouaka et al. (2008)
	CYP6P3	Dodowa	permethrin	Muller et al. (2008b)
	CYP6M2, CYP6Z2	Odumasy	permethrin	Muller et al. (2007)
	CYP4C27, CYP4H15, CYP6Z1,2, CYP12F1, CYP314A1,	ZAN/U	DDT	Vontas et al. (2005); David et al. (2005)
Anopheles stephensi	CYP325C1	DUB-R	pyrethroids	Vontas et al. (2007)
Anopheles funestus	CYP6P4, CYP6P9	FUMOZ-R	pyrethroids	Amenya et al. (2008); Wondji et al. (2009)
Aedes aegypti	CYP9J10,27,32	PMD-R, IM	pyrethroids	Strode et al. (2008)
Culex pipiens quinquefasciatus	CYP9M10	ISOP450 ISOJPAL	permethrin	Hardstone et al. (2010)
	CYP6F1			Kasai et al. (2000)
	CYP4H34, CYP6Z10, CYP9M10	Jpal-per	permethrin	Komagata et al. (2010)
	CYP4H21, H22, H23 CYP4J4, CYP4J6	RR	deltamethrin	Shen et al. (2003)
Lepidoptera	Heliothis virescens	CYP9A1	Macon Ridge	thiodicarb	Rose et al. (1997)
	Helicoverpa zea	CYP6B8, B9	Texas (field)	cypermethrin	Hopkins et al. (2010)
	Helicoverpa armigera	CYP4G8		pyrethroids	Pittendrigh et al. (1997)
		CYP6B7		pyrethroids	Ranasinghe and Hobbs (1998)
		CYP6B7, CYP9A12, CYP9A14	YGF	pyrethroids	Yang et al. (2006)
		CYP4S1, CYP337B1	AN02	fenvalerate	Wee et al. (2008)
		CYP4L5,11, CYP4M6,7, CYP6AE11, CYP9A14, CYP332A1, CYP337B1	various strains	deltamethrin	Brun-Barale et al., (2010)
	Plutella xylostella	CYP6BG1			Bautista et al. (2007)
		CYP4M20	CR	cypermethrin	Baek et al. (2009)
Hemiptera	Lygus lineolaris	CYP6X1		permethrin	Zhu and Snodgrass (2003)
	Bemisia tabaci	CYP6CM1vQ		imidacloprid	Karunker et al. (2008)
	Myzus persicae	CYP6CY3	5191A	neonicotinoids	Puinena et al. (2010)
Coleoptera	Diabrotica virgifera	CYP4		Me-parathion, carbaryl	Scharf et al. (2001)
	Tribolium castaneum	CYP6BQ8, 9, 10, CYP436B1, B2, CYP6BK2	QTC279	deltamethrin	Zhu et al. (2010)
Blattaria	Blattella germanica	P450MA		chlorpyriphos	Scharf et al. (1999)
		CYP4G19		pyrethroids	Pridgeon et al.(2003)

overexpression of one or more P450 genes is a common phenomenon of metabolic resistance, but they do not by themselves establish a causal relationship with resistance. Instead, they provide lists of candidate genes (see, for example, Brun-Barale *et al.*, 2010).

Genetic linkage between increased mRNA or protein levels for a particular P450 and resistance has been obtained to the chromosome level or closer to marker genes (Bogwitz *et al.*, 2005; Feyereisen, 2005; Wee *et al.*, 2008; Hardstone *et al.*, 2010). Linkage is just the first step in establishing a causal link between a P450 gene and resistance. QTL mapping of resistance in mosquitoes has also provided shortlists of candidate genes for further functional studies.

Functional expression of the P450 enzymes in heterologous systems is required to establish the contribution to resistance of a CYP gene reported to be overexpressed. This has been achieved in several cases (Yang *et al.*, 2008; Karunker *et al.*, 2009; and sections below). Field-collected *H. zea* moths surviving a diagnostic dose of cypermethrin were shown to overexpress *CYP6B8* (Hopkins *et al.*, 2010), thus associating this P450 (known to metabolize cypermethrin (X. Li *et al.*, 2004) with resistance. Transgenic expression in *Drosophila* can provide convincing evidence (see section 8.6.2.6), but such transgenesis can be a blunt instrument. For instance, while Bogwitz *et al.* (2005) demonstrated that *Cyp12a4* overexpression in the midgut and Malpighian tubules confers lufenuron resistance in *Drosophila*, they also showed that high-level ectopic expression of the gene was embryonic-lethal (for an unknown reason). RNAi suppression of CYP genes (Yang *et al.*, 2007; Zhu *et al.*, 2010; section 8.4.5.7) is very informative, but this technique is not equally effective in all insect species. RNAi suppression of *CYP6BG1* in a permethrin-resistant strain of *Plutella xylostella* caused a significant reduction of resistance, thus strongly supporting the importance of *CYP6BG1* overexpression in resistance (Bautista *et al.*, 2009).

The following is a discussion of selected cases of P450 genes associated with insecticide resistance. Evidence for mutations causing constitutive overexpression in *cis* and *trans*, as well as examples of P450 gene duplication/amplification, are currently available. The variety of mechanisms, even in a single species in response to the same insecticide, is striking. The paucity of available data on the molecular definition of the resistant genotype and on its causal relationship to resistance is also striking when compared to the wealth of data on target site resistance (ffrench-Constant *et al.*, 1999).

8.4.5.2. CYP6A1 and diazinon resistance in the house fly Rutgers strain

CYP6A1 is the first insect P450 cDNA to have been cloned. The gene is phenobarbital-inducible and constitutively overexpressed in the multiresistant Rutgers strain (Feyereisen *et al.*, 1989). A survey of 15 house fly strains (Cariño *et al.*, 1992) showed that *CYP6A1* is constitutively overexpressed at various degrees in eight resistant strains, but not in all resistant strains – including, notably, R-Fc, known to possess a P450-based resistance mechanism. Thus, the first survey with a P450 molecular probe confirmed the results of the first survey of house fly strains with marker P450 activities (aldrin epoxidation and naphthalene hydroxylation) (Schonbrod *et al.*, 1968): that there is no simple relationship between resistance and a molecular marker, here the level of expression of a single P450 gene. That different P450 genes would be involved in different cases of insecticide resistance was a sobering observation (Cariño *et al.*, 1992), even before the total number of P450 genes in an insect genome was known. The constitutive overexpression of *CYP6A1* in larvae and in adults is linked to a semi-dominant factor on chromosome 2 (Cariño *et al.*, 1994), but the *CYP6A1* gene maps to chromosome 5 (Cohen *et al.*, 1994), implying the existence of a chromosome 2 *trans*-acting factor(s) differentially regulating *CYP6A1* expression in the two strains (Cariño *et al.*, 1994). The gene copy number and the coding sequence of CYP6A1 are identical between Rutgers and a standard susceptible strain (*sbo*) (Cariño *et al.*, 1994; Cohen *et al.*, 1994). Competitive ELISA using purified recombinant CYP6A1 protein as standard showed that the elevated mRNA levels are indeed translated into elevated protein levels (Sabourault *et al.*, 2001) (**Table 8**). Reconstitution of recombinant CYP6A1 expressed in *E.coli* with its redox partners (Sabourault *et al.*, 2001) provided the conclusive evidence for its role in diazinon resistance, as CYP6A1 metabolizes the insecticide with a high turnover (18.7 pmol/pmol

Table 8 Comparison of a Susceptible and a Resistant Strain of the House Fly

Fly Strain	Sbo (S)	Rutgers (R)
Diazinon contact toxicity LC50 (μg/pint jar)	4.4	167.8
resistance ratio	1	37.8
Diazinon topical toxicity LD50 (μg/fly)	0.059*	7.1
resistance ratio	1	120
P450 level (nmol/mg protein)	0.14	0.29
aldrin epoxidation (pmol/min/pmol P450)	4.4*	15.6
CYP6A1 mRNA relative level	1	27.5
CYP12A1 mRNA relative level	1	15
CYP6A1 protein level (fmol/abdomen)	36	565
Diazinon metabolized by CYP6A1		
oxidative cleavage	0.5	7.7
desulfuration to oxon (pmol/min/fly)	0.2	2.9
OP oxon metabolized by mutant ali-esterase (pmol/ min/fly)**	0.0	1.5
NADPH-dep. diazinon metabolism by microsomes (pmol/min/ abdomen)	2.9*	15.5

*NAIDM susceptible strain
**estimations based on the calculations of Devonshire *et al.* (2003)

CYP6A1 per min), and a favorable ratio, 2.7, between oxidative ester cleavage and desulfuration (**Table 6**).

The nature of the chromosome 2 *trans*-acting factor and of the mutation leading to resistance in the Rutgers strain has remained enigmatic, despite considerable circumstantial evidence for a major resistance factor on chromosome 2 (Plapp, 1984). Diazinon resistance and high CYP6A1 protein levels could not be separated by recombination in the short distance between the *ar* and *car* genes (3.3–12.4 cM). This region carries an ali-esterase gene (MdαE7). A Gly137 to Asp mutation in this ali-esterase abolishes carboxylesterase activity towards model compounds such as methylthiobutyrate, and confers a low but measurable phosphotriester hydrolase activity towards an organophosphate ("P=O"), chlorfenvinphos, in both the sheep blowfly *Lucilia cuprina* LcαE7 and *M. domestica* MdαE7 enzymes (Newcomb *et al.*, 1997; Claudianos *et al.*, 1999). Chromosome 2 of the Rutgers strain carries this *MdαE7* Gly137 to Asp mutation, and low CYP6A1 protein levels are correlated with the presence of at least one wild type (Gly137) allele of *MdαE7*. Recombination in the *ar–car* region could not dissociate diazinon susceptibility, low CYP6A1 protein level, and the presence of a Gly137 allele of the ali-esterase (Sabourault *et al.*, 2001). It was therefore hypothesized that the wild type ali-esterase metabolizes an (unknown) endogenous substrate into a negative regulator of *CYP6A1* transcription. Removal of this regulator (by loss of function of the ali-esterase) would increase CYP6A1 production, and hence diazinon metabolism. House fly strains that are susceptible or that are not known to overexpress *CYP6A1* predictably carry at least one wild type Gly137 allele (Scott and Zhang, 2003). The LPR strain, which overexpresses *CYP6A1* (Cariño *et al.*, 1992) and has increased OP metabolism (Hatano and Scott, 1993), as well as other resistant strains, carry other alleles of *MdαE7* (Scott and Zhang, 2003; Gacar and Taskin, 2009). These alleles, Trp251 to Ser or Leu, also have impaired ali-esterase activity in *M. domestica* (Claudianos *et al.*, 1999; Gacar and Taskin, 2009). The hypothesis that the wild type allele of *MdαE7* is a *trans*-acting negative regulator causing low expression of *CYP6A1* in susceptible strains has implications. The pleiotropic effect of *trans* regulation is compatible with constitutive overexpression of *CYP12A1* (whose product metabolizes diazinon as well; Guzov *et al.*, 1998) and of *GST-1*, which are both controlled in the Rutgers strain by a chromosome 2 factor – possibly the same as the one controlling *CYP6A1* expression.

The diazinon resistance (*Rop-1*) and malathion resistance (*Rmal*) in *L. cuprina* are linked to the Gly137Asp and Trp251Leu mutations in *LcαE7* (Newcomb *et al.*, 1997). The Trp251Leu mutation enhances hydrolysis of dimethylorganophosphates and of the malathion carboxylesters, while the Gly137Asp mutation enhances preferentially the hydrolysis of diethylorganophosphates,

but virtually abolishes malathion carboxylesterase activity (Devonshire *et al.*, 2003). In view of these data, could the Gly137Asp mutation alone be responsible for diazinon resistance in the blowfly? The elegant calculations of Devonshire *et al.* (2003) show that, despite the very low activity of the Gly137Asp mutant esterase to hydrolyze the oxon form of the pesticide (k_{cat} ~0.05/min), the 10- to 20-fold resistance to diazinon and parathion could indeed be accounted for by LcαE7. Those calculations, based on the oxon form, do not take into account the necessary P450-dependent desulfuration of the parent OP to produce the oxon, which is invariably associated with ester cleavage (see above), so P450 may still play a role in the blowfly. Indeed, the Q strain of the sheep blowfly is more resistant to parathion than to paraoxon (Hughes and Devonshire, 1982), and indirect evidence for a P450 involvement in *L. cuprina* diazinon resistance has also been presented (Kotze, 1995; Kotze and Sales, 1995).

In the Rutgers strain of the house fly, the situation is different. The mutant ali-esterase cannot account for the P450-dependent carbamate and JHA resistance. Furthermore, the higher resistance to diazinon than in *L. cuprina* (120× vs 10×) requires a more efficient clearance of the oxon. The contribution of CYP6A1 alone accounts for over five times more than the mutant ali-esterase to the timely removal of the toxic form. Lethality of the *MdαE7* null (Sabourault *et al.*, 2001) suggests that the Gly137Asp mutation is the optimal loss-of-function mutation, as the Rutgers haplotype has swept through global populations of the house fly (Claudianos, 1999; Gacar and Taskin, 2009). The low phosphotriester hydrolase activity of the mutant ali-esterase probably helps clearing the activated form (P=O) of the insecticide (Sabourault *et al.*, 2001). The endogenous function of the wild type *MdαE7* gene remains to be elucidated.

8.4.5.3. *Drosophila Cyp6g1*, the *Rst(2)DDT* gene at 64.5 on chromosome 2

The power of *Drosophila* genetics, coupled with the tools made possible by the complete genome sequence, has provided the most detailed, yet complex, molecular genetic detail about a P450-based insecticide resistance mechanism. The resistance gene *Rst(2)DDT* has been genetically characterized for over 40 years (see reviews in Daborn *et al.*, 2001; Wilson, 2001). The position of this gene around 64.5 cM on the left arm of chromosome 2 has become almost mythical, as a number of phenotypes were linked to this locus, ranging from dominant DDT resistance to phenylthiourea susceptibility, from organophosphorus to carbamate resistance, and from various P450-dependent activities to vinyl chloride activation. EMS mutagenesis of a wild type stock and selection with imidacloprid led to two strains with moderate imidacloprid resistance and moderate cross-resistance to DDT (Daborn *et al.*, 2001). Conversely, two DDT-resistant strains (Hikone-R

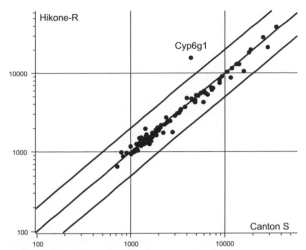

Figure 26 *Cyp6g1* overexpression in the Hikone-R insecticide-resistant strains when compared to the Canton-S susceptible strain of *Drosophila melanogaster*. DNA microarray analysis revealed overexpression of just one CYP gene (Daborn *et al.*, 2002). Adapted from ffrench-Constant *et al.*, 2004.

and Wisconsin-1) were shown to be cross-resistant to imidacloprid. Fine-scale mapping of this dominant resistance localized *Rst(2)DDT* to a region from 48D5-6 to 48F3-6 on the polytene chromosome map. Of three candidate P450 genes in this region, *Cyp6g1*, *Cyp6g2*, and *Cyp6t3*, only the first showed constitutive overexpression in the DDT and imidacloprid resistant strains tested (Daborn *et al.*, 2001). A DNA microarray comprising probes for all the *Drosophila* P450 genes was addressed with target cDNAs from susceptible strains, and from the DDT-resistant Hikone-R strain and the propoxur-resistant WC2 strain. In both cases, *Cyp6g1* was the only P450 gene showing constitutive overexpression (Daborn *et al.*, 2002) (**Figure 26**). Overexpression of *Cyp6g1* was confirmed by quantitative RT-PCR in 20 strains, and DDT, imidacloprid, nitenpyram, and lufenuron resistances were all independently mapped to the *Cyp6g1* locus in the Hikone-R and WC2 strains. The insertion of a terminal direct repeat of the transposable element *Accord* was systematically found in the 5′ UTR of 20 different resistant strains from across the globe. Phylogenetic analysis of the first intron sequence of the gene showed a unique haplotype in resistant strains vs a large diversity of susceptible haplotypes, suggesting a selective sweep had occured in global *Drosophila* populations (Daborn *et al.*, 2002). This was further demonstrated in a survey of 673 lines from 34 populations collected around the world showing perfect correlation between the presence of the *Accord* insertion and resistance, and a reduction in variability measured by microsatellite analysis in a 20-kb region downstream of *Cyp6g1* (Catania *et al.*, 2004). In some resistant lines, the presence of a *P*-element insertion into the *Accord* element was reported in that study. In fact,

the *Cyp6g1* locus is even more complex, with at least six different alleles found in nature (Schmidt *et al.*, 2010). Beyond the *Cyp6g1* ± *Accord* insertion alleles, Schmidt and colleagues reported four alleles in which the *Cyp6g1* with *Accord* insertion was duplicated, and in three of them additional insertions of *P* elements or *HMS Beagle* elements were found within the original *Accord* insertion. This allelic succession is adaptive, with higher resistance and *Cyp6g1* transcription found for the most complex allele (*Cyp6g1-[BP]*; Schmidt *et al.*, 2010).

Transgenic flies producing CYP6G1 under the control of a variety of promoters in the GAL4/UAS system (heat-shock, tubulin, or midgut/fat body/Malpighian tubules) showed increased survival to acetamiprid, imidacloprid, and nitenpyram in larvae, and to DDT in adults (Daborn *et al.*, 2002; Le Goff *et al.*, 2003; Chung *et al.*, 2007; Daborn *et al.*, 2007). Significantly, the 491-bp *Accord* sequence carries enhancer elements itself, and can direct expression of reporter GFP in the tissues (gastric ceca, midgut, Malpighian tubules, and fat body of larvae) in which *Cyp6g1* expression is localized in resistant strains (Chung *et al.*, 2007). Moreover, the Malpighian tubules of adults are critical, as overexpression of *Cyp6g1*, or its RNAi knockdown, in just this tissue can significantly shift the toxicity of DDT to lower or higher levels, respectively (Yang *et al.*, 2007). Definitive functional evidence that CYP6G1 metabolizes insecticides was provided by Joussen *et al.* (2008), who showed that the enzyme produced in tobacco-cell suspension cultures metabolizes imidacloprid and DDT, as predicted by the experiments with transgenic flies resulting in resistance, and by homology modeling (Jones *et al.*, 2010).

In most field-collected strains, DDT resistance is significant but low compared to that of strains further selected in the laboratory (e.g., 91-R, see below), suggesting that while *Cyp6g1* may constitute a first line of defense seen in field populations, further insecticide pressure in the laboratory may select additional mechanisms. In addition, mechanisms other than *Cyp6g1* overexpression (including target site resistance) can also be involved in DDT resistance. For instance, while *Cyp6g1* overexpression is observed in the DDT-resistant Wisconsin and 91R strains, the *Cyp12d1* (or *Cyp12d2*) gene is also overexpressed in both strains (Brandt *et al.*, 2002; Festucci-Buselli *et al.*, 2005), and its transgenic overexpression can confer resistance to DDT (Daborn *et al.*, 2007). Moreover, several other genes are overtranscribed in these strains as well (Pedra *et al.*, 2004). Therefore, it is hardly surprising that resistance, which has always been a relative term, is not restricted to *Cyp6g1* overexpression, especially when compared to strains that are themselves resistant (Festucci-Buselli *et al.*, 2005; Kuruganti *et al.*, 2006).

In a Brazilian strain of *Drosophila simulans* resistant to DDT, imidacloprid, and malathion, only the *Cyp6g1* ortholog is overexpressed (Le Goff *et al.*, 2003). In a

Californian population of *D. simulans*, the 5′-flanking sequence of the *Cyp6g1* ortholog is nearly fixed for a *Doc* transposable element insertion. This insertion is absent from African populations, and is associated with increased transcript abundance of *Cyp6g1* and resistance, in what appears to be a case of resistance analogous with the *Accord* case of *D. melanogaster Cyp6g1* (Schlenke and Begun, 2004).

8.4.5.4. Overexpression of *Cyp6a2* in *Drosophila*

The *Drosophila Cyp6a2* gene is constitutively overexpressed in the 91-R strain, resistant to malathion and DDT by a factor of 20–30 relative to the susceptible 91C strain (Waters *et al.*, 1992). DDT resistance maps to 56 cM on the left arm of chromosome 2 in the 91R strain (Dapkus, 1992), which is at or near the chromosomal location of *Cyp6a2* (43A1-2). Initially, the presence of a 96-bp insertion in the 3′ UTR of the gene was proposed to confer a low level of expression to the *Cyp6a2* gene (Waters *et al.*, 1992), but this insertion (or, rather, the lack of it) was neither correlated with DDT resistance (Delpuech *et al.*, 1993) nor confirmed to be linked to overexpression in resistant strains (Dombrowski *et al.*, 1998). *Cyp6a8* is also constitutively overexpressed in the 91R strain (Maitra *et al.*, 1996), and the expression of both *Cyp6a2* and *Cyp6a8* is repressed in 91R/91C hybrids (Maitra *et al.*, 1996; Dombrowski *et al.*, 1998; Maitra *et al.*, 2000). Transgenic flies (ry[506], insecticide susceptible and with constitutively low expression of *Cyp6a2*) carrying *Cyp6a2* with 985- or 1331-bp upstream DNA of the 91R strain express the transgene at higher levels than the endogenous *Cyp6a2*, but at lower levels than in their native 91R background (Dombrowski *et al.*, 1998). The expression of both *Cyp6a2* and *Cyp6a8* is also constitutively higher in the MHIII-D23 strain initially selected for malathion resistance (Maitra *et al.*, 2000). Genetic crosses and chromosome substitution experiments conclusively showed that the expression of both genes (located on the 2R chromosome) is repressed by factors on the third chromosome of the insecticide-susceptible 91C and rosy[506] strains. In contrast, the third chromosome of the MHIII-D23 and 91R strains carries a loss-of-function mutation for this negative *trans* regulator, allowing the constitutive overexpression of the two genes (Maitra *et al.*, 2000). Further careful examination of the promoter activity of the *Cyp6a8* gene by fusion with a luciferase reporter gene in transgenic flies identified a –11/–761-bp region of the *Cyp6a8* gene of the 91R strain that was sufficient to respond to the negative regulation by the rosy[506] (wild type) *trans* acting factor (Maitra *et al.*, 2002). CYP6A8 does not metabolize DDT when expressed in yeast (Helvig *et al.*, 2004b), but CYP6A2, similar to CYP6A1, appears to have a broad substrate specificity. Whether CYP6A2 of the 91R strain is capable of metabolizing DDT is unknown, as its sequence differs

from that of the baculovirus-produced CYP6A2, which does not metabolize DDT. Transgenic overexpression of *Cyp6a2* or *Cyp6a8* does not confer resistance to DDT, nitenpyram, dicyclanil, or diazinon (Daborn *et al.*, 2007), so the role of *Cyp6a2* in resistance remains conjectural. Also, transgenic overexpression of an SVL mutant of *Cyp6a2* previously reported to enable DDT metabolism (Amichot *et al.*, 2004) did not confer DDT resistance (Perry *et al.*, 2011), and the former study has not been confirmed. Nonetheless, the studies with the 91R and MHIII-D23 strains are clear indications for loss-of-function mutations in gene(s) encoding negative regulators of P450 gene expression on chromosome 3. Genetic analyses of malathion or fenitrothion resistance and of P450 expression (electrophoretic bands and marker activities) (see, for example, Hallstrom, 1985; Hallstrom and Blanck, 1985; Houpt *et al.*, 1988; Waters and Nix, 1988; Miyo *et al.*, 2002) have pointed to one or more loci between 51 cM and 61 cM on the right arm of chromosome 3, and it is interesting that this chromosome arm is thought to be orthologous to chromosome 2 of the house fly (Weller and Foster, 1993). The *Drosophila* ortholog of the *MdαE7* gene (*Est23*) is located at 84D9.

8.4.5.5. *CYP6D1*, the LPR strain, and pyrethroid resistance in the house fly

The LPR strain of the house fly is highly resistant to pyrethroids with a phenoxybenzyl moiety. This permethrin-selected strain has several resistance mechanisms, with important contributions of the genes *pen* (for reduced penetration) and *kdr* (for target site resistance) (Liu and Scott, 1995), and with P450-based detoxification as a major contributor. An abundant form of P450 (P450Lpr) was purified from abdomens of adult LPR flies, and immunological data indicated that P450Lpr represents 67% of the P450 in microsomes from LPR flies – a 10-fold (Wheelock and Scott, 1990) increase over the reference susceptible strain. The *P450Lpr* gene, *CYP6D1* (Tomita and Scott, 1995), is located on chromosome 1 of the house fly (Liu *et al.*, 1995), and is constitutively overexpressed by about 10-fold in the LPR strain. This overexpression is not caused by gene amplification, but by increased transcription. It has been claimed that increased transcription of *CYP6D1* causes insecticide resistance (Liu and Scott, 1998), but transgenic expression of *CYP6D1* in *Drosophila* (Korytko *et al.*, 2000a) has not been reported to confer resistance, and heterologously expressed CYP6D1 (Smith and Scott, 1997) has not been reported to metabolize pyrethroids. Instead, the evidence for the role of CYP6D1 in pyrethroid resistance is based on the inhibition of microsomal deltamethrin and cypermethrin metabolism by anti-P450Lpr antibodies (Wheelock and Scott, 1992a; Korytko and Scott, 1998). Deltamethrin is metabolized preferentially at the gem-dimethyl group on the acid moiety (Wheelock and Scott, 1992a), whereas cypermethrin is mainly hydroxylated at the 4′ position on the alcohol moiety, at an extremely low rate (Zhang and Scott, 1996b). Whether overexpression or point mutations of *CYP6D1*, or both, are involved in pyrethroid resistance in the LPR strain is yet unknown. Indeed, the *CYP6D1* gene sequence from 5 strains shows a high polymorphism, with 57 variable sites in the coding region alone, of which 12 are non-silent (Tomita *et al.*, 1995). Six amino acid changes are specific to the LPR strain (CYP6D1v1) when compared to pyrethroid-susceptible strains, and several of these mutations appear to align with SRS3.

The 170-fold piperonyl butoxide-suppressible resistance is conferred by a combination of the resistant chromosomes 1 and 2 from the parent LPR strain in the homozygous condition (Liu and Scott, 1995). There is no substantial resistance or CYP6D1 overproduction conferred by isolated LPR chromosomes 1 or 2, or by their subsequent combination (Liu and Scott, 1995; Liu and Scott, 1996). Thus, in the LPR strain, P450-mediated resistance requires both copies of the LPR chromosomes 1 and 2. The resistance and *CYP6D1* overexpression linked to chromosome 1 are dominant, whereas the contributions of chromosome 2 are mostly recessive (Liu and Scott, 1996; Liu and Scott, 1997). These data suggest a unique combination of the LPR strain of chromosome 2 *trans*-acting factor(s) with at least a matched *cis*-factor on chromosome 1. A key sequence difference between the 5′ UTR of the *CYP6D1v1* allele of LPR and of susceptible strains is the presence of a 15-bp insert that interrupts a binding site of the transcriptional repressor mdGfi-1, and reduces the amount of mdGfi-1 binding to the *CYP6D1* promoter in electrophoretic mobility shift assays (Gao and Scott, 2006). This strongly suggests that the 15-bp insertion is the major *cis*-mutation on chromosome 1. This insert is also found in a permethrin-resistant strain NG98, from Georgia, USA, that carries a virtually identical *CYP6D1v1* haplotype (Seifert and Scott, 2002).

However, in a strain from neighboring Alabama with high permethrin resistance (ALHF), chromosomes 1 and 2 play little role, but chromosome 5 plays a major piperonyl butoxide-suppressible role (Liu and Yue, 2001). Two P450 genes located on chromosome 5, *CYP6A36* and *CYP6A5v2* (a probable recent duplicate of *CYP6A5*), are constitutively overexpressed in that strain (Zhu and Liu, 2008; Zhu *et al.*, 2008a). A gene closely linked to *CYP6D1* on chromosome 1 codes for a similar (78% identity) P450, CYP6D3 (Kasai and Scott, 2001a). *CYP6D3* is 12-fold overexpressed in adult flies of the LPR strain, but it is also expressed in larvae (Kasai and Scott, 2001b), as opposed to *CYP6D1*, which has an adult-specific pattern of expression. *CYP6D1* is overexpressed by about 2.4-fold in the Japanese strain YPER (Kamiya *et al.*, 2001; Shono *et al.*, 2002). This strain does not carry the *CYP6D1v1* allele, and chromosome 2 has a major role in this permethrin-resistant strain. *CYP6D3v2* and *CYP6A24* are also overexpressed in Japanese strains (Kamiya *et al.*, 2001).

Pyrethroid resistance in the house fly appears thus to involve multiple mutations playing a role in the contribution of several P450s to the multifactorial resistance.

8.4.5.6. DDT and pyrethroid resistance in mosquitoes

Major efforts have followed the sequencing of the *A. gambiae* genome to identify P450 genes involved in insecticide resistance in mosquitoes, in particular pyrethroid resistance, which threatens the effectiveness of indoor residual spraying and of insecticide-treated bednets. The genome of *A. gambiae*, as well as the genomes of *Aedes aegypti* and *Culex pipiens* (**Table 1**), have among the largest CYPomes described to date, so global approaches have been most successful. These include QTL mapping in *A. gambiae* (Ranson *et al.*, 2004) and *A. funestus* (Wondji *et al.*, 2009), and microarray hybridizations. The microarray experiments have compared susceptible and resistant strains, but also field-caught mosquitoes, in *A. gambiae* (David *et al.*, 2005; Vontas *et al.*, 2005; Muller *et al.*, 2007; Djouaka *et al.*, 2008; Muller *et al.*, 2008a), *A. stephensi* (Vontas *et al.*, 2007), and *A. arabiensis* (Muller *et al.*, 2008b). Microarrays have also been used for *Aedes aegypti* (Strode *et al.*, 2008) and *Culex pipiens* (Komagata *et al.*, 2010). These and other approaches (Nikou *et al.*, 2003; Amenya *et al.*, 2008; Boonsuepsakul *et al.*, 2008; Hardstone *et al.*, 2010) have collectively identified a relatively restricted number of CYP genes overexpressed in mosquitoes and potentially involved in resistance. In most cases several CYP genes are overexpressed together, albeit to different levels. Functional expression of a few *A. gambiae* P450s has identified CYP6Z1 as metabolizing DDT and carbaryl but not pyrethroids (Chiu *et al.*, 2008), whereas CYP6Z2 metabolizes carbaryl and alkoxyresorufins, but neither DDT nor pyrethroids (Chiu *et al.*, 2008; McLaughlin *et al.*, 2008). In contrast, CYP6P3 and CYP6M2 were both shown to metabolize pyrethroids (Muller *et al.*, 2008a; Stevenson *et al.*, 2011). CYP6AA3 of *A. minimus* was also shown to metabolize deltamethrin (Boonsuepsakul *et al.*, 2008). Several important candidate genes remain to be characterized (e.g., *CYP6P4*, *CYP6P9*, *CYP9M10*, *CYP325A3*), and the molecular mechanisms of their overexpression explored.

8.4.5.7. Deltamethrin resistance in the red flour beetle

Resistance to deltamethrin in the QTC279 strain of *T. castaneum* is caused by P450-mediated metabolism that is suppressible by piperonyl butoxide. *CYP6BQ9* was identified as a major contributor to resistance (Zhu *et al.*, 2010). These authors used microarrays to identify differentially expressed genes, and confirmed overexpression by qRT-PCR for six P450 genes. One of them, *CYP6BQ9*, shows over 200-fold overexpression, and knockdown of its expression by dsRNA injection causes a substantial suppression of resistance. Heterologous expression in the baculovirus system has established that CYP6BQ9 can hydroxylate deltamethrin at the 4′ position. Remarkably, *CYP6BQ9* is expressed mainly in the brain, and overexpression in the QTC279 strain is seen mostly in the brain. Predominant transgenic expression in the brain of *Drosophila* is sufficient to confer resistance *in vivo* (Zhu *et al.*, 2010). *CYP6BQ9* is 1 of 3 overexpressed genes in a large cluster of 12 *CYP6BQ* genes, and this raises interesting questions regarding the nature of the mutation(s) that led to resistance in this experimentally tractable system.

8.4.5.8. CYP gene duplications and resistance

Copy number variations that can lead to fixed gene duplications are emerging as an important factor in genome evolution, and these are targets of selection (Emerson *et al.*, 2008). Careful analysis of several CYP genes associated with resistance has revealed cases of copy number variation or gene duplication. The *Cyp6g1* duplication in *Drosophila* (Schmidt *et al.*, 2010) has been discussed above, and the neighboring *Cyp12d1/d2* pair may also be linked to resistance. *CYP9M10* is duplicated within a 100-kb amplicon in a resistant strain of *Culex quinquefasciatus*, and perhaps linked to the presence of a MITE-like element. The duplicated haplotype is linked to the 260-fold pyrethroid resistance (Itokawa *et al.*, 2010). In *A. funestus*, *CYP6P4* and *CYP6P9* are both overexpressed in a pyrethroid-resistant strain. These genes are both present as tandem duplicates in the middle of a cluster of nine other *CYP6AA* and *CYP6P* genes (Wondji *et al.*, 2009). These duplications predate selection by insecticides, as judged by their 5% nt divergence. Perhaps caused by a more recent event, the *CYP6CY3* gene of the aphid *M. persicae* is overexpressed about 22-fold in the 5191A neonicotinoid-resistant clone. When compared to a susceptible clone, a nine-fold amplification of the gene can be measured (Puinean *et al.*, 2010). There are at least five sequence variants in the 5′ UTR region of *CYP6CY3*, and at least one amplified copy represents a pseudogene with a premature stop codon. The role of this gene in resistance and the mutational events involved are still unclear.

8.4.5.9. P450-mediated resistance: evolution

Resistance-conferring point mutations in insecticide target sites (Rdl, AchE, sodium channel) have a remarkable pattern of parallel evolution in widely divergent species. This is clearly not the case for P450-based resistance, beyond the probable example of *Cyp6g1* in *D. melanogaster* and *D. simulans* (which are only 2.5 million years apart). Rather, the multitude of P450 genes, whose expression is often inducible and therefore not strongly expressed in most developmental stages/tissues, constitutes a "reservoir" in which mutations affecting expression levels can be selected following insecticide exposure. In the field, this can lead to selective "sweeps" of these most adapted mutations, as seen for the global predominance of the

Cyp6g1/Accord haplotypes in fruit flies. In the latter case the resistance allele is actually favored, because it carries no fitness cost in the absence of insecticide (McCart *et al.*, 2005) – a rather unusual observation that may be linked to some yet unrecognized reproductive/developmental function of the CYP6G1 enzyme (Drnevich *et al.*, 2004; McCart and ffrench-Constant, 2008). Patterns of P450-dependent resistance evolution will emerge when the precise molecular basis of single or multiple CYP gene overexpression is better documented.

8.5. Regulation of P450 Gene Expression

8.5.1. Spatial and Temporal Patterns of P450 Gene Expression

P450-dependent enzyme activities have been detected in virtually all insect tissues and developmental stages (Feyereisen, 2005). With molecular probes, the expression patterns of individual P450 genes have now been presented in different tissues, developmental stages, or induction regimes with varying degrees of detail, and with probes of varying degree of specificity (Feyereisen, 2005). Quantitative RT-PCR is currently the most common method used in studies reporting a specific pattern of expression, often in confirmation of DNA microarray experiments. The FlyAtlas resource for gene expression in *Drosophila* (Chintapalli *et al.*, 2007) (http://flyatlas.org) was specifically mined for P450s (Yang *et al.*, 2007), and the analysis of the Malpighian tubule transcriptome showed that 29 CYP genes were significantly enriched in this issue (Wang *et al.*, 2004). Microarrays and other transcriptomic approaches have been used to describe spatial and developmental patterns and to analyze many experimental settings, particularly in the model insect *Drosophila* (e.g., Graveley *et al.*, 2011), but also in mosquitoes (e.g., Marinotti *et al.*, 2005; Vontas et al., 2005; Strode *et al.*, 2006; Poupardin *et al.*, 2010) and the silkworm (Xia *et al.*, 2007). Often, information on CYP genes remains to be mined from these studies. Comprehensive and specific analyses of the CYPomes in the human liver (Yang *et al.*, 2010) or in the model plant *Arabidopsis thaliana* (Ehlting *et al.*, 2008) show the power of such correlative studies.

In most studies using transcriptomics, one or more CYP genes is shown to be affected in its expression but it is easy to overstate the relationship that may be a remote effect in a cascade of events. Nonetheless, CYP genes figure regularly in transcriptomic studies related to infection and immunity, aging, oxidative stress (Girardot *et al.*, 2004; Feyereisen, 2005; Johansson *et al.*, 2005; Lai *et al.*, 2007; Felix *et al.*, 2010), and circadian rhythmicity. Profiles of circadian rhythms in *Drosophila* have revealed the co-regulation of 6 *Cyp6a* genes that are adjacent in a cluster at 51D5 on chromosome 2R (Ueda *et al.*, 2002),

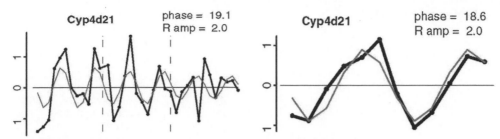

Figure 27 Circadian rhythmicity of *Cyp4d21* expression in *Drosophila melanogaster* expressed as a log ratio of change. Left: 36 time points collected over 6 days (Affymetrix microarray data); Right: 12 time points collected over 2 days. Estimated phases and log ratio amplitudes are indicated. Red lines are 24-h guidelines. Reproduced from Claridge-Chang *et al.* (2001), with permission.

as well as the control of *Cyp4e2* expression by the *Clk* gene. A number of other P450 genes are also cycling with different phases, *Cyp4d21* in particular (Claridge-Chang *et al.*, 2001; McDonald and Rosbash, 2001; Ceriani *et al.*, 2002) (**Figure 27**). Circadian variations in ethoxycoumarin dealkylase activity, and propoxur and fipronil toxicity, have been reported in *Drosophila* (Hooven *et al.*, 2009). Moreover, flies deficient in the clock element cyc and its target transcription factor Pdp1 have increased sensitivity to permethrin, and downregulated *Cyp6g1* and *Cyp6a2* expression (Beaver *et al.*, 2010). These studies suggest that great care must be exercised in studies of P450 regulation and, more generally, insect toxicology, to eliminate (or study) variability due to circadian rhythms.

Spatial and temporal patterns of CYP gene expression are also studied at a finer anatomical scale, by *in situ* hybridization (Feyereisen, 2005) or with reporter genes. For instance, the *Cyp6a2* promoter linked to a GFP reporter gene has been used in transgenic *Drosophila* to document precisely the expression of this gene (Giraudo *et al.*, 2010). The landmark study of Chung *et al.* (2009) describes by *in situ* hybridization the expression pattern of 50 CYP genes during embryogenesis, and 58 CYP genes in third instar larvae. Their study highlights the importance of the midgut, fat body, and Malpighian tubules as the tissues where at least 35 CYP genes are expressed, most probably involved in detoxification. These authors also showed that 9 of 59 P450s targeted by RNAi were lethal – a figure that may be an underestimate, but suggests many unexplored physiological functions.

8.5.2. CYP Gene Induction by Xenobiotics

Beyond the developmental and hormone regulation of CYP genes involved in precise physiological functions, one remarkable feature of P450 genes is that the transcription rate of many of them (sometimes gene batteries that also include genes other than P450s) is induced by foreign chemicals or "xenobiotics." Much of our current understanding of induction is derived from studies on the mammalian CYP genes and the receptors that regulate their expression, in particular the Ah receptor and the nuclear

receptors PXR (NR1I2) and CAR (NR1I3) (see Pascussi *et al.*, 2008, for review). These receptors play a key role at the interface between physiological responses and environmental responses. Little is known of these "xenosensors" in insects, but the major types of inducers known in the vertebrate toxicology literature, if not each specific chemical, are known to be active as inducers in *Drosophila* and in other insects (Feyereisen, 2005).

The variety of chemicals known to induce specific CYP genes in insects is shown in **Table 9**. It illustrates the fact that virtually any chemical, natural or synthetic, can at some dose induce one or many CYP genes. But how many genes? Microarray studies on *Drosophila* have started to answer this question (Zou *et al.*, 2000; Girardot *et al.*, 2004; Willoughby *et al.*, 2006, 2007; Jensen *et al.*, 2006a, 2006b; King-Jones *et al.*, 2006; Le Goff *et al.*, 2006; Sun *et al.*, 2006). A survey of these 17 datasets on chemical treatments of *Drosophila* with 10 different chemicals showed that about a third of the CYPome comprises xenobiotic-inducible genes (Giraudo *et al.*, 2010). In that survey, two broad classes were recognized. One class is thought to represent a stress response represented by tunicamycin- (8) and paraquat- (9) inducible genes. The other covers genes induced by the other chemicals, and is thought to represent the response to chemical challenge. This classification may be biased by the small number of chemicals tested. This "round up the usual suspects" approach identifies eight genes as the most likely to be included in a list of significantly induced genes. This list – *Cyp4p1*, *Cyp6a2*, *Cyp6a8*, *Cyp6d4*, *Cyp6d5*, *Cyp6w1*, *Cyp12d1*, and *Cyp28a5* (Giraudo *et al.*, 2010) – consists mainly of CYP3 clan genes, but also includes a CYP4 clan and a mitochondrial clan P450. Inducibility is therefore not restricted to a specific CYP clan, and involves cohorts of other genes often associated with detoxification (in particular, glutathione S-transferases and esterases). Baek *et al.* (2009) compared various induction conditions by cypermethrin in *P. xylostella*, and showed that low sublethal amounts and short exposures using a leaf dip method were more efficient than a topical application at high dose resulting in long exposure. They argue cogently that the onset of toxicity would affect gene expression

Table 9   Inducers of CYP Genes in Insects

Inducers	P450 Induced	Reference
Alkaloids		
nicotine	CYP4M1, 4M3	Snyder *et al.* (1995a)
senita/saguaro cactus alkaloids	CYP28A1, A2, A3, CYP4D10	Danielson *et al.* (1997,1998); Fogleman *et al.* (1998)
caffeine	Cyp6a2, 6a8 + 9 *Drosophila* genes*	Bhaskara *et al.* (2006) *Willoughby *et al.* (2006)
Piper nigrum extracts (piperamides)	6 *Drosophila* genes*	Jensen *et al.* (2006a)
Terpenoids		
monoterpenes (peppermint oil)	CYP6B2	Ranasinghe *et al.* (1997)
α-pinene	CYP6B2, 6B7	Ranasinghe *et al.* (1997); Ranasinghe and Hobbs (1999b)
menthol	CYP6B2	Ranasinghe *et al.* (1997)
limonene	Cyp6a2	Giraudo *et al.* (2010)
gossypol	CYP6B27, CYP6AE14, CYP9A12, A17	Li *et al.* (2002c); Y. B. Mao *et al.* (2007); Zhou *et al.* (2010)
Derived from phenylpropanoid pathway		
cinnamic acid	CYP6AE14	Y. B. Mao *et al.* (2007)
salicylic acid	CYP6B8, B9, B27, B28, CYP321A1	Li *et al.* (2002b); Zeng *et al.* (2009)
chlorogenic acid	CYP6B8, B9, B27, B28	Li *et al.* (2002b)
tannic acid	CYP6AE14	Y. B. Mao *et al.* (2007)
Coumarins		
coumarin	CYP6B8, CYP321A1	Li *et al.* (2002b); Zeng *et al.* (2007)
furanocoumarins	CYP6B1, B3, B4, B8, B9, B17, B27, B28, CYP6AE14 CYP9A2, A4, A5 CYP321A1	Cohen *et al.* (1992); Hung *et al.* (1995b); Stevens *et al.* (2000); Li *et al.* (2000, 2001, 2002b); Y. B. Mao *et al.*, (2007); Wen *et al.* (2009)
Flavonoids		
flavone	CYP6B8, B9, B27, B28, CYP321A1	Li *et al.* (2002b); Wen *et al.* (2009); Zhang *et al.* (2010)
quercetin	CYP6B8	Li *et al.* (2002b)
rutin	CYP6B8, B27, B28	Li *et al.* (2002b)
b-naphthoflavone	CYP6B8, CYP321A1	Wen *et al.* (2009)
Various other natural products		
ethanol	CYP6A1, Cyp6a8	Cariño *et al.* (1992); Morozova *et al.* (2006)
2-tridecanone	CYP4M3, CYP6A2, CYP6B6	Snyder *et al.* (1995a); Stevens *et al.* (2000); Liu *et al.* (2006)
2-undecanone	CYP4M1, M3, CYP9A2, A4, A5	Snyder *et al.* (1995a); Stevens *et al.* (2000)
jasmonic acid	CYP6B8, B9, B27, B28 CYP321A1	Li *et al.* (2002b); Zeng *et al.* (2009)
indole-3-carbinol	CYP6B8, B9, B27, B28, CYP9A2, CYP321A1	Stevens *et al.* (2000); Li *et al.* (2002b); Zeng *et al.* (2007)
pyrethrum	Cyp9f2, Cyp12d1/2	* Jensen *et al.* (2006b)
Synthetic chemicals		
phenobarbital	CYP4D10, CYP4L2, CYP4M1, 4M3,CYP6A1, CYP6B7 ,B8, B9, B27, B28, CYP6D1,D3,CYP9A2, A12, A17 CYP12A1, CYP28A1, A2, A3 + 11 *Drosophila* genes*	Snyder *et al.* (1995); Dunkov *et al.* (1997); Danielson *et al.*, (1997); Danielson *et al.* (1998); Guzov *et al.* (1998); Ranasinghe and Hobbs (1999a); Li *et al.* (2000a, 2002b); Kasai and Scott (2001b); * Le Goff *et al.*, (2006); *Willoughby *et al.* (2006); *King-Jones *et al.* (2006); *Sun *et al.* (2006)
barbital	Cyp6a2, a8, a9	Dombrowski *et al.* (1998); Maitra *et al.* (1996)
glyphosate	CYP6N11, N12, CYP6Z6, CYP6AG7, CYP325AA1	Riaz *et al.* (2009)
DDT	Cyp12d1/2, Cyp6a8	Brandt *et al.* (2002); Willoughby *et al.* (2006); Morra *et al.* (2010)
alkylbenzenes (pentamethyl benzene)	CYP4	Scharf *et al.* (2001)

(Continued)

Table 9 Inducers of CYP genes in insects—cont'd

Inducers	P450 Induced	Reference
aldrin	*Cyp6a2*	Giraudo *et al.* (2010)
PCBs	*Cyp6a2*	Giraudo *et al.* (2010)
butylated hydroxyanisole	*CYP6B2*	Ranasinghe *et al.* (1997)
benzo[a]pyrene	*CYP6B1, B4, CYP6Z6, Z8, CYP9M5*	Petersen Brown *et al.* (2005); McDonnell *et al.* (2004); Riaz *et al.* (2009)
fluoranthene	*CYP6M6, M11, CYP6N12, CYP6Z8, Z9, CYP6AL1, CYP9M6, M9, CYP12F8*	Poupardin *et al.* (2008, 2010)
piperonyl butoxide	*CYP6A1, CYP6B2* + 12 *Drosophila* genes*	Cariño *et al.* (1992); Ranasinghe *et al.*(1997); *Willoughby *et al.* (2007)
clofibrate	*CYP4M1, M3, CYP9A2, A4*	Snyder *et al.* (1995a); Stevens *et al.* (2000)
trans-stilbene oxide	*Cyp6a2*	Giraudo *et al.* (2010)
atrazine	*CYP4, CYP6AL1, CYP9M8, M9* + 6 *Drosophila* genes*	Londono *et al.* (2004); Poupardin *et al.* (2008)* Le Goff *et al.* (2006)
paraquat	9 *Drosophila* genes*	Girardot *et al.* (2004); Zou *et al.* (2000)
tunicamycin	7 *Drosophila* genes*	Girardot *et al.* (2004)
p-hydroxybenzoate	*CYP6B8, B9, B27, B28*	Li *et al.* (2002b)
imidacloprid	*CYP4G36, CYP6CC1*	Riaz *et al.*, (2009)
permethrin	*CYP4D4v2, CYP4G2, CYP6A38, CYP6M6, M11, CYP6N12,CYP6X1, CYP6AE13, CYP6AL1, CYP6BG1, CYP9M8, M9, CYP314A1*	Zhu and Snodgrass (2003); Bautista *et al.* (2007); Zhu *et al.* (2008); Poupardin *et al.* (2008, 2010)
fenvalerate	*CYP6B7*	Ranasinghe and Hobbs (1999b)
cypermethrin	*CYP6B7, B27, B28,* + 8 *P. xylostella* genes	Ranasinghe and Hobbs (1999b); Li *et al.* (2002b); Baek *et al.* (2009)
deltamethrin	*CYP6CE1,CE2 CYP9A12,A17*	Zhou *et al.* (2010); Jiang *et al.* (2010)

*Data from microarrays. The consensus list is given in Giraudo *et al.* (2010)

patterns in a way that is different from the more subtle gene induction. It is likely that future studies will revisit the induction of CYP genes, and distinguish more accurately the dose-dependent response to a chemical from the physiological stresses caused by impending toxicity.

8.5.2.1. Phenobarbital-inducible genes Phenobarbital-like inducers include a variety of chemicals with widely divergent physico-chemical properties. In mammals these inducers are mostly tumor promoters (phenobarbital, polychlorinated biphenyls, chlordane), but they are not genotoxic. They must affect some important homeostatic process in cells, as phenobarbital induction is observed in plants, nematodes, insects, and mammals – and some bacteria. Thus, phenobarbital may not be an ecologically relevant inducer of environmental response genes *sensu* Berenbaum (2002), but a physiologically relevant inducer *sensu* Nebert (1991). It highlights a pathway shared by many chemically unrelated compounds. Both the CAR and the PXR nuclear receptors are involved in the induction of drug-metabolizing enzymes in vertebrates, by heterodimerization with the retinoid X receptor, and, at least in part, through a very distal phenobarbital-responsive enhancer module (PBREM).

The closest *Drosophila* relative to PXR and CAR (and of the vitamin D receptor) is the nuclear receptor DHR96 (NR1J1), and its potential role in xenobiotic response was

studied in *DHR96[1]*, a strong loss of function allele (King-Jones *et al.*, 2006). This mutant is viable, more sensitive to phenobarbital in a geotaxis assay than wild type, and marginally more sensitive to DDT in a long-term (3-week) assay on DDT in the diet. Wild type and mutant adult flies with or without a 10-hour phenobarbital exposure were compared by microarray analysis, and wild type larvae vs transgenic larvae were compared 4 hours after heat shock to induce ectopic DHR96 expression. The results of these experiments paint a complex picture of genes responsive to phenobarbital, to the presence or absence of DHR96, or to its ectopic expression. While some genes respond to phenobarbital only in the presence of DHR96, some respond only in its absence. Also, basal expression of many genes is affected by DHR96, but the majority of phenobarbital-responsive genes (and in particular most CYP genes) are not affected by the DHR96 mutation, and still respond to the chemical (King-Jones *et al.*, 2006). At least one phenobarbital-type ligand, CITCO, was shown to activate a DHR96 ligand binding domain reporter *in vivo* (Palanker *et al.*, 2006). Therefore, while DHR96 certainly plays a role in the response of some genes to phenobarbital in *Drosophila*, it is not the only player in the regulatory cascade, and an exploration of its role in induction by other chemicals, or, as nuclear receptor, its DNA target and heterodimerization partner, are needed. DHR96 is reported to play a role in sterol and lipid homeostasis (Horner et al., 2009; Sieber and Thummel, 2009), so its role in the xenobiotic response may be peripheral and it may just participate in receptor crosstalk, a situation well described in vertebrates (Pascussi et al., 2008).

The definition of regulatory elements of phenobarbital induction in *Drosophila* started with transfection of *Cyp6a2*/luciferase reporter constructs in Schneider cells. Promoter elements sufficient for directing both basal and phenobarbital-inducible expression are located within 428 bp 5′ of the start codon (Dunkov et al., 1997). Elements further upstream (984 and 1328 bp) are needed for higher basal activity. *In vivo* experiments with transgenic flies showed that (pheno)barbital induction was functional with 1331 or 985 bp of 5′ upstream DNA, whereas 129 bp upstream of the start codon, while conferring a low level of basal expression, does not support induction (Dombrowski et al., 1998). Further experiments with transgenic flies carrying the luciferase reporter gene driven by promoter sequences of the *Cyp6a8* gene showed low basal activity of a −11/−199-bp construct, with higher basal activities seen with 761 and 3100 bp of upstream DNA (Maitra et al., 2002). Phenobarbital inducibility was apparent with the three types of constructs, the highest level of induced activity being achieved with the −11/−761-bp promoter region. These studies thus consistently identify a small upstream region of the promoter that is important for phenobarbital inducibility. The *Cyp6a2* gene as well as the *Cyp6a8* gene are also inducible by caffeine (Bhaskara et al., 2006). Caffeine-responsive elements were dispersed in both genes. The D-JUN transcription factor acts as a negative regulator of both genes, decreased by caffeine treatment, whereas cAMP upregulated both genes (Bhaskara et al., 2008). Caffeine induction and phenobarbital induction are mediated by different pathways, and are synergistic on the *Cyp6a8* promoter (Morra et al., 2010). Interestingly, *Cyp6a8* is still inducible by phenobarbital in *DHR96[1]* mutants (King-Jones et al., 2006).

The house fly *CYP6D1* gene induction by phenobarbital requires a short 5′ UTR sequence (−330 to −280) when measured with a luciferase reporter assay in *Drosophila* Schneider cells. Suppression of DHR96 by RNAi in those cells results in a loss of inducibility. The BR-C transcription factor plays an opposite, suppressive role in *CYP6D1* expression, consistent with the presence of a BrC-Z4 binding site in the PB-responsive promoter region that was studied (Lin et al., 2011).

8.5.2.2. Furanocoumarin-inducible genes
The inducibility of *CYP6B* genes by furanocoumarins has been studied extensively in *Papilio* species. The *CYP6B1* and *CYP6B3* genes are inducible by the linear furanocoumarin xanthotoxin, and variably induced by bergapten, angelicin, and sphondin (Prapaipong et al., 1994; Hung et al., 1995a,

1995b). The 5′ flanking sequence of the *CYP6B1v3* gene comprising nt −838 to +22 (relative to the transcription start site) was fused to the reporter chloramphenical acetyl transferase (CAT) gene, and this construct was transfected into Sf9 cells (Prapaipong et al., 1994). The promoter region of the *CYP6B1* gene had a low basal activity, and was induced (about two-fold) by xanthotoxin. Thus, at least some of the sequences required for induction are present in this upstream region, and the xanthotoxin signaling cascade is present in these cells from the generalist herbivore *S. frugiperda*. Xanthotoxin inducibility was mapped to a region between −146 and −97. Within it, the region of nt −136 to −119 was identified as the XRE-xan (for Xenobiotic Response Element-xanthotoxin) element (Petersen et al., 2003). Sequences similar to the ecdysone RE (EcRE) and an ARE (antioxidant RE) overlap on the 5′ end of the XRE-xan. Mutation of the EcRE element affects basal expression but not ecdysone repression of the gene, and more downstream sequences are also needed for promoter activity (Petersen Brown et al., 2004). A sequence similar to XRE-AhR (the vertebrate aryl hydrocarbon receptor) was identified in that region, and benzo[a]pyrene and xanthotoxin appear to act through the same response elements (McDonnell et al., 2004). Transfection with plasmids for two *Drosophila* paralogs of the Ah receptor, spineless and tango, showed that these proteins enhanced basal expression but decreased the response to benzo[a]pyrene and xanthotoxin (Petersen Brown et al., 2005). The *Papilio glaucus* CYP6B4 promoter is similar to that of CYP6B1, and also contains an essential EcRE/ARE/XRE-xan composite element, but it has a lower constitutive expression and shows higher inducibility (McDonnell et al., 2004). The promoter of the flavone-inducible *CYP321A1* gene of *H. zea* contains a different element, called XRE-Fla, that was precisely mapped and shown to bind to *H. zea* fat body nuclear extracts (Zhang et al., 2010).

8.6. Working with Insect P450 Enzymes

8.6.1. Biochemical Approaches

8.6.1.1. Subcellular fractions
The enzymology of insect P450 can be studied in different types of environments. These are enriched subcellular organelles from insects (microsomes, mitochondria) where multiple P450 interact *in situ* with their redox partners; membranes from cellular expression systems where a cloned recombinant P450 interacts with native or engineered redox partners; and ultimately a reconstituted system of purified recombinant P450 and its redox partners in a defined system devoid of biological membranes. Transgenic expression of P450 genes is another way to study P450 biochemistry, but also regulation.

The classical preparation of microsomal fractions from insect tissue homogenates by differential centrifugation

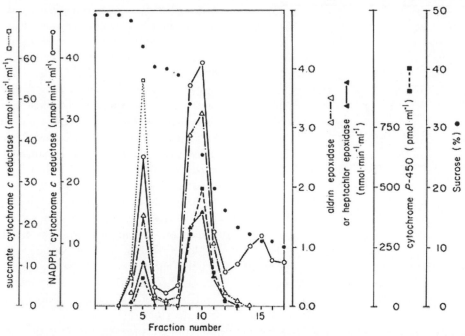

Figure 28 Sucrose density centrifugation separation of subcellular fractions of house fly larval homogenates showing the distribution of marker enzyme activities between mitochondrial and microsomal fractions. Note some P450 reductase activity in the top (soluble) fractions representing proteolytically cleaved enzyme. From Feyereisen, 1983.

has been extensively described by Hodgson (1985) and Wilkinson (1979). It remains, with minor modifications, the most widely used first step in the biochemical characterization of insect P450 enzymes. Linear or step gradients of sucrose for the centrifugal preparation of microsomes and mitochondria, or CaCl$_2$ precipitation of microsomes, have been less favored. In all approaches, the careful use of marker enzymes is critical. A technique for the rapid preparation of microsomal fractions of small tissue samples relying on centrifugation at very high speed on sucrose layers in a vertical rotor has been described (Feyereisen *et al.*, 1985).

The well-documented sedimentation of P450-associated activities at low *g* forces in many early insect studies (reviewed in Wilkinson and Brattsten, 1972; Wilkinson, 1979) has been considered a peculiar difficulty. The discovery of insect CYP12 enzymes and their characterization as mitochondrial P450 enzymes capable of metabolizing xenobiotics (Guzov *et al.*, 1998) has brought a new light to the early difficulties in sedimenting insect P450 activities in the "correct" fractions. It is quite probable that at least part of the P450 activities observed in "mitochondrial" fractions were indeed carried out by CYP12 enzymes. In house fly larvae, 15–20% of the aldrin and heptachlor epoxidase activities were associated with mitochondrial fractions after sucrose density centrifugation (Feyereisen, 1983) (**Figure 28**). The insect midgut is a particularly rich source of P450 activity (Hodgson, 1985), but the external brush border membrane is a significant source of membrane vesicles (BBMV) upon homogenization of the

tissue. Centrifugal methods to separate the BBMV fraction from the microsomes derived from the endoplasmic reticulum have been described (Neal and Reuveni, 1992).

8.6.1.2. Spectral characterization and ligand binding The analysis of P450 levels in subcellular fractions follows the original procedure of Omura and Sato (1964). A difference spectrum between reduced microsomes and reduced microsomes after gentle bubbling of CO readily displays the famous redshifted Soret peak at 450 nm (**Figure 6**). The concentration of P450 can be calculated from the ΔOD between 490 and 450 nm and Omura and Sato's extinction coefficient $\varepsilon = 91$ mM^{-1}.cm^{-1}. This measure gives the total concentration of all forms of P450 present in the preparation. Individual P450 proteins may have peaks that are one or two nm off the 450-nm norm, and when they represent a large portion of the total P450, the total P450 peak may be shifted as a consequence. The degradation of P450 to the inactive P420 form may interfere with the measurement of P450, as already reviewed by Hodgson (1985), and the respiratory chain pigments interfere with the measurement of P450 in mitochondrial fractions. The classical Omura and Sato procedure remains the technique of choice to measure the amount and purity of P450 proteins produced in heterologous systems (see below).

Ligand-induced spectral changes also follow classical procedures, detailed in a useful review (Jefcoate, 1978). Type I spectra (peak at 380–390 nm, trough at 415–425 nm) result from ligand in the substrate binding site displacing

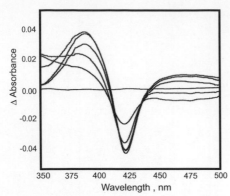

Figure 29 Type I substrate-induced difference spectrum of recombinant CYP12A1 with increasing concentrations of progesterone.

water as the sixth ligand to the heme iron (**Figure 29**). Type I spectra are concentration dependent, giving a spectral dissociation constant (K_s), and this titration is correlated with a shift of the iron from high spin to low spin. Not all type I ligands are substrates, and not all substrates are type I ligands, so this useful tool must be used with caution. A Type I spectral K_s is not an enzymatic K_d.

Type II spectra (peak at 425–435 nm, trough at 390–405 nm) result from the binding of a strong ligand to the heme iron, typically the nitrogen coordination of compounds such as pyrimidines, azoles, or n-octylamine. Type II spectral titration is correlated with a shift from high spin to low spin, and is a hallmark of strong inhibitors such as imidazoles. Other, less frequently studied spectral changes induced by ligands or their metabolism will not be discussed here (e.g., type III spectra, Hodgson, 1985; spectra of phenyl–iron complexes, Andersen *et al.*, 1997).

8.6.1.3. Assays and substrates

Measurement of P450 activity is a special challenge because of the large number of different P450 enzymes, each catalyzing the metabolism of a specific (broad or narrow) range of substrates. There are, therefore, a very large number of assays for P450 activity. Direct assays of product appearance or substrate disappearance rely on all the tools of analytical chemistry. The assay of a P450 produced in a heterologous system can be straightforward, but the assay of a P450 in its native microsomal or mitochondrial membrane, where it is mixed with an undetermined number and amount of other P450s, is more problematic. Metabolism of compound M to product N in microsomes is the sum of the contributions of all P450 enzymes that catalyze the M to N reaction (and sometimes non-P450 enzymes can catalyze the same reaction!). Selective inhibitors (chemicals or antibodies) of one P450 can, by substraction, indicate the relative contribution of that particular P450 to the reaction being measured (see, for example, Wheelock and Scott, 1992b; Hatano and Scott, 1993; Korytko

et al., 2000b). This indirect inference is only as valid as the inhibitor is selective. Substrates that are selective for one P450 and that are easily assayed have been the object of considerable research in biomedical toxicology. The relative success of this quest (e.g., nifedipine as model probe for CYP3A4) is a result of both the limited number of major P450s expressed in human liver and the heavy investment in their study. The large number of insect P450s multiplied by the large number of insect species under study divided by the investment in their research makes a similar quest seem quixotic. By default, then, a certain number of assays have taken their place in the literature as some measure of "global" P450 activity. Most authors are now fully aware that the microsomal activity of, for example, aldrin epoxidation, p-nitroanisole O-demethylation, or 7-ethoxycoumarin O-deethylation is only a measure of those P450 enzymes catalyzing these reactions. But this awareness is only as recent as our understanding that there are really many P450 enzymes, and that their individual catalytic competence may be broad or narrow, overlapping with other P450 enzymes or not. Thus, the pioneers who used aldrin epoxidation as an assay used an analytical tool then readily available in pesticide toxicology laboratories (GC with electron capture detection for the sensitive detection of organochlorine pesticide residues), but they didn't use it without a caveat – quoting the classical Krieger *et al.* study (1971), "to the extent that the rate of epoxidation of aldrin to dieldrin typifies the activity of the enzymes toward a wider range of substrates…." The current and still widespread use of a "model" substrate to explore P450 activities in insect subcellular fractions can be useful. If the metabolism of a randomly chosen P450 substrate (e.g., aldrin, aminopyrine, 7-methoxyresorufin) is quantitatively different between insect strains, in different tissues, following induction, etc., then this substrate has provided a clue that the qualitative or quantitative complement of P450 enzymes is changing. It is up to the investigator to follow up on this clue. Alkoxycoumarins and alkoxyresorufins are useful substrates for sensitive fluorometric assays, and have largely replaced organochlorines as model substrates. They can be used to "map" the catalytic competence of a heterologously expressed P450 – for example, the preference of CYP12A1 for pentoxycoumarin (Guzov *et al.*, 1998) (**Figure 30**), or the preference of CYP6Z2 for benzyloxyresorufin (McLaughlin *et al.*, 2008). The latter study also made extensive use of a model substrate (benzyloxyresorufin) to survey a large number of CYP6Z2 inhibitors. Steroids such as testosterone have multiple sites of attack by P450 enzymes, and can likewise be used to characterize the activity of subcellular fractions or of heterologously expressed P450s (for example, Jacobsen *et al.*, 2006; Murataliev *et al.*, 2008).

Another approach to the study of P450 activity is to follow the consumption of the other substrates, O_2 or

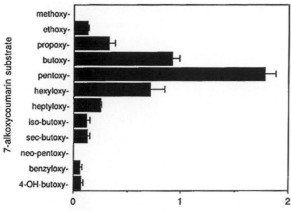

Figure 30 O-dealkylation activity of *E. coli*-expressed recombinant house fly CYP12A1 in a reconstituted system. Reproduced from Guzov *et al.* (1998), with permission.

NADPH. This approach is certainly valid when the stoichiometry of the reaction (see [1]) is under study, but its use to monitor metabolism by subcellular fractions is fraught with difficulties. Other enzymes consume O_2 and NADPH as well, and the background activity can be very high.

8.6.1.4. P450 assays in individual insects

The design of assays suitable for assessing P450 activities in single insects has accompanied the need to study variations in P450 activities from the individual to the population level. Such assays allow the presentation of frequency histograms of activity levels in field-collected samples or in laboratory-selected populations, and are ideally adapted to microtiter plate format. The NADPH-dependent conversion of *p*-nitroanisole to *p*-nitrophenol was followed in individual homogenates of *H. virescens* and *Pseudoplusia includens* larvae (Kirby *et al.*, 1994; Rose *et al.*, 1995; Thomas *et al.*, 1996). This assay has a relatively low sensitivity, but clearly distinguished individuals from susceptible and insecticide-resistant strains. Cut abdomens of adult *Drosophila* in buffer containing 7-ethoxycoumarin can be used to measure 7-hydroxycoumarin formation in a 96-well microtiter plate format (de Sousa *et al.*, 1995). This rapid technique allows, for instance, the monitoring of individual variability over the course of a selection regime (Bride *et al.*, 1997). Inceoglu *et al.* (2009) used a luminescent P450-Glo assay kit adapted to the gut of a single mosquito (after removal of gut content) and this has the potential for high-throughput, sensitive screening.

A simple assay based on the peroxidase activity of the heme group with tetramethylbenzidine was developed for use in single mosquitoes (Brogdon *et al.*, 1997). This assay, easily developed on a microtiter plate format, is an indirect assay, measuring total heme content of the insect homogenate rather than P450 activity, and therefore needs to be carefully validated.

8.6.1.5. Solubilization and purification

Solubilization and purification of insect P450 enzymes from insects has generally followed techniques developed in the purification of vertebrate P450 (Feyereisen, 2005). P450 purification from microsomes of mixed tissues (e.g., fly abdomens) can be sufficient to obtain a protein fraction suitable for antibody generation or peptide sequencing (Feyereisen *et al.*, 1989; Wheelock and Scott, 1990; Scott and Lee, 1993).

Sequential chromatography on octylamino agarose, DEAE-cellulose, and hydroxyapatite was used to purify sodium cholate-solubilized P450s from *Drosophila* (Sundseth *et al.*, 1990). The two protein fractions obtained, P450 A and B, had only a very low 7-sethoxycoumarin O-deethylase activity (0.01 nmol/nmol P450 per min), but the proteins were useful in generating monoclonal antibodies that allowed the subsequent cloning of CYP6A2 (Waters *et al.*, 1992).

An original approach was the purification of a locust P450 by affinity chromatography with type II and type I ligands (Winter *et al.*, 2001) that led to cloning of CYP6H1 (Winter *et al.*, 1999). In this approach, microsomes from larval *L. migratoria* Malpighian tubules were first treated with the detergent synperonic NP10 to solubilize P450. The extract was then chromatographed on ω-octylamino agarose and hydroxylapatite. The third step was chromatography on a triazole agarose affinity column. The affinity ligand was a derivative of the fungicide difenoconazole, which has an affinity for ecdysone 20-monoxygenase of the same level as that of the substrate ecdysone (0.5 vs 0.2 µM). This substituted triazole is a typical type II ligand (active site liganding of the heme), and its use on the affinity matrix led to the adsorption of all the P450 loaded on the column (Winter *et al.*, 2001). Elution from the affinity column was done by replacing the immobilized type II ligand with a soluble type I ligand, ecdysone. A major protein band of 60 kDa was thus obtained in 4% yield, with a P450 specific activity of 13.1 nmol/mg protein. Unfortunately, biochemical evidence that this P450 is in fact an ecdysone 20-monooxygenase was not obtained in this study (Winter *et al.*, 2001), so the nature of CYP6H1 (Winter *et al.*, 1999) remains conjectural.

Another variant on the classical purification schemes has been the use of immobilized artificial membrane high performance liquid chromatography (IAM-HPLC) of microsomal proteins (Scharf *et al.*, 1998). This technique allowed the 70-fold purification of a P450 from the German cockroach, and the subsequent production of antibodies with this 49-kDa protein as antigen.

Purification of heterologously expressed, recombinant insect P450s, especially with His-tags, is usually much easier than purification from insect tissues, and such expressions systems are reviewed below.

8.6.2. Heterologous Expression Systems

Biochemical characterization of a P450 protein and its substrate selectivity remains a *sine qua non* condition of its functional identification. Only a few P450 enzymes are characterized well enough (e.g., steroid metabolizing P450s in vertebrates) that sequence comparison can reasonably predict activity. For most other P450s, the sequence does not provide a clue to the activity, and there are now innumerable papers describing how one or a few mutations can change substrate selectivity (reviewed in Domanski and Halpert, 2001). The expression of P450 cDNAs in heterologous systems has become the standard way of characterizing insect P450 proteins. A number of such expression systems have been developed (**Table 5**), and the techniques are essentially similar to those used for the production of P450 proteins from mammalian or plant tissues.

8.6.2.1. *Escherichia coli* The host organism *E. coli* is a rare organism devoid of P450 genes of its own, while other bacteria can carry over 20. Bacterial production of insect P450s has required several modifications of the P450 sequence. At the 5′ end of the cDNA, mutations are introduced to optimize expression (Barnes *et al.*, 1991). The second codon is typically replaced by Ala to enhance expression, and silent substitutions are introduced to increase the A/T content (Sutherland *et al.*, 1998). In some cases, introduction of four to six His codons just before the stop codon directs the translational production of a C-terminal "histidine tag" useful for purification (Guzov *et al.*, 1998; Sutherland *et al.*, 1998; Helvig *et al.*, 2004a). P450 production can be enhanced by the addition of δ-aminolevulinic acid (a precursor for heme biosynthesis) to the culture broth (Sutherland *et al.*, 1998; Helvig *et al.*, 2004a). The P450 produced in bacteria is found mostly in a membrane fraction, and sometimes in a fraction of inclusion bodies that are difficult to extract. In some cases (Andersen *et al.*, 1994), a significant amount of P450 is produced as a soluble form. The *E. coli* membrane fraction carrying the recombinant P450 protein is generally suitable for analysis by difference spectroscopy for either P450 content by the Omura and Sato (1964) procedure, or for ligand binding (type I binding for potential substrates or type II binding for azoles). Although some P450 proteins (e.g., CYP17) expressed in *E. coli* can utilize an endogenous flavodoxin reductase/flavodoxin system for catalysis, none of the insect P450 proteins tested thus far have been catalytically active in *E. coli* membrane fractions in the absence of a P450 reductase. Therefore, P450 produced in bacteria needs to be solubilized and purified by classical methods. The proteins produced with a histidine tag, once solubilized, are purified by nickel chelate affinity chromatography. Extensive dialysis is needed in both procedures to remove excess detergent or imidazole used for elution from the nickel affinity column. The P450 obtained is then suitable for reconstitution with redox partners. These partners (microsomal or mitochondrial redox partners) are themselves produced in *E.coli* and purified (Guzov *et al.*, 1998). Reconstitution of a catalytically active enzyme system is then tedious or artistic, depending on one's degree of patience! It requires attention to the details of concentrations of phospholipids, detergents, and proteins, and their order of addition, mixing, and dilution (Sutherland *et al.*, 1998). The advantages of bacterial expression are the low cost of production of large amounts of P450, and the possibility of working with a precisely defined *in vitro* system with highly purified enzymes and their partners. A thorough characterization of the enzyme can be undertaken. The disadvantage of this formal biochemical approach is that purification and reconstitution are difficult and time-consuming, probably not suitable when the goal is simply a survey of the catalytic competence of the P450, or the comparison of a large number of P450s or P450 mutants.

This disadvantage can be avoided by targeting the expressed proteins to the membrane fraction of *E. coli*, and thus preventing aggregation in inclusion bodies. N-terminal fusion of the bacterial pelB signal sequence thus allows production of high amounts of active, membrane-bound P450 reductase (Andersen *et al.*, 2004). Production of several insect P450s and P450 reductase in a membrane fraction of *E. coli*, suitable for spectral measurement of P450 levels and NADPH-dependent enzyme assays, has now been reported (McLaughlin *et al.*, 2008; Muller *et al.*, 2008a; Karunker *et al.*, 2009; Stevenson *et al.*, 2011). In this successful strategy, the ompA and pelB signal sequences were used to direct the production of P450s and of the *A. gambiae* P450 reductase, respectively, to the membrane. Following cell disruption and centrifugation, this membrane preparation can be used directly as surrogate microsomal preparation in enzymatic assays. The more easily purified cytochrome b_5 can optionally be added.

8.6.2.2. Baculovirus Expression of P450 in lepidopteran cells (Sf9, Tn5, etc.) by the baculovirus system requires no modification of sequence, and is a widely used method for the production of proteins in a eukaryotic system. It has the potential of producing large amounts of P450 proteins for subsequent purification, but studies with insect P450 expressed with this system (**Table 5**) have relied instead on the advantage that the protein is present in a suitable milieu: the endoplasmic reticulum of an insect cell. Thus, cell lysates, briefly centrifuged to pellet cell debris, are used as an enzyme source. Difference spectroscopy or immunological methods (Dunkov *et al.*, 1997) can be used to assess the amount of P450 produced. The host cells provide their endogenous P450 reductase to support the activity of the heterologous P450 when the cell lysates are incubated with an NADPH regenerating system. Although this is sufficient to allow the measurement of a number of P450-dependent activities (Dunkov *et al.*, 1997), the stoichiometry of endogenous P450 reductase, and cytochrome b_5, to heterologously expressed P450 is probably not optimal. The activities measured do not represent the full potential of the P450 under study. For instance, a 30-fold increase in CYP6A2 activity was observed when purified house fly P450 reductase and cytochrome b_5 were added to lysates of cells expressing *Cyp6a2* (Dunkov *et al.*, 1997). An improvement of the baculovirus expression system has therefore been designed, wherein the cells are co-infected with a virus engineered to carry the P450, and a virus engineered to carry a P450 reductase (house fly P450 reductase; Wen *et al.*, 2003). A significant increase (33-fold) in CYP6B1 activity towards the substrate xanthotoxin was achieved with the improved P450 reductase/P450 ratio. In fact, the improved conditions allowed the measurement of angelicin metabolism that was barely detectable in the absence of additional P450 reductase. Thus, in the baculovirus system, insect P450s can be studied in an insect membrane environment, without need for purification. These are great advantages over the *E. coli* expression + solubilization/purification system. However, the interactions with redox partners are not manipulated as easily. The total amounts of P450 produced are also smaller, although addition of hemin to the culture medium can increase the amount of P450 produced (Dunkov *et al.*, 1997; Wen *et al.*, 2003). The total amount of P450 produced is less important in the baculovirus system than in the *E. coli* system, as purification is not required for most applications, and the highest activity of cell lysates is achieved at the optimal P450/P450 reductase ratio, not at the maximal P450 production level. The level of endogenous P450 in the control experiments (i.e., uninfected cells, or cells infected with a virus carrying a non-P450 "control" cDNA) are virtually undetectable. However, this needs to be systematically and carefully verified, because Sf9 cells express constitutive and inducible P450s that may be significant when the activity measured is very low or difficult to detect. P450 proteins that are poorly expressed in the baculovirus system can be engineered appropriately. For instance, the CYP6B33 N-terminal was modified, and even a single mutation V32A in the linker between the signal-anchor and the Pro/Gly domain led to proper folding and expression in this system (Mao *et al.*, 2008a).

8.6.2.3. Transfection in cell lines   Expression in transfected *Drosophila* Schneider 2 cells has been used to produce P450s (**Table 5**). The advantage of the method is its simplicity. When the expression of the P450 is coupled with a very sensitive assay, the method can rapidly provide qualitative data on the catalytic competence of the enzyme.

However, the quantitative determination of P450 levels is more difficult to achieve, and the interaction with redox partners cannot be optimized except by co-infection. It is interesting that the CYP302A1 and CYP315A1 expressed by this method are mitochondrial P450s, and the S2 cell homogenates were able to provide adequate redox partners. As used so far, it has not allowed a measurement of the amounts of P450 produced, and neither have the redox partners been characterized or optimized. Cell transfection does not have the potential of the baculovirus system for large-scale production of P450 proteins.

8.6.2.4. Yeast   Yeast expression systems have only started to be exploited for the production of insect P450 proteins. *Saccharomyces cerevisiae* has three P450 genes that are fully characterized, and are expressed at low levels, so that inducible expression of an exogenous P450 is not hindered by the endogenous P450. Co-production of CYP6A2 from *Drosophila* and of human P450 reductase in yeast (Saner *et al.*, 1996) generated a cell system capable of activating several pro-carcinogens to active metabolites that induced mitotic gene conversion or cytotoxicity. A few other P450s have been produced in yeast (**Table 5**), but this has not yet achieved the success seen with plant P450 production in yeast. P450 cDNAs may need to be engineered to recode the N-terminus of the protein. This has been done successfully with plant P450s to conform with the yeast codon usage (Hehn *et al.*, 2002). The replacement of the yeast P450 reductase gene with an insect P450 reductase gene by homologous recombination (Pompon *et al.*, 1996) should increase the usefulness of this as yet underutilized expression system. Indeed, yeast combines the advantages of *E. coli*-inducible production of large amounts of protein with the advantage of the eukaryotic cell system in which P450 enzymes can be studied in a normal membrane environment.

8.6.2.5. Plant cell suspension cultures   Transformation of tobacco (*Nicotiana tabacum*) heterotrophic cell suspensions with an *Agrobacterium tumefaciens* vector carrying *Drosophila* CYP6G1 was successfully used to study imidacloprid, DDT, and methoxychlor metabolism (Joussen *et al.*, 2008, 2010). This system allows *in vivo* metabolism in a cellular system carrying its own source of P450 reductase. Potentially, it can generate good amounts of metabolites for identification. One drawback is that the tobacco cells have their own complement of P450 enzymes, as well as other enzymes that can alternatively, or further, metabolize the substrate under study. Careful attention to metabolism by non-transformed cells is needed, as in the other heterologous systems, such as baculovirus and S2 cells.

8.6.2.6. Transgenic insects   The use of transgenic insects to study P450 function (or regulation) has until now been restricted to *Drosophila*. Heterologous expression of vertebrate P450s was achieved in studies aimed at developing *Drosophila* as a genotoxicity model organism (Feyereisen, 2005). *Drosophila* has been used further: (1) as a heterologous *in vivo* system; or (2) as a system for ectopic and/or overexpression of *Drosophila* P450s. Examples of the former include house fly CYP6D1 production leading to a significant increase in benzo[a]pyrene hydroxylation (Korytko *et al.*, 2000a), and *T. castaneum* CYP6BQ9 production leading to deltamethrin resistance (Zhu *et al.*, 2010). Examples of the latter have mostly aimed at inferring a role in pesticide detoxification by showing altered levels of susceptibility (Daborn *et al.*, 2002; Le Goff *et al.*, 2003; Bogwitz *et al.*, 2005; Chung *et al.*, 2007; Yang *et al.*, 2007). The more systematic study of Daborn *et al.* (2007) used transgenic flies to overexpress eight P450 genes and test their possible involvement in insecticide metabolism. These elegant studies do not provide the same information as the more straightforward conversion of a substrate to a product by a purified enzyme. Positive results are directly useful and are indicative of "resistance potential," as phrased by Daborn *et al.* (2007). Negative results require more caution, because a modification of the *in vivo* toxicokinetics of a compound can be difficult to interpret. Similarly, negative results in the analysis of null mutants, as for *Cyp6d4* in *Drosophila*, are not always informative (Hardstone *et al.*, 2006). Transformation of other insects, notably with the piggyBac vector as in *B. mori* (Tamura *et al.*, 2000), will undoubtedly increase the applications of transgenesis to P450 research.

8.6.3. Mechanisms and Specificity of P450 Inhibitors

The common features of electron transfer, ligand binding, and catalysis described above are the features that determine the relative success of P450 inhibitors. Compounds that act as electron sinks and are readily auto-oxidizable can inhibit P450 reactions by inhibiting electron transfer by the respective redox partners. This mechanism is typical of the eye pigment xanthommatin, which was identified as an "endogenous inhibitor" in early studies (for review, see Hodgson, 1985). Several flavonoids may act in this way, and care must be taken to distinguish P450 inhibition *per se* from inhibition of electron transfer.

Insecticide synergists (**Figure 31**) are among the most interesting inhibitors of P450 because of their widespread commercial use – in particular, piperonyl butoxide. A landmark review paper on synergists remains that of Casida (1970). Synergists (Hodgson, 1985; Bernard and Philogene, 1993) as well as P450 inhibitors in general (Ortiz de Montellano and Correia, 1995) were also covered in other insightful reviews. The synergism of carbaryl by piperonyl butoxide has been used in a survey of 54 insect species to estimate P450 activity *in vivo* (Brattsten

piperonyl butoxide

TCPPE

verbutin

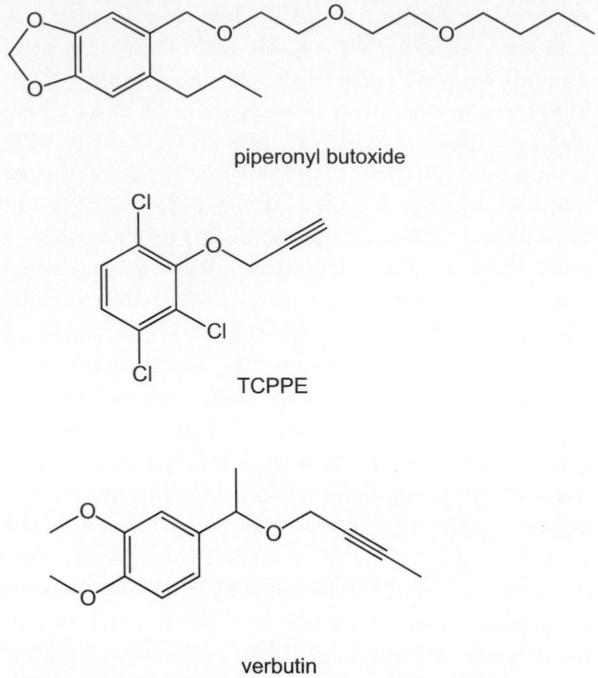

Figure 31 Structures of the synergists and P450 inhibitors piperonyl butoxide, a typical methylene dioxyphenyl (MDP) compound; TCPPE (trichlorophenylpropynyl ether); and verbutin.

and Metcalf, 1970). Although the synergistic ratio is most often presented, the usefulness of a synergistic difference has been proposed as well (Brindley, 1977). The mode of action of piperonyl butoxide and other related methylene-dioxyphenyl (MDP) compounds (or benzodioxole compounds) involves an initial metabolic activation by the P450 enzyme, leading to the formation of a carbene–iron complex that is virtually irreversible (Ortiz de Montellano and Correia, 1995). It follows that those P450 enzymes (1) with low affinity for the MDP compound and/or (2) with low capacity to metabolize it to the carbene inhibitory form will not be inhibited, and thus piperonyl butoxide and other MDP compounds are not universal inhibitors of all P450 enzymes. Some compounds with an MDP moiety can be preferentially metabolized at another site of the molecule. Myristicin is epoxidized at its distal propene site by CYP6AB3v2 (Mao *et al.*, 2008b). Binding of this natural product to the active site of CYP6AB3v2 is in the opposite orientation than that needed for inhibition by activation of the MDP group. Inhibition of CYP6AB3v2 by myristicin is therefore not an example of the classical inhibition of P450 enzymes by MDP compounds.

Inhibition of some esterase activities by piperonyl butoxide can be achieved at high concentrations (see, for example, Young *et al.*, 2005), so indiscriminate use of this synergist to infer P450-dependent metabolism is risky. Confirmation by another, non-MDP type synergist is recommended. Esterase inhibition is not a function of the MDP moiety of piperonyl butoxide, as a benzofuran analog retains the inhibitory activity towards 1-naphthyl butyrate hydrolysis (Moores *et al.*, 2009). Because the effectiveness of MDP compounds as P450 inhibitors is metabolism-dependent, they are not rapid inhibitors, and their activity *in vivo* can be delayed. Another potential drawback of piperonyl butoxide (and MDP compounds in general) is that it can induce CYP genes (Willoughby *et al.*, 2007) as well inhibiting P450 enzymes. This has long been known in vertebrates and insects (Thongsinthusak and Krieger, 1974). Long-term diet exposure to piperonyl butoxide, for instance, can have effects that are difficult to interpret in the absence of independent evidence.

Self-catalyzed destruction of P450 enzymes by terminal acetylenes or olefins and other "suicide substrates" is also, and for the same reasons as for MDP compounds, not equally effective for all P450 enzymes. The phenyl-propynyl ether synergists such as TCPPE fall into this category, as well as the newer synergist verbutin (Bertok *et al.*, 2003). This non-generality of inhibition has been well documented *in vivo* – for instance, TCPPE can be an effective synergist when piperonyl butoxide cannot (Brown *et al.*, 1996; Zhang *et al.*, 1997). In the case of 1-aminobenzotriazole (ABT), which is metabolized to an intermediate formally described as benzyne, which covalently binds to the prosthetic heme, *in vitro* P450 destruction and formation of the porphyrin adduct have been measured in the house fly (Feyereisen *et al.*, 1984). P450 protein labeling by P450 inhibitors has also been achieved (Andersen *et al.*, 1995; Cuany *et al.*, 1995). The NADPH-dependent decrease in P450 caused by ABT, TCPPE, or piperonyl butoxide differs according to the induction status and fly strain, suggesting selectivity (Feyereisen *et al.*, 1984). This selectivity can be harnessed into the design of useful "suicide" inhibitors of, for example, ecdysone biosynthesis (Luu and Werner, 1996). The synthesis of the MDP moiety seen in many plant natural products is itself dependent on a P450 activity (CYP719; Ikezawa *et al.*, 2003), and it is postulated that such compounds may be a legacy of evolutionary interactions with insect and other enemies (Berenbaum and Neal, 1987).

Another class of powerful P450 inhibitors comprises the heterocyclic compounds with an *sp2* hybridized nitrogen, as in pyridines, azoles, and imidazoles. These compounds bind simultaneously to the heme iron (type II ligands) and to a hydrophobic binding site of the P450 for its substrate. This "two-point binding" therefore has an intrinsic potential for selectivity, with the substrate mimic moiety of the inhibitor targeting the specific P450 and the type II-ligand moiety coordinating the heme, and inhibiting the enzyme. This reasoning has led to the design of potent inhibitors and photoaffinity labels (Andersen *et al.*, 1995) for methyl farnesoate epoxidase. The commercial importance of this type of P450 inhibitors is emphasized by the fungicides and CYP51 inhibitors miconazole and ketoconazole. Heterologous expression of P450 enzymes now makes it possible to screen large numbers of compounds as potential inhibitors (Wen *et al.*, 2006; McLaughlin *et al.*, 2008). Such studies can help "map" the active site of the enzyme, understand the interactions of the insect with components of its diet, as well as lead to resistance-breaking synergists.

8.7. Conclusion and Prospects

Some examples of the multiple functions of insect P450 enzymes, of their complex biochemistry, and of their toxicological and physiological importance have been presented. It is now clear that there is not one class of P450s involved in physiological processes and another distinct class involved mostly in xenobiotic metabolism. Physiological functions are not restricted to one branch of the P450 evolutionary tree, and the ramifications that seem typical of "environmental response genes" are found in both microsomal and mitochondrial P450 encoding genes. The CYP3 and CYP4 clans seem the most diverse, and the smaller CYP2 and mitochondrial CYP clans seem to have proportionally more genes found as orthologs across species. However, there are many physiological functions for P450s that just are not known or suspected. CYP4C7 was shown to be selectively expressed in the corpora allata, and to metabolize JH and its precursors to new metabolites (Sutherland *et al.*, 1998). The presence of this P450 in corpora allata and the existence of these metabolites were unexpected. Similarly, the presence of CYP6G2 in the CA portion of *Drosophila* ring glands is not yet understood. Consequently, the search for new functions of P450 enzymes should involve broad, rational screens, and the technology is now available to perform such functional screens. CYP genes "appear" prominently without obvious explanation in unexpected areas of research. This is the case for *Cyp6a20* and *Cyp4p2* in studies of aggressiveness in *Drosophila* (Dierick and Greenspan, 2006; Wang *et al.*, 2008; Edwards *et al.*, 2009). There is also a regular occurence of CYP genes in studies of caste differences in social insects (for example, Cornette *et al.*, 2006; Zhou *et al.*, 2006) that merits further exploration. New regulatory pathways and new signal molecules remain to be discovered through a better understanding of insect P450 enzymes such as CYP303A1 (Willingham and Keil, 2004), CYP310A1 (Mohit *et al.*, 2006), and many others.

When significantly more data are obtained on the catalytic competence of a wide variety of insect P450 enzymes, it will become easier to understand the way insects maintain a wide repertoire of P450 genes. If positive selection of a few P450 genes can lead to specialized enzymes in oligophagous species (Li *et al.*, 2003), is this an evolutionary dead end? Do the P450 enzymes with "broad and overlapping" specificity serve as a perpetual reservoir where some genes, because of their pattern of expression or inducibility or catalytic competence, can then serve as templates for the evolution of a new branch of specialized enzymes? Does this "primordial soup" perpetuate itself simply by a neutral process of intense gene duplication, or are new chemical insults of the environment frequent enough to positively select for a minimal number of "jack-of-all-trades" P450 enzymes? How do P450 genes become recruited into physiological networks and biosynthetic pathways? The gap between "endogenous" and "xenobiotic" is continuously being filled with new data and insights from P450 research (Nebert, 1991).

Of the 57 human P450 genes, 15 are known xenobiotic metabolizers, 14 metabolize sterols or steroids, 15 are various lipids, and 13 remain "orphans" with unknown function (Guengerich *et al.*, 2005). If these proportions are any guide – and indeed it was estimated that one-third or less of the *Drosophila* P450s are involved in the xenobiotic response (Giraudo *et al.*, 2010) – then P450 enzymes will increase in importance in insect biochemistry, physiology, and ecology. It is hoped that this review will help all new explorers of the insect CYPome.

References

Adams, M. D., Celniker, S. E., Holt, R. A., Evans, C. A., Gocayne, J. D., et al. (2000). The genome sequence of *Drosophila melanogaster*. *Science, 287*, 2185–2195.

Agosin, M. (1985). Role of microsomal oxidations in insecticide degradation. In G. A. Kerkut, & L. I. Gilbert (Eds.), *Comprehensive Insect Physiology, Biochemistry and Pharmacology* (pp. 647–712). Oxford, UK: Pergamon.

Agrawal, A. A., Vala, F., & Sabelis, M. W. (2002). Induction of preference and performance after acclimation to novel hosts in a phytophagous spider mite: adaptive plasticity? *Am. Naturalist, 159*, 553–565.

Ahmad, S. (1986). Enzymatic adaptations of herbivorous insects and mites to phytochemicals. *J. Chem. Ecol., 12*, 533–560.

Ahmad, S., Brattsten, L. B., Mullin, C. A., & Yu, S. J. (1986). Enzymes involved in the metabolism of plant allelochemicals. In L. B. Brattsten, & S. Ahmad (Eds.), *Molecular Aspects of Insect–Plant Interactions* (pp. 73–151). New York, NY: Plenum Press.

Ahmad, S., Kirkland, K. E., & Blomquist, G. J. (1987). Evidence for a sex pheromone metabolizing cytochrome P-450 mono-oxygenase in the housefly. *Arch. Insect Biochem. Physiol., 6*, 21–40.

Aigrain, L., Pompon, D., Morera, S., & Truan, G. (2009). Structure of the open conformation of a functional chimeric NADPH cytochrome P450 reductase. *EMBO Rep., 10*, 742–747.

Amenya, D. A., Naguran, R., Lo, T. C., Ranson, H., Spillings, B. L., et al. (2008). Over expression of a cytochrome P450 (CYP6P9) in a major African malaria vector, *Anopheles funestus*, resistant to pyrethroids. *Insect Mol. Biol., 17*, 19–25.

Amichot, M., Tares, S., Brun-Barale, A., Arthaud, L., Bride, J. M., & Berge, J. B. (2004). Point mutations associated with insecticide resistance in the *Drosophila* cytochrome P450 Cyp6a2 enable DDT metabolism. *Eur. J. Biochem., 271*, 1250–1257.

Anandatheerthavarada, H. K., Addya, S., Dwivedi, R. S., Biswas, G., Mullick, J., & Avadhani, N. G. (1997). Localization of multiple forms of inducible cytochromes P450 in rat liver mitochondria: Immunological characteristics and patterns of xenobiotic substrate metabolism. *Arch. Biochem. Biophys., 339*, 136–150.

Andersen, J. F., Utermohlen, J. G., & Feyereisen, R. (1994). Expression of house fly CYP6A1 and NADPH-cytochrome P450 reductase in *Escherichia coli* and reconstitution of an insecticide-metabolizing P450 system. *Biochemistry, 33*, 2171–2177.

Andersen, J. F., Ceruso, M., Unnithan, G. C., Kuwano, E., Prestwich, G. D., & Feyereisen, R. (1995). Photoaffinity labeling of methyl farnesoate epoxidase in cockroach corpora allata. *Insect Biochem. Mol. Biol., 25*, 713–719.

Andersen, J. F., Walding, J. K., Evans, P. H., Bowers, W. S., & Feyereisen, R. (1997). Substrate specificity for the epoxidation of terpenoids and active site topology of house fly cytochrome P450 6A1. *Chem. Res. Toxicol., 10*, 156–164.

Appel, H. M., & Martin, M. (1992). Significance of metabolic load in the evolution of host specificity of *Manduca sexta*. *Ecology, 73*, 216–228.

Arensburger, P., Megy, K., Waterhouse, R. M., Abrudan, J., Amedeo, P., et al. (2010). Sequencing of *Culex quinquefasciatus* establishes a platform for mosquito comparative genomics. *Science, 330*, 86–88.

Aw, T., Schlauch, K., Keeling, C. I., Young, S., Bearfield, J. C., et al. (2010). Functional genomics of mountain pine beetle (*Dendroctonus ponderosae*) midguts and fat bodies. *BMC Genomics, 11*, 215.

Baek, J. H., Clark, J. M., & Lee, S. H. (2009). Cross-strain comparison of cypermethrin-induced cytochrome P450 transcription under different induction conditions in diamonback moth. *Pestic. Biochem. Physiol., 96*, 43–50.

Baldwin, W. S., Marko, P. B., & Nelson, D. R. (2009). The cytochrome P450 (CYP) gene superfamily in *Daphnia pulex*. *BMC Genomics, 10*, 169.

Barnes, H. J., Arlotto, M. P., & Waterman, M. R. (1991). Expression and enzymatic activity of recombinant cytochrome P450 17 alpha-hydroxylase in *Escherichia coli*. *Proc. Natl. Acad. Sci. USA, 88*, 5597–5601.

Bassett, M. H., McCarthy, J. L., Waterman, M. R., & Sliter, T. J. (1997). Sequence and developmental expression of Cyp18, a member of a new cytochrome P450 family from *Drosophila*. *Mol. Cell Endocrinol., 131*, 39–49.

Baudry, J., Li, W., Pan, L., Berenbaum, M. R., & Schuler, M. A. (2003). Molecular docking of substrates and inhibitors in the catalytic site of CYP6B1, an insect cytochrome p450 monooxygenase. *Protein Eng., 16*, 577–587.

Bautista, M. A., Tanaka, T., & Miyata, T. (2007). Identification of permethrin-inducible cytochrome P450s from the diamondback moth, *Plutella xylostella* (L.), and the possibility of involvement in permethrin resistance. *Pestic. Biochem. Physiol., 87*, 85–93.

Bautista, M. A., Miyata, T., Miura, K., & Tanaka, T. (2009). RNA interference-mediated knockdown of a cytochrome P450, CYP6BG1, from the diamondback moth, *Plutella xylostella*, reduces larval resistance to permethrin. *Insect Biochem. Mol. Biol.*, *39*, 38–46.

Beaver, L. M., Hooven, L. A., Butcher, S. M., Krishnan, N., Sherman, K. A., et al. (2010). Circadian clock regulates response to pesticides in *Drosophila* via conserved Pdp1 pathway. *Toxicol. Sci.*, *115*, 513–520.

Berenbaum, M. R. (1990). Evolution of specialization in insect-umbellifer associations. *Ann. Rev. Entomol.*, *35*, 319–343.

Berenbaum, M. R. (1995). The chemistry of defense: Theory and practice. *Proc. Natl. Acad. Sci. USA*, *92*, 2–8.

Berenbaum, M. R. (2002). Postgenomic chemical ecology: From genetic code to ecological interactions. *J. Chem. Ecol.*, *28*, 873–896.

Berenbaum, M. R., & Feeny, P. (1981). Toxicity of angular furanocoumarins to swallowtails: Escalation in the coevolutionary arms race. *Science*, *212*, 927–929.

Berenbaum, M. R., & Neal, J. J. (1987). Interactions among allelochemicals and insect resistance in crop plants. *ACS Symposium Series*, *330*, 416–430.

Berenbaum, M. R., Cohen, M. B., & Schuler, M. A. (1992). Cytochrome P450 monooxygenase genes in oligophagous lepidoptera. *ACS Symposium Series*, *505*, 114–124.

Bernard, C. B., & Philogene, B. J. (1993). Insecticide synergists: Role, importance, and perspectives. *J. Toxicol. Environ. Health*, *38*, 199–223.

Bernays, E., & Graham, M. (1988). On the evolution of host specificity in phytophagous arthropods. *Ecology*, *69*, 886–892.

Bertok, B., Pap, L., Arvai, G., Bakonyvari, I., & Kuruczne Ribai, Z. (2003). Structure–activity relationship study of alkynyl ether insecticide synergists and the development of MB-599 (verbutin). *Pest. Manag. Sci.*, *59*, 377–392.

Bhaskara, S., Dean, E. D., Lam, V., & Ganguly, R. (2006). Induction of two cytochrome P450 genes, *Cyp6a2* and *Cyp6a8*, of *Drosophila melanogaster* by caffeine in adult flies and in cell culture. *Gene*, *377*, 56–64.

Bhaskara, S., Chandrasekharan, M. B., & Ganguly, R. (2008). Caffeine induction of *Cyp6a2* and *Cyp6a8* genes of *Drosophila melanogaster* is modulated by cAMP and D-JUN protein levels. *Gene*, *415*, 49–59.

Black, B. C., Hollingworth, R. M., Ahammadsahib, K. I., Kukel, C. D., & Donovan, S. (1994). Insecticidal action and mitochondrial uncoupling activity of AC303,630 and related halogenated pyrroles. *Pestic. Biochem. Physiol.*, *50*, 115–128.

Blais, C., & Lafont, R. (1986). Ecdysone 20-hydroxylation in imaginal wing discs of *Pieris brassicae* (Lepidoptera): Correlations with ecdysone and 20-hydroxyecdysone titers in pupae. *Arch. Insect Biochem. Physiol.*, *3*, 501–512.

Blomquist, G. J., Dillwith, J. W., & Pomonis, J. G. (1984). Sex pheromone of the housefly. Metabolism of (Z)-9-tricosene to (Z)-9,10-epoxytricosane and (Z)-14-tricosen-10-one. *Insect Biochem.*, *14*, 279–284.

Blomquist, G. J., Figueroa-Teran, R., Aw, M., Song, M., Gorzalski, A., et al. (2010). Pheromone production in bark beetles. *Insect Biochem. Mol. Biol.*, *40*, 699–712.

Bogwitz, M. R., Chung, H., Magoc, L., Rigby, S., Wong, W., et al. (2005). *Cyp12a4* confers lufenuron resistance in a natural population of *Drosophila melanogaster*. *Proc. Natl. Acad. Sci. USA*, *102*, 12807–12812.

Bollenbacher, W. E., Smith, S. L., Wielgus, J. J., & Gilbert, L. I. (1977). Evidence for an α-ecdysone cytochrome P-450 mixed function oxidase in insect fat body mitochondria. *Nature*, *268*, 660–663.

Bonasio, R., Zhang, G., Ye, C., Mutti, N. S., Fang, X., et al. (2010). Genomic comparison of the ants *Camponotus floridanus* and *Harpegnathos saltator*. *Science*, *329*, 1068–1071.

Bono, J. M., Matzkin, L. M., Castrezana, S., & Markow, T. A. (2008). Molecular evolution and population genetics of two *Drosophila mettleri* cytochrome P450 genes involved in host plant utilization. *Mol. Ecol.*, *17*, 3211–3221.

Boonsuepsakul, S., Luepromchai, E., & Rongnoparut, P. (2008). Characterization of *Anopheles minimus* CYP6AA3 expressed in a recombinant baculovirus system. *Arch. Insect Biochem. Physiol.*, *69*, 13–21.

Bowers, W. S., Ohta, T., Cleere, J. S., & Marsella, P. A. (1976). Discovery of insect anti-juvenile hormones in plants. *Science*, *193*, 542–547.

Bradfield, J. Y., Lee, Y. H., & Keeley, L. L. (1991). Cytochrome P450 family 4 in a cockroach: Molecular cloning and regulation by regulation by hypertrchalosemic hormone. *Proc. Natl. Acad. Sci. USA*, *88*, 4558–4562.

Brandt, A., Scharf, M., Pedra, J. H., Holmes, G., Dean, A., et al. (2002). Differential expression and induction of two *Drosophila* cytochrome P450 genes near the Rst(2)DDT locus. *Insect Mol. Biol.*, *11*, 337–341.

Brattsten, L. B. (1979a). Ecological significance of mixed-function oxidations. *Drug Metab. Rev.*, *10*, 35–58.

Brattsten, L. B. (1979b). Biochemical defense mechanisms in herbivores against plant allelochemicals. In G. A. Rosenthal, & D. H. Janzen (Eds.), *Herbivores, Their Interaction with Secondary Plant Metabolites* (pp. 199–270). New York, NY: Academic Press.

Brattsten, L. B. (1983). Cytochrome P-450 involvement in the interactions between plant terpenes and insect herbivores. *ACS Symposium Series*, *208*, 173–195.

Brattsten, L. B. (1987). Sublethal virus infection depresses cytochrome P-450 in an insect. *Experientia*, *43*, 451–454.

Brattsten, L. B., & Metcalf, R. L. (1970). The synergistic ratio of carbaryl with piperonyl butoxide as an indicator of the distribution of multifunction oxidases in the insecta. *J. Econ. Entomol.*, *63*, 101–104.

Brattsten, L. B., & Wilkinson, C. F. (1977). Insecticide solvents: Interference with insecticidal action. *Science*, *196*, 1211–1213.

Brattsten, L. B., Wilkinson, C. F., & Eisner, T. (1977). Herbivore–plant interactions: Mixed-function oxidases and secondary plant substances. *Science*, *196*, 1349–1352.

Bride, J. M., Cuany, A., Amichot, M., Brun, A., Babault, M., et al. (1997). Cytochrome P-450 field insecticide tolerance and development of laboratory resistance in grape vine populations of *Drosophila melanogaster* (Diptera: Drosophilidae). *J. Econ. Entomol.*, *90*, 1514–1520.

Brindley, W. A. (1977). Synergist differences as an alternate interpretation of carbaryl-piperonyl butoxide toxicity data. *Environ. Entomol.*, *6*, 885–888.

Brogdon, W. G., McAllister, J. C., & Vulule, J. M. (1997). Heme peroxidase activity measured in single mosquitoes identifies individuals expressing an elevated oxidase for insecticide resistance. *J. Am. Mosq. Control Assoc.*, *13*, 233–237.

Brooks, G. T. (1979). The metabolism of xenobiotics in insects. In J. W. Bridges, & L. Chasseaud (Eds.), *Progress in Drug Metabolism* (pp. 151–214). London, UK: John Wiley & Son.

Brooks, G. T., Pratt, G. E., Mace, D. W., & Cocks, J. A. (1985). Inhibitors of juvenile hormone biosynthesis in corpora allata of the cockroach *Periplaneta amaericana* (L.) *in vitro. Pestic. Sci.*, *16*, 132–142.

Brown, R. P., McDonnell, C. M., Berenbaum, M. R., & Schuler, M. A. (2005). Regulation of an insect cytochrome P450 monooxygenase gene (CYP6B1) by aryl hydrocarbon and xanthotoxin response cascades. *Gene*, *358*, 39–52.

Brown, T. M., Bryson, P. K., & Payne, G. T. (1996). Synergism by propynyl aryl ethers in permethrin-resistant tobacco budworm larvae. *Pesticide Science*, *46*, 323–331.

Brun-Barale, A., Hema, O., Martin, T., Suraporn, S., Audant, P., et al. (2010). Multiple P450 genes overexpressed in deltamethrin-resistant strains of *Helicoverpa armigera. Pest. Manag. Sci.*, *66*, 900–909.

Bull, D. L., Ivie, G. W., Beier, R. C., & Pryor, N. W. (1986). *In vitro* metabolism of a linear furanocoumarin (8-methoxypsoralen, xanthotoxin) by mixed-function oxidases of larvae of black swallowtail butterfly and fall armyworm. *J. Chem. Ecol.*, *12*, 885–892.

Capdevila, J. H., Harris, R. C., & Falck, J. R. (2002). Microsomal cytochrome P450 and eicosanoid metabolism. *Cell. Mol. Life Sci.*, *59*, 780–789.

Cariño, F., Koener, J. F., Plapp, F. W., Jr., & Feyereisen, R. (1992). Expression of the cytochrome P450 gene CYP6A1 in the housefly, *Musca domestica. ACS Symposium Series*, *505*, 31–40.

Cariño, F. A., Koener, J. F., Plapp, F. W., Jr., & Feyereisen, R. (1994). Constitutive overexpression of the cytochrome P450 gene CYP6A1 in a house fly strain with metabolic resistance to insecticides. *Insect Biochem. Mol. Biol.*, *24*, 411–418.

Casida, J. E. (1970). Mixed-function oxidase involvement in the biochemistry of insecticide synergists. *J. Agric. Food Chem.*, *18*, 753–771.

Catania, F., Kauer, M. O., Daborn, P. J., Yen, J. L., ffrench-Constant, R. H., & Schlotterer, C. (2004). World-wide survey of an Accord insertion and its association with DDT resistance in *Drosophila melanogaster. Mol. Ecol.*, *13*, 2491–2504.

Ceriani, M. F., Hogenesch, J. B., Yanovsky, M., Panda, S., Straume, M., & Kay, S. A. (2002). Genome-wide expression analysis in *Drosophila* reveals genes controlling circadian behavior. *J. Neurosci.*, *22*, 9305–9319.

Chavez, V. M., Marques, G., Delbecque, J. P., Kobayashi, K., Hollingsworth, M., et al. (2000). The *Drosophila* disembodied gene controls late embryonic morphogenesis and codes for a cytochrome P450 enzyme that regulates embryonic ecdysone levels. *Development*, *127*, 4115–4126.

Chen, C. D., Doray, B., & Kemper, B. (1998). A conserved proline-rich sequence between the N-terminal signal-anchor and catalytic domains is required for assembly of functional cytochrome P450 2C2. *Arch. Biochem. Biophys.*, *350*, 233–238.

Chen, C. K., Berry, R. E., Shokhireva, T., Murataliev, M. B., Zhang, H., & Walker, F. A. (2010). Scanning chimeragenesis: The approach used to change the substrate selectivity of fatty acid monooxygenase CYP102A1 to that of terpene omega-hydroxylase CYP4C7. *J. Biol. Inorg. Chem.*, *15*, 159–174.

Chen, J. H., Hara, T., Fisher, M. J., & Rees, H. H. (1994). Immunological analysis of developmental changes in ecdysone 20-monooxygenase expression in the cotton leafworm, *Spodoptera littoralis. Biochem. J.*, *299*, 711–717.

Chen, J. S., Berenbaum, M. R., & Schuler, M. A. (2002). Amino acids in SRS1 and SRS6 are critical for furanocoumarin metabolism by CYP6B1v1, a cytochrome P450 monooxygenase. *Insect Mol. Biol.*, *11*, 175–186.

Chintapalli, V. R., Wang, J., & Dow, J. A. (2007). Using FlyAtlas to identify better *Drosophila melanogaster* models of human disease. *Nat. Genet.*, *39*, 715–720.

Chiu, T. L., Wen, Z., Rupasinghe, S. G., & Schuler, M. A. (2008). Comparative molecular modeling of *Anopheles gambiae* CYP6Z1, a mosquito P450 capable of metabolizing DDT. *Proc. Natl. Acad. Sci. USA*, *105*, 8855–8860.

Chung, H., Bogwitz, M. R., McCart, C., Andrianopoulos, A., ffrench-Constant, R. H., et al. (2007). Cis-regulatory elements in the accord retrotransposon result in tissue-specific expression of the *Drosophila melanogaster* insecticide resistance gene *Cyp6g1. Genetics*, *175*, 1071–1077.

Chung, H., Sztal, T., Pasricha, S., Sridhar, M., Batterham, P., & Daborn, P. J. (2009). Characterization of Drosophila melanogaster cytochrome P450 genes. *Proc. Natl. Acad. Sci. USA*, *106*, 5731–5736.

Cianfrogna, J. A., Zangerl, A. R., & Berenbaum, M. R. (2002). Dietary and developmental influences on induced detoxification in an oligophage. *J. Chem. Ecol.*, *28*, 1349–1364.

Cilek, J. E., Dahlman, D. L., & Knapp, F. W. (1995). Possible mechanism of diazinon negative cross-resistance in pyrethroid-resistant horn flies (Diptera: Muscidae). *J. Econ. Entomol.*, *88*, 520–524.

Claridge-Chang, A., Wijnen, H., Naef, F., Boothroyd, C., Rajewsky, N., & Young, M. W. (2001). Circadian regulation of gene expression systems in the *Drosophila* head. *Neuron*, *32*, 657–671.

Claudianos, C. (1999). *The Evolution of a-Esterase Mediated Organophosphate Resistance in Musca domestica, Ph.D. Thesis.* Canberra, Australia: Australian National University.

Claudianos, C., Russell, R. J., & Oakeshott, J. G. (1999). The same amino acid substitution in orthologous esterases confers organophosphate resistance on the house fly and a blowfly. *Insect Biochem. Mol. Biol.*, *29*, 675–686.

Claudianos, C., Ranson, H., Johnson, R. M., Biswas, S., Schuler, M. A., et al. (2006). A deficit of detoxification enzymes: Pesticide sensitivity and environmental response in the honeybee. *Insect Mol. Biol.*, *15*, 615–636.

Cohen, M. B., & Feyereisen, R. (1995). A cluster of cytochrome P450 genes of the CYP6 family in the house fly. *DNA Cell. Biol.*, *14*, 73–82.

Cohen, M. B., Berenbaum, M. R., & Schuler, M. A. (1989). Induction of cytochome P450-mediated detoxification of xanthotoxin in the black swallowtail. *J. Chem. Ecol.*, *15*, 2347–2355.

Cohen, M. B., Schuler, M. A., & Berenbaum, M. R. (1992). A host-inducible cytochrome P-450 from a host-specific caterpillar: Molecular cloning and evolution. *Proc. Natl. Acad. Sci. USA*, *89*, 10920–10924.

Cohen, M. B., Koener, J. F., & Feyereisen, R. (1994). Structure and chromosomal localization of CYP6A1, a cytochrome P450-encoding gene from the house fly. *Gene*, *146*, 267–272.

Cornette, R., Koshikawa, S., Hojo, M., Matsumoto, T., & Miura, T. (2006). Caste-specific cytochrome P450 in the damp-wood termite *Hodotermopsis sjostedti* (Isoptera: Termopsidae). *Insect Mol. Biol.*, *15*, 235–244.

Couillaud, F., Debernard, S., Darrouzet, E., & Rossignol, F. (1996). Hidden face of juvenile hormone metabolism in the African locust. *Arch. Insect Biochem. Physiol.*, *32*, 387–397.

Crampton, A. L., Baxter, G. D., & Barker, S. C. (1999). Identification and characterisation of a cytochrome P450 gene and processed pseudogene from an arachnid: The cattle tick, *Boophilus microplus. Insect Biochem. Mol. Biol.*, *29*, 377–384.

Crankshaw, D. L., Hetnarski, K., & Wilkinson, C. F. (1981). The functional role of NADPH-cytochrome c reductase in southern armyworm (*Spodoptera eridania*) midgut microsomes. *Insect Biochemistry*, *11*, 515–522.

Cuany, A., Helvig, C., Amichot, M., Pflieger, P., Mioskowski, C., et al. (1995). Fate of a terminal olefin with *Drosophila* microsomes and its inhibitory effects on some P-450 dependent activities. *Arch. Insect Biochem. Physiol.*, *28*, 325–338.

Daborn, P., Boundy, S., Yen, J., Pittendrigh, B., & ffrench-Constant, R. (2001). DDT resistance in *Drosophila* correlates with *Cyp6g1* over-expression and confers cross-resistance to the neonicotinoid imidacloprid. *Mol. Genet. Genomics.*, *266*, 556–563.

Daborn, P. J., Yen, J. L., Bogwitz, M. R., Le Goff, G., Feil, E., et al. (2002). A single P450 allele associated with insecticide resistance in *Drosophila. Science*, *297*, 2253–2256.

Daborn, P. J., Lumb, C., Boey, A., Wong, W., ffrench-Constant, R. H., & Batterham, P. (2007). Evaluating the insecticide resistance potential of eight *Drosophila melanogaster* cytochrome P450 genes by transgenic over-expression. *Insect Biochem. Mol. Biol.*, *37*, 512–519.

d'Alencon, E., Sezutsu, H., Legeai, F., Permal, E., Bernard-Samain, S., et al. (2010). Extensive synteny conservation of holocentric chromosomes in Lepidoptera despite high rates of local genome rearrangements. *Proc. Natl. Acad. Sci. USA*, *107*, 7680–7685.

Danielson, P. B., Letman, J. A., & Fogleman, J. C. (1995). Alkaloid metabolism by cytochrome P-450 enzymes in *Drosophila melanogaster. Comp. Biochem. Physiol.*, *110B*, 683–688.

Danielson, P. B., MacIntyre, R. J., & Fogleman, J. C. (1997). Molecular cloning of a family of xenobiotic-inducible drosophilid cytochrome p450s: Evidence for involvement in host-plant allelochemical resistance. *Proc. Natl. Acad. Sci. USA*, *94*, 10797–10802.

Danielson, P. B., Foster, J. L., McMahill, M. M., Smith, M. K., & Fogleman, J. C. (1998). Induction by alkaloids and phenobarbital of Family 4 Cytochrome P450s in *Drosophila*: Evidence for involvement in host plant utilization. *Mol. Gen. Genet.*, *259*, 54–59.

Danielson, P. B., Foster, J. L., Cooper, S. K., & Fogleman, J. C. (1999). Diversity of expressed cytochrome P450 genes in the adult Mediterranean Fruit Fly, *Ceratitis capitata. Insect Mol. Biol.*, *8*, 149–159.

Dapkus, D. (1992). Genetic localization of DDT resistance in *Drosophila melanogaster* (Diptera: Drosophilidae). *J. Econ. Entomol.*, *85*, 340–347.

Darrouzet, E., Mauchamp, B., Prestwich, G. D., Kerhoas, L., Ujvary, I., & Couillaud, F. (1997). Hydroxy juvenile hormones: New putative juvenile hormones biosynthesized by locust corpora allata *in vitro. Biochem. Biophys. Res. Commun.*, *240*, 752–758.

Darvas, B., Rees, H. H., & Hoggard, N. (1993). Ecdysone 20-monooxygenase systems in flesh-flies (Diptera: Sarcophagidae), *Neobellieria bullata* and *Parasarcophaga argyrostoma. Comp. Biochem. Physiol.*, *105B*, 765–773.

David, J. P., Strode, C., Vontas, J., Nikou, D., Vaughan, A., et al. (2005). The *Anopheles gambiae* detoxification chip: A highly specific microarray to study metabolic-based insecticide resistance in malaria vectors. *Proc. Natl. Acad. Sci. USA*, *102*, 4080–4084.

David, J. P., Boyer, S., Mesneau, A., Ball, A., Ranson, H., & Dauphin-Villemant, C. (2006). Involvement of cytochrome P450 monooxygenases in the response of mosquito larvae to dietary plant xenobiotics. *Insect Biochem. Mol. Biol.*, *36*, 410–420.

Davies, L., Williams, D. R., Turner, P. C., & Rees, H. H. (2006). Characterization in relation to development of an ecdysteroid agonist-responsive cytochrome P450, CYP18A1, in Lepidoptera. *Arch. Biochem. Biophys.*, *453*, 4–12.

Davis, R. H., & Nahrstedt, A. (1987). Biosynthesis of cyanogenic glucosides in butterflies and moths. *Insect Biochemistry*, *17*, 689–693.

Delpuech, J. M., Aquadro, C. F., & Roush, R. T. (1993). Noninvolvement of the long terminal repeat of transposable element 17.6 in insecticide resistance in *Drosophila. Proc. Natl. Acad. Sci. USA*, *90*, 5643–5647.

Dethier, V. G. (1954). Evolution of feeding preferences in phytophagous insects. *Evolution*, *8*, 33–54.

de Sousa, G., Cuany, A., Brun, A., Amichot, M., Rahmani, R., & Berge, J. B. (1995). A microfluorometric method for measuring ethoxycoumarin-O-deethylase activity on individual *Drosophila melanogaster* abdomens: Interest for screening resistance in insect populations. *Anal. Biochem.*, *229*, 86–91.

Devonshire, A. L., Heidari, R., Bell, K. L., Campbell, P. M., Campbell, B. E., et al. (2003). Kinetic efficiency of mutant carboxylesterases implicated in organophosphate insecticide resistance. *Pestic. Biochem. Physiol.*, *76*, 1–13.

Dierick, H. A., & Greenspan, R. J. (2006). Molecular analysis of flies selected for aggressive behavior. *Nat. Genet.*, *38*, 1023–1031.

Djouaka, R. F., Bakare, A. A., Coulibaly, O. N., Akogbeto, M. C., Ranson, H., et al. (2008). Expression of the cytochrome P450s, CYP6P3 and CYP6M2 are significantly elevated in multiple pyrethroid resistant populations of *Anopheles gambiae* s.s. from Southern Benin and Nigeria. *BMC Genomics*, *9*, 538.

Domanski, T. L., & Halpert, J. R. (2001). Analysis of mammalian cytochrome P450 structure and function by site-directed mutagenesis. *Curr. Drug Metab.*, *2*, 117–137.

Dombrowski, S. M., Krishnan, R., Witte, M., Maitra, S., Diesing, C., et al. (1998). Constitutive and barbital-induced expression of the Cyp6a2 allele of a high producer strain of CYP6A2 in the genetic background of a low producer strain. *Gene, 221*, 69–77.

Dowd, P. F., Smith, C. M., & Sparks, T. C. (1983). Detoxification of plant toxins by insects. *Insect Biochemistry, 13*, 453–468.

Drabek, J., & Neumann, R. (1985). Proinsecticides. In D. H. Hutson, & T. R. Roberts (Eds.), *Insecticides* (pp. 35–86). Chichester, UK: John Wiley & Sons Ltd.

Drnevich, J. M., Reedy, M. M., Ruedi, E. A., Rodriguez-Zas, S., & Hughes, K. A. (2004). Quantitative evolutionary genomics: Differential gene expression and male reproductive success in *Drosophila melanogaster. Proc. R. Soc. Lond. B Biol. Sci., 271*, 2267–2273.

Dunkov, B. C., Rodriguez-Arnaiz, R., Pittendrigh, B., ffrench-Constant, R.H., & Feyereisen, R. (1996). Cytochrome P450 gene clusters in *Drosophila melanogaster. Mol. Gen. Genet., 251*, 290–297.

Dunkov, B. C., Guzov, V. M., Mocelin, G., Shotkoski, F., Brun, A., et al. (1997). The *Drosophila* cytochrome P450 gene Cyp6a2: Structure, localization, heterologous expression, and induction by phenobarbital. *DNA Cell. Biol., 16*, 1345–1356.

Durham, E. W., Siegfried, B. D., & Scharf, M. E. (2002). *In vivo* and *in vitro* metabolism of fipronil by larvae of the European corn borer *Ostrinia nubilalis. Pest. Manag. Sci., 58*, 799–804.

Edwards, A. C., Ayroles, J. F., Stone, E. A., Carbone, M. A., Lyman, R. F., & Mackay, T. F. (2009). A transcriptional network associated with natural variation in *Drosophila* aggressive behavior. *Genome Biol., 10*, R76.

Ehlting, J., Sauveplane, V., Olry, A., Ginglinger, J. F., Provart, N. J., & Werck-Reichhart, D. (2008). An extensive (co-) expression analysis tool for the cytochrome P450 superfamily in *Arabidopsis thaliana. BMC Plant Biol., 8*, 47.

Ehrlich, P. R., & Raven, P. H. (1964). Butterflies and plants: A study in coevolution. *Evolution, 18*, 586–608.

Ellis, J., Gutierrez, A., Barsukov, I. L., Huang, W. C., Grossmann, J. G., & Roberts, G. C. (2009). Domain motion in cytochrome P450 reductase: Conformational equilibria revealed by NMR and small-angle x-ray scattering. *J. Biol. Chem., 284*, 36628–36637.

Emerson, J. J., Cardoso-Moreira, M., Borevitz, J. O., & Long, M. (2008). Natural selection shapes genome-wide patterns of copy-number polymorphism in *Drosophila melanogaster. Science, 320*, 1629–1631.

Estabrook, R. W. (1996). The remarkable P450s: A historical overview of these versatile hemeprotein catalysts. *FASEB J., 10*, 202–204.

Fan, Y., Zurek, L., Dykstra, M. J., & Schal, C. (2003). Hydrocarbon synthesis by enzymatically dissociated oenocytes of the abdominal integument of the German cockroach, *Blattella germanica. Naturwissenschaften, 90*, 121–126.

Felix, R. C., Muller, P., Ribeiro, V., Ranson, H., & Silveira, H. (2010). *Plasmodium* infection alters *Anopheles gambiae* detoxification gene expression. *BMC Genomics, 11*, 312.

Feng, R., Houseman, J. G., & Downe, A. E.R. (1992). Effect of ingested meridic diet and corn leaves on midgut detoxification processes in the European corn borer, *Ostrinia nubilalis. Pestic. Biochem. Physiol., 42*, 203–210.

Festucci-Buselli, R. A., Carvalho-Dias, A. S., de Oliveira-Andrade, M., Caixeta-Nunes, C., Li, H. M., et al. (2005). Expression of Cyp6g1 and Cyp12d1 in DDT resistant and susceptible strains of *Drosophila melanogaster. Insect Mol. Biol., 14*, 69–77.

Feyereisen, R. (1977). Cytochrome P-450 et hydroxylation de l'ecdysone en ecdysterone chez *Locusta migratoria. C.R. Acad. Sc. Paris, 284*, 1831–1834.

Feyereisen, R. (1983). Polysubstrate monooxygenases (cytochrome P-450) in larvae of susceptible and resistant strains of house flies. *Pestic. Biochem. Physiol., 19*, 262–269.

Feyereisen, R. (1999). Insect P450 enzymes. *Annu. Rev. Entomol., 44*, 507–533.

Feyereisen, R. (2005). Insect cytochrome P450. In L. I. Gilbert, K. Iatrou, & S. S. Gill (Eds.), *Comprehensive Molecular Insect Science* (pp. 1–77). Oxford, UK: Elsevier.

Feyereisen, R. (2006). Evolution of insect P450. *Biochem. Soc. Trans., 34*, 1252–1255.

Feyereisen, R. (2011). Arthropod CYPomes illustrate the tempo and mode in P450 evolution. *Biochim. Biophys. Acta., 1814*, 19–28.

Feyereisen, R., & Durst, F. (1978). Ecdysterone biosynthesis: A microsomal cytochrome-P-450-linked ecdysone 20-monooxygenase from tissues of the African migratory locust. *Eur. J. Biochem., 88*, 37–47.

Feyereisen, R., & Vincent, D. R. (1984). Characterization of antibodies to house fly NADPH-cytochrome P-450 reductase. *Insect Biochemistry, 14*, 163–168.

Feyereisen, R., Pratt, G. E., & Hamnett, A. F. (1981). Enzymic synthesis of juvenile hormone in locust corpora allata: Evidence for a microsomal cytochrome P-450 linked methyl farnesoate epoxidase. *Eur. J. Biochem., 118*, 231–238.

Feyereisen, R., Langry, K. C., & Ortiz de Montellano, P. R. (1984). Self-catalyzed destruction of insect cytochrome P-450. *Insect Biochemistry, 14*, 19–26.

Feyereisen, R., Baldridge, G. D., & Farnsworth, D. E. (1985). A rapid method for preparing insect microsomes. *Comp. Biochem. Physiol. [B], 82*, 559–562.

Feyereisen, R., Koener, J. F., Farnsworth, D. E., & Nebert, D. W. (1989). Isolation and sequence of cDNA encoding a cytochrome P-450 from an insecticide-resistant strain of the house fly, *Musca domestica. Proc. Natl. Acad. Sci. USA, 86*, 1465–1469.

ffrench-Constant, R. H., Park, Y., & Feyereisen, R. (1999). Molecular biology of insecticide resistance. In A. Puga, & K. B. Wallace (Eds.), *Molecular Biology of the Toxic Response* (pp. 533–551). London, UK: Taylor & Francis.

ffrench-Constant, R. H., Daborn, P. J., & Le Goff, G. (2004). The genetics and genomics of insecticide resistance. *Trends Genet., 20*, 163–170.

Fogleman, J. C. (2000). Response of *Drosophila melanogaster* to selection for P450-mediated resistance to isoquinoline alkaloids. *Chem. Biol. Interact., 125*, 93–105.

Fogleman, J. C., & Danielson, P. B. (2000). Analysis of fragment homology among DNA sequences from cytochrome P450 families 4 and 6. *Genetica, 110*, 257–265.

Fogleman, J. C., Danielson, P. B., & MacIntyre, R. J. (1998). The molecular basis of adaptation in *Drosophila*: The role of cytochrome P450s. In M. Hecht, R. MacIntyre, & M. Clegg (Eds.), *Evolutionary Biology* (pp. 15–77). New York, NY: Plenum Press.

Fraenkel, G. S. (1959). The raison d'etre of secondary plant substances. *Science, 129*, 1466–1470.

Frank, M. R., & Fogleman, J. C. (1992). Involvement of cytochrome P450 in host-plant utilization by Sonoran Desert *Drosophila. Proc. Natl. Acad. Sci. USA, 89*, 11998–12002.

Freeman, M. R., Dobritsa, A., Gaines, P., Segraves, W. A., & Carlson, J. R. (1999). The dare gene: Steroid hormone production, olfactory behavior, and neural degeneration in *Drosophila. Development, 126*, 4591–4602.

Frolov, M. V., & Alatortsev, V. E. (1994). Cluster of cytochrome P450 genes on the X chromosome of *Drosophila melanogaster. DNA Cell. Biol., 13*, 663–668.

Fuji-Kuriyama, Y., Mizukami, Y., Kawajiri, K., Sogawa, K., & Muramatsu, M. (1982). Primary structure of a cytochrome P-450: Coding nucleotide sequence of phenobarbital-inducible cytochrome P-450 cDNA from rat liver. *Proc. Natl. Acad. Sci. USA, 79*, 2793–2797.

Fujii, S., & Amrein, H. (2002). Genes expressed in the *Drosophila* head reveal a role for fat cells in sex-specific physiology. *EMBO J., 21*, 5353–5363.

Fujii, S., Toyama, A., & Amrein, H. (2008). A male-specific fatty acid omega-hydroxylase, SXE1, is necessary for efficient male mating in *Drosophila melanogaster. Genetics, 180*, 179–190.

Gacar, F., & Taskin, V. (2009). Partial base sequence analysis of MdaE7 gene and ali-esterase enzyme activities in field collected populations of house fly (*Musca domestica* L.) from Mediterranean and Aegean regions of Turkey. *Pesticide Biochem. Physiol., 94*, 86–92.

Gandhi, R., Varak, E., & Goldberg, M. L. (1992). Molecular analysis of a cytochrome P450 gene of family 4 on the *Drosophila* X chromosome. *DNA Cell. Biol., 11*, 397–404.

Gao, J., & Scott, J. G. (2006). Role of the transcriptional repressor mdGfi-1 in CYP6D1v1-mediated insecticide resistance in the house fly, *Musca domestica. Insect Biochem. Mol. Biol., 36*, 387–395.

Genter, M. B., Clay, C. D., Dalton, T. P., Dong, H., Nebert, D. W., & Shertzer, H. G. (2006). Comparison of mouse hepatic mitochondrial versus microsomal cytochromes P450 following TCDD treatment. *Biochem. Biophys. Res. Commun., 342*, 1375–1381.

Girardot, F., Monnier, V., & Tricoire, H. (2004). Genome wide analysis of common and specific stress responses in adult *Drosophila melanogaster. BMC Genomics, 5*, 74.

Giraudo, M., Unnithan, G. C., Le Goff, G., & Feyereisen, R. (2010). Regulation of cytochrome P450 expression in *Drosophila*: Genomic insights. *Pestic. Biochem. Physiol., 97*, 115–122.

Glendinning, J. I., & Slansky, F., Jr. (1995). Consumption of a toxic food by caterpillars increases with dietary exposure: Support for a role of induced detoxification enzymes. *J. Comp. Physiol. A, 176*, 337–345.

Gotoh, O. (1992). Substrate recognition sites in cytochrome P450 family 2 (CYP2) proteins inferred from comparative analyses of amino acid and coding nucleotide sequences. *J. Biol. Chem., 267*, 83–90.

Gotoh, O. (1993). Evolution and Differentiation of P-450 genes. In T. Omura, Y. Ishimura, & Y. Fuji-Kuriyama (Eds.), *Cytochrome P-450* (2nd ed.), (pp. 255–272). Tokyo, Japan: Kodansha.

Gotoh, O. (1998). Divergent structures of *Caenorhabditis elegans* cytochrome P450 genes suggest the frequent loss and gain of introns during the evolution of nematodes. *Mol. Biol. Evol., 15*, 1447–1459.

Gould, F. (1984). Mixed function oxidases and herbivore polyphagy: The devil's advocate position. *Ecological Entomology, 9*, 29–34.

Govind, G., Mittapalli, O., Griebel, T., Allmann, S., Bocker, S., & Baldwin, I. T. (2010). Unbiased transcriptional comparisons of generalist and specialist herbivores feeding on progressively defenseless *Nicotiana attenuata* plants. *PLoS One, 5*, e8735.

Graveley, B. R., Brooks, A. N., Carlson, J. W., Duff, M. O., Landolin, J. M., et al. (2011). The developmental transcriptome of *Drosophila melanogaster. Nature, 471*, 473–479.

Greenwood, D. R., & Rees, H. H. (1984). Ecdysone 20-monooxygenase in the desert locust, *Schistocerca gregaria. Biochem. J., 223*, 837–847.

Grieneisen, M. L., Warren, J. T., & Gilbert, L. I. (1993). Early steps in ecdysteroid biosynthesis: Evidence for the involvement of cytochrome P-450 enzymes. *Insect Biochemistry and Molecular Biology, 23*, 13–23.

Guengerich, F. P. (2001). Common and uncommon cytochrome P450 reactions related to metabolism and chemical toxicity. *Chem. Res. Toxicol., 14*, 611–650.

Guengerich, F. P., Wu, Z. L., & Bartleson, C. J. (2005). Function of human cytochrome P450s: Characterization of the orphans. *Biochem. Biophys. Res. Commun., 338*, 465–469.

Guittard, E., Blais, C., Maria, A., Parvy, J. P., Pasricha, S., et al. (2011). CYP18A1, a key enzyme of *Drosophila* steroid hormone inactivation, is essential for metamorphosis. *Dev. Biol., 349*, 35–45.

Gunderson, C. A., Brattsten, L. B., & Fleming, J. T. (1986). Microsomal oxidase and glutathione transferase as factors influencing the effects of pulegone in southern and fall armyworm larvae. *Pestic. Biochem. Physiol., 26*, 238–249.

Gutierrez, E., Wiggins, D., Fielding, B., & Gould, A. P. (2007). Specialized hepatocyte-like cells regulate *Drosophila* lipid metabolism. *Nature, 445*, 275–280.

Guzov, V. M., Houston, H. L., Murataliev, M. B., Walker, F. A., & Feyereisen, R. (1996). Molecular cloning, overexpression in *Escherichia coli*, structural and functional characterization of house fly cytochrome b5. *J. Biol. Chem., 271*, 26637–26645.

Guzov, V. M., Unnithan, G. C., Chernogolov, A. A., & Feyereisen, R. (1998). CYP12A1, a mitochondrial cytochrome P450 from the house fly. *Arch. Biochem. Biophys., 359*, 231–240.

Hahn, M. W., Han, M. V., & Han, S. G. (2007). Gene family evolution across 12 *Drosophila* genomes. *PLoS Genet., 3*, e197.

Hainzl, D., & Casida, J. E. (1996). Fipronil insecticide: Novel photochemical desulfinylation with retention of neurotoxicity. *Proc. Natl. Acad. Sci. USA, 93*, 12764–12767.

Hainzl, D., Cole, L. M., & Casida, J. E. (1998). Mechanisms for selective toxicity of fipronil insecticide and its sulfone metabolite and desulfinyl photoproduct. *Chem. Res. Toxicol.*, *11*, 1529–1535.

Halliday, W. R., Farnsworth, D. E., & Feyereisen, R. (1986). Hemolymph ecdysteroid titer and midgut ecdysone 20-monooxygenase activity during the last larval stage of *Diploptera punctata*. *Insect Biochemistry*, *16*, 627–634.

Hallstrom, I. (1985). Genetic regulation of the cytochrome P-450 system in *Drosophila melanogaster*. II. Localization of some genes regulating cytochrome P-450 activity. *Chem. Biol. Interact.*, *56*, 173–184.

Hallstrom, I., & Blanck, A. (1985). Genetic regulation of the cytochrome P-450 system in *Drosophila melanogaster*. I. Chromosomal determination of some cytochrome P-450-dependent reactions. *Chem. Biol. Interact.*, *56*, 157–171.

Hamdane, D., Xia, C., Im, S. C., Zhang, H., Kim, J. J., & Waskell, L. (2009). Structure and function of an NADPH-cytochrome P450 oxidoreductase in an open conformation capable of reducing cytochrome P450. *J. Biol. Chem.*, *284*, 11374–11384.

Hammock, B. D. (1975). NADPH dependent epoxidation of methyl farnesoate to juvenile hormone in the cockroach *Blaberus giganteus* L. *Life Sci.*, *17*, 323–328.

Hammock, B. D. (1985). Regulation of juvenile hormone titer: Degradation. In G. A. Kerkut, & L. I. Gilbert (Eds.), *Comprehensive Insect Physiology, Biochemistry and Pharmacology* (pp. 431–472). Oxford, UK: Pergamon.

Hammock, B. D., & Mumby, S. M. (1978). Inhibition of epoxidation of methyl farnesoate to juvenile hormone III by cockroach corpus allatum homogenates. *Pestic. Biochem. Physiol.*, *9*, 39–47.

Hannemann, F., Bichet, A., Ewen, K. M., & Bernhardt, R. (2007). Cytochrome P450 systems-biological variations of electron transport chains. *Biochim. Biophys. Acta.*, *1770*, 330–344.

Hardstone, M. C., & Scott, J. G. (2010). Is *Apis mellifera* more sensitive to insecticides than other insects? *Pest. Manag. Sci.*, *66*, 1171–1180.

Hardstone, M. C., Baker, S. A., Gao, J., Ewer, J., & Scott, J. G. (2006). Deletion of Cyp6d4 does not alter toxicity of insecticides to *Drosophila melanogaster*. *Pesticide Biochemistry and Physiology*, *84*, 236–242.

Hardstone, M. C., Komagata, O., Kasai, S., Tomita, T., & Scott, J. G. (2010). Use of isogenic strains indicates CYP9M10 is linked to permethrin resistance in *Culex pipiens quinquefasciatus*. *Insect Mol. Biol.*, *19*, 717–726.

Harrison, P. M., Milburn, D., Zhang, Z., Bertone, P., & Gerstein, M. (2003). Identification of pseudogenes in the *Drosophila melanogaster* genome. *Nucleic Acids Res.*, *31*, 1033–1037.

Harrison, T. L., Zangerl, A. R., Schuler, M. A., & Berenbaum, M. R. (2001). Developmental variation in cytochrome P450 expression in *Papilio polyxenes* in response to xanthotoxin, a hostplant allelochemical. *Arch. Insect Biochem. Physiol.*, *48*, 179–189.

Harwood, S. H., Moldenke, A. F., & Berry, R. E. (1990). Toxicity of peppermint monoterpenes to the variegated cutworm, *Peridroma saucia* Hubner (Lepidoptera: Noctuidae). *J. Econ. Entomol.*, *83*, 1761–1767.

Hasemann, C. A., Kurumbail, R. G., Boddupalli, S. S., Peterson, J. A., & Deisenhofer, J. (1995). Structure and function of cytochromes P450: A comparative analysis of three crystal structures. *Structure*, *3*, 41–62.

Hatano, R., & Scott, J. G. (1993). Anti-P450lpr antiserum inhibits the activation of chlorpyrifos to chlorpyrifos oxon in house fly microsomes. *Pestic. Biochem. Physiol.*, *45*, 228–233.

Hatano, R., Scott, J. G., & Dennehy, T. J. (1992). Enhanced activation is the mechanism of negative cross-resistance to chlorpyriphos in the Dicofol-IR strain of *Tetranychus urticae* (Acari: Tetranychidae). *J. Econ. Entomol.*, *85*, 1088–1091.

He, H., Chen, A. C., Davey, R. B., & Ivie, G. W. (2002). Molecular cloning and nucleotide sequence of a new P450 gene, *CYP319A1*, from the cattle tick, *Boophilus microplus*. *Insect Biochem. Mol. Biol.*, *32*, 303–309.

Hehn, A., Morant, M., & Werck-Reichhart, D. (2002). Partial recoding of P450 and P450 reductase cDNAs for improved expression in yeast and plants. *Methods Enzymol.*, *357*, 343–351.

Helvig, C., Koener, J. F., Unnithan, G. C., & Feyereisen, R. (2004a). CYP15A1, the cytochrome P450 that catalyzes epoxidation of methyl farnesoate to Juvenile Hormone III in cockroach corpora allata. *Proc. Natl. Acad. Sci. USA*, *101*, 4024–4029.

Helvig, C., Tijet, N., Feyereisen, R., Walker, F. A., & Restifo, L. L. (2004b). *Drosophila melanogaster* CYP6A8, an insect P450 that catalyzes lauric acid (omega-1)-hydroxylation. *Biochem. Biophys. Res. Commun.*, *325*, 1495–1502.

Hodgson, E. (1985). Microsomal monooxygenases. In G. A. Kerkut, & L. I. Gilbert (Eds.), *Comprehensive Insect Physiology, Biochemistry and Pharmacology* (pp. 225–331). Oxford, UK: Pergamon.

Hoggard, N., & Rees, H. H. (1988). Reversible activation–inactivation of mitochondrial ecdysone 20-mono-oxygenase: A possible role for phosphorylation–dephosphorylation. *J. Insect Physiol.*, *34*, 647–653.

Hoggard, N., Fisher, M. J., & Rees, H. H. (1989). Possible role for covalent modification in the reversible activation of ecdysone 20-monooxygenase activity. *Arch. Insect Biochem. Physiol.*, *10*, 241–253.

Hold, K. M., Sirisoma, N. S., & Casida, J. E. (2001). Detoxification of alpha- and beta-Thujones (the active ingredients of absinthe): Site specificity and species differences in cytochrome P450 oxidation *in vitro* and *in vivo*. *Chem. Res. Toxicol.*, *14*, 589–595.

Hold, K. M., Sirisoma, N. S., Ikeda, T., Narahashi, T., & Casida, J. E. (2000). Alpha-thujone (the active component of absinthe): Gamma-aminobutyric acid type A receptor modulation and metabolic detoxification. *Proc. Natl. Acad. Sci. USA*, *97*, 3826–3831.

Holt, R. A., Subramanian, G. M., Halpern, A., Sutton, G. G., Charlab, R., et al. (2002). The genome sequence of the malaria mosquito *Anopheles gambiae*. *Science*, *298*, 129–149.

Holzkamp, G., & Nahrstedt, A. (1994). Biosynthesis of cyanogenic glucosides in the lepidoptera. Incorporation of [U-14C]-2-methylpropanealdoxime, 2S-[U-14C]-methylbutanealdoxime and -[U-14C]-N-hydroxyisoleucine into linamarin and lotaustralin by the larvae of *Zygaena trifolli*. *Insect Biochem. Mol. Biol.*, *24*, 161–165.

Hooven, L. A., Sherman, K. A., Butcher, S., & Giebultowicz, J. M. (2009). Does the clock make the poison? Circadian variation in response to pesticides. *PLoS One, 4,* e6469.

Hopkins, B. W., Longnecker, M. T., & Pietrantonio, P. V. (2010). Transcriptional overexpression of CYP6B8/CYP6B28 and CYP6B9 is a mechanism associated with cypermethrin survivorship in field-collected *Helicoverpa zea* (Lepidoptera: Noctuidae) moths. *Pest. Manag. Sci., 67,* 21–25.

Horike, N., & Sonobe, H. (1999). Ecdysone 20-monooxygenase in eggs of the silkworm, *Bombyx mori*: Enzymatic properties and developmental changes. *Arch. Insect Biochem. Physiol., 41,* 9–17.

Horike, N., Takemori, H., Nonaka, Y., Sonobe, H., & Okamoto, M. (2000). Molecular cloning of NADPH-cytochrome P450 oxidoreductase from silkworm eggs. Its involvement in 20-hydroxyecdysone biosynthesis during embryonic development. *Eur. J. Biochem., 267,* 6914–6920.

Horner, M. A., Pardee, K., Liu, S., King-Jones, K., Lajoie, G., et al. (2009). The *Drosophila* DHR96 nuclear receptor binds cholesterol and regulates cholesterol homeostasis. *Genes Dev., 23,* 2711–2716.

Houpt, D. R., Pursey, J. C., & Morton, R. A. (1988). Genes controlling malathion resistance in a laboratory-selected population of *Drosophila melanogaster*. *Genome, 30,* 844–853.

Hovemann, B. T., Sehlmeyer, F., & Malz, J. (1997). *Drosophila melanogaster* NADPH-cytochrome P450 oxidoreductase: Pronounced expression in antennae may be related to odorant clearance. *Gene, 189,* 213–219.

Huber, D. P., Erickson, M. L., Leutenegger, C. M., Bohlmann, J., & Seybold, S. J. (2007). Isolation and extreme sex-specific expression of cytochrome P450 genes in the bark beetle, *Ips paraconfusus*, following feeding on the phloem of host ponderosa pine, *Pinus ponderosa*. *Insect Mol. Biol., 16,* 335–349.

Hughes, P. B., & Devonshire, A. L. (1982). The biochemical basis of resistance to organophosphorus insecticides in the sheep blowfly, *Lucilia cuprina*. *Pestic. Biochem. Physiol., 18,* 289–297.

Hung, C. F., Harrison, T. L., Berenbaum, M. R., & Schuler, M. A. (1995a). CYP6B3: A second furanocoumarin-inducible cytochrome P450 expressed in *Papilio polyxenes*. *Insect Mol. Biol., 4,* 149–160.

Hung, C. F., Prapaipong, H., Berenbaum, M. R., & Schuler, M. A. (1995b). Differential induction of cytochrome P-450 transcripts in *Papilio polyxenes* by linear and angular furanocoumarins. *Insect Biochem. Mol. Biol., 25,* 89–99.

Hung, C. F., Holzmacher, R., Connolly, E., Berenbaum, M. R., & Schuler, M. A. (1996). Conserved promoter elements in the CYP6B gene family suggest common ancestry for cytochrome P450 monooxygenases mediating furanocoumarin detoxification. *Proc. Natl. Acad. Sci. USA, 93,* 12200–12205.

Hung, C. F., Berenbaum, M. R., & Schuler, M. A. (1997). Isolation and characterization of CYP6B4, a furanocoumarin-inducible cytochrome P450 from a polyphagous caterpillar (Lepidoptera: Papilionidae). *Insect Biochem. Mol. Biol., 27,* 377–385.

Hurban, P., & Thummel, C. S. (1993). Isolation and characterization of fifteen ecdysone-inducible *Drosophila* genes reveal unexpected complexities in ecdysone regulation. *Mol. Cell Biol., 13,* 7101–7111.

Ikezawa, N., Morisaki, M., & Fujimoto, Y. (1993). Sterol metabolism in insects: Dealkylation of phytosterol to cholesterol. *Acc. Chem. Res., 26,* 139–146.

Ikezawa, N., Tanaka, M., Nagayoshi, M., Shinkyo, R., Sakaki, T., et al. (2003). Molecular cloning and characterization of CYP719, a methylenedioxy bridge-forming enzyme that belongs to a novel P450 family, from cultured Coptis japonica cells. *J. Biol. Chem., 278,* 38557–38565.

Inceoglu, A. B., Waite, T. D., Christiansen, J. A., McAbee, R. D., Kamita, S. G., et al. (2009). A rapid luminescent assay for measuring cytochrome P450 activity in individual larval *Culex pipiens* complex mosquitoes (Diptera: Culicidae). *J. Med. Entomol., 46,* 83–92.

Ishikawa, C., Yoshinaga, N., Aboshi, T., Nishida, R., & Mori, N. (2009). Efficient incorporation of free oxygen into volicitin in *Spodoptera litura* common cutworm larvae. *Biosci. Biotechnol. Biochem., 73,* 1883–1885.

Itokawa, K., Komagata, O., Kasai, S., Okamura, Y., Masada, M., & Tomita, T. (2010). Genomic structures of Cyp9m10 pyrethroid-resistant and -susceptible strains of *Culex quinquefasciatus*. *Insect Biochem. Mol. Biol., 40,* 631–640.

Ivanov, A. S., Gnedenko, O. V., Molnar, A. A., Archakov, A. I., & Podust, L. M. (2010). FMN binding site of yeast NADPH-cytochrome P450 reductase exposed at the surface is highly specific. *ACS Chem. Biol., 5,* 767–776.

Ivie, G. W., Bull, D. L., Beier, R. C., Pryor, N. W., & Oertli, E. H. (1983). Metabolic detoxification: Mechanism of insect resistance to plant psoralens. *Science, 221,* 374–376.

Iyengar, S., Arnason, J. T., Philogene, B. J.R., Werstiuk, N. H., & Morand, P. (1990). Comparative metabolism of the phototoxic allelochemical α-terthienyl in three species of Lepidopterans. *Pestic. Biochem. Physiol., 37,* 154–164.

Jacobsen, N. E., Kover, K. E., Murataliev, M. B., Feyereisen, R., & Walker, F. A. (2006). Structure and stereochemistry of products of hydroxylation of human steroid hormones by a housefly cytochrome P450 (CYP6A1). *Magn. Reson. Chem., 44,* 467–474.

Jefcoate, C. F. (1978). Measurement of substrate and inhibitor binding to microsomal cytochrome P-450 by optical-difference spectroscopy. *Methods Enzymol., 52,* 258–279.

Jensen, H. R., Scott, I. M., Sims, S., Trudeau, V. L., & Arnason, J. T. (2006a). Gene expression profiles of *Drosophila melanogaster* exposed to an insecticidal extract of *Piper nigrum*. *J. Agric. Food Chem., 54,* 1289–1295.

Jensen, H. R., Scott, I. M., Sims, S. R., Trudeau, V. L., & Arnason, J. T. (2006b). The effect of a synergistic concentration of a *Piper nigrum* extract used in conjunction with pyrethrum upon gene expression in *Drosophila melanogaster*. *Insect Mol. Biol., 15,* 329–339.

Jensen, N. B., Zagrobelny, M., Hjerno, K., Olsen, C. E., Houghten-Larsen, J., et al. (2011). Convergent evolution in biosynthesis of cyanogenic defence compounds in plants and insects. *Nature communications* 2, Article number: 273, doi:10.1038/ncomms 1271.

Jermy, T. (1984). Evolution of insect/host plant relationships. *Am Naturalist, 124,* 609–630.

Jiang, H. B., Tang, P. A., Xu, Y. Q., An, F. M., & Wang, J. J. (2010). Molecular characterization of two novel deltamethrin-inducible P450 genes from *Liposcelis bostrychophila* Badonnel (Psocoptera: Liposcelididae). *Arch. Insect Biochem. Physiol., 74,* 17–37.

Johansson, K. C., Metzendorf, C., & Soderhall, K. (2005). Microarray analysis of immune challenged *Drosophila* hemocytes. *Exp. Cell Res.*, *305*, 145–155.

Johnson, P., & Rees, H. H. (1977). The mechanism of C-20 hydroxylation of α-ecdysone in the desert locust, *Schistocerca gregaria. Biochem. J.*, *168*, 513–520.

Jones, D., & Jones, G. (2007). Farnesoid secretions of dipteran ring glands: What we do know and what we can know. *Insect Biochem. Mol. Biol.*, *37*, 771–798.

Jones, R. T., Bakker, S. E., Stone, D., Shuttleworth, S. N., Boundy, S., et al. (2010). Homology modelling of *Drosophila* cytochrome P450 enzymes associated with insecticide resistance. *Pest. Manag. Sci.*, *66*, 1106–1115.

Joussen, N., Heckel, D. G., Haas, M., Schuphan, I., & Schmidt, B. (2008). Metabolism of imidacloprid and DDT by P450 CYP6G1 expressed in cell cultures of *Nicotiana tabacum* suggests detoxification of these insecticides in Cyp6g1-overexpressing strains of *Drosophila melanogaster*, leading to resistance. *Pest. Manag. Sci.*, *64*, 65–73.

Joussen, N., Schuphan, I., & Schmidt, B. (2010). Metabolism of methoxychlor by the P450-monooxygenase CYP6G1 involved in insecticide resistance of *Drosophila melanogaster* after expression in cell cultures of *Nicotiana tabacum. Chem. Biodivers.*, *7*, 722–735.

Jurenka, R. A., Subchev, M., Abad, J. L., Choi, M. Y., & Fabrias, G. (2003). Sex pheromone biosynthetic pathway for disparlure in the gypsy moth, *Lymantria dispar. Proc. Natl. Acad. Sci. USA*, *100*, 809–814.

Kabbouh, M., Kappler, C., Hetru, C., & Durst, F. (1987). Further characterization of the 2-deoxyecdysone C-2 hydroxylase from *Locusta migratoria. Insect Biochem.*, *17*, 1155–1161.

Kaewpa, D., Boonsuepsakul, S., & Rongnoparut, P. (2007). Functional expression of mosquito NADPH-cytochrome P450 reductase in *Escherichia coli. J. Econ. Entomol.*, *100*, 946–953.

Kahn, R. A., Bak, S., Svendsen, I., Halkier, B. A., & Moller, B. L. (1997). Isolation and reconstitution of cytochrome P450ox and *in vitro* reconstitution of the entire biosynthetic pathway of the cyanogenic glucoside dhurrin from sorghum. *Plant Physiol.*, *115*, 1661–1670.

Kamataki, T., & Neal, R. A. (1976). Metabolism of diethyl p-nitrophenyl phosphorothionate (parathion) by a reconstituted mixed-function oxidase enzyme system: Studies of the covalent binding of the sulfur atom. *Molecular Pharmacology*, *12*, 933–944.

Kamiya, E., Yamakawa, M., Shono, T., & Kono, Y. (2001). Molecular cloning, nucleotide sequences and gene expression of new cytochrome P450s (CYP6A24, CYP6D3v2) from the pyrethroid resistant housefly, *Musca domestica* L. (Diptera: Muscidae). *Appl. Ent. Zool.*, *36*, 225–229.

Kao, L. M., Wilkinson, C. F., & Brattsten, L. B. (1995). *In vivo* effects of 2,4-D and atrazine on cytochrome P-450 and insecticide toxicity in southern armyworm (*Spodoptera eridania*) larvae. *Pestic. Sci.*, *45*, 331–334.

Kappers, W. A., Edwards, R. J., Murray, S., & Boobis, A. R. (2001). Diazinon is activated by CYP2C19 in human liver. *Toxicol. Appl. Pharmacol.*, *177*, 68–76.

Kappler, C., Kabbouh, M., Durst, F., & Hoffmann, J. A. (1986). Studies on the C-2 hydroxylation of 2-deoxyecdysone in *Locusta migratoria. Insect Biochem.*, *16*, 25–32.

Kappler, C., Kabbouh, M., Hetru, C., Durst, F., & Hoffmann, J. A. (1988). Characterization of three hydroxylases involved in the final steps of biosynthesis of the steroid hormone ecdysone in *Locusta migratoria* (Insecta: Orthoptera). *J. Steroid Biochem.*, *31*, 891–898.

Karunker, I., Benting, J., Lueke, B., Ponge, T., Nauen, R., et al. (2008). Over-expression of cytochrome P450 CYP6CM1 is associated with high resistance to imidacloprid in the B and Q biotypes of *Bemisia tabaci* (Hemiptera: Aleyrodidae). *Insect Biochem. Mol. Biol.*, *38*, 634–644.

Karunker, I., Morou, E., Nikou, D., Nauen, R., Sertchook, R., et al. (2009). Structural model and functional characterization of the *Bemisia tabaci* CYP6CM1vQ, a cytochrome P450 associated with high levels of imidacloprid resistance. *Insect Biochem. Mol. Biol.*, *39*, 697–706.

Kasai, S., & Scott, J. G. (2000). Overexpression of cytochrome P450 CYP6D1 is associated with monooxygenase-mediated pyrethroid resistance in house flies from Georgia. *Pestic. Biochem. Physiol.*, *68*, 34–41.

Kasai, S., & Scott, J. G. (2001a). Cytochrome P450s CYP6D3 and CYP6D1 are part of a P450 gene cluster on autosome 1 in the house fly. *Insect Mol. Biol.*, *10*, 191–196.

Kasai, S., & Scott, J. G. (2001b). Expression and regulation of CYP6D3 in the house fly, *Musca domestica* (L.). *Insect Biochem. Mol. Biol.*, *32*, 1–8.

Kasai, S., Weerasinghe, I. S., & Shono, T. (1998). P450 Monooxygenases are an important mechanism of permethrin resistance in *Culex quinquefasciatus* Say larvae. *Arch. Insect Biochem. Physiol.*, *37*, 47–56.

Kasai, S., Weerashinghe, I. S., Shono, T., & Yamakawa, M. (2000). Molecular cloning, nucleotide sequence and gene expression of a cytochrome P450 (CYP6F1) from the pyrethroid-resistant mosquito, *Culex quinquefasciatus* Say. *Insect Biochem. Mol. Biol.*, *30*, 163–171.

Kayser, H., & Eilinger, P. (2001). Metabolism of diafenthiuron by microsomal oxidation: Procide activation and inactivation as mechanisms contributing to selectivity. *Pest. Management Science*, *57*, 975–980.

Kayser, H., Winkler, T., & Spindler-Barth, M. (1997). 26-hydroxylation of ecdysteroids is catalyzed by a typical cytochrome P-450-dependent oxidase and related to ecdysteroid resistance in an insect cell line. *Eur. J. Biochem.*, *248*, 707–716.

Kayser, H., Ertl, P., Eilinger, P., Spindler-Barth, M., & Winkler, T. (2002). Diastereomeric ecdysteroids with a cyclic hemiacetal in the side chain produced by cytochrome P450 in hormonally resistant insect cells. *Arch. Biochem. Biophys.*, *400*, 180–187.

Kennaugh, L., Pearce, D., Daly, J. C., & Hobbs, A. A. (1993). A piperonyl butoxide synergizable resistance to permethrin in *Helicoverpa armigera* which is not due to increased detoxification by cytochrome P450. *Pesticide Biochem. Physiol.*, *45*, 234–241.

Kennedy, G. G. (1984). 2-tridecanone, tomatoes and *Heliothis zea*: Potential incompability of plant antibiosis with insecticidal control. *Entomol. exp. appl.*, *35*, 305–311.

King-Jones, K., Horner, M. A., Lam, G., & Thummel, C. S. (2006). The DHR96 nuclear receptor regulates xenobiotic responses in *Drosophila*. *Cell Metab.*, *4*, 37–48.

Kirby, M. L., & Ottea, J. A. (1995). Multiple mechanisms for enhancement of glutathione S-transferase activities in *Spodoptera frugiperda* (Lepidopter: Noctuidae). *Insect Biochem. Mol. Biol.*, *25*, 347–353.

Kirby, M. L., Young, R. J., & Ottea, J. A. (1994). Mixed-function oxidase and glutathione S-transferase activities from field-collected larval and adult tobacco budworms, *Heliothis virescens* (F.). *Pest. Biochem. Physiol.*, *49*, 24–36.

Kirischian, N., McArthur, A. G., Jesuthasan, C., Krattenmacher, B., & Wilson, J. Y. (2011). Phylogenetic and functional analysis of the vertebrate cytochrome P450 2 family. *J. Mol. Evol.*, *72*, 56–71.

Koener, J. F., Carino, F. A., & Feyereisen, R. (1993). The cDNA and deduced protein sequence of house fly NADPH-cytochrome P450 reductase. *Insect Biochem. Mol. Biol.*, *23*, 439–447.

Komagata, O., Kasai, S., & Tomita, T. (2010). Overexpression of cytochrome P450 genes in pyrethroid-resistant *Culex quinquefasciatus*. *Insect Biochem. Mol. Biol.*, *40*, 146–152.

Konno, T., E., H., & Dauterman, W. C. (1989). Studies on methyl parathion resistance in *Heliothis virescens*. *Pestic. Biochem. Physiol.*, *33*, 189–199.

Korytko, P. J., & Scott, J. G. (1998). CYP6D1 protects thoracic ganglia of houseflies from the neurotoxic insecticide cypermethrin. *Arch. Insect Biochem. Physiol.*, *37*, 57–63.

Korytko, P. J., MacLntyre, R. J., & Scott, J. G. (2000a). Expression and activity of a house-fly cytochrome P450, CYP6D1, in *Drosophila melanogaster*. *Insect Mol. Biol.*, *9*, 441–449.

Korytko, P. J., Quimby, F. W., & Scott, J. G. (2000b). Metabolism of phenanthrene by house fly CYP6D1 and dog liver cytochrome P450. *J. Biochem. Mol. Toxicol.*, *14*, 20–25.

Kotaki, T., Shinada, T., Kaihara, K., Ohfune, Y., & Numata, H. (2009). Structure determination of a new juvenile hormone from a heteropteran insect. *Org. Lett.*, *11*, 5234–5237.

Kotze, A. C. (1995). Induced insecticide tolerance in larvae of *Lucilia cuprina* (Wiedemann) (Diptera: Calliphoridae) following dietary phenobarbital treatment. *J. Australian Entomol. Soc.*, *34*, 205–209.

Kotze, A. C., & Sales, N. (1995). Elevated *in vitro* monooxygenase activity associated with insecticide resistances in field-strain larvae of the Austalian sheep blowfly (Diptera: Calliphoridae). *J. Econ. Entomol.*, *88*, 782–787.

Krieger, R. I., Feeny, P. P., & Wilkinson, C. F. (1971). Detoxication enzymes in the guts of caterpillars: An evolutionary answer to plant defenses? *Science*, *172*, 579–581.

Kula, M. E., & Rozek, C. E. (2000). Expression and translocation of *Drosophila* nuclear encoded cytochrome b(5) proteins to mitochondria. *Insect Biochem. Mol. Biol.*, *30*, 927–935.

Kula, M. E., Allay, E. R., & Rozek, C. E. (1995). Evolutionary divergence of the cytochrome b5 gene of *Drosophila*. *J. Mol. Evol.*, *41*, 430–439.

Kuruganti, S., Lam, V., Zhou, X., Bennett, G., Pittendrigh, B. R., & Ganguly, R. (2006). High expression of Cyp6g1, a cytochrome P450 gene, does not necessarily confer DDT resistance in *Drosophila melanogaster*. *Gene*, *388*, 43–53.

Lafont, R. (2005). Ecdysteroid chemistry and biochemistry. In L. I. Gilbert, K. Iatrou, & S. S. Gill (Eds.), *Comprehensive Molecular Insect Science* (pp. 125–195). Oxford, UK: Elsevier.

Lai, C. Q., Parnell, L. D., Lyman, R. F., Ordovas, J. M., & Mackay, T. F. (2007). Candidate genes affecting *Drosophila* life span identified by integrating microarray gene expression analysis and QTL mapping. *Mech. Ageing Dev.*, *128*, 237–249.

Lamb, D. C., Kim, Y., Yermalitskaya, L. V., Yermalitsky, V. N., Lepesheva, G. I., et al. (2006). A second FMN binding site in yeast NADPH-cytochrome P450 reductase suggests a mechanism of electron transfer by diflavin reductases. *Structure*, *14*, 51–61.

Latli, B., & Prestwich, G. D. (1991). Metabolically blocked analogs of housefly sex pheromone: I. Synthesis of alternative substrates for the cuticular monooxygenases. *J. Chem. Ecol.*, *17*, 1745–1768.

Le Goff, G., Boundy, S., Daborn, P. J., Yen, J. L., Sofer, L., et al. (2003). Microarray analysis of cytochrome P450 mediated insecticide resistance in *Drosophila*. *Insect Biochem. Mol. Biol.*, *33*, 701–708.

Le Goff, G., Hilliou, F., Siegfried, B. D., Boundy, S., Wajnberg, E., et al. (2006). Xenobiotic response in *Drosophila melanogaster*: Sex dependence of P450 and GST gene induction. *Insect Biochem. Mol. Biol.*, *36*, 674–682.

Lee, S. H., Kang, J. S., Min, J. S., Yoon, K. S., Strycharz, J. P., et al. (2010). Decreased detoxification genes and genome size make the human body louse an efficient model to study xenobiotic metabolism. *Insect Mol. Biol.*, *19*, 599–615.

Li, W., Berenbaum, M. R., & Schuler, M. A. (2001). Molecular analysis of multiple CYP6B genes from polyphagous *Papilio* species. *Insect Biochem. Mol. Biol.*, *31*, 999–1011.

Li, W., Petersen, R. A., Schuler, M. A., & Berenbaum, M. R. (2002). CYP6B cytochrome P450 monooxygenases from *Papilio canadensis* and *Papilio glaucus*: Potential contributions of sequence divergence to host plant associations. *Insect Mol. Biol.*, *11*, 543–551.

Li, W., Schuler, M. A., & Berenbaum, M. R. (2003). Diversification of furanocoumarin-metabolizing cytochrome P450 monooxygenases in two papilionids: Specificity and substrate encounter rate. *Proc. Natl. Acad. Sci. USA*, 14593–14598.

Li, W., Zangerl, A. R., Schuler, M. A., & Berenbaum, M. R. (2004). Characterization and evolution of furanocoumarin-inducible cytochrome P450s in the parsnip webworm, *Depressaria pastinacella*. *Insect Mol. Biol.*, *13*, 603–613.

Li, X., Berenbaum, M. R., & Schuler, M. A. (2000a). Molecular cloning and expression of CYP6B8: A xanthotoxin-inducible cytochrome P450 cDNA from *Helicoverpa zea*. *Insect Biochem. Mol. Biol.*, *30*, 75–84.

Li, X., Zangerl, A. R., Schuler, M. A., & Berenbaum, M. (2000b). Cross-resistance to α-cypermethrin after xanthotoxin ingestion in *Helicoverpa zea* (Lepidoptera: Noctuidae). *J. Econ. Entomol.*, *93*, 18–25.

Li, X., Berenbaum, M. R., & Schuler, M. A. (2002a). Cytochrome P450 and actin genes expressed in *Helicoverpa zea* and *Helicoverpa armigera*: Paralogy/orthology identification, gene conversion and evolution. *Insect Biochem. Mol. Biol.*, *32*, 311–320.

Li, X., Schuler, M. A., & Berenbaum, M. R. (2002b). Jasmonate and salicylate induce expression of herbivore cytochrome P450 genes. *Nature, 419*, 712–715.

Li, X., Berenbaum, M. R., & Schuler, M. A. (2002c). Plant allelochemicals differentially regulate *Helicoverpa zea* cytochrome P450 genes. *Insect Mol. Biol., 11*, 343–351.

Li, X., Baudry, J., Berenbaum, M. R., & Schuler, M. A. (2004). Structural and functional divergence of insect CYP6B proteins: From specialist to generalist cytochrome P450. *Proc. Natl. Acad. Sci. USA, 101*, 2939–2944.

Li, X., Schuler, M. A., & Berenbaum, M. R. (2007). Molecular mechanisms of metabolic resistance to synthetic and natural xenobiotics. *Annu. Rev. Entomol., 52*, 231–253.

Liebrich, W., & Hoffmann, K. H. (1991). Ecdysone 20-monooxygenase in a cricket, *Gryllus bimaculatus* (Ensifera: Gryllidae): Characterization of the microsomal midgut steroid hydroxylase in adult females. *J. Comp. Physiol. B, 161*, 93–99.

Lin, G. G., Kozaki, T., & Scott, J. G. (2011). Hormone receptor-like in 96 and Broad-Complex modulate phenobarbital induced transcription of cytochrome P450 CYP6D1 in *Drosophila* S2 cells. *Insect Mol. Biol., 20*, 87–95.

Lindigkeit, R., Biller, A., Buch, M., Schiebel, H. M., Boppre, M., & Hartmann, T. (1997). The two faces of pyrrolizidine alkaloids: The role of the tertiary amine and its N-oxide in chemical defense of insects with acquired plant alkaloids. *Eur. J. Biochem., 245*, 626–636.

Lindroth, R. L. (1989). Host plant alteration of detoxication activity in *Papilio glaucus glaucus*. *Entomol. Exp. Appl., 50*, 29–35.

Lindroth, R. L. (1991). Differential toxicity of plant allelochemicals to insects: Roles of enzymatic detoxication Systems. In Bernays, E.A. (Ed.), *Insect–Plant Interactions* (pp. 1–33). Boca Raton, FL: CRC Press.

Lindsley, D. L., & Zimm, G. G. (1992). *The Genome of Drosophila melanogaster*. San Diego, CA: Academic Press, Inc.

Liu, X., Liang, P., Gao, X., & Shi, X. (2006). Induction of the cytochrome P450 activity by plant allelochemicals in the cotton bollworm, Helicoverpa amigera (Hubner). *Pesticide Biochem. Physiol., 84*, 127–134.

Liu, N., & Scott, J. G. (1995). Genetics of resistance to pyrethroid insecticides in the house fly, *Musca domestica*. *Pestic. Biochem. Physiol., 52*, 116–124.

Liu, N., & Scott, J. G. (1996). Genetic analysis of factors controlling high-level expression of cytochrome P450, CYP6D1, cytochrome b5, P450 reductase, and monooxygenase activities in LPR house flies, *Musca domestica*. *Biochem. Genet., 34*, 133–148.

Liu, N., & Scott, J. G. (1997). Inheritance of CYP6D1-mediated pyrethroid resistance in house fly (Diptera: Muscidae). *J. Econ. Entomol., 90*, 1478–1481.

Liu, N., & Scott, J. G. (1998). Increased transcription of CYP6D1 causes cytochrome P450-mediated insecticide resistance in house fly. *Insect Biochem. Mol. Biol., 28*, 531–535.

Liu, N., & Yue, X. (2001). Genetics of pyrethroid resistance in a strain (ALHF) of house flies (Diptera: Muscidae). *Pestic. Biochem. Physiol., 70*, 151–158.

Liu, N., Tomita, T., & Scott, J. G. (1995). Allele-specific PCR reveals that CYP6D1 is on chromosome 1 in the house fly, *Musca domestica*. *Experientia, 51*, 164–1677.

Lockey, K. H. (1991). Insect hydrocarbon classes: Implications for chemotaxonomy. *Insect Biochem., 21*, 91–97.

Londono, D. K., Siegfried, B. D., & Lydy, M. J. (2004). Atrazine induction of a family 4 cytochrome P450 gene in *Chironomus tentans* (Diptera: Chironomidae). *Chemosphere, 56*, 701–706.

Luu, B., & Werner, F. (1996). Sterols that modify moulting in insects. *Pesticide Science, 46*, 49–53.

Lycett, G. J., McLaughlin, L. A., Ranson, H., Hemingway, J., Kafatos, F. C., et al. (2006). *Anopheles gambiae* P450 reductase is highly expressed in oenocytes and *in vivo* knockdown increases permethrin susceptibility. *Insect Mol. Biol., 15*, 321–327.

Lynch, M., & Conery, J. S. (2000). The evolutionary fate and consequences of duplicate genes. *Science, 290*, 1151–1155.

Ma, R., Cohen, M. B., Berenbaum, M. R., & Schuler, M. A. (1994). Black swallowtail (*Papilio polyxenes*) alleles encode cytochrome P450s that selectively metabolize linear furanocoumarins. *Arch. Biochem. Biophys., 310*, 332–340.

Maeda, S., Nakashima, A., Yamada, R., Hara, N., Fujimoto, Y., et al. (2008). Molecular cloning of ecdysone 20-hydroxylase and expression pattern of the enzyme during embryonic development of silkworm *Bombyx mori*. *Comp. Biochem. Physiol. B Biochem. Mol. Biol., 149*, 507–516.

Maestro, J. L., Pascual, N., Treiblmayr, K., Lozano, J., & Belles, X. (2010). Juvenile hormone and allatostatins in the German cockroach embryo. *Insect Biochem. Mol. Biol., 40*, 660–665.

Maibeche-Coisne, M., Jacquin-Joly, E., Francois, M. C., & Nagnan-Le Meillour, P. (2002). cDNA cloning of biotransformation enzymes belonging to the cytochrome P450 family in the antennae of the noctuid moth *Mamestra brassicae*. *Insect Mol. Biol., 11*, 273–281.

Maibeche-Coisne, M., Nikonov, A. A., Ishida, Y., Jacquin-Joly, E., & Leal, W. S. (2004). Pheromone anosmia in a scarab beetle induced by *in vivo* inhibition of a pheromone-degrading enzyme. *Proc. Natl. Acad. Sci. USA, 101*, 11459–11464.

Maibeche-Coisne, M., Merlin, C., Francois, M. C., Porcheron, P., & Jacquin-Joly, E. (2005). P450 and P450 reductase cDNAs from the moth *Mamestra brassicae*: Cloning and expression patterns in male antennae. *Gene, 346*, 195–203.

Maitra, S., Dombrowski, S. M., Waters, L. C., & Ganguly, R. (1996). Three second chromosome-linked clustered Cyp6 genes show differential constitutive and barbital-induced expression in DDT-resistant and susceptible strains of *Drosophila melanogaster*. *Gene, 180*, 165–171.

Maitra, S., Dombrowski, S. M., Basu, M., Raustol, O., Waters, L. C., & Ganguly, R. (2000). Factors on the third chromosome affect the level of cyp6a2 and cyp6a8 expression in *Drosophila melanogaster*. *Gene, 248*, 147–156.

Maitra, S., Price, C., & Ganguly, R. (2002). *Cyp6a8* of *Drosophila melanogaster*: Gene structure, and sequence and functional analysis of the upstream DNA. *Insect Biochem. Mol. Biol., 32*, 859–870.

Malka, O., Karunker, I., Yeheskel, A., Morin, S., & Hefetz, A. (2009). The gene road to royalty – differential expression of hydroxylating genes in the mandibular glands of the honeybee. *FEBS J., 276*, 5481–5490.

Mansuy, D. (1998). The great diversity of reactions catalyzed by cytochrome P450. *Comp. Biochem. Physiol., 121C*, 5–14.

Mansuy, D., & Renaud, J. P. (1995). Heme-thiolate proteins different from cytochromes P450 catalyzing monooxygenations. In P. R. Ortiz de Montellano (Ed.), *Cytochrome P450* (2nd ed.), (pp. 537–574). New York, NY: Plenum Press.

Mao, W., Rupasinghe, S., Zangerl, A. R., Schuler, M. A., & Berenbaum, M. R. (2006a). Remarkable substrate-specificity of CYP6AB3 in *Depressaria pastinacella*, a highly specialized caterpillar. *Insect Mol. Biol.*, *15*, 169–179.

Mao, W., Berhow, M. A., Zangerl, A. R., McGovern, J., & Berenbaum, M. R. (2006b). Cytochrome P450-mediated metabolism of xanthotoxin by *Papilio multicaudatus*. *J. Chem. Ecol.*, *32*, 523–536.

Mao, W., Rupasinghe, S. G., Zangerl, A. R., Berenbaum, M. R., & Schuler, M. A. (2007a). Allelic variation in the *Depressaria pastinacella* CYP6AB3 protein enhances metabolism of plant allelochemicals by altering a proximal surface residue and potential interactions with cytochrome P450 reductase. *J. Biol. Chem.*, *282*, 10544–10552.

Mao, W., Schuler, M. A., & Berenbaum, M. R. (2007b). Cytochrome P450s in *Papilio multicaudatus* and the transition from oligophagy to polyphagy in the Papilionidae. *Insect Mol. Biol.*, *16*, 481–490.

Mao, W., Berenbaum, M. R., & Schuler, M. A. (2008a). Modifications in the N-terminus of an insect cytochrome P450 enhance production of catalytically active protein in baculovirus-Sf9 cell expression systems. *Insect Biochem. Mol. Biol.*, *38*, 66–75.

Mao, W., Zangerl, A. R., Berenbaum, M. R., & Schuler, M. A. (2008b). Metabolism of myristicin by *Depressaria pastinacella* CYP6AB3v2 and inhibition by its metabolite. *Insect Biochem. Mol. Biol.*, *38*, 645–651.

Mao, W., Rupasinghe, S. G., Johnson, R. M., Zangerl, A. R., Schuler, M. A., & Berenbaum, M. R. (2009). Quercetin-metabolizing CYP6AS enzymes of the pollinator *Apis mellifera* (Hymenoptera: Apidae). *Comp. Biochem. Physiol. B Biochem. Mol. Biol.*, *154*, 427–434.

Mao, Y. B., Cai, W. J., Wang, J. W., Hong, G. J., Tao, X. Y., et al. (2007). Silencing a cotton bollworm P450 monooxygenase gene by plant-mediated RNAi impairs larval tolerance of gossypol. *Nature Biotechnology*, *25*, 1307–1313.

Mao, Y. B., Tao, X. Y., Xue, X. Y., Wang, L. J., & Chen, X. Y. (2010). Cotton plants expressing CYP6AE14 double-stranded RNA show enhanced resistance to bollworms. *Transgenic Res.*, doi 10.1007/s11248-010-9450-1.

Marchal, E., Zhang, J. R., Badisco, L., Verlinden, H., Hult, E. F., et al. (2011). Final steps in juvenile hormone biosynthesis in the desert locust, *Schistocerca gregaria*. *Insect Biochem. Mol. Biol.*, *41*, 219–227.

Marinotti, O., Nguyen, Q. K., Calvo, E., James, A. A., & Ribeiro, J. M. (2005). Microarray analysis of genes showing variable expression following a blood meal in *Anopheles gambiae*. *Insect Mol. Biol.*, *14*, 365–373.

Martin, T., Ochou, O. G., Vaissayre, M., & Fournier, D. (2003). Oxidases responsible for resistance to pyrethroids sensitize *Helicoverpa armigera* (Hubner) to triazophos in West Africa. *Insect Biochem. Mol. Biol.*, *33*, 883–887.

Matsunaga, E., Umeno, M., & Gonzalez, F. J. (1990). The rat P450 IID subfamily: Complete sequences of four closely linked genes and evidence that gene conversions maintained sequence homogeneity at the heme-binding region of the cytochrome P450 active site. *J. Mol. Evol.*, *30*, 155–169.

Mauchamp, B., Darrouzet, E., Malosse, C., & Couillaud, F. (1999). 4′-OH-JH-III: An additional hydroxylated juvenile hormone produced by locust corpora allata *in vitro*. *Insect Biochem. Mol. Biol.*, *29*, 475–480.

McCart, C., & ffrench-Constant, R. H. (2008). Dissecting the insecticide-resistance-associated cytochrome P450 gene Cyp6g1. *Pest. Manag. Sci.*, *64*, 639–645.

McCart, C., Buckling, A., & ffrench-Constant, R. H. (2005). DDT resistance in flies carries no cost. *Curr. Biol.*, *15*, R587–R589.

McDonald, M. J., & Rosbash, M. (2001). Microarray analysis and organization of circadian gene expression in *Drosophila*. *Cell*, *107*, 567–578.

McDonnell, C. M., Brown, R. P., Berenbaum, M. R., & Schuler, M. A. (2004). Conserved regulatory elements in the promoters of two allelochemical-inducible cytochrome P450 genes differentially regulate transcription. *Insect Biochem. Mol. Biol.*, *34*, 1129–1139.

McLaughlin, L. A., Niazi, U., Bibby, J., David, J. P., Vontas, J., et al. (2008). Characterization of inhibitors and substrates of *Anopheles gambiae* CYP6Z2. *Insect Mol. Biol.*, *17*, 125–135.

Megias, A., Saborido, A., & Municio, A. M. (1983). Properties of the NADH-cytochrome b5 reductase from *Ceratitis capitata*. *Comp. Biochem. Physiol. B Biochem. Mol. Biol.*, *74*, 411–416.

Miota, F., Siegfried, B. D., Scharf, M. E., & Lydy, M. J. (2000). Atrazine induction of cytochrome P450 in *Chironomus tentans* larvae. *Chemosphere*, *40*, 285–291.

Miyo, T., Kono, Y., & Oguma, Y. (2002). Genetic basis of cross-resistance to three organophosphate insecticides in *Drosophila melanogaster* (Diptera: Drosophilidae). *J. Econ. Entomol.*, *95*, 871–877.

Mohit, P., Makhijani, K., Madhavi, M. B., Bharathi, V., Lal, A., et al. (2006). Modulation of AP and DV signaling pathways by the homeotic gene Ultrabithorax during haltere development in *Drosophila*. *Dev. Biol.*, *291*, 356–367.

Moldenke, A. F., Berry, R. E., Miller, J. C., Kelsey, R. G., Wernz, J. G., & Venkateswaran, S. (1992). Carbaryl susceptibility and detoxication enzymes in gypsy moth (Lepidoptera: Lymantriidae): Influence of host plant. *J. Econ. Entomol.*, *85*, 1628–1635.

Moores, G. D., Philippou, D., Borzatta, V., Trincia, P., Jewess, P., et al. (2009). An analogue of piperonyl butoxide facilitates the characterisation of metabolic resistance. *Pest. Manag. Sci.*, *65*, 150–154.

Morozova, T. V., Anholt, R. R., & Mackay, T. F. (2006). Transcriptional response to alcohol exposure in *Drosophila melanogaster*. *Genome Biol.*, *7*, R95.

Morra, R., Kuruganti, S., Lam, V., Lucchesi, J. C., & Ganguly, R. (2010). Functional analysis of the cis-acting elements responsible for the induction of the *Cyp6a8* and *Cyp6g1* genes of *Drosophila melanogaster* by DDT, phenobarbital and caffeine. *Insect Mol. Biol.*, *19*, 121–130.

Mpuru, S., Reed, J. R., Reitz, R. C., & Blomquist, G. J. (1996). Mechanism of hydrocarbon biosynthesis from aldehyde in selected insect species: Requirement for O_2 and NADPH and carbonyl group released as CO_2. *Insect Biochem. Mol. Biol.*, *26*, 203–208.

Mpuru, S., Blomquist, G. J., Schal, C., Roux, M., Kuenzli, M., et al. (2001). Effect of age and sex on the production of internal and external hydrocarbons and pheromones in the housefly, *Musca domestica. Insect Biochem. Mol. Biol., 31,* 139–155.

Muller, P., Donnelly, M. J., & Ranson, H. (2007). Transcription profiling of a recently colonised pyrethroid resistant *Anopheles gambiae* strain from Ghana. *BMC Genomics, 8,* 36.

Muller, P., Warr, E., Stevenson, B. J., Pignatelli, P. M., Morgan, J. C., et al. (2008a). Field-caught permethrin-resistant *Anopheles gambiae* overexpress CYP6P3, a P450 that metabolises pyrethroids. *PLoS Genet., 4,* e1000286.

Muller, P., Chouaibou, M., Pignatelli, P., Etang, J., Walker, E. D., et al. (2008b). Pyrethroid tolerance is associated with elevated expression of antioxidants and agricultural practice in *Anopheles arabiensis* sampled from an area of cotton fields in Northern Cameroon. *Mol. Ecol., 17,* 1145–1155.

Mullin, C. A. (1986). Adaptive divergence of chewing and sucking arthropods to plant allelochemicals. In L. B. Brattsten, & S. Ahmad (Eds.), *Molecular Aspects of Insect–Plant Interactions* (pp. 175–209). New York, NY: Plenum.

Mullin, C. A., Croft, B. A., Strickler, K., Matsumura, F., & Miller, J. R. (1982). Detoxification enzyme differences between a herbivorous and predatory mite. *Science, 217,* 1270–1272.

Munro, A. W., Girvan, H. M., & McLean, K. J. (2007). Cytochrome P450 – redox partner fusion enzymes. *Biochim. Biophys. Acta., 1770,* 345–359.

Murataliev, M. B., & Feyereisen, R. (1999). Mechanism of cytochrome P450 reductase from the house fly: Evidence for an FMN semiquinone as electron donor. *FEBS Lett., 453,* 201–204.

Murataliev, M. B., & Feyereisen, R. (2000). Interaction of NADP(H) with oxidized and reduced P450 reductase during catalysis. Studies with nucleotide analogues. *Biochemistry, 39,* 5066–5074.

Murataliev, M. B., Arino, A., Guzov, V. M., & Feyereisen, R. (1999). Kinetic mechanism of cytochrome P450 reductase from the house fly (*Musca domestica*). *Insect Biochem. Mol. Biol., 29,* 233–242.

Murataliev, M. B., Feyereisen, R., & Walker, F. A. (2004). Electron transfer by diflavin reductases. *Biochem. Biophys. Acta., 1698,* 1–26.

Murataliev, M. B., Guzov, V. M., Walker, F. A., & Feyereisen, R. (2008). P450 reductase and cytochrome b5 interactions with cytochrome P450: Effects on house fly CYP6A1 catalysis. *Insect Biochem. Mol. Biol., 38,* 1008–1015.

Mutch, E., Daly, A. K., Leathart, J. B., Blain, P. G., & Williams, F. M. (2003). Do multiple cytochrome P450 isoforms contribute to parathion metabolism in man? *Arch. Toxicol., 77,* 313–320.

Nahrstedt, A. (1988). Cyanogenesis and the role of cyanogenic compounds in insects. *CIBA Foundation Symposia, 140,* 131–150.

Namiki, T., Niwa, R., Sakudoh, T., Shirai, K., Takeuchi, H., & Kataoka, H. (2005). Cytochrome P450 CYP307A1/ Spook: A regulator for ecdysone synthesis in insects. *Biochem. Biophys. Res. Commun., 337,* 367–374.

Naumann, C., Hartmann, T., & Ober, D. (2002). Evolutionary recruitment of a flavin-dependent monooxygenase for the detoxication of host plant-acquired pyrrolizidine alkaloid-defended arctiid moth *Tyria Jacobaeae. Proc. Natl. Acad. Sci. USA, 99,* 6085–6090.

Neal, J. J. (1987). Metabolic costs of mixed-function oxidase induction in *Heliothis zea. Entomol. Exp. Appl., 43,* 175–179.

Neal, J. J., & Reuveni, M. (1992). Separation of cytochrome P450 containing vesicles from the midgut microsomal fraction of *Manduca sexta. Comp. Biochem. Physiol., 102C,* 77–82.

Neal, J. J., & Wu, D. (1994). Inhibition of insect cytochromes P450 by furanocoumarins. *Pesticide Biochem. Physiol., 50,* 43–50.

Nebert, D. W. (1991). Proposed role of drug-metabolizing enzymes: Regulation of steady state levels of the ligands that effect growth, homeostasis, differentiation, neuroendocrine functions. *Mol. Endocrinol., 5,* 1203–1214.

Nebert, D. W., Adesnik, M., Coon, M. J., Estabrook, R. W., Gonzalez, F. J., et al. (1987). The P450 gene superfamily: Recommended nomenclature. *DNA, 6,* 1–11.

Nebert, D. W., Nelson, D. R., Coon, M. J., Estabrook, R. W., Feyereisen, R., et al. (1991). The P450 superfamily: Update on new sequences, gene mapping, recommended nomenclature. [published erratum appears in *DNA Cell Biol.* 1991 Jun;10(5):397–398] *DNA Cell. Biol., 10,* 1–14.

Nelson, D. R. (1998). Metazoan cytochrome P450 evolution. *Comp. Biochem. Physiol. C Pharmacol. Toxicol. Endocrinol., 121,* 15–22.

Nelson, D. R., Kamataki, T., Waxman, D. J., Guengerich, F. P., Estabrook, R. W., et al. (1993). The P450 superfamily: Update on new sequences, gene mapping, accession numbers, early trivial names of enzymes, nomenclature. *DNA Cell. Biol., 12,* 1–51.

Nelson, D. R., Koymans, L., Kamataki, T., Stegeman, J. J., Feyereisen, R., et al. (1996). P450 superfamily: Update on new sequences, gene mapping, accession numbers and nomenclature. *Pharmacogenetics, 6,* 1–42.

Nelson, D. R., Zeldin, D. C., Hoffman, S. M., Maltais, L. J., Wain, H. M., & Nebert, D. W. (2004). Comparison of cytochrome P450 (CYP) genes from the mouse and human genomes, including nomenclature recommendations for genes, pseudogenes and alternative-splice variants. *Pharmacogenetics, 14,* 1–18.

Neve, E. P., & Ingelman-Sundberg, M. (2008). Intracellular transport and localization of microsomal cytochrome P450. *Anal. Bioanal. Chem., 392,* 1075–1084.

Newcomb, R. D., Campbell, P. M., Ollis, D. L., Cheah, E., Russell, R. J., & Oakeshott, J. G. (1997). A single amino acid substitution converts a carboxylesterase to an organophosphorus hydrolase and confers insecticide resistance on a blowfly. *Proc. Natl. Acad. Sci. USA, 94,* 7464–7468.

Nikou, D., Ranson, H., & Hemingway, J. (2003). An adult-specific CYP6 P450 gene is overexpressed in a pyrethroid-resistant strain of the malaria vector, *Anopheles gambiae. Gene, 318,* 91–102.

Nitao, J. K. (1989). Enzymatic adaptation in a specialist herbivore for feeding on furanocoumarin-containing plants. *Ecology, 70,* 629–635.

Nitao, J. K. (1990). Metabolism and excretion of the furano-coumarin xanthotoxin by parsnip webworm, *Depressaria pastinacella*. *J. Chem. Ecol.*, *16*, 417–428.

Nitao, J. K., Berhow, M., Duval, S. M., Weisleder, D., Vaughn, S. F., et al. (2003). Characterization of furanocoumarin metabolites in parsnip webworm, *Depressaria pastinacella*. *J. Chem. Ecol.*, *29*, 671–682.

Niu, G., Wen, Z., Rupasinghe, S. G., Zeng, R. S., Berenbaum, M. R., & Schuler, M. A. (2008). Aflatoxin B1 detoxification by CYP321A1 in *Helicoverpa zea*. *Arch. Insect Biochem. Physiol.*, *69*, 32–45.

Niu, G., Rupasinghe, S. G., Zangerl, A. R., Siegel, J. P., Schuler, M. A., & Berenbaum, M. R. (2011). A substrate-specific cytochrome P450 monooxygenase, CYP6AB11, from the polyphagous navel orangeworm (*Amyelois transitella*). *Insect Biochem. Mol. Biol.*, *41*, 244–253.

Niwa, R., Matsuda, T., Yoshiyama, T., Namiki, T., Mita, K., et al. (2004). CYP306A1, a cytochrome P450 enzyme, is essential for ecdysteroid biosynthesis in the prothoracic glands of *Bombyx* and *Drosophila*. *J. Biol. Chem.*, *279*, 35942–35949.

Niwa, R., Sakudoh, T., Namiki, T., Saida, K., Fujimoto, Y., & Kataoka, H. (2005). The ecdysteroidogenic P450 Cyp302a1/disembodied from the silkworm, *Bombyx mori*, is transcriptionally regulated by prothoracicotropic hormone. *Insect Mol. Biol.*, *14*, 563–571.

Niwa, R., Namiki, T., Ito, K., Shimada-Niwa, Y., Kiuchi, M., et al. (2010). Non-molting glossy/shroud encodes a short-chain dehydrogenase/reductase that functions in the "Black Box" of the ecdysteroid biosynthesis pathway. *Development*, *137*, 1991–1999.

Oakeshott, J. G., Johnson, R. M., Berenbaum, M. R., Ranson, H., Cristino, A. S., & Claudianos, C. (2010). Metabolic enzymes associated with xenobiotic and chemosensory responses in *Nasonia vitripennis*. *Insect Mol. Biol.*, *19*(Suppl. 1), 147–163.

Oi, M., Dauterman, W. C., & Motoyama, N. (1990). Biochemical factors responsble for an extremely high level of diazinon resistance in a housefly strain. *J. Pestic. Sci.*, *15*, 217–224.

Omura, T. (1993). History of cytochrome P450. In T. Omura, Y. Ishimura, & Y. Fujii-Kuriyama (Eds.). *Cytochrome P-450* (2nd ed.) (pp. 1–15). Tokyo, Japan: Kodansha.

Omura, T., & Sato, R. (1964). The carbon monoxide-binding pigment of liver microsomes I. Evidence for its hemoprotein nature. *J. Biol. Chem.*, *239*, 2370–2378.

Ono, H., Ozaki, K., & Yoshikawa, H. (2005). Identification of cytochrome P450 and glutathione-S-transferase genes preferentially expressed in chemosensory organs of the swallowtail butterfly, *Papilio xuthus* L. *Insect Biochem. Mol. Biol.*, *35*, 837–846.

Ono, H., Rewitz, K. F., Shinoda, T., Itoyama, K., Petryk, A., et al. (2006). Spook and Spookier code for stage-specific components of the ecdysone biosynthetic pathway in Diptera. *Dev. Biol.*, *298*, 555–570.

Ortiz de Montellano, P. R. (1995). Oxygen activation and reactivity. In T. Omura, Y. Ishimura, & Y. Fujii-Kuriyama (Eds.). *Cytochrome P-450* (2nd ed.). (pp. 245–303). New York, NY: Plenum Press.

Ortiz de Montellano, P. R. (2005). *Cytochrome P450, Structure, Mechanism, Biochemistry* (3rd ed.). New York, NY: Kluwer Academic/Plenum Publishers.

Ortiz de Montellano, P. R., & Correia, M. A (1995). Inhibition of cytochrome P450 enzymes. In T. Omura, Y. Ishimura, & Y. Fujii-Kuriyama (Eds.), *Cytochrome P-450* (2nd ed.). (pp. 305–364). New York, NY: Plenum Press.

Ortiz de Montellano, P. R., & Graham-Lorence, S. E. (1993). Structure of cytochrome P450: Heme-binding site and heme reactivity. In J. B. Schenkman, & H. Greim (Eds.), *Cytochrome P450* (pp. 169–181). Berlin, Germany: Springer-Verlag.

Osada, N., & Innan, H. (2008). Duplication and gene conversion in the *Drosophila melanogaster* genome. *PLoS Genet.*, *4*, e1000305.

Palanker, L., Necakov, A. S., Sampson, H. M., Ni, R., Hu, C., et al. (2006). Dynamic regulation of *Drosophila* nuclear receptor activity *in vivo*. *Development*, *133*, 3549–3562.

Pan, L., Wen, Z., Baudry, J., Berenbaum, M. R., & Schuler, M. A. (2004). Identification of variable amino acids in the SRS1 region of CYP6B1 modulating furanocoumarin metabolism. *Arch. Biochem. Biophys.*, *422*, 31–41.

Paquette, S. M., Bak, S., & Feyereisen, R. (2000). Intron–exon organization and phylogeny in a large superfamily, the paralogous cytochrome P450 genes of *Arabidopsis thaliana*. *DNA Cell Biol.*, *19*, 307–317.

Pare, P. W., Alborn, H. T., & Tumlinson, J. H. (1998). Concerted biosynthesis of an insect elicitor of plant volatiles. *Proc. Natl. Acad. Sci. USA*, *95*, 13971–13975.

Parker, R. S., & McCormick, C. C. (2005). Selective accumulation of alpha-tocopherol in *Drosophila* is associated with cytochrome P450 tocopherol-omega-hydroxylase activity but not alpha-tocopherol transfer protein. *Biochem. Biophys. Res. Commun.*, *338*, 1537–1541.

Pascussi, J. M., Gerbal-Chaloin, S., Duret, C., Daujat-Chavanieu, M., Vilarem, M. J., & Maurel, P. (2008). The tangle of nuclear receptors that controls xenobiotic metabolism and transport: Crosstalk and consequences. *Annu. Rev. Pharmacol. Toxicol.*, *48*, 1–32.

Pauchet, Y., Wilkinson, P., Vogel, H., Nelson, D. R., Reynolds, S. E., et al. (2010). Pyrosequencing the *Manduca sexta* larval midgut transcriptome: Messages for digestion, detoxification and defence. *Insect Mol. Biol.*, *19*, 61–75.

Pedra, J. H., McIntyre, L. M., Scharf, M. E., & Pittendrigh, B. R. (2004). Genome-wide transcription profile of field- and laboratory-selected dichlorodiphenyltrichloroethane (DDT)-resistant *Drosophila*. *Proc. Natl. Acad. Sci. USA*, *101*, 7034–7039.

Perry, T., Batterham, P., & Daborn, P. J. (2011). The biology of insecticidal activity and resistance. *Insect Biochem. Mol. Biol.*, *41*, doi:10.1016/j.ibmb.2011.03.003.

Petersen, R. A., Zangerl, A. R., Berenbaum, M. R., & Schuler, M. A. (2001). Expression of CYP6B1 and CYP6B3 cytochrome P450 monooxygenases and furanocoumarin metabolism in different tissues of *Papilio polyxenes* (Lepidoptera: Papilionidae). *Insect Biochem. Mol. Biol.*, *31*, 679–690.

Petersen, R. A., Niamsup, H., Berenbaum, M. R., & Schuler, M. A. (2003). Transcriptional response elements in the promoter of CYP6B1, an insect P450 gene regulated by plant chemicals. *Biochim. BiophysActa. – Gen. Subj.*, *1619*, 269–282.

Petersen Brown, R., Berenbaum, M. R., & Schuler, M. A. (2004). Transcription of a lepidopteran cytochrome P450 promoter is modulated by multiple elements in its 5′ UTR and repressed by 20-hydroxyecdysone. *Insect Mol. Biol.*, *13*, 337–347.

Petryk, A., Warren, J. T., Marques, G., Jarcho, M. P., Gilbert, L. I., et al. (2003). Shade is the *Drosophila* P450 enzyme that mediates the hydroxylation of ecdysone to the steroid insect molting hormone 20-hydroxyecdysone. *Proc. Natl. Acad. Sci. USA*, *100*, 13773–13778.

Pikuleva, I. A., Mackman, R. L., Kagawa, N., Waterman, M. R., & Ortiz de Montellano, P. R. (1995). Active-site topology of bovine cholesterol side-chain cleavage cytochrome P450 (P450scc) and evidence for interaction of tyrosine 94 with the side chain of cholesterol. *Arch. Biochem. Biophys.*, *322*, 189–197.

Pikuleva, I. A., Cao, C., & Waterman, M. R. (1999). An additional electrostatic interaction between adrenodoxin and P450c27 (CYP27A1) results in tighter binding than between adrenodoxin and P450scc (CYP11A1). *J. Biol. Chem.*, *274*, 2045–2052.

Pimprale, S. S., Besco, C. L., Bryson, P. K., & Brown, T. M. (1997). Increased susceptibility of pyrethroid-resistant tobacco budworm (Lepidoptera: Noctuidae) to chlorfenapyr. *J. Econ. Entomol.*, *90*, 49–54.

Pittendrigh, B., Aronstein, K., Zinkovsky, E., Andreev, O., Campbell, B., et al. (1997). Cytochrome P450 genes from *Helicoverpa armigera*: Expression in a pyrethroid-susceptible and -resistant strain. *Insect Biochem. Mol. Biol.*, *27*, 507–512.

Plapp, F. W., Jr. (1984). The genetic basis of insecticide resistance in the house fly: Evidence that a single locus plays a major role in metabolic resistance to insecticides. *Pest. Biochem. Physiol.*, *22*, 94–201.

Plettner, E., Slessor, K. N., & Winston, M. L. (1998). Biosynthesis of mandibular acids in honey bees (*Apis mellifera*): *De novo* synthesis, route of fatty acid hydroxylation and caste selective ω-oxidation. *Insect Biochem. Mol. Biol.*, *28*, 31–42.

Pompon, D., Louerat, B., Bronine, A., & Urban, P. (1996). Yeast expression of animal and plant P450s in optimized redox environments. *Methods Enzymol.*, 51–64.

Pondeville, E., Maria, A., Jacques, J. C., Bourgouin, C., & Dauphin-Villemant, C. (2008). *Anopheles gambiae* males produce and transfer the vitellogenic steroid hormone 20-hydroxyecdysone to females during mating. *Proc. Natl. Acad. Sci. USA*, *105*, 19631–19636.

Porter, T. D. (2002). The roles of cytochrome b5 in cytochrome P450 reactions. *J. Biochem. Mol. Toxicol.*, *16*, 311–316.

Porter, T. D., & Kasper, C. B. (1986). NADPH-cytochrome P-450 oxidoreductase: flavin mononucleotide and flavin adenine dinucleotide domains evolved from different flavoproteins. *Biochemistry*, *25*, 1682–1687.

Poulos, T. L. (2003). Cytochrome P450 flexibility. *Proc. Natl. Acad. Sci.*, *100*, 13121–13122.

Poulos, T. L., Finzel, B. C., Gunsalus, I. C., Wagner, G. C., & Kraut, J. (1985). The 2.6-A crystal structure of *Pseudomonas putida* cytochrome P-450. *J. Biol. Chem.*, *260*, 16122–16130.

Poupardin, R., Reynaud, S., Strode, C., Ranson, H., Vontas, J., & David, J. P. (2008). Cross-induction of detoxification genes by environmental xenobiotics and insecticides in the mosquito *Aedes aegypti*: Impact on larval tolerance to chemical insecticides. *Insect Biochem. Mol. Biol.*, *38*, 540–551.

Poupardin, R., Riaz, M. A., Vontas, J., David, J. P., & Reynaud, S. (2010). Transcription profiling of eleven cytochrome P450s potentially involved in xenobiotic metabolism in the mosquito *Aedes aegypti*. *Insect Mol. Biol.*, *19*, 185–193.

Prapaipong, H., Berenbaum, M. R., & Schuler, M. A. (1994). Transcriptional regulation of the *Papilio polyxenes* CYP6B1 gene. *Nucleic Acids Res.*, *22*, 3210–3217.

Pratt, G. E., Kuwano, E., Farnsworth, D. E., & Feyereisen, R. (1990). Structure/activity studies on 1,5-disubstituted imidazoles as inhibitors of juvenile hormone biosynthesis in isolated corpora allata of the cockroach, *Diploptera punctata*. *Pestic. Biochem. Physiol.*, *38*, 223–230.

Pridgeon, J. W., Zhang, L., & Liu, N. (2003). Overexpression of CYP4G19 associated with a pyrethroid-resistant strain of the German cockroach, *Blattella germanica* (L.). *Gene.*, *314*, 157–163.

Puinean, A. M., Foster, S. P., Oliphant, L., Denholm, I., Field, L. M., et al. (2010). Amplification of a cytochrome P450 gene is associated with resistance to neonicotinoid insecticides in the aphid *Myzus persicae*. *PLoS Genet.*, *6*, e1000999.

Pyke, F. M., Bogwitz, M. R., Perry, T., Monk, A., Batterham, P., & McKenzie, J. A. (2004). The genetic basis of resistance to diazinon in natural populations of *Drosophila melanogaster*. *Genetica*, *121*, 13–24.

Ranasinghe, C., & Hobbs, A. A. (1998). Isolation and characterization of two cytochrome P450 cDNA clones for CYP6B6 and CYP6B7 from *Helicoverpa armigera* (Hubner): Possible involvement of CYP6B7 in pyrethroid resistance. *Insect Biochem. Mol. Biol.*, *28*, 571–580.

Ranasinghe, C., & Hobbs, A. A. (1999a). Isolation and characterisation of a cytochrome b5 cDNA clone from *Helicoverpa armigera* (Hubner): Possible involvement of cytochrome b5 in cytochrome P450 CYP6B7 activity towards pyrethroids. *Insect Biochem. Mol. Biol.*, *29*, 145–151.

Ranasinghe, C., & Hobbs, A. A. (1999b). Induction of cytochrome P450 CYP6B7 and cytochrome b5 mRNAs from *Helicoverpa armigera* (Hubner) by pyrethroid insecticides in organ culture. *Insect Mol. Biol.*, *8*, 443–447.

Ranasinghe, C., Headlam, M., & Hobbs, A. A. (1997). Induction of the mRNA for CYP6B2, a pyrethroid inducible cytochrome P450, in *Helicoverpa armigera* (Hubner) by dietary monoterpenes. *Arch. Insect. Biochem. Physiol.*, *34*, 99–109.

Ranson, H., Claudianos, C., Ortelli, F., Abgrall, C., Hemingway, J., et al. (2002a). Evolution of supergene families associated with insecticide resistance. *Science*, *298*, 179–181.

Ranson, H., Nikou, D., Hutchinson, M., Wang, X., Roth, C. W., et al. (2002b). Molecular analysis of multiple cytochrome P450 genes from the malaria vector, *Anopheles gambiae*. *Insect Mol. Biol.*, *11*, 409–418.

Ranson, H., Paton, M. G., Jensen, B., McCarroll, L., Vaughan, A., et al. (2004). Genetic mapping of genes conferring permethrin resistance in the malaria vector, *Anopheles gambiae*. *Insect Mol. Biol.*, *13*, 379–386.

Reed, J. R., Vanderwel, D., Choi, S., Pomonis, J. G., Reitz, R. C., & Blomquist, G. J. (1994). Unusual mechanism of hydrocarbon formation in the housefly: Cytochrome P450 converts aldehyde to the sex pheromone component (Z)-9-tricosene and CO_2. *Proc. Natl. Acad. Sci. USA*, *91*, 10000–10004.

Reed, J. R., Quilici, D. R., Blomquist, G. J., & Reitz, R. C. (1995). Proposed mechanism for the cytochrome P450-catalyzed conversion of aldehydes to hydrocarbons in the house fly, *Musca domestica*. *Biochemistry, 34*, 16221–16227.

Rees, H. H. (1995). Ecdysteroid biosynthesis and inactivation in relation to function. *Eur. J. Entomol., 92*, 9–39.

Rewitz, K. F., & Gilbert, L. I. (2008). Daphnia Halloween genes that encode cytochrome P450s mediating the synthesis of the arthropod molting hormone: Evolutionary implications. *BMC Evol. Biol., 8*, 60.

Rewitz, K. F., Rybczynski, R., Warren, J. T., & Gilbert, L. I. (2006). Developmental expression of *Manduca* shade, the P450 mediating the final step in molting hormone synthesis. *Mol. Cell Endocrinol., 247*, 166–174.

Rewitz, K. F., O'Connor, M. B., & Gilbert, L. I. (2007). Molecular evolution of the insect Halloween family of cytochrome P450s: Phylogeny, gene organization and functional conservation. *Insect Biochem. Mol. Biol., 37*, 741–753.

Rewitz, K. F., Larsen, M. R., Lobner-Olesen, A., Rybczynski, R., O'Connor, M. B., & Gilbert, L. I. (2009). A phosphoproteomics approach to elucidate neuropeptide signal transduction controlling insect metamorphosis. *Insect Biochem. Mol. Biol., 39*, 475–483.

Rewitz, K. F., Yamanaka, N., & O'Connor, M. B. (2010). Steroid hormone inactivation is required during the juvenile–adult transition in *Drosophila. Dev. Cell, 19*, 895–902.

Riaz, M. A., Poupardin, R., Reynaud, S., Strode, C., Ranson, H., & David, J. P. (2009). Impact of glyphosate and benzo[a]pyrene on the tolerance of mosquito larvae to chemical insecticides. Role of detoxification genes in response to xenobiotics. *Aquat. Toxicol., 93*, 61–69.

Ringo, J., Jona, G., Rockwell, R., Segal, D., & Cohen, E. (1995). Genetic variation for resistance to chlorpyrifos in *Drosophila melanogaster* (Diptera: Drosophilidae) infesting grapes in Israel. *J. Econ. Entomol., 88*, 1158–1163.

Riskallah, M. R., Dauterman, W. C., & Hodgson, E. (1986a). Host plant induction of microsomal monooxygenases activity in relation to diazinon metabolism and toxicity in larvae of the tobacco budworm *Heliothis virescens* (F.). *Pestic. Biochem. Physiol., 25*, 233–247.

Riskallah, M. R., Dauterman, W. C., & Hodgson, E. (1986b). Nutritional effects on the induction of cytochrome P-450 and glutathione transferase in larvae of the tobacco budworm, *Heliothis virescens* (F.). *Insect Biochem., 16*, 491–499.

Rittle, J., & Green, M. T. (2010). Cytochrome P450 compound I: Capture, characterization, C-H bond activation kinetics. *Science, 330*, 933–937.

Robertson, H. M., Martos, R., Sears, C. R., Todres, E. Z., Walden, K. K., & Nardi, J. B. (1999). Diversity of odourant binding proteins revealed by an expressed sequence tag project on male *Manduca sexta* moth antennae. *Insect Mol. Biol., 8*, 501–518.

Rose, H. A. (1985). The relationship between feeding specialization and host plants to aldrin epoxidase activities of midgut homogenates in larval Lepidoptera. *Ecol. Entomol., 10*, 455–467.

Rose, R. L., Gould, F., Levi, P. E., & Hodgson, E. (1991). Differences in cytochrome P450 activities in tobacco budworm larvae as influenced by resistance to host plant allelochemicals and induction. *Comp. Biochem. Physiol. B Biochem. Mol. Biol., 99*, 535–540.

Rose, R. L., Barbhaiya, L., Roe, R. M., Rock, G. C., & Hodgson, E. (1995). Cytochrome P450-associated insecticide resistance and the development of biochemical diagnostic assays in *Heliothis virescens. Pesticide Biochem. Physiol., 51*, 178–191.

Rose, R. L., Goh, D., Thompson, D. M., Verma, K. D., Heckel, D. G., et al. (1997). Cytochrome P450 (CYP)9A1 in *Heliothis virescens*: The first member of a new CYP family. *Insect Biochem. Mol. Biol., 27*, 605–615.

Rupasinghe, S., Schuler, M. A., Kagawa, N., Yuan, H., Lei, L., et al. (2006). The cytochrome P450 gene family CYP157 does not contain EXXR in the K-helix reducing the absolute conserved P450 residues to a single cysteine. *FEBS Lett., 580*, 6338–6342.

Rupasinghe, S. G., Wen, Z., Chiu, T. L., & Schuler, M. A. (2007). *Helicoverpa zea* CYP6B8 and CYP321A1: Different molecular solutions to the problem of metabolizing plant toxins and insecticides. *Protein Eng. Des. Sel., 20*, 615–624.

Sabourault, C., Guzov, V. M., Koener, J. F., Claudianos, C., Plapp, F. W., Jr., & Feyereisen, R. (2001). Overproduction of a P450 that metabolizes diazinon is linked to a loss-of-function in the chromosome 2 ali-esterase (MdalphaE7) gene in resistant house flies. *Insect Mol. Biol., 10*, 609–618.

Salzemann, A., Nagnan, P., Tellier, F., & Jaffe, K. (1992). Leaf-cutting ant *Atta laevigata* (Formicidae: Attini) marks its territory with colony-specific Dufour gland secretion. *J. Chem. Ecol., 18*, 183.

Sandstrom, P., Welch, W. H., Blomquist, G. J., & Tittiger, C. (2006). Functional expression of a bark beetle cytochrome P450 that hydroxylates myrcene to ipsdienol. *Insect Biochem. Mol. Biol., 36*, 835–845.

Sandstrom, P., Ginzel, M. D., Bearfield, J. C., Welch, W. H., Blomquist, G. J., & Tittiger, C. (2008). Myrcene hydroxylases do not determine enantiomeric composition of pheromonal ipsdienol in *Ips* spp. *J. Chem. Ecol., 34*, 1584–1592.

Saner, C., Weibel, B., Wurgler, F. E., & Sengstag, C. (1996). Metabolism of promutagens catalyzed by *Drosophila melanogaster* CYP6A2 enzyme in *Saccharomyces cerevisiae. Environ. Mol. Mutagen., 27*, 46–58.

Sarapusit, S., Xia, C., Misra, I., Rongnoparut, P., & Kim, J. J. (2008). NADPH-cytochrome P450 oxidoreductase from the mosquito *Anopheles minimus*: Kinetic studies and the influence of Leu86 and Leu219 on cofactor binding and protein stability. *Arch. Biochem. Biophys., 477*, 53–59.

Sarapusit, S., Pethuan, S., & Rongnoparut, P. (2010). Mosquito NADPH-cytochrome P450 oxidoreductase: Kinetics and role of phenylalanine amino acid substitutions at leu86 and leu219 in CYP6AA3-mediated deltamethrin metabolism. *Arch. Insect Biochem. Physiol., 73*, 232–244.

Sasabe, M., Wen, Z., Berenbaum, M. R., & Schuler, M. A. (2004). Molecular analysis of CYP321A1, a novel cytochrome P450 involved in metabolism of plant allelochemicals (furanocoumarins) and insecticides (cypermethrin) in *Helicoverpa zea. Gene, 338*, 163–175.

Scharf, M. E., Neal, J. J., Marcus, C. B., & Bennett, G. W. (1998). Cytochrome P450 purification and immunological detection in an insecticide resistant strain of German cockroach (*Blattella germanica*, L.). *Insect Biochem. Mol. Biol., 28*, 1–9.

Scharf, M. E., Lee, C. Y., Neal, J. J., & Bennett, G. W. (1999). Cytochrome P450 MA expression in insecticide-resistant German cockroaches (Dictyoptera: Blattellidae). *J. Econ. Entomol., 92*, 788–793.

Scharf, M. E., Parimi, S., Meinke, L. J., Chandler, L. D., & Siegfried, B. D. (2001). Expression and induction of three family 4 cytochrome P450 (CYP4)* genes identified from insecticide-resistant and -susceptible western corn rootworms, *Diabrotica virgifera virgifera. Insect Mol. Biol., 10*, 139–146.

Scharf, M. E., Scharf, D. W., Bennett, G. W., & Pittendrigh, B. R. (2004). Catalytic activity and expression of two flavin-containing monooxygenases from *Drosophila melanogaster. Arch. Insect Biochem. Physiol., 57*, 28–39.

Schenkman, J. B., & Jansson, I. (2003). The many roles of cytochrome b5. *Pharmacol. Ther., 97*, 139–152.

Schlenke, T. A., & Begun, D. J. (2004). Strong selective sweep associated with a transposon insertion in *Drosophila simulans. Proc. Natl. Acad. Sci. USA, 101*, 1626–1631.

Schmidt, J. M., Good, R. T., Appleton, B., Sherrard, J., Raymant, G. C., et al. (2010). Copy number variation and transposable elements feature in recent, ongoing adaptation at the Cyp6g1 locus. *PLoS Genet., 6*, e1000998.

Schonbrod, R. D., Khan, M. A. Q., Terriere, L. C., & Plapp, F. W., Jr. (1968). Microsomal oxidases in the house fly: A survey of fourteen strains. *Life Sci., 7*, 681–688.

Schrag, M. L., & Wienkers, L. C. (2000). Topological alteration of the CYP3A4 active site by the divalent cation Mg$^{(2+)}$. *Drug Metab. Dispos., 28*, 1198–1201.

Scott, J. A., Collins, F. H., & Feyereisen, R. (1994). Diversity of cytochrome P450 genes in the mosquito, *Anopheles albimanus. Biochem. Biophys. Res. Commun., 205*, 1452–1459.

Scott, J. G., & Lee, S. S. (1993). Purification and characterization of a cytochrome P-450 from insecticide susceptible and resistant strains of housefly, *Musca domestica* L., before and after phenobarbital exposure. *Arch. Insect Biochem. Physiol., 24*, 1–19.

Scott, J. G., & Zhang, L. (2003). The house fly aliesterase gene(MdαE7) is not associated with insecticide resistance or P450 expression in three strains of house fly. *Insect Biochem. Mol. Biol., 33*, 139–144.

Scott, J. G., Foroozesh, M., Hopkins, N. E., Alefantis, T. G., & Alworth, W. L. (2000). Inhibition of cytochrome P450 6D1 by alkynylarenes, methylenedioxyarenes, and other substituted aromatics. *Pestic. Biochem. Physiol., 67*, 63–71.

Scrimshaw, S., & Kerfoot, W. C. (1987). Chemical defenses of freshwater organisms: Beetles and bugs. In W. C. Kerfoot, & A. Sih (Eds.), *Predation: Direct and Indirect Impacts on Aquatic Communities* (pp. 240–262). Hanover, NH: University Press of New England.

Sehlmeyer, S., Wang, L., Langel, D., Heckel, D. G., Mohagheghi, H., et al. (2010). Flavin-dependent monooxygenases as a detoxification mechanism in insects: New insights from the arctiids (lepidoptera). *PLoS One, 5*, e10435.

Seifert, J., & Scott, J. G. (2002). The CYP6D1v1 allele is associated with pyrethroid resistance in the house fly, *Musca domestica. Pestic. Biochem. Physiol., 72*, 40–44.

Seliskar, M., & Rozman, D. (2007). Mammalian cytochromes P450 – importance of tissue specificity. *Biochim. Biophys. Acta., 1770*, 458–466.

Shen, B., Dong, H. Q., Tian, H. S., Ma, L., Li, X. L., et al. (2003). Cytochrome P450 genes expressed in the deltamethrin-susceptible and -resistant strains of *Culex pipiens pallens. Pestic. Biochem. Physiol., 75*, 19–26.

Sheppard, D. G., & Joyce, J. A. (1998). Increased susceptibility of pyrethroid-resistant horn flies (Diptera: Muscidae) to chlorfenapyr. *J. Econ. Entomol., 91*, 398–400.

Shergill, J. K., Cammack, R., Chen, J. H., Fisher, M. J., Madden, S., & Rees, H. H. (1995). EPR spectroscopic characterization of the iron-sulphur proteins and cytochrome P-450 in mitochondria from the insect *Spodoptera littoralis* (cotton leafworm). *Biochem. J., 307*, 719–728.

Shono, T., Kasai, S., Kamiya, E., Kono, Y., & Scott, J. G. (2002). Genetics and mechanisms of permethrin resistance in the YPER strain of house fly. *Pestic. Biochem. Physiol., 73*, 27–36.

Sieber, M. H., & Thummel, C. S. (2009). The DHR96 nuclear receptor controls triacylglycerol homeostasis in *Drosophila. Cell Metab., 10*, 481–490.

Smith, F. F., & Scott, J. G. (1997). Functional expression of house fly (*Musca domestica*) cytochrome P450 CYP6D1 in yeast (*Saccharomyces cerevisiae*). *Insect Biochem. Mol. Biol., 27*, 999–1006.

Smith, S. L. (1985). Regulation of ecdysteroid titer: Synthesis. In G. A. Kerkut, & L. I. Gilbert (Eds.), *Comprehensive Insect Physiology, Biochemistry and Pharmacology* (pp. 295–341). Oxford, UK: Pergamon.

Smith, S. L., Bollenbacher, W. E., Cooper, D. Y., Schleyer, H., Wielgus, J. J., & Gilbert, L. I. (1979). Ecdysone 20-monooxygenase: Characterization of an insect cytochrome P-450 dependent steroid hydroxylase. *Mol. Cell Endocrinol., 15*, 111–133.

Snyder, M. J., & Glendinning, J. I. (1996). Causal connection between detoxification enzyme activity and consumption of a toxic plant compound. *J. Comp. Physiol. [A], 179*, 255–261.

Snyder, M. J., Hsu, E. -L., & Feyereisen, R. (1993). Induction of cytochrome P450 activities by nicotine in the tobacco hornworm *Manduca sexta. J. Chem. Ecol., 19*, 2903–2916.

Snyder, M. J., Walding, J. K., & Feyereisen, R. (1994). Metabolic fate of the allelochemical nicotine in the tobacco hornworm, *Manduca sexta. Insect Biochem. Mol. Biol., 24*, 837–846.

Snyder, M. J., Stevens, J. L., Andersen, J. F., & Feyereisen, R. (1995). Expression of cytochrome P450 genes of the CYP4 family in midgut and fat body of the tobacco hornworm, *Manduca sexta. Arch. Biochem. Biophys., 321*, 13–20.

Snyder, M. J., Scott, J. A., Andersen, J. F., & Feyereisen, R. (1996). Sampling P450 diversity by cloning polymerase chain reaction products obtained with degenerate primers. *Methods Enzymol., 272*, 304–312.

Stevens, J. L., Snyder, M. J., Koener, J. F., & Feyereisen, R. (2000). Inducible P450s of the CYP9 family from larval *Manduca sexta* midgut. *Insect Biochem. Mol. Biol., 30*, 559–568.

Stevenson, B. J., Bibby, J., Pignatelli, P., Muangnoicharoen, S., O'Neill, P. M., et al. (2011). Cytochrome P450 6M2 from the malaria vector *Anopheles gambiae* metabolizes pyrethroids: Sequential metabolism of deltamethrin revealed. *Insect Biochem. Mol. Biol., 41*, doi:10.1016/j.ibmb.2011.02.003.

Strode, C., Steen, K., Ortelli, F., & Ranson, H. (2006). Differential expression of the detoxification genes in the different life stages of the malaria vector *Anopheles gambiae. Insect Mol. Biol., 15*, 523–530.

Strode, C., Wondji, C. S., David, J. P., Hawkes, N. J., Lum-juan, N., et al. (2008). Genomic analysis of detoxification genes in the mosquito *Aedes aegypti*. *Insect Biochem. Mol. Biol.*, *38*, 113–123.

Sun, W., Margam, V. M., Sun, L., Buczkowski, G., Bennett, G. W., Schemerhorn, B., Muir, W. M., & Pittendrigh, B. R. (2006). Genome-wide analysis of phenobarbital-inducible genes in *Drosophila melanogaster*. *Insect Mol. Biol.*, *15*, 455–464.

Sundseth, S. S., Nix, C. E., & Waters, L. C. (1990). Isolation of insecticide resistance-related forms of cytochrome P-450 from *Drosophila melanogaster*. *Biochem. J.*, *265*, 213–217.

Sutherland, T. D., Unnithan, G. C., Andersen, J. F., Evans, P. H., Murataliev, M. B., et al. (1998). A cytochrome P450 terpenoid hydroxylase linked to the suppression of insect juvenile hormone synthesis. *Proc. Natl. Acad. Sci. USA*, *95*, 12884–12889.

Sutherland, T. D., Unnithan, G. C., & Feyereisen, R. (2000). Terpenoid omega-hydroxylase (CYP4C7) messenger RNA levels in the corpora allata: A marker for ovarian control of juvenile hormone synthesis in *Diploptera punctata*. *J. Insect Physiol.*, *46*, 1219–1227.

Sztal, T., Chung, H., Gramzow, L., Daborn, P. J., Batterham, P., & Robin, C. (2007). Two independent duplications forming the *Cyp307a* genes in *Drosophila*. *Insect Biochem. Mol. Biol.*, *37*, 1044–1053.

Tamura, T., Thibert, C., Royer, C., Kanda, T., Abraham, E., et al. (2000). Germline transformation of the silkworm *Bombyx mori* L. using a piggyBac transposon-derived vector. *Nat. Biotechnol.*, *18*, 81–84.

Tang, J., Cao, Y., Rose, R. L., Brimfield, A. A., Dai, D., et al. (2001). Metabolism of chlorpyrifos by human cytochrome P450 isoforms and human, mouse, and rat liver microsomes. *Drug Metab. Dispos.*, *29*, 1201–1204.

Taylor, M., & Feyereisen, R. (1996). Molecular biology and evolution of resistance of toxicants. *Mol. Biol. Evol.*, *13*, 719–734.

Thomas, J. D., Ottea, J. A., Boethel, D. J., & Ibrahim, S. (1996). Factors influencing pyrethroid resistance in a per-methrin-resistant strain of the soybean looper, *Pseudoplusia includens* (Walker). *Pestic. Biochem. Physiol.*, *55*, 1–9.

Thompson, M., Svoboda, J., Lusby, W., Rees, H., Oliver, J., et al. (1985). Biosynthesis of a C21 steroid conjugate in an insect. The conversion of [14C]cholesterol to 5-[14C]preg-nen-3 beta,20 beta-diol glucoside in the tobacco hornworm, *Manduca sexta*. *J. Biol. Chem.*, *260*, 15410–15412.

Thongsinthusak, T., & Krieger, R. I. (1974). Inhibitory and inductive effects of piperonyl butoxide on dihydroisodrin hydroxylation *in vivo* and *in vitro* in black cutworm (*Agrotis ypsilon*) larvae. *Life Sci.*, *14*, 2131–2141.

Tijet, N., Helvig, C., & Feyereisen, R. (2001). The cytochrome P450 gene superfamily in *Drosophila melanogaster*: Annotation, intron–exon organization and phylogeny. *Gene.*, *262*, 189–198.

Tomita, T., & Scott, J. G. (1995). cDNA and deduced protein sequence of CYP6D1: The putative gene for a cytochrome P450 responsible for pyrethroid resistance in house fly. *Insect Biochem. Mol. Biol.*, *25*, 275–283.

Tomita, T., Liu, N., Smith, F. F., Sridhar, P., & Scott, J. G. (1995). Molecular mechanisms involved in increased expression of a cytochrome P450 responsible for pyrethroid resistance in the housefly, *Musca domestica*. *Insect Mol. Biol.*, *4*, 135–140.

Tribolium Genome Sequencing Consortium (2008). The genome of the model beetle and pest *Tribolium castaneum*. *Nature*, *452*, 949–955.

Ueda, H. R., Matsumoto, A., Kawamura, M., Iino, M., Tan-imura, T., & Hashimoto, S. (2002). Genome-wide transcriptional orchestration of circadian rhythms in *Drosophila*. *J. Biol. Chem.*, *277*, 14048–14052.

Ugaki, M., Shono, T., & Fukami, J. I. (1985). Metabolism of fenitrothion by organophosphorous-resistant and -susceptible house flies, *Musca domestica* L. *Pestic. Biochem. Physiol.*, *23*, 33–40.

Unnithan, G. C., Andersen, J. F., Hisano, T., Kuwano, E., & Feyereisen, R. (1995). Inhibition of juvenile hormone biosynthesis and methyl farnesoate epoxidase activity by 1,5-disubstituted imidazoles in the cockroach, *Diploptera punctata*. *Pesticide Science*, *43*, 13–19.

Valles, S. M., Koehler, P. G., & Brenner, R. J. (1997). Antagonism of fipronil toxicity by piperonyl butoxide and S,S,S-tributyl phosphorotrithioate in the German cockroach (Dictyoptera: Blattelliidae). *J. Econ. Entomol.*, *90*, 1254–1258.

Vontas, J., Blass, C., Koutsos, A. C., David, J. P., Kafatos, F. C., et al. (2005). Gene expression in insecticide resistant and susceptible *Anopheles gambiae* strains constitutively or after insecticide exposure. *Insect Mol. Biol.*, *14*, 509–521.

Vontas, J., David, J. P., Nikou, D., Hemingway, J., Christophides, G. K., et al. (2007). Transcriptional analysis of insecticide resistance in *Anopheles stephensi* using cross-species microarray hybridization. *Insect Mol. Biol.*, *16*, 315–324.

Wada, A., & Waterman, M. R. (1992). Identification by site-directed mutagenesis of two lysine residues in cholesterol side chain cleavage cytochrome P450 that are essential for adrenodoxin binding. *J. Biol. Chem.*, *267*, 22877–22882.

Walsh, J. B. (1987). Sequence-dependent gene conversion: Can duplicated genes diverge fast enough to escape conversion? *Genetics*, *117*, 543–557.

Wang, J., Kean, L., Yang, J., Allan, A. K., Davies, S. A., et al. (2004). Function-informed transcriptome analysis of *Drosophila* renal tubule. *Genome Biol.*, *5*, R69.

Wang, L., Bieber Urbauer, R. J., Urbauer, J. L., & Benson, D. R. (2003). House fly cytochrome b_5 exhibits kinetically trapped hemin and selectivity in hemin binding. *Biochem. Biophys. Res. Commun.*, *305*, 840–845.

Wang, L., Cowley, A. B., Terzyan, S., Zhang, X., & Benson, D. R. (2007). Comparison of cytochromes b5 from insects and vertebrates. *Proteins*, *67*, 293–304.

Wang, L., Dankert, H., Perona, P., & Anderson, D. J. (2008). A common genetic target for environmental and heritable influences on aggressiveness in *Drosophila*. *Proc. Natl. Acad. Sci. USA*, *105*, 5657–5663.

Wang, M., Roberts, D. L., Paschke, R., Shea, T. M., Masters, B. S., & Kim, J. J. (1997). Three-dimensional structure of NADPH-cytochrome P450 reductase: Prototype for FMN- and FAD-containing enzymes. *Proc. Natl. Acad. Sci. USA*, *94*, 8411–8416.

Wang, Q., Hasan, G., & Pikielny, C. W. (1999). Preferential expression of biotransformation enzymes in the olfactory organs of *Drosophila melanogaster*, the antennae. *J. Biol. Chem.*, *274*, 10309–10315.

Warren, J. T., Petryk, A., Marques, G., Jarcho, M., Parvy, J. P., et al. (2002). Molecular and biochemical characterization of two P450 enzymes in the ecdysteroidogenic pathway of *Drosophila melanogaster*. *Proc. Natl. Acad. Sci. USA*, *99*, 11043–11048.

Warren, J. T., Petryk, A., Marques, G., Parvy, J. P., Shinoda, T., et al. (2004). Phantom encodes the 25-hydroxylase of *Drosophila melanogaster* and *Bombyx mori*: A P450 enzyme critical in ecdysone biosynthesis. *Insect Biochem. Mol. Biol.*, *34*, 991–1010.

Waters, L. C., & Nix, C. E. (1988). Regulation of insecticide resistance-related cytochrome P-450 expression in *Drosophila melanogaster*. *Pest. Biochem. Physiol.*, *30*, 214–227.

Waters, L. C., Zelhof, A. C., Shaw, B. J., & Ch'ang, L. Y. (1992). Possible involvement of the long terminal repeat of transposable element 17.6 in regulating expression of an insecticide resistance-associated P450 gene in *Drosophila*. [published erratum appears in *Proc. Natl. Acad. Sci. USA* 1992 Dec 15;89(24):12209] *Proc. Natl. Acad. Sci. USA*, *89*, 4855–4859.

Wee, C. W., Lee, S. F., Robin, C., & Heckel, D. G. (2008). Identification of candidate genes for fenvalerate resistance in *Helicoverpa armigera* using cDNA-AFLP. *Insect Mol. Biol.*, *17*, 351–360.

Weirich, G. F., Williams, V. P., & Feldlaufer, M. F. (1996). Ecdysone 20-hydroxylation in *Manduca sexta* midgut: Kinetic parameters of mitochondrial and microsomal ecdysone 20-monooxygenases. *Arch. Insect Biochem. Physiol.*, *31*, 305–312.

Weller, G. L., & Foster, G. G. (1993). Genetic maps of the sheep blowfly *Lucilia cuprina*: Linkage-group corelations with other dipteran genera. *Genome*, *36*, 495–506.

Wen, Z., Horak, C. E., & Scott, J. G. (2001). CYP9E2, CYP4C21 and related pseudogenes from German cockroaches, *Blattella germanica*: Implications for molecular evolution, expression studies and nomenclature of P450s. *Gene*, *272*, 257–266.

Wen, Z., Pan, L., Berenbaum, M. B., & Schuler, M. A. (2003). Metabolism of linear and angular furanocoumarins by *Papilio polyxenes* CYP6B1 co-expressed with NADPH cytochrome P450 reductase. *Insect Biochem. Mol. Biol.*, *33*, 937–947.

Wen, Z., Baudry, J., Berenbaum, M. R., & Schuler, M. A. (2005). Ile115Leu mutation in the SRS1 region of an insect cytochrome P450 (CYP6B1) compromises substrate turnover via changes in a predicted product release channel. *Protein Eng. Des. Sel.*, *18*, 191–199.

Wen, Z., Rupasinghe, S., Niu, G., Berenbaum, M. R., & Schuler, M. A. (2006). CYP6B1 and CYP6B3 of the black swallowtail (*Papilio polyxenes*): Adaptive evolution through subfunctionalization. *Mol. Biol. Evol.*, *23*, 2434–2443.

Wen, Z., Zeng, R. S., Niu, G., Berenbaum, M. R., & Schuler, M. A. (2009). Ecological significance of induction of broad-substrate cytochrome P450s by natural and synthetic inducers in *Helicoverpa zea*. *J. Chem. Ecol.*, *35*, 183–189.

Werck-Reichhart, D., & Feyereisen, R. (2000). Cytochromes P450: A success story. *Genome Biology*, *1*, 3003.1–3003.9.

Wheeler, G. S., Slansky, F., Jr., & Yu, S. J. (1993). Fall armyworm sensitivity to flavone: Limited role of constitutive and induced detoxifying enzyme activity. *J. Chem. Ecol.*, *19*, 645–667.

Wheelock, G. D., & Scott, J. G. (1990). Immunological detection of cytochrome P450 from insecticide resistant and susceptible house flies (*Musca domestica*). *Pestic. Biochem. Physiol.*, *38*, 130–139.

Wheelock, G. D., & Scott, J. G. (1992a). The role of cytochrome P450 in deltamethrin metabolism by pyrethroid-resistant and susceptible strains of house flies. *Pestic. Biochem. Physiol.*, *43*, 67–77.

Wheelock, G. D., & Scott, J. G. (1992b). Anti-P450lpr antiserum inhibits specific monooxygenase activities in LPR house fly microsomes. *J. Exp. Zool.*, *264*, 153–158.

White, J. R.A., Franklin, R. T., & Agosin, M. (1979). Conversion of a-pinene to a-pinene oxide by rat liver and the bark beetle *Dendroctonus terebrans* microsomal fractions. *Pesticide Biochem. Physiol.*, *10*, 233–242.

Whittaker, R. H., & Feeny, P. P. (1971). Allelochemics: Chemical interactions between species. *Science*, *171*, 757–770.

Wigglesworth, V. B. (1970). Structural lipids in the insect cuticle and the function of the oenocytes. *Tissue Cell*, *2*, 155–179.

Wilding, C. S., Weetman, D., Steen, K., & Donnelly, M. J. (2009). High, clustered, nucleotide diversity in the genome of *Anopheles gambiae* revealed through pooled-template sequencing: Implications for high-throughput genotyping protocols. *BMC Genomics*, *10*, 320.

Wilkinson, C. F. (1979). The use of insect subcellular components for studying the metabolism of xenobiotics. *ACS Symposium Series*, *97*, 249–284.

Wilkinson, C. F., & Brattsten, L. B. (1972). Microsomal drug metabolizing enzymes in insects. *Drug Metab. Rev.*, *1*, 153–228.

Williams, D. R., Chen, J. H., Fisher, M. J., & Rees, H. H. (1997). Induction of enzymes involved in molting hormone (ecdysteroid) inactivation by ecdysteroids and an agonist, 1,2-dibenzoyl-1-*tert*-butylhydrazine (RH-5849). *J. Biol. Chem.*, *272*, 8427–8432.

Williams, P. A., Cosme, J., Sridhar, V., Johnson, E. F., & McRee, D. E. (2000a). Mammalian microsomal cytochrome P450 monooxygenase: Structural adaptations for membrane binding and functional diversity. *Mol. Cell*, *5*, 121–131.

Williams, D. R., Fisher, M. J., & Rees, H. H. (2000b). Characterization of ecdysteroid 26-hydroxylase: An enzyme involved in molting hormone inactivation. *Arch. Biochem. Biophys.*, *376*, 389–398.

Willingham, A. T., & Keil, T. (2004). A tissue specific cytochrome P450 required for the structure and function of *Drosophila* sensory organs. *Mech. Dev.*, *121*, 1289–1297.

Willoughby, L., Chung, H., Lumb, C., Robin, C., Batterham, P., & Daborn, P. J. (2006). A comparison of *Drosophila melanogaster* detoxification gene induction responses for six insecticides, caffeine and phenobarbital. *Insect Biochem. Mol. Biol.*, *36*, 934–942.

Willoughby, L., Batterham, P., & Daborn, P. J. (2007). Piperonyl butoxide induces the expression of cytochrome P450 and glutathione S-transferase genes in *Drosophila melanogaster*. *Pest. Manag. Sci.*, *63*, 803–808.

Wilson, T. G. (2001). Resistance of *Drosophila* to toxins. *Annu. Rev. Entomol.*, *46*, 545–571.

Winter, J., Bilbe, G., Richener, H., Sehringer, B., & Kayser, H. (1999). Cloning of a cDNA encoding a novel cytochrome P450 from the insect *Locusta migratoria*: CYP6H1, a putative ecdysone 20-hydroxylase. *Biochem. Biophys. Res. Commun.*, *259*, 305–310.

Winter, J., Eckerskorn, C., Waditschatka, R., & Kayser, H. (2001). A microsomal ecdysone-binding cytochrome P450 from the insect *Locusta migratoria* purified by sequential use of type-II and type-I ligands. *Biol. Chem.*, *382*, 1541–1549.

Wojtasek, H., & Leal, W. S. (1999). Degradation of an alkaloid pheromone from the pale-brown chafer, *Phyllopertha diversa* (Coleoptera: Scarabaeidae), by an insect olfactory cytochrome P450. *FEBS Lett.*, *458*, 333–336.

Wondji, C. S., Irving, H., Morgan, J., Lobo, N. F., Collins, F. H., et al. (2009). Two duplicated P450 genes are associated with pyrethroid resistance in *Anopheles funestus*, a major malaria vector. *Genome Res.*, *19*, 452–459.

Xia, Q., Cheng, D., Duan, J., Wang, G., Cheng, T., et al. (2007). Microarray-based gene expression profiles in multiple tissues of the domesticated silkworm, *Bombyx mori*. *Genome Biol.*, *8*, R162.

Yamazaki, S., Sato, K., Suhara, K., Sakaguchi, M., Mihara, K., & Omura, T. (1993). Importance of the proline-rich region following signal-anchor sequence in the formation of correct conformation of microsomal cytochrome P-450s. *J. Biochem. (Tokyo)*, *114*, 652–657.

Yang, J., McCart, C., Woods, D. J., Terhzaz, S., Greenwood, K. G., et al. (2007). A *Drosophila* systems approach to xenobiotic metabolism. *Physiol. Genomics*, *30*, 223–231.

Yang, X., Zhang, B., Molony, C., Chudin, E., Hao, K., et al. (2010). Systematic genetic and genomic analysis of cytochrome P450 enzyme activities in human liver. *Genome Res.*, *20*, 1020–1036.

Yang, Y., Chen, S., Wu, S., Yue, L., & Wu, Y. (2006). Constitutive overexpression of multiple cytochrome P450 genes associated with pyrethroid resistance in *Helicoverpa armigera*. *J. Econ. Entomol.*, *99*, 1784–1789.

Yang, Y., Yue, L., Chen, S., & Wu, Y. (2008). Functional expression of *Helicoverpa armigera* CYP9A12 and CYP9A14 in *Saccharomyces cerevisiae*. *Pestic. Biochem. Physiol.*, *92*, 101–105.

Yoshinaga, N., Alborn, H. T., Nakanishi, T., Suckling, D. M., Nishida, R., et al. (2010). Fatty acid-amino acid conjugates diversification in lepidopteran caterpillars. *J. Chem. Ecol.*, *36*, 319–325.

Yoshiyama, T., Namiki, T., Mita, K., Kataoka, H., & Niwa, R. (2006). Neverland is an evolutionarily conserved Rieske-domain protein that is essential for ecdysone synthesis and insect growth. *Development*, *133*, 2565–2574.

Young, S. J., Gunning, R. V., & Moores, G. D. (2005). The effect of piperonyl butoxide on pyrethroid-resistance-associated esterases in *Helicoverpa armigera* (Hubner) (Lepidoptera: Noctuidae). *Pest. Manag. Sci.*, *61*, 397–401.

Yu, S. J. (1984). Interactions of allelochemicals with detoxication enzymes of insecticide-susceptible and resistant fall armyworms. *Pest. Biochem. Physiol.*, *22*, 60–68.

Yu, S. J. (1986). Consequences of induction of foreign compound-metabolizing enzymes in insects. In L. B. Brattsten, & S. Ahmad (Eds.), *Molecular Aspects of Insect–Plant Interactions* (pp. 211–255). New York, NY: Plenum.

Yu, S. J. (1987). Microsomal oxidation of allelochemicals in generalist (*Spodoptera frugiperda*) and semispecialist (*Anticarsia gemmatalis*) insects. *J. Chem. Ecol.*, *13*, 423–436.

Yu, S. J. (1995). Allelochemical stimulation of ecdysone 20-monooxygenase in fall armyworm larvae. *Arch. Insect Biochem. Physiol.*, *28*, 365–375.

Zagrobelny, M., Scheibye-Alsing, K., Jensen, N. B., Moller, B. L., Gorodkin, J., & Bak, S. (2009). 454 pyrosequencing based transcriptome analysis of *Zygaena filipendulae* with focus on genes involved in biosynthesis of cyanogenic glucosides. *BMC Genomics*, *10*, 574.

Zdobnov, E. M., von Mering, C., Letunic, I., Torrents, D., Suyama, M., et al. (2002). Comparative genome and proteome analysis of *Anopheles gambiae* and *Drosophila melanogaster*. *Science*, *298*, 149–159.

Zeng, R. S., Wen, Z., Niu, G., Schuler, M. A., & Berenbaum, M. R. (2007). Allelochemical induction of cytochrome P450 monooxygenases and amelioration of xenobiotic toxicity in *Helicoverpa zea*. *J. Chem. Ecol.*, *33*, 449–461.

Zeng, R. S., Wen, Z., Niu, G., Schuler, M. A., & Berenbaum, M. R. (2009). Enhanced toxicity and induction of cytochrome P450s suggest a cost of "eavesdropping" in a multitrophic interaction. *J. Chem. Ecol.*, *35*, 526–532.

Zhang, C., Luo, X., Ni, X., Zhang, Y., & Li, X. (2010). Functional characterization of cis-acting elements mediating flavone-inducible expression of CYP321A1. *Insect Biochem. Mol. Biol.*, *40*, 898–908.

Zhang, L., Harada, K., & Shono, T. (1997). Genetic analysis of pyriproxifen resistance in the housefly, *Musca domestica* L. *Appl. Ent. Zool.*, *32*, 217–226.

Zhang, L., Kasai, S., & Shono, T. (1998). *In vitro* metabolism of pyriproxyfen by microsomes from susceptible and resistant housefly larvae. *Arch. Insect Biochem. Physiol.*, *37*, 215–224.

Zhang, M., & Scott, J. G. (1994). Cytochrome b5 involvement in cytochrome P450 monooxygenase activities in house fly microsomes. *Arch. Insect Biochem. Physiol.*, *27*, 205–216.

Zhang, M., & Scott, J. G. (1996a). Purification and characterization of cytochrome b5 reductase from the house fly, *Musca domestica*. *Comp. Biochem. Physiol.*, *113B*, 175–183.

Zhang, M., & Scott, J. G. (1996b). Cytochome b5 is essential for cytochome P450 6D1-mediated cypermethrin resistance in LPR house flies. *Pestic. Biochem. Physiol.*, *55*, 150–156.

Zhou, S., Gao, Y., Jiang, W., Huang, M., Xu, A., & Paxton, J. W. (2003). Interactions of herbs with cytochrome P450. *Drug Metab. Rev.*, *35*, 35–98.

Zhou, X., Sheng, C., Li, M., Wan, H., Liu, D., & Qiu, X. (2010). Expression responses of nine cytochrome P450 genes to xenobiotics in the cotton bollworm, *Helicoverpa armigera*. *Pestic. Biochem. Physiol.*, *97*, 209–213.

Zhou, X., Song, C., Grzymala, T. L., Oi, F. M., & Scharf, M. E. (2006). Juvenile hormone and colony conditions differentially influence cytochrome P450 gene expression in the termite Reticulitesmes flavipes. *Insect Mol. Biol.*, *6*, 749–761.

Zhu, F., & Liu, N. (2008). Differential expression of CYP6A5 and CYP6A5v2 in pyrethroid-resistant house flies, *Musca domestica*. *Arch. Insect Biochem. Physiol.*, *67*, 107–119.

Zhu, F., Feng, J. N., Zhang, L., & Liu, N. (2008a). Characterization of two novel cytochrome P450 genes in insecticide-resistant house-flies. *Insect Mol. Biol.*, *17*, 27–37.

Zhu, F., Li, T., Zhang, L., & Liu, N. (2008b). Co-up-regulation of three P450 genes in response to permethrin exposure in permethrin resistant house flies, *Musca domestica*. *BMC Physiol.*, *8*, 18.

Zhu, F., Parthasarathy, R., Bai, H., Woithe, K., Kaussmann, M., et al. (2010). A brain-specific cytochrome P450 responsible for the majority of deltamethrin resistance in the QTC279 strain of *Tribolium castaneum*. *Proc. Natl. Acad. Sci. USA*, *107*, 8557–8562.

Zhu, Y. C., & Snodgrass, G. L. (2003). Cytochrome P450 CYP6X1 cDNAs and mRNA expression levels in three strains of the tarnished plant bug *Lygus lineolaris* (Heteroptera: Miridae) having different susceptibilities to pyrethroid insecticide. *Insect Mol. Biol.*, *12*, 39–49.

Ziegler, D. M. (2002). An overview of the mechanism, substrate specificities, and structure of FMOs. *Drug Metab. Rev.*, *34*, 503–511.

Zou, S., Meadows, S., Sharp, L., Yan, L. Y., & Jan, Y. N. (2000). Genome-wide study of aging and oxidative stress response in *Drosophila melanogaster*. *Proc. Natl. Acad. Sci. USA*, *97*, 13726–13731.

Zumwalt, J. G., & Neal, J. J. (1993). Cytochromes P450 from *Papilio polyxenes*: Adaptations to host plant allelochemicals. *Comp. Biochem. Physiol.*, *106C*, 111–118.

9   Lipid Transport

Dick J Van der Horst
Utrecht University, Utrecht, The Netherlands
Robert O Ryan
Children's Hospital Oakland Research Institute,
Oakland, CA, USA

9.1. Historical Perspective

9.1.1. Lipophorin Structure, Assembly, and Morphology

Lipophorin was discovered nearly 50 years ago as a major hemolymph component and key transport vehicle for water-insoluble metabolites (for reviews, see Beenakkers *et al.*, 1985; Chino, 1985). Lipophorin is generally regarded as a multifunctional carrier because it displays a broad ability to accommodate hydrophobic biomolecules. In essence, lipophorin can be described as a non-convalent assembly of lipids and proteins, organized as a largely spherical particle. The core of the particle is made up of hydrophobic lipid molecules, such as diacylglycerol (DAG), hydrocarbons, and carotenoids. DAG, which serves as the transport form of neutral glycerolipid in hemolymph, provides an energy source for various tissues through oxidative metabolism of its fatty acid constituents. Hydrocarbons, in the form of long-chain aliphatic alkanes and alkenes, are extremely hydrophobic lipid molecules that are deposited on the cuticle, where they serve to prevent desiccation and may function as semiochemicals. Carotenoids are plant-derived pigments used for coloration, and as a precursor to visual pigments (Canavoso *et al.*, 2001). Another important lipid component of lipophorin is phospholipid. In general, the major glycerophospholipids present are phosphatidylcholine and phosphatidylethanolamine (Wang *et al.*, 1992). These amphiphilic lipids exist as a monolayer at the lipophorin particle surface, positioned in such a way that their fatty acyl chains interact with the hydrophobic core of the particle while their polar head groups are presented to the aqueous milieu. In this manner, the phospholipid moieties of lipophorin serve a key structural role. The other major structural component of lipophorin is protein. All lipophorin particles possess two apolipoproteins, termed apolipophorin I (apoLp-I) and apolipophorin II (apoLp-II). ApoLp-I and apoLp-II are integral components of the lipophorin particle, and cannot be removed without destruction of lipophorin particle integrity. It is recognized that apoLp-I and apoLp-II are the product of the same gene, and that the two proteins arise through post-translational cleavage of their common precursor protein (Weers *et al.*, 1993). This finding is consistent with the fact that apoLp-I and apoLp-II are found in a 1:1 molar ratio in lipophorin particles. Their common precursor protein is arranged with apoLp-II at the N-terminal end and apoLp-I at the C-terminal end (Bogerd *et al.*, 2000), and therefore is termed apoLp-II/I. The apoLp-II/I cDNA of several insect species has been isolated and characterized or identified in genome analysis

DOI:10.1016/B978-0-12-384747-8.10009-1

projects (for review, see Van der Horst and Rodenburg, 2010a); based on sequence similarity and ancestral exon boundaries, these insect apolipoprotein precursors were revealed to belong to the large lipid transfer (LLT) protein (LLTP) superfamily that emerged from an ancestral molecule and includes vertebrate apolipoprotein B (apoB), microsomal triglyceride-transfer protein (MTP), and vitellogenin (Vg) (Babin et al., 1999). The LLT domain shared by these proteins comprises a large N-terminal domain of about 1000 amino acids; the LLT domains of apoB, MTP, and Vg contain a large lipid-binding cavity which was proposed to act to store and transfer lipids to the apolipoprotein in a coordinated manner (for reviews, see Rodenburg and Van der Horst, 2005; Smolenaars et al., 2007a). A recent model of locust (*Locusta migratoria*) apoLp-II/I, constructed on homology with the X-ray crystal structure of lamprey lipovitellin, the processed form of Vg (Thompson and Banaszak, 2002), as well as a structural model for nascent human apoB lipoprotein (Richardson et al., 2005), reveals a similar putative lipid pocket in the insect LLT domain (Smolenaars et al., 2005). The cleavage of insect apoLp-II/I into apoLp-II and apoLp-I is mediated by an insect furin, acting at a consensus substrate sequence (RQKR) between two residues (720 and 721) of the LLT module (Smolenaars et al., 2005). Since protein cleavage by furin homologs is performed late in the secretory pathway, mainly in the *trans*-Golgi network, insect lipoprotein biosynthesis was proposed to proceed by initial lipidation of apoLp-II/I to a lipoprotein, while cleavage of apoLp-II/I into apoLp-I and -II would occur at a later stage (Smolenaars et al., 2005). The uncleaved LLT domain in apoLp-II/I, comprising intimately linked regions of apoLp-I and apoLp-II, is likely to be essential to enable the first step in lipidation, as in apoB. Moreover, the occurrence of a cleavage step prior to lipidation might result in the parting of apoLp-I and apoLp-II, and thus in impairment of lipoprotein biosynthesis. Indeed, it was shown that if cleavage was impaired by a furin inhibitor or mutagenesis of the consensus substrate sequence for furin, uncleaved apoLp-II/I appeared to be lipidated and functioned as a single apolipoprotein in the formation of a lipoprotein particle with a buoyant density and molecular mass identical to wild type lipophorin (Smolenaars et al., 2005; Van der Horst and Rodenburg, 2010a); it was therefore proposed that cleavage of apoLp-II/I by insect furin is required neither for biosynthesis nor for secretion of the insect lipoprotein. Although the apparent conservation of apoLp-II/I cleavage in all insects characterized to date reveals the importance of this processing step, at present it is not known if one or the other apoLp possesses additional functions aside from its primary role in stabilizing lipophorin particle integrity, although a role in receptor interactions is implied (see below). The structural role is fulfilled by the capacity of apoLp-I and apoLp-II to interact with lipid and create an interface between the non-polar core of the particle and the external environment. In this capacity, apoLp-I and apoLp-II function in a manner similar to that proposed for apoB in vertebrate plasma. This is consistent with the finding that the genes encoding these proteins are derived from a common ancestor (Babin et al., 1999).

The structural resemblance between apoLp-II/I and apoB is not limited to their LLT modules, but extends to the entire polypeptide chains. Prediction of amphipathic clusters in apoB proposed a pentapartite structure of α-helical domains (α) and amphipathic β-strand domains (β) along the apoB polypeptide, organized as N-α_1-β_1-α_2-β_2-α_3-C (Segrest et al., 2001); the C-terminal β_1-α_2-β_2-α_3 clusters stabilize the expansion of the initial lipid core in the LLT module and accommodate most of the lipid-binding capacity. Recent data on the amphipathic clusters in apoLp-II/I propose the polypeptide to contain a similar, albeit smaller, (tripartite) structure, organized as N-α_1-β-α_2-C; recombinant expression experiments demonstrated the β cluster to accommodate the apoLp-II/I lipid-binding capacity (Smolenaars et al., 2007b). After cleavage of apoLp-II/I, the β cluster is almost entirely situated in apoLp-I, suggesting that apoLp-I, and not apoLp-II, binds the vast majority of lipids (Smolenaars et al., 2007b). This finding is consistent with lipophorin dissociation experiments in which >98% of the total lipid in lipophorin remained associated with apoLp-I (Kawooya et al., 1989). On the basis of the similar structural organization of apoLp-II/I and apoB, the pathway for lipoprotein biogenesis in insects might be assumed to show similarity with that in mammals. Lipoprotein assembly in mammals has disclosed the role of MTP in acquiring the initial binding of lipids to the amphipathic lipid-associating segment of apoB (Hussain et al., 2001; Shelness and Sellers, 2001; Ledford et al., 2006). From the discovery of an MTP homolog in the fruit fly, *Drosophila melanogaster*, which was able to promote the assembly and secretion of human apoB (Sellers et al., 2003), insect lipoprotein assembly early in the secretory pathway has been proposed to occur similarly (Smolenaars et al., 2005). The recovery of MTP homologs in all available insect genomes (Smolenaars et al., 2007a) provides significant support for the concept that an MTP-dependent mechanism for initial lipoprotein biosynthesis is also operative in the biogenesis of insect lipoproteins. Moreover, insect MTP was experimentally shown to stimulate insect lipoprotein biogenesis considerably, since co-expression of the *Drosophila* MTP homolog (dMTP) and recombinant full-length *L. migratoria* apoLp-II/I cDNA in an insect cell (Sf9) expression system resulted in a several-fold increase in the secretion of apoLp-I and -II, as well as uncleaved apoLp-II/I (Smolenaars et al., 2007b; for reviews, see Van der Horst et al., 2009; Van der Horst and Rodenburg, 2010a). Concomitant with their secretion, dMTP significantly stimulated the lipidation of the apoLp-II/I proteins, since the secreted lipoprotein particles were recovered at a decreased buoyant density compared to control cells lacking the dMTP gene. Recombinant co-expression of dMTP and a series of C-terminal truncation variants of apoLp-II/I in Sf9 cells revealed that formation of a buoyant lipoprotein particularly requires the amphipathic β cluster (Smolenaars *et al.*, 2007b). Taken together, these data support a unifying concept for lipoprotein biogenesis, and led to the conclusion that, regardless of specific modifications, the assembly of lipoproteins both in mammals and insects requires amphipathic structures in the apolipoprotein carriers, as well as MTP (Smolenaars *et al.*, 2007b; for reviews, see Van der Horst *et al.*, 2009; Van der Horst and Rodenburg, 2010a).

9.1.2. Lipophorin Subspecies

One of the hallmark features of lipophorin-mediated lipid transport relates to the dynamic nature of the particle. Lipophorin isolated from various life stages is generally of a unique density and lipid composition. For example, lipophorin from *Manduca sexta* fifth instar larvae displays a density of 1.15 g/ml, with a particle diameter in the range of 16 nm. By contrast, lipophorin isolated from adult hemolymph is of lower density and larger diameter. Indeed, in *M. sexta* a broad array of unique lipophorin subspecies has been identified, each with characteristic properties (Prasad *et al.*, 1986). On the basis of this diversity, a nomenclature system has been adopted that distinguishes various lipophorin subspecies based on their density. Since most particles fall within the density limits of 1.21 and 1.07 g/ml, the term "high-density lipophorin" (HDLp) is commonly used. Because many lipophorin subspecies are present at well-defined developmental stages, a suffix may be added to denote this. Hence, HDLp-P and HDLp-A may be used to distinguish HDLp from pupal and adult hemolymphs, respectively.

One of the features of HDLp-A is its ability to associate with a third apolipophorin, apoLp-III. In insect species that use lipid as a fuel for flight (such as *L. migratoria* and *M. sexta*), apoLp-III is present in abundance in adult hemolymph as a lipid-free protein. Whereas a small amount of apoLp-III may be associated with HDLp under resting conditions, flight activity induces association of large amounts of apoLp-III with the lipophorin particle surface (Van der Horst *et al.*, 1979). This process, which is dependent upon the uptake of DAG by the lipophorin particle, leads to the conversion of HDLp into low-density lipophorin (LDLp). LDLp has a larger diameter, a significantly increased DAG content, and a lower density. In studies of this conversion, it has been shown that apoLp-III associates with the surface of the expanding lipophorin particle as a function of DAG enrichment (for reviews, see Soulages and Wells, 1994a; Ryan and Van der Horst, 2000; Van der Horst and Ryan, 2005; Van der Horst and Rodenburg, 2010a, 2010b). Thus, it has

been hypothesized that apoLp-III serves to stabilize the DAG-enriched particle, providing an interface between surface-localized hydrophobic DAG molecules and the external aqueous medium. It is envisioned that continued DAG accumulation by HDLp results in partitioning of DAG between the hydrophobic core of the particle and the surface monolayer (Wang *et al.*, 1995). The presence of DAG in the surface monolayer exerts a destabilizing effect on the particle structure, and, if allowed to persist, would result in deleterious particle fusion and aggregation. By "sensing" the presence of DAG in the lipophorin surface monolayer, apoLp-III is attracted to the particle surface and forms a stable binding interaction. This event is fully reversible, and, upon removal of DAG from the particle, apoLp-III dissociates, leading to regeneration of HDLp. Importantly, it is recognized that lipophorin particles can then bind additional DAG, forming a cycle of transport. It is noteworthy that these concepts about apoLp-III association/dissociation from lipophorin emerged from physiological studies of flight activity in *L. migratoria* conducted in the late 1970s and early 1980s in The Netherlands and England (Mwangi and Goldsworthy, 1977, 1981; Van der Horst *et al.*, 1979, 1981). A cartoon depicting metabolic and biochemical processes related to the induction of flight-related lipophorin conversions and the accompanying increase in neutral lipid transport capacity is presented in **Figure 1**. Elaboration of various aspects of this central scheme will occur in subsequent sections. At this point, however, it should be noted that this generalized mechanism differs fundamentally from metabolic processes in vertebrates, where lipoproteins do not have a function in the transport of energy substrates during exercise (Van der Horst *et al.*, 2002; Van der Horst and Rodenburg, 2010a). That said, it is evident that novel insight into structural and functional aspects of vertebrate lipid transport processes can be gained from the study of insect lipid transport (for reviews, see Rodenburg and Van der Horst, 2005; Van der Horst *et al.*, 2009; Van der Horst and Rodenburg, 2010a).

9.2. Flight-Related Processes

9.2.1. Adipokinetic Hormone

Insect flight involves the mobilization, transport, and utilization of endogenous energy reserves at extremely high rates. In insects that engage in long-distance flight, the demand for fuel, particularly lipids, by the flight muscles can remain elevated for extended periods of time. Adipokinetic hormones (AKHs), synthesized and stored in neuroendocrine cells, play a crucial role in this process, as they integrate flight energy metabolism. Insect AKHs comprise a family of short peptides consisting of 8–11 amino acid residues. Over 40 bioanalogs of this family have been identified in representative species of

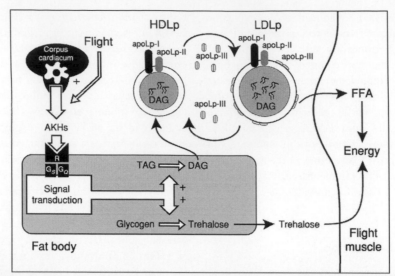

Figure 1 Molecular basis of the lipophorin lipid shuttle. AKH-controlled DAG mobilization from insect fat body during flight activity results in the reversible alternation of lipophorin from a relatively lipid-poor (HDLp) to a lipid-rich (LDLp) state, and apoLp-III from a lipid-free to a lipid-bound state. The reversible conformational change in apoLp-III induced by DAG loading of lipophorin is schematically visualized. AKHs, adipokinetic hormones; R, receptor; G, G protein; HDLp, high-density lipophorin; LDLp, low-density lipophorin; apoLp-I, -II, -III, apolipophorin I, II, III; TAG, triacylglycerol; DAG, diacylglycerol; FFA, free fatty acids.
Based on data from several insect species, particularly *Locusta migratoria* and *Manduca sexta*, reviewed in Ryan and Van der Horst (2000) and Van der Horst *et al.* (2001).
Reprinted with permission from Van der Horst D.J., Ryan R.O., 2005. Lipid transport. In: Gilbert, L.I., Iatrou, K., Gill, S.S. (eds), *Comprehensive Molecular Insect Science*, Vol. 4. Elsevier, Amsterdam, pp. 225–246.

most insect orders; in spite of considerable variation in their structures, they are clearly related (for reviews, see Gäde, 1997, 2009; Van der Horst *et al.*, 2001; Oudejans and Van der Horst, 2003). All AKHs are N-terminally blocked by a pyroglutamate (pGlu) residue, and all but one (Köllisch *et al.*, 2000) are C-terminally amidated. Initiation of flight activity induces the release of AKHs from the intrinsic AKH-producing cells (adipokinetic cells) in the glandular lobes of the corpus cardiacum, a neuroendocrine gland located caudal to the insect brain and physiologically equivalent to the pituitary of mammals. The fat body plays a fundamental role in lipid storage, as well as in the process of lipolysis controlled by the AKHs. Binding of these hormones to their G protein-coupled receptors at the fat body target cells triggers a number of coordinated signal transduction processes that ultimately result in the mobilization of carbohydrate and lipid reserves as fuels for flight activity (see **Figure 1**). Energy-yielding metabolites are transported via the hemolymph to the contracting flight muscles. Carbohydrate (trehalose) in the circulation provides energy for the initial period of flight, and is replenished from glycogen reserves. However, similar to sustained activity in many other animal species, flight activity of insects covering vast distances non-stop is powered principally by mobilization of endogenous reserves of triacylglycerol (TAG), the most concentrated form of energy available to biological tissues. As a result of TAG mobilization, the concentration of sn-1,2-DAG in the

hemolymph increases progressively, and gradually constitutes the principal fuel for flight. The mechanism for hormonal activation of glycogen phosphorylase, the enzyme determining the rate of glycogen breakdown and trehalose biosynthesis, has been well established. In contrast, the mechanism by which the pivotal enzyme TAG lipase catalyzes AKH-controlled production of the DAG on which long-distance flight is dependent is less well understood.

For a considerable part, the success of insects in long-distance flights is attributable to their system of neuropeptide AKHs integrating flight energy metabolism, involving the transfer of energy substrates, particularly lipids, to the flight muscles, as discussed above. Therefore, in the following sections, recent advances in the strategy of adipokinetic cells in hormone storage and release will be discussed, along with the effects of the AKHs on lipid mobilization.

9.2.2. Strategy of the Adipokinetic Cells

In view of their involvement in the regulation and integration of extremely intense metabolic processes, the AKH-producing cells (adipokinetic cells) of the corpus cardiacum constitute an appropriate model system for studying neuropeptide biosynthesis and processing, as well as the coherence between biosynthesis, storage, and release of these neurohormones (for reviews, see Ryan and Van der Horst, 2000; Van der Horst *et al.*, 2001; Diederen *et al.*,

2002; Van der Horst, 2003; Van der Horst and Ryan, 2005). These processes have been particularly studied in two locust species notorious for their long-range flight capacity, *L. migratoria* and *Schistocerca gregaria*, which, similarly to several other insect species, mobilize more than one AKH. The three AKHs synthesized in the adipokinetic cells of *L. migratoria* consist of a decapeptide AKH-I and two octapeptides (AKH-II and -III). AKH-I is by far the most abundant peptide; the ratio of AKH-I : -II : -III in the corpus cardiacum is approximately 14 : 2 : 1 (Oudejans *et al.*, 1993). All three AKHs are involved in the mobilization of both lipids and carbohydrates, although their action is differential (for reviews, see Vroemen *et al.*, 1998; Van der Horst and Oudejans, 2003). In addition, several other effects of AKH are known, such as inhibition of the synthesis of proteins, fatty acids, and RNA (for reviews, see Gäde, 1996, 2009; Gäde *et al.*, 1997).

The transport of these hydrophobic peptides in the circulation occurs independently of a carrier (Oudejans *et al.*, 1996). The AKHs of *L. migratoria* appear to be catabolized differentially after their release; turnover half-lives of AKH-I and -II during flight are relatively slow (35 and 37 minutes, respectively), whereas the hemolymph half-life of AKH-III is very rapid (3 minutes) (Oudejans *et al.*, 1996). Degradation of the (single) AKH in the hemolymph of adult females of the cricket *Gryllus bimaculatus*, which do not fly well, was estimated to be remarkably short (half-life approximately 3 minutes) in the resting state (Woodring *et al.*, 2002). A study in which AKH concentrations were measured by radio immunoassay shows that the hemolymph concentrations of the two AKHs from *S. gregaria* (AKH-I and -II; this species lacks AKH III (Oudejans *et al.*, 1991)) increase within 5 minutes of initiation of flight, and are maintained at approximately 15-fold (AKH-I) and 6-fold (AKH-II) the resting levels over flights of at least 60 min (Candy, 2002). The increase in hormone level preceded an increase in hemolymph lipid content. Furthermore, a rapid release of

the AKHs over the first few minutes was followed by a slower release, maintaining the elevated hormone levels.

The AKH peptides are derived from pre-prohormones that are translated from separate mRNAs and subsequently enzymatically processed. Co-translational cleavage of the signal sequences in *L. migratoria* generates the AKH-I, -II, and -III prohormones, consisting of a single copy of AKH, a GKR or GRR processing site, and an AKH-associated peptide (AAP). AKH-I and -II prohormones are structurally very similar, whereas AKH-III prohormone is remarkably different (Bogerd *et al.*, 1995) (**Figure 2**). Prior to further processing, the AKH-I and -II prohormones dimerize at random by oxidation of their (single) cysteine residues in the AAP, giving rise to two homodimers and one heterodimer. Proteolytic processing of these dimeric products at their processing sites, involving removal of the two basic amino acid residues and amidation, using glycine as the donor, yields the bioactive hormones as well as three (two homodimeric and one heterodimeric) AKH-precursor related peptides (APRPs) (for reviews, see Van der Horst *et al.*, 2001; Diederen *et al.*, 2002; Oudejans and Van der Horst, 2003). Data from capillary liquid chromatography-tandem mass spectrometry analysis indicate that these APRPs are further processed to form smaller peptides, designated AKH joining peptide 1 (AKH-JP I) and 2 (AKH-JP II), respectively (Baggerman *et al.*, 2002; for review, see Van der Horst, 2003) (**Figure 2**). The production of AKH-III results in the formation of an APRP with two disulfide bonds, a homodimer resulting from the cross-linking (in a parallel and/or antiparallel fashion) of two AKH-III prohormone molecules (Huybrechts *et al.*, 2002) (**Figure 2**). A putative cleavage site within the sequence of the latter APRP is lacking, suggesting that this APRP likely is not processed further (Huybrechts *et al.*, 2002).

In situ hybridization showed that the mRNA signals encoding the three different AKH pre-prohormones are co-localized in the cell bodies of the glandular lobes of

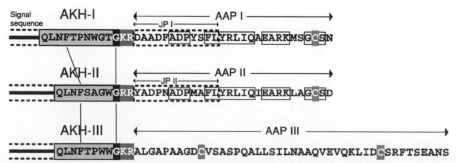

Figure 2 Sequence and proteolytic processing of *Locusta migratoria* AKH prohormones. The AKH sequence is followed by a processing site (GKR or GRR); identical residues in the AKH-associated peptides (AAPs) I and II are boxed. The cysteine residues forming disulfide bridges prior to proteolytic processing of all AKH prohormones are shown in white. JP, joining peptide. Based on data from Bogerd *et al.*, 1995; Baggerman *et al.*, 2002; Huybrechts *et al.*, 2002; Van der Horst, 2003. Reprinted with permission from Van der Horst D.J., Ryan R.O., 2005. Lipid transport. In: Gilbert, L.I., Iatrou, K., Gill, S.S. (eds), *Comprehensive Molecular Insect Science*, Vol. 4. Elsevier, Amsterdam, pp. 225–246.

the corpus cardiacum (Bogerd *et al.*, 1995). Following their synthesis in the rough endoplasmic reticulum in the cell bodies, the AKH prohormones are transported to the Golgi complex and packaged into secretory granules at the trans-Golgi network, whereas proteolytic processing of the prohormones to bioactive AKHs is presumed to take place in the secretory granules (for reviews, see Van der Horst *et al.*, 2001; Diederen *et al.*, 2002; Oudejans and Van der Horst, 2003). All three AKHs were shown to be co-localized in the same secretory granules, and are released simultaneously during flight (Harthoorn *et al.*, 1999). Since the membranes of exocytosed secretory granules fuse with the plasma membrane, the total content of the granules is released into the hemolymph. Consequently, in addition to bioactive AKHs the APRPs, and possibly other products encoded by the AKH genes, are also released during flight activity and might execute specific functions. Intriguingly, aligning of all known AKH pre-prohormone genes showed the APRP region to be better conserved in evolution (nematodes, insects, crustaceans) than that of AKH, suggesting an important biological role (De Loof *et al.*, 2009). However, although APRPs have been tested extensively in a large variety of bioassays, APRP function has not yet been uncovered (Hatle and Spring, 1999; De Loof *et al.*, 2009). A recent peptidomic survey of the locust neuroendocrine system confirmed the corpora cardiaca of both *L. migratoria* and *S. gregaria* to contain the two processing products of the APRPs, AKH-JP I and II (Clynen and Schoofs, 2009). However, whether the AKH-JPs are released is not yet clear (Baggerman *et al.*, 2002; Huybrechts *et al.*, 2002); in bioassays, AKH-JP I and II appeared neither to stimulate lipid release from the fat body nor to activate fat body glycogen phosphorylase – both key functions of the AKHs (Baggerman *et al.*, 2002).

The secretory activity of the adipokinetic cells, which has been investigated *in vitro* primarily for AKH-I, is subject to many regulatory substances, including neurogenic locustatachykinins and humoral crustacean cardioactive peptide (CCAP) as initiating factors, trehalose as an inhibitor, and several positive and negative modulators (for reviews, see Van der Horst *et al.*, 1999; Vullings *et al.*, 1999; Van der Horst, 2003; Van der Horst and Ryan, 2005). Data on the release of AKH from the corpora cardiaca *in vitro* show that regulatory substances (including CCAP) affect the release of all three AKHs in proportion to their concentration in the corpus cardiacum (Harthoorn *et al.*, 2001). However, the only natural stimulus for the release of the AKHs is flight activity, and the relative contributions of all known substances effective in the process of release of these neurohormones remain to be established *in vivo*.

The amount of AKHs released during flight represents only a few percent of the huge stores harbored in the adipokinetic cells. On the other hand, only a limited part of these AKH stores appear to be actually releasable. In studies in which young secretory granules were specifically labeled, these newly formed secretory granules were preferentially released (last in, first out) (for reviews, see Van der Horst *et al.*, 2001; Diederen *et al.*, 2002; Oudejans and Van der Horst, 2003). Following the biosynthesis of new AKH prohormones, their packaging into secretory granules and their processing to bioactive AKHs, which takes less than 1 hour, granules containing newly synthesized AKHs appeared to be available for release during a restricted period of approximately 8 hours before they are supposed to enter a pool of older secretory granules that appear to be unable to release their content upon secretory stimulation. This indicates that only a relatively small readily releasable pool of new secretory granules exists. Therefore, an important question is whether the secretory output of AKHs during flight would induce a stimulation of the rate of AKH biosynthesis. The mRNA levels of all three AKH pre-prohormones, however, did not appear to be affected by flight activity, while the rate of synthesis of AKH prohormones and AKHs was not affected either (Harthoorn *et al.*, 2001). Apparently, a coupling between release and biosynthesis of AKHs is absent. Inhibition of AKH biosynthesis *in vitro* by Brefeldin A, a specific blocker of the transport of newly synthesized secretory proteins from the endoplasmic reticulum to the Golgi complex, resulted in a considerable decrease in the release of AKHs induced by CCAP, highlighting once more that the regulated secretion of AKHs is completely dependent on the existence of a readily releasable pool of newly formed secretion granules (Harthoorn *et al.*, 2002). Therefore, we conclude that the strategy of the adipokinetic cells to cope with variations in secretory output of AKHs apparently is to rely on the continuous biosynthesis of AKHs, which produces a readily releasable pool that is sufficiently large and constantly replenished.

An important question remaining unanswered, however, is: what might be the rationale for the storage of such large quantities of hormones that are not accessible for secretory release?

9.2.3. Effect of Adipokinetic Hormones on Lipid Mobilization

9.2.3.1. Adipokinetic hormone receptors   Binding of the AKHs to their plasma membrane receptor(s) at the fat body cells is the primary step to the induction of signal transduction events that ultimately lead to the activation of target key enzymes and the mobilization of lipids as a fuel for flight. Although the AKHs constitute extensively studied neurohormones, and their actions have been shown to occur via G protein-coupled receptors (GPCRs) (for reviews, see Van Marrewijk and Van der Horst, 1998; Vroemen *et al.*, 1998), the general properties of which are remarkably well conserved during evolution (for review, see Vanden Broeck, 2001), insect AKH receptors have been identified only recently. However, in *L. migratoria*, which produces three different AKHs and may be envisaged to have (three) different AKH receptors, the receptor(s) are as yet unidentified. The first insect AKH receptors characterized at the molecular level, namely those of *D. melanogaster* and the silkworm *Bombyx mori* (Staubli *et al.*, 2002), were shown to be GPCRs structurally related to mammalian gonadotropin-releasing hormone (GnRH) receptors. No other AKH receptors were isolated until 2006, when an AKH receptor from the American cockroach *Periplaneta americana* was identified (Hansen *et al.*, 2006); the production of two intrinsic AKHs (*Periplaneta* AKH-I and -II) may suggest the presence of a second AKH receptor. A similar cockroach AKH receptor was also identified by Wicher *et al.* (2006); there are, however, differences in one amino acid residue, as well as in the response towards the two *Periplaneta* AKHs (*cf.* Hansen *et al.*, 2006). In the malaria mosquito *Anopheles gambiae* an AKH receptor has been characterized in addition to an orphan receptor, the close relationship of which to the insect AKH receptors identified thus far suggesting that this receptor is an AKH receptor as well (Belmont *et al.*, 2006). For the yellow fever mosquito *Aedes aegypti*, two splice variants of the AKH receptor gene, differing at their C-terminal ends, were reported (Kaufmann *et al.*, 2009); it was postulated that both receptor variants could selectively bind the two AKH peptides found in *Ae. aegypti*. The signaling of the AKH receptor of *B. mori* and its peptide ligands (*Bombyx* AKH1, -2 and -3) have been recently characterized at the molecular and functional levels (Zhu *et al.*, 2009). Recent cloning studies demonstrating that the GnRH receptor in the nematode *Caenorhabditis elegans* is stimulated by both a *C. elegans* AKH-GnRH-like peptide and *Drosophila* AKH suggest that the AKH-GnRH signaling system arose very early in metazoan evolution (Lindemans *et al.*, 2009).

9.2.3.2. Signal transduction of adipokinetic hormones
The signal transduction mechanism of the three locust AKHs has been studied extensively, and involves stimulation of cAMP production, which is dependent on the presence of extracellular Ca^{2+}. Additionally, the AKHs enhance the production of inositol phosphates, including inositol 1,4,5-trisphosphate (IP_3), which may mediate the mobilization of Ca^{2+} from intracellular stores. This depletion of Ca^{2+} from intracellular stores stimulates the influx of extracellular Ca^{2+}, indicative of the operation of a capacitative (store-operated) calcium entry mechanism. The interactions between the AKH signaling pathways ultimately result in mobilization of stored reserves as fuel for flight (for reviews, see Van Marrewijk and Van der Horst, 1998; Vroemen *et al.*, 1998; Ryan and Van der Horst, 2000; Van der Horst *et al.*, 2001; Van Marrewijk, 2003;

Van der Horst and Ryan, 2005). The concentration of DAG in the hemolymph increases progressively at the expense of stored TAG reserves in the fat body, which implies hormonal activation of the key enzyme, fat body TAG lipase. In a bioassay, all three AKHs are able to stimulate lipid mobilization, although their relative potencies are different. This recalls the concept of a hormonally redundant system involving multiple regulatory molecules with overlapping actions (for reviews, see Goldsworthy *et al.*, 1997; Vroemen *et al.*, 1998). Results obtained with combinations of two or three AKHs, which are likely to occur together in locust hemolymph under physiological conditions *in vivo*, revealed that the maximal responses for the lipid-mobilizing effects were much lower than the theoretically calculated responses based on dose–response curves for the individual hormones. In the lower (probably physiological) range, however, combinations of the AKHs were more effective than the theoretical values calculated from the responses elicited by the individual hormones (for review, see Van Marrewijk and Van der Horst, 1998).

The mechanism by which TAG lipase catalyzes AKH-controlled production of the DAG on which long-distance flight depends is only partly understood, mainly due to technical problems in isolating or activating the lipase. In vertebrates, hormone-sensitive lipase (HSL) and adipose TAG lipase (ATGL) are the key enzymes in the control of lipid mobilization from TAG stores in adipose tissue, and although contrary to insects, free fatty acids (FFA) are released into the blood for uptake and oxidation in muscle, there is a clear functional similarity between vertebrate adipose tissue HSL and ATGL on one hand, and insect fat body TAG lipase on the other (for reviews, see Ryan and Van der Horst, 2000; Van der Horst *et al.*, 2001; Van der Horst and Oudejans, 2003; Van der Horst and Ryan, 2005; Van der Horst and Rodenburg, 2011).

9.2.3.3. Activation of lipolysis in insect fat body
Regarding the process of lipid mobilization, recent data reveal insects to be very similar to mammals (for review, see Van der Horst and Rodenburg, 2011). For example, packaging lipid in intracellular lipid droplets and the mechanisms guiding mobilization of stored lipids are conserved between insects and mammals (Kulkarni and Perrimon, 2005; Martin and Parton, 2006; Brasaemle, 2007; Murphy *et al.*, 2009; Walther and Farese, 2009). Lipid droplets, which are progressively recognized to represent ubiquitous dynamic organelles regulating intracellular TAG metabolism, are surrounded by a phospholipid monolayer coated with specific proteins, belonging to the evolutionary ancient PAT (perilipin/ADRP/TIP47) family of proteins, that participate in the regulation of TAG storage and lipolysis (Martin and Parton, 2006; Brasaemle, 2007; Londos *et al.*, 1999; Miura *et al.*, 2002; Grönke *et al.*, 2003; Gross *et al.*, 2006; Arrese

et al., 2008a; Bickel *et al.*, 2009). Similar to mammalian adipocytes, the TAG accumulated in cytosolic lipid storage droplets of insect fat body cells provides the major long-term energy reserve of the organism, for which *Drosophila* recently emerged as a powerful system, to a large extent due to its well-developed genetics (Grönke *et al.*, 2003, 2005, 2007). Generation of loss-of-function mutants evidenced that simultaneous loss of the AKH receptor – and thus the signaling pathway for lipid mobilization, which is related to β-adrenergic signaling in mammals – and the lipid droplet-associated TAG lipase brummer (bmm), a homolog of human adipose TAG lipase (ATGL; for recent reviews, see Zechner *et al.*, 2009; Zimmermann *et al.*, 2009), caused extreme obesity and blocked acute storage fat mobilization in flies (Grönke *et al.*, 2005). Intriguingly, excessive fat storage in flies lacking *bmm* function reduced the median lifespan by only 10%, and acute TAG mobilization is impaired but not abolished in *bmm* mutants (Grönke *et al.*, 2005), suggesting that, as in mammals, mobilization of TAG in *Drosophila* is controlled by more than one TAG lipase (Grönke *et al.*, 2005; Kulkarni and Perrimon, 2005). In addition, *Akhr*null mutant flies appeared to be markedly starvation resistant, suggesting that their higher TAG content confers a survival benefit. Consequently, lipolytic mechanisms independent of the AKH pathway must exist in *Drosophila*, enabling *Akhr* mutants to mobilize TAG reserves, although they retain considerable energy stores as well when challenged with starvation (Bharucha *et al.*, 2008).

In addition to a similar TAG lipase, two lipid storage droplet (Lsd) proteins (Lsd1 and -2) belonging to the PAT protein family were identified in insects (Miura *et al.*, 2002; Grönke *et al.*, 2003; Teixera *et al.*, 2003: Arrese *et al.*, 2008a, 2008b, 2008c; Bickel *et al.*, 2009), suggesting that the overall processes of lipid storage and mobilization in insects may function similar to those in vertebrates. To further demonstrate the functional similarity between mammalian and *Drosophila* TAG lipases, the lipid droplet surface-localized bmm was shown to antagonize a perilipin-related lipid droplet surface protein (Lsd2) (Grönke *et al.*, 2005) that functions as an evolutionarily conserved modulator of lipolysis (Grönke *et al.*, 2003). Moreover, *Drosophila* key candidate genes for lipid droplet regulation were identified, the functions of which are conserved in the mouse. These include regulation of lipolysis by the vesicle-mediated Coat Protein Complex I (COPI) transport complex, required for limiting lipid storage by regulating the PAT protein composition and promoting the association of TAG lipase at the lipid droplet surface and composition (Beller *et al.*, 2008; Guo *et al.*, 2008).

In contrast to the mechanism of lipid mobilization in *Drosophila*, however, the main TAG lipase in the fat body of *M. sexta* was identified as the homolog of *D. melanogaster* GC8552. This protein, which was named triglyceride

lipase (TGL), is conserved among insects and also displays significant phospholipase A_1 activity (Arrese *et al.*, 2006; for review, see Arrese and Soulages, 2010). TGL shares significant sequence similarity with vertebrate phospholipases, but shows no homology to the main vertebrate adipose TAG lipase, ATGL.

In vertebrates, mobilization of TAG stores in adipose tissue is facilitated by the phosphorylation of several key proteins, including HSL and lipid droplet PAT proteins such as perilipin. The principal substrate for HSL is DAG, which is provided by the upstream ATGL (for reviews, see Watt and Steinberg, 2008; Zechner *et al.*, 2009). In insect fat body, AKH induces increased cAMP levels, which in turn may lead to increased PKA activity (reviewed in Van der Horst *et al.*, 2001; Gäde and Auerswald, 2003). Although the resulting PKA-mediated protein phosphorylation is considered a major factor in the activation of lipolysis (Arrese and Wells, 1994; for reviews, see Van der Horst *et al.*, 2001; Van der Horst and Ryan, 2005), *in vitro* studies showed the phosphorylation level of TGL in *M. sexta* fat body to be unchanged by AKH (Patel *et al.*, 2006). Instead, activation of lipid droplets by phosphorylation of Lsd1 was identified to mediate AKH-induced lipolysis (Arrese *et al.*, 2008b; for review, see Arrese and Soulages, 2010). Also in mammalian adipocytes, the PKA-mediated phosphorylation of perilipin at the surface of the lipid droplets is directly involved in the activation of lipolysis (Londos *et al.*, 2005) as mentioned above, and the phosphorylation of perilipin mediates the translocation of the likewise phosphorylated HSL to the surface of perilipin-coated lipid droplets (Sztalryd *et al.*, 2003; Wang *et al.*, 2009; for reviews, see Martin and Parton, 2006; Brasaemle, 2007; Walther and Farese, 2009; Bickel *et al.*, 2009).

In spite of the similarities in overall processes of lipid storage and mobilization in insects and mammals, however, both the transport form and the transport vehicle of the lipid substrate mobilized from the TAG stored in lipid droplets are different. During prolonged exercise of mammals, long-chain FFAs are mobilized from adipose tissue TAG stores and transported in the circulation bound to the abundant serum protein, albumin, for uptake and oxidation in the working muscles. However, in the locust and other insect species recruiting fat body TAG depots to power their flight muscles during migratory flight, the TAG-derived lipid is released as DAG into the hemolymph, as indicated above, and transported to the flight muscles in LDLp particles as discussed earlier (see section 9.1.2. and **Figure 1**).

It is interesting to note that in mammalian adipocytes ATGL is the predominant TAG lipase, whereas HSL and monoacylglycerol (MAG) lipase are the major lipases responsible for the hydrolysis of DAG and MAG, respectively. The net result of the consecutive actions of these three enzymes is the hydrolysis of a fatty acyl side chain from TAG, DAG, and MAG, and the release of the liberated FFAs and glycerol from the cells. The efflux of DAG from insect fat body cells following bmm action would suggest a lack of (the net activity of) the other downstream lipases found in adipocytes (for review, see Van der Horst and Rodenburg, 2010a).

L. migratoria DAG were shown to be stereospecific, revealing the *sn*-1,2 configuration, thus demonstrating stereospecific conversions to be involved in their production from TAG (for reviews, see Beenakkers *et al.*, 1985; Van der Horst, 1990). Data on the (nonapeptide) AKH-stimulated synthesis of *sn*-1,2-DAG in the fat body of *M. sexta* support the hypothesis of stereospecific hydrolysis of fat body TAG stores (Arrese and Wells, 1997; for reviews, see Gibbons *et al.*, 2000; Arrese *et al.*, 2001).

9.3. Apolipophorin III

9.3.1. Lipid-Free Helix Bundle Structure

ApoLp-III was discovered in the late 1970s and early 1980s by research groups in Europe and North America (for reviews, see Blacklock and Ryan, 1994; Ryan and Van der Horst, 2000; Weers and Ryan, 2006). ApoLp-III was first isolated from hemolymph of *L. migratoria* (Van der Horst *et al.*, 1984) and the tobacco hawkmoth, *M. sexta* (Kawooya *et al.*, 1984). *M. sexta* apoLp-III is a 166-aa protein that lacks tryptophan and cysteine (Cole *et al.*, 1987). However, the well-characterized apoLp-III from *L. migratoria* is 164 residues long, and lacks cysteine, methionine, and tyrosine (Kanost *et al.*, 1988; Smith *et al.*, 1994). *M. sexta* apoLp-III is non-glycosylated, while *L. migratoria* apoLp-III contains two complex carbohydrate chains (Hård *et al.*, 1993). The sequences of numerous apoLp-III have been reported, and important aspects are summarized in **Table 1**. Sequence analysis predicts that all apoLp-IIIs are composed of a predominantly amphipathic α-helix secondary structure, consistent with far ultraviolet circular dichroism (CD) studies (Ryan *et al.*, 1993; Weers *et al.*, 1998). An important breakthrough in our understanding of the structure of apoLp-III occurred with the determination of the X-ray crystal structure of *L. migratoria* apoLp-III (Breiter *et al.*, 1991). These authors showed that apoLp-III exists as a globular, up-and-down amphipathic α-helix bundle in the absence of lipid. The molecule is composed of five discrete α-helix segments that orient their hydrophobic faces toward the center of the bundle. Using a convenient method for bacterial overexpression, recombinant *M. sexta* apoLp-III was enriched with stable isotopes (Ryan *et al.*, 1995; Wang *et al.*, 1997a). Application of heteronuclear multidimensional nuclear magnetic resonance (NMR) techniques to isotopically enriched *M. sexta* apoLp-III yielded a complete assignment of this protein (Wang *et al.*, 1997b). Structure calculations revealed a five-helix bundle molecular architecture,

Table 1 Insect Species from which ApoLp-III has been Identified and Characterized*

ApoLp-III	Number of Residues, Mass	Glycosylation
Orthoptera		
Locusta migratoria	162–164, 20 kDa	14%
Gastrimargus africanus	20 kDa	5.3%
Acheta domesticus	161, 17.2 kDa	–
Barytettix psolus	20 kDa	5%
Melanoplus differentialis	20 kDa	5%
Lepidoptera		
Manduca sexta	166, 18.4 kDa	–
Bombyx mori	164, 18.3 kDa	–
Bombyx mandarina	164	–
Galleria mellonella	18.1 kDa	–
Spodoptera litura	166, 18.3 kDa	–
Acherontia atropos	20 kDa	–
Diatraea grandiosella	17 kDa	–
Heliothis virescens	18.0 kDa	–
Hyphantria cunea	165, 18.3 kDa	–
Hyalophora cecropia	18 kDa	–
Coleoptera		
Derobrachus geminatus	18 kDa	+ (% NA)
Hemiptera		
Lethocerus medius	19 kDa	–
Thasus acutangulus	20 kDa	–

*See Weers and Ryan (2006) for individual references.

representing the first full-length apolipoprotein whose high resolution solution structure has been determined in the absence of detergent (Wang *et al.*, 2002) (**Figure 3**). In keeping with the X-ray structure of *L. migratoria* apoLp-III, this structure also reveals an up-and-down bundle of five amphipathic α-helices. Interestingly, however, Wang and coworkers identified a distinct short segment of α-helix that connects helices 3 and 4 in the bundle (termed helix 3′). This sequence segment (P95DVEKE100) aligns perpendicular to the long axis of the bundle, and, as discussed below, has been shown to play a role in the initiation of apoLp-III lipid interaction. More recently, Fan *et al.* (2001, 2003) employed multidimensional NMR techniques to obtain a complete assignment and solution structure determination for *L. migratoria* apoLp-III. This work is significant in that it permits direct comparison between the X-ray crystal structure and the NMR structure. Interestingly, Fan and colleagues provide previously unreported structural evidence for a solvent exposed short helix that is positioned perpendicular to the long axis of the helix bundle. These authors propose that this short helix can serve as a recognition helix for initiation of apoLp-III lipid interaction, leading to conformational opening of the helix bundle in a manner that is different from the original proposal on the basis of the X-ray crystal structure of this protein. Another important aspect of this

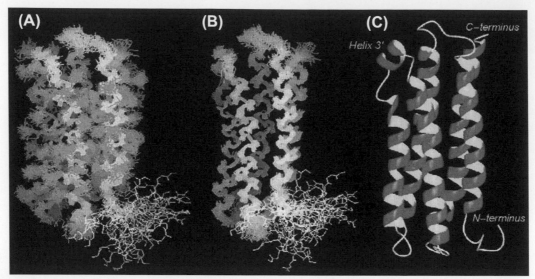

Figure 3 Nuclear magnetic resonance (NMR) visualizations of structure of lipid-free *Manduca sexta* apoLp-III. (A, B) Superposition of 40 NMR-derived structures of apoLp-III, with backbone atoms displayed in white and side chain heavy atoms displayed in green. (C) Ribbon representation of an energy-minimized, average structure of apoLp-III (PDB code 1EQ1). Reproduced with permission from Wang, J., Sykes, B.D., Ryan, R.O., 2002. Structural basis for the conformational adaptability of apolipophorin III, a helix bundle exchangeable apolipoprotein. *Proc. Natl. Acad. Sci. USA* 99, 1188–1193; ©National Academy of Sciences of the United States of America.

work is that buried interhelical H-bonds provide a driving force for the helix bundle recovery of apoLp-III from the lipid-bound open conformation.

9.3.2. Lipid-Induced Conformation Change

The up-and-down antiparallel organization of helical segments in apoLp-III allows for a simple opening of the bundle about putative "hinge" loops that connect the helices as originally proposed by Breiter *et al.* (1991). The model suggests that apoLp-III initiates contact with lipid surfaces via one end of the helix bundle. Conformational opening could then occur with retention of helix boundaries present in the bundle configuration. Such an event would result in substitution of helix–helix interactions in the bundle conformation for helix–lipid interactions. Current evidence suggests that this conformational transition is triggered by availability of a suitable lipid surface, and is reversible (Singh *et al.*, 1992; Liu *et al.*, 1993; Soulages and Wells, 1994b; Soulages *et al.*, 1995, 1996). Thus, it is conceivable that helices 3 and 4 move away from helices 1, 2, and 5 in concert as the bundle opens about the loop segments connecting helices 2 and 3, and helices 4 and 5 (Breiter *et al.*, 1991; Narayanaswami *et al.*, 1996a).

A well-known property of amphipathic exchangeable apolipoproteins in general is an ability to disrupt phospholipid bilayer vesicles and transform them into apolipoprotein–phospholipid disc complexes (Pownall *et al.*, 1978). This property provides a useful method to investigate aspects of the proposed lipid-induced helix bundle molecular switch process. The disc-shaped complexes

formed between apoLp-III and dimyristoylphosphatidylcholine (DMPC) are of uniform size and composition, permitting detailed analysis of their structural organization (Wientzek *et al.*, 1994). Attenuated total reflectance Fourier transformed infrared spectroscopy has been employed to characterize helix orientation in apoLp-III–DMPC disc complexes (Raussens *et al.*, 1995, 1996). This analysis, and more recent studies (Soulages and Arrese, 2001), reveal that apoLp-III helical segments interact with phospholipid fatty acyl chains around the perimeter of the disc complex.

Several independent studies have provided convincing evidence that apoLp-III undergoes a significant conformational change upon association with lipid. Kawooya *et al.* (1986) used a monolayer balance to investigate apoLp-III behavior at the air–water interface, while Narayanaswami *et al.* (1996b) studied the unique fluorescence properties of the lone tyrosine in *M. sexta* apoLp-III. Near-ultraviolet CD analysis of *L. migratoria* apoLp-III indicates that helix realignment and reorientation occurs upon interaction with phospholipid vesicles (Weers *et al.*, 1994). Sahoo *et al.* (2000) used pyrene excimer fluorescence spectroscopy to investigate lipid-binding induced realignment of helix 2 and helix 3 in *M. sexta* apoLp-III. In this study, cysteine residues were introduced into the protein by site-directed mutagenesis (N40C and L90C). These sites were selected for introduction of cysteine residues based on the fact that they reside in close proximity in the helix bundle conformation. Covalent modification of the cysteine thiol groups with pyrene maleimide yielded a double pyrene labeled apoLp-III. In the absence of lipid, pyrene labeled apoLp-III adopts a helix bundle conformation. Fluorescence spectroscopy experiments revealed normal pyrene emission at 375 and 395 nm (excitation 345 nm) as well as excimer (excited state dimer) fluorescence at longer wavelengths (460 nm). Control experiments verified that the excimer peak arose from intramolecular pyrene–pyrene interactions in the labeled protein, and was not due to intermolecular interactions. Because it is known that excimer fluorescence is manifest only when pyrene moieties are within 10 Å of one another, this property was used to assess the effect of lipid binding. The observation that excimer fluorescence was greatly reduced when apoLp-III was complexed with DMPC was taken as evidence for a conformational change in the protein upon lipid binding that results in relocation of helix 2 away from helix 3.

In fluorescence studies of apoLp-III, carried out by Soulages and Arrese (2000a, 2000b), site-directed mutagenesis was used to create various mutant apoLp-IIIs with a single tryptophen residue in each of the five helical segments of the protein. Data obtained in this study suggest that apoLp-III undergoes a conformational change that brings helices 1, 4, and 5 into contact with the lipid surface, while others (helices 2 and 3) appear to behave differently. In other studies, Soulages et al. (2001) used disulfide bond engineering to show that conformational flexibility of helices 1 and 5 of *L. migratoria* apoLp-III plays an important role in the lipid-binding process. Dettloff et al. (2001a) reported that a C-terminal truncated apoLp-III from the wax moth *Galleria mellonella*, comprising the first three helical segments of the protein, retains structural integrity and an ability to interact with lipid surfaces. More recently, Dettloff et al. (2002)

expanded this work to encompass two additional three-helix mutants derived from *G. mellonella* apoLp-III, a C-terminal fragment comprising helices 3–5, and a core fragment comprising helices 2–4. All three truncation mutants retained their ability to solubilize bilayer vesicles of DMPC – an event that led to large increases in their α-helix content. The N-terminal and core fragment, but not the C-terminal fragment, were able to interact with phospholipase C modified human low-density lipoprotein, thereby preventing its aggregation. This result suggests that impairment of the lipid interaction properties of the C-terminal fragment has occurred as a result of removal of N-terminal helix segments. Taken together, it appears that the minimal essential elements required for apoLp-III lipid-binding function are less than the intact five-helix bundle. Recent experiments have provided evidence that opening of the helix bundle is even more dramatic than originally postulated. It is now proposed that the protein adopts a fully extended belt-like conformation (Garda et al., 2002; Sahoo et al., 2002) (**Figure 4**). Garda et al. (2002) employed fluorescence resonance energy transfer methods, while Sahoo et al. (2002) used pyrene excimer fluorescence to probe aspects of helix repositioning upon interaction with DMPC. In both of these studies, knowledge of the three-dimensional structure of the apoLp-III in the absence of lipid (i.e., the helix bundle conformation) allowed for structure-guided site-directed mutagenesis to introduce strategically placed cysteine residues to which fluorescent reporter groups could be covalently attached. Subsequent characterization studies yielded a unifying model of apoLp-III conformation on disc complexes wherein the resulting structure resembles

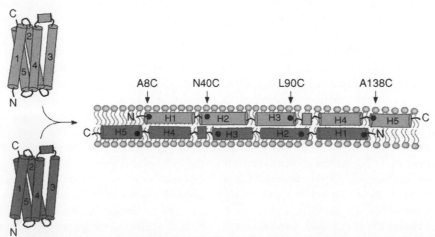

Figure 4 Model of apoLp-III bound to phospholipid discoidal complexes. ApoLp-III complexes with phospholipids on a discoidal particle adopting an extended α-helical conformation. Lipid-triggered association involves extension of H1 away from H5, helix bundle opening and repositioning of H2 and H3. The positions of cysteine substitution mutations employed in this and previous analyses are indicated: A8C, N40C, L90C, and A138C; H1, H2, H3, and H5. Apolp-III adopts an extended helical conformation around the periphery of discoidal phospholipid bilayer complexes, with neighboring molecules aligned antiparallel with respect to each other, and shifted by one helix.

Reprinted with permission from Sahoo, D., Weers, P.M.M., Ryan, R.O., Narayanaswami, V., 2002. Lipid-triggered conformational switch of apolipophorin III helix bundle to an extended helix organization. *J. Mol. Biol.* 321, 201–214; ©Elsevier.)

concepts and models that describe the organization of human apolipoprotein A-I on nascent high-density lipoproteins (Klon *et al.*, 2002).

Recently, Vasquez *et al.* (2009) employed *L. migratoria* apoLp-III as a model to study apolipoprotein lipid-binding interactions. To investigate the role of positive charges on lipid binding, lysine residues in apoLp-III were acetylated. Modified apoLp-III possessed a reduced amount of α-helix, and a slight increase in protein stability. While the ability to solubilize DMPC vesicles was unchanged, the rate of anionic dimyristoylphosphatidylglycerol (DMPG) vesicle solubilization was reduced twofold. These results indicate that the eight lysine residues in *L. migratoria* apoLp-III are not required for the protein's ability to bind zwitterionic phospholipids, but are required for optimal binding to anionic lipid surfaces, likely through an electrostatic effect, and this may provide a means of modulating apoLp-III interaction with complex membrane environments *in vivo*. Chiu *et al.* (2009) employed differential scanning calorimetry to measure the binding interaction of *L. migratoria* apoLp-III with liposomes composed of mixtures of lipids. Association of apoLp-III with multilamellar liposomes occurred over a temperature range around the liquid crystalline phase transition of the the lipid in question. Thus, surface defects arising from non-ideal packing at the lipid phase transition temperature influence apoLp-III binding properties.

9.3.3. Initiation of ApoLp-III Lipid Binding

Analysis of the structure of *L. migratoria* apoLp-III indicates the presence of solvent-exposed leucine residues at one end of the protein (Breiter *et al.*, 1991). These authors proposed that this region of the molecule functions as a "hydrophobic sensor" that recognizes potential lipid surface binding sites. Surface plasmon resonance spectroscopy studies revealed that small amounts of DAG induce binding of apoLp-III to a phospholipid bilayer with its long molecular axis normal to the lipid surface (Soulages *et al.*, 1995). This interaction is proposed to be the first step in formation of a stable binding interaction. Site-directed mutagenesis was performed to determine whether alteration in the hydrophobicity of the putative sensor region of *L. migratoria* apoLp-III affects its ability to initiate contact with lipid surfaces (Weers *et al.*, 1999). In this study three partially exposed leucine residues, located at the end of the protein containing the loop segments that connect helices 1 and 2 and helices 3 and 4, were mutated to arginine. Three single arginine-to-leucine substitution mutants and a triple mutant were expressed in *Escherichia coli*, and characterized in terms of their structural and stability properties. The effect of these mutations on phospholipid bilayer vesicle transformation into disc complexes versus lipoprotein binding suggests that the former binding interaction has an electrostatic

component. Taken together, the data support the view that the end of the molecule bearing Leu 32, 34, and 95 is responsible for initiating contact with potential lipid surface binding sites.

The solution structure of *M. sexta* apoLp-III revealed the presence of helix 3′ at one end of the protein globule (Wang *et al.*, 1997b, 2002). One possibility is that helix 3′ reorientation facilitates contact with a lipid surface by exposing the hydrophobic interior of the helix bundle. The lipid surface could then trigger a molecular switch to induce conformational opening of the helix bundle and formation of a stable binding interaction. To investigate this, protein engineering was employed to remove helix 3′ and replace it with a sequence that has a high probability of forming a β-turn (Narayanaswami *et al.*, 1999). Characterization of the lipid-binding properties of this "helix-to-turn" mutant apoLp-III revealed defective lipid-binding properties. In more refined site-directed mutagenesis studies, it was determined that Val 97, located in the center of helix 3′, is a critical residue for initiation of apoLp-III lipid binding. As described above, a similar short helix was identified in *L. migratoria* apoLp-III based on its solution structure (Fan *et al.*, 2003). This helix, however, is present at the opposite end of the apoLp-III helix bundle, suggesting that, if it is a recognition helix, bundle opening is different from that proposed for *M. sexta* apoLp-III by Narayanaswami *et al.* (1999) and Wang *et al.* (2002).

The role of the conformational flexibility of helices and loops in *L. migratoria* apoLp-III in lipid-binding activity was investigated by disulfide bond engineering experiments (Chetty *et al.*, 2003). The ability of helix-tethered apoLp-III mutants to interact with phospholipid vesicles, mixed micelles, and spherical lipoprotein particles was studied. The authors determined that: (1) opening of the helix bundle does not require the separation of loops 2 and 4; (2) α-helices 3 and/or 4 are involved in the insertion of apoLp-III in both phospholipid bilayers and monolayers; and (3) interaction of helices 1 and/or 5 with the lipid surface is required for the formation of stable lipoprotein complexes. In *L. migratoria* apoLp-III, most hydrophobic residues are buried in the protein interior. However, it was postulated that the presence of polar residues in the hydrophobic protein interior may contribute to protein instability and lipid-binding induced conformational opening (Weers *et al.*, 2005). To test this, Thr-31 was changed to alanine by site-directed mutagenesis. Lipid-binding studies using phospholipid vesicles showed that Thr31Ala apoLp-III was able to transform phospholipid vesicles into discoidal particles, but at a three-fold reduced rate compared to wild type apoLp-III. In contrast, less stable apoLp-III mutants displayed an increased ability to transform phospholipid vesicles. Thus, an inverse correlation exists between protein stability and phospholipid vesicle transformation activity. Furthermore, these data suggest that Thr-31 is a key determinant of apoLp-III lipid-binding activity.

Studies of the effect of the glycosyl moieties of *L. migratoria* apoLp-III on its lipid-binding properties have also been investigated. Soulages *et al.* (1998) showed that recombinant apoLp-III, which lacks covalently bound carbohydrate, displayed a much stronger interaction with phospholipid vesicles than natural insect-derived apoLp-III. From the X-ray structure of *L. migratoria* apoLp-III in the absence of lipid, it is known that both glycosylation sites (at residues 18 and 85) are localized in the central region of the long axis of the bundle. Further study of this phenomenon revealed that apoLp-III sugar moieties interfere with helix bundle penetration into the bilayer surface during disruption and transformation into disc complexes (Weers *et al.*, 2000). Thus, it is apparent that structural aspects of the helix bundle as well as the composition of the lipid surface influence the ability of apoLp-III to initiate and form a stable lipid-binding interaction.

9.3.4. ApoLp-III Alternate Functions

Based on the developmentally timed upregulation of its mRNA, apoLp-III has been implicated in muscle and neuron programmed cell death (Sun *et al.*, 1995). When considered in light of its known lipid interaction properties, it is conceivable that apoLp-III functions in membrane dissolution and/or lipid reabsorption during metamorphosis. ApoLp-III has also been identified as a hemagglutinating agent in larval hemolymph of *G. mellonella* (Ishikawa *et al.*, 1998). It interacts with lipoteichoic acids and with surface components of Gram-positive bacteria (Halwani *et al.*, 2000), binds to fungal conidia and β-1,3-glucan (Whitten *et al.*, 2004), and has been implicated in the detoxification of lipopolysaccharide (LPS) (Dunphy and Halwani, 1997). Given its recognition and binding properties, apoLp-III may serve a surveillance role as a pattern recognition receptor (Kanost *et al.*, 2004). Since its function as a lipid transport protein is required only when large amounts of lipids are transported, a large pool of apoLp-III may be available for immediate protection against foreign invaders. Indeed, such a role could explain the presence of apoLp-III in hemolymph during life stages wherein its association with lipophorin is not required or does not occur.

It has been reported that apoLp-III functions in insect immunity (Wiesner *et al.*, 1997). Indeed, results suggest that lipid-associated apoLp-III manifests this biological activity (Dettloff *et al.*, 2001b, 2001c). These authors hypothesized that LDLp serves as an endogenous signal for immune activation, specifically mediated by apoLp-III interaction with hemocyte membrane receptors. From a structural standpoint, truncated variants of *G. mellonella* apoLp-III (see above) that retain function represent useful tools to probe the structural and physiological roles of

apoLp-III in innate immunity. Support for this general concept has emerged from studies of *G. mellonella* apoLp-III variants wherein point mutations were introduced at residues 66 and 68 (Niere *et al.*, 2001). The observation that mutation-induced decreases in apoLp-III lipid interaction properties correlate with decreased immune-inducing activity is consistent with the hypothesis that apoLp-III immune activation is related to the conformational change that accompanies lipid interaction of this protein.

Fluorescence studies, exploiting the unique tyrosine residue in *G. mellonella* apoLp-III (Tyr-142), provided additional evidence for LPS interaction (Pratt and Weers, 2004). In the absence of lipid, Tyr-142 fluorescence is quenched. However, upon incubation with LPS or detoxified LPS, Tyr-142 fluorescence intensity is enhanced, indicating relocation of this residue to a new environment. Dissociation constants (K_d) measured by apoLp-III titration were estimated at approximately 1 μM (Leon *et al.*, 2006a). Increasing the ionic strength had no effect on the K_d, and nor did removal of LPS phosphate moieties. Interestingly, a truncated apoLp-III variant missing two complete helices retained LPS-binding activity. To further investigate the structural requirements for LPS binding, Leon *et al.* (2006b) used a protein engineering approach. These authors introduced two cysteine residues in close spatial proximity, resulting in disulfide bond formation between the residues. Tethering the helix bundle in this manner abolished the ability of apoLp-III to undergo conformational opening. Furthermore, tethered apoLp-III was unable to bind LPS. Disulfide bond reduction, however, restored helix bundle opening and LPS-binding capability. It may be concluded that helix bundle opening is required for LPS binding, and, since the interior of the bundle is hydrophobic, it appears that apoLp-III–LPS interactions are governed by hydrophobic interactions.

Although they lack an adaptive immune system, insects possess an innate immune system that recognizes and destroys intruding microorganisms. Its operation under natural conditions has not been well studied, as most studies have introduced microbes to laboratory-reared insects via artificial mechanical wounding. One of the most common routes of natural exposure and infection, however, is via food. Freitak *et al.* (2007) examined the immune system response to non-infectious microorganisms via simple oral consumption. Eight proteins were highly expressed in the hemolymph of the larvae fed bacteria including apoLp-III. Studies in the fall webworm, *Hyphantria cunea*, also demonstrated an immune activation role for apoLp-III, including induction of antimicrobial peptide expression and *E. coli* membrane binding (Kim *et al.*, 2004). Changes in apoLp-III concentration and lipophorin interconversions after immune challenge are known (Dettloff *et al.*, 2001a; Mullen and Goldsworthy, 2003), and it has been suggested that lipid-bound apoLp-III acts as the immune activator, while lipid-free apoLp-III does not display this activity (Dettloff *et al.*, 2001b; Niere *et al.*, 2001; Whitten *et al.*, 2004). The reason for this is unclear, but a ligand-induced conformational change may create new structural features on the surface of apoLp-III that may form a signal for secondary immune responses.

Giannoulis and colleagues (2007) studied the effects of LPS and lipoteichoic acid on the immune response of *Malacosoma disstria*. LPS induced an increase in the number of damaged hemocytes, and limited removal of the entomopathogenic bacterium *Xenorhabdus nematophila* from hemolymph. Similar effects were observed with the lipid A moiety of LPS. At the same time, the effects of LPS and lipid A on hemocyte were abrogated by polymyxin B, an antibiotic that binds to lipid A, confirming that LPS is the hemocytotoxin. Lipoteichoic acid elicited nodulation and enhanced phenoloxidase activation. Importantly, apoLp-III interfered with the effects of LPS, lipid A, and lipoteichoic acid on hemocytes and prophenoloxidase until a critical threshold was exceeded.

The potential for two roles for apoLp-III raises several issues. For example, Adamo *et al.* (2008) tested the hypothesis that competition between apoLp-III immune function and lipid transport causes transient immunosuppression in crickets. Both flying and an immune challenge reduced the amount of free apoLp-III in hemolymph. The authors showed that immune function is sensitive to the amount of free apoLp-III in hemolymph. Reducing the amount of free apoLp-III in hemolymph, using AKH, produced immunosuppression. Likewise, increasing apoLp-III levels after flight by pre-loading insects with trehalose, or by injecting isolated apoLp-III, reduced immunosuppression. Thus, it appears that competition between lipid transport and immune function can lead to transient immunosuppression.

Seo *et al.* (2008) studied the antioxidant properties of apoLp-III and compared them to its mammalian counterpart, apolipoprotein A-I (apoA-I). In order to compare the antioxidant abilities of apoLp-III and apoA-I in the lipid-free and lipid-bound state, both proteins were purified and synthesized individually as a palmitoyloleoyl phosphatidylcholine (POPC)-reconstituted high-density lipoprotein (rHDL) using the same molar ratio. In the lipid-bound state, apoLp-III and apoA-I showed good antioxidant activities against copper-mediated low density lipoprotein (LDL) oxidation. However, while lipid-free apoA-I treatment prevented cellular uptake of oxLDL in macrophages, lipid-free apoLp-III did not. These results indicate that the putative conformational change of apoLp-III during lipid association is critical for the maintenance of antioxidant activity, and that the physiologic role of apoLp-III may differ when it is in the lipid-free state and the lipid-bound state.

On a broader scale, it is important to understand the molecular details of this family of proteins (Narayanaswami

and Ryan, 2000) because their property of reversible interconversion between water-soluble and lipid-bound states could have application beyond their natural biological settings. Indeed, as work on this system continues, it is evident that apoLp-III and analogous helix bundle apolipoproteins represent novel biosurfactants with potentially useful properties, including biodegradability. An example of such an application is the use of apoLp-III to generate functional nanolipoprotein particles (NLPs; Chromy *et al.*, 2007). These authors made NLPs from DMPC in combination with apoLp-III or mammalian apolipoproteins. Predominately discoidal in shape, these particles have diameters of 10–20 nm, and a uniform height of 4.5–5 nm. The apolipoprotein employed, the lipid-to-lipoprotein ratio, and the assembly parameters affect the size and homogeneity of NLPs generated. These particles have myriad potential uses, including membrane protein solubilization (Bayburt and Sligar, 2010), drug delivery (Ryan, 2008), and diagnostic imaging (Cormode *et al.*, 2009).

9.4. Lipophorin Receptor Interactions

9.4.1. Receptor-Mediated Endocytosis of Lipophorin

In the concept of lipid transport during intense lipid utilization in insects, a major difference between the functioning of lipoproteins of mammals and insects is the selective mechanism by which insect lipoproteins transfer their hydrophobic cargo. Circulating HDLp particles may serve as a DAG donor or acceptor, dependent on the physiological situation, and function as a reusable lipid shuttle without additional synthesis or increased degradation of the apolipoprotein matrix, as discussed above. In apparent contrast to this concept, in fat body tissue of larval and young adult locusts, receptor-mediated uptake of HDLp was demonstrated (Dantuma *et al.*, 1997). A receptor has been cloned and sequenced from locust fat body cDNA, and identified as a novel member of the LDL receptor (LDLR) family, which is particularly expressed in fat body, oocytes, midgut, and the brain (Dantuma *et al.*, 1999). When stably transfected in an LDLR-deficient Chinese hamster ovary (CHO) cell line, the locust lipophorin receptor (LpR) mediated endocytic uptake of fluorescently labeled HDLp that was absent in mock-transfected cells, suggesting that the receptor may function *in vivo* as an endocytic receptor for HDLp (Dantuma *et al.*, 1999). To date, the LpR sequences of several other insect species have been elucidated (Cheon *et al.*, 2001; Lee *et al.*, 2003a, 2003b; Seo *et al.*, 2003; Gopalapillai *et al.*, 2006; Ciudad *et al.*, 2007; Guidugli-Lazzarini *et al.*, 2008; Tufail *et al.*, 2009; for review, see Tufail and Takeda, 2009). Domain organization of LpR is identical to that of mammalian LDLR (Dantuma *et al.*, 1999). However, the ligand-binding domain of LpR contains one additional cysteine-rich repeat compared to the seven repeats in LDLR, and is therefore identical to that of the human very low-density lipoprotein (VLDL) receptor (VLDLR), which also contains eight consecutive cysteine-rich repeats in this domain (schematically depicted in **Figure 5**). The amino acid sequence of the longer cytoplasmic tail of LpR is unique for insect lipophorin receptors: the 12 C-terminal amino acid residues of LDLR are completely different from those of LpR, whereas the C-terminal tail of LpR contains an additional 10 amino acid residues (Van Hoof *et al.*, 2002; Rodenburg *et al.*, 2006). Three-dimensional models of the elements representing both the ligand-binding domain and the epidermal growth factor precursor homology domain of locust LpR bear a striking resemblance to those of mammalian LDLR (Van der Horst *et al.*, 2002). Despite their pronounced structural similarity, however, the ligand specificity of LpR and LDLR for lipophorin and LDL, respectively, is mutually exclusive (Van Hoof *et al.*, 2002). Additionally, the functioning of both receptors in lipid transport in insects and mammals

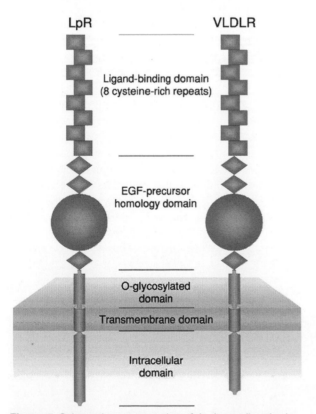

Figure 5 Schematic representation of the insect lipophorin receptor (LpR) and the mammalian VLDL receptor (VLDLR), indicating the identical domain organization. The mammalian LDL receptor has the same organization, but one less ligand-binding repeat. EGF, epidermal growth factor.
Based on data from Dantuma *et al.*, 1999.
Reprinted with permission from Van der Horst D.J., Ryan R.O., 2005. Lipid transport. In: Gilbert, L.I., Iatrou, K., Gill, S.S. (eds), *Comprehensive Molecular Insect Science*, Vol. 4. Elsevier, Amsterdam, pp. 225–246.

appears to be intriguingly different (ligand recycling *versus* ligand degradation), as discussed below in more detail. Possibly, these specific properties may be attributable to relatively small structural differences governing different properties of ligand binding and/or release.

Interaction of HDLp with a specific high-affinity binding site or receptor in the cell membrane of the fat body and other tissues of several insect species has been well documented (for reviews, see Ryan and Van der Horst, 2000; Van der Horst and Ryan, 2005). In the binding of human LDL to its receptor, the most C-terminal 1000 amino acids in apoB are involved (Borén *et al.*, 1998). Remarkably, although both the sequence and domain structure of insect apoLp-II/I resemble that of apoB100 (Babin *et al.*, 1999; Smolenaars *et al.*, 2007b; for reviews, see Van der Horst *et al.*, 2002, 2009; Van der Horst and Rodenburg, 2010a), apoLp-II/I does not show homology to this C-terminal part of apoB100, leaving the receptor-binding domain of apoLp-II/I to be disclosed.

Immunocytochemical localization of HDLp has demonstrated the presence of the lipoprotein in endosomes of fat body of the larval dragonfly *Aeshna cyanea* (Bauerfeind and Komnick, 1992) and in developing mosquito oocytes (Sun *et al.*, 2000), suggesting endocytosis of circulating HDLp. In addition, the uptake of HDLp in the fat body of young adult locusts was shown to be receptor mediated (Dantuma *et al.*, 1997). Van Hoof *et al.* (2003) presented evidence for the involvement of LpR in the endocytic uptake mechanism for HDLp in the locust that is temporally present during specific periods of development. Shortly after ecdysis, when lipid reserves are depleted, LpR is expressed in fat body tissue of young adult locusts as well as larvae, and fat body cells are able to endocytose the complete HDLp particle. On the fourth day after (larval or imaginal) ecdysis, however, expression of LpR is downregulated and drops below detectable levels; consequently, HDLp is no longer internalized. Downregulation of LpR was postponed by experimental starvation of adult locusts immediately after ecdysis. Moreover, by starving adult locusts after downregulation of LpR, expression of the receptor was re-induced. These data suggest that LpR expression is regulated by a deficiency of lipid components in fat body tissue (Van Hoof *et al.*, 2003). Receptor-mediated endocytosis of HDLp might therefore provide a mechanism for uptake of specific lipid components, independent of the mechanism of selective unloading of the lipid cargo of circulatory HDLp particles at the cell surface.

On the other hand, in contrast to specific uptake of lipid components by receptor-mediated endocytosis of HDLp, experiments using HDLp that was partially delipidated *in vitro*, yielding a particle of buoyant density 1.17 g/ml, indicated that LpR favors the binding of this lipid-unloaded HDLp over HDLp of normal density. The latter data would suggest a preferential mechanism for the intracellular loading of specific fat body lipid components onto relatively lipid-poor HDLp, while the lipid loading of the particle additionally results in decreased affinity for LpR (Roosendaal *et al.*, 2009), and thus facilitates the process of HDLp recycling (discussed in the following section).

9.4.2. Lipophorin Receptor-Mediated Ligand Recycling

Receptor-mediated uptake of HDLp in newly ecdysed adult and larval locusts may provide a mechanism for the uptake of specific lipid components, in addition to the mechanism of selective unloading of HDLp lipid cargo at the cell surface. However, the downregulation of LpR expression in fat body cells after these developmental periods suggests that this receptor is not involved in the lipophorin shuttle mechanism operative in the flying insect. Nevertheless, an endocytic uptake of HDLp seems to conflict with the selective process of lipid transport between HDLp and fat body cells without degradation of the lipophorin matrix. However, the pathway followed by the internalized HDLp appears to be different from the classical receptor-mediated lysosomal pathway typical of LDLR-internalized ligands.

In mammalian cells, LDL and diferric transferrin have been used extensively to study intracellular transport of ligands that are internalized by receptor-mediated endocytosis (Goldstein *et al.*, 1985; Brown and Goldstein, 1986; Mellman, 1996; Mukherjee *et al.*, 1997). Whereas LDL dissociates from its receptor upon delivery to the low pH milieu of the endosome and is completely degraded in lysosomes (for reviews, see Jeon and Blacklow, 2005; Beglova and Blacklow, 2005), transferrin remains attached to its receptor and, following the unloading of its two

iron ions, is eventually re-secreted from the cells (Ghosh *et al.*, 1994; Maxfield and McGraw, 2004). The endocytic uptake and intracellular trafficking of fluorescently labeled locust HDLp were studied simultaneously with fluorescently labeled human LDL or transferrin in LDLR-expressing CHO cells transfected with LpR cDNA, by multicolor confocal laser scanning microscopy, and provided evidence for different intracellular routes followed by the mammalian and insect lipoproteins (Van Hoof *et al.*, 2002) (**Figure 6**). Both HDLp and LDL appeared to co-localize to the same early endocytic vesicle structures. However, whereas LDL was eventually degraded in lysosomes after dissociating from its receptor, HDLp remained coupled to LpR and was transported to a non-lysosomal juxtanuclear compartment. Co-localization of HDLp with transferrin (Van Hoof *et al.*, 2002) (**Figure 6**) identified this organelle as the endocytic recycling compartment (ERC), from which internalized HDLp was eventually resecreted (half-life ~13 minutes) in a manner similar to that operative in the transferrin recycling pathway, thus escaping from lysosomal degradation.

The above data indicate that, in mammalian cells, endocytosed insect HDLp, in contrast to human LDL, follows a recycling pathway mediated by LpR. Although this behavior of LpR in mammalian cells proposes a novel function of an LDLR family member, recycling of endocytosed HDLp in insect fat body cells remained to be shown. Since a locust fat body cell line is not available, fat body tissue from young adults after ecdysis, endogenously expressing LpR, was used for tracking the intracellular pathway of fluorescently labeled HDLp. The lipoprotein appeared not to be transported to a recognizable ERC-like compartment, but remained in vesicles in the periphery of the cell from which, during a chase, the labeled HDLp disappeared almost completely (Van Hoof *et al.*, 2005a),

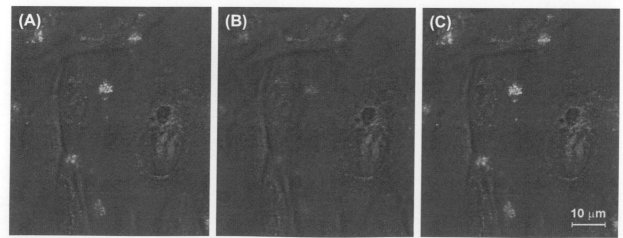

Figure 6 Confocal laser microscopic digital image of Chinese hamster ovary cells incubated with fluorescently labeled HDLp (A) and transferrin (B) after a chase period of 20 min. Co-localization of both ligands is visualized in yellow when images (A) and (B) are merged (C).

Reproduced with permission from Van der Horst, D.J., Van Hoof, D., Van Marrewijk, W.J.A., Rodenburg, K.W., 2002. Alternative lipid mobilization: the insect shuttle system. *Mol. Cell. Biochem.* 239, 113–119; ©Kluwer Academic Publishers.

indicative of re-secretion of the ligand and thus supporting the concept of ligand recycling that was demonstrated for LpR-transfected mammalian cells (Van Hoof *et al.*, 2002).

The above concept, which implies that lipophorin is recycled following endocytosis by an LDLR family member, conflicts with the generally accepted fate of ligands endocytosed by all other LDLR family members. Binding assays using flow cytometry demonstrated that, in contrast to the LDL–LDLR complex, HDLp and LpR remain in the complex at endosomal pH (Roosendaal *et al.*, 2008). Since, in addition to pH lowering, the Ca^{2+} concentration in the early endosome is also lowered to the low micromolar range (Gerasimenko *et al.*, 1998), the HDLp–LpR complex was treated with an EDTA-containing buffer to mimic the effect of the low Ca^{2+} concentration in the endosome. This treatment did not induce complex dissociation either, once more in contrast to the effect of EDTA treatment on the LDL–LDLR complex (Roosendaal *et al.*, 2008). These results indicate that endocytic conditions fail to induce dissociation of the complex, and imply that HDLp and LpR remain in complex throughout the itinerary from the early endosome to the ERC (Roosendaal *et al.*, 2008). This remarkable stability of the ligand–receptor complex is likely to provide a crucial key to the recycling mechanism.

Extensive studies have proposed that LDLR releases LDL at endosomal pH by undergoing a conformational change, in which the β-propeller of LDLR interacts with the ligand-binding domain, resulting in displacement of LDL (Herz, 2001; Rudenko *et al.*, 2002; Innerarity, 2002; Beglova *et al.*, 2004a). Sequence alignment of the amino acid sequence of LDLR with that of LpR revealed that a number of residues crucial for LDL release by LDLR (Beglova *et al.*, 2004a, 2004b; Boswell *et al.*, 2004), notably Gln540, His 562, Glu581, and Lys582, are not conserved in LpR. Changing the complete ligand-binding domain of LpR for that of LDLR ($LDLR_{1–292}LpR_{343–850}$) (Van Hoof *et al.*, 2005b) resulted in a hybrid receptor that was able to bind LDL but unable to release this ligand at endosomal pH, suggesting that the lack of Gln540, His 562, Glu581, and Lys582 renders the β-propeller of LpR indeed incapable of inducing LDL release and causes the lack of HDLp release by LpR. However, the inverse hybrid, in which the β-propeller of LDLR was introduced into LpR, did not lead to release of HDLp by this hybrid receptor either, implying that other domains produce the remarkable stability of the complex (Van Hoof *et al.*, 2005b). In LDLR, the interface between the most C-terminal LA-repeat (LA-7) and the adjacent cysteine-rich repeat of the EGF domain (EGF-A), the hinge region, has additionally been proposed to play an important role in LDL release by functioning as a rigid scaffold that allows the β-propeller to fold over the ligand-binding domain (Beglova *et al.*, 2004a, 2004b; Jeon and Blacklow, 2005). Potentially crucial residues in the hinge region of LDLR

(His264, Ser265, and Ile313) are not conserved in LpR, and might abolish ligand release by increasing the flexibility of the hinge region. However, a hybrid LpR in which both the hinge region and β-propeller of LDLR were introduced ($LpR_{1–301}LDLR_{252–839}$) failed once more to release HDLp, in spite of the fact that this hybrid contains all the domains that LDLR brings into action for LDL release. Consequently, since these functional LDLR domains appeared unable to evoke HDLp release, the lack of dissociation of the HDLp–LpR complex was proposed to result from the specific binding interaction of the ligand-binding domain of LpR with HDLp, which may be different from that used by other LDLR family members for the interaction with their ligands (Roosendaal *et al.*, 2008). However, the molecular mechanism for the stability of the HDLp–LpR complex awaits disclosure.

Additionally, even though the acidic endosomal environment of endocytosed HDLp has been postulated to facilitate the transfer of lipid components other than DAG or cholesterol (Dantuma *et al.*, 1997), both the precise function of the process of receptor-mediated endocytosis and the rationale for its occurrence during specific stages of insect development remain to be elucidated.

9.5. Other Lipid-Binding Proteins

9.5.1. Lipid Transfer Particle

In mammals, specialized proteins function in redistribution of hydrophobic lipid molecules. A wide variety of distinct lipid transfer proteins have been characterized, and their metabolic roles investigated. In 1986, a lipid transfer particle (LTP) was isolated from *M. sexta* larvae and shown to facilitate vectorial redistribution of lipids among plasma lipophorin subspecies (Ryan *et al.*, 1986a, 1986b). In subsequent studies, LTP was implicated in the formation of LDLp from HDLp in response to AKH (Van Heusden and Law, 1989). The concept that LTP functions in flight-related lipophorin conversions correlates well with observed increases in LTP concentration in adult hemolymph compared with other developmental stages (Van Heusden *et al.*, 1996; Tsuchida *et al.*, 1998). When compared to other lipid transfer proteins, however, LTP displays unique structural characteristics. For example, it exists as a high molecular weight complex of three apoproteins (apoLTP-I, 320,000 kDa; apoLTP-II, 85,000 kDa; and apoLTP-III, 55,000 kDa) and 14% non-covalently associated lipid (Ryan *et al.*, 1988). LTPs exhibiting similar structural properties have been isolated from *L. migratoria*, *P. americana*, *B. mori*, and *Rhodnius prolixus* hemolymph (Hirayama and Chino, 1990; Takeuchi and Chino, 1993; Tsuchida *et al.*, 1997; Golodne *et al.*, 2001). The large size of LTP permitted examination of its structural properties by negative stain electron microscopy (Ryan *et al.*, 1990a; Takeuchi and Chino, 1993). LTP from distinct species displays a highly asymmetric morphology with two major structural features: a quasispherical head region, and an elongated cylindrical tail that appears to possess a central hinge. The lipid component resembles that of lipophorin in that it contains predominantly phospholipid and DAG (Ryan, 1990). An important question arising from these physical characteristics relates to the requirement of the lipid component as a structural entity and/or its involvement in catalysis of lipid transfer. Studies employing lipoproteins containing radiolabeled lipids in incubations with LTP have revealed that the lipid component of the particle is in dynamic equilibrium with that of lipoprotein substrates (Ryan *et al.*, 1988). Thus, it is evident that the lipid moiety is not merely a static structural component of LTP; rather, it can be considered as a functional element in the mechanism of lipid transfer.

9.5.1.1. Lipid substrate specificity

Experiments have been conducted to examine the ability of LTP to utilize various substrate lipids. As reviewed earlier (Ryan and Van der Horst, 2000; Arrese *et al.*, 2001), LTP catalyzes the exchange and net transfer of DAG, in keeping with its proposed role in lipophorin interconversions *in vivo*. Extending this concept, Canavoso and Wells (2001) incubated radiolabeled midgut sacs with lipophorin-containing medium. These authors found that transfer of DAG from the midgut sacs to lipophorin was blocked by preincubation with antibody against LTP, supporting the view that LTP functions in DAG export from the midgut to lipophorin. In a similar manner, LTP was shown by Jouni *et al.* (2003) to be required for DAG transfer from lipophorin to *B. mori* ovarioles.

In studies of other potential lipid substrates, Singh and Ryan (1991) used [^{14}C]acetate to label the DAG and hydrocarbon moiety of lipophorin *in vivo*. Subsequent lipid transfer experiments revealed that LTP is capable of facilitating transfer of hydrocarbon among lipoprotein substrates, suggesting that LTP plays a role in the movement of these extremely hydrophobic, specialized lipids from their site of synthesis to their site of deposition at the cuticle (Takeuchi and Chino, 1993). Interestingly, the rate of LTP-mediated hydrocarbon transfer was slower than DAG transfer. In other work, *B. mori* LTP was employed in studies of LTP-mediated carotene transfer among lipophorin particles (Tsuchida *et al.*, 1998). Again, compared to DAG transfer, the rate of LTP-mediated carotene redistribution was much slower. Taken together, these results suggest that LTP may have a preference for DAG versus hydrocarbon or carotenes. Alternatively, the observed preference for DAG may be a function of the relative accessibility of the substrates within the donor lipoprotein. The ability of LTP to facilitate phospholipid transfer was studied by Golodne *et al.* (2001). These authors observed that LTP-mediated phospholipid transfer is non-selective. In contrast to the requirement for LTP to mediate transfer of

DAG, hydrocarbon, carotenoids, and phospholipids, Yun *et al.* (2002) provided evidence that LTP does not function in cholesterol transfer or redistribution in *M. sexta*. Rather, cholesterol is proposed to diffuse among tissues via mass action, freely transferring between lipophorin and tissues, depending on the physiological need. In keeping with this interpretation, Jouni *et al.* (2003) found that cholesterol transfer from lipophorin to *B. mori* ovarioles was unaffected by antibodies directed against LTP, whereas DAG transfer was inhibited.

9.5.1.2. Mechanism of facilitated lipid transfer

In general, lipid transfer catalysts may act as carriers of lipid between donor and acceptor lipoproteins, or transfer may require formation of a ternary complex between donor, acceptor, and LTP. Based on the observed LTP-mediated net transfer of DAG from HDLp to human LDL (Ryan *et al.*, 1990b), a strategy was developed to address this question experimentally (Blacklock *et al.*, 1992). [^{3}H]-DAG–HDLp and unlabeled LDL were covalently bound to Sepharose matrices and packed into separate columns connected in series, followed by circulation of LTP or buffer. Circulation of LTP, but not buffer, resulted in a concentration-dependent increase in the amount of radiolabeled DAG recovered in the LDL fraction, revealing that LTP facilitates net lipid transfer via a carrier-mediated mechanism.

Blacklock and Ryan (1995) employed LTP apolipoprotein-specific antibodies to probe the structure and catalytic properties of *M. sexta* LTP, obtaining evidence that apoLTP-II is a catalytically important apoprotein. In a similar manner, Van Heusden *et al.* (1996) employed LTP antibody inhibition experiments to demonstrate that all three LTP apoproteins are important for lipid transfer activity. These authors found that, unlike apoLp-III, apoLTP-III is not found as a free protein in hemolymph (Van Heusden *et al.*, 1996), despite the fact that it dissociates from the complex following exposure to non-ionic detergent (Blacklock and Ryan, 1995).

9.5.2. Carotenoid-Binding Proteins

In insects, the involvement of lipophorin in the hemolymph transport of dietary carotenoids is well documented (Tsuchida *et al.*, 1998). In keeping with its general function to accept and deliver hydrophobic lipid cargo, lipophorin may be anticipated to selectively deposit these isoprenoids at specific tissues. The mechanism involved is not fully understood, but likely involves LTP, the lipophorin receptor, and other proteins. Carotenoids fulfill several important roles in insects. Certain carotenoids are provitamins for vitamin A, which is required as the visual pigment chromophore (Giovannucci and Stephenson, 1999). Studies in *Drosophila* have shown that mutations in the ninaD gene result in blindness due to a defect in cellular uptake of carotenoids (Kiefer *et al.*, 2002). The ninaD gene encodes a protein that possesses significant sequence identity with mammalian class B scavenger receptors (i.e., SR-BI and CD36). In mammals, SR-BI is involved in cholesterol homeostasis, and mediates cholesterol flux between target cells and lipoproteins (Jian *et al.*, 1998; Yancey *et al.*, 2000). Insofar as lipophorin is structurally related to mammalian lipoproteins, it is conceivable that ninaD mediates transfer of carotenoids from lipophorin in a mechanistically similar manner (Kiefer *et al.*, 2002). Consistent with this concept, ninaD mRNA levels are particularly high in pupae, suggesting a role in the transport/ delivery of zeaxanthin from fat body to the developing eye during pupation (Giovannucci and Stephenson, 1999). In a screen for mutations that affect the biosynthesis of rhodopsin in *Drosophila*, Wang *et al.* (2007) identified a second class-B scavenger receptor, named Santa Maria. Subsequent studies revealed that Santa Maria functions upstream of vitamin A formation in neurons and glia. The protein is co-expressed, and functionally coupled, with the β, β-carotene-15, 15′-monooxygenase (NinaB), which converts β-carotene to all-trans-retinal.

Another vivid example of carotenoid transport is the production of a yellow cocoon in the silkworm, *B. mori* (Tabunoki *et al.*, 2002). A carotenoid-binding protein (CBP) from silk glands of *B. mori* larvae was identified (Tabunoki *et al.*, 2002). The function of this 33-kDa protein was investigated using *B. mori* mutants. Interestingly, only in larvae carrying the dominant *Y* (yellow hemolymph) gene, was CBP present in the villi of the midgut epithelium, suggesting that CBP may be involved in absorption of carotenoids. Tsuchida *et al.* (2004a) fed radiolabeled triolein to *B. mori* mutants, and found no defects in fatty acid uptake from midgut or delivery to fat body and silk glands. However, analysis revealed that yellow coloration of hemolymph was attributable to its carotenoid content. Lipophorin from the *Y+I* mutant exhibited the highest concentration of total carotenoids (55.8 μg/mg lipophorin) compared to 3.1 μg/mg in the +*Y+I* mutant, 1.2 μg/mg in the *YI* mutant, and 0.5 μg/mg in the +*YI* mutant. Thus, although lipid metabolism in the mutants is normal, defective lutein uptake was associated with the recessive *Y*-gene. Tsuchida *et al.* (2004b) studied the expression of CBP in the larval midgut and silk glands from wild type and four naturally occurring carotenoid transport mutants. CBP was expressed throughout the fifth stadium, with the highest expression on days 4–5 in the silk gland and days 3–5 in the midgut. Microscopic immunocytochemistry demonstrated uniform expression along the brush border of columnar cells in the midgut epithelium, consistent with a role in absorption of dietary carotenoids. CBP was also highly expressed along the distal membrane of the middle part of the silk gland, indicating a function in the uptake of carotenoids from lipophorin. When the middle silk glands and midguts of the four mutant strains were incubated with rabbit anti-CBP antibody, only proteins of the *Y*-gene dominant mutants cross reacted with the antibody, supporting the hypothesis that CBP is a *Y*-gene dependent protein.

Tabunoki *et al.* (2004) reported that CBP expression in the silk glands of larvae matched the period of carotenoid uptake into the silk gland. When these authors provided double-stranded CBP RNA to *B. mori* larvae, reduced expression of CBP in the silk gland was observed, along with a decrease in yellow pigmentation in the cocoon. Thus, CBP modulates cocoon pigmentation via its role in the uptake of carotenoids by silk gland tissue. The deduced amino acid sequence of CBP indicates it is a member of the steroidogenic acute regulatory (StAR) protein family, despite the fact that CBP binds carotenoids rather than cholesterol. In 2005, Sakudoh and colleagues identified a novel isoform of CBP, Start1, in *B. mori* that is comprised of a membrane-spanning N-terminal "MENTAL" domain together with a C-terminal lipid-binding "START" domain. This molecular architecture is identical to mammalian MLN64 and *Drosophila* Start1 proteins that are implicated in cholesterol transport and regulation of steroidogenesis. Interestingly, *B. mori* Start1 is expressed in both white and yellow cocoon strains, while CBP is only detected in the yellow cocoon strain. *B. mori* Start1 mRNA abundance in the prothoracic gland, the main ecdysteroidogenic tissue, positively correlates with changes in the hemolymph ecdysteroid level. Genomic analysis revealed that *B. mori* Start1 and CBP are generated from the same gene locus by alternative splicing. Thus, alternative splicing of the Start1/CBP gene generates unique protein isoforms whose endogenous ligands, sterol or carotenoid, are structurally different. Sakudoh and coworkers (2007) showed that, in the Y-recessive strain, the absence of a specific exon generates a non-functional CBP mRNA. Germ-line transformation with a wild type CBP gene, however, induced carotenoid uptake and normal cocoon coloration. More recently, Sakudoh *et al.* (2010) used positional cloning and transgenic rescue experiments to identify Cameo2, a transmembrane protein gene belonging to the CD36 family. In mutant larvae, Cameo2 expression was strongly repressed in the silk gland, resulting in colorless silk glands and white cocoons. The developmental profile of Cameo2 expression, CBP expression, and lutein pigmentation in the silk gland of the yellow cocoon strain are correlated. Thus, it may be considered that selective delivery of lutein requires the combination of two components: CBP as a carotenoid transporter in cytosol; and Cameo2 as the cell surface receptor.

9.5.3. Fatty Acid-Binding Proteins

Hydrolysis of LDLp-carried DAG by a lipophorin lipase at the flight muscles (for review, see Van der Horst *et al.*, 2001, 2010b) results in the extracellular production of FFAs. After uptake by the flight muscle cells, these FFAs are oxidized for energy generation. The mechanism by which the extracellularly liberated FFAs are translocated across the plasma membrane is as yet unknown, but may involve membrane fatty acid transporter proteins similar to those identified in mammals, i.e., fatty acid translocase (FAT)/CD36, a family of fatty acid transport proteins (FATP1-6), and plasma membrane-associated fatty acid-binding protein (FABP$_{pm}$) that not only facilitate but also regulate cellular fatty acid uptake, for instance through their inducible rapid (and reversible) translocation from intracellular storage pools to the cell membrane (for reviews, see Bonen et al., 2007; Glatz et al., 2010; Schwenk et al., 2010). The intracellular transport of FFAs in insect flight muscle cells closely resembles that in mammalian skeletal red muscle cells, and is mediated by a fatty acid-binding protein (FABP) (for review, see Haunerland, 1997). This insect FABP belongs to the cytoplasmic FABPs that comprise a family of 14- to 15-kDa proteins that bind fatty acid ligands with high affinity and are involved in shuttling fatty acids to cellular compartments, modulating intracellular lipid metabolism, and regulating gene expression (for reviews, see Boord et al., 2002; Storch and McDermott, 2009). Intriguingly, in contrast to FABP in mammals, locust FABP is an adult-specific protein, the expression of which is directly linked to metamorphosis; the concentration of FABP in locust flight muscle cytosol is over three times that in the mammalian heart (for review, see Haunerland, 1997), suggesting an adaptation to the extremely high metabolic rate of fatty acid oxidation for energy generation during migratory flight (Haunerland et al., 1992; Van der Horst et al., 1993; Qu et al., 2007). The flight muscle tissue of migratory birds such as the Western sandpiper also contains unusually high amounts of FABP (Guglielmo et al., 2002), supporting a role of FABP in migratory flight as suggested above.

The high amino acid sequence similarity (82%) between the FABP of L. migratoria flight muscle and that of human skeletal muscles (Maatman et al., 1994) is reflected in a strong similarity in their three-dimensional structures (for review, see Van der Horst et al., 2002). The structure of L. migratoria FABP complexed with a fatty acid, however, although resembling the closely related mammalian heart- and adipocyte-type FABPs, is characterized by binding features that differ significantly from the typical hairpin-turn ligand shapes of the latter forms. As a result of these evolutionary variations, insect FABPs may display a much greater diversity in intracellular lipid binding than that observed for the mammalian transport proteins (Lücke et al., 2006).

9.5.4. Vitellogenin

Oocyte development in adult females involves the accumulation of large amounts of lipid, most of which is extraovarian in origin and is delivered by lipophorin. Another lipid-binding protein that serves a role in lipid transport to the oocyte is Vg; although its overall contribution to the oocyte lipid accumulation is relatively modest (about 5%) (Sun et al., 2000), it is by far the most abundant yolk protein precursor accumulated in all insect species (for review, see Tufail and Takeda, 2009). While the structural properties of insect Vgs are diverse (for review see Tufail and Takeda, 2008), they generally possess 10% lipid, primarily phospholipid and glycerolipid. Vg is synthesized and assembled in the fat body, secreted into hemolymph, and taken up by oocytes. Vg uptake is facilitated by members of a subfamily of the LDLR family that have been characterized so far in Drosophila (Schonbaum et al., 1995), Ae. aegypti (Sappington et al., 1996), Solenopsis invicta (Chen et al., 2004), P. americana (Tufail and Takeda, 2005), Blattella germanica (Ciudad et al., 2006), and Leucophaea maderae (Tufail and Takeda, 2007; for review, see Tufail and Takeda, 2009). Cheon et al. (2001) demonstrated that the ovarian vitellogenin receptor (VgR) is only distantly related to LpR, another ovarian LDLR homolog with a different ligand; a recent phylogenetic analysis places insect VgRs and LpRs in separate groups, and reveals that insect LpRs are more closely related to vertebrate VLDLRs/VgRs and LDLRs than to insect VgRs (Tufail and Takeda, 2009). These data imply that the receptor-mediated mechanisms involved in the uptake of lipid and the accumulation of yolk protein precursors (which provide a key nutrient source for the developing oocyte) utilize two separate receptors (VgR and LpR), which are specific for their respective ligands, Vg and HDLp. Considerable early work was performed to characterize the lipid transport properties of Vg (for review, see Kunkel and Nordin, 1985), while more recent work has focused on molecular and evolutionary aspects of vitellogenin proteins and receptor-mediated endocytosis. These aspects, which are beyond the scope of the present chapter, have been comprehensively reviewed elsewhere (Sappington and Raikhel, 1998; Raikhel et al., 2002; Tufail and Takeda, 2008, 2009).

With respect to HDLp internalization, it is important to note that immunocytochemical data in the mosquito (Sun et al., 2000) revealed that only a small amount of HDLp accumulates in developing oocytes as yolk protein, comprising 3% of total ovarian proteins upon completion of protein internalization. Since lipid accounts for 35–40% of the insect egg dry weight (Kawooya and Law, 1988), Sun et al. (2000) proposed that internalization of HDLp is unlikely to be the major route of lipid delivery to the developing oocyte (for review, see Ziegler and Van Antwerpen, 2006). A dual mechanism for lipophorin-mediated lipid delivery to oocytes (a lipophorin shuttle mechanism involving LDLp and internalization of HDLp, with stripping of most of its lipid) had been demonstrated earlier (Kawooya and Law, 1988; Kawooya et al., 1988; Liu and Ryan, 1991; Van Antwerpen et al., 2005). In addition, recently LDLp formation in the eggs of *B. mori* was reported, in which the apoLp-III associated with LDLp was apparently synthesized in the eggs (Tsuchida *et al.*, 2010). However, considering that the LpR involved in uptake of HDLp by mosquito oocytes bears a high structural similarity to the LpR discovered in locust fat body cell membranes (Dantuma *et al.*, 1999; Cheon *et al.*, 2001), the precise mechanism of LpR-mediated endocytosis in the oocyte, and the fate of HDLp, remain open questions. In light of the ligand recycling mechanism discussed earlier for lipid delivery to the fat body, the possibility exists that LpR recycles its ligand after intracellular trafficking, providing another mechanism for the uptake of specific lipid components by the oocyte.

References

Adamo, S. A., Roberts, J. L., Easy, R. H., & Ross, N. W. (2008). Competition between immune function and lipid transport for the protein apolipophorin III leads to stress-induced immunosuppression in crickets. *J. Exp. Biol.*, *211*, 531–538.

Arrese, E. L., & Soulages, J. L. (2010). Insect fat body: Energy, metabolism, and regulation. *Annu. Rev. Entomol.*, *55*, 207–225.

Arrese, E. L., & Wells, M. A. (1994). Purification and properties of a phosphorylatable triacylglycerol lipase from the fat body of an insect. *Manduca sexta. J. Lipid Res.*, *35*, 1652–1660.

Arrese, E. L., & Wells, M. A. (1997). Adipokinetic hormone-induced lipolysis in the fat body of an insect, *Manduca sexta*: Synthesis of sn-1,2-diacylglycerols. *J. Lipid Res.*, *38*, 68–76.

Arrese, E. L., Canavoso, L. E., Jouni, Z. E., Pennington, J. E., Tsuchida, K., et al. (2001). Lipid storage and mobilization in insects: Current status and future directions. *Insect Biochem. Mol. Biol.*, *31*, 7–17.

Arrese, E. L., Patel, R. T., & Soulages, J. L. (2006). The main triglyceride-lipase from the insect fat body is an active phospholipase A(1): Identification and characterization. *J. Lipid Res.*, *47*, 2656–2667.

Arrese, E. L., Rivera, L., Hamada, M., & Soulages, J. L. (2008a). Purification and characterization of recombinant lipid storage protein-2 from *Drosophila melanogaster. Protein Pept. Lett.*, *15*, 1027–1032.

Arrese, E. L., Rivera, L., Hamada, M., Mirza, S., Hartson, S. D., et al. (2008b). Function and structure of lipid storage droplet protein 1 studied in lipoprotein complexes. *Arch. Biochem. Biophys.*, *473*, 42–47.

Arrese, E. L., Mirza, S., Rivera, L., Howard, A. D., Chetty, P. S., et al. (2008c). Expression of lipid storage droplet protein-1 may define the role of AKH as a lipid mobilizing hormone in *Manduca sexta. Insect Biochem. Mol. Biol.*, *38*, 993–1000.

Babin, P. J., Bogerd, J., Kooiman, F. P., Van Marrewijk, W. J. A., & Van der Horst, D. J. (1999). Apolipophorin II/I, apolipoprotein B, vitellogenin, and microsomal triglyceride transfer protein genes are derived from a common ancestor. *J. Mol. Evol.*, *49*, 150–160.

Baggerman, G., Huybrechts, J., Clynen, E., Hens, K., Harthoorn, L., et al. (2002). New insights in adipokinetic hormone (AKH) precursor processing in *Locusta migratoria* obtained by capillary liquid chromatography-tandem mass spectrometry. *Peptides*, *23*, 635–644.

Bauerfeind, R., & Komnick, H. (1992). Immunocytochemical localization of lipophorin in the fat body of dragonfly larvae (*Aeshna cyanea*). *J. Insect Physiol.*, *38*, 185–198.

Bayburt, T. H., & Sligar, S. G. (2010). Membrane protein assembly into Nanodiscs. *FEBS Lett.*, *584*, 1721–1727.

Beenakkers, A. M. T., Van der Horst, D. J., & Van Marrewijk, W. J. A. (1985). Insect lipids and lipoproteins, and their role in physiological processes. *Prog. Lipid. Res.*, *24*, 19–67.

Beglova, N., & Blacklow, S. C. (2005). The LDL receptor: How acid pulls the trigger. *Trends Biochem. Sci.*, *30*, 309–317.

Beglova, N., Jeon, H., Fisher, C., & Blacklow, S. C. (2004a). Cooperation between fixed and low pH-inducible interfaces controls lipoprotein release by the LDL receptor. *Mol. Cell*, *16*, 281–292.

Beglova, N., Jeon, H., Fisher, C., & Blacklow, S. C. (2004b). Structural features of the low-density lipoprotein receptor facilitating ligand binding and release. *Biochem. Soc. Trans.*, *32*, 721–723.

Beller, M., Sztalryd, C., Southall, N., Bell, M., Jäckle, H., et al. (2008). COPI complex is a regulator of lipid homeostasis. *PLoS Biol.*, *6*, e292, doi:10.1371/journal.pbio.0060292.

Belmont, M., Cazzamali, G., Williamson, M., Hauser, F., & Grimmelikhuijzen, C. J. (2006). Identification of four evolutionarily related G protein-coupled receptors from the malaria mosquito *Anopheles gambiae. Biochem. Biophys. Res. Commun.*, *344*, 160–165.

Bharucha, K. N., Tarr, P., & Zipursky, S. L. (2008). A glucagon-like endocrine pathway in *Drosophila* modulates both lipid and carbohydrate homeostasis. *J. Exp. Biol.*, *211*, 3103–3110.

Bickel, P. E., Tansey, J. T., & Welte, M. A. (2009). PAT proteins, an ancient family of lipid droplet proteins that regulate cellular lipid stores. *Biochim. Biophys. Acta*, *1791*, 419–440.

Blacklock, B. J., & Ryan, R. O. (1994). Hemolymph lipid transport. *Insect Biochem. Mol. Biol.*, *24*, 855–873.

Blacklock, B. J., & Ryan, R. O. (1995). Structural studies of *Manduca sexta* lipid transfer particle with apolipoprotein-specific antibodies. *J. Lipid. Res.*, *35*, 108–116.

Blacklock, B. J., Smillie, M., & Ryan, R. O. (1992). *Manduca sexta* lipid transfer particle can facilitate lipid transfer via a carrier-mediated mechanism. *J. Biol. Chem.*, *267*, 14033–14037.

Bogerd, J., Kooiman, F. P., Pijnenburg, M. A. P., Hekking, L. H. P., Oudejans, R. C. H. M., et al. (1995). Molecular cloning of three distinct cDNAs, each encoding a different adipokinetic hormone precursor, of the migratory locust, *Locusta migratoria*: Differential expression of the distinct adipokinetic hormone precursor genes during flight activity. *J. Biol. Chem.*, *270*, 23038–23043.

Bogerd, J., Babin, P. J., Kooiman, F. P., André, M., Ballagny, C., et al. (2000). Molecular characterization and gene expression in the eye of the apolipophorin II/I precursor from *Locusta migratoria. J. Comp. Neurol.*, *427*, 546–558.

Bonen, A., Chabowski, A., Luiken, J. J. F. P., & Glatz, J. F. C. (2007). Is membrane transport of FFA mediated by lipid, protein, or both? Mechanisms and regulation of protein-mediated cellular fatty acid uptake: Molecular, biochemical, and physiological evidence. *Physiol. (Bethesda)*, *22*, 15–29.

Boord, J. B., Fazio, S., & Linton, M. F. (2002). Cytoplasmic fatty acid-binding proteins: Emerging roles in metabolism and atherosclerosis. *Curr. Opin. Lipidol.*, *13*, 141–147.

Borén, J., Lee, I., Zhu, W., Arnold, K., Taylor, S., et al. (1998). Identification of the low density lipoprotein receptor-binding site in apolipoprotein B100 and the modulation of its binding activity by the carboxyl terminus in familial defective apo-B100. *J. Clin. Invest.*, *101*, 1084–1093.

Boswell, E. J., Jeon, H., Blacklow, S. C., & Downing, A. K. (2004). Global defects in the expression and function of the low density lipoprotein receptor (LDLR) associated with two familial hypercholesterolemia mutations resulting in misfolding of the LDLR epidermal growth factor–AB pair. *J. Biol. Chem.*, *279*, 30611–30621.

Brasaemle, D. L. (2007). The perilipin family of structural lipid droplet proteins: Stabilization of lipid droplets and control of lipolysis. *J. Lipid Res.*, *48*, 2547–2559.

Breiter, D. R., Kanost, M. R., Benning, M. M., Wesenberg, G., Law, J. H., et al. (1991). Molecular structure of an apolipoprotein determined at 2.5-Å resolution. *Biochemistry*, *30*, 603–608.

Brown, M. S., & Goldstein, J. L. (1986). A receptor-mediated pathway for cholesterol homeostasis. *Science*, *232*, 34–47.

Canavoso, L. E., & Wells, M. A. (2001). Role of lipid transfer particle in delivery of diacylglycerol from midgut to lipophorin in larval *Manduca sexta*. *Insect Biochem. Mol. Biol.*, *31*, 783–790.

Canavoso, L. E., Jouni, Z. E., Karnas, K. J., Pennington, J. E., & Wells, M. A. (2001). Fat metabolism in insects. *Annu. Rev. Nutr.*, *21*, 23–46.

Candy, D. J. (2002). Adipokinetic hormone concentrations in the haemolymph of *Schistocerca gregaria*, measured by radio-immunoassay. *Insect Biochem. Mol. Biol.*, *32*, 1361–1367.

Chen, M. -E., Lewis, D. K., Keeley, L. L., & Pietrantonio, P. V. (2004). cDNA cloning and transcriptional regulation of the vitellogenin receptor from the fire ant, *Solenopsis invicta* Buren (Hymenoptera: Formicidae). *Insect Molecular Biology*, *13*, 195–204.

Cheon, H. -M., Seo, S. -J., Sun, J., Sappington, T. W., & Raikhel, A. S. (2001). Molecular characterization of the VLDL receptor homolog mediating uptake of lipophorin in oocytes of the mosquito *Aedes aegypti*. *Insect Biochem. Mol. Biol.*, *31*, 753–760.

Chetty, P. S., Arrese, E. L., Rodriguez, V., & Soulages, J. L. (2003). Role of helices and loops in the ability of apolipophorin-III to interact with native lipoproteins and form discoidal lipoprotein complexes. *Biochemistry*, *42*, 15061–15067.

Chino, H. (1985). Lipid transport: Biochemistry of hemolymph lipophorin. In G. A. Kerkut, & L. I. Gilbert (Eds.), *Comprehensive Insect Physiology, Biochemistry, and Pharmacology* Vol. 10, (pp. 115–134). Oxford, UK: Pergamon Press.

Chiu, M. H., Wan, C. P., Weers, P. M. M., & Prenner, E. J. (2009). Apolipophorin III interaction with model membranes composed of phosphatidylcholine and sphingomyelin using differential scanning calorimetry. *Biochim. Biophys. Acta*, *1788*, 2160–2168.

Chromy, B. A., Arroyo, E., Blanchette, C. D., Bench, G., Benner, H., et al. (2007). Different apolipoproteins impact nanolipoprotein particle formation. *J. Am. Chem. Soc.*, *129*, 14348–14354.

Ciudad, L., Bellés, X., & Piulachs, M. -D. (2007). Structural and RNAi characterization of the German cockroach lipophorin receptor, and the evolutionary relationships of lipoprotein receptors. *BMC Mol. Biol.*, *8*, 53.

Clynen, E., & Schoofs, L. (2009). Peptidomic survey of the locust neuroendocrine system. *Insect Biochem. Mol. Biol.*, *39*, 491–507.

Cole, K. D., Fernando-Warnakulasuriya, G. J. P., Boguski, M. S., Freeman, M., Gordon, J. I., et al. (1987). Primary structure and comparative sequence analysis of an insect apolipoprotein: Apolipophorin III from *Manduca sexta*. *J. Biol. Chem.*, *262*, 11794–11800.

Cormode, D. P., Frias, J. C., Ma, Y., Chen, W., Skajaa, T., et al. (2009). HDL as a contrast agent for medical imaging. *Clin. Lipidol.*, *4*, 493–500.

Dantuma, N. P., Pijnenburg, M. A. P., Diederen, J. H. B., & Van der Horst, D. J. (1997). Developmental down-regulation of receptor-mediated endocytosis of an insect lipoprotein. *J. Lipid. Res.*, *38*, 254–265.

Dantuma, N. P., Potters, M., De Winther, M. P. J., Tensen, C. P., Kooiman, F. P., et al. (1999). An insect homolog of the vertebrate very low density lipoprotein receptor mediates endocytosis of lipophorins. *J. Lipid. Res.*, *40*, 973–978.

De Loof, A., Vandersmissen, T., Huybrechts, J., Landuyt, B., Baggerman, G., et al. (2009). APRP, the second peptide encoded by the adipokinetic hormone gene(s), is highly conserved in evolution: A role in control of ecdysteroidogenesis? *Ann. NY Acad. Sci.*, *1163*, 376–378.

Dettloff, M., Weers, P. M. M., Niere, M., Kay, C. M., Ryan, R. O., et al. (2001a). An N-terminal three-helix fragment of an exchangeable insect apolipoprotein, apolipophorin III, conserves the lipid binding properties of the full length protein. *Biochemistry*, *40*, 3150–3157.

Dettloff, M., Kaiser, B., & Wiesner, A. (2001b). Localization of injected apolipophorin III *in vivo*: New insights into the immune activation process directed by this protein. *J. Insect Physiol.*, *47*, 789–797.

Dettloff, M., Wittwer, D., Weise, C., & Wiesner, A. (2001c). Lipophorin of a lower density is formed during immune responses in the lepidopteran insect *Galleria mellonella*. *Cell Tissue Res.*, *306*, 449–458.

Dettloff, M., Niere, M., Ryan, R. O., Kay, C. M., Wiesner, A., et al. (2002). Differential lipid binding of truncation mutants of *Galleria mellonella* apolipophorin III. *Biochemistry*, *41*, 9688–9695.

Diederen, J. H. B., Oudejans, R. C. H. M., Harthoorn, L. F., & Van der Horst, D. J. (2002). Cell biology of the adipokinetic hormone-producing neurosecretory cells in the locust corpus cardiacum. *Microsc. Res. Tech.*, *56*, 227–236.

Dunphy, G., & Halwani, A. (1997). Haemolymph proteins of larvae of *Galleria mellonella* detoxify endotoxins of the insect pathogenic bacteria *Xenorhabdus nematophilus* (Enterobacteriaceae). *J. Insect Physiol., 43,* 1023–1029.

Fan, D., Reese, L., Ren, X., Weers, P. M. M., Ryan, R. O., et al. (2001). Complete ^{1}H, ^{15}N, and ^{13}C assignments of an exchangeable apolipoprotein, *Locusta migratoria* apolipophorin III. *J. Biomol. NMR, 19,* 83–84.

Fan, D., Zheng, Y., Yang, D., & Wang, J. (2003). NMR solution structure and dynamics of an exchangeable apolipoprotein, *Locusta migratoria* apolipophorin III. *J. Biol. Chem., 278,* 21210–21220.

Freitak, D., Wheat, C. W., Heckel, D. G., & Vogel, H. (2007). Immune system responses and fitness costs associated with consumption of bacteria in larvae of *Trichoplusia ni. BMC Biol., 5,* 56.

Gäde, G. (1996). The revolution in insect neuropeptides illustrated by the adipokinetic hormone/red pigment-concentrating hormone family of peptides. *Z. Naturforsch, 51C,* 607–617.

Gäde, G. (1997). The explosion of structural information on insect neuropeptides. *Prog. Chem. Organ. Nat. Prod., 71,* 1–128.

Gäde, G. (2009). Peptides of the adipokinetic hormone/red pigment-concentrating hormone family: A new take on biodiversity. *Ann. NY Acad. Sci., 1163,* 125–136.

Gäde, G., & Auerswald, L. (2003). Mode of action of neuropeptides from the adipokinetic hormone family. *Gen. Comp. Endocrinol., 132,* 10–20.

Gäde, G., Hoffmann, K. H., & Spring, J. H. (1997). Hormonal regulation in insects: Facts, gaps and future directions. *Physiol. Rev., 77,* 963–1032.

Garda, H. A., Arrese, E. L., & Soulages, J. L. (2002). Structure of apolipophorin III in discoidal lipoproteins: Inter-helical distances in the lipid-bound state and conformational change upon binding to lipid. *J. Biol. Chem., 262,* 11794–11800.

Gerasimenko, J. V., Tepikin, A. V., Petersen, O. H., & Gerasimenko, O. V. (1998). Calcium uptake via endocytosis with rapid release from acidifying endosomes. *Curr. Biol., 8,* 1335–1338.

Ghosh, R. N., Gelman, D. L., & Maxfield, F. R. (1994). Quantification of low density lipoprotein and transferrin endocytic sorting HEp2 cells using confocal microscopy. *J. Cell Sci., 107,* 2177–2189.

Giannoulis, P., Brooks, C. L., Dunphy, G. B., Niven, D. F., & Mandato, C. A. (2008). Surface antigens of *Xenorhabdus nematophila* (F. Enterobacteriaceae) and *Bacillus subtilis* (F. Bacillaceae) react with antibacterial factors of *Malacosoma disstria* (C. Insecta: O. Lepidoptera) hemolymph. *J. Invertebr. Pathol., 97,* 211–222.

Gibbons, G. F., Islam, K., & Pease, R. J. (2000). Mobilisation of triacylglycerol stores. *Biochim. Biophys. Acta, 1483,* 37–57.

Giovannucci, D. R., & Stephenson, R. S. (1999). Identification and distribution of dietary precursors of the *Drosophila* visual pigment chromophore: Analysis of carotenoids in wild type and ninaD mutants by HPLC. *Vision Res., 39,* 219–229.

Glatz, J. F. C., Luiken, J. J. F. P., & Bonen, A. (2010). Membrane fatty acid transporters as regulators of lipid metabolism: Implications for metabolic disease. *Physiol. Rev., 90,* 367–417.

Goldstein, J. L., Brown, M. S., Anderson, R. G. W., Russell, D. W., & Schneider, W. J. (1985). Receptor-mediated endocytosis: Concepts emerging from the LDL receptor system. *Annu. Rev. Cell Biol., 1,* 1–39.

Goldsworthy, G. J., Lee, M. J., Luswata, R., Drake, A. F., & Hyde, D. (1997). Structures, assays and receptors for locust adipokinetic hormones. *Comp. Biochem. Physiol. B, 117,* 483–496.

Golodne, D. M., Van Heusden, M. C., Gondim, K. C., Masuda, H., & Atella, G. C. (2001). Purification and characterization of a lipid transfer particle in *Rhodnius prolixus*: Phospholipid transfer. *Insect Biochem. Mol. Biol., 31,* 563–571.

Gopalapillai, R., Kadono-Okuda, K., Tsuchida, K., Yamamoto, K., Nohata, J., et al. (2006). Lipophorin receptor of *Bombyx mori*: cDNA cloning, genomic structure, alternative splicing, and isolation of a new isoform. *J. Lipid Res., 47,* 1005–1013.

Grönke, S., Beller, M., Fellert, S., Ramakrishnan, H., Jäckle, H., et al. (2003). Control of fat storage by a *Drosophila* PAT domain protein. *Curr. Biol., 13,* 603–606.

Grönke, S., Mildner, A., Fellert, S., Tennagels, N., Petry, S., et al. (2005). Brummer lipase is an evolutionary conserved fat storage regulator in *Drosophila. Cell Metabol., 1,* 323–330.

Grönke, S., Müller, G., Hirsch, J., Fellert, S., Andreou, A., et al. (2007). Dual lipolytic control of body fat storage and mobilization in *Drosophila. PLoS Biol., 5,* e137, doi:10.1371/journal.pbio.0050137.

Gross, D. N., Miyoshi, H., Hosaka, T., Zhang, H. -H., Pino, E. C., et al. (2006). Dynamics of lipid droplet-associated proteins during hormonally stimulated lipolysis in engineered adipocytes: Stabilization and lipid droplet binding of adipocyte differentiation-related protein/adipophilin. *Mol. Endocrinol., 20,* 459–466.

Guglielmo, C. G., Haunerland, N. H., Hochachka, P. W., & Williams, T. D. (2002). Seasonal dynamics of flight muscle fatty acid binding protein and catabolic enzymes in a migratory shorebird. *Am. J. Physiol., 282,* R1405–R1413.

Guidugli-Lazzarini, K. R., do Nascimento, A. M., Tanaka, E. D., Piulachs, M. -D., Hartfelder, K., et al. (2008). Expression analysis of putative vitellogenin and lipophorin receptors in honey bee (*Apis mellifera* L.) queens and workers. *J. Insect Physiol., 54,* 1138–1147.

Guo, Y., Walther, T. C., Rao, M., Stuurman, N., Goshima, G., et al. (2008). Functional genomic screen reveals genes involved in lipid-droplet formation and utilization. *Nature, 453,* 657–661.

Halwani, A. E., Niven, D. F., & Dunphy, G. B. (2000). Apolipophorin-III and the interactions of lipoteichoic acids with the immediate immune responses of *Galleria mellonella. J. Invertebr. Pathol., 76,* 233–241.

Hansen, K. K., Hauser, F., Cazzamali, G., Williamson, M., & Grimmelikhuijzen, C. J. (2006). Cloning and characterization of the adipokinetic hormone receptor from the cockroach *Periplaneta americana. Biochem. Biophys. Res. Commun., 343,* 638–643.

Hård, K., Van Doorn, J. M., Thomas-Oates, J. E., Kamerling, J. P., & Van der Horst, D. J. (1993). Structure of the Asn-linked oligosaccharides of apolipophorin III from the insect *Locusta migratoria*: Carbohydrate-linked 2-amino-ethylphosphonate as a constituent of a glycoprotein. *Biochemistry, 32*, 766–775.

Harthoorn, L. F., Diederen, J. H. B., Oudejans, R. C. H. M., & Van der Horst, D. J. (1999). Differential location of peptide hormones in the secretory pathway of insect adipokinetic cells. *Cell Tissue Res., 298*, 361–369.

Harthoorn, L. F., Oudejans, R. C. H. M., Diederen, J. H. B., Van de Wijngaart, D. J., & Van der Horst, D. J. (2001). Absence of coupling between release and biosynthesis of peptide hormones in insect neuroendocrine cells. *Eur. J. Cell Biol., 80*, 451–457.

Harthoorn, L. F., Oudejans, R. C. H. M., Diederen, J. H. B., & Van der Horst, D. J. (2002). Coherence between biosynthesis and secretion of insect adipokinetic hormones. *Peptides, 23*, 629–634.

Hatle, J. D., & Spring, J. H. (1999). Tests of potential adipokinetic hormone precursor-related peptide (APRP) functions: Lack of responses. *Arch. Insect Biochem. Physiol., 42*, 163–166.

Haunerland, N. H. (1997). Transport and utilization of lipids in insect flight muscle. *Comp. Biochem. Physiol. B, 117*, 475–482.

Haunerland, N. H., Andolfatto, P., Chisholm, J. M., Wang, Z., & Chen, X. (1992). Fatty-acid-binding protein in locust flight muscle. Developmental changes of expression, concentration and intracellular distribution. *Eur. J. Biochem., 210*, 1045–1051.

Herz, J. (2002). Deconstructing the LDL receptor – a rhapsody in pieces. *Nat. Struct. Biol., 8*, 476–478.

Hirayama, Y., & Chino, H. (1990). Lipid transfer particle in locust hemolymph: Purification and characterization. *J. Lipid Res., 31*, 793–799.

Hussain, M. M., Kedees, M. H., Singh, K., Athar, H., & Jamali, N. Z. (2001). Signposts in the assembly of chylomicrons. *Front. Biosci., 6*, D320–D321.

Huybrechts, J., Clynen, E., Baggerman, G., De Loof, A., & Schoofs, L. (2002). Isolation and identification of the AKH III precursor-related peptide from *Locusta migratoria*. *Biochem. Biophys. Res. Commun., 296*, 1112–1117.

Innerarity, T. L. (2002). Structural biology: LDL receptor's beta-propeller displaces LDL. *Science, 298*, 2337–2339.

Ishikawa, H., Yamamoto, K., & Sehnal, F. (1998). Hemagglutinating properties of apolipophorin III from the hemolymph of *Galleria mellonella* larvae. *Arch. Insect Biochem. Physiol., 38*, 119–125.

Jeon, H., & Blacklow, S. C. (2005). Structure and physiologic function of the low-density lipoprotein receptor. *Annu. Rev. Biochem., 74*, 535–562.

Jian, B., de la Llera-Moya, M., Ji, Y., Wang, N., Phillips, M. C., et al. (1998). Scavenger receptor class B type I as mediator of cellular cholesterol efflux to lipoproteins and phospholipid acceptors. *J. Biol. Chem., 273*, 5599–5606.

Jouni, Z. E., Takada, N., Gazard, J., Maekawa, H., Wells, M. A., et al. (2003). Transfer of cholesterol and diacylglycerol from lipophorin to *Bombyx mori* ovarioles *in vitro*: Role of lipid transfer particle. *Insect Biochem. Mol. Biol., 33*, 145–153.

Kanost, M. R., Boguski, M. S., Freeman, M., Gordon, J. I., Wyatt, G. R., et al. (1988). Primary structure of apolipophorin-III from the migratory locust, *Locusta migratoria*. *J. Biol. Chem., 263*, 10568–10573.

Kaufmann, C., Merzendorfer, H., & Gäde, G. (2009). The adipokinetic hormone system in Culicinae (Diptera: Culicidae): Molecular identification and characterization of two adipokinetic hormone (AKH) precursors from *Aedes aegypti* and *Culex pipiens* and two putative AKH receptor variants from *Ae. aegypti. Insect Biochem. Mol. Biol., 39*, 770–781.

Kawooya, J. K., & Law, J. H. (1988). Role of lipophorin in lipid transport to the insect egg. *J. Biol. Chem., 263*, 8748–8753.

Kawooya, J. K., Keim, P. S., Ryan, R. O., Shapiro, J. P., Samaraweera, P., et al. (1984). Insect apolipophorin III: Purification and properties. *J. Biol. Chem., 259*, 10733–10737.

Kawooya, J. K., Meredith, S. C., Wells, M. A., Kézdy, F. J., & Law, J. H. (1986). Physical and surface properties of insect apolipophorin III. *J. Biol. Chem., 261*, 13588–13591.

Kawooya, J. K., Osir, E. O., & Law, J. H. (1988). Uptake of the major lipoprotein and its transformation in the insect egg. *J. Biol. Chem., 263*, 8740–8747.

Kawooya, J. K., Wells, M. A., & Law, J. H. (1989). A strategy for solubilizing delipidated apolipoprotein with lysophosphatidylcholine and reconstitution with phosphatidylcholine. *Biochemistry, 28*, 6658–6667.

Kiefer, C., Sumser, E., Wernet, M. F., & von Lintig, J. (2002). A class B scavenger receptor mediates the cullular uptake of carotenoids in *Drosophila. Proc. Natl. Acad. Sci. USA, 99*, 10581–10586.

Kim, H. J., Je, H. J., Park, S. Y., Lee, I. H., Jin, B. R., et al. (2004). Immune activation of apolipophorin-III and its distribution in hemocytes from *Hyphantria cunea. Insect Biochem. Mol. Biol., 34*, 1011–1023.

Klon, A. E., Segrest, J. P., & Harvey, S. C. (2002). Comparative models for human apolipoprotein A-I bound to lipid in discoidal high-density lipoprotein particles. *Biochemistry, 41*, 10895–10905.

Köllisch, G. V., Lorenz, M. W., Kellner, R., Verhaert, P. D., & Hoffman, K. H. (2000). Structure elucidation and biological activity of an unusual adipokinetic hormone from corpora cardiaca of the butterfly, *Vanessa cardui. Eur. J. Biochem., 267*, 5502–5508.

Kulkarni, M. M., & Perrimon, N. (2005). Super-size flies. *Cell Metabol., 1*, 288–290.

Kunkel, J. G., & Nordin, J. H. (1985). Yolk proteins. In G. A. Kerkut, & L. I. Gilbert (Eds.), *Comprehensive Insect Physiology, Biochemistry, and Pharmacology* (Vol. 1, pp. 83–111). Oxford, UK: Pergamon Press.

Ledford, A. S., Weinberg, R. B., Cook, V. R., Hantgan, R. R., & Shelness, G. S. (2006). Self-association and lipid binding properties of the lipoprotein initiating domain of apolipoprotein B. *J. Biol. Chem., 281*, 8871–8876.

Lee, C. S., Han, J. H., Lee, S. M., Hwang, J. S., Kang, S. W., et al. (2003a). Wax moth, *Galleria mellonella*, fat body receptor for high-density lipophorin (HDLp). *Arch. Insect Biochem. Physiol., 54*, 14–24.

Lee, C. S., Han, J. H., Kim, B. S., Lee, S. M., Hwang, J. S., et al. (2003b). Wax moth, *Galleria mellonella*, high density lipophorin receptor: Alternative splicing, tissue-specific expression, and developmental regulation. *Insect Biochem. Mol. Biol., 33*, 761–771.

Leon, L. J., Pratt, C. C., Vasquez, L. J., & Weers, P. M. M. (2006a). Tyrosine fluorescence analysis of apolipophorin III–lipopolysaccharide interaction. *Arch. Biochem. Biophys.*, *452*, 38–45.

Leon, L. J., Idangodage, H., Wan, C. P., & Weers, P. M. M. (2006b). Apolipophorin III: Lipopolysaccharide binding requires helix bundle opening. *Biochem. Biophys. Res. Commun.*, *348*, 1328–1333.

Lindemans, M., Liu, F., Janssen, T., Husson, S. J., Mertens, I., et al. (2009). Adipokinetic hormone signaling through the gonadotropin-releasing hormone receptor modulates egg-laying in *Caenorhabditis elegans*. *Proc. Natl. Acad. Sci. USA*, *106*, 1642–1647.

Liu, H., & Ryan, R. O. (1991). Role of lipid transfer particle in transformation of lipophorin in insect oocytes. *Biochim. Biophys. Acta*, *1085*, 112–118.

Liu, H., Scraba, D. G., & Ryan, R. O. (1993). Prevention of phospholipase-C induced aggregation of low density lipoprotein by amphipathic apolipoproteins. *FEBS Lett.*, *316*, 27–33.

Londos, C., Brasaemle, D. L., Schultz, C. J., Segrest, J. P., & Kimmel, A. R. (1999). Perilipins, ADRP, and other proteins that associate with intracellular neutral lipid droplets in animal cells. *Semin. Cell Dev. Biol.*, *10*, 51–58.

Londos, C., Sztalryd, C., Tansey, J. T., & Kimmel, A. R. (2005). Role of PAT proteins in lipid metabolism. *Biochimie*, *87*, 45–49.

Lücke, C., Qiao, Y., Van Moerkerk, H. T., Veerkamp, J. H., & Hamilton, J. A. (2006). Fatty-acid-binding protein from the flight muscle of *Locusta migratoria*: Evolutionary variations in fatty acid binding. *Biochemistry*, *45*, 6296–6305.

Maatman, R. G. H. J., Degano, M., Van Moerkerk, H. T. B., Van Marrewijk, W. J. A., Van der Horst, D. J., et al. (1994). Primary structure and binding characteristics of locust and human muscle fatty-acid-binding proteins. *Eur. J. Biochem.*, *221*, 801–810.

Martin, S., & Parton, R. G. (2006). Lipid droplets: A unified view of a dynamic organelle. *Nat. Rev. Mol. Cell Biol.*, *7*, 373–378.

Maxfield, F. R., & McGraw, T. E. (2004). Endocytic recycling. *Nat. Rev. Mol. Cell Biol.*, *5*, 121–132.

Mellman, I. (1996). Endocytosis and molecular sorting. *Annu. Rev. Cell Devel. Biol.*, *12*, 575–625.

Miura, S., Gan, J. W., Brzostowski, J., Parisi, M. J., Schultz, C. J., et al. (2002). Functional conservation for lipid storage droplet association among Perilipin, ADRP, and TIP47 (PAT)-related proteins in mammals, *Drosophila*, and *Dictyostelium*. *J. Biol. Chem.*, *277*, 32253–32257.

Mukherjee, S., Ghosh, R. N., & Maxfield, F. R. (1997). Endocytosis. *Physiol. Rev.*, *77*, 759–803.

Mullen, L., & Goldsworthy, G. (2003). Changes in lipophorins are related to the activation of phenoloxidase in the haemolymph of *Locusta migratoria* in response to injection of immunogens. *Insect Biochem. Mol. Biol.*, *33*, 661–670.

Murphy, S., Martin, S., & Parton, R. G. (2009). Lipid droplet–organelle interactions: Sharing the fats. *Biochim. Biophys. Acta*, *1791*, 441–447.

Mwangi, R. W., & Goldsworthy, G. J. (1977). Diglyceride-transporting lipoproteins in *Locusta*. *J. Comp. Physiol.*, *114*, 177–190.

Mwangi, R. W., & Goldsworthy, G. J. (1981). Diacylglycerol-transporting lipoproteins and flight in *Locusta*. *J. Insect Physiol.*, *27*, 47–50.

Narayanaswami, V., & Ryan, R. O. (2000). The molecular basis of exchangeable apolipoprotein function. *Biochim. Biophys. Acta*, *1483*, 15–36.

Narayanaswami, V., Wang, J., Kay, C. M., Scraba, D. G., & Ryan, R. O. (1996a). Disulfide bond engineering to monitor conformational opening of apolipophorin III during lipid binding. *J. Biol. Chem.*, *271*, 26855–26862.

Narayanaswami, V., Frolov, A., Schroeder, F., Oikawa, K., Kay, C. M., et al. (1996b). Fluorescence studies of lipid-association induced conformational adaptations of an exchangeable amphipathic apolipoprotein. *Arch. Biochem. Biophys.*, *334*, 143–150.

Narayanaswami, V., Wang, J., Schieve, D., Kay, C. M., & Ryan, R. O. (1999). A molecular trigger of lipid-binding induced opening of a helix bundle exchangeable apolipoprotein. *Proc. Natl. Acad. Sci. USA*, *96*, 4366–4371.

Niere, M., Dettloff, M., Maier, T., Ziegler, M., & Wiesner, A. (2001). Insect immune activation by apolipophorin III is correlated with the lipid-binding properties of this protein. *Biochemistry*, *40*, 11502–11508.

Oudejans, R. C. H. M., Kooiman, F. P., Heerma, W., Versluis, C., Slotboom, A. J., et al. (1991). Isolation and structure elucidation of a novel adipokinetic hormone (Lom-AKH-III) from the glandular lobes of the corpus cardiacum of the migratory locust, *Locusta migratoria*. *Eur. J. Biochem.*, *195*, 351–359.

Oudejans, R. C. H. M., Mes, T. H. M., Kooiman, F. P., & Van der Horst, D. J. (1993). Adipokinetic peptide hormone content and biosynthesis during locust development. *Peptides*, *14*, 877–881.

Oudejans, R. C. H. M., Vroemen, S. F., Jansen, R. F. R., & Van der Horst, D. J. (1996). Locust adipokinetic hormones: Carrier-independent transport and differential inactivation at physiological concentrations during rest and flight. *Proc. Natl. Acad. Sci. USA*, *93*, 8654–8659.

Oudejans, R. C. H. M., & Van der Horst, D. J. (2003). Adipokinetic hormones: Structure and biosynthesis. In H. L. Henry, & A. W. Norman (Eds.). *Encyclopedia of Hormones* (vol. 1, pp. 38–42). London: Academic Press.

Patel, R. T., Soulages, J. L., & Arrese, E. L. (2006). Adipokinetic hormone-induced mobilization of fat body triglyceride stores in *Manduca sexta*: Role of TG-lipase and lipid droplets. *Arch. Insect Biochem. Physiol.*, *63*, 73–81.

Pownall, H. J., Massey, J. B., Kusserow, S. K., & Gotto, A. M., Jr. (1978). Kinetics of lipid–protein interactions: Interaction of apolipoprotein A-I from human plasma high density lipoproteins with phosphatidylcholines. *Biochemistry*, *17*, 1183–1188.

Prasad, S. V., Ryan, R. O., Wells, M. A., & Law, J. H. (1986). Changes in lipoprotein composition during larval–pupal metamorphosis of an insect, *Manduca sexta*. *J. Biol. Chem.*, *261*, 558–562.

Pratt, C. C., & Weers, P. M. M. (2004). Lipopolysaccharide binding of an exchangeable apolipoprotein, apolipophorin III, from *Galleria mellonella*. *Biol. Chem.*, *385*, 1113–1119.

Qu, H., Cui, L., Rickers-Haunerland, J., & Haunerland, N. H. (2007). Fatty acid-dependent expression of the muscle FABP gene – comparative analysis of gene control in functionally related, but evolutionary distant animal systems. *Mol. Cell. Biochem.*, *299*, 45–53.

Raikhel, A. S., Kokoza, V. A., Zhu, J., Martin, D., Wang, S. F., et al. (2002). Molecular biology of mosquito vitellogenesis: From basic studies to genetic engineering of antipathogen immunity. *Insect Biochem. Mol. Biol., 32,* 1275–1286.

Raussens, V., Goormaghtigh, E., Narayanaswami, V., Ryan, R. O., & Ruysschaert, J. M. (1995). Alignment of apolipophorin III α-helices in complex with dimyristoylphosphatidylcholine: A unique spatial orientation. *J. Biol. Chem., 270,* 12542–12547.

Raussens, V., Narayanaswami, V., Goormaghtigh, E., Ryan, R. O., & Ruysschaert, J. M. (1996). Hydrogen/deuterium exchange kinetics of apolipophorin III in lipid free and phospholipid bound states: An analysis by Fourier transform infrared spectroscopy. *J. Biol. Chem., 271,* 23089–23095.

Richardson, P. E., Manchekar, M., Dashti, N., Jones, M. K., Beigneux, A., et al. (2005). Assembly of lipoprotein particles containing apolipoprotein-B: Structural model for the native lipoprotein particle. *Biophys. J., 88,* 2789–2800.

Rodenburg, K. W., & Van der Horst, D. J. (2005). Lipoprotein-mediated lipid transport in insects: Analogy to the mammalian lipid carrier system and novel concepts for the functioning of LDL receptor family members. *Biochim. Biophys. Acta, 1736,* 10–29.

Rodenburg, K. W., Smolenaars, M. M. W., Van Hoof, D., & Van der Horst, D. J. (2006). Sequence analysis of the nonrecurring C-terminal domains shows that insect lipoprotein receptors constitute a distinct group of LDL receptor family members. *Insect Biochem. Mol. Biol., 36,* 250–263.

Roosendaal, S. D., Kerver, J., Schipper, M., Rodenburg, K. W., & Van der Horst, D. J. (2008). The complex of the insect LDL receptor homologue, LpR, and its lipoprotein ligand does not dissociate at endosomal conditions. *FEBS J., 275,* 1751–1766.

Roosendaal, S. D., Van Doorn, J. M., Valentijn, K. M., Van der Horst, D. J., & Rodenburg, K. W. (2009). Delipidation of insect lipoprotein, lipophorin, affects its binding to the lipophorin receptor, LpR: Implications for the role of LpR-mediated endocytosis. *Insect Biochem. Mol. Biol., 39,* 135–144.

Rudenko, G., Henry, L., Henderson, K., Ichtchenko, K., Brown, M. S., et al. (2002). Structure of the LDL receptor extracellular domain at endosomal pH. *Science, 298,* 2353–2358.

Ryan, R. O. (1990). The dynamics of insect lipophorin metabolism. *J. Lipid Res., 31,* 1725–1739.

Ryan, R. O. (2008). Nanodisks: Hydrophobic drug delivery vehicles. *Expert Opin. Drug Deliv, 3,* 343–351.

Ryan, R. O., & Van der Horst, D. J. (2000). Lipid transport biochemistry and its role in energy production. *Annu. Rev. Entomol., 45,* 233–260.

Ryan, R. O., Prasad, S. V., Henriksen, E. J., Wells, M. A., & Law, J. H. (1986a). Lipoprotein interconversions in an insect, *Manduca sexta*: Evidence for a lipid transfer factor in the hemolymph. *J. Biol. Chem., 261,* 563–568.

Ryan, R. O., Wells, M. A., & Law, J. H. (1986b). Lipid transfer protein from *Manduca sexta* hemolymph. *Biochem. Biophys. Res. Commun., 136,* 260–265.

Ryan, R. O., Senthilathipan, K. R., Wells, M. A., & Law, J. H. (1988). Facilitated diacylglycerol exchange between insect hemolymph lipophorins: Properties of *Manduca sexta* lipid transfer particle. *J. Biol. Chem., 263,* 14140–14145.

Ryan, R. O., Howe, A., & Scraba, D. G. (1990a). Studies of the morphology and structure of the plasma lipid transfer particle from the tobacco hornworm. *Manduca sexta. J. Lipid Res., 31,* 871–879.

Ryan, R. O., Wessler, A. N., Ando, S., Price, H. M., & Yokoyama, S. (1990b). Insect lipid transfer particle catalyzes bidirectional vectorial transfer of diacylglycerol from lipophorin to human low density lipoprotein. *J. Biol. Chem., 265,* 10551–10555.

Ryan, R. O., Oikawa, K., & Kay, C. M. (1993). Conformational, thermodynamic and stability properties of *Manduca sexta* apolipophorin III. *J. Biol. Chem., 268,* 1525–1530.

Ryan, R. O., Schieve, D. S., Wientzek, M., Narayanaswami, V., Oikawa, K., et al. (1995). Bacterial expression and site directed mutagenesis of a functional recombinant apolipoprotein. *J. Lipid Res., 36,* 1066–1072.

Sahoo, D., Narayanaswami, V., Kay, C. M., & Ryan, R. O. (2000). Pyrene excimer fluorescence: A spatially sensitive probe to monitor lipid induced rearrangement of apolipophorin III. *Biochemistry, 39,* 6594–6601.

Sahoo, D., Weers, P. M. M., Ryan, R. O., & Narayanaswami, V. (2002). Lipid-triggered conformational switch of apolipophorin III helix bundle to an extended helix organization. *J. Mol. Biol., 321,* 201–214.

Sakudoh, T., Tsuchida, K., & Kataoka, H. (2005). BmStart1, a novel carotenoid-binding protein isoform from *Bombyx mori*, is orthologous to MLN64, a mammalian cholesterol transporter. *Biochem. Biophys. Res. Commun, 336,* 1125–1135.

Sakudoh, T., Sezutsu, H., Nakashima, T., Kobayashi, I., Fujimoto, H., et al. (2007). Carotenoid silk coloration is controlled by a carotenoid-binding protein, a product of the *Yellow blood* gene. *Proc. Natl. Acad. Sci. USA, 104,* 8941–8946.

Sakudoh, T., Iizuka, T., Narukawa, J., Sezutsu, H., Kobayashi, I., et al. (2010). A CD36-related transmembrane protein is coordinated with an intracellular lipid-binding protein in selective carotenoid transport for cocoon coloration. *J. Biol. Chem., 285,* 7739–7751.

Sappington, T. W., & Raikhel, A. S. (1998). Molecular characteristics of insect vitellogenins and vitellogenin receptors. *Insect Biochem. Mol. Biol., 28,* 277–300.

Sappington, T. W., Kokoza, V. A., Cho, W. L., & Raikhel, A. S. (1996). Molecular characterization of the mosquito vitellogenin receptor reveals unexpected high homology to the *Drosophila* yolk protein receptor. *Proc. Natl. Acad. Sci. USA, 93,* 8934–8939.

Schwenk, R. W., Holloway, G. P., Luiken, J. J. F. P., Bonen, A., & Glatz, J. F. C. (2010). Fatty acid transport across the cell membrane: Regulation by fatty acid transporters. *Prostaglandins Leukot. Essent. Fatty Acids, 82,* 149–154.

Schonbaum, C. P., Lee, S., & Mahowald, A. P. (1995). The *Drosophila* yolkless gene encodes a vitellogenin receptor belonging to the low-density lipoprotein receptor superfamily. *Proc. Natl. Acad. Sci. USA, 92,* 1485–1489.

Segrest, J. P., Jones, M. K., De Loof, H., & Dashti, N. (2001). Structure of apolipoprotein B-100 in low density lipoproteins. *J. Lipid Res., 42,* 1346–1367.

Sellers, J. A., Hou, L., Athar, H., Hussain, M. M., & Shelness, G. S. (2003). A *Drosophila* microsomal triglyceride transfer protein homolog promotes the assembly and secretion of human apolipoprotein B: Implications for human and insect lipid transport and metabolism. *J. Biol. Chem., 278,* 20367–20373.

Seo, S. -J., Cheon, H. -M., Sun, J., Sappington, T. W., & Raikhel, A. S. (2003). Tissue- and stage-specific expression of two lipophorin receptor variants with seven and eight ligand-binding repeats in the adult mosquito. *J. Biol. Chem., 278,* 41954–41962.

Seo, S. J., Park, K. H., & Cho, K. H. (2008). Apolipophorin III from *Hyphantria cunea* shows different anti-oxidant ability against LDL oxidation in the lipid-free and lipid-bound state. *Comp. Biochem. Physiol. B, 151,* 433–439.

Shelness, G. S., & Sellers, J. A. (2001). Very-low-density lipoprotein assembly and secretion. *Curr. Opin. Lipidol., 12,* 151–157.

Singh, T. K. A., & Ryan, R. O. (1991). Lipid transfer particle catalyzed transfer of lipoprotein-associated diacylglycerol and long chain aliphatic hydrocarbons. *Arch. Biochem. Biophys., 286,* 376–382.

Singh, T. K. A., Scraba, D. G., & Ryan, R. O. (1992). Conversion of human low density lipoprotein into a very low density lipoprotein like particle *in vitro. J. Biol. Chem., 267,* 9275–9280.

Smith, A. F., Owen, L. M., Strobel, L. M., Chen, H., Kanost, M. R., et al. (1994). Exchangeable apolipoproteins of insects share a common structural motif. *J. Lipid Res., 35,* 1976–1984.

Smolenaars, M. M. W., Kasperaitis, M. A. M., Richardson, P. E., Rodenburg, K. W., & Van der Horst, D. J. (2005). Biosynthesis and secretion of insect lipoprotein: Involvement of furin in cleavage of the apoB homolog, apolipophorin-II/I. *J. Lipid Res., 46,* 412–421.

Smolenaars, M. M. W., Madsen, O., Rodenburg, K. W., & Van der Horst, D. J. (2007a). Molecular diversity and evolution of the large lipid transfer protein superfamily. *J. Lipid Res., 48,* 489–502.

Smolenaars, M. M. W., De Morrée, A., Kerver, J., Van der Horst, D. J., & Rodenburg, K. W. (2007b). Insect lipoprotein biogenesis depends on an amphipathic β cluster in apolipophorin-II/I and is stimulated by microsomal triglyceride transfer protein. *J. Lipid Res., 48,* 1955–1965.

Soulages, J. L., & Arrese, E. L. (2000a). Dynamics and hydration of the α-helices of apolipophorin III. *J. Biol. Chem., 275,* 17501–17509.

Soulages, J. L., & Arrese, E. L. (2000b). Fluorescence spectroscopy of single tryptophan mutants of apolipophorin-III in discoidal lipoproteins of dymyristoylphosphatidylcholine. *Biochemistry, 39,* 10574–10580.

Soulages, J. L., & Arrese, E. L. (2001). Interaction of the α-helices of apolipophorin III with phospholipid acyl chains in discoidal lipoprotein particle: a fluorescence quenching study. *Biochemistry, 40,* 14279–14290.

Soulages, J. L., & Wells, M. A. (1994a). Lipophorin: The structure of an insect lipoprotein and its role in lipid transport in insects. *Adv. Protein Chem., 45,* 371–415.

Soulages, J. L., & Wells, M. A. (1994b). Effect of diacylglycerol content on some physicochemical properties of the insect lipoprotein, lipophorin: Correlation with the binding of apolipophorin-III. *Biochemistry, 33,* 2356–2362.

Soulages, J. L., Salamon, Z., Wells, M. A., & Tollin, G. (1995). Low concentrations of diacylglycerol promote the binding of apolipophorin III to a phospholipid bilayer: A surface plasmon resonance spectroscopy study. *Proc. Natl. Acad. Sci. USA, 92,* 5650–5654.

Soulages, J. L., Van Antwerpen, R., & Wells, M. A. (1996). Role of diacylglycerol and apolipophorin-III in regulation of physicochemical properties of the lipophorin surface: Metabolic implications. *Biochemistry, 35,* 5191–5198.

Soulages, J. L., Pennington, J., Bendavid, O., & Wells, M. A. (1998). Role of glycosylation in the lipid-binding activity of the exchangeable apolipoprotein, apolipophorin-III. *Biochem. Biophys. Res. Commun, 243,* 372–376.

Soulages, J. L., Arrese, E. L., Chetty, P. S., & Rodriguez, V. (2001). Essential role of the conformational flexibility of helices 1 and 5 on the lipid binding activity of apolipophorin III. *J. Biol. Chem., 276,* 34162–34166.

Staubli, F., Jørgensen, T. J. D., Cazzamali, G., Williamson, M., Lenz, C., et al. (2002). Molecular identification of the insect adipokinetic hormone receptors. *Proc. Natl. Acad. Sci. USA, 99,* 3446–3451.

Storch, J., & McDermott, L. (2009). Structural and functional analysis of fatty acid-binding proteins. *J. Lipid Res., 50*(Suppl.), 126–131.

Sun, D., Ziegler, R., Milligan, C. E., Fahrbach, S., & Schwartz, L. M. (1995). Apolipophorin III is dramatically up-regulated during the programmed cell death of insect skeletal muscle and neurons. *J. Neurobiol., 26,* 119–129.

Sun, J., Hiraoka, T., Dittmer, N. T., Cho, K. -H., & Raikhel, A. S. (2000). Lipophorin as a yolk protein in the mosquito, *Aedes aegypti. Insect Biochem. Mol. Biol., 30,* 1161–1171.

Sztalryd, C., Xu, G., Dorward, H., Tansey, J. T., Contreras, J. A., et al. (2003). Perilipin A is essential for the translocation of hormone-sensitive lipase during lipolytic activation. *J. Cell Biol., 161,* 1093–1103.

Tabunoki, H., Sugiyama, H., Tanaka, Y., Fujii, H., Banno, Y., et al. (2002). Isolation, characterization, and cDNA sequence of a carotenoid binding protein from the silk gland of *Bombyx mori* larvae. *J. Biol. Chem., 277,* 32133–32140.

Tabunoki, H., Higurashi, S., Ninagi, O., Fujii, H., Banno, Y., et al. (2004). A carotenoid-binding protein (CBP) plays a crucial role in cocoon pigmentation of silkworm (*Bombyx mori*) larvae. *FEBS Lett., 567,* 175–178.

Takeuchi, N., & Chino, H. (1993). Lipid transfer particle in the hemolymph of the American cockroach: Evidence for its capacity to transfer hydrocarbons between lipophorin particles. *J. Lipid Res., 34,* 543–551.

Teixeira, L., Rabouille, C., Rorth, P., Ephrussi, A., & Vanzo, N. F. (2003). *Drosophila* Perilipin/ADRP homologue Lsd2 regulates lipid metabolism. *Mech. Devel., 120,* 1071–1081.

Thompson, J. R., & Banaszak, L. J. (2002). Lipid–protein interactions in lipovitellin. *Biochemistry, 41,* 9398–9409.

Tsuchida, K., Soulages, J. L., Moribayashi, A., Suzuki, K., Maekawa, H., et al. (1997). Purification and properties of a lipid transfer particle from *Bombyx mori*: Comparison to the lipid transfer particle from *Manduca sexta. Biochim. Biophys. Acta, 1337,* 57–65.

Tsuchida, K., Arai, M., Tanaka, Y., Ishihara, R., Ryan, R. O., et al. (1998). Lipid transfer particle catalyzes transfer of carotenoids between lipophorins of *Bombyx mori*. *Insect Biochem. Mol. Biol.*, *28*, 927–934.

Tsuchida, K., Katagiri, C., Tanaka, Y., Tabunoki, H., Sato, R., et al. (2004a). The basis for colorless hemolymph and cocoons in the *Y*-gene recessive *Bombyx mori* mutants: A defect in the cellular uptake of carotenoids. *J. Insect Physiol.*, *50*, 975–983.

Tsuchida, K., Jouni, Z. E., Gardetto, J., Kobayashi, Y., Tabunoki, H., et al. (2004b). Characterization of the carotenoid-binding protein of the *Y*-gene dominant mutants of *Bombyx mori*. *J. Insect Physiol.*, *50*, 363–372.

Tufail, M., & Takeda, M. (2008). Molecular characteristics of insect vitellogenins. *J. Insect Physiol.*, *54*, 1447–1458.

Tufail, M., & Takeda, M. (2009). Insect vitellogenin/lipophorin receptors: Molecular structures, role in oogenesis, and regulatory mechanisms. *J. Insect Physiol.*, *55*, 87–103.

Tufail, M., Elmogy, M., Ali Fouda, M. M., Elgendy, A. M., Bembenek, J., et al. (2009). Molecular cloning, characterization, expression pattern and cellular distribution of an ovarian lipophorin receptor in the cockroach, *Leucophaea maderae*. *Insect Mol. Biol.*, *18*, 281–294.

Van Antwerpen, R., Pham, D. Q. -D., & Ziegler, R. (2005). Accumulation of lipids in insect oocytes. In A. S. Raikhel, & T. W. Sappington (Eds.). *Progress in Vitellogenesis: Reproductive Biology of Invertebrates, Vol. XII, Part B* (pp. 265–288). Enfield, NH: Science Publishers Inc.

Vanden Broeck, J. (2001). Insect G protein-coupled receptors and signal transduction. *Arch. Insect Biochem. Physiol.*, *48*, 1–12.

Van der Horst, D. J. (1990). Lipid transport function of lipoproteins in flying insects. *Biochim. Biophys. Acta*, *1047*, 195–211.

Van der Horst, D. J. (2003). Insect adipokinetic hormones: Release and integration of flight energy metabolism. *Comp. Biochem. Physiol. B*, *136*, 217–226.

Van der Horst, D. J., & Oudejans, R. C. H. M. (2003). Adipokinetic hormones and lipid mobilization. In H. L. Henry, & A. W. Norman (Eds.). *Encyclopedia of Hormones* (Vol. 1, pp. 34–38). London, UK: Academic Press.

Van der Horst, D. J., & Rodenburg, K. W. (2010a). Lipoprotein assembly and function in an evolutionary perspective. *BioMol. Concepts*, *1*, 165–183.

Van der Horst, D. J., & Rodenburg, K. W. (2010b). Locust flight activity as a model for hormonal regulation of lipid mobilization and transport. *J. Insect Physiol.*, *56*, 844–853.

Van der Horst, D. J., & Rodenburg, K. W. (2011). Adipokinetic hormones and their role in lipid mobilization in insects. In M. Tufail, & M. Takeda (Eds.). *Hemolymph Proteins and Functional Peptides: Recent Advances in Insects and Other Arthropods*. Oak Park, IL: Bentham Science Publishers, in press.

Van der Horst, D. J., & Ryan, R. O. (2005). Lipid transport. In L. I. Gilbert, K. Iatrou, & S. S. Gill (Eds.). *Comprehensive Molecular Insect Science* (Vol. 4, pp. 225–246). Amsterdam: Elsevier.

Van der Horst, D. J., Van Doorn, J. M., & Beenakkers, A. M. T. (1979). Effects of the adipokinetic hormone on the release and turnover of hemolymph diglycerides and on the function of the diglyceride-transporting lipoprotein system during locust flight. *Insect Biochem.*, *9*, 627–635.

Van der Horst, D. J., Van Doorn, J. M., De Keijzer, A. N., & Beenakkers, A. M. T. (1981). Interconversions of diacylglycerol-transporting lipoproteins in the haemolymph of *Locusta migratoria*. *Insect Biochem.*, *11*, 717–723.

Van der Horst, D. J., Van Doorn, J. M., & Beenakkers, A. M. T. (1984). Hormone-induced rearrangements of locust hemolymph lipopoproteins: the involvement of glycoprotein C_2. *Insect Biochem.*, *14*, 495–504.

Van der Horst, D. J., Van Doorn, J. M., Passier, P. C. C. M., Vork, M. M., & Glatz, J. F. C. (1993). Role of fatty acid-binding protein in lipid metabolism of insect flight muscle. *Mol. Cell. Biochem.*, *123*, 145–152.

Van der Horst, D. J., Van Marrewijk, W. J. A., Vullings, H. G. B., & Diederen, J. H. B. (1999). Metabolic neurohormones: Release, signal transduction and physiological responses of adipokinetic hormones in insects. *Eur. J. Entomol.*, *96*, 299–308.

Van der Horst, D. J., Van Marrewijk, W. J. A., & Diederen, J. H. B. (2001). Adipokinetic hormones of insect: Release, signal transduction, and responses. *Int. Rev. Cytol.*, *211*, 179–240.

Van der Horst, D. J., Van Hoof, D., Van Marrewijk, W. J. A., & Rodenburg, K. W. (2002). Alternative lipid mobilization: The insect shuttle system. *Mol. Cell. Biochem.*, *239*, 113–119.

Van der Horst, D. J., Roosendaal, S. D., & Rodenburg, K. W. (2009). Circulatory lipid transport: Lipoprotein assembly and function from an evolutionary perspective. *Mol. Cell. Biochem.*, *326*, 105–119.

Van Heusden, M. C., & Law, J. H. (1989). An insect lipid transfer particle promotes lipid loading from fat body to lipoprotein. *J. Biol. Chem.*, *264*, 17287–17292.

Van Heusden, M. C., Yepiz-Plascencia, G. M., Walker, A. M., & Law, J. H. (1996). *Manduca sexta* lipid transfer particle: Synthesis by fat body and occurrence in hemolymph. *Arch. Insect Biochem. Physiol.*, *31*, 39–51.

Van Hoof, D., Rodenburg, K. W., & Van der Horst, D. J. (2002). Insect lipoprotein follows a transferrin-like recycling route that is mediated by the insect LDL receptor homologue. *J. Cell Sci.*, *115*, 4001–4012.

Van Hoof, D., Rodenburg, K. W., & Van der Horst, D. J. (2003). Lipophorin receptor-mediated lipoprotein endocytosis in insect fat body cells. *J. Lipid Res.*, *44*, 1431–1440.

Van Hoof, D., Rodenburg, K. W., & Van der Horst, D. J. (2005a). Receptor-mediated endocytosis and intracellular trafficking of lipoproteins and transferrin in insect cells. *Insect Biochem. Mol. Biol.*, *35*, 117–128.

Van Hoof, D., Rodenburg, K. W., & Van der Horst, D. J. (2005b). Intracellular fate of LDL receptor family members depends on the cooperation between their ligand-binding and EGF domains. *J. Cell Sci.*, *118*, 1309–1320.

Van Marrewijk, W. J. A. (2003). Adipokinetic hormones and carbohydrate metabolism. In H. L. Henry, & A. W. Norman (Eds.). *Encyclopedia of Hormones* (Vol. 1, pp. 29–34). London, UK: Academic Press.

Van Marrewijk, W. J. A., & Van der Horst, D. J. (1998). Signal transduction of adipokinetic hormone. In G. M. Coast, & S. G. Webster (Eds.). *Recent Advances in Arthropod Endocrinology* (pp. 172–188). Cambridge, UK: Cambridge University Press.

Vasquez, L. J., Abdullahi, G. E., Wan, C. P., & Weers, P. M. M. (2009). Apolipophorin III lysine modification: Effect on structure and lipid binding. *Biochim. Biophys. Acta, 1788,* 1901–1906.

Vroemen, S. F., Van der Horst, D. J., & Van Marrewijk, W. J. A. (1998). New insights into adipokinetic hormone signaling. *Mol. Cell. Endocrinol., 141,* 7–12.

Vullings, H. G. B., Diederen, J. H. B., Veelaert, D., & Van der Horst, D. J. (1999). Multifactorial control of the release of hormones from the locust retrocerebral complex. *Microsc. Res. Tech., 45,* 142–153.

Walther, T. C., & Farese, R. V., Jr. (2009). The life of lipid droplets. *Biochim. Biophys. Acta, 1791,* 459–466.

Wang, H., Hu, L., Dalen, K., Dorward, H., Marcinkiewicz, A., et al. (2009). Activation of hormone-sensitive lipase requires two steps, protein phosphorylation and binding to the PAT-1 domain of lipid droplet coat proteins. *J. Biol. Chem., 284,* 32116–32125.

Wang, J., Liu, H., Sykes, B. D., & Ryan, R. O. (1992). P-NMR study of the phospholipid moiety of lipophorin subspecies. *Biochemistry, 31,* 8706–8712.

Wang, J., Liu, H., Sykes, B. D., & Ryan, R. O. (1995). Localization of two distinct microenvironments for the diacylglycerol component of lipophorin particles by ^{13}C NMR. *Biochemistry, 34,* 6755–6761.

Wang, J., Sahoo, D., Schieve, D., Gagne, S., Sykes, B. D., et al. (1997a). Multidimensional NMR studies of an exchangeable apolipoprotein and its interaction with lipids. *Tech. Protein Chem., 8,* 427–438.

Wang, J., Gagne, S., Sykes, B. D., & Ryan, R. O. (1997b). Insight into lipid surface recognition and reversible conformational adaptations of an exchangeable apolipoprotein by multidimensional heteronuclear NMR techniques. *J. Biol. Chem., 272,* 17912–17920.

Wang, J., Sykes, B. D., & Ryan, R. O. (2002). Structural basis for the conformational adaptability of apolipophorin III, a helix bundle exchangeable apolipoprotein. *Proc. Natl. Acad. Sci. USA, 99,* 1188–1193.

Wang, T., Jiao, Y., & Montell, C. (2007). Dissection of the pathway required for generation of vitamin A and for *Drosophila* phototransduction. *J. Cell Biol., 177,* 305–316.

Watt, M. J., & Steinberg, G. R. (2008). Regulation and function of triacylglycerol lipases in cellular metabolism. *Biochem. J., 414,* 313–325.

Weers, P. M. M., & Ryan, R. O. (2006). Apolipophorin III: Role model apolipoprotein. *Insect Biochem. Mol. Biol., 36,* 231–240.

Weers, P. M. M., Van Marrewijk, W. J. A., Beenakkers, A. M. T., & Van der Horst, D. J. (1993). Biosynthesis of locust lipophorin: Apolipophorins I and II originate from a common precursor. *J. Biol. Chem., 268,* 4300–4303.

Weers, P. M. M., Kay, C. M., Oikawa, K., Wientzek, M., Van der Horst, D. J., et al. (1994). Factors affecting the stability and conformation of *Locusta migratoria* apolipophorin III. *Biochemistry, 33,* 3617–3624.

Weers, P. M. M., Wang, J., Van der Horst, D. J., Kay, C. M., Sykes, B. D., et al. (1998). Recombinant locust apolipophorin III: Characterization and NMR spectroscopy. *Biochim. Biophys. Acta, 1393,* 99–107.

Weers, P. M. M., Narayanaswami, V., Kay, C. M., & Ryan, R. O. (1999). Interaction of an exchangeable apolipoprotein with phospholipid vesicles and lipoprotein particles: Role of leucines 32, 34, and 95 in *Locusta migratoria* apolipophorin III. *J. Biol. Chem., 274,* 21804–21810.

Weers, P. M. M., Van der Horst, D. J., & Ryan, R. O. (2000). Interaction of locust apolipophorin III with lipoproteins and phospholipid vesicles: Effect of glycosylation. *J. Lipid Res., 41,* 416–423.

Weers, P. M. M., Abdullahi, W. E., Cabrera, J. M., & Hsu, T. C. (2005). Role of buried polar residues in helix bundle stability and lipid binding of apolipophorin III: Destabilization by threonine 31. *Biochemistry, 44,* 8810–8816.

Whitten, M. M. A., Tew, I. F., Lee, B. L., & Ratcliffe, N. A. (2004). A novel role for an insect apolipoprotein (apolipophorin III) in β-1,3-glucan pattern recognition and cellular encapsulation reactions. *J. Immunol., 172,* 2177–2185.

Wicher, D., Agricola, H. J., Söhler, S., Gundel, M., Heinemann, S. H., et al. (2006). Differential receptor activation by cockroach adipokinetic hormones produces differential effects on ion currents, neuronal activity, and locomotion. *J. Neurophysiol., 95,* 2314–2325.

Wientzek, M., Kay, C. M., Oikawa, K., & Ryan, R. O. (1994). Binding of insect apolipophorin III to dimyristoylphosphatidylcholine vesicles: Evidence for a conformational change. *J. Biol. Chem., 269,* 4605–4612.

Wiesner, A., Losen, S., Kopacek, P., Weise, C., & Gotz, P. (1997). Isolated apolipophorin III from *Galleria mellonella* stimulates the immune reactions of this insect. *J. Insect Physiol., 43,* 383–391.

Woodring, J., Lorenz, M. W., & Hoffmann, K. H. (2002). Sensitivity of larval and adult crickets (*Gryllus bimaculatus*) to adipokinetic hormone. *Comp. Biochem. Physiol. A, 133,* 637–644.

Yancey, P. G., de la Llera-Moya, M., Swarnakar, S., Monzo, P., Klein, S. M., et al. (2000). High density lipoprotein phospholipid composition is a major determinant of the bi-directional flux and net movement of cellular free cholesterol mediated by scavenger receptor BI. *J. Biol. Chem., 275,* 36596–36604.

Yun, H. K., Jouni, Z. E., & Wells, M. A. (2002). Characterization of cholesterol transport from midgut to fat body in *Manduca sexta* larvae. *Insect Biochem. Mol. Biol., 32,* 1151–1158.

Zechner, R., Kienesberger, P. C., Haemmerle, G., Zimmermann, R., & Lass., A. (2009). Adipose triglyceride lipase and the lipolytic catabolism of cellular fat stores. *J. Lipid Res., 50,* 3–21.

Zhu, C., Huang, H., Hua, R., Li, G., Yang, D., et al. (2009). Molecular and functional characterization of adipokinetic hormone receptor and its peptide ligands in *Bombyx mori*. *FEBS Lett., 583,* 1463–1468.

Ziegler, R., & Van Antwerpen, R. (2006). Lipid uptake by insect oocytes. *Insect Biochem. Mol. Biol., 36,* 264–272.

Zimmermann, R., Lass, A., Haemmerle, G., & Zechner, R. (2009). Fate of fat: The role of adipose triglyceride lipase in lipolysis. *Biochim. Biophys. Acta, 1791,* 494–500.

10   Insect Proteases

Michael R Kanost
Department of Biochemistry, Kansas State University,
Manhattan, KS, USA
Rollie J Clem
Division of Biology, Kansas State University,
Manhattan, KS, USA

10.1. Introduction and History	346
10.2. Proteases in Eggs and Embryos	347
10.2.1. Proteases that Digest Egg Yolk Proteins	347
10.2.2. The Dorsal Pathway in Embryonic Development	347
10.3. Hemolymph Plasma Proteases	348
10.3.1. Serine Proteases	348
10.3.2. Protease Inhibitors	351
10.4. Cellular Proteases	352
10.4.1. Cathepsin-Type Cysteine Proteases	352
10.4.2. Caspases	352
10.4.3. Metalloproteases	354
10.4.4. Aspartic Acid Proteases	355
10.4.5. Proteasomes	355
10.5. Conclusions and Future Prospects	355

10.1. Introduction and History

Proteases (peptidases) are enzymes that hydrolyze peptide bonds in proteins. Exopeptidases cleave a terminal amino acid residue at the end of a polypeptide; endopeptidases cleave internal peptide bonds. Hooper (2002) provides a useful introduction to the general properties of proteases. Proteases can be classified based on the chemical groups that function in catalysis. In serine proteases, the hydroxyl group in the side chain of a serine residue in the active site acts as a nucleophile in the reaction that hydrolyzes a peptide bond, whereas in cysteine proteases the sulfhydryl group of a cysteine side chain performs this function. In aspartic acid proteases and metalloproteases, a water molecule in the active site (positioned by interacting with an aspartyl group or a metal ion, respectively) functions as the nucleophile that attacks the peptide bond. Proteases are classified on this basis of catalytic mechanism in a system developed by the Nomenclature Committee of the International Union of Biochemistry and Molecular Biology (http://www.chem.qmul.ac.uk/iubmb/enzyme/EC3/4/). However, proteases can have the same catalytic mechanism but be unrelated in amino acid sequence, as products of convergent evolution. The MEROPS classification system groups proteases into families based on sequence similarity (Rawlings et al. 2010) (http://merops.sanger.ac.uk).

A protease cleaves a peptide bond, called the scissile bond, between two amino acid residues named P1 and P1′ (Schechter and Berger, 1967). Residues on the amino-terminal side of the scissile bond are numbered in the C to N direction, whereas residues on the carboxyl-terminal side of the scissile bond (the "prime" side) are numbered in the N to C direction (**Figure 1**). The substrate specificity of most endopeptidases is highly dependent on the nature of the side chain of the P1 residue, but the sequence of other residues near the scissile bond can also affect binding of the substrate to the active site, and thus influence substrate specificity.

Insects produce abundant proteases that function in the digestion of dietary proteins in the gut. Such proteases are thoroughly discussed in Chapter 11 of this volume. This chapter focuses on non-digestive proteases, which have many diverse roles in insect biology. These proteases often function in cascade pathways, in which one protease activates the zymogen form of another protease, leading to amplification of an initial signal that may involve a few molecules, and finally generating a very large number of effector molecules at the end of the pathway. The complement and blood coagulation pathways in mammals are well understood examples of this type of protease cascade, which also occurs in insect embryonic development and insect immune responses (Krem and DiCerra, 2002; Cerenius *et al.*, 2010). Details of the organization and regulation of such pathways in insect biology are beginning to be understood in a few species. Intracellular cysteine protease cascades leading to apoptosis represent

DOI:10.1016/B978-0-12-384747-8.10010-8

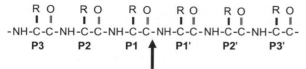

Figure 1 The Schechter and Berger (1967) notation for protease cleavage sites. The arrow designates the scissile peptide bond between amino acid residues P1 and P1′.

another important regulatory pathway involving proteases (Feinstein-Rotkopf and Arama, 2009). This chapter will include an emphasis on the current state of knowledge in these rapidly developing areas.

Insect proteases have previously been reviewed by Law *et al.* (1977), Applebaum (1985), Terra *et al.* (1996), and Reeck *et al.* (1999). These reviews deal primarily with proteases as they function in the digestion of food. Only recently has much detailed information appeared about proteases with other functions in insect biology. An exception is cocoonase, the first insect protease that was purified and well characterized biochemically. Cocoonase is a serine protease from silkmoths that functions to hydrolyze silk proteins in the cocoon, enabling the adult moth to emerge (Kafatos *et al.*, 1967a, 1967b). It digests sericin, the silk protein that cements fibroin threads together. A specialized tissue called the galea, derived from modified mouthparts, synthesizes and secretes the zymogen form, procicoonase (Kafatos, 1972). On the surface of the galea, procicoonase is activated by cleavage at a specific site by an unknown protease in the molting fluid (Berger *et al.*, 1971). Sequencing of an amino terminal fragment and the peptide containing the active site Ser residue indicated that the activation and catalytic mechanisms of cocoonase were quite similar to those of mammalian trypsin (Felsted *et al.*, 1973; Kramer *et al.*, 1973). It is surprising that the gene encoding this historically important insect protease has not yet been identified.

10.2. Proteases in Eggs and Embryos

10.2.1. Proteases that Digest Egg Yolk Proteins

Vitellin and a few other egg-specific proteins stored in yolk granules of insect eggs are digested by proteases to release amino acids for use in embryonic development (Raikhel and Dhadialla, 1992). Such enzymes in eggs represent several different protease families and mechanistic classes. A serine protease that degrades vitellin was purified from *Bombyx mori* eggs (Indrasith *et al.*, 1988), and its cDNA was cloned (Ikeda *et al.*, 1991). This protease cleaves after Arg or Lys P1 residues, and is a member of the S1 (chymotrypsin-like) family of serine proteases. It is synthesized in ovaries as a zymogen, and is activated during embryogenesis. A second serine protease from the S1 family specifically degrades the 30-kDa yolk proteins present in *B. mori*

eggs (Maki and Yamashita, 1997, 2001). This protease, which is synthesized at the end of embryogenesis, has elastase-like specificity, cleaving after P1 residues with small side chains. A serine carboxypeptidase is synthesized in the fat body of a mosquito, *Aedes aegypti*, transported through the hemolymph, and taken up by oocytes (Cho *et al.* 1991). This protease is synthesized as a zymogen, and activated within eggs during embryogenesis.

Cysteine proteases have been characterized from eggs of several insect species. Those that have been sequenced are from the C1 (papain-like) family of cysteine proteases. They typically have acidic pH optima, and have biochemical properties similar to mammalian cysteine proteases known as cathepsins (although not all proteases called cathepsins are cysteine proteases). A 47-kDa cysteine protease that can digest vitellin has been purified from *B. mori* eggs (Kageyama and Takahashi, 1990; Yamamoto and Takahashi, 1993), and its cDNA has been cloned (Yamamoto *et al.*, 1994). It has sequence similarity to mammalian cathepsin L, and a preference for cleaving at sites that have hydrophobic residues at the P2 and P3 positions. It is synthesized as a zymogen in ovary and fat body as a maternal product, and taken up into oocytes (Yamamoto *et al.*, 2000). This cysteine protease is self-activated at low pH by proteolytic processing, apparently by a weak activity of the proenzyme under acidic conditions (Takahashi *et al.*, 1993a).

Cysteine proteases with sequence similarity to mammalian cathepsin B have also been identified as enzymes that digest insect egg yolk proteins. In *Drosophila melanogaster*, a cysteine protease is associated with yolk granules (Medina *et al.*, 1988). Its zymogen is apparently activated by a serine protease during embryonic development, and the active cathepsin-B then digests yolk proteins. A cysteine protease that digests yolk proteins has also been identified in another higher dipteran, *Musca domestica* (Ribolla and De Bianchi, 1995). In *Ae. aegypti*, a "vitellogenic cathepsin B" is synthesized in adult female fat body after the insect has taken a blood meal, and the zymogen is transported through the hemolymph and taken up by oocytes (Cho *et al.*, 1999). The enzyme is activated by proteolytic processing when embryonic development begins, and then probably functions to digest vitellin. A cathepsin B-like protease that can digest vitellin is also synthesized in fat body and ovaries of a lepidopteran insect, *Helicoverpa armigera* (Zhao *et al.*, 2002). However, its gene is expressed in fat body of males and females, and in larvae and pupae, and thus is not coordinated with vitellogenesis as is the mosquito cathepsin-B.

10.2.2. The Dorsal Pathway in Embryonic Development

A signal transduction system that regulates dorsal/ventral patterning in *D. melanogaster* embryonic development is activated by an extracellular serine protease cascade

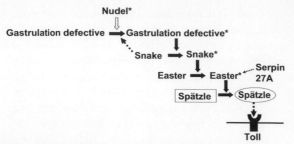

Figure 2 A model of the protease cascade that activates the Dorsal signal transduction pathway in *D. melanogaster* embryonic development. Nudel, gastrulation defective, snake, and easter are serine proteases that are synthesized as zymogens. The active forms of the proteases are indicated with an asterisk. Solid arrows indicate proteolytic activation steps that have been demonstrated by biochemical studies. A dotted arrow indicates that interaction between the snake and gastrulation defective zymogens can lead to activation of gastrulation defective. Easter* cleaves spätzle to produce an active ligand that binds to Toll, a transmembrane receptor. Easter is negatively regulated by interaction with an inhibitor from the serpin family.

(Morisato and Anderson, 1995; LeMosy *et al.*, 1999; Moussian and Roth, 2005). The members of this cascade are produced maternally, and deposited in the space between the vitellin membrane and the embryo. The pathway was elucidated by genetic analysis, and recently the recombinant forms of the proteases have been studied. This pathway involves a serine protease cascade (**Figure 2**) that eventually cleaves an inactive protein called spätzle, making it competent to bind to a transmembrane receptor named Toll. Binding of spätzle to Toll initiates a signal transduction pathway that leads to activation of a transcription factor from the rel family named Dorsal.

A large (350-kDa), multi-domain protein called nudel, containing a serine protease domain, regions of LDL receptor repeats, and an amino-terminal glycosaminoglycan modification, is secreted by the ovarian follicle cells into the perivitelline space (Hong and Hashimoto, 1995; Turcotte and Hashimoto, 2002). The nudel protease is autoactivated by a mechanism not yet understood (LeMosy *et al.*, 1998, 2000), and is thought to activate the second protease in the pathway, "gastrulation defective," by specific proteolysis. Mutations in nudel's protease domain produce a dorsalizing phenotype and can also result in fragile eggshells, suggesting an additional function for the protease activity (Hong and Hashimoto, 1996; LeMosy *et al.*, 1998, 2000; LeMosy and Hashimoto, 2000).

Gastrulation defective is a serine protease (Konrad *et al.*, 1998; Han *et al.*, 2000), with sequence similarity to mammalian complement factors C2 and B (DeLotto, 2001). Experiments with recombinant proteins have demonstrated that the gastrulation defective zymogen can be autoactivated when it interacts with the zymogen form of a protease named snake, and that active gastrulation defective can in turn proteolytically activate snake

(Dissing *et al.*, 2001; LeMosy *et al.*, 2001). Computer modeling studies indicate that the zymogen activation site of gastrulation defective is a good fit in the active site of nudel and that the snake zymogen activation site can dock in the active site of gastrulation defective, consistent with the proposed functions of these enzymes in the cascade pathway (Rose *et al.*, 2003). A potential lower affinity interaction of the gastrulation defective active site with its own zymogen activation sequence may explain the autoactivation of gastrulation defective in the absence of nudel when it is overexpressed in embryos or at high concentration *in vitro*.

The final two proteases in this cascade, snake and easter, contain carboxyl-terminal serine protease domains and amino-terminal clip domains similar to horseshoe crab proclotting enzyme (DeLotto and Spierer, 1986; Chasan and Anderson, 1989; Gay and Keith, 1992; Smith and DeLotto, 1992, 1994). Clip domains, thought to function in protein–protein interactions, are also present in some hemolymph proteases of insects (Jiang and Kanost, 2000) (see section 10.3.1 below). Mutations that eliminate the protease activity of snake (Smith *et al.*, 1994) or easter (Jin and Anderson, 1990) have abnormal dorsoventral phenotypes, indicating that a functional protease domain is essential for their roles in embryonic development. *In vitro* experiments with recombinant snake and easter zymogens confirm their order in the cascade indicated by genetic analysis: snake cleaves and activates easter, which cleaves prospätzle (Smith *et al.*, 1995; DeLotto and DeLotto, 1998; Dissing *et al.*, 2001; LeMosy *et al.*, 2001). These results are consistent with predictions based on computer modeling of the snake and easter three-dimensional structures and substrate interactions sites (Rose *et al.*, 2003). Active easter is converted *in vivo* to a high molecular mass form which is probably a complex with a protease inhibitor that regulates its activity (Misra *et al.*, 1998; Chang and Morisato, 2002). Female flies with a mutation in the gene for a serine protease inhibitor, serpin 27A, produce embryos that show defects in dorsal–ventral polarity, suggesting that this inhibitor is a maternal product that regulates at least one of the proteases in the pathway (Hashimoto *et al.*, 2003; Ligoxygakis *et al.*, 2003).

10.3. Hemolymph Plasma Proteases

10.3.1. Serine Proteases

Serine proteases in hemolymph have several types of physiological functions in defense against infection or wounding. An unusual phenomenon, perhaps related to protection against predation, involves serine proteases in hemolymph of South American Saturniid caterpillars of the genus *Lonomia*, which are toxic to mammals. Contact with these caterpillars can result in acquired bleeding disorders due to the potent fibrinolytic activity of these hemolymph proteases (Amarant *et al.*, 1991; Arocha-Pinango *et al.*, 2000; Pinto *et al.*, 2006).

Extracellular serine protease cascades mediate rapid responses to infection and wounding in vertebrate and invertebrate animals. Biochemical and genetic evidence indicates that activation of serine proteases in arthropod hemolymph is a component of several immune responses, including coagulation, melanotic encapsulation, activation of antimicrobial peptide synthesis, and modulation of hemocyte function (Barillas-Mury, 2007; Kanost and Gorman, 2008; Cerenius *et al.*, 2010; Jiang *et al.*, 2010). Most hemolymph proteases are expressed in fat body or hemocytes, but a bacteria-induced protease, scolexin, from *Manduca sexta* is expressed in epidermis (Finnerty *et al.*, 1999). Another novel serine protease expressed in pupal yellow body of *Sarcophaga peregrina* has antibacterial activity distinct from its protease activity (Nakajima *et al.*, 1997; Tsuji *et al.*, 1998).

Serine proteases that contain a carboxyl-terminal protease domain and an amino-terminal clip domain are known to act in cascade pathways in arthropod hemolymph (Jiang and Kanost, 2000; Cerenius *et al.*, 2010). Among clip domain proteases with known function are horseshoe crab proclotting enzyme and clotting factor B (Kawabata *et al.*, 1996), *D. melanogaster* snake and easter (see section 10.2.2), and proteases involved in activating phenoloxidase and the cytokine Spätzle during immune responses, as described below. Clip domains are 35- to 55-amino acid residue sequences that contain three conserved disulfide bonds. Structures of clip domains from two proteins have been solved, and they contain a conserved pattern of a three-stranded antiparallel β-sheet flanked by two α-helices (Piao *et al.*, 2005; Huang *et al.*, 2007). They may function to mediate interactions of members of protease cascade pathways. Proteases may contain one or more amino-terminal clip domains, followed by a 20- to 100-residue linking sequence connecting them to an S1 family protease domain.

Insects that have been investigated in some detail are known to contain a large number of genes for clip domain proteases. Among the 204 genes with homology to the S1 serine protease family in the *D. melanogaster* genome, 24 are clip domain proteases, most of whose functions are unknown (Ross *et al.*, 2003). Numbers of clip domain proteases in some other insect genomes include 18 in the honeybee *Apis mellifera* (Zou *et al.*, 2006) and 31 in *Tribolium castaneum* (Zou *et al.*, 2007), while the genomes of mosquitoes are extremely rich in genes encoding clip domain proteases, with 55 in *Anopheles gambiae* and 71 in *Ae. aegypti* (Waterhouse *et al.*, 2007). Among Lepidopterans, the *B. mori* genome contains 15 clip domain proteases (Tanaka *et al.* 2008), whereas more than 20 clip domain proteases expressed in fat body or hemocytes have been identified in *M. sexta* (Jiang *et al.*, 2005; Zou *et al.*, 2008).

Melanization, a response to wounding and infection in insects and crustaceans, involves activation of a cascade of serine protease zymogens (**Figure 3**). This pathway leads to rapid activation of a protease which then activates a phenoloxidase zymogen (prophenoloxidase; proPO) (Kanost and Gorman, 2008). Oxidation of phenols by phenoloxidase leads to production of quinones that polymerize to form melanin. Melanization of encapsulated parasites is believed to be an important defensive response in insects, including insect vectors of human diseases (Cerenius *et al.*, 2008). Serine proteases demonstrated to activate prophenoloxidase have been characterized from two lepidopteran insects, *M. sexta* (Jiang *et al.*, 1998, 2003a, 2003b) and *B. mori* (Satoh *et al.*, 1999); two beetles, *Holotrichia diomphalia* (Lee *et al.*, 1998a, 1998b) and *Tenebrio molitor* (Kan *et al.*, 2008); a mosquito, *A. gambiae* (An *et al.*, 2011a); and a crayfish (Wang *et al.*, 2001). All of these enzymes contain a carboxyl-terminal serine protease catalytic domain and one or two amino-terminal clip domains, and they are synthesized as zymogens which must be activated by a protease upstream in the pathway. The pathways are initiated by modular proteins that contain multiple small domains and a carboxyl-terminal serine protease domain (Wang and Jiang, 2006, 2007; Kan *et al.*, 2008). The amino terminal modules interact with hemolymph pattern recognition proteins and microbial polysaccharides at a microbial surface (see Chapter 14), and become self-activated by mechanisms not yet understood (Wang and Jiang, 2007, 2010; Kan *et al.*, 2008).

Similar pathways for activation of proPO have been characterized in *M. sexta* and in *T. molitor* (**Figure 3**). In *M. sexta*, a modular protease, hemolymph protease 14 (HP14) becomes active upon exposure to fungi or Gram-positive bacteria and interaction with a β-1,3-glucan recognition protein. HP14 activates a clip domain protease, HP21, which then activates prophenoloxidase-activating proteases 2 and 3 (PAP2, PAP3) (Gorman *et al.*, 2007; Wang and Jiang, 2007). These proteases then cleave and activate proPO (Jiang *et al.*, 1998, 2003a, 2003b). In an alternative pathway, a clip domain protease HP6 activates PAP1, which then can activate proPO (An *et al.*, 2009). In *T. molitor*, interactions with microbial surfaces and hemolymph pattern recognition proteins stimulate activation of a modular protease homologous to *M. sexta* HP14 (Roh *et al.*, 2009), which then activates a clip domain protease, SAE, and SAE activates a clip domain protease (SPE), which can cleave and activate proPO (Kan *et al.*, 2008; Roh *et al.*, 2009). In *D. melanogaster*, genetic evidence indicates that two clip domain proteases named MP1 and MP2 are involved in a proPO pathway (Tang *et al.*, 2006), but their order in the pathway and the identity of the protease that directly activates proPO are not yet known.

Initial characterization of a proPO-activating proteinase in *M. sexta* indicated that the purified protease could

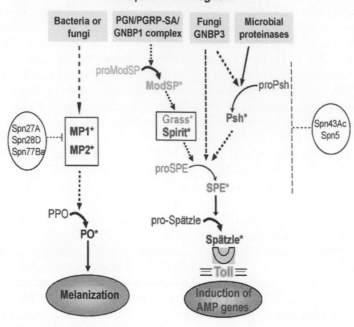

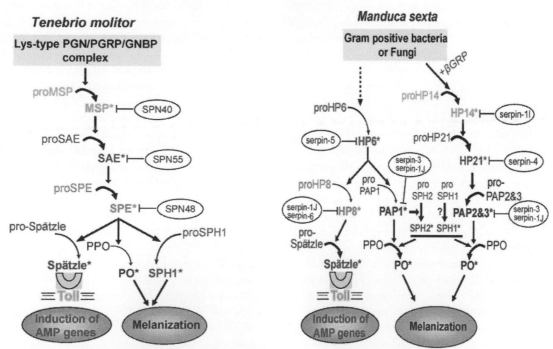

Figure 3 A model of hemolymph protease cascades in immune systems of species from three insect orders. Hemolymph plasma proteins known as pattern recognition proteins bind to polysaccharides on the surface of microorganisms. This interaction leads to activation of modular serine proteases by a mechanism not yet understood, which triggers cascade pathways formed from clip domain proteases and protease homologs. The final protease in the pathway cleaves and activates prophenoloxidase (ProPO) to form active phenoloxidase (PO) or pro-spätzle to produce an active ligand that binds to Toll, a transmembrane receptor, stimulating synthesis of antimicrobial peptides. PO catalyzes the oxidation of hemolymph catecholic phenols to corresponding quinones, which can undergo further reactions to form melanin. Proteases in the pathway are regulated by serine protease inhibitors known as serpins. For protease names shown in boxes, genetic evidence indicates participation in an immune pathway, but the activating protease and the protease's substrate are not yet known. Dashed arrows indicate putative steps that have not been verified experimentally. Serpins connected to proteases by dashed lines indicate serpins known from genetic evidence to be involved in regulating a pathway, but the proteases they inhibit have not been determined.

not efficiently activate proPO, but required participation of a non-proteolytic protein fraction (Jiang *et al.*, 1998). This protein cofactor was identified in *H. diomphalia* (Kwon *et al.*, 2000) and *M. sexta* (Yu *et al.*, 2003) as a protein with a clip domain and a serine protease domain, in which the active site serine residue is changed to glycine. Such clip domain serine protease homolog genes are also abundant in insect genomes (Ross *et al.*, 2003; Waterhouse *et al.*, 2007; Zou *et al.*, 2007). The serine protease homologs lack protease activity due to the incomplete catalytic triad, and must therefore have other functions. They may promote a structural change in the proPO substrate or the activating protease, and may function to cause proper spatial interaction of proPO and its activating protease (Wang and Jiang, 2004a; Gupta *et al.*, 2005; Piao, *et al.*, 2005). In *A. gambiae*, different SPH proteins may activate or suppress the melanization response (Barillas-Mury *et al.*, 2007). One *D. melanogaster* serine protease homolog, masquerade, functions in nerve and muscle development (Morugasu-Oei *et al.*, 1995, 1996). A serine protease homolog in crayfish hemolymph has a role in immune responses, indicating evolutionary conservation of function in these unusual proteins (Lee and Söderhäll, 2001). The active forms of the serine protease homologs that function as cofactors for proPO activation are themselves activated through specific cleavage by a serine protease in hemolymph (Kim *et al.*, 2002; Lee *et al.*, 2002; Yu *et al.*, 2003; Kan *et al.*, 2008), adding additional complexity to this pathway. The serine protease homologs from *M. sexta* that stimulate proPO activation bind to a hemolymph lectin that is a recognition protein for lipopolysaccharides from Gram-negative bacteria, and to proPO and prophenoloxidase-activating protease (Yu *et al.*, 2003). The interaction between the lectin and a proPO activation complex may serve to localize melanin synthesis to the surface of invading bacteria.

Serine proteases are also involved in other insect immune responses. The signal transduction system that regulates dorsal–ventral development in *D. melanogaster* embryos also regulates expression of the gene for an antifungal peptide in larvae and adults (Ferrandon *et al.*, 2007). In embryonic development, this pathway involves an extracellular serine protease cascade that eventually cleaves an inactive protein, spätzle, making it competent to bind to a transmembrane receptor named Toll (section 10.2.2). This same receptor–ligand interaction activates a signal pathway that leads to activation of rel family transcription factors that stimulate expression of drosomycin, an antifungal peptide synthesized by the fat body after microbial challenge. However, mutants of gastrulation defective, snake, or easter do not have an impaired antifungal response, indicating that a different set of proteases functions in the immune response protease cascade. Like prophenoloxidase activation, this pathway is initiated by interactions of pattern recognition proteins

with microbial surface polysaccharides (Ferrandon *et al.*, 2007), stimulating activation of a modular serine protease as the initiating protease of the cascade (Buchon *et al.*, 2009). Clip domain proteases known as Persephone, Grass, and Spirit participate in this pathway (Ligoxygakis *et al.*, 2002; Kambris *et al.*, 2006), but their positions in the cascade and their substrates are not yet known. Persephone may be a "danger sensing" protease, activated by microbial proteases and thereby stimulating a protective antimicrobial response (El Chamy *et al.*, 2008). A clip domain protease that cleaves and activates proSpätzle, Spätzle-processing enzyme, has been characterized (Jang *et al.*, 2006) and shown to be required for the Toll signaling immune response.

A hemolymph protease cascade leading to Spätzle activation, and presumably the Toll signaling pathway, has been fully defined in *T. molitor* (Kim *et al.*, 2008; Roh *et al.*, 2009). This pathway includes the same proteases that lead to proPO activation, including the same modular initiating protease which activates SPE-activating enzyme (SAE), which then activates Spätzle-processing enzyme (SPE) (**Figure 3**). SPE can activate both proSpätzle and proPO. In contrast, in *M. sexta* the proPO-activating proteases cannot activate proSpätzle, and instead a different clip domain protease, HP8, cleaves and activates this cytokine (An *et al.*, 2010). HP8 is activated by the clip domain protease HP6, which is a putative ortholog of *D. melanogaster* Persephone (An *et al.*, 2009).

10.3.2. Protease Inhibitors

Insect hemolymph contains high concentrations of serine protease inhibitors from several different gene families (Kanost, 1999). Protease cascade pathways in mammalian blood are regulated by ~45-kDa protease inhibitors known as serpins (Gettins, 2002). Serpins also function in arthropod hemolymph to regulate protease cascades, preventing detrimental effects of uncontrolled immune responses. Serpin gene families present in sequenced insect genomes have been analyzed in *B. mori* (Zou *et al.*, 2009), *A. gambiae* (Suwanchaichinda and Kanost, 2009), *A. mellifera* (Zou *et al.*, 2006), and *T. castaneum* (Zou *et al.*, 2007).

The reactive site in a serpin protein that interacts with the target protease is part of an exposed loop near the carboxyl-terminal end of the serpin sequence. Some insect serpin genes have a unique structure in which mutually exclusive alternate splicing of an exon that encodes the reactive site loop results in production of several inhibitors with different specificity. This was first observed in the gene for *M. sexta* serpin-1, which contains 12 copies of its ninth exon. Each version of exon 9 encodes a different reactive site loop sequence and inhibits a different spectrum of proteases (Jiang *et al.*, 1996; Jiang and Kanost, 1997; Ragan *et al.*, 2010). Structures of two of the *M. sexta* serpin-1 variants have been determined by X-ray crystallography (Li *et al.*, 1999; Ye *et al.*, 2001). Serpin genes with alternate exons in the same position as in *M sexta* serpin-1 have been identified in other insect species, including the lepidopterans *B. mori* (Sasaki, 1991; Narumi *et al.*, 1993; Zou *et al.*, 2009) and *Mamestra configurata* (Chamankhah *et al.*, 2003), the dipterans *D. melanogaster* (Kruger *et al.*, 2002) and *A. gambiae* (Danielli *et al.*, 2003), and the cat flea, *Ctenocephalides felis* (Brandt *et al.*, 2004).

Phenoloxidase activation is normally regulated *in vivo* as a local reaction of brief duration. This regulation involves serine protease inhibitors in plasma (Kanost, 1999). Serpins from several insect species have been demonstrated to regulate proPO activation. In *M. sexta*, serpin-1J, serpin-3, serpin-4, serpin-5, and serpin-6 disrupt proPO activation when added as recombinant proteins to plasma. Serpins-1J, 3, and 6 directly inhibit proPO activating proteases (Jiang *et al.*, 2003a, 2003b; Zhu *et al.*, 2003a; Wang and Jiang, 2004b), serpin-4 inhibits HP21 (Tong and Kanost, 2005; Tong *et al.*, 2005), and serpin-5 inhibits HP6 (An and Kanost, 2010).

The reactive center loop of serpin-3 contains a sequence very similar to the conserved activation site in proPO (Zhu *et al.*, 2003a). This sequence is likely an excellent fit in the active site of PAPs, and it is probable that serpin-3 is a physiologically important regulator of PAP activity. A variety of experiments have shown that *D. melanogaster* serpin 27A (a putative ortholog of *M. sexta* serpin-3) regulates melanization, and it can inhibit a proPO activating protease from a beetle, *H. diomphalia* (De Gregorio *et al.*, 2002; Ligoxygakis *et al.*, 2002; Nappi *et al.*, 2005). In *A. gambiae*, decreased expression of a serpin-3 ortholog (serpin-2) results in formation of melanotic pseudotumors (Michel *et al.*, 2005). Recombinant *A. gambiae* serpin-2 can block proPO activation in *M. sexta* plasma, and it can inhibit *M. sexta* PAP-3, suggesting that serpin-2 is likely to inhibit an *A. gambiae* proPO-activating protease (Michel *et al.*, 2006). This was demonstrated to be correct by data showing that recombinant serpin-2 inhibits *A. gambiae* clip domain protease CLIPB9, which can activate *M. sexta* proPO and is required for the melanizing phenotype caused by serpin-2 expression knockdown (An *et al.*, 2011a).

Serpins also regulate proteases in the Toll activation cascade (**Figure 3**). In *T. molitor*, three serpins have been demonstrated to specifically inhibit the three proteases in the pathway leading to activation of spätzle and proPO (Jiang *et al.*, 2009). The structure of one of these serpins, serpin-48, which inhibits spätzle processing enzyme, has interesting structural and functional similarity to human antithrombin (Park *et al.*, 2010). In *M. sexta*, Spätzle activation is regulated through inhibition of HP6 by serpin 5 (An and Kanost, 2010) and inhibition of HP8 by serpin-1J (Ragan *et al.*, 2010; An *et al.*, 2011b). In

D. melanogaster, mutation in serpin 43Ac (Necrotic) leads to constitutive expression of drosomycin, indicating that this serpin regulates a protease in the cascade that processes spätzle (Levashina *et al.*, 1999; Green *et al.*, 2000). It is not yet known which protease is inhibited by serpin 43Ac.

In addition to serpins, lower molecular weight inhibitors from the Kunitz family (Sugumaran *et al.*, 1985; Saul *et al.*, 1986; Aso *et al.*, 1994) and a family of 4-kDa inhibitors from locusts (Boigegrain *et al.*, 1992) can interfere with proPO activation (reviewed in Kanost, 1999), although it is not yet known which proteases in the pathway they can inhibit.

10.4. Cellular Proteases

10.4.1. Cathepsin-Type Cysteine Proteases

Cysteine proteases related to cathepsin B and cathepsin L have been identified as proteins produced by hemocytes that participate in tissue remodeling in the metamorphosis of several insects. In *S. peregrina*, a 26/29-kDa protease synthesized in hemocytes was identified as a cathepsin B (Kurata *et al.*, 1992a; Saito *et al.*, 1992; Takahashi *et al.*, 1993b; Fujimoto *et al.*, 1999). This protease may be released from pupal hemocytes to cause dissociation of fat body at metamorphosis (Kurata *et al.*, 1992b). A cathepsin B from hemocytes of *B. mori* may also function in tissue degradation during metamorphosis, including histolysis of silk glands (Shiba *et al.*, 2001; Xu and Kawasaki, 2001). Cathepsin B expressed in fat body of *B. mori* and *H. armigera* also appears to have a role in tissue remodeling during metamorphosis (Lee *et al.*, 2009; Yang *et al.*, 2006).

Cysteine proteases classified as cathepsin L have also been identified as participants in tissue remodeling at metamorphosis. A cathepsin L from *S. peregrina* appears to function in differentiation of imaginal discs (Homma *et al.*, 1994; Homma and Natori, 1996). A similar cathepsin L from another dipteran, *Delia radicum*, is expressed highly in midgut beginning in late third instar, and may function in metamorphosis of the midgut (Hegedus *et al.*, 2002). A cathepsin L-like protease expressed in a *D. melanogaster* hemocyte cell line is present in granules, and may be a lysosomal enzyme functioning to degrade phagocytosed material (Tryselius and Hultmark, 1997). Cathepsin L expressed in hemocytes of a lepidopteran, *H. armigera*, appears to have a similar function (Wang *et al.*, 2010).

10.4.2. Caspases

The caspases (MEROPS family C14) are cysteine proteases that are best known for their roles in apoptosis (Steller, 2008), but are also involved in other functions in insects, including immunity, gamete development, and basal lamina remodeling (Cooper *et al.*, 2009; Feinstein-Rotkopf and Arama, 2009). In fact, the first caspase identified was not involved in apoptosis; this was mammalian caspase-1, which is responsible for proteolytic maturation of interleukin-1β (hence its original name, IL-1β converting enzyme, or ICE) (Cerretti *et al.*, 1992). It was not until the subsequent discovery that the product of a gene required for apoptosis in nematodes (CED-3) was homologous to ICE that caspases were first implicated in apoptosis (Yuan *et al.*, 1993). Caspases were initially known as "ICE-like proteases" until 1996, when their present name was proposed (Alnemri *et al.*, 1996). The word "caspase" is derived from "cysteinyl aspartate-specific proteases," since caspases have strict substrate specificity for Asp in the P1 position (although one insect caspase, Dronc, can also cleave substrates after Glu residues). However, since residues surrounding the cleavage site are also involved in determining substrate recognition, caspases do not indiscriminately cleave after any Asp residue. Individual caspases differ in their substrate specificity, and numerous small molecule inhibitors and fluorometric or colorimetric substrates are available which are recognized (with varying degrees of specificity) by different caspases.

Caspases are synthesized as inactive zymogens, which are themselves activated by specific caspase cleavage after Asp residues, yielding a small and a large subunit that heterodimerize and together form an active site. Two of these heterodimers further dimerize to form a tetramer containing two active sites (Shi, 2002). Two types of caspases are recognized, initiator and effector. Initiator caspases have relatively long amino-terminal sequences containing protein–protein interaction domains which allow for binding to various adaptor proteins. Binding to these adaptors allows for dimerization and autoactivation of initiator caspases. It was initially thought that activation of initiator caspases required autocatalytic cleavage, but it is now clear that dimerization itself is sufficient to activate at least some initiator caspases (Renatus *et al.*, 2001; Snipas *et al.*, 2008) (although as a consequence of dimerization, autocatalytic cleavage normally occurs). Activated initiator caspases then cleave and activate effector caspases, which in turn recognize and cleave various cellular substrates.

Caspases have been identified by genome sequencing projects in a number of insects, but, as might be expected, most of what is known about insect caspases comes from studies done in *D. melanogaster*. The genome of *D. melanogaster* contains seven caspase genes, including three initiator (Dronc, Dredd, and Dream/Strica) and four effector (Decay, Damm, Drice, and DCP-1) caspases. The three mosquito genomes that have been sequenced to date (*A. gambiae*, *Ae. aegypti*, and *Culex quinquefasciatus*) contain larger numbers of caspase genes (13, 11, and 16, respectively (Bryant *et al.*, 2010), which initially led to suggestions that the expansion of caspases in mosquitoes may be related to their hemophagous life history. However, many gene families are expanded in mosquitoes

compared to *D. melanogaster* (Waterhouse *et al.*, 2008). In addition, recent phylogenetic analysis of caspase genes in 12 sequenced *Drosophila* species showed that there have been numerous caspase gene duplications even within the *Drosophila* genus (Bryant *et al.*, 2010).

10.4.2.1. Caspases involved in apoptosis The two caspases that are most important in apoptosis in *D. melanogaster* are the initiator caspase Dronc and the effector caspase Drice (**Figure 4**). Like other initiator caspases, activation of Dronc requires binding to an adaptor protein. Binding to this protein, known as Ark, results in a multimeric complex known as the apoptosome, and promotes Dronc dimerization and activation (Yu *et al.*, 2006). Unlike in mammalian cells, where apoptosome formation depends on release of cytochrome c from mitochondria, Dronc appears to bind to Ark and become activated constitutively in most *D. melanogaster* cell types (Igaki *et al.*, 2002; Muro *et al.*, 2002; Zimmermann *et al.*, 2002). Excessive Dronc activation is prevented by a ubiquitin ligase protein called DIAP1, which promotes ubiquitination of Dronc (Steller, 2008). Upon receiving an apoptotic signal, IAP antagonist proteins (including Reaper, Hid, and Grim) are upregulated, and these allow accumulation of active Dronc, and subsequent cleavage and activation of the effector caspase Drice. Activated Drice then cleaves numerous substrates, leading to apoptosis. A second effector caspase, DCP-1, is also activated by

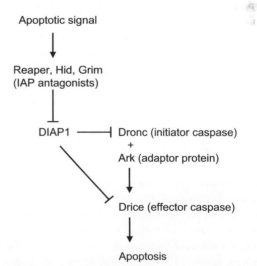

Figure 4 The involvement of caspases in the core apoptosis pathway in *D. melanogaster*. In unstimulated cells, the initiator caspase Dronc undergoes autoactivation due to its interaction with the adaptor protein Ark. However, the ubiquitin ligase DIAP1 prevents excessive caspase activity by inhibiting both the initiator caspase Dronc and the effector caspase Drice. When cells receive an apoptotic signal, IAP antagonists Reaper, Hid, and Grim are upregulated and inhibit the ability of DIAP1 to inhibit caspases. Activated Dronc is then able to cleave and activate Drice, which then cleaves numerous substrates, leading to apoptosis.

Dronc but does not appear to be required for apoptosis in most cell types (Xu *et al.*, 2006), although it is required for nurse cell death during oogenesis (McCall and Steller, 1998), along with either Dronc or Dream/Strica, which act redundantly in this tissue (Baum *et al.*, 2007).

Dronc expression is stimulated by ecdysteroids in both *D. melanogaster* and *Ae. aegypti* (Cooper *et al.*, 2007; Dorstyn *et al.*, 1999), and Dronc is required for most if not all programmed cell deaths that occur during *Drosophila* development (Chew *et al.*, 2004; Daish *et al.*, 2004; Xu *et al.*, 2005). Although Dronc is not required for death of the larval midgut during metamorphosis, the death of this tissue was later shown to be due to autophagy rather than apoptosis (Denton *et al.*, 2009). Drice expression is also stimulated by ecdysteroids (Kilpatrick *et al.*, 2005), and Drice plays an important role in developmentally programmed cell death (Muro *et al.*, 2006; Xu *et al.*, 2006).

Studies in other insects have also demonstrated important roles for caspases in apoptosis. In the lepidopteran *Spodoptera frugiperda*, an effector caspase called Sf-caspase-1 is activated upon apoptotic signaling (Seshagiri and Miller, 1997; LaCount *et al.*, 2000), and silencing Sf-caspase-1 reduces apoptosis (Lin *et al.*, 2007). In *Ae. aegypti*, homologs of Dronc and Ark are required for apoptosis, while silencing the expression of either of two effector caspases related to Drice (CASPS7 or CASPS8) also reduces apoptosis (Liu and Clem, 2011).

10.4.2.2. Caspases involved in non-apoptotic processes

In addition to apoptosis, caspases are also important in other cellular processes in insects. The *Drosophila* initiator caspase Dredd and its adaptor protein FADD are both required for innate immunity signaling through the IMD pathway (Leulier *et al.*, 2000, 2002). Dredd appears to be required at two steps in the pathway, due to its requirement for cleaving both the IMD protein and the inhibitor protein Relish (Stoven *et al.*, 2003; Paquette *et al.*, 2010). Multiple caspases also play non-apoptotic roles in *Drosophila* spermatogenesis. Dronc, Dredd, Drice, and DCP-1 are all involved in the process of spermatid individualization, where individual spermatids are formed from a syncitial cell mass, degrading unneeded cytoplasm in the process (Huh *et al.*, 2004; Muro *et al.*, 2006). Caspases have also been shown to be required for cell proliferation in response to injury in *Drosophila*. When cells are lost from a tissue due to injury, they can be replenished through stimulated mitosis of the surviving cells. This response, known as compensatory proliferation, requires caspases, which apparently trigger the release of mitogenic cytokines from dying cells (Bergmann and Steller, 2010). Finally, a role for caspases in basal lamina remodeling was recently identified in the lepidopteran *Trichoplusia ni* (Means and Passarelli, 2010). In response to fibroblast growth factor signaling, matrix metalloproteases activate one or more unidentified effector caspase(s), which then cleave laminin and collagen in the basal lamina.

Caspase-like decoy molecules are enzymatically inactive versions of caspases that have arisen through gene duplication and have the ability to either positively or negatively regulate the activity of closely related caspases at the post-translational level (Lamkanfi *et al.*, 2007). The first putative decoy caspases in insects were recently identified in the genomes of four *Drosophila* species and two species of mosquitoes (Bryant *et al.*, 2010). These eight putative decoy caspases each contain amino acid substitutions in one or more highly conserved residues that are required for substrate recognition and catalysis. The function of one of these putative decoy caspases, CASPS18 from *Ae. aegypti*, was examined, and CASPS18 was shown to stimulate the activity of a closely related caspase, CASPS19 (Bryant *et al.*, 2010).

10.4.3. Metalloproteases

Three families of metalloproteases that are not involved in the digestion of food have been targets of fairly limited investigation in insects. They appear to function in remodeling of the extracellular matrix, or in degradation of peptide hormones.

Matrix metalloproteases (MMP) are integral membrane proteins, present on the outer surface of cells. They are multi-domain proteins which include a catalytic domain that incorporates a zinc ion in the active site. In mammals, these enzymes regulate processes involving morphogenesis in development and tissue remodeling by digesting protein components of the extracellular matrix (Nagase and Woessner, 1999). In *D. melanogaster*, there are two matrix metalloprotease genes. Dm1-MMP is expressed strongly in embryos, and may have a role in remodeling of the extracellular matrix in development of the central nervous system (Llano *et al.*, 2000) and tracheae (Guha *et al.*, 2009; Glasheen *et al.*, 2010; Wang *et al.*, 2010). Dm2-MMP is expressed at a low level at all developmental stages, but with strong expression in regions of the brain and the eye imaginal discs (Llano *et al.*, 2002). Dm2-MMP is required for dendrite reshaping after adult eclosion, probably through digestion of extracellular matrix (Yasunaga *et al.*, 2010). Mutants in Dm1-MMP have defects in larval tracheal development and pupal head eversion, whereas mutants in Dm2-MMP do not undergo proper tissue remodeling during metamorphosis. However, normal embryonic development was observed even in double mutants lacking both MMPs (Page-McCaw *et al.*, 2003). It appears that these proteases are required primarily for remodeling of extracellular matrix in metamorphosis (Page-McCaw, 2007). However, in *T. castaneum*, knockdown of MMP expression resulted in defects in both embryonic development and metamorphosis (Knorr *et al.*, 2009).

A different class of metalloproteases known as ADAMs (because they contain a disintegrin and metalloprotease domain) are also integral membrane proteins, with the zinc-containing protease domain on the extracellular surface. In mammals, ADAMs participate in growth factor processing, cell adhesion, cell fusion, and tissue remodeling processes, although little is known about their physiological functions in insects. A *D. melanogaster* ADAM called kuzbanian has been shown to function in axon extension in nervous system development (Fambrough *et al.*, 1996; Rooke *et al.*, 1996). Kuzbanian is involved in initiation of a signal transduction pathway by a transmembrane receptor called Notch (Sotillos *et al.*, 1997). The metalloprotease domain of Kuzbanian may exert its physiological effect through cleavage of Notch (Lieber *et al.*, 2002; Delwig and Rand, 2008) or the Notch ligand, Delta (Bland *et al.*, 2003).

A third family of zinc metalloproteases identified in insects contains members that may function in degradation of peptide hormones. These transmembrane proteins are similar to mammalian neprilysins (also called neutral endopeptidases), which cleave oligopeptides on the amino side of hydrophobic residues. They cleave physiologically active signaling peptides with functions in nervous, cardiovascular, inflammatory, and immune systems, and have a wide tissue distribution (Turner *et al.*, 2001; Isaac *et al.*, 2009). Metalloprotease activities with properties similar to neprylins have been identified as enzymes that can degrade tachykinin-related peptides in a cockroach, *Leucophaea maderae*, a locust, *Locusta migratoria*, a dipteran, *D. melanogaster*, and a lepidopteran, *Lacanobia oleracea* (Isaac *et al.*, 2002; Isaac and Nassel, 2003). Similar activities that degrade adipokinetic hormone have been identified in lepidopterans *Lymantria dispar* (Masler *et al.*, 1996) and *M. sexta* (Fox and Reynolds, 1991), and a dipteran, *M. domestica* (Lamango and Isaac, 1993). cDNA clones for proteases with sequence similarity to neprilysin have been described in *B. mori* (Zhao *et al.*, 2001), *M. sexta* (Zhu *et al.*, 2003b), and *L. migratoria* (Macours *et al.*, 2003). The *D. melanogaster* genome contains 24 neprilysin-like genes (Coates *et al.*, 2000). Further study is needed to determine the substrates and physiological roles of these proteases, but it seems likely that they function as negative regulators of peptide signaling molecules hormones.

10.4.4. Aspartic Acid Proteases

The aspartic acid proteases from the MEROPS families A1 and A2 (similar to human pepsin) are a group that has received little study in insects. A cathepsin D-like aspartic acid protease from *Ae. aegypti* has been identified as a lysosomal enzyme, which accumulates in fat body during vitellogenin synthesis (Cho and Raikhel, 1992; Dittmer and Raikhel, 1997). A cathepsin D-like protease may function in fat body histolysis during metamorphosis of

Ceratitis capitata (Rabossi *et al.*, 2004). A cathepsin D may also have a role in metamorphosis in *B. mori* (Gui *et al.*, 2006).

An enzymatically inactive protein from the A1 family is an important allergen from a cockroach (*Blattella germanica*) that triggers asthmatic responses in humans (Pomés *et al.*, 2002). It is present at highest concentration in the gut (Arruda *et al.*, 1995), but since it apparently lacks proteolytic activity, its function in the insect is unknown.

10.4.5. Proteasomes

Proteasomes are complex intracellular proteases that function in regulated degradation of cellular proteins. Turnover of proteins by the proteasome regulates many processes, including the cell cycle, circadian cycles, transcription, growth, development, as well as removal of abnormal proteins. Proteins are targeted for degradation by the proteasome by attachment of polyubiquitin chains to an amino group on a lysine side chain. Eukaryotic proteasomes are composed of four stacked heptameric rings that form a cylinder with multiple protease catalytic sites in its interior. This structure, the 20S proteasome, is composed of 28 subunits from multiple homologous gene products, and has a mass of ~700 kDa. The 20S proteasome has little activity unless it is activated by another 700-kDa, 20-subunit protein, PA700, that can bind to one or both ends of the cylinder. When both ends of the 20S proteasome are capped by a PA700, the resulting complex is the 26S proteasome, which is active in degrading ubiquitinated proteins, an ATP-dependent process (DeMartino and Slaughter, 1999).

Proteasomes in insects have been studied primarily in *D. melanogaster* and *M. sexta* (Mykles, 1997, 1999). They have physical and catalytic properties similar to those of proteasomes from other eukaryotic species (Uvardy, 1993; Haire *et al.*, 1995; Walz *et al.*, 1998). Mutants in genes for subunits of *D. melanogaster* proteasomes or proteins that regulate proteasome activity have lethal or otherwise complex phenotypes involving disruption of multiple aspects of physiology and development (Schweisguth, 1999; Ma *et al.*, 2002; Watts *et al.*, 2003). Some mutants alter subunit composition or disrupt proteasome assembly (Covi *et al.*, 1999; Smyth and Belote, 1999; Szlanka *et al.*, 2003) In *M. sexta*, programmed cell death of intersegmental muscles at metamorophosis is accompanied by marked changes in proteasome activity and subunit composition (Dawson *et al.*, 1995; Jones *et al.*, 1995; Takayanagi *et al.*, 1996; Low *et al.*, 1997, 2000, 2001; Hastings *et al.*, 1999).

10.5. Conclusions and Future Prospects

Tremendous advances have been made in the past 10 years in our knowledge of the existence, structure, and function of insect proteases that have biological roles unrelated to the digestion of food. It is apparent that proteolysis is an important mechanism for post-translational modification of proteins, involved in many regulatory pathways. Much more remains to be discovered. Through examination of genome sequences, we can see that insects have an enormous number of genes encoding proteases, and that many of them are unstudied and have unknown functions. Nearly all of the research in this area has focused on a few holometabolous model species. It is to be expected that detailed investigation of a broader range of species will reveal complex and diverse functions for proteases in regulating intracellular and extracellular processes. A common feature of proteases is that they are synthesized as inactive zymogens, activated by proteolysis when the time is right, and then rapidly inhibited by specific inhibitors. Better understanding of molecular mechanisms of this tight regulation of multiple and varied protease cascade pathways will impact many areas of insect biology.

Acknowledgments

We thank Chunju An for help with figure preparation.

References

Alnemri, E. S., Livingston, D. J., Nicholson, D. W., Salvesen, G., Thornberry, N. A., et al. (1996). Human ICE/CED-3 protease nomenclature. *Cell, 87,* 171.

Amarant, T., Burkhart, W., LeVine, H., III, Arocha-Pinango, C. L., & Parikh, I. (1991). Isolation and complete amino acid sequence of two fibrinolytic proteinases from the toxic Saturnid caterpillar *Lonomia achelous. Biochem. Biophys. Acta, 1079,* 214–221.

An, C., & Kanost, M. R. (2010). *Manduca sexta* serpin-5 regulates prophenoloxidase activation and the Toll signaling pathway by inhibiting hemolymph proteinase HP6. *Insect Biochem. Mol. Biol., 40,* 683–689.

An, C., Ishibashi, J., Ragan, E. J., Jiang, H., & Kanost, M. R. (2009). Functions of *Manduca sexta* hemolymph proteinases HP6 and HP8 in two innate immune pathways. *J. Biol. Chem., 284,* 19716–19726.

An, C., Jiang, H., & Kanost, M. R. (2010). Proteolytic activation and function of the cytokine Spätzle in innate immune response of a lepidopteran insect, *Manduca sexta. FEBS J., 277,* 148–162.

An, C., Budd, A., Kanost, M. R., & Michel, K. (2011a). Characterization of a regulatory unit that controls melanization and affects longevity of mosquitoes. *Cell Mol. Life Sci.,* in press.

An, C., Ragan, E. J., & Kanost, M. R. (2011b). Serpin-1 splicing isoform J inhibits the proSpätzle-activating proteinase HP8 to regulate expression of antimicrobial hemolymph proteins in *Manduca sexta. Dev. Comp. Immunol., 35,* 135–141.

Applebaum, S. W. (1985). Biochemistry of digestion. In G. A. Kerkut, & L. I. Gilbert (Eds.). *Comprehensive Insect Physiology Biochemistry and Pharmacology* (Vol. 4, pp. 279–311). Oxford, UK: Pergamon Press.

Arocha-Pinango, C. L., & Guerrero, B. (2000). *Lonomia* genus caterpillar toxins: Biochemical aspects. *Biochemie, 82,* 937–942.

Arruda, L. K., Vailes, L. D., Mann, B. J., Shannon, J., Fox, J. W., et al. (1995). Molecular cloning of a major cockroach (*Blattella germanica*) allergen, Bla g 2. Sequence homology to the aspartic proteases. *J. Biol. Chem., 270,* 19563–19568.

Aso, Y., Yamashita, T., Meno, K., & Murakami, M. (1994). Inhibition of prophenoloxidase-activating enzyme from *Bombyx mori* by endogenous chymotrypsin inhibitors. *Biochem. Mol. Biol. International, 33,* 751–758.

Barillas-Mury, C. (2007). CLIP proteases and *Plasmodium* melanization in *Anopheles gambiae. Trends Parasitol., 23,* 297–299.

Baum, J. S., Arama, E., Steller, H., & McCall, K. (2007). The *Drosophila* caspases Strica and Dronc function redundantly in programmed cell death during oogenesis. *Cell Death Differ, 14,* 1508–1517.

Berger, E., Kafatos, F. C., Felsted, R. L., & Law, J. H. (1971). Cocoonase III. Purification, preliminary characterization, and activation of the zymogen of an insect protease. *J. Biol. Chem., 246,* 4131–4137.

Bergmann, A., & Steller, H. (2010). Apoptosis, stem cells, and tissue regeneration. *Sci. Signal, 3,* re8.

Bland, C. E., Kimberly, P., & Rand, M. D. (2003). Notch-induced proteolysis and nuclear localization of the Delta ligand. *J. Biol. Chem., 278,* 13607–13610.

Boigegrain, R., Mattras, H., Brehélin, M., Paroutaud, P., & Coletti-Previero, M. (1992). Insect immunity: Two proteinase inhibitors from hemolymph of *Locusta migratoria. Biochem. Biophys. Res. Commun., 189,* 790–793.

Brandt, K. S., Silver, G. M., Becher, A. M., Gaines, P. J., Maddux, J. D., et al. (2004). Isolation, characterization, and recombinant expression of multiple serpins from the cat flea, *Ctenocephalides felis. Arch. Insect Biochem. Physiol., 55,* 200–214.

Bryant, B., Ungerer, M. C., Liu, Q., Waterhouse, R. M., & Clem, R. J. (2010). A caspase-like decoy molecule enhances the activity of a paralogous caspase in the yellow fever mosquito, *Aedes aegypti. Insect Biochem. Mol. Biol., 40,* 516–523.

Buchon, N., Poidevin, M., Kwon, H. M., Guillou, A., Sottas, V., et al. (2009). A single modular serine protease integrates signals from pattern-recognition receptors upstream of the *Drosophila* Toll pathway. *Proc. Natl. Acad. Sci. USA, 106,* 12442–12447.

Cerenius, L., Lee, B. L., & Söderhäll, K. (2008). The proPO-system: Pros and cons for its role in invertebrate immunity. *Trends Immunol., 29,* 263–271.

Cerenius, L., Kawabata, S., Lee, B. L., Nonaka, M., & Söderhäll, K. (2010). Proteolytic cascades and their involvement in invertebrate immunity. *Trends Biochem. Sci., 35,* 575–583.

Cerretti, D. P., Kozlosky, C. J., Mosley, B., Nelson, N., Van Ness, K., et al. (1992). Molecular cloning of the interleukin-1 beta converting enzyme. *Science, 256,* 97–100.

Chamankhah, M., Braun, L., Visal-Shah, S., O'Grady, M., Baldwin, D., et al. (2003). *Mamestra configurata* serpin-1 homologues: Cloning, localization and developmental regulation. *Insect Biochem. Mol. Biol., 33,* 355–369.

Chang, A. J., & Morisato, D. (2002). Regulation of Easter activity is required for shaping the Dorsal gradient in the *Drosophila* embryo. *Development, 129,* 5635–5645.

Chasan, R., & Anderson, K. (1989). The role of easter, an apparent serine protease, in organizing the dorsal–ventral pattern of the *Drosophila* embryo. *Cell, 56,* 391–400.

Chew, S. K., Akdemir, F., Chen, P., Lu, W. J., Mills, K., et al. (2004). The apical caspase *dronc* governs programmed and unprogrammed cell death in *Drosophila. Dev. Cell, 7,* 897–907.

Cho, W. -L., & Raikhel, A. (1992). Cloning of cDNA for mosquito lysosomal aspartic protease. *J. Biol. Chem., 267,* 21823–21829.

Cho, W. L., Deitsch, K. W., & Raikhel, A. S. (1991). An extra-ovarian protein accumulated in mosquito oocytes is a carboxypeptidase activated in embryos. *Proc. Natl. Acad. Sci. USA, 88,* 10821–10824.

Cho, W. -L., Tsao, S. -M., Hays, A. R., Walter, R., Chen, J. -S., et al. (1999). Mosquito cathepsin B-like protease involved in embryonic degradation of vitellin is produced as a latent extraovarian precursor. *J. Biol. Chem., 274,* 13311–13321.

Coates, D., Siviter, R., & Isaac, R. E. (2000). Exploring the *Caenorhabditis elegans* and *Drosophila melanogaster* genomes to understand neuropeptide and peptidase function. *Biochem. Soc. Trans., 28,* 464–469.

Cooper, D. M., Thi, E. P., Chamberlain, C. M., Pio, F., & Lowenberger, C. (2007). *Aedes* Dronc: A novel ecdysone-inducible caspase in the yellow fever mosquito, *Aedes aegypti. Insect Mol. Biol., 16,* 563–572.

Cooper, D. M., Granville, D. J., & Lowenberger, C. (2009). The insect caspases. *Apoptosis, 14,* 247–256.

Covi, J. A., Belote, J. M., & Mykles, D. L. (1999). Subunit compositions and catalytic properties of proteasomes from developmental temperature-sensitive mutants of *Drosophila melanogaster. Arch. Biochem. Biophys., 368,* 85–97.

Daish, T. J., Mills, K., & Kumar, S. (2004). *Drosophila* caspase DRONC is required for specific developmental cell death pathways and stress-induced apoptosis. *Develop. Cell, 7,* 909–915.

Danielli, A., Kafatos, F. C., & Loukeris, T. G. (2003). Cloning and characterization of four *Anopheles gambiae* serpin isoforms, differentially induced in the midgut by *Plasmodium berghei* invasion. *J. Biol. Chem., 278,* 4184–4193.

Dawson, S. P., Arnold, J. E., Mayer, N. J., Reynolds, S. E., Billett, M. A., et al. (1995). Developmental changes of the 26 S proteasome in abdominal intersegmental muscles of *Manduca sexta* during programmed cell death. *J. Biol. Chem., 270,* 1850–1858.

De Gregorio, E., Han, S. J., Lee, W. J., Baek, M. J., Osaki, T., et al. (2002). An immune-responsive Serpin regulates the melanization cascade in *Drosophila. Dev. Cell, 3,* 581–592.

Delwig, A., & Rand, M. D. (2008). Kuz and TACE can activate Notch independent of ligand. *Cell Mol. Life Sci., 65,* 2232–2243.

DeLotto, R. (2001). Gastrulation defective, a complement factor C2/B-like protease interprets a ventral prepattern in *Drosophila. EMBO Rep., 2,* 721–726.

DeLotto, Y., & DeLotto, R. (1998). Proteolytic processing of the *Drosophila* Spätzle protein by easter generates a dimeric NGF-like molecule with ventralising activity. *Mech. Dev., 72,* 141–148.

DeLotto, R., & Spierer, P. (1986). A gene required for the specification of dorsal–ventral pattern in *Drosophila* appears to encode a serine protease. *Nature, 323,* 688–692.

DeMartino, G. N., & Slaughter, C. A. (1999). The proteasome, a novel protease regulated by multiple mechanisms. *J. Biol. Chem., 274,* 22123–22126.

Denton, D., Shravage, B., Simin, R., Mills, K., Berry, D. L., et al. (2009). Autophagy, not apoptosis, is essential for midgut cell death in *Drosophila. Curr. Biol., 19,* 1741–1746.

Dissing, M., Giordano, H., & DeLotto, R. (2001). Autoproteolysis and feedback in a protease cascade directing *Drosophila* dorsal-ventral cell fate. *EMBO J., 20,* 2387–2393.

Dittmer, N. T., & Raikhel, A. S. (1997). Analysis of the mosquito lysosomal aspartic protease gene: An insect housekeeping gene with fat body-enhanced expression. *Insect Biochem. Mol. Biol., 27,* 323–335.

Dorstyn, L., Colussi, P. A., Quinn, L. M., Richardson, H., & Kumar, S. (1999a). DRONC, an ecdysone-inducible *Drosophila* caspase. *Proc. Natl. Acad. Sci.USA, 96,* 4307–4312.

El Chamy, L., Leclerc, V., Caldelari, I., & Reichhart, J. M. (2008). Sensing of "danger signals" and pathogen-associated molecular patterns defines binary signaling pathways "upstream" of Toll. *Nat. Immunol., 9,* 1165–1170.

Fambrough, D., Pan, D., Rubin, G. M., & Goodman, C. S. (1996). The cell surface metalloprotease/disintegrin Kuzbanian is required for axonal extension in *Drosophila. Proc. Natl. Acad. Sci. USA, 93,* 13233–13238.

Feinstein-Rotkopf, Y., & Arama, E. (2009). Can't live without them, can live with them: Roles of caspases during vital cellular processes. *Apoptosis, 14,* 980–995.

Felsted, R. L., Kramer, K. J., Law, J. H., Berger, E., & Kafatos, F. C. (1973). Cocoonase IV. Mechanism of activation of procococonase from *Antheraea polyphemus. J. Biol. Chem., 248,* 3021–3028.

Ferrandon, D., Imler, J. L., Hetru, C., & Hoffmann, J. A. (2007). The *Drosophila* systemic immune response: Sensing and signalling during bacterial and fungal infections. *Nat. Rev. Immunol., 7,* 862–874.

Finnerty, C. M., Karplus, P. A., & Granados, R. R. (1999). The insect immune protein scolexin is a novel serine proteinase homolog. *Protein Sci., 8,* 242–248.

Fox, A. M., & Reynods, S. E. (1991). Degradation of adipokinetic hormone family peptides by a circulating endopeptidase in the insect *Manduca sexta. Peptides, 12,* 937–944.

Fujimoto, Y., Kobayashi, A., Kurata, S., & Natori, S. (1999). Two subunits of the insect 26/29-kDa proteinase are probably derived from a common precursor protein. *J. Biochem. (Tokyo), 125,* 566–573.

Gay, N. J., & Keith, F. J. (1992). Regulation of translation and proteolysis during the development of embryonic dorso-ventral polarity in *Drosophila.* Homology of easter proteinase with *Limulus* proclotting enzyme and translational activation of Toll receptor synthesis. *Biochim. Biophys. Acta, 1132,* 290–296.

Gettins, P. G. (2002). Serpin structure, mechanism, and function. *Chem. Rev., 102,* 4751–4804.

Glasheen, B. M., Robbins, R. M., Piette, C., Beitel, G. J., & Page-McCaw, A. (2010). A matrix metalloproteinase mediates airway remodeling in *Drosophila. Dev. Biol., 44,* 772–783.

Gorman, M. J., Wang, Y., Jiang, H., & Kanost, M. R. (2007). *Manduca sexta* hemolymph proteinase 21 activates pro-phenoloxidase activating proteinase 3 in an insect innate immune response proteinase cascade. *J. Biol. Chem.*, *282*, 11742–11749.

Green, C., Levashina, E., McKimmie, C., Dafforn, T., Reichhart, J. -M., & Gubb, D. (2000). The necrotic gene in *Drosophila* corresponds to one of a cluster of three serpin transcripts mapping at 43A1.2. *Genetics*, *156*, 1117–1127.

Guha, A., Lin, L., & Kornberg, T. B. (2009). Regulation of *Drosophila* matrix metalloprotease Mmp2 is essential for wing imaginal disc:trachea association and air sac tubulogenesis. *Dev. Biol.*, *335*, 317–326.

Gui, Z. Z., Lee, K. S., Kim, B. Y., Choi, Y. S., Wei, Y. D., et al. (2006). Functional role of aspartic proteinase cathepsin D in insect metamorphosis. *BMC Dev. Biol.*, *6*, 49.

Gupta, S., Wang, Y., & Jiang, H. (2005). *Manduca sexta* pro-phenoloxidase (proPO) activation requires proPO-activating proteinase (PAP) and serine proteinase homologs (SPHs) simultaneously. *Insect Biochem. Mol. Biol.*, *35*, 241–248.

Haire, M. F., Clark, J. J., Jones, M. E., Hendil, K. B., Schwartz, L. M., & Mykles, D. L. (1995). The multicatalytic protease (proteasome) of the hawkmoth, *Manduca sexta*: Catalytic properties and immunological comparison with the lobster enzyme complex. *Arch. Biochem. Biophys.*, *318*, 15–24.

Han, J. -H., Lee, S. H., Tan, Y. -Q., Lemosy, E. K., & Hashimoto, C. (2000). Gastrulation defective is a serine protease involved in activating the receptor Toll to polarize the *Drosophila* embryo. *Proc. Natl. Acad. Sci.*, *97*, 9093–9097.

Hashimoto, C., Kim, D. R., Weiss, L. A., Miller, J. W., & Morisato, D. (2003). Spatial regulation of developmental signaling by a serpin. *Dev. Cell.*, *5*, 945–950.

Hastings, R. A., Eyheralde, I., Dawson, S. P., Walker, G., Reynolds, S. E., et al. (1999). A 220-kDa activator complex of the 26 S proteasome in insects and humans. A role in type II programmed insect muscle cell death and cross-activation of proteasomes from different species. *J. Biol. Chem.*, *274*, 25691–25700.

Hegedus, D., O'Grady, M., Chamankhah, M., Baldwin, D., Gleddie, S., et al. (2002). Changes in cysteine protease activity and localization during midgut metamorphosis in the crucifer root maggot (*Delia radicum*). *Insect Biochem. Molec. Biol.*, *32*, 1585–1596.

Homma, H., & Natori, N. (1996). Identification of substrate proteins for cathepsin L that are selectively hydrolyzed during the differentiation of imaginal discs of *Sarcophaga peregrina*. *Eur. J. Biochem.*, *240*, 443–447.

Homma, K., Kurata, S., & Natori, S. (1994). Purification, characterization, and cDNA cloning of procathepsin L from the culture medium of NIH-SAPE-4, an embryonic cell line of *Sarcophaga peregrina* (flesh fly), and its involvement in the differentiation of imaginal discs. *J. Biol. Chem.*, *269*, 15258–15264.

Hong, C., & Hashimoto, C. (1995). An unusual mosaic protein with a protease domain, encoded by the nudel gene, is involved in defining embryonic dorsoventral polarity in *Drosophila*. *Cell*, *82*, 785–794.

Hong, C., & Hashimoto, C. (1996). The maternal nudel protein of *Drosophila* has two distinct roles important for embryogenesis. *Genetics*, *143*, 1653–1661.

Hooper, N. M. (2002). Proteases: A primer. In N. M. Hooper (Ed.). *Essays in Biochemistry: Proteases in Biology and Medicine* (pp. 1–8). London, UK: Portland Press.

Huang, R., Lu, Z., Dai, H., Velde, D. V., & Prakash, O. (2007). The solution structure of clip domains from *Manduca sexta* prophenoloxidase activating proteinase-2. *Biochemistry*, *46*, 11431–11439.

Huh, J. R., Vernooy, S. Y., Yu, H., Yan, N., Shi, Y., et al. (2004). Multiple apoptotic caspase cascades are required in nonapoptotic roles for *Drosophila* spermatid individualization. *PLoS Biol.*, *2*, E15.

Igaki, T., Yamamoto-Goto, Y., Tokushige, N., Kanda, H., & Miura, M. (2002). Down-regulation of DIAP1 triggers a novel *Drosophila* cell death pathway mediated by Dark and DRONC. *J. Biol. Chem.*, *277*, 23103–23106.

Ikeda, M., Yaginuma, T., Kobayashi, M., & Yamashita, O. (1991). cDNA cloning, sequencing and temporal expression of the protease responsible for vitellin degradation in the silkworm *Bombyx mori*. *Comp. Biochem. Physiol.*, *B 99*, 405–411.

Indrasith, L. S., Sasaki, T., & Yamashita, O. (1988). A unique protease responsible for selective degradation of a yolk protein in *Bombyx mori*. Purification, characterization, and cleavage profile. *J. Biol. Chem.*, *263*, 1045–1051.

Isaac, R. E., & Nassel, D. R. (2003). Identification and localization of a neprilysin-like activity that degrades tachykinin-related peptides in the brain of the cockroach, *Leucophaea maderae*, and locust, *Locusta migratoria*. *J. Comp. Neurol.*, *457*, 57–66.

Isaac, R. E., Parkin, E. T., Keen, J. N., Nassel, D. R., Siviter, R. J., & Shirras, A. D. (2002). Inactivation of a tachykinin-related peptide: identification of four neuropeptide-degrading enzymes in neuronal membranes of insects from four different orders. *Peptides*, *23*, 725–733.

Isaac, R. E., Bland, N. D., & Shirras, A. D. (2009). Neuropeptidases and the metabolic inactivation of insect neuropeptides. *Gen. Comp. Endocrinol.*, *162*, 8–17.

Jang, I. H., Chosa, N., Kim, S. H., Nam, H. J., Lemaitre, B., et al. (2006). A spätzle-processing enzyme required for toll signaling activation in *Drosophila* innate immunity. *Developmental Cell*, *10*, 45–55.

Jiang, H., & Kanost, M. R. (1997). Characterization and functional analysis of twelve naturally occurring reactive site variants of serpin-1 from *Manduca sexta*. *J. Biol. Chem.*, *272*, 1082–1087.

Jiang, H., & Kanost, M. R. (2000). The clip-domain family of serine proteinases in arthropods. *Insect Biochem. Molec. Biol.*, *30*, 95–105.

Jiang, H., Wang, Y., Huang, Y., Mulnix, A. B., Kadel, J., et al. (1996). Organization of serpin gene-1 from *Manduca sexta*: Evolution of a family of alternate exons encoding the reactive site loop. *J. Biol. Chem.*, *271*, 28017–28023.

Jiang, H., Wang, Y., & Kanost, M. R. (1998). Pro-phenol oxidase activating proteinase from an insect, *Manduca sexta*: A bacteria-inducible protein similar to *Drosophila* easter. *Proc. Natl. Acad. Sci. USA*, *95*, 12220–12225.

Jiang, H., Wang, Y., Yu, X. -Q., & Kanost, M. R. (2003a). Prophenoloxidase-activating proteinase-2 from hemolymph of *Manduca sexta*. A bacteria-inducible serine proteinase containing two clip domains. *J. Biol. Chem.*, *278*, 3552–3561.

Jiang, H., Wang, Y., Yu, X. -Q., Zhu, Y., & Kanost, M. R. (2003b). Prophenoloxidase-activating proteinase-3 (PAP-3) from *Manduca sexta* hemolymph: A clip-domain serine proteinase regulated by serpin-1J and serine proteinase homologs. *Insect Biochem. Mol. Biol., 33*, 1049–1060.

Jiang, H., Wang, Y., Gu, Y., Guo, X., Zou, Z., et al. (2005). Molecular identification of a bevy of serine proteinases in *Manduca sexta* hemolymph. *Insect Biochem. Molec. Biol., 35*, 931–943.

Jiang, R., Kim, E. H., Gong, J. H., Kwon, H. M., Kim, C. H., et al. (2009). Three pairs of protease-serpin complexes cooperatively regulate the insect innate immune responses. *J. Biol. Chem., 284*, 35652–35658.

Jiang, H., Vilcinskas, A., & Kanost, M. R. (2010). Immunity in Lepidopteran insects. In K. Söderhäll (Ed.). *Invertebrate Immunity*: Landes Bsciencehttp://www.landesbioscience.com/curie/chapter/4692/2010.

Jin, Y., & Anderson., K. V. (1990). Dominant and recessive alleles of the *Drosophila* easter gene are point mutations at conserved sites in the serine protease catalytic domain. *Cell, 60*, 873–881.

Jones, M. E., Haire, M. F., Kloetzel, P. M., Mykles, D. L., & Schwartz, L. M. (1995). Changes in the structure and function of the multicatalytic proteinase (proteasome) during programmed cell death in the intersegmental muscles of the hawkmoth, *Manduca sexta. Dev. Biol., 169*, 436–447.

Kafatos, F. C. (1972). The cocoonase zymogen cells of silk moths: A model of terminal cell differentiation for specific protein synthesis. *Curr. Top. Dev. Biol., 7*, 125–191.

Kafatos, F. C., Tartakoff, A. M., & Law, J. H. (1967a). Cocoonase I. Preliminary characterization of a proteolytic enzyme from silk moths. *J. Biol. Chem., 242*, 1477–1487.

Kafatos, F. C., Law, J. H., & Tartakoff, A. M. (1967b). Cocoonase II. Substrate specificity, inhibitors, and classification of the enzyme. *J. Biol. Chem., 424*, 1488–1494.

Kageyama, T., & Takahashi, S. Y. (1990). Purification and characterization of a cysteine proteinase from silkworm eggs. *Eur. J. Biochem., 193*, 203–210.

Kambris, Z., Brun, S., Jang, I. H., Nam, H. J., Romeo, Y., et al. (2006). *Drosophila* immunity: A large-scale *in vivo* RNAi screen identifies five serine proteases required for Toll activation. *Curr. Biol., 16*, 808–813.

Kan, H., Kim, C. H., Kwon, H. M., Park, J. W., Roh, K. B., et al. (2008). Molecular control of phenoloxidase-induced melanin synthesis in an insect. *J. Biol. Chem., 283*, 25316–25323.

Kanost, M. R. (1999). Serine proteinase inhibitors in arthropod immunity. *Dev. Comp. Immunol., 23*, 291–301.

Kanost, M. R., & Gorman, M. G. (2008). Phenoloxidases in insect immunity. In N. Beckage (Ed.). *Insect Immunology* (pp. 69–96). San Diego, CA: Academic Press/Elsevier.

Kawabata, S., Muta, T., & Iwanaga, S. (1996). The clotting cascade and defense molecules found in the hemolymph of the horseshoe crab. In K. Söderhäll, S. Iwanaga, & G. Vasta (Eds.). *New Directions in Invertebrate Immunology* (pp. 255–283). Cambridge, UK: SOS Publications.

Kilpatrick, Z. E., Cakouros, D., & Kumar, S. (2005). Ecdysone-mediated up-regulation of the effector caspase DRICE is required for hormone-dependent apoptosis in *Drosophila* cells. *J. Biol. Chem., 280*, 11981–11986.

Kim, C. H., Kim, S. J., Kan, H., Kwon, H. M., Roh, K. B., et al. (2008). A three-step proteolytic cascade mediates the activation of the peptidoglycan-induced toll pathway in an insect. *J. Biol. Chem., 283*, 7599–7607.

Kim, M. S., Baek, M. J., Lee, M. H., Park, J. W., Lee, S. Y., et al. (2002). A new easter-type serine protease cleaves a masquerade-like protein during prophenoloxidase activation in *Holotrichia diomphalia* larvae. *J. Biol. Chem., 277*, 39999–40004.

Knorr, E., Schmidtberg, H., Vilcinskas, A., & Altincicek, B. (2009). MMPs regulate both development and immunity in the *Tribolium* model insect. *PLoS One, 4*(3), e4751.

Konrad, K. D., Goralski, T. J., Mahowald, A. P., & Marsh, J. L. (1998). The gastrulation defective gene of *Drosophila melanogaster* is a member of the serine protease superfamily. *Proc. Natl. Acac. Sci. USA, 95*, 6819–6824.

Kramer, K. J., Felsted, R. L., & Law, J. H. (1973). Cocoonase V. Structural studies on an insect serine protease. *J. Biol. Chem., 248*, 3021–3028.

Krem, M. M., & Di Cerra, E. (2002). Evolution of enzyme cascades from embryonic development to blood coagulation. *Trends Biochem. Sci., 27*, 67–74.

Kruger, O., Ladewig, J., Koster, K., & Ragg, H. (2002). Widespread occurrence of serpin genes with multiple reactive centre-containing exon cassettes in insects and nematodes. *Gene, 293*, 97–105.

Kurata, S., Saito, H., & Natori, S. (1992a). Purification of a 29-kDa proteinase of *Sarcophaga peregrina. Eur. J. Biochem., 204*, 911–914.

Kurata, S., Saito, H., & Natori, S. (1992b). The 29-kDa hemocyte proteinase dissociates fat body at metamorphosis of *Sarcophaga. Dev. Biol., 153*, 115–121.

Kwon, T. H., Kim, M. S., Choi, H. W., Loo, C. H., Cho, M. Y., & Lee, B. L. (2000). A masquerade-like serine proteinase homologue is necessary for phenoloxidase activity in the coleopteran insect, *Holotrichia diomphalia* larvae. *Eur. J. Biochem., 267*, 6188–6196.

LaCount, D. J., Hanson, S. F., Schneider, C. L., & Friesen, P. D. (2000). Caspase inhibitor P35 and inhibitor of apoptosis Op-IAP block *in vivo* proteolytic activation of an effector caspase at different steps. *J. Biol. Chem., 275*, 15657–15664.

Lamango, N. S., & Isaac, R. E. (1993). Metabolism of insect neuropeptides: Properties of a membrane-bound endopeptidase from heads of Musca domestica. *Insect Biochem. Mol. Biol., 23*, 801–808.

Lamkanfi, M., Festjens, N., Declercq, W., Vanden Berghe, T., & Vandenabeele, P. (2007). Caspases in cell survival, proliferation and differentiation. *Cell Death Diff., 14*, 44–55.

Law, J. H., Dunn, P. E., & Kramer, K. J. (1977). Insect proteases and peptidases. *Adv. Enzymol. Relat. Areas Mol. Biol., 45*, 389–425.

Lee, K. S., Kim, B. Y., Choo, Y. M., Yoon, H. J., Kang, P. D., et al. (2009). Expression profile of cathepsin B in the fat body of *Bombyx mori* during metamorphosis. *Comp. Biochem. Physiol. B, 154*, 188–194.

Lee, K. Y., Zhang, R., Kim, M. S., Park, J. W., Park, H. Y., et al. (2002). A zymogen form of masquerade-like serine proteinase homologue is cleaved during pro-phenoloxidase activation by Ca²⁺ in coleopteran and *Tenebrio molitor* larvae. *Eur. J. Biochem., 269*, 4375–4383.

Lee, S. Y., & Söderhäll, K. (2001). Characterization of a pattern recognition protein, a masquerade-like protein, in the freshwater crayfish *Pacifastacus leniusculus*. *J. Immunol.*, *166*, 7319–7326.

Lee, S. Y., Kwon, T. H., Hyun, J. H., Choi, J. S., Kawabata, S. I., et al. (1998a). *In vitro* activation of pro-phenol-oxidase by two kinds of pro-phenol-oxidase activating factors isolated from hemolymph of coleopteran *Holotrichia diomphalia* larvae. *Eur. J. Biochem.*, *15*, 50–57.

Lee, S. Y., Cho, M. Y., Hyun, J. H., Lee, K. M., Homma, K., et al. (1998b). Molecular cloning of cDNA for pro-phenol-oxidase-activating factor I, a serine proteinase is induced by lipopolysaccharide or 1,3-β-glucan in a coleopteran insect, *Holotrichia diomphalia* larvae. *Eur. J. Biochem.*, *257*, 615–621.

LeMosy, E. K., & Hashimoto, C. (2000). The nudel protease of *Drosophila* is required for eggshell biogenesis in addition to embryonic patterning. *Dev. Biol.*, *217*, 352–361.

LeMosy, E. K., Kemler, D., & Hashimoto, C. (1998). Role of Nudel protease activation in triggering dorsoventral polarization of the *Drosophila* embryo. *Development*, *125*, 4045–4053.

LeMosy, E. K., Hong, C. C., & Hashimoto, C. (1999). Signal transduction by a protease cascade. *Trends Cell Biol.*, *9*, 102–107.

LeMosy, E. K., Leclerc, C. L., & Hashimoto, C. (2000). Biochemical defects of mutant *nudel* alleles causing early developmental arrest or dorsalization of the *Drosophila* embryo. *Genetics*, *154*, 247–257.

LeMosy, E. K., Tan, Y. -Q., & Hashimoto, C. (2001). Activation of a protease cascade involved in patterning the *Drosophila* embryo. *Proc. Natl. Acad. Sci.*, *98*, 5055–5060.

Leulier, F., Rodriguez, A., Khush, R. S., Abrams, J. M., & Lemaitre, B. (2000). The *Drosophila* caspase Dredd is required to resist gram-negative bacterial infection. *EMBO Rep.*, *1*, 353–358.

Leulier, F., Vidal, S., Saigo, K., Ueda, R., & Lemaitre, B. (2002). Inducible expression of double-stranded RNA reveals a role for dFADD in the regulation of the antibacterial response in *Drosophila* adults. *Curr. Biol.*, *12*, 996–1000.

Levashina, E. A., Langley, E., Green, C., Gubb, D., Ashburner, M., et al. (1999). Constitutive activation of Toll-mediated antifungal defense in serpin-deficient *Drosophila*. *Science*, *285*, 1917–1919.

Li, J., Wang, Z., Canagarajah, B., Jiang, H., Kanost, M. R., & Goldsmith, E. J. (1999). The structure of active serpin K from *Manduca sexta* and a model for serpin–protease complex formation. *Structure*, *7*, 103–109.

Lieber, T., Kidd, S., & Young, M. W. (2002). Kuzbanian-mediated cleavage of *Drosophila* Notch. *Genes. Dev.*, *16*, 209–221.

Ligoxygakis, P., Pelte, N., Ji, C., Leclerc, V., Duvic, B., et al. (2002). A serpin mutant links Toll activation to melanization in the host defence of *Drosophila*. *EMBO J.*, *21*, 6330–6337.

Ligoxygakis, P., Roth, S., & Rechhart, J. M. (2003). A serpin regulates dorsal–ventral axis formation in the *Drosophila* embryo. *Curr. Biol.*, *13*, 2097–2102.

Lin, C. C., Hsu, J. T., Huang, K. L., Tang, H. K., Shu, C. W., & Lai, Y. K. (2007). Sf-caspase-1-repressed stable cells: Resistance to apoptosis and augmentation of recombinant protein production. *Biotechnol. Appl. Biochem.*, *48*, 11–19.

Liu, Q., & Clem, R. J. (2011). Defining the core apoptosis pathway in the mosquito disease vector *Aedes aegypti*: The roles of *iap1*, *ark*, *dronc*, and effector caspases. *Apoptosis*, *16*, 105–113.

Llano, E., Pendas, A. M., Aza-Blanc, P., Kornberg, T. B., & Lopez-Otin, C. (2000). Dm1-MMP, a matrix metalloproteinase from *Drosophila* with a potential role in extracellular matrix remodeling during neural development. *J. Biol. Chem.*, *275*, 35978–35985.

Llano, E., Adam, G., Pendas, A. M., Quesada, V., Sanchez, L. M., et al. (2002). Structural and enzymatic characterization of *Drosophila* Dm2-MMP, a membrane-bound matrix metalloproteinase with tissue-specific expression. *J. Biol. Chem.*, *277*, 23321–23329.

Low, P., Bussell, K., Dawson, S. P., Billett, M. A., Mayer, R. J., & Reynolds, S. E. (1997). Expression of a 26S proteasome ATPase subunit, M73, in muscles that undergo developmentally programmed cell death, and its control by ecdysteroid hormones in the insect *Manduca sexta*. *FEBS Lett.*, *400*, 345–349.

Low, P., Hastings, R. A., Dawson, S. P., Sass, M., Billett, M. A., et al. (2000). Localisation of 26S proteasomes with different subunit composition in insect muscles undergoing programmed cell death. *Cell Death Differ.*, *7*, 1210–1217.

Low, P., Reynolds, S. E., & Sass, M. (2001). Proteolytic activity of 26s proteasomes isolated from muscles of the tobacco hornworm, *Manduca sexta*: Differences between surviving muscles and those undergoing developmentally programmed cell death. *Acta Biol. Hung.*, *52*, 435–442.

Ma, J., Katz, E., & Belote, J. M. (2002). Expression of proteasome subunit isoforms during spermatogenesis in *Drosophila melanogaster*. *Insect Mol. Biol.*, *11*, 627–639.

Macours, N., Poels, J., Hens, K., Luciani, N., De Loof, A., & Huybrechts, R. (2003). An endothelin-converting enzyme homologue in the locust, *Locusta migratoria*: Functional activity, molecular cloning and tissue distribution. *Insect Mol. Biol.*, *12*, 233–240.

Maki, N., & Yamashita, O. (1997). Purification and characterization of a protease degrading 30-kDa yolk proteins of the silkworm, *Bombyx mori*. *Insect Biochem. Mol. Biol.*, *27*, 721–728.

Maki, N., & Yamashita, O. (2001). The 30k protease A responsible for 30-kDa yolk protein degradation of the silkworm, *Bombyx mori*: cDNA structure, developmental change and regulation by feeding. *Insect Biochem. Mol. Biol.*, *31*, 407–413.

Masler, E. P., Wagner, R. M., & Kovaleva, E. S. (1996). *In vitro* metabolism of an insect neuropeptide by neural membrane preparations from Lymantria dispar. *Peptides*, *17*, 321–326.

McCall, K., & Steller, H. (1998). Requirement for DCP-1 caspase during *Drosophila* oogenesis. *Science*, *279*, 230–234.

Means, J. C., & Passarelli, A. L. (2010). Viral fibroblast growth factor, matrix metalloproteases, and caspases are associated with enhancing systemic infection by baculoviruses. *Proc. Natl. Acad. Sci. USA*, *107*, 9825–9830.

Medina, M., Leon, P., & Vallejo, C. G. (1988). *Drosophila* cathepsin B-like proteinase: A suggested role in yolk degradation. *Arch. Biochem. Biophys.*, *263*, 355–363.

Michel, K., Budd, A., Pinto, S., Gibson, T. J., & Kafatos, F. C. (2005). *Anopheles gambiae* SRPN2 facilitates midgut invasion by the malaria parasite Plasmodium berghei. *EMBO Rep.*, *6*, 891–897.

Michel, K., Suwanchaichinda, C., Morlais, I., Lambrechts, L., Cohuet, A., et al. (2006). Increased melanizing activity in *Anopheles gambiae* does not affect development of *Plasmodium falciparum*. *Proc. Natl. Acad. Sci. USA*, *103*, 16858–16863.

Misra, S., Hecht, P., Maeda, R., & Anderson, K. V. (1998). Positive and negative regulation of Easter, a member of the serine protease family that controls dorsal–ventral patterning in the *Drosophila* embryo. *Development*, *125*, 1261–1267.

Morisato, D., & Anderson, K. V. (1995). Signaling pathways that establish the dorsal–ventral pattern of the *Drosophila* embryo. *Annu. Rev. Genet.*, *29*, 371v399.

Morugasu-Oei, B., Rodrigues, V., Yang, X., & Chia, W. (1995). Masquerade: A novel secreted serine protease-like molecule is required for somatic muscle attachment in the *Drosophila* embryo. *Genes. Dev.*, *9*, 139–154.

Morugasu-Oei, B., Balakrishnan, R., Yang, X., Chia, W., & Rodrigues, V. (1996). Mutations in masquerade, a novel serine-protease-like molecule, affect axonal guidance and taste behavior in *Drosophila*. *Mech. Dev.*, *57*, 91–101.

Moussian, B., & Roth, S. (2005). Dorsoventral axis formation in the *Drosophila* embryo – Shaping and transducing a morphogen gradient. *Current Biol.*, *15*, R887–R899.

Muro, I., Hay, B. A., & Clem, R. J. (2002). The *Drosophila* DIAP1 protein is required to prevent accumulation of a continuously generated, processed form of the apical caspase DRONC. *J. Biol. Chem.*, *277*, 49644–49650.

Muro, I., Berry, D. L., Huh, J. R., Chen, C. H., Huang, H., et al. (2006). The *Drosophila* caspase Ice is important for many apoptotic cell deaths and for spermatid individualization, a nonapoptotic process. *Development*, *133*, 3305–3315.

Mykles, D. L. (1997). Biochemical properties of insect and crustacean proteasomes. *Mol. Biol. Rep.*, *24*, 133–138.

Mykles, D. L. (1999). Structure and functions of arthropod proteasomes. *Mol. Biol. Rep.*, *26*, 103–111.

Nagase, H., & Woessner, F. J. (1999). Matrix metalloproteinases. *J. Biol. Chem.*, *274*, 21491–21494.

Nakajima, Y., Tsuji, Y., Homma, K., & Natori, S. (1997). A novel protease in the pupal yellow body of *Sarcophaga peregrina* (flesh fly). Its purification and cDNA cloning. *J. Biol. Chem.*, *272*, 23805–23810.

Nappi, A. J., Frey, F., & Carton, Y. (2005). *Drosophila* serpin 27A is a likely target for immune suppression of the blood cell-mediated melanotic encapsulation response. *J. Insect Physiol.*, *51*, 197–205.

Narumi, H., Hishida, T., Sasaki, T., Feng, D. F., & Doolittle, R. F. (1993). Molecular cloning of silkworm (*Bombyx mori*) antichymotrypsin. A new member of the serpin superfamily of proteins from insects. *Eur. J. Biochem.*, *214*, 181–187.

Page-McCaw, A. (2007). Remodeling the model organism: Matrix metalloproteinase functions in invertebrates. *Semin. Cell Dev. Biol.*, *19*, 14–23.

Page-McCaw, A., Serano, J., Sante, J. M., & Rubin, G. M. (2003). *Drosophila* matrix metalloproteinases are required for tissue remodeling, but not embryonic development. *Dev. Cell*, *4*, 95–106.

Paquette, N., Broemer, M., Aggarwal, K., Chen, L., Husson, M., et al. (2010). Caspase-mediated cleavage, IAP binding, and ubiquitination: linking three mechanisms crucial for *Drosophila* NF-κB signaling. *Mol. Cell*, *37*, 172–182.

Park, S.H., Jiang, R., Piao, S., Zhang, B., Kim, E.H., et al., 2010. Structural and functional characterization of a highly specific serpin in the insect innate immunity, in press.

Piao, S., Song, Y. L., Kim, J. H., Park, S. Y., Park, J. W., et al. (2005). Crystal structure of a clip-domain serine protease and functional roles of the clip domains. *EMBO J.*, *24*, 4404–4414.

Pinto, A. F. M., Silva, K. R. L. M., & Guimaraes, J. A. (2006). Proteases from *Lonomia obliqua* venomous secretions: Comparison of procoagulant, fibrin(ogen)olytic and amidolytic activities. *Toxicon*, *47*, 113–121.

Pomés, A., Chapman, M. D., Vailes, L. D., Blundell, T. L., & Dhanaraj, V. (2002). Cockroach allergen Bla g2: Structure, function, and implications for allergic sensitization. *Am. J. Respir. Crit. Care Med.*, *165*, 391–397.

Rabossi, A., Stoka, V., Puizdar, V., Turk, V., & Quesada-Allué, L. A. (2004). Novel aspartyl proteinase associated to fat body histolysis during *Ceratitis capitata* early metamorphosis. *Arch. Insect Biochem. Physiol.*, *57*, 51–67.

Ragan, E. J., An, C., Yang, C., & Kanost, M. R. (2010). Analysis of mutually-exclusive alternatively spliced serpin-1 isoforms and identification of serpin-1 proteinase complexes in *Manduca sexta* hemolymph. *J. Biol. Chem.*, *285*, 29642–29650.

Raikhel, A. S., & Dhadialla, T. S. (1992). Accumulation of yolk proteins in insect oocytes. *Annu. Rev. Entomol.*, *37*, 217–251.

Rawlings, N. D., Barrett, A. J., & Bateman, A. (2010). *MEROPS*: the peptidase database. *Nucleic Acids Res.*, *38*, D227–D233.

Reeck, G., Oppert, B., Denton, M., Kanost, M., Baker, J., & Kramer, K. (1999). Insect proteinases. In V. Turk (Ed.). *Proteases: New Perspectives* (pp. 125–148). Basel, Switzerland: Birkhäuser.

Renatus, M., Stennicke, H. R., Scott, F. L., Liddington, R. C., & Salvesen, G. S. (2001). Dimer formation drives the activation of the cell death protease caspase 9. *Proc. Natl. Acad. Sci. USA*, *98*, 14250–14255.

Ribolla, P. E., & De Bianchi, A. G. (1995). Processing of procathepsin from *Musca domestica* eggs. *Insect Biochem. Mol. Biol.*, *25*, 1011–1017.

Roh, K. B., Kim, C. H., Lee, H., Kwon, H. M., Park, J. W., et al. (2009). Proteolytic cascade for the activation of the insect toll pathway induced by the fungal cell wall component. *J. Biol. Chem.*, *284*, 19474–19481.

Rooke, J., Pan, D., Xu, T., & Rubin, G. M. (1996). KUZ, a conserved metalloprotease-disintegrin protein with two roles in *Drosophila* neurogensis. *Science*, *273*, 1227–1231.

Rose, T., LeMosy, E. K., Cantwell, A. M., Banerjee-Roy, D., Skeath, J. B., & Di Cera, E. (2003). Three-dimensional models of proteases involved in patterning of the *Drosophila* embryo. Crucial role of predicted cation binding sites. *J. Biol. Chem.*, *278*, 11320–11330.

Ross, J., Jiang, H., Kanost, M. R., & Wang, Y. (2003). Serine proteases and their homologs in the *Drosophila melanogaster* genome: An initial analysis of sequence conservation and phylogenetic relationships. *Gene*, *304*, 117–131.

Saito, H., Kurata, S., & Natori, S. (1992). Purification and characterization of a hemocyte proteinase of *Sarcophaga*, possibly participating in elimination of foreign substances. *Eur. J. Biochem.*, *209*, 939–944.

Sasaki, T. (1991). Patchwork-structure serpins from silkworm (*Bombyx mori*) larval hemolymph. *Eur. J. Biochem.*, *202*, 255–161.

Satoh, D., Horii, A., Ochiai, M., & Ashida, M. (1999). Prophenoloxidase-activating enzyme of the silkworm, *Bombyx mori*: Purification, characterization and cDNA cloning. *J. Biol. Chem.*, *274*, 7441–7453.

Saul, S., & Sugumaran, M. (1986). Protease inhibitor controls prophenoloxidase activation in *Manduca sexta*. *FEBS Lett.*, *208*, 113–116.

Schechter, I., & Berger, A. (1967). On the size of the active site in proteases. I. Papain. *Biochem. Biophys. Res. Commun.*, *27*, 157–162.

Schweisguth, F. (1999). Dominant-negative mutation in the β2 and β6 proteasome subunit genes affect alternate cell fate decisions in the *Drosophila* sense organ lineage. *Proc. Natl. Acad. Sci. USA*, *96*, 11382–11386.

Seshagiri, S., & Miller, L. K. (1997). Baculovirus inhibitors of apoptosis (IAPs) block activation of Sf-caspase-1. *Proc. Natl. Acad. Sci. USA*, *94*, 13606–13611.

Shi, Y. (2002). Mechanisms of caspase activation and inhibition during apoptosis. *Mol. Cell*, *9*, 459–470.

Shiba, H., Uchida, D., Kobayashi, H., & Natori, M. (2001). Involvement of cathepsin B- and L-like proteinases in silk gland histolysis during metamorphosis of *Bombyx mori*. *Arch. Biochem. Biophys.*, *390*, 28–34.

Smith, C., & DeLotto, R. (1992). A common domain within the proenzyme regions of the *Drosophila* snake and easter proteins and *Tachypleus* proclotting enzyme defines a new subfamily of serine proteases. *Protein Sci.*, *11*, 1225–1226.

Smith, C., & DeLotto, R. (1994). Ventralizing signal determined by protease activation in *Drosophila* embryogenesis. *Nature*, *368*, 548–551.

Smith, C., Giordano, H., & DeLotto, R. (1994). Mutational analysis of the *Drosophila* snake protease: An essential rule for domains within the proenzyme polypeptide chain. *Genetics*, *136*, 1355–1365.

Smith, C. L., Giordano, H., Schwartz, M., & DeLotto, R. (1995). Spatial regulation of *Drosophila* snake protease activity in the generation of dorsal–ventral polarity. *Development*, *121*, 4127–4135.

Smyth, K. A., & Belote, J. M. (1999). The dominant temperature-sensitive lethal *DTS7* of *Drosophila melanogaster* encodes an altered 20S proteasome β-type subunit. *Genetics*, *151*, 211–220.

Snipas, S. J., Drag, M., Stennicke, H. R., & Salvesen, G. S. (2008). Activation mechanism and substrate specificity of the *Drosophila* initiator caspase DRONC. *Cell Death Differ.*, *15*, 938–945.

Sotillos, S., Roch, F., & Campuzano, S. (1997). The metalloprotease-disintintegrin Kuzbanian participates in Notch activation during growth and patterning of *Drosophila* imaginal discs. *Development*, *124*, 4769–4779.

Steller, H. (2008). Regulation of apoptosis in *Drosophila*. *Cell Death Diff.*, *15*, 1132–1138.

Stoven, S., Silverman, N., Junell, A., Hedengren-Olcott, M., Erturk, D., et al. (2003). Caspase-mediated processing of the *Drosophila* NF-kappaB factor Relish. *Proc. Natl. Acad. Sci. USA*, *100*, 5991–5996.

Sugumaran, M., Saul, S., & Ramesh, N. (1985). Endogenous protease inhibitors prevent undesired activation of prophenolase in insect hemolymph. *Biochem. Biophys. Res. Commun.*, *132*, 1124–1129.

Suwanchaichinda, C., & Kanost, M. R. (2009). The serpin gene family in *Anopheles gambiae*. *Gene*, *442*, 47–54.

Szlanka, T., Haracska, L., Kiss, I., Deak, P., Kurucz, E., et al. (2003). Deletion of proteasomal subunit S5a/Rpn10/p54 causes lethality, multiple mitotic defects and overexpression of proteasomal genes in *Drosophila melanogaster*. *J. Cell Sci.*, *116*, 1023–1033.

Takahashi, S. Y., Yamamoto, Y., Shionoya, Y., & Kageyama, T. (1993a). Cysteine proteinase from the eggs of the silkmoth, *Bombyx mori*: Identification of a latent enzyme and characterization of activation and proteolytic processing *in vivo* and *in vitro*. *J. Biochem. (Tokyo)*, *114*, 267–272.

Takahashi, N., Kurata, S., & Natori, S. (1993b). Molecular cloning of cDNA for the 29-kDa proteinase participating on decomposition of the larval fat body during metamorphosis of *Sarcophaga peregrina* (flesh fly). *FEBS Lett.*, *334*, 153–157.

Takayanagi, K., Dawson, S., Reynolds, S. E., & Mayer, R. J. (1996). Specific developmental changes in the regulatory subunits of the 26 S proteasome in intersegmental muscles preceding eclosion in *Manduca sexta*. *Biochem. Biophys. Res. Commun.*, *228*, 517–523.

Tanaka, H., Ishibashi, J., Fujita, K., Nakajima, Y., Sagisaka, A., et al. (2008). A genome-wide analysis of genes and gene families involved in innate immunity of *Bombyx mori*. *Insect Biochem. Mol. Biol.*, *38*, 1087–1110.

Tang, H., Kambris, Z., Lemaitre, B., & Hashimoto, C. (2006). Two proteases defining a melanization cascade in the immune system of *Drosophila*. *J. Biol. Chem.*, *281*, 28097–28104.

Terra, W. R., Ferreira, C., Jordão, B. P., & Dillon, R. J. (1996). Digestive enzymes. In M. J. Lehane, & P. F. Billingsley (Eds.). *Biology of the Insect Midgut* (pp. 153–194). London, UK: Chapman and Hall.

Tong, Y., & Kanost, M. R. (2005). *Manduca sexta* serpin-4 and serpin-5 inhibit the prophenoloxidase activation pathway. cDNA cloning, protein expression, and characterization. *J. Biol. Chem.*, *280*, 14923–14931.

Tong, Y., Jiang, H., & Kanost, M. R. (2005). Identification of plasma proteases inhibited by *Manduca sexta* serpin-4 and serpin-5 and their association with components of the prophenol oxidase activation pathway. *J. Biol. Chem.*, *280*, 14932–14942.

Tryselius, Y., & Hultmark, D. (1997). Cysteine proteinase 1 (CP1), a cathespin L-like enzyme expressed in the *Drosophila melanogaster* haemocyte cell line mbn-2. *Insect Mol. Biol.*, *6*, 173–181.

Tsuji, Y., Nakajima, Y., Homma, K. -I., & Natori, S. (1998). Antibacterial activity of a novel 26-kDa serine protease in the yellow body of *Sarcophaga peregrina* (flesh fly) pupae. *FEBS Lett.*, *425*, 131–133.

Turcotte, C. L., & Hashimoto, C. (2002). Evidence for a glycosaminoglycan on the nudel protein important for dorsoventral patterning of the *Drosophila* embryo. *Dev. Dyn.*, *224*, 51–57.

Turner, A. J., Isaac, R. E., & Coates, D. (2001). The neprilysin (NEP) family of zinc metalloendopeptidases, genomics and function. *BioEssays*, *23*, 261–269.

Uvardy, A. (1993). Purification and characterization of a multiprotein component of the *Drosophila* 26S (1500 kDa) proteolytic complex. *J. Biol. Chem.*, *268*, 9055–9062.

Walz, J., Erdmann, A., Kania, M., Typke, D., Koster, A. J., & Baumeister, W. (1998). 26S proteasome structure revealed by three-dimensional electron microscopy. *J. Struct. Biol.*, *121*, 19–29.

Wang, L. F., Chai, L. Q., He, H. J., Wang, Q., Wang, J. X., & Zhao, X. F. (2010). A cathepsin L-like proteinase is involved in moulting and metamorphosis in *Helicoverpa armigera*. *Insect Mol. Biol.*, *19*, 99–111.

Wang, R., Lee, S. Y., Cerenius, L., & Söderhall, K. (2001). Properties of the prophenoloxidase activating enzyme of the freshwater crayfish, *Pacifastacus leniusculus*. *Eur. J. Biochem.*, *268*, 895–902.

Wang, Y., & Jiang, H. (2004a). Prophenoloxidase (proPO) activation in *Manduca sexta*: An analysis of molecular interactions among proPO, proPO-activating proteinase-3, and a cofactor. *Insect Biochem. Mol. Biol.*, *34*, 731–742.

Wang, Y., & Jiang, H. (2004b). Purification and characterization of *Manduca sexta* serpin-6: A serine proteinase inhibitor that selectively inhibits prophenoloxidase-activating proteinase-3. *Insect Biochem. Mol. Biol.*, *34*, 387–395.

Wang, Y., & Jiang, H. (2006). Interaction of β-1,3-glucan with its recognition protein activates hemolymph proteinase 14, an initiation enzyme of the prophenoloxidase activation system in *Manduca sexta*. *J. Biol. Chem.*, *281*, 9271–9278.

Wang, Y., & Jiang, H. (2007). Reconstitution of a branch of the *Manduca sexta* prophenoloxidase activation cascade *in vitro*: snake-like hemolymph proteinase 21 (HP21) cleaved by HP14 activates prophenoloxidase-activating proteinase-2 precursor. *Insect Biochem. Mol. Biol.*, *37*, 1015–1025.

Wang, Y., & Jiang, H. (2010). Binding properties of the regulatory domains in *Manduca sexta* hemolymph proteinase-14, an initiation enzyme of the prophenoloxidase activation system. *Dev. Comp. Immunol.*, *34*, 316–322.

Waterhouse, R. M., Xi, Z. Y., Kriventseva, E., Meister, S., Alvarez, K. S., et al. (2007). Evolutionary dynamics of immune-related genes and pathways in disease vector mosquitoes. *Science*, *316*, 1738–1743.

Waterhouse, R. M., Wyder, S., & Zdobnov, E. M. (2008). The *Aedes aegypti* genome: A comparative perspective. *Insect Mol. Biol.*, *17*, 1–8.

Watts, R. J., Hoopfer, E. D., & Luo, L. (2003). Axon pruning during *Drosophila* metamorphosis: Evidence for local degeration and requirement of the ubiquitin-proteasome system. *Neuron*, *38*, 871–887.

Xu, D., Li, Y., Arcaro, M., Lackey, M., & Bergmann, A. (2005). The CARD-carrying caspase Dronc is essential for most, but not all, developmental cell death in *Drosophila*. *Development*, *132*, 2125–2134.

Xu, D., Wang, Y., Willecke, R., Chen, Z., Ding, T., & Bergmann, A. (2006). The effector caspases *drICE* and *dcp-1* have partially overlapping functions in the apoptotic pathway in *Drosophila*. *Cell Death Diff.*, *13*, 1697–1706.

Xu, Y., & Kawasaki, H. (2001). Isolation and expression of cathepsin B cDNA in hemocytes during metamorphosis of *Bombyx mori*. *Comp. Biochem. Physiol.*, *130B*, 393–399.

Yamamoto, Y., & Takahashi, S. Y. (1993). Cysteine proteinase from *Bombyx* eggs: Role in programmed degradation of yolk proteins during embryogenesis. *Comp. Biochem. Physiol. B*, *106*, 35–45.

Yamamoto, Y., Takimoto, K., Izumi, S., Toriyama-Sakurai, M., Kageyama, T., & Takahashi, S. Y. (1994). Molecular cloning and sequencing of cDNA that encodes cysteine proteinase in the eggs of the silkmoth, *Bombyx mori. J. Biochem. (Tokyo)*, *116*, 1330–1335.

Yamamoto, Y., Yamahama, Y., Katou, K., Watabe, S., & Takahashi, S. Y. (2000). *Bombyx* acid cysteine protease (BCP): Hormonal regulation of biosynthesis and accumulation in the ovary. *J. Insect Physiol.*, *46*, 783–791.

Yang, X. M., Hou, L. J., Dong, D. J., Shao, H. L., Wang, J. X., & Zhao, X. F. (2006). Cathepsin B-like proteinase is involved in the decomposition of the adult fat body of *Helicoverpa armigera*. *Arch. Insect Biochem. Physiol.*, *62*, 1–10.

Yasunaga, K., Kanamori, T., Morikawa, R., Suzuki, E., & Emoto, K. (2010). Dendrite reshaping of adult *Drosophila* sensory neurons requires matrix metalloproteinase-mediated modification of the basement membranes. *Dev. Cell*, *8*, 621–632.

Ye, S., Cech, A. L., Belmares, R., Bergstrom, R. C., Tong, Y., et al. (2001). The structure of a Michaelis serpin–protease complex. *Nature Struct. Biol.*, *8*, 979–983.

Yu, X., Wang, L., Acehan, D., Wang, X., & Akey, C. W. (2006). Three-dimensional structure of a double apoptosome formed by the *Drosophila* Apaf-1 related killer. *J. Mol. Biol.*, *355*, 577–589.

Yu, X. -Q., Jiang, H., Wang, Y., & Kanost, M. R. (2003). Nonproteolytic serine proteinase homologs are involved in prophenoloxidase activation in the tobacco hornworm, *Manduca sexta. Insect Biochem. Mol. Biol.*, *33*, 197–208.

Yuan, J., Shaham, S., Ledoux, S., Ellis, H. M., & Horvitz, H. R. (1993). The *C. elegans* cell death gene ced-3 encodes a protein similar to mammalian interleukin-1beta-converting enzyme. *Cell*, *75*, 641–652.

Zhao, X., Mita, K., Shimada, T., Okano, K., Quan, G. X., et al. (2001). Isolation and expression of an ecdysteroid-inducible neutral endopeptidase 24.11-like gene in wing discs of *Bombyx mori. Insect Biochem. Mol. Biol.*, *31*, 1213–1219.

Zhao, X. F., Wang, J. X., Xu, X. L., Schid, R., & Wieczorek, H. (2002). Molecular cloning and characterization of the cathepsin B-like proteinase from the cotton boll worm, *Helicoverpa armigera. Insect Mol. Biol.*, *11*, 567–575.

Zhu, Y., Wang, Y., Gorman, M. J., Jiang, H., & Kanost, M. R. (2003a). *Manduca sexta* serpin-3 regulates prophenoloxidase activation in response to infection by inhibiting prophenoloxidase-activating proteinases. *J. Biol. Chem.*, *278*, 46556–46564.

Zhu, Y., Johnson, T., Myers, A., & Kanost, M. R. (2003b). Identification by subtractive suppression hybridization of bacteria-induced genes expressed in *Manduca sexta* fat body. *Insect Biochem. Molec. Biol.*, *33*, 541–559.

Zimmermann, K. C., Ricci, J. -E., Droin, N. M., & Green, D. R. (2002). The role of ARK in stress-induced apoptosis in *Drosophila* cells. *J. Cell Biol.*, *156*, 1077–1087.

Zou, Z., Lopez, D. L., Kanost, M. R., Evans, J. D., & Jiang, H. B. (2006). Comparative analysis of serine protease-related genes in the honey bee genome: Possible involvement in embryonic development and innate immunity. *Insect Molecular Biology, 2006*(15), 603–614.

Zou, Z, Evans, J. D., Lu, Z., Zhao, P., Williams, M., et al. (2007). Comparative genomic analysis of the *Tribolium* immune system. *Genome Biol., 8,* R177.

Zou, Z., Najar, F., Wang, Y., Roe, B., & Jiang, H. (2008). Pyrosequence analysis of expressed sequence tags for *Manduca sexta* hemolymph proteins involved in immune responses. *Insect Biochem. Mol. Biol., 38,* 677–682.

Zou, Z., Picheng, Z., Weng, H., Mita, K., & Jiang, H. (2009). A comparative analysis of serpin genes in the silkworm genome. *Genomics, 93,* 367–375.

11 Biochemistry and Molecular Biology of Digestion

Walter R Terra
University of São Paulo, São Paulo, Brazil
Clélia Ferreira
University of São Paulo, São Paulo, Brazil

DOI:10.1016/B978-0-12-384747-8.10011-X

11.1. Introduction

The growth of knowledge in the biochemistry of insect digestion had a bright start during the first decades of the last century, but slowed down after the development of synthetic chemical insecticides in the 1940s. Later on, with the environmental problems caused by chemical insecticides, new approaches for insect control were investigated. Midgut studies were particularly stimulated after the realization that the gut is a very large and relatively unprotected interface between the insect and its environment. Hence, an understanding of gut function was thought to be essential when developing methods of control that act through the gut, such as the use of transgenic plants to control phytophagous insects.

Applebaum (1985), in his review on the biochemistry of digestion, described the beginning of the renewed growth of the field. He discussed contemporary research showing that most insect digestive enzymes are similar to their mammalian counterparts, but that insect exotic diets require specific enzymes. In the next decade it became apparent that even enzymes similar to those of mammals have distinct characteristics, because each insect taxon deals with food in a special way (Terra and Ferreira, 1994). Since then, the field of digestive physiology and biochemistry has progressed dramatically at the molecular level (Terra and Ferreira, 2005).

The aim of this chapter is to review the recent and spectacular progress in the study of insect digestive biochemistry. To provide a broad coverage while keeping the chapter within reasonable size limits, only a brief account with key references is given for work done prior to 2000. Papers after 2000 have been selected from those richer in molecular details, and, when they were too numerous, representative papers were chosen, especially when abundant in references to other papers. Throughout, the focus is on providing a coherent picture of phenomena and highlighting further research areas. Amino acid residues are denoted by the one-letter code, if in peptides, for the sake of brevity. When mentioned in text with a position number, amino acid residues are denoted by the three-letter code to avoid ambiguity. For consistency, traditional abbreviations, like BAPA for benzoyl-arginine p-nitroanilide, have been changed, in the example to B-R-pNA, because the one-letter code for arginine is R.

The chapter is organized into four parts. The first part (sections 11.2 and 11.3) tries to establish uniform parameters for studying insect digestive enzymes, providing an overview of the biochemistry of insect digestion, and discusses factors affecting digestive enzymes *in vivo*. The second part (sections 11.4–11.7) reviews digestive enzymes and microvillar proteins, with the emphasis on molecular aspects, whereas the third part (sections 11.8 and 11.9) describes the details of the digestive biochemical process alongside insect evolution. Finally, the fourth part (section 11.10) discusses data on digestive enzyme secretion mechanisms.

11.2. Overview of the Digestive Process

11.2.1. Initial Considerations

Digestion is the process by which food molecules are broken down into smaller molecules that are absorbed by cells in the gut tissue. This process is controlled by digestive enzymes, and is dependent on their localization in the insect gut.

11.2.2. Characterization of Digestive Enzymes

Enzyme kinetic parameters are meaningless unless assays are performed in conditions in which enzymes are stable. If researchers adopt uniform parameters and methods, comparisons among similar and different insect species will be more meaningful. A rectilinear plot of product formation (or substrate disappearance) versus time will ensure that enzymes are stable in a given condition. Activities (velocities) calculated from this plot are reliable parameters. According to the International Union of Biochemistry and Molecular Biology, the assay temperature should be 30°C, except when the enzyme is unstable at this temperature or altered for specific purposes. Owing to partial inactivation, the optimum temperature is not a true property of enzymes, and therefore should not be included in the characterization. Optimum enzyme pH should be determined using different buffers to discount the effects of chemical constituents of the buffers and their ionic strength on enzyme activity. The number of molecular forms of a given enzyme should be evaluated by submitting the enzyme preparation to a separation process (gel permeation, ion-exchange chromatography, hydrophobic chromatography, electrophoresis, gradient ultracentrifugation, etc.), followed by assays of the resulting fractions. Substrate specificity of each molecular form of a given enzyme should be evaluated and substrate preference quantified by determining V_m/K_m ratios for each substrate, keeping the amount of each enzyme form constant. Substrate preference expressed as the percentage activity towards a given substrate in relation to the activity upon a reference substrate may be misleading because, in this condition, enzyme activities are determined at different substrate saturations. The isoelectric points of many enzymes can be determined after staining with specific substrates following the separation of the native enzymes on isoelectrofocusing gels.

If enzyme characterization is performed as part of a digestive physiology study, emphasis should be given to enzyme compartmentalization, substrate specificity, and substrate preference, in order to discover the sequential action of enzymes during the digestive process.

Knowledge of the effect of pH on enzyme activity is useful in evaluating enzyme action in gut compartments (**Figure 1**) with different pH values. Finally, the determination of the molecular masses of digestive enzymes, associated with the ability of enzymes to pass through the peritrophic membrane, allows estimation of the pore sizes of the peritrophic membrane. Molecular masses determined in non-denaturing conditions are preferred, since in these conditions the enzymes should maintain their *in vivo* aggregation states (not only their quaternary structures if present). The method of choice in this case is gradient ultracentrifugation.

Complete enzymological characterization requires purification to homogeneity, and sequencing. Furthermore,

details of the catalytic mechanisms, including involvement of amino acid residues in catalysis and substrate specificity, should be determined. This permits the classification of insect digestive enzymes into catalytic families, and discloses the structural basis of substrate specificities; it will also enable us to establish evolutionary relationships with enzymes from other organisms.

Cloning cDNA sequences encoding digestive enzymes enables the expression of large amounts of recombinant enzymes that may be crystallized or used for the production of antibodies. Antibodies are used in Western blots to identify a specific enzyme in protein mixtures, or to localize the enzyme in tissue sections in a light or electron microscope. Enzyme crystals used for resolving

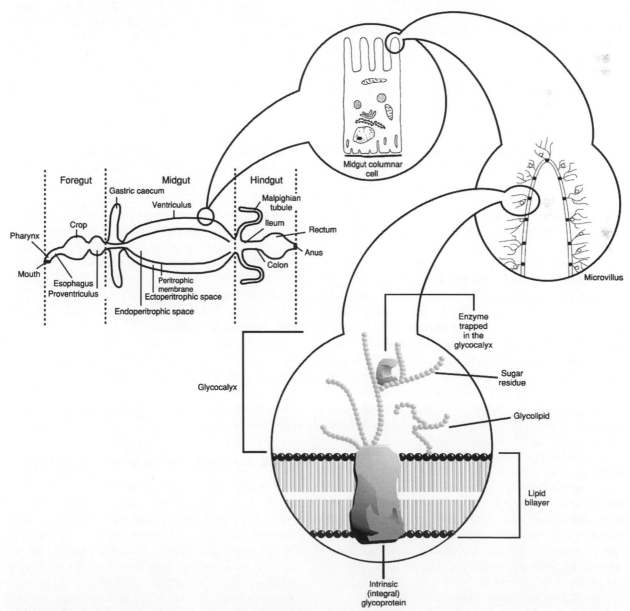

Figure 1 Diagrammatic representation of insect gut compartments. Glycocalyx: the carbohydrate moiety of intrinsic proteins and glycolipids occurring in the luminal face of microvillar membranes.

three-dimensional (3D) structures (via X-ray diffraction or nuclear magnetic resonance, NMR) need amounts of purified enzymes that frequently are difficult to isolate from insects by conventional separation procedures. However, detailed 3D structures are necessary to understand enzyme mechanisms and the binding of inhibitors to enzyme molecules. Alternatively, cDNA may be amenable to site-directed mutagenesis for structure–function studies. Site-directed mutagenesis tests the role of individual amino acid residues in enzyme function or structure. Such knowledge is a prerequisite in developing new biotechnological approaches to control insects via the gut. Finally, interference RNA may be used to suppress the expression of one enzyme, in order to test hypotheses regarding its physiological role (see Chapter 2).

11.2.3. Classification of Digestive Enzymes

Digestive enzymes are hydrolases. The enzyme classification and numbering system used here is that recommended by the Nomenclature Committee of the International Union of Biochemistry and Molecular Biology (Enzyme Commission).

Peptidases (peptide hydrolases, EC 3.4) are enzymes that act on peptide bonds and include the proteinases (endopeptidases, EC 3.4.21–24) and the exopeptidases (EC 3.2.4.11–19; see also Chapter 10 in this volume). Proteinases are divided into subclasses on the basis of catalytic mechanism, as shown with specific reagents or effect of pH. Specificity is only used to identify individual enzymes within subclasses. Serine proteinases (EC 3.4.21) have a serine and a histidine in the active site. Cysteine proteinases (EC 3.4.22) possess a cysteine in the active site, and are inhibited by mercurial compounds. Aspartic proteinases (EC 3.4.23) have a pH optimum below 5, due to the involvement of a carboxyl residue in catalysis. Metalloproteinases (EC 2.3.24) need a metal ion in the catalytic process. Exopeptidases include enzymes that hydrolyze single amino acids from the N-terminus (aminopeptidases, EC 3.4.11) or from the C-terminus (carboxypeptidases, EC 3.4.16–18) of the peptide chain, and those enzymes specific for dipeptides (dipeptide hydrolases, EC 3.4.13) (**Figure 2**).

Glycosidases (EC 3.2) are classified according to their substrate specificities. They include the enzymes that cleave internal bonds in polysaccharides and are usually named from their substrates – for example, amylase, cellulase, pectinase, and chitinase. They also include the enzymes that hydrolyze oligosaccharides and disaccharides. Oligosaccharidases and disaccharidases are usually named based on the monosaccharide that gives its reducing group to the glycosidic bond, and on the configuration (α or β) of this bond (**Figure 2**).

Lipids are a large and heterogeneous group of substances that are relatively insoluble in water but readily soluble in apolar solvents. Some contain fatty acids (fats, phospholipids, glycolipids, and waxes), while others lack them (terpenes, steroids, and carotenoids). Ester bonds are hydrolyzed in lipids containing fatty acids before they are absorbed. The enzymes that hydrolyze ester bonds comprise: (1) carboxylic ester hydrolases (EC 3.1.1), such as lipases, esterases, and phospholipases A and B; (2) phosphoric monoester hydrolases (EC 3.1.3), which are the phosphatases; and (3) phosphoric diester hydrolases (EC 3.1.4), including phospholipases C and D (**Figure 2**).

11.2.4. Phases of Digestion and Their Compartmentalization in the Insect Gut

Most food molecules to be digested are polymers, such as proteins and starch, and are digested sequentially in three phases. Primary digestion is the dispersion and reduction in molecular size of the polymers, and results in oligomers. During intermediate digestion, these undergo a further reduction in molecular size to dimers; in final digestion, they become monomers. Digestion usually occurs under the action of digestive enzymes from the midgut, with little or no participation of salivary enzymes.

Any description of the spatial organization of digestion in an insect must relate the midgut compartments (cell, ecto-, and endoperitrophic spaces) to each phase of digestion, and hence to the corresponding enzymes. To accomplish this, enzyme determinations must be performed in each midgut luminal compartment and in the corresponding tissue. Techniques of sampling enzymes from midgut luminal compartments and enzymes trapped in cell glycocalyx have been reviewed elsewhere (Terra and Ferreira, 1994). Microvillar enzymes are discussed in detail in section 11.7.

Frequently, initial digestion occurs inside the peritrophic membrane (see sections 11.8.1 and 11.8.2), intermediate digestion in the ectoperitrophic space, and final digestion at the surface of midgut cells by integral microvillar enzymes or by enzymes trapped into the glycocalyx (**Figure 1**). Exceptions to this rule, and the procedures for studying the organization of the digestive process, will be detailed below.

11.2.5. Role of Microorganisms in Digestion

Most insects harbor a substantial microbiota, including bacteria, yeast, and protozoa. Microorganisms might be symbiotic or fortuitous contaminants from the external environment. They are found in the lumen, adhering to the peritrophic membrane, attached to the midgut surface, or within cells. Intracellular bacteria are usually found in special cells, mycetocytes, which may be organized in groups, mycetomes. Microorganisms produce and secrete their own hydrolases, and cell death will result in the release of enzymes into the intestinal milieu. Any

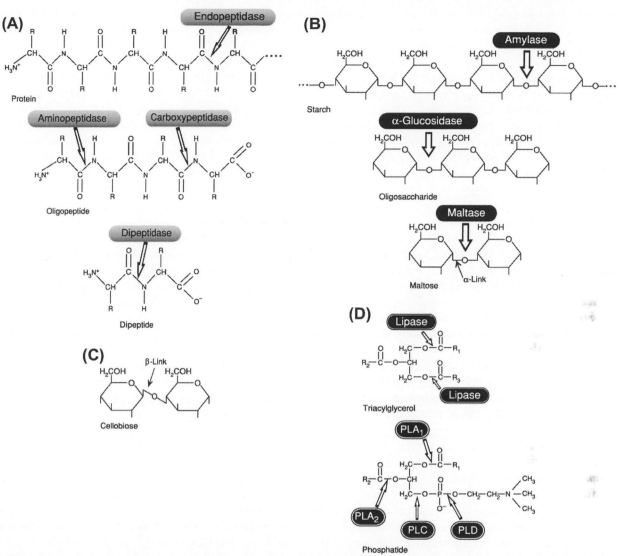

Figure 2 Digestion of important nutrient classes. Arrows point to bonds cleaved by enzymes. (A) Protein digestion; R, different amino acid moieties; (B) starch digestion; (C) β-linked glucoside; (D) lipid digestion; PL, phospholipase; R, fatty acyl moieties. Reprinted with permission from Terra, W.R., 2009. Digestion. In: Resh, V.H., Cardé, R.T. (eds), *Encyclopedia of Insects*, 2nd edn, Academic Press, San Diego, CA, pp. 271–273; ©Elsevier.)

consideration of the spectrum of hydrolase activity in the midgut must include the possibility that some of the activity may derive from microorganisms. Despite the fact that digestive enzymes of some insects are thought to be derived from the microbiota, there are relatively few studies that show an unambiguous contribution of microbial hydrolases. Best examples are found among wood- and humus-feeding insects like termites, tipulid fly larvae, and scarabid beetle larvae. Although these insects may have their own cellulases (see section 11.4.3), only fungi and certain filamentous bacteria developed a strategy for the chemical breakdown of lignin. Lignin is a phenolic polymer that forms an amorphous resin in which the polysaccharides of the secondary plant cell wall are embedded, thus becoming protected from enzymatic attack (Terra *et al.*, 1996; Brune, 1998; Dillon and Dillon, 2004).

Microorganisms play a limited role in digestion, but they may enable phytophagous insects to overcome biochemical barriers to herbivory – for example, detoxifying flavonoid alkaloids and the phenolic aglycones of plant glycosides. They may also provide complex-B vitamins for blood-feeders and essential amino acids for phloem feeders, produce pheromone components, or withstand the colonization of the gut by non-indigenous species (including pathogens) (Dillon and Dillon, 2004; Genta *et al.*, 2006a).

11.3. Midgut Conditions Affecting Enzyme Activity

The pH of the contents of the midgut is one of the important internal environmental properties that affect digestive enzymes. Although midgut pH is hypothesized to result from adaptation of an ancestral insect to a particular diet, its descendants may diverge, feeding on different diets, while still retaining the ancestral midgut pH condition. Thus, there is not necessarily a correlation between midgut pH and diet. In fact, midgut pH correlates well with insect phylogeny (Terra and Ferreira, 1994; Clark, 1999).

The pH of insect midgut contents is usually in the 6–7.5 range. Major exceptions are the very alkaline midgut contents (pH 9–12) of Lepidoptera, scarab beetles, and nematoceran Diptera larvae; the very acid (pH 3.1–3.4) middle region of the midgut of cyclorrhaphous Diptera; and the acid posterior region of the midgut of heteropteran Hemiptera (Terra and Ferreira, 1994; Clark, 1999). pH values may not be equally buffered along the midgut. Thus, midgut contents are acidic in the anterior midgut and nearly neutral or alkaline in the posterior midgut in Dictyoptera, Orthoptera, and most families of Coleoptera. Cyclorrhaphan Diptera midguts have nearly neutral contents in the anterior and posterior regions, whereas in middle midgut the contents are very acid (Terra and Ferreira, 1994).

A pH in the ectoperitrophic space lower than in the midgut lumen was reported in some lepidopteran larvae. This is an artefact caused by a halt in alkaline secretion by the isolated midgut tissue (Grigorten et al., 1993). Nevertheless, the pH in the immediate neighborhood of the negatively-charged microvillar glycocalyx is expected to be lower than in the bulk solution, because of proton retention (Quina et al., 1980).

The high alkanity of lepidopteran midgut contents is thought to allow these insects to feed on plant material rich in tannins, which bind to proteins at lower pH, reducing the efficiency of digestion (Berenbaum, 1980). This explanation may also hold for scarab beetles and for detritus-feeding nematoceran Diptera larvae that usually feed on refractory materials such as humus. Nevertheless, mechanisms other than high gut pH must account for the resistance to tannin displayed by some locusts (Bernays et al., 1981) and beetles (Fox and Macauley, 1977). One possibility is the effect of surfactants such as lysolecithin, which is formed in insect fluids due to the action of phospholipase A on cell membranes (**Figure 2**), and which occurs widely in insect digestive fluids (De Veau and Schultz, 1992). Surfactants are known to prevent the precipitation of proteins by tannins even at pH as low as 6.5 (Martin and Martin, 1984). Present knowledge is not sufficient to relate midgut detergency to diet or phylogeny, or to both.

Tannins may have deleterious effects other than precipitating proteins. Tannic acid is frequently oxidized in the midgut lumen, generating peroxides, including hydrogen peroxide, which readily diffuses across cell membranes and is a powerful cytotoxin. In some insects (e.g., *Orgyia leucostigma*), tannic acid oxidation and the generation of peroxides are suppressed by the presence of high concentrations of ascorbate and glutathione in the midgut lumen (Barbehenn et al., 2003). Dihydroxy phenolics in an alkaline medium are converted to quinones that react with lysine ε-amino groups. This leads to protein aggregation and a decrease in lysine availability for the insect. Other compounds (e.g., oleuropein, alkylate lysine residues in proteins) cause the same problems as dihydroxy phenolics. These phenomena are inhibited in larvae of several lepidopteran species by secreting glycine into the midgut lumen. Glycine competes with lysine residues in the denaturing reaction (Konno et al., 2001). In some insects, tannins reduce the overall efficiency of conversion of ingested matter to body mass by an unknown mechanism. Nevertheless, the performance of these insects remains unchanged, because of compensatory feeding (Barbehenn et al., 2009).

A high midgut pH may also be of importance, in addition to its role in preventing tannin binding to proteins, in freeing hemicelluloses from plant cell walls ingested by insects. Hemicelluloses are usually extracted in alkaline solutions for analytical purposes (Blake et al., 1971), and insects such as the caterpillar *Erinnyis ello* are able to digest hemicelluloses efficiently without affecting the cellulose from the leaves they ingest to any degree (Terra, 1988). This explanation is better than the previous one in accounting for the very high pH observed in several insects, since a pH of about 8 is sufficient to prevent tannin binding to proteins (Terra, 1988).

The acid region in the cyclorrhaphous Diptera midgut is assumed to be involved in the process of killing and digesting bacteria, which may be an important food for maggots. This region is retained in Muscidae that have not diverged from the putative ancestral bacteria-feeding habit, as well as in the flesh-feeding Calliphoridae and the fruit-feeding Tephritidae (Terra and Ferreira, 1994). The acid posterior midgut of Hemiptera may be related to their lysosome-like digestive enzymes (cysteine and aspartic proteinase) (see sections 11.5.3 and 11.5.4).

Few papers have dealt with midgut pH buffering mechanisms. Dow (1992) described a carbonate secretion system, which may be responsible for the high pH found in Lepidoptera midguts (**Figure 3**). Phosphorus NMR microscopy has been used to show that valinomycin leads to a loss of alkalinization in the midgut of *Spodoptera litura* (Skibbe et al., 1996). As valinomycin is known to transport K^+ down its concentration gradient, this result gives further support to the model described in **Figure 3**. Midgut alkalinization in nematoceran Diptera occurs by mechanisms similar to those of lepidopteran larvae (Okech et al., 2008), whereas no data are available for scarab beetles. Terra and Regel (1995) determined pH values and concentrations of ammonia, chloride, and phosphate in the presence or absence of ouabain and vanadate in *Musca*

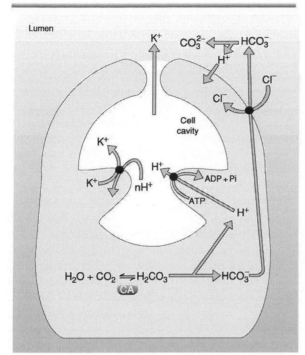

Figure 3 A model for generation of high gut pH by the goblet cells of lepidopteran larvae. Carbonic anhydrase (CA) produces carbonic acid that dissociates into bicarbonate and a proton. The proton is pumped by a V-ATPase into the goblet cell cavity, from where it is removed in exchange with K^+ that eventually diffuses into lumen. Bicarbonate is secreted in exchange with chloride and loses a proton due to the intense field near the membrane, forming carbonate and raising the gut pH. Data from Dow (1992).

domestica midguts. From the results, they proposed that middle midgut acidification is accomplished by a proton pump of mammalian-like oxyntic cells, whereas the neutralization of posterior midgut contents depends on ammonia secretion (**Figure 4**).

Redox conditions in the midgut are regulated and may be the result of phylogeny, although data are scarce. Reducing conditions are observed in clothes moth, sphinx moths, owlet moths, and dermestid beetles (Appel and Martin, 1990), and in Hemiptera (Silva and Terra, 1994). Reducing conditions are important to open disulfide bonds in keratin ingested by some insects (clothes moths, dermestid beetles) (Appel and Martin, 1990), to maintain the activity of the major proteinase in Hemiptera (see section 11.5.3), and to reduce the impact of some plant allelochemicals, such as phenol, in some herbivores (Appel and Martin, 1990). In spite of this, the artificial lowering of *in vivo* redox potentials did not significantly impact digestive efficiency of the herbivore *Helicoverpa zea*, although the reducing agent used (dithiothreitol) inhibited some proteinases *in vitro* (Johnson and Felton, 2000). Midgut antioxidant enzymes in *Spodoptera littoralis* are upregulated in response to increased oxidative stress caused by oxidizable allelochemicals (Krishnan and Kodrik, 2006).

Although several allelochemicals other than phenols may be present in the insect gut lumen, including alkaloids, terpene aldehydes, saponins, and hydroxamic acids (Appel, 1994), data on their effect on digestion are lacking.

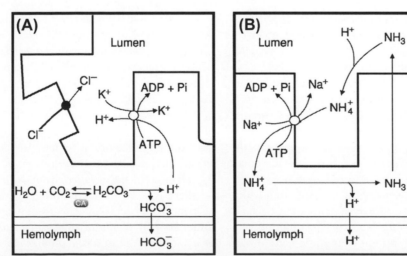

Figure 4 Diagrammatic representation of ion movements, proposed as being responsible for maintenance of pH in the larval midgut contents of *Musca domestica*. Carbonic anhydrase (CA) in cup-shaped oxyntic cells in the middle of the midgut (A) produces carbonic acid which dissociates into bicarbonate and a proton. Bicarbonate is transported into the hemolymph, whereas the proton is actively translocated into the midgut lumen acidifying its contents to pH 3.2. Chloride ions follow the movement of protons. NH_3 diffuses from anterior and posterior midgut cells (B) into the midgut lumen, becoming protonated and neutralizing their contents to pH 6.1–6.8. NH_4^+ is then exchanged for Na^+ by a microvillar Na^+/K^+-ATPase. Inside the cells, NH_4^+ forms NH_3, which diffuses into midgut lumen, and proton that is transferred into the hemolymph. Reprinted with permission from Terra, W.R., Regel, R., 1995. pH buffering in *Musca domestica* midguts. *Comp. Biochem. Physiol. A* 112, 559–564; ©Elsevier.

11.4. Digestion of Carbohydrates

11.4.1. Initial Considerations

Polysaccharides are major constituents of cell walls and energy reserves, such as starch granules within plant cells and glycogen within animal cells. For phytophagous insects, disruption of plant cell walls is necessary in order to expose storage polymers in cell contents to polymer hydrolases. Cell wall breakdown may be achieved by mastication, but more frequently it is the result of the action of digestive enzymes. Thus, even insects unable to obtain nourishment from the cellulosic and non-cellulosic cell wall biochemicals would profit from having enzymes active against these structural components. Cell walls are disrupted by β-glucanases, xylanases, and pectinases (plant cells), lysozyme (bacterial cells), or chitinase and β-1,3-glucanase (fungal cells). The carbohydrates associated with cellulose are frequently named "hemicelluloses" and the enzymes that attack them "hemicellulases." Thus, the hemicellulases include β-glucanases other than cellulases, xylanases, and pectinases. Following the loss of cell wall integrity, starch digestion is initiated by amylases. A complex of carbohydrases converts the oligomers resulting from the action of the polymer hydrolyzing enzymes into dimers (such as sucrose, cellobiose, and maltose, which also occur free in some cells), and finally into monosaccharides like glucose and fructose.

11.4.2. Amylases

α-Amylases (EC 3.2.1.1) catalyze the endohydrolysis of long α-1,4-glucan chains such as starch and glycogen. Amylases are usually purified by glycogen–amylase complex formation followed by precipitation in cold ethanol, or, alternatively, by affinity chromatography in a gel matrix linked to a protein amylase inhibitor. In sequence, isoamylases can be resolved by anion exchange chromatography (Terra and Ferreira, 1994).

Most insect amylases have molecular weights in the range 48–60 kDa, pI values of 3.5–4.0, and K_m values with soluble starch around 0.1%. pH optima generally correspond to the pH prevailing in midguts from which the amylases were isolated. Insect amylases are calcium-dependent enzymes, and are activated by chloride with displacement of the pH optimum. Activation also occurs with anions other than chloride, such as bromide and nitrate, and it seems to depend upon the ionic size (Terra and Ferreira, 1994).

The best-known insect α-amylase, and the only one whose 3D structure has been resolved, is the midgut α-amylase of *Tenebrio molitor* larvae. The enzyme has three domains. The central domain (domain A) is a $(\beta/\alpha)_8$-barrel that comprises the core of the molecule and includes the catalytic amino acid residues (Asp 185, Glu 222, and Asp 287) (*T. molitor* α-amylase numbering throughout). Domains B and C are almost opposite each other, on each side of domain A. The substrate-binding site is located in a long V-shaped cleft between domains A and B. There, six saccharide units can be accommodated, with the sugar chain being cleaved between the third and fourth bound glucose residues. A calcium ion is placed in domain B, and is coordinated by Asn 98, Arg 146, and Asp 155. This ion is important for the structural integrity of the enzyme, and also seems to be relevant due its contact with His 189. This residue interacts with the fourth sugar of the substrate bound in the active site, forming a hinge between the catalytic site and the Ca^{2+}-binding site. A chloride ion is coordinated by Arg 183, Asn 285, and Arg 321 in domain A (Strobl *et al.*, 1998a). Domain C is placed in the C-terminal part of the enzyme, contains the so-called "Greek key" motif, and has no clear function (**Figure 5**). These structural features are shared by all known α-amylases (Nielsen and Borchert, 2000), although not all α-amylases have a chloride-binding site (Strobl *et al.*, 1998a). The most striking difference between mammalian and insect α-amylases is the presence of additional loops in the vicinity of the active site of the mammalian enzymes (Strobl *et al.*, 1998a).

Asp 287 is conserved in all α-amylases. Comparative studies have shown that Glu 222 is the proton donor and Asp 185 the nucleophile, and that Asp 287 is important, but not a direct participant in catalysis. It is proposed that its role is to elevate the pK_a of the proton donor (Nielsen and Borchert, 2000).

Chloride ion is an allosteric activator of α-amylases, leading to a conformation change in the enzyme that changes the environment of the proton donor. This change causes an increase in the pK_a of the proton donor, thus displacing the pH optimum of the enzyme and increasing its V_{max} (Levitzki and Steer, 1974). According to Strobl *et al.* (1998a), the increase in V_{max} is a consequence of the chloride ion being close to the water molecule that has been suggested to initiate the cleavage of the substrate chain. The nucleophilicity of this water molecule might be enhanced by the negative charge of the ion.

There is a large and growing number of complete sequences of insect α-amylases registered in the GenBank. Examples may be found among Hymenoptera (Da Lage *et al.*, 2002), Coleoptera (Strobl *et al.*, 1997; Titarenko and Chrispeels, 2000), Diptera (Grossmann *et al.*, 1997; Charlab *et al.*, 1999), and Lepidoptera (Da Lage *et al.*, 2002; Pytelkova *et al.*, 2009). All the sequences have the catalytic triad (Asp 185, Glu 222, and Asp 287), the substrate-binding histidine residues (His 99, His 189, and His 286), and the Ca^{2+}-coordinating residues (Asn 98, Arg 146, and Asp 155) (**Figure 5**). From the residues found to be involved in chloride binding, Arg 183

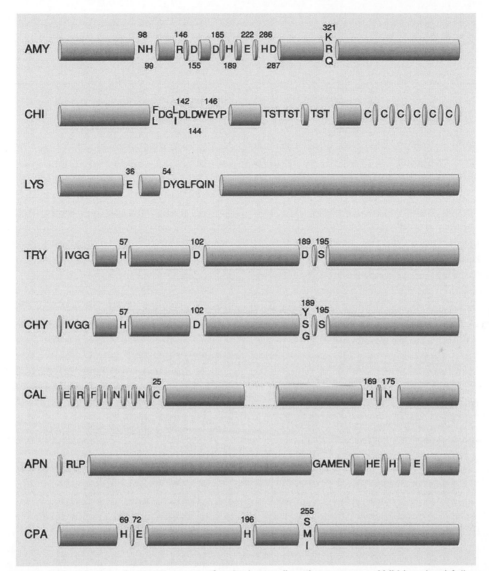

Figure 5 Conserved residues in the primary structures of major insect digestive enzymes. AMY (amylase) follows *Tenebrio molitor* amylase numbering; CHI (chitinase), molting-fluid *Manduca sexta* chitinase numbering. TRY (trypsin) and CHY (chymotrypsin) follow the bovine chymotrypsin numbering; CAL (cathepsin L), papain numbering; APN (aminopeptidase N) does not have a consensual numbering; CPA (carboxypeptidase A), mammalian CPA numbering.

and Asn 285 are conserved, whereas position 321 varies. According to D'Amico *et al.* (2000), all known chloride-activated α-amylases have an arginine or lysine residue at position 321. Insect α-amylase sequences have arginine at position 321, except those of *Zabrotes subfasciatus* and *Anthonomus grandis*, which have lysine, and the lepidopteran α-amylases, which have glutamine. This agrees with the observation that most insect α-amylases are activated by chloride, with the remarkable exception of lepidopteran amylases (Terra and Ferreira, 1994). The few coleopteran and hymenopteran α-amylases reported not to be affected by chloride (Terra and Ferreira, 1994) deserve reinvestigation. It is possible that another anion is replacing chloride as an activator, as shown for some hemipteran amylases (Hori, 1972). It is worth noting

that there is more Arg + Lys in the lepidopteran amylase than in *T. molitor* amylase (Pytelkowa *et al.*, 2009). This is thought to be an adaptation to the higher pH milieu of lepidopteran midguts.

The "action pattern" refers to the number of bonds hydrolyzed during the lifetime of a particular enzyme–substrate complex. If more than one bond is hydrolyzed after the first hydrolytic step, the action pattern is said to be processive. The degree of multiple attack is the average number of hydrolyzed bonds after the first bond is broken. *Rhynchosciara americana* amylase has a degree of multiple attack between that of the amylase of *Bacillus subtilis* (1.7) and porcine pancreas (6). Amylases from larvae and adults of *Sitophilus zeamais*, *S. granarius*, and *S. oryzae*, and larvae of *Bombyx mori*, have action patterns similar to that of porcine pancreas amylase (Terra and Ferreira, 1994). These studies need to be taken further, including the determination of the affinities corresponding to each subsite in the active center. Such studies, especially if combined with crystallographic data, may describe in molecular detail the reasons why amylases act differently toward starches of distinct origins.

There is a variety of natural compounds that affect amylases, including many plant protein inhibitors (Franco *et al.*, 2002). Crystallographic data have shown that these protein inhibitors always occupy the amylase active site (Strobl *et al.*, 1998b; Payan, 2004). In the case of the *Amaranth* α-amylase inhibitor, a comparison of *T. molitor* amylase–inhibitor complex with a modeled complex between porcine pancreatic α-amylase and the inhibitor identified six hydrogen bonds that can be formed only in the *T. molitor* amylase–inhibitor complex (Pereira *et al.*, 1999). This was the first successful explanation of how a protein inhibitor specifically inhibits α-amylases from insects, but not from mammalian sources. As will be discussed with details for trypsins (see section 11.5.2.1), specific amylases are induced when insect larvae are fed with α-amylase inhibitor-containing diets (Silva *et al.*, 2001; Pytelkova *et al.*, 2009). The mechanisms underlying this induction are unknown.

11.4.3. β-Glucanases

β-Glucanases are enzymes acting on β-glucans. These are major polysaccharide components of plant cell walls, and include β-1,4-glucans (cellulose), β-1,3-glucans (callose), and β-1,3;1,4-glucans (cereal β-glucans) (Bacic *et al.*, 1988). The cell walls of certain groups of fungi have β-1,3;1,6-glucans (Bacic *et al.*, 1988).

11.4.3.1. Cellulases Cellulose is by far the most important β-glucan. It occurs in the form of β-1,4-glucan chains packed in an ordered manner to form compact aggregates which are stabilized by hydrogen bonds. The resulting structure is crystalline and not soluble. According to work done with microbial systems, cellulose is digested by a combined action of two enzymes. An endo-β-1,4-glucanase (EC 3.2.1.4), with an open substrate-binding cleft, cleaves bonds located within chains in the amorphous regions of cellulose, creating new chain ends. An exo-β-1,4-glucanase (EC 3.2.1.91) processively releases cellobiose from the ends of cellulose chains in a tunnel-like active site. Surface loops in cellobiohydrolase prevent the dislodged cellulose chains from readhering to the crystal surface, as the enzyme progresses into crystalline cellulose (Rouvinen *et al.*, 1990; Kleywegt *et al.*, 1997). Cellobiohydrolase structure is modular, comprising a catalytic domain linked to a distinct cellulose-binding domain, which enhances the activity of the enzyme towards insoluble cellulose (Linder and Teeri, 1997).

Cellulose digestion in insects is rare, presumably because the dietary factor that usually limits growth in plant feeders is protein quality. Thus, cellulose digestion is unlikely to be advantageous to an insect that can meet its dietary requirements using more easily digested constituents. This is exemplified by lepidopterans that, even ingesting plant material, lack cellulase genes (Watanabe and Tokuda, 2010). Nevertheless, cellulose digestion occurs in several insects that have, as a rule, nutritionally poor diets such as wood or humus (Terra and Ferreira, 1994). The role of symbiotic organisms in insect cellulose digestion is less important than initially believed (Slaytor, 1992), although symbiotic nitrogen-fixing organisms are certainly involved in increasing the nutritive value of diets of many insects (Terra, 1990).

Insect cellulases are known mainly from the lower termite *Reticulitermes speratus*, the higher termite *Nasutitermes takasagoensis*, and the woodroach *Panesthia cribrata* (Lo *et al.*, 2000; Watanabe and Tokuda, 2010). Alignments of the sequences of termite and woodroach endoglucanases from data banks showed that they belong to family 9 of glycoside hydrolases (Coutinho and Henrissat, 1999). The paradigm of this family is the endo/exocellulase from the bacteria *Thermomonospora fusca*, whose catalytic center binds a cellopentaose residue and cleaves it into cellotetraose plus glucose or cellotriose plus cellobiose, and has Asp 55 as a nucleophile and Glu 424 as a proton donor (Sakon *et al.*, 1997).

The active site groups are conserved in the termite and woodroach endoglucanases, although these enzymes lack the cellulose-binding domains that improve the binding and facilitate the activity of the catalytic domain on crystalline cellulose (Linder and Teeri, 1997). The conclusions drawn from sequence alignments were confirmed by the 3D structure resolution of the *N. takasagoensis* endoglucanase, which also revealed a Ca^{2+}-binding site near its substrate binding cleft (Khadeni *et al.*, 2002). According to Slaytor (1992), the large production of endoglucanases in termites and woodroaches would compensate for their low efficiency on crystalline cellulose.

The phytophagous beetle *Phaedon cochleariae* has cellulase activity in its midgut (Girard and Jouanin, 1999a). A cDNA that encodes one cellulase was cloned, and was shown to belong to the glycoside hydrolase family 45 and to consist only of a catalytic domain. Similar enzymes occur in the midgut of the beetles *Psacothea hilaris* (Sugimura *et al.*, 2003) and *Apriona germari* (Wei *et al.*, 2006). The cellulase from the cricket *Teleogryllus emma*, however, is from family 9 (Kim *et al.*, 2008).

11.4.3.2. Laminarinases and licheninases Licheninases (EC 3.2.1.73) digest only β-1,3;1,4-glucans, whereas laminarinases may hydrolyze β-1,3;1,4-glucans and also β-1,3-glucans (EC 3.2.1.6), or only the last polymer (EC 3.2.1.39). In spite of laminarinase activities being widespread among insects (Terra and Ferreira, 1994), few of these enzymes were purified and characterized. Three laminarinases produced by the salivary glands were isolated from *Periplaneta americana* (LAM, LIC1 and LIC2; Genta *et al.*, 2003). LIC1 and LIC2 hydrolyze laminarin and lichenin, whereas LAM is active only against laminarin. *Abracris flavolineata* (Genta *et al.*, 2007), *T. molitor* (Genta *et al.*, 2009), and *Spodoptera frugiperda* (Bragatto *et al.*, 2010) have only one major laminarinase (ALAM, TLAM, and SLAM, respectively) produced by the midgut. ALAM and SLAM have only laminarin as substrate, and TLAM hydrolyzes, besides laminarin, yeast β-1,3-β-1,6-glucan. Except for SLAM and LIC2 (not determined) all the mentioned enzymes lyse *Saccharomyces cerevisae* cells. The enzymes have pH optima from 5.0 to 6.5, except for SLAM (9.0), K_m values for laminarin hydrolysis from 0.074 to 0.36%, and molecular masses from 23.4 to 46.2 kDa.

Some laminarinases are processive – that is, they perform multiple rounds of catalysis when the enzyme remains attached to the substrate. The exo-β-1,3-glucanase of *A. flavolineata* has a high-affinity accessory site that, on substrate binding, causes active site exposure, followed by the transference of the substrate to the active site. Processivity results in this case from consecutive transferences of substrate between accessory and active site (Genta *et al.*, 2007).

Insect laminarinases belong to glycoside hydrolase family 16, and are evolutionarily related and strikingly similar in sequence to Gram-negative bacteria-binding proteins (GNBPs) and other glucan-binding proteins that are active in the insect innate immune system. Because of this, many true laminarinases had been wrongly annotated as glucan-binding proteins in sequence databases.

Laminarinases and GNBPs are derived from the laminarinase of the ancestor of mollusks and arthropods. The insect laminarinases lost an extended N-terminal region of the ancestral laminarinase, whereas the β-glucan binding proteins lost the catalytical residues (**Figure 6**) (Bragatto *et al.*, 2010).

The role of laminarinases in the insect gut is not yet clear. Insects eating contaminated food, such as *T. molitor* and *P. americana*, may use β-1,3 glucanases to digest fungi from their diet. In Lepidoptera that eat a less contaminated food, β-1,3- glucanase may be important to digest callose. Since the larvae eats on and in the same plant, callose deposition can impair nutrient digestibility in the absence of a β-1,3- glucanase. Pauchet *et al.* (2009a) correlate the increase in β-1,3-glucanase activity in *Helicoverpa armigera* after fungi or bacteria ingestion with a role of the enzyme in immunity. It cannot be ruled out, however, that the enzyme activity increase may be due to a higher substrate concentration, and not because of a role in immunity. Further studies are necessary to settle the subject.

11.4.4. Xylanases and Pectinases

Xylans constitute the major non-cellulosic polysaccharides (hemicelluloses) of primary walls of grasses and secondary walls of all angiosperms, accounting for one-third of all renewable organic carbon available on earth (Bacic *et al.*, 1988). Chemically, xylans are a family of linear β-1,4-xylans with a few branches. Endo-β-1,4-xylanase activities (EC 3.2.1.8) were found in several insects (Terra and Ferreira, 1994). One of these enzymes was cloned from a beetle and shown to correspond to a protein of 22 kDa, with high sequence identity to fungal xylanases, and conserving the usual two catalytic regions (Girard and Jouanin, 1999a). An exo-β-1,4-xylanase (EC 3.2.1.37) was partially purified from termites (Matoub and Rouland, 1995), and was thought to act synergistically with an endo-β-1,4-xylanase originating from fungus ingested by the termites. Recent work showed that xylanase genes were expressed in the hindguts of termites by symbiotic flagellates (Arakawa *et al.*, 2009). Much more work is needed on this class of enzymes that may be important mainly for detritivorous insects.

Pectin is a linear chain of a D-galacturonic acid units with α-1,4-linkages in which varying proportions of the acid groups are present as methyl esters. It is the main component of the rhamnogalacturonan backbone of the structure formed by the pectin polysaccharides. Pectin is hydrolyzed by pectinases (polygalacturonases, EC 3.2.1.15) described in many insects (Terra and Ferreira, 1994).

Pectinases are thought to be important for hemipterans, as they would facilitate penetration of their stylets through plant tissues into sap-conducting structures, and for insects boring plant parts. Accordingly, pectinases have been found in hemipteran saliva (Vonk and Western, 1984), and have been isolated and characterized from two weevils (Shen *et al.*, 1996; Dootsdar *et al.*, 1997) and cloned from a phytophagous beetle (Girard and Jouanin, 1999a).

The pectinases from the weevils *S. oryzae* (Shen *et al.*, 1996) and *Diaprepes abbreviatus* (Dootsdar *et al.*, 1997) were purified to electrophoretical homogeneity from whole-body extracts and gut homogenates, respectively. Purification of the pectinases was achieved by affinity chromatography through cross-linked pectate in addition to two ion-exchange chromatographic steps. The enzymes have molecular masses of 35–45 kDa, pH optimum 5.5, and K_m values of 1–4 mg/ml for pectic acid. The *D. abbreviatus* pectinase is inhibited by a polygalacturonase-inhibitor protein that may be associated with plant resistance to insects (Dootsdar *et al.*, 1997). Although the weevil pectinases may originate from endosymbiotic bacteria (Campbell *et al.*, 1992), the finding that the cDNA-coding pectinase of the beetle *P. cochleariae* has a poly(A) tail (Girard and Jouanin,

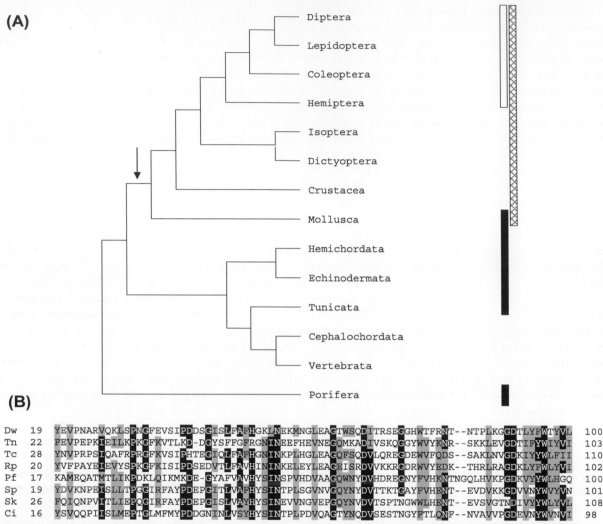

(A)

Diptera
Lepidoptera
Coleoptera
Hemiptera
Isoptera
Dictyoptera
Crustacea
Mollusca
Hemichordata
Echinodermata
Tunicata
Cephalochordata
Vertebrata
Porifera

(B)

```
Dw  19  YEVPNARVQKLSPNGFEVSIPDDSGISLFAFHGKLNEKMNGLEAGTWSQDITRSEGGHWTFRNT--NTPLKGGDTLYFWTYVL  100
Tn  22  FEVPEPKIEILKPKGFKVTLKD-DGYSFFGFRGNINEEFHEVNEGQMKADIVSKQGGYWVYKNR--SKKLEVGDTIBYWIYVI  103
Tc  28  YNVPRPSIQAFRPRGFKVSIPHTEGIQLFAFHGNINKPLHGLEAGQFSQDVLQREGDEWVFQDS--SAKLNVGDKIYYWLFII  110
Rp  20  YVFPAYEIEVYSPKGFKISIPDSEDVTLFAVHININKELEYLEAGEISRDVVKKRGDRWVYEDK--THRLRAGDKLYFWLYVI  102
Pf  17  KAMEQATMTLIKPDKLQIKMKDE-GYAFVAVHYSINSPVHDVAAGQWNYDVHDREGNYFVHKNTNGQLHVKPGDKVYYWLHGQ  100
Sp  19  YDVKNPEISLLTPGGIRFAYPDEPGITLVAFHYSINTPLSGVNVGQYNYDVTTKTGAYFVHENT--EVDVKKGDVVNYWVYVN  101
Sk  26  FQIQNPVITLIEPQGIRFAYPDEPGISLVAYHYSINEVVNGVEPGQYNVDVTSPTNGWWLHENT--EVSVGTNDIVYYWLYVL  108
Ci  16  YSVQQPIISLMEPTGLMFMYPDDGNINLVSYHYSINTPLPDVQAGTYNQDVSESTNGYFTLQNF--NVAVVPGDEVNYWVNVI   98
```

Figure 6 Occurrence of GNBPs and β-1,3-glucanases in metazoans. (A) Phylogeny tree of some metazoans according to Halanych (2004). Sidebars indicate the occurrence of insect GNBPs (white), β-1,3-glucanases lacking (crosshatched) and featuring (black) the extended N-terminal region. The arrow indicates the point where possible gene duplication occurred. (B). Partial alignment of the extended N-terminal region sequences of selected GNBPs and β-1,3-glucanases. Strongly conserved residues are shaded in black. Organisms and nucleotide GenBank Access number: Dw, *Drosophila willistoni* (**XM_002061567**); Tn, *Trichoplusia ni* (**EU770373**); Tc, *Tribolium castaneum* (**XM_966587**); Rp, *Rhodnius prolixus* (**EF634459**); Pf, *Pinctata fucata* (**FJ775601**); Sp, *Strongylocentrotus purpuratus* (**XR_025993**); Sk, *Saccoglossus kowalevskii* (**XM_002740940**); Ci, *Ciona intestinalis* (**XM_002126654**). Reprinted with permission from Bragatto, I., Genta, F.A., Ribeiro, A.F., Terra, W.R., Ferreira, C., 2010. Characterization of a β-1,3-glucanase active in the alkaline midgut of *Spodoptera frugiperda* larvae and its relation to β-glucan-binding proteins. *Insect Biochem. Mol. Biol.*, 40, 861–872.

1999a) argues against this hypothesis. The beetle pectinase belongs to family 28 of the glycoside hydrolases and is most related to fungal endopolygalacturonases, conserving the active site signature centered on the His 223 catalytic residue (Girard and Jouanin, 1999a; Markovic and Janecek, 2001).

11.4.5. Chitinases and Lysozymes

Chitin, the simplest of the glycosaminoglycans, is a β-1,4-homopolymer of N-acetylglucosamine (see Chapters 5–7 in this volume). Chitinolytic enzymes include chitinase

(EC 3.2.1.14), which catalyzes the random hydrolysis of internal bonds in chitin forming smaller oligosaccharides, and β-N-acetyl-D-glucosaminidase (EC 3.2.1.52), which liberates N-acetylglucosamine from the non-reducing end of oligosaccharides (Arakane and Muthukrishnan, 2010). Lysozyme, as described below, also has some chitinase activity, whereas chitinase rarely has lysozyme activity. These enzymes, besides being active in the ecdysial cycle, may also have a digestive role. Chitinase assays with midguts of several insects showed that there is a correlation between the presence of chitinase and a diet rich in chitin (Terra and Ferreira, 1994).

The best-known insect chitinase is the molting fluid chitinase from the lepidopteran *Manduca sexta* (**Figure 5**), which pertains to family 18 of glycosyl hydrolases. The enzyme has a multidomain architecture that includes a signal peptide, an N-terminal catalytic domain with the consensus sequence (F/L)DG(L/I)DLDWEYP, and a C-terminal cysteine-rich chitin-binding domain, which are connected by the interdomain serine/threonine-rich O-glycosylated linker. The residues Asp 142, Asp 144, Trp 145, and Glu 146 of the consensus sequence have been shown by site-directed mutagenesis to be involved in catalysis. Glu 146 functions as a proton donor in the hydrolysis like homologous residues in other glycoside hydrolases. Asp 144 apparently functions as an electrostatic stabilizer of the positively charged transition state, whereas Asp 142, and perhaps also Trp 145, influences the pK_a values of Asp 144 and Glu 146. The chitin-binding domains have six cysteines (with the consensus sequence CXnCXnCXnCXnCXnC, where Xn stands for a variable number of residues), and include several highly conserved aromatic residues (Tellam *et al.*, 1999). The three disulfide bonds in the domain may constrain the polypeptide to present the aromatic amino acids on the protein surface for interactions with the ring structures of sugars. Thus, the chitin-binding domains enhance activity toward the insoluble polymer, whereas the linker region facilitates secretion from the cell and helps to stabilize the enzyme in the presence of proteolytic enzymes (Lu *et al.*, 2002; Arakane *et al.*, 2003). In some cases, proteins with chitin-binding domains but lacking catalytic activity aid in the degradation of chitin (Vaaje-Kolstad *et al.*, 2005).

The *Anopheles gambiae* gut chitinase is secreted upon blood-feeding as an inactive proenzyme that is later activated by trypsin. Sequencing a cDNA coding for the gut chitinase showed that the enzyme comprises a putative catalytic domain at the N-terminus, a putative chitin-binding domain at the C-terminus, and a serine/threonine/proline-rich amino acid stretch between them (Shen and Jacobs-Lorena, 1997). The mosquito chitinase seems to modulate the thickness and permeability of the chitin-containing peritrophic membrane (see section 11.8.1). Supporting this conjecture, the authors found that the peritrophic membrane is stronger and persisted longer when the mosquitoes were fed diets containing chitinase inhibitor.

The beetle *P. cochleariae* has one group of chitinases of 40–70 kDa active at pH 5.0 and detected in guts. This enzyme has an active site centered on the catalytic residues Asp 146 and Glu 150 (*M. sexta* chitinase numbering), but lacks the C-terminal chitin-binding domain and the serine/threonine-rich interdomain (Girard and Jouanin, 1999b). *T. molitor* midgut chitinase (Genta *et al.*, 2006b) is similar to that of *P.cochleariae*, and both pertain to group IV chitinases (Arakane and Muthukrishnan, 2010) that as a rule lack a chitin-binding domain. The function of the *T. molitor* enzyme may be digestion of the cell walls of fungi usually present in its food, without damaging the peritrophic membrane.

Lysozyme (EC 3.2.1.17) catalyzes the hydrolysis of the 1,4-β-glycosidic linkage between N-acetyl-muramic acid and N-acetylglucosamine of the peptidoglycan present in the cell wall of many bacteria, causing cell lysis. Lysozyme is part of an immune defense mechanism against bacteria, and has been described in most animals, including insects (see Chapter 14 in this volume). Lysozyme has also been implicated in the midgut digestion of bacteria by organisms which ingest large amounts of them, such as marine bivalves, or that harbor a bacterial culture in their guts (exemplified by Callewaert and Michiels, 2010).

Among insects, the capacity for digesting bacteria in the midgut seems to be an ancestral trait of Diptera Cyclorrhapha (Lemos and Terra, 1991a; Regel *et al.*, 1998), which agrees with the fact that most Diptera Cyclorrhapha larvae are saprophagous, feeding largely on bacteria (Terra, 1990). These insects have midgut lysozymes (Lemos *et al.*, 1993; Regel *et al.*, 1998) similar to those of vertebrate fermenters. Thus, these enzymes have low pI values, are more active at pH values 3–4 and when present in media with physiological ionic strengths, and are resistant to the cathepsin D-like aspartic proteinase present in midguts (Lemos *et al.*, 1993; Regel *et al.*, 1998; Fujita, 2004; Cançado *et al.*, 2008).

Sequence analyses (Daffre *et al.*, 1994; Cançado *et al.*, 2008) showed that cyclorrhaphan (*Drosophila melanogaster* and *Musca domestica*) digestive lysozymes have the same conserved residues as vertebrate lysozymes (Imoto *et al.*, 1972) (numbering according to Regel *et al.*, 1998): positions 55–61, Glu 36, and Asp 54. Glu 36 is believed to act as a general acid in catalysis, whereas Asp 54 is postulated to stabilize the resulting metastable oxocarbonium intermediate (Imoto *et al.*, 1972). More recently, Asp 54 has been implicated more strongly in catalysis of the hydrolysis of chitin-derived substrates (Matsumura and Kirsch, 1996). The ability of *D. melanogaster* and *M. domestica* purified lysozymes in hydrolyzing chitosan favors this view. The 3D structures of two *M. domestica* lysozymes were resolved (Marana *et al.*, 2006; Cançado *et al.*, 2007), and site-directed mutagenesis was performed (Cançado *et al.*, 2010). The results supported the hypothesis that the acidic pH optimum with synthetic substrates is determined by the presence of N46, S106, and T107 around the catalytic residues, which favors pK_a reduction. Furthermore, the acid pH optimum upon bacterial walls is caused by a decrease in surface positive charges.

Lysozyme is also found in the salivary glands of *Reticulitermes speratus*. This insect is a termite that feeds mainly on dead wood, which tends to lack nitrogen. Fujita *et al.* (2001) suggested, on the basis of the distribution and activity of lysozyme in this termite, that wood-feeding termites use lysozyme secreted from the salivary gland to digest their hindgut bacteria, which are transferred by proctodeal trophallaxis. The termite lysozyme is active in neutral pH (Fujita *et al.*, 2002), thus differing from the digestive cyclorrhaphan lysozymes.

11.4.6. α-Glucosidases

α-Glucosidases (EC 3.2.1.20) catalyze the hydrolysis of terminal, non-reducing α-1,4-linked glucose residues from aryl (or alkyl)-glucosides, disaccharides, or oligosaccharides. α-Glucosidases are frequently named maltases, although some of them may have weak activity on maltose. Insect α-glucosidases occur as soluble forms in the midgut lumen, or are trapped in the midgut cell glycocalyx. They are also bound to microvillar membranes (Terra and Ferreira, 1994), to perimicrovillar membranes (lipoprotein membranes ensheathing the midgut cell microvillar membranes in most hemipterans) (Silva and Terra, 1995), or to the modified perimicrovillar membranes of aphid midgut cells (Cristofoletti *et al.*, 2003). The last two membrane-bound α-glucosidases, as well as the soluble enzyme from bee midguts (Nishimoto *et al.*, 2001), were purified to electrophoretic homogeneity. *Culex pipiens* microvillar α-glucosidase is the primary target of the binary toxin of *Bacillus sphericus*, and, although not purified, cDNA sequencing data suggest it is bound by a glycosyl phosphatidyl inositol anchor (Darboux *et al.*, 2001). α-Glucosidase is a major protein in dipteran midgut microvilli (Terra and Ferreira, 1994), and probably because of that it is the receptor of endotoxins, similar to that observed with aminopeptidase N in lepidopteran midgut cells (see section 11.5.5).

Although biochemical properties of many crude, partially or completely purified gut α-glucosidases are known, including molecular masses (range 60–80 kDa or a multiple of these values), pH optima (range 5–6.5, irrespective of the corresponding midgut pH value), pI values (range 5.0–7.2), and inhibition by tris(hydroxylmethyl) aminomethane (Tris), only a few studies report on α-glucosidases specificities. These studies showed that insect α-glucosidases hydrolyze oligosaccharides up to at least maltopentaose (Terra and Ferreira, 1994), although there are exceptions. The perimicrovillar α-glucosidase from *Dysdercus peruvianus* (Silva and Terra, 1995) and *Quesada gigas* (Fonseca *et al.*, 2010) prefers oligosaccharides up to maltotetraose, and the midgut bee α-glucosidase, up to maltotriose (Nishimoto *et al.*, 2001). The purified midgut α-glucosidase of the pea aphid *Acyrthosiphon pisum* catalyzes transglycosylation reactions in the presence of excess sucrose, thus freeing glucose from sucrose without increasing the osmolarity of the medium (Cristofoletti *et al.*, 2003). This phenomenon, associated with a quick fructose absorption (Ashford *et al.*, 2000), explains why the midgut luminal osmolarity decreases as the ingested sucrose-containing phloem sap passes along

the aphid midgut. *A. pisum* α-glucosidase sequence does not contain any consensus sequences for membrane association, such as transmembrane helices or anchorage by C-terminal glycosylphosphatidyl inositol (GPI) moiety. It is argued that the enzyme may associate with the perimicrovillar membrane through its C-terminal region, which is predominantly hydrophobic (Price *et al.*, 2007).

A. pisum α-glucosidase, like other insect midgut α-glicosidases (e.g., *C.pipiens* midgut α-glucosidase, Darboux *et al.*, 2001) pertains to family 13 glycosidase. All the sequences have the invariant residues: Asp 123, His 128, Asp 206, Arg 221, Glu 271, His 296, and Asp 297 (numbers are relative to the positions in the sequence of *C. pipiens* α-glucosidases) (Darboux *et al.*, 2001) that are involved in the active site of the α-amylase family of enzymes, and the three residues Gly 69, Pro 77, and Gly 323, that have a structural role for some α-glucosidases (Janecek, 1997).

11.4.7. β-Glucosidases, β-Galactosidases, and Myrosinases

β-Glycosidases (EC 3.2.1) catalyze the hydrolysis of terminal, non-reducing β-linked monosaccharide residues from the corresponding glycoside. Depending on the monosaccharide that is removed, the β-glycosidase is named β-glucosidase (glucose), β-galactosidase (galactose), β-xylosidase (xylose), and so on. Frequently, the same β-glycosidase is able to hydrolyze several different monosaccharide residues from glycosides. In this case, β-glucosidase (EC 3.2.1.21) is used to name all enzymes that remove glucose efficiently. The active site of these enzymes is formed by subsites numbered from the glycosidic linkage to be broken, with negative (towards the non-reducing end of the substrate) or positive (towards the reducing end of the substrate) integers (Davies *et al.*, 1997). The non-reducing monosaccharide residue binds at the glycone (–1) subsite, whereas the rest of the molecule accommodates at the aglycone subsite, which actually may correspond to several monosaccharide residue-binding subsites.

Some insects have three or four digestive β-glycosidases with different substrate specificity. In others, only two of these enzymes are found, which are able to hydrolyze as many different β-glycosides as the other three or four enzymes together (Ferreira *et al.*, 1998; Azevedo *et al.*, 2003).

Insect β-glycosidases best characterized have molecular masses of 30–150 kDa, pH optima of 4.5–6.5, and pI values of 3.7–6.8, whereas K_m values, with cellobiose or p-nitrophenyl β-glucoside (NpβGlu) as substrates, are in the range of 0.2–2 mM. Although hydrolyzing several similar substances, insect digestive β-glycosidases have different specificities, preferring β-glucosides or β-galactosides as substrates, with hydrophobic or hydrophilic moieties in the aglycone part of the substrate (Terra and Ferreira, 1994; Azevedo *et al.*, 2003).

Based on relative catalytic efficiency on several substrates, insect β-glycosidases can be divided into two classes. Class A includes the enzymes that efficiently hydrolyze substrates with hydrophilic aglycones, such as disaccharides and oligosaccharides. Class B comprises enzymes that have high activity only on substrates with hydrophobic aglycones, such as alkyl-, p-nitrophenyl-, and methylumbelliferyl-glycosides. Enzymes from class A are more abundant than β-glycosidases from class B. Class A β-glycosidases hydrolyze di- and oligosaccharides, and have four subsites for glucose binding in the active site: one in the glycone (–1) and three in the aglycone (+1, +2, +3) positions (Ferreira *et al.*, 2001, 2003; Marana *et al.*, 2001; Azevedo *et al.*, 2003). Some enzymes seem to be adapted to use disaccharides as well as oligosaccharides as substrates. Optimal hydrolysis of disaccharides relies on high affinities to glucose moieties in –1 and primarily in the +1 subsite (Ferreira *et al.*, 2003). The enzymes highly active against oligosaccharides have subsites –1, +1, and +2, with similar affinities to glucose moieties (Ferreira *et al.*, 2001, 2003).

Class A β-glycosidases are able to hydrolyze β-1,3, β-1,4, and β-1,6 glycoside bonds from di- and oligosaccharides. These enzymes are likely to be involved in the intermediate and terminal digestion of cellulose, hemicellulose, and glycoproteins present in food.

Class B β-glycosidases (or active site) with high activity against hydrophobic substrates may have the physiological role of hydrolyzing glycolipids, mainly galactolipids that are found in high amounts in vegetal tissues. The main galactolipids in plants are 2,3-diacyl β-galactoside D-glycerol (mono galactosyl diglyceride) and 2,3-diacyl 1-(α-galactosyl 1,6 β-galactosyl)-D-glycerol (digalactosyl diglyceride) (Harwood, 1980). These enzymes may act directly against the monogalactosyl diglyceride, or on digalactosyl diglyceride after the removal of one of the galactose residues by α-galactosidase. The activation by amphiphatic substances (as Triton X-100; see above) may be a mechanism to maintain high enzyme activity only in the neighborhood of plant cell membranes undergoing digestion in the insect midgut. These membranes are the source of glycolipid substrates and activating detergent-like molecules. Distant from membranes, the β-glycosidase would be less active, thus hydrolyzing plant glucosides (see below) ingested by the insect with decreased efficiency.

In agreement with the hypothesis presented above, βGly47 from *S. frugiperda* can hydrolyze glycosylceramide, although with low activity (Marana *et al.*, 2000). In mammals, sphingosine and ceramide hydrolysis are dependent on enzyme activation by proteins called saposins (Harzer *et al.*, 2001). Given that genome sequences similar to saposins were found in *D. melanogaster*, insect

β-glycosidases active on glycolipids may also need the same kind of activators, and their absence in the assay reaction may explain why the activity against ceramides is low or not detected at all.

A few digestive β-glycosidases were purified and had their substrate specificity determined using k_{cat}/K_m values. Regarding natural substrates, two β-glycosidases from *T. molitor* (Ferreira *et al.*, 2001, 2003) and three β-glycosidases from *Diatraea saccharalis* (Azevedo *et al.*, 2003) are more active on laminaribiose (glucose β-1,3-glucose) than on cellobiose. On the other hand, the termite *Reticulitermes flavipes* β-glycosidase prefers cellobiose (glucose β-1,4-glucose) (Scharf *et al.*, 2010), and apparently the enzyme from *Neotermes koshunensis* has almost the same activity on both substrates (Tokuda *et al.*, 2002). This difference in specificity may be due to the fact that the β-glycosidases finish the digestion of the β-1,4-glucan chains of cellulose (termites, see section 11.4.3) or the β-1,3-glucan chains of hemicelluloses (lepidopterans, Terra *et al.*, 1987). It would be interesting to identify which amino acid residues are responsible for this difference in specificity. Free energy relationships (Withers and Rupitz, 1990) were used to compare the specificity of insect β-glycosidase subsites (Azevedo *et al.*, 2003). Since each of the three *D. saccharalis* β-glycosidases has a counterpart in *T. molitor*, Azevedo *et al.* (2003) speculated that insects with the same number of β-glycosidases could have similar enzymes.

The amino acid sequence was determined for a few β-glycosidases produced by the salivary glands of the termite *N. koshunensis* (Tokuda *et al.*, 2002) and *B.mori* (Byeon *et al.*, 2005), or produced by the midgut of *S. frugiperda* (Marana *et al.*, 2001), *T. molitor* (Ferreira *et al.*, 2001), and the termites *R. flavipes* (Scharf *et al.*, 2010) and *N. takasagoensis* (Tokuda *et al.*, 2009). All these enzymes pertain to family 1 of glycoside hydrolases.

The overall 3D structure of the *N. koshunensis* β-glucosidase (Jeng *et al.*, 2010) is similar to that of other glycosyl hydrolase family 1 enzymes, which have a classical $(\alpha/\beta)_8$-TIM barrel fold. The active site, containing the catalytic residues Glu 193 and Glu 402, forms a deep slot-like cleft and is located on connecting loops at the C-terminal end of the β-sheets of the TIM barrel. In the βGly50 from *S. frugiperda*, the pK_a of the nucleophile (Glu 399) is 4.9 and of the proton donor (Glu 187) is 7.5. In this enzyme, residue Glu 451 seems to be a key residue in determining the enzyme preference for glucosides versus galactosides. This is due to its interaction, in the glycone site, with substrate equatorial or axial hydroxyl 4, which is the only position where glucose differs from galactose. The steric hindrance of the same residue with hydroxyl 6 probably also explains why fucosides are the best substrate for many β-glycosidases (Marana *et al.*, 2002a).

Besides having a role in digestion, β-glycosidases are important in insect–plant relationships. To avoid herbivory, plants synthesize a large number of toxic glucosides (**Figure 7**), which may be present in concentrations from 0.5% to 1% (Spencer, 1988). The presence of these glucosides in some insect diets may explain the variable number of β-glycosidases with different specificities present in their guts. Most plant glucosides have a hydrophobic aglycone and are β-linked O-glycosyl compounds. Since aglycones are usually more toxic than the glycosides themselves, intoxication may be avoided by decreasing the activity of the enzyme most active on toxic glucosides, without affecting the final digestion of hemicellulose and cellulose carried out by the other enzymes. This is exemplified by *D. saccharalis* larvae, which have three β-glycosidases in their midgut, feeding on diets containing the cyanogenic glucoside amygdalin. In this condition, the activity of the enzyme responsible for the hydrolysis of prunasin is decreased (Ferreira *et al.*, 1997). Prunasin is the saccharide resulting after the removal from amygdalin, and that forms the cyanogenic mandelonitrille upon hydrolysis (**Figure 7**). Resistance to toxic glucosides may also be achieved by the lack of enzymes able to hydrolyze toxic β-glucosides, as observed with *S. frugiperda* larvae, which have two β-glycosidases unable to efficiently hydrolyze prunasin (Marana *et al.*, 2001; S.R. Marana, personal communication). Progress in this field will require disclosing the mechanisms by which the presence of toxic β-glucosides differentially affects the midgut β-glycosidases, and knowing the details of the active site architecture responsible for the specificity of these enzymes.

Myrosinase (EC 3.2.147) is a member of the glycosyl hydrolase family 1 that hydrolyzes glucosinolates, which are β-D-thioglucosides. The aphid *Brevicoryne brassicae* has a midgut myrosinase for which coding cDNA was cloned and sequenced (Jones *et al.*, 2002), and its 3D structure resolved (Husebye *et al.*, 2005). The enzyme has two catalytic Glu residues (plant myrosinase has only one), and its sequence is more similar to animal β-O-glucosidases than to plant myrosinases. The data suggest that myrosinase has twice arisen from β-glucosidases in plants and animals. Further structural data are necessary to clarify the features responsible for the hydrolysis of the thioglucoside bond.

11.4.8. Trehalases

Trehalase (EC 3.2.1.28) hydrolyzes α,α'-trehalose into two glucose molecules, and is one of the most widespread carbohydrases in insects, occurring in most tissues. Trehalase is very important for insect metabolism, since trehalose is the main circulating sugar in these organisms. There are many insect midgut trehalase sequences deposited in GenBank, but the single trehalase with known

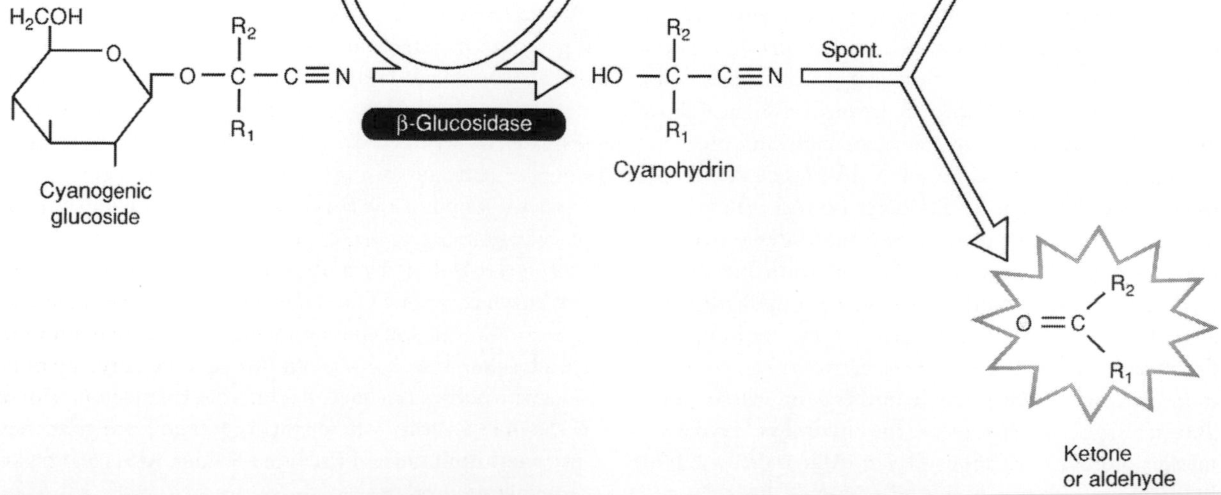

Figure 7 β-Glucosidase acting on a cyanogenic glucoside releases glucose and cyanohydrin that spontaneously decomposes, producing a ketone (or an aldehyde) and hydrogen cyanide. If $R_1 = R_2 = CH_3$, the glucoside is linamarin and the resulting ketone is acetone. If $R_1 = H$ and $R_2 = $ phenyl, the glucoside is prunasin and the resulting aldehyde is benzaldehyde (see more examples in Vetter, 2000).

3D structure resolved is a periplasmic enzyme from *Escherichia coli* (Gibson *et al.*, 2007)

Apical and basal trehalases can be distinguished in insect midguts. The apical trehalase may be soluble (glycocalyx-associated or secreted into the midgut lumen) or microvillar, whereas the midgut basal trehalase is an integral protein of the basal plasma membrane. The apical midgut trehalase is a true digestive enzyme. The midgut basal trehalase probably plays a role in the midgut utilization of hemolymph trehalose (Terra and Ferreira, 1994). There is a report that localizes the soluble trehalase at the cavity of the goblet cell, and the membrane bound trehalase at the visceral muscles (Mitsumatsu *et al.*, 2005). This should be reinvestigated, because they may be artefacts (Silva *et al.*, 2009). Trehalases, partially or completely purified from insect guts, have pH optima from 4.8 to 6.0, K_m from 0.33 to 1.1 mM, pI around 4.6, and molecular masses from 60 to 138 kDa (Terra and Ferreira, 1994).

There have been a few attempts to identify important groups in the midgut trehalase active site. Terra *et al.* (1978, 1979, 1983) determined the pK_a values of the catalytical groups of the *R. americana* midgut trehalase. The pK_a value of the nucleophile was 5.0 (kinetic data) or 5.3 (carbodiimide modification results), whereas the pK_a value of the proton donor was 8.3 (kinetic data) or 7.7 (carbodiimide modification). Since there was a disagreement between the pK_a values determined for the proton donor, and taking into account that carbodiimide modification is only partially protected by trehalose, the authors suggested that the proton donor is near, but not at, the active site, and that it participates in the reaction through another amino acid residue, like histidine (Terra *et al.*, 1979). Lee *et al.* (2001), with the same approach as Terra *et al.* (1979), found pK_a values of 5.3 and 8.5 for the *Apis mellifera* trehalase. These authors and Valaitis and Bowers (1993), who worked with *Lymantria dispar* trehalase, showed that the trehalases were inactivated by diethyl pyrocarbonate (DPC), but the work did not progress further.

The active site of the *S. frugiperda* soluble midgut trehalase was modeled based on the 3D structure of *E. coli* trehalase. The model guided the choice of the trehalase residues to be mutated. Site-directed mutagenesis confirmed that D322 and E520 are the basic and acid catalysts, respectively, and showed that three Arg residues (R169, R222, and R287) are also essential for enzyme activity (Silva *et al.*, 2010). As the phenylglyoxal modified R222 has a pK_a value that is affected by a His residue in a similar way as the enzyme proton donor (Silva *et al.*, 2004), this explains the earlier implication of His residues in assisting a carboxyl group acting as a proton donor (Terra *et al.*, 1978, 1979; Lee *et al.*, 2001).

Plant toxic β-glucosides and their aglycones can inhibit, with varied efficiency, some or all trehalases from Malpighian tubules, fat body, midgut, and body wall of *P. americana*, *M. domestica*, *S. frugiperda*, and *D. saccharalis* (Silva *et al.*, 2006). Toxic β-glucosides are produced by many plant species, and are present in high concentrations (see section 11.4.7). It is not known whether those glucosides or aglycones are absorbed by the insect gut and interact with trehalases in tissues other than the midgut. It would be interesting to know if this happens, and how insects resistant to toxic β-glucosides avoid the damage they can cause.

11.4.9. Acetylhexosaminidases, β-Fructosidases, and α-Galactosidases

An enzyme related to chitinolytic enzymes is β-N-acetyl-D-hexosaminidase (EC 3.2.1.52), which differs from β-N-acetyl-D-glucosaminidase in having a rather wide substrate specificity. The enzyme is found in many insects, and its presumed physiological role is the hydrolysis of N-acetylglucosamine β-linked compounds such as glycoproteins (Terra and Ferreira, 1994). Detailed studies of this digestive enzyme are lacking.

Sucrose hydrolysis is catalyzed by enzymes that are specific for the α-glucosyl (α-glucosidase, EC 3.2.1.20; see above) or for the β-fructosyl residue (β-fructosidase, EC 3.2.1.26) of the substrate. β-Fructosidase is characterized by its activity toward sucrose and raffinose, and lack of activity upon maltose and melibiose. In insect midguts, sucrose hydrolysis generally occurs by action of the conspicuous α-glucosidase rather than by β-fructosidase. However, larvae and adults of the moth *Erinnyis ello* have a midgut β-fructosidase with pH optimum 6.0, K_m 30 mM (sucrose), pI 5.2, and molecular mass of 78 kDa. The physiological role of this enzyme is to hydrolyze sucrose, the major leaf (larvae) or nectar (adults) carbohydrate, which is not efficiently digested by *E. ello* midgut α-glucosidase (Santos and Terra, 1986). After this first report, the presence of β-fructosidase was described in other Lepidoptera (Sumida *et al.*, 1994; Carneiro *et al.*, 2004). The presence of this enzyme is thought to be an adaptation to the presence of alkaloid sugars in latex that inhibits the α-glucosidase but not the β-fructosidase (Daimon *et al.*, 2008). As there seem to be exceptions for this inhibition (Hirayama *et al.*, 2007), more data should be gathered to settle the hypothesis.

The first β-fructosidase cDNA sequence was found in a proteomic survey of the larval gut lumen of *H. armigera* (Pauchet *et al.*, 2008). Later on, other genes were found in a midgut transcriptome and in lepidopteran-specific public EST data sets. The results indicated that β-fructosidase genes are widespread among Lepidoptera, and that they could be acquired from bacteria via horizontal gene transfer (Pauchet *et al.*, 2010).

α-Galactosidase (α-D-galactoside galactohydrolase, EC 3.2.1.22) catalyzes the hydrolysis of α-D-galactosidic linkages in the non-reducing end of oligosaccharides, galactomannans, and galactolipids, and is widely distributed in nature (Dey and Pridham, 1972). Galactooligosaccharides, such as melibiose, raffinose, and stachiose, are common in plants, mainly in lipid-rich seeds (Shiroya, 1963), whereas galactolipids are widespread among leaves. The major lipids in chloroplast membranes are mono-galactosyldiglyceride and digalactosyldiglyceride (Harwood, 1980).

There have been few attempts to resolve insect midgut α-galactosidases. Gel filtration and heat inactivation suggested that there is a single α-galactosidase activity (30 kDa, pH optimum 5.0) in *D. peruvianus* midgut that is more efficient on raffinose than on melibiose and NPαGal (Silva and Terra, 1997). There are two α-galactosidases in *A. flavolineata* midguts: the major (112 kDa, pH optimum 5.4) is active on melibiose and raffinose in addition to NPαGal, whereas the minor (70 kDa, pH optimum 5.7) hydrolyzes only NPαGal (Ferreira *et al.*, 1999). In the case of *Psacothea hilaris*, gel filtration gave evidence of the presence of multiple overlapping α-galactosidases more active on NPαGal than on melibiose (Scrivener *et al.*, 1997). There are three midgut luminal α-galactosidases (TG1, TG2, and TG3) from *T. molitor* larvae that are partially resolved by ion-exchange chromatography (Grossmann and Terra, 2001). The enzymes have approximately the same pH optimum (5.0), pI value (4.6), and molecular mass (46–49 kDa). TG2 hydrolyzes α-1,6-galactosaccharides, exemplified by raffinose, whereas TG3 acts on melibiose and apparently also on digalactosyldiglyceride, the most important compound in thylacoid membranes of chloroplast, converting it into monogalactosyldiglyceride. *S. frugiperda* larvae have three midgut α-galactosidases (SG1, SG2, and SG3) partially resolved by ion-exchange chromatography (Grossmann and Terra, 2001). The enzymes have similar pH optimum (5.8), pI value (7.2), and molecular mass (46–52 kDa). SG1 and SG3 hydrolyze melibiose, and SG3 digests raffinose and, perhaps, digalactosyldiglyceride.

11.5. Digestion of Proteins

11.5.1. Initial Considerations

The initial digestion of proteins is carried out by proteinases (endopeptidases) that break internal bonds in proteins. Different proteinases are necessary to do this because the amino acid residues vary along the peptide chain. There are three subclasses of proteinases involved in digestion, classified according to their active site group (and hence by their mechanism): serine, cysteine, and aspartic proteinases. In each of the mentioned subclasses, there are several proteinases differing in substrate specificities. The oligopeptides resulting from proteinase action are attacked from the N-terminal end by aminopeptidases and from the C-terminal end by carboxypeptidases, both

enzymes liberating one amino acid residue at each catalytic step. Finally, the dipeptides formed are hydrolyzed by dipeptidases.

11.5.2. Serine Proteinases

Serine proteinases (EC 3.4.21) (MEROPS) have serine, histidine, and aspartic acid residues (called the catalytic triad) in the active site. The family of enzymes homologous to chymotrypsin (Barrett *et al.*, 1998) includes the major digestive enzymes trypsin, chymotrypsin, and elastase. These enzymes differ in structural features that are associated with their different substrate specificities, as detailed below. The numbering of residues in enzyme polypeptide chains is referred to that of bovine chymotrypsin.

11.5.2.1. Trypsins   Trypsins (EC 3.4.21.4) preferentially cleave protein chains on the carboxyl side of basic L-amino acids such as arginine and lysine. Most insect trypsins have molecular masses in the range 20–35 kDa, pI values 4–5, and pH optima 8–10. These enzymes occur in the majority of insects, with the remarkable exception of hemipteran species and some taxa belonging to the series Cucujiformia of Coleoptera like Curculionidae (Terra and Ferreira, 1994). Nevertheless, some heteropteran Hemiptera have trypsin in the salivary glands (Zeng *et al.*, 2002).

Trypsin is usually identified in insect midgut homogenates using benzoyl-arginine p-nitroanilide (B-R-pNA, often referred to as BApNA) or benzoylarginine 7-amino-4-methyl coumarin (B-R-MCA) as substrates, and with N-α-tosyl-l-lysine chloromethyl ketone (TLCK), phenylmethylsulfonyl fluoride (PMSF), or diisopropylfluorophosphate (DFP) as inactivating compounds. The substrates of choice for assaying insect trypsins are shown in **Figure 8**. Trypsins from Orthoptera, Dictyoptera, and Coleoptera are usually purified by ion-exchange chromatography, and those from Diptera and Lepidoptera by affinity chromatography – either in benzamidine-agarose (elution with benzamidine or by change in pH) or in soybean trypsin inhibitor (SBTI)-Sepharose (elution by change in pH). Due to significant autolysis, lepidopteran trypsins are more frequently purified by chromatography on benzamidine-Agarose with elution with benzamidine.

There are a great number of insect trypsin sequences registered in GenBank. Examples may be found among Hemiptera (Zeng *et al.*, 2002), Coleoptera (Zhu and Baker, 1999), Diptera (Ramalho-Ortigão *et al.*, 2003), Siphonaptera (Gaines *et al.*, 1999), and Lepidoptera (Peterson *et al.*, 1994; Zhu *et al.*, 2000). The complete sequences have signal and activation peptides, and the features typical of trypsin-like enzymes, including the conserved N-terminal residues IVGG, the catalytic amino acid triad of serine proteinase active sites (His 57, Asp 102, and Ser 195), three pairs of conserved cysteine residues for

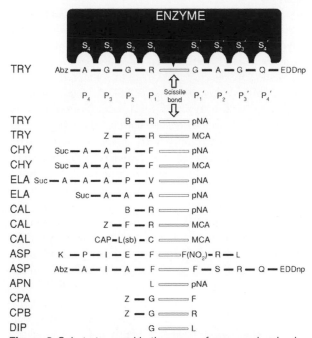

Figure 8 Substrates used in the assay of enzymes involved in protein digestion. Sn are subsites in the enzymes, and Pn are amino acid residues in substrates. The arrows point to bonds cleaved by the different enzymes. Abz-Xn-EDDnp is a class of peptides with quenching (EDDnp) and fluorescent (Abz) groups at the C- and N-terminal ends, respectively, so that after hydrolysis the peptides become fluorescent. Substrates with C-terminal MCA are fluorescent, and those with pNA are colorimetric. Contrary to GL, LG is also hydrolyzed by APN in addition to dipeptidase. For further details, see text. TRY, trypsin; CHY, chymotrypsin; ELA, elastase; CAL, cathepsin L-like enzyme; ASP, aspartic proteinase; APN, aminopeptidase N; CPA, carboxypeptidase A; CPB carboxypeptidase B; DIP, dipeptidase.

disulfide bonds, and the residue Asp 189 that determines specificity in trypsin-like enzymes (see **Figure 5**). In spite of having structural features resembling vertebrate trypsins, insect trypsins differ from these because they are not activated or stabilized by calcium ions, and frequently are unstable in acidic pH (Terra and Ferreira, 1994). Finally, other differences between vertebrate and insect trypsins include their substrate specificities and their interaction with protein inhibitors.

Amino acyl residues in proteinase substrates are numbered from the hydrolyzed peptide bonds as P1, P2, P3, …, Pn in the direction of the N-terminus, and P′1, P′2, P′3, …, P′n in the direction of the peptide C-terminus, whereas the corresponding enzyme subsites are numbered S1, S2, S3, …, Sn and S′1, S′2, S′3, …, S′ n (Schechter and Berger, 1967) (**Figure 8**). Mammalian trypsin preferably cleaves substrates having arginine rather than lysine at P1 (primary specificity) (Craik *et al.*, 1985). The same primary specificity was found for insect trypsins, except those from lepidopterans, which prefer lysine at P1 (Lopes *et al.*, 2004). This will be discussed below in relation to

trypsin insensitivity to protein inhibitors. In order to characterize the trypsin specificity at subsites other than S1, quenched fluorescent substrates such as o-aminobenzoyl-GGRGAGQ-2,4-dinitrophenyl-ethylene diamine (where R stands for arginine at P1 position) were synthesized with 15 amino acid replacements at each of the positions P′1 , P′2, P2, and P3. The results suggested that trypsin subsites are more hydrophobic in trypsins from the more evolved insects (Lopes *et al.*, 2006). Trypsins from different insects also differ in the strength with which their subsites bind the substrate or the transition state (high-energy intermediate of the reaction). In other words, trypsin subsites differ in how they favor substrate binding or catalysis (Marana *et al.*, 2002b).

Plants have protein inhibitors (PIs) of insect midgut serine proteinases that affect insect development (Ryan, 1990). Insects may adapt to the presence of PIs in the diet by overexpressing proteinases (Bown *et al.*, 1997; Broadway, 1997; Gatehouse *et al.*, 1997), by proteolytical inactivation of PIs mediated by the insect's own proteinases (Giri *et al.*, 1998), or by expressing new proteinases that are resistant to the inhibitor (Mazumdar-Leighton and Broadway, 2001a, 2001b). Current research is investigating the molecular basis of the difference between sensitive and inhibitor-insensitive trypsins, as well as the regulation of these enzymes.

PIs produced by plants have a region, named the reactive site, that interacts with the active site of their target enzymes. The reactive sites of many PIs are hydrophilic loops with a lysine residue at P₁ (Lopes *et al.*, 2004). As lepidopteran trypsins have hydrophobic subsites and prefer lysine rather than arginine at P1 (see above), they are usually more resistant to PIs than the other insect trypsins. In this respect, it is interesting to note that PI-insensitive trypsins from *Heliothis virescens* bind more tightly to a hydrophobic chromatographic column than do sensitive trypsins (Brito *et al.*, 2001). These observations led to the hypothesis that the molecular differences between sensitive and insensitive trypsins must rely on the interactions of PIs with residues in and around the enzyme active site.

An interesting approach to studying insect–PI interactions was introduced by Volpicella *et al.* (2003), who compared the sequence of a sensitive trypsin from *Helicoverpa armigera* with the insensitive trypsin from the closely related species *Helicoverpa zea*. The 57 different amino acids observed between the two enzymes were superimposed on the porcine trypsin crystal structure, where the residues known to be in contact with a Kunitz-type inhibitor (Song and Suh, 1998) were identified. The residues at positions (chymotrypsin numbering) 41, 57, 60, 95, 99, 151, 175, 213, 217, and 220 were considered by Volpicella *et al.* (2003) to be important in *H. zea* trypsin–PI interaction. However, some of the interacting residues may have been misidentified, because trypsins from different species were compared. In a similar approach,

Lopes *et al.* (2004) aligned all available trypsin sequences characterized as sensitive or insensitive to Kunitz-type inhibitor (Bown *et al.*, 1997; Mazumdar-Leighton and Broadway, 2001a) with porcine trypsin. After discounting conserved positions and positions not typical of sensitive or insensitive trypsins, the remaining positions that agreed with those involved in porcine trypsin–PI (Bowman-Birk type, Lin *et al.*, 1993; Kunitz type, Song and Suh, 1998) or substrate (Koepke *et al.*, 2000) interactions were: 60, 94, 97, 98, 99, 188, 190, 213, 215, 217, 219, 228. These positions support the tree branches in a neighbor-joining analysis of sensitive (I, III) and insensitive (II) trypsin sequences (Lopes *et al.*, 2004). Site-directed mutagenesis of trypsin, followed by the determination of the binding constants of mutated trypsins with PIs, may help to resolve the discrepancy.

The mechanism by which PIs in the diet induce the synthesis of insensitive trypsin in responsive insects remains unknown, although it was found that the first step in the process is the expression of the whole set of midgut trypsins (Brioschi *et al.*, 2007). The evolutionary "arms race" between plants and insects regarding evolving new digestive proteinases and new PIs are reviewed in Christeller (2005).

11.5.2.2. Chymotrypsins Chymotrypsin (EC 3.4.21.1) preferentially cleaves protein chains at the carboxyl side of aromatic amino acids. Insect chymotrypsins usually have molecular masses of 20–30 kDa and pH optima of 8–11, and they differ from their vertebrate counterparts in their instability at acidic pH, inhibition pattern with SBTI (Terra and Ferreira, 1994), and, finally, in reacting with N-α-tosyl-l-phenylalanine chloromethyl ketone (TPCK) (see below). The distribution of chymotrypsin among insect taxa is similar to that of trypsin (Terra and Ferreira, 1994), including the occurrence in the salivary glands of some heteropteran bugs (Colebatch *et al.*, 2002). The earlier failure to detect chymotrypsin activity in insect midguts was a consequence of using mammalian chymotrypsin substrates, such as benzoyl-tyrosine p-nitroanilide (B-Y-pNA) or benzoyl-tyrosine ethyl ester (B-Y-ee), in the assays. Insect chymotrypsins prefer Phe at P_1, and are almost inactive if Tyr is at that position (Lopes *et al.*, 2009). Furthermore, insect chymotrypsins have an extended active site (as in mammalian chymotrypsin; see Lopes *et al.*, 2009) and larger substrates, like succinyl-AAPF-p-nitroanilide (Suc-AAPF-pNA), are usually necessary for their detection (Lee and Anstee, 1995; Lopes *et al.*, 2009) (**Figure 8**). Insect chymotrypsins are usually purified by affinity chromatography in phenyl butylamina-Sepharose (elution with phenyl butylamina) or in SBTI-Sepharose (elution with benzamidine) for enzymes from lepidopterans, and by ion-exchange chromatography for those from dictyopterans, orthopterans, hymenopterans, and dipterans. They have been purified from several

sources: Hemiptera (Colebatch *et al.*, 2002), Coleoptera (Oliveira-Neto *et al.*, 2004; Elpidina *et al.*, 2005), Hymenoptera (Whitworth *et al.*, 1998), Siphonaptera (Gaines *et al.*, 1999), Diptera (de Almeida *et al.*, 2003; Ramalho-Ortigão *et al.*, 2003), and Lepidoptera (Peterson *et al.*, 1995; Volpicella *et al.*, 2006). A large number of sequences is also available for insect chymotrypsins at GenBank. All the sequences have a signal peptide, an activation peptide (ending with an arginine residue), the catalytic triad (His 57, Asp 102, and Ser 195), three pairs of conserved cysteine residues, conserved N-terminal sequence IVGG, and Ser/Gly/Tyr 189, which confers specificity to chymotrypsin-like enzymes (**Figure 5**).

The substrate preferences of chymotrypsins from insects of three different orders were studied with quenched fluorescent substrates. The result showed that although substrate preferences vary among the different chymotrypsins, no evolutionary trend as described for trypsins was observed. In spite of those differences, the data suggested that in lepidopteran chymotrypsins S_2 and S_1' bind the substrate ground state, whereas only S_1' binds the transition state, supporting aspects of the present accepted mechanism of catalysis (Sato *et al.*, 2008).

The insect digestive chymotrypsin that has been most thoroughly studied is that of *M. sexta* (Lepidoptera: Sphingidae) (Peterson *et al.*, 1995). In this enzyme, the activation peptide is longer and has a net charge different from that of bovine chymotrypsinogen, leading the authors to suggest that the insect enzyme is activated by a peculiar mechanism. The mammalian chymotrypsin has a pH optimum around 8, with two catalytic important pK_as of 6.8 and 9.5, corresponding to the active-site histidine and N-terminal leucine, respectively. In contrast, the *M. sexta* chymotrypsin has pH optimum 10.5–11, and a single kinetically significant pK_a at pH 9.2. This pK_a may represent the active-site histidine in an appropriate environment, although several other hypotheses were discussed (Peterson *et al.*, 1995). It is not clear whether the insect chymotrypsin active-site changes associated with TPCK resistance (see below) may also be the cause of the putative histidine pK_a displacement.

The resolution of the 3D structure of the fire-ant digestive chymotrypsin led to the conclusion that it is strikingly similar to mammalian chymotrypsin, but has differences beyond those found among homologs from different mammalian systems (Botos *et al.*, 2000). The similarities include a conserved backbone scaffold and structural domains. Differences include the activation mechanism and substitutions in the subsite S_1 and mainly in the other subsites (S_4–S_4') that suggest different substrate specificities and interactions with PIs. In agreement with this, different insect chymotrypsins are sensitive to distinct PIs and, like trypsins, PI-insensitive chymotrypsins may be induced in insects ingesting PI-containing diets (Bown *et al.*, 1997; Mazumdar-Leighton and Broadway, 2001b).

The molecular mechanism of chymotrypsin PI inhibition was investigated. Two chymotrypsins were purified from the midgut of *Helicoverpa punctigera*, one PI-sensitive and the other PI-insensitive. After their corresponding cDNAs were cloned and sequenced, molecular modeling revealed that a Phe→Leu substitution at position 37 in the chymotrypsin results in the loss of important contacts with the PI. This was confirmed by site-directed mutagenesis of chymotrypsin molecules, followed by inhibition tests (Dunse *et al.*, 2010). Chymotrypsins from insects that routinely ingest ketone-releasing compounds (like several plant glycosides) (see **Figures** 7 and **9**) are not affected much by these compounds and others that react with His 57. Thus, in comparison with bovine chymotrypsin, the chymotrypsin from polyphagous lepidopteran insects reacts slowly with chloromethyl ketones, whereas those of oligophagous pyralid insects react rapidly (Lopes *et al.*, 2009). Modeling *Spodoptera frugiperda* (Noctuidae) chymotrypsin, based on its sequence and on crystallographic data of bovine chymotrypsin, showed that the neighborhood of His 57 differs from bovine chymotrypsin, thus affecting His reactivity (Lopes *et al.*, 2009). These adaptations are new examples of the interplay between insects and plants during their evolutionary arms race, and deserve more attention through site-directed mutagenesis of recombinant chymotrypsins.

11.5.2.3. Elastases Since Christeller *et al.* (1990) described an elastase (EC 3.4.21.36)-like enzyme in the cricket *Teleogryllus commodus*, this enzyme has been described in many other insects, including in homogeneous form (Terra and Ferreira, 1994; Whitworth *et al.*, 1998). Usually elastase is identified with the substrate Suc-AAPL-pNA (**Figure 8**), combined with the observation of lack of activity on B-Y-pNA or B-Y-ee and resistance to TPCK. Since the mentioned substrate may also be hydrolyzed by chymotrypsin, and lack of activity on B-Y-pNA and/or resistance to TPCK are usual among chymotrypsins (see section 11.5.2.2), most described elastases may actually be chymotrypsins. True elastases

were isolated from gypsy moth midguts (Valaitis, 1995) and from whole larvae of *Solenopsis invicta* (Whitworth *et al.*, 1998). The last-mentioned enzymes hydrolyze Suc-APA-pNA, but not substrates with phenylalanine at P_1. Although the specific substrate for elastase (Suc-AAA-pNA) (Bieth *et al.*, 1974) was not tested, the hydrolysis of Suc-AAAPV-pNA and the lack of hydrolysis of substrates with phenylalanine in P1 discount a chymotrypsin. One of the *S. invicta* elastases (E2) was cloned, sequenced, and shown to be more similar to chymotrypsin than to elastase (Whitworth *et al.*, 1999). This work confirms the occurrence of elastase in insect midgut. Further work is necessary to evaluate the extent of this enzyme in insect midguts, and its importance in digestion.

11.5.3. Cysteine Proteinases

Cysteine proteinase is usually assayed in insect midgut contents or midgut homogenates at pH 5–6 with B-R-pNA, B-R-NA, casein, or hemoglobin as substrate. Activation by sulfhydryl agents (dithiothreitol (DTT) or cysteine) and inhibition by trans-epoxysuccinyl-l-leucyl-amido (4-guanidinobutane) (E-64) are usually indicative of the presence of the enzyme. The observation of inhibition of hydrolytic activity on any of the mentioned substrates by E-64 is insufficient for a positive identification of cysteine proteinase. Trypsin hydrolyzes the same substrates, and may be reversibly inhibited by E-64 (Novillo *et al.*, 1997). The identification of cysteine proteinase was made easier with the substrate e-amino-caproyl-leucyl-(S-benzyl)-cysteinyl-MCA, which is hydrolyzed by cysteine proteinase but not by serine proteinases (Alves *et al.*, 1996). Using such criteria, cysteine proteinases were described in Hemiptera Heteroptera and in species belonging to the series Cucujiformia of Coleoptera (Terra and Ferreira, 1994). Exceptions to this rule are the identification of cysteine proteinase in Hemiptera Stermorrhyncha (aphids) (Cristofoletti *et al.*, 2003; Deraison *et al.*, 2004), and the lack of this enzyme in cucujiform cerambycid beetles (Johnson and Rabosky, 2000).

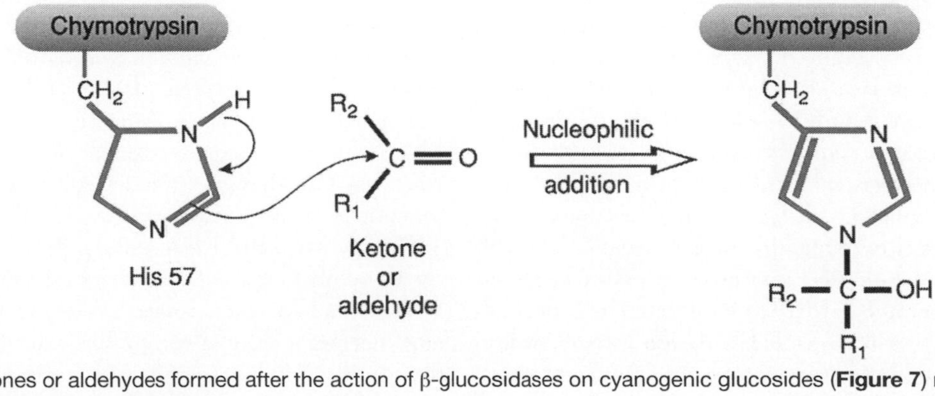

Figure 9 Ketones or aldehydes formed after the action of β-glucosidases on cyanogenic glucosides (**Figure 7**) may react with His 57 of chymotrypsin, inactivating it.

Insect midgut cysteine proteinases were at first denoted as cathepsin B (EC 3.4.22.1)-like enzymes, because cathepsin B was the first animal cysteine proteinase described. Later on it became known that cathepsin B is more important as a peptidyl dipeptidase, rather than as an endopeptidase, because of the existence of an extended loop that forms a cap to the active-site cleft, and carries a pair of histidine residues that are thought to bind to the C-terminal carboxylate of the substrate (Barrett *et al.*, 1998). Cathepsin L (EC 3.4.22.15) is a true endopeptidase that preferentially cleaves peptide bonds with hydrophobic amino acid residues in P_2 (cathepsin B prefers arginine at the same position) (Barrett *et al.*, 1998). Thus, by using substrates like carbobenzoxy (Z)-FR-MCA and Z-RR-MCA it is possible to distinguish between the two enzymes (see **Figure 8**). Current research has revealed that cathepsin L-like enzymes are the only insect midgut cysteine proteinase that is quantitatively important. Much more difficult to ascertain is that a cathepsin L-like enzyme assayed in insect midguts has been secreted into midgut contents, and hence may be considered as a truly digestive enzyme. As in other animals (Barrett *et al.*, 1998), cathepsin L-like enzymes in insects are expected to occur in lysosomes and never leave the cells. The same difficulties arise in trying to relate digestion to cathepsin L-like enzymes encoded by cDNAs cloned from midgut cells.

The problems that may arise during cathepsin L-like enzyme characterization are well illustrated in a study with *T. molitor* larvae. Three cathepsin L-like sequences were recognized in a cDNA library prepared from *T. molitor* midguts. One sequence, after being expressed and used to raise antibodies, was found to correspond to a lysosomal cathepsin L immunolocalized mainly at hemocytes and fat body cells. The second sequence was not related yet with any enzyme active in midgut. Finally, the third sequence corresponds to a cathepsin L-like enzyme purified from midgut contents (Cristofoletti *et al.*, 2005).

Digestive cathepsin L-like enzymes have been purified to homogeneity only from *Diabrotica virgifera* (Coleoptera: Cucujiformia) (Koiwa *et al.*, 2000), *Acyrthosiphon pisum* (Hemiptera: Sternorrhyncha) (Cristofoletti *et al.*, 2003), and *T. molitor* (Coleoptera: Cucujiformia) (Cristofoletti *et al.*, 2005). The *A. pisum* enzyme is cell membrane-bound and faces the luminal contents, whereas those in *D. virgifera* and *T. molitor* are soluble enzymes secreted into midgut contents. The complete purification of these enzymes was achieved by affinity chromatography with soyacystatin as a ligand, or by a combination of ion-exchange chromatographies. The enzymes have pH optima of 5–6, molecular masses of 20–40 kDa, prefer Z-FR-MCA over Z-RR-MCA, are inhibited by E-64, and are activated by cysteine or DTT. At least, the *A. pisum* enzyme is also inhibited by chymostatin (completely) and elastatinal (partly) (Cristofoletti *et al.*, 2003).

There are many cathepsin L-like sequences corresponding to coleopterans and hemipterans (those known to have digestive cathepsins) registered in GenBank. The sequences demonstrate the features characteristic of family 1 of cysteine proteinases (Barrett *et al.*, 1998): an N-terminal propeptide that must be removed to activate the enzyme and the catalytic triad, Cys 25, His 169, and Asn 175 (papain numbering), and the ERFNIN motif (**Figure 5**). The sequences form a monophyletic grouping, and a polyphyletic array seems to correspound to lysosomal and digestive enzymes, respectively (Cristofoletti *et al.*, 2005). The data suggest that the Hemiptera digestive cathepsin is close to the lysosomal one from *T. molitor* (CAL 1), whereas those from the coleopterans (except from *Sitophilis oryzae*) are similar to the *T. molitor* digestive enzyme (CAL 3). It is probable that the *S. oryzae* enzymes are not true digestive but lysosomal, like *T. molitor* CAL (Cristofoletti *et al.*, 2005). More work is needed to clarify the role of cathepsin L-like enzymes in insect digestion.

11.5.4. Aspartic Proteinases

Aspartic proteinases are active at acid pH, hydrolyze internal peptide bonds in proteins, and attack some synthetic substrates – either chromophoric (Dunn *et al.*, 1986) or internally quenched fluorescent substrates (Pimenta *et al.*, 2001) (**Figure 8**). Mammalian cathepsin D has a substrate-binding cleft that can accommodate up to seven amino acids, and prefers to cleave between two hydrophobic residues (Barrett *et al.*, 1998).

The first report of aspartic proteinases in insects was made by Greenberg and Paretsky (1955), who found a strong proteolytic activity at pH 2.5–3.0 in homogenates of whole bodies of *Musca domestica*. Lemos and Terra (1991b) showed that the enzyme occurs in midguts, and is cathepsin D-like. An aspartic proteinase similar to cathepsin D was found in families of Hemiptera and Heteroptera, and in several families belonging to the cucujiform series of Coleoptera (Terra and Ferreira, 1994). Thus, it is possible that aspartic proteinases occur together with cysteine proteinase in Hemiptera and in most Coleoptera. The aspartic proteinase isolated from *Callosobruchus maculatus* (pH optimum 3.3, 62 kDa) (Silva and Xavier-Filho, 1991) and *Tribolium castaneum* (pH optimum 3.0, 22 kDa) (Blanco-Labra *et al.*, 1996) were partially purified and shown to be similar to cathepsin D.

M. domestica midguts express 3 cathepsin D-like enzymes (CAD1, CAD2, and CAD3). All CADs have catalytic Asp 33 (together with T34 and G35) and Asp 229 (together with T230 and G231) (bovine cathepsin D numbering), and the conserved substrate binding pockets (Padilha *et al.*, 2009). CAD3 is a luminal digestive enzyme, and lacks the proline loop (as defined by the motif: DxPxPx (G/A)P, thus resembling vertebrate pepsin). CAD2 also lacks the proline loop and may be another digestive enzyme. CAD1 is expressed in all *M. domestica* tissues (not only in midgut as CAD2 and CAD3), and shows the proline loop as the vertebrate lysosomal cathepsin D (Padilha *et al.*, 2009). The data suggest that on adapting to deal with a bacteria-rich food in an acid midgut region, *M. domestica* CAD resulted from the same archetypical gene as the intracellular cathepsin D, paralleling what happened with vertebrates. The lack of the proline loop may be somehow associated with the extracellular role of both pepsin and digestive CAD3. Further work is necessary to clarify this point.

11.5.5. Aminopeptidases

Aminopeptidases sequentially remove amino acids from the N-terminus of peptides, and are classified on the basis of their dependence on metal ions (usually Zn^{2+} or Mn^{2+}) and substrate specificity. Aminopeptidase N (EC 3.4.11.2) has a broad specificity, although it preferentially removes alanine and leucine residues from peptides, whereas aminopeptidase A (EC 3.4.11.7) prefers aspartyl (or glutamyl) peptides as substrates. Both are metalloenzymes (Norén *et al.*, 1986).

In insect midguts, major amounts of soluble aminopeptidases are found in less evolved insects (e.g., Orthoptera, Hemiptera, Coleoptera Adephaga), whereas in more evolved insects (e.g., Coleoptera Polyphaga, Diptera, and Lepidoptera) aminopeptidase is found mainly bound to the microvillar membranes of midgut cells (Terra and Ferreira, 1994). Insect midgut aminopeptidases are metalloenzymes (ethylenediaminetetraacetic acid (EDTA) inhibition) and have pH optima of 7.2–9.0, irrespective of the pH of the midgut lumen from the different species, K_m values (L-pNA) of 0.13–0.78 mM, and molecular masses of 90–130 kDa. With a single exception (see below), all known insect aminopeptidases have a broad specificity, hydrolyzing a variety of amino acyl β-naphthylamides (except acidic amino acyl β-naphthylamides), indicating they are aminopeptidases N (Terra and Ferreira, 1994) (**Figure 8**). The exception is a soluble glycocalyx-associated midgut aminopeptidase from *R. americana*. This enzyme is an aminopeptidase removing N-terminal aspartic acid or glutamic acid residues from peptides that are not efficiently attacked by the other aminopeptidases (Klinkowstrom *et al.*, 1994).

In addition to a midgut aminopeptidase A, the dipteran *R. americana* has three midgut aminopeptidase Ns (one soluble and two membrane-bound). The soluble aminopeptidase N (115.7 kDa) prefers tetrapeptides over tripeptides (Ferreira and Terra, 1984), like the minor 107-kDa membrane-bound enzyme, whereas the contrary is true for the major 169-kDa membrane-bound aminopeptidase (Ferreira and Terra, 1985, 1986a, 1986b). The single midgut aminopeptidase N of the coleopterans *Attagenus megatoma* (Baker and Woo, 1981) and *T. molitor* (Cristofoletti and Terra, 1999) resemble the 115.7-kDa

and 107-kDa aminopeptidases of *R. americana*. Approximately the same substrate specificity was observed with the two midgut aminopeptidases of the lepidopteran *Tineola bisselliella* (Ward, 1975a, 1975b), that from the aphid *A. pisum* (Cristofoletti *et al.*, 2006), and that of the cerambycid beetle *Morimus funereus* (Bozic *et al.*, 2008). The data suggest that panorpoid insects (Diptera and Lepidoptera) present multiple aminopeptidases with different substrate specificities, in contrast with the single aminopeptidase of coleopterans. However, considerably more data are needed to support this hypothesis.

There have been few attempts to characterize the active site of insect midgut aminopeptidases. Using multiple inhibition analysis and observing the protection against EDTA inactivation that different competitive inhibitors conferred to the enzyme, two subsites were proposed to occur in the active center of *R. americana* microvillar aminopeptidase: a hydrophobic subsite, to which isoamyl alcohol binds, exposing the metal ion, and a polar subsite, to which hydroxylamine binds. Exposure of the metal ion after isoamyl alcohol binding may be analogous to the situation that results when part of the substrate occupies the hydrophobic subsite, causing conformational changes associated with the catalytic step (Ferreira and Terra, 1986b). The effect of pH at different temperatures on kinetic parameters of *T. molitor* midgut aminopeptidase and its inactivation by different compounds were studied (Cristofoletti and Terra, 2000). The data showed that *T. molitor* aminopeptidase catalysis depends on a metal ion, a carboxylate, and a protonated imidazole group, and is somehow influenced by an arginine residue in the neighborhood of the active site. The catalytic metal binding depends on at least a deprotonated imidazole. In addition to the above-mentioned groups involved in catalysis, at least one phenol group and one carboxylate are associated with substrate binding. Thus, *T. molitor* aminopeptidase shares common features with those of other zinc metallopeptidases, especially with mammalian aminopeptidase N, but it differs in some details. An imidazole group seems to be involved in catalysis in *T. molitor* aminopeptidase; this is not observed in mammalian aminopeptidase N, which has an imidazole group participating in substrate binding.

Sequences of aminopeptidase N from insect midgut are available from different insects. Some examples are *Trichoplusia ni* (Wang *et al.*, 2005), *A. pisum* (Cristofoletti *et al.*, 2006), *Aedes aegypti* (Chen *et al.*, 2009), *H. armigera* (Angelucci *et al.*, 2008), *Ostrinia nubilalis*, and *B. mori* (Crava *et al.*, 2010). The sequences have a signal peptide, a conserved RLP motif near the N-terminal, a zinc binding/gluzincin motif $HEXXHX_{18}E$, a GAMEN conserved motif, and a long hydrophobic C-terminal containing a glycosyl phosphatidyl inositol anchor (**Figure 5**). Based on the crystal structure of leukotriene A4 hydrolase, the two histidine residues and the distant glutamic acid residue of the gluzincin motif are zinc ligands, the glutamic acid residue between the histidine residues is involved in catalysis (Hooper, 1994; Rawlings and Barrett, 1995), and the glutamic acid residue of the GAMEN motif binds the substrate N-terminal amino acid (Luciani *et al.*, 1998). In contrast to the situation in mammals, insect aminopeptidase N is membrane bound at the C-terminal. No soluble insect aminopeptidase N has been sequenced.

Dendrograms derived from alignments of lepidopteran midgut aminopeptidases suggest that there are at least four major groups of lepidopteran aminopeptidases, with the isoforms of the same animal distributed among the groups (Wang *et al.*, 2005; Cristofoletti *et al.*, 2006; Crava *et al.*, 2010). The existence of a number of different aminopeptidases in lepidopterans could be explained by the need for enzymes with different substrate specificities (as shown above for *R. americana*) or different susceptibilities to inhibitors, similar to serine proteinases (see section 11.5.2).

Probably associated with the fact that aminopeptidases are major proteins in some microvillar membranes (55% of *T. molitor* midgut microvillar proteins) (Cristofoletti and Terra, 1999), they are targets of insecticidal *Bacillus thuringiensis* crystal δ-endotoxins. These toxins, after binding to aminopeptidases and receptor molecules called cadherins, form channels through which cell contents leak, leading to the death of the insect (Knight *et al.*, 1995). Although data on substrate specificity for lepidopteran aminopeptidase isoforms are lacking, there is evidence that the isoforms may have differences in toxin binding (Nakanishi *et al.*, 2002; Rajagopal *et al.*, 2003).

Cloning and sequencing dipteran aminopeptidases, for which differences in substrate specificity are known, and a study of substrate specificities of lepidopteran aminopeptidases, may clarify the selective advantages of the evolution of aminopeptidase groups. Furthermore, this study may support the hypothesis that aminopeptidase gene duplications have occurred in the panorpoid ancestor, before differentiation between dipterans and lepidopterans.

11.5.6. Carboxypeptidases and Dipeptidases

Carboxypeptidases hydrolyze single amino acids from the C-terminus of the peptide chain, and are divided into classes on the basis of their catalytic mechanism. There are two digestive metallocarboxypeptidases in mammals: carboxypeptidase A (EC 3.4.17.1), which hydrolyzes (in alkaline medium) C-terminal amino acids, except arginine, lysine, and proline; and carboxypeptidase B (EC 3.4.1.7.2), which releases (in alkaline conditions) C-terminal lysine and arginine preferentially. Insect digestive carboxypeptidases have been classified as carboxypeptidase A or B depending on activity in alkaline medium against Z-GF (or hippuryl β-phenyllactic acid) or Z-GR (or hippuryl-l-arginine), respectively (**Figure 8**). Digestive insect carboxypeptidase A-like enzymes are widespread among insects, and most of them have pH optima

of 7.5–9.0 and molecular masses of 20–50 kDa (Terra and Ferreira, 1994). They have been cloned and sequenced from Diptera (Ramos *et al.*, 1993; Edwards *et al.*, 1997) and Lepidoptera (Bown *et al.*, 1998), and the enzyme from the lepidopteran *H. armigera* was also submitted to crystallographic studies (Estébanez-Perpiña *et al.*, 2001). The sequences have signal and activation peptides, and the features typical of carboxypeptidase As, including the residues His 69, Glu 72, and His 196, which bind the catalytic zinc ion; Arg 71, Asn 144, Arg 145, and Tyr 248, responsible for substrate binding; and Arg 127 and Glu 270, responsible for catalysis. In spite of the overall similarity of *H. armigera* procarboxypeptidase with human procarboxypeptidase A2, there are differences in the loops between the conserved secondary structures, including the loop where the activation processing occurs. Another important difference is the residue 255 (bottom of the S′1 pocket) that defines the enzyme specificity. In mammalian sequences, Asp 255 is found in carboxypeptidase B and Ile 255 in carboxypeptidase A. In insect carboxypeptidase As, this residue varies (but never is an acid residue) (**Figure 5**).

Carboxypeptidase B-like enzymes have been detected in insect midguts (Terra and Ferreira, 1994), and the enzyme from *H. zea* (CPBz) has been described in 3D detail (Bayés *et al.*, 2005). CPBhz has Glu 255 in the specificity pocket and differs from other carboxypeptidase Bs in some loops, which renders it insensitive to the potato carboxypeptidade inhibitor. Furthermore, the enzyme is unable to hydrolase substrates ending in Arg because of its bulky size, whereas it cleaves quite well substrates ending in Lys (Bayés *et al.*, 2005).

Dipeptidases hydrolyze dipeptides, and are classified according to their substrate specificities. Dipeptidases comprise the least known of the insect peptide hydrolases. There have been few studies in which dipeptidase assays were performed, and even fewer attempts to characterize the enzymes (Terra and Ferreira, 1994). The larval midgut of *R. americana* has three dipeptidases (two soluble, of 63 kDa and 73 kDa, respectively, and one membrane-bound) that hydrolyze Gly-Leu, resembling dipeptide hydrolase (dipeptidase, E.C. 3.4.13.18), although, in contrast to the mammalian enzyme, they are very active upon Pro-Gly (**Figure 8**). *R. americana* also seems to have an amino acyl-histidine dipeptidase (carnosinase, EC 3.4.13.3) (Klinkowstrom *et al.*, 1995). More work on insect digestive dipeptidases is urgently needed.

11.6. Digestion of Lipids and Phosphates

11.6.1. Overview

Lipids that contain fatty acids comprise storage lipids and membrane lipids. Storage lipids, such as oils present in seeds, and fats in adipose tissue of animals, are triacylglycerols (triglycerides), and are hydrolyzed by lipases. Membrane lipids include phospholipids and glycolipids, such as mono- and digalactosyldiglycerides (see section 11.4.7). Phospholipids are digested by phospholipases. A combination of α- and β-galactosidases may remove galactose residues from mono- and digalactosyldiglyceride to leave a diacylglycerol which may be hydrolyzed by a triacylglycerol lipase.

Phosphate moieties need to be removed from phosphorylated compounds prior to absorption. This is accomplished by non-specific phosphatases. The phosphatases may be active in an alkaline (alkaline phosphatase, EC 3.1.3.1) or acid (acid phosphates, EC 3.1.3.2) medium.

11.6.2. Lipases

Triacylglycerol lipases (EC 3.1.1.3) are enzymes that preferentially hydrolyze the outer links of triacylglycerols and act only on the water–lipid interface. Activity of the lipase is increased as the interface becomes larger due to lipid emulsification caused by emulsifiers (surfactants). Insects lack emulsifiers comparable to the bile salts of vertebrates, but surfactant phospholipids, including lysolecithin, occur in their midguts in sufficient concentration to alter the surface tension of midgut contents (De Veau and Schultz, 1992). Lysolecithin, and other surfactants, may be formed by the action of phospholipase A on ingested phospholipids (see below, and **Figure 2**).

Insect midgut triacylglycerol lipases have been studied in few insects, and only in crude preparations. The data suggest that the enzyme preferentially releases fatty acids from the α-positions, prefers unsaturated fatty acids, and is activated by calcium ions, thus resembling the action of mammalian pancreatic lipase. The resulting 2-monoacylglycerol may be absorbed or hydrolyzed (Terra *et al.*, 1996). Hydrolysis of 2-monoacylglycerol may be accomplished by triacylglycerol lipase, following migration of the fatty acid to the 1-position, which seems to be favored by the alkaline midgut pH, at least in *M. sexta* (Terra *et al.*, 1996).

Current research on insect midgut lipases is focused on classifying the different enzymes in the families already described for mammals. Thus, a large group of proteins is identified as neutral (or pancreatic) lipase. They have the catalytic triad (Ser 152, Asp 176, His 263, human pancreatic lipase numbering) and two important features that are associated with triacylglycerol hydrolysis; the β9 loop and lid, which cover the active site and are implicated in substrate recognition (Horne *et al.*, 2009). The C-terminal domain is absent in insect neutral lipases, which is consistent with the lack of a colipase, the protein associated with the vertebrate pancreatic lipases. Also differing from the pancreatic lipases, insect neutral lipases may have the lid structure severely reduced and the β9 loop partially deleted, resembling phospholipases (Christeller

et al., 2010). In accordance with this, the major triacylglycerol lipase from *Manduca sexta* fat body is also an active phospholipase A1 (Arrese *et al.*, 2006).

Another large group of lipases correspond to the acid (gastric) lipases. They have conserved catalytic triad and lid similar to those of the neutral lipases (Horne *et al.*, 2009). Alignment of the sequences of insect midgut lipases with those of mammalian lipases, for which substrate specificity information is available, suggests that the lepidopteran neutral lipases are galactolipases and phospholipases *in vivo*, whereas the lepidopteran acid lipases are triacylglycerol lipases (Christeller *et al.*, 2010).

Insect midgut lipases may have a primary role in the acquisition of dietary lipid, but may also have strong antiviral activity by disrupting viral envelopes, as observed in *B. mori* larvae (Ponnuvel *et al.*, 2003).

Esterases, which are usually named the carboxylesterases (ali-esterases, EC 3.1.1.1) catalyze the hydrolysis of carboxyl ester into alcohol and carboxylate. This enzyme, in contrast to lipases, attacks molecules that are completely dissolved in water. It also hydrolyzes water-insoluble long-chain fatty acid esters in the presence of surfactants, but at a rate much slower than that of triacylglycerol lipase. A role for esterases in digestion is unclear, and because of this they are not reviewed in detail here.

In spite of the fact that a requirement for essential fatty acids is probably universal in insects, progress has been limited in the study of lipid digestion. Presumably, the lack of comparatively simple, sensitive assays, and the complexities of digestion related to lipid solubilization, have hindered work in this area. Another reason to study enzymes associated with lipid digestion is that they might be important in limiting the development of pathogens and parasites. Hydrolysis of membrane lipids might cause cellular lysis, and fatty acid products of digestion may possess antibiotic effects.

11.6.3. Phospholipases

Phospholipase A2 (EC 3.1.1.4) and phospholipase A1 (EC 3.1.1.32) remove the fatty acids from phosphatides attached to the 2- and 1-positions, respectively, resulting in a lysophosphatide (**Figure 2**). Lysophosphatide is more stable in micellar aggregates than on membranes. Thus, the action of phospholipase A on the membrane phosphatides causes the solubilization of cell membranes, rendering the cell contents free to be acted upon by the appropriate digestive enzymes. Phospholipase is widespread among insects (Terra *et al.*, 1996). Phospholipase A2 partially purified from the midgut of adult beetle *Cincindella circumpicta* has a molecular mass of 22 kDa and pH optimum 9.0, is calcium dependent, and is inhibited by the site-specific inhibitor oleyoxyethyl phosphorylcholine. Unfed beetles did not express the phospholipase in the midgut contents (Uscian *et al.*, 1995).

Since no ESTs characteristic of classical phospholipases have been reported from lepidopteran midgut, it is likely the phospholipase A$_2$ activities that have been characterized from the midgut and salivary gland of *M. sexta* (Rana and Stanley, 1999; Tunaz and Stanley, 2004) are actually neutral lipases.

Although lysophosphatide may be further hydrolyzed by a lysophospholipase (phospholipase B, EC 3.1.1.5), evidence suggests it is absorbed intact by insects (Terra *et al.*, 1996). Phosphatides may also be hydrolyzed by phospholipase C (EC 3.1.4.3), yielding the phosphoryl base moiety and diacylglycerol, or by phospholipase D (EC 3.1.4.4), resulting in phosphatidate and the base (**Figure 2**). Both enzymes have been found in insect midgut (Terra *et al.*, 1996), but have not been studied in detail.

11.6.4. Phosphatases

Alkaline phosphatase is usually a midgut microvillar membrane marker in dipteran and lepidopteran species, although it may also occur in midgut basolateral membranes, and even as a secretory enzyme. Acid phosphatase is usually soluble in the cytosol of midgut cells in many insects, and may also appear in midgut contents or be found membrane-bound in midgut cells (Terra and Ferreira, 1994).

The best-known alkaline phosphatases are those from *B. mori* (Lepidoptera: Bombycidae) larval midgut. The major membrane-bound and the minor soluble alkaline phosphatases were purified and shown to be monomeric enzymes with the following properties: (1) soluble enzyme, molecular mass of 61 kDa, pH optimum 9.8; (2) membrane-bound enzyme, molecular mass of 58 kDa, pH optimum 10.9. Both enzymes have wide substrate specificity and are inhibited by cysteine. The membrane-bound alkaline phosphatase occurs in the microvillar membranes of columnar cells, whereas the soluble enzyme is loosely attached to the goblet cell apical membrane facing the cell cavity (Eguchi, 1995). The determination of the complete sequence of the membrane-bound alkaline phosphatase led to the finding of putative regions for phosphatidylinositol anchoring and zinc-binding site, but not for N-glycosylation, despite the fact that the enzyme contains N-linked oligosaccharides (Itoh *et al.*, 1991). The sequence of the soluble alkaline phosphatase was also determined, and has high identity with the membrane-bound enzyme (Itoh *et al.*, 1999).

Acid phosphatases have been characterized in some detail only in *Rhodnius prolixus* (Hemiptera: Reduvidae). The major enzyme activity is soluble, and has the following properties: wide specificity, a molecular mass of 82 kDa, K$_m$ for p-nitrophenyl phosphate 0.7 mM, and is inhibited by fluoride, tartrate, and molybdate. The minor enzyme activity is membrane-bound, and is resolved into two enzymes (123 and 164 kDa) that are resistant to fluoride and tartrate (Terra *et al.*, 1988).

11.7. Microvillar Membranes

11.7.1. Isolation, Chemistry, and Enzymology

Midgut cells are associated with one another by junctions that separate the plasma cell membranes into an apical and a basolateral domain. The apical domain is usually modified into finger-like projections, the microvilli. The insect midgut cell microvillus is homologous to that described in vertebrates and reviewed by Bement and Mooseker (1996). Thus, a bundle of parallel actin filaments crosslinked by actin-bundling proteins like fimbrin and villin form the core of a microvillus. Lateral side arms (composed of myosin I and calmodulin) connect the sides of the actin bundle to the overlying plasma membrane.

It has been known for a long time that insect midgut cell apices are involved in the transport of water (Wigglesworth, 1933) and organic compounds (Treherne, 1959). Nevertheless, only since 1980 has it been recognized that insect midgut apical cell membrane plays a role in digestive events. Before 1980 all insect digestive enzymes were considered to be secreted into the lumen, as with mammals, where, in 1961, Miller and Crane had provided cell fractionation data showing that disaccharidases are firmly bound to cell membrane covering the enterocyte microvilli.

Insect midgut microvilli were isolated for the first time by Ferreira and Terra (1980) from *R. americana* (a lower dipteran) larvae using a differential calcium (magnesium) precipitation technique (Schmitz *et al.*, 1973) developed for mammals. Digestive enzymes performing final digestion were found associated with microvilli. A few months later, Hanozet *et al.* (1980) used the same technique to isolate microvilli from the columnar (principal) cell of the midgut of lepidopteran larvae, in order to study *in vitro* amino acid transport. After this paper, and complementary data (Terra *et al.*, 2006), differential precipitation became the method of choice to prepare microvilli from columnar cells of lepidopteran midguts. In addition to lower Diptera and Lepidoptera, differential precipitation has been used to isolate microvilli from midgut cells from other insect taxa, such as Dictyoptera (Parenti *et al.* 1986), Coleoptera (Ferreira *et al.*, 1990; Reuveni *et al.*, 1983), and higher Diptera (Lemos and Terra, 1992).

Microvilli prepared by differential precipitation methods are free from contaminants from other cells (such as goblet cells in the case of lepidopterans), but still contain most of the microvillus skeleton. A successful procedure was developed for cytoskeleton removal from insect midgut microvilli by treatment with hyperosmotic Tris, followed by pelleting the purified membranes with negligible amounts of cytoskeleton and slight contamination by basolateral membranes (Coleoptera and Diptera, see Jordão *et al.*, 1995; Lepidoptera, see Capella *et al.*, 1997).

Insect midgut microvillar membrane densities vary widely, with insects appearing later in evolution (more derived insects) having denser membranes (and hence a higher protein content) than insects appearing earlier in evolution (less derived insects) (Terra *et al.*, 2006). Although higher protein content does not necessarily mean a richer variety of proteins, there is evidence supporting this. SDS-PAGE of midgut microvillar proteins of a coleopteran (membrane density 1.08–1.10) resolves fewer clearly visible bands than in the case of a lepidopteran or a dipteran (membrane density 1.14–1.16) (Jordão *et al.*, 1995; Capella *et al.*, 1997). The observed range of protein–lipid mass ratio of insect microvillar membranes is 1.41–3.13, which is wider than that found among mammalian enterocytes (Terra *et al.*, 2006).

Apparently there is an inverse relationship between the protein–lipid mass ratio (or membrane density) and the cholesterol and carbohydrate content in insect microvillar membranes. Thus, protein–lipid mass ratio, carbohydrate (µg/mg protein) and cholesterol (µg/mg protein) contents are, respectively: 1.4–1.7, 400–700, and 110–140 for Coleoptera; 2.0–2.6, 240–410, and 40–59 for higher Diptera; and 2.6–3.8, 0–80, 17–28 for Lepidoptera (Jordão *et al.*, 1995; Capella *et al.*, 1997). A detailed study of the chemical composition of microvilli (microvillar membranes plus contaminant cytoskeleton) from *B. mori* midgut cells (Leonardi *et al.*, 2001) confirms previous data. Thus, the protein–lipid ratio is smaller (1.85) in anterior plus middle in comparison to posterior midgut (2.3). Phospholipids account for 77% (phosphatides make up 62%) of total lipids, with glycolipids summing 8%.

11.7.2. Microvillar Proteins

The physiological role of midgut microvillar membranes may change along the midgut and among insect taxa, and should include surface (terminal) digestion, absorption, ion homeostasis, signaling, and unique digestive enzyme secretion mechanisms. At first, microvilli preparations revealed that microvillar integral digestive enzymes vary among different taxa. Most frequently, they are aminopeptidase, alkaline phosphatase, carboxypeptidase, dipeptidase, and α-glucosidase (Terra and Ferreira, 1994). Many of those proteins have been characterized as detailed before in the corresponding headings.

To date, the comprehensive studies of microvillar proteins are performed with microvilli preparations according to two approaches. The proteomics approach is based on the resolution of the microvillar proteins and mass spectrometry for identification. The resolution of the proteins usually uses two-dimensional gel electrophoresis (2-D PAGE), with electrofocusing in the first-dimension, followed by SDS-PAGE in the second dimension

(Candas *et al.*, 2003; McNall and Adang, 2003; Krishnamoorthy *et al.*, 2007; Bayyaredy *et al.*, 2009). Another variant of the proteomics approach is the shotgun method of Popova-Butler and Dean (2009). In this method, a microvilli preparation is subjected to trypsin and chymotrypsin digestion. The combined digesta is then submitted to two-dimensional liquid chromatography coupled with tandem spectrometry. Pauchet *et al.* (2009b) used two different separation procedures: separation of microvillar Triton X-100 (or digitonin)-solubilized microvilli proteins by anion-exchange chromatography or native electrophoresis in one dimension, followed by SDS-PAGE in the second dimension.

The immunoscreening approach to identify microvillar proteins is based on immunoscreening a midgut-specfic cDNA library using antibodies raised against cytoskeleton-free microvillar proteins, followed by sequencing the positive clones and searching for similarities in databases (Ferreira *et al.*, 2007). The proteomics approach is limited by solubility problems affecting many proteins, by occasional failures of protein bands in originating useful mass spectra, and by the quality of peptide mass fingerprints obtained when the sequences of the specific organism under study are not abundant in the databases (frequent among insects). Furthermore, all the reports based on the proteomic approach use microvilli preparations instead of cytoskeleton-free microvillar membranes, resulting in the recovery of proteins deriving from cytoskeleton, mitochondria, and cytosol among the microvillar proteins. The immunoscreening method also has some possible sources of errors: (1) undetected proteins because of a lack of reacting antibodies caused by extremely low amounts of antigens, or because they were not immunogenic enough; (2) contaminants detected because they are highly immunogenic; (3) non-microvillar proteins detected because they share epitopes or were accidentally associated with microvillar proteins; (4) failure of inserted-cDNA-phage expression.

In spite of the limitations discussed above, both methods allowed the characterization of a substantial number of midgut microvillar proteins of different taxa. Thus, three groups of predicted microvillar proteins were recognized in *T. molitor* (Coleoptera) (Ferreira *et al.*, 2007): digestive enzymes (aminopeptidase and α-mannosidase), putative peritrophic membrane ancillary protein (PMAP)-like proteins; and peritrophic membrane proteins (peritrophins). The aminopeptidase corresponds to that described previously (Cristofoletti and Terra, 1999, 2000). PMAP may be involved in PM formation, and is homologous to proteins having insect-allergen repeat-related domains (Ferreira *et al.*, 2008) (see section 11.8.1).

Lepidopteran midgut microvillar proteins are grouped into six classes (McNall and Adang, 2003; Ferreira *et al.*, 2007; Krishnamoorthy *et al.*, 2007; Pauchet *et al.*, 2009b): (1) digestive enzymes (aminopeptidases with GPI-anchors, carboxypeptidase, alkaline phosphatase, astacin-like protein, dipeptidyl peptidase A, maltase-like protein; (2) midgut protection (thioredoxin peroxidase; protein disulfide isomerase; aldehyde dehydrogenase, serpin); (3) peritrophic membrane formation (peritrophins); (4) membrane-tight-bound cytoskeleton proteins (fimbrin, actin, afadin, desmocollin, cadherin-like proteins); (5) proteins associated with microapocrine secretion (annexin, calmodulin, gelsolin); and (6) others (V-ATPase, chlorophylide A binding protein, ABC transporters). Microapocrine secretion is discussed in section 11.10. V-ATPase is thought to be a contamination by goblet cells.

Larval mosquito midgut microvillar proteins fall into four categories (Bayyareddy *et al.*, 2009; Popova-Butler and Dean, 2009): (1) digestive enzymes (alkaline phosphatase, trypsin, serine proteinase, zinc-metalloprotease, α-amylase); (2) midgut protection (protein disulfide isomerase, aldehyde dehydrogenase, peroxiredoxin); (3) membrane-tightly-bound cytoskeleton proteins (flotillins, prohibitin, actin); and (4) others (V-ATPase, calmodulin).

Lepidopteran and dipteran midgut microvillar membranes are denser and show a greater variety of proteins than those of coleopterans, because there are a host of microvillar proteins in lepidopterans and dipterans that assist the larvae in dealing quickly with huge amounts of food derived from a variety of plants. There are proteins involved in counteracting plant chemical defenses, in protecting the midgut surface against the larval serine proteinases, and in promoting peculiar secretory mechanisms. In addition to those proteins, there are still others that lepidopterans and dipterans share with coleopterans; namely, those forming the peritrophic membrane, a few digestive enzymes, and those not found among predicted proteins, like receptors and ion and organic compound transporters. Coleopterans attack a variety of plants, but they usually deal with their food more slowly than lepidopterans and dipterans. Perhaps because of this, their digestive physiology strategy seems to rely less on midgut microvillar proteins than is the case in lepidopterans and dipterans.

11.8. The Peritrophic Membrane

11.8.1. The Origin, Structure, and Formation of the Peritrophic Membrane

There is a film surrounding the food bolus in most insects that is occasionally fluid (peritrophic gel), but is more frequently membranous (peritrophic membrane, PM). The PM is made up of a matrix of proteins (mainly peritrophins) and chitin to which other components (e.g., enzymes, food molecules) may associate. This anatomical structure is sometimes called the peritrophic matrix, but this term should be avoided because it does not convey the idea of a film and rather suggests that it is the

fundamental substance of some structure, usually filling a space, as in the mitochondrial matrix. The argument that "membrane" means a lipid bilayer is not valid, because the PM is not a cell part but an anatomical structure, like the nictitating membrane of birds and reptiles.

PM proteins have been classified by Tellam *et al.* (1999) according to the ease with which they can be extracted: class 1 proteins can be removed with physiological buffers, class 2 with mild detergents, and class 3 with strong denaturants (e.g., urea), while class 4 proteins are the remaining residue. Class 1 proteins are thought to be digestive enzymes and food proteins loosely adsorbed at the PM surface. Class 2 proteins, at least in lepidopteran PMs, may correspond to proteins enclosed in membrane vesicles entrapped between PM sheets. These membrane vesicles bud off from the cell microvilli (Ferreira *et al.*, 1994; Bolognesi *et al.*, 2001; see also 11.8.2.2.7). The numerous hydrolases found in the PM proteomes (Campbell *et al.*, 2008; Dinglasan *et al.*, 2009) may correspond to those entrapped enzymes. The same might be true for the so-called "GNBP-like" proteins (β-glucan binding proteins) (Campbell *et al.*, 2008), which might correspond to active laminarinases (see 11.4.3.2).

Class 3 proteins are the integral proteins of PM. Class 4 proteins are likely the same as those in class 3, since Campbell *et al.* (2008) were able to completely solubilise PM with anhydrous trifluoromethanesulfonic acid, and found the same proteins previously decribed as class 3. The major class 3 proteins are named peritrophins, are characterized by the presence of chitin-binding domains (CBDs) (called peritrophin, pfam 01607, CBM_14) (CDART database) (Geer *et al.*, 2002), and may also have mucin-like domains (Tellam *et al.*, 1999).

Peritrophins are made of several domains. The major domain (peritrophin A-domain) is a cysteine-rich domain with chitin-binding properties having the consensus sequence $CX_{13-20}CX_{5-6}CX_{9-19}CX_{10-14}CX_{4-14}C$ (where X is any amino acid except cysteine), which includes several conserved aromatic amino acids. Variations of this chitin-binding domain are the peritrophin-B and peritrophin-C domains, with consensus sequences $CX_{12-13}CX_{20-21}CX_{10}CX_{12}CX_2CX_8CX_{7-12}C$ and $CX_{8-9}CX_{17-21}CX_{10-11}CX_{12-13}CX_{11}C$, respectively. These variations may be absent from many insects, exemplified by the predicted sequences of *T. castaneum* proteins (Jasrapuria *et al.*, 2010). The mucin-like domains occurring in peritrophins are proline/threonine-rich domains that are heavily glycosylated and similar to mucins (Lang *et al.*, 2007). Peritrophins may have one (e.g., Cb-peritrophin-15 from *Lucilia cuprina*) to several (e.g., peritrophin-44 from *L. cuprina*) chitin-binding domains or chitin-binding domains with small (e.g., Ag-AperI from *Anopheles gambiae*) or very large mucin-like domains (e.g., IIM from *Trichoplusia ni*) (Wang and Granados, 1997; Shen and Jacobs-Lorena, 1998; Tellam *et al.*, 1999, 2003).

The availability of genome data pemitted the description of the complete sets of peritrophins of *Ae. aegypti* (65 peritrophins), *T. castaneum* (25 peritrophins), and *D. melanogaster* (65 peritrophins) (Venancio *et al.*, 2009). It is not necessary for all the peritrophins to be true components of PM. The presence of those peritrophins in PM needs to be experimentally confirmed in the light that peritrophin-like proteins may have functions other than to form PM (Jasrapuria *et al.*, 2010).

The 3D structure of PM is thought to result from chitin fibrils being interlocked with the chitin-binding domains of peritrophins. Mucin-like domains of peritrophins are thought to face the ectoperitrophic and endoperitrophic sides of the PM. As these domains are highly hydrated, they lubricate the surface of the PM, easing the movement of food inside the PM and of the ectoperitrophic fluid outside the PM. Furthermore, the glucan chains associated with peritrophin mucin-like domains may ensure high proteinase resistance to PM (see **Figure 9** in Schorderet *et al.*, 1998; **Figure 5** in Wang and Granados, 2001; **Figure 2** in Hegedus *et al.*, 2008).

The structure of peritrophins prompted Terra (2001) to develop a speculative model of the origin and evolution of the PM. As mucins have a very early origin among animals (confirmed by Lang *et al.*, 2007), Terra (2001) proposed that PM derived from the ancestral mucus. According to this hypothesis, the peritrophins, the major PM proteins, evolved from mucins by acquiring chitin-binding domains. The concomitant evolution of chitin secretion by midgut cells permitted the formation of the chitin–protein network characteristic of the PM structure, described above. Later in evolution, some peritrophins lost their mucin-like domains. If the hypothesis that the PM is derived from the gastrointestinal mucus is correct, it should have originally been synthesized by midgut cells along the whole midgut, and should have had the properties of the mucus. Later in evolution, insect species would have appeared with a chitin–protein network, resulting in PM formation. Therefore, the formation of the PM by the whole midgut epithelium is the ancestral condition, whereas the restriction of PM production to midgut sections, or the lack of a PM and its replacement by the peritrophic gel, are derived conditions.

PMs are classified into two types (Peters, 1992). Type I PM is found in cockroaches (Dictyoptera), grasshoppers (Orthoptera), beetles (Coleoptera), bees, wasps, and ants (Hymenoptera), moths and butterflies (Lepidoptera), and in hematophagous adult mosquitoes (Diptera). Type II PM occurs in larval and adult (except hematophagous ones) mosquitoes and flies (Diptera), and in a few adult Lepidoptera.

Other PM integral proteins are chitin deacetylases (CDA) (Dixit *et al.*, 2008). This kind of protein was first found in the PM of *Trichoplusia ni* by Guo *et al.* (2005). The authors did not detect any chitin deacetylase activity, and suggested that binding might be the only interaction of those proteins with chitin. The interaction was suggested by its extraction from PM with calcofluor. Taking into account that some CDA-like proteins found associated with PM are active in producing chitosan from chitin (e.g., in *Mamestra configurata* PM; Toprak *et al.*, 2008), Campbell *et al.* (2008) suggested several functions for chitosan: (1) it might be present to modify the flexibility of chitin fibers; (2) it may regulate redox condition, as it has antioxidant properties; (3) it may have antimicrobial properties; and, finally, (4) modification of chitin may be required for the binding of other proteins.

Type I PM is formed either by the whole midgut epithelium, or by part of it (anterior or posterior regions). The formation of these PMs is frequently induced by the distension of the gut caused by food ingestion. PM produced by the whole or anterior midgut epithelium envelops the food along the whole midgut. When PM is produced only by the posterior part, the anterior midgut epithelium is usually covered with a viscous material, the peritrophic gel, as observed in carabid beetles and bees. This gel is also observed in the anteriorly placed midgut ceca of some insects, and along the whole midgut of others (Terra, 2001).

During formation of type I PM, chitin is synthesized outside the cells by a chitin synthase bound to microvillar membranes using precursors formed inside the cells (Zimoch and Merzendorfer, 2002; Arakane *et al.*, 2005). Once chitin is self-organized in chitin fibers (Hegedus *et al.*, 2008), it interlocks with peritrophin molecules that are released by exocytosis (Bolognesi *et al.*, 2001). The micrographs of Harper and Hopkins (1997) show that, during the formation of Type I PM, a fibrous material appears first at the tips of the microvilli of anterior midgut cells and then is rapidly included in a thin PM surrounding the food bolus. Thus, the crucial events in PM formation appear to take place among microvilli.

Peritrophins are soluble proteins that are extensively immobilized at the surface of *T. molitor* cells due to cell glycocalyx association (Ferreira *et al.*, 2007). This arguably facilitates chitin–peritrophin association, and it is conceivable that ancillary proteins help this process. *T. molitor* PMAP may be one of those ancillary proteins. PMAP is homologous to proteins having insect allergen-related repeat domains like AEG12 (Shao *et al.*, 2005). The role of PMAP is based on circumstancial evidence: (1) PMAP is a soluble protein that, like peritrophins, occurs partially immobilized at the cell surface; and (2) PMAP is found associated with PM, but, in contrast with peritrophins, it seems to be removed from the PM as it moves along the midgut. A search of EST databanks led to the suggestion that PMAP-like proteins concur in the formation of type I, but not type II, PM. Nevertheless, lepidopterans do not have PMAP-like proteins, and may have other ancillary proteins.

Type II PM is secreted by a few rows of cells at the entrance of the midgut (cardia), and is usually found in insects irrespective of food ingestion. Peritrophins are secreted by exocytosis (Eisemann *et al.*, 2001).

A comparison of the complete sets of peritrophins from PM from larval *Ae. aegypti* and *D. melanogaster* (type II PM) showed they have more complex domain structures than from adult *Ae. aegypti* and *T. castaneum* (type I PM). Furthermore, mucin-like domains of peritrophins from *T. castaneum* (feeding on rough food) are lengthier than those of adult *Ae. aegypti* (blood-feeding). This suggested that type I and type II PMs may have variable architectures determined by different peritrophins and/or ancillary proteins, which may be partly modulated by diet (Venâncio *et al.*, 2009).

Although a PM is found in most insects, it does not occur in Hemiptera and Thysanoptera, which have perimicrovillar membranes in their cells (see below). The other insects that apparently do not have a PM are adult Lepidoptera, Phthiraptera, Psocoptera, Zoraptera, Strepsiptera, Raphidioptera, Megaloptera, adult Siphonaptera, bruchid beetles, and some adult ants (Hymenoptera) (Peters, 1992). These insects may have a peritrophic gel instead of PM, one example being bruchid beetles (Terra, 2001), or may have had their PMs overlooked because the insects were unfed.

Another possibility is that minute hematophagous insects (like Siphonaptera and Phthiraptera) have lost their PM because the blood clot ensures countercurrent flows (see section 11.8.2.2.3), and their small size makes for easy and efficient diffusion of digestion products up to the midgut surface.

The PM may have a wide range of pore sizes, with some being small or very large, but most of them in the middle range. The average pore sizes of PM may be determined by comparing molecular masses of enzymes restricted to the ectoperitrophic fluid (**Figure 1**) with those of enzymes present inside PM. This method of pore size estimation is probably the most accurate, since it reflects *in vivo* conditions. Pore sizes have also been determined by feeding insects with colloidal gold particles or fluorescent dextran molecules of known molecular masses, and recording their passage through the PM *in vivo*, or using PM mounted as a sac and measuring diffusing rates. Determinations performed with these techniques by different authors found pores in the range 7–9 nm for insects pertaining to different orders. Other authors described pores in the range 17–36 nm (Terra, 2001). Pore sizes in the range 17–36 nm were obtained with fluorescent dextran molecules in conditions able to detect very small amounts of substances traversing the PM. Those pores probably correspond to the large pores occurring at low frequency in PMs. Although these large pores are supposed to be of no importance regarding digestive events, they set the size limits for an infecting particle to successfully pass through the PM.

As a consequence of its small pores, the PM hinders the free movement of molecules, dividing the midgut lumen into two compartments (**Figure 1**) with different molecules. The functions of this structure include those of the mucus (protection against food abrasion and invasion by microorganisms), and several roles associated with the compartmentalization of the midgut. These roles result in improvements in digestive physiology efficiency, thereby leading to decreased digestive enzyme excretion, and restrict the production of the final products of digestion close to their transporters, thus facilitating absorption. These roles will be detailed below (see section 11.8.2).

Major points needing clarification are how the chemical nature of peritrophic gel and PM define their strength, elasticity, and porosity, and how these structures are self-assembled in the midgut lumen.

11.8.2. The Physiological Role of the Peritrophic Membrane

11.8.2.1. Protection against food abrasion and invasion by microorganisms.
As mentioned before, gut cells in most animals are covered with a gel-like coating of mucus, which has been most thoroughly studied in mammals (Lang *et al.*, 2007). In these animals, the mucus is supposed to lubricate the mucosa, protecting it from mechanical damage, and to trap bacteria and parasites. Since the insect midgut epithelium lacks a mucus coating, PM functions were supposed to be analogous to that of mucus. Thus, insects deprived of PM may have the midgut cells damaged by coarse food, and may be liable to microorganism invasion (see Peters, 1992; Tellam, 1996; Lehane, 1997).

The PM as a barrier against invasion by microorganisms has particular relevance in insects that transmit viruses and parasites to human beings, as these microorganisms may have specific developmental phases in insect tissues (Tellam, 1996; Lehane, 1997). Microorganisms invade the insect midgut cells after disrupting the PM with the use of chitinase (Shahabuddin, 1995), or by using a proteinase such as enhancin that specifically affects the peritrophins (Peng *et al.*, 1999; Ivanova *et al.*, 2003).

A barrier against microorganism invasion is probably less important for the majority of insects that feed on plants, as exemplified by observations carried out with the moth *T. ni*. Larvae of this insect deprived of PM by Calcofluor treatment show high mortality. Examination of dead larvae showed no signs of microbial infection or cell damage by Calcofluor, although these larvae were more susceptible to experimental infection (Wang and Granados, 2000). The results may be interpreted as Calcofluor killing larvae by affecting PM functions in digestion.

Larvae deprived of PM by Calcofluor treatment show a five-fold decrease in the growth rate (GR) that cannot be explained by only a two-fold reduction in the consumption rate (CR). The GR decrease should be a consequence of the increase in the metabolic costs associated with the conversion of food into body mass, rather than the reduction of consumption. One of the presumed causes of the increase in metabolic costs is the increase in digestive enzyme excretion (see section 11.8.2.2.3) (Bolognesi *et al.*, 2008).

Similarly, it has been observed that some plants respond to herbivorous insect attack by producing a unique 33-kDa cysteine proteinase with chitin-binding activity. This proteinase damages the PM, resulting in significant reduction in caterpillar growth caused by impaired nutrient utilization (Pechan *et al.*, 2002).

11.8.2.2. Enhancing digestive efficiency

11.8.2.2.1. Overview The proposal of roles for the PM in digestion has benefited from studies on the organization of the digestive process. These studies (for reviews, see Terra, 1990; Terra and Ferreira, 1994, 2009) revealed that in most insects, initial digestion occurs in the endoperitrophic space (**Figure 10**), intermediate digestion in the ectoperitrophic space, and final digestion at the surface of midgut cells. Such studies led to the formulation of the hypothesis of the endo–ectoperitrophic circulation of digestive enzymes. It was suggested that there is a recycling mechanism (**Figure 10**) whereby food flows inside the PM from the anterior midgut to the posterior, whereas in the ectoperitrophic space water flows from the posterior midgut to the ceca. When the polymeric food molecules become sufficiently small to pass through the PM (with the accompanying polymer hydrolases), the flow patterns result in carriage towards the ceca or the anterior midgut, where intermediate and final digestion occurs.

Terra and colleagues (for reviews, see Terra and Ferreira, 1994; Terra, 2001) hypothesized that, as a consequence of the compartmentalization of digestive events, there is an increase in the efficiency of digestion of polymeric food by allowing the removal of the oligomeric molecules from the endoperitrophic space, which is powered by the recycling mechanism associated with the midgut fluxes. Because oligomers may be substrates or inhibitors for some polymer hydrolases, their presence should decrease the rate of polymer degradation. Fast polymer degradation ensures that polymers are not excreted, and hence increases their digestibility. Another possible consequence of compartmentalization is an increase in the efficiency of oligomeric food hydrolysis due to the transference of oligomeric molecules to the ectoperitrophic space and restriction of oligomer hydrolases to this compartment. In these conditions, oligomer hydrolysis occurs in the absence of probable partial inhibition (because of nonproductive binding) by polymer food and presumed nonspecific binding by non-dispersed undigested food. This process should lead to the production of food monomers

in the vicinity of midgut cell surface, causing an increase in the concentration of the final products of digestion close to their transporters, thus facilitating absorption. Experimental evidence supporting the adaptations for increasing digestive efficiency proposals are discussed in the following sections.

11.8.2.2.2. Prevention of non-specific binding A model system was used to test the hypothesis that the PM prevents non-specific binding of undigested material onto midgut cell surface, with beneficial results (Terra, 2001). For this, purified microvillar membranes were isolated and added to peritrophic membrane contents. The activity of the microvillar enzymes decreased in these conditions, thus supporting the hypothesis (Bolognesi *et al.*, 2008).

11.8.2.2.3. Prevention of enzyme excretion This function was at first proposed based on results obtained with dipteran larvae (for reviews, see Terra and Ferreira, 1994; Terra, 2001). Both *R. americana* and *M. domestica* present a decreasing trypsin gradient along midgut contents (putatively generated by the recycling mechanism), and excreted less than 15% of midgut luminal trypsin after each gut emptying. The findings have now been extended to include Lepidoptera (Borhegyi *et al.*, 1999; Bolognesi *et al.*, 2001, 2008), Coleoptera (Ferreira *et al.*, 2002; Caldeira *et al.*, 2007), and Orthoptera (Biagio *et al.*, 2009). Furthermore, a theoretical model has been developed that is able to calculate the distribution of digestive enzymes along the midgut contents, given the site of water secretion and absorption, or to identify accurately the enzyme secretory site, given the other variables (Bolognesi *et al.*, 2008).

11.8.2.2.4. Increase in the efficiency of digestion of polymeric food favored by oligomer removal A model system was used to test this hypothesis (Bolognesi *et al.*, 2008). Midgut contents from *S. frugiperda* larvae were placed in dialysis bags suspended in stirred and unstirred media. Trypsin activities in stirred and unstirred bags were 210% and 160%, respectively, over the activities of similar samples maintained in a test tube (Bolognesi *et al.*, 2008).

11.8.2.2.5. Increase in the efficiency of oligomeric food hydrolysis The hypothesis that the efficiency of oligomer digestion (intermediate digestion) increases if separated from the initial digestion (Terra, 2001) was confirmed by the experiments of Bolognesi *et al.* (2008). These authors collected ectoperitrophic fluid from the large midgut ceca of *R. americana*, and assayed several digestive enzymes restricted to the fluid. When those enzymes were put in the presence of PM contents, their activities decreased in relation to controls. These decreases in activity probably result from oligomer hydrolase competitive inhibition by luminal polymers.

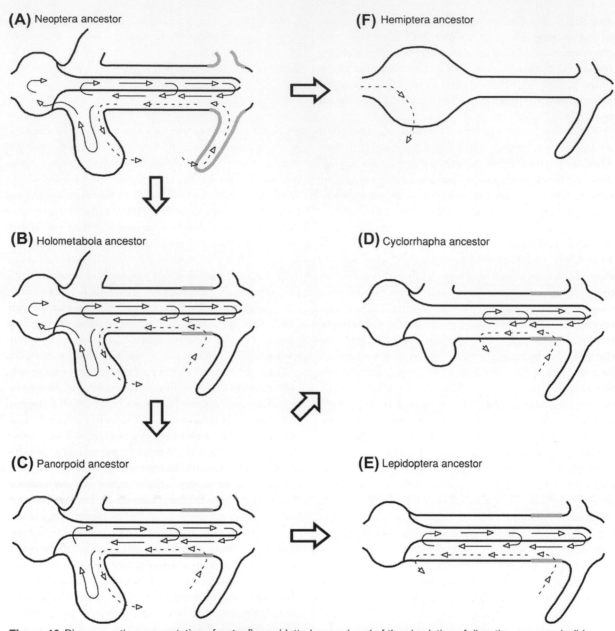

Figure 10 Diagrammatic representation of water fluxes (dotted arrows) and of the circulation of digestive enzymes (solid arrows) in putative insect ancestors that correspond to the major basic gut plans.

In Neoptera ancestors (A), midgut digestive enzymes pass into the crop. Countercurrent fluxes depend on the secretion of fluid by the Malpighian tubules and its absorption by the ceca. Enzymes involved in initial, intermediate, and final digestion circulate freely among gut compartments. Holometabola ancestors (B) are similar except that secretion of fluid occurs in posterior ventriculus. The ancestors of hymenopteran and panorpoid (Lepidoptera and Diptera assemblage) insects (C) display countercurrent fluxes like Holometabola ancestors, midgut enzymes are not found in the crop, and only the enzymes involved in initial digestion pass through the peritrophic membrane. Enzymes involved in intermediate digestion are restricted to the ectoperitrophic space and those responsible for terminal digestion are immobilized at the surface of midgut cells. Cyclorrhapha ancestors (D) have a reduction in ceca, absorption of fluid in middle midgut, and anterior midgut playing a storage role. Lepidoptera ancestors (E) are similar to panorpoid ancestors, except that the anterior midgut replaces the ceca in fluid absorption. Hemiptera ancestors (F) have lost crop, ceca, and fluid-secreting regions. Fluid is absorbed in anterior midgut. Reprinted with permission from Terra, W.R., Ferreira, C., 2009. Digestive system. In: Resh, V.H., Cardé, R.T. (eds), *Encyclopedia of Insects*, 2nd edn. Academic Press, San Diego, CA, pp. 273–281. ©Elsevier.

11.8.2.2.6. Restriction of food monomer production at the cell surface This is a consequence of restricting oligomer hydrolases to the ectoperitrophic space (see section 11.8.2.2.5), and causes an increase in the concentration of the final products of digestion close to the carriers responsible for their absorption. A model system should be developed to test this hypothesis.

11.8.2.2.7. Enzyme immobilization Midgut luminal enzymes, in addition to occurring in the endoperitrophic and ectoperitrophic spaces, may be associated with the PM. For example, results obtained with *S. frugiperda* larvae showed that PM may contain up to 13% and 18% of the midgut luminal activity of amylase and trypsin, respectively (Ferreira *et al.*, 1994). Hence, enzyme immobilization may play a role in digestion, although a minor one. The attachment mechanism of enzymes in PM is not well known. Nevertheless, there is evidence, at least in *S. frugiperda,* that trypsin, amylase, and microvillar enzymes are incorporated into the jelly-like substance associated with PM when the enzymes, still bound to membranes, are released from midgut cells by a microaprocrine process (Jordão *et al.*, 1999; Bolognesi *et al.*, 2001).

11.8.2.2.8. Toxin binding Although potentially toxic dietary tannins are attached to and excreted with *Schistocerca gregaria* PM (Bernays and Chamberlain, 1980), toxin binding by the PM seems to be a less widespread phenomenon than previously suggested. Thus, tannins in *M. sexta* (Barbehenn and Martin, 1998) and lipophilic and amphiphilic noxious substances in *Melanoplus sanguinipes* (Barbehenn, 1999) are maintained in the endoperitrophic space because they form high molecular weight complexes, not because of PM binding. However there is evidence that PM may bind heme in blood-feeding insects (Devenport *et al.*, 2006).

11.8.2.2.9. Peritrophic membrane functions and insect phylogeny Current data detailed below suggest that PMs of all insects have functions (see sections 11.8.2.1, 11.8.2.2.2–11.8.2.2.4), whereas functions (see sections 11.8.2.2.5 and 11.8.2.2.6) are demonstrable only in PMs of Panorpodea (the taxon that includes Diptera and Lepidoptera) and of the hymenopteran sawflies. PM function (see section 11.8.2.2.7) may occur in all insects, but this needs further confirmation. Function (see section 11.8.2.2.8), although it may be important for some insects, should be viewed as opportunistic. In other words, the PM probably evolved from a protective role (see section 11.8.2.1) to more sophisticated functions (see sections 11.8.2.2.2–11.8.2.2.7) under selective pressures and, due to the chemical properties of their constituents, the PM also developed the ability to bind different compounds including toxins.

11.9. Organization of the Digestive Process

11.9.1. Evolutionary Trends of Insect Digestive Systems

After studying the spatial organization of the digestive events in insects of different taxa and diets, it was realized that insects may be grouped relative to their digestive physiology, assuming they have common ancestors. Those putative ancestors correspond to basic gut plans from which groups of insects may have evolved by adapting to different diets (Terra and Ferreira, 1994, 2009).

The basic plan of digestive physiology for most winged insects (Neoptera ancestors) is summarized in **Figure 10**. In these ancestors, the major part of digestion is carried out in the crop by digestive enzymes propelled by antiperistalsis forward from the midgut. Saliva plays a variable role in carbohydrate digestion. Shortly after ingestion, the crop contracts, transferring digestive enzymes and partly digested food into the ventriculus. The anterior ventriculus is acid and has high carbohydrase activity, whereas the posterior ventriculus is alkaline and has high proteinase activity. This differentiation along the midgut may be an adaptation to the instability of ancestral carbohydrases in the presence of proteinases. The food bolus moves backward in the midgut of the insect by peristalsis. As soon as the polymeric food molecules are digested and become sufficiently small to pass through the peritrophic membrane, they diffuse with the digestive enzymes into the ectoperitrophic space (**Figure 1**). The enzymes and nutrients are then displaced toward the ceca with a countercurrent flux caused by secretion of fluid at the Malpighian tubules and its absorption back by cells at the ceca (**Figure 10**), where final digestion is completed and nutrient absorption occurs. When the insect starts a new meal, the ceca contents are moved into the crop. As a consequence of the countercurrent flux, digestive enzymes occur at a decreasing gradient in the midgut, and their excretion is lowered.

The Neoptera basic plan gave origin to that of the Polyneoptera orders, which include Dictyoptera (cockroaches, termites and mantids), Orthoptera and Phasmatodea, and evolved to the basic plans of Paraneoptera and Holometabola. The characteristics of the Paraneoptera ancestors cannot be inferred, because midgut function data are available only for Hemiptera.

The basic gut plan of the Holometabola (**Figure 10B**) (which include Coleoptera, Megaloptera, Hymenoptera, Diptera, and Lepidoptera) is similar to that of Neoptera, except that fluid secretion occurs by the posterior ventriculus, instead of by the Malpighian tubules. Because the posterior midgut fluid does not contain wastes, as is the case for Malpighian tubular fluid, the accumulation of wastes in the ceca is decreased. Ceca loss probably further decreases the accumulation of noxious substances in the midgut, which would be more serious in insects that have high relative food consumption rates, as is common among Holometabola.

The basic plan of Coleoptera did not evolve dramatically from the holometabolan ancestor, whereas the basic plan of the Hymenoptera, Diptera, and Lepidoptera ancestor (hymenopteran–panorpoid ancestor, **Figure 10C**) presents important differences. Thus, hymenopteran–panorpoid ancestors have countercurrent fluxes like holometabolan ancestors, but differ from them in the lack of crop digestion, midgut differentiation in luminal pH, and in which compartment is responsible for each phase of digestion. In holometabolan ancestors all phases of digestion occur in the endoperitrophic space (**Figure 1**), whereas in hymenopteran–panorpoid ancestors only initial digestion occurs in that region. In the latter ancestors, intermediate digestion is carried out by free enzymes in the ectoperitrophic space, and final digestion occurs at the midgut cell surface by immobilized enzymes. The free digestive enzymes do not pass through the PM, because they are larger than the PM's pores. As a consequence of the compartmentalization of digestive events in hymenopteran and panorpoid insects, there is an increase in the efficiency of digestion of polymeric food, as discussed previously.

The evolution of insect digestive systems summarized above and in **Figure 10** was proposed, as discussed above, from studies carried out in 12 species pertaining to 4 insect orders. To give further support to the hypothesis that the characteristics of gut function and morphology depend more on phylogeny than on diet, another approach was used. A total of 29 gut morphology and digestive physiology characteristics (e.g., luminal pH, ratio of gut section volumes, type of peritrophic membrane, presence of special gut cells, distribution of digestive enzymes along the gut, major proteinase) were identified in 23 species from 8 different insect orders. Making use of these characteristics, a cladogram was constructed, putting together the data from studied species (**Figure 11**). The data confirmed that the morphological and functional traits associated with the digestive system are more dependent on taxon than on dietary habits of the different insects (Dias, Vanin, Marques, and Terra, unpublished data). There are two insect species that do not apparently fit the model: *Anopheles* spp. and *Themos malaisei*. *Anopheles* spp. is an adult, whereas the other Diptera is larval. *T. malaisei* is an unexpected finding that will be discussed below (see section 11.9.2.7).

11.9.2. Digestion in the Major Insect Orders

The organization of the digestive process in the different insect orders has been reviewed several times (Terra, 1988, 1990; Terra and Ferreira, 1994, 2009). The following section is therefore an abridged version of those texts, highlighting new findings and trying to identify points that

deserve more research, especially in relation to molecular aspects. Only key references before 2000 are cited, and the reader will find more references in the above-mentioned reviews.

11.9.2.1. Dictyoptera   Dictyoptera comprises two suborders: Blattaria (cockroaches) and Mantodea (mantids). After extensive molecular phylogenetic analyses, Inward *et al.* (2007) showed that termites are social cockroaches, no longer deserving classification as a separate order (Isoptera) from cockroaches. Actually, termites pertain to a family (Termitidae) that is sister to that of the woodroach *Cryptocercus* (Cryptocercidae). The branch Cryptocercidae–Termitidae is a sister of Blattidae, forming Blattoidea that is a sister of Blaberoidea (Blattelidae plus Blaberidae), which in addition to Polyphagoidea form the Blattaria. Cockroaches are usually omnivorous. It is thought that digestion in cockroaches occurs as described for the Neoptera ancestor (**Figure 10A**), except that part of the final digestion of proteins occurs on the surface of midgut cells (Terra and Ferreira, 1994). This was confirmed by the finding in *P. americana* that most trypsin, maltase, and amylase is found in the crop, whereas aminopeptidase predominates in the microvillar membranes of posterior midgut. There is a decreasing gradient of trypsin, maltase, and amylase along

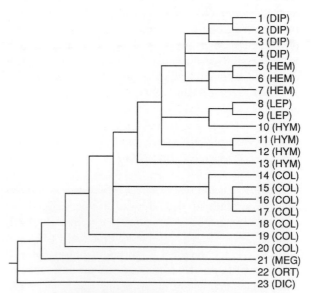

Figure 11 Cladogram of representative insects based on 29 gut morphology and digestive physiology characteristics. Insects: 1, *Trichosia pubescens*; 2, *Rhynchosciara americana*; 3, *Musca domestica*; 4, *Anopheles* spp.; 5, *Rhodnius prolixus*; 6, *Dysdercus peruvianus*; 7, *Acyrthosiphon pisum*; 8, *Erinnyis ello*; 9, *Spodoptera frugiperda*; 10, *Themos malaisei*; 11, *Camponotus rufipes*; 12, *Scaptotrigona bipunctata*; 13, *Bracon hebetor*; 14, *Tenebrio molitor*; 15, *Migdolus fryanus*; 16, *Sphenophorus levis*; 17, *Cyrtomon solana*; 18, *Dermestes maculatus*; 19, *Pyrearinus termitilluminans*; 20, *Pheropsophus aequinoctialis*; 21, *Corydalus* spp.; 22, *Abracris flavolineata*; 23, *Periplaneta americana*. Courtesy of A.B. Dias.

the midgut contents, and less than 5% of trypsin and maltase (amylase, 27%) is excreted during each midgut emptying. This suggests the existence of midgut digestive enzyme recycling, with amylase excretion increased probably due to excess dietary starch. The recycling mechanism is thought to be powered by water fluxes, as in the Neoptera ancestor, although there are no data supporting this. Major digestive proteinases are trypsin and chymotrypsin (Dias and Terra, unpublished data).

The differentiation of pH along the midgut (acid anterior midgut and alkaline posterior midgut) is not conserved among some cockroaches, like *P. americana*, (Blattidae) and *Blatella germanica* (Blatellidae), but is maintained in others, exemplified by the blaberid *Nauphoeta cinerea* (Elpidina *et al.*, 2001a; Vinokurov *et al.*, 2007). The organization of digestion in this cockroach seems similar to that in *P. americana*, although data on enzyme excretion are lacking. At least, blaberoid cockroaches possess proteinase inhibitory proteins active in the anterior midgut (Elpidina *et al.*, 2001b; Vinokurov *et al.*, 2007). These inhibitors are thought to be a primitive device to decrease the proteolytic inactivation of the animal's own carbohydrases, which are thus expected to be more active in the anterior midgut. The digestive carbohydrases from more evolved insects are stable in the presence of their own proteinases (Terra, 1988). Recently, several proteinase inhibitors have been partially purified from *N. cinerea* (Elpidina *et al.*, 2001b).

Another difference between cockroaches and the Neoptera ancestor is the enlargement of hindgut structures, noted mainly in wood-feeding cockroaches. These hindgut structures harbor bacteria producing acetate and butyrate from ingested wood or other cellulose-containing materials. Acetate and butyrate are absorbed by the hindgut of all cockroaches, but this is more remarkable with woodroaches (Terra and Ferreira, 1994). Cellulose digestion may be partly accomplished by bacteria in the hindgut of *P. americana* or protozoa in *Cryptocercus punctulatus*. Nevertheless, now it is clear that *P. americana* saliva contains two cellulases and three laminarinases that may open plant cells and lyse fungal cells (Genta *et al.*, 2003). This agrees with the omnivorous detritus-feeding habit of the insect. The woodroach *Panestria cribrata* also has its own cellulase (Scrivener *et al.*, 1989).

Termites may be seen as insects derived from woodroaches, and more adapted to dealing with refractory material such as wood and humus. Associated with this specialization, they lost the crop and midgut ceca, and enlarged their hindgut structures. Both lower and higher termites digest cellulose with their own cellulase, despite the occurrence of cellulose-producing protozoa in the paunch – an enlarged region of the anterior hindgut in lower termites. The products of cellulose digestion pass from the midgut into the hindgut, where they are converted into acetate and butyrate by hindgut bacteria, as in woodroaches. Symbiotic bacteria are also responsible for nitrogen fixation in the hindgut, resulting in bacterial protein. This is incorporated into the termite body mass after being expelled in feces by one individual, and being ingested and digested by another. This explains the ability of termites to develop successfully on diets very poor in protein. Both lower- and higher-feeding termites seem to have an endo–ectoperitrophic circulation of digestive enzymes (Terra and Ferreira, 1994; Nakashima *et al.*, 2002; see also section 11.4.3.1).

11.9.2.2. Orthoptera Grasshoppers feed mainly on grasses, and their digestive physiology has clearly evolved from the neopteran ancestor. Carbohydrate digestion occurs mainly in the crop, under the action of midgut enzymes, whereas protein digestion and final carbohydrate digestion take place at the anterior midgut ceca. The abundant saliva (devoid of significant enzymes) produced by grasshoppers saturates the absorbing sites in the midgut ceca, thus hindering the countercurrent flux of fluid. This probably avoids excessive accumulation of noxious wastes, coming from Malpighian tubule secretion, in the ceca, and makes possible the high relative food consumption observed among locusts in their migratory phases. Starving grasshoppers present midgut countercurrent fluxes. Cellulase found in some grasshoppers is believed to facilitate the access of digestive enzymes to the plant cells ingested by the insects, by degrading the cellulose framework of cell walls (Dow, 1986; Terra and Ferreira, 1994; Marana *et al.*, 1997).

Crickets are omnivorous or predatory insects with initial starch digestion occurring in their capacious crop and ending in ceca lumina. Regarding protein, initial trypsin digestion occurs mostly in ceca lumina, whereas final aminopeptidase digestion takes place in ceca and ventriculus. The emptying of ceca in some crickets is propelled by peristalsis, whereas in others it depends on the relative pressure produced by proventriculus and ceca. Differing from grasshoppers, the final digestion of both protein and carbohydrates depends on membrane-bound enzymes in addition to soluble ones. Both starving and feeding crickets present midgut countercurrent fluxes (Woodring and Lorenz, 2007; Biagio *et al.*, 2009).

11.9.2.3. Phasmatodea The stick and leaf insects are remarkable mimics of the stems and leaves on which they feed (Grimaldi and Engel, 2005). The Phasmatodea ancestors derived from the Neoptera ancestors. The stick insect lacks ceca, which were replaced in *Phibalosoma phyllinum* (Phasmatidae) by the anterior midgut as the site of fluid absorption. Countercurrent flux of fluid resembles that in grasshoppers in being detected only in starving animals. Initial digestion of protein and carbohydrates takes place in the crop. The final digestion of proteins and carbohydrates occurs along the ventriculus, under the action of a microvillar aminopeptidase and a soluble-glycocalyx-associated maltase, respectively (Monteiro, Tamaki, Terra, Ribeiro, unpublished data).

11.9.2.4. Hemiptera The Hemiptera comprise insects of the major suborders Auchenorrhyncha (cicadas, spittlebugs, leafhoppers, and planthoppers) and Sternorrhyncha (aphids and white flies), which feed almost exclusively on plant sap, and Heteroptera (e.g., assassin bugs, plant bugs, stink bugs, and lygaeid bugs), which are adapted to different diets.

The ancestor of the entire order Hemiptera is supposed to have been a cell-feeder similar to the present day phloem-feeder Auchenorrhyncha. Evolutionary shifts to phloem feeding or predation were not associated with a marked increase in body size. In contrast, xylem feeders are large to be able to extract xylem fluid from a host plant (Novotny and Wilson, 1997). Phloem and xylem sap have a very low protein content (with the exception of a few phloem saps; see below) and carbohydrate polymer content, and are relatively poor in free essential amino acids. In contrast to xylem sap, phloem sap is very rich in sucrose (Douglas, 2006). Thus, except for dimer (sucrose) hydrolysis, no food digestion is usually necessary in sap-suckers. Upon adapting to dilute phloem and/or xylem sap, hemipteran ancestors would lose the enzymes involved in initial and intermediate digestion, and lose the peritrophic membrane (**Figure 10F**). These changes are associated with the lack of luminal digestion.

The major problem facing a sap-sucking insect (especially on dilute phloem or xylem sap) is to absorb nutrients, such as essential amino acids, that are present in very low concentrations in sap. Whichever mechanism is used, xylem feeders may absorb as much as 99% of dietary amino acids and carbohydrate (Andersen *et al.*, 1989). Amino acids may be absorbed according to a hypothesized mechanism that depends on perimicrovillar membranes, which are membranes ensheathing the midgut microvilli with a dead end (**Figure 12**). A role in midgut amino acid absorption depends on the presence of amino acid–K^+ symports on the surface of the perimicrovillar membranes, and of amino acid carriers and potassium pumps on the microvillar membranes. Although amino acid carriers have been found in the microvillar membranes of several insects (Wolfersberger, 2000), no attempts have been made to study the other postulated proteins. Thus, in spite of the model providing an explanation for the occurrence of these peculiar cell structures in Hemiptera, it is supported only by: (1) evidence that amino acids are absorbed with potassium ions in *D. peruvianus* (Silva and Terra, 1994); and (2) occurrence of particles studying the cytoplasmic face of the midgut microvillar membranes of *D. peruvianus*. These might be ion pumps responsible for the putative potassium ion transport, like similar structures in several epithelia (Silva *et al.*, 1995).

Another problem that deserves more attention regarding perimicrovillar membranes is their origin. Immunolocalization of the perimicrovillar enzyme marker, α-glucosidase, suggests that these membranes are formed when double membrane vesicles fuse their outer membranes with the microvillar membranes and their inner membranes with the perimicrovillar membranes. A double-membrane Golgi cisterna (on budding) forms the double-membrane vesicles (Silva *et al.*, 1995).

Organic compounds in xylem sap need to be concentrated before they can be absorbed by the perimicrovillar system. This occurs in the filter chamber of Cicadoidea and Cercopoidea, which concentrates xylem sap 10-fold. The filter chamber consists of a thin-walled, dilated anterior midgut in close contact with the posterior midgut and the proximal ends of the Malpighian tubules. This arrangement enables water to pass directly from the anterior midgut to the Malpighian tubules, concentrating food in the midgut and eliminating excess water. The high permeability of the filter chamber membrane to water results from the occurrence of specific channels formed by proteins named aquaporins. These were characterized as 15- to 26-kDa membrane proteins, and were immunolocalized in the filter chamber of several xylem sap feeders (Le Cahérec *et al.*, 1997).

Sternorrhyncha, as exemplified by aphids, may suck, more or less continuously, phloem sap of sucrose concentration up to 1.0 M and osmolarity up to three times that of the insect hemolymph. This results in a considerable hydrostatic pressure caused by the tendency of water to move from the hemolymph into midgut lumen. To withstand these high hydrostatic pressures, aphids have developed several adaptations. Midgut stretching resistance is helped by the existence of links between apical lamellae (replacing usual midgut cell microvilli) that become less conspicuous along the midgut. As a consequence of the links between the lamellae, the perimicrovillar membranes could no longer exist and were replaced by membranes seen associated with the tips of the lamellae, the modified perimicrovillar membranes (Ponsen, 1991; Cristofoletti *et al.*, 2003). A modified perimicrovillar membrane-associated α-glucosidase frees fructose from sucrose without increasing the osmolarity by promoting transglycosylations. As the fructose is quickly absorbed the osmolarity decreases, resulting in a honeydew iso-osmotic with hemolymph (Ashford *et al.*, 2000; Cristofoletti *et al.*, 2003). Another interesting adaptation is observed in whiteflies, where a trehalulose synthase forms trehalulose from sucrose, thus making available less substrate for an α-glucosidase that otherwise would increase the osmolarity of ingested fluid on hydrolyzing sucrose (Salvucci, 2003).

A cathepsin L (see section 11.5.3) bound to the modified perimicrovillar membranes of *A. pisum* (Cristofoletti *et al.*, 2003) may explain the capacity of some phloem sap feeders to rely on protein found in some phloem saps

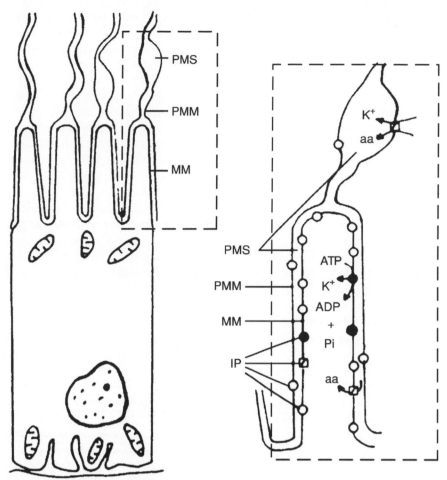

Figure 12 Model for the structure and physiological role of the microvillar border of midgut cells from Hemiptera. The left figure is a diagrammatic representation of a typical Hemiptera midgut cell, and the right figure details its apex. The microvillar membrane (MM) is ensheathed by the perimicrovillar membrane (PMM), which extends toward the luminal compartment with a dead end. The microvillar and perimicrovillar membranes delimit a closed compartment, i.e., the perimicrovillar space (PMS). The microvillar membrane is rich, and the perimicrovillar membrane poor, in integral proteins (IP). Microvillar membranes actively transport potassium ions (the most important ion in sap) from PMS into the midgut cells, generating a concentration gradient between the gut luminal sap and the PMS. This concentration gradient may be a driving force for the active absorption of organic compounds (amino acids, for example) by appropriate carriers present in the PMM. Organic compounds, once in the PMS, may diffuse up to specific carriers on the microvillar surface. This movement is probably enhanced by a transfer of water from midgut lumen to midgut cells, following (as solvation water) the transmembrane transport of compounds and ions by the putative carriers. Reprinted with permission from Terra, W.R., Ferreira, C., 1994. Insect digestive enzymes: properties, compartmentalization and function. *Comp. Biochem. Physiol*. B 109, 1–62; ©Elsevier.

(Salvucci *et al.*, 1998), and the failure of other authors to find an active proteinase in sap feeders. These other authors worked with homogenate supernatants or supernatants of Triton X-100-treated samples, under which conditions the cathepsin L would remain in the pellet.

An aminopeptidase, also bound to the modified perimicrovillar membranes, is the major binding site of the lectin Concanavalin A. On binding, the lectin impairs aphid development, in spite of the fact that the lectin does not affect aminopeptidase activity. It is thought that the aminopeptidase is located near the amino acid carriers responsible for amino acid absorption, and that these are inhibited when the lectin binds to the aminopeptidase (Cristofoletti *et al.*, 2006).

Amino acid absorption in *A. pisum* midguts is influenced by the presence of the bacteria *Buchnera* in the mycetocytes of the mycetomes occurring in the aphid hemocoel (Douglas, 2006). The molecular mechanisms underlying this phenomenon are not known, in spite of the fact that there is strong evidence showing that *Buchnera* uses the non-essential amino acids absorbed by the host in the synthesis of essential amino acids (Shigenobu *et al.*, 2000; Douglas, 2006).

The evolution of Heteroptera was associated with regaining the ability to digest polymers. Because the appropriate digestive enzymes were lost, they instead used enzymes derived from lysosomes. Lysosomes are cell organelles involved in intracellular digestion carried out by special proteinases referred to as cathepsins. Compartmentalization of digestion was maintained by the perimicrovillar membranes, as a substitute for the absent peritrophic membrane. Digestion in the two major Heteroptera taxa, Cimicomorpha (exemplified by the blood-feeder *R. prolixus*), and Pentatomorpha (exemplified by the seed-sucker *D. peruvianus*), are similar. The dilated anterior midgut stores food and absorbs water, and, at least in *D. peruvianus*, also absorbs glucose, which is transported with the aid of a uniporter (GLUT) and a K^+-glucose symporter (SGLT) (Bifano *et al.*, 2010). Digestion of proteins and absorption of amino acids occurs in the posterior ventriculus. Most protein digestion occurs in the lumen with the aid of a cysteine proteinase, and ends in the perimicrovillar space under the action of aminopeptidases and dipeptidases (Terra and Ferreira, 1994). Symbiont bacteria may occur in blood-feeders, putatively to provide vitamins (see section 11.2.5). At least in *R. prolixus*, the neuroendocrine system has factors important for maintaining the ultrastructural organization of the midgut epithelial cells (Gonzales *et al.*, 1998).

11.9.2.5. Megaloptera

The Megaloptera includes alderflies and dobsonflies, and is often considered to be the most primitive group of insects with complete metamorphosis. All their larvae are aquatic predators feeding on invertebrates (Theischinger, 1991). Megaloptera ancestors are like Holometabola ancestors, except for the anterior midgut ceca, which were lost and replaced in function by the anterior midgut. Thus, in *Corydalus* spp. larvae, most digestion occurs in the crop under the action of soluble amylase, maltase, aminopeptidase, and trypsin (major proteinase). Digestive enzyme recycling should occur, as less than 3.3% of midgut amylase, maltase, and aminopeptidase are lost at each midgut emptying. The higher excretory rate of trypsin (27%) probably results from excess dietary protein (Dias and Terra, unpublished data).

11.9.2.6. Coleoptera

Coleoptera ancestors are like those of Megaloptera. Nevertheless, there are evolutionary trends leading to a great reduction or loss of the crop, and, as in the panorpoid orders, occurrence of at least final digestion of proteins at the surface of midgut cells. Thus, in predatory Carabidae most of the digestive phases occur in the crop by means of midgut enzymes, whereas in predatory larvae of Elateridae initial digestion occurs extra-orally by the action of enzymes regurgitated onto their prey. The preliquified material is then ingested by the larvae, and its digestion is finished at the surface of midgut cells (Terra and Ferreira, 1994).

Initial digestion of glycogen and proteins occurs in the dermestid larval endoperitrophic space, which is limited by a peritrophic gel in the anterior midgut and a peritrophic membrane in the posterior midgut. Final digestion takes place at the midgut cell surface, in the anterior and posterior midgut in the case of glycogen and proteins, respectively. There is a decreasing gradient along the midgut of amylase and trypsin (major proteinase), suggesting the occurrence of digestive enzyme recycling (Terra and Ferreira, 1994; Caldeira *et al.*, 2007). Thus, dermestid beetles digest keratin with serine proteinases. Keratins are the major protein components of wool, hair, horn, and feathers, and have peptide chains cross-linked by disulfide bonds to give a very stable structure. These bonds are reduced only in a very reducing (−100 to −300 mV) environment, characteristic of the midgut of these insects (Vonk and Western, 1984).

Like dermestid beetles, the larva of *Sphenophorus levis* (Curculionidae) has a peritrophic gel and a peritrophic membrane in the anterior and posterior midgut, respectively, and a decreasing gradient of amylase, maltase, and proteinase (cathepsin L) along the midgut. A microvillar protein carries out the final digestion of proteins at the posterior midgut (Soares-Costa *et al.*, 2011). The spatial organization of digestion in the larvae of *Migdolus fryanus* (Cerambicidae) is similar to that of *S. levis*, except that their major digestive proteinase is trypsin (Dias and Terra, unpublished data). Tenebrionid larvae also have aminopeptidase as a microvillar enzyme, and the distribution of enzymes in gut regions of adults is similar to that in the larvae (Terra and Ferreira, 1994). This suggests that the overall pattern of digestion in larvae and adults of Coleoptera is similar, despite the fact that, in contrast to adults, beetle larvae usually lack a crop.

Insects of the series Cucujiformia (which includes Tenebrionidae, Chrysomelidae, Bruchidae, and Curculionidae) have cysteine proteinases (see section 11.5.3) in addition to (or in place of) serine proteinases as digestive enzymes, suggesting that the ancestors of the whole taxon were insects adapted to feed on seeds rich in serine proteinase inhibitors (Terra and Ferreira, 1994). The occurrence of trypsin as the major proteinase in *M. fryanus* (Dias and Terra, unpublished data) confirmed the preliminary work (Murdock *et al.*, 1987), according to which cerambycid larvae reacquired digestive serine proteinases.

Scarabaeidae and several related families are relatively isolated in the series Elateriformia, and evolved considerably from the Coleoptera ancestor. Scarabid larvae, exemplified by dung beetles, usually feed on cellulose materials undergoing degradation by a fungus-rich flora. Digestion occurs in the midgut, which has three rows of ceca, with a ventral groove between the middle and posterior row. The alkalinity of gut contents increases to almost pH 12 along the midgut ventral groove. This high pH probably enhances cellulose digestion, which occurs mainly in the hindgut fermentation chamber, through the action of endogenous and bacterial cell-bound enzymes. The final product of cellulose degradation is mainly acetic acid, which is absorbed through the hindgut wall (Terra and Ferreira, 1994; Huang *et al.*, 2010). There is controversy as to whether scarabid larvae ingest feces to obtain nitrogen compounds, as described above for termites. Nevertheless, this is highly probable on the grounds that the microbial biomass in the fermentation chamber is incorporated into the larval biomass (Li and Brune, 2005).

Keratin beetles, as *Trox sp.* (Scarabaeoidea: Trogidae), digest keratin with serine proteinases (Hughes and Vogler, 2006) like dermestids.

11.9.2.7. Hymenoptera Organization of the digestive process is variable among hymenopterans, and to understand its peculiarities it is necessary to review briefly their evolution. The hymenopteran basal lineages are phytophagous as larvae, feeding both ecto- and endophytically, and include several superfamilies such as Xyeloidea and Tenthredinoidea, all known as sawflies. Close to these are the Siricoidea (wood wasps), which are adapted to ingest fungus-infected wood. Wood wasp-like ancestors gave rise to the Apocrita (wasp-waisted Hymenoptera), which are parasitoids of insects. They use their ovipositor to injure or kill their host, which represents a single meal for their complete development. A taxon sister of Ichneumonoidea in Apocrita gave rise to Aculeata (bees, ants, and wasps with thin waists) (Quicke, 2003).

The digestive systems of Hymenoptera ancestors are like the panorpoid ancestors (**Figure 10C**). However, there are evolutionary trends leading to the loss of midgut ceca (replaced in function by the anterior midgut) and changes in midgut enzyme compartmentalization. These trends appear to be associated with the development of parasitoid habits, and were maintained in Aculeata, as described below.

The sawfly *T. malaisei* (Tenthredinoidea: Argidae) larva has a midgut with a ring of anterior ceca that forms a U at the ventral side. Luminal pH is above 9.5 in the first two-thirds of the midgut. Trypsin (major proteinase) and amylase have a decreasing activity along the endoperitrophic space, suggesting enzyme recycling. Maltase predominates in the anterior midgut tissue as a soluble glycocalyx-associated enzyme, whereas aminopeptidase is a microvillar enzyme in the posterior midgut (Dias, Ribeiro, and Terra, unpublished data). These characteristics (except the presence of ceca) are similar to those of lepidopteran larvae (see section 11.9.2.10), and explain the fact that this insect is close to the lepidopterans in **Figure 11**. Otherwise, Aculeata, with their less sophisticated midgut (see below) branches closer to coleopterans (**Figure 11**).

Wood wasp larvae of the genus *Sirex* are believed to be able to digest and assimilate wood constituents by acquiring cellulase and xylanase, and possibly other enzymes, from fungi present in the wood on which they feed (Martin, 1987). The larvae of Apocrita present a midgut which is closed at its rear end, and which remains unconnected with the hindgut until the time of pupation. It is probable that this condition evolved as an adaptation of

endoparasitoid Apocrita ancestors to avoid the release of toxic compounds into the host in which they lived (Terra, 1988).

In larval bees, most digestion occurs in the endoperitrophic space. Countercurrent fluxes seem to occur, but there is no midgut luminal pH gradient. Adult bees ingest nectar and pollen. Sucrose from nectar is hydrolyzed in the crop by the action of a sucrase from the hypopharyngeal glands. After ingestion, pollen grains extrude their protoplasm in the ventriculus, where digestion occurs. As in larvae, there is also evidence of an endo–ectoperitrophic circulation of digestive enzymes (Terra and Ferreira, 1994).

Although many authors favor the view that pollen grains are digested in bees after their extrusion by osmotic shock, this subject is controversial not only in bees, but also among pollen-feeder beetles (Human and Nicholson, 2003).

Digestion in larval ants is similar to that in larval bees (Erthal *et al.*, 2007), whereas worker ants feed on nectar, honeydew, and plant exudates, acquiring the necessary amino acids for growth with the aid of microsymbionts (Cook and Davidson, 2006). Worker ants may also feed on partly digested food regurgitated by their larva. Thus, they have frequently been said to lack digestive enzymes or display only those enzymes associated with intermediate and (or) final digestion (Terra and Ferreira, 1994). Although this seems true for leaf-cutting ants (Erthal *et al.*, 2007), which appear to rely only on monosaccharides produced by fungal enzymes acting on plant polysaccharides (Silva *et al.*, 2003), it is not widespread. Thus, adult *Camponotus rufipes* (Formicinae) have soluble amylase, trypsin (major proteinase), maltase, and aminopeptidase enclosed in a type I PM in their midguts. As only 14% of amylase and less than 7% of the other digestive enzymes are excreted during the midgut emptying, these insects may have a digestive enzyme recycling mechanism (Dias and Terra, unpublished data).

11.9.2.8. Diptera The Diptera evolved along two major lines: an assemblage (early Nematocera) of suborders corresponding to the mosquitoes, including the basal Diptera, and the suborder Brachycera, which includes the most evolved flies (Cyclorrhapha). The dipteran ancestor is similar to the panorpoid ancestor (**Figure 10C**) in having the enzymes involved in intermediate digestion free in the ectoperitrophic fluid (mainly in the large ceca), whereas the enzymes of terminal digestion are membrane bound at the midgut cell microvilli (Terra and Ferreira, 1994). These characteristics correspond to those of a nematoceran sciarid larvae, *R. americana* (Terra and Ferreira, 1994). As expected, the digestive process in larval mosquitoes follows a similar pattern, according to a microarray-based analysis of *A. gambiae* midgut transcripts (Oviedo *et al.*, 2008).

Non-hematophagous adults store liquid food (nectar or decay products) in their crops. Digestion occurs in their midgut, as in larvae. Nectar ingested by mosquitoes (males and females) is stored in the crop, and is digested and absorbed at the anterior midgut. Blood, which is ingested only by females, passes to the posterior midgut, where it is digested and absorbed (Billingsley, 1990; Terra and Ferreira, 1994).

The adult *Ae. aegypti* midgut surface is covered, in large part, by tubular bilayers with a diameter four-fold smaller than the microvilli. They fuse and branch, forming bundles that seem to originate in the intercellular crypts and appear to be fused with the microvillar surface (Ziegler *et al.*, 2000). These structures are not related with the perimicrovillar membranes of Hemiptera. The latter envelop the microvilli before extending into the lumen in structures that may resemble the tubular membrane bilayers of *Ae. aegypti* (see section 11.9.2.4 and **Figure 12**). The puzzling structures of *Ae. aegypti* should be further studied to discover their relationships with digestion.

Gall midges (Cecidomyiidae) are interesting because they manipulate plant growth, arguably with small secretory proteins that lack matches in gene sequence databanks (Zhang *et al.*, 2010).

The Cyclorrhapha ancestor (**Figure 10D**) evolved dramatically from the panorpoid ancestor (**Figure 10C**), apparently as a result of adaptations to a diet consisting mainly of bacteria. Digestive events in Cyclorrhapha larvae are exemplified by larvae of the housefly *M. domestica*. These ingest food rich in bacteria. In the anterior midgut there is a decrease in the starch content of the food bolus, facilitating bacterial death. The bolus now passes into the middle midgut, where bacteria are killed by the combined action of low pH, a special lysozyme (see section 11.4.5), and a cathepsin D-like proteinase (see section 11.5.4). Finally, the material released by bacteria is digested in the posterior midgut, as is observed in the whole midgut of insects of other taxa. Countercurrent fluxes occur in the posterior midgut, powered by secretion of fluid in the distal part of the posterior midgut and its absorption back into the middle midgut. The middle midgut has specialized cells for buffering the luminal contents in the acidic zone (**Figure 4**), in addition to those functioning in fluid absorption. Cyclorrhaphan adults, except for a few blood-suckers, feed mainly on liquids associated with decaying material (rich in bacteria), in a way similar to housefly *M. domestica* adults. These salivate (or regurgitate their crop contents) onto their food. After the dispersed material is ingested, starch digestion is accomplished primarily in the crop by the action of salivary amylase. Digestion follows in the midgut, essentially as described for larvae (Terra and Ferreira, 1994).

The stable fly *Stomoxys calcitrans* stores and concentrates the blood meal in the anterior midgut, and gradually passes it to the posterior midgut, where digestion takes place, resembling what occurs in larvae. These adults lack the characteristic cyclorrhaphan middle midgut and the associated luminal low pH. Stable flies occasionally take nectar (Jordão *et al.*, 1996a).

11.9.2.9. Lepidoptera   Lepidopteran ancestors (**Figure 10E**) differ from panorpoid ancestors because they lack midgut ceca, have all their digestive enzymes (except those of initial digestion) immobilized at the midgut cell surface, and present long-necked goblet cells and stalked goblet-cells in the anterior and posterior larval midgut regions, respectively. Goblet cells excrete K^+ ions that are absorbed from leaves ingested by larvae. Goblet cells also seem to assist anterior columnar cells in water absorption, and posterior columnar cells in water secretion (Terra and Ferreira, 1994; Ortego *et al.*, 1996).

Although most lepidopteran larvae have a common pattern of digestion, species that feed on unique diets generally display some adaptations. *Tineola bisselliella* (Tineidae) larvae feed on wool, and display a highly reducing midgut for cleaving the disulfide bonds in keratin to facilitate proteolytic hydrolysis of this otherwise insoluble protein (Terra and Ferreira, 1994). Similar results were obtained with *Hofmannophila pseudospretella* (Christeller, 1996). Wax moths (*Galleria mellonella*) infest beehives, and digest and absorb wax. The participation of symbiotic bacteria in this process is controversial. Another adaptation has apparently occurred in lepidopteran adults, which feed solely on nectar. Digestion of nectar only requires the action of an α-glucosidase (or a β-fructosidase) to hydrolyze sucrose, the major component present. Many nectar-feeding lepidopteran adults have amylase in salivary glands, and several glycosidases and peptidases in the midgut (Terra and Ferreira, 1994).

Woods and Kingsolver (1999) developed a chemical reactor model of the caterpillar midgut, and used the model as a framework for generating hypotheses about the relationship between feeding responses to variable dietary proteins, and the physical and biochemical events in the midgut and body. They concluded that absorption (or post-absorptive processes) was limiting in a caterpillar maintained on artificial diets. Caterpillars eating leaves may not have the same limiting step, and this deserves a similar detailed study. Another interesting study would be the development of a model for the beetle midgut. This would determine whether beetles have digestion or consumption as the limiting step to compensate for their less sophisticated midguts.

11.10. Digestive Enzyme Secretion Mechanisms

Digestive enzyme secretory mechanisms and control are probably the least understood areas in insect digestion. Studies of secretory mechanisms have only described major differences, which seem to include unique aspects not seen in other animals.

Insects are continuous (e.g., Lepidoptera and Diptera larvae) or discontinuous (e.g., predators and many hematophagous insects) feeders. The synthesis and secretion of digestive enzymes in continuous feeders seem to be constitutive – that is, they occur continuously (at least between molts) – whereas in discontinuous feeders they are regulated (Lehane *et al.*, 1996). Digestive enzymes, as with all animal proteins, are synthesized in the rough endoplasmic reticulum and processed in the Golgi complex, and are packed into secretory vesicles (**Figure 13**). There are several mechanisms by which the contents of the secretory vesicles are freed in the midgut lumen. In holocrine secretion, secretory vesicles are stored in the cytoplasm until they are released, at which time the whole secretory cell is lost to the extracellular space. During exocytic secretion, secretory vesicles fuse with the midgut cell apical membrane, emptying their contents without any loss of cytoplasm (**Figure 13A**). In contrast, apocrine secretion involves the loss of at least 10% of the apical cytoplasm following the release of secretory vesicles (**Figure 13B**). These have previously undergone fusions, leading to larger vesicles that, after release, eventually free their contents by solubilization (**Figure 13B**). When the loss of cytoplasm is very small, the secretory mechanism is called microapocrine. Microapocrine secretion consists in releasing budding double-membrane vesicles (**Figure 13C**) or, at least in insect midguts, pinched-off vesicles that may contain a single or several secretory vesicles (**Figure 13D**). In both cases, the secretory vesicle contents are released by membrane fusion and/or by membrane solubilization caused by high pH contents or by luminal detergents.

The secretory mechanisms of insect midgut cells reviewed below are based on immunocytolocalization data, or on data combining biochemical procedures and electron micrographs. Studies based only on traditional cytology have been reviewed elsewhere (Terra and Ferreira, 1994; Lehane *et al.*, 1996).

Holocrine secretion is usually described on histological grounds mainly in midgut of insects other than higher Holometabola. These insects have large numbers of regenerative cells in their midguts. Thus, it is probable that cell renewal in these insects is being misinterpreted as holocrine secretion (Terra and Ferreira, 1994). In spite of this, immunocytochemical data showed that trypsin-containing vesicles along with cell organelles are discharged by opaque zone cells of adult stable flies, suggesting holocrine secretion (Jordão *et al.*, 1996a).

Exocytic, apocrine, and microaprocrine secretory mechanisms depend largely on midgut regions. Digestive

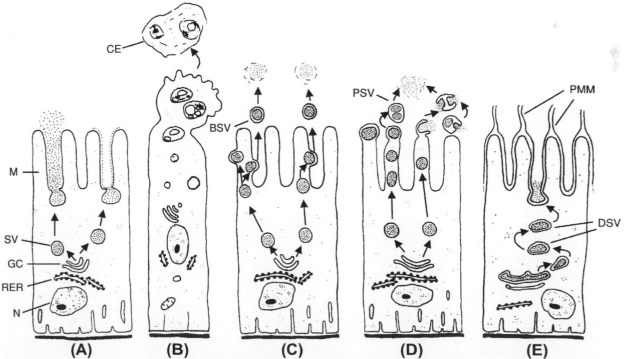

Figure 13 Models for secretory processes of insect digestive enzymes. (A) Exocytic secretion; (B) apocrine secretion; (C) microapocrine secretion with budding vesicles; (D) microapocrine secretion with pinched-off vesicles; (E) modified exocytic secretion in hemipteran midgut cell. BSV, budding secretory vesicle; CE, cellular extrusion; DSV, double-membrane secretory vesicle; GC, Golgi complex; M, microvilli; N, nucleus; PMM, perimicrovillar membrane; PSV, pinched-off secretory vesicle; RER, rough endoplasmatic reticulum; SV, secretory vesicle. Reprinted with permission from Terra, W.R., Ferreira, C., 2009. Digestive system. In: Resh, V.H., Cardé, R.T. (eds.), *Encyclopedia of Insects*, 2nd edn. Academic Press, San Diego, CA, pp. 273–281; ©Elsevier.

enzymes are usually secreted by exocytosis in the posterior midgut, whereas alternate mechanisms may be observed in anterior midgut. Thus, trypsin is secreted by the posterior midgut of adult mosquitoes (Graf *et al.*, 1986), larval flies (Jordão *et al.*, 1996b), and caterpillars (Jordão *et al.*, 1999) by exocytosis, as well as β-glycosidase by *T. molitor* middle midguts (Ferreira *et al.*, 2002). Trypsin is secreted by the anterior midgut of caterpillars using a microapocrine route (Santos *et al.*, 1986; Jordão *et al.*, 1999), whereas in the anterior midgut of *T. molitor* amylase secretion occurs by an apocrine mechanism (Cristofoletti *et al.*, 2001). Based only on morphological evidence, it appears that, in addition to *E. ello* and *Spodoptera frugiperda*, microapocrine secretion occurs in other lepidopteran species, such as *Manduca sexta*, whereas apocrine secretion is observed in some Orthoptera and in many coleopteran species other than *T. molitor* (Terra and Ferreira, 1994).

Immunocytolocalization data (Silva *et al.*, 1995) showed that secretion by hemipteran midgut cells displays special features, as the cells have perimicrovillar membranes in addition to microvillar ones (**Figure 13E**). In this case, double-membrane vesicles bud from modified (double membrane) Golgi structures (**Figure 13E**). The double-membrane vesicles move to the cell apex, their outer membranes fuse with the microvillar membrane, and their inner membranes fuse with the perimicrovillar membranes, emptying their contents (**Figure 13E**).

The molecular mechanisms underlying the insect midgut secretory processes are unknown. Nevertheless, there is suggestive evidence involving calmodulin, annexin, and midgut-specific gelsolin in the unique microapocrine process (Ferreira *et al.*, 2007). This area of research deserves more effort, because it may provide insights regarding new insect control procedures.

11.11. Concluding Remarks

In spite of numerous gaps demanding further research, already indicated in this chapter, it is clear that insect digestive biochemistry is becoming a developed science and that its methods are powerful enough to lead to steady progress. It is conceivable that, in the next few decades, knowledge of the structural biology and function of digestive enzymes, and of the control of expression of alternate digestive enzymes and their secretory mechanisms, as well as on microvillar biochemistry, will support the development of more effective and specific methods of insect control. According to a Brazilian proverb, "*Quem viver, verá*" – "Whoever is alive, will see."

Acknowledgments

Our work was supported by Brazilian research agencies Fundação de Amparo a Pesquisa do Estado de São Paulo (FAPESP) (Temático program) and CNPq. The authors are staff members of the Biochemistry Department, research fellows of CNPq, and members of the INCT-Entomologia molecular.

References

Alves, L. C., Almeida, P. C., Franzoni, L., Juliano, L., & Juliano, M. A. (1996). Synthesis of N-a-protected aminoacyl 7-amino-4-methyl-coumarin amide by phosphorus oxychloride and preparation of specific fluorogenic substrates for papain. *Peptide Res.*, *9*, 92–96.

Andersen, P. C., Brodbeck, B. V., & Mizell, R. F. (1989). Metabolism of aminoacids, organic acids and sugars extracted from the xylem fluid of four host plants by adult *Homalodisca coagulata*. *Entomol. Exp. Appl.*, *50*, 149–159.

Angelucci, C., Barrett-Wilt, G. A., Hunt, D. F., Akhurst, R. J., East, P. D., Gordon, K. H. J., & Campbell, P. M. (2008). Diversity of aminopeptidases derived from four lepidopteran gene duplications, and polycalins expressed in the midgut of *Helicoverpa armigera*: Identification of proteins binding the δ-endotoxin, Cry 1AC of *Bacillus thuringiensis*. *Insect Biochem. Mol. Biol.*, *38*, 685–696.

Appel, H. M. (1994). The chewing herbivore gut lumen: Physicochemical conditions and their impact on plant nutrients, allelochemicals, and insect pathogens. In E. A. Bernays (Ed.), *Insect–Plant Interactions* (Vol. 5, pp. 203–223). Boca Raton, FL: CRC Press.

Appel, H. M., & Martin, M. M. (1990). Gut redox conditions in herbivorous lepidopteran larvae. *J. Chem. Ecol.*, *16*, 3277–3290.

Applebaum, S. W. (1985). Biochemistry of digestion. In G. A. Kerkut, & L. I. Gilbert (Eds.), *Comprehensive Insect Physiology, Biochemistry, and Pharmacology* (Vol. 4, pp. 279–312). Oxford, UK: Pergamon.

Arakane, Y., & Muthukrishan, S. (2010). Insect chitinase and chitinase-like proteins. *Cell. Mol. Life Sci.*, *67*, 201–216.

Arakane, Y., Zhu, Q., Matsumiya, M., Muthukrishnan, S., & Kramer, K. J. (2003). Properties of catalytic, linker and chitin-binding domains of insect chitinase. *Insect Biochem. Mol. Biol.*, *33*, 631–648.

Arakane, Y., Muthukrishnan, S., Kramer, K. J., Specht, C. A., Tomoyasu, Y., et al. (2005). The *Tribolium* chitin synthase genes TcCHS1 and TcCHS2 are specialized for synthesis of epidermal cuticle and midgut peritrophic matrix. *Insect Molec. Biol.*, *14*, 453–463.

Arakawa, G., Watanabe, H., Yamasaki, H. Y., Meekawa, H., & Tokuda, G. (2009). Purification and molecular cloning of xylanases from the wood-feeding termite, *Coptotermes formosanus* Shiraki. *Biosci. Biotechnol. Biochem.*, *73*, 710–718.

Arrese, E. L., Patel, R. T., & Soulages, J. L. (2006). The main triglyceride-lipase from the insect fat body is an active phospholipase A1: Identification and characterization. *J. Lipid Res.*, *47*, 2656–2667.

Ashford, D. A., Smith, W. A., & Douglas, A. E. (2000). Living on a high sugar diet: The fate of sucrose ingested by a phloem-feeding insect, the pea aphid *Acyrthosiphon pisum*. *J. Insect Physiol.*, *46*, 335–341.

Azevedo, T. M., Terra, W. R., & Ferreira, C. (2003). Purification and characterization of three β-glycosidases from midgut of the sugar cane borer, *Diathraea saccharalis*. *Insect Biochem. Mol. Biol.*, *33*, 81–92.

Bacic, A., Harris, P. J., & Stone, B. A. (1988). Structure and function of plant cell walls. In P. K. Stumpf, & E. E. Conn (Eds.), *Biochemistry of Plants* (Vol. 14, pp. 297–371). New York, NY: Academic Press.

Baker, J. E., & Woo, S. M. (1981). Properties and specificities of a digestive aminopeptidase from larvae of *Attagenus megatoma* (Coleoptera: Dermestidae). *Comp. Biochem. Physiol. B*, *69*, 189–193.

Barbehenn, R. V. (1999). Non-absorption of ingested lipophilic and amphiphilic allelochemicals by generalist grasshoppers: The role of extractive ultrafiltration by the peritrophic envelope. *Arch. Insect Biochem. Physiol.*, *42*, 130–137.

Barbehenn, R. V., & Martin, M. M. (1998). Formation of insoluble and colloidally dispersed tannic acid complexes in the midgut fluid of *Manduca sexta* (Lepidoptera: Sphingidae): An explanation for the failure of tannic acid to cross the peritrophic envelopes of Lepidopteran larvae. *Arch. Insect Biochem. Physiol.*, *39*, 109–117.

Barbehenn, R. V., Poopat, U., & Spencer, B. (2003). Semiquinone and ascorbyl radicals in the gut fluids of caterpillars with EPR spectrometry. *Insect Biochem. Mol. Biol.*, *33*, 125–130.

Barbehenn, R. V., Jaros, A., Lee, G., Mozola, C., Weir, Q., & Salminen, J. -P. (2009). Hydrolyzable tannins as "qualitative defences": Limited impact against *Lymantria dispar* caterpillars on hybrid poplar. *J. Insect Physiol.*, *55*, 297–304.

Barrett, A. J., Rawlings, N. D., & Wasner, J. F. (1998). *Handbook of Proteolytic Enzymes*. London, UK: Academic Press.

Bayés, A., Comellas-Bigler, M., de la Vega, M. R., Maskos, K., Bode, W., et al. (2005). Structural basis of the resistance of an insect carboxypeptidase to plant protease inhibitors. *Proc. Natl. Acad. Sci., USA*, *102*, 16602–16607.

Bayyareddy, K., Andacht, T. M., Abdullah, M. A., & Adang, M. J. (2009). Proteomic identification of *Bacillus thuringiensis* subsp. *israelensis* toxin Cry4Ba binding proteins in midgut membranes from *Aedes* (*Stegomyia*) *aegypti* Linnaeus (Diptera: Culicidae) larvae. *Insect Biochem. Mol. Biol.*, *39*, 279–286.

Bement, W. M., & Mooseker, M. S. (1996). The cytoskeleton of the intestinal epithelium: Components, assembly, and dynamic rearrangements. In J. E. Hesketh, & J. F. Pryme (Eds.), *The Cytoskeleton: A Multi-volume Treatise* (3, pp. 359–404). Greenwich, CT: JAI Press.

Berenbaum, M. (1980). Adaptive significance of midgut pH in larval Lepidoptera. *Am. Natural.*, *115*, 138–146.

Bernays, E. A., & Chamberlain, D. J. (1980). A study of tolerance of ingested tannin in *Schistocerca gregaria*. *J. Insect Physiol.*, *26*, 415–420.

Bernays, E. A., Chamberlain, D. J., & Leather, E. M. (1981). Tolerance of acridids to ingested condensed tannin. *J. Chem. Ecol.*, *7*, 247–256.

Biagio, F. P., Tawaki, F. K., Terra, W. R., & Ribeiro, A. F. (2009). Digestive morphophysiology of *Gryllodes sigillatus* (Orthoptera: Gryllidae). *J. Insect Physiol.*, *55*, 1125–1133.

Bieth, J., Spiess, B., & Wermuth, C. G. (1974). The synthesis and analytical use of a highly sensitive and convenient substrate of elastase. *Biochem. Med.*, *11*, 350–357.

Bifano, T. D., Alegria, T. G.P., & Terra, W. R. (2010). Transporters involved in glucose absorption in the *Dysdercus peruvianus* (Hemiptera: Pyrrhocoridae) anterior midgut. *Comp. Biochem. Physiol.*, *157 B*, 1–9.

Billingsley, P. F. (1990). The midgut ultrastructure of hematophagous insects. *Annu. Rev. Entomol.*, *35*, 219–248.

Blake, J. D., Murphy, P. T., & Richards, G. N. (1971). Isolation and A/B classification of hemicelluloses. *Carbohydr. Res.*, *16*, 49–57.

Blanco-Labra, A., Martinez-Gallardo, N. A., Sandoval-Cardoso, L., & Delano-Frier, J. (1996). Purification and characterization of a digestive cathepsin D proteinase isolated from *Tribolium castaneum* larvae (Herbst). *Insect Biochem. Mol. Biol.*, *26*, 95–100.

Bolognesi, R., Ribeiro, A. F., Terra, W. R., & Ferreira, C. (2001). The peritrophic membrane of *Spodoptera frugiperda*: Secretion of peritrophins and role in immobilization and recycling digestive enzymes. *Arch. Insect Biochem. Physiol.*, *47*, 62–75.

Bolognesi, R., Terra, W. R., & Ferreira, C. (2008). Peritrophic membrane role in enhancing digestive efficiency. Theoretical and experimental models. *J. Insect Physiol.*, *54*, 1413–1422.

Borhegyi, N. H., Molnár, K., Csikós, G., & Sass, M. (1999). Isolation and characterization of an apically sorted 41-kDa protein from the midgut of tobacco hornworm (*Manduca sexta*). *Cell Tissue Res.*, *297*, 513–525.

Botos, I., Meyer, E., Nguyen, M., Swanson, S. M., Koomen, J. M., et al. (2000). The structure of an insect chymotrypsin. *J. Mol. Biol.*, *298*, 895–901.

Bown, D. P., Wilkinson, H. S., & Gatehouse, J. A. (1997). Differentially regulated inhibitor-sensitive and insensitive protease genes from the phytophagous insect pest, *Helicoverpa armigera*, are members of complex multigene families. *Insect Biochem. Mol. Biol.*, *27*, 625–638.

Bown, D. P., Wilkinson, H. S., & Gatehouse, J. A. (1998). Midgut carboxypeptidase from *Helicoverpa armigera* (Lepidoptera: Noctuidae) larvae: Enzyme characterization, cDNA cloning and expression. *Insect Biochem. Mol. Biol.*, *28*, 739–749.

Bozic, N., Ivanovic, J., Nenadovic, V., Berdstrom, J., Larsson, T., & Vujcic, Z. (2008). Purification and properties of a major leucyl aminopeptidase of *Morimus funereus* (Coleoptera: Cerambycidae) larvae. *Comp. Biochem. Physiol. B*, *149*, 454–462.

Bragato, I., Genta, F. A., Ribeiro, A. F., Terra, W. R., & Ferreira, C. (2010). Characterization of a β-1,3-glucanase active in the alkaline midgut of *Spodoptera frugiperda* larvae and its relation to β-glucan-binding proteins. *Insect Biochem. Mol. Biol.*, *40*, 861–872.

Brioschi, D., Nadalini, L. D., Bengtson, M. H., Sogayar, M. C., Moura, D. S., & Silva-Filho, M. C. (2007). General upregulation of *Spodoptera frugiperda* trypsins and chymotrypsins allows its adaptation to soybean proteinase inhibitor. *Insect. Biochem. Mol. Biol.*, *37*, 1283–1290.

Brito, L. O., Lopes, A. R., Parra, J. R.P., Terra, W. R., & Silva-Filho, M. C. (2001). Adaptation of tobacco budworm *Heliothis virescens* to proteinase inhibitor may be mediated by the synthesis of new proteinases. *Comp. Biochem. Physiol. B*, *128*, 365–375.

Broadway, P. (1997). Dietary regulation of serine proteinases that are resistant to serine proteinase inhibitors. *J. Insect Physiol.*, *43*, 855–874.

Brune, A. (1998). Termite guts: The world's smallest bioreactors. *Trends Biotechnol.*, *16*, 16–21.

Byeon, M. G., Lee, K. S., Gui, Z. Z., Kim, I., Kamg, P. D., et al. (2005). A digestive β-glucosidase from the silkworm, *Bombyx mori*: cDNA cloning, expression and enzymatic characterization. *Comp. Biochem. Physiol. B*, 418–427.

Caldeira, W., Dias, A. B., Terra, W. R., & Ribeiro, A. F. (2007). Digestive enzyme compartmentalization and recycling and sites of absorption and secretion along the midgut of *Dermestes maculatus* (Coleoptera) larvae. *Arch. Insect Biochem. Physiol.*, *64*, 1–18.

Callewaert, L., & Michiels, C. W. (2010). Lysozymes in the animal kingdom. *J. Biosci.*, *35*, 127–160.

Campbell, B. C., Bragg, T. S., & Turner, C. E. (1992). Phylogeny of symbiotic bacteria of four weevil species (Coleoptera: Curculionidae) based on analysis of 16s ribosomal DNA. *Insect Biochem. Mol. Biol.*, *22*, 415–421.

Campbell, P. M., Cao, A. T., Hines, E. R., East, P. D., & Gordon, K. H.J. (2008). Proteomic analysis of the peritrophic matrix from the gut of the caterpillar, *Helicoverpa armigera*. *Insect Biochem. Mol. Biol.*, *38*, 950–958.

Cançado, F. C., Valério, A. A., Marana, S. R., & Barbosa, J. A. (2007). The crystal structure of a lysozyme c from housefly *Musca domestica*, the first structure of a digestive lysozyme. *J. Struct Biol.*, *160*, 83–92.

Cançado, F. C., Chimoy Effio, P., Terra, W. R., & Marana, S. R. (2008). Cloning, purification and comparative characterization of two digestive lysozymes from *Musca domestica* larvae. *Braz. J. Med. Biol. Res.*, *41*, 969–977.

Cançado, F. C., Barbosa, J. A. G., & Marana, S. R. (2010). Role of the triad N46, S106 and T107 and the surface changes in the determination of the acidic pH optimum of the digestive lysozymes from *Musca domestica*. *Comp. Biochem. Physiol.*, *155B*, 387–395.

Candas, M., Loseva, O., Oppert, B., Kosaraju, P., & Bulla, L. A., Jr. (2003). Insect resistence to *Bacillus thuringiensis*: Alterations in the Indian-meal moth larval gut proteome. *Molec. Cell Proteomics.*, *2*, 19–28.

Capella, A. N., Terra, W. R., Ribeiro, A. F., & Ferreira, C. (1997). Cytoskeleton removal and characterization of the microvillar membranes isolated from two midgut regions of *Spodoptera frugiperda* (Lepidoptera). *Insect Biochem. Mol. Biol.*, *27*, 793–801.

Carneiro, C. N., Isejima, E. M., Samuels, R. I., & Silva, C. P. (2004). Sucrose hydrolases from the midgut of the sugarcane stalk borer *Diatraea saccharalis*, *J. Insect Physiol.*, *50*, 1093–1101.

Charlab, R., Valenzuela, J. G., Rowton, E. D., & Ribeiro, J. M. (1999). Toward an understanding of the biochemical and pharmacological complexity of the saliva of a hematophagous sand fly *Lutzomyia longipalpis*. *Proc. Natl. Acad. Sci. USA*, *96*, 15155–15160.

Chen, J., Aimanova, K. G., Pan, S., & Gill, S. S. (2009). Identification and characterization of *Aedes aegypti* aminopeptidase N as a putative receptor of *Bacillus thuringiensis* Cry11A toxin. *Insect Biochem. Mol. Biol.*, *39*, 688–696.

Christeller, J. T. (1996). Degradation of wool by *Hofmannophila pseudospretella* (Lepidoptera: Oecophoridae) larval midgut extracts under conditions simulating the midgut environment. *Arch. Insect Biochem. Physiol.*, *33*, 99–119.

Christeller, J. T. (2005). Evolutionary mechanisms acting on proteinase inhibitor variability. *FEBS J.*, *272*, 5710–5222.

Christeller, J. T., Laing, W. A., Shaw, B. D., & Burgess, E. P.J. (1990). Characterization and partial purification of the digestive proteases of the black field cricket, *Teleogryllus commodus* (Walker): Elastase is a major component. *Insect Biochem.*, *20*, 157–164.

Christeller, J. T., Poulton, J., Markwick, N. M., & Simpson, R. M. (2010). The effect of diet on the expression of lipase genes in the midgut of the lightbrown apple moth (*Epiphyas postuittana* Walker; Tortricidade). *Insect Mol. Biol.*, *19*, 9–25.

Clark, T. M. (1999). Evolution and adaptive significance of larval midgut alkalinization in the insect superorder Mecopterida. *J. Chem. Ecol.*, *25*, 1945–1960.

Colebatch, G., Cooper, P., & East, P. (2002). cDNA cloning of a salivary chymotrypsin-like protease and the identification of six additional cDNAs encoding putative digestive proteases from the green mirid, *Creontiades dilutus* (Hemiptera: Miridae). *Insect Biochem. Mol. Biol.*, *32*, 1065–1075.

Cook, S. C., & Davidson, D. W. (2006). Nutritional and functional biology of exudate-feeding ants. *Ent. Exp. Appl.*, *118*, 1–10.

Coutinho, P.M., Henrissat, B., 1999. Carbohydrate-active enzymes server. http://www.afmb.cnrs-mrs.fr/~cazy/CAZY/index.html.

Craik, C. S., Largman, C., Fletcher, T., Roczmiak, S., Barr, P. J., et al. (1985). Redesigning trypsin: Alteration of substrate specificity. *Science*, *228*, 291–297.

Crava, C. M., Bel, Y., Lee, S. F., Manachini, B., Heckel, D. G., & Esviche, B. (2010). Study of the aminopeptidase N gene family in the lepidopterans *Ostrinia nubinalis* (Hubner) and *Bombyx mori* (L.): Sequence mapping and expression. *Insect Biochem. Mol. Biol.*, *40*, 506–515.

Cristofoletti, P. T., & Terra, W. R. (1999). Specificity, anchoring, and subsites in the active center of a microvillar aminopeptidase purified from *Tenebrio molitor* (Coleoptera) midgut cells. *Insect Biochem. Mol. Biol.*, *29*, 807–819.

Cristofoletti, P. T., & Terra, W. R. (2000). The role of amino acid residues in the active site of a midgut microvillar aminopeptidase from the beetle *Tenebrio molitor*. *Biochim. Biophys. Acta*, *1479*, 185–195.

Cristofoletti, P. T., Ribeiro, A. F., & Terra, W. R. (2001). Apocrine secretion of amylase and exocytosis of trypsin along the midgut of *Tenebrio molitor*. *J. Insect Physiol.*, *47*, 143–155.

Cristofoletti, P. T., Ribeiro, A. F., Deraison, C., Rahbé, Y., & Terra, W. R. (2003). Midgut adaptation and digestive enzyme distribution in a phloem feeding insect, the pea aphid *Acyrthosiphon pisum*. *J. Insect Physiol.*, *49*, 11–24.

Cristofoletti, P. T., Ribeiro, A. F., & Terra, W. R. (2005). The cathepsin L-like proteinases from the midgut of *Tenebrio molitor* larvae: Sequence, properties, immunocytochemical localization and function. *Insect Biochem. Mol. Biol.*, *35*, 883–901.

Cristofoletti, P. T., Mendonça-de-Sousa, F. A., Rahbé, Y., & Terra, W. R. (2006). Characterization of a membrane-bound aminopeptidase purified from *Acyrtosiphon pisum* midgut cells. A major binding site for toxic mannose lectins. *FEBS J., 273*, 5574–5588.

Daffre, S., Kylsten, P., Samakovlis, C., & Hultmark, D. (1994). The lysozyme locus in *Drosophila melanogaster*: An expanded gene family adapted for expression in the digestive tract. *Mol. Gen. Genet., 242*, 152–162.

Daimon, T., Taguchi, T., Meng, Y., Katsuma, S., Mita, K., & Shimoda, T. (2008). Beta-fructofuranosidase genes of the silkworm, *Bombyx mori*: Insights into enzymatic adaptation of *B.mori* to toxic alkaloids in mulberry latex. *J. Biol. Chem., 283*, 15271–15279.

Da Lage, J. L., Von Wormhoudt, A., & Cariou, M. L. (2002). Diversity and evolution of the alpha-amylase genes in animals. *Biologia, 57*, 181–189.

D'Amico, S., Gerday, C., & Feller, G. (2000). Structural similarities and evolutionary relationships in chloride-dependent α-amylases. *Gene, 253*, 95–105.

Darboux, I., Nielsen-LeRoux, C., Charles, J. F., & Pauron, D. (2001). The receptor of *Bacillus sphaericus* binary toxin in *Culex pipiens* (Diptera: Culicidae) midgut: Molecular cloning and expression. *Insect Biochem. Mol. Biol., 31*, 981–990.

Davies, G. J., Wilson, K. S., & Henrisat, B. (1997). Nomenclature for sugar-binding subsites in glycosyl hydrolases. *Biochem. J., 321*, 557–559.

de Almeida, R. W., Tovar, F. J., Ferreira, I. I., & Leoncini, O. (2003). Chymotrypsin genes in the malaria mosquitoes, *Anopheles aquasalis* and *Anopheles darlingi*. *Insect Biochem. Mol. Biol., 33*, 307–315.

Deraison, C., Darboux, I., Duportets, L., Gorojankina, T., Rahbé, Y., & Jouanin, L. (2004). Cloning and characterization of a gut-specific cathepsin L from the aphid *Aphis gossypii*. *Insect Mol. Biol., 13*, 165–177.

De Veau, E. J.I., & Schultz, J. C. (1992). Reassessment of interaction between gut detergents and tannins in Lepidoptera and significance for gypsy moth larvae. *J. Chem. Ecol., 18*, 1437–1453.

Devenport, M., Alvarenga, P. H., Shao, L., Fujioka, H., Bianconi, M. L., Oliveira, P. L., & Jacobs-Lorena, M. (2006). Identification of the *Aedes aegypti* matrix protein AeIMUCI as a heme-binding protein. *Biochemistry, 45*, 9540–9549.

Dey, P. M., & Pridham, J. B. (1972). Biochemistry of α-galactosidases. *Adv. Enzymol, 36*, 91–130.

Dillon, R. J., & Dillon, V. M. (2004). The insect gut bacteria: An overview. *Annu. Rev. Entomol., 49*, 71–92.

Dinglasan, R. R., Devenport, M., Florens, L., Johnson, J. R., McHugh, C. A., et al. (2009). The *Anopheles gambiae* adult midgut peritrophic matrix proteome. *Insect Biochem. Mol. Biol., 39*, 125–134.

Dixit, R., Arakane, Y., Specht, C. A., Richard, C., Kramer, K. J., et al. (2008). Domain organization and phylogenetic analysis of proteins from the chitin deacetylase gene family of *Tribolium castaneum* and three other species of insects. *Insect Biochem. Mol. Biol., 38*, 440–451.

Dootsdar, H., McCollum, T. G., & Mayer, R. T. (1997). Purification and characterization of an endo-polygalacturonase from the gut of West Indies sugarcane rootstalk borer weevil (*Diaprepes abbreviatus* L.) larvae. *Comp. Biochem. Physiol. B, 118*, 861–867.

Douglas, A. E. (2006). Phloem-sap feeding by animals: Problems and solutions. *J. Exp. Bot., 57*, 747–754.

Dow, J. A. T. (1986). Insect midgut function. *Adv. Insect Physiol., 19*, 187–328.

Dow, J. A.T. (1992). pH gradients in lepidopteran midgut. *J. Exp.Biol., 172*, 355–375.

Dunn, B. M., Jimenez, M., Parten, B. F., Valler, M. J., Rolph, C. E., et al. (1986). A systematic series of synthetic chromophoric substrates for aspartic proteinases. *Biochem. J., 237*, 899–906.

Dunse, K. M., Kaas., Q., Guarini, R. F., Barton, P. A., Craik, D. J., & Anderson, M. A. (2010). Molecular basis for the resistance of an insect chymotrypsin to a potato type II proteinase inhibitor. *Proc. Natl. Acad. Sci. USA, 107*, 15016–15021.

Edwards, M. J., Lemos, F. J. A., Donnelly-Doman, M., & Jacobs-Lorena, M. (1997). Rapid induction by a blood meal of a carboxypeptidase gene in the gut of the mosquito *Anopheles gambiae*. *Insect Biochem. Mol. Biol., 27*, 1063–1072.

Eguchi, M. (1995). Alkaline phosphatase isozymes in insects and comparison with mammalian enzyme. *Comp. Biochem. Physiol. B, 111*, 151–162.

Eisemann, C., Wijffels, G., & Tellam, R. L. (2001). Secretion of type 2 peritrophic matrix protein, peritrophin-15, from the cardia. *Arch. Insect Biochem. Physiol., 472*, 76–85.

Elpidina, E. N., Vinokurov, K. S., Gromenko, V. A., Rudenskoya, Y. A., Dunaevsky, Y. E., et al. (2001a). Compartmentalization of proteinases and amylases in *Nauphoeta cinerea* midgut. *Arch. Insect Biochem. Physiol., 48*, 206–216.

Elpidina, E. N., Vinokurov, K. S., Rudenskaya, Y. A., Dunoevsky, Y. E., & Zhuzhikov, D. P. (2001b). Proteinase inhibitors in *Nauphoeta cinerea* midgut. *Arch. Insect Biochem. Physiol., 48*, 217–222.

Elpidina, E. N., Tsybina, T. A., Dunaevsky, Y. E., Belozersky, M. A., Zhuzhikov, D. P., & Oppert, B. (2005). A chymotrypsin-like proteinase from the midgut of *Tenebrio molitor* larvae. *Biochimie, 87*, 771–779.

Erthal, M., Jr., Silva, C. P., & Samuels, R. I. (2007). Digestive enzymes in larvae of the leaf cutting ant, *Acromyrmex subterraneous* (Hymenoptera: Formicidae: Attinae). *J. Insect Physiol., 53*, 1101–1111.

Estébanez-Perpiña, E., Bayés, A., Vendrell, J., Jongsma, M. A., Bown, D. P., et al. (2001). Crystal structure of a novel midgut procarboxypeptidase from the cotton pest *Helicoverpa armigera*. *J. Mol. Biol., 313*, 629–638.

Ferreira, A. H. P., Marana, S. R., Terra, W. R., & Ferreira, C. (2001). Purification, molecular cloning, and properties of a β-glycosidase isolated from midgut lumen of *Tenebrio molitor* (Coleoptera) larvae. *Insect Biochem. Mol. Biol., 31*, 1065–1076.

Ferreira, A. H. P., Ribeiro, A. F., Terra, W. R., & Ferreira, C. (2002). Secretion of β-glycosidase by middle midgut cells and its recycling in the midgut of *Tenebrio molitor* larvae. *J. Insect Physiol., 48*, 113–118.

Ferreira, A. H. P., Terra, W. R., & Ferreira, C. (2003). Characterization of a β-glycosidase highly active on disaccharides and of a β-galactosidase from *Tenebrio molitor* midgut lumen. *Insect Biochem. Mol. Biol., 33*, 253–265.

Ferreira, A. H. P., Cristofoletti, P. T., Lorenzini, D. M., Guerra, L. O., Paiva, P. B., et al. (2007). Identification of midgut microvillar proteins from *Tenebrio molitor* and *Spodoptera frugiperda* by cDNA library screenings with antibodies. *J. Insect Physiol., 53*, 1112–1124.

Ferreira, A. H. P., Cristofoletti, P. T., Pimenta, D. C., Ribeiro, A. F., Terra, W. R., & Ferreira, C. (2008). Structure, processing and midgut secretion of putative peritrophic membrane ancillary protein (PMAP) from *Tenebrio molitor* larvae. *Insect Biochem. Mol. Biol.*, *38*, 233–243.

Ferreira, C., & Terra, W. R. (1980). Intracelluar distribution of hydrolases in midgut caeca cells from an insect with emphasis on plasma membrane-bound enzymes. *Comp. Biochem. Physiol. B*, *66*, 467–473.

Ferreira, C., & Terra, W. R. (1984). Soluble aminopeptidases from cytosol and luminal contents of *Rhynchosciara americana* midgut caeca: Properties and phenanthroline inhibition. *Insect Biochem.*, *14*, 145–150.

Ferreira, C., & Terra, W. R. (1985). Minor aminopeptidases purified from the plasma membrane of midgut caeca cells of an insect (*Rhynchosciara americana*) larva. *Insect Biochem.*, *15*, 619–625.

Ferreira, C., & Terra, W. R. (1986a). The detergent form of the major aminopeptidase from the plasma membrane of midgut caeca cells of *Rhynchosciara americana* (Diptera) larva. *Comp. Biochem. Physiol. B*, *84*, 373–376.

Ferreira, C., & Terra, W. R. (1986b). Substrate specificity and binding loci for inhibitors in an aminopeptidase purified from the plasma membrane of midgut cells of an insect (*Rhynchosciara americana*) larva. *Arch. Biochem. Biophys.*, *244*, 478–485.

Ferreira, C., Bellinello, G. L., Ribeiro, A. F., & Terra, W. R. (1990). Digestive enzymes associated with the glycocalix, microvillar membranes and secretory vesicles from midgut cells of *Tenebrio molitor* larvae. *Insect Biochem.*, *20*, 839–847.

Ferreira, C., Capella, A. N., Sitnik, R., & Terra, W. R. (1994). Digestive enzymes in midgut cells, endo- and ectoperitrophic contents and peritrophic membranes of *Spodoptera frugiperda* (Lepidoptera) larvae. *Arch. Insect Biochem. Physiol.*, *26*, 299–313.

Ferreira, C., Parra, J. R. P., & Terra, W. R. (1997). The effect of dietary plant glycosides on larval midgut β-glycosidases from *Spodoptera frugiperda* and *Diatraea saccharalis*. *Insect Biochem. Mol. Biol.*, *27*, 55–59.

Ferreira, C., Torres, B. B., & Terra, W. R. (1998). Substrate specificities of midgut β-glycosidases from insects of different orders. *Comp. Biochem. Physiol. B*, *119*, 219–225.

Ferreira, C., Marana, S. R., Silva, C., & Terra, W. R. (1999). Properties of digestive glycosidases and peptidases and the permeability of the peritrophic membranes of *Abracris flavolineata* (Orthoptera: Acrididae). *Comp. Biochem. Physiol. B*, *123*, 241–250.

Fonseca, F. V., Silva, J. R., Samuels, R. I., DaMatta, R. A., Terra, W. R., & Silva, C. P. (2010). Purification and partial characterization of a midgut membrane-bound α-glucosidase from *Quesada gigas* (Hemiptera: Cicadidae). *Comp. Biochem. Physiol.*, *155B*, 20–25.

Fox, L. R., & Macauley, B. J. (1977). Insect grazing on *Eucalyptus* in response to variation in leaf tannins and nitrogen. *Oecologia*, *29*, 145–162.

Franco, O. L., Rigden, D. J., Melo, F. R., & Grossi-de-Sa′, M. F. (2002). Plant α-amylase inhibitors and their interaction with insect α-amylases: Structure, function and potential for crop protection. *Eur. J. Biochem.*, *269*, 397–412.

Fujita, A. (2004). Lysozymes in insects: What role do they play in nitrogen metabolism? *Physiol. Entomol.*, *29*, 305–310.

Fujita, A., Shimizu, I., & Abe, T. (2001). Distribution of lysozyme and protease, and amino acid concentration in guts of a wood-feeding termite, *Reticulitermes speratus* (Kolbe): Possible digestion of symbiont bacteria transferred by trophallaxis. *Physiol. Entomol.*, *26*, 116–123.

Fujita, A., Minamoto, T., Shimizu, I., & Abe, T. (2002). Molecular cloning of lysozyme-encoding cDNAs expressed in the salivary gland of a wood-feeding termite, *Reticulitermes speratus*. *Insect Biochem. Mol. Biol.*, *32*, 1615–1624.

Gaines, P. J., Sampson, C. M., Rushlow, K. E., & Stiegler, G. L. (1999). Cloning of a family of serine protease genes from the cat flea *Ctenocephalides felis*. *Insect Mol. Biol.*, *81*, 11–22.

Gatehouse, L. N., Shannon, A. L., Burgess, E. P. J., & Christeller, J. T. (1997). Characterization of major mid-gut proteinase cDNAs from *Helicoverpa armigera* larvae and changes in gene expression in response to four proteinase inhibitors in the diet. *Insect Biochem. Mol. Biol.*, *27*, 929–944.

Geer, L. Y., Domrachev, M., Lipman, D. J., & Bryant, S. H. (2002). CDART: Protein homology by domain architecture. *Genome Res.*, *12*, 1619–1623.

Genta, F. A., Terra, W. R., & Ferreira, C. (2003). Action pattern, specificity, lytic activities, and physiological role of five digestive β-glucanases isolated from *Periplaneta americana*. *Insect Biochem. Mol. Biol.*, *33*, 1085–1097.

Genta, F. A., Dillon, R. J., Terra, W. R., & Ferreira, C. (2006a). Potential role for gut microbiota in cell wall digestion and glucoside detoxification in *Tenebrio molitor* larvae. *J. Insect Physyiol.*, *52*, 593–601.

Genta, F. A., Blanes, L., Cristofoletti, P. T., do Lago, C. L., Terra, W. R., & Ferreira, C. (2006b). Purification, characterization and molecular cloning of the major chitinase from *Tenebrio molitor* larval midgut. *Insect Biochem. Mol. Biol.*, *36*, 789–800.

Genta, F. A., Dumont, A. F., Marana, S. R., Terra, W. R., & Ferreira, C. (2007). The interplay of processivity, substrate inhibition and a secondary substrate binding site of an insect exo-beta-1,3-glucanase. *Biochim. Biophys. Acta.*, *1774*, 1079–1091.

Genta, F. A., Bragatto, I., Terra, W. R., & Ferreira, C. (2009). Purification, characterization and sequencing of the major beta-1,3-glucanase from the midgut of *Tenebrio molitor* larvae. *Insect Biochem. Mol. Biol.*, *39*, 861–874.

Gibson, R. P., Gloster, T. M., Roberts, S., Warren, R. A., Storch de Gracia, I., Garcia, A., et al. (2007). Molecular basis for trehalase inhibition revealed by the structure of trehalase in complex with potent inhibitors. *Angew Chem. Int. Ed. Engl.*, *46*, 4115–4119.

Girard, C., & Jouanin, L. (1999a). Molecular cloning of cDNAs encoding a range of digestive enzymes from a phytophagous beetle *Phaedon cochleariae*. *Insect Biochem. Mol. Biol.*, *29*, 1129–1142.

Girard, G., & Jouanin, L. (1999b). Molecular cloning of a gut-specific chitinase cDNA from the beetle *Phaedon cochleariae*. *Insect Biochem. Mol. Biol.*, *29*, 549–556.

Giri, A. P., Harsulkar, A. M., Deshpande, V. V., Sainani, M. N., Gupta, V. S., et al. (1998). Chickpea defensive proteinase inhibitors can be inactivated by podborer gut proteinases. *Plant Physiol.*, *116*, 393–401.

Gonzales, M. S., Nogueira, N. F. S., Feder, D., de Souza, W., Azambuja, P., et al. (1998). Role of the head in the ultrastructural midgut organization in *Rhodnius prolixus* larvae: Evidence from head transplantation experiments and acdysone therapy. *J. Insect Physiol., 44*, 553–560.

Graf, R., Raikhel, A. S., Brown, M. R., Lea, A. O., & Briegel, H. (1986). Mosquito trypsin: Immunocytochemical localization in the midgut of blood-fed *Aedes aegyti* (L.). *Cell Tissue Res., 245*, 19–27.

Greenberg, B., & Paretsky, D. (1955). Proteolytic enzymes in the housefly, *Musca domestica* (L.). *Ann. Entomol. Soc. America, 48*, 46–50.

Grigorten, J. L., Cawford, D. N., & Harvey, W. R. (1993). High pH in the ectoperitrophic space of the larval lepidopteran midgut. *J. Exp. Biol., 183*, 353–359.

Grimaldi, D., & Engel, M. S. (2005). *Evolution of Insects*. New York, NY: Cambridge University Press.

Grossmann, G. A., & Terra, W. R. (2001). Alpha-galactosidases from the larval midgut of *Tenebrio molitor* (Coleoptera) and *Spodoptera frugiperda* (Lepidoptera). *Comp. Biochem. Physiol. B, 128*, 109–122.

Grossmann, G. L., Campos, Y., Senerson, D. W., & James, A. A. (1997). Evidence for two distinct members of the amylase gene family in the yellow fever mosquito, *Aedes aegypti*. *Insect Biochem. Mol. Biol., 27*, 769–781.

Guo, W., Li, G., Pang, Y., & Wang, P. (2005). A novel chitin-binding protein identification from the peritrophic membrane of the cabbage looper, *Trichoplusia ni*. *Insect Biochem. Mol. Biol., 35*, 1224–1234.

Halanych, K. M. (2004). The new view of animal phylogeny. *Ann. Rev. Ecol. Evol. Syst., 35*, 229–256.

Hanozet, G. M., Giordana, B., & Sacchi, J. F. (1980). K+-dependent phenylalanine uptake in membrane vesicles isolated from the midgut of *Philosamia cynthia*. *Biochim. Biophys. Acta., 596*, 481–486.

Harper, M. S., & Hopkins, T. L. (1997). Peritrophic membrane structure and secretion in European corn borer larvae (*Ostrinia nubilalis*). *Tissue Cell, 29*, 461–475.

Harwood, J. L. (1980). Plant acyl lipids: Structure, distribution and analysis. In P. K. Stumpf (Ed.), *The Biochemistry of Plants: A Comprehensive Treatise*, Vol. 4, *Lipids: Structure and Function* (pp. 1–56). New York, NY: Academic Press.

Harzer, K., Hiraiwa, M., & Paton, B. C. (2001). Saposins (sap) A and C activate the degradation of galactosylsphingosine. *FEBS Lett., 508*, 107–110.

Hegedus, D., Erlandson, M., Gillot, C., & Toprak, U. (2008). New insights into peritrophic matrix synthesis, architecture, and function. *Annu. Rev. Entomol., 54*, 285–302.

Hirayama, C., Konno, K., Wasano, N., & Nakamura, M. (2007). Differential effects of sugar-mimic alkaloids in mulberry latex on sugar metabolism and disaccharidases of Eri and domesticated silkworms: Enzymatic adaptation of *Bombyx mori* to mulberry defense. *Insect Biochem. Mol. Biol., 37*, 1348–1358.

Hooper, N. M. (1994). Families of zinc metalloproteases. *FEBS Lett., 354*, 1–6.

Hori, K. (1972). Comparative study of a property of salivary amylase among various heteropterous insects. *Comp. Biochem. Physiol. B, 42*, 501–508.

Horne, I., Haritos, V. S., & Oakeshott, J. G. (2009). Comparative and functional genomics of lipases in holometabolous insects. *Insect Biochem. Mol. Biol., 39*, 547–567.

Huang, S. -W., Zhang, H. -Y., Marshall, S., & Jackson, T. A. (2010). The scarab gut: A potential bioreactor for bio-fuel production. *Insect Science, 17*, 175–183.

Hughes, J., & Vogler, A. P. (2006). Gene expression in the gut of keratin-feeding clothes moths (*Tineola*) and keratin beetles (*Trox*) revealed by subtracted cDNA libraries. *Insect Biochem. Mol. Biol., 36*, 584–592.

Human, H., & Nicolson, S. W. (2003). Digestion of maize and sunflower pollen by the spotted maize *Astylus atromaculatus* (Melyridae): Is there a role for osmotic shock? *J. Insect Physiol., 49*, 633–643.

Husebye, H., Artz, S., Burmeister, W. P., Härtel, F. V., Brandt, A., et al. (2005). Crystal structure at 1.1-Å resolution of an insect myrosinase from *Brevicorine brassicae* shows its close relationship to β-glucosidases. *Insect Biochem. Mol. Biol., 35*, 1311–1320.

Imoto, T., Johnson, L. N., North, A. C. T., Phillips, D. C., & Ruppley, J. A. (1972). Vertebrate lysozymes. In P. D. Boyer (Ed.), *The Enzymes* (Vol. 7, pp. 665–868). New York, NY: Academic Press.

Inward, D., Beccaloni, G., & Eggleton, P. (2007). Death of an order: A comprehensive molecular phylogenetic study confirms that termites are eusocial cockroaches. *Biol. Lett., 3*, 331–335.

Itoh, M., Takeda, S., Yamamoto, H., Izumi, S., Tomino, S., et al. (1991). Cloning and sequence analysis of membrane-bound alkaline phosphatase cDNA of the silkworm, *Bombyx mori*. *Biochim. Biophys. Acta, 1129*, 135–138.

Itoh, M., Kanamori, Y., Takao, M., & Eguchi, M. (1999). Cloning of soluble alkaline phosphatase cDNA and molecular basis of the polymorphic nature in alkaline phosphatase isozymes of *Bombyx mori* midgut. *Insect Biochem. Mol. Biol., 29*, 121–129.

Ivanova, N., Sorokin, A., Anderson, L., Galleron, N., Candellon, B., et al. (2003). Genome sequence of *Bacillus cereus* and comparative analysis with *Bacillus anthracis*. *Nature, 423*, 87–91.

Janecek, S. (1997). Alpha-amylase family: Molecular biology and evolution. *Progr. Biophys. Mol. Biol., 67*, 67–97.

Jasrapuria, S., Arakane, Y., Osman, G., Kraes, K. J., Beerman, R. W., & Muthukrishman, S. (2010). Genes encoding proteins with peritrophin A-type chitin-binding domains in *Tribolium castaneum* are grouped into three distinct families based on phylogeny, expression and function. *Insect Biochem. Molec. Biol., 40*, 214–227.

Jeng, W. -Y., Wang, N. -C., Lin, C. -T., Liaw, Y. -C., Chang, W. -J., et al. (2010). Structural and functional analysis of three β-glucosidases from bacterium *Clostridium cellulovorans*, fungus *Trichoderma reesei* and termite *Neotermes koshunensis*. *J. Struct. Biol*, in press.

Johnson, K. S., & Felton, G. W. (2000). Digestive proteinase activity in corn earworm (*Helicoverpa zea*) after molting and in response to lowered redox potential. *Arch. Insect Biochem. Physiol., 44*, 151–161.

Johnson, K. S., & Rabosky, D. (2000). Phylogenetic distribution of cysteine proteinases in beetles: Evidence for an evolutionary shift to an alkaline digestive strategy in Cerambycidae. *Comp. Biochem. Physiol. B, 126*, 609–619.

Jones, A. M. E., Winge, P., Bones, A. M., Cole, R., & Rossiter, J. T. (2002). Characterization and evolution of a myrosinase from the cabbage aphid *Brevicorine brassicae*. *Insect Biochem. Mol. Biol.*, *32*, 275–284.

Jordão, B. P., Terra, W. R., & Ferreira, C. (1995). Chemical determinations in microvillar membranes purified from brush borders isolated from the larval midgut from one Coleoptera and two Diptera species. *Insect Biochem. Mol. Biol.*, *25*, 417–426.

Jordão, B. P., Lehane, M. J., Terra, W. R., Ribeiro, A. F., & Ferreira, C. (1996a). An immunocytochemical investigation of trypsin secretion in the midgut of the stablefly *Stomoxys calcitrans*. *Insect Biochem. Mol. Biol.*, *26*, 445–453.

Jordão, B. P., Terra, W. R., Ribeiro, A. F., Lehane, M. J., & Ferreira, C. (1996b). Trypsin secretion in *Musca domestica* larval midguts: A biochemical and immunocytochemical study. *Insect Biochem. Mol. Biol.*, *26*, 337–346.

Jordão, B. P., Capella, A. N., Terra, W. R., Ribeiro, A. F., & Ferreira, C. (1999). Nature of the anchors of membrane-bound aminopeptidase, amylase, and trypsin and secretory mechanisms in *Spodoptera frugiperda* (Lepidoptera) midgut cells. *J. Insect Physiol.*, *45*, 29–37.

Khadeni, S., Guarino, L. A., Watenabe, H., Tokuda, G., & Meyer, E. F. (2002). Structure of an endoglucanase from termite, *Nasutitermes takasagoensis*. *Acta Cryst.*, *D 58*, 653–658.

Kim, N., Choo, Y. M., Lee, K. S., Hong, S. J., Seol, K. Y., et al. (2008). Molecular cloning and characterization of a glycosyl family 9 cellulase distributed throughout the digestive tract of the cricket *Teleogryllus emma*. *Comp. Biochem. Physiol.*, *150B*, 368–376.

Kleywegt, G. J., Zou, J. Y., Divne, C., Davies, G. J., Sinning, I., et al. (1997). The crystal structure of the catalytic core domain of endoglucanase I from *Trichoderma reesei* at 3.6 angstrom resolution, and a comparison with related enzymes. *J. Mol. Biol.*, *272*, 393–397.

Klinkowstrom, A. M., Terra, W. R., & Ferreira, C. (1994). Aminopeptidase A from *Rhynchosciara americana* (Diptera) larval midguts: Properties and midgut distribution. *Arch. Insect Biochem. Physiol.*, *27*, 301–305.

Klinkowstrom, A. M., Terra, W. R., & Ferreira, C. (1995). Midgut dipeptidases from *Rhynchosciara americana* (Diptera) larvae: Properties of soluble and membrane bound forms. *Insect Biochem. Mol. Biol.*, *25*, 303–310.

Knight, P. J. K., Knowles, B. H., & Eller, D. J. (1995). Molecular cloning of an insect aminopeptidase N that serves as a receptor of *Bacillus thuringiensis* Cry 1a (c) toxin. *J. Biol. Chem.*, *270*, 17765–17770.

Koepke, J., Ermler, V., Warkentin, E., Wenzi, G., & Flecker, P. (2000). Crystal structure of cancer chemopreventive Bowman–Birk inhibition in ternary complex with bovine trypsin at 2–3-Å resolution: Structural basis of Janus-faced serine protease inhibitor specificity. *J. Mol. Biol.*, *298*, 477–491.

Koiwa, H., Shade, R. E., Zhu-Salzman, K., D'Urzo, M. P., Murdock, L. L., et al. (2000). A plant defensive cystatin (soyacystatin) targets cathepsin L-like digestive cysteine proteinases (DuCALs) in the larval midgut of the western corn rootworm (*Diabrotica virgifera virgifera*). *FEBS Lett.*, *471*, 67–70.

Konno, K., Okada, S., & Hirayama, C. (2001). Selective secretion of free glycine, a neutralizer against a plant defense chemical, in the digestive juice of the privet moth larvae. *J. Insect Physiol.*, *47*, 1451–1457.

Krishnamoorthy, M., Jurat-Fuentes, J. L., McNall, R. J., Andacht, T., & Adang, M. J. (2007). Identification of novel CryIAc binding proteins in midgut membranes from *Heliothis virescens* using proteomic analyses. *Insect Biochem. Mol. Biol.*, *37*, 189–201.

Krishnan, N., & Kodrik, D. (2006). Antioxidant enzymes in *Spodoptera littoralis* (Boisduval): Are they enhanced to protect gut tissues during oxidative stress? *J. Insect Physiol.*, *52*, 11–20.

Lang, T., Hansson, G. C., & Samuelsson, T. (2007). Gel-forming mucins appeared early in metazoan evolution. *Proc. Natl. Acad. Sci. USA*, *104*, 16209–16214.

Le Cahérec, F., Guillam, M. T., Beuron, F., Cavalier, A., Thomas, D., et al. (1997). Aquaporin-related proteins in the filter chamber of homopteran insects. *Cell Tissue Res.*, *290*, 143–151.

Lee, J. H., Tsuji, M., Nakamura, M., Nishimoto, M., Okuyama, M., et al. (2001). Purification and identification of the essential ionizable groups of honeybee, *Apis mellifera* L., trehalase. *Biosci. Biotechnol. Biochem.*, *65*, 2657–2665.

Lee, M. J., & Anstee, J. H. (1995). Endoproteases from the midgut of *Spodoptera littoralis* include a chymotrypsin-like enzyme with an extended binding site. *Insect Biochem. Mol. Biol.*, *25*, 49–61.

Lehane, M. J. (1997). Peritrophic matrix structure and function. *Annu. Rev. Entomol.*, *42*, 525–550.

Lehane, M. J., Müller, H. M., & Crisanti, A. (1996). Mechanisms controlling the synthesis and secretion of digestive enzymes in insects. In M. J. Lehane, & P. F. Billingsley (Eds.), *Biology of the Insect Midgut* (pp. 195–205). London, UK: Chapman and Hall.

Lemos, F. J.A., & Terra, W. R. (1991a). Digestion of bacteria and the role of midgut lysozyme in some insect larvae. *Comp. Biochem. Physiol. B*, *100*, 265–268.

Lemos, F. J.A., & Terra, W. R. (1991b). Properties and intracellular distribution of a cathepsin D-like proteinase active at the acid region of *Musca domestica* midgut. *Insect Biochem.*, *21*, 457–465.

Lemos, F. J.A., & Terra, W. R. (1992). A high yield preparation of *Musca domestica* larval midgut microvilli and the subcellular distribution of amylase and trypsin. *Insect Biochem. Mol. Biol.*, *22*, 433–438.

Lemos, F. J.A., Ribeiro, A. F., & Terra, W. R. (1993). A bacteria-digesting midgut lysozyme from *Musca domestica* (Diptera) larvae: Purification, properties and secretory mechanism. *Insect Biochem. Mol. Biol.*, *23*, 533–541.

Leonardi, M. G., Marciani, P., Montorfano, P. G., Cappelloza, S., & Giordana, B. (2001). Effect of fenoxycarb on leucine uptake and lipid composition of midgut brush border membrane in the silkworm, *Bombyx mori* (Lepidoptera: Bombycidae). *Pest. Biochem. Physiol.*, *70*, 42–51.

Levitzki, A., & Steer, M. L. (1974). The allosteric activation of mammalian α-amylase by chloride. *Eur. J. Biochem.*, *41*, 171–180.

Li, X., & Brune, A. (2005). Digestion of microbial mass, structural polysaccharides, and protein by the humivorous larva of *Pachnoda ephippiata* (Coleoptera: Scarabaeidae). *Soil Biol. Biochem.*, *37*, 107–116.

Lin, G., Bode, W., Huber, R., Chi, C., & Engh, R. A. (1993). The 0.25-nm X-ray structure of Bowman–Birk-type inhibitor from mung bean in ternary complex with porcine trypsin. *Eur. J. Biochem.*, *212*, 549–555.

Linder, M., & Teeri, T. T. (1997). The roles and function of cellulose-binding domains. *J. Biotechnol.*, *57*, 15–28.

Lo, N., Tokuda, G., Watanabe, H., Rose, H., Slaytor, M., et al. (2000). Evidence from multiple gene sequences indicates that termites evolved from wood-feeding cockroaches. *Current Biol. 10*, 801–804.

Lopes, A. R., Juliano, M. A., Juliano, L., & Terra, W. R. (2004). Coevolution of insect trypsins and inhibitors. *Arch. Insect Biochem. Physiol.*, *55*, 140–152.

Lopes, A. R., Juliano, M. A., Marana, S. R., Juliano, L., & Terra, W. R. (2006). Substrate specificity of insect trypsins and the role of their subsites in catalysis. *Insect Biochem. Mol. Biol.*, *36*, 130–140.

Lopes, A. R., Sato, P. M., & Terra, W. R. (2009). Insect chymotrypsins: Chloromethyl ketone inactivation and substrate specificity relative to possible coevolutional adaptation of insects and plants. *Arch. Insect Biochem. Physiol.*, *70*, 188–203.

Lu, Y., Zen, K. -C., Muthukrishnan, S., & Kramer, K. J. (2002). Site-directed mutagenesis and functional analysis of active site acidic amino acid residues D142, D144 and E146 in *Manduca sexta* (tobacco hornworm) chitinase. *Insect Biochem. Mol. Biol.*, *32*, 1369–1382.

Luciani, V., Marie-Claire, C., Ruffet, E., Beaumont, A., Roques, B. P., et al. (1998). Characterization of Glu^{350} as a critical residue involved in the N-terminal amine binding site of aminopeptidase N (EC 3.4.11.2): Insight into its mechanism of action. *Biochemistry*, *37*, 686–692.

Marana, S. R., Ribeiro, A. F., Terra, W. R., & Ferreira, C. (1997). Ultrastructure and secretory activity of *Abracris flavolineata* (Orthoptera: Acrididae) midguts. *J. Insect Physiol.*, *43*, 465–473.

Marana, S. R., Terra, W. R., & Ferreira, C. (2000). Purification and properties of a β-glycosidase purified from midgut cells of *Spodoptera frugiperda* (Lepidoptera) larvae. *Insect Biochem. Mol. Biol.*, *30*, 1139–1146.

Marana, S. R., Jacobs-Lorena, M., Terra, W. R., & Ferreira, C. (2001). Amino acid residues involved in substrate binding and catalysis in an insect digestive β-glycosidase. *Biochim. Biophys. Acta*, *1545*, 41–52.

Marana, S. R., Terra, W. R., & Ferreira, C. (2002a). The role of amino acid residues Q39 and E451 in the determination of substrate specificity of the *Spodoptera frugiperda* β-glycosidase. *Eur. J. Biochem.*, *269*, 3705–3714.

Marana, S. R., Lopes, A. R., Juliano, L., Juliano, M. A., Ferreira, C., et al. (2002b). Subsites of trypsin active site favor catalysis or substrate binding. *Biochem. Biophys. Res. Commun.*, *290*, 494–497.

Marana, S. R., Cançado, F. C., Valerio, A. A., Ferreira, C., Terra, W. R., & Barbosa, J. A. R. G. (2006). Crystallization, data collection and phasing of two digestive lysozymes from *Musca domestica*. *Acta Cryst. F62*, 750–752.

Markovic, O., & Janecek, S. (2001). Pectin degrading glycoside hydrolases of family 28: Sequence-structural features, specificities and evolution. *Protein Eng.*, *14*, 615–631.

Martin, M. M. (1987). *Invertebrate–Microbial Interactions: Ingested Fungal Enzymes in Anthropod Biology*. Ithaca, NY: Cornell University Press.

Martin, M. M., & Martin, J. S. (1984). Surfactants: Their role in preventing the precipitation of proteins by tannins in insect guts. *Oecologia*, *61*, 342–345.

Matoub, M., & Rouland, C. (1995). Purification and properties of the xylanases from the termite *Macrotermes bellicosus* and its symbiotic fungus *Termitomyces* sp. *Comp. Biochem. Physiol. B*, *112*, 629–635.

Matsumura, I., & Kirsch, J. F. (1996). Is aspartate 52 essential for catalysis by chicken egg white lysozyme? The role of natural substrate-assisted hydrolysis. *Biochemistry*, *35*, 1881–1889.

Mazumdar-Leighton, S., & Broadway, R. M. (2001a). Transcriptional induction of diverse midgut trypsins in larval *Agrotis ipsilon* and *Helicoverpa zea* feeding on the soybean trypsin inhibitor. *Insect Biochem. Mol. Biol.*, *31*, 645–657.

Mazumdar-Leighton, S., & Broadway, R. M. (2001b). Identification of six chymotrypsin cDNAs from larval midgut of *Helicoverpa zea* and *Agrotis ipsilon* feeding on the soybean (Kunitz) trypsin inhibitor. *Insect Biochem. Mol. Biol.*, *31*, 633–644.

McNall, R. J., & Adang, M. J. (2003). Identification of novel *Bacillus thuringiensis* Cry1 Ac binding proteins in *Manduca sexta* midgut through proteomic analysis. *Insect Biochem. Mol. Biol.*, *33*, 999–1010.

Miller, D., & Crane, R. K. (1961). The digestive function of the epithelium of the small intestine. II. Localization of disaccharide hydrolysis in the isolated brush border portion of intestinal epithelial cells. *Biochim. Biophys. Acta*, *52*, 293–298.

Mitsumasu, K., Azuma, M., Niimi, T., Yamashita, O., & Yaginuma, T. (2005). Membrane-penetrating trehalase from silkworm *Bombyx mori*. Molecular cloning and localization in larval midgut. *Insect Mol Biol.*, *14*, 501–508.

Murdock, L. L., Brookhart, G., Dunn, P. E., Foard, D. E., Kelley, S., et al. (1987). Cysteine digestive proteinases in Coleoptera. *Comp. Biochem. Physiol. B*, *87*, 783–787.

Nakanishi, K., Yaoi, K., Nagino, Y., Hara, H., Kitami, M., et al. (2002). Aminopeptidase N isoforms from the mid-gut of *Bombyx mori* and *Plutella xylostella*: Their classification and the factors that determine their binding specificity to *Bacillus thuringiensis* Cry 1A toxin. *FEBS Lett.*, *519*, 215–220.

Nakashima, K., Watanabe, H., Saitoh, H., Tokuda, G., & Azuma, J. I. (2002). Dual cellulose-digesting system of the wood-feeding termite, *Coptotermes formosanus* Shiraki. *Insect Biochem. Mol. Biol.*, *32*, 777–784.

Nielsen, J. E., & Borchert, T. U. (2000). Protein engineering of bacterial α-amylases. *Biochim. Biophys. Acta*, *1543*, 253–274.

Nishimoto, M., Kubota, M., Tsuji, M., Mori, H., Kimura, A., et al. (2001). Purification and substrate specificity of honeybee, *Apis mellifera* L., alpha-glucosidase III. *Biosci. Biotechnol. Biochem.*, *65*, 1610–1616.

Norén, O., Sjostrom, H., Danielsen, E. M., Cowell, G. M., & Skovbjerg, H. (1986). The enzymes of the enterocyte plasma membrane. In P. Desnuelle, H. Sjostrom, & O. Norén (Eds.), *Molecular and Cellular Basis of Digestion* (pp. 355–365). Amsterdam: Elsevier.

Novillo, C., Castanera, P., & Ortego, F. (1997). Inhibition of digestive trypsin-like proteases from larvae of several lepidopteran species by the diagnostic cysteine protease inhibitor E-64. *Insect Biochem. Mol. Biol.*, 27, 247–254.

Novotny, V., & Wilson, M. R. (1997). Why are there no small species among xylem-sucking insects? *Evol. Ecol.*, 11, 419–437.

Okech, B. A., Boudko, D. Y., Linser, P. J., & Harvey, W. R. (2008). Cationic pathway of pH regulation in larvae of *Anopheles gambiae*. *J. Exp. Biol.*, 211, 957–968.

Oliveira-Neto, O. B., Batista, J. A. N., Rigden, D. J., Fragoso, R. R., Silva, R. O., et al. (2004). A diverse family of serine proteinase genes expressed in cotton boll weevil (*Anthonomus grandis*): Implications for the design of pest-resistant transgenic cotton. *Insect Biochem. Mol. Biol.*, 34, 903–918.

Ortego, F., Novillo, C., & Catañera, P. (1996). Characterization and distribution of digestive proteases of the stalk corn borer, *Sesamia nonagrioides* Lef. (Lepidoptera: Noctuidae). *Arch. Insect Biochem. Physiol.*, 33, 136–180.

Oviedo, M. N., VanEkeris, L., Corena-Mcleod, M. D. P., & Linser, P. J. (2008). A microarray-based analysis of transcriptional compartmentalization in the alimentary canal of *Anopheles gambiae* (Diptera: Culicidae) larvae. *Insect Mol. Biol.*, 17, 61–72.

Padilha, M. H. P., Pimentel, A. C., Ribeiro, A. F., & Terra, W. R. (2009). Sequence and function of lysosomal and digestive cathepsin D-Like proteinases of *Musca domestica* midgut. *Insect Biochem. Mol. Biol.*, 39, 782–791.

Parenti, P., Sacchi, F. V., Hanozet, G. M., & Giordana, B. (1986). Na-dependent uptake of phenylalanine in the midgut of a cockroach (*Blabera gigantean*). *J. Comp. Physiol. B*, 156, 549–556.

Pauchet, Y., Muck, A., Svatos, A., Heckel, D. G., & Preiss, S. (2008). Mapping the larval midgut lumen proteome of *Helicoverpa armigera*, a generalist herbivorous insect. *J. Proteom. Res.*, 7, 1629–1639.

Pauchet, Y., Freitak, D., Heidel-Fischer, H. M., Heckel, D. G., & Vogel, H. (2009a). Immunity or digestion: Glucanase activity in a glucan-binding protein family from Lepidoptera. *J. Biol. Chem.*, 284, 2214–2224.

Pauchet, Y., Muck, A., Svatos, A., & Heckel, D. G. (2009b). Chromatographic and eletrophoretic resolution of proteins and protein complexes from the larval midgut microvilli of *Manduca sexta*. *Insect Biochem. Mol. Biol.*, 39, 467–474.

Pauchet, Y., Wilkinson, P., Vogel, H., Nelson, D. R., Reynolds, S. E., et al. (2010). Pyrosequencing the *Manduca sexta* larval midgut transcriptome: Messages for digestion, detoxification and defense. *Insect Biochem. Mol. Biol.*, 19, 61–75.

Payan, F. (2004). Structural basis for the inhibition of mammalian and insect α-amylases by plant protein inhibitors. *Biochim. Biophys. Acta*, 1696, 171–180.

Pechan, T., Cohen, A., Williams, W. P., & Luthe, D. S. (2002). Insect feeding mobilizes a unique defense protease that disrupts the peritrophic matrix of caterpillars. *Proc. Natl. Acad. Sci. USA*, 99, 13319–13323.

Peng, J., Zhong, J., & Granados, R. R. (1999). A baculovirus enhancin alters the permeability of a mucosal midgut peritrophic matrix from lepidopteran larvae. *J. Insect Physiol.*, 45, 159–166.

Pereira, P. J.B., Lozanov, V., Patthy, A., Huber, R., Bode, W., et al. (1999). Specific inhibition of insect α-amylases: Yellow mealworm α-amylase in complex with the Amaranth at 2.0 Å resolution. *Structure*, 7, 1079–1088.

Peters, W. (1992). *Peritrophic Membranes*. Berlin: Springer.

Peterson, A. M., Barillas-Mury, C. V., & Wells, M. A. (1994). Sequence of three cDNAs encoding an alkaline midgut trypsin from *Manduca sexta*. *Insect Biochem. Mol. Biol.*, 24, 463–471.

Peterson, A. M., Fernando, G. J. P., & Wells, M. (1995). Purification, characterization and cDNA sequence of an alkaline chymotrypsin from the midgut of *Manduca sexta*. *Insect Biochem. Mol. Biol.*, 25, 765–774.

Pimenta, D. C., Oliveira, A., Juliano, M. A., & Juliano, L. (2001). Substrate specificity of human cathepsin D using internally quenched fluorescent peptides derived from reactive site loop of kallistatin. *Biochim. Biophys. Acta*, 1544, 113–122.

Ponnuvel, K. M., Nakazawa, H., Furukawa, S., Asaoka, A., Ishibashi, J., Tanaka, H., & Yamakawa, M. (2003). A lipase isolated from the silkworm *Bombyx mori* shows antiviral activity against nucleopolyhedrovirus. *J. Virol.*, 77, 10725–10729.

Ponsen, M. B. (1991). *Structure of the digestive system of aphids, in particular* Hyalopterus *and* Coloradoa, *and its bearing on the evolution of filter chambers in Aphidoidea*. Wageningen: Wageningen Agricultural University Press, 91–95, pp. 3–61.

Popova-Butler, A., & Dean, D. H. (2009). Proteomic analysis of the mosquito *Aedes aegypti* midgut brush border membrane vesicles. *J. Insect Physiol.*, 55, 264–272.

Price, D. R. G., Karley, A. J., Ashford, D. A., Isaacs, H. V., Pownall, M. E., et al. (2007). Molecular characterization of a candidate gut sucrose in the pea aphid, *Acyrtosiphon pisum*. *Insect Biochem. Mol. Biol.*, 37, 307–317.

Pytelkova, J., Hubert, J., Lepsik, M., Sobotnik, J., Sindelka, R., et al. (2009). Digestive α-amylases from *Ephestia kuehniella* – adaptation to alkaline environment and plant inhibitors. *FEBS J.*, 276, 3531–3546.

Quicke, D. L. J. (2003). Hymenoptera (ants, bees, wasps). In V. H. Resh, & R. T. Cardé (Eds.). *Encyclopedia of Insects* (pp. 534–546). San Diego, CA: Academic Press.

Quina, F. H., Politi, M. J., Cuccovia, I. M., Baumgarten, E., Martins-Franchetti, S. M., & Chaimovich, H. (1980). Ion exchange in micellar solutions. 4. "Buffered" systems. *J. Phys. Chem.*, 84, 361–365.

Rajagopal, R., Agrawal, N., Selvapandiyan, A., Sivakumar, S., Ahmad, S., et al. (2003). Recombinantly expressed isoenzymic aminopeptidases from *Helicoverpa armigera* (American cotton bollworm) midgut display differential interaction with closely related *Bacillus thuringiensis* insecticidal proteins. *Biochem. J.*, 379, 971–978.

Ramalho-Ortigão, J. M., Kamhawi, S., Rowton, E. D., Ribeiro, J. M. C., & Valenzuela, J. G. (2003). Cloning and characterization of trypsin- and chymotrypsin-like proteases from the midgut of the sand fly vector *Phletomus papatasi*. *Insect Biochem. Mol. Biol.*, 33, 163–171.

Ramos, A., Mahowald, A., & Jacobs-Lorena, M. (1993). Gut-specific genes from the black-fly *Simulium vittatum* encoding trypsin-like and carboxypeptidase-like proteins. *Insect Mol. Biol.*, 1, 149–163.

Rana, R. L., & Stanley, D. W. (1999). *In vitro* secretion of digestive phospholipase A2 by midgut isolated from tobacco hornworm, *Manduca sexta. Arch. Insect Biochem. Physiol., 42*, 179–187.

Rawlings, N. D., & Barrett, A. J. (1995). Evolutionary families of metallopeptidases. *Methods Enzymol., 248*, 183–228.

Regel, R., Matioli, S. R., & Terra, W. R. (1998). Molecular adaptation of *Drosophila melanogaster* lysozymes to a digestive function. *Insect Biochem. Mol. Biol., 28*, 309–319.

Reuveni, M., Hong, Y. S., Dunn, P. E., & Neal, J. J. (1993). Leucine transport into brush border membrane vesicles from guts of *Leptinotarsa decemlineata* and *Manduca sexta. Comp. Biochem. Physiol., 104A*, 267–272.

Rouvinen, J., Bergfors, T., Teeri, T., Knowles, J. K. C., & Jones, T. A. (1990). Three-dimensional structure of cellobiohydrolase II from *Trichoderma reesei. Science, 249*, 380–386.

Ryan, C. A. (1990). Proteinase inhibitors in plants: Genes for improving defense against insects and pathogens. *Annu. Rev. Phytopathol., 28*, 425–449.

Sakon, J., Irwin, D., Wilson, D. B., & Karplus, P. A. (1997). Structure and mechanism of endo/exocellulase E4 from *Thermomonospora fusca. Nature Struct. Biol., 4*, 810–818.

Salvucci, M. E. (2003). Distinct sucrose isomerases catalyze trehalulose synthesis in whiteflies, *Bemisia argentifolii*, and *Erwinia rhapontici. Comp. Biochem. Physiol. B, 135*, 385–395.

Salvucci, M. E., Rosell, R. C., & Brown, J. K. (1998). Uptake and metabolism of leaf proteins by the silverleaf whitefly. *Arch. Insect Biochem. Physiol., 39*, 155–165.

Santos, C. D., & Terra, W. R. (1986). Midgut alpha-glucosidase and beta-fructosidase from *Erinnyis ello* larvae and imagoes: Physical and kinetic-properties. *Insect Biochem., 16*, 819–824.

Santos, C. D., Ribeiro, A. F., & Terra, W. R. (1986). Differential centrifugation, calcium precipitation and ultrasonic disruption of midgut cells of *Erinnyis ello* caterpillars: Purification of cell microvilli and inferences concerning secretory mechanisms. *Can. J. Zool., 64*, 490–500.

Sato, P. M., Lopes, A. R., Juliano, L., Juliano, M. A., & Terra, W. R. (2008). Subsite substrate specificity of midgut insect chymotrypsins. *Insect Biochem. Mol. Biol., 38*, 628–633.

Scharf, M. E., Kovaleva, E. S., Jadhao, S., Campbell, J. H., Buchman, G. W., & Boucias, D. G. (2010). Functional and translational analyses of a beta-glucosidase gene (glycosyl hydrolase family 1) isolated from the gut of the lower termite *Reticulitermes flavipes. Insect Biochem. Mol. Biol., 40*, 611–620.

Schechter, I., & Berger, A. (1967). On the size of active site in proteases. *Biochim. Biophys. Res. Commun., 27*, 157–162.

Schmitz, J., Preiser, H., Maestracci, D., Ghosh, B. K., Cerda, J., & Crane, R. K. (1973). Purification of the human intestinal brush border membrane. *Biochim. Biophys. Acta, 323*, 98–112.

Schorderet, S., Pearson, R. D., Vuocolo, T., Eisemann, C., Riding, G. A., et al. (1998). cDNA and deduced amino acid sequences of a peritrophic membrane glycoprotein, "Peritrophin-48," from the larvae of *Lucilia cuprina. Insect Biochem. Mol. Biol., 28*, 99–111.

Scrivener, A. M., Slaytor, M., & Rose, H. A. (1989). Symbiont-independent digestion of cellulose and starch in *Panesthia cribrata* Saussure, an Australian wood-eating cockroach. *J. Insect Physiol., 35*, 935–941.

Scrivener, A. M., Watanabe, H., & Noda, H. (1997). Diet and carbohydrate digestion in the yellow-spotted longicorn beetle *Psacothea hilaris. J. Insect Physiol., 43*, 1039–1052.

Shahabuddin, M. (1995). Chitinase as a vaccine. *Parasitol. Today, 11*, 46–47.

Shao, L., Devenport, M., Fujioka, H., Ghosh, A., & Jacobs-Lorena, M. (2005). Identification and characterization of a novel peritrophic matrix protein, Ae-Aper50, and the microvillar membrane protein, AEG12, from the mosquito, *Aedes aegypti. Insect Biochem. Mol. Biol., 35*, 947–959.

Shen, Z., & Jacobs-Lorena, M. (1997). Characterization of a novel gut-specific chitinase gene from the human malaria vector *Anopheles gambiae. J. Biol. Chem., 272*, 28895–28900.

Shen, Z., & Jacobs-Lorena, M. (1998). A type I peritrophic matrix protein from the malaria vector *Anopheles gambiae* binds to chitin: Cloning, expression and characterization. *J. Biol. Chem., 273*, 17665–17670.

Shen, Z., Reese, J. C., & Reeck, G. R. (1996). Purification and characterization of polygalacturonase from the rice weevil, *Sitophilus oryzae* (Coleoptera: Curculionidae). *Insect Biochem. Mol. Biol., 26*, 427–433.

Shigenobu, S., Watanabe, H., Hattori, M., Sasaki, Y., & Ishikawa, H. (2000). Genome sequence of the endocellular bacterial symbiont of aphids *Buchnera* sp. APS. *Nature, 407*, 81–86.

Shiroya, T. (1963). Metabolism of raffinose in cotton seeds. *Phytochemistry, 2*, 23–46.

Silva, A., Bacci, M., Jr., Siqueira, C. Q., Bueno, O. L., Pagnoca, F. C., et al. (2003). Survival of *Atta sexdens* workers on different food sources. *J. Insect Physiol., 49*, 307–313.

Silva, C. P., & Terra, W. R. (1994). Digestive and absorptive sites along the midgut of the cotton seed sucker bug *Dysdercus peruvianus* (Hemiptera: Pyrrhocoridae). *Insect Biochem. Mol. Biol., 24*, 493–505.

Silva, C. P., & Terra, W. R. (1995). An α-glucosidase from perimicrovillar membranes of *Dysdercus peruvianus* (Hemiptera: Pyrrhocoridae) midgut cells: Purification and properties. *Insect Biochem. Mol. Biol., 25*, 487–494.

Silva, C. P., & Terra, W. R. (1997). Alpha-galactosidase activity in ingested seeds and in the midgut of *Dysdercus peruvianus* (Hemiptera: Pyrrhocoridae). *Arch. Insect Biochem. Physiol., 34*, 443–460.

Silva, C. P., & Xavier-Filho, J. (1991). Comparison between the levels of aspartic and cysteine proteinase of the larval midguts of *Callosobruchus maculatus* (F.) and *Zabrotes subfasciatus* (Boh.) (Coleoptera: Bruchidae). *Comp. Biochem. Physiol. B, 99*, 529–533.

Silva, C. P., Ribeiro, A. F., Gulbenkian, S., & Terra, W. R. (1995). Organization, origin and function of the outer microvillar (perimicrovillar) membranes of *Dysdercus peruvianus* (Hemiptera) midgut cells. *J. Insect Physiol., 41*, 1093–1103.

Silva, C. P., Terra, W. R., de Sá, M. F. G., Samuels, R. I., Isejima, E. M., et al. (2001). Induction of digestive α-amylases in larvae of *Zabrotes subfasciatus* (Coleoptera: Bruchidae) in response to ingestion of common bean α-amylase inhibitor 1. *J. Insect Physiol., 47*, 1283–1290.

Silva, M. C. P., Terra, W. R., & Ferreira, C. (2004). The role of carboxyl, guanidine and imidazole groups in catalysis by a midgut trehalase purified from an insect larvae. *Insect Biochem. Mol. Biol., 34*, 1089–1099.

Silva, M. C. P., Terra, W. R., & Ferreira, C. (2006). Absorption of toxic beta-glucosidses produced by insects and their effect on tissue trehalases from insects. *Comp. Biochem. Physiol*, *143 B.*, 367–373.

Silva, M. C. P., Ribeiro, A. F., Terra, W. R., & Ferreira, C. (2009). Sequencing of *Spodoptera frugiperda* midgut trehalases and demonstration of secretion of soluble trehalase by midgut columnar cells. *Insect Mol. Biol.*, *18*, 769–784.

Silva, M. C. P., Terra, W. R., & Ferreira, C. (2010). The catalytic and other residues essential for the activity of the midgut trehalase from *Spodoptera frugiperda*. *Insect Biochem. Mol. Biol.*, *40*, 733–741.

Skibbe, U., Christeller, J. T., Callaghan, P. T., Eccles, C. D., & Laing, W. A. (1996). Visualization of pH gradients in the larval midgut of *Spodoptera litura* using ^{31}P-NMR microscopy. *J. Insect Physiol.*, *42*, 777–790.

Slaytor, M. (1992). Cellulose digestion in termites and cockroaches: What role do symbionts play? *Comp. Biochem. Physiol. B*, *103*, 775–784.

Soares-Costa, A., Dias, A. B., Dellamano, M., de Paula, F. F. P., Carmona, A. K., et al. (2011). Digestive physiology and characterization of a digestive cathepsin L-like proteinase from the sugar cane weevil *Sphenophorus levis*. *J. Insect Physiol.*, *57*, 462–468.

Song, H. K., & Suh, S. W. (1998). Kunitz-type soybean trypsin inhibitor revisited: Refined structure of its complex with porcine trypsin reveals an insight into the interaction between a homologous inhibitor from *Erythrina caffra* and tissue-type plasminogen activator. *J. Mol. Biol.*, *275*, 347–363.

Spencer, K. C. (1988). Chemical mediation of coevolution in the *Passiflora–Heliconius* interaction. In K. C. Spencer (Ed.). *Chemical Mediation of Coevolution* (pp. 167–240). London, UK: Academic Press.

Strobl, S., Gomis-Ruth, F. X., Maskos, K., Frank, G., Huber, R., et al. (1997). The alpha-amylase from the yellow mealworm: Complete primary structure, crystallization and preliminary X-ray analysis. *FEBS Lett.*, *409*, 109–114.

Strobl, S., Maskos, K., Betz, M., Wiegand, G., Huber, R., et al. (1998a). Crystal structure of yellow meal worm α-amylase at 1.64 A resolution. *J. Mol. Biol.*, *278*, 617–628.

Strobl, S., Maskos, K., Wiegand, G., Huber, R., GomisRuth, F. X., et al. (1998b). A novel strategy for inhibition of α-amylases: Yellow mealworm α-amylases in complex with the Ragi bifunctional inhibitor at 2.5 A resolution. *Structure*, *6*, 911–921.

Sugimura, M., Watanabe, H., Lo, N., & Saito, H. (2003). Purification, characterization, cDNA cloning and nucleotide sequencing of a cellulase from the yellow-spotted longicorn beetle, *Psacothea vilaris*. *Eur. J. Biochem.*, *270*, 3455–3460.

Sumida, M., Yuan, X. L., & Matsubara, F. (1994). Purification and some properties of soluble of soluble beta-fructofuranosidase from larval midgut of the silkworm, *Bombyx mori*. *Comp. Biochem. Physiol. B. Biochem. Molec. Biol.*, *107*, 273–284.

Tellam, R. L. (1996). The peritrophic matrix. In M. J. Lehane, & P. F. Billingsley (Eds.). *Biology of the Insect Midgut* (pp. 86–114). London, UK: Chapman and Hall.

Tellam, R. L., Wijffels, G., & Willadsen, P. (1999). Peritrophic matrix proteins. *Insect Biochem. Mol. Biol.*, *29*, 87–101.

Tellam, R. L., Vuocolo, T., Eisemann, C., Briscoe, S., Riding, G., et al. (2003). Identification of an immunoprotective mucin-like protein, peritrophin-55, from the peritrophic matrix of *Lucilia cuprina* larvae. *Insect Biochem. Mol. Biol.*, *33*, 239–252.

Terra, W. R. (1988). Physiology and biochemistry of insect digestion: An evolutionary perspective. *Braz. J. Med. Biol. Res.*, *21*, 675–734.

Terra, W. R. (1990). Evolution of digestive systems of insects. *Annu. Rev. Entomol.*, *35*, 181–200.

Terra, W. R. (2001). The origin and functions of the insect peritrophic membrane and peritrophic gel. *Arch. Insect Biochem. Physiol.*, *47*, 47–61.

Terra, W. R. (2009). Digestion. In V. H. Resh, & R. T. Cardé (Eds.), *Encyclopedia of Insects* (2nd edn, pp. 271–273). San Diego, CA: Academic Press.

Terra, W. R., & Ferreira, C. (1994). Insect digestive enzymes: Properties, compartmentalization and function. *Comp. Biochem. Physiol. B*, *109*, 1–62.

Terra, W. R., & Ferreira, C. (2005). Biochemistry of digestion. In L. I. Gilbert, K. Iatrou, & S. S. Gill (Eds.), *Comprehensive Molecular Insect Science, Vol. 4, Biochemistry and Molecular Biology* (pp. 171–224). Oxford, UK: Elsevier.

Terra, W. R., & Ferreira, C. (2009). Digestive system. In V. H. Resh, & R. T. Cardé (Eds.), *Encyclopedia of Insects* (2nd edn, pp. 273–281). San Diego, CA: Academic Press.

Terra, W. R., & Regel, R. (1995). pH buffering in *Musca domestica* midguts. *Comp. Biochem. Physiol. A*, *112*, 559–564.

Terra, W. R., Ferreira, C., & De Bianchi, A. G. (1978). Physical properties and Tris inhibition of an insect trehalase and a thermodynamic approach to the nature of its active site. *Biochim. Biophys. Acta*, *524*, 131–141.

Terra, W. R., Terra, I. C. M., Ferreira, C., & de Bianchi, A. G. (1979). Carbodiimide-reactive carboxyl groups at the active site of an insect midgut trehalase. *Biochim. Biophys. Acta*, *571*, 79–85.

Terra, W. R., Terra, I. C. M., & Ferreira, C. (1983). Inhibition of an insect midgut trehalase by dioxane and d-gluconolactone: Enzyme pKa values and geometric relationships at the active site. *Intl. J. Biochem.*, *15*, 143–146.

Terra, W. R., Ferreira, C., & Garcia, E. S. (1988). Origin, distribution, properties and functions of the major *Rhodnius prolixus* midgut hydrolases. *Insect Biochem.*, *18*, 423–434.

Terra, W. R., Valentin, A., & Santos, C. D. (1987). Utilization of sugars, hemicellulose, starch, protein, fat and minerals by *Erinnyis ello* larvae and digestive role of their midgut hydrolases. *Insect Biochem.*, *17*, 1143–1147.

Terra, W. R., Ferreira, C., Jordão, B. P., & Dillon, R. J. (1996). Digestive enzmes. In M. J. Lehane, & P. F. Billingsley (Eds.), *Biology of the Insect Midgut* (pp. 153–194). London, UK: Chapman and Hall.

Terra, W. R., Costa, R. H., & Ferreira, C. (2006). Plasma membranes from insect midgut cells. *An. Acad. Bras. Cien.*, *86*, 1–15.

Theischinger, G. (1991). Megaloptera. In I. D. Naumann, P. B. Carne, J. F. Lawrence, E. S. Nielsen, J. P. Spradbery, et al. (Eds.), *The Insects of Australia* (2nd edn., pp. 516–520). Melbourne, Australia: Melbourne University Press.

Titarenko, E., & Chrispeels, M. J. (2000). cDNA cloning, biochemical characterization and inhibition by plant inhibitors of the alpha-amylases of the Western corn rootworm, *Diabrotica virgifera virgifera*. *Insect Biochem. Mol. Biol.*, *30*, 979–990.

Tokuda, G., Saito, H., & Watanabe, H. (2002). A digestive β-glucosidase from the salivary glands of the termite, *Neotermes koshunensis* (Shiraki): Distribution, characterization and isolation of its presursor cDNA by 5′- and 3′-RACE amplifications with degenerate primers. *Insect Biochem. Mol. Biol.*, *32*, 1681–1689.

Tokuda, G., Miyagi, M., Makiya, H., Watanabe, H., & Arakawa, G. (2009). Digestive β-glucosidases from the wood-feeding termite, *Nasutitermes takasagoensis*: Intestine distribution, molecular characterization, and alteration in sites of expression. *Insect Biochem. Mol. Biol.*, *39*, 931–937.

Toprak, U., Baldwin, D., Erlandson, M., Gillott, C., Houx, et al. (2008). A chitin diacetylase and putative insect intestinal lipases are components of the *Mamestra configurata* (Lepidoptera: Noctuidae) peritrophic matrix. *Insect Mol. Biol.*, *17*, 573–585.

Treherne, J. E. (1959). Amino acid absorption in the locust (*Schistocerca gregaria* Forsk). *J. Exp. Biol.*, *36*, 533–545.

Tunaz, H., & Stanley, D. W. (2004). Phospholipase A2 in salivary glands isolated from tobacco hornworms, *Manduca sexta*. *Comp. Biochem. Physiol.*, *139B*, 27–33.

Uscian, J. M., Miller, J. S., Sarath, G., & Stanley-Samuelson, D. W. (1995). A digestive phospholipase A2 in the tiger beetle *Cincindella circumpicta*. *J. Insect Physiol*, *41.*, 135–141.

Vaaje-Kolstad, G., Horn, S. J., van Aalten, D. M., Synstad, B., & Eijsink, V. G. (2005). The non-catalytic chitin-binding protein CBP21 from *Serratia marcescens* is essential for chitin degradation. *J. Biol. Chem.*, *280*, 28492–28497.

Valaitis, A. P. (1995). Gypsy moth midgut proteinases: Purification and characterization of luminal trypsin, elastase and the brush-border membrane leucine aminopeptidase. *Insect Biochem. Mol. Biol.*, *25*, 139–149.

Valaitis, A. P., & Bowers, D. F. (1993). Purification and properties of the soluble midgut trehalase from the gypsy moth, *Lymantria dispar*. *Insect Biochem. Mol. Biol.*, *23*, 599–606.

Venâncio, T. M., Cristofolletti, P. T., Ferreira, C., Verjovski-Almeida, S., & Terra, W. R. (2009). The *Aedes aegypti* larval transcriptome: A comparative perspective with emphasis on trypsins and the domain structure of peritrophins. *Insect Mol. Biol.*, *18*, 33–44.

Vetter, J. (2000). Plant cyanogenic glycosides. *Toxicon*, *38*, 11–36.

Vinokurov, K., Taranushenko, Y., Krishnan, N., & Sehnal, F. (2007). Proteinase, amylase, and proteinase-inhibitor activities in the gut of six cockroach species. *J. Insect Physiol.*, *53*, 794–802.

Volpicella, M., Ceci, L. R., Cordewenwe, J., America, T., Gallerani, R., et al. (2003). Properties of purified gut trypsin from *Helicoverpa zea*, adapted to proteinase inhibitors. *Eur. J. Biochem.*, *270*, 10–19.

Volpicella, M., Cordewener, J., Jongsma, M. A., Gallerani, R., Ceci, J. R., & Beekwilder, J. (2006). Identification and characterization of digestive serine proteases from inhibitor-resistant *Helicoverpa zea* larval midgut. *J. Chromatogr. B*, *833*, 26–32.

Vonk, H. J., & Western, J. R. H. (1984). *Comparative Biochemistry and Physiology of Enzymatic Digestion*. New York, NY: Academic Press.

Wang, P., & Granados, R. R. (1997). Molecular cloning and sequencing of a novel invertebrate intestinal mucin. *J. Biol. Chem.*, *272*, 16663–16669.

Wang, P., & Granados, R. R. (2000). Calcofluor disrupts the midgut defense system in insects. *Insect Biochem. Mol. Biol.*, *30*, 135–143.

Wang, P., & Granados, R. R. (2001). Molecular structure of the peritrophic membrane (PM): Identification of potential PM target sites for insect control. *Arch. Insect Biochem. Physiol.*, *47*, 110–118.

Wang, P., Zhang, X., & Zhang, J. (2005). Molecular characterization of four midgut aminopeptidase N isoenzymes from the cabbage looper, *Trichoplusia ni*. *Insect Biochem. Mol. Biol.*, *35*, 611–620.

Ward, C. W. (1975a). Aminopeptidases in webbing clothes moth larvae: Properties and specificities of the enzymes of intermediate electrophoretic mobility. *Biochim. Biophys. Acta*, *410*, 361–369.

Ward, C. W. (1975b). Aminopeptidases in webbing clothes moth larvae: Properties and specificities of enzymes of highest electrophoretic mobility. *Austral. J. Biol. Sci.*, *28*, 447–455.

Watanabe, H., & Tokuda, G. (2010). Cellulolytic systems in insects. *Annu. Rev. Entomol.*, *55*, 609–632.

Wei, Y. D., Lee, K. S., Gui, Z. Z., Yoon, H. J., Kim, I., et al. (2006). Molecular cloning, expression, and enzymatic activity of a novel endogenous cellulose from the mulberry longicorn beetle, *Apriona germari*. *Comp. Biochem. Physiol.*, *145 B*, 220–229.

Whitworth, S. T., Blum, M. S., & Travis, J. (1998). Proteolytic enzymes from larvae of the fire ant, *Solenopsis invicta*: Isolation and characterization of four serine endopeptidases. *J. Biol. Chem.*, *273*, 14430–14434.

Whitworth, S. T., Kordula, T., & Travis, J. (1999). Molecular cloning of Soli EC: An elastase-like serine proteinase from the imported red fire ant (*Solenopsis invicta*). *Insect Biochem. Mol. Biol.*, *29*, 249–254.

Wigglesworth, V. B. (1933). The function of the anal gills of the mosquito larva. *J. Exp. Biol.*, *10*, 16–26.

Withers, S. G., & Rupitz, K. (1990). Measurement of active-site homology between potato and rabbit muscle α-glucan phosphorylases through use of a free energy relationship. *Biochemistry*, *29*, 6405–6409.

Woodring, J., & Lorenz, M. W. (2007). Feeding, nutrient flow, and functional gut morphology in the cricket, *Gryllus maculatus*. *J. Morphol.*, *268*, 815–825.

Wolfersberger, M. G. (2000). Amino acid transport in insects. *Annu. Rev. Entomol.*, *45*, 111–120.

Woods, H. A., & Kingsolver, J. G. (1999). Feeding rate and the structure of protein digestion and absorption in Lepidoptera midguts. *Arch. Insect Biochem. Physiol.*, *42*, 74–87.

Zeng, F., Zhu, Y. C., & Cohen, A. C. (2002). Molecular cloning and partial characterization of a trypsin-like protein in salivary glands of *Lygus hesperus* (Hemiptera: Miridae). *Insect Biochem. Mol. Biol.*, *32*, 455–464.

Zhang, S., Shukle, R., Mittapalli, O., Zhu, Y. C., Reese, J. C., et al. (2010). The gut transcriptome of a gall midge, *Mayetiola destructor*. *J. Insect Physiol.*, *56*, 1198–1206.

Zhu, Y. C., & Baker, J. E. (1999). Characterization of midgut trypsin-like enzymes and three trypsinogen cDNAs from the lesser grain borer, *Rhyzopertha dominica* (Coleoptera: Bostrichidae). *Insect Biochem. Mol. Biol., 29*, 1053–1063.

Zhu, Y. C., Kramer, K. K., Oppert, B., & Dowdy, A. K. (2000). cDNAs of aminopeptidase-like protein genes from *Plodia interpunctella* strains with different susceptibilities to *Bacillus thuringiensis* toxins. *Insect Biochem. Mol. Biol., 30*, 215–224.

Ziegler, H., Garon, C. F., Fischer, E. R., & Shahabuddin, M. (2000). A tubular network associated with the brush-border surface of the *Aedes aegypti* midgut: Implications for pathogen transmission by mosquitoes. *J. Exp. Biol., 203*, 1599–1611.

Zimoch, L., & Merzendorfer, H. (2002). Immunolocalization of chitin synthase in the tabacco hornworm. *Cell Tissue Res., 308*, 287–297.

12   Programmed Cell Death in Insects

Susan E Fahrbach
Department of Biology, Wake Forest University,
Winston-Salem, NC, USA
John R Nambu
Department of Biological Sciences,
Charles E. Schmidt College of Science,
Florida Atlantic University, Boca Raton, FL, USA
Lawrence M Schwartz
Department of Biology, 221 Morrill Science Center,
University of Massachusetts, Amherst, MA, USA

Summary

Programmed cell death (PCD) is a normal component of development and homeostasis in animals, plants, and even some single-celled organisms. While there appear to be multiple forms of PCD, the best characterized are apoptosis and autophagy. In insects, PCD has been observed in diverse tissues, and is required for the normal completion of metamorphosis. This chapter reviews the history of studies of PCD in two key models, the hawkmoth *Manduca sexta* and the fruit fly *Drosophila melanogaster*. This highly active field of research is built on a sturdy foundation of decades of studies of hormonally-regulated PCD in neuromuscular systems in these two species. Major discoveries based on insect research include identification of the RHG protein apoptosis activators and IAP family proteins as well as the first demonstration of the role of ubiquitination in muscle PCD. Contemporary studies of PCD in neuromuscular systems and dying larval tissues (salivary gland, midgut) have demonstrated the co-occurrence of apopotic and autophagic gene expression in individual cells fated to die. The study of PCD during metamorphosis in insects is a mature field of inquiry that offers numerous opportunities for study of mechanisms related both to insect development and human disease.

DOI:10.1016/B978-0-12-384747-8.10012-1

12.1. Introduction

Programmed cell death (PCD) is a fundamental component of the post-embryonic development of insects, and has been the subject of at least 3000 PubMed-indexed publications as of this writing. PCD is particularly evident during metamorphosis of the neuromuscular systems of holometabolous insects, and these systems will be emphasized in this chapter. Nerve and muscle cells born during embryogenesis are deleted in a segment- and cell-specific fashion at both the larval–pupal and pupal–adult transitions. Although such deaths presumably occur in all insects with complete metamorphosis, almost all of the information on this topic has been obtained from two species: the fruit fly *Drosophila melanogaster* (Diptera: Drosophilidae), and the tobacco hawkmoth *Manduca sexta* (Lepidoptera: Sphingidae). The sophistication of the genetic tools and genomic resources available in *Drosophila* has allowed many key regulatory pathways in insect PCD to be identified. In addition, the success of large-scale genetic screens for embryonic pattern formation as well as detailed analysis of *Hox* gene clusters have provided insight into mechanisms underlying elaboration of body axes, specification of cell populations, and organogenesis. These insights into early development provide a strong base for an understanding of metamorphosis that is now supplemented by large-scale analyses of gene expression using tools such as DNA micro- and tiling arrays. The small size of fruit flies, however, limits the utility of this species for physiological and biochemical studies involving hormone manipulations, tissue transplantation, and surgical interventions. The lepidopteran *Manduca sexta* is an alternative model uniquely suited for these experimental approaches, serving as the insect equivalent of the laboratory white rat (Fahrbach, 1997). Experimental results from these species provide the foundation of this chapter, but the explosion in availability of sequenced insect genomes that characterized the first decade of the 21st century opens the door to true comparative studies of PCD.

12.2. PCD, Apoptosis, Autophagy, or Necrosis?

It is now recognized that cell death is a normal part of life in animals, plants, and even some single-celled organisms (Ameisen, 2002; Segovia *et al.*, 2003; Reape and McCabe, 2008; Shemarova, 2010). For example, parasitic protists may undergo PCD to minimize their impact on the immune system of a mammalian host (Deponte, 2008), and several types of bacteria exhibit PCD in response to stress or nutrient deprivation (Engelberg-Kulka *et al.*, 2006). In multicellular organisms, the capacity to remove selected cells during development provides organisms with a plastic response to developmental contingencies. In vertebrates, PCD has been demonstrated to:

1   Match sizes of interacting populations of cells, such as oligodendrocytes, to the axons they myelinate (Barres *et al.*, 1992).
2   Remove deleterious cells, such as self-reactive T cells (Smith *et al.*, 1989; Opferman, 2008).
3   Sculpt the body, such as in the loss of interdigital cells in the limb bud of the developing embryo (Zuzarte-Luís and Hurlé, 2002; Montero and Hurlé, 2010).
4   Remove developmentally obsolete cells, such as the tail of the tadpole and other larval organs in amphibians (Yoshizato, 1996; Ishizuya-Oka *et al.*, 2010).

In insects, dramatic examples of PCD take place during embryogenesis and during metamorphosis, when larval tissues are destroyed to allow the formation of new adult structures. The predictable events of insect metamorphosis provide accessible experimental systems for defining the regulatory mechanisms that mediate PCD, and which can be extended to other organisms.

12.2.1. Apoptosis

The best-characterized morphology associated with both developmental and pathological cell death is apoptosis, a term coined by Kerr, Currie, and Wyllie (Kerr *et al.*, 1972). At its introduction, apoptosis was a morphological term that implied neither mechanism nor specific developmental context. During the process of apoptosis, cells shrink and display plasma membrane zeosis, defined as the formation of numerous protuberances or blebs. Time-lapse photographs of cells undergoing apoptosis in culture look like drops of water skittering on a hot skillet. Under *in vivo* conditions, apoptotic cells are typically rapidly phagocytosed either by neighboring cells or by phagocytes (Franc, 2002; Geske *et al.*, 2002; Hart *et al.*, 2008). An informative video is available at: http://www.youtube.com/watch?v=V-NsR-krKME&feature=related.

The nucleus of apoptotic cells condenses, and the chromatin becomes electrondense and marginated along the inner aspect of the nuclear envelope. These morphological changes are physical manifestations of a massive cleavage of genomic DNA that occurs when endogenous nucleases become activated and cleave the linker DNA between individual nucleosomes (Wyllie *et al.*, 1984; Enari *et al.*, 1998). The fragmentation of the genome can be visualized when DNA is extracted and fractionated by size in agarose (Eastman, 1995). It is sometimes difficult to detect clear apoptotic ladders because, in many tissues, dying cells are intermingled with healthy ones. In addition, because tissues are homogenized during DNA isolation, it is impossible to determine which subpopulation of cells contributed the degraded DNA. An alternative strategy for detecting fragmented genomic DNA is to employ *in situ* labeling techniques on tissue sections, such as terminal deoxynucleotidyl transferase (TdT)-mediated dUTP-biotin nick end labeling, referred to as the TUNEL method (Gavrieli *et al.*, 1992; Ben-Sasson *et al.*, 1995). The TUNEL method is now supplemented by other *in situ* methods, including use of monoclonal antibodies to detect single-stranded DNA (Frankfurt *et al.*, 1996). Other *in situ* methods rely on antibodies targeted to proteins that form the molecular machinery of apoptosis or to cleavage products resulting from the action of proteolytic enzymes on their protein targets during the execution phase of apoptosis (Huerta *et al.*, 2007). Antibodies that recognize activated, cleaved caspase-3 can be used to detect apoptosis in *Drosophila* (Xu *et al.*, 2006; Shapiro *et al.*, 2008), although there has been some controversy regarding the identity of endogenous proteins recognized by these antisera (Fan and Bergmann, 2010). New computational tools have been developed for high throughput analysis of apoptotic cells *in vivo*, which will facilitate genetic screens (Forero *et al.*, 2009).

12.2.2. Autophagy

While apoptosis is the morphology most commonly observed during PCD, particularly during embryogenesis, it is not the only one. Other suicidal cell deaths are designated autophagic (Bursch, 2001; Ryoo and Baehrecke, 2010). Like apoptosis, autophagy appears to be an ancient cellular process with counterparts in unicellular organisms (King and Gottlieb, 2009). Cells undergoing autophagy lack, in general, initial condensation of chromatin, and instead form autophagic bodies (autophagosomes) – double-membrane bounded portions of cytoplasm that eventually fuse with a lysosome (Xie and Klionsky, 2007). The identification and analysis of autophagosomes was originally dependent upon transmission electron microscopy, and there was a lag in development of useful cellular markers for autophagic cells. This lag initially hindered analysis of the developmental significance of autophagy.

One of the most common triggers for autophagy is nutrient limitation (King and Gottlieb, 2009), and in this context autophagy represents a cell survival mechanism (because the net effect is to recycle cellular materials into metabolic pathways) rather than a mediator of PCD. However, autophagy plays an active role in PCD in several tissues in developing *Drosophila*, including the salivary gland (Berry and Baehrecke, 2007) and ovary (Nezis *et al.*, 2009). There is evidence for cross-talk between autophagic and apoptotic death pathways (Hou *et al.*, 2008; Nezis *et al.*, 2010).

Recent interest in the contribution of autophagy to human diseases such as cancer, Crohn's disease, and neurodegeneration associated with aging has led to development of several novel markers for autophagy, including homologs of yeast genes essential for autophagy, such as LC3 in mice and *Draut1* in *Drosophila melanogaster* (Juhász *et al.*, 2003; Tanida *et al.*, 2008). As a result, rapid progress in understanding autophagy in insects can be expected (Barth *et al.*, 2010).

12.2.3. PCD is an Inclusive Term

In this chapter, we employ the term PCD to describe cell loss that occurs in a temporally and spatially predictable manner. PCD encompasses both apoptotic and autophagic cell deaths that occur as normal components of development. Current evidence argues strongly that, even within a single cell, multiple molecular mechanisms of PCD can be simultaneously active (Schwartz *et al.*, 1993a; Jones and Schwartz, 2001; Thummel, 2001; Muppidi *et al.*, 2004; Stoica and Faden, 2010). For example, death of the larval salivary glands during metamorphosis in *Drosophila* requires a combination of apoptosis and autophagy, both of which are activated by pulses of the steroid 20-hydroxyecdysone (20E) secreted by the ring gland at pupation (Jiang *et al.*, 2000; Berry and Baehrecke, 2007; Conradt, 2009). A trend in the current literature is to refer to specific examples of PCD in terms of the underlying biochemical pathways (e.g., caspase-dependent, autophagy (Atg)-dependent) when these have been experimentally determined.

12.2.4. Necrosis

Apoptosis and autophagy are orchestrated developmental decisions, but all cells can be induced to die by necrosis, a form of cell death that occurs in response to external insult such as heat, salt, abrasion, toxin exposure, etc. (Dive *et al.*, 1992; Kroemer *et al.* 1998). Necrosis typically follows disruption of plasma membrane integrity, which in turn results in an influx of water and ions, most notably calcium, which leads to subsequent cellular swelling and lysis. In vertebrates, which have an adaptive immune response, necrosis provides a valuable warning to the immune system that focal injury has occurred, because the cellular constituents that are liberated during necrosis are highly inflammatory (MacDonald and Stoodley, 1998; Fadok *et al.*, 2001). Endogenous responses to necrosis are responsible for much of the secondary tissue damage that accompanies injury (Whelan *et al.*, 2010).

12.2.5. Hybrid Forms of Cell Death

Novel cell death pathways, including hybrids that incorporate features of established pathways, have been described, but their relevance to insect systems in general, and insect neuromuscular systems in particular, has not yet been established. A recent example from mammalian cells is necroptosis, an orderly, programmed form of necrosis. Necroptosis can be triggered in cells lacking caspase activity by a death domain containing kinase RIP1 (Christofferson and Yuan, 2010) as a consequence of the binding of TNF (tumor necrotic factor) to its receptor, TNFR1 (Vandenabeele *et al.*, 2010). It is associated with ischemic cell death following stroke (Degterev *et al.*, 2005). One lesson from the broader cell death literature for investigators of PCD in insects is that cross-talk across death/survival pathways is to be expected. One consequence of such cross-talk is that simple, tried and true histological or single marker methods are not always reliable gauges of cellular status (Mohseni *et al.*, 2009).

12.3. Historical Overview and Current Trends

The first morphological description of PCD during insect metamorphosis we have found in the literature is a description of the death of intersegmental muscles (ISMs) in the silkmoth *Bombyx mori* (Kuwana, 1936). The phenomenon of ISM death during insect metamorphosis was largely unknown outside of Asia until independently described by Finlayson, who also studied development in Lepidoptera (Finlayson, 1956). With the exception of a modest number of descriptive papers documenting examples of PCD in other insect taxa, little was published about cell loss during arthropod development for most of the subsequent decade. A compendium of the early studies of cell death in insects can be found in Glücksmann's early review of the field (Glücksmann, 1951). A comprehensive investigation of PCD based on analysis of the ISMs was initiated in the mid-1960s by Lockshin and Williams (1964, 1965a, 1965b, 1965c). In fact, it was in R. A. Lockshin's 1963 doctoral dissertation on ISM development that the term programmed cell death was first introduced.

The first description of neuronal death during insect metamorphosis detailed the post-embryonic changes in neuronal populations observed in the abdominal ganglia of *Manduca sexta* (Taylor and Truman, 1974). Truman and his colleagues exploited their ability to count motoneurons in individual ganglia by backfilling cut ends of peripheral nerves with cobalt; they reported that motoneurons were lost from the abdominal ganglia at both the larval–pupal and pupal–adult transitions. Many of these motoneurons were later determined to be uniquely identifiable cells, which permitted the precisely regulated nature of metamorphic neuron death to be established (Truman, 1983; Levine and Truman, 1985).

Studies by Schwartz and Truman subsequently demonstrated that changes associated with ISMs at the end of metamorphosis are controlled by hormonal signals, particularly by the steroid 20E (Schwartz and Truman, 1982; 1983). The principle of steroid control of nerve and muscle fate at metamorphic transitions was extended to the death of abdominal neurons that occurs after adult eclosion in *Manduca* (Truman and Schwartz, 1984), and to the death of *Manduca* proleg motoneurons (and their associated muscles) at pupation (Weeks and Truman, 1985; Weeks, 1987, 1999). Using the ISMs from the moth *Manduca sexta*, Schwartz and colleagues were the first to clone death-associated transcripts from any organism (Schwartz *et al.*, 1990a). The results of these influential studies are detailed below.

Descriptive studies of the death of neurons and muscles after adult eclosion in *Drosophila melanogaster* (Kimura and Truman, 1990) foreshadowed the initiation in the 1990s of an intensive genetic analysis of PCD in fruit flies. Efforts focused on identifying the molecular mechanisms that mediate PCD, and the extracellular signals that trigger it. Subsequently, interest grew in the development of *Drosophila* models of human neurodegenerative disease (e.g., Mutsuddi and Nambu, 1998; Driscoll and Gerstbrein, 2003; Lu, 2009). The underpinning of this work is the consensus that mechanisms of cell death are broadly conserved across phylogeny. Studies of PCD in insects are well positioned to address specific questions concerning metamorphosis, basic questions in cell biology, and topics of relevance to human health.

12.3.1. Earlier Reviews of PCD Relevant to Insects

The field of PCD in insects has been extensively reviewed, although many reviews focus on narrow subtopics within the field. For early reviews, see Wing and Nambu (1998), Abrams (1999), Bangs and White (2000), and Lee and Baehrecke (2000). Recent reviews covering the discovery of the first genes associated with PCD in *Drosophila melanogaster* include those by Hay and Guo (2006); Steller (2008); and Xu *et al.* (2009). Autophagic and apoptotic cell deaths during *Drosophila* metamorphosis were first compared by Thummel (2001). Notable reviews of PCD during animal development in general (including insects) were provided by Jacobson *et al.* (1997), Milligan and Schwartz (1997), Meier *et al.* (2000), and Baehrecke (2002). Kornbluth and White (2005) and Conradt (2009) compared the molecular basis of PCD during development in three model organisms: the nematode *Caenorhabditis elegans*, the fruit fly *Drosophila melanogaster*, and the mouse *Mus musculus*.

12.3.2. Current Trends in PCD Research and Publication

Persistent research interest in the general field of cell death studies is indicated by the success of print and online journals dedicated to PCD. These include *Cell Death and Differentiation*, *Autophagy*, *Journal of Cell Death*, *Cell Death and Disease*, and *Apoptosis*, all of which publish studies based on insect models.

12.4. The *Manduca* Model

12.4.1. Choice of *Manduca sexta* as a Model System for the Study of PCD of Nerve and Muscle Cells

Insect endocrinologists and neurobiologists favor *Manduca sexta* because of its large size and ease of rearing in culture on an artificial diet (Bell and Joachim, 1976; Arnett, 1993; Fahrbach, 1997). This species has a facultative rather than an obligatory diapause, and therefore all life stages can be produced in the laboratory at any time of the year. Development from egg to adult requires roughly 40 days. There are five larval stages (each referred to as an instar, so that the final larval stage is the fifth instar) and a single pupal stage. The adult emerges after about 18 days, and lives approximately 10 days. The large size of this insect permits extensive surgical and endocrine manipulations. The central nervous system (CNS) consists of a dorsal brain and a ventral nerve cord. The abdominal portion of the ventral nerve cord retains its segmental organization of discrete thoracic and abdominal ganglia throughout the post-embryonic period – a feature that facilitates anatomical, electrophysiological, and functional analyses.

The thoracic and abdominal musculature of *Manduca sexta* has been described (Eaton, 1988). Many aspects of neuromuscular metamorphosis first described in *Manduca* were subsequently identified in metamorphosing fruit flies (e.g., Kimura and Truman, 1990; Truman *et al.*, 1994), despite evidence that the insect orders of Lepidoptera and Diptera have been evolving independently for at least 200, and possibly 300, million years (Hoy, 2003). This conservation suggests that many aspects of neuromuscular metamorphosis in insects, including PCD, are ancient in origin. This relationship permits phenomena first observed in *Manduca* to be subjected to molecular genetic analysis by shifting to *Drosophila* for follow-up studies (e.g., Robinow *et al.,* 1993). Conversely, orthologs of genes first identified in *Drosophila* can be cloned in *Manduca*, and their patterns of expression followed throughout development in identified neurons and muscles (e.g., Nagy *et al.*, 1991; Kraft and Jäckle, 1994).

The striking changes in the organization of the nervous and muscular systems that accompany lepidopteran metamorphosis result from a combination of post-embryonic cell proliferation, modification of structures formed initially in the embryo, and PCD. In the CNS, while a large proportion of developing neurons and glia undergo PCD, these deaths occur in specific, isolated cells. In contrast, in the developing musculature there is wholesale loss of entire bundles of muscle fibers (Schwartz, 2008). The coordinated loss of the ISMs following adult eclosion in

Manduca offers a robust system for the study of PCD, because these cells are exceptionally large, easily accessible throughout the entire process of degeneration, and uncontaminated by persisting muscle fibers (**Figure 1**).

The ISMs of *Manduca* are composed of giant syncytial fibers approximately 5 mm long and 1 mm in diameter. These embryonically-derived fibers are used by the larva for locomotion, and by the pupa for defensive and respiratory behaviors. The ISMs also perform the major abdominal movements required for the eclosion behavior of the adult moth. The ISMs begin to atrophy 3 days prior to adult eclosion. This period of atrophy results in a loss of 40% of muscle mass; despite this atrophy, the ability to contract is maintained during this early phase (Schwartz and Ruff, 2002). Once the ISMs have participated in eclosion, they are no longer required for any adult behavior. They die over the course of the subsequent 30 hours (Lockshin and Williams, 1965a; Schwartz, 2008).

12.4.2. Hormonal Regulation of Metamorphosis

As in all insects, metamorphosis in *Manduca* is regulated by two categories of developmental hormones: ecdysteroids and juvenoids (reviewed by Nijhout, 1994; Truman and Riddiford 2002). The ecdysteroids are steroid hormones that exert their primary actions through members of the nuclear receptor superfamily of proteins (Robinson-Rechavi *et al.,* 2003). The juvenile hormones are terpinoids (Riddiford, 2008). At present, the cellular mode of the action of the juvenile hormones in metamorphosis remains to be defined, and likely involves more than one signaling pathway (Gilbert *et al.*, 2000; Jones and Jones, 2000; Wilson *et al.*, 2006; Bitra and Palli, 2009). Experiments on the hormonal regulation of neuromuscular metamorphosis in *Manduca* have clearly established that the changes in cell populations and connectivity that occur during neuromuscular metamorphosis are controlled by direct actions of ecdysteroids and juvenile hormones on neurons, glial cells, and muscle (Bennett and Truman 1985; Streichert *et al.*, 1997). Receptors for ecdysteroids have been localized to neuronal, glial cell, and muscle cell nuclei in *Manduca* (Bidmon and Koolman, 1989; Fahrbach and Truman, 1989; Bidmon and Sliter, 1990; Fahrbach, 1992; Hegstrom *et al.*, 1998).

12.4.3. PCD of Neurons during Metamorphosis

12.4.3.1. Background and overview Neuronal death during metamorphosis in *Manduca* involves motoneurons, interneurons, and identified peptidergic neurons. Most studies have focused on the death of motoneurons not only because of the greater ease of identification of specific neurons across individuals, but also because the highly visible degeneration of muscles during post-embryonic development often suggests that innervating

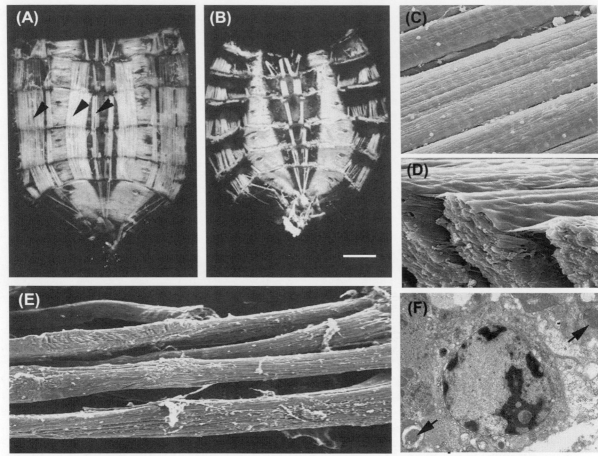

Figure 1 The morphology of the ISMs at the light (A, B), scanning electron microscopic (C–E), and transmission electron microscopic (F) levels. ISMs were examined in the abdomens of newly eclosed adults (A) and 30-hour-old adults (B). Three sets of bilaterally symmetric ISMs span four abdominal segments (arrowheads) in the newly ecdysed adult, but they disappear during the subsequent 30 hours. Other abdominal muscle groups are spared, and persist throughout adulthood. The ISMs are composed of large, well-defined muscle fibers (C, D) that rapidly lose mass as they die (E). This death is cell autonomous, and the dying fibers are not phagocytosed (F). Ultrastructural analysis of the ISMs during death reveals pyknotic nuclei and numerous autophagic vesicles (arrows) (F). Scale bar approximately 5 mm (A, B), 10 μm (C), 4 μm (D), 40 μm (E), and 2 μm m (F). Adapted from Schwartz (2008).

motoneurons will be lost, so that investigators in effect "know where to look." The presence of sexually dimorphic structures in the adult, such as the oviduct (Giebultowicz and Truman, 1984; Thorn and Truman, 1989) and brain (Kalberer *et al.*, 2010), also often suggests where neuronal PCD will be found. Dying peptidergic neurons can be identified after the detection of stage-specific patterns of antibody labeling (Ewer *et al.*, 1998). Cell counts and the presence of small pycnotic (shrunken) profiles often provide the only anatomical clues that interneurons have died. The observation that the corpses of many dying neurons are not immediately phagocytosed and therefore persist in the ganglia facilitates the detection of PCD well after the initiation of the PCD program.

Patterns of neuronal death during metamorphosis in *Manduca* have been most fully described for the motoneurons of the abdominal ganglia, although histological surveys of thoracic ganglia reveal that PCD of neurons is

also found in this tissue at the metamorphic transitions (S. E. Fahrbach, unpublished data) (**Figure 2**). At the larval–pupal transition, the best-studied examples of dying neurons are the proleg motoneurons (Weeks and Truman, 1985; Weeks, 1999). Studies of neuronal death associated with adult eclosion have focused primarily on the death of motoneurons (including the death of proleg motoneurons that persist after pupation) in the unfused abdominal ganglia, A3 through A5. In addition to the deaths of fully differentiated neurons, PCD of undifferentiated neurons occurs in the imaginal nests of the segmental ganglia, both as the larvae feed and at the onset of metamorphosis (Giebultowicz and Truman, 1984; Booker and Truman 1987; Thorn and Truman, 1989; Booker *et al.*, 1996). PCD of neuroblasts that are active during post-embryonic life terminate proliferation in specific lineages in both the segmental ganglia and the brain: in *Manduca*, the death of neuroblasts has been most comprehensively studied in

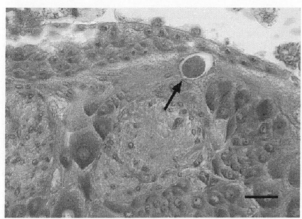

Figure 2 Transverse sections (8 μm) through the first thoracic (prothoracic or T1) ganglion of the ventral nerve cord of *Manduca sexta* fixed in alcoholic Bouins 72 hours after pupal ecdysis. The ganglion was dehydrated, embedded in a paraffin-based medium, sectioned, and stained with hematoxylin and eosin. The arrow points to the shrunken, condensed profile of a motoneuron that has undergone PCD. Because of the absence of phagocytosis in the ventral nerve cord during metamorphosis, the persisting fragments of such neurons are easily detected in insect nervous tissue prepared using routine histological techniques. Scale bar = 30 μm.

the segmental ganglia (Booker and Truman, 1987; Booker *et al.*, 1996).

The emphasis on PCD of neurons at the larval–pupal and pupal–adult transitions in *Manduca* should not obscure the fact that the majority of larval neurons survive metamorphosis and become integral components of the adult nervous system.

12.4.3.2. PCD of abdominal motoneurons during metamorphosis in *Manduca sexta*

With rare exceptions, the abdominal motoneurons of *Manduca* are born during embryonic life. At both the larval–pupal and the pupal–adult metamorphic transitions, some of the larval motoneurons lose their muscle targets as a consequence of changes in the organization of the skeletal musculature (Levine and Truman, 1985). At the larval–pupal transition, some of these now targetless motoneurons die, including motoneurons that innervate muscles associated with the five pairs of abdominal prolegs, the locomotory appendages of the caterpillar that disappear at pupation (Weeks and Truman, 1985). In some segments, however, the proleg motoneurons survive this round of PCD, and, after a period of dendritic regression, acquire new muscle targets in the developing adult (Weeks and Ernst-Utzschneider, 1989; Weeks *et al.*, 1992). Targetless motoneurons therefore can either undergo PCD or become respecified. The death of these motoneurons appears to be triggered by endocrine cues independent of contact with target muscles (Bennett and Truman, 1985; Weeks and Truman, 1985). Abdominal motoneurons that lose their muscle targets after adult eclosion all die within 48 hours

of emergence; possibly the short (approximately 10-day) life of the adult may not provide sufficient opportunity for respecification. It should be noted that unpublished studies on the giant silkmoth *Antheraea pernyi* (J. W. Truman) and *Hyalophora cecropia* (S. E. Fahrbach) suggest that, in some species, motoneurons persist after the death of the ISMs, neither respecifying nor dying. The survival of these motoneurons therefore is not dependent on target-derived trophic signals, in contrast to vertebrate motoneurons. Comparative studies on a broader range of insect taxa are needed to complete our understanding of neuronal PCD following adult eclosion.

12.4.3.3. Sex-specific PCD of abdominal motoneurons during metamorphosis

The populations of abdominal motoneurons are the same in male and female larvae, but during metamorphosis the genital segments undergo sex-specific morphological changes that are accompanied by a wave of sex-specific PCD in the ganglia that innervate these segments (Giebultowicz and Truman, 1984). This is an example of equal opportunity sexual differentiation, as some motoneurons persist in males but degenerate in females, while others show the opposite pattern (Thorn and Truman, 1989). There is a sex difference in the timing of sex-specific PCD relative to pupation, with most neurons in females dying during the first 2 days following pupation, while most neurons in males die during the third to the sixth days after pupation (Thorn and Truman, 1994a).

An interesting example of sex-specific neuronal death in *Manduca* is the case of the imaginal midline neurons (IMNs). These are unusual motoneurons that are born post-embryonically during the fourth (penultimate) larval instar. These motoneurons innervate visceral rather than skeletal musculature, and a subset can be tracked during metamorphosis because they are immunoreactive with an antibody against molluscan small cardioactive peptide b (Thorn and Truman, 1994b). IMNs that innervate the sperm duct in males are absent from the terminal ganglia of females, while IMNs that innervate the oviduct in females are absent from the terminal ganglia of males. There is some evidence that contact with an appropriate target enhances the survival of the IMNs that innervate the sperm duct in males, another atypical (but vertebrate-like) aspect of their physiology.

12.4.3.4. PCD of identified peptidergic neurons during metamorphosis

Two identified interneurons (INs) that contain crustacean cardioactive peptide (CCAP), cell 27 and IN 704, undergo PCD within 36 hours of adult eclosion (Ewer *et al.*, 1998). Both cell 27 and IN 704 display increases in cGMP immunoreactivity during larval ecdyses. Cell 27 also shows this response at pupal ecdysis and adult eclosion (Ewer and Truman, 1997). Because application of CCAP to the isolated CNS can trigger the motor patterns of ecdysis, this set of neurons appears to be central to the control of molting behavior (Gammie and Truman, 1997; Mesce and Fahrbach, 2002). The death of peptidergic neurons involved in the control of ecdysis behavior, after the moth's molting career is finished, supports the hypothesis that obsolescent neurons are actively eliminated from the *Manduca* nervous system.

12.4.3.5. PCD in the brain during metamorphosis

Detailed studies of specific regions of the *Manduca* brain have clearly demonstrated that newly-generated neurons die during adult development. PCD is a prominent feature of the development of both the medulla and lamina cortices in the optic lobes (Monsma and Booker, 1996a). Both developmental hormones and retinal afferents appear to regulate this process (Monsma and Booker, 1996b). The extent of PCD in other regions of the developing *Manduca* brain is largely unstudied. A general role for ingrowing sensory afferents in the regulation of neuronal survival is suggested by the sex differences that arise in populations of antennal lobe interneurons as a result of sexual dimorphisms in the antennae (Schneiderman *et al.*, 1982; Kalberer *et al.*, 2010). Recent studies attempting 3D reconstructions of the developing *Manduca* brain suggest a revival of interest in brain metamorphosis that may lead to new data on PCD in the brain (Huetteroth *et al.*, 2010). Several peptidergic neurons in the brain that might be expected to undergo PCD persist through the adult stage of life. These are the protocerebral prothoracicotropic (PTTH) neurons, which regulate the synthesis and secretion of ecdysteroids by the prothoracic glands. PTTH activity and, presumably, PTTH neurons persist in the adult brain (Rybczynski *et al.*, 2009), despite the PCD of the prothoracic glands in the pharate adult (Dai and Gilbert, 1997).

12.4.4. Regulation of PCD in the Nervous System during Metamorphosis

The ecdysteroids couple neuronal PCD with other metamorphic changes. It is important to note that the ecdysteroid cue for triggering metamorphic neuronal death can be either a rising or a falling titer, depending upon developmental stage. For example, the decline in circulating levels of 20E that occurs at the end of adult development prior to adult eclosion is the cue for PCD of abdominal motoneurons at this time, and treatment with exogenous 20E blocks this PCD (Truman and Schwartz, 1984). By contrast, it is the pre-pupal rise in circulating ecdysteroids that is responsible for the larval–pupal transition death of the proleg motoneurons in abdominal ganglia A5 and A6 (Weeks and Ernst-Utzschneider, 1989; Weeks *et al.*, 1992). The response of the proleg motoneurons to the steroid signal, however, is segment-specific. Homologous neurons in abdominal ganglia A3 and A4 persist through the pupal stage and adult development, but then

undergo PCD within 24 hours of adult eclosion (Zee and Weeks, 2001). These responses to 20E, as well as the segment specificity of response at different stages in development, are retained when individual proleg motoneurons are cultured *in vitro*, providing evidence for the cell-autonomous, target-independent nature of these PCDs (Streichert *et al.*, 1997; Hoffman and Weeks, 1998).

While ecdysteroids regulate gene expression in the *Manduca* nervous system (see, for example, Garrison and Witten, 2010) the target genes that mediate neuronal PCD have not been identified. In *Drosophila*, the expression of the ecdysteroid receptor A (EcR-A) has been directly correlated with the occurrence of post-eclosion neuronal death in the CNS and transcriptional activation of the *reaper* and *hid* death genes (Robinow *et al.*, 1993; Jiang *et al.*, 2000). It is not known if a similar relationship prevails in *Manduca* neurons, although autoradiographic evidence has demonstrated that *Manduca* motoneurons fated to die at the start of adult life display nuclear concentration of radiolabeled ecdysteroids (Fahrbach and Truman, 1989).

Evidence that other signals fine-tune timing of the death of neurons during metamorphosis comes from several sources. Adult *Manduca sexta* emerge from their pupal cuticle in an underground pupation chamber, and then must dig to the surface before inflating their wings. Adult moths forced to continue digging for hours beyond the time this behavior would normally cease exhibited delayed death of abdominal motoneurons (Truman, 1983). In addition, transection of the ventral nerve cord prior to adult eclosion blocks the death of specific motoneurons in ganglia posterior to the point of transection, even in moths in which the levels of 20E undergo a normal decline (Fahrbach and Truman, 1987).

A well-studied example of a spared abdominal motoneuron is MN-12. Subsequent to ventral nerve cord transection, this supernumerary member of the adult abdominal ganglion maintains its normal central arborizations and electrophysiological properties, implying that the cell death program has been completely blocked in the absence of a descending signal (Fahrbach *et al.*, 1995; DeLorme and Mesce, 1999). Treatment of cultured abdominal ganglia with extracts prepared from ventral nerve cord restores the normal pattern of cell death to MN-12, but the active factor in the extracts remains to be identified (Choi and Fahrbach, 1995). Other examples of motoneuron death, such as the death of the accessory planta retractors (APRs) at the larval–pupal transition, however, are unaffected by cutting of the connectives prior to the normal time of death (Weeks and Davidson, 1994). This suggests that the phenomenon of interganglionic cell death signaling affects only a subset of neurons.

Because of the scattered and episodic nature of neuronal death during metamorphosis, and the unavailability of transgenic *Manduca* for analysis, little is known about the molecular mechanisms of neuronal PCD in this species.

Hormone-dependent neuronal PCD is blocked by treatment with inhibitors of transcription or translation (Weeks *et al.*, 1993; Fahrbach *et al.*, 1994; Ewer *et al.*, 1998; Hoffman and Weeks, 1998). In support of the hypothesis that PCD of *Manduca* neurons requires *de novo* protein synthesis, a two-dimensional gel electrophoresis analysis of the *Manduca* abdominal ganglia revealed changes in protein expression patterns associated with newly-eclosed adults, a period of massive PCD. These changes included expression of novel proteins (Montemayor *et al.*, 1990). In cultured proleg motoneurons, inhibition of caspase activity blocked PCD (Hoffman and Weeks, 2001), but ultrastructural studies of the APR motoneurons and the motoneurons that innervate the ISMs indicate that neuronal death in *Manduca* during metamorphosis may be autophagic rather than apoptotic, or combine features of both PCD programs (Stocker *et al.*, 1978; Kinch *et al.*, 2003).

Immunolabeling studies, in which the distribution of several death-associated gene products (initially identified from a screen of dying moth muscle; see section 12.5.4.3 and **Table 1**) was examined in the segmental ganglia, failed to reveal a reliable correlation of enhanced ubiquitination- or multicatalytic proteinase-immunoreactivity within dying neurons (Fahrbach and Schwartz, 1994; Hashimoto *et al.*, 1996), despite association of these gene products with PCD in insect skeletal muscles (Haas *et al.*,

1995; Jones *et al.*, 1995). This neuron–muscle discrepancy may reflect that the basal levels of these components are sufficient to mediate neuronal PCD, but are inadequate for the destruction of the giant muscle fibers. By contrast, apolipophorin III is upregulated both by dying neurons and by degenerating muscles in *Manduca*, a finding that suggests that this molecule has functions in PCD in addition to its role in lipid transport (Sun *et al.*, 1995).

12.4.5. PCD of Muscles during Metamorphosis

The ISMs of *Manduca* (**Figure 1**) are the major abdominal muscles of the larva, pupa, and pharate adult. The ISMs are divided into separate pairs of bilaterally symmetric bundles, each of which attaches to the cuticle at the intersegmental boundaries. These muscles form in the embryo, and span eight of the abdominal segments in the larva. The ISMs provide the propulsive force for hatching and subsequent larval locomotion. Following pupation, the muscles in the first two and last two abdominal segments die and rapidly disappear. The muscles in the middle four segments persist throughout metamorphosis, and are used for the defensive and respiratory movements of the pupa. Following adult eclosion, the remaining ISMs undergo PCD during the subsequent 30 hours. While the basis for this segmental fate determination has not been examined, presumably it is established early in embryogenesis as a

Table 1 Genes Differentially Expressed in Condemned *Manduca* ISMs

Process	Gene	Response	Reference
Proteolysis	Polyubiquitin	Induced	Schwartz *et al.* (1990b)
	14-kDa E2 ubiquitin conjugase	Induced	Haas *et al.* (1995)
	18–56 (Sug1) 26S proteasome ATPase	Induced	Sun *et al.* (1996)
	28.1-kDa subunit catalytic subunit of 20S proteasome	Repressed	Löw *et al.* (2000)
	S6 (TBP7, MS73) 26S proteasome ATPase	Induced	Jones *et al.* (1995)
	S6′ (TBP1) 26S proteasome ATPase	Induced	Löw *et al.* (2000)
	S7 (MSS1) 26S proteasome ATPase	Induced	Löw *et al.* (2000)
	S10b (SUG2) 26S proteasome ATPase	Induced	Löw *et al.* (2000)
Transcription	E75B	Repressed	Löw *et al.* (2005)
Translation	Acheron (putative RNA binding protein)	Induced	Valavanis *et al.* (2007)
	eIF1A Translation-Initiation Factor	Induced	Löw *et al.* (2005)
	Oskar (maternal effect protein)	Repressed	Zhang *et al.* (2007)
Signal transduction	Small Cytoplasmic Repeat Protein SCLP)	Induced	Kuelzer *et al.* (1999)
	G coupled receptor GPR85	Induced	Zhang *et al.* (2007)
	Calmodulin-dependent calcineurin A1 subunit	Induced	Zhang *et al.* (2007)
	Death Associated LIM-Only Protein (DALP) (insect ortholog of Hic-5)	Induced	Hu *et al.* (1999)
Metabolism	Apolipoprotein III	Induced	Sun *et al.* (1995)
	Low MW lipoprotein PBMHPC-23	Induced	Zhang *et al.* (2007)
	Hydroxy acid oxidase 1	Induced	Zhang *et al.* (2007)
Contractile protein	Actin	Repressed	Schwartz *et al.* (1993b)
	Myosin heavy chain	Repressed	Schwartz *et al.* (1993b)
	Myosin light chain	Repressed	Zhang *et al.* (2007)
	Myosin Regulatory Light Chain isoforms 1 and 2	Repressed	Zhang *et al.* (2007)
	Calponin 1	Repressed	Zhang *et al.* (2007)
	Troponin 1	Repressed	Zhang *et al.* (2007)

result of the actions of segmentation and homeotic genes (Bejsovsc and Wieschaus, 1993; DiNardo *et al.*, 1994; French, 2001; Sanson, 2001).

The nuclear changes that accompany ISM death display none of the features of apoptosis (Schwartz *et al.*, 1993a) (**Figure 1**). The chromatin does not become electron-dense, but remains dispersed throughout the nucleoplasm. In addition, agarose gel electrophoresis of ISM genomic DNA fails to reveal apoptotic ladders. Ultrastructurally, there is an increase in autophagic vesicles, and the death of these cells is accompanied by autophagy (Lockshin and Beaulaton, 1974, 1979).

Following PCD of muscles in many animals, the cell corpse is phagocytosed by neighboring cells or circulating macrophage-like cells (Hart *et al.*, 2008). A classic example of this phenomenon is found in amphibian metamorphosis, where the massive tail musculature is lost during the transition from larva to adult (Weber, 1964; Watanabe and Sassaki, 1974). During this process, muscle fibers become decorated with macrophages that contain identifiable remnants of skeletal muscle debris (Metchnikoff, 1892; Nishikawa *et al.*, 1998). While dying muscles in insects are sometimes phagocytosed by circulating hemocytes (Crossley, 1968), this is not universally so (Jones *et al.*, 1978). In particular, the death of the ISMs following adult eclosion in moths does not attract macrophage-like cells, or rely on phagocytosis for resolution (Beaulaton and Lockshin, 1977) (**Figure 1**). In fact, estimates of ISM volumes and hemocyte numbers in adult *Manduca* suggest that removal of dying cells in these animals would require at least an order of magnitude greater number of phagocytic cells than has been shown to reside in the hemolymph (Jones and Schwartz, 2001).

While the ISMs of adult moths are not phagocytosed, Rheuben (1992) observed an intimate association between phagocytic hemocytes and the sarcolemma during the death of mesothoracic muscles in pupae. The phagocytes were well spaced along the fibers, and appeared to degrade the basal lamina. One difference between the ISMs and the mesothoracic fibers is that the latter are not completely degraded during development. Instead, they act as scaffolds for myoblasts that remodel the fibers during formation of adult muscle fibers. Phagocytes may play a more significant role in tissue remodeling rather than cell death.

12.4.5.1. Endocrine control of ISM death
Timing of ISM death must be coordinated with other metamorphic events, or the animal might suffer disastrous consequences. For example, premature loss of the ISMs in moths would leave the animal trapped within the pupal cuticle and locked in either a cocoon or an underground chamber. Delays in ISM death might have other deleterious consequences, including depriving the adult of nutrients required for gametogenesis. As described in section 12.4.4, the titer of ecdysteroids serves as an endogenous developmental time reference that can be used by the different organs of the pupa to coordinate developmental decisions, including the timing of ISM death.

Early reports suggested that the cessation of motoneuron activity was the proximal trigger for ISM death (Lockshin and Williams, 1965b). Subsequent studies demonstrated that the timing of ISM death in *Antheraea polyphemus* was not altered by silencing motoneuron activity with the sodium channel blocker tetrodotoxin or by removal of the entire ventral nerve cord (Schwartz and Truman, 1983, 1984a). Instead, in this species, the trigger for cell death is the peptide eclosion hormone (EH) (Schwartz and Truman, 1984a, 1984b). EH acts via cGMP; the description of its role in ISM death represented the first study demonstrating that cGMP met all of Earl Sutherland's requirements for identifying a second messenger for action of a hormone (Sutherland, 1972; Schwartz and Truman, 1984b). The capacity of EH to act on the ISMs is itself under the control of circulating ecdysteroids, as a decline in 20E regulates both the timing of EH release and the capacity of the ISMs to respond to this trigger (Truman, 1984; Schwartz and Truman, 1984a). The possible role of other insect peptides, such as ecdysis-triggering hormone (ETH), in PCD has not been explored.

12.4.5.2. Physiology of ISM death
The size of the ISMs, and the coordinated nature of the developmental changes that take place in this tissue during metamorphosis, facilitate examination of the physiological changes that accompany naturally occurring muscle atrophy and PCD. Under laboratory conditions, metamorphosis in *Manduca* is complete in 18 days, with adult eclosion taking place late on day 18. On day 15 of adult development the mass of the ISMs begins to decline, and during the next 3 days ISMs lose 40% of their mass. This pre-eclosion program of atrophy is non-pathological, and the muscles retain almost all of their normal physiological responses, including force/cross-sectional area and sensitivity to calcium ions in skinned fiber preparations (Schwartz and Ruff, 2002). These observations suggest that the reduction in muscle mass observed during the atrophy phase reflects a generalized enhancement of protein turnover rather than selective destructive of contractile proteins, despite the fact that entire contractile bundles are lost during this phase (Lockshin and Beaulaton, 1979).

The ISMs of *Manduca* begin PCD coincident with adult eclosion. At this time, the ISMs begin to lose mass at a rate of approximately 4% per hour (Schwartz and Ruff, 2002). By 24 h post-eclosion, reliable resting potentials can no longer be recorded (Lockshin, 1973). While there are few changes in the organization of the contractile apparatus during the atrophy phase, the post-emergence period is marked by profound sarcomere disruption (Lockshin and Beaulaton, 1979). Whole filaments disappear rapidly, with a preferential loss of thick filaments relative to thin filaments (Beaulaton and Lockshin, 1977). During this same period mitochondria are lost, autophagic vacuoles form, and the T tubule system swells. As a consequence, the muscle fibers rapidly weaken, even when force is normalized to cross-sectional area. This is true for twitches, and for tetanus and caffeine-induced contractions (Schwartz and Ruff, 2002). There are also defects in the ability of the contractile apparatus to respond to free calcium in both intact muscles and skinned fiber preparations.

12.4.5.3. Patterns of gene expression during PCD of ISMs

The primary biochemical mechanism that mediates the atrophy phase appears to be an increase in the ubiquitin–proteasome pathway, which allows protein catabolism to outstrip synthesis (Haas *et al.*, 1995). There is a transient increase in polyubiquitin expression in the ISMs on days 15 and 16 of pupal–adult development during the early phases of atrophy that is controlled by the falling ecdysteroid titer (Schwartz *et al.*, 1990b).

Lockshin (1969) demonstrated that ISM PCD in silkmoths is blocked by inhibitors of RNA or protein synthesis, suggesting a requirement for *de novo* gene expression. These results are similar to those reported for PCD in other tissues, including the insect nervous system, as well as in other metamorphosing taxa such as amphibians (Weber, 1965).

To identify genes that may mediate ISM death, Schwartz and colleagues utilized a differential screening approach using cDNA libraries constructed from Day 18 *Manduca* ISM mRNA (Schwartz *et al.*, 1990). Even though the ISMs are dying, the abundance of most transcripts was unchanged during the last days of pupation. However, a few transcripts were found that dramatically induced or repressed with the commitment of the cells to die. The cloning of these differentially expressed genes resulted in the first identification of death-associated gene expression from any organism (Schwartz *et al.*, 1990b). Among the genes that are repressed when the ISMs become committed to die are actin and myosin heavy chain (Schwartz *et al.*, 1993b) (**Table 1; Figure 3**). These transcripts are among the most abundant in the muscle during larval and pupal life, but begin to disappear late on day 17 (the day before adult eclosion), and are almost undetectable by late day 18 when the animals eclose.

One mechanism for reducing transcript abundance is transcriptional repression. A complementary mechanism is enhancement of transcript degradation. In this regard, it was found that there is a transient increase in endogenous ISM RNase activity on day 17, which facilitates removal of transcripts prior to induction of new gene expression (Cascone and Schwartz, 2001). Coordinated control of transcription and degradation may allow the muscles to rapidly shift developmental programs from homeostasis to death. While the molecular mechanism

that mediates this global change in transcript abundance has not been determined, microRNAs are potential regulators. MicroRNAs bind to sequences within the 3′ untranslated region (UTR) of target mRNAs and regulate transcript stability and translatability (Fabian *et al.*, 2010). In the case of *Manduca*, the stability of ISM transcripts could be transferred to ectopic reporter mRNAs by swapping the 3′ UTRs (Cascone and Schwartz, 2001). In other models, such as *Drosophila*, microRNAs can exert a profound effect on developmental processes (Jones and Newbury, 2010). Specific microRNAs, such as Let-7C, control the timing of intersegmental muscle cell death following adult eclosion (Sokol *et al.*, 2008).

A small number of induced cell death-associated genes were identified from the ISM screen (Schwartz *et al.*, 1990b). Some encoded known proteins, including polyubiquitin (Schwartz *et al.*, 1990b) and several proteasome subunits (Sun *et al.*, 1996), while others encoded novel proteins (Hu *et al.*, 1999; Kuelzer *et al.*, 1999; Valavanis *et al.*, 2007). More recently, additional PCD-associated genes expressed in ISMs have been identified either via

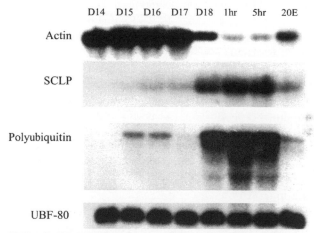

Figure 3 The ISMs begin to atrophy on day 15 of pupal–adult development and then initiate programmed cell death coincident with adult eclosion late on day 18. By 5 hours post-eclosion, the muscles have lost significant mass and physiological function. Most genes are constitutively expressed, like the ubiquitin fusion 80 (UBF-80) gene, which plays a role in ribosome biogenesis. While most genes are constitutively expressed within the ISMs, independent of developmental stage, a small number are induced or repressed with the commitment to die. Actin mRNA goes from being one of the most abundant transcripts within the ISMs prior to eclosion to all but disappearing in the dying cells. Conversely, Small Cytoplasmic Leucine Rich Repeat Protein (SCLP) is almost undetectable in the muscles prior to day 18 and is then dramatically induced with the commitment to die. Polyubiquitin is transiently induced on day 15, coincident with the onset of atrophy, and is then expressed at very high levels on day 18. All of the death-associated changes in transcript abundance can be prevented with injection of 25 μg of 20-hydroxyecdysone (20E) on day 17. D, day of pupal–adult development; h, hours post-eclosion; 20E, 20-hydroxyecdysone. Adapted from Schwartz (2008).

proteomics (Löw et al., 2000, 2005; Zhang et al., 2007) or by direct examination of proposed candidates in both Manduca and Bombyx (Table 1). The following section will focus on the genes identified from the molecular screen.

Ubiquitin is a 76 amino-acid peptide that is the most highly conserved protein present in all eukaryotes. At the protein level, insect and human ubiquitins are identical (Rechsteiner, 1988). The post-translational covalent attachment of ubiquitin to selected lysine residues on substrate proteins serves as a molecular tag to target proteins to specific fates within the cell (Salomons et al., 2010). The addition of single ubiquitin moieties directs proteins to specific subcellular locations, while addition of multiple head-to-tail ubiquitin chains promotes binding to the 26S proteasome. This multisubunit protease then releases the ubiquitin, unfolds the substrate, and rapidly degrades it to small peptides.

As mentioned above, there is a transient increase in polyubiquitin expression that correlates with ISM atrophy. Polyubiquitin mRNA then accumulates to prodigious levels on day 18, along with the coordinated expression of both 20S and 26S proteasome subunits (Schwartz et al., 1990b; Dawson et al., 1995; Haas et al., 1995; Jones et al., 1995; Takayanagi et al., 1996; Löw et al., 1997). This enhancement in ubiquitin-dependent proteolysis is presumably adaptive, because the ISMs are not phagocytosed, and therefore require a cell-autonomous mechanism for the liberation of cellular constituents (Jones and Schwartz, 2001). The ubiquitin–proteasome pathway is presumably serving roles in both cell death signal transduction and large-scale protein turnover (Broemer and Meier, 2009). In Manduca, injection of day-18 pharate adults with proteasome inhibitors (hemin and N-acetyl-leu-leu-norleucinal) delayed ISM death (Bayline et al., 2005).

Other known genes are induced in the dying ISMs, but their role in PCD is unknown. For example, the abundance of apolipoprotein III (apoLp-III) is dramatically induced, at both the RNA and protein levels, in both the ISMs and a subpopulation of neurons undergoing PCD (Sun et al., 1995). ApoLp-III is synthesized predominantly in the fat body, and normally facilitates lipid transport in the hemolymph by associating with lipophorin. Given that the ISMs do not express lipophorin, the role of ApoLp-III in PCD is currently mysterious.

The majority of cDNAs isolated in the ISM screen encoded novel proteins of unknown function. An example is SCLP (small cytoplasmic leucine-rich repeat protein), which is induced at both the RNA and protein levels in condemned ISMs expressed on day 18 (Kuelzer et al., 1999). This small protein is composed of multiple leucine-rich repeat protein–protein interaction motifs, and likely serves as a signal transduction protein. Ectopic expression of SCLP in different tissues in Drosophila did not result in an overt phenotype (Kuelzer et al., 1999).

Two of the novel genes identified in the ISM screen do have vertebrate homologs. DALP (death-associated LIM-only protein) contains one perfect and two imperfect LIM domains (Hu et al., 1999), structural motifs that consist of paired zinc fingers that mediate protein–protein interaction (Rétaux and Bachy, 2002). DALP is induced on day 17, well in advance of the other death-associated cDNAs from Manduca ISMs. Expression of the DALP protein is likely restricted to the ISMs, as it was not detected in flight muscle, fat body, Malpighian tubules, the ovary, oocytes, or the male sexual accessory gland. As with SCLP, the function of DALP was explored using transgenic flies (Hu et al., 1999). Ectopic expression of Manduca DALP in the abdominal ISMs of fly pupae resulted in the disorganization of the contractile apparatus and subsequent muscle atrophy. Targeted mutations in the LIM domain blocked muscle atrophy, suggesting that the observed effects of ectopic DALP expression were dependent on expression of the intact functional protein.

Further insights into the function of DALP were gained by examining the effects of expressing Manduca DALP in the mouse myoblast C_2C_{12} line (Hu et al., 1999). This muscle satellite cell line has been extensively used as a model for examining muscle differentiation and PCD in mammals (Schwartz et al., 2009). C_2C_{12} cells can be maintained as a stable, non-transformed line that, when incubated in a low serum medium, ceases cycling and differentiates into multinucleated myotubes (Yaffe and Saxel, 1977). Expression of DALP blocked the differentiation of C_2C_{12} cells into myotubes by blocking induction of MyoD, a basic helix–loop–helix muscle transcription factor required for differentiation. Effects of DALP were overcome by co-transfecting cells with an expression vector driving production of MyoD. In addition to blocking differentiation, DALP enhanced the probability of cell death. Identical results were obtained with C_2C_{12} cells transformed to express Hic-5 (hydrogen peroxide-inducible clone-5), the mammalian ortholog of Manduca DALP. These data show that DALP and Hic-5 are likely conserved proteins that function as negative regulators of muscle differentiation and survival in insect and mammalian cells.

Another of the novel cell death associated genes from Manduca may play a role in human pathogenesis. Acheron (Achn) contains three Lupus antigen (La) repeats, nuclear localization and export (NLS and NES) signals, and an RNA recognition motif (Valavanis et al., 2007). In fact, Achn defines a new subfamily of Lupus antigen (La) proteins that appears to have branched from authentic La protein relatively late in metazoan evolution. In mammalian cells, Achn (also known as La related protein 6, or LARP6), binds to the 5′ untranslated region of collagen mRNA and facilitates translation (Cai et al., 2010). While its role in ISM death has not been explored, Achn plays an essential role in myogenesis in zebrafish (Wang et al., 2009).

In C_2C_{12} myoblasts, Achn acts upstream of MyoD and is required for these cells to either differentiate or undergo apoptosis following loss of growth factors (Wang *et al.*, 2009). Other studies have explored the role of Achn in regulating integrin–extracellular matrix interactions required for myogenesis. Both control C_2C_{12} myoblasts and those engineered to express ectopic Achn expressed the fibronectin receptor integrin α5β1 in the presence of growth factors and the laminin receptor α7β1 following growth factor withdrawal. Expression of the laminin receptor was blocked in cells expressing either Achn antisense or dominant-negative Achn. Control cells and those expressing ectopic Achn undergo sequential and transient increases in both substrate adhesion and migration before cell fusion. Blockade of Achn expression reduced these effects on laminin but not on fibronectin. Taken together, these data suggest that Achn may influence differentiation in part via its control of cell adhesion dynamics (Glenn *et al.*, 2010).

A recent study has demonstrated that Achn is expressed in the myoepithelial cells of the mammary gland (Shao *et al.*, 2011). Microarray and immunohistochemical analysis of tissues from patients with breast cancer have demonstrated that Achn expression is significantly elevated in some basal-like tumors, the most aggressive of the breast cancers. Ectopic expression of Achn in MDA-MB-231 breast cancer cells induced a number of phenotypic changes that are associated with malignancy and metastasis, including enhanced cell proliferation, lamellipodia formation, greater invasive activity, and elevated expression of the metastasis-associated proteins MMP-9 and VEGF. In xenograph studies using athymic mice, MDA-MB-231 cells expressing ectopic Achn displayed enhanced angiogenesis and an approximately five-fold increase in tumor size relative to control cells. These data support the hypothesis that Achn enhances human breast tumor growth and vascularization, and may represent a target for diagnostics and therapeutics.

12.5. The *Drosophila* Model

12.5.1. Choice of *Drosophila melanogaster* as a Model System for the Study of PCD of Nerve and Muscle Cells

Like moths, fruit flies are holometabolous insects. Three larval (feeding) stages are followed by pupation. The pupa forms inside the pupal case (puparium), the hardened cuticle of the final larval stage. *Drosophila* development is rapid and the entire life cycle is complete within 10 days at 25°C. While *Drosophila* have provided a powerful genetic model system for a century, it arrived relatively late to the modern study of PCD. Thus, it was not until 1990 that Kimura and Truman (1990) extended classic observations of muscle death during *Drosophila* metamorphosis

by systematically documenting neuronal death in the fused ventral ganglia following adult eclosion. Injection of living flies with a toluidine blue solution revealed numerous examples of dying neurons in the dorsal and lateral regions of the abdominal and metathoracic neuromeres. These investigators also used bi-refringence to map degeneration of head and abdominal musculature during metamorphosis and adult eclosion. Several studies described below have made use of these baseline observations to explore ecdysteroid regulation and molecular mechanisms of PCD in the nervous system of newly-eclosed adult flies (Robinow *et al.*, 1993, 1997; Draizen *et al.*, 1999).

Interestingly, the first insight into the genetic basis of PCD in flies did not come from studies of metamorphosis but, rather, of embryogenesis. The laboratory of Hermann Steller used a classic histological technique, acridine orange staining, to identify dying cells with an apoptotic phenotype in wild type embryos (Abrams *et al.*, 1993). A subsequent genetic screen using a set of 129 chromosomal deficiency strains identified a small region at 75C1,2 on chromosome 3L that is essential for developmental and X-irradiation induced apoptosis (White *et al.*, 1994). Subsequent studies led to the identification of four related pro-apoptotic genes that reside within a 300-kb interval in this region: *reaper* (*rpr*), *head involution defective* (*hid*), *grim*, and *sickle* (*skl*) (White *et al.*, 1994; Grether *et al.*, 1995; Chen *et al.*, 1996; Christich *et al.*, 2002; Srinivasula *et al.*, 2002; Wing *et al.*, 2002) (**Figure 4**). A genetic modifier screen resulted in the identification of *Drosophila* Inhibitor of Apoptosis Proteins, DIAP1 and DIAP2, which are crucial inhibitors of caspase activities (Hay *et al.*, 1995). DIAP1 mutants exhibit embryonic lethality associated with massive ectopic PCD, indicating it is essential for survival of many cell types (Hay, 2000).

The Reaper, Grim, and Sickle proteins are small (65, 138, and 108 amino acids, respectively), while Hid is substantially larger (410 amino acids). Each possesses a related 14-aa region at the N-terminus, designated the RHG (Reaper, Hid, Grim) motif or IBM (inhibitor of apoptosis (IAP)-binding motif). This motif has potent pro-apoptotic activities, and can bind to and repress the caspase-inhibiting activities of IAPs (Vaux and Silke, 2005; Steller, 2008; Orme and Meier, 2009; also discussed below). The RHG/IBM motifs of Reaper and Grim share 71% identity, with three of the four amino-acid differences conservative substitutions. However, despite this strong similarity, *in vivo* studies have shown that these domains possess distinct death-inducing activities (Wing *et al.*, 1998). More recently, two unlinked *Drosophila* genes, *omi/htr2A* (a homolog of the mammalian Htr2A serine protease) and *jafrac2* (a thiodoxin peroxidase), were identified that encode proteins with similar RHG domains (see, for example, Challa *et al.*, 2007), but for this review we will focus on the four linked RHG genes, *reaper*, *hid*, *grim*, and *sickle*. Reaper, Grim, and

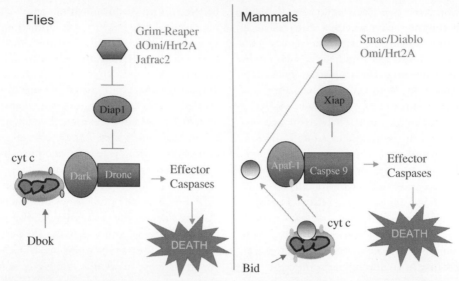

Figure 4 The *grim-reaper* genes encode related pro-apoptotic proteins. (A) The *hid*, *grim*, *reaper*, and *sickle* genes all reside within a 300-kb interval in the 75C region of chromosome 3L of *Drosophila melanogaster*. The four genes are transcribed in the same direction (arrows above genes) and are expressed predominantly in doomed and dying cells. No genes are predicted to reside between *grim* and *reaper*, or *reaper* and *sickle*, while four unrelated genes (orange blocks) reside between *hid* and *grim*. (B) *reaper*, *sickle*, and *grim* encode small proteins that contain an N-terminal RHG motif/IBM as well as a C-terminal Trp block/GH3 domain. Hid is a substantially larger protein that also contains an RHG motif but does not exhibit strong similarity to the Trp block/GH3 domain.

Sickle also share a second region of sequence similarity, a 15-aa Grim helix 1, 2, and 3 (GH3) domain or Trp block (**Figure 4**), which has distinct pro-apoptotic functions. This domain is necessary for mitochondrial association of Reaper and Grim during apoptosis, and activates PCD independently of the IAP antagonizing functions of the NH2 RHG domain/IBM (see, for example, Abdelwahid *et al.*, 2007). Hid possesses four regions that resemble a Trp block/GH3 domain, although the level of similarity is much lower (Wing *et al.*, 2001). In addition, Hid also contains a mitochondrial localization signal and can form a heterodimer with Reaper that translocates to mitochondria to promote cell death (Sandu *et al.*, 2010). Reaper also forms homodimers and heterodimers with Grim but not Sickle. Formation of these dimers appears essential for strong pro-apoptotic activity of RHG proteins (Sandu *et al.*, 2010). Finally, Reaper and Grim, but not Hid, can induce apoptosis via general translation inhibition (Holley *et al.*, 2002; Yoo *et al.*, 2002). The related RHG proteins are multifunctional apoptosis activators that have distinct and overlapping modes of action.

Insights into the molecular mechanisms of PCD gained by further studies of RHG genes during *Drosophila* embryonic development and metamorphosis are described in the following sections. Significantly, targeted ectopic expression of these genes in cells that are normally fated to live has provided an invaluable tool for producing cell-specific lesions of the fly CNS, particularly of peptidergic neurons (see, for example, McNabb *et al.*, 1997;

Zhou *et al.*, 1997; Renn *et al.*, 1999; Rulifson *et al.*, 2002; Park *et al.*, 2003; Zhao *et al.*, 2010).

12.5.2. Hormonal Regulation of Metamorphosis in *Drosophila melanogaster*

Hormonal regulation of metamorphosis in *Drosophila* is similar to that described for other insects, with post-embryonic development wholly dependent upon exposure of tissues to coordinated pulses of 20E, which in turn produce a coordinated cascade of gene expression (Riddiford, 1993; Truman and Riddiford, 2002). Cloning of the *Drosophila* ecdysone receptor (*EcR*) gene led to studies of the expression of EcR in tissues including neurons and muscle (Koelle *et al.*, 1991; Robinow *et al.*, 1993; Talbot *et al.*, 1993; Truman *et al.*, 1994). These studies support the view, based on earlier studies of *Manduca*, that ecdysteroid regulation of PCD in insects is a result of direct action of the steroid on the cells that are fated to die.

12.5.3. PCD of Neurons during Early Development and Metamorphosis

Four periods of PCD have been described in the nervous system of *Drosophila*: during mid-to-late embryogenesis, in late third instar larvae, in pupae during metamorphosis, and in the newly-eclosed adult. The PCD that occurs during these stages generates in turn the larval and adult nervous systems by eliminating cells that are produced in excess, such as the embryonic midline glia, and cells with transient functions, such as larval abdominal ganglion neuroblasts (Kimura and Truman, 1990; Truman *et al.*, 1993; Sonnenfeld and Jacobs, 1995; Zhou *et al.*, 1995). As in mammalian development, the extent of cell death in the embryonic *Drosophila* nervous system is profound; approximately two-thirds of all cells born die before larval hatching (Abrams *et al.*, 1993; White *et al.*, 1994).

The extent of cell death in *Drosophila* embryos suggests that this process is critical for development of the larval nervous system. However, homozygous Df(3L)H99 mutant embryos that lack cell death exhibit relatively normal organization of the ventral nerve cord despite having a large excess of surviving cells (White *et al.*, 1994; Zhou *et al.*, 1995). Nonetheless, mutations in several genes that are important for PCD result in hypertrophy of the embryonic, larval, or adult CNS, and these mutants often exhibit stage-specific lethality or sterility (Grether *et al.*, 1995; Song *et al.*, 1997; Peterson *et al.*, 2002; Rogulja-Ortmann *et al.*, 2007; Kumar *et al.*, 2009). Removal of dead cells from the CNS also appears to be important for nervous system development, as mutants lacking functional macrophages exhibit disruptions in the normal architecture of the embryonic axon scaffold (Sears *et al.*, 2003).

The proportion of neurons that die during the postembryonic period is lower than during embryogenesis, and the widespread degeneration of larval tissues during metamorphosis in flies does not extend to the CNS. As in *Manduca*, most neurons in the larva persist rather than die (Truman, 1990). After pupariation, however, many dying neurons can be observed in the ventral CNS in both thoracic and abdominal neuromeres (Truman *et al.*, 1993). The identity of most of these neurons has not been determined, with the exception of several motoneurons identified by retrograde fills of the T2 mesothoracic nerve (Consoulas *et al.*, 2002). In addition, PCD of neuroblasts after pupariation marks the termination of post-embryonic neurogenesis in the ventral nervous system (Truman and Bate, 1988; Truman *et al.*, 1993; Kumar *et al.*, 2009). As in *Manduca*, identified peptidergic neurons provide a population that can be tracked across metamorphic transitions. Corazonin is an 11-amino acid peptide expressed by 8 pairs of neurons in the ventral nerve cord of larval fruit flies. Detailed studies of the time-course of corazonin disappearance revealed that these neurons undergo PCD 2–6 hours after puparium formation. Because their fate can be readily tracked by anti-corazonin immunolabeling, they were chosen to explore the ecdysteroid-dependent and cell-specific mechanisms of PCD (Choi *et al.*, 2006). These studies revealed that the pulses of ecdysteroids that drive the onset of metamorphosis trigger the death of the corazonin neurons via activation of the EcR nuclear receptor. The *EcR* gene of *Drosophila* encodes three ecdysone receptor subunits: EcR-A, EcR-B1, and Ecr-B2 (Talbot *et al.*, 1993). These receptor subunits share common DNA and ligand-binding domains, but have different N-terminal regions. Analysis of loss-of-function mutants for specific EcR isoforms revealed that either EcR-B1 or EcR-B2 can activate the cascade of transcriptional events that results in the PCD of these neurons (Choi *et al.*, 2006).

The use of EcR isoform-specific antibodies also revealed that a high level of expression of EcR-A is correlated with post-eclosion PCD (Robinow *et al.*, 1993). Prior to eclosion, nearly 300 neurons in the ventral ganglia displayed high levels of EcR-A immunoreactivity. During the first 24 hours after adult eclosion, essentially the EcR-A-immunopositive neurons in the ventral ganglia displayed characteristic features of PCD and were basically all absent by the end of this first day. In contrast to the death of the corazonin neurons at the end of larval life described above (Choi *et al.*, 2006), it appears that it is EcR-A expression during the period of ecdysteroid decline at the end of metamorphosis which is critical to the decision of these neurons to commit suicide. A subsequent analysis of EcR-A mutants supported the view that the distinct EcR-isoforms have specific functions during development, and linked EcR-A receptors to PCD in an additional tissue, the larval salivary glands (Davis *et al.*, 2005). Analysis of isoform-specific activities of EcR is an active area of research with a current focus on the regulatory role of heterodimerization partners (Braun *et al.*, 2009).

12.5.4. Molecular Mechanisms of Neuronal Death

PCD research in *Drosophila* is a large and active field that can no longer be adequately summarized in a single chapter. We focus here on selected topics related to the death of neurons and glial cells, covering both apoptosis and autophagy. Accounts of the control of neuron number by neuroblast apoptosis in the ventral nerve cord and the brain can be found in Peterson *et al.* (2002), Bello *et al.* (2003, 2007), Kumar *et al.* (2009), and Siegrist *et al.* (2010). Descriptions of the regulation of cell survival by cell intrinsic mechanisms in sensory neuron lineages can be found in Spana and Doe (1996) and Orgogozo *et al.* (2002). These studies have provided insights into the regulation of apoptosis and other cell fates by asymmetric cell division (Hatzold and Conradt, 2008; Zhong, 2008).

12.5.4.1. The *Drosophila* cell death "machinery" for apoptosis
The basic molecular machinery of apoptosis includes both initiator and executioner caspases, a Ced (cell death abnormal)-4/Apaf (apoptosome associated factor)-1 ortholog, and Ced-9/Bcl-2 protein family members (Tittel and Steller, 2000; Vernooy *et al.*, 2000) (**Table 2**). These proteins all have critical functions in apoptosis that were initially defined by studies in the nematode *Caenorhabditis elegans* (Horvitz *et al.*, 1994; Liu and Hengartner, 1999). Caspases are considered the

Table 2 *Drosophila* Cell Death Regulators

Family	Proteins
Caspases	Dronc
	Dcp-1
	Drice
	Dredd
	Strica/Dream
	Decay
	Damm
Caspase inhibitor	DIAP1/Thread
	DIAP2
Bcl-2 family	Drob-1/Debcl/Dborg-1/Dbok
	Buffy/Dborg-2
IAP inhibitors	Reaper
	Hid
	Grim
	Sickle
	Jafrac2
Apaf-1/Ded 4 family	Dark/HAC-1/Dapaf-1

cellular executioners because they activate proteolytic enzymes that digest cellular structural elements; Apaf-1 factors form complexes called apoptosomes that regulate caspase activity, and Bcl-2 family proteins regulate the integrity of mitochondrial membranes. An important difference between apoptosis in *C. elegans*, *Drosophila*, and mammals is the role of mitochondria (and, hence, an important role for Bcl-2 family members). In mammals, mitochondria undergo membrane permeability changes in dying cells and serve as critical sources of several pro-apoptotic factors, including cytochrome *c* and Smac/Diablo. However, while mitochondria have clearly been implicated in apoptotic pathways in flies and worms, the importance and extent of their contribution is less certain (see, for example, Colin *et al.*, 2009; Krieser and White, 2009). In flies, the major arbiter and point of regulation for cell survival decisions appears to be DIAP1 (Hay, 2000; Vaux and Silke, 2005; Orme and Meier, 2009).

In flies, worms, and mammals, caspases serve as the effectors of PCD. These specialized cysteine proteases cleave at aspartate or glutamate residues within enzymatic or structural substrate proteins to promote the dismantling of a cell (Cooper *et al.*, 2009; Feinstein-Rotkopf and Arama, 2009). Caspases are initially synthesized as inactive zymogens that contain either a long or a short prodomain, and a large and small subunit. Proteolytic cleavage of caspases in dying cells results in formation of an active heterotetramer comprised of two large and two small subnits. Heterotetramers derived from long prodomain zyomogens correspond to initiator caspases, and these cleave short prodomain zymogens to generate active effector caspases that ultimately dismantle diverse cellular substrates. Thus, a cascade of caspase activities defines apoptotic cell deaths. The *Drosophila* genome encodes seven caspases (**Table 2**), including three long prodomain initiator caspases such as Dronc (*Drosophila* nedd2-like caspase,

similar to caspase 9), and four short prodomain effector caspases such as Drice (similar to caspase-3) and DCP-1 (death caspase 1; similar to caspase-7). Dronc and Drice are now recognized as key elements of the core of the *Drosophila* caspase-dependent cell death machinery (Hay and Guo, 2006). Mutant *Drosophila* embryos lacking zygotic *dronc* gene product display significantly reduced levels of apoptosis in many cell populations, including the developing ventral nerve cord (Chew *et al.*, 2004). During metamorphosis, targeted expression of the baculovirus pan-caspase inhibitor p35 blocks the death of the larval corazonin neurons of the ventral nerve cord (Choi *et al.*, 2006). Interestingly, a recent report links the regulation of Dronc activity (via phosphorylation at S130) to cellular metabolism by demonstration that increases in NADPH inhibit this caspase while inhibition of NADPH production triggers apoptosis (Yang *et al.*, 2010); this relationship between NADPH and Dronc activity was first demonstrated in *Drosophila* S2 cells, but was also found in neurons in the developing CNS. Focal activation of caspases plays a role in neurite pruning in *Drosophila*, which provides a surprising and elegant tool for sculpting the nervous system and refining synaptic connections (Williams *et al.*, 2006).

The fly genome encodes two Bcl-2 family members, Drob-1/Debcl/dBorg-1/dBok and Buffy/dBorg-2, that possess the Bcl-2 homology (BH) domains BH1, BH2, and BH3 (Brachmann *et al.*, 2000; Colussi *et al.*, 2000; Zhang *et al.*, 2000; Quinn *et al.*, 2003). Both of these proteins resemble the pro-apoptotic mammalian Bok protein, and act to promote apoptosis (Igaki and Miura, 2004). This difference supports the view that flies are less dependent upon mitochondrial disruption for the activation of caspases than are mammals (Wang and Youle, 2009). This observation provokes speculation concerning the evolution of cell death mechanisms, and raises the question: what factors inhibit death in insect cells?

12.5.4.2. Role of IAP and RHG family proteins in apoptosis The major anti-apoptotic proteins in *Drosophila* are the IAPs (inhibitors of apoptosis proteins), which function by binding to and inactivating caspases (Hay, 2000; Vaux and Silke, 2005; Steller, 2008; Orme and Meier, 2009). IAPs were first identified from the *Cydia pomonella* granulosis virus, and found to inhibit apoptosis in infected host cells (Crook *et al.*, 1993). Subsequently, a large number of viral and cellular IAPs have been identified in divergent species that function as potent caspase repressors and inhibitors of PCD. IAPs all contain one or more copies of a 70-aa baculovirus IAP repeat (BIR) domain, typically located in the central or N-terminal region of the protein. In addition, they contain a 50-aa RING (Really Interesting New Gene) domain, generally situated towards the C-terminal region of the protein. The sequences of the BIR and RING domains both resemble zinc fingers, and serve as critical protein–protein interaction domains. The BIR domains can directly associate with procaspases and caspases, and inhibit their activation or activities (Vaux and Silke, 2005; Steller, 2008; Orme and Meier, 2009). The RING domain possesses ubiquitin E3 ligase activity that can target bound caspases for polyubiquitination and degradation via the 26S proteasome (reviewed by Bergmann, 2010). In addition, IAPs bound to caspases are themselves targeted to proteasome-dependent proteolysis via the N-end rule pathway of protein turnover (Ditzel *et al.*, 2003). The IAPs therefore can efficiently reduce cellular caspase levels, and the interplay between caspases and IAPs is a key determinant of cell survival. The levels of IAPS themselves are also under rigid regulation. In particular, the RHG proteins bind not only to the BIR domains of IAPs and displace bound caspases, but also promote IAP auto-ubiquitination (Yang *et al.*, 2000; Yoo *et al.*, 2002). The release and de-repression of bound caspases as well as increased turnover of IAPs act together to strongly promote PCD.

The dependence on IAP proteins for cellular survival provides a key regulatory point for the regulation of apoptosis. The BIR domains of DIAP1 bind both caspases and the Grim-Reaper proteins. In particular, the BIR2 domain of DIAP1 associates both with Dronc and Reaper, Hid, Grim, and Sickle. The BIR1 domain binds to Drice and DCP-1 and also to Reaper and Grim, and, to some extent, Hid. Thus, the two BIR domains exhibit distinct abilities to bind to both specific caspases and RHG proteins (see, for example, Zachariou *et al.*, 2003). In surviving cells, DIAP1 binds and inhibits caspases, thereby repressing cell death (**Figure 5**). In contrast, in cells that receive

signals to die, Grim-Reaper proteins are expressed and compete with caspases for DIAP1 binding. The displacement of caspases from DIAP1 results in increased levels of proteolytically active enzymes that promote cell death. Cell survival decisions, therefore, are determined by the interactions among the pro-apoptotic RHG proteins, the anti-apoptotic IAPs, and caspases.

IAP regulation is a conserved mechanism for controlling cell survival (**Figure 5**). In mammals, Smac/Diablo and Omi/Htr2A are mitochondrial proteins released with cytochrome *c* in dying cells (Du *et al.*, 2000; Suzuki, 2001; Verhagen and Vaux, 2002). Cytoplasmic Smac/Diablo and Omi/Htr2A associates with X-linked inhibitor of apoptosis (XIAP) to prevent it from binding and thereby inhibiting caspase-9. Strikingly, the binding of Smac/Diablo and Omi/Htr2A to IAPs is mediated through an N-terminal tetrapeptide sequence that is conserved in the Grim-Reaper RHG motif, implying conserved modes of action for the fly and mammalian proteins (Chai *et al.*, 2000; Srinivasula *et al.*, 2000, 2002). As expression of IAPs is upregulated in many types of tumors, interest is intense regarding the possibility that either these endogenous inhibitors or related synthetic compounds can be exploited to develop new therapies for cancer and other diseases (Fulda, 2007; Gyrd-Hansen and Meier, 2010).

12.5.4.3. Regulation of embryonic glial cell survival in *Drosophila*

One important mechanism for regulating the survival of neurons and glia within the vertebrate nervous system involves the actions of trophic factors (Raff *et al.*, 1993). These pro-survival molecules are synthesized in restricted amounts by target tissues, and permit winnowing of innervating cells to ensure matching

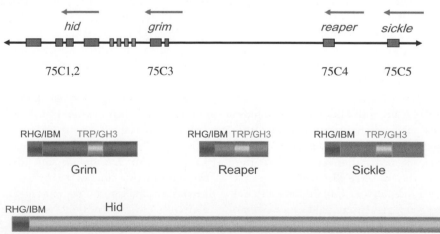

Figure 5 IAP inhibition is a conserved mechanism of regulation of apoptosis. In *Drosophila*, expression of the *grim-reaper*, *jafrac2, or dOmi/Hrt2A* genes is activated in doomed cells, and the corresponding proteins bind to DIAP1 and block its ability to inhibit caspases. This results in active caspases that degrade cellular proteins to mediate apoptosis. In mammals, pro-apoptotic stimuli induce the release of Smac/Diabl, Omi/Hrt2A, as well as other pro-apoptotic factors, including cytochrome c, from the mitochondria. Cytoplasmic Smac/Diablo or Omi/Hrt2A binds XIAP and represses its caspase-inhibitory actions, thereby promoting apoptosis. In both flies and mammals, the orthologous Dark and Apaf-1 proteins as well as pro-apoptotic members of the Ced-9/Bcl-2 family promote activation of inhibitor caspases.

of interacting cell populations. Such molecules appear to be involved in regulating glial cell survival in *Drosophila* embryos.

Approximately 10% of all cells within the *Drosophila* nervous system are glial cells. Neurons and glia display complex signaling relationships during development. For example, the PCD of neurons de-represses division of glial cells, leading to glial cell proliferation (Kato *et al.*, 2009). Among the best-characterized glial cells are the embryonic longitudinal and midline glia of the CNS. In both these lineages, glial cells undergo extensive apoptosis during embryogenesis (Sonnenfeld and Jacobs, 1995; Zhou *et al.*, 1995; Kinrade *et al.*, 2001). The survival of subsets of these cell populations is governed by trophic actions of EGF-related ligands and activation of the Ras/MAP kinase pathway via the EGF receptor homolog (EGFR).

During embryogenesis, a single, laterally positioned glioblast precursor gives rise to approximately 10 longitudinal glia in each hemisegment of the ventral nerve cord. These glial cells migrate medially, contact pioneer longitudinal axons, and ultimately ensheathe the longitudinal nerve bundles (Hidalgo and Booth, 2000). Coincident with the onset of axon/glial contact, many of these longitudinal glia undergo apoptosis, suggesting that as the glia contact the axons, they become dependent upon them for survival (Kinrade *et al.*, 2001). Consistent with this notion, ablation of pioneer and other neurons results in a decrease in longitudinal glia (Hidalgo *et al.*, 2001). Thus, apoptosis determines the final numbers of longitudinal glia during embryogenesis, and axon-derived factors are required for longitudinal glial cell survival. One of these factors is Vein (Vn), a *Drosophila* neuregulin homolog that contains both an IgG domain and an EGF domain (Schnepp *et al.*, 1996). The *vein* gene is expressed in a subset of neurons within the embryonic CNS including the midline precursor 2 (MP2) pioneer neurons and the Ventral Unpaired Median (VUM) neurons of the CNS midline (Hidalgo *et al.*, 2001). *Vein* mutants exhibit ectopic apoptosis of longitudinal glial cells, which is also observed for RNAi-mediated knockdown of *vein* gene product in either all neurons or the MP2 neurons. Vein is a secreted ligand for *Drosophila* EGFR, and EGFR-mediated activation of the Ras/MAP kinase pathway is essential for longitudinal glial cell survival. EGFR is transiently expressed in a subset of these glia, suggesting that Vein is secreted from axons to promote survival of these glial cells.

Similar to the Vein-dependent longitudinal glia, survival of the midline glia is also promoted by an EGF family member, the TGF-α homolog Spitz (Bergmann *et al.*, 2002). Midline glia are essential for proper formation of the axon scaffold, as migrating midline glia contact, separate, and ultimately ensheathe the anterior and posterior axon commissures (Klambt *et al.*, 1991). The midline glia are normally in close contact with commissural axons, suggesting that glial–axon contact may be important for midline glial cell survival. Observations made in *commissureless* mutants, where the commissures fail to form, are consistent with this notion; in these mutants, the isolated midline glial cells that fail to contact axons undergo apoptosis (Sonnenfeld and Jacobs, 1995). Ultrastructural and genetic analyses indicate that the midline glia undergo apoptosis, which reduces their number from an initial set of nine cells to three cells in each segment of the mature ventral nerve cord (Sonnenfeld and Jacobs, 1995; Zhou *et al.*, 1995). This apoptosis is dependent upon caspases and the actions of multiple Grim-Reaper proteins. Simultaneous loss of *reaper*, *hid*, and *grim* gene expression blocks all midline glial cell death, and results in the survival of the nine midline glia per segment. The loss of *hid* and *grim* results in approximately seven to eight midline glia per segment, while the loss of *hid* alone results in six midline glia (Zhou *et al.*, 1997; Bergmann *et al.*, 2002). Thus, *hid* expression is required for the death of three midline glia, and *reaper* and *grim* are together essential for the death of the other three midline glia. Both *hid* gene transcription and activity of Hid protein in the midline glia are regulated by proteins in the EGF signaling pathway, including Spitz and EGFR (Kurada and White, 1998). The loss of midline glia in *spitz* mutants is rescued in *spitz;hid* double mutants, implying that the pro-apototic functions of Hid are normally opposed by the prosurvival functions of Spitz.

The Spitz protein appears to be an axon-derived trophic factor required for midline glial survival (Bergmann *et al.*, 2002), as the loss of midline glia in *spitz* mutants is rescued by targeted expression of the transmembrane Spitz precursor protein in commissural neurons, and the number of surviving midline glia can be modulated by controlling the levels of Spitz activity. The model that has emerged is that axon-derived Spitz protein signals the midline glia via EGFR, resulting in activation of the Ras/MAP kinase pathway. Hid activity is subsequently downregulated via phosphorylation, which permits midline glial survival. Taken together, the actions of Vein in the longitudinal glia and Spitz in the midline glia indicate that trophic survival mechanisms are utilized to match the sizes of interacting populations of neurons and glial cells in invertebrates as well as vertebrates.

12.5.4.4. The *Drosophila* cell death "machinery" for autophagy Studies in *Drosophila* are now contributing significantly to our understanding of the molecular mechanisms of autophagy (Chang and Neufeld, 2010; Ryoo and Baehrecke, 2010; **Figure 6**). In animal cells, autophagy is regulated by the class I and class III phosphatidylinositol 3-kinase pathways (PI3K). Metabolic signals such as the binding of insulin to its receptor activate PI3K, which ultimately releases an autophagy inhibitor, TOR (target of rapamycin), from inhibition. The activation of TOR in turn activates

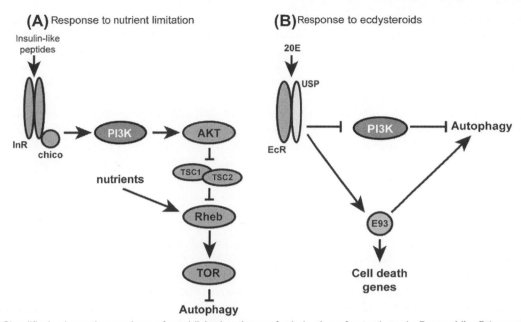

Figure 6 Simplified schematic overviews of established pathways for induction of autophagy in *Drosophila*. Other pathways have also been described. (A) Autophagy can be induced by nutrient limitation through inhibition of TOR, which negatively regulates autophagy in well-fed organisms. Rheb and TSC1/2 control the activity of TOR, both acting downstream of PI3K and AKT. (B) Developmental autophagy as described in this chapter is triggered by ecdysone signaling that it is time to eliminate tissues or organs. Note that in this scenario activation of EcR by 20E inhibits PI3K, providing an opportunity for cross-talk between these signaling pathways. (For clarity, the steps in the pathway linking P13K with autophagy are omitted in panel (B)). This form of autophagy can involve activation of known cell death genes, including caspases in some cell types. 20E, 20-hydroxyecdysone; AKT, a serine/threonine protein kinase; chico, *Drosophila* homolog of mammalian IRS1-IRS4, the insulin receptor substrate adaptor proteins; E93, an early ecdysone response gene that encodes a pipsqueak domain transcription factor; EcR, ecdysone receptor (nuclear hormone receptor); InR, *Drosophila* homolog of the mammalian insulin receptor; PI3K, phosphoinositide 3-kinase; Rheb, "Ras homolog enriched in brain," a GTP-binding protein (GTPase); TOR, "target of rapamycin," a serine/threonine kinase that belongs to the phosphoinositide-3-kinase-related kinase (PIKK) family of checkpoint kinases and is a part of a signaling pathway that regulates growth; TSC1/2, tuberous sclerosis complex proteins that function as GTPase activating proteins against Rheb; USP, *Drosophila* homolog of RXR that forms a heterodimer with EcR to form the functional insect ecdysteroid receptor. Adapted from Chang and Neufeld (2010).

Atg-family proteins that regulate the formation of the autophagosome (Ryoo and Baehrecke, 2010). This same pathway functions in *Drosophila*, with insulin-like peptides and insulin-like receptor corresponding to the mammalian insulin and insulin receptor, respectively. Other autophagy pathways have also been described in *Drosophila*. The first is a pathway in which autophagy results from an ecdysteroid signal (**Figure 6**); the second is part of the cellular response to oxidative stress in which Atg-family genes are transcriptionally activated by Jun-N-terminal kinase (JNK) signaling (Chang and Neufeld, 2010).

12.5.4.5. Autophagy and metamorphosis of neuromuscular systems A surprising phenotype is associated with mutations in *Drosophila* Atg-family genes (Shen and Ganetzky, 2010). In *Drosophila*, the number of boutons associated with each neuromuscular junction can be counted by immunolabeling synapses with markers such as anti-Synaptotagmin or anti-Nervous wreck. When bouton numbers were compared between wild type flies and flies with loss-of-function Atg mutations, the

mutants displayed significantly reduced bouton numbers, by up to 50%. Bouton numbers can be increased by pro-autophagic genetic manipulations. The site at which autophagic processes regulate normal development of the neuromuscular junction appears to be the motoneuron, rather than the muscle target (Shen and Ganetzky, 2009). Reduced expression of the Highwire protein appears to be instrumental in the regulation of synaptic development via autophagy. Highwire is an E3 ubiquitin ligase previously shown to reduce the formation of synapses (Wan *et al.*, 2000). The substrate specificity of autophagic pathways in motoneurons is astonishing. Whether this is a singular mechanism for the regulation of synaptic plasticity or the first of many "selective autophagy" targets remains to be discovered.

12.5.5. PCD of Muscles during Metamorphosis

There is extensive loss of the embryonically-derived musculature within the abdomen in *Drosophila* following adult eclosion (Miller, 1950; Kimura and Truman, 1990). Although some of the persisting muscles continue to function during pupal life and adult emergence, others provide a template for adult myogenesis (Broadie and Bate, 1991; Bate, 1993). Some of these new muscles, including male-specific muscles, depend upon innervation for differentiation and survival (Bate, 1993). Phagocytosis of degenerated (also referred to as histolyzed) muscle is a feature of the larval–pupal transition in *Drosophila* (Crossley, 1978; Bate, 1993), but not of muscles that die following adult eclosion (Kimura and Truman, 1990; Jones and Schwartz, 2001).

In contrast with *Manduca*, relatively little is known about the genes involved in the death of *Drosophila* muscles during metamorphosis. A recent study has demonstrated that the nuclear spindle matrix scaffold proteins East and Chromator play antagonistic roles in the death of the dorsal oblique external muscles during metamorphosis in flies (Wasser *et al.*, 2007). The Chromator protein appears to act late in histolysis, and is required for the complete destruction of dying muscles. EAST, which physically interacts with Chromator, appears to delay the loss of structural integrity in the muscles. The post-eclosion death of abdominal muscles in *Drosophila* is dependent on the microRNA Let7C (Sokol *et al.*, 2008). Deletion of Let7C ablates muscle death, which can be rescued with ectopic expression. The mRNA targets for Let7C that are essential for muscle death have not yet been determined.

12.6. Insights from Other Tissues

Study of PCD during embryogenesis and metamorphosis outside of the neuromuscular system is a growth area in cell death research. Many tissues have been studied, including the fat body and the germ cell lineages. Of particular relevance are two tissues that undergo PCD during the early hours of metamorphosis: the larval salivary glands, and the larval midgut. This is because they have been used effectively to explore the coupling of ecdysteroid signals and PCD. They also illustrate the co-occurrence of apoptosis and autophagy, and the cell-specificity of PCD pathways.

12.6.1. Larval Salivary Glands

The salivary glands of *Drosophila* form during embryogenesis, and constitute the largest secretory organ in the fly body. During larval life the salivary glands secrete digestive enzymes, while at the end of larval life they secrete the "glue proteins" that attach the pupa to the substrate – typically, the wall of the culture vial. In response to the pre-pupal pulse of ecdysone secretion at 10–12h post-pupariation, the larval salivary glands undergo PCD (Jiang *et al.*, 1997). This hormonal cue serves to upregulate the expression of a number of multiple pro-apoptotic proteins, including Reaper, HID, Dronc, and Dark (Jiang *et al.*, 1997). Some

of this transcriptional regulation is direct: for example, EcR/USP response elements have been identified in both the *rpr* and *dronc* promoters (Jiang *et al.*, 2000; Cakorous *et al.*, 2004). Much of this regulation, however, reflects the activation of a cascade of early response genes by 20E, including E74, Broad Complex (BRC), and E93 (Lee and Baehrecke, 2001; Yin and Thummel, 2005). The importance of this indirect regulation is shown by the fact that E93 loss-of-function mutants have severe deficits in salivary gland PCD (Lee *et al.*, 2000). Notably, ectopic Reaper expression does not kill the larval salivary glands before the middle of the third and final larval instar, as prior to this PCD is blocked by high levels of DIAP1 (Yin *et al.*, 2007). Resistance to Reaper is removed during the second half of the third instar by a CREB-binding protein (CBP) mediated downregulation of DIAP1 in response to a mid-third instar ecdysteroid pulse (Yin *et al.*, 2007). The role of CBP in steroid-regulated nerve and muscle PCD has not yet been assessed.

Detailed analyses of the morphology of the dying larval salivary gland cells and the accompanying gene cascades provided a compelling example of the co-existence of caspase activation and the formation of autophagosomes. Expression of p35 in salivary glands keeps the cells alive, but does not block the formation of autophagosomes (Lee and Baehrecke, 2001). Confirmation of the simultaneous co-activation of multiple PCD-related programs comes from microarray and SAGE-based studies of gene expression in dying larval salivary glands (Gorski *et al.*, 2003; Lee *et al.*, 2003). Among the several hundred transcripts identified in these studies were known apoptosis-related transcripts, genes associated with autophagy, and non-caspase proteases (Yin and Thummel, 2005).

PCD of larval salivary glands differs from the PCD observed in the embryonic and metamorphosing nervous system, in that the process in the salivary gland destroys an entire organ, whereas in the CNS selected individual cells die while their neighbors survive. PCD in salivary glands is more similar to the phenomenon observed in the ISMs. Whether the distinct selective nature of PCD activation in one case but not the others represents fundamentally or subtly different forms of regulation is unknown.

12.6.2. Larval Midgut

By the time the larval salivary glands begin to degenerate in response to the pre-pupal ecdysone pulse, the death of the larval midgut in response to the late third larval instar ecdysone pulse is already well underway (Lee *et al.*, 2002). However, in contrast to the salivary gland, midgut death occurs in a caspase-independent manner (Denton *et al.*, 2009). Instead, it is completely dependent on the autophagic pathway (Denton *et al.*, 2010). This model may represent the first clear demonstration of an autophagic cell death in the absence of a caspase-based component. Functional analyses of previously identified larval salivary gland death genes were in turn identified as acting as 20E-dependent pro-survival or pro-death factors for the midgut (Chittaranjan *et al.*, 2009).

12.7. Summary and Conclusions

12.7.1. A Mature Field

The extent of research described in this chapter, much of it based on detailed genetic analyses of PCD in insect systems, indicates how thoroughly insect systems are embedded into the broader field of cell death studies. In addition to gaining a better understanding of how PCD matches insect neuromuscular systems to the stage of development, investigators are now linking the molecular machinery of PCD to their mechanistic analyses of the regulation of organ size, aging, and human neurodegenerative disease. Important pathways to watch in the future include the Hippo kinase pathway, which regulates the balance of cell proliferation and apoptosis during organogenesis (Badouel *et al.*, 2009); TOR-dependent autophagy and its relationship to lifespan (Bjedov *et al.*, 2010); and the regulation of cell death related genes by the forkhead box transcription factor FoxO1, which is in turn regulated by phosphorylation by the Parkinson's Disease-associated leucine-rich repeat kinase 2 (LRRK2) (Gandhi *et al.*, 2009; Kanao *et al.*, 2010).

12.7.2. Broad Phylogenetic Comparisons

Rapid strides in genome sequencing will ultimately result in numerous interesting comparative studies of the molecular machinery of all forms in cell death. There is already a wide range of insect genomes that have been sequenced, including 12 different *Drosophila* species and several other dipterans, as well as members of the Lepidoptera, Hymenoptera, Coleoptera, Hemiptera, and Phthiraptera. In addition, the genome of *Daphnia pulex*, a member of the crustacean sister group, has also been sequenced. These data provide a rich resource for comparative analysis of PCD pathways and PCD evolution. Studies of PCD in insects have already raised several pertinent questions, including the basis for distinct molecular palettes and pathways for regulating common cellular outcomes in divergent organisms.

12.7.2.1. Evolutionary scenarios for the *grim-reaper* gene complex
The small Reaper, Grim, and Sickle proteins are clearly related, and the much larger Hid protein is also likely to share a common origin. Why are four related proteins needed to control DIAP1 inhibition and apoptosis, and how similar are the functions of these proteins? Gene complexes can provide enhanced capabilities for a cell to control specific biological processes by providing gene products that exhibit overlapping yet distinct expression patterns and functions. Presumably,

the existence of the *reaper*, *hid*, *grim*, and *sickle* genes in *Drosophila* enables cells to control apoptosis and other processes with the exquisite specificity and efficiency that are required for development. It will be fascinating to discover the functional differences that distinguish these proteins. Already, gene expression studies, as well as loss-of-function and gain-of-function mutants, have indicated that these genes exhibit distinct patterns of expression and non-redundant cell killing activities (see, for example, Grether *et al.*, 1995; Robinow *et al.*, 1997; Zhou *et al.*, 1997; Wing *et al.*, 1998, 2002).

While the *grim-reaper* genes are essential cell death activators in *Drosophila*, *bona fide* structural homologs are not found in *C. elegans* or vertebrates. Indeed, until recently, RHG genes had only been identified in other dipterans, including several *Drosophila* species, the blowfly *Lucilia cuprina*, and the mosquitoes *Anopheles gambiae* and *Aedes aegypti* (White *et al.*, 1994; Chen *et al.*, 2004; Zhou *et al.*, 2005). However, a recent bioinformatic anlysis identified an RHG gene from the lepidopteran *Bombyx mori* (Bryant *et al.*, 2009). The predicted proteins share a highly conserved NH2-terminal RHG domain, and most also possess a GH3 domain/Trp block. These RHG proteins are likely to serve as PCD-promoting IAP antagonists. Interestingly, the mosquitoes and silkworm genomes each appear to possess a single RHG family member, as opposed to the four related, linked genes in *Drosophila*. This suggests that gene duplication events leading to the four linked RHG genes in *Drosophila* occurred recently, and that, in comparison with the caspases and Bcl-2 family proteins, there has been an extremely rapid divergence of the coding sequences of IAP-antagonists. It will be of interest to clarify the phylogenetic distribution of RHG genes.

12.7.3. Future Directions

It can be confidently predicted that studies of the diverse mechanisms of PCD in insect neuromuscular systems across the lifespan will lead to greater understanding of the molecular mechanisms of metamorphosis, particularly the ways in which developmental hormone signals are coupled to neuronal and muscle survival. Although studies of developmental PCD in insects now encompass other tissues, studies of PCD in the CNS will remain important because they offer the opportunity to study the life (and death) history of individual identified neurons. Studies of PCD in ISMs will continue to offer the challenge of segment-specific death phenotypes. Studies driven by investigators eager to identify the molecular pathologies involved in human neurodegenerative disease and human myopathies are likely to become an ever-increasing force in the field of insect PCD, particularly as researchers become ever more adept at transitioning seamlessly across mammalian and insect model systems (see, for example, Godin *et al.*, 2010). The focus on translational research, however, should not obscure the many fascinating basic questions that remain unanswered concerning PCD and insect neurons and muscle. A partial list includes the following:

- Why are there so many PCD pathways? Can we develop a model that allows us to predict the PCD program a cell will select under specific developmental and environmental conditions?

- What are the downstream effects of PCD in neuromuscular systems? That is, do dying cells in turn produce signals that influence the phenotype and/or survival of neighboring cells? For example, it has been shown in fly wing that dying cells produce a caspase-dependent signal (wingless) that induces neighboring cells to proliferate. This phenomenon, referred to as compensatory cell proliferation, is thought to help maintain homeostasis in developing tissues (see, for example, Ryoo *et al.*, 2004). The relevance of this phenomenon to other organs and tissues is unknown.

- Is our view of insect PCD skewed because we rely so heavily on a single insect model? For example, there is growing interest in epigenetic regulation of apoptotic genes (Mazin, 2009; Hajji and Joseph, 2010; Jazirehi, 2010), yet the *Drosophila melanogaster* genome does not encode any methyltransferases – thus one key component of epigenetic DNA modification cannot be studied in this species, although other insect species such as the honey bee have methyltransferase enzymes comparable to those of mammals (Gabor Miklos and Maleszka, 2010).

- Do vertebrate-like neurotrophins function to regulate PCD in insect embryogenesis? During metamorphosis? The recent report describing *Drosophila* Neurotrophin 1 (DNT-1) should spark new interest in this possibility (Zhu *et al.*, 2008).

- What is the relationship of cell energy budgets to PCD in insect neurons and muscles?

- What role, if any, do microRNAs play in the regulation of nerve and muscle survival?

- Can genetic tools (RNAi, transfection, etc.) be applied to the study PCD in *Manduca* to facilitate the analysis of the novel cell death genes isolated from this species?

A persisting question in the field is the nature of the relationship between pathological cell death and PCD that is necessary for normal development.

References

Abdelwahid, E., Yokokura, T., Krieser, R. J., Balasundaram, S., Fowle, W. H., & White, K. (2007). Mitochondrial disruption in *Drosophila* apoptosis. *Dev. Cell*, *12*, 793–806.

Abrams, J. M. (1999). An emerging blueprint for apoptosis in *Drosophila*. *Trends Cell Biol.*, *9*, 435–440.

Abrams, J. M., White, K., Fessler, L. I., & Steller, H. (1993). Programmed cell death during *Drosophila* embryogenesis. *Development*, *117*, 29–43.

Ameisen, J. C. (2002). On the origin, evolution, and nature of programmed cell death: A timeline of four billion years. *Cell Death Differ.*, *9*, 367–393.

Arnett, R. H.J. (1993). *American Insects. A Handbook of the Insects of America North of Mexico*. Gainesville, FL: Sandhill Crane Press, Inc.

Badouel, C., Garg, A., & McNeill, H. (2009). Herding hippos: Regulating growth in flies and man. *Curr. Opin. Cell Biol.*, *21*, 837–843.

Baehrecke, E. H. (2002). How death shapes life during development. *Nat. Rev. Mol. Cell Biol.*, *3*, 779–787.

Bangs, P., & White, K. (2000). Regulation and execution of apoptosis during *Drosophila* development. *Dev. Dyn.*, *218*, 68–79.

Barres, B. A., Hart, I. K., Coles, H. S., Burne, J. F., Voyvodic, J. T., *et al.* (1992). Cell death and control of cell survival in the oligodendrocyte lineage. *Cell*, *70*, 31–46.

Barth, S., Glick, D., & Macleod, K. F. (2010). Autophagy: Assays and artifacts. *J. Pathol.*, *221*, 117–124.

Bate, M. (1993). The mesoderm and its derivatives. In M. Bate, & A. Arias (Eds.), *The Development of Drosophila melanogaster* (pp. 1013–1090). Plainview, NY: Academic Press.

Bayline, R. J., Dean, D. M., & Booker, R. (2005). Inhibitors of ubiquitin-dependent proteolysis can delay programmed cell death of adult intersegmental muscles in the moth *Manduca sexta*. *Dev. Dyn.*, *233*, 445–455.

Beaulaton, J., & Lockshin, R. A. (1977). Ultrastructural study of the normal degeneration of the intersegmental muscles of *Antheraea polyphemus* and *Manduca sexta* (Insecta: Lepidoptera) with particular reference to autophagy. *J. Morphol.*, *154*, 39–58.

Bejsovsc, A., & Wieschaus, E. (1993). Segment polarity gene interactions modulate epidermal patterning in *Drosophila* embryos. *Development*, *119*, 501–517.

Bell, R. A., & Joachim, F. G. (1976). Techniques for rearing laboratory cultures of tobacco hornworms and pink bollworms. *Ann. Entomol. Soc. Am.*, *69*, 365–373.

Bello, B., Holbro, N., & Reichert, H. (2007). Polycomb group genes are required for neural stem cell survival in postembryonic neurogenesis of *Drosophila*. *Development*, *134*, 1091–1099.

Bello, B. C., Hirth, F., & Gould, A. P. (2003). A pulse of the *Drosophila* Hox protein Abdominal-A schedules the end of neural proliferation via neuroblast apoptosis. *Neuron*, *37*, 209–219.

Bennett, K. L., & Truman, J. W. (1985). Steroid-dependent survival of identifiable neurons in cultured ganglia of the moth *Manduca sexta*. *Science*, *229*, 58–60.

Ben-Sasson, S. A., Sherman, Y., & Gavrieli, Y. (1995). Identification of dying cells – in situ staining. In L. M. Schwartz, & B. A. Osborne (Eds.), *Cell Death, Methods in Cell Biology* (pp. 29–39). San Diego, CA: Academic Press.

Bergmann, A. (2010). The role of ubiquitylation for the control of cell death in *Drosophila*. *Cell Death Differ.*, *17*, 61–67.

Bergmann, A., Tugentman, M., Shilo, B. Z., & Steller, H. (2002). Regulation of cell number by MAPK-dependent control of apoptosis: A mechanism for trophic survival signaling. *Dev. Cell*, *2*, 159–170.

Berry, D. L., & Baehrecke, E. H. (2007). Growth arrest and autophagy are required for salivary gland cell degradation in *Drosophila*. *Cell*, *131*, 1137–1148.

Bidmon, H. J., & Koolman, J. (1989). Ecdysteroid receptors located in the central nervous system of an insect. *Experientia*, *45*, 106–109.

Bidmon, H. J., & Sliter, T. (1990). The ecdysteroid receptor. *Intl. J. Invertebr. Reprod. Dev.*, *18*, 13–27.

Bitra, K., & Palli, S. R. (2009). Interaction of proteins involved in ecdysone and juvenile hormone signal transduction. *Arch. Insect Biochem. Physiol.*, *70*, 90–105.

Bjedov, I., Toivonen, J. M., Kerr, F., Slack, C., Jacobson, J., et al. (2010). Mechanisms of life span extension by rapamycin in the fruit fly *Drosophila melanogaster*. *Cell Metab.*, *11*, 35–46.

Booker, R., & Truman, J. W. (1987). Postembryonic neurogenesis in the CNS of the tobacco hornworm, *Manduca sexta*. I. Neuroblast arrays and the fate of their progeny during metamorphosis. *J. Comp. Neurol.*, *255*, 548–559.

Booker, R., Babashak, J., & Kim, J. B. (1996). Postembryonic neurogenesis in the central nervous system of the tobacco hornworm, *Manduca sexta*. III. Spatial and temporal patterns of proliferation. *J. Neurobiol.*, *29*, 233–248.

Brachmann, C. B., Jassim, O. W., Wachsmuth, B. D., & Cagan, R. L. (2000). The *Drosophila* Bcl-2 family member dBorg-1 functions in the apoptotic response to UV-irradiation. *Curr. Biol.*, *10*, 547–550.

Braun, S., Azoitei, A., & Spindler-Barth, M. (2009). DNA-binding properties of *Drosophila* ecdysone receptor isoforms and their modification by the heterodimerization partner ultraspiracle. *Arch. Insect Biochem. Physiol.*, *72*, 172–191.

Broadie, K. S., & Bate, M. (1991). Development of adult muscles in *Drosophila*: Ablation of identified muscle precursor cells. *Development*, *113*, 103–118.

Broemer, M., & Meier, P. (2009). Ubiquitin-mediated regulation of apoptosis. *Trends Cell Biol.*, *19*, 130–140.

Bryant, B., Zhang, Y., Zhang, C., Santos, C. P., Clem, R. J., & Zhou, L. (2009). A lepidopteran ortholog of *reaper* reveals functional conservation and evolution of IAP antagonists. *Insect Mol. Biol.*, *18*, 341–351.

Bursch, W. (2001). The autophagosomal–lysosomal compartment in programmed cell death. *Cell Death Differ.*, *8*, 569–581.

Cai, L., Fritz, D., Stefanovic, L., & Stefanovic, B. (2010). Binding of LARP6 to the conserved 5′ stem-loop regulates translation of mRNAs encoding type I collagen. *J. Mol. Biol.*, *395*, 309–326.

Cakouros, D., Daish, T. J., & Kumar, S. (2004). Ecdysone receptor directly binds the promoter of the *Drosophila* caspase *dronc*, regulating its expression in specific tissue. *J. Cell Biol.*, *165*, 631–640.

Cascone, P. J., & Schwartz, L. M. (2001). Role of the 3′ UTR in regulating the stability and translatability of death-associated mRNAs in moth skeletal muscle. *Dev. Genes. Evol.*, *211*, 397–405.

Chai, J., Du, C., Wu, J. W., Kyin, S., Wang, X., & Shi, Y. (2000). Structural and biochemical basis of apoptotic activation by Smac/DIABLO. *Nature*, *406*, 855–862.

Challa, M., Malladi, S., Pellock, B. J., Dresneck, D., Varadarajan, S., et al. (2007). *Drosophila Omi*, a mitochondrial-localized IAP antagonist and proapoptotic serine protease. *EMBO J.*, *26*, 3144–3156.

Chang, Y., & Neufeld, T. P. (2010). Autophagy takes flight in *Drosophila*. *FEBS Lett.*, *584*, 1342–1349.

Chen, P., Nordstrom, W., Gish, B., & Abrams, J. M. (1996). *Grim*, a novel cell death gene in *Drosophila*. *Genes Dev.*, *10*, 1773–1782.

Chen, P., Ho, S., Shi, Z., & Abrams, J. M. (2004). Bifunctional killing activity encoded by conserved reaper proteins. *Cell Death Differ.*, *11*, 704–713.

Chew, S. K., Akdemir, F., Chen, P., Lu, W., Mills, K., et al. (2004). The apical caspase *dronc* governs programmed and unprogrammed cell death in *Drosophila*. *Dev. Cell*, *7*, 897–907.

Chittaranjan, S., McConechy, M., Hou, Y. C., Freeman, J. D., Devorkin, L., & Gorski, S. M. (2009). Steroid hormone control of cell death and cell survival: Molecular insights using RNAi. *PLoS Genet.*, *5*, e1000379.

Choi, M. K., & Fahrbach, S. E. (1995). Evidence for an endogenous neurocidin in the *Manduca sexta* ventral nerve cord. *Arch. Insect Biochem. Physiol.*, *28*, 273–289.

Choi, Y., Lee, G., & Park, J. H. (2006). Programmed cell death mechanisms of identifiable peptidergic neurons in *Drosophila melanogaster*. *Development*, *133*, 2223–2232.

Christich, A., Kauppila, S., Chen, P., Sogame, N., Ho, S. I., & Abrams, J. M. (2002). The damage-responsive *Drosophila* gene *sickle* encodes a novel IAP binding protein similar to but distinct from *reaper, grim*, and *hid*. *Curr. Biol.*, *12*, 137–140.

Christofferson, D. E., & Yuan, J. (2010). Necroptosis as an alternative form of programmed cell death. *Curr. Opin. Cell Biol.*, *22*, 263–268.

Colin, J., Gaumer, S., Guenal, I., & Mignotte, B. (2009). Mitochondria, Bcl-2 family proteins and apoptosomes: Of worms, flies and men. *Front. Biosci.*, *14*, 4127–4137.

Colussi, P. A., Quinn, L. M., Huang, D. C., Coombe, M., Read, S. H., et al. (2000). Debcl, a proapoptotic Bcl-2 homologue, is a component of the *Drosophila melanogaster* cell death machinery. *J. Cell Biol.*, *148*, 703–714.

Conradt, B. (2009). Genetic control of programmed cell death during animal development. *Annu. Rev. Genet.*, *43*, 493–523.

Consoulas, C., Restifo, L. L., & Levine, R. B. (2002). Dendritic remodeling and growth of motoneurons during metamorphosis of *Drosophila melanogaster*. *J. Neurosci.*, *22*, 4906–4917.

Cooper, D. M., Granville, D. J., & Lowenberger, C. (2009). The insect caspases. *Apoptosis*, *14*, 247–256.

Crook, N. E., Clem, R. J., & Miller, L. K. (1993). An apoptosis-inhibiting baculovirus gene with a zinc finger-like motif. *J. Virol.*, *67*, 2168–2174.

Crossley, A. C. (1968). The fine structure and mechanism of breakdown of larval intersegmental muscles in the blowfly *Calliphora erythrocephala*. *J. Insect Physiol.*, *14*, 1389–1407.

Crossley, A. C. (1978). The morphology and development of the *Drosophila* muscular system. In M. Ashburner, & T. Wright (Eds.), *The Genetics and Biology of Drosophila* (pp. 499–560). New York, NY: Academic Press.

Dai, J. D., & Gilbert, L. I. (1997). Programmed cell death of the prothoracic glands of *Manduca sexta* during pupal-adult metamorphosis. *Insect Biochem. Mol. Biol.*, *27*, 69–78.

Davis, M. B., Carney, G. E., Robertson, A. E., & Bender, M. (2005). Phenotypic analysis of EcR-A mutants suggests that EcR isoforms have unique functions during *Drosophila* development. *Dev. Biol., 282*, 385–396.

Dawson, S. P., Arnold, J. E., Mayer, N. J., Reynolds, S. F., Billett, M. A., et al. (1995). Developmental changes of the 26S proteasome in abdominal intersegmental muscles of *Manduca sexta* during programmed cell death. *J. Biol. Chem., 270*, 1850–1858.

Degterev, A., Huang, Z., Boyce, M., Li, Y., Jagtap, P., et al. (2005). Chemical inhibitor of nonapoptotic cell death with therapeutic potential for ischemic brain injury. *Nat. Chem. Biol., 1*, 112–119.

DeLorme, A. W., & Mesce, K. A. (1999). Programmed cell death of an identified motoneuron examined *in vivo*: Electrophysiological and morphological correlates. *J. Neurobiol., 39*, 307–322.

Denton, D., Shravage, B., Simin, R., Mills, K., Berry, D. L., et al. (2009). Autophagy, not apoptosis, is essential for midgut cell death in *Drosophila*. *Curr. Biol., 19*, 1741–1746.

Denton, D., Shravage, B., Simin, R., Baehrecke, E. H., & Kumar, S. (2010). Larval midgut destruction in *Drosophila*: not dependent on caspases but suppressed by the loss of autophagy. *Autophagy, 6*, 163–165.

Deponte, M. (2008). Programmed death in protists. *Biochim. Biophys. Acta, 1783*, 1396–1405.

DiNardo, S., Heemskerk, J., Dougan, S., & O'Farrell, P. (1994). The making of a maggot: Patterning the *Drosophila* embryonic epidermis. *Curr. Opin. Genet. Dev., 4*, 529–534.

Ditzel, M., Wilson, R., Tenev, T., Zachariou, A., Paul, A., et al. (2003). Degradation of DIAP1 by the N-end rule pathway is essential for regulating apoptosis. *Nat. Cell Biol., 5*, 467–473.

Dive, C., Gregory, C. D., Phipps, D. J., Evans, D. L., Milner, A. E., & Wyllie, A. H. (1992). Analysis and discrimination of necrosis and apoptosis (programmed cell death) by multiparameter flow cytometry. *Biochim. Biophys. Acta, 1133*, 275–285.

Draizen, T. A., Ewer, J., & Robinow, S. (1999). Genetic and hormonal regulation of the death of peptidergic neurons in the *Drosophila* central nervous system. *J. Neurobiol., 38*, 455–465.

Driscoll, M., & Gerstbrein, B. (2003). Dying for a cause: Invertebrate genetics takes on human neurodegeneration. *Nat. Rev. Genet., 4*, 181–194.

Du, C., Fang, M., Li, Y., & Wang, X. (2000). Smac, a mitochondrial protein that promotes cytochrome c-dependent caspase activation by eliminating IAP inhibition. *Cell, 102*, 33–42.

Eastman, A. (1995). Assays for DNA fragmentation, endonucleases, and intracellular pH and Ca^{2+} associated with apoptosis. In L. M. Schwartz, & B. A. Osborne (Eds.), *Cell Death, Methods in Cell Biology* (pp. 41–55). San Diego, CA: Academic Press.

Eaton, J. L. (1988). *Lepidopteran Anatomy*. New York, NY: Wiley.

Enari, M., Sakahira, H., Yokoyama, H., Okawa, K., Iwamatsu, A., & Nagata, S. (1998). A caspase-activated DNase that degrades DNA during apoptosis, and its inhibitor ICAD. *Nature, 391*, 43–50.

Engelberg-Kulka, H., Amitai, S., Kolodkin-Gal, I., & Hazan, R. (2006). Bacterial programmed cell death and multicellular behavior in bacteria. *PLoS Genet., 2*, e135.

Ewer, J., & Truman, J. W. (1997). Invariant association of ecdysis with increases in cyclic 3′, 5′-guanosine monophosphate immunoreactivity in a small network of peptidergic neurons in the hornworm, *Manduca sexta*. *J. Comp. Physiol. A, 181*, 319–330.

Ewer, J., Wang, C. M., Klukas, K. A., Mesce, K. A., Truman, J. W., & Fahrbach, S. E. (1998). Programmed cell death of identified peptidergic neurons involved in ecdysis behavior in the moth, *Manduca sexta*. *J. Neurobiol., 37*, 265–280.

Fabian, M. R., Sonenberg, N., & Filipowicz, W. (2010). Regulation of mRNA translation and stability by micro RNAs. *Annu. Rev. Biochem., 79*, 351–379.

Fadok, V. A., Bratton, D. L., Guthrie, L., & Henson, P. M. (2001). Differential effects of apoptotic versus lysed cells on macrophage production of cytokines: Role of proteases. *J. Immunol., 166*, 6847–6854.

Fahrbach, S. E. (1992). Developmental regulation of ecdysteroid receptors in the nervous system of *Manduca sexta*. *J. Exp. Zool., 261*, 245–253.

Fahrbach, S. E. (1997). The regulation of neuronal death during insect metamorphosis. *BioScience, 47*, 77–85.

Fahrbach, S. E., & Schwartz, L. M. (1994). Localization of immunoreactive ubiquitin in the nervous system of the *Manduca sexta* moth. *J. Comp. Neurol., 343*, 464–482.

Fahrbach, S. E., & Truman, J. W. (1987). Possible interactions of a steroid hormone and neural inputs in controlling the death of an identified neuron in the moth *Manduca sexta*. *J. Neurobiol., 18*, 497–508.

Fahrbach, S. E., & Truman, J. W. (1989). Autoradiographic identification of ecdysteroid-binding cells in the nervous system of the moth *Manduca sexta*. *J. Neurobiol., 20*, 681–702.

Fahrbach, S. E., Choi, M. K., & Truman, J. W. (1994). Inhibitory effects of actinomycin D and cycloheximide on neuronal death in adult *Manduca sexta*. *J. Neurobiol., 25*, 59–69.

Fahrbach, S. E., DeLorme, A. W., Klukas, K. A., & Mesce, K. A. (1995). A motoneuron spared from steroid-activated developmental death by removal of descending neural inputs exhibits stable electrophysiological properties and morphology. *J. Neurobiol., 26*, 511–522.

Fan, Y., & Bergmann, A. (2010). The cleaved-Caspase-3 antibody is a marker of Caspase-9-like DRONC activity in *Drosophila*. *Cell Death Differ., 17*, 534–539.

Feinstein-Rotkopf, Y., & Arama, E. (2009). Can't live without them, can live with them: Roles of caspases during vital cellular processes. *Apoptosis, 14*, 980–995.

Finlayson, L. H. (1956). Normal and induced degeneration of abdominal muscles during metamorphosis in the Lepidoptera. *J. Cell Sci.* (*formerly Q.J. Microsc. Sci.*), *97*, 215–233.

Forero, M. G., Pennack, J. A., Learte, A. R., & Hidalgo, A. (2009). Dead easy caspase: Automatic counting of apoptotic cells in *Drosophila*. *PLoS One, 4*, e5441.

Franc, N. C. (2002). Phagocytosis of apoptotic cells in mammals, *Caenorhabditis elegans*, and *Drosophila melanogaster*:

Molecular mechanisms and physiological consequences. *Front. Biosci., 1*, 1298–1313.

Frankfurt, O. S., Robb, J. A., Sugarbaker, E. V., & Villa, L. (1996). Monoclonal antibody to single-stranded DNA is a specific and sensitive cellular marker of apoptosis. *Exp. Cell Res., 226*, 387–397.

French, V. (2001). Insect segmentation: genes, stripes and segments in "Hoppers." *Curr. Biol., 11*, R910–R913.

Fulda, S. (2007). Inhibitor of apoptosis proteins as targets for anticancer therapy. *Expert Rev. Anticancer Ther., 7*, 1255–1264.

Gabor Miklos, G. L., & Maleszka, R. (2010). Epigenomic communication systems in humans and honey bees: From molecules to behavior. *Horm. Behav.*, in press.

Gammie, S. C., & Truman, J. W. (1997). Neuropeptide hierarchies and the activation of sequential motor behaviors in the hawkmoth, *Manduca sexta. J. Neurosci., 17*, 4389–4397.

Gandhi, P. N., Chen, S. G., & Wilson-Delfosse, A. L. (2009). Leucine-rich repeat kinase 2 (LRRK2): A key player in the pathogenesis of Parkinson's disease. *J. Neurosci. Res., 87*, 1283–1295.

Garrison, S. L., & Witten, J. L. (2010). Steroid hormone regulation of the voltage-gated, calcium activated potassium channel expression in developing muscular and neural systems. *Dev. Neurobiol*, Jul. 28. (epub ahead of print).

Gavrieli, Y., Sherman, Y., & Ben-Sasson, S. A. (1992). Identification of programmed cell death *in situ* via specific labeling of nuclear DNA fragmentation. *J. Cell Biol., 119*, 493–501.

Geske, F. J., Monks, J., Lehman, L., & Fadok, V. A. (2002). The role of the macrophage in apoptosis: Hunter, gatherer, and regulator. *Intl. J. Hematol., 76*, 16–26.

Giebultowicz, J. M., & Truman, J. W. (1984). Sexual differentiation in the terminal ganglion of the moth *Manduca sexta*: Role of sex-specific neuronal death. *J. Comp. Neurol., 226*, 87–95.

Gilbert, L. I., Granger, N. A., & Roe, R. M. (2000). The juvenile hormones: Historical facts and speculations on future research directions. *Insect Biochem. Mol. Biol., 30*, 617–644.

Glenn, H. L., Wang, Z., & Schwartz, L. M. (2010). Acheron, a Lupus antigen family member, regulates integrin expression, adhesion, and motility in differentiating myoblasts. *Am. J. Physiol. Cell Physiol., 298*, C46–55.

Glücksmann, A. (1951). Cell deaths in normal vertebrate ontogeny. *Biol. Rev., 26*, 59–86.

Godin, J. D., Poizat, G., Hickey, M. A., Maschat, F., & Humbert, S. (2010). Mutant huntingtin-impaired degradation of beta-catenin causes neurotoxicity in Huntington's disease. *EMBO J., 29*, 2433–2445.

Gorski, S. M., Chittaranjan, S., Pleasance, E. D., Freeman, J. D., Anderson, C. L., et al. (2003). A SAGE approach to discovery of genes involved in autophagic cell death. *Curr. Biol., 13*, 358–363.

Grether, M., Abrams, J. M., Agapite, J., White, K., & Steller, H. (1995). The *head involution defective* gene of *Drosophila melanogaster* functions in programmed cell death. *Genes Dev., 9*, 1694–1708.

Gyrd-Hansen, M., & Meier, P. (2010). IAPs: From caspase inhibitors to modulators of NF-kappaB, inflammation and cancer. *Nat. Rev. Cancer, 10*, 561–574.

Haas, A., Baboshina, O., Williams, B., & Schwartz, L. M. (1995). Coordinated induction of the ubiquitin pathway accompanies the developmentally programmed death of insect skeletal muscle. *J. Biol. Chem., 270*, 9407–9412.

Hajji, N., & Joseph, B. (2010). Epigenetic regulation of cell life and death decisions and deregulation in cancer. *Essays Biochem., 48*, 121–146.

Hart, S. P., Dransfield, I., & Rossi, A. G. (2008). Phagocytosis of apoptotic cells. *Methods, 44*, 280–285.

Hashimoto, M. K., Mykles, D. L., Schwartz, L. M., & Fahrbach, S. E. (1996). Imaginal cell-specific accumulation of the multicatalytic proteinase complex (proteasome) during post-embryonic development in the tobacco hornworm, *Manduca sexta. J. Comp. Neurol., 365*, 329–341.

Hatzold, J., & Conradt, B. (2008). Control of apoptosis by asymmetric cell division. *PLoS Biol., 6*, e84.

Hay, B. A. (2000). Understanding IAP function and regulation: A view from *Drosophila. Cell Death Differ., 7*, 1045–1056.

Hay, B. A., & Guo, M. (2006). Caspase-dependent cell death in *Drosophila. Annu. Rev. Cell Dev. Biol., 22*, 623–650.

Hay, B. A., Wassarman, D. A., & Rubin, G. M. (1995). *Drosophila* homologs of baculovirus inhibitor of apoptosis proteins function to block cell death. *Cell, 83*, 1253–1262.

Hegstrom, C. D., Riddiford, L. M., & Truman, J. W. (1998). Steroid and neuronal regulation of ecdysone receptor expression during metamorphosis of muscle in the moth, *Manduca sexta. J. Neurosci., 18*, 1786–1794.

Hidalgo, A., & Booth, G. E. (2000). Glia dictate pioneer axon trajectories in the *Drosophila* embryonic CNS. *Development, 127*, 393–402.

Hidalgo, A., Kinrade, E. F., & Georgiou, M. (2001). The *Drosophila* neuregulin vein maintains glial survival during axon guidance in the CNS. *Dev. Cell, 1*, 679–690.

Hoffman, K. L., & Weeks, J. C. (1998). Programmed cell death of an identified motoneuron *in vitro*: Temporal requirements for steroid exposure and protein synthesis. *J. Neurobiol., 35*, 300–322.

Hoffman, K. L., & Weeks, J. C. (2001). Steroid-mediated programmed cell death of motoneurons: Required role of caspases and mitochondrial events. *Dev. Biol., 229*, 517–536.

Holley, C. L., Olson, M. R., Colón-Ramos, D. A., & Kornbluth, S. (2002). Reaper eliminates IAP proteins through stimulated IAP degradation and generalized translational inhibition. *Nat. Cell Biol., 4*, 439–444.

Horvitz, H. R., Shaham, S., & Hengartner, M. O. (1994). The genetics of programmed cell death in the nematode *Caenorhabditis elegans. Cold Spring Harb. Symp. Quant. Biol., 59*, 377–385.

Hou, Y. C., Chittaranjan, S., Barbosa, S. G., McCall, K., & Gorski, S. M. (2008). Effector caspase Dcp-1 and IAP protein Bruce regulate starvation-induced autophagy during *Drosophila melanogaster* oogenesis. *J. Cell Biol., 182*, 1127–1139.

Hoy, M. A. (2003). *Insect Molecular Genetics* (2nd ed.). New York, NY: Academic Press.

Hu, Y., Cascone, P., Cheng, L., Sun, D., Nambu, J. R., & Schwartz, L. M. (1999). Lepidopteran DALP, and its mammalian ortholog Hic-5, function as negative regulators of muscle differentiation. *Proc. Natl. Acad. Sci. USA, 96*, 10218–10223.

Huerta, S., Goulet, E. J., Huerta-Yepez, S., & Livingston, E. H. (2007). Screening and detection of apoptosis. *J. Surg. Res.,* *139*, 143–156.

Huetteroth, W., El Jundi, B., El Jundi, S., & Schachtner, J. (2010). 3D-reconstructions and virtual 4D-visualization to study metamorphic brain development in the sphinx moth *Manduca sexta. Front. Syst. Neurosci., 4,* 1–15.

Igaki, T., & Miura, M. (2004). Role of Bcl-2 family members in invertebrates. *Biochim. Biophys. Acta, 1644,* 73–81.

Ishizuya-Oka, A., Hasebe, T., & Shi, Y. (2010). Apoptosis in amphibian organs during metamorphosis. *Apoptosis, 15,* 350–364.

Jacobson, M., Weil, M., & Raff, M. (1997). Programmed cell death in animal development. *Cell, 88,* 347–354.

Jazirehi, A. R. (2010). Regulation of apoptosis-associated genes by histone deacetylase inhibitors: Implications in cancer therapy. *Anticancer Drugs, 21,* 805–813.

Jiang, C., Baehrecke, E. H., & Thummel, C. S. (1997). Steroid regulated programmed cell death during *Drosophila* metamorphosis. *Development, 124,* 4673–4683.

Jiang, C., Lamblin, A. F., Steller, H., & Thummel, C. S. (2000). A steroid-triggered transcriptional hierarchy controls salivary gland cell death during *Drosophila* metamorphosis. *Mol. Cell, 5,* 445–455.

Jones, G., & Jones, D. (2000). Considerations on the structural evidence of a ligand-binding function of ultraspiracle, an insect homolog of vertebrate RXR. *Insect Biochem. Mol. Biol., 30,* 671–679.

Jones, C. I., & Newbury, S. F. (2010). Functions of microRNAs in *Drosophila* development. *Biochem. Soc. Trans., 38,* 1137–1143.

Jones, M. E. E., & Schwartz, L. M. (2001). Not all muscles meet the same fate when they die. *Cell Biol. Intl., 25,* 539–545.

Jones, M. E. E., Haire, M. F., Kloetzel, P. M., Mykles, D. L., & Schwartz, L. M. (1995). Changes in the structure and function of the multicatalytic proteinase (proteasome) during programmed cell death in the intersegmental muscles of the hawkmoth, *Manduca sexta. Dev. Biol., 169,* 437–447.

Jones, R. G., Davis, W. L., Hung, A. C., & Vinson, S. B. (1978). Insemination-induced histolysis of the flight musculature in fire ants (*Solenopsis,* spp). *Am. J. Anat., 151,* 603–610.

Juhász, G., Csikós, G., Sinka, R., Erdélyi, M., & Sass, M. (2003). The *Drosophila* homolog of Aut1 is essential for autophagy and development. *FEBS Lett., 543,* 154–158.

Kalberer, N. M., Reisenman, C. E., & Hildebrand, J. G. (2010). Male moths bearing transplanted female antennae express characteristically female behaviour and central neural activity. *J. Exp. Biol., 213,* 1272–1280.

Kanao, T., Venderova, K., Park, D. S., Unterman, T., Lu, B., & Imai, Y. (2010). Activation of FoxO by LRRK2 induces expression of proapoptotic proteins and alters survival of postmitotic dopaminergic neuron in *Drosophila. Hum. Mol. Genet., 19,* 3747–3758.

Kato, K., Awasaki, T., & Ito, K. (2009). Neuronal programmed cell death induces glial cell division in the adult *Drosophila* brain. *Development, 136,* 51–59.

Kerr, J. F. R., Wyllie, A. H., & Currie, A. R. (1972). Apoptosis: A basic biological process with wide-ranging implications in tissue kinetics. *Br. J. Cancer, 26,* 239–257.

Kimura, K., & Truman, J. W. (1990). Postmetamorphic cell death in the nervous and muscular systems of *Drosophila melanogaster. J. Neurosci., 10,* 403–411.

Kinch, G., Hoffman, K. L., Rodrigues, E. M., Zee, M. C., & Weeks, J. C. (2003). Steroid-triggered programmed cell death of a motoneuron is autophagic and involves structural changes in mitochondria. *J. Comp. Neurol., 457,* 384–403.

King, A., & Gottlieb, E. (2009). Glucose metabolism and programmed cell death: An evolutionary and mechanistic perspective. *Curr. Opin. Cell Biol., 21,* 885–893.

Kinrade, E. F., Brates, T., Tear, G., & Hidalgo, A. (2001). Roundabout signaling, cell contact and trophic support confine longitudinal glia and axons in the *Drosophila* CNS. *Development, 128,* 207–216.

Klambt, C., Jacobs, R., & Goodman, C. S. (1991). The midline of the *Drosophila* central nervous system: A model for the genetic analysis of cell fate, cell migration, and growth cone guidance. *Cell, 64,* 801–815.

Koelle, M. R., Talbot, W. S., Segraves, W. A., Bender, M. T., Cherbas, P., & Hogness, D. S. (1991). The *Drosophila EcR* gene encodes an ecdysone receptor, a new member of the steroid receptor superfamily. *Cell, 67,* 59–77.

Kornbluth, S., & White, K. (2005). Apoptosis in *Drosophila:* Neither fish nor fowl (nor man, nor worm). *J. Cell. Sci., 118,* 1779–1787.

Kraft, R., & Jäckle, H. (1994). *Drosophila* mode of metamerization in the embryogenesis of the lepidopteran insect, *Manduca sexta. Proc. Natl. Acad. Sci. USA, 91,* 6634–6638.

Krieser, R. J., & White, K. (2009). Inside an enigma: Do mitochondria contribute to cell death in *Drosophila? Apoptosis, 14,* 961–968.

Kroemer, G., Dallaporta, B., & Resche-Rigon, M. (1998). The mitochondrial death/life regulator in apoptosis and necrosis. *Annu. Rev. Physiol., 60,* 619–642.

Kuelzer, F., Kuah, K., Bishoff, S. J., Cheng, L., Nambu, J., & Schwartz, L. M. (1999). SCLP (small cytoplasmic leucine-rich repeat protein) encodes a novel protein that is dramatically up-regulated during the programmed death of moth skeletal muscle. *J. Neurobiol., 41,* 482–494.

Kumar, A., Bello, B., & Reichert, H. (2009). Lineage-specific cell death in postembryonic brain development of *Drosophila. Development, 136,* 3433–3442.

Kurada, P., & White, K. (1998). Ras promotes cell survival in *Drosophila* by down regulating hid expression. *Cell, 95,* 319–329.

Kuwana. (1936). Degeneration of muscles in the silkworm moth. *Zool. Mag. Tokyo, 48,* 881–884.

Lee, C. Y., & Baehrecke, E. H. (2000). Genetic regulation of programmed cell death in *Drosophila. Cell Res., 10,* 193–204.

Lee, C. Y., & Baehrecke, E. H. (2001). Steroid regulation of autophagic programmed cell death during development. *Development, 128,* 1443–1455.

Lee, C. Y., Wendel, D. P., Reid, P., Lam, G., Thummel, C. S., & Baehrecke, E. H. (2000). E93 directs steroid-triggered programmed cell death in *Drosophila. Mol. Cell, 6,* 433–443.

Lee, C. Y., Cooksey, B., & Baehrecke, E. H. (2002). Steroid regulation of midgut cell death during *Drosophila* development. *Dev. Biol., 250,* 101–111.

Lee, C. Y., Clough, E. A., Yellon, P., Teslovich, T. M., Stephan, D. A., & Baehrecke, E. H. (2003). Genome-wide analyses of steroid- and radiation-triggered programmed cell death in *Drosophila*. *Curr. Biol.*, *13*, 350–357.

Levine, R. B., & Truman, J. W. (1985). Dendritic reorganization of abdominal motoneurons during metamorphosis of the moth, *Manduca sexta*. *J. Neurosci.*, *5*, 2424–2431.

Liu, Q. A., & Hengartner, M. O. (1999). The molecular mechanism of programmed cell death in *C. elegans*. *Ann. NY Acad. Sci.*, *887*, 92–104.

Lockshin, R. A. (1969). Programmed cell death. Activation of lysis by a mechanism involving the synthesis of protein. *J. Insect Physiol.*, *15*, 1505–1516.

Lockshin, R. A. (1973). Degeneration of insect intersegmental muscles: Electrophysiological studies of populations of fibres. *J. Insect Physiol.*, *19*, 2359–2372.

Lockshin, R. A., & Beaulaton, J. (1974). Programmed cell death. Cytochemical evidence for lysosomes during the normal breakdown of the intersegmental muscles. *J. Ultrastruct. Res.*, *46*, 43–62.

Lockshin, R. A., & Beaulaton, J. (1979). Programmed cell death. Electrophysiological and ultrastructural correlations in metamorphosing muscles of lepidopteran insects. *Tissue Cell*, *11*, 803–819.

Lockshin, R. A., & Williams, C. A. (1964). Programmed cell death. II. Endocrine potentiation of the breakdown of the intersegmental muscles of silkmoths. *J. Insect Physiol.*, *10*, 643–649.

Lockshin, R. A., & Williams, C. A. (1965a). Programmed cell death. I. Histology and cytology of the breakdown of the intersegmental muscles in Saturniid moths. *J. Insect Physiol.*, *11*, 123–133.

Lockshin, R. A., & Williams, C. A. (1965b). Programmed cell death. III. Neural control of the breakdown of the intersegmental muscles of silkmoths. *J. Insect Physiol.*, *11*, 601–610.

Lockshin, R. A., & Williams, C. A. (1965c). Programmed cell death. IV. The influence of drugs on the breakdown of the intersegmental muscles of silkmoths. *J. Insect Physiol.*, *11*, 803–809.

Löw, P., Bussell, K., Dawson, S., Billett, M., Mayer, R., & Reynolds, S. E. (1997). Expression of a 26S proteasome ATPase subunit, MS73, in muscles that undergo developmentally programmed cell death, and its control by ecdysteroid hormones in the insect *Manduca sexta*. *FEBS Lett.*, *400*, 345–349.

Löw, P., Hastings, R. A., Dawson, S. P., Sass, M., Billett, M. A., et al. (2000). Localisation of 26S proteasomes with different subunit composition in insect muscles undergoing programmed cell death. *Cell Death Differ.*, *7*, 1210–1217.

Löw, P., Talián, G. C., & Sass, M. (2005). Up- and down-regulated genes in muscles that undergo developmentally programmed cell death in the insect *Manduca sexta*. *FEBS Lett.*, *579*, 4943–4948.

Lu, B. (2009). Recent advances in using *Drosophila* to model neurodegenerative diseases. *Apoptosis*, *14*, 1008–1020.

MacDonald, R. L., & Stoodley, M. (1998). Pathophysiology of cerebral ischemia. *Neurol. Med. Chir.* (*Tokyo*), *38*, 1–11.

Mazin, A. L. (2009). Suicidal function of DNA methylation in age-related genome disintegration. *Ageing Res. Rev.*, *8*, 314–327.

McNabb, S. L., Baker, J. D., Agapite, J., Steller, H., Riddiford, L. M., & Truman, J. W. (1997). Disruption of a behavioral sequence by targeted death of peptidergic neurons in *Drosophila*. *Neuron*, *19*, 813–823.

Meier, P., Finch, A., & Evan, G. (2000). Apoptosis in development. *Nature*, *407*, 796–801.

Mesce, K. A., & Fahrbach, S. E. (2002). Integration of endocrine signals that regulate insect ecdysis. *Front. Neuroendocrinol.*, *23*, 179–199.

Metchnikoff, E. (1892). La phagocytose musculaire. I Atrophie des muscles pendant la transformation des bateaciens. *Ann. Inst. Pasteur*, *6*, 1–12.

Miller, A. (1950). The internal anatomy and histology of the imago of *Drosophila melanogaster*. In M. Demerec (Ed.), *Biology of Drosophila* (pp. 420–534). New York, NY: Wiley.

Milligan, C. M., & Schwartz, L. M. (1997). Programmed cell death during animal development. *Br. Med. Bull.*, *53*, 570–590.

Mohseni, N., McMillan, S. C., Chaudhary, R., Mok, J., & Reed, B. H. (2009). Autophagy promotes caspase-dependent cell death during *Drosophila* development. *Autophagy*, *5*, 329–338.

Monsma, S. A., & Booker, R. (1996a). Genesis of the adult retina and outer optic lobes of the moth, *Manduca sexta*. I. Patterns of proliferation and cell death. *J. Comp. Neurol.*, *367*, 10–20.

Monsma, S. A., & Booker, R. (1996b). Genesis of the adult retina and outer optic lobes of the moth, *Manduca sexta*. II. Effects of deafferentation and developmental hormone manipulation. *J. Comp. Neurol.*, *367*, 21–35.

Montemayor, M. E., Fahrbach, S. E., Giometti, C. S., & Roy, E. J. (1990). Characterization of a protein that appears in the nervous system of the moth *Manduca sexta* coincident with neuronal death. *FEBS Lett.*, *276*, 219–222.

Montero, J. A., & Hurlé, J. M. (2010). Sculpturing digit shape by cell death. *Apoptosis*, *15*, 365–375.

Muppidi, J., Porter, M., & Siegel, R. M. (2004). Measurement of apoptosis and other forms of cell death. *Curr. Protoc. Immunol.*, Chapter 3, Unit 3.17.

Mutsuddi, M., & Nambu, J. R. (1998). Neural disease: *Drosophila* degenerates for a good cause. *Curr. Biol.*, *8*, R809–R811.

Nagy, L. M., Booker, R., & Riddiford, L. M. (1991). Isolation and embryonic expression of an abdominal-A-like gene from the lepidopteran, *Manduca sexta*. *Development*, *112*, 119–129.

Nezis, I. P., Lamark, T., Velentzas, A. D., Rusten, T. E., Bjørkøy, G., Johansen, T., Papassideri, I. S., Stravopodis, D. J., Margaritis, L. H., Stenmark, H., & Brech, A. (2009). Cell death during *Drosophila melanogaster* early oogenesis is mediated through autophagy. *Autophagy*, *5*, 298–302.

Nezis, I. P., Shravage, B. V., Sagona, A. P., Lamark, T., Bjørkøy, G., et al. (2010). Autophagic degradation of dBruce controls DNA fragmentation in nurse cells during late *Drosophila melanogaster* oogenesis. *J. Cell Biol.*, *190*, 523–531.

Nijhout, H. F. (1994). *Insect Hormones*. Princeton, NJ: Princeton University Press.

Nishikawa, A., Murata, E., Akita, M., Kaneko, K., Moriya, O., et al. (1998). Role of macrophages in programmed cell death and remodeling of tail and body muscle of *Xenopus laevis* during metamorphosis. *Histochem. Cell Biol.*, *109*, 11–17.

Opferman, J. T. (2008). Apoptosis in the development of the immune system. *Cell Death Differ.*, *15*, 234–242.

Orgogozo, V., Schweisguth, F., & Bellaïche, Y. (2002). Binary cell death decision regulated by unequal partitioning of Numb at mitosis. *Development*, *129*, 4677–4684.

Orme, M., & Meier, P. (2009). Inhibitor of apoptosis proteins in *Drosophila*: Gatekeepers of death. *Apoptosis*, *14*, 950–960.

Park, J. H., Schroeder, A. J., Helfrich-Forster, C., Jackson, F. R., & Ewer, J. (2003). Targeted ablation of CCAP neuropeptide-containing neurons of *Drosophila* causes specific defects in execution and timing of ecdysis behavior. *Development*, *130*, 2645–2656.

Peterson, C., Carney, G. E., Taylor, B. J., & White, K. (2002). Reaper is required for neuroblast apoptosis during *Drosophila* development. *Development*, *129*, 1467–1476.

Quinn, L., Coombe, M., Mills, K., Daish, T., Colussi, P., et al. (2003). Buffy, a *Drosophila* Bcl-2 protein, has antiapoptotic and cell cycle inhibitory functions. *EMBO J.*, *22*, 3568–3579.

Raff, M. C., Barres, B. A., Burne, J. F., Coles, H. S., Ishizaki, Y., & Jacobson, M. D. (1993). Programmed cell death and the control of cell survival: Lessons from the nervous system. *Science*, *262*, 695–700.

Reape, T. J., & McCabe, P. F. (2008). Apoptotic-like programmed cell death in plants. *New Phytol.*, *180*, 13–26.

Rechsteiner, M. (1988). *Ubiquitin*. New York, NY: Plenum Press.

Renn, S. C.P., Park, J. H., Rosbash, M., Hall, J. C., & Taghert, P. H. (1999). A *pdf* neuropeptide gene mutation and ablation of PDF neurons each cause severe abnormalities of behavioral circadian rhythms in *Drosophila*. *Cell*, *99*, 781–802.

Rétaux, S., & Bachy, I. (2002). A short history of LIM domains (1993–2002): From protein interaction to degradation. *Mol. Neurobiol.*, *26*, 269–281.

Rheuben, M. B. (1992). Degenerative changes in the muscle fibers of *Manduca sexta* during metamorphosis. *J. Exp. Biol.*, *167*, 91–117.

Riddiford, L. M. (1993). Hormones and *Drosophila* development. In M. Bate, & A. Arias (Eds.). *The Development of Drosophila melanogaster* (pp. 899–939). Plainview, NY: Cold Spring Harbor Laboratory Press.

Riddiford, L. M. (2008). Juvenile hormone action: A 2007 perspective. *J. Insect Physiol.*, *54*, 895–901.

Robinow, S., Talbot, W. S., Hogness, D. S., & Truman, J. W. (1993). Programmed cell death in the *Drosophila* CNS is ecdysone-regulated and coupled with a specific ecdysone receptor isoform. *Development*, *119*, 1251–1259.

Robinow, S., Draizen, T. A., & Truman, J. W. (1997). Genes that induce apoptosis: Transcriptional regulation in identified, doomed neurons of the *Drosophila* CNS. *Dev. Biol.*, *190*, 206–213.

Robinson-Rechavi, M., Garcia, H. E., & Laudet, V. (2003). The nuclear receptor superfamily. *J. Cell. Sci.*, *116*, 585–586.

Rogulja-Ortmann, A., Lüer, K., Seibert, J., Rickert, C., & Technau, G. M. (2007). Programmed cell death in the embryonic central nervous system of *Drosophila melanogaster*. *Development*, *134*, 105–116.

Rulifson, E. J., Kim, S. K., & Nusse, R. (2002). Ablation of insulin-producing neurons in flies: Growth and diabetic phenotypes. *Science*, *296*, 1118–1120.

Rybczynski, R., Snyder, C. A., Hartmann, J., Gilbert, L. I., & Sakurai, S. (2009). *Manduca sexta* prothoracicotropic hormone: Evidence for a role beyond steroidogenesis. *Arch. Insect Biochem. Physiol.*, *70*, 217–229.

Ryoo, H. D., & Baehrecke, E. H. (2010). Distinct death mechanisms in *Drosophila* development. *Curr. Opin. Cell Biol.*, *22*, 889–995.

Ryoo, H. D., Gorenc, T., & Steller, H. (2004). Apoptotic cells can induce compensatory cell proliferation through the JNK and the Wingless signaling pathways. *Dev. Cell*, *7*, 491–501.

Salomons, F. A., Acs, K., & Dantuma, N. P. (2010). Illuminating the ubiquitin/proteasome system. *Exp. Cell Res.*, *316*, 1289–1295.

Sandu, C., Ryoo, H. D., & Steller, H. (2010). *Drosphila* IAP antagonists form multimeric complexes to promote cell death. *J. Cell Bio.*, *190*, 1039–1052.

Sanson, B. (2001). Generating patterns from fields of cells. Examples from *Drosophila* segmentation. *EMBO Rep.*, *2*, 1083–1088.

Schneiderman, A., Matsumoto, S. G., & Hildebrand, J. G. (1982). Trans-sexually grafted antennae influence development of sexually dimorphic neurons in moth brain. *Nature*, *298*, 844–846.

Schnepp, B., Grumbling, G., Donaldson, T., & Simcox, A. (1996). Vein is a novel component in the *Drosophila* epidermal growth factor receptor pathway with similarity to the neuregulins. *Genes Dev.*, *10*, 2302–2313.

Schwartz, L. M. (2008). Atrophy and programmed cell death of skeletal muscle. *Cell Death Differ.*, *15*, 1163–1169.

Schwartz, L. M., & Ruff, R. L. (2002). Changes in contractile properties of skeletal muscle during developmentally programmed atrophy and death. *Am. J. Physiol. Cell Physiol.*, *282*, C1270–C1277.

Schwartz, L. M., & Truman, J. W. (1982). Peptide and steroid regulation of muscle degeneration in an insect. *Science*, *215*, 1420–1421.

Schwartz, L. M., & Truman, J. W. (1983). Hormonal control of rates of metamorphic development in the tobacco hornworm, *Manduca sexta*. *Dev. Biol.*, *99*, 103–114.

Schwartz, L., & Truman, J. W. (1984a). Cyclic GMP may serve as a second messenger in peptide-induced muscle degeneration in an insect. *Proc. Natl. Acad. Sci. USA*, *81*, 6718–6722.

Schwartz, L. M., & Truman, J. (1984b). Hormonal control of muscle atrophy and degeneration in the moth *Antheraea polyphemus*. *J. Exp. Biol.*, *111*, 13–30.

Schwartz, L. M., Myer, A., Kosz, L., Engelstein, M., & Maier, C. (1990). Activation of polyubiquitin gene expression during developmentally programmed cell death. *Neuron*, *5*, 411–419.

Schwartz, L. M., Kosz, L., & Kay, B. K. (1990a). Gene activation is required for developmentally programmed cell death. *Proc. Natl. Acad. Sci. USA*, *87*, 6594–6598.

Schwartz, L. M., Myer, A., Kosz, L., Engelstein, M., & Maier, C. (1990b). Activation of polyubiquitin gene expression during developmentally programmed cell death. *Neuron*, *5*, 411–419.

Schwartz, L. M., Smith, S. W., Jones, M. E. E., & Osborne, B. A. (1993a). Do all programmed cell deaths occur via apoptosis? *Proc. Natl. Acad. Sci. USA, 90,* 980–984.

Schwartz, L. M., Jones, M. E., Kosz, L., & Kuah, K. (1993b). Selective repression of actin and myosin heavy chain expression during the programmed death of insect skeletal muscle. *Dev. Biol., 158,* 448–455.

Schwartz, L. M., Gao, Z., Brown, C., Parelkar, S. S., & Glenn, H. (2009). Cell death in myoblasts and muscles. *Methods Mol. Biol., 559,* 313–332.

Sears, H. C., Kennedy, C. J., & Garrity, P. A. (2003). Macrophage-mediated corpse engulfment is required for normal *Drosophila* CNS morphogenesis. *Development, 130,* 3557–3565.

Segovia, M., Haramaty, L., Berges, J. A., & Falkowski, P. G. (2003). Cell death in the unicellular chlorophyte *Dunaliella tertiolecta*. A hypothesis on the evolution of apoptosis in higher plants and metazoans. *Plant Physiol., 132,* 99–105.

Shao, R., Scully, S. J., Jr., Yan, W., Bentley, B., Mueller, J., et al. (2011). The novel lupus antigen related protein Acheron enhances the development and growth of human breast cancer. *Intl. J. Cancer*, in press.

Shapiro, P. J., Hsu, H. H., Jung, H., Robbins, E. S., & Ryoo, H. D. (2008). Regulation of the *Drosophila* apoptosome through feedback inhibition. *Nat. Cell Biol., 10,* 1440 1446.

Shemarova, I. V. (2010). Signaling mechanisms of apoptosis-like programmed cell death in unicellular eukaryotes. *Comp. Biochem. Physiol. B, Biochem. Mol. Biol., 155,* 341–353.

Shen, W., & Ganetzky, B. (2009). Autophagy promotes synapse development in *Drosophila*. *J. Cell Biol., 187,* 71–79.

Shen, W., & Ganetzky, B. (2010). Nibbling away at synaptic development. *Autophagy, 6,* 168–169.

Siegrist, S. E., Haque, N. S., Chen, C., Hay, B. A., & Hariharan, I. K. (2010). Inactivation of both Foxo and reaper promotes long-term adult neurogenesis in *Drosophila*. *Curr. Biol., 20,* 643–648.

Smith, C. A., Williams, G. T., Kingston, R., Jenkinson, E. J., & Owen, J. J. (1989). Antibodies to CD3/T-cell receptor complex induce death by apoptosis in immature T cells in thymic cultures. *Nature, 337,* 181–184.

Sokol, N. S., Xu, P., Jan, Y. N., & Ambros, V. (2008). *Drosophila* let-7 microRNA is required for remodeling of the neuromusculature during metamorphosis. *Genes Dev., 22,* 1591–1596.

Song, Z., McCall, K., & Steller, H. (1997). DCP-1, a *Drosophila* cell death protease essential for development. *Science, 275,* 536–540.

Sonnenfeld, M. J., & Jacobs, J. R. (1995). Apoptosis of the midline glia during *Drosophila* embryogenesis: A correlation with axon contact. *Development, 121,* 569–578.

Spana, E. P., & Doe, C. Q. (1996). Numb antagonizes Notch signaling to specify sibling neuron cell fates. *Neuron, 17,* 21–26.

Srinivasula, S. M., Datta, P., Fan, X. J., Fernandes-Alnemri, T., Huang, Z., & Alnemri, E. S. (2000). Molecular determinants of the caspase-promoting activity of Smac/DIABLO and its role in the death receptor pathway. *J. Biol. Chem., 275,* 36152–36157.

Srinivasula, S. M., Datta, P., Kobayashi, M., Wu, J. W., Fujioka, M., et al. (2002). *Sickle*, a novel *Drosophila* death gene in the *reaper/hid/grim* region, encodes an IAP-inhibitory protein. *Curr. Biol., 12,* 125–130.

Steller, H. (2008). Regulation of apoptosis in *Drosophila*. *Cell Death Differ., 15,* 1132–1138.

Stocker, R. F., Edwards, J. S., & Truman, J. W. (1978). Fine structure of degenerating moth abdominal motorneurons after eclosion. *J. Cell Tissue Res., 191,* 317–331.

Stoica, B. A., & Faden, A. I. (2010). Cell death mechanisms and modulation in traumatic brain injury. *Neurotherapeutics, 7,* 3–12.

Streichert, L. C., Pierce, J. T., & Weeks, J. C. (1997). Steroid hormones act directly to trigger segment-specific programmed cell death of identified neurons *in vitro*. *Dev. Biol., 183,* 95–107.

Sun, D., Ziegler, R., Milligan, C. E., Fahrbach, S. E., & Schwartz, L. M. (1995). Apoliphorin III is dramatically up-regulated during the programmed death of insect skeletal muscle and neurons. *J. Neurobiol., 26,* 119–129.

Sun, D., Sathyanarayana, U. G., Johnston, S. A., & Schwartz, L. M. (1996). A member of the phylogenetically conserved CAD family of transcriptional regulators is dramatically up-regulated during the programmed cell death of skeletal muscle in the tobacco hawkmoth *Manduca sexta*. *Dev. Biol., 173,* 499–509.

Sutherland, E. W. (1972). Studies on the mechanism of hormone action. *Science, 177,* 401–408.

Suzuki, Y., Imai, Y., Nakayama, H., Takahashi, K., Takio, K., & Takahashi, R. (2001). A serine protease, HtrA2, is released from the mitochondria and interacts with XIAP, inducing cell death. *Mol. Cell, 8,* 613–621.

Takayanagi, K., Dawson, S., Reynolds, S. E., & Mayer, R. J. (1996). Specific developmental changes in the regulatory subunits of the 26S proteasome in intersegmental muscles preceding eclosion in *Manduca sexta*. *Biochem. Biophys. Res. Commun., 228,* 517–523.

Talbot, W. S., Swyryd, E. A., & Hogness, D. S. (1993). *Drosophila* tissues with different metamorphic responses to ecdysone express different ecdysone receptor isoforms. *Cell, 73,* 1323–1337.

Tanida, I., Ueno, T., & Kominami, E. (2008). LC3 and autophagy. *Methods Mol. Biol., 445,* 77–88.

Taylor, H. M., & Truman, J. W. (1974). Metamorphosis of the abdominal ganglia of the tobacco hornworm, *Manduca sexta*: Changes in population of identified motor neurons. *J. Comp. Physiol., 90,* 367–388.

Thorn, R. S., & Truman, J. W. (1989). Sex-specific neuronal respecification during the metamorphosis of the genital segments of the tobacco hornworm moth *Manduca sexta*. *J. Comp. Neurol., 284,* 489–503.

Thorn, R. S., & Truman, J. W. (1994a). Sexual differentiation in the central nervous system of the moth *Manduca sexta*. I. Sex and segment-specificity in production, differentiation, and survival of the imaginal midline neurons. *J. Neurobiol., 25,* 1039–1044.

Thorn, R. S., & Truman, J. W. (1994b). Sexual differentiation in the central nervous system of the moth, *Manduca sexta*. II. Target dependence for the survival of the imaginal midline neurons. *J. Neurobiol., 10,* 1054–1066.

Thummel, C. (2001). Steroid-triggered death by autophagy. *Bioessays, 23,* 677–682.

Tittel, J. N., & Steller, H. (2000). A comparison of programmed cell death between species. *Genome Biol.*, *1*, reviews0003.

Truman, J. W. (1983). Programmed cell death in the nervous system of an adult insect. *J. Comp. Neurol.*, *216*, 445–452.

Truman, J. W. (1984). Ecdysteroids regulate the release and action of eclosion hormone in the moth *Manduca sexta*. In J. Hoffman, & M. Porchet (Eds.), *Biosynthesis, Metabolism, and Mode of Action of Invertebrate Hormones* (pp. 136–144). Berlin: Springer-Verlag.

Truman, J. W. (1990). Metamorphosis of the central nervous system of *Drosophila*. *J. Neurobiol.*, *21*, 1072–1084.

Truman, J. W., & Bate, M. (1988). Spatial and temporal patterns of neurogenesis in the CNS of *Drosophila melanogaster*. *Dev. Biol.*, *125*, 146–157.

Truman, J. W., & Riddiford, L. M. (2002). Insect developmental hormones and their mechanism of action. In D. W. Pfaff (Ed.), *Hormones, Brain and Behavior* (pp. 841–873). San Diego, CA: Academic Press.

Truman, J. W., & Schwartz, L. M. (1984). Steroid regulation of neuronal death in the moth nervous system. *J. Neurosci.*, *4*, 274–280.

Truman, J. W., Taylor, B. J., & Awad, T. A. (1993). Formation of the adult nervous system. In M. Bate, & A. M. Arias (Eds.), *The Development of Drosophila melanogaster* (pp. 1245–1275). Plainview, NY: Cold Spring Harbor Laboratory Press.

Truman, J. W., Talbot, W. S., Fahrbach, S. E., & Hogness, D. S. (1994). Ecdysone receptor expression in the CNS correlates with stage-specific responses to ecdysteroids during *Drosophila* and *Manduca* development. *Development*, *120*, 219–234.

Valavanis, C., Wang, Z., Sun, D., Vaine, M., & Schwartz, L. M. (2007). Acheron, a novel member of the Lupus Antigen family, is induced during the programmed cell death of skeletal muscles in the moth *Manduca sexta*. *Gene*, *393*, 101–109.

Vandenabeele, P., Galluzzi, L., Vanden Berghe, T., & Kroemer, G. (2010). Molecular mechanisms of necroptosis: An ordered cellular explosion. *Nat. Rev. Mol. Cell Biol.*, *11*, 700–714.

Vaux, D. L., & Silke, J. (2005). IAPs, RINGs and ubiquitylation. *Nat. Rev. Mol. Cell Biol.*, *6*, 287–297.

Verhagen, A. M., & Vaux, D. L. (2002). Cell death regulation by the mammalian IAP antagonist Diablo/Smac. *Apoptosis*, *7*, 163–166.

Vernooy, S. Y., Copeland, J., Ghaboosi, N., Griffin, E. E., Yoo, S. J., & Hay, B. A. (2000). Cell death regulation in *Drosophila*: Conservation of mechanism and unique insights. *J. Cell Biol.*, *150*, F69–76.

Wan, H. I., DiAntonio, A., Fetter, R. D., Bergstrom, K., Strauss, R., & Goodman, C. S. (2000). Highwire regulates synaptic growth in *Drosophila*. *Neuron*, *26*, 313–329.

Wang, C., & Youle, R. J. (2009). The role of mitochondria in apoptosis. *Annu. Rev. Genet.*, *43*, 95–118.

Wang, Z., Glenn, H., Brown, C., Valavanis, C., Liu, J. X., et al. (2009). Regulation of muscle differentiation and survival by Acheron. *Mech. Dev.*, *126*, 700–709.

Wasser, M., Bte Osman, Z., & Chia, W. (2007). EAST and Chromator control the destruction and remodeling of muscles during *Drosophila* metamorphosis. *Dev. Biol.*, *307*, 380–393.

Watanabe, K., & Sassaki, F. (1974). Ultrastructural changes in regressing tail muscles of anuran tadpoles during metamorphosis. *Cell Tissue Res.*, *155*, 321–336.

Weber, R. (1964). Ultrastructural changes in regressing tail muscles of *Xenopus* larvae at metamorphosis. *J. Cell Biol.*, *68*, 251–306.

Weber, R. (1965). Inhibitory effect of actinomycin D on tail atrophy in *Xenopus* larvae at metamorphosis. *Experientia*, *21*, 665–666.

Weeks, J. C. (1987). Time course of hormonal independence for developmental events in neurons and other cell types during insect metamorphosis. *Dev. Biol.*, *124*, 163–176.

Weeks, J. C. (1999). Steroid hormones, dendritic remodeling and neuronal death: Insights from insect metamorphosis. *Brain Behav. Evol.*, *54*, 51–60.

Weeks, J. C., & Davidson, S. K. (1994). Influence of interganglionic interactions on steroid-mediated dendritic reorganization and death of proleg motor neurons during metamorphosis in *Manduca sexta*. *J. Neurobiol.*, *25*, 535–554.

Weeks, J. C., & Ernst-Utzschneider, K. (1989). Respecification of larval proleg motoneurons during metamorphosis of the tobacco hornworm, *Manduca sexta*: Segmental dependence and hormonal regulation. *J. Neurobiol.*, *20*, 569–592.

Weeks, J. C., & Truman, J. W. (1985). Independent steroid control of the fates of motoneurons and their muscles during insect metamorphosis. *J. Neurosci.*, *5*, 2290–2300.

Weeks, J. C., Roberts, W. M., & Trimble, D. L. (1992). Hormonal regulation and segmental specificity of motoneuron phenotype during metamorphosis of the tobacco hornworm, *Manduca sexta*. *Dev. Biol.*, *149*, 185–196.

Weeks, J. C., Davidson, S. K., & Debu, B. H.G. (1993). Effects of a protein synthesis inhibitor on the hormonally mediated regression and death of motoneurons in the tobacco hornworm, *Manduca sexta*. *J. Neurobiol.*, *24*, 125–140.

Whelan, R. S., Kaplinskiy, V., & Kitsis, R. N. (2010). Cell death in the pathogenesis of heart disease: Mechanisms and significance. *Annu. Rev. Physiol.*, *72*, 19–44.

White, K., Grether, M. E., Abrams, J. M., Young, L., Farrell, K., & Steller, H. (1994). Genetic control of programmed cell death in *Drosophila*. *Science*, *264*, 677–683.

Williams, D. W., Kondo, S., Krzyzanowska, A., Hiromi, Y., & Truman, J. W. (2006). Local caspase activity directs engulfment of dendrites during pruning. *Nat. Neurosci.*, *9*, 1234–1306.

Wilson, T. G., Yerushalmi, Y., Donnell, D. M., & Restifo, L. L. (2006). Interaction between hormonal signaling pathways in *Drosophila melanogaster* as revealed by genetic interaction between methoprene-tolerant and broad-complex. *Genetics*, *172*, 253–264.

Wing, J. P., & Nambu, J. (1998). Apoptosis in *Drosophila*. In C. Potten, C. Booth, & J. W. Wilson (Eds.), *Apoptosis Regulatory Genes*. London, UK: Chapman and Hall.

Wing, J. P., Zhou, L., Schwartz, L. M., & Nambu, J. R. (1998). Distinct cell killing properties of the Drosophila *reaper, head involution defective*, and *grim* genes. *Cell Death Differ.*, *5*, 930–939.

Wing, J. P., Schwartz, L. M., & Nambu, J. R. (2001). The RHG motifs of *Drosophila* reaper and grim are important for their distinct cell death-inducing abilities. *Mech. Dev.*, *102*, 193–203.

Wing, J. P., Karres, J. S., Ogdahl, J. L., Zhou, L., Schwartz, L. M., & Nambu, J. R. (2002). *Drosophila sickle* is a novel *grim-reaper* cell death activator. *Curr. Biol., 12*, 131–135.

Wyllie, A. H., Morris, R. G., Smith, A. L., & Dunlop, D. (1984). Chromatin cleavage in apoptosis: Association with condensed chromatin morphology and dependence on macromolecular synthesis. *J. Pathol., 142*, 67–77.

Xie, Z., & Klionsky, D. J. (2007). Autophagosome formation: Core machinery and adaptations. *Nat. Cell Biol., 9*, 1102–1109.

Xu, D., Wang, Y., Willecke, R., Chen, Z., Ding, T., & Bergmann, A. (2006). The effector caspases drICE and dcp-1 have partially overlapping functions in the apoptotic pathway in *Drosophila. Cell Death Differ., 13*, 1697–1706.

Xu, D., Woodfield, S. E., Lee, T. V., Fan, Y., Antonio, C., & Bergmann, A. (2009). Genetic control of programmed cell death (apoptosis) in *Drosophila. Fly (Austin), 3*, 78–90.

Yaffe, D., & Saxel, O. (1977). Serial passaging and differentiation of myogenic cells isolated from dystrophic mouse muscle. *Nature, 270*, 725–727.

Yang, C., Thomenius, M. J., Gan, E. C., Tang, W., Freel, C. D., et al. (2010). Metabolic regulation of *Drosophila* apoptosis through inhibitory phosphorylation of Dronc. *EMBO J., 29*, 3196–3207.

Yang, Y., Fang, S., Jensen, J. P., Weissman, A. M., & Ashwell, J. D. (2000). Ubiquitin protein ligase activity of IAPs and their degradation in proteasomes in response to apoptotic stimuli. *Science, 288*, 874–877.

Yin, V. P., & Thummel, C. S. (2005). Mechanisms of steroid-triggered programmed cell death in *Drosophila. Semin. Cell Dev. Biol., 16*, 237–243.

Yin, V. P., Thummel, C. S., & Bashirullah, A. (2007). Down-regulation of inhibitor of apoptosis levels provides competence for steroid-triggered cell death. *J. Cell Biol., 178*, 85–92.

Yoo, S. J., Huh, J. R., Muro, I., Yu, H., Wang, L., et al. (2002). Hid, Rpr and Grim negatively regulate DIAP1 levels through distinct mechanisms. *Nat. Cell Biol., 4*, 416–424.

Yoshizato, K. (1996). Cell death and histolysis in amphibian tail during metamorphosis. In L. Gilbert, J. Tata, & B. Atkinson (Eds.), *Metamorphosis. Postembryonic Reprogramming of Gene Expression in Amphibian and Insect Cells* (pp. 647–671). San Diego, CA: Academic Press.

Zachariou, A., Tenev, T., Goyal, L., Agapite, J., Steller, H., & Meier, P. (2003). IAP antagonists exhibit non-redundant modes of action through differential DIAP1 binding. *EMBO J., 22*, 6642–6652.

Zee, M. C., & Weeks, J. C. (2001). Developmental change in the steroid hormone signal for cell-autonomous, segment-specific programmed cell death of a motoneuron. *Dev. Biol., 235*, 45–61.

Zhang, H., Huang, Q., Ke, N., Matsuyama, S., Hammock, B., et al. (2000). *Drosophila* pro-apoptotic Bcl-2/Bax homologue reveals evolutionary conservation of cell death mechanisms. *J. Biol. Chem., 275*, 27303–27306.

Zhang, P., Aso, Y., Jikuya, H., Kusakabe, T., Lee, J. M., et al. (2007). Proteomic profiling of the silkworm skeletal muscle proteins during larval-pupal metamorphosis. *J. Proteome Res., 6*, 2295–2303.

Zhao, Y., Bretz, C. A., Hawksworth, S. A., Hirsh, J., & Johnson, E. C. (2010). Corazonin neurons function in sexually dimorphic circuitry that shape behavioral responses to stress in *Drosophila. PLoS One, 5*, e9141.

Zhong, W. (2008). Timing cell-fate determination during asymmetric cell divisions. *Curr. Opin. Neurobiol., 18*, 472–478.

Zhou, L., Hashimi, H., Schwartz, L. M., & Nambu, J. R. (1995). Programmed cell death in the *Drosophila* central nervous system midline. *Curr. Biol., 5*, 784–790.

Zhou, L., Schnitzler, A., Agapite, J., Schwartz, L. M., Steller, H., & Nambu, J. R. (1997). Cooperative functions of the *reaper* and *head involution defective* genes in the programmed cell death of *Drosophila* central nervous system midline cells. *Proc. Natl. Acad. Sci. USA, 94*, 5131–5136.

Zhou, L., Jiang, G., Chan, G., Santos, C. P., Severson, D. W., & Xiao, L. (2005). Michelob x is the missing inhibitor of apoptosis protein antagonist in mosquito genomes. *EMBO Rep., 6*, 769–774.

Zhu, B., Pennack, J. A., McQuilton, P., Forero, M. G., Mizuguchi, K., et al. (2008). *Drosophila* neurotrophins reveal a common mechanism for nervous system formation. *PLoS Biol., 6*, e284.

Zuzarte-Luís, V., & Hurlé, J. M. (2002). Programmed cell death in the developing limb. *Intl J. Dev. Biol., 46*, 871–876.

13 Regulation of Insect Development by TGF-β Signaling

Philip A Jensen
Department of Biology, Rocky Mountain College,
Billings, MT, USA

Summary

The Transforming Growth Factor-β (TGF-β) signaling family influences the differentiation and growth of many tissues during the development of insects. This chapter focuses on the TGF-β system of the well-studied insect *Drosophila melanogaster*. The components and functions of both the canonical Bone Morphogenic Protein (BMP) and Activin branches of TGF-β family in *Drosophila* are discussed, as are instances where the two pathways interact or signal non-canonically. Regulators of the pathway are also presented in several contexts where TGF-β signaling is necessary for proper development. Finally, evolution of TGF-β signaling in other insects and arthropods is discussed briefly.

13.1. Overview and Components

The Transforming Growth Factor-beta (TGF-β) signaling family uses a relatively simple signal transduction pathway in development of many organisms across the animal kingdom, including insects. TGF-β signaling utilizes extracellular, secreted signaling molecules that bind to receptors on the surface of cells. These receptors then activate a family of proteins inside the cell that ultimately alter transcription of target genes in the nucleus. This general signaling cascade – from outside the cell through the cell membrane and into the nucleus – and the classes of molecules that compose it are strongly conserved throughout the animal kingdom. This simple scheme is utilized by many insect tissues to control growth and patterning during development. This chapter describes some of the mechanisms by which this control is achieved. This section of the chapter outlines some of the mechanistic details of the TGF-β signaling cascade, and introduces some of the primary signaling molecules found in the TGF-β pathways of the fruit fly *Drosophila melanogaster*.

In this chapter, most of the referenced work will apply to the *Drosophila* system because of its role as a well-studied experimental system in molecular biology. As such, this chapter is laden with whimsical names of genes and proteins that are typical of fly genetics. These gene names are often derived either from the phenotypes conferred by mutation of the gene, or from the gene's evolutionary relationships to homologs that have been characterized in other systems. As a matter of nomenclature, these names are in lower-case italicized text when referring to the gene itself or the mutation of that gene; proteins encoded by a gene bear the same name but start with a capital letter and are not in italics. As an additional reference guide,

DOI:10.1016/B978-0-12-384747-8.10013-3

Table 1 Core Components (Ligands, Receptors, R-Smads, and Co-Smad) of the TGF-β Pathway in *Drosophila*, Categorized by Function in the BMP or Activin Family*

	Name in Drosophila	*Synonym*(s)	*Molecular Function*	*Key Reference*(s)	*Human Homolog*
BMP Components	Decapentaplegic (Dpp)		Ligand (BMP)	Spencer *et al.*, 1982, among others	BMP-2/4
	Glass bottom boat (Gbb)	60A	Ligand (BMP)	Wharton *et al.*, 1991; Haerry *et al.*, 1998; Khalsa *et al.*, 1998	BMP-5/6/7/8
	Mother against Decapentaplegic (Mad)		R-Smad (BMP)	Sekelsky *et al.*, 1995; Newfeld *et al.*, 1997	SMAD-1/5/8
	Saxophone (Sax)		Type I Receptor (BMP)	Xie *et al.*, 1994	ALK-1/2
	Screw (Scw)		Ligand (BMP)	Arora *et al.*, 1994; Shimmi *et al.*, 2005a	–
	Thickveins (Tkv)		Type I Receptor (BMP)	Okano *et al.*, 1994; Terracol and Lengyel, 1994	ALK-3/6
Both BMP and Activin	Medea		Co-Smad	Raftery *et al.*, 1995; Inoue *et al.*, 1998	SMAD-4
	Punt	Atr-II	Type II Receptor	Childs *et al.*, 1993	ActR-IIA/B
	Wishful thinking (Wit)		Type II Receptor	Aberle *et al.*, 2002; Marqués *et al.*, 2002	BMPR-II
Activin Components	Activin-β (Act-β)	dActivin (dAct)	Ligand (Activin)	Kutty *et al.*, 1998	
	Baboon (Babo)	Atr-I	Type I Receptor (Activin)	Wrana *et al.*, 1994; Brummel *et al.*, 1999; Das *et al.*, 1999	ALK-4/5
	Dawdle (Daw)	Alp23B	Ligand (Activin)	Parker *et al.*, 2006; Serpe *et al.*, 2006	–
	Myoglianin (Myg)	Myo	Ligand (Activin)	Lo and Frasch, 1999	Myostatin
	Maverick (Mav)		Ligand (Activin)	Nguyen *et al.*, 2000	BMP-3/3a ?
	Smad on X (Smox)	dSmad2	R-Smad (Activin)	Brummel *et al.*, 1999; Das *et al.*, 1999	SMAD-2/3

*Components listed in the gray cells near the center of the table function in both families. Published synonyms and molecular functions of each pathway member are also listed. Key references indicate papers that report the discovery of the component and/or elucidate its molecular function in *Drosophila*. For more thorough referencing, please see the text. Because much of the mechanistic detail of the TGF-β signaling pathway has come from work in the human system, human homologs are given to aid further literature research of pathway members.

many of the molecules that are discussed in this chapter can be found in **Tables 1** and **2**, as can their key references, abbreviations, and synonyms. Further, because this section provides only a cursory overview of TGF-β signaling, it will not contain references to the processes or genes described therein; those references can be found throughout the main body of this chapter. For other general references about the TGF-β signaling pathway, see reviews by Shi and Massague (2003), Parker *et al.* (2004), or many others.

13.1.1. General Mechanism and Overview

The canonical (TGF-β) signaling cascade initiates when an extracellular ligand binds two types of transmembrane receptor kinases: one Type I and one Type II. Upon formation of this ligand–Type I– Type II complex, the Type II receptor phosphorylates and activates the Type I receptor, which then phosphorylates a cytoplasmic transcription factor called a Response (R-)Smad. Phosphorylated R-Smad proteins bind to a related protein called a Co-Smad, and this complex accumulates in the nucleus and can affect transcription of diverse target genes (**Figure 1A**).

Many animals utilize two related but separable branches of the canonical TGF-β pathway: the Bone Morphogenic Protein (BMP) and Activin/TGF-β branches (**Figure 1B**). The BMP and Activin pathways both operate using the mechanism described above, but do so using separate ligands, Type I receptors, and R-Smads. However, these two signaling cascades both utilize the same Type II receptors and Co-Smad. This section highlights the components and mechanisms of both the BMP and Activin pathways in the well characterized TGF-β family of the fruit fly *Drosophila melanogaster*.

13.1.2. Components of the BMP and Activin Pathways

13.1.2.1. Ligands The TGF-β signaling pathway in *Drosophila melanogaster* includes seven extracellular signaling ligands, each with an N-terminal prodomain and a C-terminal ligand domain. The prodomain contains signals for secretion, trafficking, and processing, and aids in the folding of the ligand domain in the secretory system of the cell. After secretion from the cell, the ligand

Table 2 Selected Regulators of the TGF-β Signaling Pathway in *Drosophila**

Name in Drosophila	Molecular Function	Key Reference(s)	Human Homolog
Brinker (Brk)	Inhibitor of Dpp target genes	Campbell and Tomlinson, 1999; Jazwinska et al., 1999a, 1999b; Minami et al., 1999	
Crossveinless (Cv)	BMP inhibitor/shuttle	Shimmi et al., 2005b; Vilmos et al., 2005	
Crossveinless-2 (Cv-2)	Binds BMP ligands extracellularly	Conley et al., 2000; Serpe et al., 2008	Cv-2
Daughters against deca-pentaplegic (Dad)	Inhibitory Smad	Tsuneizumi et al., 1997; Kamiya et al., 2008	SMAD-6/7
dSmurf	E3 Ubiquitin Ligase, targets Mad	Podos et al., 2001; Liang et al., 2003	Smurf1, 2
dSno	Medea cofactor, influences R-Smad binding	Takaesu et al., 2006; Barrio et al., 2007	Ski, Sno
Eukaryotic Initiation Factor 4A (eIF4A)	Influences Mad/Medea proteolysis	Li et al., 2005; Li and Li, 2006	eIF4A
Highwire (Hiw)	E3 Ubiquitin Ligase, targets Medea	Wan et al., 2000; McCabe et al., 2004	
MAN1	Nuclear lamin-binding protein, inhibits BMP signaling	Wagner et al., 2010	MAN1
Moleskin (Msk)	Nuclear Importin, imports Mad to nucleus	Xu et al., 2007	Imp-7,8
Nemo	Kinase, phosphorylates Mad MH1 domain	Zeng et al., 2007; Merino et al., 2009	Nemo
Otefin (Ote)	Nuclear lamin-binding protein, binds Medea	Jiang et al., 2008	(LEM family members)
Protein Phosphatase 2A (PP2A)	Phosphatase	Batut et al., 2008	PP2A
Schnurri (Shn)	Cofactor for transcription	Arora et al., 1995; Grieder et al., 1995; Staehling-Hampton et al., 1995	PDRDII
Smad Anchor for Receptor Activation (SARA)	Recruits Smad and Phosphatase to Receptor	Bennet and Alphey, 2002; Bökel et al., 2006	SARA
Short gastrulation (Sog)	BMP inhibitor/shuttle	Ferguson and Anderson, 1992a, 1992b	Chordin
Tolkin (Tok; also Tolloid-related)	Protease, cleaves Sog and Activin Prodomains	Nguyen et al., 1994; Serpe and O'Connor, 2006	BMP-1
Tolloid (Tld)	Protease, cleaves Sog	Shimell et al., 1991; Marqués et al., 1997	BMP-1
Twisted gastrulation (Tsg)	BMP inhibitor/shuttle	Mason et al., 1994; Ross et al., 2001	Twisted Gastrulation
Type 1 Protein Phospha-tase c (PP1c)	Phosphatase, dephosphorylates Tkv	Bennet and Alphey, 2002	PPM1A

*Key references indicate papers that report the discovery of the regulator and/or elucidate its molecular function in *Drosophila*. For more thorough referencing, please see the text. Because much of the mechanistic detail of the TGF-β signaling pathway has come from work in the human system, human homologs are given to aid further literature research of pathway members.

domain is cleaved from its adjoined prodomain at a site of highly charged amino acids. The removal of the prodomain is necessary for ligand maturation, and for the ligand's interaction with receptors on the surface of the cell.

Cells secrete TGF-β ligands as covalently linked dimers. These dimers are tethered together by a disulfide bond between two cysteine residues, one from each monomer. Other disulfide bonds between cysteines within each monomer give the ligands their three-dimensional structure. Ligands can either homo- or hetero-dimerize via this process, and the identities of the ligand monomers that comprise the dimer pair determine the composition of the subsequent receptor complex that is recruited at the cell surface. Because each ligand monomer physically interacts with one Type II and one Type I receptor, ligand dimer pairs recruit tetrameric complexes of two

Type II and two Type I receptors at the cell surface. The formation of this six-protein complex – two conjoined ligand monomers, two Type II receptors, and two Type I receptors – is necessary for initiation of canonical TGF-β signaling.

Ligands in the BMP and Activin families have different numbers of cysteine residues in their ligand domains. In the fruit fly *Drosophila melanogaster*, members of the BMP family – Decapentaplegic (Dpp), Screw (Scw), and Glass bottom boat (Gbb) – have seven cysteines, while members of the Activin family – Activin-β (Act-β) and Dawdle (Daw) – have nine. Interestingly, the last two TGF-β ligands that were discovered in *Drosophila*, Myoglianin (Myg) and Maverick (Mav), have nine cysteines like Activins, but their amino acid sequences somewhat resemble BMPs. For the former reason, they are often classified in the Activin/TGF-β family (**Figure 1B**).

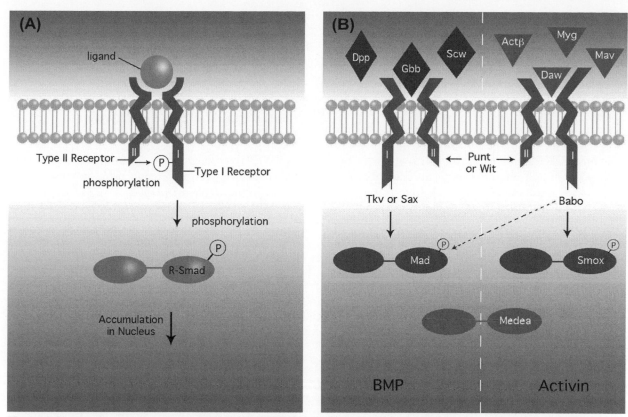

Figure 1 Mechanism and components of TGF-β signaling in *Drosophila*. (A) The canonical signaling cascade of the *Drosophila* TGF-β pathway. An extracellular ligand binds a Type II and a Type I transmembrane receptor kinase. Upon formation of this signaling complex, the Type II receptor phosphorylates the Type I receptor, which is then able to phosphorylate a cytosolic R-Smad. Phosphorylated R-Smads can accumulate in the nucleus after binding to a Co-Smad. In **Figures 1(A)** and **1(B)**, ligands are represented as single units that attract single Type II and Type I receptors. This depiction represents half of the actual signaling complex, where a ligand dimer attracts a tetrameric complex of two Type II and two Type I receptors. (B) The Bone Morphogenic Protein (BMP) and Activin pathways in *Drosophila*. In the BMP pathway, the three extracellular BMP ligands – Decapentaplegic (Dpp), Glass bottom boat (Gbb), Screw (Scw) – attract a Type II receptor and one of the two BMP-specific Type I receptors, Thickveins (Tkv) or Saxophone (Sax). Formation of this complex results in the phosphorylation of the R-Smad of the BMP pathway, Mothers against decapentaplegic (Mad). In the Activin pathway there are four extracellular ligands: Activin-β (Actβ), Dawdle (Daw), Myoglianin (Myg), and Maverick (Mav). An Activin ligand binds a Type II receptor and the Type I receptor Baboon (Babo), which phosphorylates the Activin R-Smad, Smad on X (Smox). Babo may also phosphorylate Mad. The two Type II receptors – Punt (Put) and Wishful thinking (Wit) – can function in both the BMP and Activin pathways, as can the Co-Smad, Medea (Med). Image courtesy of Leah Moak.

13.1.2.2. Receptors Each Type I and Type II TGF-β receptor is a single-pass transmembrane receptor kinase (**Figure 1A**). The N-terminus of the protein forms an extracellular ligand-binding domain whose structure is determined in part by intramolecular disulfide bonds between cysteine residues. This domain is adjacent to a single transmembrane motif composed of hydrophobic amino acids, which separates it from a C-terminal, intracellular serine/threonine kinase domain. Upon formation of the ligand–receptor complex, the Type II receptor phosphorylates serine residues in the Type I receptor. These serines reside in a glycine- and serine-rich stretch of amino acids just inside the cell membrane called the G-S box. Phosphorylation at this site activates the Type I receptor, which then phosphorylates cytosolic R-Smads (**Figure 1A**).

In the fruit fly *Drosophila melanogaster*, there are two Type II and three Type I TGF-β receptors (**Figure 1B**). Both of the Type II receptors – Punt and Wishful thinking (Wit) – function in both the BMP and Activin pathways. In contrast, the Type I receptors can be separated into distinct pathways (**Figure 1B**). The Type I receptors Thickveins (Tkv) and Saxphone (Sax) can interact with the BMP ligands (Dpp, Gbb, and Scw), while the Type I receptor Baboon (Babo) interacts with the Activin ligands (dAct, Daw, Myo, and perhaps Mav).

13.1.2.3. Smads Smads operate downstream of receptors and transduce the TGF-β signal into the nucleus of the cell. Originally, the term "Smad" resulted from a fusion of two different names for this class of proteins in two model organisms: the "Sma" proteins in the nematode *C. elegans* and the "Mad" family in *Drosophila*. There are three forms of Smads in TGF-β signaling generally: Response (R-) Smads, Co-Smads, and Inhibitory (I-) Smads. All of these proteins are structurally related but play distinct roles downstream of receptor activation inside of the cell.

R-Smads comprise three domains: an N-terminal MH1 domain, a C-terminal MH2 domain, and a "linker region" between them. The MH1 domain is highly conserved and contains a DNA-binding domain that is necessary for the R-Smad's role as a transcription factor. The linker region of R-Smads is the site of several post-translational modifications that can influence the stability, subcellular trafficking, or protein-binding properties of the Smad. Also, the linker serves as a binding site for some of the proteins that regulate the activities of R-Smads inside of the cell.

The MH2 domain of the R-Smad is also modified post-translationally and interacts with other proteins. When a Type I receptor is activated by a Type II receptor by phosphorylation, it is the MH2 domain of the R-Smad that binds the Type I receptor and becomes phosphorylated. The target sequence for this phosphorylation is an S–X–S motif at the extreme C-terminus that is conserved in R-Smads across the animal kingdom. Once phosphorylated, the MH2 domain of the R-Smad interacts with the Co-Smad that will shuttle it into the nucleus, and this complex can alter the transcriptional profile of the cell. Once inside the nucleus, the MH2 domain of an R-Smad is also often responsible for binding cofactors for transcription. Importantly, the phosphorylation states of the C-terminal serines in the MH2 domain can be monitored experimentally by utilizing antibodies that recognize only the phosphorylated form of an R-Smad. This technique is commonly used to measure if, or to what degree, the signaling pathway upstream of the R-Smad has been stimulated.

In *Drosophila* there are two R-Smads: Mothers against decapentaplegic (Mad), and Smad on X (Smox) (**Figure 1B**). These proteins are largely thought to be pathway specific: Mad is considered the BMP Smad because it is activated by the BMP Type I receptors Tkv and Sax, while Smox is considered the Smad of the Activin family because of its roles downstream of the Activin Type I receptor Baboon. However, there is evidence that Baboon can also phosphorylate Mad (dotted line in **Figure 1B**; also see section 13.4.2).

The Co-Smad in *Drosophila*, Medea, binds R-Smads after they have been phosphorylated by Type I receptors and shuttles them to the nucleus. Co-Smads have MH1, linker, and MH2 domains, the latter of which is responsible for binding R-Smads after stimulation. These Co-Smad/R-Smad complexes are trimeric, with one Co-Smad and two R-Smads. The cell can modify the Co-Smad post-translationally to regulate the subcellular trafficking of these trimeric signaling complexes.

Drosophila has one I-Smad, Daughters against decapentaplegic (Dad), that has an MH2 and a "linker" domain,

but is missing most of the MH1 domain. I-Smads interfere with BMP signaling by several mechanisms, including binding to receptors to prevent interactions with R-Smads, and binding to R-Smads to prevent their interactions with other proteins.

13.2. Dpp, the BMP Pathway, and Gradients

The TGF-β signaling pathway is involved in the development of many tissues. This section highlights the discoveries of many members of the *Drosophila* BMP pathway and their roles in development, with emphasis given to three developmental contexts that involve gradients of BMP activity: dorsoventral patterning in the embryo, the formation of the posterior crossvein, and development of the wing.

13.2.1. *dpp* and Patterning in *Drosophila*

Much of the early focus on TGF-β signaling in *Drosophila melanogaster* centered on the ligand Decapentaplegic (Dpp) and its myriad roles in spatial patterning during development. Among the first structures linked to Dpp signaling were the appendages of the adult animal, and the tissues that gave rise to them in the larva. Indeed, the name of the *decapentaplegic* gene refers to the 15 imaginal tissues that had pattern deficiencies or duplications during development in *dpp* mutants (Spencer *et al.*, 1982). Studies of some of the adult structures that arose from these tissues demonstrated that *dpp* mutants had massive cell death in the distal portions of their limbs (Bryant, 1988). These results, coupled with transplantation experiments, suggested that Dpp was a secreted signal that conveyed positional information to pattern imaginal tissues during development (Bryant, 1988). The mechanism of this patterning was uncovered in part by later studies that showed that Dpp regulated expression of tissue-specific transcription factors in the limbs. Dpp regulated the homeobox gene *aristaless* at the presumptive tips of the leg and wing (Campbell *et al.*, 1993), and also induced expression of *distal-less* in primordia of imaginal tissues of the thorax (Cohen *et al.*, 1993). Further evidence for Dpp's role as a secreted signal came from the organ that gives rise to the wing – the imaginal wing disc – where Dpp appeared to pattern the whole tissue despite its expression in only a subset of cells near the anterior–posterior boundary (Posakony *et al.*, 1990). Also, veins in the adult wings of some *dpp* mutants were disrupted (Diaz-Benjumea and Garcia-Bellido, 1990). Taken together, these experiments demonstrated that the *dpp* gene encoded a secreted signal that patterned many tissues during *Drosophila* development.

In addition to linking Dpp to the patterning of the appendages of the fly, early findings also implicated Dpp in the development of internal organs. Prominent among these is the developing gut (Masucci and Hoffman, 1993, and others). In the embryo, Dpp induces expression of the homeotic gene *labial* to ensure proper midgut development (Tremml and Bienz, 1992). *labial* is expressed in the endodermal cells of the midgut, but Dpp is expressed in the neighboring visceral mesoderm, demonstrating induction by signaling across germ layers (Immergluck *et al.*, 1990; Panganiban *et al.*, 1990). These findings exhibited yet again that Dpp is a secreted factor. Expression of *dpp* in the visceral mesoderm is itself regulated by several homeotic transcription factors, including Ultrabithorax (Ubx) and Abdominal-A (Abd-A). Ubx positively regulates *dpp* expression via a feedback loop (Hursh *et al.*, 1993; Thuringer and Bienz, 1993; Thuringer *et al.*, 1993), and Abd-A negatively regulates *dpp* expression (Capovilla *et al.*, 1994; Manak *et al.*, 1994). Together these findings began to uncover both the complex regulation of the Dpp signal in the gut and the diverse roles of Dpp during development.

In addition to the adult appendages and the midgut, Dpp signaling has been implicated in the development of structures as diverse as the salivary gland (Panzer *et al.*, 1992), the retina of the eye (Heberlein *et al.*, 1993), the foregut (Pankratz and Hoch, 1995), tracheal cells (Vincent *et al.*, 1997), and even the eggshell (Twombly *et al.*, 1996). In short, Dpp has been found to pattern many tissues – some in multiple ways – during development. However, one developmental context in particular has helped uncover the mechanism by which Dpp acts on its target tissues. The study of the patterning of the dorso–ventral (D–V) axis during early embryonic development identified and characterized the functions of downstream TGF-β pathway members in *Drosophila*.

13.2.2. The BMP Pathway and Extracellular Regulators

Dpp was implicated in dorsal–ventral patterning when *dpp* mRNA was discovered only on the dorsal side of the early embryo (St Johnston and Gelbart, 1987) and when *dpp* mutant embryos were found completely ventralized (Irish and Gelbart, 1987). That same year Dpp was identified as a putative TGF-β signal (Padgett *et al.*, 1987), and subsequent work highlighted another TGF-β signal, Screw (Scw), that was also necessary for patterning dorsal tissues during embryonic development (Arora *et al.*, 1994). Furthermore, several coincident studies reported the discovery of two putative TGF-β receptors, Thickveins (Tkv; Okano *et al.*, 1994; Terracol and Lengyel, 1994) and Saxophone (Sax; Xie *et al.*, 1994), as well as their roles in D–V patterning and their apparent functions as downstream modulators of Dpp and Scw (Nellen *et al.*, 1994; Brummel *et al.*, 1994; Ruberte *et al.*, 1995). Also, Punt was implicated as the Type II receptor in the pathway (Letsou *et al.*, 1995). Together these experiments identified ligands and receptors in what would be called

the Bone Morphogenic Protein (BMP) signaling pathway, and established the importance of the pathway in dorso–ventral patterning in *Drosophila*.

The discoveries of extracellular BMP ligands and transmembrane receptors left unanswered questions about how the signaling pathway continued into the interior of the cell. In particular, several studies concurrent with discoveries of the receptors Tkv and Sax reported the initiation of transcription of target genes downstream of Dpp activity, but the mechanism(s) of their activation in the nucleus was unclear (Staehling-Hampton and Hoffman, 1994; Staehling-Hampton *et al.*, 1994a). Clues about how Dpp, Scw, Tkv, and Sax ultimately induced transcription came from a genetic screen and subsequent studies that uncovered a gene dubbed *mothers against decapentaplegic* (*mad*). Phenotypes of *mad* mutants were similar to *dpp* mutant phenotypes, and the phenotypes of double mutants were synergistically severe (Sekelsky *et al.*, 1995). Mad was later shown to be a cytosolic protein that was a required downstream component of Dpp signaling (Newfeld *et al.*, 1996; Wiersdorff *et al.*, 1996). Finally, a report demonstrated that Mad accumulated in the nucleus after it was phosphorylated by the Type I receptor Thickveins in response to stimulation by Dpp (Newfeld *et al.*, 1997). Together, these experiments recognized Mad as a critical cytosolic component of the Dpp signaling pathway and linked Dpp signaling events at the plasma membrane to a subsequent signaling event in the nucleus.

Mad, however, did not act alone inside of the cell. Other genetic experiments uncovered a gene that also acted downstream of *dpp* and enhanced its mutant phenotypes (Raftery *et al.*, 1995). Because this genetic interaction with *dpp* was due to a maternal contribution, the gene was named *medea* after the notorious mother in Greek mythology. Subsequently, several contemporaneous studies demonstrated the molecular details of Medea's inclusion in the BMP pathway. Medea localized to the nucleus after stimulation of Mad by Tkv (Das *et al.*, 1998), and this nuclear translocation was the result of a physical interaction with phosphorylated Mad (Inoue *et al.*, 1998). Further, Medea is a homolog of the mammalian Co-Smad, Smad4, and injection of human Smad4 mRNA into *medea* mutant embryos rescued the ventralized phenotype of *medea* mutants (Hudson *et al.*, 1998). These studies established Medea as another intracellular component of the TGF-β pathway in *Drosophila*.

The ligands Dpp and Scw; the receptors Tkv, Sax, and Punt; the R-Smad Mad; and the Co-Smad Medea outlined the components of the BMP signaling cascade in dorso–ventral patterning. However, mutations in other genes also gave similar D–V phenotypes in the embryo, and seemed to interact genetically with the genes that encoded the signaling molecules themselves. While the workings of the pathway itself were elucidated, studies also determined the roles of several genes that did not encode signaling molecules *per se*, but that influenced signaling. The molecular details of these regulators of BMP signaling eventually uncovered the nature of their interactions with BMP pathway members. This section highlights the discovery and functions of some of the first-known regulators of the BMP pathway during dorsoventral patterning in *Drosophila*.

The gene *short gastrulation* (*sog*) was implicated in embryonic morphogenesis before it was linked to TGF-β signaling (Wieschaus *et al.*, 1984; Eberl and Hilliker, 1988; Zusman *et al.*, 1988), but *sog* was later associated with the TGF-β pathway because of its genetic interactions with *dpp*. Ferguson and Anderson (1992a) demonstrated that Sog repressed patterning of dorsal tissues of the embryo by Dpp, and that increased levels of *dpp* strengthened the dorsalization of *sog* mutants. Subsequent studies also demonstrated repression of Dpp by Sog, and described Sog as a diffusible extracellular protein that contained four cysteine-rich (CR) domains (Francois *et al.*, 1994; Biehs *et al.*, 1996; Yu *et al.*, 1996.) These domains enabled Sog to bind BMP ligands, and prevented their interaction with receptors at the cell surface (**Figure 2**). These studies demonstrated that Short gastrulation was important for dorsoventral patterning, and could antagonize Dpp signaling.

Like *sog*, *twisted gastrulation* (*tsg*) was first characterized independent of any interaction with TGF-β signaling; *tsg* was described as a maternal gene that was necessary for embryonic development (Wieschaus *et al.*, 1984; Zusman and Wieschaus, 1985, Perrimon *et al.*, 1989). Also like *sog*, *tsg* influenced *dpp* phenotypes in dorso–ventral patterning, and when mutated resulted in aberrant specification of dorsal tissues (Mason *et al.*, 1994, 1997; Yu *et al.*, 2000). Tsg is a secreted protein that contains several cysteine-rich domains that bind to TGF-β ligands. During embryonic development, Tsg increases Sog's affinity for Dpp, resulting in stronger repression of signaling via the formation of a Dpp–Sog–Tsg complex (**Figure 2**). This repression of Dpp signaling was first demonstrated in cell-based signaling assays, and as a result of these experiments Tsg appeared to join Sog as a negative regulator of Dpp signaling in the early embryo (Ross *et al.*, 2001).

The *tolloid* (*tld*) gene also interacted genetically with the *dpp* pathway, but, unlike Sog and Tsg, the *tolloid* gene product appeared to enhance Dpp signaling. For example, some dorsal embryonic tissues that are patterned by Dpp were missing in *tld* mutant embryos: they lacked amnioserosa at the dorsal midline, and had less tissue fated as dorsal ectoderm (Shimell *et al.*, 1991; Arora and Nüsslein-Volhard, 1992; Ferguson and Anderson, 1992a). These results suggested that Tld somehow facilitated Dpp signaling. That *tld* encoded a metalloprotease was the first clue about the mechanism of this facilitation (Shimell *et al.*, 1991). Further work demonstrated that Tolloid cleaved the Dpp inhibitor Sog, and did so in a Dpp-dependent fashion (Marqués *et al.*, 1997). That

is, Tld appeared to cut Sog only when Sog was bound to Dpp, which then allowed Dpp to interact with cell-surface receptors (**Figure 2**). This proteolytic cleavage was later demonstrated *in vivo* (Yu *et al.*, 2000). Furthermore, the cutting of Sog by Tld dissociated the entire Sog–Tsg–Dpp complex (Shimmi and O'Connor, 2003). These experiments outlined a molecular mechanism of Tolloid's enhancement of Dpp signaling, and explained many of the phenotypes seen in genetic experiments involving *dpp*, *sog*, *tsg*, and *tld*. However, not all embryonic phentoypes could be explained by this interaction.

Despite Sog and Tsg's established roles as inhibitors of Dpp signaling in some contexts, they also appeared to enhance Dpp activity *in vivo*. For example, *tsg* embryos lacked the patterning of the dorsal-most tissues of the early embryo, which is a phenotype similar to *dpp* and *scw* mutants (Mason *et al.*, 1994). This phenotype was not consistent with Tsg's role as strictly an inhibitor of Dpp signaling. Also paradoxically, though Sog was clearly established as a repressor of Dpp signaling *in vitro* (Marqués *et al.*, 1997; Ross *et al.*, 2001), it too appeared to facilitate Dpp signaling *in vivo* (Ashe and Levine, 1999). Namely, though *sog* mutant embryos did have an expanded domain of dorsal ectoderm, which was consistent with Sog's role as a repressor of Dpp, they also lacked patterning of the amnioserosa at the dorsal midline, just like Dpp mutants (Francois *et al.*, 1994). These unintuitive findings – collectively termed the "Sog paradox" – suggested that Sog and Tsg could also facilitate the very signaling process that they inhibited. This apparent inconsistency can be explained by the spatial expression patterns of *dpp*, *sog*, *tsg*, *tld*, and the *screw* gene, which encodes another BMP ligand, during early embryonic development. These five species of secreted molecules diffuse and interact within the perivitelline (PV) space around the embryo to establish a gradient of BMP signaling that is necessary for proper dorso–ventral patterning of the embryo.

13.2.3. Dorso–Ventral Patterning in the *Drosophila* Embryo

In dorso–ventral patterning Dpp acts as a morphogen, or a molecule that confers spatial information during development. Morphogens are often molecules that diffuse from a source throughout a developing tissue to form a gradient, and they often affect cells in a concentration-dependent manner. As a result, the distance between a cell and the source of the morphogen determines the position of the cell in the concentration gradient, which often greatly influences the cell's fate. BMP molecules can pattern several tissues by forming concentration gradients during development. That is, though BMP molecules will stimulate identical downstream signaling pathways in each cell within a developing tissue, the cells will often respond differently depending on the level of the signal.

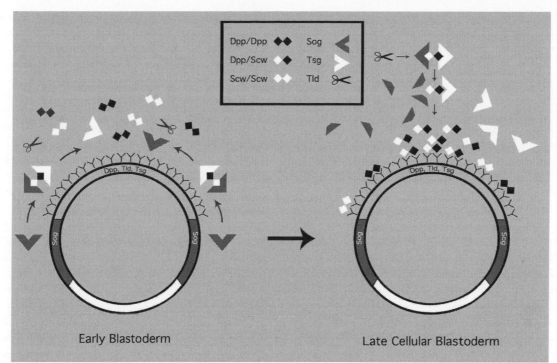

Figure 2 Spatial organization of BMP signaling components and regulators during dorso–ventral patterning of the *Drosophila* embryo. Each circle represents a transverse section of the embryo, with dorsal at the top and ventral at the bottom. A gradient of BMP activity patterns the dorsal tissues of the embryo, with the highest levels of signaling patterning the amnioserosa at the dorsal midline. The gradient of low-level signaling in the early blastoderm embryo (left) is broad, but refines to a stripe of high-level signaling at the dorsal midline in the late cellular blastoderm embryo. The distinct signaling and binding properties of the Dpp/Scw heterodimers (conjoined white and black diamonds) enable this dynamic gradient. Early in development, the inhibitor Sog is expressed ventrolaterally and diffuses towards the dorsal midline. Sog (gray "V") and Tsg (white "V") preferentially bind the Dpp/Scw heterodimers, shuttling them – and relatively few homodimers – towards the dorsal midline. In the dorsal region, Tld (represented by scissors) cuts Sog, irreversibly dissociating the Dpp/Scw–Sog–Tsg complex. Repeated binding by Sog–Tsg and proteolysis by Tld results in concentration of Dpp/Scw heterodimers, which signal more strongly than homodimers, at the dorsal midline in the late cellular blastoderm embryo (right). Image courtesy of Leah Moak.

That Dpp was a morphogen in dorso–ventral patterning was apparent early during its study (Arora and Nusslein-Volhard, 1992; Ferguson and Anderson, 1992b; Wharton *et al.*, 1993). A gradient of Dpp forms around the early embryo, and depending on where a cell resides along this gradient, the cell will adopt a particular cell fate. In general, exposure to higher concentrations of Dpp results in the adoption of more dorsal cell fates. This concept is clearly represented by the lack of all dorsal tissues in *dpp* mutants.

Dpp is secreted with its activating protease, Tld, by cells on the dorsal side of the embryo (**Figure 2**; St Johnston and Gelbart, 1987; Shimmel *et al.*, 1991). This expression pattern is intuitively consistent with diffusion of Dpp from the dorsal-most regions and the subsequent formation of an activity gradient with highest concentration at the dorsal midline. That *sog* is expressed in ventro–lateral regions of the embryo (**Figure 2**) also fits this model: diffusion of Sog might lead to inhibition of Dpp laterally and to no inhibition at the dorsal midline, where signaling is greatest. However, this simple model does not account for several experimental observations; in particular, it does not explain the mutant phenotypes that present the "Sog paradox" described above, and it does not account for the involvement of Scw, which is also necessary for patterning dorsal tissues. Perhaps most intriguingly, in the embryo the gradient of Dpp activity changes its intensity, domain, and shape throughout early development. This phenomenon relies on a series of molecular interactions to pattern the tissue that forms at the dorsal midline, the amnioserosa (**Figure 2**).

Near the end of early embryonic development, a high level of phosphorylated Mad (P-Mad) is found in a stripe approximately 10 cells wide at the dorsal midline of the embryo. However, this stripe of P-Mad – indicative of high levels of BMP signaling – is not found at earlier stages of development, but is the result of the narrowing and strengthening of a broad gradient over the dorsal side of the embryo. That is, during earlier stages of development a more diffuse gradient of weak signaling covers much of the dorsal half of the embryo, but this pattern gradually sharpens and results in a thin stripe of strong signal at the dorsal midline. This dynamic gradient can be explained in part by the expression patterns and diffusion of Sog and Scw.

As mentioned above, the regulator Sog is secreted by cells in the ventro–lateral region of the embryo, resulting in a gradient of high Sog concentration in the ventro–lateral regions with low levels at the dorsal midline. That is, the expression pattern of Sog results in a net diffusion towards the dorsal midline of the embryo (**Figure 2**). The sharpening of the gradient of BMP activity relies both on this flux of Sog towards the dorsal midline and on Sog's strong interaction with one specific type of BMP signal, the Dpp/Screw heterodimer. Compared with Dpp/Dpp and Scw/Scw homodimers, Dpp/Scw heterodimers more readily form complexes with Sog and Tsg, and they also signal more strongly via cell surface receptors (Shimmi *et al.*, 2005a). As a result, these potent Dpp/Scw dimers that are secreted from the dorsal portion of the embryo are more likely to bind Sog and to diffuse as an inert complex towards the dorsal midline, where Tld will cleave Sog and liberate the ligand to transduce a strong signal. Dpp/Dpp or Scw/Scw homodimers that are secreted from the same cells, on the other hand, are less likely to be shepherded to the midline, and as a result will more often transduce their weaker signals in more lateral regions. Iterative rounds of this selective transport of the strongest signaling pairs towards the midline – via repeated selective binding of Dpp/Scw by Sog-Tsg and subsequent cleavage by Tld – are critical to forming the dynamic gradient of BMP signaling seen in the *Drosophila* embryo. This system demonstrates a solution to the aforementioned "Sog paradox;" Sog and Tsg can inhibit ligand–receptor interactions locally, but can also sequester and concentrate the most potent ligand pairs to enable high-level signaling at the dorsal midline (**Figure 2**). This dual inhibitor/enhancer function is also consistent with findings from the cell-based signaling assays that labeled Sog and Tsg as strictly inhibitors of Dpp signaling: cell-based assays would demonstrate the inhibition of Dpp by Sog and Tsg, but their enhancement of Dpp signaling would only become apparent in the unique spatial arrangements of expression patterns in the early *Drosophila* embryo.

13.2.4. The Posterior Crossvein

Other molecules can also both inhibit and enhance BMP signaling. One such molecule in *Drosophila*, Crossveinless-2 (Cv-2), acts in the dorso–ventral patterning system highlighted above and in the patterning of veins in the *Drosophila* wing. Interestingly, both of these systems involve the sharpening of a BMP signaling gradient. As its name suggests, Cv-2 helps form the posterior crossvein (PCV) of the wing, a process dependent on high levels of BMP signaling (Conley *et al.*, 2000; Ralston and Blair, 2005; Serpe *et al.*, 2008). *cv-2* mutants lack this strong signal during development, suggesting that Cv-2 helps facilitate BMPs. However, *in vitro* signaling assays show that at high concentrations Cv-2 can also inhibit

signaling. While the debate concerning the mechanism for this discrepancy is not settled, it aids our understanding of gradients and how multiple molecules interact to maintain them (reviewed in O'Connor *et al.*, 2006; Umulis *et al.*, 2009).

Cv-2 has two general domains: an N-terminal, cysteine-rich domain responsible for binding BMPs, and a C-terminal domain that is thought to aid interactions with the cell surface (Conley *et al.*, 2000; Serpe *et al.*, 2008). One proposed mechanism for enhancement of BMP signaling suggests that Cv-2 acts as a kind of co-receptor by facilitating an interaction between ligand and receptor, while another model involves the binding of Cv-2 to extracellular inhibitors, such as Sog (reviewed in Umulis *et al.*, 2009). In the first case, Cv-2 is thought to bind both a BMP ligand using its N-terminal, cysteine-rich domain, and the cell surface with its C-terminal domain. Such a configuration would bring the ligand into close proximity with the receptor, especially since Cv-2 can also recruit BMP receptors themselves. In the second case, because Cv-2 can also bind Sog using its N-terminal domain, Cv-2 is thought to bind Sog as well as the cell surface and/or its receptors. Such a model can also be consistent with the concept of a co-receptor, as Cv-2's interaction with Sog might free the ligand in the presence of a cell-surface receptor.

In either case, *Drosophila* Cv-2 can enhance BMP signaling at low concentrations, but it inhibits signaling at high concentrations, perhaps because it simply sequesters BMP ligands without interacting with receptors or the cell surface in a way that promotes signaling. Interestingly, Cv-2's role as an inhibitor is not solely dependent on its concentration, but also depends on the identity of the BMP ligand (Serpe *et al.*, 2008). While the agonist/antagonist scenario above aptly describes the interplay between Cv-2 and the BMP ligand Glass bottom boat (Gbb), the relationship between Cv-2 and Dpp could be described as simply antagonistic. That is, while Cv-2 facilitates Gbb signaling at low concentrations but inhibits it at high concentrations, Cv-2 inhibits Dpp signaling at all concentrations. Such selectivity suggests that Cv-2 can bias a cell's reception of various BMP ligands; such a feature might allow a cell to fine-tune its level of signaling when confronted with several signals that otherwise would use the same cell-surface receptors. In a developing organism where multiple ligands are used simultaneously to pattern a field of cells of different fates, such a trait might provide important specificity of signaling.

The posterior crossvein is not the only developmental context in which Gbb plays a role that is distinct from that of Dpp. The gene *glass bottom boat* was named in reference to two of its other functions during development. First, unlike wild type larvae, which appear white due to the presence of large fat bodies, the aptly named *glass bottom boat* mutant larvae are translucent. Indeed, a recent report has implicated Gbb signaling in metabolism and energy homeostasis during larval growth (Ballard *et al.*, 2010). Second, Gbb has a complex genetic interaction with Dpp, especially in the *Drosophila* wing. Though Gbb is a BMP signal that is related to Dpp (Wharton *et al.*, 1991), the two do not function interchangeably in the embryo (Staehling-Hampton *et al.*, 1994b) or in the wing (Haerry *et al.*, 1998; Khalsa *et al.*, 1998; Ray and Wharton, 2001; Bangi and Wharton, 2006a, 2006b). Gbb and Dpp diffuse differently in the developing wing imaginal disc, in part because of their different affinities for the two BMP Type I receptors, Thickveins (Tkv) and Saxophone (Sax). In general, Gbb and Dpp have higher affinities for Sax and Tkv, respectively, and as a result serve almost complementary roles during development of the organ. In reference to this relationship, the name *glass bottom boat* was chosen because the letters in its abbreviation – *gbb* – are nearly a vertical mirror image of the letters in *dpp*. Specifically, Gbb and Dpp activate different receptors with different transcriptional consequences during development of the wing (Haerry *et al.*, 1998; Khalsa *et al.*, 1998). Indeed, a recent report suggests that this may be due to the ligands' affinities for different receptors: Tkv and Sax differ slightly in their signaling activities inside the cell, especially with regard to the transcriptional activities of the Smads they phosphorylate (Haerry, 2010). In short, Dpp and Gbb can collaborate during gradient formation in both the larval wing imaginal disc and the pupal wing at the posterior crossvein.

The gradient of BMP activity at the eventual posterior crossvein – like the gradient in the early embryo – sharpens and intensifies over time, and Cv-2 might provide interesting insight into the molecular mechanisms of that phenomenon in both contexts (reviewed in O'Connor *et al.*, 2008). BMP signaling initiates transcription of *cv-2*, highlighting a potential positive feedback loop where cells that receive more signal could sequester even more ligands using newly produced Cv-2 at the cell surface, leaving fewer for neighboring cells that initially received less signal, and subsequently express less Cv-2. This "rich get richer" scenario fits nicely with observed sharpening of gradients in the embryo and the wing, and may highlight a general mechanism used in developmental patterning.

However, while the BMP signaling systems in the embryo and at the posterior crossvein both utilize the sharpening of gradients to determine their patterns, they occur on remarkably different timescales: the gradient in the embryo sharpens in minutes, but the gradient in the crossvein sharpens over days. The systems are similar in that they involve shuttling of BMPs using extracellular regulators (Sog and Tsg in the embryo, Sog and the Tsg-like molecule Crossveinless (Cv) in the wing (Bridges, 1920; Shimmi *et al.*, 2005b; Vilmos *et al.*, 2005), but a significant difference involves the rates at which ligands are released from their inhibitory complexes. This discrepancy is due in part to the proteases that destroy the inhibitory complexes in each case. In the embryo, Tolloid (Tld) cleaves Sog, but a similar enzyme called Tolkin (Tok, also Tolloid-related or Tlr; Nguyen *et al.*, 1994; Finelli *et al.*, 1995) cuts Sog at the crossvein, only much more slowly (Serpe and O'Connor, 2006). This matching of catalytic activity to time-sensitive developmental processes demonstrates one way that distinct members of the BMP signaling pathway can be utilized to achieve similar results in different tissues during development.

13.2.5. The *Drosophila* Wing

Genetic studies uncovered another important component of the BMP pathway in *Drosophila*. The *schnurri* (*shn*) gene was necessary to activate transcription of many Dpp target genes in many tissues; cells with no *shn* activity could not respond to even hyperactivated Type I receptors (Arora *et al.*, 1995; Grieder *et al.*, 1995; Staehling-Hampton *et al.*, 1995; Marty *et al.*, 2000; Muller *et al.*, 2003). These results, in conjunction with reports showing that Shn is a zinc-finger transcription factor, suggested that the *schnurri* gene product was a necessary downstream component of the BMP pathway, much like Mad and Medea. However, unlike *mad* and *medea*, *shn* mutants did not display the embryonic dorso–ventral patterning phenotype that is typical of BMP pathway members. Instead, like some *punt* and *tkv* mutants, *shn* mutants displayed incomplete dorsal closure late in embryogenesis (Affolter *et al.*, 1994; Brummel *et al.*, 1994; Nellen *et al.*, 1994; Penton *et al.*, 1994; Arora *et al.*, 1995; Grieder *et al.*, 1995). Together these results suggested that Schnurri played a role in transcriptional activation downstream of the BMP ligand–receptor complex, but was not a component of the pathway that was necessary for BMP signaling in all contexts.

Schnurri's connection to the pathway was uncovered with the discovery of another gene, *brinker* (*brk*), that also affected cells' responses to BMPs. *brk* encodes a DNA-binding protein that represses many gene targets of BMP signaling (Campbell and Tomlinson, 1999; Jazwinska *et al.*, 1999a, 1999b; Minami *et al.*, 1999). As a result, initiation of transcription of these BMP target genes requires not only nuclear accumulation of activated complexes of Co- and R-Smads, but also downregulation of Brk activity. Interestingly, this downregulation is achieved at the transcriptional level by BMP signaling and requires the Schnurri protein, which acts as a co-repressor by forming a complex with Mad and Medea (Torres-Vaquez *et al.*, 2000, 2001; Muller *et al.*, 2003; Pyrowolakis *et al.*, 2004). These results explained the observation that cells lacking *shn* were unresponsive to activated Type I receptors: in those cells, *brk* was expressed at high levels and repressed expression of target genes (Muller *et al.*, 2003). Together, Schnurri and Brinker influence the transcriptional consequences of BMP signaling in many, but not all, contexts during *Drosophila* development. Several of these contexts involve the interpretation of a gradient of BMP signaling in a tissue.

Gradients of BMP signaling activity, the transcriptional repressor Schnurri, and the transcriptional repressor Brinker greatly influence the development of the wing primordium in *Drosophila* (**Figure 3**). Expression of *dpp* is initiated by Hedgehog signaling at the anterior–posterior (A/P) compartment boundary of the imaginal wing disc (Basler and Struhl, 1994). This pattern of expression and subsequent diffusion results in a gradient of Dpp across the wing disc that is most concentrated at the A/P border. As a result of Dpp's signaling via cell-surface receptors, a matching, shallow gradient of nuclearly accumulated phosphorylated Mad (P-Mad) is established throughout the tissue, with high nuclear P-Mad at the A/P border. This activated Mad forms complexes with Medea and Schnurri to downregulate *brk* expression in the nucleus, which results in an inverse gradient of Brinker activity in the wing disc, with high expression at the lateral edges and little expression at the A/P boundary (Muller *et al.*, 2003; Pyrowolakis *et al.*, 2004). The combination of these inverted P-Mad and Brinker gradients affects the spatial expression of several genes in the *Drosophila* wing.

The transcriptional regulation of two genes – *optomoter blind* (*omb*) and *spalt* (*sal*) – by BMP signaling and Brinker are well documented (reviewed in Raftery *et al.*, 2006; Pyrowolakis *et al.*, 2008). These two genes require different mechanisms of transcriptional activation, and their regulation demonstrates the complexity that can be achieved by the relatively simple system of a signaling gradient and differential transcription factor-binding. *omb* is expressed in a broad stripe that is centered around the A/P boundary, while *sal* is expressed only in a narrow stripe of cells near the A/P boundary. The borders of both of these expression domains are defined by the Brinker concentration gradient, demonstrating that Brinker represses expression of both genes (Muller *et al.*, 2003). However, these genes do not respond to Brinker levels identically. Expression of the *sal* gene is sensitive to Brk: even the low levels of Brk near the A/P boundary repress transcriptional activity. In contrast, *omb* is relatively insensitive to binding by Brinker, responding only to high levels of the repressor in the lateral portions of the wing disc. These differential binding affinities by Brinker to the *sal* and *omb* genes account for the breadths of the genes' patterns of expression.

However, Brk is not the only input for expression. First, *omb* is only an indirect target of Dpp signaling because *omb* expression requires repression of the Brinker repressor itself; in wild type cells, P-Mad, Medea, and Schnurri form a complex that represses expression of Brinker, and the lack of Brinker results in *omb* expression. In this relatively simple case, BMP signaling only de-represses *omb* expression, which is activated by another source. This

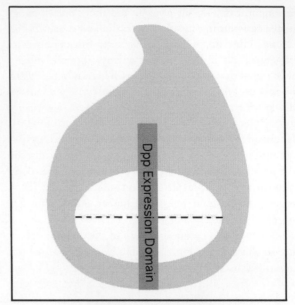

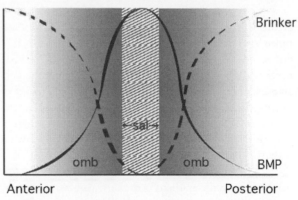

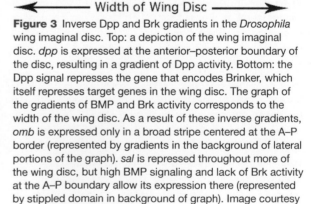

Figure 3 Inverse Dpp and Brk gradients in the *Drosophila* wing imaginal disc. Top: a depiction of the wing imaginal disc. *dpp* is expressed at the anterior–posterior boundary of the disc, resulting in a gradient of Dpp activity. Bottom: the Dpp signal represses the gene that encodes Brinker, which itself represses target genes in the wing disc. The graph of the gradients of BMP and Brk activity corresponds to the width of the wing disc. As a result of these inverse gradients, *omb* is expressed only in a broad stripe centered at the A–P border (represented by gradients in the background of lateral portions of the graph). *sal* is repressed throughout more of the wing disc, but high BMP signaling and lack of Brk activity at the A–P boundary allow its expression there (represented by stippled domain in background of graph). Image courtesy of Leah Moak.

relationship is demonstrated by experiments that show a lack of *omb* expression in cells that are mutant for BMP receptors, but that show restoration of that expression in cells mutant both for BMP receptors and for *brk*. In contrast to *omb*, *sal* expression is both de-repressed and actively induced by BMP signaling. That is, P-Mad, Medea, and Schnurri form a complex that represses *brinker* and relieves repression, but P-Mad, Medea, and other cofactors also form a complex to activate transcription themselves.

This joint de-repression and activation are highlighted by genetic experiments similar to those done with *omb*. As is the case with *omb*, cells mutant for BMP receptors do not express *sal* due to high levels of Brinker. However, unlike with *omb*, cells mutant for both BMP receptors and Brinker also lack *sal* transcription, demonstrating a requirement for BMP signaling for *sal* expression beyond repression of Brinker. These interactions between gradients, transcriptional targets, and transcriptional cofactors result in the nested expression of *sal* at the A/P boundary of the wing disc, and of *omb* in an overlapping but broader domain of the same tissue.

13.3. Other Developmental Contexts and Regulation of BMPs

13.3.1. The Neuromuscular Junction

Neurons communicate with other cells via structures called synapses, which form at the termini of neuronal axons. Synapses allow neurons to communicate with other neurons or with other cells of different types, such as muscle cells. In embryos of *Drosophila melanogaster*, synapses form between motoneurons and muscle fibers to form neuromuscular junctions (NMJ). During larval development the surface area of *Drosophila* muscles grows over 100-fold, and as a result the synapses at the NMJ must be remodeled to ensure proper communication between motoneurons and muscles. This maintenance of communication at the synapse – termed *synaptic homeostasis* – requires molecular signals both from the motoneuron to the muscle (or anterograde) and from the muscle to the motoneuron (or retrograde). In recent years some of these signals have been identified as classical morphogens, and among these the signals of the TGF-β pathway have been identified as key regulators of synaptic homeostasis. Because synapse formation and remodeling is thought to be the basis for learning and memory, and because the *Drosophila* system is amenable to powerful genetic manipulation, the *Drosophila* NMJ and the TGF-β signals within it have attracted much attention as a model system of developmental biology and neuroscience (as reviewed by Marqués (2005), among others).

To accommodate the increase in muscle size during development, a *Drosophila* neuromuscular junction grows by increasing its number of boutons, which are presynaptic structures that, in this context, house the neurotransmitter glutamate before its release. Improperly low numbers of boutons during development are a hallmark phenotype of several BMP pathway mutants, including the ligand Glass bottom boat (Gbb; McCabe *et al.*, 2003), the Type I receptors Thickveins (Tkv; McCabe *et al.*, 2004) and Saxophone (Sax; Rawson *et al.*, 2003),

the Type II receptor Wishful Thinking (Wit; Aberle *et al.*, 2002; Marqués *et al.*, 2002), the BMP R-Smad Mothers Against Decapentaplegic (Mad; Rawson *et al.*, 2003), and the Co-Smad Medea (McCabe *et al.*, 2004). These studies implicate every necessary member of a working BMP pathway in the phenomenon of synaptic homeostasis. However, BMP mutants display other neuronal phenotypes in addition to decreased size of the NMJ caused by insufficient bouton number. BMP signaling is necessary not only for growth of the synapse, but also for its stability and its regulation of both the actin and microtubule cytoskeletons (Eaton and Davis, 2005; Wang *et al.*, 2007). Also, BMP mutants have disrupted synapse function, as evidenced by their decreased synaptic transmission. This compromised function may be due in part to the altered ultrastructure of the synapses, particularly the detachment of both pre- and post-synaptic membranes (Aberle *et al.*, 2002; Marqués *et al.*, 2002). Perhaps more predictably, BMP mutants also show lack of phosphorylated Mad (P-Mad), which displays a lack of BMP activity.

Experiments that measured levels of P-Mad at the larval neuromuscular junction in BMP mutants have demonstrated that synaptic homeostasis largely requires only retrograde BMP signaling. For example, the Type II receptor Wishful thinking (Wit) is expressed only at the presynapse, and its loss-of-function results in decreased P-Mad in the motoneuron (Marqués *et al.*, 2002). Furthermore, the decreased size of the NMJ in *wit* mutants can be rescued by presynaptic overexpression of *wit* cDNA (McCabe *et al.*, 2004). Similar functional complementation can be achieved by neuronal expression of *tkv*, *sax*, *mad*, or *medea* in the relevant mutant background. Furthermore, blockage of Gbb secretion from the post-synaptic side of the NMJ mimicked the small synapses, compromised synaptic transmission, and decreased pre-synaptic P-Mad levels of *wit* mutants (McCabe *et al.*, 2003). These results suggest that the BMP signal delivered by the ligand Gbb is retrograde, and is necessary presynaptically at the NMJ.

This retrograde, pre-synaptic BMP signaling is important for proper synapse development and function, but it does not explain all BMP mutant phenotypes at the NMJ. While post-synaptic expression of Gbb largely rescues *gbb* mutant phenotypes, the rescue is more complete when Gbb is expressed in both muscles and motoneurons (McCabe *et al.*, 2003). Further, Tkv and Mad localize to the post-synapse, and nuclear shuttling of Mad in muscle cells increases upon muscle stimulation (Dudu *et al.*, 2006). These results suggest that BMP signaling is required in both the pre-synapse and post-synapse for proper development and function of the NMJ. Such post-synaptic signaling could be stimulated by either an anterograde or an autocrine ligand.

The neuromuscular junction is not the only context in the *Drosophila* nervous system in which Gbb acts as a retrograde signal. The neurosecretory Tv neurons in the ventral nerve cord produce the small, physiologically important neuropeptide FMRFamide. Tv neurons release these peptides into the hemolymph of the animal via the dorsal neurohemal organ (NHO), which they innervate. Expression of FMRFamide by Tv neurons requires BMP signaling, which involves Wit and ligand Gbb (Allan *et al.*, 2003; Marqués *et al.*, 2003), and this signaling appears to continue to regulate FMRFamide expression in mature animals (Eade and Allan, 2009). As in the context of the NMJ, Wit is only necessary pre-synaptically, or in the Tv neurons, and Gbb is delivered postsynaptically, or from the NHO. Further, BMP-dependent expression of FMRFamide in Tv neurons apparently relies on an interaction between activated BMP Smads and the transcription cofactors Eyes Absent (Eya) and Dachsund (Dac) in those cells (Miguel-Aliaga *et al.*, 2004). In the context of FMRFamide expression, Gbb again regulates development via retrograde signaling and via the Type II receptor Wishful Thinking.

The levels of BMP signaling in pre-synaptic motoneuron are regulated by two independent mechanisms. The first involves the E3 ubiquitin ligase Highwire (Hiw; Wan *et al.*, 2000), which is capable of attenuating BMP signaling in the presynaptic neuron by targeting Medea, the Co-Smad, for proteosomal degradation (McCabe *et al.*, 2004). The second mechanism involves Spinster, which is a transmembrane protein that localizes to the late endosome (Nakano *et al.*, 2001). Spinster is responsible for downregulating BMP signaling via lysosomal degradation (Sweeney and Davis, 2002). Removal of either of these mechanisms of downregulation via mutation of either of the *hiw* or *spin* genes results in the predictable phenotype of synaptic overgrowth at the NMJ. Both Hiw and Spin degrade components of the BMP pathway to maintain synaptic homeostasis.

Another regulator of BMP signaling is critical for proper development of the NMJ. Animals mutant for the *spichthyin* (*spict*) gene displayed significant overgrowth at the larval NMJ, having nearly double the wild type number of synaptic boutons (Wang *et al.*, 2007). This phenotype can be fully suppressed by loss-of-function of other members of the BMP pathway, suggesting a role downstream of BMP signaling. Spichthyin is an early endosome protein that binds the Type II receptor Wit and that co-localizes with the Type I receptor Thickveins. These subcellular localization and binding data suggest that Spict regulates signaling by affecting trafficking of TGF-β receptors (Wang *et al.*, 2007).

13.3.2. Germ-line Stem Cells

The germ-line stem cells (GSCs) in the *Drosophila* testis and ovary provide a powerful *in vivo* system for studying stem cells and the signals that regulate their behaviors. These two niches have been useful in elucidating several

of the mechanisms that determine stem cell fate. Studies of GSCs in *Drosophila* have demonstrated that the BMP pathway is a central regulator of their maintenance, division, and differentiation. This section will highlight the role of the BMP pathway and regulators in the biology of germ-line stem cells (reviewed in Gilboa and Lehmann, 2004; Xie *et al.*, 2005; Yamashita *et al.*, 2005).

At the anterior tip of the *Drosophila* testis lies a structure called the hub, which is composed of somatic cells (for a more detailed description of germline structures, see Gilboa and Lehmann, 2004; Yamashita *et al.*, 2005). Approximately nine germ-line stem cells surround the hub, and the hub cells can send signals to the GSCs that provide cues for maintenance and division. GSCs are adjacent to somatic stem cells (SSCs) that also can influence GSCs by supplying signals for differentiation and division. After a germ-line stem cell divides in the male, one daughter cell remains in contact with the hub and continues to receive signals that maintain its undifferentiated state, while the other moves away from the hub and begins to differentiate. This gradual posterior progression results in gradual differentiation into a gonialblast, which is still surrounded by its partner SSCs. During differentiation, a gonialblast will undergo 4 consecutive mitotic divisions without cytokinesis to form a cyst; all 16 cells in this cyst will undergo meiosis and eventually become 64 mature sperm cells.

The ovary in the *Drosophila* female is composed of up to 20 units called ovarioles. At the anterior of an ovariole is a region called the germarium, which houses the stem cells that are responsible for generating oocytes. At the anterior tip of the germarium are the somatic cap cells, which are connected at their posterior to the germ-line stem cells. Cap cells signal to the GSCs, and can alter their behavior. Each GSC in the germarium is ensheathed by other somatic cells called inner-sheath cells, which can also influence GSC division and differentiation. In females, the division of a GSC results in the formation of two daughter cells. One cell remains in contact with the cap cell at the anterior tip of the germarium, and the other moves posteriorly, where it starts to differentiate into a cystoblast. As in males, this cystoblast undergoes 4 rounds of mitotic divisions to give rise to a 16-cell cyst. However, unlike in males, only 1 of these cyst cells undergoes meiosis to generate an oocyte – the other 15 cells become nurse cells that nourish the oocyte during oogenesis.

BMP signaling is critical for proper germ-line stem cell maintenance and division in the *Drosophila* ovary (**Figure 4**). The ligand Dpp is expressed in the cap and inner-sheath cells of the germarium, two tissues that regulate behavior of GSCs (Xie and Spradling, 1998). Gbb is also expressed in somatic tissues in the anterior of the ovariole (Song *et al.*, 2004). *dpp* and *gbb* mutant ovaries display loss of cells in the germ-line, and mutants for intracellular BMP pathway members *tkv* and *medea* have GSCs with short half-lives and reduced rates of division

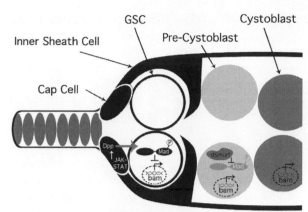

Figure 4 BMP signaling in maintenance of germ-line stem cells in the *Drosophila* ovary. BMP signals are expressed in somatic cells near the anterior of the ovariole (left). In cap cells, Dpp expression is regulated by the JAK-STAT pathway. Dpp signals to germ-line stem cells (GSCs), inducing phosphorylation of P-Mad, which negatively regulates transcription of the *bag of marbles* (*bam*) gene. Repression of *bam* discourages differentiation. Upon cell division of GSCs, one daughter cell moves posteriorly in the ovariole, increasing its distance from the Dpp-secreting cap cells. As a result, in these daughter cells levels of P-Mad decrease, *bam* transcription initiates, cells begin differentiating. dSmurf also actively degrades P-Mad, facilitating this differentiation process. Increased states of differentiation are denoted by increasingly dark shades of gray. Image courtesy of Leah Moak.

(Xie and Spradling, 1998, 2000). Moreover, overexpression of Dpp but not Gbb results in an increase in the number of GSCs (Xie and Spradling, 1998; Song, *et al.*, 2004). These results demonstrate that BMP signals from somatic tissues directly regulate division and maintenance of germ-line stem cells in the ovary.

BMP signaling is also necessary for the maintenance of GSCs in the *Drosophila* testis. As in the ovary, *gbb* and *dpp* are expressed by somatic tissues in the male niche, although in the male Gbb appears to play a more significant role than Dpp in regulating stem cell renewal (Shivdasani and Ingham, 2003; Kawase *et al.*, 2004). Also as in females, loss of downstream BMP signaling components (i.e., *put*, *tkv*, *mad*, *med*, and *shn*) results in decreased GSC number (Matunis *et al.*, 1997; Kawase *et al.*, 2004). In both sexes, BMP signals expressed by somatic tissues regulate GSC maintenance and division.

The niche controls maintenance and differentiation of germ-line stem cells in males and females in large part by regulating expression of the gene *bag of marbles* (*bam*). Bam expression in differentiating cystoblasts is a critical determinant of differentiation (McKearin and Spradling, 1990; McKearin and Ohlstein, 1995). Expression of *bam* is repressed by BMP signaling: mutants for *punt* and *medea* have high levels of *bam* mRNA, and overexpression of Dpp blocks *bam* expression in GSCs (Xie and Spradling, 1998; Chen and McKearin, 2003a; Schulz *et al.*, 2004; Song *et al.*, 2004). Further, Mad and Medea bind

a repressor element in the *bam* locus that is necessary for downregulation of *bam* expression (Chen and McKearin, 2003a, 2003b). Consistent with the role of BMP signaling as a repressor of *bam* expression and subsequent differentiation, cystoblast cells in the *Drosophila* germarium actively repress BMP signaling by several mechanisms inside the cells (Casanueva and Ferguson, 2004). P-Mad in these cells decreases as cells move posteriorly and differentiate, which is coincident with an increase in *bam* expression. One mechanism of this downregulation involves the E3 ubiquitin ligase dSmurf, which can target Mad for proteasomal degradation in a signal-dependent manner (see section 13.3.3). In addition to the regulation of BMP signaling inside GSCs, expression of the BMP ligand Dpp is positively regulated in somatic tissues of the ovarian niche by the Janus kinase/signal transducer and activator of transcription (JAK/STAT) pathway, which is also critical for proper maintenance of GSCs in the male (López-Onieva *et al.*, 2008; Wang *et al.*, 2008). The requirement for JAK/STAT signaling in both sexes demonstrates a functionally conserved mechanism for regulating BMP signaling in the male and female germ-line stem cell niches (Singh *et al.*, 2005; see also Fuller and Spradling, 2007). By regulating the level of BMP signaling in the niche, the JAK/STAT pathway can control the size of the ovarian niche and the fates of the GSCs within it.

13.3.3. Intracellular Regulation of the BMP Pathway

TGF-β signaling can be regulated by several mechanisms downstream of ligand–receptor binding. In this section, we will highlight some additional regulators of TGF-β signaling inside the cell. Interactions highlighted in this section are depicted in **Figure 5**, and molecules discussed in this section are summarized in **Table 2**.

13.3.3.1. Dad   The gene *daughters against decapentaplegic* (*dad*) was discovered as a downstream antagonist of Dpp signaling: expression of *dad* was induced by Dpp and subsequently downregulated Dpp activity. Further, ectopic overexpression of *dad* could suppress phenotypes caused by BMP gain-of-function (Tsuneizumi *et al.*, 1997). Dpp and Dad form an inducible negative feedback loop of BMP signaling in many tissues during development, and Dad is often used to suppress BMP signaling in genetic assays. *dad* encodes a protein that is weakly structurally similar to the BMP R-Smad Mad, but, unlike Mad, it does not have a DNA-binding MH1 domain and does not transduce a signal downstream of TGF-β receptors. Instead, Dad is an inhibitory Smad (I-Smad), and interferes with signaling by physically interacting with either receptors or other Smads. For instance, Dad inhibits BMP signaling by binding the Type I receptors Thickveins and Saxophone and preventing binding of Mad, but

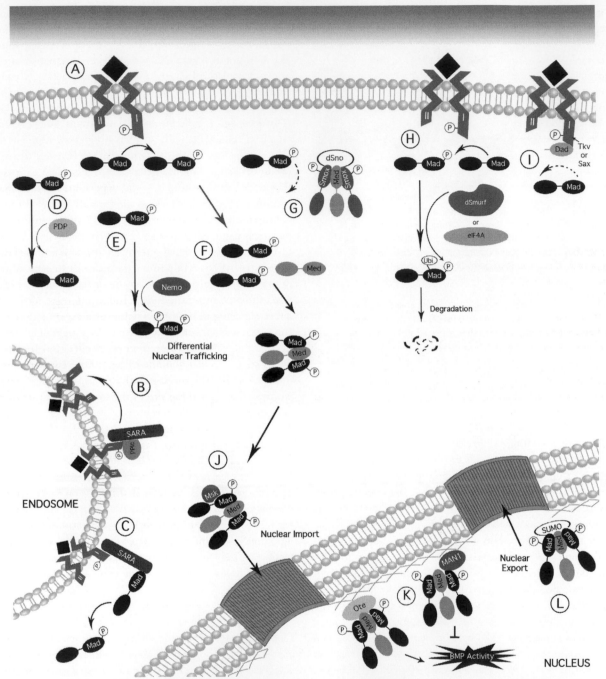

Figure 5 Selected intracellular regulators of BMP signaling. (A) Formation of a BMP ligand–Type II–Type I signaling complex results in the phosphorylation of the R-Smad Mad at its C-terminus. (B) PP1c is recruited to the signaling complex in the endosome by SARA and dephosphorylates the Type I receptor, inactivating it. (C) SARA recruits unphosphorylated Mad to the signaling complex in the endosome, both facilitating BMP signaling and preventing nuclear localization of unphosphorylated Mad. (D) Phosphorylated Mad (P-Mad) is dephosphorylated by PDP. (E) P-Mad is phosphorylated in the linker region by the Nemo kinase, which can affect the rate of nuclear accumulation of P-Mad. (F & G) P-Mad will form a complex with the Co-Smad Medea that will eventually enter the nucleus (F), but in the presence of dSno, Medea preferentially binds the Activin R-Smad Smox, inhibiting nuclear translocation of the BMP signal (G). (H) dSmurf and eIF4A can induce ubiquitination and proteasomal degradation of Mad after stimulation by the signaling complex. (I) The I-Smad Dad can prevent the interaction between the signaling complex and Mad, inhibiting BMP signaling. Dad binds the BMP Type I receptors Tkv and Sax, but not the Activin Type I receptor Baboon. (J) The importin family member Msk binds Mad and shuttles it into the nucleus. (K) The nuclear lamin-binding proteins Ote and MAN1 interact with Medea and Mad, respectively, at the nuclear periphery. The former interaction facilitates BMP signaling, the latter inhibits it. (L) SUMOylation of Medea results in its export from the nucleus. Image courtesy of Leah Moak.

it does not interact similarly with the Activin Type I receptor Baboon (Inoue *et al.*, 1998; Kamiya *et al.*, 2008). In this way Dad is functionally more similar to one of two homologous I-Smads in humans, Smad6, which also primarily inhibits the BMP pathway (Imamura *et al.*, 1997; Hata *et al.*, 1997). The other human I-Smad, Smad7, predominantly antagonizes the TGF-β/Activin pathway, though both Smad6 and Smad7 can inhibit both signaling families at high levels (Hayashi, 1997; Nakao *et al.*, 1997). In *Drosophila*, Dad also interacts with Mad directly and prevents its translocation to the nucleus in the presence of activated receptors (Inoue *et al.*, 1998). Dad is the lone *Drosophila* I-Smad, and inhibits BMP signaling via several mechanisms.

13.3.3.2. dSARA and PP1c

Like Dad, the protein Smad Anchor for Receptor Activation (SARA) can regulate interactions between Smads and receptors. SARA positively regulates TGF-β signaling in the vertebrate system by recruiting Smads to activated receptor complexes (Tsukazaki *et al.*, 1998; Xu *et al.*, 2000). SARA localizes to the endosomal compartment via a FYVE domain and recruits unphosphorylated Smad2, which is a homolog of the Smad in the *Drosophila* Activin family, Smox. This binding both inhibits nuclear localization of unstimulated Smads and presents activated receptors with phosphorylation targets. The *Drosophila melanogaster* genome includes a *SARA* homolog, *dSARA*, that appears to positively regulate the BMP pathway in a similar manner (Bökel *et al.*, 2006). In addition to apparently attracting Smads to activated receptor complexes like its vertebrate counterpart, however, dSARA also negatively regulates the BMP pathway in *Drosophila* (Bennet and Alphey, 2002). dSARA binds the catalytic subunit of a phosphatase, type 1 serine/threonine protein phosphatase (PP1c), and recruits PP1c to the activated receptor complex, where it apparently dephosphorylates the Type I receptor. PP1c antagonizes BMP signaling in the wing under overexpressed conditions, demonstrating a role for both it and dSARA as negative regulators of BMP signaling.

13.3.3.3. PDP

PP1c is not the only phosphatase that can downregulate TGF-β signaling in *Drosophila*. *Drosophila* S2 cells respond to Dpp by increasing levels of phosphorylated Mad (P-Mad), but high levels of P-Mad return to unstimulated levels soon after removal of Dpp. An RNAi-based screen in S2 cells identified a serine/threonine phosphatase – pyruvate dehydrogenase phosphatase (PDP) – that was responsible for the rapid decline of P-Mad levels post-stimulation (Chen *et al.*, 2006). Further, PDP was shown to physically interact with and dephosphorylate activated Mad, and *pdp* mutant embryos showed increased P-Mad staining. These results implicated PDP as an antagonist of BMP signaling. The mammalian PDP homologs also dephosphorylate Smad1,

a BMP R-Smad, suggesting evolutionary conservation of this mechanism of downregulation (Chen *et al.*, 2006). In addition to PDP homologs there are several other phosphatases in the vertebrate system that regulate phosphorylation of Smads at the C-terminal serines and in the linker region, but many of their homologs in *Drosophila* have not been identified or characterized (Duan *et al.*, 2006; Knockaert, *et al.*, 2006; Sapkota *et al.*, 2006; and others).

13.3.3.4. PP2A

Another phosphatase that influences TGF-β signaling in *Drosophila* is Protein Phosphatase 2A (PP2A), a conserved multi-subunit phosphatase that has been implicated in several biological processes. One of the subunits of PP2A is the regulatory subunit, which influences the activities of the catalytic phosphatase itself. The lone PP2A regulatory subunit in *Drosophila* is encoded by the *twins* gene, which was named for its role in patterning imaginal discs (Mayer-Jaekel *et al.*, 1993; Uemura *et al.*, 1993). PP2A was recently implicated in the regulation of *Drosophila* TGF-β signaling because of a genetic interaction between *twins* and members of the Activin pathway (Batut *et al.*, 2008). Overexpression of *twins* in the wing disc resulted in small, wrinkled, blistered adult wings, and this phenotype was suppressed by co-expression of wild type *smox* and, to some extent, by co-expression of one isoform of the Activin receptor Baboon. These interactions hinted that Twins may antagonize Activin signaling, though the mechanism of the interaction between Twins and the Activin pathway members is unclear. Vertebrate PP2A and its multiple regulatory subunits influence several members of the TGF-β signaling network by distinct mechanisms, including stabilization of levels of receptor by dephosphorylation and dephosphorylation of a BMP Smad (Batut *et al.*, 2008; Bengtsson *et al.*, 2009). Whether PP2A and its regulatory subunit Twins downregulate Activin signaling via stabilizing Activin receptors, interacting with Smox, enhancing BMP signaling, or a combination thereof has yet to be determined.

13.3.3.5. Nemo

Like dephosphorylation by phosphatases, phosphorylation by non-receptor kinases can profoundly affect the functions of Smads inside the cell. Phosphorylation at sites other than the C-terminal SSXS motifs can affect cellular trafficking and stability of Smads after stimulation by traditional receptor kinases. The serine/threonine kinase Nemo in *Drosophila* is the founding member and namesake of a family of proteins related to MAP kinases (Choi and Benzer, 1994). Nemo plays a role in patterning of imaginal discs, in part because of an interaction with the BMP pathway. *nemo* mutants display increased BMP signaling in their wing discs, and overexpression of *nemo* results in decreased BMP signaling, suggesting that Nemo is a BMP antagonist. This antagonism results from phosphorylation of Mad by Nemo at a serine residue in the MH1 domain, which can inhibit nuclear localization of Mad by promoting nuclear export (Zeng *et al.*, 2007; Merino *et al.*, 2009). This interaction between Nemo and BMP signaling can also be observed in the growth of the neuromuscular junction, though the nuclear trafficking as a result of this phosphorylation is less clear in this context (see section 13.3.1 above; Merino *et al.*, 1999). Other kinases phosphorylate the linker regions of vertebrate Smads, but many homologous inter-actions have not been found in flies; however, some of the phosphorylation sites have been found in *Drosophila* and play important roles in regulating BMP signaling during development (Eivers *et al.*, 2009).

13.3.3.6. dSno

The *Drosophila* dSno protein is another intracellular mediator of TGF-β signaling (Takaesu *et al.*, 2006; Barrio *et al.*, 2007; Ramel *et al.*, 2007). dSno inter-acts with the Co-Smad Medea and biases its binding of R-Smads. dSno-bound Medea has increased affinity for Smox, the Smad of the *Drosophila* Activin pathway, and as a result may function as an enhancer of Activin signal-ing. Indeed, *dSno* mutants have mutant phenotypes that mimic the small larval brains found in some Activin-path-way mutants (Takaesu *et al.*, 2006). Moreover, the ability of dSno to increase the affinity of Medea for Smox results in a functional inhibition of BMP signaling by excluding Mad from the nucleus of stimulated cells (Takaesu *et al.*, 2006; Barrio *et al.*, 2007). This dSno-mediated interplay between the BMP and Activin pathways is reminiscent of the functions of the related Ski and namesake Sno onco-proteins in humans, which can also antagonize BMP sig-naling (reviewed in Massagué *et al.*, 2005). In *Drosophila*, dSno may function as a type of switch between the BMP and Activin pathways in cells that receive both forms of signal. That *dSno* is a downstream target of Dpp and EGF signaling suggests the presence of an inducible negative feedback loop (Shravage *et al.*, 2007).

13.3.3.7. dSmurf

In addition to regulating the traf-ficking of activated Smads to control signaling levels in the nucleus, the cell can also destroy the Smads via the ubiquitin/proteasome pathway. The E3-ubiquitin ligase Highwire and its effects on BMP signaling at the larval neuromuscular junction have been discussed previously (see section 13.3.1 above). Unlike Highwire, which tar-gets the Co-Smad Medea for degradation, the HECT domain family E3-ubiquitin ligase dSmurf (also known as Lack) antagonizes BMP signaling by interacting with Mad (Podos *et al.*, 2001). This interaction regulates P-Mad levels in many developmental contexts, including dorso–ventral patterning of the early embryo, the development of the embryonic hindgut, and the maintenance of germ-line stem cells in the ovary (Podos *et al.*, 2001; Casanueva and Ferguson, 2004). dSmurf adds the small protein ubiquitin to Mad, targeting it for proteasomal degradation, and this interaction is dependent on activation of Mad by a Type I

receptor (Liang *et al.*, 2003). This degradation of activated BMP Smads is conserved in humans, where the dSmurf homolog Smurf1 can degrade Smad1 (Zhu *et al.*, 1999). This degradation can be enhanced by phosphorylation of the Smad linker region by non-receptor kinases (Sapkota *et al.*, 2007). Interestingly, like its vertebrate homolog Smurf2, dSmurf can bind both BMP- and Activin-family R-Smads in a signal-dependent manner, though destruc-tion of Smox by dSmurf has not been demonstrated (Lin *et al.*, 2000; Zhang *et al.*, 2001; Liang *et al.*, 2003). Unlike its vertebrate relatives, dSmurf does not appear to interact with the *Drosophila* I-Smad, Dad (Ebisawa *et al.*, 2001; Liang *et al.*, 2003).

13.3.3.8. eIF4A

A genetic screen for regulators of Dpp unexpectedly implicated Eukaryotic Initiation Factor 4A (eIF4A), the well-established regulator of protein transla-tion, as an inhibitor of BMP signaling (Li *et al.*, 2005). Mutation of eIF4A phenocopied *dpp* gain-of-function, and its overexpression mimicked *dpp* loss-of-function, sug-gesting that eIF4A is a *dpp* antagonist (Li and Li, 2006). Surprisingly, eIF4A physically interacts with Mad and Medea upon stimulation by Type I receptors. This interac-tion results in degradation of the BMP pathway members, and does so independently of dSmurf (see above). This regulation of downstream BMP signaling components by eIF4A appears to function in several developmental contexts, and occurs independently of eIF4A's role in cap-dependent translation (Li and Li, 2006). Whether this mechanism of signal-dependent regulation also applies to the Activin pathway in *Drosophila* has yet to be reported.

13.3.3.9. SUMOylation of Medea

Unlike ubiquitina-tion, addition of the SUMO protein to other proteins does not target them for degradation, but instead merely alters their trafficking or function inside the cell. Recent genetic experiments demonstrated that mutants for com-ponents of the SUMOylation cascade had expanded expression of Dpp target genes in the early embryo (Miles *et al.*, 2008). Further analysis demonstrated that the Co-Smad Medea is SUMOylated both *in vitro* and *in vivo*, and overexpression of SUMO–Medea fusion proteins decreased transcriptional activation of Dpp target genes. The mechanism for this downregulation of BMP signal-ing appears to involve the export of SUMO-modified Medea from the nucleus. Such a mechanism evidently provides cells with another level of regulation in reading or fine-tuning extracellular morphogen gradients: rather than altering the amount of ligand, the level of receptor activation, or the levels of phosphorylated Smad, cells can instead regulate the nuclear localization of the Co-Smad. Further, a model has been suggested where changing the nuclear localization of the Smads may alter their stabil-ity – and therefore the stability of the signal – in the cell (Miles *et al.*, 2008). In addition to its effects on BMP signaling, SUMOylation of Medea may also impact the Activin signaling pathway, though this has not yet been addressed experimentally.

13.3.3.10. Moleskin In most contexts, a cellular response to BMP signaling requires accumulation of Smads in the nucleus. A recent cell-based RNAi screen isolated the nuclear importin family member Moleskin (Msk; Lorenzen *et al.*, 2001) as a key molecule in the nuclear translocation of the *Drosophila* BMP Smad Mad (Xu *et al.*, 2007). *moleskin* mutant eye imaginal discs also lacked nuclear accumulation of phosphorylated Mad (P-Mad), which is consistent with Msk's role as an importin family member. Moleskin physically interacts with both unstimulated and stimulated Mad, and imports it to the nucleus (Xu *et al.*, 2007). That this process occurs independent of BMP stimulation intimates the roles of other factors that retain phosphorylated Mad in the nucleus. Further, the human homologs of Msk, Importin 7 and Importin 8, also bind the human BMP Smad Smad1 and shuttle it to the nucleus, demonstrating conservation of this mechanism of spatial regulation of downstream BMP signaling molecules in the cell (Xu *et al.*, 2007).

13.3.3.11. Otefin and MAN1 Once inside nuclei, Smads can interact with proteins associated with the nuclear membrane. Otefin (Ote), a nuclear lamin-binding protein of the LEM family, is required for proper maintenance of germ-line stem cells (GSCs) in *Drosophila*: *ote* mutants display germ-line defects, and *ote* overexpression increases the number of GSCs (Jiang *et al.*, 2008). Otefin interfaces with BMP signaling by forming a complex with the Co-Smad Medea, and this complex represses *bam* transcription by binding silencer elements of the *bam* locus (Jiang *et al.*, 2008). In this way, Ote positively regulates BMP signaling at the nuclear periphery.

However, another LEM family member in *Drosophila*, MAN1, antagonizes BMP signaling. MAN1 contains two putative Mad-binding domains, and insects lacking one of these sites showed increased levels of phosphorylated Mad and increased expression of BMP target genes in the wing. Moreover, overexpression of MAN1 phenocopied BMP loss-of-function (Wagner *et al.*, 2010). MAN1 also inhibits BMP signaling at the larval neuromuscular junction. In vertebrates, several LEM family members antagonize TGF-β signaling by interacting with both BMP and Activin/TGF-β Smads (Raju *et al.*, 2003, Lin *et al.*, 2005). Interestingly, MAN1 expression is apparently upregulated by BMP signaling in *Drosophila*, demonstrating a possible negative feedback loop during wing development (Wagner *et al.*, 2010). In this way, Otefin and MAN1 demonstrate how the nuclear membrane and the nuclear periphery can both enhance and inhibit BMP signaling in different developmental contexts.

13.4. Activins and Non-Canonical TGF-β Signaling

13.4.1. The Activin Family

The Activin pathway in *Drosophila melanogaster* has been implicated in several biological processes, most of which involve cellular proliferation and neuronal development. The discovery of the Activin pathway in *Drosophila* relied heavily upon work that described the BMP pathway in flies and the Activin pathway in vertebrates. Using the sequences of these other pathway members and the recently acquired understanding of their molecular functions, researchers could search not only for members of the Activin subfamily in insects, but also for the biological roles that they played.

The first clues for the existence of an Activin/TGF-β pathway in *Drosophila* came with the discoveries of the Type II receptor Punt (then called Atr-II; Childs *et al.*, 1993) and the Type I Activin receptor Baboon (then called Atr-I; Wrana *et al.*, 1994). No *Drosophila* Activin ligands or R-Smads were known at that time, but biochemical and cell-culture assays demonstrated that these receptors could interact predictably with human Activin ligands and Smads, pointing to the existence of Activin signaling in insects. The gene encoding the R-Smad in the *Drosophila* Activin pathway, Smox (for **Sm**ad **on X**, formerly "dSmad2"), was subsequently found on the X-chromosome (Henderson and Andrew, 1998), and this R-Smad functioned downstream of Baboon (Das *et al.*, 1999). Importantly, known BMP receptors could not activate this newfound R-Smad, marking Activins as a distinct branch of the TGF-β pathway. Finally, a biological role for the *Drosophila* Activin pathway emerged when Baboon and Smox were linked to cell growth and proliferation (Brummel *et al.*, 1999). Furthermore, the Activin pathway seemed to differ from the BMP pathway in that it affected proliferation of cells, but not necessarily the patterning of the tissues themselves.

Activin-β (Act-β) in *Drosophila melanogaster* was the first Activin ligand described in an invertebrate organism (Kutty *et al.*, 1998), and it serves several roles in the developing nervous system. Activin-β signals via Baboon and Smox to enable the rewiring and pruning of neuronal axons during development of the third-instar larval brain, and does so by positively regulating expression of the *EcR-B1* gene, which encodes a receptor for ecdysone, an ecdysteroid (Zheng *et al.*, 2003). Because ecdysone is a principal cue for developmental transitions in insects, this role for the Activin pathway has implicated the TGF-β pathway in the regulation of developmental timing. In addition to regulating expression of a hormone receptor that aids neuronal remodeling, Act-β also serves other roles in the nervous system. For example, visual detection of ultraviolet light requires Act-β-dependent Activin signaling to ensure proper axon guidance in the developing visual system (Ting *et al.*, 2007). In this context Act-β functions as an autocrine signal in photoreceptor axons, where activated Smox is shuttled into the nucleus after binding to Importin-α3. Proper Activin signaling in these photoreceptor axons prevents their aberrant innervation of the optic lobe. In yet another developmental context, Activin-β cooperates with the Activin family member Dawdle to ensure proper brain size (Zhu *et al.*, 2008). Act-β and Dawdle signal via Babo and Smox to redundantly stimulate proliferation rates of neuroblasts during larval brain development; lack of any of these Activin pathway members results in decreased brain size.

The *Drosophila* Activin ligand Dawdle (Daw) is expressed in many tissues, and plays several roles in the development of the nervous system in addition to its role in neuroblast proliferation. During embryonic development, Dawdle signals via the Type I receptor Baboon and the Type II receptor Punt to ensure proper innervation of motoneurons; the slow, wandering axons in *dawdle* mutants give the gene its name (Parker *et al.*, 2006; Serpe and O'Connor, 2006). During axon guidance the Daw signal is received by the motoneuron itself, but another developmental context – the formation of the neuromuscular junction (NMJ) – requires Daw signaling post-synaptically, in the muscle. Post-synaptic, Daw-dependent Activin signaling controls growth of the NMJ in part by regulating expression of Gbb in muscle cells (Ellis *et al.*, 2010). In turn, this muscle-derived Gbb delivers a retrograde signal to the pre-synaptic neuron via the BMP receptors Tkv, Sax, and Wit to regulate bouton number and synaptic homeostasis during development of the NMJ (see section 13.3.1 above). This interplay between Dawdle and Glass bottom boat – where the Activin and the BMP deliver anterograde and retrograde signals, respectively – highlights a convergence of the BMP and Activin signaling pathways during development.

Prior to signaling, the prodomain of the Dawdle ligand is cleaved by the extracellular protease Tolloid-related (Tlr, also Tolkin) (Serpe and O'Connor, 2006). Unlike activation of BMPs during embryonic development, where the protease Tolloid cleaves inhibitory molecules that are bound to the ligands, Tlr cleaves the prodomains of Activin ligands themselves. This cleavage is distinguishable from the separation of the ligand and the prodomain at the basic site between the domains during maturation. Cleavage of the Daw prodomain by Tlr increases the ligand's ability to signal in cell-culture assays and *in vivo*. Tlr similarly cleaves the prodomains of the ligands Act-β and Myoglianin, though any effect on signaling has not been reported.

The roles of the Activin ligands Myoglianin (Myg) and Maverick (Mav) during the development of *Drosophila melanogaster* are poorly understood. However, Myoglianin – whose name stems from its strong expression in glial and muscle cells (Lo and Frasch, 1999) – binds Baboon and the Type II receptor Wishful thinking (Wit) in biochemical assays, though robust signaling via the *Drosophila* Activin pathway has not been reported (Lee-Hoeflich *et al.*, 2005). Maverick, named for its divergent amino acid sequence (Nguyen *et al.*, 2000), remains enigmatic: it has not demonstrated a capacity to signal using the Activin or BMP pathways, nor has it been implicated in any biological process.

Recent work has uncovered a potential mechanism of spatial regulation of Activin signaling in *Drosophila* by focusing on the potential specificity of four potential Activin ligands for the single Activin receptor, Baboon (Jensen *et al.*, 2009). The *baboon* gene encodes three differentially spliced transcripts that differ by only one exon, which in each case codes for the cysteine-rich region in the ligand-binding domain (Wrana *et al.*, 1994). As a result, the transcripts encode three proteins – $Babo_A$, $Babo_B$, and $Babo_C$ – that differ only in their ligand-binding domains. The inclusion of these different exons is consequential for ligand-binding: Dawdle demonstrates ligand-receptor specificity by signaling exclusively via $Babo_C$. Whether other Activin ligands discriminate between other receptor isoforms for signaling has not been reported. Because the isoforms of the Type I receptor are expressed in different tissues during development, this mechanism of discriminate ligand–receptor interactions may enable ligands to target specific cell types depending on which isoform(s) of Baboon that they express. This mechanism may explain how several Activin ligands can apparently signal via the same receptor to achieve distinct results during development.

13.4.2. Cross-Talk between the Activin and BMP Pathways

Recent evidence demonstrates that the Activin and BMP pathways physically interact inside of the cell, or downstream of ligand–receptor binding. In *Drosophila*, both BMP and Activin signaling are responsible for regulating growth of the wing (Capdevila *et al.*, 1994; Zecca *et al.*, 1995; Nellen *et al.*, 1996; Brummel *et al.*, 1999; Martin-Castellanos and Edgar, 2002). Specifically, Activin signaling appears to regulate wing growth but not patterning (Brummel *et al.*, 1999; Gesualdi and Haerry, 2007), and BMP signaling regulates both growth of the wing and patterning of veins within it (for review, see Affolter and Basler, 2007). Several lines of evidence suggest that these two pathways do not act independently in this developmental context.

First, altering levels of Activin signaling in the wing gave phenotypes similar to those caused by aberrant BMP signaling (Sander *et al.*, 2010). Impairing Activin signaling in the wing disc by expressing *smox* RNAi constructs mimicked the large discs typical of BMP gain-of-function.

Expression of *smox* RNAi also caused formation of extra vein tissue in adult wings, another BMP gain-of-function phenotype. Indeed, these ectopic vein tissues had increased levels of phosphorylated Mad (P-Mad), suggesting that Activin signaling antagonizes BMP signaling in the *Drosophila* wing. This antagonism was shown to be dependent on *medea*, suggesting that the mechanism for the interaction was competition for binding of the Co-Smad inside the cell. That is, increased levels of phosphorylated Smox (P-Smox) would bind Medea, preventing its binding to (P-Mad) (Sander *et al.*, 2010). Such a mechanism has been proposed for the development of *Xenopus* embryos, suggesting strong evolutionary conservation (Candia *et al.*, 1997).

Downstream interactions between BMP and Activin signaling in *Drosophila* are not limited to competition for the Co-Smad Medea and for subsequent entrance to the nucleus. Recent cell culture and genetic assays demonstrate that the Activin Type I receptor, Baboon, is capable of phosphorylating the BMP R-Smad Mad *in vitro*, and under overexpressed conditions *in vivo* (Gesualdi and Haerry *et al.*, 2007). Constitutively activated Baboon (Babo*) phosphorylated both Smox and Mad in S2 cells, and overexpression of the Mad with either Baboon or selected Activin family ligands in the wing phenocopied BMP overexpression. Such a result is consistent with the finding mentioned above where wing cells with depleted *smox* levels have increased P-Mad accumulation (Sander *et al.*, 2010). Also, similar signaling events between Activin family Type I receptors and BMP family R-Smads have been reported in the vertebrate system, and support a new model of interaction between BMP and Activin signaling (Sander *et al.*, 2010, and others – see below). Instead of simple competition between BMP and Activin R-Smads for Medea binding and subsequent accumulation in the nucleus, perhaps there exist Smad complexes with two species of R-Smad. Because the activated Smad complex is trimeric, with one Co-Smad and two R-Smads, two R-Smads from different signaling subfamilies may simultaneously bind the Co-Smad and enter the nucleus (Shi *et al.*, 1997; Kawabata *et al.*, 1998). Such complexes may demonstrate unique biological properties and transcriptional activities (Goumans *et al.*, 2002, 2003; Bharathy *et al.*, 2008; Daly *et al.*, 2008; Liu *et al.*, 2009).

Interestingly, recent genetic experiments also demonstrate similar phenotypes of *smox* and *saxophone* mutants in dividing spermatogonia in the *Drosophila* testis (Li *et al.*, 2007). These mutant phenotypes suggest a reciprocal cross-talk signaling event between a BMP Type I receptor and the Activin R-Smad. However, biochemical or cell culture signaling data that support this finding have not yet been reported.

The discovery of physical interactions between Activin family Type I receptors and BMP family Smads raises questions regarding the relationship between the two branches of the TGF-β signaling family. In particular, the original classification of TGF-β superfamily ligands, receptors, and Smads into BMP and Activin branches comes into question. That is, if the *Drosophila* Activin receptor can phosphorylate the BMP Smad, why was this interaction initially not detected, resulting in assignment of the molecules into two different signaling groups? One potential answer comes from a recent report in the human TGF-β system that demonstrated that, indeed, Activin/ TGF-β family Type I receptors are capable of phosphorylating BMP R-Smads, but have greater affinity for Activin family R-Smads (Liu *et al.*, 2009). Type I receptors affinities for Smads are at least in part determined by a region of the intracellular kinase domain of the receptor, the L45 loop. Experiments that "swapped" L45 domains of receptors from the Activin and BMP signaling families resulted in the receptors' binding of different Smads (Feng and Derynck, 1997; Chen *et al.*, 1998). In the case of Activin-type receptors, the L45 loop was thought to bind only Activin-type Smads (Feng and Derynck, 1997; Chen *et al.*, 1998; Itoh *et al.*, 2003). However, it is also capable of binding BMP Smads, albeit more weakly and perhaps only after the pool of available Activin Smads has been depleted (Liu *et al.*, 2009). Such discrimination between Smads of different families by the kinase domain may explain the initial classification of members of the Activin family. For instance, early biochemical experiments that tested for putative interactions between Baboon and Mad were done in cells with endogenous levels of Activin-type Smads (Brummel *et al.*, 1999). No interaction was found, perhaps because of the combination of a weak interaction between Baboon and Mad and the presence of Activin-type Smads in the cells.

13.4.3. Non-Canonical TGF-β Signaling

Not all TGF-β signals require activation of Smad proteins. In the canonical model of BMP and Activin signaling in *Drosophila*, a ligand dimer forms a complex with Type II and Type I receptors at the cell membrane, the Type II receptor phosphorylates the Type I receptor at the GS domain (Wrana *et al.*, 1994), and the Type I receptor phosphorylates an intracellular Smad. Several reports in *Drosophila melanogaster* have indicated that ligand-stimulated receptors have downstream targets in addition to Smads (Eaton and Davis, 2005; Ng, 2008). Such pathways indicate the presence of non-canonical TGF-β signaling pathways in *Drosophila*.

The first report of non-canonical TGF-β signaling in *Drosophila* involved signals that stabilize the neuromuscular junction (NMJ). Presynaptic BMP signaling – delivered by Gbb, Tkv, Sax, Wit, and Mad – is critical for proper development of the NMJ; the requirement of a Smad in this pathway indicates the necessity of canonical BMP signaling (see section 13.3.1 above). In addition to its role in the canonical signaling pathway, the Type II receptor Wit also activates the kinase LIMK1, and does so independently of Mad (Eaton and Davis, 2005). LIMK1 influences development in the nervous system in part by phosphorylating Twinstar, the *Drosophila* homolog of cofilin, and this signal inhibits polymerization of the actin cytoskeleton (Ohashi, 2000; Ng and Luo, 2004). LIMK1 is also associated with other aspects of the cytoskeleton in synapses (Eaton and Davis, 2005).

The activation of LIMK1 by Wit does not require Wit's kinase activity. Assays for synapse stability showed that *wit* mutants had more severe phenotypes than animals mutant for other members of the canonical BMP pathway, suggesting that Wit had other functions in the motoneuron outside of the Smad pathway. Unlike other TGF-β receptors in *Drosophila*, Wit has a long C-terminal tail, and this domain is present in the mammalian homolog of Wit (Aberle *et al.*, 2002; Marqués *et al.*, 2002). Viability of *wit* mutants is restored by expression of a transgene that lacks this domain, demonstrating that the C-terminal tail is not necessary for much of Wit function (Marqués *et al.*, 2002). However, restoration of synaptic stability in *wit* mutants requires expression of the full-length *wit* cDNA: the C-terminal tail of Wit regulates stability by interacting with LIMK1 and controlling its activity at the presynapse (Eaton and Davis, 2005). Vertebrate homologs of these molecules interact similarly (Foletta *et al.*, 2003; Lee-Hoeflich *et al.*, 2004). As a result, Wit affects synapse development via both canonical, Smad-dependent and non-canonical, Smad-independent BMP signaling pathways, both of which are strongly conserved throughout animal evolution.

The Activin pathway in *Drosophila* can also function Smad-independently. Activin signaling is critical for proper axon extension in mushroom body neurons: *baboon* mutants have overextended axons. However, this phenotype is not found in *smox*, *medea*, or *mad* mutants, suggesting that Baboon functions Smad-independently in this developmental context (Ng, 2008). Also, overexpression of activated *baboon* (*babo**) results in truncated axons, or the opposite of the loss-of-function phenotype. Interestingly, misexpression of LIMK1 mimics this gain-of-function phenotype, and it can be suppressed by *babo* or *wit* loss-of-function. That is, ectopic LIMK1 activity is rescued by loss of Babo and Wit, suggesting that LIMK1 acts downstream of these receptors, and does so Smad-independently (Ng, 2008).

The mechanism of LIMK1 activation in this context is atypical not only because it is Smad-independent, but also because it apparently challenges a well-established model of receptor activation. As mentioned above, in canonical TGF-β signaling the formation of the ligand–receptor complex stimulates phosphorylation of Type I receptors by Type II receptors. That is, Type I receptors are downstream of Type II receptors. In the non-canonical system

of Activin signaling during development of mushroom body neurons, there is evidence that this relationship is reversed: the Type I receptor Baboon may act upstream of the Type II receptors Punt and Wit. Specifically, in mushroom body neurons, loss of one copy of *punt* or *wit* suppresses phenotypes that result from overexpression of Babo*, suggesting that Type II receptors are either necessary for or downstream of Babo function. Also, ectopic expression of Type II receptors in *babo* mutant neurons rescued Smad-independent – but not Smad-dependent – phenotypes, which appears to support the model of downstream Type II receptors (Ng, 2008). In any case, these genetic experiments may provide a foundation for the exploration of biochemical and cell culture-based signaling experiments that further elucidate the mechanisms of non-canonical TGF-β signaling in *Drosophila*.

13.5. Evolution of TGF-β Signaling in Insects

While the *Drosophila* system has contributed profoundly to our understanding of TGF-β signaling generally, it might not be a representative system for TGF-β signaling in other insects, especially in some developmental contexts. One such context is early embryonic development. Drosophilids belong to class of higher Diptera whose embryos have an amnioserosa instead of a distinct amnion and serosa, among other differences from other insects (Stauber *et al.*, 2002; summarized in van der Zee *et al.*, 2006, 2008). As such, the signals and signaling mechanisms utilized in early *Drosophila* embryonic development are in some ways different from those in other insects and arthropods who have a more ancestral developmental program. This section briefly highlights recent work that characterizes some of the disparities between TGF-β systems in *Drosophila* and other insects, as well as spiders.

Analysis of the TGF-β signaling family and its regulators in the genome of the red flour beetle *Tribolium castaneum* reveals many similarities to the *Drosophila* system (van der Zee, 2008). The beetle genome houses genes that are similar to all fly Activin ligands, Type I receptors, Type II receptors, R-, Co-, and I-Smads, and most BMP ligands. These findings are consistent with the observation that TGF-β signaling plays myriad roles in numerous tissues in all stages of life in many animals. However, there is a notable absence in the beetle genome: there is no apparent gene that is homologous to *screw* in *Drosophila*. As discussed in section 13.2.3, Screw is a BMP signal that is critical to dorso–ventral patterning in flies, and perhaps interestingly, unlike other more versatile TGF-β signals, it seemingly does little else during development. Screw is expressed ubiquitously early in development, and it is essential to the constriction and intensification of the BMP signaling domain on the dorsal side of the late blastoderm embryo. Screw's role in this phenomenon involves heterodimerization with Dpp and subsequent shuttling to the dorsal midline by the Sog–Tsg complex (**Figure 2**; see also Shimmi *et al.*, 2005a). This mechanism of dorso–ventral patterning is not shared by all insects and arthropods, including *Tribolium* and the honeybee *Apis mellifera* (see above) and spiders (see below), which have more ancestral modes of early development. That *screw* is absent from both the beetle and bee genomes suggests rapid functional divergence of a Gbb-like BMP pathway member in higher Diptera that coincides with the evolution of such a specialized role in embryogenesis.

Other disparities between the beetle and fly genomes suggest that *Tribolium* indeed has a more ancestral TGF-β signaling system: *Tribolium* has several signaling components and regulators that *Drosophila* lacks, but that are found in vertebrates. For example, beetles have a BMP homolog that is not found in *Drosophila*, but is found in many other animals, including bees (van der Zee, 2008). Furthermore, beetles express several vertebrate-like regulators of TGF-β signaling that fruit flies do not express, including DAN, Gremlin, and BAMBI. That these signaling components are found in vertebrates and some insects, but not *Drosophila*, raises questions about biological processes unique to the *Drosophila* lineage.

Some biological differences are seen even between the functions of related molecules during the development of *Drosophila* and *Tribolium*. For instance, unlike in flies, where Short Gastrulation (Sog) and Twisted gastrulation (Tsg) function synergistically and somewhat redundantly during embryonic development, the Tsg homolog in beetles functions largely independently of Sog (Nunes da Fonseca *et al.*, 2010). Moreover, while *Drosophila* has three Tsg-like molecules – Tsg, Crossveinless (Cv), and Shrew (Srw; Bonds *et al.*, 2007) – *Tribolium* only has a single ortholog. Similar redundancy in the fly system is absent regarding the beetle Tolloid homolog: unlike flies, which express both Tolloid and Tolloid-related (see sections 13.2.3 and 13.2.4), beetles have only one Tolloid-like protease (Nunes da Fonseca *et al.*, 2010). The observation that these two species have so few differences between their TGF-β systems, and that those differences are so prominently associated with biological processes that the species do not share, has uncovered an exciting area of future study.

Roles of TGF-β signals and their regulators also differ between fruit flies and another arthropod, the spider (reviewed in Oda and Akiyama-Oda, 2008). Dpp-like molecules pattern the dorsal side of the embryo in both *Drosophila* and *Achaearanea*, but the mechanisms by which the signals are localized differ. As discussed in section 13.2.3, in *Drosophila* Dpp molecules are secreted by cells on the dorsal side of the embryo, and these signals are eventually concentrated at the dorsal midline by repeated rounds of shuttling by the extracellular regulators Sog and Tsg (**Figure 2**). In this context, the domain of the embryo that receives Dpp signaling constricts over time to eventually pattern the amnioserosa at the dorsal midline. Lack of Dpp in this system results in a ventralized embryo, demonstrating that ventral cell fates are induced by a mechanism independent of Dpp signaling. In spiders, however, a nearly opposite phenomenon occurs. Instead of a broad domain of Dpp expression giving rise to a narrowing, intensifying gradient of dorsal signaling as in *Drosophila*, *Achaearanea* Dpp is expressed by a small cluster of cells that sits at the dorsal pole, and continued secretion of Dpp induces the expansion of a BMP signaling gradient throughout the embryo (Akiyama-Oda and Oda, 2003, 2006). Also unlike fly development, this signaling domain in spiders is necessary for the patterning of both dorsal and ventral cell fates. That is, while Dpp in *Drosophila* patterns only dorsal tissue in the embryo, in spiders Dpp is responsible for both dorsal and ventral cell fates.

The balance of dorsal and ventral domains in *Achaearanea* depends largely on Short Gastrulation (Sog). Unlike in fruit flies, where *sog* mutants still have most of their ventral tissues intact, spiders lacking Sog are almost completely dorsalized (Akiyama-Oda and Oda, 2006). This observation is similar to *sog* loss-of-function phenotypes in *Tribolium* (van der Zee, 2006). Interestingly, the *sog* expression domain shrinks in the spider embryo during embryogenesis, resulting in only a stripe of *sog* expression at the ventral midline. This constricting domain complements the growing domain of phosphorylated Mad (P-Mad) in the embryo, suggesting that Dpp signaling inhibits Sog expression (Akiyama-Oda and Oda, 2006). In spiders, the *sog* expression domain appears to retreat ventrally and "protect" ventral cells in the embryo from BMP signaling. In flies, Sog expression remains constant, and Sog diffusion "attacks" the domain of BMP signaling by facilitating P-Mad's retreat to the dorsal midline. Such disparate mechanisms of dorso–ventral patterning raise interesting questions about the evolution of D–V patterning in arthropods.

Acknowledgments

Thank you to Leah Moak, who constructed the figures, to Michael O'Connor for help in outlining the chapter, and to Lisa Bofenkamp for preparation of reference materials. Thanks also to Mark Osterlund for comments on the manuscript.

References

Aberle, H., Haghighi, A. P., Fetter, R. D., McCabe, B. D., Magalhães, T. R., & Goodman, C. S. (2002). *wishful thinking* encodes a BMP type II receptor that regulates synaptic growth in *Drosophila*. *Neuron*, *33*, 545–558.

Affolter, M., & Basler, K. (2007). The Decapentaplegic morphogen gradient: From pattern formation to growth regulation. *Nat. Rev. Genet.*, *8*, 663–674.

Affolter, M., Nellen, D., Nussbaumer, U., & Basler, K. (1994). Multiple requirements for the receptor serine/threonine kinase *thickveins* reveal novel functions of TGF beta homologs during *Drosophila* embryogenesis. *Development, 120,* 3105–3117.

Akiyama-Oda, Y., & Oda, H. (2003). Early patterning of the spider embryo: A cluster of mesenchymal cells at the cumulus produces Dpp signals received by germ disc epithelial cells. *Development, 130,* 1735–1747.

Akiyama-Oda, Y., & Oda, H. (2006). Axis specification in the spider embryo: Dpp is required for radial-to-axial symmetry transformation and sog for ventral patterning. *Development, 133,* 2347–2357.

Allan, D. W., St Pierre, S. E., Miguel-Aliaga, I., & Thor, S. (2003). Specification of neuropeptide cell identity by the integration of retrograde BMP signaling and a combinatorial transcription factor code. *Cell, 113,* 73–86.

Arora, K., & Nüsslein-Volhard, C. (1992). Altered mitotic domains reveal fate map changes in *Drosophila* embryos mutant for zygotic dorsoventral patterning genes. *Development, 114,* 1003–1024.

Arora, K., Levine, M. S., & O'Connor, M. B. (1994). The *screw* gene encodes a ubiquitously expressed member of the TGF-beta family required for specification of dorsal cell fates in the *Drosophila* embryo. *Genes Dev., 8,* 2588–2601.

Arora, K., Dai, H., Kazuko, S. G., Jamal, J., O'Connor, et al. (1995). The *Drosophila schnurri* gene acts in the Dpp/TGF signaling pathway and encodes a transcription factor homologous to the human MBP family. *Cell, 81,* 781–790.

Ashe, H. L., & Levine, M. (1999). Local inhibition and long-range enhancement of Dpp signal transduction by Sog. *Nature, 398,* 427–431.

Ballard, S. L., Jarolimova, J., & Wharton, K. A. (2010). Gbb/BMP signaling is required to maintain energy homeostasis in *Drosophila*. *Dev. Biol., 337,* 375–385.

Bangi, E., & Wharton, K. (2006a). Dpp and Gbb exhibit different effective ranges in the establishment of the BMP activity gradient critical for *Drosophila* wing patterning. *Dev. Biol., 295,* 178–193.

Bangi, E., & Wharton, K. (2006b). Dual function of the *Drosophila* Alk1/Alk2 ortholog Saxophone shapes the Bmp activity gradient in the wing imaginal disc. *Development, 133,* 3295–3303.

Barrio, R., López-Varea, A., Casado, M., & de Celis, J. F. (2007). Characterization of dSnoN and its relationship to Decapentaplegic signaling in *Drosophila*. *Dev. Biol., 306,* 66–81.

Basler, K., & Struhl, G. (1994). Compartment boundaries and the control of *Drosophila* limb pattern by hedgehog protein. *Nature, 368,* 208–214.

Batut, J., Schmierer, B., Cao, J., Raftery, L. A., Hill, C. S., & Howell, M. (2008). Two highly related regulatory subunits of PP2A exert opposite effects on TGF-beta/Activin/Nodal signalling. *Development, 135,* 2927–2937.

Bengtsson, L., Schwappacher, R., Roth, M., Boergermann, J. H., Hassel, S., & Knaus, P. (2009). PP2A regulates BMP signalling by interacting with BMP receptor complexes and by dephosphorylating both the C-terminus and the linker region of Smad1. *J. Cell Sci., 122,* 1248–1257.

Bennett, D., & Alphey, L. (2002). PP1 binds Sara and negatively regulates Dpp signaling in *Drosophila melanogaster*. *Nat. Genet., 31,* 419–423.

Bharathy, S., Xie, W., Yingling, J. M., & Reiss, M. (2008). Cancer-associated transforming growth factor beta type II receptor gene mutant causes activation of bone morphogenic protein-Smads and invasive phenotype. *Cancer Res., 68,* 1656–1666.

Biehs, B., Francois, V., & Bier, E. (1996). The *Drosophila short gastrulation* gene prevents Dpp from autoactivating and suppressing neurogenesis in the neuroectoderm. *Genes Dev., 10,* 2922–2934.

Bökel, C., Schwabedissen, A., Entchev, E., Renaud, O., & Gonzalez-Gaitan, M. (2006). Sara endosomes and the maintenance of Dpp signaling levels across mitosis. *Science, 314,* 1135–1139.

Bonds, M., Sands, J., Poulson, W., Harvey, C., & Von Ohlen, T. (2007). Genetic screen for regulators of *ind* expression identifies *shrew* as encoding a novel twisted gastrulation-like protein involved in Dpp signaling. *Dev. Dyn., 236,* 3524–3531.

Bridges, C. B. (1920). The mutant *crossveinless* in *Drosophila melanogaster*. *Proc. Natl. Acad. Sci. USA, 6,* 660–663.

Brummel, T., Twombly, V., Marqués, G., Wrana, J. L., Newfeld, S. J., et al. (1994). Characterization and relationship of dpp receptors encoded by the *saxophone* and *thickveins* genes in *Drosophila*. *Cell, 78,* 251–261.

Brummel, T., Abdollah, S., Haerry, T. E., Shimell, M. J., Merriam, J., et al. (1999). The *Drosophila* activin receptor baboon signals through dSmad2 and controls cell proliferation but not patterning during larval development. *Genes Dev., 13,* 98–111.

Bryant, P. J. (1988). Localized cell death caused by mutations in a *Drosophila* gene coding for a transforming factor-beta homolog. *Dev. Biol., 128,* 386–395.

Campbell, G., & Tomlinson, A. (1999). Transducing the Dpp morphogen gradient in the wing of *Drosophila*: Regulation of Dpp targets by brinker. *Cell, 96,* 553–562.

Campbell, G., Weaver, T., & Tomlinson, A. (1993). Axis specification in the developing *Drosophila* appendage: The role of *wingless, decapentaplegic*, the homeobox gene aristaless. *Cell, 74,* 1113–1123.

Candia, A. F., Watabe, T., Hawley, S. H., Onichtchouk, D., Zhang, Y., et al. (1997). Cellular interpretation of multiple TGF-beta signals: Intracellular antagonism between activin/BVg1 and BMP-2/4 signaling mediated by Smads. *Development, 124,* 4467–4480.

Capdevila, J., Estrada, M. P., Sanchez-Herrero, E., & Guerrero, I. (1994). The *Drosophila* segment polarity gene *patched* interacts with *decapentaplegic* in wing development. *EMBO J., 13,* 71–82.

Capovilla, M., Brandt, M., & Botas, J. (1994). Direct regulation of *decapentaplegic* by Ultrabithorax and its role in *Drosophila* midgut morphogenesis. *Cell, 76,* 461–475.

Casanueva, M. O., & Ferguson, E. L. (2004). Germline stem cell number in the *Drosophila* ovary is regulated by redundant mechanisms that control Dpp signaling. *Development, 131,* 1881–1890.

Chen, D., & McKearin, D. (2003a). Dpp signaling silences bam transcription directly to establish asymmetric divisions of germline stem cells. *Curr. Biol., 13,* 1786–1791.

Chen, D., & McKearin, D. M. (2003b). A discrete transcriptional silencer in the *bam* gene determines asymmetric division of the *Drosophila* germline stem cell. *Development, 130,* 1159–1170.

Chen, H. B., Shen, J., Ip, Y. T., & Xu, L. (2006). Identification of phosphatases for Smad in the BMP/DPP pathway. *Genes Dev., 20,* 648–653.

Chen, Y. G., Hata, A., Lo, R. S., Wotton, D., Shi, Y., et al. (1998). Determinants of specificity in TGF-beta signal transduction. *Genes Dev., 12,* 2144–2152.

Childs, S. R., Wrana, J. L., Arora, K., Attisano, L., O'Connor, M. B., & Massagué, J. (1993). Identification of a *Drosophila* activin receptor. *Proc. Natl. Acad. Sci. USA, 90,* 9475–9479.

Choi, K. W., & Benzer, S. (1994). Rotation of photoreceptor clusters in the developing *Drosophila* eye requires the *nemo* gene. *Cell, 78,* 125–136.

Cohen, B., Simcox, A. A., & Cohen, S. M. (1993). Allocation of the thoracic imaginal primordia in the *Drosophila* embryo. *Development, 117,* 597–608.

Conley, C. A., Silburn, R., Singer, M. A., Ralston, A., Rohwer-Nutter, D., et al. (2000). Crossveinless 2 contains cysteine-rich domains and is required for high levels of BMP-like activity during the formation of the cross veins in *Drosophila. Development, 127,* 3947–3959.

Daly, A. C., Randall, R. A., & Hill, C. S. (2008). Transforming growth factor beta-induced Smad1/5 phosphorylation in epithelial cells is mediated by novel receptor complexes and is essential for anchorage-independent growth. *Mol. Cell Biol., 28,* 6889–6902.

Das, P., Maduzia, L. L., Wang, H., Finelli, A. L., Cho, S. H., et al. (1998). The *Drosophila* gene Medea demonstrates the requirement for different classes of Smads in dpp signaling. *Development, 125,* 1519–1528.

Das, P., Inoue, H., Baker, J. C., Beppu, H., Kawabata, M., et al. (1999). *Drosophila* dSmad2 and Atr-I transmit activin/TGFbeta signals. *Genes Cells, 4,* 123–134.

Diaz-Benjumea, F. J., & Garcia-Bellido, A. (1990). Genetic analysis of the wing vein pattern of *Drosophila. Rouxs Arch. Dev. Biol., 198,* 336–354.

Duan, X., Liang, Y. Y., Feng, X. H., & Lin, X. (2006). Protein serine/threonine phosphatase PPM1A dephosphorylates Smad1 in the bone morphogenetic protein signaling pathway. *J. Biol. Chem., 281,* 36526–36532.

Dudu, V., Bittig, T., Entchev, E., Kicheva, A., Julicher, F., & Gonzalez-Gaitan, M. (2006). Postsynaptic mad signaling at the *Drosophila* neuromuscular junction. *Curr. Biol., 16,* 625–635.

Eade, K. T., & Allan, D. W. (2009). Neuronal phenotype in the mature nervous system is maintained by persistent retrograde bone morphogenetic protein signaling. *J. Neurosci., 29,* 3852–3864.

Eaton, B. A., & Davis, G. W. (2005). LIM Kinase1 controls synaptic stability downstream of the type II BMP receptor. *Neuron, 47,* 695–708.

Eberl, D. F., & Hilliker, A. J. (1988). Characterization of X-linked recessive lethal mutations affecting embryonic morphogenesis in *Drosophila melanogaster. Genetics, 118,* 109–120.

Ebisawa, T., Fukuchi, M., Murakami, G., Chiba, T., Tanaka, K., et al. (2001). Smurf1 interacts with transforming growth factor-beta type I receptor through Smad7 and induces receptor degradation. *J. Biol. Chem., 276,* 12477–12480.

Eivers, E., Fuentealba, L. C., Sander, V., Clemens, J. C., Hartnett, L., & De Robertis, E. M. (2009). Mad is required for wingless signaling in wing development and segment patterning in *Drosophila. PLoS ONE, 4,* e6543.

Ellis, J. E., Parker, L., Cho, J., & Arora, K. (2010). Activin signaling functions upstream of Gbb to regulate synaptic growth at the *Drosophila* neuromuscular junction. *Dev. Biol., 342,* 121–133.

Feng, X. H., & Derynck, R. (1997). A kinase subdomain of transforming growth factor-beta (TGF-beta) type I receptor determines the TGFbeta intracellular signaling specificity. *EMBO J., 16,* 3912–3923.

Ferguson, E. L., & Anderson, K. V. (1992a). Localized enhancement and repression of the activity of the TGF-β family member, decapentaplegic, is necessary for dorsal–ventral pattern formation in the *Drosophila* embryo. *Development, 114,* 583–597.

Ferguson, E. L., & Anderson, K. V. (1992b). Decapentaplegic acts as a morphogen to organize dorsal–ventral pattern in the *Drosophila* embryo. *Cell, 71,* 451–461.

Finelli, A. L., Xie, T., Bossie, C. A., Blackman, R. K., & Padgett, R. W. (1995). The *tolkin* gene is a tolloid BMP-1 homologue that is essential for *Drosophila* development. *Genetics, 141,* 271–281.

Foletta, V. C., Lim, M. A., Soosairajah, J., Kelly, A. P., Stanley, E. G., et al. (2003). Direct signaling by the BMP type II receptor via the cytoskeletal regulator LIMK1. *J. Cell Biol., 162,* 1089–1098.

Francois, V., Solloway, M., O'Neill, J. W., Emery, J., & Bier, E. (1994). Dorsal–ventral patterning of the *Drosophila* embryo depends on a putative negative growth factor encoded by the *short gastrulation* gene. *Genes Dev., 8,* 2602–2616.

Fuller, M. T., & Spradling, A. C. (2007). Male and female *Drosophila* germline stem cells: Two versions of immortality. *Science, 316,* 402–404.

Gesualdi, S. C., & Haerry, T. E. (2007). Distinct signaling of *Drosophila* Activin/TGF-beta family members. *Fly, 1*(4), 212–221.

Gilboa, L., & Lehmann, R. (2004). How different is Venus from Mars? The genetics of germ-line stem cells in *Drosophila* females and males. *Development, 131,* 4895–4905.

Goumans, M. J., Valdimarsdottir, G., Itoh, S., Rosendahl, A., Sideras, P., & ten Dijke, P. (2002). Balancing the activation state of the endothelium via two distinct TGF-beta type I receptors. *EMBO J., 21,* 1743–1753.

Goumans, M. J., Valdimarsdottir, G., Itoh, S., Lebrin, F., Larsson, J., et al. (2003). Activin receptor-like kinase (ALK)1 is an antagonistic mediator of lateral TGFbeta/ALK5 signaling. *Mol. Cell, 12,* 817–828.

Grieder, N. C., Nellen, D., Burke, R., Basler, K., & Affolter, M. (1995). Schnurri is required for *Drosophila* dpp signaling and encodes a zinc finger protein similar to the mammalian transcription factor PRDII-BF1. *Cell, 81,* 791–800.

Haerry, T. E. (2010). The interaction between two TGF-beta type I receptors plays important roles in ligand binding, SMAD activation, gradient formation. *Mech. Dev., 127,* 358–370.

Haerry, T. E., Khalsa, O., O'Connor, M. B., & Wharton, K. A. (1998). Synergistic signaling by two BMP ligands through the SAX and TKV receptors controls wing growth and patterning in *Drosophila. Development, 125,* 3977–3987.

Hata, A., Lo, R. S., Wotton, D., Lagna, G., & Massagué, J. (1997). Mutations increasing autoinhibition inactivate tumour suppressors Smad2 and Smad4. *Nature, 388*, 82–87.

Hayashi, H., Abdollah, S., Qiu, Y., Cai, J., Xu, Y. Y., et al. (1997). The MAD-related protein Smad7 associates with the TGFβ receptor and functions as an antagonist of TGFβ signaling. *Cell, 89*, 1165–1173.

Heberlein, U., Wolff, T., & Rubin, G. M. (1993). The TGF homolog dpp and the segment polarity gene hedgehog are required for propagation of a morphogenetic wave in the *Drosophila* retina. *Cell, 75*, 913–926.

Henderson, K. D., & Andrew, D. J. (1998). Identification of a novel *Drosophila* SMAD on the X chromosome. *Biochem. Biophys. Res. Commun., 252*, 195–201.

Hudson, J. B., Podos, S. D., Keith, K., Simpson, S. L., & Ferguson, E. L. (1998). The *Drosophila* Medea gene is required downstream of *dpp* and encodes a functional homolog of human Smad4. *Development, 125*, 1407–1420.

Hursh, D. A., Padgett, R. W., & Gelbart, W. M. (1993). Cross-regulation of *decapentaplegic* and *Ultrabithorax* transcription in the embryonic visceral mesoderm of *Drosophila*. *Development, 117*, 1211–1222.

Imamura, T., Takase, M., Nishihara, A., Oeda, E., Hanai, J., et al. (1997). Smad6 inhibits signalling by the TGF-β superfamily. *Nature, 389*, 622–626.

Immergluck, K., Lawrence, P. A., & Bienz, M. (1990). Induction across germ layers in *Drosophila* mediated by a genetic cascade. *Cell, 62*, 261–268.

Inoue, H., Imamura, T., Ishidou, Y., Takase, M., Udagawa, Y., et al. (1998). Interplay of signal mediators of decapentaplegic (Dpp): Molecular characterization of mothers against dpp, Medea, daughters against dpp. *Mol. Biol. Cell, 9*, 2145–2156.

Irish, V. F., & Gelbart, W. M. (1987). The *decapentaplegic* gene is required for dorsal–ventral patterning of the *Drosophila* embryo. *Genes Dev., 1*, 868–879.

Itoh, S., Thorikay, M., Kowanetz, M., Moustakas, A., Itoh, F., et al. (2003). Elucidation of Smad requirement in transforming growth factor-beta type I receptor-induced responses. *J. Biol. Chem., 278*, 3751–3761.

Jazwinska, A., Rushlow, C., & Roth, S. (1999a). The role of *brinker* in mediating the graded response to Dpp in early *Drosophila* embryos. *Development, 126*, 3323–3334.

Jazwinska, A., Kirov, N., Wieschaus, E., & Roth, S., Rushlow, C. (1999b). The *Drosophila* gene *brinker* reveals a novel mechanism of Dpp target gene regulation. *Cell, 96*(4), 563–573.

Jensen, P. A., Zheng, X., Lee, T., & O'Connor, M. B. (2009). The *Drosophila* Activin-like ligand Dawdle signals preferentially through one isoform of the Type-I receptor Baboon. *Mech. Dev., 126*, 950–957.

Jiang, X., Xia, L., Chen, D., Yang, Y., Huang, H., et al. (2008). Otefin, a nuclear membrane protein, determines the fate of germline stem cells in *Drosophila* via interaction with Smad complexes. *Dev. Cell, 14*, 494–506.

Kamiya, Y., Miyazono, K., & Miyazawa, K. (2008). Specificity of the inhibitory effects of Dad on TGF-beta family type I receptors, Thickveins, Saxophone, Baboon in *Drosophila*. *FEBS Lett., 582*, 2496–2500.

Kawabata, M., Inoue, H., Hanyu, A., Imamura, T., & Miyazono, K. (1998). Smad proteins exist as monomers *in vivo* and undergo homo- and hetero-oligomerization upon activation by serine/threonine kinase receptors. *EMBO J., 17*, 4056–4065.

Kawase, E., Wong, M. D., Ding, B. C., & Xie, T. (2004). Gbb/Bmp signaling is essential for maintaining germline stem cells and for repressing *bam* transcription in the *Drosophila* testis. *Development, 131*, 1365–1375.

Khalsa, O., Yoon, J., Torres-Schumann, S., & Wharton, K. A. (1998). TGF-β/BMP superfamily members, gbb-60A and dpp, cooperate to provide pattern information and establish cell identity in the *Drosophila* wing. *Development, 125*, 2723–2734.

Knockaert, M., Sapkota, G., Alarcón, C., Massagué, J., & Brivanlou, A. H. (2006). Unique players in the BMP pathway: Small C-terminal domain phosphatases dephosphorylate Smad1 to attenuate BMP signaling. *Proc. Natl. Acad. Sci. USA, 103*, 11940–11945.

Kutty, G., Kutty, R. K., Samuel, W., Duncan, T., Jaworski, C., & Wiggert, B. (1998). Identification of a new member of transforming growth factor-beta superfamily in *Drosophila*: the first invertebrate activin gene. *Biochem. Biophys. Res. Commun., 246*, 644–649.

Lee-Hoeflich, S. T., Causing, C. G., Podkowa, M., Zhao, X., Wrana, J. L., & Attisano, L. (2004). Activation of LIMK1 by binding to the BMP receptor, BMPRII, regulates BMP-dependent dendritogenesis. *EMBO J., 23*, 4792–4801.

Lee-Hoeflich, S. T., Zhao, X., Mehra, A., & Attisano, L. (2005). The *Drosophila* type II receptor, Wishful thinking, binds BMP and myoglianin to activate multiple TGFbeta family signaling pathways. *FEBS Lett., 579*, 4615–4621.

Letsou, A., Arora, K., Wrana, J. L., Simin, K., Twombly, V., et al. (1995). *Drosophila* Dpp signaling is mediated by the *punt* gene product: A dual ligand-binding type II receptor of the TGF beta receptor family. *Cell, 80*, 899–908.

Li, C. Y., Guo, Z., & Wang, Z. (2007). TGFbeta receptor saxophone non-autonomously regulates germline proliferation in a Smox/dSmad2-dependent manner in *Drosophila* testis. *Dev. Biol., 309*, 70–77.

Li, J., & Li, W. X. (2006). A novel function of *Drosophila* eIF4A as a negative regulator of Dpp/BMP signalling that mediates SMAD degradation. *Nat. Cell Biol., 8*, 1407–1414.

Li, J., Li, W. X., & Gelbart, W. M. (2005). A genetic screen for maternal-effect suppressors of *decapentaplegic* identifies the eukaryotic translation initiation factor 4A in *Drosophila*. *Genetics, 171*, 1629–1641.

Liang, Y. Y., Lin, X., Liang, M., Brunicardi, F. C., ten Dijke, P., et al. (2003). dSmurf selectively degrades *decapentaplegic*-activated MAD, its overexpression disrupts imaginal disc development. *J. Biol. Chem., 278*, 26307–26310.

Lin, F., Morrison, J. M., Wu, W., & Worman, H. J. (2005). MAN1, an integral protein of the inner nuclear membrane, binds Smad2 and Smad3 and antagonizes transforming growth factor-beta signaling. *Hum. Mol. Genet., 14*, 437–445.

Lin, X., Liang, M., & Feng, X. H. (2000). Smurf2 is a ubiquitin E3 ligase mediating proteasome-dependent degradation of Smad2 in transforming growth factor-beta signaling. *J. Biol. Chem., 275*, 36818–36822.

Liu, I. M., Schilling, S. H., Knouse, K. A., Choy, L., Derynck, R., & Wang, X. F. (2009). TGFbeta-stimulated Smad1/5 phosphorylation requires the ALK5 L45 loop and mediates the pro-migratory TGFbeta switch. *EMBO J.*, *28*, 88–98.

Lo, P. C., & Frasch, M. (1999). Sequence and expression of myoglianin, a novel *Drosophila* gene of the TGF-beta superfamily. *Mech. Dev.*, *86*, 171–175.

López-Onieva, L., Fernández-Miñán, A., & González-Reyes, A. (2008). Jak/Stat signalling in niche support cells regulates *dpp* transcription to control germline stem cell maintenance in the *Drosophila* ovary. *Development*, *135*, 533–540.

Lorenzen, J. A., Baker, S. E., Denhez, F., Melnick, M. B., Brower, D. L., & Perkins, L. A. (2001). Nuclear import of activated D-ERK by DIM-7, an importin family member encoded by the gene *moleskin*. *Development*, *128*, 1403–1414.

Manak, J. R., Mathies, L. D., & Scott, M. P. (1994). Regulation of a *decapentaplegic* midgut enhancer by homeotic proteins. *Development*, *120*, 3605–3619.

Marqués, G. (2005). Morphogens and synaptogenesis in *Drosophila*. *J. Neurobiol.*, *64*, 417–434.

Marqués, G., Musacchio, M., Shimell, M. J., Wunnenberg-Stapleton, K., Cho, K. W., & O'Connor, M. B. (1997). Production of a DPP activity gradient in the early *Drosophila* embryo through the opposing actions of the SOG and TLD proteins. *Cell*, *91*, 417–426.

Marqués, G., Bao, H., Haerry, T. E., Shimell, M. J., Duchek, P., et al. (2002). The *Drosophila* BMP type II receptor Wishful Thinking regulates neuromuscular synapse morphology and function. *Neuron*, *33*, 529–543.

Marqués, G., Haerry, T. E., Crotty, M. L., Xue, M., et al. (2003). Retrograde Gbb signaling through the Bmp type 2 receptor wishful thinking regulates systemic FMRFa expression in *Drosophila*. *Development*, *130*, 5457–5470.

Martín-Castellanos, C., & Edgar, B. A. (2002). A characterization of the effects of Dpp signaling on cell growth and proliferation in the *Drosophila* wing. *Development*, *129*, 1003–1013.

Marty, T., Muller, B., Basler, K., & Affolter, M. (2000). Schnurri mediates Dpp-dependent repression of *brinker* transcription. *Nat. Cell Biol.*, *2*, 745–749.

Mason, E. D., Konrad, K. D., Webb, C. D., & Marsh, J. L. (1994). Dorsal midline fate in *Drosophila* embryos requires *twisted gastrulation*, a gene encoding a secreted protein related to human connective tissue growth factor. *Genes Dev.*, *8*, 1489–1501.

Mason, E. D., Williams, S., Grotendorst, G. R., & Marsh, J. L. (1997). Combinatorial signaling by twisted gastrulation and decapentaplegic. *Mech. Dev.*, *64*, 61–75.

Massagué, J., Seoane, J., & Wotton, D. (2005). Smad transcription factors. *Genes Dev.*, *19*, 2783–2810.

Masucci, J. D., & Hoffmann, F. M. (1993). Identification of two regions from the *Drosophila* decapentaplegic gene required for embryonic midgut development and larval viability. *Dev. Biol.*, *159*, 276–287.

Matunis, E., Tran, J., Gonczy, P., Caldwell, K., & DiNardo, S. (1997). *punt* and *schnurri* regulate a somatically derived signal that restricts proliferation of committed progenitors in the germline. *Development*, *124*, 4383–4391.

Mayer-Jaekel, R. E., Ohkura, H., Gomes, R., Sunkel, C. E., Baumgartner, S., Hemmings, B. A., & Glover, D. M. (1993). The 55 kd regulatory subunit of *Drosophila* protein phosphatase 2A is required for anaphase. *Cell*, *72*, 621–633.

McCabe, B. D., Marqués, G., Haghighi, A. P., Fetter, R. D., Crotty, M. L., et al. (2003). The BMP homolog Gbb provides a retrograde signal that regulates synaptic growth at the *Drosophila* neuromuscular junction. *Neuron*, *39*, 241–254.

McCabe, B. D., Hom, S., Aberle, H., Fetter, R. D., Marqués, G., et al. (2004). Highwire regulates presynaptic BMP signaling essential for synaptic growth. *Neuron*, *41*, 891–905.

McKearin, D., & Spradling, A. C. (1990). *bag-of-marbles*: A *Drosophila* gene required to initiate both male and female gametogenesis. *Genes Dev.*, *4*, 2242–2251.

McKearin, D., & Ohlstein, B. (1995). A role for the *Drosophila* bag-of-marbles protein in the differentiation of cystoblasts from germline stem cells. *Development*, *121*, 2937–2947.

Merino, C., Penney, J., González, M., Tsurudome, K., Moujahidine, M., et al. (2009). Nemo kinase interacts with Mad to coordinate synaptic growth at the *Drosophila* neuromuscular junction. *J. Cell Biol.*, *185*, 713–725.

Miguel-Aliaga, I., Allan, D. W., & Thor, S. (2004). Independent roles of the dachshund and eyes absent genes in BMP signaling, axon pathfinding and neuronal specification. *Development*, *131*, 5837–5848.

Miles, W. O., Jaffray, F., Campbell, S. G., Takeda, S., Bayston, L. J., et al. (2008). Medea SUMOylation restricts the signaling range of the Dpp morphogen in the *Drosophila* embryo. *Genes Dev.*, *22*, 2578–2590.

Minami, M., Kinoshita, N., Kamoshida, Y., Tanimoto, H., & Tabata, T. (1999). *brinker* is a target of Dpp in *Drosophila* that negatively regulates Dpp-dependent genes. *Nature*, *398*, 242–246.

Muller, B., Hartmann, B., Pyrowolakis, G., Affolter, M., & Basler, K. (2003). Conversion of an extracellular Dpp/BMP morphogen gradient into an inverse transcriptional gradient. *Cell*, *113*, 221–233.

Nakano, Y., Fujitani, K., Kurihara, J., Ragan, J., Usui-Aoki, K., et al. (2001). Mutations in the novel membrane protein spinster interfere with programmed cell death and cause neural degeneration in *Drosophila melanogaster*. *Mol. Cell Biol.*, *21*, 3775–3788.

Nakao, A., Afrakhte, M., Morén, A., Nakayama, T., Christian, J. L., et al. (1997). Identification of Smad7, a TGFβ-inducible antagonist of TGF-β signaling. *Nature*, *389*, 631–635.

Nellen, D., Affolter, M., & Basler, K. (1994). Receptor serine/threonine kinases implicated in the control of *Drosophila* body pattern by decapentaplegic. *Cell*, *78*, 225–237.

Nellen, D., Burke, R., Struhl, G., & Basler, K. (1996). Direct and long-range action of a DPP morphogen gradient. *Cell*, *85*, 357–368.

Newfeld, S. J., Chartoff, E. H., Graff, J. M., Melton, D. A., & Gelbart, W. M. (1996). *Mothers against dpp* encodes a conserved cytoplasmic protein required in DPP/TGF-beta responsive cells. *Development*, *122*, 2099–2108.

Newfeld, S. J., Mehra, A., Singer, M. A., Wrana, J. L., Attisano, L., & Gelbart, W. M. (1997). Mothers against dpp participates in a DDP/TGF-beta responsive serine-threonine kinase signal transduction cascade. *Development*, *124*, 3167–3176.

Ng, J. (2008). TGF-β signals regulate axonal development through distinct Smad-independent mechanisms. *Development*, *135*, 4025–4035.

Ng, J., & Luo, L. (2004). Rho GTPases regulate axon growth through convergent and divergent signaling pathways. *Neuron*, *44*, 779–793.

Nguyen, M., Parker, L., & Arora, K. (2000). Identification of maverick, a novel member of the TGF-beta superfamily in *Drosophila*. *Mech. Dev.*, *95*, 201–206.

Nguyen, T., Jamal, J., Shimell, M. J., Arora, K., & O'Connor, M. B. (1994). Characterization of tolloid-related-1: A BMP-1-like product that is required during larval and pupal stages of *Drosophila* development. *Dev. Biol.*, *166*, 569–586.

Nunes da Fonseca, R., van der Zee, M., & Roth, S. (2010). Evolution of extracellular Dpp modulators in insects: The roles of tolloid and twisted-gastrulation in dorsoventral patterning of the *Tribolium* embryo. *Dev. Biol.*, *345*, 80–93.

O'Connor, M. B., Umulis, D., Othmer, H. G., & Blair, S. S. (2006). Shaping BMP morphogen gradients in the *Drosophila* embryo and pupal wing. *Development*, *133*, 183–193.

Oda, H., & Akiyama-Oda, Y. (2008). Differing strategies for forming the arthropod body plan: Lessons from Dpp, Sog and Delta in the fly *Drosophila* and spider *Achaearanea*. *Dev. Growth Differ.*, *50*, 203–214.

Ohashi, K., Hosoya, T., Takahashi, K., Hing, H., & Mizuno, K. (2000). A *Drosophila* homolog of LIM-kinase phosphorylates cofilin and induces actin cytoskeletal reorganization. *Biochem. Biophys. Res. Commun*, *276*, 1178–1185.

Okano, H., Yoshikawa, S., Suzuki, A., Ueno, N., Kaizu, M., et al. (1994). Cloning of a *Drosophila melanogaster* homologue of the mouse type-I bone morphogenetic proteins-2/-4 receptor: A potential decapentaplegic receptor. *Gene.*, *148*, 203–209.

Padgett, R. W., St Johnston, R. D., & Gelbart, W. M. (1987). A transcript from a *Drosophila* pattern gene predicts a protein homologous to the transforming growth factor-β family. *Nature*, *325*, 81–84.

Panganiban, G., Reuter, R., Scott, M. P., & Hoffmann, F. M. (1990). A *Drosophila* growth factor homolog, decapentaplegic, regulates homeotic gene expression within and across germ layers during midgut morphogenesis. *Development*, *110*, 1041–1050.

Pankratz, M. J., & Hoch, M. (1995). Control of epithelial morphogenesis by cell signaling and integrin molecules in the *Drosophila* foregut. *Development*, *121*, 1885–1898.

Panzer, S., Weigel, D., & Beckendorf, S. K. (1992). Organogenesis in *Drosophila melanogaster*: Embryonic salivary gland determination is controlled by homeotic and dorsoventral patterning genes. *Development*, *114*, 49–57.

Parker, L., Stathakis, D. G., & Arora, K. (2004). Regulation of BMP and activin signaling in *Drosophila*. *Prog. Mol. Subcell. Biol.*, *34*, 73–101.

Parker, L., Ellis, J. E., Nguyen, M. Q., & Arora, K. (2006). The divergent TGF-beta ligand Dawdle utilizes an activin pathway to influence axon guidance in *Drosophila*. *Development*, *133*, 4981–4991.

Penton, A., Chen, Y., Staehling-Hampton, K., Wrana, J. L., Attisano, L., et al. (1994). Identification of two bone morphogenetic protein type I receptors in *Drosophila* and evidence that Brk25D is a decapentaplegic receptor. *Cell*, *78*, 239–250.

Perrimon, N., Engstrom, L., & Mahowald, A. P. (1989). Zygotic lethals with specific maternal effect phenotypes in *Drosophila melanogaster*. I. Loci on the X chromosome. *Genetics*, *121*, 333–352.

Podos, S. D., Hanson, K. K., Wang, Y. C., & Ferguson, E. L. (2001). The DSmurf ubiquitin-protein ligase restricts BMP signaling spatially and temporally during *Drosophila* embryogenesis. *Dev. Cell*, *1*, 567–578.

Posakony, L. G., Raftery, L. A., & Gelbart, W. M. (1990). Wing formation in *Drosophila melanogaster* requires *decapentaplegic* gene function along the anterior–posterior compartment boundary. *Mech. Dev.*, *33*, 69–82.

Pyrowolakis, G., Hartmann, B., Mueller, B., Basler, K., & Affolter, M. (2004). A simple molecular complex mediates widespread BMP-induced repression during *Drosophila* development. *Dev. Cell*, *7*, 229–240.

Pyrowolakis, G., Hartmann, B., & Affolter, M. (2008). 17 TGF-β family signaling in *Drosophila*. In R. Derynk, & M. Miyazono (Eds.). *The TGF-β Family. Cold Spring Harbor Monograph Archive* (50, pp. 493–526).

Raftery, L. A., Twombly, V., Wharton, K. A., & Gelbart, W. M. (1995). Genetic screens to identify elements of the decapentaplegic signaling pathway in *Drosophila*. *Genetics*, *139*, 241–254.

Raftery, L. A., Korochkina, S., & Cao, J. (2006). Smads in *Drosophila* – interpretation of graded signals *in vivo*. In P. ten Dijke, & C. -H. Heldin (Eds.), *"Smad Signal Transduction" Proteins and Cell Regulation* (5, pp. 55–73).

Raju, G. P., Dimova, N., Klein, P. S., & Huang, H. C. (2003). SANE, a novel LEM domain protein, regulates bone morphogenetic protein signaling through interaction with Smad1. *J Biol Chem.*, *278*, 428–437.

Ralston, A., & Blair, S. S. (2005). Long-range Dpp signaling is regulated to restrict BMP signaling to a crossvein competent zone. *Dev. Biol.*, *280*, 187–200.

Ramel, M. C., Emery, C. S., Foulger, R., Goberdhan, D. C.I., van den Heuvel, M., & Wilson, C. (2007). *Drosophila* SnoN modulates growth and patterning by antagonizing TGF-beta signalling. *Mech. Dev.*, *124*, 304–317.

Rawson, J. M., Lee, M., Kennedy, E. L., & Selleck, S. B. (2003). *Drosophila* neuromuscular synapse assembly and function require the TGF-beta type I receptor saxophone and the transcription factor Mad. *J. Neurobiol.*, *55*, 134–150.

Ray, R. P., & Wharton, K. A. (2001). Context-dependent relationships between the BMPs gbb and dpp during development of the *Drosophila* wing imaginal disc. *Development*, *128*, 3913–3925.

Ross, J. J., Shimmi, O., Vilmos, P., Petryk, A., Kim, H., et al. (2001). Twisted gastrulation is a conserved extracellular BMP antagonist. *Nature*, *410*, 479–483.

Ruberte, E., Marty, T., Nellen, D., Affolter, M., & Basler, K. (1995). An absolute requirement for both the Type II and Type I receptors, punt and thickveins, for Dpp signaling *in vivo*. *Cell*, *80*, 889–897.

Sander, V., Eivers, E., Choi, R. H., & De Robertis, E. M. (2010). *Drosophila* Smad2 opposes *mad* signaling during wing vein development. *PLoS ONE 5*, e10383.

Sapkota, G., Knockaert, M., Alarcón, C., Montalvo, E., Brivanlou, A. H., & Massagué, J. (2006). Dephosphorylation of the linker regions of Smad1 and Smad2/3 by small C-terminal domain phosphatases has distinct outcomes for bone morphogenetic protein and transforming factor-beta pathways. *J. Biol. Chem., 281*, 40412–40419.

Sapkota, G., Alarcón, C., Spagnoli, F. M., Brivanlou, A. H., & Massagué, J. (2007). Balancing BMP signaling through integrated inputs into the Smad1 linker. *Mol. Cell, 25*, 441–454.

Schulz, C., Kiger, A. A., Tazuke, S. I., Yamashita, Y. M., Pantalena-Filho, L. C., et al. (2004). A misexpression screen reveals effects of *bag-of-marbles* and TGF beta class signaling on the *Drosophila* male germ-line stem cell lineage. *Genetics, 167*, 707–723.

Sekelsky, J. J., Newfeld, S. J., Raftery, L. A., Chartoff, E. H., & Gelbart, W. G. (1995). Genetic characterization and cloning of *Mothers against dpp*, a gene required for *decapentaplegic* function in *Drosophila melanogaster*. *Genetics, 139*, 1347–1358.

Serpe, M., & O'Connor, M. B. (2006). The metalloprotease tolloid-related and its TGF-beta-like substrate Dawdle regulate *Drosophila* motoneuron axon guidance. *Development, 133*, 4969–4679.

Serpe, M., Umulis, D., Ralston, A., Chen, J., Olson, D. J., et al. (2008). The BMP-binding protein Crossveinless 2 is a short-range, concentration-dependent, biphasic modulator of BMP signaling in *Drosophila*. *Dev. Cell, 14*, 940–953.

Shi, Y., & Massague, J. (2003). Mechanisms of TGF-beta signaling from cell membrane to the nucleus. *Cell, 113*, 685–700.

Shi, Y., Hata, A., Lo, R. S., Massagué, J., & Pavletich, N. P. (1997). A structural basis for mutational inactivation of the tumour suppressor Smad4. *Nature, 388*, 87–93.

Shimell, M. J., Ferguson, E. L., Childs, S. R., & O'Connor, M. B. (1991). The *Drosophila* dorsal–ventral patterning gene *tolloid* is related to human bone morphogenetic protein 1. *Cell, 67*, 469–481.

Shimmi, O., & O'Connor, M. B. (2003). Physical properties of Tld, Sog, Tsg and Dpp protein interactions are predicted to help create a sharp boundary in Bmp signals during dorso–ventral patterning of the *Drosophila* embryo. *Development, 130*, 4673–4682.

Shimmi, O., Umulis, D., Othmer, H., & O'Connor, M. B. (2005a). Facilitated transport of a dpp/scw heterodimer by sog/tsg leads to robust patterning of the *Drosophila* blastoderm embryo. *Cell, 120*, 873–886.

Shimmi, O., Ralston, A., Blair, S. S., & O'Connor, M. B. (2005b). The *crossveinless* gene encodes a new member of the Twisted gastrulation family of BMP-binding proteins which, with Short gastrulation, promotes BMP signaling in the crossveins of the *Drosophila* wing. *Dev. Biol., 282*, 70–83.

Shivdasani, A. A., & Ingham, P. W. (2003). Regulation of stem cell maintenance and transit amplifying cell proliferation by tgf-β signaling in *Drosophila* spermatogenesis. *Curr. Biol., 13*, 2065–2072.

Shravage, B. V., Altmann, G., Technau, M., & Roth, S. (2007). The role of Dpp and its inhibitors during eggshell patterning in *Drosophila*. *Development, 134*, 2261–2271.

Singh, S. R., Chen, X., & Hou, S. X. (2005). JAK/STAT signaling regulates tissue outgrowth and male germline stem cell fate in *Drosophila*. *Cell Res., 15*, 1–5.

Song, X., Wong, M. D., Kawase, E., Xi, R., Ding, B. C., McCarthy, J. J., & Xie, T. (2004). Bmp signals from niche cells directly repress transcription of a differentiation-promoting gene, *bag of marbles*, in germline stem cells in the *Drosophila* ovary. *Development, 131*, 1353–1364.

Spencer, F. A., Hoffmann, F. M., & Gelbart, W. M. (1982). Decapentaplegic: A gene complex affecting morphogenesis in *Drosophila melanogaster*. *Cell, 28*, 451–461.

Staehling-Hampton, K., & Hoffmann, F. M. (1994). Ectopic decapentaplegic in the *Drosophila* midgut alters the expression of five homeotic genes, *dpp, wingless*, causing specific morphological defects. *Dev. Biol., 164*, 502–512.

Staehling-Hampton, K., Hoffmann, F. M., Baylies, M. K., Rushton, E., & Bate, M. (1994a). *dpp* induces mesodermal gene expression in *Drosophila*. *Nature, 372*, 783–786.

Staehling-Hampton, K., Jackson, P. D., Clark, M. J., Brand, A. H., & Hoffmann, F. M. (1994b). Specificity of bone morphogenetic protein-related factors: Cell fate and gene expression changes in *Drosophila* embryos induced by decapentaplegic but not 60A. *Cell Growth Diffn., 5*, 585–593.

Staehling-Hampton, K., Laughon, A. S., & Hoffmann, F. M. (1995). A *Drosophila* protein related to the human zinc finger transcription factor PRDII/MBPI/HIV-EP1 is required for dpp signaling. *Development, 121*, 3393–3403.

Stauber, M., Prell, A., & Schmidt-Ott, U. (2002). A single Hox3 gene with composite *bicoid* and *zerknullt* expression characteristics in non-Cyclorrhaphan flies. *Proc. Natl. Acad. Sci. USA, 99*, 274–279.

St Johnston, R. D., & Gelbart, W. M. (1987). Decapentaplegic transcripts are localized along the dorsal–ventral axis of the *Drosophila* embryo. *EMBO J., 6*, 2785–2791.

Sweeney, S. T., & Davis, G. W. (2002). Unrestricted synaptic growth in spinster – a late endosomal protein implicated in TGF-beta-mediated synaptic growth regulation. *Neuron, 36*, 403–416.

Takaesu, N. T., Hyman-Walsh, C., Ye, Y., Wisotzkey, R. G., Stinchfield, M. J., et al. (2006). dSno facilitates *baboon* signaling in the *Drosophila* brain by switching the affinity of Medea away from Mad and toward dSmad2. *Genetics, 174*, 1299–1313.

Terracol, R., & Lengyel, J. A. (1994). The *thickveins* gene of *Drosophila* is required for dorsoventral polarity of the embryo. *Genetics, 138*, 165–178.

Thuringer, F., & Bienz, M. (1993). Indirect autoregulation of a homeotic *Drosophila* gene mediated by extracellular signaling. *Proc. Natl. Acad. Sci. USA, 90*, 3899–3903.

Thuringer, F., Cohen, S. M., & Bienz, M. (1993). Dissection of an indirect autoregulatory response of a homeotic *Drosophila* gene. *EMBO J., 12*, 2419–2430.

Ting, C. Y., Herman, T., Yonekura, S., Gao, S., Wang, J., et al. (2007). Tiling of r7 axons in the *Drosophila* visual system is mediated both by transduction of an activin signal to the nucleus and by mutual repulsion. *Neuron, 56*, 793–806.

Torres-Vazquez, J., Warrior, R., & Arora, K. (2000). *schnurri* is required for dpp-dependent patterning of the *Drosophila* wing. *Dev. Biol., 227*, 388–402.

Torres-Vazquez, J., Park, S., Warrior, R., & Arora, K. (2001). The transcription factor Schnurri plays a dual role in mediating Dpp signaling during embryogenesis. *Development, 128*, 1657–1670.

Tremml, G., & Bienz, M. (1992). Induction of labial expression in the *Drosophila* endoderm: Response elements for dpp signalling and for autoregulation. *Development, 116*, 447–456.

Tsukazaki, T., Chiang, T. A., Davison, A. F., Attisano, L., & Wrana, J. L. (1998). SARA, a FYVE domain protein that recruits Smad2 to the TGFbeta receptor. *Cell, 95*, 779–791.

Tsuneizumi, K., Nakayama, T., Kamoshida, Y., Kornberg, T. B., Christian, J. L., & Tabata, T. (1997). *Daughters against dpp* modulates *dpp* organizing activity in *Drosophila* wing development. *Nature, 389*, 627–631.

Twombly, V., Blackman, R. K., Jin, H., Graff, J. M., Padgett, R. W., & Gelbart, W. M. (1996). The TGF-beta signaling pathway is essential for *Drosophila* oogenesis. *Development, 122*, 1555–1565.

Uemura, T., Shiomi, K., Togashi, S., & Takeichi, M. (1993). Mutation of *twins* encoding a regulator of protein Phosphatase 2A leads to pattern duplication in *Drosophila* imaginal discs. *Genes Dev., 7*, 429–440.

Umulis, D., O'Connor, M. B., & Blair, S. S. (2009). The extracellular regulation of bone morphogenetic protein signaling. *Development, 136*, 3715–3728.

van der Zee, M., Stockhammer, O., von Levetzow, C., Nunes da Fonseca, R., & Roth, S. (2006). Sog/Chordin is required for ventral-to-dorsal Dpp/BMP transport and head formation in a short germ insect. *Proc. Natl. Acad. Sci. USA, 103*, 16307–16312.

van der Zee, M., Nunes da Fonseca, R., & Roth, S. (2008). TGFbeta signaling in *Tribolium*: Vertebrate-like components in a beetle. *Dev. Genes Evol., 218*, 203–213.

Vilmos, P., Sousa-Neves, R., Lukacsovich, T., & Marsh, J. L. (2005). *crossveinless* defines a new family of Twisted-gastrulation-like modulators of bone morphogenetic protein signalling. *EMBO Rep., 6*, 262–267.

Vincent, S., Ruberte, E., Grieder, N. C., Chen, C. K., Haerry, T., et al. (1997). DPP controls tracheal cell migration along the dorsoventral body axis of the *Drosophila* embryo. *Development, 124*, 2741–2750.

Wagner, N., Weyhersmüller, A., Blauth, A., Schuhmann, T., Heckmann, M., et al. (2010). The *Drosophila* LEM-domain protein MAN1 antagonizes BMP signaling at the neuromuscular junction and the wing crossveins. *Dev. Biol., 339*, 1–13.

Wan, H. I., DiAntonio, A., Fetter, R. D., Bergstrom, K., Strauss, R., & Goodman, C. S. (2000). Highwire regulates synaptic growth in *Drosophila*. *Neuron, 26*, 313–329.

Wang, L., Li, Z., & Cai, Y. (2008). The JAK/STAT pathway positively regulates DPP signaling in the *Drosophila* germline stem cell niche. *J. Cell Biol., 180*, 721–728.

Wang, X., Shaw, R., Tsang, H. T. H., Reid, E., & O.Kane, C. J. (2007). *Drosophila* spichthyin inhibits BMP signaling and regulates synaptic growth and axonal microtubules. *Nat. Neurosci., 10*, 177–185.

Wharton, K. A., Thomsen, G. H., & Gelbart, W. M. (1991). *Drosophila 60A* gene, another transforming growth factor family member is closely related to human bone morphogenetic proteins. *Proc. Natl. Acad. Sci. USA, 88*, 9214–9218.

Wharton, K. A., Ray, R. P., & Gelbart, W. M. (1993). An activity gradient of *decapentaplegic* is necessary for the specification of dorsal pattern elements in the *Drosophila* embryo. *Development, 117*, 807–822.

Wiersdorff, V., Lecuit, T., Cohen, S. M., & Mlodzik, M. (1996). Mad acts downstream of Dpp receptors, revealing a differential requirement for dpp signaling in initiation and propagation of morphogenesis in the *Drosophila* eye. *Development, 122*, 2153–2162.

Wieschaus, E., Nusslein-Volhard, C., & Jurgens, G. (1984). Mutations affecting the pattern of the larval cuticle in *Drosophila melanogaster*. III. Zygotic loci on the X-chromosome and fourth chromosome. *Rouxs Arch. Dev. Biol., 193*, 296–307.

Wrana, J. L., Tran, H., Attisano, L., Arora, K., Childs, S. R., et al. (1994). Two distinct transmembrane serine/threonine kinases from *Drosophila melanogaster* form an activin receptor complex. *Mol. Cell Biol., 14*, 944–950.

Xie, T., & Spradling, A. C. (1998). *decapentaplegic* is essential for the maintenance and division of germline stem cells in the *Drosophila* ovary. *Cell, 94*, 251–260.

Xie, T., & Spradling, A. C. (2000). A niche maintaining germ line stem cells in the *Drosophila* ovary. *Science, 290*, 328–330.

Xie, T., Finelli, A., & Padgett, R. W. (1994). The *Drosophila saxophone* gene: A serine-threonine kinase receptor of the TGF-β superfamily. *Science, 263*, 1756–1759.

Xie, T., Kawase, E., Kirilly, D., & Wong, M. D. (2005). Intimate relationships with their neighbors: Tales of stem cells in *Drosophila* reproductive systems. *Dev. Dyn., 232*, 775–790.

Xu, L., Chen, Y. G., & Massagué, J. (2000). The nuclear import function of Smad2 is masked by SARA and unmasked by TGFbeta-dependent phosphorylation. *Nat. Cell Biol., 2*, 559–562.

Xu, L., Yao, X., Chen, X., Lu, P., Zhang, B., & Ip, Y. T. (2007). Msk is required for nuclear import of TGF-beta/BMP-activated Smads. *J. Cell Biol., 178*, 981–994.

Yamashita, Y. M., Fuller, M. T., & Jones, D. L. (2005). Signaling in stem cell niches: Lessons from the *Drosophila* germline. *J. Cell Sci., 118*, 665–672.

Yu, K., Sturtevant, M. A., Biehs, B., Francois, V., Padgett, R. W., et al. (1996). The *Drosophila decapentaplegic* and *short gastrulation* genes function antagonistically during adult wing vein development. *Development, 122*, 4033–4044.

Yu, K., Srinivasan, S., Shimmi, O., Biehs, B., Rashka, K. E., et al. (2000). Processing of the *Drosophila* Sog protein creates a novel BMP inhibitory activity. *Development, 127*, 2143–2154.

Zecca, M., Basler, K., & Struhl, G. (1995). Sequential organizing activities of *engrailed, hedgehog* and *decapentaplegic* in the *Drosophila* wing. *Development, 121*, 2265–2278.

Zeng, Y. A., Rahnama, M., Wang, S., Sosu-Sedzorme, W., & Verheyen, E. M. (2007). *Drosophila* Nemo antagonizes BMP signaling by phosphorylation of Mad and inhibition of its nuclear accumulation. *Development, 134*, 2061–2071.

Zhang, Y., Chang, C., Gehling, D. J., Hemmati-Brivanlou, A., & Derynck, R. (2001). Regulation of Smad degradation and activity by Smurf2, an E3 ubiquitin ligase. *Proc. Natl. Acad. Sci. USA, 98*, 974–979.

Zheng, X., Wang, J., Haerry, T. E., Wu, A. Y., Martin, J., et al. (2003). Baboon/Smad2-mediated TGF-β signaling activates ecdysone receptor B1 expression during neuronal remodeling in the *Drosophila* brain. *Cell, 112*, 303–315.

Zhu, C. C., Boone, J. Q., Jensen, P. A., Hanna, S., Podemski, L., et al. (2008). *Drosophila* Activin-β and the Activin-like product Dawdle function redundantly to regulate proliferation in the larval brain. *Development, 135*, 513–521.

Zhu, H., Kavsak, P., Abdollah, S., Wrana, J. L., & Thomsen, G. H. (1999). A SMAD ubiquitin ligase targets the BMP pathway and affects embryonic pattern formation. *Nature, 400*, 687–693.

Zusman, S., & Wieschaus, E. (1985). Requirements for zygotic gene activity during gastrulation in *Drosophila melanogaster*. *Dev. Biol., 111*, 359–371.

Zusman, S., Sweeton, D., & Wieschaus, E. (1988). *short gastrulation*, a mutation causing delays in stage-specific cell shape changes during gastrulation in *Drosophila melanogaster*. *Dev. Biol., 129*, 417–427.

14 Insect Immunology

Ji Won Park
Pusan National University, Busan, Korea
Bok Luel Lee
Pusan National University, Busan, Korea

14.1. Introduction

Although insects do not have the adaptive immunity that defines the antigen-specific immune responses of vertebrates, insects do have innate immunity, which is comprised of cellular and humoral immune responses. The cellular innate immune responses, including phagocytosis, nodulation, and encapsulation, are mediated by hemocytes (insect blood cells). Clotting, melanin synthesis, and antimicrobial peptide (AMP) production, known collectively as humoral innate immunity, are mediated by soluble plasma proteins or fat bodies (corresponding to mammalian liver). Recent accumulated experimental data from insects, which are infected with pathogenic bacteria, fungi, parasites or viruses that specifically elicit their innate immune responses, provide strong clues that insects are useful model systems to study the innate immunity of invertebrates. As a result of drastic new technique developments in molecular cellular biology and genetics in the fruit fly, mosquito, and other large insects over the past 20 years, a dramatic explosion of knowledge of insect immunology has enabled us to understand its basic mechanisms. In this chapter, we will overview recent progress in elucidating these molecular mechanisms, and current knowledge about the biological significance of insect immunology.

14.2. Insect Immunology Background

14.2.1. Pioneering Studies of Insect Immunology

In 1884, Elie Metchnikoff, a Russian scientist, introduced the word "phagocyte" as a name for the invertebrate circulating cells responsible for ingesting foreign bodies, and first discovered the phagocytotic immune response in insects. Another Russian scientist, Serge Metalnikov, intensively explored the mechanism of phagocytosis in insects. He began by describing the various types of insect hemocytes and their roles in phagocytosis and nodule formation. He found granular hemocytes (granulocytes) to be the major phagocytic cells. Metalnikov spent nearly three decades studying the phagocytosis of bacteria in two species of lepidopteran larvae: *Galleria mellonella* and *Pyrausta nubilarisa*. His numerous observations led him to believe that phagocytosis was not one type of response but an extremely complex

DOI:10.1016/B978-0-12-384747-8.10014-5

phenomenon with degrees of intensity. Our molecular understanding of the cellular innate reaction in insects has advanced dramatically during the past two decades. Several receptors involved in phagocytosis and its regulatory mechanisms have recently been determined using genetic and molecular biology techniques (Stuart and Ezekowitz, 2008). Several molecules related to encapsulation have also been identified in large insect and mosquito systems.

In 1918, Rudolf W. Glaser first reported the existence of immunity in the grasshopper (Glaser, 1918). Next, June M. Stevens reported the identification of inducible bactericidal activity in the hemolymph (insect blood) of the wax moth *G. mellonella* (Stevens, 1962). True humoral insect immunology research was started by the Hans Boman group at Stockholm University, in an attempt to answer the question of how holometabolous insects survive against pathogenic microbial infection if their circulating hemolymph contains no antibodies and no immune cells capable of mounting an adaptive immune response. By injecting several different Gram-positive bacteria, they demonstrated that the fruit fly *Drosophila melanogaster* also had an inducible bactericidal effect in hemolymph (Boman *et al.*, 1972). The group realized that the purification of these bactericidal molecules from *Drosophila* would be exceedingly difficult, and Boman's colleagues turned instead to the giant silkmoth *Hyalophora cecropia*, from which about 1 ml of hemolymph could be collected per pupa. After accumulating knowledge about inducible bactericidal effects, they finally reported the amino acid sequences of two novel antibacterial peptides called cecropins that are involved in insect innate immunity (Steiner *et al.*, 1981). This seminal discovery opened up a whole new field of insect immunology, leading to the isolation of AMPs from many different insects. Also, α-defensins were isolated from mammalian neutrophils only a few years after the identification and sequencing of cecropins (Selsted *et al.*, 1985), leading to the identification of a large number of AMPs found in vertebrates, and providing the essential knowledge for understanding immune defense in mammalian hosts (Zasloff, 2002). A variety of additional advances over the next 25 years resulted in the discovery of *Drosophila* Toll and IMD signaling pathways (Lemaitre *et al.*, 1996; Choe *et al.*, 2002; Gottar *et al.*, 2002), and human Toll-like receptors and their signaling pathways (Kawai and Akira, 2010).

14.2.2. Basic Concept and Development of Innate Immunity

In 1989, Charles A. Janeway first proposed the basic concept of innate immunity as the first-line host defense responses that serve to limit infection in the early hours after exposure to microorganisms (Janeway, 1989). He suggested that this system was activated by a group of germ-line-encoded receptors and soluble proteins, termed pattern-recognition receptors and proteins, respectively. These pattern-recognition molecules were suggested to recognize microbial cell wall components that are conserved among microbes but absent in the host. These conserved motifs, called pathogen-associated molecular patterns (PAMPs), include lipopolysaccharide (LPS) of Gram-negative bacteria, peptidoglycan (PGN) of Gram-positive bacteria, and β-1,3-glucan of fungi (Medzhitov and Janeway, 2002). Upon recognition, these pattern-recognition receptors activate distinct signaling cascades leading to the expression of genes that participate in innate immune response, such as inflammatory cytokines or AMPs. Janeway also suggested that pattern recognition might be critically important not only in the early phase of infection but also in the initiation of adaptive immunity in vertebrates. His hypothesis has been proved by recent innate immunity studies in invertebrates and vertebrates. In 1996, Hoffmann and his colleagues reported that *Drosophila* Toll receptor controls the antifugal immune reponse in the fruit fly, and that mutations in the Toll signaling pathway dramatically reduce survival after fungal infection (Lemaitre *et al.*, 1996). In 1997, Janeway and his colleagues first identified a human homolog of the *Drosophila* Toll protein, and proved that this Toll-like receptor induced activation of adaptive immunity signals (Medzhitov *et al.*, 1997). This report opened up new research in mammalian innate immunity. Dramatic developments in the field of vertebrate and invertebrate innate immunity have contributed invaluable knowledge toward new drug development against infectious, inflammatory, and immune diseases that seriously affect modern human society (Hennessy *et al.*, 2010).

14.3. PAMP-Recognition Proteins in Insect Immunology

14.3.1. Peptidoglycan Recognition Proteins

PGN is a polymer consisting of glycan strands of alternating *N*-acetylglucosamine (GlcNAc) and *N*-acetylmuramic acid (MurNAc) that are cross-linked to each other by short-peptide bridges (Schleifer and Kandler, 1972). PGNs from Gram-negative bacteria and some *Bacillus* species differ from other Gram-positive PGNs by the replacement of lysine (Lys) residue with *meso*-diaminopimelic acid (DAP) at the third amino acid in the stem-peptide chain (**Figure 1**). Insect PGN recognition receptors (PGRPs) were first characterized in the *Bombyx mori* moth (Yoshida *et al.*, 1996), and proposed to be a trigger of the innate immune response in the moth. At present, many PGRPs have been purified or identified from different vertebrates and invertebrates, including *Drosophila* (Kang *et al.*, 1998; Royet and Dziarski, 2007). All PGRPs have so-called PGRP domains of about 160 amino acid residues that show high similarity to those of the bacteriophage

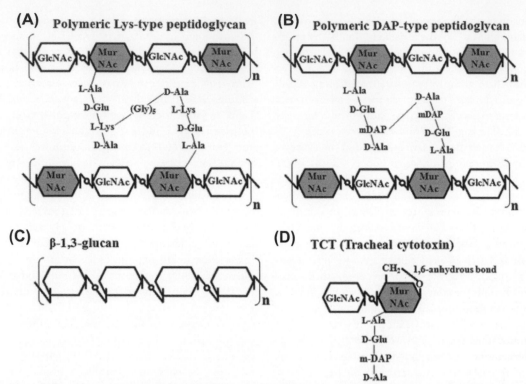

Figure 1 The primary structures of peptidoglycan, TCT and β-1,3-glucan. Peptidoglycan is a stem-peptide cross-linked polymer consisting of N-acetylglucosamine (GlcNAc) and N-acetylmuramic acid (MurNAc). (A) Structure of *S. aureus* polymeric Lys-type peptidoglycan. The third residue of the stem-peptide is L-Lys. (B) Polymeric DAP-type peptidoglycan structure of *E. coli* containing a *meso*-diaminopimelic acid (DAP) residue at the third residue of the stem-peptide. (C) A polymer structure of β-1,3-glucan of yeast and some fungi. (D) Structure of monomeric DAP-type peptidoglycan containing an internal 1,6-anhydro bond at the MurNAc residue functioning as a ligand molecule of the IMD pathway. These molecules function as pathogen-associated molecular patterns (PAMPs).

T7 lysozyme, which is known to have a typical zinc-dependent *N*-acetylmurarmoyl-L-alanine amidase. This amidase can cleave the amide bond connecting the stem-peptide to the carbohydrate backbone of PGN. The tertiary structures of several PGRPs, such as PGRP-LB, PGRP-LE, or PGRP-LC, show an extended surface groove domain that may be involved in the recognition of downstream molecules during the transfer of PGN signals, and in the interaction between the stem-peptide of PGN and the PGRP domain (Royet and Dziarski, 2007).

Drosophila has 13 PGRP genes that encode 17 PGRP proteins through alternative splicing. *Drosophila* PGRPs are classified into catalytic and non-catalytic PGRPs. PGRP-LB, -SB, and -SC belong to the catalytic PGRPs. Members of the second group, such as PGRP-SA, -SD, and -LC, lack the zinc-binding residues required for amidase activity, but still retain the ability to bind and recognize PGN. *Drosophila* PGRPs can also be categorized based on their size. The short-form proteins PGRP-SA, -SB, -SC, and -SD are secreted hemolymph proteins with PGRP domains and signal sequences, while the long-form proteins PGRP-LA, -LB, -LC, -LD, and -LE contain PGRP domains, signal sequences, and additional domains. PGRP-LA, -LC and -LD are membrane-associated

proteins, while others are assumed to be cytosolic (Werner *et al.*, 2000). Recent *Drosophila* genetic studies have demonstrated that two non-catalytic short-form PGRPs, -SA and -SD, are involved in the recognition of Lys-type-PGN and in the activation of the Toll signaling pathway. Two long-form receptors, PGRP-LC and PGRP-LE, are necessary for the recognition of DAP-type PGN and for the activation of the IMD signaling pathway (Lemaitre and Hoffmann, 2007) (see section 14.4.2).

14.3.2. β-1,3-Glucan Recognition Proteins

The β-1,3-glucans are polysaccharides of D-glucose monomers linked by β-1,3-glycosidic bonds that are important cell wall components of baker's yeast, mushrooms, and other fungi (**Figure 1C**). The first β-1,3-glucan recognition protein (βGRP) was purified from the hemolymph of the silkworm *B. mori* as a 50-kDa protein with a strong affinity for the cell wall of Gram-negative bacteria. The cDNA encoding this protein was isolated from an immunized silkworm fat body cDNA library, sequenced, and named the Gram-negative bacteria-binding protein (GNBP) (Lee *et al.*, 1996). Unexpectedly, the amino acid sequence of this GNBP protein contained a region with significant homology to the putative catalytic region of a group of bacterial β-1,3-glucanases and β-1,4-glucanases. At the time, it was unknown whether this GNBP could bind to β-1,3-glucans. In 2010, the Ashida group determined the cDNA of the real βGRP, which has an ability to activate the β-1,3-glucan-mediated silkworm prophenoloxidase cascade (Ochiai and Ashida, 2000). The deduced amino acid sequence of βGRP was not coincident with that of GNBP even though they were purified from the same insect, suggesting that *Bombyx* GNBP and βGRP may recognize different ligands and may regulate different immune responses in the silkworm. In 2003, the Hoffman group published a series of breakthrough *Drosophila* genetic studies. The *Drosophila* genome encodes three GNBPs (GNBP-1, -2, and -3) and two related proteins. Gobert *et al.* (2003) reported that *Drosophila* GNBP-1 functions as an essential molecule for the activation of the Gram-positive-mediated Toll signaling pathway. Furthermore, in 2006, Gottar and colleagues demonstrated that *Drosophila* GNBP-3 is involved in sensing fungal β-1, 3-glucan and functions as an upstream β-1,3-glucan recognition receptor of the Toll pathway (Gottar *et al.*, 2006). The exact biological function of *Drosophila* GNBP-2 has not yet been determined. Based on these pioneer works, *Drosophila* GNBP-1 and GNBP-3 homologs have been identified from several different insects, such as *Anopheles gambiae*, *Manduca sexta*, and *Tenebrio molitor*, and their biological functions characterized (see below). Recently, the tertiary structures of the N-terminal domain of *Drosophila* GNBP-3 have been solved, showing that this domain contains an immunoglobulin-like fold in which the glucan-binding site is masked by a loop that is highly conserved among glucan-binding proteins (Mishima *et al.*, 2009).

14.3.3. Lipopolysaccharide Recognition Proteins

Hemolin, identified as a bacteria-inducible immune gene in the giant silkmoth *H. cecropia*, is a member of the immunoglobulin gene superfamily comprised of four immunoglobulin domains (Sun *et al.*, 1990). When bacteria were injected into the pupal diapause, the hemolin concentration increased 18-fold. It was shown to specifically interact with the lipid A part of LPS. An X-ray structure was also determined, and it revealed immunoglobulin domains 2 and 3 as a phosphate-binding site, supporting the idea that the specificity of the bacterial interaction occurs through the diphosphate lipid A (Su *et al.*, 1998). The cellular defense effects of hemolin are reflected by its capacity to both inhibit hemocyte aggregation and enhance phagocytosis *in vitro*. Based on these findings, hemolin is considered a pattern-recognition molecule in the insect immune response. Several hemolins from different insects, including *M. sexta*, have also been purified and show similar bacterial binding specificities and

biological functions to those of giant silkmoth hemolin (Yu and Kanost, 2002). Another LPS recognition protein (LRP), a novel 40-kDa protein, was purified to homogeneity from the large beetle *Holotrichia diomphalia*. LRP exhibited agglutinating activity on *Escherichia coli*, but not on *Staphylococcus aureus* or *Candida albicans*. This *E. coli*-agglutinating activity was preferentially inhibited by rough-type LPS with a complete oligosaccharide core. Interestingly, LRP consists of six repeats of an epidermal growth factor (EGF)-like domain. Furthermore, *E. coli* coated with purified LRP was cleared more rapidly in the *H. larvae* than naked *E. coli*, indicating that the protein participates in the clearance of *E. coli in vivo*. Three amino-terminal EGF-like domains of LRP, but not the three carboxyl epidermal growth factor-like domains, are involved in the LPS-binding activity, suggesting that LRP functions as a pattern-recognition protein for LPS and plays a role as an innate immune protein (Ju *et al.*, 2006).

14.4. Humoral Innate Immune Responses

14.4.1. Biochemical Properties of Insect Antimicrobial Peptides

Over the past two decades, many AMPs have been isolated from insects and plants (Broekaert *et al.*, 1995; Bulet *et al.*, 1999; Imler and Bulet, 2005). Most of these insect AMPs are classified into two groups: inducible AMPs that are secreted into the hemolymph when bacteria or fungi are injected into insects, and constitutive AMPs that exist in the naïve hemolymph. More than 250 insect AMPs have been isolated since the discovery of cecropin by the Boman group. In spite of great differences in size, amino acid composition, and structure, insect AMPs can be grouped into three categories. The largest AMPs belong to the first category: peptides with intramolecular disulfide bonds forming an α-helical-β-sheet, a hairpin-like β-sheet, or mixed structures. The second group comprises linear peptides forming amphipathic α-helices. The last category contains peptides with Pro- and/or Gly-rich residues.

Because the molecular activation and regulation mechanisms of insect AMP production are mainly studied in the *Drosophila* system, the biochemical characteristics of *Drosophila* AMPs will be summarized as a way to easily explain AMP induction signaling pathways, such as the Toll and IMD pathways (Lemaitre and Hoffmann, 2007). Many insect AMPs similar to those of *Drosophila* have been identified from a wide variety of different insects, and the relationships between their structures and activities are reviewed in other excellent papers (Brown and Hancock, 2006). The technical advances of analytical biochemistry and mass spectrometry enabled the purification of eight different AMPs using *Drosophila* hemolymph. These AMPs are grouped into three families based on their antimicrobial activities. Specifically, defensin showed antibacterial activity against Gram-positive bacteria. Cecropin, drosocin, attacins, diptericin, and MPAC (matured pro-domain of attacin C) have antibacterial activity against Gram-negative bacteria. Drosomycin and metchnikowin are known to have antifungal activity. These peptides are inducible AMPs that are synthesized by the fat body in response to bacterial or fungal infections and then secreted into the hemolymph. The biochemical characteristics and functions of the eight *Drosophila* AMPs can be summarized as follows.

Defensin contains a 92-residue precursor containing an N-terminal signal peptide (20 residues) followed by a pro-domain (32 residues) and the mature defensin (40 residues). It consists of an α-helical domain linked to antiparallel β-strands with three internal disulfide linkages: Cys_1–Cys_4, Cys_2–Cys_5, and Cys_3–Cys_6. This disulfide bridge pattern is also observed in plant defensins but not in mammalian defensins, suggesting a difference in tertiary structures between invertebrate and vertebrate defensins. This AMP showed high homology with many invertebrate defensin-like AMPs, such as *Sarcophaga* sapecin A and *Tenebrio* tenecin 1. These insect defensins showed antibacterial activity against many Gram-positive bacteria, a limited number of Gram-negative bacteria, and some filamentous fungi. Insect defensins are reported to disrupt the permeability of the cytoplasmic membrane of bacteria, leading to the leakage of cytoplasmic potassium and partial depolarization of the inner membrane (Cociancich *et al.*, 1993). The antibacterial activity decreased in the presence of high salt concentration, as shown in the vertebrate and plant defensins.

Cecropins are synthesized as 63-residue precursors, with a putative signal peptide of 23 residues followed by mature cecropin (40 residues), and ending in a glycine residue that is amidated. Cecropin-like AMP is the first purified animal-inducible AMP generated in response to an experimental infection in diapausing pupae of the *H. cecropia* moth (Lepidoptera) (Steiner *et al.*, 1981). The three *Drosophila* cecropin genes (*cecropin* A–C) are also the first genes to be cloned. Synthetic *Drosophila* cecropin is highly efficacious on many Gram-negative bacteria at concentrations below 10 μM, while most of the Gram-positive strains tested remained insensitive even at higher concentrations. It has been speculated that helix-forming cecropins induce the disintegration of bacterial membrane structures and lysis of bacteria.

Drosocin, which shows antibacterial activity against Gram-negative bacteria, is synthesized as a 64-residue precursor corresponding to a signal peptide (21 residues) followed by the mature drosocin (19 residues), and ending in a prodomain of 24 residues. This peptide has two biochemical characteristics: an *o*-glycosylated Thr residue, and three repeats of a Pro–Arg–Pro motif in the mature sequence. Apidaecins purified from honeybee *Apis*

mellifera are homologs of drosocin-like AMP (Casteels *et al.*, 1989). Unlike other α-helical-β-sheet- or hairpin-like β-sheet-forming AMPs, which work in a matter of minutes, drosocin takes several hours to kill bacteria. Interestingly, synthetic drosocin created with only D-type amino acid residues was totally inactive against Gram-negative bacteria, suggesting that drosocin has stereoselective bactericidal effects and is not only a pore-forming AMP.

Diptericin, which also displays antibacterial activity against Gram-negative bacteria, is synthesized as a 106-residue precursor containing signal peptide (23 residues) followed by the mature diptericin (83 residues). Similar peptides have been identified in other dipteran species, such as the blow fly *Protophormia terraenovae*, the flesh fly *Sarcophaga peregrina* (Natori *et al.*, 1999), and the tsetse fly *Glossina marsitans* (Bulet *et al.*, 1999). The presence of *o*-glycosylation on the Thr residue within the N-terminal Pro-rich domain of this peptide was confirmed by biochemical methods. This *o*-glycosylation modification was also observed in drosocin and MPAC (see below). When full-size synthetic unglycosylated and *o*-glycosylated 82 mer *Prototophormia* diptericins were prepared by solid-phase synthesis (Cudic *et al.*, 1999) or expressed in an *E. coli* system (Winans *et al.*, 1999) and tested for antibacterial activity, the full-size synthetic *Protophormia* diptericin showed antibacterial activity against only a limited number of Gram-negative bacteria. These bacteria were killed by increasing the permeability of the outer and inner membranes of *E. coli*.

Attacin homologs, at more than 190 amino acid residues in length, are the largest AMPs. These Cys-free attacin-like AMPs have no particular post-translational modification, and were initially reported in lepidopteran species with antibacterial activity against Gram-negative bacteria (Engström *et al.*, 1984). *Hyalophora* attacin has been shown to interfere with transcription of the *omp* gene of *E. coli*. The *omp* gene is involved in the synthesis of porines, which form protein channels in bacterial membranes (Carlsson *et al.*, 1991). MPAC (mature prodomain of attacin C) is synthesized as a 241-residue precursor containing a signal peptide (21 residues) followed by MPAC (23 residues). When a dibasic cleavage site of attacin C is cleaved, mature attacin C containing amidated Gly residue at the C-terminal is generated. MPAC is a Cys-free, Pro-rich 23 residue peptide containing N-terminus pyroglutamic acid and an *o*-glycosylation motif on the Thr residue similar to the Pro-rich drosocin. Amino acid comparisons showed that MPAC has high homology (37%) with the prodomains of dipteran diptericin, sarcotoxin IIA, and lepidopteran diptericins. Because MAPC showed high similarity to drosocin and Pro-rich AMPs from other insects, functional studies of MPAC were performed *in vitro* using synthetic non-glycosylated and glycosylated forms against several bacteria. Unexpectedly, none of the strains appeared to be sensitive to MPAC, while a significant synergy was observed between MPAC and cecropin (Bulet *et al.*, 2003).

Drosomycin, which shows antifungal activity, is synthesized as a 70-residue precursor polypeptide containing a signal (26 residues) followed by the mature drosomycin (44 residues). Compared to insect defensin, which contains three disulfide linkages, drosomycin contains an additional disulfide linkage; the bonds are formed in the pattern $Cys_1–Cys_8$, $Cys_2–Cys_5$, $Cys_3–Cys_6$, $Cys_4–Cys_7$, suggesting the presence of an additional β-strand at the N-terminus of the molecule. Even though the six amino acid sequences of the *Drosophila* drosomycins do not show any similarities to those of the insect defensins, insect drosomycins are very similar to the plant defensins and r-thionins, two classes of defense molecules in plants (Padovan *et al.*, 2010). Drosomycins are potent antifungal peptides that inhibit the growth of filamentous fungi, which are pathogenic to humans and plants. Unlike insect defensins, drosomycins retain their biological activities at a high salt concentration.

Metchnikowin, another Cys-free and Pro-rich *Drosophila* antifungal peptide, is synthesized as a 52-residue precursor peptide containing a signal peptide (26 residues) followed by the mature metchnikowin (26 residues). The primary sequence of this AMP shows similarity to the C-terminal domains of dosocin, abaecin, and the lebocins, which are Pro-rich AMPs that have been purified from the fruit fly *D. melanogaster*, the bumblebee *Bombus pascuorum*, and the silkmoth *B. mori*, respectively (Chowdhury *et al.*, 1995; Rees *et al.*, 1997). Recent studies have demonstrated that synthetic 26 amino acid metchnikowin caused strong growth inhibition of the pathogenic fungus *Fusarium graminearum in vitro*. Transgenic barley expressing the metchnikowin gene in its 52-aa form also displayed enhanced resistance to plant fungi, suggesting that antifungal peptides from insects can be a valuable source of crop plant improvements, and that insect peptides can be used as selective compounds against specific plant diseases (Rahnamaeian *et al.*, 2009). Finally, two Gly- and His-rich antifungal peptides named *Sarcophaga* antifungal protein (AFP) and holotricin-3 are purified from the hemolymph of the flesh fly *S. peregrina* (Iijima *et al.*, 1993) and the larvae of the coleopteran beetle, *H. diomphalia* (Lee *et al.*, 1995). Both peptides showed fungicidal activity against human *C. albicans*. However, the exact mode of action of these novel antifungal peptides is not yet clear.

14.4.2. Toll and IMD Signaling Pathways after Bacterial or Fungal Infection

14.4.2.1. Activation and regulation of *Drosophila* Toll and IMD signaling pathways Classical deletion-mapping studies on the cecropin A1 and diptericin genes began in 1990, to answer the question of how insect AMP proteins are strongly induced or upregulated by microbial infection. The unique DNA sequences (GGGGATTYYT) identified in the cecropin A1 and diptericin promoter regions are specifically recognized by the Relish family of transcription factors, which belong to the NF-κB protein family. This discovery led to the detection of this DNA motif in all other AMP genes (Engström *et al.*, 1993). Based on these *Drosophila* genome sequences, three different Relish family genes, Dorsal, Dif (Dorsal-related immunity factor), and Relish, were targeted for mutagenesis screening (Hetru and Hoffmann, 2009). Dif was shown to be essential for the induction of several *Drosophila* genes when flies were challenged with Gram-positive bacteria and fungi. Namely, *Dif* mutant flies strongly reduced the expression of drosomycin and defensin genes after microbial infection. The *Relish* gene was also induced in infected flies. The Relish protein contains two domains, an N-terminal Rel domain and a C-terminal IκB-like ankyrin repeat domain. Upon infection, Relish was rapidly processed by the caspase Dredd molecule into two peptides: the Relish homology domain, which is translocated to the nucleus, and the ankyrin repeat domain, which remains cytosolic. This processing is essential for the expression of *Drosophila* AMPs. Dorsal-mutant larvae and flies exhibited a normal response to bacterial infection challenges. These basic studies provided important clues regarding the key roles the Relish family transcription factors play in the regulation of insect AMP genes via the Toll and IMD pathways.

The Toll pathway, named for the Toll transmembrane-associated receptor, was first genetically characterized for its role in the establishment of dorso–ventral polarity during *Drosophila* embryo development (Belvin and Anderson, 1996). The breakthrough genetic evidence of this pathway in insect immunology was reported between 2001 and 2003 by the group of Jules Hoffmann, which screened flies with mutant *Drosophila* PGRP-SA encoded by the gene *semmelweis*, and β-1,3-glucan recognition protein (βGRP)/Gram-negative binding protein-1 (GNBP1) encoded by the gene *osiris* (Michel *et al.*, 2001; Gobert *et al.*, 2003). Unexpectedly, loss-of-function mutations in either the PRGP-SA or the GNBP1 gene showed very similar phenotypes of compromised survival to Gram-positive bacterial infection, indicating that these two hemolymph proteins cooperate to sense Gram-positive bacteria and are essential for the activation of proteolytic enzymes(s) that cleave Spätzle, a Toll-receptor ligand. Conversely, when these two genes were overexpressed together, flies induced the activation of the Toll pathway even in the absence of a bacterial challenge (Gobert *et al.*, 2003). Hoffman's group also demonstrated that activation of Toll by fungal infection is independent of the *semmelweis* and *osiris* genes. A further genetic screen has identified the *persephone* gene, which encodes a hemolymph trypsin-like serine protease that mediates the fungal-dependent cleavage of Spätzle and the activation of Toll. Overexpression of the *persephone* gene is sufficient to lead to Spätzle-dependent induction of the Toll pathway in the absence of an immune challenge (Ligoxygakis *et al.*, 2002a). Lately, Ferrandon and his colleagues have studied flies with mutant GNBP3, a protein encoded by the gene *hades*. They found that GNBP3 is a pattern-recognition receptor that is required for the detection of β-1,3-glucan, a fungal cell wall component. They also found a parallel pathway that acts jointly with GNBP3 (Gottar *et al.*, 2006). Specifically, when *Drosophila* persephone protease is proteolytically matured by the secreted fungal virulence protease PR1, it activates the Toll pathway. Thus, the detection of fungal infections in *Drosophila* relies both on the recognition of β-1,3-glucan, an invariant microbial cell wall component, and on the effects of virulence factors such as PR1 protease (**Figure 2**).

As shown in **Figure 2**, the *Drosophila* Toll signaling pathway is divided into three steps: (1) extracellular recognition of invading Gram-positive bacteria or fungi by the PGRP-SA/GNBP1 complex or GNBP3, respectively, and signaling amplification step by several serine proteases, leading to the cleavage of pro-Spätzle; (2) activation via binding between a Toll ligand, processed Spätzle and a Toll receptor; and (3) intracellular activation and the expression of effector molecules. Specifically, upon recognition of invading Gram-positive bacteria or fungi, recognition signals are amplified by the serine protease proteolytic cascade, as in mammalian blood coagulation or complement activation cascades (Krem and Di Cera, 2002). It has been suggested that the activation of pro-Spätzle is achieved by a set of serine proteases distinct from those involved in Toll activation during embryonic development. Recently, Lemaitre and colleagues identified modular serine protease (ModSP), the most upstream serine protease of *Drosophila* Toll cascade (Buchon *et al.*, 2009). They demonstrated that *Drosophila* ModSP integrates signals originating from the GNBP3 and PGRP-SA/GNBP1 complex, and connects them to the downstream serine proteases. The terminal serine protease that processes Spätzle has been identified, and named *Drosophila* Spätzle-processing enzyme (SPE) (Jang *et al.*, 2006). Even though several *Drosophila* clip-domain-containing serine proteases, such as Grass and Spirit, which are suggested to act between ModSP and SPE, have been identified, the biochemical functional studies of these serine proteases are not yet complete (Kambris *et al.*, 2006).

Upon binding of the cleaved Spätzle to the Toll receptor, the production of AMPs is induced from the fat body (Lemaitre and Hoffmann, 2007; Weber *et al.*, 2003). The Spätzle–Toll complex recruits a set of downstream molecules. First, Toll/IL-1 Receptor (TIR) and/or the death-domain-containing adaptor molecules dMyD88

and Tube lead to the activation of the kinase Pelle. Pelle induces the degradation of Cactus, an inhibitor that maintains the cytoplasmic localization of Dif and Dorsal, two transactivators of the NF-κB family. The translocation of these transcription factors induces the expression of several immune genes, including AMP genes, through binding to the κB DNA motifs of their promoter.

The IMD pathway is named after the first *Drosophila* mutant flies, called *immune deficiency* (*IMD*), which are susceptible to Gram-negative bacteria infection, but are more resistant to fungi and Gram-positive bacteria infection. The *imd* gene encodes a death-domain-containing protein similar to that of the receptor interacting protein (RIP) of the mammalian tumor necrosis factor receptor (TNFR) pathway (Georgel *et al.*, 2001). In 2002, three groups simultaneously reported the identity of the upstream receptor molecule of the *Drosophila* IMD pathway: PGRP-LC, a putative transmembrane protein that is required for the activation of the Gram-negative-mediated IMD pathway (Choe *et al.*, 2002; Gottar *et al.*, 2002; Rämet *et al.*, 2002). Compared to the IMD null mutant, loss-of-function mutants of PGRP-LC showed less severe phenotypes, suggesting that PGRP-LC is only one of several receptors that sense Gram-negative bacteria. Kurata and colleagues reported that *Drosophila* PGRP-LE, another member of the PGRP family, activates the IMD pathway when this gene is overexpressed (Takehana *et al.*, 2002). Taken together, these studies showed that PGRP-LC and PGRP-LE function as receptor molecules for the activation of the Gram-negative bacteria-mediated IMD pathway (**Figure 2**).

The intracellular signaling mechanism of the IMD pathway, mostly determined by genetic studies, is summarized as follows (Lemaitre and Hoffmann, 2007) (**Figure 2**): upon Gram-negative bacterial infection, PGRP-LC recruits the IMD protein, which interacts with the adaptor molecule dFADD via death-domain interaction. The dFADD then recruits the Dredd caspase, which is proposed to associate with Relish. After cleavage of the Relish protein, the Relish transactivator domain translocates to the nucleus, while the inhibitory domain of Relish remains in the cytoplasm. Relish that was phosphorylated by the inhibitory κB-kinase signaling complex is proposed to be activated by Tak1 and its adaptor TAB2 in the IMD pathway. The molecular mechanisms behind the dimerization of PGRP-LC and the activation of the downstream signal pathway by the ligand–receptor complex have not yet been determined.

The serpins belong to a superfamily of **ser**ine protease **in**hibitors that act as suicide substrates by binding covalently to their target proteases. Serpins are known to regulate various physiological processes and defense reactions in mammals and invetebrates (Reichhart, 2005; Silverman *et al.*, 2010; Whisstock *et al.*, 2010). To date, four *Drosophila* serpins related to innate immunity

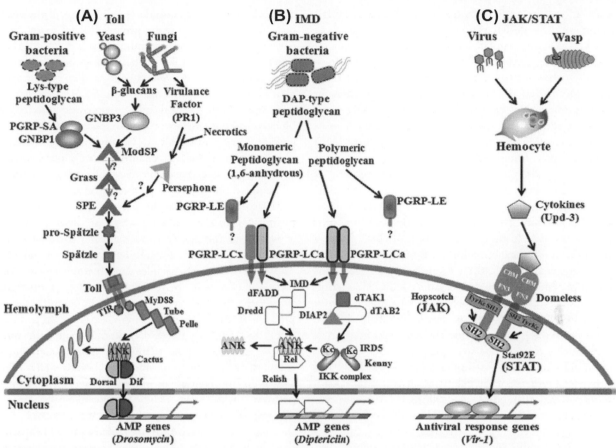

Figure 2 Three major *Drosophila* signaling pathways that regulate systemic immune responses against bacterial, fungal and viral infection. (A) The Toll pathway is activated by the Lys-type peptidoglycan of Gram-positive bacteria and β-1,3-glucan of yeast and some fungi through two different pattern recognition proteins, such as PGRP-SA/GNBP1 and GNBP3, respectively. These recognition signals initiate the activation of the proteolytic serine protease cascade via ModSP, Grass, and Spätzle-processing enzyme (SPE), which are clip-domains containing serine protease, leading to cleavage of pro-Spätzle to the processed Spätzle by activated SPE. Upon binding processed Spätzle to Toll, dimerized Toll recruits dMyD88, Tube, and Pelle, resulting in the phosphorylation and proteosomal degradation of Catus. Catus degradation induces the translocation of Rel transcriptional factors, Dif and Dosal, to the nucleus. These factors bind to NF-κB-response elements and induce activation of transcription of AMP genes, such as Drosomycin. The virulence factor of fungi, such as PR1 protease, is also suggested to activate pro-SPE to activated SPE, leading to the cleavage of pro-Spätzle to the processed Spätzle. (B) The polymeric and monomeric DAP-type peptidoglycans of Gram-negative bacteria are recognized by PGRP-LE and PGRP-LCs. These recognition signals are transferred to the IMD, which is localized in cytoplasm. Upon binding of monomeric DAP-type peptidoglycan to PGRP-Lcx/Lca heterodimer or polymeric DAP-type peptidoglycan to PGRP-Lcx homodimer, IMD is recruited by the intracellular domain of PGRP-LCs. IMD then recruits dFADD and caspase Dredd, which cleave the phosphorylated Relish. leading to translocation of Rel domain to nucleus, where the Rel domain binds to NF-κB response elements and activates the transcription of AMP genes, such as Diptericin. The phosphorylation of Relish is suggested to be mediated by an inhibitory κB (IKK) complex (containing IRD5 and Kenny). dTAK1 activation is induced by adaptor dTAB2 nd DIAP2. (C) Upon binding of cytokine Upd3, which is secreted from hemocytes into hemolymph by viral and wasp-parasitoid infection, to the dimerized Domelss receptors, JAK (Hopscotch) is activated. The JAK recruits the STATs (Start92E), which are phosphorylated and dimerized, leading to translocation to the nucleus to activate transcription of target genes, such as *Vir-1*.

(SPN43Ac, SPN27A, SPN77Ba, and SPN28D) have been analyzed in detail by genetic approaches. *SPN43Ac* mutant flies accumulated cleaved Spätzle, resulting in constitutive activation of the Toll pathway and the expression of AMPs (Levashina *et al.*, 1999). *SPN27A* and *SPN28D* are known to regulate the Toll pathway during early development (Hashimoto *et al.*, 2003; Ligoxygakis *et al.*, 2003; Scherfer *et al.*, 2008), and are also involved in the melanin biosynthesis reaction (De Gregorio *et al.*,

2002; Ligoxygakis *et al.*, 2002b). However, the molecular identities of the serpin target serine proteases and the biochemical regulatory mechanisms of these serpins have not been clearly demonstrated, leading to a lack of molecular understanding of the roles of serpins in the Toll signaling cascade.

The regulatory mechanisms of the *Drosophila* IMD pathway have also been studied; four negative regulators of the IMD pathway have been identified and characterized.

PGRP-LF, a membrane-bound non-catalytic PGRP containing two PGRP domains, was demonstrated to be a key negative regulator of the PGRP-LC-mediated IMD signaling pathway (Maillet *et al.*, 2008). However, the inhibition of the IMD pathway by PGRP-LF was induced even in the absence of infection, allowing the prevention of aberrant activation of the IMD pathway by residual PGN fragments that are ingested from food or released by indigenous microbes. Pirk (**p**oor **I**MD **r**esponse upon **k**nock-in), a protein interacting with PGRP-LC, is also known as Rutra (a new regulator of the IMD pathway); PIMS (**P**GRP-LC-interacting **i**nhibitor of **IMD s**ignaling) is another negative regulator of the IMD pathway. The biological functions of these proteins are involved in the precise control of IMD pathway induction, but the exact biological functions are still not clear (Aggarwal *et al.*, 2008; Kleino *et al.*, 2008; Lhocine *et al.*, 2008). Furthermore, catalytic PGRPs such as PGRP-SC1 and PGRP-LB, which show amidase activity against PGN, were reported as another family of negative regulators of the IMD pathway. These PGRPs function as scavengers by cleaving the stem-peptide of PGN, thereby eliminating the immune-stimulating activity of PGN and leading to shutdown of the PGRP-LC-mediated IMD signaling pathway (Bischoff *et al.*, 2006).

Although we have observed how *Drosophila* Toll and IMD pathways are activated through microbe-recognition proteins and their extracellular and intracellular adaptor molecules after recognition of Gram-positive, Gram-negative bacteria or fungi, it is necessary to answer the question of which PAMP molecules of these microbes can activate these signaling pathways. In 2003, Lemaitre and his colleagues first reported the determination of ligand molecules of the *Drosophila* Toll and IMD pathways, demonstrating that DAP-type PGN of Gram-negative bacteria and certain *bacilli* species function as the most potent activators of the IMD pathway, while the Toll pathway is predominantly activated by Lys-type PGN of Gram-positive bacteria (Leulier *et al.*, 2003). These results clearly demonstrated that the discrimination between Gram-positive and Gram-negative bacteria in *Drosophila* relies on the recognition of specific forms of PGN but not other bacterial cell wall components, such as LPS, wall teichoic acid, lipoteichoic acid, and lipoprotein.

14.4.2.2. Activation and regulation of *Tenebrio* Toll signaling pathway

As shown above, *Drosophila* genetics are very powerful tools for characterizing and ordering the components in the *Drosophila* Toll and IMD signaling pathways. However, this system is still limited in terms of determining the biochemical mechanisms involved in regulating the proteolytic Toll cascade. Since *Drosophila* has several alternative routes to the Toll pathway, used at various developmental stages and infection protocols, it is difficult to determine the clear activation mechanism

of the Toll signaling cascade. For instance, *Drosophila* persephone is another serine protease linked to the Toll pathway and antifungal immunity, yet the biological functions of this molecule have only been partially characterized by *Drosophila* genetic studies. The proper identification of downstream factor(s) of persephone still awaits further investigation. To provide compelling biochemical data on how the Lys-type PGN and β-1,3-glucan recognition signals can be sequentially transferred to Spätzle during the Toll signaling pathway, it is necessary to use a larger insect that enables us to collect larger amounts of hemolymph. The coleopteran larvae *T. molitor* was used for intensive biochemical studies, resulting in the purification of nine proteins, which are involved in the activation of the *Tenebrio* Toll signaling pathway (Park *et al.*, 2007; Kim *et al.*, 2008; Roh *et al.*, 2009) (**Figure 3**). The nine molecules include three pattern-recognition proteins (*Tenebrio* PGRP-SA, GNBP-1, GNBP-3), three serine protease zymogens (*Tenebrio* modular serine protease (MSP); Spätzle processing enzyme (SPE), and SPE-activating enzyme (SAE)), and recombinant pro-Spätzle and recombinant Toll-ecto domain-containing proteins, which were purified to homogeneity. The activation mechanism of the *Tenebrio* Toll signaling pathway was then studied biochemically using *in vitro* reconstitution experiments. We proposed that the *Tenebrio* PGRP-SA/GNBP1/MSP/SAE/SPE/Spätzle cascade is an essential unit that triggers the Lys-type PGN recognition signaling pathway in response to Gram-positive bacterial infection in the *Tenebrio* system (Kim *et al.*, 2008).

The β-1,3-glucan recognition signal is also transferred via the sequential activation of the three *Tenebrio* serine proteases, MSP, SAE, and SPE, which leads to the processing of pro-Spätzle to its mature form Spätzle, demonstrating that a three-step proteolytic cascade is essential for Toll pathway activation by fungal β-1,3-glucan in *Tenebrio* larvae. This cascade is shared with Lys-type PGN-induced Toll pathway activation (Roh *et al.*, 2009). Furthermore, we demonstrated that β-1,3-glucan and Lys-type PGN activate the Toll signaling cascade using the same three-step proteolytic cascade that results in the production of two *Tenebrio* AMPs, tenecin 1 and tenecin 2. The amino acid sequence of tenecin 1 and its disulfide bond arrangement were found to be very similar to *Drosophila* defensin, while tenecin 2 showed high sequence identity (65% and 36%) with coleoptericin and holotricin-2, respectively. Holotricin-2, previously identified by our group, is also an inducible antibacterial peptide purified from coleopteran *H. diomphalia* larvae (Lee *et al.*, 1994). The molecular mechanisms behind the recognition of Gram-positive bacteria or fungi by *Tenebrio* larvae *in vivo*, the activation of the Toll pathway, and the type of AMPs induced after the activation of the bacteria- or fungi-mediated Toll signaling cascade are clearly determined. Finally, the upstream

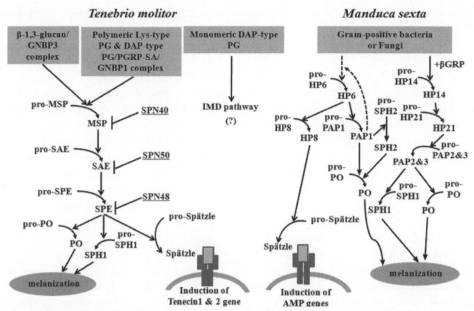

Figure 3 Comparison of *Tenebrio* and *Manduca* extracellular proteolytic Toll signaling pathways. (A) The *Tenebrio* Toll signaling pathway shares a three-step proteolytic cascade, which consists of three serine proteases, such as modular serine protease (MSP), Spätzle-processing enzyme (SPE), and SPE activating enzyme (SAE). Lys-type peptidoglycan and β-1,3-glucan are recognized by the *Tenebrio* PGRP–SA/GNBP1 complex and GNBP3, as in the *Drosophila* system. The recognition signals induce three serine proteases downstream. The activated SPE leads to the cleavage of pro-Spätzle to processed Spätzle, leading to the activation of the Toll cascade and, subsequently, production of the *Tenebrio* AMPs, tenecin 1 and 2. Three serpins (SPN 40, SPN 55, and SPN 45) make specific three serpin–serine protease complexes, and inhibit the processing of pro-Spätzle and phenoloxidase-mediated melanin synthesis. Polymeric DAP-type peptidoglycan of Gram-negative bacteria in *Tenebrio* system also uses the same cascade as that of the polymeric Lys-type peptidoglycan-mediated Toll signaling pathway. (B) The production of AMP and melanin synthesis in *Manduca* system is separated by two branches. *Manduca* AMP production begins with hemolymph protease 6 (HP6). The activated HP6 activates proHP8 to the active form of HP8, which can cleave pro-Spätzle to Spätzle, resulting in the induction of *Manduca* AMPs. The Michael Kanost group of Kansas State University proposed that activated HP6 can also activate prophenoloxidase activating proteinases-1 (PAP-1), which can cleave *Manduca* prophenoloxidase to phenoloxidase, leading to melanin synthesis in the presence of serine protease homolog 2. Also, *Manduca* prophenoloxidase is activated by a β-1,3-glucan recognition protein (βGRP)-mediated three-step proteolytic cascade (HP14/HP21/PAP2&3), leading to melanin synthesis in the presence of SPH1.

pathogen recognition features of the *Tenebrio* Toll cascade are reminiscent of the complement activation by the lectin pathway in mammals in which the recognition of carbohydrates by mannose-binding lectin (MBL) leads to the auto-activation of MBL-associated serine proteases (MASPs) (Matsushita and Fujita, 1992). The domain organization of MASPs is similar to those of insect MSPs.

Recently, the recognition mechanism of DAP-type PGN in the *Tenebrio* system has also been proposed (Yu *et al.*, 2010), although the *Drosophila* IMD pathway responds to polymeric and monomeric DAP-type PGNs of Gram-negative bacteria and certain Gram-positive *Bacillus* species through PGRP-LC or PGRP-LE receptors. Unexpectedly, polymeric (but not monomeric) DAP-type PGN formed a complex with *Tenebrio* PGRP-SA, and this complex activated the three-step proteolytic cascade to produce processed Spätzle, leading to the induction of tenecin 1 in *Tenebrio* larvae, as in the Lys-type PGN-mediated Toll pathway (**Figure 3**). In addition, both polymeric and monomeric DAP-type PGNs induced the expression

of *Tenebrio* PGRP-SC2, which is a DAP-type PGN-selective *N*-acetylmuramyl-L-alanine amidase that functions as a DAP-type PGN scavenger. PGRP-SC2 appears to function as a negative regulator of DAP-type PGN signaling by cleaving DAP-type PGN, rendering it incapable of inducing AMPs in *Tenebrio* larvae. These results demonstrated that the molecular recognition mechanism for polymeric DAP-type PGN differs between *Tenebrio* larvae and *Drosophila* adults, providing biochemical evidence of biological diversity in the innate immune responses of insects.

Because three serine proteases that are directly involved in the activation of the *Tenebrio* Toll cascade were identified, *Tenebrio* larvae were assumed to be a useful system for identifying and characterizing novel target serpins that directly regulate the Toll proteolytic cascade. As described above, because the molecular identities of the *Drosophila* target serine proteases of the four *Drosophila* serpins and the biochemical regulatory mechanisms of these serpins were not clearly demonstrated, we tried to understand the molecular regulation mechanism biochemically using the *Tenebrio* system. Three novel serpins (SPN40, SPN55, and SPN48) from the hemolymph of *T. molitor* larvae were purified (Jiang *et al.*, 2009). These serpins made specific serpin–serine protease pairs with three Toll cascade-activating serine proteases (MSP, SAE, and SPE), and cooperatively blocked the Toll signaling cascade and β-1,3-glucan-mediated melanin biosynthesis. In addition, the levels of SPN40 and SPN55 were dramatically increased *in vivo* by the injection of the processed Spätzle into *Tenebrio* larvae. This increase in SPN40 and SPN55 levels indicates that these serpins function as inducible negative feedback inhibitors. In addition, SPN55 and SPN48 were cleaved at Tyr and Glu residues in reactive center loops, respectively, despite being targeted by trypsin-like SAE and SPE serine proteases. These cleavage patterns are also highly similar to those of the unusual mammalian serpins involved in blood coagulation and blood pressure regulation, and they may contribute to highly specific and timely inactivation of detrimental serine proteases during innate immune responses. It had been thought that the *Drosophila* Toll cascade was regulated by the activity of a single "bottle-neck" protease inhibitor, but our data presented the first indication that each individual protease in a cascade may be regulated by a specific serpin (**Figure 3**).

14.4.2.3. *Manduca* Toll signaling pathway

The Toll cascade of another large insect was also studied biochemically. The Kanost group characterized more than 20 clip-domain-containing serine proteases in the hemolymph of the tobacco hornworm, *M. sexta* (Jiang and Kanost, 2000). Recently, they reported the function of two *Manduca* serine proteases, hemolymph proteases 6 and 8 (HP6 and HP8; An *et al.*, 2009; see also **Figure 3**). HP6 and HP8 are each composed of an N-terminal clip domain and a C-terminal serine protease domain. HP6 was an apparent ortholog of *Drosophila* persephone, whereas HP8 was most similar to *Drosophila* and *Tenebrio* SPE, all of which activate the Toll pathway. Recombinant HP6 was found to activate prophenoloxidase-activating proteinase (proPAP1) *in vitro* and induce prophenoloxidase activation in plasma. HP6 was also determined to activate proHP8. Active HP6 or HP8 injected into larvae induced the expression of AMPs such as attacin, cecropin, and lysozyme. These results suggest that proHP6 becomes activated in response to microbial infection, and participates in two immune pathways: activation of PAP1, which leads to prophenoloxidase activation and melanin synthesis, and activation of HP8, which stimulates the *Manduca* Toll-like signaling pathway. However, the most upstream receptors of the Toll pathway, such as *Manduca* PGRP-SA and GNBP1, have not yet been characterized in the *Manduca* system.

The Kanost group has done pioneering work in the biochemical characterization of insect serpins. They first described the *M. sexta* serpin-1, which has 12 different copies of exon 9 that undergo mutually exclusive alternative splicing to produce 12 putative protein isoforms. These isoforms differ in their carboxyl-terminal 39–46 residues, including the P1 residue, and inhibit serine proteases with different specificities (Kanost and Jiang, 1997). Recently, they also reported the biological function of the *Manduca* serpins: for example, serpin-1 isoforms can inhibit HP8, which activates pro-Spätzle, suggesting that serpin-1 isoforms may be involved in regulation of the *Manduca* Toll cascade (Ragan *et al.*, 2010). The proposed model of activation and regulation of the *Manduca* Toll cascades is shown in **Figure 3**.

14.4.3. Lectin Induction

Lectins are defined as a protein family capable of recognizing specific oligosaccharides, which were initially purified from various plant seeds. Similar proteins have also been isolated from a number of organs in a wide range of vertebrates and invertebrates (Barondes, 1984). Because the diverse biological roles of purified insect lectins cannot all be mentioned in this chapter, we will focus on the insect lectins that mainly participate in insect immunology. The pioneering work was performed by Natori and his colleagues using *S. peregrina* flesh fly larvae (Natori *et al.*, 1999). *Sarcophaga* lectins functioning as defense proteins are synthesized by the fat body and secreted into the hemolymph when the larval body wall is pricked with a hypodermic needle (Okada and Natori, 1983). This 260-aa lectin is a typical C-type lectin that requires calcium ions for agglutinating activity. C-type lectins, a superfamily of calcium-dependent carbohydrate-binding proteins, are known to function in pathogen recognition, cell–cell interactions, and innate immunity in mammals (Weis *et al.*, 1998). This lectin is rapidly induced when sheep red cells are injected, but is not present in the hemolymph of naïve larvae. Elegant biochemical studies performed by the Natori group demonstrated that this lectin has dual functions in defense and development, and provided evidence that the *Sarcophaga* lectin activates insect hemocytes in some way, resulting in the activation of hemocyte-mediated digestion of non-self foreign cells (Nakajima *et al.*, 1982; Komano *et al.*, 1983). They showed that this lectin was also needed for the differentiation of imaginal discs in the pupal stage. When antibodies against *Sarcophaga* lectin were added to the culture medium of *Sarcophaga* imaginal discs, differentiation was strongly inhibited; none of the discs reached the stage of terminal differentiation (Kawaguchi *et al.*, 1991). These results suggest that *Sarcophaga* lectin is not simply a defense molecule that is needed for the elimination of the invading non-self cells, but also a regulatory molecule in the development of imaginal discs.

Kanost and his colleagues also purified four soluble C-type lectins with two carbohydrate-recognition domains from the tobacco hornworm *M. sexta*, and named them immulectin-1 through -4 (Yu *et al.*, 1999). Expression of all four immulectins is upregulated in the fat body upon bacterial challenge. Immulectin-1 and -4 agglutinate Gram-negative and Gram-positive bacteria and yeast. However, immulectin-2 binds to a wide range of microbial cell wall components such as lipoteichoic acid, laminarin (branched β-1,3-glucan), mannose, and LPS. Furthermore, knockdown of the immulectin-2 gene at the level of both mRNA and protein by RNA interference (RNAi) markedly decreased the ability of *M. sexta* to defend bacterial infection when exposed to either species of the insect pathogen *Photorhabdus*, suggesting that the *Manduca* lectin is not the only one that recognizes *Photorhabdus*; the lectin-mediated insect immune system also plays an essential role in defending bacterial infection (Eleftherianos *et al.*, 2006). Similar tandem-domain C-type lectins are identified in other lepidopterans, and are involved in bacterial binding and hemocyte aggregation (Koizumi *et al.*, 1999). However, further studies are necessary to determine the structure of the ligand molecules and the biological functions of these C-type insect lectins during activation of insect innate immunity.

Insect galectin homologs, which have the ability to bind β-galactoside sugars, are identified from two insects, *D. melanogaster* (Pace *et al.*, 2002) and *A. gambiae* (Dimopoulos *et al.*, 1997). Fourteen galectins have been identified in mammals (Rabinovich *et al.*, 2002). However, the identification of precise biological functions for these mammalian galectins is difficult because of the redundancy in tissue expression and the complexity of recognition mechanisms in the target cells. A benefit of working with insect systems such as *Drosophila* and *Anopheles* is the ease of genetic manipulation and the rapid generation time. In addition, there are relatively small numbers of putative galectins in the *Drosophila* and *Anopheles* genomes; lower organisms such as insects are therefore useful in deciphering the precise biological functions of galectins. The Kafatos group demonstrated that a putative galectin homolog was upregulated in the salivary and gut of *A. gambiae* when mosquitoes were infected with malaria and bacteria. The *Anopheles* galectin was suggested to function as a pattern-recognition protein by binding saccharide ligands on the microbial cell wall surface to trigger a host innate immune response (Dimopoulos *et al.*, 2001). However, *Drosophila* galectin was expressed in naïve hemocytes, but not by the fat body or larval hemolymph (Pace *et al.*, 2002). Mammalian galectins were suggested to participate in the innate immune response by facilitating microbial recognition and/or lectin-mediated phagocytosis (Fradin *et al.*, 2000; Rahnamaeian *et al.*, 2009). The elucidation of Toll receptor-mediated immune response in insects has led to rapid progress in the understanding of innate immune function in mammals; the determination of biological functions of insect galectins may similarly provide insight into the novel biological functions of mammalian galectins.

14.4.4. Melanin Synthesis

The melanization reaction is a major humoral immune response in insects and arthropods. Biochemical studies using large insects, which enable the collection of large amounts of hemolymph for biochemical studies, have defined the molecular basis of melanization activation and its regulation upon pathogenic microbe infection. These studies, along with molecular genetic analysis of melanization in the fruit fly and mosquito, have provided new insight into its biological role in fighting microbe infections (Cerenius *et al.*, 2008, 2010).

14.4.4.1. Melanin synthesis by bacterial and fungal infection Melanization of the insect cuticle was observed by Pasteur in 1870 while studying silkworm disease. In 1927, Metalnikov observed the hemolymph melanization of the wax moth *G. mellonella* and assumed that the formation of the dark pigment was the result of the action of a special enzyme on a chromogenic substrate found in the insect blood. In 1953, Ohnishi discovered the inactive zymogen form of phenoloxidase and its proteinaceous activator in *Drosophila* hemolymph (Ohnishi, 1953). Then, in 1955, Mason defined phenoloxidase as a copper-containing enzyme that catalyzes two reactions: oxygenation of monophenols to *o*-diphenols and oxidation of *o*-diphenols to *o*-quinones. These reactions are key steps in the synthesis of the black melanin pigment (Mason, 1955). In 1974, Pye reported that bacteria (*Pseudomonas aeruginosa*) and zymosan (the yeast cell wall components mainly composed from β-1,3-glucan and mannan) trigger the activation of prophenoloxidase in *G. mellonella* hemolymph (Pye, 1974). At the time, it was unknown whether bacteria or zymosan directly activated prophenoloxidase, or if a proteolytic activation cascade was involved. Subsequently, the Ashida group intensively studied the basic activation mechanisms of the prophenoloxidase cascade system using the silkworm *B. mori* (Ashida *et al.*, 1974).

The Ashida group demonstrated that *Bombyx* prophenoloxidase is activated when silkworm plasma is incubated with β-1,3-glucan from fungi or PGN from bacteria (Ashida *et al.*, 1983; Yoshida *et al.*, 1986), leading to the purification and cDNA cloning of the so-called insect PGRP and β-1,3-glucan recognition protein (βGRP) from silkworm hemolymph (Ochiai and Ashida, 2000). These pioneering works provided important biochemical evidence that the invertebrate prophenoloxidase activation system may be activated by elicitors derived from microbial cell walls such as PGN and β-1,3-glucan. The prophenoloxidase system, like the complementary vertebrate system, is a proteolytic cascade containing several serine proteases and their inhibitors, and terminates with the prophenoloxidase zymogen. Microbial carbohydrates such as PGN or β-1,3-glucan first react with pattern-recognition proteins, such as PGRP or βGRP, which then induce activation of several serine proteases within the prophenoloxidase system.

In 1986, Collins *et al.* (1986) reported the important discovery that refractoriness to malaria parasites in the mosquito *A. gambiae* is correlated with melanization of the malaria ookinetes before they develop to the oocyst stage. This unique internal defense mechanism has received considerable attention, because this melanization reaction could be exploited to control and treat mosquito-borne diseases. Following this report, major research efforts have been focused on the identification of genes and gene products that are associated with the melanization response, with the intention of manipulating susceptible mosquito strains to be more resistant to malaria. The detailed results are mentioned below.

In 1995, the complete amino acid sequences of prophenoloxidase from three insect species (silkworm *B. mori*, tobacco hornworm *M. sexta*, and fruit fly *D. melanogaster*) were deduced from the respective cDNA sequences by three independent groups (Fujimoto *et al.*, 1995; Hall *et al.*, 1995; Kawabata *et al.*, 1995). Surprisingly, the deduced sequences clearly indicated that insect prophenoloxidase is a protein homolog to arthropod hemocyanin; the conserved histidine residues of the arthropod hemocyanins perfectly aligned with those in the insect prophenoloxidases and sequence homology ranged from 30% to 40%. A recent crystal structure of *M. sexta* prophenoloxidase shows that this unique insect oxidase makes a heterodimer consisting of two similar polypeptide subunit chains, prophenoloxidase-1 and -2. The enzyme active site of each subunit contains a typical type-3 dinuclear copper center; each copper ion coordinates with three structurally conserved histidine residues. The acidic glutamic acid residue located at the active site of prophenoloxidase-2 serves as a general base that deprotonates monophenolic substrates. This step is necessary for the *o*-hydroxylase activity of the generated phenoloxidase, and allows us to propose a model for localized prophenoloxidase activation in insects (Li *et al.*, 2009).

After the purification of the first prophenoloxidase-activating enzyme (PPAE) from the cuticle of silkworm (Dohke, 1973), the entire domain structure of insect PPAE, which is a clip-domain-containing serine protease, was determined in 1998 (Jiang *et al.*, 1998; Lee *et al.*, 1998). The clip-domain-containing serine protease-mediated proteolytic cascade was previously studied in the hemolymph coagulation cascade of the horseshoe crab *Tachypleus tridentatus* (Iwanaga *et al.*, 1992). Clip-domains have also been found in the *Drosophila* snake and easter precursor proteins, which are both indispensable proteins for normal embryonic development (Chasan and Anderson, 1989). These serine proteases are synthesized and secreted into the hemolymph as inactive zymogens, which are then activated at a specific cleavage site by another upstream protease. The activation of most prophenoloxidases by PPAE induces the cleavage of a conserved Arg-Phe bond in prophenoloxidase, resulting in the removal of the N-terminal fragment consisting of about 50 amino acid residues (Fujimoto *et al.*, 1995; Kim *et al.*, 2002). The generated phenoloxidase was assumed to make a polymer with melanin synthesis activity. An additional serine protease homolog (SPH), a non-catalytic enzyme with an active Ser to Gly mutation within the N-terminal clip-domain, was required for the activation of prophenoloxidase and for phenoloxidase activity (Kwon *et al.*, 2000). Similar SPHs are identified in other insects, and are involved in the activation of the prophenoloxidase cascade (Yu *et al.*, 2003). The crystal structures and the functional roles of one clip-domain-containing PPAE and one SPH during the prophenoloxidase activation cascade were resolved (Piao *et al.*, 2005, 2007), demonstrating that insect SPH forms a homooligomer upon cleavage by the upstream protease, and that the clip-domain of SPH functions as a module for binding phenoloxidase through the central cleft, while the clip-domain of a serine protease plays an essential role in the rapid activation of its protease domain.

Recently, our lab, as well as the Kanost group, has intensively studied the molecular activation mechanisms of the insect prophenoloxidase cascade at the molecular level using biochemical methods with the meal worm *T. molitor* and with *M. sexta* larvae (Kan *et al.*, 2008; An *et al.*, 2009). As shown in **Figure 3**, we provided clear biochemical evidence of the *Tenebrio* melanin synthesis cascade: *Tenebrio* Spätzle processing enzyme (SPE) cleaves both the 79-kDa *Tenebrio* prophenoloxidase and the *Tenebrio* clip-domain SPH zymogen to an active melanization complex. This complex, consisting of the 76-kDa *Tenebrio*-activated phenoloxidase and an active form of *Tenebrio* clip-domain SPH, efficiently produces melanin pigment on the surface of bacteria, to strong bactericidal effect. Additionally, we found the phenoloxidase-induced melanization reaction to be tightly regulated by *Tenebrio* prophenoloxidase itself, which functions as a competitive inhibitor of melanization complex formation. Our results demonstrate that the *Tenebrio* Toll signaling pathway and the melanization cascade share a common serine protease cascade for the regulation of these two major humoral innate immune responses. In the *Manduca* melanin synthesis cascade, an initiation proteinase precursor, proHP14, is autoactivated in response to Gram-positive bacterial or fungal infection. HP14 activates proHP21; HP21 activates proPAP2 or proPAP3; PAP2 or PAP3 then cleaves prophenoloxidase to form active phenoloxidase in the presence of SPH1 and

SPH2. Activation of prophenoloxidase can also be catalyzed by PAP1 when the high molecular SPH complex is also present. PAP1 also activates proSPH2 directly, and can indirectly lead to proHP6 activation (**Figure 3**).

Melanization in insects is crucial for defense and development, but must be tightly controlled. Systemic hyperactivation of the prophenoloxidase system, excessive formation of quinones, and inappropriate excessive melanin synthesis are deleterious to the hosts, suggesting that the prophenoloxidase activation system and melanin formation should be tightly regulated by serpins or melanization-regulatory molecules or inhibitors. Three different serpins (*Drosophila* Spn27A, Spn28D, and Spn77Ba) that regulate *Drosophila* melanization have been identified (Silverman *et al.*, 2010). Spn27A was the first serpin identified to regulate activation of phenolxoidase and melanization in the hemolymph. A loss-of-function mutation in Spn27A resulted in uncontrolled melanization as well as in semi-lethality, while overexpression of Spn27A suppressed phenoloxidase activation induced by microorganisms (De Gregorio *et al.*, 2002; Ligoxygakis *et al.*, 2002b). The target protease of Spn27A in the *Drosophila* melanization cascade has not yet been determined, but in an *in vitro* experiment, recombinant Spn27A was able to make a protease–serpin complex with the PPAE from the beetle *H. diomphalia* (De Gregorio *et al.*, 2002). Another serpin, Spn28D, inhibited phenoloxidase activation at a different level than Spn27A (Scherfer *et* al., 2008). Knockdown of Spn28D by RNAi led to constitutive melanization in various tissues. Spn28D may function to prevent premature activation of phenoloxidase at an early stage, and confine its general availability. The serpin Spn77Ba was discovered to be a regulator of melanization in the tracheal respiratory system of *Drosophila* (Tang *et al.*, 2008). Spn77Ba was expressed in the tracheal epithelium, and knockdown of Spn77Ba specifically in trachea led to constitutive tracheal melanization and associated lethality.

Several serpins have been purified from the large insects *M. sexta* and *T. molitor*, and characterized as negative regulators of the insect prophenoloxidase system (Jiang and Kanost, 1997; Jiang *et al.*, 2009). In the *Manduca* system, the Kanost group reported elegant results regarding serpin splicing: 12 different copies of *Manduca* serpin 1 undergo mutually exclusive alternative splicing to produce 12 putative protein isoforms, which differ in the carboxyl-terminal residues 39–46, including the P1 residue. These serpins inhibited *Manduca* serine proteases with different specificity (Ragan *et al.*, 2010). These serpins were characterized, and suggested as negative regulators of the prophenoloxidase and Toll signaling cascades (Kanost *et al.*, 2004). Three novel serpins (SPN40, SPN55, and SPN48) were purified from the hemolymph of *T. molitor* and characterized (Jiang *et al.*, 2009) (**Figure 3**). These *Tenebrio* serpins

made specific serpin–serine protease complexes with the three Toll signaling cascade-activating serine proteases MSP, SAE, and SPE, and cooperatively blocked the *Tenebrio* Toll-signaling cascade and β-1,3-glucan-mediated melanin biosynthesis. Inhibitors that are directly involved in the inhibition of the enzymatic activity of phenoloxidase are also identified as regulators of melanization. Phenoloxidase inhibitor (POI) was purified biochemically in the housefly *Musca domestica* (Daquinag *et al.*, 1995). POI homologs also have been identified in *M. sexta* (Lu and Jiang, 2007), and in the mosquito *A. gambiae* (Shi *et al.*, 2006). When *Anopheles* POI was knocked down using RNAi, melanin synthesis was suppressed. In *T. molitor*, a novel 43-kDa melanization-inhibiting protein (MIP) was found to specifically inhibit melanin synthesis (Zhao *et al.*, 2005). The biological function of MIP was proposed to negatively regulate phenoloxidase activation, and thus prevent possible tissue damage. No similar protein showing significant sequence homology to the *Tenebrio* MIP has been identified in *Drosophila*, except for Gp150, a protein involved in Notch signaling that has an aspartic acid-rich domain similar to that found in MIP.

In summary, microbial cell wall components of Gram-positive bacteria and fungi, such as PGN or β-1,3-glucan, are first recognized by specific recognition proteins, such as PGRP or β-1,3-glucan-recognition protein. Subsequently, the recognition signals trigger the activation of a serine protease cascade in which serine proteases are present as inactive zymogens, finally leading to the conversion of prophenoloxidase to phenoloxidase, resulting in the synthesis of melanin where melanization is required. The advantage of this insect prophenoloxidase cascade is that a minute microbial recognition signal is amplified by multiple activation steps. Until now, the *in vivo* importance of melanization during insect host defense has been intensively studied using classical biochemistry and molecular genetics. Currently, as the complete insect prophenoloxidase activation cascade gradually emerges at the molecular level, the predictions of pioneering researchers like Ohnishi and Ashida are turning out to be correct.

14.4.4.2. Melanin synthesis by parasite and parasitoid wasps As described above, melanization is also essential for host defense against parasites and parasitoid wasps. In the mosquito *A. gambiae*, resistance to malaria parasites is deeply related to melanization of the malaria ookinete (Collins *et al.*, 1986). Following this discovery, many research groups, including the Kafatos group, intensively screened and identified a large set of regulators of melanization using microarray analysis and reverse genetics via RNAi; their findings suggest that melanization reactions are fine-tuned in response to malaria parasite infection (Sinden, 2004). Interestingly, Volz *et al.* (2006) reported that a refractory *A. gambiae* strain easily synthesized melanin while under a chronic state of oxidative stress, explaining how this process leads to resistance to parasites. Now, using a highly diverse body of vector species and strains, many researchers are trying to understand the balance between parasite invasion and elimination by the mosquito innate immune system.

In early studies, biochemical evidence of dramatically reduced phenoloxidase activity was reported in lepidopteran larvae parasitized with different parasitoids (Sroka and Vinson, 1978). Beckage *et al.* (1993) first demonstrated that this inhibition effect could be mimicked by injection of *Cotesia congregata* bracovirus into non-parasitized larvae, but that inactivated virus had no effect, suggesting that downregulated phenoloxidase activity was caused by the bracovirus. Shelby *et al.* (2000) observed that the immunosuppressive *Campoletis sonorensis* ichnovirus (CsIV) reduces the protein level of several key enzymes in the melanin synthesis pathway, such as phenoloxidase, dopachrome isomerase and DOPA decarboxylase in parasitized fifth lepitopteran larvae, providing more evidence for the reduction of melanization in parasitized insects.

In the *Drosophila* system, phenoloxidase-deficient or DOPA-decarboxylase-deficient mutant flies are significantly compromised in their cellular immune responses, such as encapsulation of the endoparasite wasp *Leptopilina boulardi*, suggesting that melanization is important for host defense (Rizki and Rizki, 1990). Additionally, flies that were incapable of melanin synthesis because of the overexpression of serpin Spn27A in the hemolymph demonstrated compromised encapsulation cellular immunity and lowered ability to kill *L. boulardi* (Nappi *et al.*, 2005); this experiment demonstrated that the melanization reaction could be a target for innate immune suppression by parasitoid wasps. A recent interesting paper demonstrated that the parasitic wasp *Microplitis demolitor* injects a *Microplitis demolitor* bracovirus (MdBV) while laying its eggs in *M. sexta* larvae. MdBV secrete the serine protease inhibitor Egf1.0, which blocks melanization by competitively inhibiting *Manduca* PPAE-1 and -3, providing evidence of the importance of melanization for both the virus and its host wasps (Beck and Strand, 2007).

14.4.5. Immune Responses to Viral Infection

Insects are susceptible to highly diverse families of DNA and RNA viruses. Recently, many insect viruses have caused substantial damage to agriculture. For example, some insect viruses kill and threaten beneficial insects, such as the honeybee and silkworm. Some viral infections of insects, such as yellow fever virus, West Nile virus, and Dengue virus, can be transmitted to humans with severe side effects, including encephalitis (Mackenzie *et al.*, 2004). Clearly, insect models for studying host–virus interactions based on genetic and molecular biology will benefit society in many ways.

Genetic studies using *Drosophila* as a virus–host model system provide two types of innate immune responses to virus infection: a constitutive defense system that induces degradation of viral RNA by RNA interference (RNAi), and an inducible immune response that prevents viral infection by inducing a large number of genes after detecting viral particles (Kemp and Imler, 2009).

In 2005, the Imler group published pioneering work on *Drosophila* inducible responses to viral infection (Dostert *et al.*, 2005). They focused on one gene, *vir-1* (virus-induced RNA 1), which contains a consensus binding site for the transcription factor STAT92E. *Drosophila* C virus (DCV) infection triggers induction of STAT DNA-binding activity in fly nuclear extracts, indicating that the JAK (Janus Kinase)/STAT pathway may be involved in the induction of the *vir-1* gene. In *Drosophila*, the JAK/STAT pathway is composed of a single JAK kinase and a single STAT factor (known as STAT92E). The JAK kinase is also regulated by the cytokine receptor Domeless, which shows similarity to the gp130 subunit of the interleukin-6 receptor in mammals (Agaisse *et al.*, 2003). Analysis of the flies carrying a JAK kinase gene mutation showed a higher viral load than wild type flies, and succumbed more rapidly to DCV infection, indicating that some genes induced by DCV infection participate in the control of viral amplification. Based on these results, the Imler group proposed a model in which one cytokine of the Unpaired (Upd) family (Upd-1, -2, and -3) is induced by *Drosophila* C virus infection, and triggers an antiviral response to infected cells through activation of the JAK/STAT pathway (Sabin *et al.*, 2010) (**Figure 2**).

In the mosquito system, the Dimopoulos group demonstrated that the JAK/STAT pathway also restricts infection of Dengue virus, a medically important arbovirus (Souza-Neto *et al.*, 2009). They identified several genes that were upregulated by Dengue infection via the JAK/STAT pathway. Among them, the authors identified and partially characterized two JAK-STAT pathway-regulated and infection-responsive Dengue virus restriction factors (DVRFs) that contain putative STAT-binding sites in their promoter regions, suggesting that the JAK-STAT pathway is part of the *A. aegypti* mosquito's anti-dengue defense and acts independently of the Toll pathway and RNAi-mediated antiviral defenses.

In 1998, Fire and Mello discovered RNA interference (RNAi) (Fire *et al.*, 1998), revealing that double-stranded RNAs (dsRNAs) can be used as a tool to knock down specific genes. RNAi is a defense reaction based on the specific base-pairing between small host RNAs and invading pathogenic nucleic acids. RNAi is now an efficient biochemical tool for gene knockdowns and for examining important physiological functions in diverse biological processes. The first evidence of RNAi as an important antiviral defense mechanism came from plant studies (Li and Ding, 2001; Vance and Vaucheret, 2001).

Further studies showed that RNAi depends on small RNAs that are 21–30 nucleotides in length and are divided into three classes: small interfering RNAs (siRNAs), microRNAs (miRNAs), and Piwi-associated interfering RNAs (piRNAs). Both siRNAs and miRNAs are known to be processed as duplexes from dsRNA precursors by an RNaseIII enzyme called Dicer (Hammond, 2005), but only after the primary miRNA transcripts (pri-miRNA) that contain imperfect intramolecular stem-loops are first processed within the nucleus. The resulting precursor miRNAs (pre-miRNA) are then converted into a single mature miRNA species in the cytoplasm (Bartel, 2004). In contrast to siRNAs and miRNAs, piRNAs are about 30 nucleotides in length and are found in the germ-line of flies and vertebrates, which are Dicer-independent (Zamore, 2007). The *Drosophila* genome has two Dicer enzymes, Dicer-1 and Dicer-2, which produce two types of regulatory RNA: microRNAs (miRNAs) and small interfering RNAs (siRNAs). Dicer-1 induces the processing of the precursor miRNAs to mature miRNAs, whereas Dicer-2 recognizes long dsRNAs and cleaves them into siRNAs (Lee *et al.*, 2004).

Drosophila studies related to RNAi immune responses against viral infection began with analysis of Dicer-2 mutant flies. These flies showed increased susceptibility and lethality to RNA virus infections from DCV, Cricket paralysis virus, Flock House virus, and *Drosophila* X virus (Galiana-Arnoux *et al.*, 2006; Wang *et al.*, 2006). An increased viral load and high levels of viral RNA were observed in these virally infected mutant flies, suggesting that RNAi is a crucial defense mechanism for viral infection. Additionally, the detection of siRNAs corresponding to viral sequences in infected cells strongly supported a mechanism whereby dsRNA, corresponding either to the viral genome or to replication intermediate forms, is detected by Dicer-2 and cleaved into siRNAs. These data also demonstrated that the Dicer-2-mediated siRNA pathway is a highly specific antiviral defense system based on the pairing of complementary nucleic acids. While Dicer-2 was suggested to play an essential role in the RNAi pathway, Imler recently reported that Dicer-2 also induces triggering of a downstream antiviral signaling cascade upon binding and recognition of viral dsRNA. This cascade results in the induction of the *Vago* gene, an antiviral gene that is required to restrict viral replication in flies (Deddouche *et al.*, 2008). *Vago* expression was shown to be dependent upon Dicer-2. Since Dicer-2 belongs to the DExD/H-box helicase family, as do the RIG-I-like receptors (Takeuchi and Akira, 2008) involved in sensing viral infection and mediating interferon induction in mammals, the authors proposed that this family represents an evolutionarily conserved sensor molecule that can detect viral nucleic acids and regulate antiviral responses.

Completely unique molecules are also identified in the hemolymph of virus-infected flies (Sabatier *et al.*, 2003). After injection of DCV into the *Drosophila* thorax, the induced molecules were analyzed by MALDI-TOF mass spectrometry and compared with those produced after the injection of bacteria or fungal spores. Interestingly, pherokine-2 (Phk-2), which was previously characterized as a putative odor/pheromone binding protein, was specifically induced, suggesting that host-defense mechanisms against viral infection are different from the mechanisms operating against bacterial or fungal infections in flies.

Taken together, these recent studies highlight the role of RNAi and innate immune signaling in antiviral defense in insects. However, many questions remain. For example, the real biological functions of only a few virus-specific inducible molecules, such as Vago and DVRF, have been elucidated. It is also unknown how invading viruses are recognized by the host. The RNAi machinery recognizes dsRNA, but the molecular mechanism is unclear. Finally, the known antiviral defense responses do not yet span the whole of host antiviral defense; loss-of-function mutation studies showed only modest impact on survival and viral replication, suggesting the presence of unidentified antiviral responses in the host. Further studies are required to identify new effector molecules and signaling pathways in insect antiviral defense.

14.4.6. Immune Responses to Malaria Infection

Malaria is a widespread and devastating vector-borne disease. *Plasmodium*, the parasite responsible for malaria, is transmitted to human beings through the blood of female *Anopheles* mosquitoes. Similar to other vector parasitic systems, *Plasmodium* induces multi-step developmental transformations inside the mosquito vector during infection of the host. The life cycle of parasites in mosquitoes can be summarized as follows. After infecting the blood, the fusion of male and female gametes generates a diploid motile zygote (ookinete), which rapidly invades epithelial cells and, upon reaching the basal side of the midgut, transforms into an oocyst. Two weeks later, the sporogonic oocyst releases thousands of newly formed sporozoites that migrate and invade the salivary glands. The life cycle of *Plasmodium* parasites in mosquitoes ends when the salivary gland sporozoites are injected via the saliva fluid into the next host during feeding (Sinden, 2002).

Two methods are commonly used to control malaria: (1) anti-malarial agents, and (2) controlling mosquito populations using entomological approaches. Anti-malarial drug treatments are only useful for malaria-infected patients. However, a mosquito population can be reduced using insecticides or water management. In addition, prevention of mosquito exposure using bed nets and repellent chemicals greatly limits the spread of malaria. As a new approach, vector control methods have been developed from molecular studies of mosquito innate immune responses to *Plasmodium* parasites. Here, we summarize the current knowledge of the mosquito innate immune responses to *Plasmodium* infection.

14.4.6.1. Toll, IMD, and JAK/STAT signaling pathways after *Plasmodiun* infection

In the mosquito, the sequence complexity and redundancy of the Toll pathway make it difficult to predict the exact biological function of mosquito Toll orthologs using comparative genomics. For example, the *Anopheles* genome encodes 10 Toll receptors, four of which are orthologs of *Drosophila* Toll receptors. However, four other genes are orthologs of the *Drosophila* Toll-1 and Toll-5 receptors. In addition, at least six *Anopheles* Spätzle genes are similar to those of *Drosophila*, but the phylogenetic relationship of these 12 genes is complex. Another pathway that is different from that of *Drosophila* is the Toll receptor-dependent regulation of NF-κB/Rel transcription factors. Dorsal is a major regulator of *Drosophila* developmental processes, and Dif is essential for innate immune responses. However, *Anopheles* do not encode the Dif gene. The mosquito homolog of Dorsal is Gambif1 (also referred to as Rel1). Upon bacterial infection, Rel1 translocates to the nucleus within the mosquito fat body, and may regulate the immune system (Barillas-Mury *et al.*, 1996).

The intracellular counterparts involved in the *Anopheles* Toll receptor signaling display high homology with those of *Drosophila*, such as MyD88, Tube, Pelle, and Cactus. *Drosophila* MyD88 and Tube contain Toll and IL-1R (TIR) domains, suggesting that these proteins may interact. Pelle contains an additional Ser-Thr kinase domain within its death domain. Cactus contains an ankyrin-repeat domain, and is a negative regulator of the Toll pathway, inhibiting the nuclear translocation of NF-κB/Rel1 proteins. The biological functions of these *Anopheles* counterparts were primarily elucidated by RNAi experiments.

In contrast, the *Anopheles* IMD pathway splits into two signaling branches. One branch is similar to the mammalian C-Jun/JNK pathway and activates the transcription factor AP-1, whereas the other branch activates NF-κB, leading to the expression of the transcription factor Rel2 (corresponding to *Drosophila* Relish). Rel2 exists as two forms in mosquito cells; Rel2-S lacks the inhibitory ankyrin domain, and full-length Rel2-F is inactive until an immune response is triggered. Similar to Cactus in the Toll pathway, Caspar functions as a negative regulator of the *Anopheles* IMD pathway. Caspar, a Fas-associated homolog, was initially identified as a negative regulator of Relish activation in *Drosophila* (Kim *et al.*, 2006). Recent studies had indicated that the *Anopheles* IMD pathway is more specific to malarial infection in comparison to the Toll pathway (see below).

As the third major immune pathway, JAK/STAT signaling is important for antiviral immunity in *Drosophila* (see above), and *Drosophila* gut immunity (see below). Two STAT transcription factors (STAT-A and STAT-B) have been identified in the *A. gambiae* genome, whereas only one STAT is present in *Drosophila*. STAT-B regulates the transcription of STAT-A, which is the predominant form expressed in adult mosquitoes. Although translocation of STAT-A into the nucleus is suggested to upregulate the expression of anti-plasmodium effectors, the detailed molecular mechanism is still obscure.

14.4.6.2. Effector molecules against *Plasmodiun* infection

Many *A. gambiae* immune-responsive genes have been identified by intensive examination of expression profiles from RNAi-mediated gene silenced mutant mosquitoes, which are infected with bacteria, fungi, and malaria parasites (Dimopoulos *et al.*, 1998). Among them, thioester-containing protein1 (TEP1), leucine-rich repeat immune protein 1 (LRIM1), *Anopheles Plasmodium*-responsive leucine rich repeat 1 (APL1), C-type lectin 4 (CTL4), Serpin 2 (SRPN2), and fibrinogen-related proteins (FREPs) are suggested to be major regulatory molecules affecting *Plasmodium* development (Cirimotich *et al.*, 2010).

TEP1 belongs to a family of thioester-containing proteins, and is homologous to the vertebrate factors C3/C4/C5 and to members of the α2-macroglobulin family. In *A. gambiae*, TEP1 is constitutively secreted from mosquito hemocytes and is present in the hemolymph as a 165-kDa zymogen. TEP1 is processed into an 80-kDa fragment upon parasite infection (Levashina *et al.*, 2001). Levashina and colleagues have thoroughly studied this molecule, and suggest that TEP1 functions as an opsonin on the surface of parasites that triggers immune responses.

LRIM1 and APL1 are leucine rich repeat (LRR) domain-containing proteins. Recent studies demonstrated that LRIM1 and APL1C form a disulfide-linked, high molecular weight complex that is secreted into the *Anopheles* hemolymph. This heterodimeric complex interacts with TEP1, leading to the cleavage and activation of TEP1. Therefore, LRR domain-containing proteins play a key role in mediating anti-*Plasmodium* immunity in mosquitoes. More than 20 LRIM1-like proteins have been identified from several mosquito genomes, but no orthologs have been identified in other organisms, indicating that these genes are mosquito-specific (Povelones *et al.*, 2009). In addition, Osta *et al.* (2004) demonstrated that LRIM1 and APL1 play an important role in parasite melanization and killing during early-stage *P. berghei* infection.

The CTL family is the most diverse animal lectin family. CTLs bind carbohydrates in a Ca^{2+}-dependent manner through the C-terminal carbohydrate recognition domain. In the *A. gambiae* genome, 23 genes that encode C-type lectin domains have been identified. CTL4 and

CTLMA2 are agonist molecules of the rodent *Plasmodium* parasite, and silencing of either of these genes induces massive melanization of *P. berghei* ookinetes in the mosquito midgut epithelium, inhibiting mosquito development at the pre-oocyst stage (Osta *et al.*, 2004). However, CTLs do not seem to be involved in defense responses to human *Plasmodium* parasites.

SRPN is a member of a large family of serine protease inhibitors that are present in all higher eukaryotes, as well as in some viruses. In *Anopheles*, Michel *et al.* (2005) reported that knockdown of SRPN2 triggers spontaneous melanization of mosquito blood cells and reduces the life-span of adults. Depletion of SRPN2 increased parasite killing and stimulated melanization.

Fibrinogen-related proteins (FREPs) contain a fibrinogen-like domain, which is evolutionarily conserved from invertebrates to mammals. FREP genes are present in mosquitoes, and are significantly more expressed in *A. gambiae* (58 genes) in comparison to *Ae. aegypti* (37 genes) and *D. melanogaster* (14 genes). RNAi-mediated gene silencing showed that FREPs in *Anopheles* are involved in anti-*Plasmodium* defense. For instance, FBN39 specifically protected mosquitoes against *P. falciparum* infection (Christophides *et al.*, 2002).

14.4.6.3. Regulation of anti-malarial immune responses in mosquitoes

Elucidation of the molecular mechanisms of the mosquito innate immune response to *Plasmodium* infection is expected to help discover novel malaria control strategies. Many researchers are trying to determine how mosquito Toll and IMD immune signaling pathways mediate anti-*Plasmodium* responses.

In this respect, two breakthrough studies have been performed using RNAi-mediated Cactus-silenced and RNAi-mediated Caspar-silenced adult female mosquitoes (Frolet *et al.*, 2006; Garver *et al.*, 2009). These studies enabled us to understand mosquito resistance mechanisms and how malarial ookinetes are killed by host mosquitoes. *Anopheles* Cactus and Caspar proteins are specific negative regulators of *Anopheles* Toll and IMD signaling pathways, respectively. In the first report, Frolet *et al.* (2006) introduced a novel concept of mosquito-malaria immunity referring to a "pre-invasion or basal immunity" phase and "post-invasion or induced immunity" phase. Basal immunity is constitutive and induces the production of immune factors, including TEP1, which is dependent on the two NF-κB factors, Rel1 (Dif-like) and Rel2 (Relish-like). When TEP1 and other hemocyte-derived immune factors are depleted after a 24-h infection, immune responses are activated and induce the transcription of genes encoding multiple immune factors. It was also noted that the replenishment of depleted factors was both NF-κB-dependent and -independent. Another remarkable discovery is that boosting basal immunity by RNAi-mediated silencing of Cactus was sufficient to completely block *P. berghei* development to the oocyst stage, suggesting that Cactus is capable of complete elimination of malarial parasites in mosquitoes. These authors developed a powerful system to determine unique mosquito proteins involved in parasite killing, which can be used to identify resistant mosquito strains.

In the second report, Garver *et al.* (2009) focused on the biological effects of Caspar during malarial infection and transcriptional regulation of three antiplasmodium genes, TEP1, LIRM1, and FREP9, using an RNAi-mediated Caspar-silenced mosquito. The authors obtained several intriguing results: (1) Caspar-silenced *A. gambiae* is refractory to the natural virulent human malaria parasite, *P. falciparum*; (2) depletion of Cactus or Caspar reduced the number of oocysts in the midgut, indicating that both the Toll and IMD immune pathways are involved in the defense against *P. falciparum* and *P. berghei*, albeit to varying degrees depending on the parasite species; and (3) silencing of *A. albimanus* and *A. stephensi* Caspar genes generated a refractory phenotype regardless of the *Anopheles* species.

These studies are significant because they implied the possibility for the development of malaria control strategies based on mosquito innate immunity. Namely, rather than targeting the malaria parasite, the mosquito host can be targeted by upregulating conserved molecules to enhance basal immunity. In addition, the differences between immunity-based resistance to rodent and human malaria parasites are dependent on the transcription factor Rel. For example, Rel2-based immunity is most efficient against *P. falciparum*, whereas Rel1-based immunity is most efficient against *P. berghei*.

14.5. Cellular Innate Immune Responses

14.5.1. Phagocytosis

Phagocytosis is a major cellular innate immune response in both mammals and insects. This process is essential for various biological events, including the elimination of pathogenic microbes, removal of apoptotic cells, tissue remodeling, and induction of innate and adaptive immune responses. Major insect hemocytes are classified as crystal cells, lamellocytes, and plasmatocytes. Plasmatocytes control phagocytic elimination of invading microbes. During the past two decades, three major questions have been addressed with regard to the molecular mechanism of insect phagocytosis: (1) What kind(s) of receptors are involved in the induction of insect phagocytosis? (2) What is the ligand molecule that activates phagocytosis? and (3) What humoral factors in the insect plasma promote the phagocytosis?

Intensive biochemical and functional studies of phagocytic receptors, ligands, and intracellular signaling pathways have been performed in mammalian systems (Jutras and Desjardins, 2005). However, in *Drosophila*, six receptor molecules involved in phagocytosis have been characterized: Croquemort, *Drosophila* scavenger receptor (dSR-CI), PGRP-LC, Draper, Eater, and Nimrod C1 (Stuart and Ezekowitz, 2008).

Croquemort (catcher of death) is a member of the CD36 family of proteins, which is specifically expressed in *Drosophila* macrophage-like hemocytes and is involved in the clearance of apoptotic cells in *Drosophila* embryos (Franc *et al.*, 1999). Human CD36 is a macrophage scavenger receptor that is able to bind a subset of polyanionic ligand molecules. Genetic analysis demonstrated that Croquemort is essential for apoptosis but is not required for the engulfment of bacteria. This suggests that signals generated by dying cells increase the expression of Croquemort, which could facilitate the clearance of cell corpses. *Drosophila* Peste, which was identified by an RNAi screen, is a class B scavenger receptor and is required for the uptake of mycobacteria, but not *E. coli* or *S. aureus* (Philips *et al.*, 2005), suggesting that mammalian class B scavenger receptors may be involved in the recognition of mycobacteria by insects.

Drosophila scavenger receptor C1 (dSR-C1) is a multidomain modular protein consisting of a 609-residue membrane protein containing several well-known sequence motifs, including two complement control protein (CCP) domains, a somatomedin B domain, a MAM (a domain found in Meprin, A5 antigen) domain, and RPTP Mu and mucin-like domains. Rämet *et al.* (2001) demonstrated that dSR-C1 is a receptor that recognizes both Gram-negative and Gram-positive bacteria, but not yeast. This receptor binds a wide variety of ligands. In addition, dSR-CI does not appear to be required for the *Drosophila* AMP response. Even though the ligand molecules in these bacteria have not been determined, scavenger receptors are primordial pattern recognition molecules that mediate evolutionarily conserved innate immunity.

Drosophila PGRP-LC regulates DAP-type PGN-mediated IMD signaling. Initially, it was discovered as a phagocytosis receptor by Rämet *et al.* (2002). Using a dsRNAi-based screen in *Drosophila* macrophage-like cells, 34 genes involved in phagocytosis were identified. Of them, PGRP-LC is involved in phagocytosis of Gram-negative but not Gram-positive bacteria, demonstrating that *Drosophila* PGRP-LC is an essential molecule for the recognition and signaling of Gram-negative bacteria during the insect innate immune response.

Draper (Manaka *et al.*, 2004), Eater (Kocks *et al.*, 2005), and Nimrod C1 (Kurucz *et al.*, 2007) are receptor molecules containing epidermal growth factor (EGF) repeats. Similar proteins are found in silkmoths and the beetle *H. diomphalia* (Ju *et al.*, 2006), in which they function as secreted opsonins and are involved in bacterial clearance. The EGF-like motifs have a conserved consensus sequence CxPxCxxx-CxNGxCxx PxxCxCxxGY, and are separated by variable loops of typically 6–11 residues. This motif differs significantly from the typical EGF repeat (xxxxCx2-7Cx1-(G/A)xCx1-13ttaxCx- CxxGax1-6GxxCx).

Draper was identified as a *Drosophila* homolog of the *C. elegans* CED-1 protein, and has 15 EGF repeats. It is involved in hemocyte phagocytosis of apoptotic cells, but not of zymosans. Draper acts as a receptor in the phagocytosis of apoptotic cells by hemocytes and glial cells in *Drosophila* embryos. Although apoptotic *Drosophila* cells express the membrane phospholipid phosphatidylserine, the best characterized eat-me signal found in mammals, Draper, does not recognize phosphatidylserine. Further studies into the intracellular signaling and identification of Draper ligand molecules have been performed (see below). Eater, a cell surface receptor containing a 32 EGF repeats motif, regulates the phagocytosis of *E. coli* and *S. aureus*, suggesting that Eater is a major phagocytic receptor for a broad range of bacterial pathogens. As expected, Eater was expressed in the plasmatocytes and the lymph glands, but not in the crystal cells, lamellocytes, or the fat body, which are cells and organs not involved in phagocytosis. Biochemical data demonstrated that the N-terminal domain of Eater directly binds to acetylated and oxidized low-density lipoproteins, which act as mammalian scavenger receptor ligands.

Nimrod C1 (NimC1) is also involved in the phagocytosis of bacteria. This receptor is a 90- to 100-kDa transmembrane protein with 10 EGF repeats, similar to Draper and Eater. Furthermore, suppression of NimC1 expression by RNAi inhibited the phagocytosis of *S. aureus* in plasmatocytes. However, when NimC1 was overexpressed in S2 cells, phagocytosis was induced in response to *S. aureus* and *E. coli*. Interestingly, the *NimC1* gene is part of a cluster of 10 related *Nimrod* genes on chromosome 2 in *Drosophila*, and similar clusters of *Nimrod*-like genes were conserved in other insects, such as *Anopheles* and *Apis*. Although these five receptors were discovered as *Drosophila* phagocytosis receptors, how these receptors regulate intracellular pathways directing cytoskeletal remodeling and membrane trafficking was not addressed. In addition, the identification of phagocytic receptor ligands is essential to understanding the molecular mechanism of phagocytosis.

In 2009, the Nakanishi group identified an endoplasmic reticulum protein, Pretaporter, as a ligand for Draper (Kuraishi *et al.*, 2009). Pretaporter appeared to relocate from the endoplasmic reticulum to the cell surface during apoptosis to serve as an eat-me signal. Pretaporter-bound Draper was tyrosine phosphorylated, and required the adaptor Ced-6 and the small G-protein Rac to signal, suggesting that Pretaporter activates the engulfment pathway involving Draper/Ced-6/Rac, which is reminiscent of the CED-1/CED-6/CED-1 pathway in *C. elegans*. In addition, Calreticulin, another endoplasmic reticulum protein in *Drosophila*, acts as an eat-me signal (Kuraishi *et al.*, 2007);

however, a corresponding phagocytosis receptor has yet to be identified. Subsequently, the Nakanishi group attempted to identify the ligand molecule that is recognized by *Drosophila* hemocytes. Analysis of the cellular immune response using a series of *S. aureus* mutants that are defective in cell wall synthesis and *Drosophila* phagocytosis identified lipoteichoic acid as the ligand for Draper (Hashimoto *et al.*, 2009). Phagocytic receptors activate uptake machinery during phagosome maturation. Recent genome-wide RNAi screening studies in *Drosophila* S2 cells have identified proteins required for microbe uptake (Rämet *et al.*, 2002; Stuart *et al.*, 2007). After recognition of the ligand by phagocytic receptors, curvature of the membrane and extension of pseudopod tips are necessary for phagocytosis. RNAi-based analysis revealed that membrane curvature is regulated by coat proteins containing Clathrin and a coatomer protein complex. The properties of the coatomer protein complex are not discussed in detail here, as this is extensively covered in other reviews (for example, Stuart *et al.*, 2007). To internalize target microbes more efficiently, it is necessary for phagocytes to recruit additional membranes from intracellular components, such as endosomes and the endoplasmic reticulum. During this process, the exocyst, an octodimeric complex, tethers the recruited endosomes to the phagocytotic cup. These exocyst and coatomer protein complexes are thought to regulate phagosome maturation. However, because *Drosophila* phagocytosis is complicated by many cell surface receptors and key machinery components are partially redundant, studies dissecting the complexity of ligand recognition and intracellular signaling are needed. In addition, because the biological significance of insect phagocytosis is unclear, further studies addressing the diverse cellular fate of internalized targets and the biological functions of matured phagosomes are also needed.

14.5.2. Hemocyte Hematopoiesis during Wasp Parasitism

As described above, early studies demonstrated that *Drosophila* plasmocytes are involved in phagocytosis, and crystal cells are required for melanization. However, lamellocytes are responsible for encapsulation (Meister and Lagueux, 2003). Encapsulation, another cellular defense response, occurs when foreign bodies that are too large to be phagocytosed are surrounded. Morphological analysis of encapsulation suggested that blood cells attach to the invading foreign body, such as wasp eggs, and establish a multi-layered cellular capsule. When parasitoid wasps lay their eggs in *Drosophila* larvae, differentiated lamellocytes, which do not exist in naïve conditions, induce encapsulation. This observation has led to questions about whether the blood cells differentiate from larval hematopoietic progenitors (prohemocytes) in response to wasp

infection, and how blood cell homeostasis is regulated at the molecular level. These questions are fundamental to the elucidation of insect encapsulation during cellular immune responses.

Similar to the recent development of mammalian stem cell research, *Drosophila* hemocyte hematopoiesis research has quickly developed (Crozatier and Meister, 2007). Here, we overview studies defining the relationship between hemocyte hematopoiesis and wasp infection. In 1984, Rizki reported that massive proliferation of lamellocytes occurred in the lymph upon parasitism of wasp parasitoid, while larvae maintained a balance of 95% plasmatocytes and 5% crystal cells (Rizki and Rizki, 1984). Subsequently, Sorrentino *et al.* (2004) observed that proliferation of lamellocytes was dependent on JAK/STAT activity, and Krzemien *et al.* (2007) reported that the JAK/STAT pathway is required for maintenance of the lymph gland and the control of hemocyte production in third instar larvae. In addition, JAK/STAT signaling is turned off upon wasp infestation, resulting in differentiation of prohemocytes into lamellocytes. The JAK/STAT signaling pathway is evolutionarily conserved, and mammalian genomes encode multiple components of the JAK/STAT pathway, including multiple receptor subunits. Therefore, it is not easy to define the biological functions of each component and receptor. In contrast to mammals, only one receptor (Domeless), one JAK (Hopscotch), one STAT (Stat92E), and three cytokines – Unpaired 1 (Upd1), Upd2, and Upd3 – have been identified and characterized in *Drosophila*, indicating that it is easy to determine the molecular mechanism because of its simplicity.

In 2010, Makki *et al.* (2010) reported intriguing results regarding the regulatory mechanism of the JAK/STAT pathway. A novel negative regulator that functions as a switch-off molecule in this pathway was identified. They noticed one gene, *CG14225/lat*, coded for a protein structurally related to the Domeless receptor, that had extracellular domains similar to the cytokine-binding domain (CBM) and Lat-Domeless Homology Region (LDHR). The *lat* gene is selectively expressed in hematopoietic progenitors in the *Drosophila* lymph gland. Several novel findings regarding *lat* gene function were made: (1) *lat* mutant flies had a dysfunctional cellular immune response to wasp parasitism; (2) Lat protein functioned as a switch-off molecule of the JAK/STAT signaling pathway by heterodimerizing with Domeless *in vivo*; (3) Upd3, but not Udp1 and Udp2, activated the JAK/STAT signaling pathway; and (4) wasp parasitism increased the Lat/Dome protein ratio, decreasing JAK/STAT signaling and differentiating prohemocytes into lamellocytes. These results suggest that the *Drosophila* cellular immune response to wasp infection is regulated by two molecules: Lat and Domeless. In addition, the JAK/STAT pathway is inactivated by an increased Lat/

Domeless protein ratio, upon wasp parasitism *in vivo*. Therefore, this study identifies the need to determine whether the Lat-like short, non-signaling receptor homolog can function as an antagonist of the mammalian JAK/STAT signaling pathway.

14.6. Newly Emerging Topics in Insect Immunology

14.6.1. Gut Insect Immunology

The insect gut contains various microorganisms, such as symbiotic commensal and food-borne pathogenic bacteria. The survival of symbiotic gut microbes is achieved by evolutionary positive selection. It is unknown how the insect gut epithelium tolerates commensal bacteria while maintaining the ability to trigger an immune response upon pathogenic microbe infection. The basis for the symbiotic relationship between the insect gut epithelia and symbiotic microbes is also unknown. To answer these questions, *Drosophila* was used as a model system because the *Drosophila* gut is known to harbor approximately 10–20 bacteria species in the midgut region, whereas the human gut is thought to host 500–1000 bacteria species. Taking advantage of this simplicity, several groups have begun to elucidate the molecular mechanisms underlying innate immunity homeostasis in the *Drosophila* gut. These studies concluded that *Drosophila* gut immunity primarily relies on two effector molecules, AMPs and reactive oxygen species (ROS), and that these two molecules are tightly regulated to prevent the growth and proliferation of pathogenic bacteria. However, AMPs and ROS preserve commensal bacteria within the gut. Lee and colleagues reported seminal work regarding insect gut immunity (Ha *et al.*, 2005a; 2005b, 2009a, 2009b; Ryu *et al.*, 2008). Here, we summarize recent research progress in insect gut immunity.

14.6.1.1. NF-κB-AMP-mediated gut immunity
An intriguing aspect of gut immunity is that constant direct contact between gut epithelia and commensal bacteria induces the activation of a constitutive basal immune response, suggesting that AMPs are always produced, even in the presence of commensal bacteria. Several groups reported that the basal level of AMP in *Drosophila* epithelia is dependent on the IMD pathway and independent of the Toll pathway. However, it was unknown how insects accommodate commensal bacteria in spite of the presence of AMP, and how insects inhibit the growth and proliferation of pathogenic bacteria in gut.

Recently, Lee and colleagues elegantly demonstrated that *Drosophila* control gut homeostasis through Caudal (Cad). Cad, a homeobox transcription factor, was originally discovered as a regulator of the antero-posterior body axis in *Drosophila* (Dearolf *et al.*, 1989). Cad expression is tightly regulated in response to developmental signals during embryogenesis. However, Cad expression is altered in the intestine and Malpighian tubules after embryogenesis. Lee and colleagues found that *Drosophila* AMPs, such as diptericin and cecropin, which are expressed by a Relish-dependent IMD pathway in the fat body, were strongly inhibited at the transcriptional level by Cad. This occurred in spite of the basal level of IMD signaling (Ryu *et al.*, 2008), suggesting that Cad acts as a gut-specific repressor of AMP gene expression. Based on these results, they hypothesized that repression of NF-κB-dependent AMP production is required to preserve and tolerate commensal bacteria, and that regulation of constitutive AMP expression leads to modification of the commensal community structure in gut (**Figure 4**).

To address these hypotheses, changes in the bacteria population after manipulation of AMPs were studied. When AMP genes were overexpressed, the expression of major commensal bacterium (A911 strain of *Acetobacteraceae*) decreased; however, the amount of minor commensal bacterium (G707 strain of *Gluconoacetobacter*) increased. These surprising results were confirmed by *in vitro* analysis; the A911 strain was very sensitive to synthetic cecropin, whereas the G707 strain was relatively resistant. These results proposed that, in the presence of the A911 strain, AMP activation antagonizes the dominance of the G707 strain under normal conditions. However, upon AMP overexpression, the A911 population is increased more than that of the G707 strain. In addition, the G707 strain modified the commensal bacterial community found in Cad-knockdown flies, leading to severe apoptosis in the gut (Ryu *et al.*, 2008). Furthermore, gut pathology of Cad-knockdown flies was attenuated in the absence of commensal bacteria. Finally, when the G707 strain was introduced into the gut of germ-free flies in the absence of the A911 strain, apoptosis of intestine cells and host mortality were induced, suggesting that outgrowth of the G707 strain is deleterious to the host.

Along with Lee's studies, several other research groups identified several novel molecules that function as negative regulators of IMD-mediated Relish activation in gut epithelia. PGRP domain-containing proteins, such as PGRP-SC and -LB, degrade DAP-type PGN. These PGRPs are thought to prevent excessive IMD pathway activation by sustaining a low level of PGNs in the gut (Bischoff *et al.*, 2006; Zaidman-Remy *et al.*, 2006). Another group identified an antagonist of the PGRP-LC receptor, PIMS (Rudra; PIRK), which inhibits IMD signaling (Lhocine *et al.*, 2008). These studies identified negative regulatory molecules that work together to regulate Relish activity in gut epithelial cells. In addition, because IMD-mediated Relish activation modifies the basal expression of negative regulators, such as PGRP-SC, -LB and PIMS, commensal bacteria-mediated and IMD-mediated Relish activation is suggested to

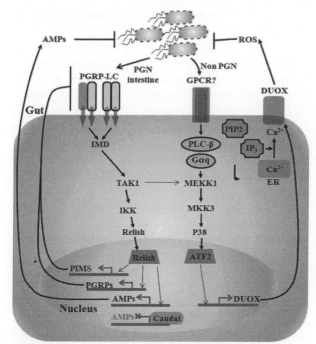

Figure 4 Proposed *Drosophila* gut immune signaling network. Two different effector molecules, AMP and ROS, are produced by activation of the Relish-mediated IMD pathway and DUOX-mediated signaling pathway, respectively. When excessive bacterial PGNs are released onto the gut epithelial cell surface following serious bacterial infection, large amounts of AMPs are produced after strong activation of the Relish-mediated IMD pathway. The excessive AMP production is negatively regulated by Caudal expression. Simultaneously, non-PGNs-mediated DUOX transcription is upregulated via p38-mediated ATF2 activation. This p38 activation is regulated by a PGNs-independent PLC-β pathway and a PGNs-dependent IMD pathway that merge into MEKK1. The production of large amounts of DUOX enzyme after upregulation of DUOX gene produces large amounts of ROS, which can kill bacteria in the gut. In contrast, small amounts of PGNs are released from commensal bacteria under uninfected conditions. In these conditions, p38 is in an inactive state via a PLC-β-dependent feedback loop in order to prevent excessive *Duox* expression. Also, PGRP-LC-mediated AMP production is regulated by Caudal, resulting in proper maintenance of the gut microbial population. Activation of the IMD pathway is negatively regulated by catalytic PGRPs and PIMS. Abbreviations: ER, endoplasmic reticulum; GPCR, G protein-coupled receptor; IKK, IκB kinase; IP3, inositol-1,4,5-trisphosphate; PIP2, phosphatidylinositol-4,5-isphosphate.

play a pivotal role in a negative regulatory feedback loop (**Figure 4**). Taken together, gut immunity studies using *Drosophila* as an animal model enabled the study of the normal flora community, which prevents colonization of potentially pathogenic microbes. Clinically, it was suggested that an imbalance in the equilibrated gut flora of humans induced by antibiotics, or anti-cancer chemotherapies or radiation, can lead to pathogenic infections (Sommer *et al.*, 2009).

14.6.1.2. Duox- and ROS-mediated gut immunity

As described above, NF-κB-AMP-mediated gut immunity is primarily regulated by DAP-type PGNs, which are secreted from commensal or pathogenic bacteria. However, recently it has been discovered that PGN-independent generation of ROS regulates insect gut immunity homeostasis. ROS are a by-product of the metabolism of oxygen, and have biologically important roles in immunity and cell signaling. They are known to have bactericidal effects due to their chemically reactive properties. The bactericidal activity of ROS in innate immunity has been well studied in mammalian phagocytes (Leto and Geiszt, 2006). NADPH oxidase within the phagosome (NOX2) produces superoxide from molecular oxygen. In a study of ROS-mediated insect gut immunity, Ha *et al.* (2005b) showed that flies lacking immune-regulated catalase (IRC) are highly susceptible to gut infection. In addition, they demonstrated that the mortality of IRC-deficient flies was due to oxidative stress after exposure to microbial components, rather than to the overproliferation of ingested microbes, suggesting that the insect gut may have a novel ROS generating system that is responsive to microbial components, but not PGN. They hypothesized that ROS generated from insect gut epithelia cells, which are not traditional phagocytic cells, act as bactericidal effector molecules in gut immunity.

Subsequently, Ha *et al.* (2005a) discovered that Duox, an oxidase, is essential for the regulation of gut immunity. *Drosophila* have one NOX and one Duox gene in their genome. Genetic analysis demonstrated that *Drosophila* Duox, but not NOX, is involved in microbial clearance in the gut, and that Duox knockdown flies are highly susceptible to gut infection. This susceptibility is due to the overproliferation of ingested microorganisms. How Duox is regulated after recognition of non-PGN molecules of gut microbes is unknown. As shown in **Figure 4**, Ha *et al.* (2009a, 2009b) provided evidence that unidentified non-PGN components of bacteria are involved in Duox-mediated ROS generation *in vitro* and *in vivo*. Subsequently, they determined how Duox is regulated. A non-PGN ligand molecule was recognized by an unknown membrane-associated G protein-coupled receptor (GPCR) and the ligand recognition signal was transmitted to the intracellular downstream molecules, Gαq and phospholipase Cβb (PLCβ), leading to the mobilization of intracellular calcium via generation of inositol 1,4,5-trisphosphate (IP3). This PLCβ/IP3-mediated calcium mobilization is sufficient for the spontaneous activation of Duox and subsequent ROS generation to promote bactericidal activity. Furthermore, Gαq and PLCβ-deficient mutant flies were unable to control gut microbes, and were highly susceptible to infection under conventional rearing conditions. However, this susceptibility was rescued under germ-free rearing conditions, confirming that regulation of Duox activity is a crucial factor in insect host defense.

Next, these authors asked how *Duox* gene expression is transcriptionally regulated in response to infectious conditions, and what kinds of intracellular molecules are involved in the regulation of *duox* gene expression. They found that *Duox* gene expression was regulated by a MEKK1–MKK3–p38 pathway in the presence of a high concentration of microbial ligands, which contain both PGN and non-PGN molecules. This result was unexpected, because PGN induced *Duox* gene expression but not Duox enzyme activity. These results clearly demonstrate that activation of the MEKK1–MKK3–p38 pathway occurs both in a PGN-independent manner via PLCβ and a PGN-dependent manner involving the PGRP-LC and IMD pathways (**Figure 4**).

Because ROS is also harmful to host cells, ROS generation is tightly regulated. Lee and colleagues demonstrated that Duox expression was precisely regulated by cross-talk between a PGN-PGRP-LC-dependent IMD–MEKK1–MKK3–p38–ATF2 signaling pathway and a non-PGN-GPCR-dependent PLCβ-signaling pathway. Namely, in the presence of the normal commensal gut population, Duox enzyme-mediated constitutive activation resulted in low PLCβ activity and downregulated PGN-dependent activation of p38. This suggests that the basal expression of Duox is required to reduce oxidative damage during the commensal–gut epithelia interaction. Conversely, upon infection by pathogenic bacteria, cooperation of Duox enzyme activity is required, and leads to ROS production. In this situation, PGN-independent PLCβ signaling and PGN-dependent PGRP-LC-mediated IMD signaling activate MEKK1 to induce p38-dependent Duox induction. At the same time, activation of PLCβ by high concentrations of non-PGN ligands increases Duox enzymatic activity (**Figure 4**). These data demonstrate how the host achieves gut microbe homeostasis by efficiently combating pathogenic bacteria while at the same time tolerating commensal microbes in the gut.

Finally, even though recent data have elucidated how insects regulate gut immunity, there are several questions that are the focus of future studies. First, how do insect gut cells distinguish between commensal bacteria and pathogenic bacteria? Second, what is the ligand structure capable of activating Duox? Future studies will enable a better understanding of the molecular mechanism of insect gut immunity.

14.6.2. Immune Priming of Insect Immunology

Insects are considered to lack both immune specificity and immune memory (immune priming) because they do not possess an adaptive immune system. However, recent molecular and genetic studies of fruit flies and mosquitoes

have identified diverse immune receptors in invertebrates (Dong *et al.*, 2006; Watson *et al.*, 2005), providing an unexpectedly high degree of specificity and immune memory within invertebrate systems (Schmid-Hempel, 2005). In addition, these studies demonstrated that a specific immune memory that is functionally similar to vertebrate immune memory may be present in invertebrates.

The phenomena of memory-like immune responses are termed immune priming, and examples of pathogen-specific and specific immune priming have been observed in numerous insects. Kurtz and Franz (2003) provided phenomenological evidence for specific memory in the invertebrate immune system by infecting a copepod, *Macrocyclops albidus*, with the tapeworm *Schistocephalus solidus*. Their results showed that the innate defense system of copepods reacted more efficiently during secondary infection after hosts had previously encountered similar parasites. These results demonstrated that the innate immune defense system in invertebrates may have an ability to discriminate amongst antigenic molecules, and that a specific immune memory exists.

In 2006, Sadd and Schmid-Hempel (2006) demonstrated the presence of specific immune priming in the bumblebee (*Bombus terrestris*). They used three different bacterial pathogens to determine whether prior homologous pathogen exposure induces long-term immune protection. As expected, they observed that bees have enhanced protection and specificity during secondary bacterial exposure even if it is several weeks after the initial exposure, suggesting that the invertebrate immune system is capable of specific immune memory. Subsequently, Pham *et al.* (2007) showed that *Streptococcus pneumoniae*-primed flies were protected against a subsequent lethal challenge with *S. pneumoniae*. This response was specific for *S. pneumoniae*, and persisted through the lifetime of the fly. In addition, they suggested that the *Drosophila* Toll pathway, but not the IMD pathway, is required for the primed response.

In 2010, Rodrigues *et al.* (2010) reported that a primed mosquito immune system has a 2- to 3.2-fold enhanced macrophage-like insect cell response upon exposure to a *Plasmodium* malaria parasite, demonstrating that the mosquito immune system can modulate macrophage-like hemocytes. Surprisingly, a strong priming response was also shown when malaria ookinetes breached the mosquito gut barrier and injured epithelial cells upon contact with bacteria, indicating a systemic immune surveillance. Upon re-exposure to a similar insult, mosquitoes mounted a more effective antibacterial response to kill *Plasmodium* parasites. This work is significant in regard to malaria control and the understanding of immune memory in invertebrates. In addition, unraveling the molecular mechanism of insect immune responses, especially the differentiation of mosquito macrophage-like hemocytes, will provide further insight into the evolution of immune memory in insects.

Three important aspects of invertebrate immune specificity have been surmised (Schulenburg *et al.*, 2007): (1) immune specificity of insects relies on the genetic diversity of pathogen recognition receptors and/or immune effectors; (2) diverse pathogen recognition receptors and/or immune effectors interact synergistically to enhance immune specificity; and (3) specific recognition of pathogens can be greatly enhanced by increasing the concentration of the relevant receptors and/or immune effectors.

Alternative splicing of exons within Dscam immunoglobulin produces diversification, with more than 30,000 isoforms in both *D. melanogaster* and *A. gambiae* (Watson *et al.*, 2005; Dong *et al.*, 2006). Different splice forms show varied affinity with pathogens, as evidence of mammalian immune specificity. However, the molecular mechanism of diversification of Dscams has not yet been determined. In addition, *Drosophila* PGRP-LC, a transmembrane protein, exists as three splice variants (PGRP-LCa, -LCx, and -LCy). LCx functions as a *bona fide* recognition receptor; however, LCa contributes to the transference of immune signaling of monomeric DAP-type PGN for recognition in the IMD pathway. Similarly, LCx homodimers are involved in the polymeric DAP-type PGN recognition pathway, providing yet another piece of evidence that synergistic interaction of pathogen recognition receptors enhances the specificity of pathogen recognition. Finally, many pathogen recognition receptors, such as PGRPs, GNBPs, and lectins, are upregulated when hosts are exposed to pathogens, increasing the efficiency of pathogen detection. However, these preliminary observations are quite different from mammalian immune specificity. Further study of immune priming as a consequence of the amount of pathogen is necessary. In addition, because the molecular mechanisms of these observations are entirely unknown, it remains to be demonstrated exactly how the specificity of insect innate immunity is enhanced. In summary, even though immune specificity and immune memory in insect innate immunity are not as developed as adaptive immunity, recent studies suggest that insect innate immunity may be different from that of vertebrates.

14.6.3. Hemimetabolous Insect Immunology

The difference between a holometabolous and hemimetabolous insect is that the former undergoes a pupal stage, whereas the latter does not. Holometabolism, also termed complete metamorphosis, is a development pattern that consists of four life stages: egg, larva, pupa, and adult. Hemimetabolism, termed incomplete metamorphosis, is a developmental process involving three distinct stages: egg, nymph, and adult. The nymph stage is simply an adult that lacks wings and functional reproductive organs.

Until now, most research regarding insect immunology has been performed using holometabolous insects, such as flies, mosquitoes, silkworms, and beetles. In addition, annotation analysis of insect immune-related genes has been based on information from holometabolous insects, including flies (*Drosophila spp*) (Sackton *et al.*, 2007), mosquitoes (*Aedes aegypti, A. gambiae*; Christophides *et al.*, 2002; Waterhouse *et al.*, 2007), bees (*A. mellifera*; Evans *et al.*, 2006), and beetles (*Tribolium castaneum*; Zou *et al.*, 2007). However, genomic information on hemimetabolous insects became available when the International Aphid Genomics Consortium (IAGC) provided results of sequencing of the pea aphid genome (IAGC, 2010). Both the genomes of the pea aphid and its primary symbiont, *Buchnera aphidicola* (Shigenobu *et al.*, 2000), are now available, providing information with regard to the nature of the interdependency between host and symbionts. In addition, this unique genomic resource allows functional genomic studies to better understand the regulatory networks underlying innate immunity and symbiosis.

Recently, Gerardo *et al.* (2010) conducted an annotation analysis of immune- and defense-related genes in the pea aphid genome. They first analyzed the presence or absence of well-known immune gene homologs that have been reported in holometabolous insects. The authors then measured the production of mRNA and protein to gain insight into the pea aphid response to the various pathogenic microbial challenges. We now summarize some interesting results of their study. Several genes related with the Toll pathway, including multiple Toll receptors and Spätzle genes, serine proteases, and serpins, were identified in aphids. However, PGRP genes, well-characterized receptor genes involved in Toll and IMD signaling in holometabolous insects, were not identified, even though two GNBP genes may function as bacterial sensing molecules. Surprisingly, pea aphids do not express many crucial components of the IMD signaling pathway. This pathway is essential for *Drosophila* protection against Gram-negative bacteria. Absence of the IMD pathway-associated genes suggested that pea aphids may express orthologs of the JNK pathway. In *Drosophila*, the JNK pathway is known to regulate developmental processes as well as wound healing. Therefore, the JNK pathway may be essential for the aphid immune response. As described above, the JAK/STAT pathway in holometabolous insects is thought to play an important role in the overproliferation of hemocytes, upregulation of TEP genes, and induction of antiviral responses. Pea aphids express homologs of the core JAK/STAT genes, including genes encoding for the cytokine receptor Domeless, JAK tyrosine kinase, and STAT92E transcription factor. However, an ortholog of the *Drosophila*

JAK/STAT ligand molecule, *upd* (unpaired), was not identified.

As for other immune-related genes, pea aphids do not express many of the AMPs common to other holometabolous insects. For example, whereas all annotated insect genomes have genes that encode defensins, defensin-like AMP genes were not found in pea aphids. In addition, insect Cecropins, Drosocin, Diptericin, Drosomycin, and Metchnikowin-like AMP genes, which are expressed in *Drosophila*, were not identified. However, surprisingly, six Thaumatin homolog genes, which are well-known AMP genes in plants, were identified. Thaumatin-family proteins contain disulfide-linked polypeptides of approximately 200 residues in length. Recently, some Thaumatin-like proteins were purified from the beetle *T. castaneum*, and were shown to have antifungal activity against filamentous fungi *Beauveria bassiana* and *Fusarium culmorum* (Altincicek *et al.*, 2008). However, why aphids have lost so many common insect AMPs while plant-originated Thaumatin-like proteins are conserved is unknown. Additionally, aphids have five C-type Lectin paralogs and two Galectin paralogs. Insects also express the scavenger receptors, Nimrod and Dscam. In conclusion, pea aphids do not have many genes that function in insect immunology; however, pea aphids have some other genes that may function in the immune response to microbial challenges.

Although symbiosis is outside the scope of this chapter, annotation of the pea aphid and *Buchnera* genomes (Shigenobu *et al.*, 2000) has provided invaluable information on symbiosis. Namely, *Buchnera* has a dramatically small genome, and it was revealed that the *Buchnera* genome has specific genes for the biosynthesis of the nine essential amino acids, histidine, isoleucine, leucine, lysine, methionine, phenylalanine, threonine, tryptophan, and valine. These amino acids are essential for pea aphid growth; however, aphids do not produce these amino acids. Indeed, the pea aphid genome does not have the genetic machinery required for amino acid synthesis. These data provide invaluable insight into aphid development, symbiotic interactions, and co-evolution with obligate and facultative symbiotic bacteria. In addition, the genomic information of these two species will enable us to characterize novel genes that are involved in the pea aphid immune and defense systems.

14.7. Conclusion

Using multiple approaches, including genetics, cell biology, molecular biology, proteomics, RNAi, and system biology, fruitful research has relied on insect model systems. In addition, the availability of many *Drosophila* mutants that are deficient in immune responses against bacterial and fungal infection has enabled the discovery of several innate immune signaling pathways, such as Toll, IMD, JNK, and JAK/STAT signaling. However, much has yet to be done regarding the molecular study of insect immunology. For example, the activation mechanism of the microbial proteinase-mediated danger signaling pathway is unclear. How is pro-persephone protease cleaved by microbial proteinases, and what is the protease directly downstream of this signaling pathway? In addition, the structures of ligands that affect these pathways are unknown, whereas the ligand structures that affect the Toll and IMD pathways have been determined. The elucidation of molecular mechanisms and virulence factors involved in the regulation of insect immune pathways will increase our understanding of insect immunology.

In addition, the DAP-type PGN-mediated IMD signal pathway has not been fully elucidated. The identities of the pattern recognition receptors in the *Drosophila* IMD pathway and the recognition mechanism between tracheal cytotoxin (TCT, **Figure 1D**) and PGRP-LE or PGRP-LC have been determined (Chang *et al.*, 2006; Lim *et al.*, 2006). However, TCT is not produced by all Gram-negative bacteria and *bacilli* species. Therefore, a different natural ligand molecule for the IMD pathway from Gram-negative bacteria must exist. Also unknown is how mycoplasma species, which are deficient in PGN in their cell membrane, are recognized by insects, and how the innate immune response after recognition of mycoplasma is activated.

A particular challenge to the insect immunity research community is the determination of the recognition and activation mechanisms after infection with intracellular pathogenic microbes, such as *Listeria monocytogenes* and *Mycobacterum marium*. Recently, Kurata and colleagues discussed the relationship between autophagy and insect innate immunity (Yano *et al.*, 2008; Kurata, 2010). *Drosophila* PGRP-LC and -LE sense the DAP-type PGN of extracellular- and intracellular-infective bacteria and induce several innate immune responses, such as AMP production, via activation of the IMD pathway. Yano *et al.* (2008) observed PGRP-LE-25-dependent autophagy, when *Drosophila* adults were infected with *L. monocytogenes*, which was independent of the IMD pathway. It will be worthwhile to identify the ligand molecules within these intracellular microbes, and to determine the molecular mechanism of DAP-type PGN-dependent autophagy and downstream signaling.

Even though the study of insect antiviral defense mechanisms has made significant progress, we are still far from completely understanding antiviral resistance mechanisms. Insect antiviral mechanisms are quite different from those triggered by bacterial and fungal infections. Future intensive studies regarding the interaction between virus and insect will help unravel antiviral response mechanisms in mammals. In addition, intensive research to understand innate immune responses in diverse insect species is needed.

Our knowledge and understanding of mosquito immunology has greatly advanced during the past 15 years due to the complete genome sequencing of *A. gambiae*. Future studies need to focus on the molecular mechanisms of mosquito immune reactions, and clarify the natural inconsistencies between responses to vectors and parasites. These studies will allow us to develop universal antiparasitic agents, which would be useful for combating malaria. Finally, many viruses, bacteria, protozoans, and metazoans are known to use insects as vectors. Study of *Anopheles* and *Plasmodium* interactions will provide valuable information regarding vector immunity to various human pathogens.

Over the next 10 years, the newly emerging fields reviewed in this chapter, such as gut immunity, immune priming, and hemimetabolous insect immunology, will be studied extensively in diverse research fields. New genomic information from different insect species will help further our understanding of the basic concepts of self and non-self discrimination in the gut, immune priming, immune specificity, and host-symbiont interactions. In particular, determination of the molecular mechanism of symbiosis will be of benefit in understanding mammalian gut immunity.

Finally, basic research of insect immunology will provide a chance to develop new drugs and clinically valuable diagnostic kits. For example, 5-S-GAD (*N*-β-Alanyl-5-S-glutathionyl-3,4-dihydroxyphenylalanine) was originally isolated as an antibacterial substance from *S. peregrina* adults (Leem *et al.*, 1996). Natori and colleagues recently reported that 5-S-GAD prevents acute lens opacity development due to 5-S-GAD antioxidant activity. Treatment with 5-S-GAD-containing eye drops delayed the progression of UVB-induced cataracts in rats. These studies resulted in lead compounds for drug development, and provide an example of drug discovery based on insect host-defense molecules (Akiyama *et al.*, 2009; Kawada *et al.*, 2009). In addition, because PGN and β-1,3-glucan activates the Toll signaling pathway in insects, which is amplified by a serine protease cascade (Kim *et al.*, 2008), the molecules involved in both PGN recognition and signal amplification are useful for development of novel diagnostic kits. These would be useful for rapid identification of bacteria-contaminated biomaterials, such as platelets and human plasma. A commercially available LPS detection kit has been developed based on the molecular mechanism of the hemolymph clotting system in *Limulus* (Iwanaga *et al.*, 1992). In addition, the components involved in β-1,3-glucan recognition signaling have facilitated the development of sensitive novel kits for the rapid identification of fungi in clinical samples and food products. Recently, a new bacterial contamination detection kit has been developed based on the molecular mechanism of *Tenebrio* Toll signaling pathway.

Acknowledgment

Works in the authors' laboratory are supported by National Research Foundation of Korea grants 2009-0322-000 and 2010-061-3026.

References

Agaisse, H., Petersen, U. M., Boutros, M., Mathey-Prevot, B., & Perrimon, N. (2003). Signaling role of hemocytes in *Drosophila* JAK/STAT-dependent response to septic injury. *Dev. Cell*, *5*, 441–450.

Aggarwal, K., Rus, F., Vriesema-Magnuson, C., Ertürk-Hasdemir, D., Paquette, N., et al. (2008). Rudra interrupts receptor signaling complexes to negatively regulate the IMD pathway. *PLoS Pathog.*, *4*, e1000120.

Akiyama, N., Umeda, I. O., Sogo, S., Nishigori, H., Tsujimoto, M., et al. (2009). 5-S-GAD, a novel radical scavenging compound, prevents lens opacity development. *Free Radic. Biol. Med.*, *46*, 511–519.

Altincicek, B., Knorr, E., & Vilcinskas, A. (2008). Beetle immunity: Identification of immune-inducible genes from the model insect *Tribolium castaneum*. *Dev. Comp. Immunol.*, *32*, 585–595.

An, C., Ishibashi, J., Ragan, E. J., Jiang, H., & Kanost, M. R. (2009). Functions of *Manduca sexta* hemolymph proteinases HP6 and HP8 in two innate immune pathways. *J. Biol. Chem.*, *284*, 19716–19726.

Ashida, M., Doke, K., & Onishi, E. (1974). Activation of prephenoloxidase. 3. Release of a peptide from prephenoloxidase by the activating enzyme. *Biochem. Biophys. Res. Commun.*, *57*, 1089–1095.

Ashida, M., Ishizaki, Y., & Iwahana, H. (1983). Activation of pro-phenoloxidase by bacterial cell walls or beta-1,3- glucans in plasma of the silkworm, *Bombyx mori*. *Biochem. Biophys. Res. Commun.*, *113*, 562–568.

Barillas-Mury, C., Charlesworth, A., Gross, I., Richman, A., Hoffmann, J. A., et al. (1996). Immune factor Gambif1, a new rel family member from the human malaria vector, *Anopheles gambiae*. *EMBO J.*, *15*, 4691–4701.

Barondes, S. H. (1984). Soluble lectins: A new class of extracellular proteins. *Science*, *223*, 1259–1264.

Bartel, D. P. (2004). MicroRNAs: Genomics, biogenesis, mechanism, and function. *Cell*, *116*, 281–297.

Beck, M. H., & Strand, M. R. (2007). A novel polydnavirus protein inhibits the insect prophenoloxidase activation pathway. *Proc. Natl. Acad. Sci. USA*, *104*, 19267–19272.

Beckage, N. E., & Kanost, M. R. (1993). Effects of parasitism by the braconid wasp *Cotesia congregata* on host hemolymph proteins of the tobacco hornworm, *Manduca sexta*. *Insect Biochem.*, *20*, 285–294.

Belvin, M. P., & Anderson, K. V. (1996). A conserved signaling pathway: The *Drosophila* toll-dorsal pathway. *Annu. Rev. Cell Dev. Biol.*, *12*, 393–416.

Bischoff, V., Vignal, C., Duvic, B., Boneca, I. G., Hoffmann, J. A., et al. (2006). Downregulation of the *Drosophila* immune response by peptidoglycan-recognition proteins SC1 and SC2. *PLoS Pathog.*, *2*, e14.

Boman, H. G., Nilsson, I., & Rasmuson, B. (1972). Inducible antibacterial defence system in *Drosophila*. *Nature*, *237*, 232–235.

Broekaert, W. F., Terras, F. R., Cammue, B. P., & Osborn, R. W. (1995). Plant defensins: Novel antimicrobial peptides as components of the host defense system. *Plant Physiol., 108*, 1353–1358.

Brown, K. L., & Hancock, R. E. (2006). Cationic host defense (antimicrobial) peptides. *Curr. Opin. Immunol., 18*, 24–30.

Buchon, N., Poidevin, M., Kwon, H. M., Guillou, A., Sottas, V., et al. (2009). A single modular serine protease integrates signals from pattern-recognition receptors upstream of the *Drosophila* Toll pathway. *Proc. Natl. Acad. Sci. USA, 106*, 12442–12447.

Bulet, P. (2003). Antimicrobial peptides in insect immunity. In R. A. Ezekowitz, & J. A. Hoffmann (Eds.). *Innate Immunity. Infectious Diseases* (pp. 89–107). Totowa, NJ: Humana Press.

Bulet, P., Hetru, C., Dimarcq, J. L., & Hoffmann, D. (1999). Antimicrobial peptides in insects; Structure and function. *Dev. Comp. Immunol., 23*, 329–344.

Carlsson, A., Engström, P., Palva, E. T., & Bennich, H. (1991). Attacin, an antibacterial protein from *Hyalophora cecropia*, inhibits synthesis of outer membrane proteins in *Escherichia coli* by interfering with omp gene transcription. *Infect. Immun., 59*, 3040–3045.

Casteels, P., Ampe, C., Jacobs, F., Vaeck, M., & Tempst, P. (1989). Antibacterial peptides from honeybees. *EMBO J., 8*, 2387–2391. Apidaecins: Antimic.

Cerenius, L., Lee, B. L., & Söderhäll, K. (2008). The proPO-system: Pros and cons for its role in invertebrate immunity. *Trends Immunol., 29*, 263–271.

Cerenius, L., Kawabata, S., Lee, B. L., Nonaka, M., & Söderhäll, K. (2010). Proteolytic cascades and their involvement in invertebrate immunity. *Trends Biochem. Sci., 35*, 575–583.

Chang, C. I., Chelliah, Y., Borek, D., Mengin-Lecreulx, D., & Deisenhofer, J. (2006). Structure of tracheal cytotoxin in complex with a heterodimeric pattern-recognition receptor. *Science, 311*, 1761–1764.

Chasan, R., & Anderson, K. V. (1989). The role of easter, an apparent serine protease, in organizing the dorsal–ventral pattern of the *Drosophila* embryo. *Cell, 56*, 391–400.

Choe, K. M., Werner, T., Stöven, S., Hultmark, D., & Anderson, K. V. (2002). Requirement for a peptidoglycan recognition protein (PGRP) in Relish activation and antibacterial immune responses in *Drosophila*. *Science, 296*, 359–362.

Chowdhury, S., Taniai, K., Hara, S., Kadono-Okuda, K., Kato, Y., et al. (1995). cDNA cloning and gene expression of lebocin, a novel member of antibacterial peptides from the silkworm, *Bombyx mori*. *Biochem. Biophys. Res. Commun., 214*, 271–278.

Christophides, G. K., Zdobnov, E., Barillas-Mury, C., Birney, E., Blandin, S., et al. (2002). Immunity-related genes and gene families in *Anopheles gambiae*. *Science, 298*, 159–165.

Cirimotich, C. M., Dong, Y., Garver, L. S., Sim, S., & Dimopoulos, G. (2010). Mosquito immune defenses against *Plasmodium* infection. *Dev. Comp. Immunol., 34*, 387–395.

Cociancich, S., Ghazi, A., Hetru, C., Hoffmann, J. A., & Letellier, L. (1993). Insect defensin, an inducible antibacterial peptide, forms voltage-dependent channels in *Micrococcus luteus*. *J. Biol. Chem., 268*, 19239–19245.

Collins, F. H., Sakai, R. K., Vernick, K. D., Paskewitz, S., Seeley, D. C., et al. (1986). Genetic selection of a *Plasmodium*-refractory strain of the malaria vector *Anopheles gambiae*. *Science, 234*, 607–610.

Crozatier, M., & Meister, M. (2007). *Drosophila* haematopoiesis. *Cell Microbiol., 9*, 1117–1126.

Cudic, M., Bulet, P., Hoffmann, R., Craik, D. J., & Otvos, L., Jr. (1999). Chemical synthesis, antibacterial activity and conformation of diptericin, an 82-mer peptide originally isolated from insects. *Eur. J. Biochem., 266*, 549–558.

Daquinag, A. C., Nakamura, S., Takao, T., Shimonishi, Y., & Tsukamoto, T. (1995). Primary structure of a potent endogenous dopa-containing inhibitor of phenol oxidase from *Musca domestica*. *Proc. Natl. Acad. Sci. USA, 92*, 2964–2968.

Dearolf, C. R., Topol, J., & Parker, C. S. (1989). The caudal gene product is a direct activator of fushi tarazu transcription during *Drosophila* embryogenesis. *Nature, 341*, 340–343.

Deddouche, S., Matt, N., Budd, A., Mueller, S., Kemp, C., et al. (2008). The DExD/H-box helicase Dicer-2 mediates the induction of antiviral activity in *Drosophila*. *Nat. Immunol., 9*, 1425–1432.

De Gregorio, E., Han, S. J., Lee, W. J., Baek, M. J., Osaki, T., et al. (2002). An immune-responsive Serpin regulates the melanization cascade in *Drosophila*. *Dev. Cell, 3*, 581–592.

Dimopoulos, G., Richman, A., Müller, H. M., & Kafatos, F. C. (1997). Molecular immune responses of the mosquito *Anopheles gambiae* to bacteria and malaria parasites. *Proc. Natl. Acad. Sci. USA, 94*, 11508–11513.

Dimopoulos, G., Seeley, D., Wolf, A., & Kafatos, F. C. (1998). Malaria infection of the mosquito *Anopheles gambiae* activates immune-responsive genes during critical transition stages of the parasite life cycle. *EMBO J., 17*, 6115–6123.

Dimopoulos, G., Müller, H. M., Levashina, E. A., & Kafatos, F. C. (2001). Innate immune defense against malaria infection in the mosquito. *Curr. Opin. Immunol., 13*, 79–88.

Dohke, K. (1973). Studies on prephenoloxidase-activating enzyme from cuticle of the silkworm *Bombyx mori*. II. Purification and characterization of the enzyme. *Arch. Biochem. Biophys., 157*, 210–221.

Dong, Y., Taylor, H. E., Dimopoulos, G., & AgDscam (2006). AgDscam, a hypervariable immunoglobulin domaincontaining receptor of the *Anopheles gambiae* innate immune system. *PLoS Biol., 4*, e229.

Dostert, C., Jouanguy, E., Irving, P., Troxler, L., Galiana-Arnoux, D., et al. (2005). The Jak-STAT signaling pathway is required but not sufficient for the antiviral response of *Drosophila*. *Nat. Immunol., 6*, 946–953.

Eleftherianos, I., Millichap, P. J., ffrench-Constant, R. H., & Reynolds, S. E. (2006). RNAi suppression of recognition protein mediated immune responses in the tobacco hornworm *Manduca sexta* causes increased susceptibility to the insect pathogen *Photorhabdus*. *Dev. Comp. Immunol., 30*, 1099–1107.

Engström, A., Engström, P., Tao, Z. J., Carlsson, A., & Bennich, H. (1984). Insect immunity. The primary structure of the antibacterial protein attacin F and its relation to two native attacins from *Hyalophora cecropia*. *EMBO J., 3*, 2065–2070.

Engström, Y., Kadalayil, L., Sun, S. C., Samakovlis, C., & Hultmark, D. (1993). Kappa B-like motifs regulate the induction of immune genes in *Drosophila. J. Mol. Biol., 232,* 327–333.

Evans, J. D., Aronstein, K., Chen, Y. P., Hetru, C., Imler, J. L., et al. (2006). Immune pathways and defence mechanisms in honey bees *Apis mellifera. Insect Mol. Biol., 15,* 645–656.

Fire, A., Xu, S., Montgomery, M. K., Kostas, S. A., Driver, S. E., et al. (1998). Potent and specific genetic interference by double-stranded RNA in *Caenorhabditis elegans. Nature, 391,* 806–811.

Fradin, C., Poulain, D., & Jouault, T. (2000). Beta-1,2-linked oligomannosides from *Candida albicans* bind to a 32- kilodalton macrophage membrane protein homologous to the mammalian lectin galectin-3. *Infect. Immun., 68,* 4391–4398.

Franc, N. C., Heitzler, P., Ezekowitz, R. A., & White, K. (1999). Requirement for croquemort in phagocytosis of apoptotic cells in *Drosophila. Science, 284,* 1991–1994.

Frolet, C., Thoma, M., Blandin, S., Hoffmann, J. A., & Levashina, E. A. (2006). Boosting NF-kappaB-dependent basal immunity of *Anopheles gambiae* aborts development of *Plasmodium berghei. Immunity, 25,* 677–685.

Fujimoto, K., Okino, N., Kawabata, S., Iwanaga, S., & Ohnishi, E. (1995). Nucleotide sequence of the cDNA encoding the proenzyme of phenol oxidase A1 of *Drosophila melanogaster. Proc. Natl. Acad. Sci. USA, 92,* 7769–7773.

Galiana-Arnoux, D., Dostert, C., Schneemann, A., Hoffmann, J. A., & Imler, J. L. (2006). Essential function *in vivo* for Dicer-2 in host defense against RNA viruses in *Drosophila. Nat. Immunol., 7,* 590–597.

Garver, L. S., Dong, Y., & Dimopoulos, G. (2009). Caspar controls resistance to *Plasmodium falciparum* in diverse anopheline species. *PLoS Pathog., 5,* e1000335.

Georgel, P., Naitza., S., Kappler., C., Ferrandon., D., Zachary., D., et al. (2001). *Drosophila* immune deficiency (IMD) is a death domain protein that activates antibacterial defense and can promote apoptosis. *Dev. Cell, 1,* 503–514.

Gerardo, N. M., Altincicek, B., Anselme., C., Atamian., H., Barribeau., S. M., et al. (2010). Immunity and other defenses in pea aphids, *Acyrthosiphon pisum. Genome Biol., 11,* R21.

Glaser, R. W. (1918). On the existence of immunity principles in insects. *Psyche, 25,* 38–46.

Gobert., V., Gottar., M., Matskevich., A. A., Rutschmann., S., Royet., J., et al. (2003). Dual activation of the *Drosophila* toll pathway by two pattern recognition receptors. *Science, 302,* 2126–2130.

Gottar, M., Gobert, V., Michel, T., Belvin., M., Duyk., G., et al. (2002). The *Drosophila* immune response against Gram-negative bacteria is mediated by a peptidoglycan recognition protein. *Nature, 416,* 640–644.

Gottar., M., Gobert., V., Matskevich., A. A., Reichhart., J. M., Wang., C., et al. (2006). Dual detection of fungal infections in *Drosophila* via recognition of glucans and sensing of virulence factors. *Cell, 127,* 1425–1437.

Ha, E. M., Oh, C. T., Bae, Y. S., & Lee, W. J. (2005a). A direct role for dual oxidase in *Drosophila* gut immunity. *Science, 310,* 847–850.

Ha, E. M., Oh, C. T., Ryu, J. H., Bae, Y. S., Kang, S. W., et al. (2005b). An antioxidant system required for host protection against gut infection in *Drosophila. Dev. Cell, 8,* 125–132.

Ha, E. M., Lee, K. A., Park, S. H., Kim, S. H., Nam, H. J., et al. (2009a). Regulation of DUOX by the galphaqphospholipase C beta-Ca2⁺ pathway in *Drosophila* gut immunity. *Dev. Cell, 16,* 386–397.

Ha, E. M., Lee, K. A., Seo, Y. Y., Kim, S. H., Lim, J. H., et al. (2009b). Coordination of multiple dual oxidase regulatory pathways in responses to commensal and infectious microbes in *Drosophila* gut. *Nat. Immunol., 10,* 949–957.

Hall, M., Scott, T., Sugumaran, M., Söderhäll, K., Law, J. H., et al. (1995). Proenzyme of *Manduca sexta* phenol oxidase: Purification, activation, substrate specificity of the active enzyme, and molecular cloning. *Proc. Natl. Acad. Sci. USA, 92,* 7764–7768.

Hammond, S. M. (2005). Dicing and slicing: The core machinery of the RNA interference pathway. *FEBS Lett., 579,* 5822–5829.

Hashimoto, C., Kim, D. R., Weiss, L. A., Miller, J. W., Morisato, D., et al. (2003). Spatial regulation of developmental signaling by a serpin. *Dev. Cell, 5,* 945–950.

Hashimoto, Y., Tabuchi, Y., Sakurai, K., Kutsuna, M., Kurokawa, K., et al. (2009). Identification of lipoteichoic acid as a ligand for draper in the phagocytosis of *Staphylococcus aureus* by *Drosophila* hemocytes. *J. Immunol., 183,* 7451–7460.

Hennessy, E. J., Parker, A. E., & O'Neill, L. A. (2010). Targeting Toll-like receptors: Emerging therapeutics? *Nat. Rev. Drug Discov., 9,* 293–307.

Hetru, C., & Hoffmann, J. A. (2009). NF-kappaB in the immune response of. *Drosophila. Cold Spring Harb. Perspect Biol., 1,* a000232.

IAGC (International Aphid genomics Consortium) (2010). Genome sequence of the pea aphid *Acyrthosiphon pisum. PLoS Biol., 8,* e1000313.

Iijima, R., Kurata, S., & Natori, S. (1993). Purification, characterization, and cDNA cloning of an antifungal protein from the hemolymph of *Sarcophaga peregrina* (flesh fly) larvae. *J. Biol. Chem., 268,* 12055–12061.

Imler, J. L., & Bulet, P. (2005). Antimicrobial peptides in *Drosophila*: Structures, activities and gene regulation. *Chem. Immunol. Allergy, 86,* 1–21.

Iwanaga, S., Miyata, T., Tokunaga, F., & Muta, T. (1992). Molecular mechanism of hemolymph clotting system in *Limulus. Thromb. Res., 68,* 1–32.

Janeway, C. A., Jr. (1989). Approaching the asymptote? Evolution and revolution in immunology. *Cold Spring Harb. Symp. Quant. Biol., 54*(Pt. 1), 1–13.

Jang, I. H., Chosa, N., Kim, S. H., Nam, H. J., Lemaitre, B., et al. (2006). A Spätzle-processing enzyme required for toll signaling activation in *Drosophila* innate immunity. *Dev. Cell, 10,* 45–55.

Jiang, H., & Kanost, M. R. (1997). Characterization and functional analysis of 12 naturally occurring reactive site variants of serpin-1 from *Manduca sexta. J. Biol. Chem., 272,* 1082–1087.

Jiang, H., & Kanost, M. R. (2000). The clip-domain family of serine proteinases in arthropods. *Insect Biochem. Mol. Biol., 30,* 95–105.

Jiang, H., Wang, Y., & Kanost, M. R. (1998). Pro-phenol oxidase activating proteinase from an insect, *Manduca sexta*: A bacteria-inducible protein similar to *Drosophila* easter. *Proc. Natl. Acad. Sci. USA, 95,* 12220–12225.

Jiang, R., Kim, E. H., Gong, J. H., Kwon, H. M., Kim, C. H., et al. (2009). Three pairs of protease-serpin complexes cooperatively regulate the insect innate immune responses. *J. Biol. Chem.*, *284*, 35652–35658.

Ju, J. S., Cho, M. H., Brade, L., Kim, J. H., Park, J. W., et al. (2006). A novel 40-kDa protein containing six repeats of an epidermal growth factor-like domain functions as a pattern recognition protein for lipopolysaccharide. *J. Immunol.*, *177*, 1838–1845.

Jutras, I., & Desjardins, M. (2005). Phagocytosis: At the cross-roads of innate and adaptive immunity. *Annu. Rev. Cell Dev. Biol.*, *21*, 511–527.

Kambris, Z., Brun, S., Jang, I. H., Nam, H. J., Romeo, Y., et al. (2006). *Drosophila* immunity: A large-scale *in vivo* RNAi screen identifies five serine proteases required for Toll activation. *Curr. Biol.*, *16*, 808–813.

Kan, H., Kim, C. H., Kwon, H. M., Park, J. W., Roh, K. B., et al. (2008). Molecular control of phenoloxidase-induced melanin synthesis in an insect. *J. Biol. Chem.*, *283*, 25316–25323.

Kang, D., Liu, G., Lundström, A., Gelius, E., & Steiner, H. (1998). A peptidoglycan recognition protein in innate immunity conserved from insects to humans. *Proc. Natl. Acad. Sci. USA*, *95*, 10078–10082.

Kanost, M. R., & Jiang, H. (1997). Serpins from an insect, *Manduca sexta*. *Adv. Exp. Med. Biol.*, *425*, 155–161.

Kanost, M. R., Jiang, H., & Yu, X. Q. (2004). Innate immune responses of a lepidopteran insect, *Manduca sexta*. *Immunol. Rev.*, *198*, 97–105.

Kawabata, T., Yasuhara, Y., Ochiai, M., Matsuura, S., & Ashida, M. (1995). Molecular cloning of insect pro-phenol oxidase: A copper-containing protein homologous to arthropod hemocyanin. *Proc. Natl. Acad. Sci. USA*, *92*, 7774–7778.

Kawada, H., Kojima, M., Kimura, T., Natori, S., Sasaki, K., et al. (2009). Effect of 5-S-GAD on UV-B-induced cataracts in rats. *Jpn J. Ophthalmol.*, *53*, 531–535.

Kawaguchi, N., Komano, H., & Natori, S. (1991). Involvement of *Sarcophaga* lectin in the development of imaginal discs of *Sarcophaga peregrina* in an autocrine manner. *Dev. Biol.*, *144*, 86–93.

Kawai, T., & Akira, S. (2010). The role of pattern-recognition receptors in innate immunity: Update on Toll-like receptors. *Nat. Immunol.*, *11*, 373–384.

Kemp, C., & Imler, J. L. (2009). Antiviral immunity in *Drosophila*. *Curr. Opin. Immunol.*, *21*, 3–9.

Kim, C. H., Kim, S. J., Kan, H., Kwon, H. M., Roh, K. B., et al. (2008). A three-step proteolytic cascade mediates the activation of the peptidoglycan-induced Toll pathway in an insect. *J. Biol. Chem.*, *283*, 7599–7607.

Kim, M., Lee, J. H., Lee, S. Y., Kim, E., & Chung, J. (2006). Caspar, a suppressor of antibacterial immunity in *Drosophila*. *Proc. Natl. Acad. Sci. USA*, *103*, 16358–16363.

Kim, M. S., Baek, M. J., Lee, M. H., Park, J. W., Lee, S. Y., et al. (2002). A new easter-type serine protease cleaves a masquerade-like protein during prophenoloxidase activation in *Holotrichia diomphalia* larvae. *J. Biol. Chem.*, *277*, 39999–40004.

Kleino, A., Myllymäki, H., Kallio, J., Vanha-aho, L. M., Oksanen, K., et al. (2008). Pirk is a negative regulator of the *Drosophila* Imd pathway. *J. Immunol.*, *180*, 5413–5422.

Kocks, C., Cho, J. H., Nehme, N., Ulvila, J., Pearson, A. M., et al. (2005). Eater, a transmembrane protein mediating phagocytosis of bacterial pathogens in *Drosophila*. *Cell*, *123*, 335–346.

Koizumi, N., Imai, Y., Morozumi, A., Imamura, M., Kadotani, T., et al. (1999). Lipopolysaccharide-binding protein of *Bombyx mori* participates in a hemocyte-mediated defense reaction against gram-negative bacteria. *J. Insect. Physiol.*, *45*, 853–859.

Komano, H., Nozawa, R., Mizuno, D., & Natori, S. (1983). Measurement of *Sarcophaga peregrina* lectin under various physiological conditions by radioimmunoassay. *J. Biol. Chem.*, *258*, 2143–2147.

Krem, M. M., & Di Cera, E. (2002). Evolution of enzyme cascades from embryonic development to blood coagulation. *Trends Biochem. Sci.*, *27*, 67–74.

Krzemień, J., Dubois, L., Makki, R., Meister, M., Vincent, A., et al. (2007). Control of blood cell homeostasis in *Drosophila* larvae by the posterior signalling centre. *Nature*, *446*, 325–328.

Kuraishi, T., Manaka, J., Kono, M., Ishii, H., Yamamoto, N., et al. (2007). Identification of calreticulin as a marker for phagocytosis of apoptotic cells in *Drosophila*. *Exp. Cell Res.*, *313*, 500–510.

Kuraishi, T., Manaka, J., Kono, M., Ishii, H., Yamamoto, N., et al. (2009). Pretaporter, a *Drosophila* protein serving as a ligand for Draper in the phagocytosis of apoptotic cells. *EMBO J.*, *28*, 3868–3878.

Kurata, S. (2010). Extracellular and intracellular pathogen recognition by *Drosophila* PGRP-LE and PGRP-LC. *Intl. Immunol.*, *22*, 143–148.

Kurtz, J., & Franz, K. (2003). Innate defence: Evidence for memory in invertebrate immunity. *Nature*, *425*, 37–38.

Kurucz, E., Márkus, R., Zsámboki, J., Folkl-Medzihradszky, K., Darula, Z., et al. (2007). Nimrod, a putative phagocytosis receptor with EGF repeats in *Drosophila* plasmatocytes. *Curr. Biol.*, *17*, 649–654.

Kwon, T. H., Kim, M. S., Choi, H. W., Joo, C. H., Cho, M. Y., et al. (2000). A masquerade-like serine proteinase homologue is necessary for phenoloxidase activity in the coleopteran insect, *Holotrichia diomphalia* larvae. *Eur. J. Biochem.*, *267*, 6188–6196.

Levashina, E. A., Langley, E., Green, C., Gubb, D., Ashburner, M., Hoffmann, J. A., & Reichhart, J. M. (1999). Constitutive activation of toll-mediated antifungal defense in serpin-deficient *Drosophila*. *Science*, *285*, 1917–1919.

Lee, S. Y., Moon, H. J., Kurata, S., Kurama, T., Natori, S., et al. (1994). Purification and molecular cloning of cDNA for an inducible antibacterial protein of larvae of a coleopteran insect, *Holotrichia diomphalia*. *J. Biochem.*, *115*, 82–86.

Lee, S. Y., Moon, H. J., Kurata, S., Natori, S., & Lee, B. L. (1995). Purification and cDNA cloning of an antifungal protein from the hemolymph of *Holotrichia diomphalia* larvae. *Biol. Pharm. Bull.*, *18*, 1049–1052.

Lee, S. Y., Cho, M. Y., Hyun, J. H., Lee, K. M., Homma, K. I., et al. (1998). Molecular cloning of cDNA for prophenol-oxidase-activating factor I, a serine protease is induced by lipopolysaccharide or 1,3-beta-glucan in coleopteran insect, *Holotrichia diomphalia* larvae. *Eur. J. Biochem*, *257*, 615–621.

Lee, W. J., Lee, J. D., Kravchenko, V. V., Ulevitch, R. J., & Brey, P. T. (1996). Purification and molecular cloning of an inducible gram-negative bacteria-binding protein from the silkworm, *Bombyx mori*. *Proc. Natl. Acad. Sci. USA*, *93*, 7888–7893.

Lee, Y. S., Nakahara, K., Pham, J. W., Kim, K., He, Z., et al. (2004). Distinct roles for *Drosophila* Dicer-1 and Dicer-2 in the siRNA/miRNA silencing pathways. *Cell*, *117*, 69–81.

Leem, J. Y., Nishimura, C., Kurata, S., Shimada, I., Kobayashi, A., et al. (1996). Purification and characterization of N-beta-alanyl-5-S-glutathionyl-3,4-dihydroxyphenylalanine, a novel antibacterial substance of *Sarcophaga peregrina* (flesh fly). *J. Biol. Chem.*, *271*, 13573–13577.

Lemaitre, B., & Hoffmann, J. (2007). The host defense of. *Drosophila melanogaster*. *Annu. Rev. Immunol.*, *25*, 697–743.

Lemaitre, B., Nicolas, E., Michaut, L., Reichhart, J. M., & Hoffmann, J. A. (1996). The dorsoventral regulatory gene cassette spätzle/Toll/cactus controls the potent antifungal response in *Drosophila* adults. *Cell*, *86*, 973–983.

Leto, T. L., & Geiszt, M. (2006). Role of Nox family NADPH oxidases in host defense. *Antioxid. Redox Signal*, *8*, 1549–1561.

Leulier, F., Parquet, C., Pili-Floury, S., Ryu, J. H., Caroff, M., et al. (2003). The *Drosophila* immune system detects bacteria through specific peptidoglycan recognition. *Nat. Immunol.*, *4*, 478–484.

Levashina, E. A., Langley, E., Green, C., Gub, D., Ashburner, M., et al. (1999). Constitutive activation of toll-mediated antifungal defense in serpin-deficient *Drosophila*. *Science*, *285*, 1917–1919.

Levashina, E. A., Moita, L. F., Blandin, S., Vriend, G., Lagueux, M., et al. (2001). Conserved role of a complement-like protein in phagocytosis revealed by dsRNA knockout in cultured cells of the mosquito, *Anopheles gambiae*. *Cell*, *104*, 709–718.

Lhocine, N., Ribeiro, P. S., Buchon, N., Wepf, A., Wilson, R., et al. (2008). PIMS modulates immune tolerance by negatively regulating *Drosophila* innate immune signaling. *Cell Host Microbe*, *4*, 147–158.

Li, W. X., & Ding, S. W. (2001). Viral suppressors of RNA silencing. *Curr. Opin. Biotechnol.*, *12*, 150–154.

Li, Y., Wang, Y., Jiang, H., & Deng, J. (2009). Crystal structure of *Manduca sexta* prophenoloxidase provides insights into the mechanism of type 3 copper enzymes. *Proc. Natl. Acad. Sci. USA*, *106*, 17002–17006.

Ligoxygakis, P., Pelte, N., Hoffmann, J. A., & Reichhart, J. M. (2002a). Activation of *Drosophila* Toll during fungal infection by a blood serine protease. *Science*, *297*, 114–116.

Ligoxygakis, P., Pelte, N., Ji, C., Leclerc, V., Duvic, B., et al. (2002b). A serpin mutant links Toll activation to melanization in the host defence of *Drosophila*. *EMBO J.*, *21*, 6330–6337.

Ligoxygakis, P., Roth, S., & Reichhart, J. M. (2003). A serpin regulates dorsal–ventral axis formation in the *Drosophila* embryo. *Curr. Biol.*, *13*, 2097–2102.

Lim, J. H., Kim, M. S., Kim, H. E., Yano, T., Oshima, Y., et al. (2006). Structural basis for preferential recognition of diaminopimelic acid-type peptidoglycan by a subset of peptidoglycan recognition proteins. *J. Biol. Chem.*, *281*, 8286–8295.

Lu, Z., & Jiang, H. (2007). Regulation of phenoloxidase activity by high- and low-molecular-weight inhibitors from the larval hemolymph of *Manduca sexta*. *Insect Biochem. Mol. Biol.*, *37*, 478–485.

Mackenzie, J. S., Gubler, D. J., & Petersen, L. R. (2004). Emerging flaviviruses: The spread and resurgence of Japanese encephalitis, *West Nile* and *Dengue* viruses. *Nat. Med.*, *10*, 98–109.

Maillet, F., Bischoff, V., Vignal, C., Hoffmann, J., & Royet, J. (2008). The *Drosophila* peptidoglycan recognition protein PGRP-LF blocks PGRP-LC and IMD/JNK pathway activation. *Cell Host Microbe*, *3*, 293–303.

Makki, R., Meister, M., Pennetier, D., Ubeda, J. M., Braun, A., et al. (2010). A short receptor downregulates Jak/STAT signalling to control the *Drosophila* cellular immune response. *PLoS Biol.*, *8*, e1000441.

Manaka, J., Kuraishi, T., Shiratsuchi, A., Nakai, Y., Higashida, H., et al. (2004). Draper-mediated and phosphatidylserine-independent phagocytosis of apoptotic cells by *Drosophila* hemocytes/macrophages. *J. Biol. Chem.*, *279*, 48466–48476.

Mason, H. S. (1955). Comparative biochemistry of the phenoloxidase complex. *Adv. Enzymol.*, *16*, 105–184.

Matsushita, M., & Fujita, T. (1992). Activation of the classical complement pathway by mannose-binding protein in association with a novel C1s-like serine protease. *J. Exp. Med.*, *176*, 1497–1502.

Medzhitov, R., & Janeway, C. A. (2002). Decoding the patterns of self and nonself by the innate immune system. *Science*, *296*, 298–300.

Medzhitov, R., Preston-Hurlburt, P., & Janeway, C. A. (1997). A human homologue of the *Drosophila* Toll protein signals activation of adaptive immunity. *Nature*, *388*, 394–397.

Meister, cM., & Lagueux, M. (2003). *Drosophila* blood cells. *Cell Microbiol.*, *5*, 573–580.

Michel, K., Budd, A., Pinto, S., Gibson, T. J., & Kafatos, F. C. (2005). *Anopheles gambiae* SRPN2 facilitates midgut invasion by the malaria parasite *Plasmodium berghei*. *EMBO Rep.*, *6*, 891–897.

Michel, T., Reichhart, J. M., Hoffmann, J. A., & Royet, J. (2001). *Drosophila* Toll is activated by Gram-positive bacteria through a circulating peptidoglycan recognition protein. *Nature*, *414*, 756–759.

Mishima, Y., Quintin, J., Aimanianda, V., Kellenberger, C., Coste, F., et al. (2009). The N-terminal domain of *Drosophila* Gram-negative binding protein 3 (GNBP3) defines a novel family of fungal pattern recognition receptors. *J. Biol. Chem.*, *284*, 28687–28697.

Nakajima, H., Komano, H., Esumi-Kurisu, M., Abe, S., Yamazaki, M., et al. (1982). Induction of macrophage-mediated tumor lysis by an animal lectin, *Sarcophaga peregrina* agglutinin. *Gann.*, *73*, 627–632.

Nappi, A. J., Frey, F., & Carton, Y. (2005). *Drosophila* serpin 27A is a likely target for immune suppression of the blood cell-mediated melanotic encapsulation response. *J. Insect Physiol.*, *51*, 197–205.

Natori, S., Shiraishi, H., Hori, S., & Kobayashi, A. (1999). The roles of *Sarcophaga* defense molecules in immunity and metamorphosis. *Dev. Comp. Immunol.*, *23*, 317–328.

Ochiai, M., & Ashida, M. (2000). A pattern-recognition protein for beta-1,3-glucan. The binding domain and the cDNA cloning of beta-1,3-glucan recognition protein from the silkworm, *Bombyx mori*. *J. Biol. Chem.*, *275*, 4995–5002.

Ohnishi, M. (1953). Tyrosinase activity during puparium formation in *Drosophila melanogaster*. *Jpn. J. Zool.*, *11*, 69–74.

Okada, M., & Natori, S. (1983). Purification and characterization of an antibacterial protein from haemolymph of *Sarcophaga peregrina* (flesh-fly) larvae. *Biochem. J.*, *211*, 727–734.

Osta, M. A., Christophides, G. K., & Kafatos, F. C. (2004). Effects of mosquito genes on *Plasmodium* development. *Science*, *303*, 2030–2032.

Pace, K. E., Lebestky, T., Hummel, T., Arnoux, P., Kwan, K., et al. (2002). Characterization of a novel *Drosophila melanogaster* galectin. Expression in developing immune, neural, and muscle tissues. *J. Biol. Chem.*, *277*, 13091–13098.

Padovan, L., Scocchi, M., & Tossi, A. (2010). Structural aspects of plant antimicrobial peptides. *Curr. Protein Pept. Sci.*, *11*, 210–219.

Park, J. W., Kim, C. H., Kim, J. H., Je, B. R., Roh, K. B., et al. (2007). Clustering of peptidoglycan recognition protein-SA is required for sensing lysine-type peptidoglycan in insects. *Proc. Natl. Acad. Sci. USA*, *104*, 6602–6607.

Pham, L. N., Dionne, M. S., Shirasu-Hiza, M., & Schneider, D. S. (2007). A specific primed immune response in *Drosophila* is dependent on phagocytes. *PLoS Pathog.*, *3*, e26.

Philips, J. A., Rubin, E. J., & Perrimon, N. (2005). *Drosophila* RNAi screen reveals CD36 family member required for mycobacterial infection. *Science*, *309*, 1251–1253.

Piao, S., Song, Y. L., Kim, J. H., Park, S. Y., Park, J. W., et al. (2005). Crystal structure of a clip-domain serine protease and functional roles of the clip domains. *EMBO J.*, *24*, 4404–4414.

Piao, S., Kim, S., Kim, J. H., Park, J. W., Lee, B. L., et al. (2007). Crystal structure of the serine protease domain of prophenoloxidase activating factor-I. *J. Biol. Chem.*, *282*, 10783–10791.

Povelones, M., Waterhouse, R. M., Kafatos, F. C., & Christophides, G. K. (2009). Leucine-rich repeat protein complex activates mosquito complement in defense against *Plasmodium parasites*. *Science*, *324*, 258–261.

Pye, A. E. (1974). Microbial activation of prophenoloxidase from immune insect larvae. *Nature*, *251*, 610–613.

Rabinovich, G. A., Baum, L. G., Tinari, N., Paganelli, R., Natoli, C., et al. (2002). Galectins and their ligands: Amplifiers, silencers or tuners of the inflammatory response?. *Trends Immunol.*, *23*, 313–320.

Ragan, E. J., An, C., Yang, C. T., & Kanost, M. R. (2010). Analysis of mutually exclusive alternatively spliced serpin-1 isoforms and identification of serpin-1 proteinase complexes in *Manduca sexta* hemolymph. *J. Biol. Chem.*, *285*, 29642–29650.

Rahnamaeian, M., Langen, G., Imani, J., Khalifa, W., Altincicek, B., et al. (2009). Insect peptide metchnikowin confers on barley a selective capacity for resistance to fungal ascomycetes pathogens. *J. Exp. Bot.*, *60*, 4105–4114.

Rämet, M., Pearson, A., Manfruelli, P., Li, X., Koziel, H., et al. (2001). *Drosophila* scavenger receptor CI is a pattern recognition receptor for bacteria. *Immunity*, *15*, 1027–1038.

Rämet, M., Manfruelli, P., Pearson, A., Mathey-Prevot, B., & Ezekowitz, R. A. (2002). Functional genomic analysis of phagocytosis and identification of a *Drosophila* receptor for. *E. coli*. *Nature*, *416*, 644–648.

Rees, J. A., Moniatte, M., & Bulet, P. (1997). Novel antibacterial peptides isolated from a European bumblebee, *Bombus pascuorum* (Hymenoptera, Apoidea). *Insect Biochem. Mol. Biol.*, *27*, 413–422.

Reichhart, J. M. (2005). Tip of another iceberg: *Drosophila* serpins. *Trends Cell Biol.*, *15*, 659–665.

Rizki, R. M., & Rizki, T. M. (1984). Selective destruction of a host blood cell type by a parasitoid wasp. *Proc. Natl. Acad. Sci. USA*, *81*, 6154–6158.

Rizki, T., & Rizki, R. (1990). Encapsulation of parasitoid eggs in phenoloxidase-deficient mutants of *Drosophila melanogaster*. *J. Insect. Physiol.*, *36*, 523–529.

Rodrigues, J., Brayner, F. A., Alves, L. C., Dixit, R., & Barillas-Mury, C. (2010). Hemocyte differentiation mediates innate immune memory in *Anopheles gambiae* mosquitoes. *Science*, *329*, 1353–1355.

Roh, K. B., Kim, C. H., Lee, H., Kwon, H. M., Park, J. W., et al. (2009). Proteolytic cascade for the activation of the insect toll pathway induced by the fungal cell wall component. *J. Biol. Chem.*, *284*, 19474–19481.

Royet, J., & Dziarski, R. (2007). Peptidoglycan recognition proteins: Pleiotropic sensors and effectors of antimicrobial defences. *Nat. Rev. Microbiol.*, *5*, 264–277.

Ryu, J. H., Kim, S. H., Lee, H. Y., Bai, J. Y., Nam, Y. D., et al. (2008). Innate immune homeostasis by the homeobox gene caudal and commensal-gut mutualism in *Drosophila*. *Science*, *319*, 777–782.

Sabatier, L., Jouanguy, E., Dostert, C., Zachary, D., Dimarcq, J. L., et al. (2003). Pherokine-2 and -3. *Eur. J. Biochem.*, *270*, 3398–3407.

Sabin, L. R., Hanna, S. L., & Cherry, S. (2010). Innate antiviral immunity in *Drosophila*. *Curr. Opin. Immunol.*, *22*, 4–9.

Sackton, T. B., Lazzaro, B. P., Schlenke, T. A., Evans, J. D., Hultmark, D., et al. (2007). Dynamic evolution of the innate immune system in *Drosophila*. *Nat. Genet.*, *39*, 1461–1468.

Sadd, B. M., & Schmid-Hempel, P. (2006). Insect immunity shows specificity in protection upon secondary pathogen exposure. *Curr. Biol.*, *16*, 1206–1210.

Scherfer, C., Tang, H., Kambris, Z., Lhocine, N., Hashimoto, C., et al. (2008). *Drosophila* Serpin-28D regulates hemolymph phenoloxidase activity and adult pigmentation. *Dev. Biol.*, *323*, 189–196.

Schleifer, K. H., & Kandler, O. (1972). Peptidoglycan types of bacterial cell walls and their taxonomic implications. *Bacteriol. Rev.*, *36*, 407–477.

Schmid-Hempel, P. (2005). Natural insect host-parasite systems show immune priming and specificity: Puzzles to be solved. *Bioessays*, *27*, 1026–1034.

Schulenburg, H., Boehnisch, C., & Michiels, N. K. (2007). How do invertebrates generate a highly specific innate immune response? *Mol. Immunol.*, *44*, 3338–3344.

Selsted, M. E., Harwig, S. S., Ganz, T., Schilling, J. W., & Lehrer, R. I. (1985). Primary structures of three human neutrophil defensins. *J. Clin. Invest.*, *76*, 1436–1439.

Shelby, K. S., Adeyeye, O. A., Okot-Kotber, B. M., & Webb, B. A. (2000). Parasitism-linked block of host plasma melanization. *J. Invertebr. Pathol.*, *75*, 218–225.

Shi, L., Li, B., & Paskewitz, S. M. (2006). Cloning and characterization of a putative inhibitor of melanization from *Anopheles gambiae. Insect Mol. Biol.*, *15*, 313–320.

Shigenobu, S., Watanabe, H., Hattori, M., Sakaki, Y., & Ishikawa, H. (2000). Genome sequence of the endocellular bacterial symbiont of aphids *Buchnera* sp. APS. *Nature*, *407*, 81–86.

Silverman, G. A., Whisstock, J. C., Bottomley, S. P., Huntington, J. A., Kaiserman, D., et al. (2010). Serpins flex their muscle: I. Putting the clamps on proteolysis in diverse biological systems. *J. Biol. Chem.*, *285*, 24299–24305.

Sinden, R. E. (2002). Molecular interactions between *Plasmodium* and its insect vectors. *Cell Microbiol.*, *4*, 713–724

Sinden, R. E. (2004). A proteomic analysis of malaria biology: Integration of old literature and new technologies. *Intl. J. Parasitol.*, *34*, 1441–1450.

Sommer, M. O., Dantas, G., & Church, G. M. (2009). Functional characterization of the antibiotic resistance reservoir in the human microflora. *Science*, *325*, 1128–1131.

Sorrentino, R. P., Melk, J. P., & Govind, S. (2004). Genetic analysis of contributions of dorsal group and JAK-Stat92E pathway genes to larval hemocyte concentration and the egg encapsulation response in *Drosophila. Genetics*, *166*, 1343–1356.

Souza-Neto, J. A., Sim, S., & Dimopoulos, G. (2009). An evolutionary conserved function of the JAK-STAT pathway in anti-dengue defense. *Proc. Natl. Acad. Sci. USA*, *106*, 17841–17846.

Sroka, P., & Vinson, S. B. (1978). Phenoloxidase activity in the hemolymph of parasitized and unparasitized *Heliothis virescens. Insect Biochem.*, *8*, 399–402.

Steiner, H., Hultmark, D., Engström, A., Bennich, H., & Boman, H. G. (1981). Sequence and specificity of two antibacterial proteins involved in insect immunity. *Nature*, *292*, 246–248.

Stevens, J. M. (1962). Bactericidal activity of the blood of actively immunized wax moth larvae. *Can. J. Microbiol.*, *8*, 491–499.

Stuart, L. M., & Ezekowitz, R. A. (2008). Phagocytosis and comparative innate immunity: Learning on the fly. *Nat. Rev. Immunol.*, *8*, 131–141.

Stuart, L. M., Boulais, J., Charriere, G. M., Hennessy, E. J., Brunet, S., et al. (2007). A systems biology analysis of the *Drosophila* phagosome. *Nature*, *445*, 95–101.

Su, X. D., Gastinel, L. N., Vaughn, D. E., Faye, I., Poon, P., et al. (1998). Crystal structure of hemolin: A horseshoe shape with implications for homophilic adhesion. *Science*, *281*, 991–995.

Sun, S. C., Lindström, I., Boman, H. G., Faye, I., & Schmidt, O. (1990). Hemolin: An insect-immune protein belonging to the immunoglobulin superfamily. *Science*, *250*, 1729–1732.

Takehana, A., Katsuyama, T., Yano, T., Oshima, Y., Takada, H., et al. (2002). Overexpression of a pattern-recognition receptor, peptidoglycan-recognition protein-LE, activates imd/relish-mediated antibacterial defense and the prophenoloxidase cascade in *Drosophila* larvae. *Proc. Natl. Acad. Sci. USA*, *99*, 13705–13710.

Takeuchi, O., & Akira, S. (2008). MDA5/RIG-I and virus recognition. *Curr. Opin. Immunol.*, *20*, 17–22.

Tang, H., Kambris, Z., Lemaitre, B., & Hashimoto, C. (2008). A serpin that regulates immune melanization in the respiratory system of *Drosophila. Dev. Cell.*, *15*, 617–626.

Vance, V., & Vaucheret, H. (2001). RNA silencing in plants – defense and counterdefense. *Science*, *292*, 2277–2280.

Volz, J., Müller, H. M., Zdanowicz, A., Kafatos, F. C., & Osta, M. A. (2006). A genetic module regulates the melanization response of *Anopheles* to *Plasmodium. Cell Microbiol.*, *8*, 1392–1405.

Wang, X. H., Aliyari, R., Li, W. X., Li, H. W., Kim, K., et al. (2006). RNA interference directs innate immunity against viruses in adult *Drosophila. Science*, *312*, 452–454.

Waterhouse, R. M., Kriventseva, E. V., Meister, S., Xi, Z., Alvarez, K. S., et al. (2007). Evolutionary dynamics of immune-related genes and pathways in disease-vector mosquitoes. *Science*, *316*, 1738–1743.

Watson, F. L., Püttmann-Holgado, R., Thomas, F., Lamar, D. L., Hughes, M., et al. (2005). Extensive diversity of Ig33 superfamily proteins in the immune system of insects. *Science*, *309*, 1874–1878.

Weber, A. N., Tauszig-Delamasure, S., Hoffmann, J. A., Lelièvre, E., Gascan, H., et al. (2003). Binding of the *Drosophila* cytokine Spätzle to Toll is direct and establishes signaling. *Nat. Immunol.*, *4*, 794–800.

Weis, W. I., Taylor, M. E., & Drickamer, K. (1998). The C-type lectin superfamily in the immune system. *Immunol. Rev.*, *163*, 19–34.

Werner, T., Liu, G., Kang, D., Ekengren, S., Steiner, H., et al. (2000). A family of peptidoglycan recognition proteins in the fruit fly *Drosophila melanogaster. Proc. Natl. Acad. Sci. USA*, *97*, 13772–13777.

Whisstock, J. C., Silverman, G. A., Bird, P. I., Bottomley, S. P., Kaiserman, D., et al. (2010). Serpins flex their muscle: II. Structural insights into target peptidase recognition, polymerization, and transport functions. *J. Biol. Chem.*, *285*, 24307–24312.

Winans, K. A., King, D. S., Rao, V. R., & Bertozzi, C. R. (1999). A chemically synthesized version of the insect antibacterial glycopeptide, diptericin, disrupts bacterial membrane integrity. *Biochemistry*, *38*, 11700–11710.

Yano, T., Mita, S., Ohmori, H., Oshima, Y., Fujimoto, Y., et al. (2008). Autophagic control of listeria through intracellular innate immune recognition in *Drosophila. Nat. Immunol.*, *9*, 908–916.

Yoshida, H., Ochiai, M., & Ashida, M. (1986). Beta-1,3-glucan receptor and peptidoglycan receptor are present as separate entities within insect prophenoloxidase activating system. *Biochem. Biophys. Res. Commun.*, *141*, 1177–1184.

Yoshida, H., Kinoshita, K., & Ashida, M. (1996). Purification of a peptidoglycan recognition protein from hemolymph of the silkworm, *Bombyx mori. J. Biol. Chem.*, *271*, 13854–13860.

Yu, X. Q., & Kanost, M. R. (2002). Binding of hemolin to bacterial lipopolysaccharide and lipoteichoic acid. An immunoglobulin superfamily member from insects as a pattern-recognition receptor. *Eur. J. Biochem.*, *269*, 1827–1834.

Yu, X. Q., Gan, H., & Kanost, M. R. (1999). Immulectin, an inducible C-type lectin from an insect, *Manduca sexta*, stimulates activation of plasma prophenol oxidase. *Insect Biochem. Mol. Biol.*, *29*, 585–597.

Yu, X. Q., Jiang, H., Wang, Y., & Kanost, M. R. (2003). Nonproteolytic serine proteinase homologs are involved in prophenoloxidase activation in the tobacco hornworm, *Manduca sexta*. *Insect Biochem. Mol. Biol.*, *33*, 197–208.

Yu, Y., Park, J. W., Kwon, H. M., Hwang, H. O., Jang, I. H., et al. (2010). Diversity of innate immune recognition mechanism for bacterial polymeric meso-diaminopimelic acid-type peptidoglycan in insects. *J. Biol. Chem.*, *285*, 32937–32845.

Zaidman-Rémy, A., Hervé, M., Poidevin, M., Pili-Floury, S., Kim, M. S., et al. (2006). The *Drosophila* amidase PGRPLB modulates the immune response to bacterial infection. *Immunity*, *24*, 463–473.

Zamore, P. D. (2007). RNA silencing: Genomic defence with a slice of pi. *Nature*, *446*, 864–865.

Zasloff, M. (2002). Antimicrobial peptides of multicellular organisms. *Nature*, *415*, 389–395.

Zhao, M., Söderhäll, I., Park, J. W., Ma, Y. G., Osaki, T., et al. (2005). A novel 43-kDa protein as a negative regulatory component of phenoloxidase-induced melanin synthesis. *J. Biol. Chem.*, *280*, 24744–24751.

Zou, Z., Evans, J. D., Lu, Z., Zhao, P., Williams, M., et al. (2007). Comparative genomic analysis of the *Tribolium* immune system. *Genome Biol.*, *8*, R177.

15 Molecular and Neural Control of Insect Circadian Rhythms

Yong Zhang
University of Massachusetts Medical School,
Department of Neurobiology, Worcester, MA, USA
Patrick Emery
University of Massachusetts Medical School,
Department of Neurobiology, Worcester, MA, USA

15.1. Introduction

15.1.1. Circadian Rhythms: Function and Fundamental Properties

The environment on Earth is constantly changing. The day/night cycle imposes dramatic variations in light intensity and temperature on most organisms. Moreover, yearly cycles in day length and weather patterns add a layer of complexity to these daily rhythms. The ecological surroundings of most organisms are also rhythmic: food availability, predator activity, and mating opportunities usually show both daily and yearly variations. To cope with the day/night cycle, cyanobacteria, fungi, plants, and animals use circadian clocks (Dunlap, 1999). The word

"circadian" is derived from the Latin *circa diem*, which means "about a day." Circadian clocks help organisms to optimize gene expression, protein activity, cell biochemistry, body physiology, and behavior with the day/night cycle to maximize their chances of survival. There are four fundamental properties that define circadian rhythms. Perhaps their most striking property is their "free running" nature, which means that they persist under constant conditions, without any environmental cues. A second canonical property is that these circadian rhythms have a period of approximately, but not exactly, 24 hours. The third property is that they can be synchronized by various environmental inputs, called Zeitgebers (a German term meaning "time-giver"). Light intensity is arguably

DOI:10.1016/B978-0-12-384747-8.10015-7

the most important Zeitgeber for most organisms. Other Zeitgebers include temperature cycles, food availability, and social interactions. The fourth fundamental property of circadian rhythms is called temperature compensation. The period of circadian rhythms stays very stable under a wide range of ambient temperatures, although the kinetics of most biochemical reactions increase with higher temperature. These four canonical properties define endogenous circadian rhythms (Pittendrigh, 1960).

Insects are the most widely distributed and diverse group of animals on the planet. They can be found in nearly all environments, including those exposed to extremely cold and warm temperatures. Insects are thus powerful models to study circadian rhythms at the cellular, physiological, behavioral, ecological, and evolutionary levels.

15.1.2. Circadian Behavioral Rhythms

Many insect behaviors are under circadian regulation. Locomotor activity is the most intensively studied circadian behavior in insects (see, for example, Konopka and Benzer, 1971; Saunders and Thomson, 1977; Kenny and Saunders, 1991; Page, 1982; Fleury *et al.*, 1995; Cymborowski *et al.*, 1996; Clark *et al.*, 1997). *Drosophila melanogaster*, for example, shows two peaks of locomotor activity around dawn and dusk, which are frequently called morning and evening peaks, respectively (**Figure 1**). These two peaks persist under constant darkness and constant temperature, and are thus circadian, although the morning

peak tends to progressively weaken, for unclear reasons. We will discuss in detail the neural bases for these two peaks in section 15.4, and the molecular mechanisms underlying these rhythms in section 15.2.

Another well-studied insect circadian behavior is feeding (Abraham and Muraleedharan, 1990; Khan *et al.*, 1997; Sothern *et al.*, 1998; Das and Dimopoulos, 2008; Xu *et al.*, 2008). Among feeding rhythms, blood-sucking might be the first that comes to mind. Indeed, female mosquitoes bite us preferentially at certain times of the day, and are vectors of devastating diseases (malaria and Dengue fever, for example). These rhythmic biting activities have long been reported for various mosquitoes (see, for example, Haddow, 1956; Standfast, 1967), and were more recently proven to be controlled by circadian clocks (Pandian, 1994; Das and Dimopoulos, 2008). Related to feeding behavior, foraging behavior can also be circadianly controlled – for example, in honeybees and bumblebees (Moore *et al.*, 1989; Bloch, 2010; Stelzer *et al.*, 2010).

Circadian clocks also influence reproduction. Timing of mating behavior might even contribute to reproductive isolation, since closely related *Drosophila* species prefer to mate at different times (Sakai and Ishida, 2001a, 2001b). In *Drosophila melanogaster*, male courtship is more frequent during the night and seems to drive rhythmic mating behavior (Fujii *et al.*, 2007), although earlier studies indicate that the female circadian clock also affects mating rhythms (Sakai and Ishida, 2001a). Mating rhythms are also found in cockroaches and moths,

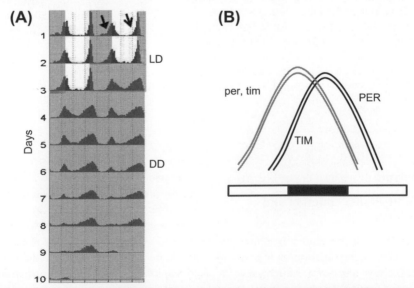

Figure 1 *Drosophila* circadian locomotor behavior and gene expression. (A) Actogram showing the average locomotor activity of a group of *Drosophila* males. Individual flies are placed in small glass tubes with sugar food and monitored under a 12:12 light:dark (LD) cycle for 3 days, and then released in constant darkness (DD). Gray shading indicates the dark phase of LD or DD. The actogram is double-plotted: each day (except the first) is plotted twice, first on the right half of the actogram, and then on the left half on the next line. Full and empty arrows indicate the morning and evening peaks of activity, respectively, which are controlled by the circadian clock. Zeitgeber Time (ZT) 0 corresponds to the beginning of the light phase. (B) Schematic representation of the expression pattern of two key circadian genes (*per* and *tim*) and their protein products. Gray lines represent mRNA levels, and black lines protein levels. The white and black bars indicate day and night, respectively.

for example (Silvegren *et al.*, 2005; Rymer *et al.*, 2007). Interestingly, the production and release of, and response to, sex pheromones can be regulated by the circadian clock (Silvegren *et al.*, 2005; Merlin *et al.*, 2007; Krupp *et al.*, 2008). Another well-characterized circadian reproductive behavior is oviposition (egg-laying rhythms; for review, see Howlader and Sharma, 2006). Interestingly, the neural substrate regulating oviposition rhythms in *Drosophila melanogaster* is distinct from those controlling circadian eclosion (see below) and adult locomotor rhythms (Howlader *et al.*, 2006).

After egg-laying, important behaviors related to the transition from one developmental stage to another can be driven by the circadian clock. For instance, egg hatching is gated by the circadian clock in many insects (see, for example, Minis and Pittendrigh, 1968; Lazzari, 1991; Sauman and Reppert, 1996). Metamorphosis is also frequently circadian. During insect metamorphosis, ecdysteroid hormones are synthesized by the prothoracic gland and released into the hemolymph to trigger adult ecdysis. The prothoracicotropic hormone (PTTH) plays an important role in this process by stimulating the prothoracic gland. Circadian control of PTTH release, PTTH responsiveness, ecdysteroid synthesis, hemolymph ecdysteroid titer, and ecdysteroid receptor expression have been described in *Rhodnius prolixus* and other insects (Cymborowski *et al.*, 1991; Vafopoulou and Steel, 1991, 1996; Pelc and Steel, 1997; Richter, 2001; for review, see Steel and Vafopoulou, 2006). Interestingly, in *Drosophila* the prothoracic gland shows rhythmic autonomous expressions of clock proteins (Emery *et al.*, 1997). *Drosophila* eclosion occurs preferentially around dawn, and this eclosion rhythm is absent in classic clock gene mutants (Konopka and Benzer, 1971; Sehgal *et al.*, 1994). *Drosophila* eclosion rhythms require both brain pacemaker neurons (the pigment dispersing factor (PDF) positive ventral lateral neurons – see section 15.4) and a functional clock in the prothoracic gland (Myers *et al.*, 2003).

Another behavior that is modulated by the circadian clock in *Drosophila* is larval photophobicity (Mazzoni *et al.*, 2005). *Drosophila* larvae (all stages except for late third-instar larvae) prefer to avoid exposure to light if possible. The clock modulates the photosensitivity of this behavior: larvae are more sensitive to light in the early morning.

Sophisticated behaviors such as time-compensated sun compass navigation require the circadian clock to correct the direction of flight as a function of time, because the position of the sun changes during the course of the day. This plays an essential role in long-distance foraging and migratory behavior – for example, in bees (Lindauer, 1960; Bloch, 2010) and Monarch butterflies (Mouritsen and Frost, 2002; Froy *et al.*, 2003; Reppert *et al.*, 2010). Strikingly, the waggle dance performed by forager honeybees to indicate the location of food to other foragers is also time-compensated to correct for changes in the solar azimuth (Lindauer, 1954).

15.1.3. Physiological and Metabolic Rhythms

Circadian clocks are found not only in the brain but also in inner organs such as the Malpighian tubules and the gut, and in external sensory organs (e.g., antennae, eyes, and proboscis) (Zeng *et al.*, 1994; Emery *et al.*, 1997; Giebultowicz and Hege, 1997; Plautz *et al.*, 1997; Krishnan *et al.*, 1999; Merlin *et al.*, 2007, 2009; Ito *et al.*, 2008; Uryu and Tomioka, 2010). Interestingly, these peripheral clocks function independently of the brain pacemaker neurons, since isolated organs show self-sustained molecular circadian rhythms in culture (Emery *et al.*, 1997; Giebultowicz and Hege, 1997; Plautz *et al.*, 1997; Merlin *et al.*, 2009). Peripheral clocks affect local physiology. For example, *Drosophila melanogaster* shows robust rhythmic physiological responses to odor and taste stimuli (Krishnan *et al.*, 1999; Krishnan *et al.*, 2008; Chatterjee *et al.*, 2010). Flies are more sensitive to odors at night, and to gustatory stimulation at dawn. Electrophysiological recordings reveal that spike amplitude is modulated by the circadian clock in gustatory and olfactory receptors (Krishnan *et al.*, 2008; Chatterjee *et al.*, 2010). The circadian clock also controls spike duration and frequency in gustatory neurons. The rhythmically expressed G Protein Receptors Kinase 2 (GPRK2) kinase plays a particularly important role in these processes (Tanoue *et al.*, 2008; Chatterjee *et al.*, 2010). In antennae, it regulates the localization of olfactory receptors to the dendritic membrane of the olfactory sensory neurons, and probably works similarly for the gustatory receptors as well.

Drosophila's innate immune response to pathogens is also under circadian regulation (Shirasu-Hiza *et al.*, 2007; Lee and Edery, 2008; see also Chapter 14 in this volume). *Drosophila*'s rate of survival is higher if an injection of pathogenic bacteria is performed during the night than during the day. Interestingly, this greater survival rate is correlated with a stronger induction of specific antimicrobial peptides at night (Lee and Edery, 2008).

In many pest insects, resistance to insecticides has been found to be dependent on the time of day (Sullivan *et al.*, 1970; Pszczolkowski and Dobrowolski, 1999). Interestingly, genes involved in detoxification were found to be rhythmically expressed in whole-genome expression profiling studies (reviewed in Wijnen and Young, 2006). The activities of specific xenobiotic metabolizing enzymes were confirmed to be expressed rhythmically, which probably explains why the efficiency of some insecticides varies significantly during the day (Hooven *et al.*, 2009).

These examples illustrate the profound effect that circadian clocks have on the physiology of insects.

Whole-genome expression studies have identified other physiological and metabolic processes that might be under circadian regulation in *Drosophila* (Claridge-Chang *et al.*, 2001; McDonald and Rosbash, 2001; Ceriani *et al.*, 2002; Y. Lin *et al.*, 2002; Ueda *et al.*, 2002; Boothroyd *et al.*, 2007; Keegan *et al.*, 2007). These studies have for the most part focused on head tissues. It would thus be interesting to extend these studies to specific organs such as the Malpighian tubules to better understand the role played by peripheral circadian clocks.

15.2. The *Drosophila* Circadian Pacemaker

This chapter focuses heavily on *Drosophila* circadian rhythms. For more than half a century, *Drosophila* has played a particularly important role in the field of chronobiology. In the early 1950s, Pittendrigh and colleagues began to use *Drosophila* eclosion (adult ecdysis) as a readout to understand the fundamental properties of circadian clocks, such as their free-running nature, their entrainment, and the phenomenon of temperature compensation (for reviews, see Pittendrigh, 1960, 1993). In the late 1960s, Konopka and Benzer had the groundbreaking idea of screening for single gene mutants affecting *Drosophila melanogaster*'s timing of eclosion, with the conviction that such screens would ultimately lead to a mechanistic dissection of the circadian clock. In 1971, they reported the identification of three mutants affecting a single locus, *period* (*per*) (Konopka and Benzer, 1971). The *per^S*, *per^L*, and *per^0* mutations accelerated, slowed down, or completely eliminated rhythms of eclosion. Adult locomotor rhythms were also affected in a similar way, which indicated that the *per* gene was probably a central element of a circadian pacemaker responsible for many, if not all, circadian rhythms in the fruit fly. Adult locomotor activity progressively replaced eclosion as a read-out for circadian clocks in genetic screens, because it allows monitoring of individual flies, while eclosion can only be monitored at a population level. Moreover, sophisticated forward genetic approaches based on the use of the *luciferase* reporter gene controlled by *per* regulatory sequences (Brandes *et al.*, 1996) have also successfully identified circadian genes. Thus, many of the genes that will be discussed below were identified through forward genetics. In the past decade, the completion of the *Drosophila* genome sequence and the development of methods to monitor whole-genome expression patterns have allowed reverse genetic approaches to become important alternatives for circadian gene discovery. Another major development over this period has been the design of genetic tools to manipulate specific circadian tissues or cell types. This has particularly impacted our understanding of the neural circuits controlling *Drosophila* circadian locomotor behavior (see Chapter 4).

15.2.1. The PER Transcriptional Feedback Loop

15.2.1.1. PERIOD (PER) At the end of 1984, two teams, led by Michael Young at Rockefeller University and Michael Rosbash and Jeffrey C. Hall at Brandeis University, reported the cloning of *per* (Bargiello *et al.*, 1984; Reddy *et al.*, 1984; Zehring *et al.*, 1984). This was a heroic effort: fragments of DNA from the region of the X-chromosome to which the *per* gene had been mapped were systematically integrated into the *per^0* mutant fly genome until normal circadian rhythms were restored. The sequence of the *per* gene was at first uninformative, and PER function remained unclear for another decade. It was only 9 years later that a domain common to PER and two transcription factors (the PAS domain) was identified, indicating that PER might also be a transcription factor (Huang *et al.*, 1993). Interestingly, it was evident by that time that PER was somehow involved in a transcriptional feedback loop (Hardin *et al.*, 1990, 1992) (**Figure 2**). This conclusion was based on three fundamental observations: first, *per* mRNA levels oscillate robustly during the course of the light/dark (LD) cycle, and also under constant darkness (DD) (**Figure 1**); second, the period of *per* mRNA oscillations, or their absence, parallels the circadian behavior phenotypes of *per^S*, *per^L* and *per^0* mutants; and third, this regulation is in great part transcriptional, since rhythmic reporter gene expression is obtained with *per* promoter sequences. In addition, PER acts as its own gene repressor, since *per* mRNA levels are low when PER is overexpressed (Zeng *et al.*, 1994). This basic feedback mechanism of a key circadian gene on its own gene expression has proven to be crucial for all known eukaryotic circadian clocks (Dunlap, 1999).

Critically, PER protein levels, its phosphorylation state, and its nuclear localization are rhythmic as well (Siwicki *et al.*, 1988; Zerr *et al.*, 1990; Zwiebel *et al.*, 1991; Edery *et al.*, 1994a; Vosshall *et al.*, 1994; Curtin *et al.*, 1995)

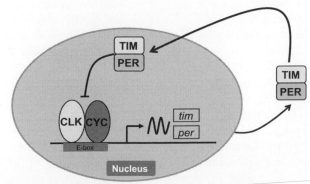

Figure 2 The PER circadian feedback loop in *Drosophila*. CLOCK (CLK) and CYCLE (CYC) are two transactivators that bind to the E-box of the *tim* and *per* promoter. PER and TIM first accumulate in the cytoplasm and then enter into the nucleus to block their own gene transcription. See main text for details.

(**Figure 1**). These rhythms determine when and how efficiently PER can function as a repressor. This is discussed in detail below.

15.2.1.2. TIMELESS (TIM)

timeless (*tim*) – the second circadian gene identified in *Drosophila* – is also an integral part of the *per* negative transcriptional feedback loop (Sehgal *et al.*, 1994, 1995; Myers *et al.*, 1995; see also **Figure 2**). *tim* and *per* mRNA levels cycle with a similar phase and amplitude (**Figure 1**). In addition, *tim* mRNA cycling is disrupted by *per* mutations, and *tim* mutations affect *per* mRNA cycling. TIM actually promotes PER function: it binds to PER, stabilizes it, and contributes to PER's translocation into the nucleus (Vosshall *et al.*, 1994; Gekakis *et al.*, 1995). In TIM's absence, flies are arrhythmic. TIM is also an important target for light-input pathways (Hunter-Ensor *et al.*, 1996; Lee *et al.*, 1996; Myers *et al.*, 1996; Zeng *et al.*, 1996; Mazzoni *et al.*, 2005), and thus connects the circadian pacemaker with the outside world (see section 15.3).

15.2.1.3. CLOCK (CLK) and CYCLE (CYC)

The promoter elements that drive *per* and *tim* mRNA cycles contain E-box binding sites, which can be recognized by bHLH/PAS family members (Sogawa *et al.*, 1995; Swanson *et al.*, 1995; Hao *et al.*, 1997; McDonald *et al.*, 2001). However, although PER carries a PAS domain, it has no identifiable bHLH DNA binding domain (Huang *et al.*, 1993), suggesting that it does not regulate transcription through direct binding to the *per* and *tim* promoters. The proteins that bind to these E-boxes are actually two bHLH-containing members of the PAS family: CLOCK (CLK) and CYCLE (CYC) (Allada *et al.*, 1998; Darlington *et al.*, 1998; Rutila *et al.*, 1998) (**Figure 2**). In the dominant-negative *Clk^jrk* and the null *cyc^0* mutants, *per* and *tim* mRNA levels are dramatically low, indicating that both CLK and CYC function as positive regulators of PER and TIM. CLK and CYC can indeed bind as heterodimers to E-boxes present in the *per* and *tim* genes (Lee *et al.*, 1999), and transactivate a luciferase reporter under the control of a *per* or *tim*

promoter in S2 cells (Darlington *et al.*, 1998). (The S2 cell line is a commonly used embryonic *Drosophila* cell line, which has proved immensely useful to understand the molecular mechanisms underlying *Drosophila* circadian rhythms; see also below.) Furthermore, CLK/CYC transactivation potential is strongly inhibited by PER and TIM. The PER feedback loop is therefore closed (Darlington *et al.*, 1998). Interestingly, in mammals CLK, CYC, and PER homologs are also key components of the circadian pacemaker (Dunlap, 1999). The mammalian and fly clocks are thus based on similar mechanisms, although mammals and flies are separated by at least 600 million years of evolution.

15.2.1.4. Mechanisms of PER repression

The PAS domain is responsible for homo and heterodimerization between proteins containing these domains (e.g., CLK/CYC) (Huang *et al.*, 1993). Since PER contains a PAS domain but no DNA binding domain, it would appear likely that PER interferes with CLK/CYC transcriptional activity by interacting with the PAS domains of these proteins. However, when the repression domain for PER was mapped using a transcriptional reporter assay in S2 cells, it was found to lie in the C-terminus of PER (Chang and Reppert, 2003) (**Figure 3**). This domain was called CCID (CLK/CYC inhibitory domain), and it has recently been shown that it contains a small domain required for PER interactions with CLK (Sun *et al.*, 2010). The PAS domain is completely unnecessary for repression, at least in the context of this cell culture assay. What, then, is the function of the PAS domain? First, the PAS region and the C-domain located C-terminal of it are responsible for interactions with TIM, which is important for PER stability (Vosshall *et al.*, 1994; Gekakis *et al.*, 1995). Of note, the *per^L* mutation is located near the N-terminus of the PAS domain. Second, PER can form homodimers through its PAS domain and the C-domain (Huang *et al.*, 1993; Yildiz *et al.*, 2005; Landskron *et al.*, 2009), although *in vivo* most PER is associated with TIM (Zeng *et al.*, 1996). It has recently been proposed that these homodimers have an important functional role

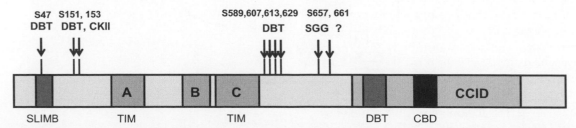

Figure 3 PER functional domains and phosphorylation sites. The red box shows the SLIMB binding domain of PER, the grey box the DBT binding domain, and the black box the CLK binding domain (CBD). Three orange boxes indicate the PAS-A, PAS-B, and PAS-C domains, necessary for TIM binding. The blue box indicates the CCID (CLK-CYC inhibitory domain). Arrows indicate the phosphorylated amino acids with demonstrated function for circadian time-keeping; the kinases that modify these sites are also indicated.

in the circadian clock (Landskron *et al.*, 2009). Single amino acid mutants were specifically designed to disrupt PER homodimer formation, based on crystallographic studies. The M560D mutant was studied in great detail. Flies expressing only this mutant PER showed very poor PER protein cycling, and were frequently behaviorally arrhythmic (about 60% of flies) or showed a slight period lengthening. Importantly, while PER homodimer formation was disrupted, PER/TIM heterodimerization was unaffected by the M560D mutation. These results support the idea that PER homodimers are important for circadian rhythms. However, there is no way to be certain that only PER homodimer formation was affected by the M560D mutation. Interactions with kinases (see below) could for example have been altered. It is also not clear how PER homodimers would mechanistically affect the circadian pacemaker. Further studies are thus necessary to clarify the role of PER homodimers.

How, then, does PER repress CLK? It could somehow interfere with CLK/CYC activation, either by blocking the recruitment of co-activators or by recruiting co-repressors, or it could displace CLK from the E-box. Electrophoretic mobility shift assays (EMSAs) gave an early indication that the second hypothesis was correct. Indeed, CLK/CYC affinity for the E-box is reduced in the presence of PER/TIM *in vitro* (Lee *et al.*, 1999). Furthermore, occupancy of the E-box was more recently shown by chromatin immunoprecipitation (Chip) assays to be rhythmic, with CLK/CYC being more strongly bound to the *per* E-box when *per* gene transcription is high, and being displaced from DNA when PER is in the nucleus and represses transcription (Yu *et al.*, 2006). A recent study actually indicates that the two mechanisms proposed above might be correct (Menet *et al.*, 2010). Indeed, Chip assays show that PER is first associated with chromatin along with CLK/CYC at a time when *per* transcription decreases. It must thus somehow interfere with the transactivation of DNA-bound CLK/CYC. Then, PER remains associated with CLK/CYC when it is detached from DNA (**Figure 4**).

In vivo, TIM contributes to PER repression, since it stabilizes PER, as discussed above. Does it also play an active role in PER repression? In cell culture, PER repression does not require TIM (Rothenfluh *et al.*, 2000; Nawathean and Rosbash, 2004), although it has been

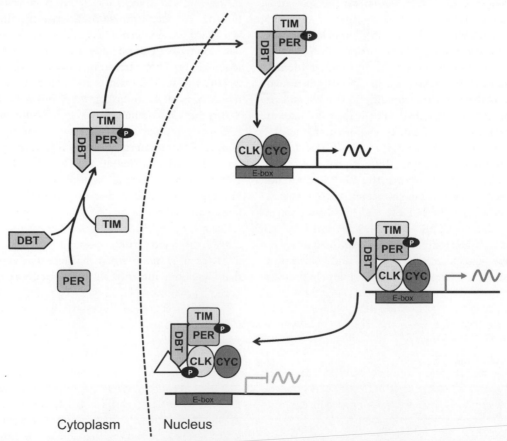

Cytoplasm | Nucleus

Figure 4 Mechanisms of PER repression. DBT binds to PER and phosphorylates it in the cytoplasm. TIM binds to the complex and stabilizes PER. DBT, PER, and TIM then enter into the nucleus (although PER and TIM probably do not translocate together; see text for details). After binding to CLK, PER starts to repress its own transcription. CLK and CYC are subsequently removed from the E-box, and *per* and *tim* transcription is shut down. CLK is phosphorylated by an unknown kinase (empty triangle), recruited by DBT, and degraded.

reported that TIM can promote PER repression (Darlington *et al.*, 1998). TIM might also be important for the timing of PER entry into the nucleus, and therefore for its repressive potential. Initial studies indicated that PER and TIM probably translocate together in the nucleus (Vosshall *et al.*, 1994; Hunter-Ensor *et al.*, 1996; Saez and Young, 1996). However, under certain circumstances, TIM is dispensable for PER nuclear entry – for example, in a $tim^0:dbt^P$ double mutant in which PER levels are quite high (Cyran *et al.*, 2005). Moreover, Shafer *et al.* showed that in circadian neurons that control circadian behavior – the ventral lateral neurons (LN_vs; see section 15.4) – PER enters into the nucleus before TIM (Shafer *et al.*, 2002). However, these results do not exclude the possibility that a small fraction of TIM shuttles between the nucleus and the cytoplasm to promote PER localization. TIM could also promote PER nuclear entry from a cytoplasmic localization. Strikingly, recent results using YFP and CFP tagged PER and TIM and FRET (Foerster (or fluorescence) resonance energy transfer) to visualize in cell culture the formation of PER and TIM dimers, and their nuclear entry, actually confirm that PER and TIM enter into the nucleus independently of each other, rather than together (Meyer *et al.*, 2006). However, prior to PER nuclear translocation, a PER/TIM dimer is observed in perinuclear foci, which indicates that TIM is indeed probably important as a cytoplasmic catalyst of PER nuclear entry. PER might then form homodimers after dissociating from TIM, since the aforementioned homodimerization mutant M560D shows defective nuclear entry (Landskron *et al.*, 2009). But if TIM enters into the nucleus independently of PER, then what is its nuclear function? One possibility is that it could still help stabilize PER late in the circadian cycle, when it becomes hyperphosphorylated. Recent results indicate that TIM might also help to stabilize PER–CLK interactions. Indeed, if TIM is present, a mutant PER protein missing the CLK interaction domain can still repress CLK/CYC in S2 cells. The same mutant protein associates with CLK *in vivo* in the dark, but when light is present this complex is dissociated and *per/tim* transcription increases (Sun *et al.*, 2010). This is explained by the light-dependent degradation of TIM (see section 15.3). TIM cannot, however, entirely compensate for the absence of the CLK binding domain of PER, since flies expressing this mutant protein have a long-period phenotype.

As will be discussed in detail in section 15.2.2.1, PER repression requires the kinase DOUBLETIME (DBT) – with which it is tightly associated – and probably additional kinases as well. A crucial role of PER in repression is thus to bring these modifying enzymes into the vicinity of CLK/CYC to trigger repression.

15.2.1.5. CLOCKWORK ORANGE (CWO) As discussed above, DNA microarrays probing the expression

of the entire *Drosophila* genome were used to identify genes regulated by the circadian clock (Claridge-Chang *et al.*, 2001; McDonald and Rosbash, 2001; Ceriani *et al.*, 2002; Y. Lin *et al.*, 2002; Ueda *et al.*, 2002). One of them was *cg17100*. With the use of an inducible CLK–Glucocorticoid receptor fusion, it was found in cell culture and in fly heads to be a primary target of CLK (Kadener *et al.*, 2007). Indeed, its promoter contains E-boxes similar to those of *per* and *tim*. CG17100 was renamed CLOCKWORK ORANGE (CWO) (**Figure 5**) because it is an Orange domain containing bHLH protein (Lim *et al.*, 2007). CWO binds E-boxes, and represses both basal E-box-containing promoter activity and CLK-mediated transactivation in cell culture (Kadener *et al.*, 2007; Lim *et al.*, 2007; Matsumoto *et al.*, 2007). Surprisingly, however, mRNA oscillations of most CLK/CYC regulated genes show a blunted peak in *cwo* loss-of-function mutants (Kadener *et al.*, 2007; Lim *et al.*, 2007; Matsumoto *et al.*, 2007; Richier *et al.*, 2008). Peaks of mRNA cycling are reduced for *per* and *tim*, and for two other CLK/CYC targets that will be discussed in section 15.2.4 (*pdp1ε* and *vri*). For some of these genes, trough levels are nevertheless increased as well. In sharp contrast, *cwo* mRNA is constantly high in *cwo* mutant flies. Taken together these results indicate that different CLK/CYC target genes are differentially sensitive to CWO. CWO might be predominantly a repressor in the case of its own gene regulation, while on other circadian genes it might both inhibit transcription at certain times of the cycle, and promote full activation at other times (**Figure 5**). It is conceivable that CWO contributes to the recruitment of activators that will help in opening chromatin, and promote the return of CLK/CYC on the promoter. However, how CWO is being displaced from chromatin is unclear. CWO protein levels do not cycle robustly (only about two-fold in the circadian neurons regulating behavior), but CWO could be modified by kinases, for example. Alternatively, it might be the state of CLK phosphorylation – regulated by PER and DBT (see section 15.2.2.1) – that determines its DNA binding affinity, and thus whether it is CLK/CYC or CWO that is bound to the promoter.

Behaviorally, severe hypomorphic *cwo* mutants, or even null mutants, are rhythmic. However, rhythms have a long period and amplitude is affected to diverse degrees, depending on the genetic manipulations that were used to reduce CWO levels (RNA interference (RNAi), P-element insertion, null mutation) (Kadener *et al.*, 2007; Lim *et al.*, 2007; Matsumoto *et al.*, 2007; Richier *et al.*, 2008). Most significantly, in null mutants rhythms were actually only slightly reduced in amplitudes, and persisted for at least 10 days (Richier *et al.*, 2008). CWO therefore acts as a modulator of circadian rhythms, important for setting the proper pace of the circadian clock, but it is not required for circadian rhythms to occur and to persist.

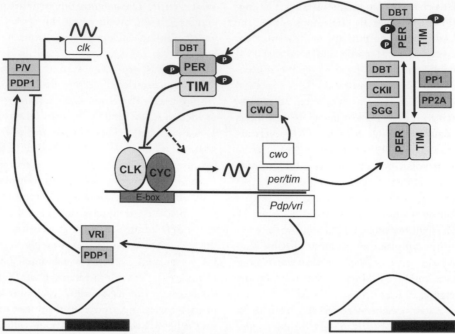

Figure 5 Interlocked feedback loops. The PER feedback loop is modulated by *cwo*, which is itself regulated by this loop. CWO is a transcriptional repressor, but it also appears to be necessary for full activation of a subset of CLK/CYC regulated genes (dotted arrow). A second loop is attached to the PER loop. VRI and PDP1 generate circadian transcriptional rhythms peaking early during the light phase (left), in antiphase of those generated by the PER loop (right).

It should be noted that there are other members of the bHLH/Orange family. It is thus possible that one of them could partially substitute for CWO (Matsumoto *et al.*, 2007), but whether this indeed happens has not yet been reported.

15.2.2. Circadian Kinases and Phosphatases

The end result of the PER transcriptional feedback loop described above is the robust cycling of PER and TIM (**Figure 1**), which is crucial for the rhythmic expression of output genes, and for behavioral and physiological rhythms. However, when one compares *per* and *tim* mRNA cycles with the oscillations of the proteins they encode, there is a striking phase delay between the accumulation of protein and mRNA (**Figure 1**). This is indicative of post-transcriptional control. Moreover, PER and TIM become progressively phosphorylated during the circadian cycle (Edery *et al.*, 1994a; Zeng *et al.*, 1996). It is now clear that the regulation of PER and TIM phosphorylation determines the pace of the circadian pacemaker.

15.2.2.1. DOUBLETIME (DBT)
DBT (also called DCO) was the first kinase to be identified in circadian genetic screens (Kloss *et al.*, 1998; Price *et al.*, 1998). DBT is the homolog of CK-Iδ/ε; these are also important kinases for the circadian pacemaker of mammals (Lowrey *et al.*, 2000; Xu *et al.*, 2005; Etchegaray *et al.*, 2009). DBT function has been closely studied. The first striking

molecular observation that was made with a null *dbt* allele (*dbt*ᵖ) was that PER is much more stable than in wild type flies, and hypophosphorylated (Price *et al.*, 1998). DBT would thus phosphorylate PER and trigger its degradation (**Figures 3** and **5**). Using candidate gene approaches, two groups identified SLIMB as a critical subunit of an SCF (Skp1/Cul1/F-box) complex that functions as a PER E3-ubiquitin ligase (Grima *et al.*, 2002; Ko *et al.*, 2002). Indeed, SLIMB is an F-box protein known to recognize phosphorylated substrates. In severe *slimb* loss-of-function mutants, which are behaviorally arrhythmic, PER oscillations are disrupted, and PER hyperphosphorylated isoforms are abnormally stable (Grima *et al.*, 2002). This phenomenon could be recapitulated in cell culture, and it was shown with this system that TIM stabilizes PER by protecting it from DBT phosphorylation and subsequent SLIMB-mediated degradation (Ko *et al.*, 2002).

Interestingly, the short- and long-period phenotypes observed in *dbt*ˢ and *dbt*ᴸ mutants do not correlate with DBT kinase activity measured *in vitro*; indeed, both mutations reduce kinase activity (Preuss *et al.*, 2004). This could be because DBT phosphorylates multiple PER amino acid residues (see below and **Figure 3**), with the *dbt* mutations having differential effects on the phosphorylation of these residues. An alternative explanation is that DBT has multiple targets in the circadian pacemaker (see below). In any case, understanding how DBT binds to PER, and which residues are phosphorylated, is essential.

DBT binds to a specific domain of PER that was mapped to residues 755–809 by the Edery group, and to residues 762–788 by the Rosbash group (Kim *et al.*, 2007; Nawathean *et al.*, 2007) (**Figure 3**). PER mutant proteins with a deletion of the DBT-binding domain cannot restore circadian rhythms in *per⁰* flies. They are very stable, hypophosphorylated, but inefficient at repressing transcription *in vivo*. This would fit with the observation made in cell culture that DBT promotes PER repression (Nawathean and Rosbash, 2004).

Mass spectrometry has been used to identify residues that are phosphorylated by DBT *in vitro* or in a DBT-dependent manner in S2 cells (Chiu *et al.*, 2008; Kivimae *et al.*, 2008). Serine 47 has been identified as a crucial site for DBT phosphorylation, and, interestingly, its temporally regulated modification creates an atypical SLIMB binding site (**Figure 3**). This would explain how DBT triggers PER degradation (Chiu *et al.*, 2008). DBT-mediated phosphorylation of Ser44, 45, and 48 probably also contributes to PER degradation. Importantly, the circadian behavior period is shortened in *per⁰* flies expressing only a S47D PER mutant protein, which would mimic constitutive S47 phosphorylation, while the period is long with an S47A mutation.

In addition, DBT can target three Ser clusters *in vitro* (Kivimae *et al.*, 2008). Mutation of the N- and C-terminal clusters had little effect on PER repressive activity in a transcriptional cell culture assay. This, however, does not mean that these residues have no function *in vivo*. Indeed, two of them (Ser 151 and 153) are actually important for PER function in flies (see next subsection) (Lin *et al.*, 2005). The third cluster is located in the middle of the PER protein (**Figure 3**). It includes the residue mutated in *perˢ* mutants. Studies performed in cell culture and *in vivo* reveal an interaction between the *perˢ* residue (Ser 589) and the other Serine residues of the cluster targeted by DBT. It appears that Ser589 phosphorylation represses phosphorylation at the other residues, and it was proposed that in order for the full program of PER phosphorylation, nuclear entry, and repressive activity to occur, dephosphorylation of this residue should first occur (Kivimae *et al.*, 2008).

Interestingly, DBT forms a very stable complex with PER (Kloss *et al.*, 2001). DBT is actually translocated with PER into the nucleus, suggesting that it might keep phosphorylating PER in that subcellular compartment. Recent studies, however, point to a novel role for DBT: regulating CLK phosphorylation. Indeed, like PER, CLK becomes progressively hyperphosphorylated, although overall CLK levels do not change much because the hyper- and hypophosphorylated isoforms cycle in antiphase (Lee *et al.*, 1998; Yu *et al.*, 2006). Interestingly, the time at which hyperphosphorylated CLK is the highest corresponds to the time at which CLK is "off DNA" (Kim and Edery, 2006; Yu *et al.*, 2006; Menet

et al., 2010). It is therefore possible that phosphorylation influences CLK's affinity for the E-box. What phosphorylation definitely regulates is CLK stability; as for PER, CLK is destabilized when hyperphosphorylated (Kim and Edery, 2006; Yu *et al.*, 2006). This destabilization requires DBT both *in vivo* and in S2 cell culture, and the presence of PER to dock DBT near CLK (Yu *et al.*, 2009). Moreover, a highly phosphorylated CLK isoform is not detectable in *dbtᴬᴿ* mutant flies (Kim and Edery, 2006). It would therefore appear that DBT phosphorylates not only PER, but also CLK to regulate circadian transcription. Unexpectedly, however, this does not seem to be the case. In flies expressing only DBT proteins with compromised kinase activity, PER phosphorylation is largely eliminated, but CLK is still clearly phosphorylated (Yu *et al.*, 2006, 2009). In addition, *per* mRNAs are near trough levels, strengthening the notion that CLK phosphorylation inhibits transcription. Taken together, these results indicate that DBT recruits other kinases for CLK phosphorylation, but does not function catalytically to repress CLK/CYC-mediated transcription. They also indicate that it is DBT-regulated CLK phosphorylation, rather than DBT-controlled PER phosphorylation, that determines CLK/CYC transcriptional activity. Moreover, the DBT-binding domain of PER is part of the CCID (Chang and Reppert, 2003; Kim *et al.*, 2007; Nawathean *et al.*, 2007), as is the CLK binding domain (Sun *et al.*, 2010). This therefore explains how the CCID region can be sufficient for transcriptional repression *in vitro*.

In summary, DBT plays at least two roles in the circadian clock: first, it phosphorylates PER and thus controls its timing of nuclear entry and its degradation; second, it recruits other kinases that will phosphorylate CLK and regulate its ability to activate transcription.

15.2.2.2. CASEIN KINASE II (CKII) CASEIN KINASE II (CKII) is a tetrameric kinase comprised of two catalytic α and two modulatory β subunits. The *Tik* and *Andante* mutation affect the α and β encoding genes, respectively (J. M. Lin *et al.*, 2002; Akten *et al.*, 2003). *Tik* is a dominant mutation that results in a long-period (26.4-h) phenotype in heterozygous flies (homozygous *Tik* flies are lethal, like *ckIIα* null mutants). *And* is a hypomorphic mutation also with a long-period phenotype (25.8 h in hemizygous *And* mutant males). More severe phenotypes were observed with overexpression of the dominant negative TIK mutant protein, with very long-period rhythms (33 h) or arrhythmicity (Smith *et al.*, 2008). The PER phosphorylation cycle was found to be altered in *Tik* mutant flies (J. M. Lin *et al.*, 2002), but not to a detectable level in *And* flies (Akten *et al.*, 2003). However, PER nuclear entry is delayed in circadian neurons of both mutants. CKII is capable of phosphorylating PER *in vitro*. Moreover, in cell culture, CKII, like DBT, promotes PER repressor activity, independently of nuclear entry (Nawathean and Rosbash, 2004). PER Ser 151 and 153 might be crucial targets for CKII. *In vivo*, mutagenesis of these residues results in phenotypes reminiscent of the *Tik* or *And* mutations, which supports the idea that CKII regulates circadian rhythms through phosphorylation of PER's Ser 151 and 153 (Lin *et al.*, 2005). However, it should be noted that DBT is also capable of phosphorylating these sites, at least *in vitro* (Kivimae *et al.*, 2008). In addition, as mentioned above, more severe suppression of CKII activity results in a more severe phenotype than those found in *Tik* heterozygotes or *And* hemizygous males, and in *per⁰* rescued flies expressing the Ser151–153 mutants (Smith *et al.*, 2008). Thus, CKII should have additional targets in the circadian pacemaker.

TIM might be one of them. CKII can also phosphorylate TIM *in vitro* (Zeng *et al.*, 1996; J. M. Lin *et al.*, 2002). In TIK-overexpressing flies and in *And* flies, TIM translocation into the nucleus is altered (J. M. Lin *et al.*, 2002; Akten *et al.*, 2003). In TIK-overexpressing flies, TIM is also unusually stable, particularly during the subjective day (Meissner *et al.*, 2008). Interestingly, the deletion of a TIM domain with four putative CKII phosphorylation sites considerably lengthens the circadian period (Ousley *et al.*, 1998). This TIM mutant protein is again unusually stable, and appears hypophosphorylated (Meissner *et al.*, 2008). Similar behavioral and biochemical observations were made with the *tim^{ultralong}* mutation, which affects a residue (Glu260) that belongs to the domain containing the four potential CKII phosphorylation sites (Rothenfluh *et al.*, 2000). The *tim^{UL}* mutation can partially suppress the period-lengthening effect of the *Tik* mutation (Meissner *et al.*, 2008). All these data support the notion that CKII also phosphorylates TIM.

15.2.2.3. p90 RIBOSOMAL S6 KINASE (S6KII)

A third kinase that can influence PER stability and repression potential is p90 ribosomal S6 kinase (S6KII) (Akten *et al.*, 2009). A null *s6kII* mutant shows a period shortening of about 1 hour. PER protein is elevated, while *per* mRNA is reduced, particularly during the repression phase of the *per* mRNA cycle. It therefore appears that PER repression is increased in the absence of S6KII. This observation was recapitulated in S2 cells. *s6kII* mRNA levels cycle, but this does not translate in any detectable protein rhythms, probably because that protein is very stable. Interestingly, S6KII and CKIIβ were shown to physically interact in a yeast two-hybrid assay, and in cell culture (Kusk *et al.*, 1999; Akten *et al.*, 2009). Consistent with this notion, the *And* mutation suppresses the period lengthening caused by the *s6kII* null mutation. Since S6KII and CKII mutants have opposite effects on circadian period, S6KII might inhibit CKII function through direct protein–protein interaction.

15.2.2.4. SHAGGY (SGG)

SHAGGY was isolated in a misexpression screen (Martinek *et al.*, 2001). A collection of flies containing *P* elements with UAS binding sites inserted pseudo-randomly in the genome (*P* elements prefer to land near the 5′-end of genes) (Rorth *et al.*, 1998) were crossed to a *tim-(UAS)-GAL4* driver. Most flies thus overexpressed the genes targeted by the *P* element. SGG-overexpressing flies showed a strikingly short phenotype (*ca.* 21 h). In SGG-overexpressing flies, PER and TIM nuclear entry is advanced in clock neurons, consistent with the short-period behavior phenotype. *sgg* null mutants are homozygous lethal, because SGG is crucial during embryonic development. However, viability can be rescued with an *sgg* cDNA controlled by the *hsp70* promoter whose expression is induced once a day by a heat pulse during development. The rescued adult flies express very low SGG levels, and have a long-period phenotype (Martinek *et al.*, 2001). In these flies, TIM cycles poorly, and is constantly hypophosphorylated. On the contrary, SGG overexpression increases TIM phosphorylation. Moreover, the mammalian homolog of SGG, GSK3β, can phosphorylate TIM *in vitro*. These results suggest that SGG phosphorylates TIM, but there is currently no direct evidence that this is indeed the case. The enzymes that would promote TIM degradation after its phosphorylation by SGG or CKII are not yet known.

SGG might also regulate PER phosphorylation (Ko *et al.*, 2010). Indeed, in S2 cells, deletion mapping and mass spectrometry analysis show that SGG can phosphorylate PER at Ser657 (**Figure 3**). This phosphorylation is dependent on prior phosphorylation at Ser661, by an as yet unidentified kinase. The presence of a conserved Proline residue adjacent to Ser661 indicates that this kinase is a member of the Proline-directed kinase family (e.g., MAPK). Phosphorylation of this residue also appears to affect the extent of DBT-dependent PER phosphorylation.

Flies expressing PER proteins with mutagenized Ser661 show a 2-hour period-lengthening phenotype that is correlated with decreased PER phosphorylation. Consistent with the long-period behavior rhythms, PER nuclear entry is delayed in Ser661 mutants. However, there is apparently no effect on PER transcriptional repression activity, based on *per* mRNA levels. Finally, phosphorylation at Ser661 is not temporally regulated. In summary, constitutive phosphorylation at Ser661 is important to organize subsequent PER phosphorylation by SGG, DBT, and possibly other kinases that regulate the timing of PER nuclear entry.

Mutations at residues 657 have much weaker effects on circadian behavioral rhythms (0.5- to 1-hour period lengthening). Thus, regulation of this residue's phosphorylation is not sufficient to explain the *sgg* loss-of-function phenotype (2.5-hour period lengthening) (Martinek *et al.*, 2001). However, additional SGG-dependent PER phosphorylation sites were identified in S2 cells (Ko *et al.*,

2010). Whether these sites are important for SGG's circadian function *in vivo* is not yet known. SGG's effect on the clock could also be a combination of its kinase activity on PER, TIM, and possibly other targets such as CRY (see section 15.3) (Ko *et al.*, 2010; Martinek *et al.*, 2001; Stoleru *et al.*, 2007).

15.2.2.5. PROTEIN PHOSPHATASE 2A (PP2A)

In most biological pathways, kinase activities are balanced by phosphatase activities. Circadian pacemakers are no exception. PP2A plays an important role in regulating the degree of PER phosphorylation, its stability and nuclear entry (Sathyanarayanan *et al.*, 2004). In cell culture, PP2A activity knockdown (by RNAi) decreases PER stability, and, to a lesser degree, that of TIM. PP2A is comprised of two subunits: a catalytic subunit (called *mts* in flies), and a regulatory subunit important for substrate recognition. There are four genes encoding PP2A regulatory subunits in flies, and RNAi knockdown in cell culture indicated that both *wdb* and *tws* could affect PER stability (Sathyanarayanan *et al.*, 2004). Interestingly, both subunits are expressed rhythmically, at least at the mRNA level, with *tws* cycling more robustly (fourfold for *tws*, two-fold for *wdb*). WDB overexpression resulted in a slight period lengthening, and combining this overexpression with long-period *dbt* mutants (*dbt^L* and *dbt^G*) resulted in period lengthening that was more than additive compared to the single mutant. This synergistic interaction suggests that DBT and PP2A act antagonistically on PER phosphorylation, which fits well with the observations made in cell culture (DBT promotes PER degradation, PP2A protects PER from it). TWS overexpression had a much more severe phenotype (arrhythmicity).

Complete *mts*, *tws*, or *wdb* loss-of-functions are lethal. However, a severe *tws* mutant survives until the third instar larvae. PER cycling was thus studied in the brain of these mutant larvae by immunohistochemistry. PER still cycled, but cycling and nuclear entry appeared significantly delayed. Moreover, expression of a dominant-negative mutant of *mts* specifically in circadian neurons (LN_vs) gave a consistent phenotype in adult flies: long-period rhythms, and very low PER levels in circadian neurons. The aforementioned period lengthening observed with WDB overexpression is, however, difficult to reconcile with these loss-of-function data, since the directionality of the phenotypes is identical. One possibility is that WDB is less efficient at recognizing PER than TWS, and thus acts as a weak dominant negative. Alternatively, it could be selective to specific PER isoforms.

What also appears surprising is that loss-of-function in both kinases and phosphatases results in period lengthening, even though the molecular effects on PER levels are opposite. This actually indicates that PP2A does not simply remove all the phosphate groups added by DBT,

CKII, and other kinases. *In vivo* at least, the activity of PP2A must be selective to certain phospho-residues, and function in concert with DBT and CKII to remove these specific residues in a temporally regulated manner. We have mentioned above that Ser589 might need to be dephosphorylated for other residues to be modified (Kivimae *et al.*, 2008). The fact that *wdb* and *tws* mRNA cycle with different phases could actually contribute to the choreography of PER phosphorylation and dephosphorylation. The period paradox could also be explained by the fact that PP2A might target not only PER, but also CLK (Kim and Edery, 2006). At least in cell culture, knockdown of the two circadianly regulated regulatory subunits of PP2A resulted in reduction of CLK levels. It would be important to determine whether PP2A indeed regulates CLK phosphorylation *in vivo*.

15.2.2.6. PROTEIN PHOSPHATASE 1 (PP1)

PP1 is a second phosphatase that plays an important role in the control of circadian rhythms (Fang *et al.*, 2007). PP1 appears predominantly to target TIM, although through its interaction with TIM it is also capable of influencing the PER phosphorylation state. There are four catalytic subunits for PP1 (PPic) that appear to be redundant. In cell culture, only a combination of dsRNAs directed against all four PP1cs results in significant TIM level decrease. To obtain *in vivo* evidence for a role of PP1, a peptidic inhibitor was overexpressed in circadian tissues. Flies with this inhibitor had a long-period phenotype, decreased TIM level, and delayed PER/TIM nuclear entry. This behavior suggests that PP1 does not simply counterbalance SGG or CKII phosphorylation, since SGG overexpression shortens period length, and SGG and CKII loss-of-function lengthens it. There is, however, a genetic interaction between SGG and PP1, since the magnitude of the period lengthening caused by overexpression of the PP1 peptidic inhibitor is reduced if SGG is overexpressed. Thus, either PP1 dephosphorylates a subset of SGG Ser/Thr targets, or SGG influences PP1's ability to regulate the phosphorylation state of residues targeted by other kinases, such as CKII.

15.2.3. Additional Levels of Control in the PER Transcriptional Feedback Loop

Beside transcriptional and post-translational controls, there is evidence for additional levels of regulation in the PER feedback loop. When transcriptional and mRNA expression profiles are compared for *per* and *tim*, there is a notable difference for *per*, while for *tim* the mRNA profile closely follows the transcriptional profile, with a short delay (So and Rosbash, 1997). Thus, *tim* mRNA must be short-lived throughout the whole circadian cycle. However, there is a significantly longer delay in *per* mRNA accumulation when transcriptional activity increases, while *per* mRNA levels closely follow transcriptional decline. This means that *per* mRNA is more stable during the late day than during the mid–late night. Moreover, low amplitude *per* mRNA rhythms in the absence of rhythmic transcription are obtained from a promoterless *per* transgene (Frisch *et al.*, 1994; So and Rosbash, 1997). The mechanism regulating *per* mRNA stability is not yet clear, but it might involve both PER and TIM. Indeed, these proteins are capable of stabilizing *per* mRNAs when TIM expression is induced in an arrhythmic *tim⁰* background (Suri *et al.*, 1999). This could be a function of the newly synthesized, still cytoplasmic PER/TIM dimer. It has been proposed that sequences located in *per* first intron might be responsible for *per* mRNA stability regulation. Indeed, expression of reporter genes fits well with *per* mRNA cycling if this intron is present, but fits better with transcriptional rhythms when it is absent (Stanewsky *et al.*, 1997, 2002). It is, however, unclear how this intronic sequence would affect mRNA stability

A role for miRNAs has also been recently proposed. miRNAs are small RNAs that usually bind specific sequences in the 3′ UTRs of genes, and induce RNA degradation and block translation (Bartel, 2004). The *Clk* 3′ UTR contains four binding sites for the microRNA *bantam*. When the hypomorphic, behaviorally arrhythmic *Clk^{AR}* mutant expresses a *Clk* transgene with mutagenized *bantam* binding sites, behavioral rhythms are poorly rescued (Kadener *et al.*, 2009). *bantam* thus contributes to the robustness of circadian rhythms. It might have additional functions, since its overexpression also lengthens the period of circadian behavioral rhythms by almost 3 hours. *bantam* is probably not the only miRNA involved in the control of circadian rhythms. Rhythmically expressed miRNAs have been identified, as well as miRNAs enriched in circadian tissues, but their function is not yet known (Yang *et al.*, 2008; Kadener *et al.*, 2009).

15.2.4. Interconnected Transcriptional Feedback Loops

The PER feedback loop is essential for circadian rhythms. We have already seen that CWO is a modulator of this loop, and that its own gene is part of the same regulatory network as the *per* and *tim* genes. In addition, the *Clk* gene is also under circadian transcriptional regulation (Bae *et al.*, 1998; Darlington *et al.*, 1998) (**Figure 5**). Interestingly, the phase of *Clk* mRNA oscillation is essentially anti-phase of *per*, *tim*, and *cwo*. *Clk* mRNA levels peak in the early morning. The *Clk* promoter is repressed by VRI (**Figure 5**), which was identified before the advent of DNA microarray by differential display as rhythmically expressed gene (Blau and Young, 1999; Cyran *et al.*, 2003; Glossop *et al.*, 2003). VRI binds specific DNA sequences found in the *Clk* promoter. PDP1 belongs to the same subfamily of Leucine Zipper DNA binding proteins, and its

ε isoform functions as a *Clk* activator (Cyran *et al.*, 2003) (**Figure 5**). Both *Pdp1* and *vri* are direct targets of CLK/CYC. However, their mRNA and protein oscillations are phased differently, with VRI oscillations preceding those of PDP1ε. The model is thus that when PDP1ε concentration increases, it replaces VRI on the *Clk* promoter. The VRI/PDP1ε and PER loops are interconnected through the CLK/CYC dimer (Cyran *et al.*, 2003; Glossop *et al.*, 1999, 2003).

How important is this loop for circadian rhythms? If *Clk* mRNA cycling is disrupted, the effects on circadian rhythms (molecular and behavioral) are actually minimal (Kim *et al.*, 2002). Thus, *Clk* mRNA cycling is not essential for circadian rhythmicity. Nevertheless, a mutation that eliminates specifically the ε isoform of PDP1 (which is the only isoform expressed in circadian tissues) compromises CLK and PER expression. As a result, flies are arrhythmic (Zheng *et al.*, 2009). Interestingly, molecular circadian rhythms can be rescued by forcing CLK expression in circadian tissues of *Pdp1ε* mutant flies, but behavioral arrhythmicity persists. This is reminiscent of an earlier study in which *Pdp1ε* dsRNAs were expressed in circadian tissues (Benito *et al.*, 2007). Molecular rhythms were more or less normal (even, surprisingly, *Clk* mRNA cycles, presumably because of residual PDP1ε expression), but behavior was arrhythmic. Thus, by generating transcriptional rhythms that are anti-phase to the PER loop, the PDP1/VRI loop plays a very important role in the regulation of output genes necessary for behavioral rhythms. Strikingly, both in flies with VRI overexpression and in PDP1ε mutant flies, expression of the neuropeptide PDF – essential for behavioral rhythms (see section 15.4.2.1) – is low (Blau and Young, 1999; Zheng *et al.*, 2009).

15.2.5. Relative Importance of Transcriptional and Post-Translational Regulation

In summary, the molecular pacemaker of *Drosophila* is based on a transcriptional feedback loop, with extremely sophisticated and complex post-transcriptional levels of control. The most prominent of these are phosphorylation modifications of PER, TIM, and CLK. It has been proposed that a key function for these protein modifications is to generate a delay between transcriptional and protein cycles, so that the circadian pacemaker has a period of 24 hours. The fact that kinase mutants have profound effects on period length fits with this idea. However, a delay between mRNA and protein accumulation is not necessary for circadian rhythmicity, since in two long-period rhythm *dbt* mutants (*dbt^G* and *dbt^H*) there is no such delay (Suri *et al.*, 2000). However, PER/TIM nuclear entry was not measured in these mutants, and the delay between transcription and nuclear entry might actually be a critical step for the self-sustained nature of circadian oscillations.

Interestingly, studies in cyanobacteria have shown that transcription is not required for temperature compensated free-running 24-hour molecular rhythms to happen. In fact, they can even happen in a test tube with only three proteins (KaiA, B, C), ATP, and Mg^{2+} (Nakajima et al., 2005)! What is looked at here is a rhythmic pattern of KaiC phosphorylation, which is generated by the KaiC protein itself. KaiC has both kinase and phosphatase activities. KaiC forms hexamers, and its biochemical activities are modulated by KaiA, KaiB, and KaiC's own phosphorylation state. This amazing recent development in chronobiology has raised questions about the need for rhythmic transcription within eukaryotic circadian molecular pacemakers (Rosbash, 2009). Studies in mammalian fibroblasts show that partially compromised Pol-II transcription shortens the circadian rhythm period (Dibner et al., 2009). In flies, it is possible to obtain 24-hour period behavioral rhythms when transcription of both per and tim is kept constant (Yang and Sehgal, 2001). Of course, other genes might still be rhythmically expressed (Clk, tws, wdb) and this might be sufficient to keep the molecular pacemaker oscillating. It should also be noted that rhythms observed in the absence of per and tim mRNA cyclings are not robust (see Hall et al., 2007, for additional data on this issue of robustness), and have abnormal properties in terms of responses to light. Importantly, it has also been recently shown that increasing CLK/CYC-mediated transcription by adding a VP16 activation domain to CYC results in a 2-hour short circadian period (Kadener et al., 2008). This short period is largely due to an increase in per transcription, and actually requires per's own promoter. Therefore, in flies, transcriptional activity in the PER loop, and particularly on the per promoter, determines the circadian period, along with the activity of several kinases and phosphatases.

15.2.6. Peripheral and Brain Molecular Pacemakers

As mentioned in the introduction, circadian rhythms are not only found in the brain, but also in peripheral organs such as the eyes, the antennae, or the Malpighian tubules (Emery et al., 1997; Giebultowicz and Hege, 1997; Plautz et al., 1997). As far as we can tell, these rhythms are generated by the same circadian proteins, and obey the same mechanisms. In fact, much of what we know about mRNA cycling, protein cycling, and protein modification is based on retinal circadian clocks, and not on the brain clocks that regulate circadian behavior. Indeed, 75% of PER or TIM signals in the head extracts that are usually used to follow these proteins and their encoding RNAs come from the eyes (Zeng et al., 1994). The other 25% come from a mixture of glial cells, neurons, fat body tissues, and possibly other tissues with a circadian clock. There are, however, tissue-specific differences between circadian pacemakers, which are probably at least in part reflective of the specific roles they are playing in different tissues. First, phase differences can be detected between tissues, and even between circadian neurons of the fly brain (Kaneko et al., 1997; Levine et al., 2002a). Second, in specific tissues, the photoreceptor CRY appears to affect the robustness of circadian molecular rhythms in DD (see also section 15.3) (Ivanchenko et al., 2001; Krishnan et al., 2001; Levine et al., 2002a). Third, circadian oscillations in peripheral tissues tend to dampen and ultimately disappear after a few days in constant darkness (DD), while circadian behavior, and therefore the circadian neurons that drive it, remain rhythmic for as long as behavioral monitoring is possible. Why are oscillations progressively disappearing in peripheral tissues? It is possible that their oscillators are not as robust as those of pacemaker neurons, or that each cell in a tissue actually oscillates, but with different periods that ultimately result in a dispersal of phase (this phenomenon was observed in mammalian fibroblast; Nagoshi et al., 2004). The other possibility is that all oscillators in flies tend to dampen, but communication between circadian neurons improves, the amplitude of their rhythms. Neural circuits are discussed in section 15.4.

15.3. Input Pathways to the *Drosophila* Circadian Pacemaker

So far, we have discussed exclusively the mechanisms that allow the *Drosophila* circadian pacemaker to keep oscillating in constant conditions. However, DD and constant temperature are only very rarely found in a natural environment (in deep caves, for example). Insects thus have to be able to adapt their circadian rhythms to the environment. The input pathways that synchronize circadian clocks with light/dark and temperature cycles therefore play a critical role.

Before describing the molecular mechanisms of these pathways, we need to define entrainment and some of its basic properties. Entrainment refers to the synchronization of a circadian pacemaker to an environmental cycle. The most obvious of these cycles is the light/dark (LD) cycle, but the temperature (or thermophase/cryophase; TC) cycle, and, for some animals, social interactions or food availability can also serve as a Zeitgeber, or timegiver. Particularly when looking at behavior (but even at the level of the molecular oscillator), it is important to distinguish entrainment from a non-circadian response to the Zeitgeber (masking). For example, under an LD cycle, arrhythmic mutants such as per^0 show sharp peaks of activity after the light-on and light-off transition, but these peaks do not persist under DD. These peaks, also observed in wild type flies, are startle responses to the changes in light intensity (a form of positive masking). The potential contribution of masking must always be carefully assessed when environmental cycles are used.

Entrainment does not necessarily require long exposure to an input. For example, flies are actually remarkably sensitive to light. Their response to a light pulse can be measured by comparing the phase of free-running behavior (in DD) between flies that have been pulsed or left in the dark. A 1-minute light pulse is sufficient to elicit a 4-hour phase response when administered at the correct time (Egan *et al.*, 1999). Actually, a phase–response curve (PRC) can be generated by plotting the amplitude of a phase shift as a function of the time of light pulse (Pittendrigh, 1967; Levine *et al.*, 1994; Suri *et al.*, 1998). During the night, phase delays are observed with early-night light pulses, while phase advances are seen with late-night light pulses. During the subjective day (what would be day if the LD cycle were still present), there is no phase shift.

15.3.1. Light-Input Pathways

15.3.1.1. The cell-autonomous photoreceptor CRYP-TOCHROME (CRY) Early studies on eclosion indicated that visual photoreception is not necessary for circadian entrainment. Vitamin-A deprived flies can entrain to LD cycles, and blue light is more efficient than any other wavelength detectable by photoreceptors to elicit phase shifts after short light pulses (Frank and Zimmerman, 1969; Zimmerman and Goldsmith, 1971; Klemm and Ninnemann, 1976). The latter observation was later extended to adult locomotor behavior (Suri *et al.*, 1998). In addition, mutant blind flies can entrain to LD cycles (see, for example, Wheeler *et al.*, 1993; Helfrich-Forster *et al.*, 2001).

Blue light was also found to be most efficient at triggering acute light-dependent TIM degradation (Suri *et al.*, 1998). This proteasome-dependent degradation is believed to play a crucial role in circadian clock resetting (Hunter-Ensor *et al.*, 1996; Lee *et al.*, 1996; Myers *et al.*, 1996; Zeng *et al.*, 1996; Suri *et al.*, 1998; Yang *et al.*, 1998; Naidoo *et al.*, 1999). Indeed, once TIM is degraded, PER is no longer protected, and thus becomes more sensitive to kinases and hence to its own proteasomal degradation (see section 15.2.2.1). Short pulses of light are sufficient to commit TIM to proteasomal degradation (Busza *et al.*, 2004). The molecular pacemaker is thus rapidly reset by light, like eclosion or locomotor rhythms. Moreover, a *tim* mutation (*tim^{SL}*) increases the sensitivity of the circadian clock to light, further supporting the idea that a specific light-input pathway triggers TIM degradation to reset circadian rhythms (Suri *et al.*, 1998). Importantly, cultured isolated body parts can be entrained by LD cycles; this indicates the existence of a cell-autonomous light sensor (Plautz *et al.*, 1997).

In *Drosophila*, the light sensor for TIM degradation and circadian photoresponses such as the PRC is CRYPTO-CHROME (CRY) (Emery *et al.*, 1998; Stanewsky *et al.*,

1998). CRY belongs to a family of proteins sensitive to UV and blue light that includes DNA repair enzymes called photolyases (Kanai *et al.*, 1997; Cashmore, 2003; Ozturk *et al.*, 2007). While photolyases can repair thymidine dimers generated by UV exposure, cryptochromes do not have that property, and were thus proposed to function as blue-light photoreceptors. They were actually first genetically identified as blue-light photoreceptors in plants, and later found in animals, including insects and mammals (Cashmore, 2003). Mammalian-type CRYs, called type II CRYs, probably do not function as circadian photoreceptors, although there is evidence indicating that they are able to sense light (Hattar *et al.*, 2003; Panda *et al.*, 2003; Tu *et al.*, 2004; Hoang *et al.*, 2008). What is clear is that they are an integral part of the molecular pacemaker: they are strong repressors of Clk/BmalI-driven transcription (Kume *et al.*, 1999; van der Horst *et al.*, 1999). We will come back to this notion in section 15.5.2, because type II CRYs are found in most insects, but not in drosophilid species.

Drosophila CRY is a type I cryptochrome (Yuan *et al.*, 2007). Evidence that it functions as a dedicated circadian photoreceptor is overwhelming. First, CRY overexpression increases the light-sensitivity of the PRC (Emery *et al.*, 1998). Second, severely hypomorphic (*cry^{baby}* [*cry^{b}*]) or null *cry* (e.g. *cry^{01}*) mutations have profound effects on circadian photoresponses (Stanewsky *et al.*, 1998; Dolezelova *et al.*, 2007). Rapid TIM degradation is eliminated. The light-pulse PRC is completely abolished. Interestingly, too, *cry* mutant flies remain rhythmic under constant light (LL), as if they were in constant darkness (Emery *et al.*, 2000a), while wild type flies become arrhythmic after a day or two in LL, because TIM is constantly degraded. Moreover, CRY is – as predicted from its similarity with photolyase – more sensitive to blue light and violet light, as is the PRC or TIM degradation (Suri *et al.*, 1998; Busza *et al.*, 2004; VanVickle-Chavez and Van Gelder, 2007; Hoang *et al.*, 2008). The spectral sensitivity of CRY was determined by exposing it to light of different wavelengths and measuring its own rapid light-dependent degradation, which can be observed in S2 cell culture (Lin *et al.*, 2001). The function of this degradation is not known.

CRY binds directly to TIM, even in the absence of PER (Ceriani *et al.*, 1999; Busza *et al.*, 2004). Strikingly, this CRY–TIM interaction is strongly promoted by light exposure, in head protein extracts, in cell culture, and even in yeast. This provides further evidence that CRY is a cell-autonomous circadian photoreceptor. We should also mention here that in yeast and cell culture, CRY has been shown to bind to PER (Rosato *et al.*, 2001). In head extracts, however, light-dependent CRY–PER interaction was only observed in the presence of TIM, suggesting that *in vivo* CRY does not bind to PER directly, or possibly binds only weakly (Busza *et al.*, 2004).

How does light allow CRY to bind TIM? As mentioned above, CRY is related to photolyases, which have been studied biochemically and biophysically in great detail (Sancar, 2003, 2008). Photolyases bind two chromophores: a flavin, and usually a pterin. Both chromophores can absorb photons, with the pterin functioning as the main photon-harvesting chromophore. Upon activation, it can transmit its activation energy to the flavin. The flavin is the catalytic chromophore, transferring, after activation, an electron to the thymidine dimer to repair it. The electron is returned to the flavin during repair. The redox state of the flavin is thus essential for its catalytic activity. If the flavin happens to be in an inactive oxidized state, it can be reactivated by photoreduction involving a chain of three Trp residues that transport an electron (Sancar, 2003). The role of this Trp triad has been studied in type I CRYs. Although results obtained from a first study with *Drosophila* CRY were interpreted as evidence of a need for the Trp triad for CRY function (Froy *et al.*, 2002), more recent studies with type I CRY show that although the Trp triad can help flavin to be reduced *in vitro* (after purification from bacteria), it is not necessary for CRY degradation in cell culture (Hoang *et al.*, 2008; Ozturk *et al.*, 2008). The explanation for this discrepancy would be that the FAD is in a different redox state within S2 cells than after purification in aerobic conditions, and thus does not require the Trp triad for photoreactivation. However, to complicate matters further, when the redox state of CRY's flavin is followed directly within cells, it appears that at low light intensities the Trp triad mutations have an effect on CRY's ability to be photoreduced, but this is not the case at high light intensities (Hoang *et al.*, 2008). This would mean that there are multiple intramolecular pathways for flavin photoreduction in type I CRYs.

Given these results, all obtained *in vitro* or in cell culture, does the Trp triad actually play a role *in vivo*? This was addressed in a study that focused on another function of CRY: its role in magnetoreception. This phenomenon is poorly understood. The Earth's magnetic field can be detected by many animals in a light-dependent manner (Ritz *et al.*, 2002). CRY is one of the rare proteins that is believed to have the biophysical properties to function as a light-sensor and as a magnetosensor, because of the predicted formation of magnetosensitive radical pairs during CRY's photocycle (Ritz *et al.*, 2002). Flies can sense magnetic fields (Wehner and Labhart, 1970; Phillips and Sayeed, 1993; Gegear *et al.*, 2008). For example, in a T-maze where one side is exposed to a strong magnetic field, flies prefer to avoid the side with the magnetic field. They can, however, be trained to prefer that side if it is associated with a food reward. In this assay, both the naïve and trained responses are eliminated in *cry* mutant flies (Gegear *et al.*, 2008). These CRY-dependent responses are independent of CRY's circadian function. However,

magnetic fields can impact CRY-dependent circadian photoresponses (Yoshii *et al.*, 2009a).

The *cry* magnetoreception defects can be rescued by re-expressing *Drosophila* CRY, but also by type I and, surprisingly, type II CRYs from Monarch butterflies (see section 15.5.2) (Gegear *et al.*, 2010). The effect of mutagenizing the terminal Trp residue of the Trp triad did not impair type I CRY function in magnetoreception, although for unclear reasons it did eliminate the rescue with dpCRY2 (Gegear *et al.*, 2010). It therefore appears that the triad is not required for CRY function in magnetoreception. For circadian photoresponses such as the PRC, the mutant type I CRYs were still able to rescue *cry*[b] defects, but the rescue was only partial.

Taken together, all these data suggest that CRY undergoes a light-dependent redox cycle of its flavin chromophore that allows it to function as a circadian photoreceptor or in magnetoreception. Although the Trp triad is not required for CRY function, it might nevertheless have a contribution to CRY's photocycle, and to optimal CRY function.

Once CRY is in its activated state, after electron transfer to the flavin, it has presumably changed conformation. Its affinity for TIM increases, and it also becomes much more prone to proteasomal degradation (Ceriani *et al.*, 1999; Lin *et al.*, 2001; Busza *et al.*, 2004; see also **Figure 6**). This conformational change appears to involve its short C-terminal domain (Busza *et al.*, 2004; Dissel *et al.*, 2004). Indeed, if CRY's C-terminus is missing, CRY binds to TIM, and is degraded by the proteasome even in the dark. Interestingly, overexpressing a C-terminal domain-ablated CRY (CRYΔ) lengthens the period of circadian behavior under constant darkness, and TIM levels are decreased (Dissel *et al.*, 2004). This has been interpreted as an indication that CRYΔ is constantly active, because this phenotype is reminiscent of that of flies under constant low light intensities. However,

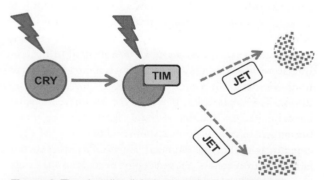

Figure 6 The circadian light input pathway in *Drosophila*. CRYPTOCHROME (CRY) is a key circadian photoreceptor of *Drosophila*. Upon blue-light photon absorption, CRY changes its conformation and binds to TIM. It then triggers TIM ubiquitination by JETLAG (JET), a step that still requires light activation. CRY also undergoes degradation, which can also be mediated by JET. Speckled shapes show the proteins undergoing proteasomal degradation.

studies of a *cry* mutant (*cry^m*) isolated in a constant light screen, in which a point mutation generates a truncated CRY protein also missing the C-terminal domain, give a different picture (Busza *et al.*, 2004). Although CRY levels are very low in *cry^m* flies, these flies can still respond to light. They still phase shift their behavior after short light pulses, and TIM levels still cycle in a light-dependent manner, although the amplitude of these CRY-dependent photoresponses is clearly reduced, presumably because of very low CRY levels. If *cry^m* flies can respond to light, then it is clear that the CRY^M protein is not constantly in a fully active state. CRY^M and CRYΔ only differ by one amino acid. It would thus seem surprising that these two molecules behave differently. What might be happening is that both CRY^M and CRYΔ have a slightly increased activity in the dark compared to wild type CRY, and that this activity becomes obvious if CRYΔ is overexpressed. What both studies unequivocally indicate, however, is that the photolyase homology domain is the active domain of the CRY protein, while the C-terminal tail is important for modulating its activity. This is a striking reversal of function compared to plant CRYs, which are involved in various blue-light dependent photoresponses, including the entrainment of the circadian clock. Indeed, in plants the photolyase domain modulates the activity of the C-terminal domain, which is the active domain that triggers light responses (Yang *et al.*, 2000). From the analysis of the *cry^m* phenotype, we can draw an additional conclusion about the mechanism of CRY photoreception. It is a two-step process requiring light activation (**Figure 6**). First, CRY changes conformation and binds to TIM, and then it triggers its degradation. This second step still requires light. Indeed, CRY binding to TIM is not sufficient to trigger circadian photoresponses, since CRY^M is constitutively interacting with TIM but can still respond to light (Busza *et al.*, 2004).

In conclusion, CRY can function as a cell-autonomous photoreceptor for circadian clocks by directly binding to TIM. This cell-autonomous property was demonstrated with tissue-specific rescues of *cry^b* mutant flies (Emery *et al.*, 2000b). For example, if CRY is rescued in the eyes, then local TIM cycling is rescued, but not circadian behavioral photoresponses such as the PRC. On the contrary, driving CRY expression in the LN_vs that drive circadian behavior in DD (section 15.4.2.1) is sufficient to restore behavioral circadian photoresponses, but not TIM light-dependent cycling in the eyes. Thus, CRY can function as a cell-autonomous photoreceptor even deep inside the fruit fly brain. Recent evidence, however, indicates that circadian behavioral photoresponses are also dependent on circadian neural network interaction (Shang *et al.*, 2008; Tang *et al.*, 2010; see also section 15.4.5).

15.3.1.2. JETLAG (JET) Light-dependent TIM degradation is proteasome-dependent (Naidoo *et al.*,

1999). Two mutants affecting a subunit of an SCF-type E3 ubiquitin ligase called JETLAG (JET) were found to have very similar phenotypes to *cry* mutant flies: these flies remain rhythmic under constant light, rather than being arrhythmic. Moreover, their PRC amplitude is reduced, and TIM degradation is attenuated in whole-head extract (Koh *et al.*, 2006; Peschel *et al.*, 2006). CRY interacts in a light-dependent manner with JET, and co-expressing CRY, JET, and TIM is sufficient for TIM's light-dependent ubiquitination and degradation to be recapitulated in S2 cells (Koh *et al.*, 2006; Peschel *et al.*, 2009). The *jet^c* and *jet^r* mutations affect a leucine-rich repeat (LRR) domain of JET, which also contains an F-box. Thus, JET is part of an SCF complex that ubiquitinates TIM (Koh *et al.*, 2006). Interestingly, recent results indicate that JET can also target CRY for proteasomal degradation (Peschel *et al.*, 2009) (**Figure 6**).

While JET definitely participates in CRY photoresponses, it is not yet clear whether it is required for them. First, TIM light-dependent oscillations in peripheral oscillators and the PRC are only partially disrupted by JET mutations (Koh *et al.*, 2006; Peschel *et al.*, 2006). Second, all molecular and behavioral phenotypes are dependent on a *tim* variant called *ls-tim* (Peschel *et al.*, 2006). Both in lab stocks and in the wild, two *tim* variants can be found (Rosato *et al.*, 1997; Tauber *et al.*, 2007). The so-called *ls-tim* variant can produce both a "long" and a "short" *tim* isoform through the use of alternative ATGs. The *s-tim* variant produces only the short isoform. Interestingly, *s-tim* flies are much more sensitive to light than the *ls-tim* flies, because CRY binds much more efficiently to the short isoform (Sandrelli *et al.*, 2007). Thus, it is only in the genetically sensitized *ls-tim* background that the *jet* mutations have strong effects on circadian light responses (Peschel *et al.*, 2006). More severe JET mutations need to be generated to know whether JET is strictly required for both TIM and CRY degradation, and for behavioral photoresponses, or if other E3 ubiquitin complexes can (at least partially) substitute for it.

Interestingly, the L-TIM/S-TIM protein variants also influence the sequences of CRY and TIM light-dependent degradation (Peschel *et al.*, 2009). Indeed, in the presence of S-TIM, which is strongly bound to CRY under light, S-TIM will be preferentially degraded. However, in the presence of L-TIM, CRY is more rapidly degraded in a JET-dependent manner because it is less efficient at targeting TIM- to JET-mediated degradation. This intimate relationship between CRY, TIM, and JET interaction, and the role of different variants of TIM, might have profound implications for the efficiency of circadian photoresponses. It has also been proposed that the *ls-tim/s-tim* variants affect ovarian diapause in *Drosophila*, which is regulated by the length of the day (Tauber *et al.*, 2007). However, the results from Tauber and colleagues actually seem to indicate that it is the probability of entering diapause that is affected by the presence of one or the other *tim* variant, rather than the critical photoperiod at which diapause occurs (for further discussion, see Emerson *et al.*, 2009). There is clearly more work needed here to establish whether TIM and the circadian clock are truly involved in photoperiodic ovarian diapause in *Drosophila*. In any case, evidence for a selective pressure on the *s/ls-tim* variants has been observed in natural populations of fruit flies, suggesting that the properties of the CRY/TIM/JET interactions provide specific selective advantages in the wild (Tauber *et al.*, 2007).

15.3.1.3. COP9 Signalosome (CSN)
SCF complexes such as those containing JET and SLIMB are comprised of a Skp protein (Skp1), a Cullin (Cul1), an Rbx protein, and an F-box protein that determines the specificity of the enzyme. Cullins are activated by neddylation (Nedd8 is a ubiquitin-like peptide). CSN5 – a subunit of the CSN, a large, evolutionary conserved eight-subunit complex – can deneddylate Cullins, and thus repress SCF activity (Cope and Deshaies, 2003). At least in clock neurons, repressing the activity of the CSN, using *csn4* and *csn5* loss-of-function mutants, or overexpressing a dominant negative CSN5 subunit, suppresses light-dependent TIM degradation (Knowles *et al.*, 2009). Dominant-negative CSN5 also reduces the amplitude of the PRC, and makes flies rhythmic in constant light. These phenotypes are similar to those observed with *cry* or *jet* mutants. These results therefore suggest that JET and the CSN are in the same pathway, with the CSN presumably functioning upstream of JET. However, there is a paradox here: since the CSN should repress JET activity, why would CSN mutants show circadian photoresponse defects? They should be hypersensitive to light. It was proposed that JET might be itself targeted to proteasomal degradation by excessive cullin neddylation (Knowles *et al.*, 2009). Indeed, this was observed in Neurospora for FWD-1, which is responsible for the SCF-dependent ubiquitination of the circadian protein FRQ (He *et al.*, 2005). In agreement with this idea, overexpression of JET corrects the phenotypes observed with the overexpression of CSN5-dominant negative mutant protein. Interestingly, the circadian pacemaker itself is not affected by decreased CSN activity, even though SLIMB containing SCF complex is, as discussed above, necessary for PER degradation and pacemaker function. One explanation could be that SLIMB is less limiting than JET in clock neurons (Knowles *et al.*, 2009).

15.3.1.4. SHAGGY (SGG)
SGG regulates TIM and PER phosphorylation (see section 15.2.2.4). Thus, it could impact circadian photoresponses, since hyperphosphorylated TIM is more sensitive to light-dependent degradation (Martinek *et al.*, 2001). Surprisingly, SGG also regulates CRY degradation

(Stoleru *et al.*, 2007). Even in the absence of TIM, SGG stabilizes CRY in cell culture, and it does so *in vivo* as well. Mechanistically, how SGG stabilizes CRY is not known. Unexpectedly, the period phenotype of SGG overexpression in DD (period shortening) is dependent on CRY, since it almost entirely disappears in *cry*^{b} mutant flies (Stoleru *et al.*, 2007). Does this mean that CRY influences the pace of the circadian pacemaker? Actually, CRY is clearly dispensable for 24-hour period behavioral rhythms, and does not appear to be required in peripheral tissues such as the eyes during and after entrainment with a temperature cycle (Stanewsky *et al.*, 1998). However, amplitude rhythms are altered in antennae and Malpighian tubules of *cry*^{b} flies, which indicates that in specific tissues or under certain circumstances, CRY might affect the function of the circadian pacemaker (Krishnan *et al.*, 1999; Ivanchenko *et al.*, 2001). Maybe SGG overexpression is a situation in which CRY's action on the circadian pacemaker is revealed. It has also been noted that *cry*^{0} flies have weaker behavioral rhythms under low ambient temperature (Dolezelova *et al.*, 2007), suggesting again that the circadian pacemaker might be sensitive to perturbation under certain conditions when CRY is missing or not functional. What would be the mechanism for CRY activity on the circadian pacemaker? It is not clear yet, but evidence for a PER-dependent repressive role for CRY in CLK/CYC transcription has been reported in peripheral tissues and in cell culture (Collins *et al.*, 2006).

15.3.1.5. Other genes that influence CRY photoresponses
A genome-wide dsRNA screen in S2 cells has identified candidate genes that influence CRY degradation by light (Sathyanarayanan *et al.*, 2008). Three genes were followed up *in vivo*: *shh*, a protein phosphatase; and two genes, *Bruce* and *cg17735*, that encode E3 ubiquitin ligase components. Weak effects on CRY degradation were found *in vivo* in hypomorphic *shh* mutants. CRY degradation and levels were more dramatically altered in *Bruce* mutants, and to a much lesser extent in flies expressing siRNAs targeting *cg17735*. However, for *Bruce* and *cg17735* these claims are disputed, since no clear differences between mutant and wild type flies were observed in a more recent study in which the color of the eye was controlled for (the eye pigments protect CRY from degradation) (Peschel *et al.*, 2009). This does not necessarily mean that *Bruce* or *cg17735* has no role in CRY degradation *in vivo*. There might be redundancies. Indeed, JET was not identified in the dsRNA screen, which again raises the question of whether JET is essential for CRY degradation, or whether other degradation pathways can substitute for JET *in vivo*. *cg17735* might also have a role in the circadian pacemaker, since flies expressing dsRNAs against this gene show a long-period phenotype.

Drosophila also possesses a second *timeless* gene, called *tim2*. This gene is important for chromosome stability, as is its mammalian homolog, called mTIM in mice (McFarlane *et al.*, 2010). In a recent study, it was proposed that TIM2 modulates CRY circadian photoresponses (Benna *et al.*, 2010). Indeed, knockdown of TIM2 expression specifically in a subset of optic lobe neurons called T1 cells – whose functions are not yet clear, but which appear to be involved in motion detection – resulted in weak increases in the amplitude of the phase shift triggered by a late-night light pulse. How T1 cells would modulate circadian light responses is unknown at the present time.

A constant light screen, performed with EP lines (see section 15.2.2.4) and hence essentially an overexpression screen, has also identified several candidate genes that might influence CRY photoreception (Dubruille *et al.*, 2009). Among them was the chromatin remodeling protein KISMET (KIS). KIS downregulation results in flies that are strongly rhythmic under constant light. KISMET interacts genetically with CRY. Indeed, overexpressing CRY can correct the light-response defect of KIS knockdown flies, while a genetic interaction with the hypomorphic mutation *cry^m* reveals that KIS affects TIM degradation. The target (or targets) of KIS in the CRY input pathway is (are) not yet identified.

Finally, whole-genome expression profiling in specific circadian neurons has identified *nocturnin* as a gene that also modulates CRY-dependent photoresponses (Nagoshi *et al.*, 2009). This gene, discussed in section 15.4.7, is specifically expressed in a subset of brain circadian neurons.

15.3.1.6. Visual inputs to circadian clocks *cry* mutant flies can entrain their behavior to LD cycles. Indeed, the brain pacemaker neurons also receive input from visual photoreceptors (Helfrich-Forster *et al.*, 2001). All three types of visual photoreceptors (eyes, ocelli, Hofbauer-Buchner eyelet) entrain circadian behavior (Rieger *et al.*, 2003; Veleri *et al.*, 2007). If *cry* and all visual photoreceptors are missing, then circadian behavior is completely blind to light (Helfrich-Forster *et al.*, 2001). It will free-run with a random phase in an LD cycle. Molecularly, there is almost nothing known about the mechanisms underlying light entrainment by visual photoreceptors. TIM could be the pacemaker target of at least the Bolwig's Organ, the larval eye that later becomes the adult Hofbauer-Buchner eyelet (Mazzoni *et al.*, 2005). Indeed, TIM light-dependent degradation kinetics is slowed down in the absence of the Bolwig's Organ. TIM might thus be the target of both CRY-dependent and CRY-independent circadian photoreception.

15.3.1.7. Evidence for additional circadian light-input pathways As mentioned above, behavior cannot be entrained to LD cycles if CRY and vision are missing. However, there is evidence for additional circadian light-input pathways. Indeed, isolated Malpighian tubules and antennae can be synchronized in the absence of CRY (Krishnan *et al.*, 1999; Ivanchenko *et al.*, 2001; Dolezelova *et al.*, 2007). The nature of this light-sensing mechanism has not been studied. Also unknown is the light-input pathway that allows a subset of dorsal neurons to be synchronized to LD cycles even in *gl cry^b* double-mutant flies (Veleri *et al.*, 2003).

15.3.2. Temperature-Input Pathways

Insects are poikilotherms, and are therefore very sensitive to environmental temperature. In the introduction to this chapter, we mentioned that an essential property of circadian clocks is their ability to function with the same period irrespective of ambient temperature (at least over a reasonable range). However, temperature cycles are also important inputs to circadian clocks in many organisms, including *Drosophila* (Pittendrigh, 1954; Wheeler *et al.*, 1993; Glaser and Stanewsky, 2005; Yoshii *et al.*, 2005; Busza *et al.*, 2007; Currie *et al.*, 2009). Temperature and light cycles actually appear to function together to synchronize circadian rhythms (Boothroyd *et al.*, 2007; Yoshii *et al.*, 2009b). The phases of molecular or behavioral rhythms obtained with an LD or a thermophase/cryophase (TC) cycle are similar, if one factors in the fact that TC cycles usually lag behind LD cycles by a few hours (Boothroyd *et al.*, 2007). Like light, temperature entrainment is tissue-autonomous: isolated body parts are entrained by temperature cycles (Glaser and Stanewsky, 2005). This suggests the existence of cell-autonomous temperature sensors that synchronize circadian clocks, or the existence of mechanisms within the pacemaker that are sensitive to temperature fluctuation. However, recent evidence also points at non-autonomous processes (Sehadova *et al.*, 2009).

15.3.2.1. NO CIRCADIAN TEMPERATURE ENTRAINMENT (NOCTE) The first mutant with a specific defect in temperature entrainment was isolated recently, and named *no circadian temperature entrainment* (*nocte*) (Glaser and Stanewsky, 2005). Flies carrying this mutation have severe defects in their ability to synchronize to a TC cycle, both behaviorally (at least under constant light) and molecularly (when luciferase activity is measured from whole flies carrying a *per*-luciferase fusion gene). The *nocte* gene encodes a protein of unknown function with relatively broad expression in the brain and in peripheral organs, which makes it a good candidate for being either a circadian thermosensor or part of a circadian temperature-sensing pathway (Sehadova *et al.*, 2009). Surprisingly, though, it was found that the circadian pacemakers of fly brains are not sensitive to TC cycles when the brain is isolated from the rest of the body (Sehadova *et al.*, 2009). Interestingly, this is also the case in mammals:

while peripheral tissues are very sensitive to the TC cycle, the suprachiasmatic nucleus (the mammalian brain pacemaker) is resistant (Buhr *et al.*, 2010). Consistent with the notion that *Drosophila* brains are insensitive to temperature, reducing NOCTE expression specifically in sensory tissues (in particular the chordotonal organs) is sufficient to disrupt thermal behavioral entrainment. NOCTE is actually expressed in chordotonal organs, and structural defects can be found in these tissues of *nocte* mutants. Other mutations that affect chordotonal organ integrity also disrupt TC entrainment. Thus, chordotonal organs are important for temperature entrainment of circadian behavior (and for other temperature dependent behaviors; Kwon *et al.*, 2010). Whether NOCTE is a structural protein or whether it is directly involved in temperature sensing is not known. Also unknown is its function in the brain, which surprisingly includes clock neurons as well.

It should be noted that most of the results presented above were obtained under constant light conditions (LL). Since LL is disruptive for circadian clocks, it is possible that using LL sensitized the clock to entrainment defects. It would thus be important to replicate the above experiments in the dark, to determine whether the chordotonal organs are really harboring the only temperature-input pathway that can entrain circadian behavior.

15.3.2.2. NO RECEPTOR POTENTIAL A (NORPA)

no receptor potential A (*norpA*) encodes a phospholipase C critical for visual phototransduction. NORPA mutants have a very similar phenotype to *nocte* mutants, with peripheral desynchronization under TC, and absence of entrainment under TC, and LL conditions (Glaser and Stanewsky, 2005). Whether NORPA is expressed in chordotonal organs is not, to our knowledge, currently known, but given its role in phototransduction, it could be part of a thermal signal transduction pathway as well.

15.3.2.3. CRY

CRY is, as discussed above, a circadian light sensor that can trigger circadian behavioral responses to short light pulses. Flies are also capable of responding to relatively short 37°C temperature pulses lasting 30 minutes (at temperatures lower than 34°C there is no measurable response to those short pulses) (Edery *et al.*, 1994b; Kaushik *et al.*, 2007). They will delay their behavior by as much as 2 hours with early night temperature pulses, but they are not able to advance their clock with later pulses. This response is unexpectedly dependent on CRY, which can interact with the PER/TIM dimer after 37°C pulses (Kaushik *et al.*, 2007). Strikingly, only early-night light pulses lead to CRY/TIM/PER interactions, which fits with the absence of a phase response with late-night temperature pulses (Edery *et al.*, 1994b). Interestingly, *per^L* mutant flies have a lower temperature threshold for phase

responses (30°C) and can also advance their phase. This is correlated with a lower temperature threshold for CRY/PER/TIM interaction. There is thus very strong evidence that this interaction mediates temperature pulse phase responses. Moreover, temperature pulses in wild type flies result in TIM degradation in about the same number of circadian neurons as a moderate intensity light pulse (Tang *et al.*, 2010). Thus, temperature can somehow change CRY's affinity for the PER/TIM dimer, by acting either on CRY's conformation or on the PER/TIM dimer, since the *per^L* mutation lowers the temperature threshold for its interaction with CRY (Kaushik *et al.*, 2007).

However, CRY is not required for temperature entrainment at temperatures that are more natural to the flies (17–29°C). Both behavioral and molecular rhythms are entrained in *cry^b* flies under their natural temperature range (Stanewsky *et al.*, 1998; Glaser and Stanewsky, 2005; Busza *et al.*, 2007). So does the CRY/TIM/PER interaction affect behavior in the physiological range of temperature? An interesting observation was made in LL. Wild type flies are, as discussed above, arrhythmic under LL conditions, and this is usually measured at 25°C. The arrhythmicity is mediated by CRY. Interestingly, though, at lower temperatures rhythmicity increases, suggesting that physiological temperature can affect CRY function, and hence its interaction with the PER/TIM dimer (Kaushik *et al.*, 2007). This interaction might thus help to integrate light and temperature inputs. It would also be important to test *nocte:cry* double mutants to determine whether these pathways might be redundant at physiological temperatures.

A final interesting observation has been made regarding CRY's role in temperature responses. It has long been known that *per^L* flies do not compensate normally for changes in ambient temperature. Their clock runs more slowly as temperature increases (hypercompensation). Surprisingly, the *cry^b* mutation can suppress this temperature compensation defect, which was thus presumably due to the lowered threshold for CRY/PER/TIM interaction (Kaushik *et al.*, 2007). However, *cry^b* flies are perfectly normal for temperature compensation, and the *per^L* temperature compensation phenotype thus reflects a gain-of-function effect. Therefore, the actual mechanism of temperature compensation remains unknown in *Drosophila*.

15.3.2.4. *per*'s temperature sensitive splicing

We have mentioned above the phenomenon of temperature compensation and the entrainment to TC cycles. Temperature has an additional well-characterized effect on circadian behavior and molecular rhythms: at different constant ambient temperatures, the circadian phase changes (Majercak *et al.*, 1999). PER/TIM oscillations are phased later as temperature increases. Behavior phase is also influenced by ambient temperature, with the phase of the evening peak of activity progressively moving to a later time of day with higher temperatures. The morning peak, however, behaves differently, becoming earlier as the temperature increases. These phase changes would make sense in the wild. Flies would probably want to avoid being active during the middle of the day, to avoid the risk of desiccation when temperatures are high. The molecular mechanism underlying temperature-dependent phase adjustment is known. Two alternative *per* transcripts can be produced, which differ only by the presence or absence of intron 8 (Cheng *et al.*, 1998) This intron is located in the 3′ UTR, and thus is non-coding. This splicing is more likely to happen at lower temperatures, because its base pair content is suboptimal for efficient splicing (Majercak *et al.*, 1999; Low *et al.*, 2008). Colder temperatures must thus facilitate binding of the splicing machinery to the splice sites. If flies are expressing only a functional *per* transgene that cannot splice this intron, or that does not contain this intron, they cannot adjust PER/TIM cycling and behavioral phases. Thus, through as yet unclear mechanisms, splicing of *per* intron 8 advances *per* mRNA and PER accumulation. What this interesting phenomenon does not explain is how the morning peak is advanced at warmer temperatures. As will be discussed in section 15.4.4, morning and evening peaks of activity can be driven by separate groups of circadian neurons. There might thus be tissue-specific regulation of the effects of PER splicing, or of the splicing itself. Alternatively, other mechanisms that are preferentially used in specific neurons could be at play. Of note, there is also a temperature sensitive splicing in the *tim* gene (Boothroyd *et al.*, 2007). Its function is not yet known.

Interestingly, *per* intron 8 splicing is also affected by light, suggesting that it could serve as a site of input integration (Collins *et al.*, 2004; Majercak *et al.*, 2004). Indeed, *per* splicing efficiency is acutely suppressed by light, and is thus higher at night. The circadian clock also controls *per* splicing and promotes it during the subjective night in DD. Finally, this oscillation is advanced if the photoperiod (length of the light phase) is short. This would further help flies to deal with winter-type environmental conditions, and advance the phase of their behavior under a short photoperiod characteristic of cold days (Majercak *et al.*, 2004). The photic mechanisms regulating *per* splicing are not yet well understood (for further discussion, see Dubruille and Emery, 2008).

Given its role in helping fruit flies to cope with seasonal variations in day length and temperature, as well as presumably helping them to respond to daily changes in temperature, it would be expected that the *per* temperature sensitive splicing is under selective pressure. Interestingly, this splicing is temperature insensitive in tropical species closely related to *Drosophila melanogaster*: *D. yakuba* and *D. santomea*. These tropical species do not adjust circadian behavior phase with temperature. Moreover, when a *D. yakuba per* transgene is used to rescue *D. melanogaster* per^0 mutants, the rescued flies show suboptimal adaptation to ambient temperature (Low *et al.*, 2008). Swapping splicing signals demonstrates that it is indeed these residues that determine thermal phase responses. Since *D. melanogaster* migrated north from a tropical environment with early human populations, these results suggest that suboptimal splice sites were selected during *D. melanogaster* northern migration. In agreement with this model, *D. simulans*, which is also closely related to *D. melanogaster* and is found across the world, has weak splicing sites in intron 8 as well. It would be interesting to determine whether this splicing is still under active selection in *D. melanogaster* and *D. simulans* populations in different parts of the world.

15.3.3. Social/Olfactory Cues as Zeitgebers

In vertebrates, non-photic, non-thermal inputs are known to be able to synchronize circadian clocks. This includes social interactions and food availability. If food is restricted to a certain time of the day, mammals will change the phase of their circadian behavior to be able to feed, even in the presence of an LD cycle (Damiola *et al.*, 2000). Flies do not appear to be able to do so, suggesting that food availability is not a strong Zeitgeber for flies, at least in the presence of an LD cycle (Oishi *et al.*, 2004).

Flies are, however, sensitive to social cues (Levine *et al.*, 2002b). If a small number of flies with a defined circadian phase are mixed with a larger population having a different phase, they can influence the phase of the larger population. Surprisingly, the contrary is not true, the large population does not seem to be able to phase-shift the smaller. This social phase-shifting is dependent on olfactory cues (Levine *et al.*, 2002b), and motivated the search for pheromones that might be rhythmically produced. Pheromones are produced by oenocytes. These oenocytes have circadian clocks, and the production of at least one key enzyme for pheromone production (DESAT1) is rhythmic (Krupp *et al.*, 2008). The amplitude is quite low (1.5- to 2-fold) at the mRNA and protein levels. However, this is nevertheless probably functionally meaningful, since mutants with DESAT1 levels reduced by *ca.* 50% have an altered hydrocarbon production profile. Moreover, the concentrations of at least four cuticular hydrocarbons involved in courtship are circadianly regulated. Whether these pheromones actually are responsible for the phase shift observed after social interaction is not known. Interestingly, social interactions have effects on clock gene expression in oenocytes, and on the pattern of expression of *desat1* and cuticular hydrocarbons (Krupp *et al.*, 2008). There is therefore a complex reciprocal relationship between the oneocyte circadian clock, the production of cuticular hydrocarbons, and social interaction. What could be the purpose of these interactions?

Synchronization of male and female behavior to maximize the chances of finding a mating partner comes to mind. Courtship behavior is influenced by the circadian clock, and the interaction of males and females profoundly alters the pattern of circadian behavior (Sakai and Ishida, 2001a; Fujii *et al.*, 2007). Oenocyte pacemakers have not yet been specifically disrupted, but the data presented above strongly indicate that they play an important role in social/olfactory synchronization of circadian behavior.

15.4. Neural Control of *Drosophila* Circadian Behavior

As discussed in the introduction, circadian clocks regulate various behaviors. The best studied at the neural level is *Drosophila*'s circadian locomotor rhythms. This section will mostly focus on the neural network that regulates this rhythmic behavior.

15.4.1. Anatomy of the *Drosophila* Circadian Neural Circuit

Immunohistochemical staining with antibodies directed against PER and other clock proteins, as well as circadian GAL4 drivers (*per-GAL4*, *tim-GAL4*) or *P*-element enhancer traps driving reporter genes (*gfp*, *lacZ*), were used to identify circadian neurons in *Drosophila* (see, for example, Helfrich-Forster, 1995; Kaneko *et al.*, 1997; Kaneko and Hall, 2000; Shafer *et al.*, 2006, reviewed in Helfrich-Forster, 2005; Nitabach and Taghert, 2008).

There are about 75–80 circadian neurons per hemisphere in the adult fly brain (all the neuron numbers below are given per brain hemisphere; see **Figure 7**). They are divided into several groups, named by the anatomical positions of their cell bodies in the brain; these groups therefore do not necessarily represent functional units.

There are three types of ventrolateral neurons (LN$_v$s). The four large LN$_v$s (l-LN$_v$s) and four small LN$_v$s (s-LN$_v$s) are pigment dispersing factor (PDF) positive. As discussed in the next section, PDF is a neuropeptide that plays an important role in communication between circadian neurons. In addition, there is a PDF negative s-LN$_v$ usually referred to as the "fifth s-LN"$_v$. There are six dorsolateral neurons (LN$_d$s), three lateral posterior neurons (LPNs), and three groups of dorsal neurons (17 DN1s and 2 DN2s, while the number of DN3s is believed to be between 30 and 40). Most of these groups can be further divided in subgroups, based on their size, neurotransmitter content, specific GAL4 driver expression or expression of other markers (for example the presence of absence of CRY). This extraordinary diversity suggests a complex division of function among circadian neurons. For a more detailed description of circadian neuron subtypes, see, for example, Kaneko and Hall, (2000), Shafer *et al.* (2006), Helfrich-Forster *et al.* (2007a, 2007b), Nitabach and Taghert (2008), Yoshii *et al.* (2008), Johard *et al.* (2009), and references therein.

The study of circadian neuron projections reveals that most circadian neurons send projections to the dorsal protocerebrum, which is believed to play an important role in locomotor activity control (Kaneko and Hall, 2000; Helfrich-Forster, 2005). One notable exception is the l-LN$_v$ group. In addition, projection patterns suggest that circadian neural groups communicate with each other. Contralateral projection of l-LN$_v$s and LN$_d$s might help circadian neurons from both hemispheres to communicate with each other (Helfrich-Forster *et al.*, 2007a). Furthermore, dorsal ipsilateral projections from s-LN$_v$s and ventral ipsilateral projections from LN$_d$s and DNs probably mutually connect ventrally and dorsally located circadian neurons. An important center of communication appears to be the accessory medulla, where many

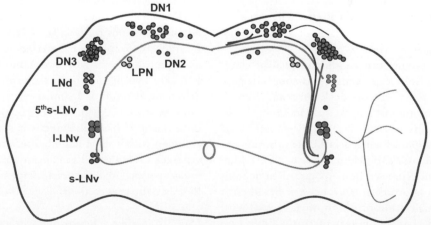

Figure 7 The circadian neurons and their projections in the *Drosophila* brain. Different groups of circadian neurons are labeled with different colors. The three groups of dorsal neurons (DN1, DN2, and DN3) are all labeled in gray. Projections are of the same color as the cell bodies. For simplicity, only the projections from the neurons of the right hemisphere are shown. Note that the l-LN$_v$s and LN$_d$s send contralateral projections.

circadian neuron projections converge, and where visual input might be received (Helfrich-Forster *et al.*, 2007a). Thus, there is a complex interconnection between different circadian neurons, which might contribute to the robustness of circadian behavior and to their ability to respond to environmental inputs (see sections 15.4.2 and 15.4.5).

15.4.2. Communication between Circadian Neurons

15.4.2.1. PIGMENT DISPERSING FACTOR (PDF) and its receptor
To understand how circadian behavior is generated, it is necessary to determine how circadian neurons communicate with each other. An important circadian neurotransmitter – and the best characterized – is pigment dispersing factor (PDF). This neuropeptide has a very restricted expression pattern in *Drosophila*, being expressed only in the four s-LN$_v$s and the four l-LN$_v$s, in four to six abdominal ganglionic neurons (non-circadian), and in three neurons in the tritocerebellum that undergo apoptosis soon after eclosion (Renn *et al.*, 1999). Wild type flies exhibit daily cycling of PDF immunoreactivity within the s-LN$_v$s dorsal projections. These rhythms are self-sustained, and are affected by *per* mutations (Park *et al.*, 2000). This rhythmic PDF immunoreactivity is interpreted as reflecting rhythmic PDF release.

The *Pdf*^{01} mutation reveals the crucial role played by PDF in the control of circadian rhythms (Renn *et al.*, 1999). Most *Pdf*^{01} mutants are arrhythmic in DD, and the rare flies that remain weakly rhythmic have a short-period behavior. In LD, the morning peak of activity is lost (see below) and the evening activity phase is advanced. The behavioral arrhythmicity in DD is correlated with desynchronization between circadian neurons, and a loss of amplitude of molecular rhythms (Peng *et al.*, 2003; Lin *et al.*, 2004). More recent studies demonstrate that PDF has different effects on different circadian neurons; for some of them PDF is necessary for rhythmicity, while for others it accelerates or slows down their pacemaker (Choi *et al.*, 2009; Yoshii *et al.*, 2009c).

The *Pdf*^{01} phenotypes can be phenocopied by eliminating the PDF cells by expressing a pro-apoptotic gene (*rpr* or *hid*) with a *Pdf-GAL4* driver (Renn *et al.*, 1999). It is also replicated in mutants affecting a class II G protein-coupled receptor (Hyun *et al.*, 2005; Lear *et al.*, 2005a; Mertens *et al.*, 2005), which has been demonstrated to respond specifically to PDF (Mertens *et al.*, 2005). PDF receptor (PDFR) is expressed in many circadian neurons and possibly in non-circadian neurons as well, although direct reliable visualization of PDFR with specific antibodies has proved challenging (Hyun *et al.*, 2005; Lear *et al.*, 2005a, 2009; Mertens *et al.*, 2005; Shafer *et al.*, 2008). The most recent study relied on the use of a large MYC-tagged transgene encompassing the whole PDFR

genomic region (Im and Taghert, 2010). It is thus likely to precisely reflect endogenous PDFR expression. PDFR is expressed in specific circadian neurons: the s-LN$_v$s, and a subset of l-LN$_v$s, LN$_d$s, DN1s, and DN3s. The presence of PDFR on s-LN$_v$s suggests the existence of an autocrine feedback and communication between large and small LN$_v$s. Interestingly, it appears that a broader range of circadian neurons rapidly responds to bath-applied PDF, based on a PDF-induced signal visualized in dissected brain with a FRET-based cAMP sensor (Shafer *et al.*, 2008). PDFR-expressing neurons might thus communicate with PDFR-negative neurons to relay PDF signals. Recent studies demonstrate that PDFR expression in a subset of DN1s plays an important role for rhythmicity in DD, and for morning activity (Zhang *et al.*, 2010a). In addition, PDF signaling collaborates with CRY to synchronize the so-called E oscillators that control the evening peak of activity (see section 15.4.4) with the LD cycle (Cusumano *et al.*, 2009; Zhang *et al.*, 2009).

Taken together, these results support the idea that rhythmic release of PDF synchronizes most *Drosophila* circadian neurons, and thus maintains circadian behavior in DD, and properly phases circadian behavior in LD. It should be noted that a recent study has challenged the notion that rhythmic PDF release is needed for circadian rhythms in DD. Indeed, in flies expressing a modified PDF in the LN$_v$s, no rhythms in PDF signals were found in the dorsal termini, yet flies were normally rhythmic (Kula *et al.*, 2006).

Interestingly, the expression of a PDF peptide that is tethered to plasma membranes in dorsally located circadian neurons can improve rhythmicity in *Pdf*^{01} flies, although the rhythms thus obtained are not normal (Choi *et al.*, 2009). Most rescued flies exhibit a complex behavior, with simultaneous complex rhythms of 22-hour and 26-hour periods. Thus, hyperactivation of the PDFR with highly expressed membrane-tethered PDF results in internal de-synchronization, with some circadian neurons running fast and some slow, probably based on the way their pacemakers respond to PDF signaling (phase advances or delays). Interestingly, when the LN$_v$s are hyperexcited with the expression of NachBac driven by *Pdf-GAL4*, a similar complex behavior arises (Sheeba *et al.*, 2008a). If this genetic manipulation is done in *Pdf*^{01} flies, rhythmicity is surprisingly rescued, although with a short period. This strongly suggests the existence of additional LN$_v$ neurotransmitters that can, under specific circumstances, compensate for the lack of PDF (see the next section).

15.4.2.2. Other neurotransmitters of circadian neurons
It has recently been found that PDF-positive s-LN$_v$s do indeed express an additional neuropeptide: short neuropeptide F (sNPF) (Johard *et al.*, 2009).

sNPF is also found in two LN$_d$s. In addition, three other neuropeptides are expressed in some circadian neurons: ion transport peptide (ITP) in the fifth s-LN$_v$ and in one LN$_d$, neuropeptide F (NPF) in the ITP-positive LN$_d$ and two other LN$_d$s, and IPNamide in two DN1s (named DN1$_a$s for their anterior position) (Shafer *et al.*, 2006; Johard *et al.*, 2009). The function of these neuropeptides in the control of circadian behavior is not yet known. However, NPF might be responsible for the difference between male and female circadian behaviors. In LD, females are much more active during the middle of the day than males (Helfrich-Forster, 2000). Interestingly, NPF is only expressed in the LN$_d$s in males, and not in females (Lee *et al.*, 2006).

Classical neurotransmitters have also been found in specific circadian neurons. A *Cha-GAL4* driver shows that the fifth s-LN$_v$ and the two sNPF-positive LN$_d$s are cholinergic.

Two DN1$_a$s, some DN1$_p$s, and several DN3s are found to express the vesicular glutamate transporter (DvGluT), which indicates that they are glutamatergic neurons (Hamasaka *et al.*, 2007). Interestingly, the s-LN$_v$s express the metabotropic glutamate receptor DmGluRA. Furthermore, in DmGluRA mutants or in flies with RNAi-mediated DmGluRA knockdown, morning activity – which is controlled by the s-LN$_v$s (see section 15.4.4) – is increased. In addition, DmGluRA mutant larvae show increased photophobic behavior, which is positively controlled by the s-LN$_v$s (Mazzoni *et al.*, 2005; Hamasaka *et al.*, 2007). These data indicate that the LN$_v$s receive inhibitory glutamate signals from glutamatergic DNs through DmGluRA (Hamasaka *et al.*, 2007). This is confirmed by the observation that glutamate application leads to decreased intracellular calcium in larval s-LN$_v$s, which is associated with decreased neuronal activity.

15.4.3. Inputs from Non-Circadian Neurons

The s-LN$_v$s are probably integrating multiple inputs from various neuronal cell types. Indeed, the larval s-LN$_v$s have been found to respond to application of GABA and serotonin (5HT), with decreased intracellular calcium (Hamasaka *et al.*, 2005; Hamasaka and Nassel, 2006). GABA and 5HT staining showed that some GABAergic and serotoninergic processes terminate near the dendrites of larval and adult LN$_v$s. Interestingly, 5HT receptors are expressed in adult LN$_v$s, and serotonin appears to modulate CRY-dependent circadian photoresponses (Yuan *et al.*, 2005). In addition, recent studies show that GABA inhibition of LN$_v$s induces sleep (Parisky *et al.*, 2008). Indeed, the large LN$_v$s promote arousal (Shang *et al.*, 2008; Sheeba *et al.*, 2008b). Histaminergic projections from the Hofbauer-Buchner eyelet are found near the LN$_v$s, and thus presumably participate in light entrainment (Hamasaka and Nassel, 2006; Veleri *et al.*, 2007).

15.4.4. Neural Control of Morning and Evening Activity

As already discussed, what is relevant for flies in the wild is to be able to cope with day/night cycles. Thus, they have to be able to integrate temperature and light inputs, and generate properly phased circadian rhythms. Under a standard LD cycle (12 hours of light and 12 hours of darkness at 25°C), flies show a bimodal activity – with a morning peak (M peak) before light on and an evening peak (E peak) before lights off (**Figure 1**). There are, in addition, non-circadian acute, transient responses to the light transitions (positive masking). Similar M and E peaks are observed under temperature cycles.

Two complementary approaches based on the GAL4/UAS system (Brand *et al.*, 1994) were used to identify the neural substrates generating the M and E peaks (Blanchardon *et al.*, 2001; Grima *et al.*, 2004; Stoleru *et al.*, 2004). The first was the ablation of specific circadian neural groups with targeted expression of a pro-apoptotic gene (*hid*). The second was to rescue PER expression in a restricted number of circadian neurons of *per⁰* flies. To achieve a good level of cellular resolution, different GAL4 drivers were used. GAL80 repressor transgenes were also introduced as tools to restrict GAL4-mediated expression (Lee and Luo, 1999; Stoleru *et al.*, 2004).

These studies concluded that the PDF-positive s-LN$_v$s control the M peak and are thus also referred to as M oscillators, while the fifth s-LN$_v$ and a subset of LN$_d$s (with possibly a few DN1s) control the E peak and are thus called E oscillators. Indeed, ablation of PDF cells – as mentioned in section 15.4.2.1 – results in the loss of the M peak, while *per⁰* rescue with *Pdf-GAL4* restores specifically the M peak (Renn *et al.*, 1999; Grima *et al.*, 2004; Stoleru *et al.*, 2004). The M peak is not rescued with a driver targeting the l-LN$_v$s, but can be rescued when the s-LN$_v$s are targeted (Grima *et al.*, 2004; Cusumano *et al.*, 2009). Ablation of all *cry13-GAL4*-positive circadian neurons (LN$_v$s, LN$_d$s, and a few DN1s) results in the loss of both M and E peaks, but if *Pdf-GAL80* is added to protect the LN$_v$s from HID expression, the morning peak is preserved (Stoleru *et al.*, 2004). Finally, rescue in a subset of LN$_d$s and in the fifth s-LN$_v$ is sufficient to restore evening activity in *per⁰* flies (Grima *et al.*, 2004; Rieger *et al.*, 2009). Surprisingly, an M peak is observed in flies that have no PER in both large and small PDF-positive LN$_v$s, but PER in all other circadian neurons. This seems at odds with the fact that the PDF-positive LN$_v$s are required for M activity. The proposed explanation is that circadian neurons driving the M and E peaks are coupled oscillators (Stoleru *et al.*, 2004). This is supported by the observation that PDF-negative cells send projections toward the s-LN$_v$s (see above). The PDF-negative circadian neurons would thus be able to activate PDF-positive neurons to generate the M peak even in the absence of a circadian clock in

PDF-positive circadian neurons. It should be noted that although PDF-positive s-LN$_v$s are sufficient for morning activity, they might not be required. Indeed, it is possible to produce flies without functional s-LN$_v$s by expressing a poly-Q containing fragment of the Huntingtin protein in PDF-positive neurons (Sheeba *et al.*, 2010). This either kills the s-LN$_v$s or completely blocks PDF expression, but leaves the l-LN$_v$s intact. Morning anticipation is preserved in these flies, but flies are arrhythmic in DD, as expected. Taken together with the absence of morning anticipation in flies without PDF cells, this suggests that the l-LN$_v$s can substitute for the s-LN$_v$s. There might therefore be a certain degree of plasticity in the circadian system (see also below).

Are the same cells responsible for the morning and evening peaks under a TC cycle? To a large extend, yes (Busza *et al.*, 2007). However, when the cells driving the M and E peaks in LD are ablated, a low-amplitude circadian evening peak can still be detected. Moreover, rescue in *per*^{0} flies restricted to these cells does not entirely restore normal behavior under a TC cycle. Hence, additional circadian neurons respond specifically to temperature cycles, and can influence circadian behavior in adult flies (Busza *et al.*, 2007). The identity of these temperature-sensitive circadian neurons is not yet known, because of lack of proper circadian drivers. However, the DN2s and LPNs are likely candidates. Indeed, when flies are exposed simultaneously to an LD and a TC cycle with an abnormal phase relationship, most circadian neurons follow the LD cycles, but the DN2s and the LPNs choose the TC cycle (Miyasako *et al.*, 2007). Moreover, the DN2s and LPNs express low CRY levels, or possibly none (Klarsfeld *et al.*, 2004; Yoshii *et al.*, 2008). This could explain their preference for TC cycles. Finally, in larvae, in which the circadian system is considerably simpler, strong evidence indicates that the DN2s (and/or maybe the fifth s-LN$_v$) can synchronize the LN$_v$s with TC cycles (Picot *et al.*, 2009).

Interestingly, while adult flies can quickly synchronize their circadian behavior with LD cycles, they are slow to entrain to TC cycles (Busza *et al.*, 2007; Currie *et al.*, 2009). This slow pace is determined by the PDF-positive circadian neurons, since in their absence flies are very sensitive to temperature input. Also, weakening the circadian molecular pacemaker in PDF-positive LN$_v$s accelerates the rate of TC entrainment (weaker oscillators are more sensitive to inputs; Pittendrigh *et al.*, 1991). The conclusion is that the robust pacemaker in the s-LN$_v$s slows down entrainment, while PDF-negative circadian neurons are very sensitive to temperature cycles. The balance between TC-resistant and -sensitive neurons could be crucial for flies to be responsive to temperature cycles, without excessively responding to a Zeitgeber that is not as reliable as the LD cycle (Busza *et al.*, 2007).

15.4.5. Plasticity in the Circadian Neural Network

The different sensitivity of circadian neurons to specific Zeitgebers is a first neural mechanism that might help flies to adapt their behavior to their ever-changing environment. Another is the modulation of the output of specific circadian neurons by environmental input. A striking example is the photic modulation of the respective role of PDF-positive and PDF-negative circadian neurons in the control of circadian locomotor rhythms (reviewed in Dubruille and Emery, 2008; Nitabach and Taghert, 2008; Sheeba *et al.*, 2008c). Under constant darkness, as described above, the PDF-positive s-LN$_v$s drive circadian locomotor behavior; they are necessary for these rhythms, and determine their pace (Renn *et al.*, 1999; Stoleru *et al.*, 2005). In constant light (LL), as we have seen, flies are usually arrhythmic, but if the CRY input pathway is compromised, flies are rhythmic (Emery *et al.*, 2000a). Interestingly, these LL rhythms are not driven by the PDF-positive circadian neurons, but by the E oscillator (three to four LN$_d$s + fifth sLN$_v$) or a subset of DN1s (Murad *et al.*, 2007; Picot *et al.*, 2007; Stoleru *et al.*, 2007). Moreover, PDF is not necessary for LL rhythms (Murad *et al.*, 2007; Picot *et al.*, 2007). The presence or absence of light thus modulates the role played by a subset of PDF-negative and PDF-positive circadian neurons. Evidence indicates that this modulation is dependent on visual photoreceptors, and regulates the output of the relevant circadian neurons (Picot *et al.*, 2007). Indeed, *cry*^{b} flies with the E oscillators being the only cells with a functional circadian clock are rhythmic in LL but not in DD, even though the circadian pacemaker of the E oscillator keeps running under DD conditions as well. The reverse is also true: *cry*^{b} flies with a functional clock restricted to the PDF-positive circadian neurons (M oscillator) are rhythmic only in DD but not in LL, even though under the latter conditions the molecular pacemaker free-runs. This light-dependent contribution of circadian neurons is probably important for dealing with days of different photoperiod. Indeed, under long day conditions, the PDF-negative circadian neurons (the E oscillators and/or the DN1s, presumably) play a more important role in determining the phase of circadian behavior, while under short photoperiod the PDF-positive LN$_v$s (M oscillators) dominate (Stoleru *et al.*, 2007). Fitting with this idea, the PDF-positive LN$_v$s (the large ones, probably) appear to be the dominant contributors to the advance zone of the PRC (see section 15.3), while PDF-negative circadian neurons dominate in the delay zone (Stoleru *et al.*, 2007; Shang *et al.*, 2008; Tang *et al.*, 2010) (for further discussion, see Stoleru *et al.*, 2007; Dubruille and Emery, 2008).

This specific contribution of different circadian neurons to the PRC has an interesting implication. CRY photoresponses are not entirely cell-autonomous, but are dependent on circuit properties. Again, the idea is that specific circadian neurons promote either phase advances or phase delays (Shang et al., 2008; Tang et al., 2010). CRY being necessary for these responses, CRY presumably functions in these circadian neurons to reset their local clock through TIM degradation. However, with the PRC, it is the phase of the s-LN$_v$s that ultimately has to be reset, because these cells are the ones driving behavior in DD – the condition in which phase is measured after the light pulse. So is CRY also degrading TIM in the s-LN$_v$s to contribute to behavior resetting? Surprisingly, at least in the delay zone of the PRC, recent evidence indicates that TIM degradation might be neither necessary nor sufficient for a delay to occur (Tang et al., 2010). This suggests that the s-LN$_v$s reset their clock only through network signaling. Interestingly, the amplitude of the phase delay can be correlated with the number of DN1s that show TIM degradation (for example, by using low light intensity pulses) (Tang et al., 2010). The number of l-LN$_v$s that are present in genetically manipulated flies also determines the amplitude of phase advance (Shang et al., 2008). This indicates that the s-LN$_v$s integrate neuronal signals coming from other circadian neurons. The origin of these signals would encode directionality of the phase shift, while the number of neurons signaling the shift would encode the magnitude of the phase changes. Interneuronal variations in CRY levels (Benito et al., 2008; Yoshii et al., 2008) could explain how light intensity determines phase amplitude: it would take greater light intensity to trigger TIM degradation and reset local clocks in circadian neurons with low CRY levels. Similar principles probably apply to temperature input as well. Indeed, as discussed above, the PDF-positive LN$_v$s are poorly sensitive to temperature input, but PDF-negative circadian neurons are much more sensitive and can entrain them to a temperature cycle (Busza et al., 2007; Miyasako et al., 2007; Picot et al., 2009; Tang et al., 2010). We can thus hypothesize that the s-LN$_v$s are poorly sensitive to inputs so that they can serve as a center of integration from input-sensing neurons that detect different information about the environment (temperature increases and decreases, time of sunset or sunlight, light intensity, etc.). In apparent contradiction to this hypothesis is the fact that CRY is expressed in the s-LN$_v$s (Emery et al., 2000b; Klarsfeld et al., 2004; Benito et al., 2008; Yoshii et al., 2008), and is required in these cells for the phase-shifting effect of light pulses (Tang et al., 2010). Maybe CRY functions very differently in the s-LN$_v$s – for example, by participating in signaling cascades that reset the circadian pacemaker in response to synaptic inputs from the l-LN$_v$s or from the DN1s.

We started this section with an example of output modulation by light. Another example is the effects light and temperature have on output from a subset of DN1s targeted by a GAL4 driver that contains a fragment of the Clk promoter. Using this driver in various rescue experiments

(per^0, pdf^{han}, na^1; narrow abdomen (na) is an important channel in circadian neurons; Lear et al., 2005b), it was found that these cells play an important role in light-induced positive masking at light-on (startle response), for morning anticipation under standard LD conditions, and for DD rhythms as a target of PDF signaling (Zhang et al., 2010a). Interestingly, their contribution to LD behavior is affected by the light intensity, as well as by temperature (Zhang et al., 2010b). In per^0 rescue experiments, they were shown to generate predominantly an M peak or an E peak, or to promote both peaks, depending on light and temperature conditions. Light intensity did not affect the amplitude or phase of molecular rhythms in the DN1s, which indicates that it is the DN1 output pathway that is under photic and thermal control. Under the same environmental conditions, output from the E oscillators was unaffected, which means that the integration of photic and thermal inputs is specific to DN1 output. However, under other specific light conditions (moonlight used during the night instead of complete darkness) the E oscillators can also generate morning activity, while under a standard LD cycle they do not (Rieger et al., 2009).

All these results thus indicate that specific circadian neurons can function as both M and E oscillators, depending on environmental conditions. It is not yet possible, however, to exclude that within the E oscillators and the clk-GAL4-positive DN1s there are different cell types, some specific to morning activity and some specific to evening activity. Indeed, these cell groups are heterogeneous (Johard et al., 2009; Zhang et al., 2010b). It seems more likely, though, that the role of these circadian neurons is plastic, and that this plasticity helps Drosophila to cope with daily changes in its environment.

15.4.6. The Contribution of Glia to Circadian Behavior

There is solid evidence for a role of glial cells in the control of circadian behavior. PER is rhythmically expressed in brain glial cells (Siwicki et al., 1988). Moreover, in gynandromorph studies (male/female mosaics), it was observed that PER expression restricted to glial cells could sustain weak behavioral rhythms (Ewer et al., 1992). In a more recent study, the glia-specific gene ebony (e) was found to be expressed in a rhythmic fashion in glial cells (Suh and Jackson, 2007). Interestingly, ebony mutants show arrhythmic or weakly rhythmic locomotor rhythms, but normal eclosion rhythms (Newby and Jackson, 1991). Restricted expression of EBONY in glial cells is sufficient to rescue robust rhythmic behavior in ebony mutants (Suh and Jackson, 2007). It should be noted, though, that repeated backcrossing of the e^1 mutation into a wild type background also restored normal behavioral rhythms (Hall et al., 2007). This indicates that multiple genetic factors contribute to the ebony behavioral phenotype; not just the mutation in the *ebony* gene. Molecular rhythms in adult circadian neurons persist in e^1 mutant flies; it is therefore the circadian behavioral output that is affected. How EBONY regulates circadian behavior is not yet known, but since it is an enzyme that can conjugate β-Alanine to various biogenic amines, it is probably significant for the metabolism of a neurotransmitter important for circadian locomotor output. Studying the circadian role of glia will certainly reveal novel mechanisms that shape circadian behavior in the future.

15.4.7. Bridging Neural Circuitry and Genomic Studies

There has been remarkable progress in methods to profile whole-genome expression in specific cell types. This technological development has allowed the study of genomic expression in subsets of *Drosophila* circadian neurons, such as the small and large LN$_v$s, as well as PDF-negative circadian neurons (Nagoshi *et al.*, 2009; Kula-Eversole *et al.*, 2010). In brief, specific subsets of circadian neurons can be GFP labeled with the tools we have described in section 15.4.4, and then manually sorted after dissociation. Comparing gene expression between non-circadian neurons and circadian neurons, or between different types of circadian neurons, should ultimately reveal which genes are critical for the function of the different circadian neurons. Already, the transcription factor FER2, whose expression is highly enriched in PDF-positive neurons, has been found to be essential for their proper development (Nagoshi *et al.*, 2009). As expected, *Fer2* mutant flies are arrhythmic. A specific isoform of NOCTURNIN is specifically expressed in dorsal neurons (Nagoshi *et al.*, 2009). Flies with reduced expression of this specific isoform are rhythmic in LL and have an attenuated PRC, indicating that it is regulating CRY circadian photoresponses. The mechanism is not yet known, but vertebrate NOCTURNIN homologs regulate mRNA stability through polyA removal (Baggs and Green, 2003). It will be very interesting to identify the RNAs targeted by NOCTURNIN in the DNs. These two examples validate cell-specific gene profiling as a method to understand circadian neural circuits both molecularly and functionally.

Interestingly, among the genes enriched in PDF positive circadian neurons, only half are shared by the large and small subtypes (Kula-Eversole *et al.*, 2010). Studying those differentially expressed genes should reveal how the l-LN$_v$s regulate sleep, light-mediated arousal, and phase advances (Parisky *et al.*, 2008; Shang *et al.*, 2008; Sheeba *et al.*, 2008b), while s-LN$_v$s control morning anticipation and DD behavior (Grima *et al.*, 2004; Stoleru *et al.*, 2005; Cusumano *et al.*, 2009). With improved tools to target specific circadian neurons, and advances in next generation sequencing technology, much will soon be learned about the gene expression patterns that define the function of different circadian neurons.

15.5. Control of Circadian Rhythms in Non-Drosophilid Insects

Most of this chapter has been dedicated to *Drosophila melanogaster* circadian rhythms, because it in this insect that we know by far the most about these adaptive mechanisms. However, the study of circadian clocks in non-drosophilid insects reveals an unexpected diversity in the molecular mechanisms and neural substrate controlling circadian rhythms. This section will illustrate this point through a few examples.

15.5.1. Neural Substrates Regulating Circadian Behavior in Non-Drosophilid Insects

To identify pacemaker neurons in the absence of genetics, a classical approach is a combination of surgical brain lesions and brain fragment transplants. Antibodies targeted against clock proteins can then be used to identify more precisely candidate pacemaker neurons. A very successful mapping of circadian pacemaker neurons using these approaches was achieved in the cockroach *Leucophaea maderae* (Page, 1982; Stengl and Homberg, 1994; Reischig and Stengl, 2003). As in *Drosophila*, the pacemaker neurons are localized in the accessory medulla of the optic lobes, and express PDF. Moreover, PDF injection can phase-shift circadian locomotor behavior (Petri and Stengl, 1997). This suggests a significant conservation of the circadian neural mechanism between flies and cockroaches. However, in Lepidoptera for example, pacemaker neurons are not located in the accessory medulla, but most likely in the dorsal lateral protocerebrum (Truman, 1972, 1974; Sauman and Reppert, 1996; Zhu *et al.*, 2008). In the Monarch butterfly, only four neurons in the pars lateralis show rhythmic nuclear entry of the critical repressor CRY2 (see next section), and are thus believed to be the brain pacemaker neurons (Zhu *et al.*, 2008). There has thus been considerable evolution of the circadian neural network in insects, which is also revealed by clock gene expression studies (reviewed in Sandrelli *et al.*, 2008; Tomioka and Matsumoto, 2010).

Surprisingly, at least in the case of the Monarch butterfly (*Danaus plexippus*), peripheral oscillators have recently been found to regulate a behavior dependent on circadian clocks. Monarch butterflies use the sun as a compass during their annual migration to Mexico (Reppert *et al.*, 2010). Since the sun moves during the day, the butterflies need to compensate their direction of migration as a function of time of day. This is dependent on the circadian clock (Mouritsen and Frost, 2002; Froy *et al.*, 2003). Unexpectedly, the antenna and its local circadian pacemakers are necessary for time-compensated sun-compass navigation (Merlin *et al.*, 2009). Whether brain pacemaker neurons are also important for this phenomenon and interact with antennal clock neurons is not yet known.

15.5.2. Surprising Diversity of Insect Circadian Molecular Pacemakers

As discussed above, *Drosophila* and mammalian pacemakers share many common principles, and homologous proteins are involved in both systems. It is worth briefly comparing the two systems before describing circadian clocks in non-drosophilid insects. First, the PER loop is very well conserved, although there are notable differences. There are two redundant functional mammalian homologs of *Drosophila* CLK: Clk and NPAS2 (Antoch *et al.*, 1997; DeBruyne *et al.*, 2007). Bmal1 is the CYC homolog (Bunger *et al.*, 2000). Interestingly, in mammals Bmal1 bears a transcriptional activation domain, while CYC does not. Two of the three mammalian PER homologs are essential for circadian rhythms (Bae *et al.*, 2001). They are degraded by the SLIMB homolog β-TRCP (Reischl *et al.*, 2007). The mammalian TIM homolog is actually closer to *Drosophila* TIM2, a gene that might play a role in light responses, in addition to its role in maintenance of chromosomal integrity (Benna *et al.*, 2010). Whether mammalian TIM might have a role in the circadian pacemaker is controversial (Gotter *et al.*, 2000; Tischkau and Gillette, 2005). It might actually connect the circadian clock with the cell cycle and DNA damage checkpoints through its interaction with CRY2 (Unsal-Kacmaz *et al.*, 2005). In fact, CRY1 and CRY2 interact with mammalian PERs (Griffin *et al.*, 1999; Kume *et al.*, 1999). Unlike *Drosophila* TIM, which is probably not directly active in transcriptional repression of CLK/CYC, mammalian CRYs are the main repressors of Clk/Bmal1

transactivation (Kume *et al.*, 1999). This is a major distinction between the two systems. Another important distinction is that CRYs are probably not circadian photoreceptors (Hattar *et al.*, 2003; Panda *et al.*, 2003). Photoreception is dependent on retinal opsins: rod and cone opsins, and melanopsin. Homologs of DBT and CKII are also involved in the mammalian pacemakers (Lowrey *et al.*, 2000; Xu *et al.*, 2005; Etchegaray *et al.*, 2009; Maier *et al.*, 2009).

The circadian pacemaker has been studied in several non-drosophilid species (for a detailed review, see Sandrelli *et al.*, 2008). Interestingly, these comparative and mechanistic studies reveal different molecular organizations. For example, there is apparently no type 1 (photoreceptive) CRY and no *Drosophila* TIM homolog in the honeybee genome (Rubin *et al.*, 2006; Yuan *et al.*, 2007). The architecture of the PER loop is probably very similar to that of mammals, because a type II CRY is a potent repressor of CLK/CYC transcription (Yuan *et al.*, 2007). Since the honeybee does not have type I CRY, it would be very interesting to determine whether isolated organs are nevertheless directly light sensitive, or if all circadian photoresponses are dependent on visual inputs.

Between this "mammalian" type of pacemaker and *Drosophila*, there are hybrid clocks. This is the case of the Monarch (*Danaus plexippus*) clock (**Figure 8**), which might be the best-understood circadian clock, at the molecular level, in a non-drosophilid insect. One of the main reasons for this deeper understanding of the Monarch clock is the existence of a Monarch cell line (Dpn1) that expresses most clock genes (Zhu *et al.*, 2008). PER, TIM,

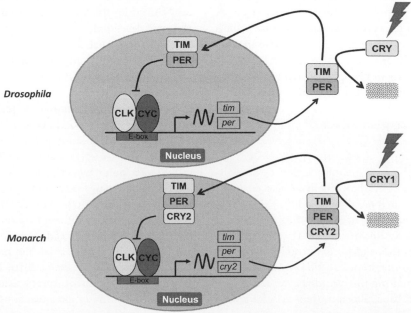

Figure 8 Comparison between the *Drosophila* and Monarch butterfly circadian pacemaker. Simplified models of the circadian pacemaker and CRY input pathway of *Drosophila* (top) and Monarch butterflies (bottom). The major difference is the presence of a type II CRY in Monarch, which functions as the main element of the negative limb of the PER feedback loop.

and CRY2 cycle with phases and amplitudes reminiscent of those observed *in vivo* under LD conditions. However, Dpn1 rhythms are not self-sustained, which means that an important element is missing in these cells for the clock to keep running under constant conditions. Nevertheless, results obtained in this cell line, in which RNAi can easily be performed, combined with *in vivo* data and *Drosophila* S2 cell transactivation assays, give us a detailed view of the functioning of the Monarch circadian clock (Zhu *et al.*, 2005, 2008). The clock truly works like a mammalian–*Drosophila* hybrid. Light is perceived by CRY1 (Type I CRY), and triggers a cascade of degradation: first TIM, then PER, and ultimately CRY2. All three proteins (TIM, PER, and CRY2) are part of a complex, with CRY2 being, as in mammals, the primary repressor of CLK/CYC activity (Zhu *et al.*, 2005, 2008). As in mammals, CYC has an activation domain, a characteristic found in most insects (Chang *et al.*, 2003; Zhu *et al.*, 2005; Rubin *et al.*, 2006; Sandrelli *et al.*, 2008). The *Tribolium castaneum* clock is probably located somewhere between the Monarch and the bee clock, since the *Tribolium* genome does not contain a type I CRY, but does bear both a type II CRY and a *Drosophila*-type TIM (Yuan *et al.*, 2007).

Of course, functional analysis of non-drosophilid circadian genes has mostly been performed in cell culture, using transcriptional assays to determine which molecules are activators or repressors in the circadian pacemaker. What is missing here is a genetic demonstration of the role of these molecules for circadian time-keeping. RNAi techniques such as the injection of dsRNAs have had limited success in insects, but appear to be working remarkably well in the cricket *Grillus bimaculatus*. In this insect, dsRNA-mediated knockdown of PER expression to trough level is sufficient to disrupt circadian behavior (Moriyama *et al.*, 2008, 2009). TIM knockdown did not abolish rhythms, but period was short, suggesting that TIM is also involved in the cricket clock (Danbara *et al.*, 2010). Whether it is essential for rhythmicity in this insect remains to be determined. These recent developments strongly suggest that it will be possible in the close future to test circadian gene function *in vivo* in many more species.

15.6. Conclusions

Our understanding of circadian clocks in insects, and in the animal kingdom in general, owes a lot to the studies in *Drosophila melanogaster*. There is no doubt that much still has to be learned in this organism. The details of the regulation of PER, TIM, and CLK phosphorylation are still far from being understood, and there are undoubtedly more genes regulating the circadian pacemaker that will have to be identified. For example, little is known about the role of Histone modifications in *Drosophila*. Also, the role of miRNAs in the control of circadian rhythms has just begun to be investigated. We have

only a very vague understanding of how circadian neurons talk to each other in the circadian neural network, and the function of at least two-thirds of adult circadian neurons has yet to be found. Finding a function for these neurons will probably require the design of novel behavioral paradigms. Indeed, adult circadian behavior has mostly been synonymous with locomotor rhythms measured under specific experimental conditions. What flies are trying to do in the small test tubes in which they are confined with a little bit of sugar food is unclear: are they trying to escape to find mates, or a more nutritious source of food? The finding that the contribution of a subset of DN1s is dependent on ambient temperature and light intensity, while the contribution of other circadian neurons does not appear to be modulated by these inputs, further raises these questions (Zhang *et al.*, 2010b). Which activity needs to be restricted to favorable environmental conditions, and which should be carried over anyway? Interestingly, the very same DN1 neurons that integrate light and temperature inputs play an important role in the circadian timing of male courtship (Fujii and Amrein, 2010).

There are, of course, very important questions on the role of insect circadian clocks that cannot be addressed with *Drosophila* – or can be addressed only with much greater difficulty. We are, for example, thinking of photoperiodism. Flies show a shallow photoperiodic ovarian diapause (Saunders *et al.*, 1989; Saunders, 1990). Thus, assessing the role of the circadian clock in this process is difficult. Many insects demonstrate much more robust photoperiodic behavior. For example, the cricket *Modicogrillus siamensis*' nymphal development is accelerated under a long photoperiod. Interestingly, injection of *per* dsRNAs entirely disrupts the circadian clock and affects nymphal development, which now occurs under the same kinetics as in animals kept in constant darkness (Sakamoto *et al.*, 2009). PER therefore appears to be critical for the timing of photoperiodic nymphal development, although this does not yet demonstrate that the circadian clock itself is indeed the timer required for photoperiodism in *M. siamensis*. (Emerson *et al.*, 2009; Sakamoto *et al.*, 2009). PER function in photoperiodism could be independent of its circadian role. This example illustrates clearly that we are at a point where gene function analysis can be performed in non-drosophilid insects. Transgenesis has been developed for many insects, RNA interference can be obtained through different approaches (virus, dsRNA injection, transgenesis), and even site-directed mutagenesis is now a possibility with the development of zinc-finger nucleases (Remy *et al.*, 2010; Takasu *et al.*, 2010). This means that behaviors such as time-compensated sun-compass navigation or circadian foraging behavior might soon be accessible to genetic dissection. We might thus be at the dawn of an explosion of knowledge in the field of insect circadian rhythms.

Acknowledgments

We would like to thank Drs Ralf Stanewsky and Isaac Edery for their critical reading of our manuscript, and their insightful comments. Patrick Emery's work is supported by two NIH R01 grants from the National Institute of General Medical Sciences (GM066777 and GM079182).

References

Abraham, G., & Muraleedharan, D. (1990). Role of corpus allatum on the modulation of feeding rhythm in the semilooper caterpillar, *Achaea janata* (L). *Chronobiol. Intl.*, *7*, 419–424.

Akten, B., Jauch, E., Genova, G. K., Kim, E. Y., Edery, I., et al. (2003). A role for CK2 in the *Drosophila* circadian oscillator. *Nat. Neurosci.*, *6*, 251–257.

Akten, B., Tangredi, M. M., Jauch, E., Roberts, M. A., Ng, F., et al. (2009). Ribosomal s6 kinase cooperates with casein kinase 2 to modulate the *Drosophila* circadian molecular oscillator. *J. Neurosci.*, *29*, 466–475.

Allada, R., White, N. E., So, W. V., Hall, J. C., & Rosbash, M. (1998). A mutant *Drosophila* homolog of mammalian Clock disrupts circadian rhythms and transcription of period and timeless. *Cell*, *93*, 791–804.

Antoch, M. P., Song, E. -J., Chang, A. -M., Vitaterna, M. H., Zhao, Y., et al. (1997). Functional identification of the mouse circadian clock gene by transgenic BAC rescue. *Cell*, *89*, 655–667.

Bae, K., Lee, C., Sidote, D., Chuang, K. Y., & Edery, I. (1998). Circadian regulation of a *Drosophila* homolog of the mammalian Clock gene: PER and TIM function as positive regulators. *Mol. Cell Biol.*, *18*, 6142–6151.

Bae, K., Jin, X., Maywood, E. S., Hastings, M. H., Reppert, S. M., & Weaver, D. R. (2001). Differential functions of mPer1, mPer2, and mPer3 in the SCN Circadian Clock. *Neuron*, *30*, 525–536.

Baggs, J. E., & Green, C. B. (2003). Nocturnin, a deadenylase in *Xenopus laevis* retina: A mechanism for posttranscriptional control of circadian-related mRNA. *Curr. Biol.*, *13*, 189–198.

Bargiello, T. A., Jackson, F. R., & Young, M. W. (1984). Restoration of circadian behavioural rhythms by gene transfer in *Drosophila*. *Nature*, *312*, 752–754.

Bartel, D. P. (2004). MicroRNAs: Genomics, biogenesis, mechanism, and function. *Cell*, *116*, 281–297.

Benito, J., Zheng, H., & Hardin, P. E. (2007). PDP1epsilon functions downstream of the circadian oscillator to mediate behavioral rhythms. *J. Neurosci.*, *27*, 2539–2547.

Benito, J., Houl, J. H., Roman, G. W., & Hardin, P. E. (2008). The blue-light photoreceptor CRYPTOCHROME is expressed in a subset of circadian oscillator neurons in the *Drosophila* CNS. *J. Biol. Rhythms.*, *23*, 296–307.

Benna, C., Bonaccorsi, S., Wulbeck, C., Helfrich-Forster, C., Gatti, M., et al. (2010). *Drosophila* timeless2 is required for chromosome stability and circadian photoreception. *Curr. Biol.*, *20*, 346–352.

Blanchardon, E., Grima, B., Klarsfeld, A., Chelot, E., Hardin, P. E., et al. (2001). Defining the role of *Drosophila* lateral neurons in the control of circadian rhythms in motor activity and eclosion by targeted genetic ablation and PERIOD protein overexpression. *Eur. J. Neurosci.*, *13*, 871–888.

Blau, J., & Young, M. W. (1999). Cycling vrille expression is required for a functional *Drosophila* clock. *Cell*, *99*, 661–671.

Bloch, G. (2010). The social clock of the honeybee. *J. Biol. Rhythms*, *25*, 307–317.

Boothroyd, C. E., Wijnen, H., Naef, F., Saez, L., & Young, M. W. (2007). Integration of light and temperature in the regulation of circadian gene expression in *Drosophila*. *PLoS Genet.*, *3*, e54.

Brand, A. H. M., A. S., & Perrimon, N. (1994). Ectopic expression in *Drosophila*. In L. S. B. Goldstein, & E. A. Fyrberg (Eds.), *Drosophila melanogaster: Practical Uses in Cell and Molecular Biology* (pp. 635–654). San Diego, CA: Academic Press.

Brandes, C., Plautz, J. D., Stanewsky, R., Jamison, C. F., Straume, M., et al. (1996). Novel features of *Drosophila* period transcription revealed by real-time luciferase reporting. *Neuron*, *16*, 687–692.

Buhr, E. D., Yoo, S. H., & Takahashi, J. S. (2010). Temperature as a universal resetting cue for mammalian circadian oscillators. *Science*, *330*, 379–385.

Bunger, M. K., Wilsbacher, L. D., Moran, S. M., Clendenin, C., Radcliffe, L. A., et al. (2000). Mop3 is an essential component of the master circadian pacemaker in mammals. *Cell*, *103*, 1009–1017.

Busza, A., Emery-Le, M., Rosbash, M., & Emery, P. (2004). Roles of the two *Drosophila* CRYPTOCHROME structural domains in circadian photoreception. *Science*, *304*, 1503–1506.

Busza, A., Murad, A., & Emery, P. (2007). Interactions between circadian neurons control temperature synchronization of *Drosophila* behavior. *J. Neurosci.*, *27*, 10722–10733.

Cashmore, A. R. (2003). Cryptochromes: Enabling plants and animals to determine circadian time. *Cell*, *114*, 537–543.

Ceriani, M. F., Darlington, T. K., Staknis, D., Mas, P., Petti, A. A., et al. (1999). Light-dependent sequestration of TIMELESS by CRYPTOCHROME. *Science*, *285*, 553–556.

Ceriani, M. F., Hogenesch, J. B., Yanovsky, M., Panda, S., Straume, M., & Kay, S. A. (2002). Genome-wide expression analysis in *Drosophila* reveals genes controlling circadian behavior. *J. Neurosci.*, *22*, 9305–9319.

Chang, D. C., & Reppert, S. M. (2003). A novel C-terminal domain of drosophila PERIOD inhibits dCLOCK: CYCLE-mediated transcription. *Curr. Biol.*, *13*, 758–762.

Chang, D. C., McWatters, H. G., Williams, J. A., Gotter, A. L., Levine, J. D., & Reppert, S. M. (2003). Constructing a feedback loop with circadian clock molecules from the silkmoth, *Antheraea pernyi*. *J. Biol. Chem.*, *278*, 38149–38158.

Chatterjee, A., Tanoue, S., Houl, J. H., & Hardin, P. E. (2010). Regulation of gustatory physiology and appetitive behavior by the *Drosophila* circadian clock. *Curr. Biol.*, *20*, 300–309.

Cheng, Y., Gvakharia, B., & Hardin, P. (1998). Two alternatively spliced transcripts from the *Drosophila* period gene rescue rhythms having different molecular and behavioral characteristics. *Mol. Cell Biol.*, *18*, 6505–6514.

Chiu, J. C., Vanselow, J. T., Kramer, A., & Edery, I. (2008). The phospho-occupancy of an atypical SLIMB-binding site on PERIOD that is phosphorylated by DOUBLETIME controls the pace of the clock. *Genes. Dev.*, *22*, 1758–1772.

Choi, C., Fortin, J. P., McCarthy, E., Oksman, L., Kopin, A. S., & Nitabach, M. N. (2009). Cellular dissection of circadian peptide signals with genetically encoded membrane-tethered ligands. *Curr. Biol.*, *19*, 1167–1175.

Claridge-Chang, A., Wijnen, H., Naef, F., Boothroyd, C., Rajewsky, N., & Young, M. (2001). Circadian regulation of gene expression systems in the *Drosophila* head. *Neuron, 32,* 657–671.

Clark, F., Deadman, D., Greenwood, M., & Larsen, K. S. (1997). A circadian rhythm of locomotor activity in newly emerged *Ceratophyllus sciurorum. Med. Vet. Entomol., 11,* 213–216.

Collins, B. H., Rosato, E., & Kyriacou, C. P. (2004). Seasonal behavior in *Drosophila melanogaster* requires the photoreceptors, the circadian clock, and phospholipase C. *Proc. Natl. Acad. Sci. USA, 101,* 1945–1950.

Collins, B., Mazzoni, E. O., Stanewsky, R., & Blau, J. (2006). *Drosophila* CRYPTOCHROME is a circadian transcriptional repressor. *Curr. Biol., 16,* 441–449.

Cope, G. A., & Deshaies, R. J. (2003). COP9 signalosome: A multifunctional regulator of SCF and other cullin-based ubiquitin ligases. *Cell, 114,* 663–671.

Currie, J., Goda, T., & Wijnen, H. (2009). Selective entrainment of the *Drosophila* circadian clock to daily gradients in environmental temperature. *BMC Biol., 7,* 49.

Curtin, K., Huang, Z. J., & Rosbash, M. (1995). Temporally regulated nuclear entry of the *Drosophila* period protein contributes to the circadian clock. *Neuron, 14,* 365–372.

Cusumano, P., Klarsfeld, A., Chelot, E., Picot, M., Richier, B., & Rouyer, F. (2009). PDF-modulated visual inputs and cryptochrome define diurnal behavior in *Drosophila. Nat. Neurosci., 12,* 1431–1437.

Cymborowski, B., Muszynska-Pytel, M., Porcheron, P., & Cassier, P. (1991). Hemo-lymph ecdysteroid titers controlled by a circadian clock mechanism in larvae of the wax moth, *Galleria mellonella. J. Insect. Physiol, 37,* 35–40.

Cymborowski, B., Hong, S. F., McWatters, H. G., & Saunders, D. S. (1996). S-antigen antibody partially blocks entrainment and the effects of constant light on the circadian rhythm of locomotor activity in the adult blow fly, *Calliphora vicina. J. Biol. Rhythms., 11,* 68–74.

Cyran, S. A., Buchsbaum, A. M., Reddy, K. L., Lin, M. C., Glossop, N. R., et al. (2003). vrille, Pdp1, and dClock form a second feedback loop in the *Drosophila* circadian clock. *Cell, 112,* 329–341.

Cyran, S. A., Yiannoulos, G., Buchsbaum, A. M., Saez, L., Young, M. W., & Blau, J. (2005). The double-time protein kinase regulates the subcellular localization of the *Drosophila* clock protein period. *J. Neurosci., 25,* 5430–5437.

Damiola, F., Le Minh, N., Preitner, N., Kornmann, B., Fleury-Olela, F., & Schibler, U. (2000). Restricted feeding uncouples circadian oscillators in peripheral tissues from the central pacemaker in the suprachiasmatic nucleus. *Genes. Dev., 14,* 2950–2961.

Danbara, Y., Sakamoto, T., Uryu, O., & Tomioka, K. (2010). RNA interference of timeless gene does not disrupt circadian locomotor rhythms in the cricket *Gryllus bimaculatus. J. Insect. Physiol.,* 1738–1745.

Darlington, T. K., Wager-Smith, K., Ceriani, M. F., Staknis, D., Gekakis, N., et al. (1998). Closing the circadian loop: CLOCK-induced transcription of its own inhibitors per and tim. *Science, 280,* 1599–1603.

Das, S., & Dimopoulos, G. (2008). Molecular analysis of photic inhibition of blood-feeding in *Anopheles gambiae. BMC Physiol., 8,* 23.

DeBruyne, J. P., Weaver, D. R., & Reppert, S. M. (2007). CLOCK and NPAS2 have overlapping roles in the suprachiasmatic circadian clock. *Nat. Neurosci., 10,* 543–545.

Dibner, C., Sage, D., Unser, M., Bauer, C., d'Eysmond, T., et al. (2009). Circadian gene expression is resilient to large fluctuations in overall transcription rates. *EMBO J., 28,* 123–134.

Dissel, S., Codd, V., Fedic, R., Garner, K. J., Costa, R., et al. (2004). A constitutively active cryptochrome in *Drosophila melanogaster. Nat. Neurosci., 7,* 834–840.

Dolezelova, E., Dolezel, D., & Hall, J. C. (2007). Rhythm defects caused by newly engineered null mutations in *Drosophila's* cryptochrome gene. *Genetics, 177,* 329–345.

Dubruille, R., & Emery, P. (2008). A plastic clock: How circadian rhythms respond to environmental cues in *Drosophila. Mol. Neurobiol., 38,* 129–145.

Dubruille, R., Murad, A., Rosbash, M., & Emery, P. (2009). A constant light-genetic screen identifies KISMET as a regulator of circadian photoresponses. *PLoS Genet., 5,* e1000787.

Dunlap, J. C. (1999). Molecular bases for circadian clocks. *Cell, 96,* 271–290.

Edery, I., Zwiebel, L. J., Dembinska, M. E., Rosbash, M. (1994a). Temporal phosphorylation of the *Drosophila* period protein. *Proc. Natl. Acad. Sci. USA, 91(6),* 2260–2264.

Edery, I., Rutila, J. E., & Rosbash, M. (1994b). Phase shifting of the circadian clock by induction of the *Drosophila* period protein. *Science, 263,* 237–240.

Egan, E. S., Franklin, T. M., Hilderbrand-Chae, M. J., McNeil, G. P., Roberts, M. A., et al. (1999). An extraretinally expressed insect cryptochrome with similarity to the blue-light photoreceptors of mammals and plants. *J. Neurosci., 19,* 3665–3673.

Emerson, K. J., Bradshaw, W. E., & Holzapfel, C. M. (2009). Complications of complexity: Integrating environmental, genetic and hormonal control of insect diapause. *Trends Genet., 25,* 217–225.

Emery, I. F., Noveral, J. M., Jamison, C. F., & Siwicki, K. K. (1997). Rhythms of *Drosophila* period gene expression in culture. *Proc. Natl. Acad. Sci. USA, 94,* 4092–4096.

Emery, P., So, W. V., Kaneko, M., Hall, J. C., & Rosbash, M. (1998). CRY, a *Drosophila* clock and light-regulated cryptochrome, is a major contributor to circadian rhythm resetting and photosensitivity. *Cell, 95,* 669–679.

Emery, P., Stanewsky, R., Hall, J. C., & Rosbash, M. (2000a). A unique circadian-rhythm photoreceptor. *Nature, 404,* 456–457.

Emery, P., Stanewsky, R., Helfrich-Forster, C., Emery-Le, M., Hall, J. C., & Rosbash, M. (2000b). *Drosophila* CRY is a deep-brain circadian photoreceptor. *Neuron, 26,* 493–504.

Etchegaray, J. P., Machida, K. K., Noton, E., Constance, C. M., Dallmann, R., et al. (2009). Casein kinase 1 delta regulates the pace of the mammalian circadian clock. *Mol. Cell. Biol., 29,* 3853–3866.

Ewer, J., Frisch, B., Hamblen-Coyle, M. J., Rosbash, M., & Hall, J. C. (1992). Expression of the period clock gene within different cell types in the brain of *Drosophila* adults and mosaic analysis of these cells' influence on circadian behavioral rhythms. *J.Neurosci., 12,* 3321–3349.

Fang, Y., Sathyanarayanan, S., & Sehgal, A. (2007). Post-translational regulation of the *Drosophila* circadian clock requires protein phosphatase 1 (PP1). *Genes Dev., 21,* 1506–1518.

Fleury, F., Allemand, R., Fouillet, P., & Bouletreau, M. (1995). Genetic variation in locomotor activity rhythm among populations of *Leptopilina heterotoma* (Hymenoptera: Eucoilidae), a larval parasitoid of *Drosophila* species. *Behav. Genet., 25*, 81–89.

Frank, K. D., & Zimmerman, W. F. (1969). Action spectra for phase shifts of a circadian rhythm in *Drosophila. Science, 163*, 688–689.

Frisch, B., Hardin, P. E., Hamblen-Coyle, M. J., Rosbash, M., & Hall, J. C. (1994). A promoterless period gene mediates behavioral rhythmicity and cyclical per expression in a restricted subset of the *Drosophila* nervous system. *Neuron, 12*, 555–570.

Froy, O., Chang, D., & Reppert, S. (2002). Redox potential differential roles in dCRY and mCRY1 functions. *Curr. Biol., 12*, 147–152.

Froy, O., Gotter, A. L., Casselman, A. L., & Reppert, S. M. (2003). Illuminating the circadian clock in monarch butterfly migration. *Science, 300*, 1303–1305.

Fujii, S., & Amrein, H. (2010). Ventral lateral and DN1 clock neurons mediate distinct properties of male sex drive rhythm in *Drosophila. Proc. Natl. Acad. Sci. USA, 107*, 10590–10595.

Fujii, S., Krishnan, P., Hardin, P., & Amrein, H. (2007). Nocturnal male sex drive in *Drosophila. Curr. Biol., 17*, 244–251.

Gegear, R. J., Casselman, A., Waddell, S., & Reppert, S. M. (2008). Cryptochrome mediates light-dependent magnetosensitivity in *Drosophila. Nature, 454*, 1014–1018.

Gegear, R. J., Foley, L. E., Casselman, A., & Reppert, S. M. (2010). Animal cryptochromes mediate magnetoreception by an unconventional photochemical mechanism. *Nature, 463*, 804–807.

Gekakis, N., Saez, L., Delahaye-Brown, A. -M., Myers, M. P., Sehgal, A., et al. (1995). Isolation of timeless by PER protein interactions: Defective interaction between timeless protein and long-period mutant PERL. *Science, 270*, 811–815.

Giebultowicz, J. M., & Hege, D. M. (1997). Circadian clock in malphigian tubules. *Nature, 386*, 664.

Glaser, F. T., & Stanewsky, R. (2005). Temperature synchronization of the *Drosophila* circadian clock. *Curr. Biol., 15*, 1352–1356.

Glossop, N. R., Lyons, L. C., & Hardin, P. E. (1999). Interlocked feedback loops within the *Drosophila* circadian pacemaker. *Science, 286*, 766–768.

Glossop, N. R., Houl, J. H., Zheng, H., Ng, F. S., Dudek, S. M., & Hardin, P. E. (2003). VRILLE feeds back to control circadian transcription of clock in the *Drosophila* circadian oscillator. *Neuron, 37*, 249–261.

Gotter, A. L., Manganaro, T., Weaver, D. R., Kolakowski, L. F., Jr., Possidente, B., Sriram, S., et al. (2000). A timeless function for mouse timeless, *Nat. Neurosci., 3*, 755–756.

Griffin, E. A. J., Staknis, D., & Weitz, C. J. (1999). Light-independent role of CRY1 and CRY2 in the mammalian circadian clock. *Science, 286*, 768–771.

Grima, B., Lamouroux, A., Chelot, E., Papin, C., Limbourg-Bouchon, B., & Rouyer, F. (2002). The F-box protein slimb controls the levels of clock proteins period and timeless. *Nature, 420*, 178–182.

Grima, B., Chelot, E., Xia, R., & Rouyer, F. (2004). Morning and evening peaks of activity rely on different clock neurons of the *Drosophila* brain. *Nature, 431*, 869–873.

Haddow, A. J. (1956). Rhythmic biting activity of certain East African mosquitoes. *Nature, 177*, 531–532.

Hall, J. C., Chang, D. C., & Dolezelova, E. (2007). Principles and problems revolving around rhythm-related genetic variants. *Cold Spring Harb Symp. Quant. Biol., 72*, 215–232.

Hamasaka, Y., & Nassel, D. R. (2006). Mapping of serotonin, dopamine, and histamine in relation to different clock neurons in the brain of *Drosophila. J. Comp. Neurol., 494*, 314–330.

Hamasaka, Y., Wegener, C., & Nassel, D. R. (2005). GABA modulates *Drosophila* circadian clock neurons via GABAB receptors and decreases in calcium. *J. Neurobiol., 65*, 225–240.

Hamasaka, Y., Rieger, D., Parmentier, M. L., Grau, Y., Helfrich-Forster, C., & Nassel, D. R. (2007). Glutamate and its metabotropic receptor in *Drosophila* clock neuron circuits. *J. Comp. Neurol., 505*, 32–45.

Hao, H., Allen, D. L., & Hardin, P. E. (1997). A circadian enhancer mediates PER-dependent mRNA cycling in *Drosophila melanogaster. Mol. Cell Biol., 17*, 3687–3693.

Hardin, P. E., Hall, J. C., & Rosbash, M. (1990). Feedback of the *Drosophila* period gene product on circadian cycling of its messenger RNA levels. *Nature, 343*, 536–540.

Hardin, P. E., Hall, J. C., & Rosbash, M. (1992). Circadian oscillations in period gene mRNA levels are transcriptionally regulated. *Proc. Natl. Acad. Sci. USA, 89*, 11711–11715.

Hattar, S., Lucas, R. J., Mrosovsky, N., Thompson, S., Douglas, R. H., et al. (2003). Melanopsin and rod-cone photoreceptive systems account for all major accessory visual functions in mice. *Nature, 424*, 76–81.

He, Q., Cheng, P., & Liu, Y. (2005). The COP9 signalosome regulates the *Neurospora* circadian clock by controlling the stability of the SCFFWD-1 complex. *Genes. Dev., 19*, 1518–1531.

Helfrich-Forster, C. (1995). The period clock gene is expressed in central nervous system neurons which also produce a neuropeptide that reveals the projections of circadian pacemaker cells within the brain of *Drosophila melanogaster. Proc. Natl. Acad. Sci. USA, 92*, 612–616.

Helfrich-Forster, C. (2000). Differential control of morning and evening components in the activity rhythm of *Drosophila melanogaster* – sex-specific differences suggest a different quality of activity. *J. Biol. Rhythms, 15*, 135–154.

Helfrich-Forster, C. (2005). Neurobiology of the fruit fly's circadian clock. *Genes Brain Behav., 4*, 65–76.

Helfrich-Forster, C., Winter, C., Hofbauer, A., Hall, J. C., & Stanewsky, R. (2001). The circadian clock of fruit flies is blind after elimination of all known photoreceptors. *Neuron, 30*, 249–261.

Helfrich-Forster, C., Shafer, O. T., Wulbeck, C., Grieshaber, E., Rieger, D., & Taghert, P. (2007a). Development and morphology of the clock-gene-expressing lateral neurons of *Drosophila melanogaster. J. Comp. Neurol., 500*, 47–70.

Helfrich-Forster, C., Yoshii, T., Wulbeck, C., Grieshaber, E., Rieger, D., et al. (2007b). The lateral and dorsal neurons of *Drosophila melanogaster*: New insights about their morphology and function. *Cold Spring Harb. Symp. Quant. Biol., 72*, 517–525.

Hoang, N., Schleicher, E., Kacprzak, S., Bouly, J. P., Picot, M., et al. (2008). Human and *Drosophila* cryptochromes are light activated by flavin photoreduction in living cells. *PLoS Biol., 6*, e160.

Hooven, L. A., Sherman, K. A., Butcher, S., & Giebultowicz, J. M. (2009). Does the clock make the poison? Circadian variation in response to pesticides. *PLoS One, 4*, e6469.

Howlader, G., & Sharma, V. K. (2006). Circadian regulation of egg-laying behavior in fruit flies *Drosophila melanogaster*. *J. Insect Physiol., 52*, 779–785.

Howlader, G., Paranjpe, D. A., & Sharma, V. K. (2006). Non-ventral lateral neuron-based, non-PDF-mediated clocks control circadian egg-laying rhythm in *Drosophila melanogaster*. *J. Biol. Rhythms, 21*, 13–20.

Huang, Z. J., Edery, I., & Rosbash, M. (1993). PAS is a dimerization domain common to *Drosophila* Period and several transcription factors. *Nature, 364*, 259–262.

Hunter-Ensor, M., Ousley, A., & Sehgal, A. (1996). Regulation of the *Drosophila* protein timeless suggests a mechanism for resetting the circadian clock by light. *Cell, 84*, 677–685.

Hyun, S., Lee, Y., Hong, S. T., Bang, S., Paik, D., et al. (2005). *Drosophila* GPCR Han is a receptor for the circadian clock neuropeptide PDF. *Neuron, 48*, 267–278.

Im, S. H., & Taghert, P. H. (2010). PDF receptor expression reveals direct interactions between circadian oscillators in *Drosophila*. *J. Comp. Neurol., 518*, 1925–1945.

Ito, C., Goto, S. G., Shiga, S., Tomioka, K., & Numata, H. (2008). Peripheral circadian clock for the cuticle deposition rhythm in *Drosophila melanogaster*. *Proc. Natl. Acad. Sci. USA, 105*, 8446–8451.

Ivanchenko, M., Stanewsky, R., & Giebultowicz, J. M. (2001). Circadian photoreception in *Drosophila*: Functions of cryptochrome in peripheral and central clocks. *J. Biol. Rhythms, 16*, 205–215.

Johard, H. A., Yoishii, T., Dircksen, H., Cusumano, P., Rouyer, F., et al. (2009). Peptidergic clock neurons in *Drosophila*: Ion transport peptide and short neuropeptide F in subsets of dorsal and ventral lateral neurons. *J. Comp. Neurol., 516*, 59–73.

Kadener, S., Stoleru, D., McDonald, M., Nawathean, P., & Rosbash, M. (2007). Clockwork Orange is a transcriptional repressor and a new *Drosophila* circadian pacemaker component. *Genes. Dev., 21*, 1675–1686.

Kadener, S., Menet, J. S., Schoer, R., & Rosbash, M. (2008). Circadian transcription contributes to core period determination in *Drosophila*. *PLoS Biol., 6*, e119.

Kadener, S., Menet, J. S., Sugino, K., Horwich, M. D., Weissbein, U., et al. (2009). A role for microRNAs in the *Drosophila* circadian clock. *Genes. Dev., 23*, 2179–2191.

Kanai, S., Kikuno, R., Toh, H., Ryo, H., & Todo, T. (1997). Molecular evolution of the photolyase-blue-light photoreceptor family. *J. Mol. Evol., 45*, 535–548.

Kaneko, M., & Hall, J. C. (2000). Neuroanatomy of cells expressing clock genes in *Drosophila*: Transgenic manipulation of the period and timeless genes to mark the perikarya of circadian pacemaker neurons and their projections. *J. Comp. Neurol., 422*, 66–94.

Kaneko, M., Helfrich-Forster, C., & Hall, J. C. (1997). Spatial and temporal expression of the period and timeless genes in the developing nervous system of *Drosophila*: Newly identified pacemakers candidates and novel features of clock gene product cycling. *J. Neurosci., 17*, 6745–6760.

Kaushik, R., Nawathean, P., Busza, A., Murad, A., Emery, P., & Rosbash, M. (2007). PER-TIM interactions with the photoreceptor cryptochrome mediate circadian temperature responses in *Drosophila*. *PLoS Biol., 5*, e146.

Keegan, K. P., Pradhan, S., Wang, J. P., & Allada, R. (2007). Meta-analysis of *Drosophila* circadian microarray studies identifies a novel set of rhythmically expressed genes. *PLoS Comput. Biol., 3*, e208.

Kenny, N. A., & Saunders, D. S. (1991). Adult locomotor rhythmicity as "hands" of the maternal photoperiodic clock regulating larval diapause in the blowfly, *Calliphora vicina*. *J. Biol. Rhythms, 6*, 217–233.

Khan, S. A., Narain, K., Dutta, O., Handique, R., Srivastava, V. K., & Mahanta, J. (1997). Biting behaviour and biting rhythm of potential Japanese encephalitis vectors in Assam. *J. Commun. Dis., 29*, 109–120.

Kim, E. Y., & Edery, I. (2006). Balance between DBT/CKIepsilon kinase and protein phosphatase activities regulate phosphorylation and stability of *Drosophila* CLOCK protein. *Proc. Natl. Acad. Sci. USA, 103*, 6178–6183.

Kim, E. Y., Bae, K., Ng, F. S., Glossop, N. R., Hardin, P. E., & Edery, I. (2002). *Drosophila* CLOCK protein is under post-transcriptional control and influences light-induced activity. *Neuron, 34*, 69–81.

Kim, E. Y., Ko, H. W., Yu, W., Hardin, P. E., & Edery, I. (2007). A DOUBLETIME kinase binding domain on the *Drosophila* PERIOD protein is essential for its hyperphosphorylation, transcriptional repression, and circadian clock function. *Mol. Cell Biol., 27*, 5014–5028.

Kivimae, S., Saez, L., & Young, M. W. (2008). Activating PER repressor through a DBT-directed phosphorylation switch. *PLoS Biol., 6*, e183.

Klarsfeld, A., Malpel, S., Michard-Vanhee, C., Picot, M., Chelot, E., & Rouyer, F. (2004). Novel features of cryptochrome-mediated photoreception in the brain circadian clock of *Drosophila*. *J. Neurosci., 24*, 1468–1477.

Klemm, E., & Ninnemann, H. (1976). Detailed action spectrum for the daily shift in pupal emergence of *Drosophila pseudoobscura*. *Photochem. Photobiol., 24*, 371.

Kloss, B., Price, J. L., Saez, L., Blau, J., Rothenfluh-Hilfiker, A., et al. (1998). The *Drosophila* clock gene double-time encodes a protein closely related to human casein kinase Iε. *Cell, 94*, 97–107.

Kloss, B., Rothenfluh, A., Young, M. W., & Saez, L. (2001). Phosphorylation of period is influenced by cycling physical associations of double-time, period, and timeless in the *Drosophila* clock. *Neuron, 30*, 699–706.

Knowles, A., Koh, K., Wu, J. T., Chien, C. T., Chamovitz, D. A., & Blau, J. (2009). The COP9 signalosome is required for light-dependent timeless degradation and *Drosophila* clock resetting. *J. Neurosci., 29*, 1152–1162.

Ko, H. W., Jiang, J., & Edery, I. (2002). Role for Slimb in the degradation of *Drosophila* Period protein phosphorylated by Doubletime. *Nature, 420*, 673–678.

Ko, H. W., Kim, E. Y., Chiu, J., Vanselow, J. T., Kramer, A., & Edery, I. (2010). A hierarchical phosphorylation cascade that regulates the timing of PERIOD nuclear entry reveals novel roles for proline-directed kinases and GSK-3beta/SGG in circadian clocks. *J. Neurosci., 30*, 12664–12675.

Koh, K., Zheng, X., & Sehgal, A. (2006). JETLAG resets the *Drosophila* circadian clock by promoting light-induced degradation of TIMELESS. *Science, 312*, 1809–1812.

Konopka, R. J., & Benzer, S. (1971). Clock mutants of *Drosophila melanogaster. Proc. Natl. Acad. Sci. USA, 68*, 2112–2116.

Krishnan, B., Dryer, S. E., & Hardin, P. E. (1999). Circadian rhythms in olfactory responses of *Drosophila melanogaster. Nature, 400*, 375–378.

Krishnan, B., Levine, J. D., Lynch, M. K., Dowse, H. B., Funes, P., et al. (2001). A new role for cryptochrome in a *Drosophila* circadian oscillator. *Nature, 411*, 313–317.

Krishnan, P., Chatterjee, A., Tanoue, S., & Hardin, P. E. (2008). Spike amplitude of single unit responses in antennal sensillae is controlled by the *Drosophila* circadian clock. *Curr. Biol.*, 803–807.

Krupp, J. J., Kent, C., Billeter, J. C., Azanchi, R., So, A. K., et al. (2008). Social experience modifies pheromone expression and mating behavior in male *Drosophila melanogaster. Curr. Biol., 18*, 1373–1383.

Kula, E., Levitan, E. S., Pyza, E., & Rosbash, M. (2006). PDF cycling in the dorsal protocerebrum of the *Drosophila* brain is not necessary for circadian clock function. *J. Biol. Rhythms, 21*, 104–117.

Kula-Eversole, E., Nagoshi, E., Shang, Y., Rodriguez, J., Allada, R., & Rosbash, M. (2010). Surprising gene expression patterns within and between PDF-containing circadian neurons in *Drosophila. Proc. Natl. Acad. Sci. USA, 107*, 13497–13502.

Kume, K., Zylka, M. J., Sriram, S., Shearman, L. P., Weaver, D. R., et al. (1999). mCRY1 and mCRY2 are essential components of the negative limb of the circadian clock feedback loop. *Cell, 98*, 193–205.

Kusk, M., Ahmed, R., Thomsen, B., Bendixen, C., Issinger, O. G., & Boldyreff, B. (1999). Interactions of protein kinase CK2beta subunit within the holoenzyme and with other proteins. *Mol. Cell Biochem., 191*, 51–58.

Kwon, Y., Shen, W. L., Shim, H. S., & Montell, C. (2010). Fine thermotactic discrimination between the optimal and slightly cooler temperatures via a TRPV channel in chordotonal neurons. *J. Neurosci., 30*, 10465–10471.

Landskron, J., Chen, K. F., Wolf, E., & Stanewsky, R. (2009). A role for the PERIOD: PERIOD homodimer in the *Drosophila* circadian clock. *PLoS Biol., 7*, e3.

Lazzari, C. R. (1991). Circadian rhythm of egg hatching in *Triatoma infestans* (Hemiptera: Reduviidae). *J. Med. Entomol., 28*, 740–741.

Lear, B. C., Merrill, C. E., Lin, J. M., Schroeder, A., Zhang, L., & Allada, R. (2005a). A G protein-coupled receptor, groom-of-PDF, is required for PDF neuron action in circadian behavior. *Neuron, 48*, 221–227.

Lear, B. C., Lin, J. M., Keath, J. R., McGill, J. J., Raman, I. M., & Allada, R. (2005b). The ion channel narrow abdomen is critical for neural output of the *Drosophila* circadian pacemaker. *Neuron, 48*, 965–976.

Lear, B. C., Zhang, L., & Allada, R. (2009). The neuropeptide PDF acts directly on evening pacemaker neurons to regulate multiple features of circadian behavior. *PLoS Biol., 7*, e1000154.

Lee, C., Parikh, V., Itsukaichi, T., Bae, K., & Edery, I. (1996). Resetting the *Drosophila* clock by photic regulation of PER and a PER-TIM complex. *Science, 271*, 1740–1744.

Lee, C., Bae, K., & Edery, I. (1998). The *Drosophila* CLOCK protein undergoes daily rhythms in abundance, phosphorylation and interactions with the PER-TIM complex. *Neuron, 4*, 857–867.

Lee, C., Bae, K., & Edery, I. (1999). PER and TIM inhibit the DNA binding activity of a *Drosophila* CLOCK-CYC/dBMAL1 heterodimer without disrupting formation of the heterodimer: A basis for circadian transcription. *Mol. Cell Biol., 19*, 5316–5325.

Lee, G., Bahn, J. H., & Park, J. H. (2006). Sex- and clock-controlled expression of the neuropeptide F gene in *Drosophila. Proc. Natl. Acad. Sci. USA, 103*, 12580–12585.

Lee, J. E., & Edery, I. (2008). Circadian regulation in the ability of *Drosophila* to combat pathogenic infections. *Curr. Biol., 18*, 195–199.

Lee, T., & Luo, L. (1999). Mosaic analysis with a repressible cell marker for studies of gene function in neuronal morphogenesis. *Neuron, 22*, 451–461.

Levine, J. D., Casey, C. I., Kalderon, D. D., & Jackson, F. R. (1994). Altered circadian pacemaker functions and cyclic AMP rhythms in the *Drosophila* learning mutant dunce. *Neuron, 13*, 967–974.

Levine, J. D., Funes, P., Dowse, H. B., & Hall, J. C. (2002a). Advanced analysis of a cryptochrome mutation's effects on the robustness and phase of molecular cycles in isolated peripheral tissues of *Drosophila. BMC Neurosci., 3*.

Levine, J. D., Funes, P., Dowse, H. B., & Hall, J. C. (2002b). Resetting the circadian clock by social experience in *Drosophila melanogaster. Science, 298*, 2010–2012.

Lim, C., Chung, B. Y., Pitman, J. L., McGill, J. J., Pradhan, S., et al. (2007). Clockwork orange encodes a transcriptional repressor important for circadian-clock amplitude in *Drosophila. Curr. Biol., 17*, 1082–1089.

Lin, F. J., Song, W., Meyer-Bernstein, E., Naidoo, N., & Sehgal, A. (2001). Photic signaling by cryptochrome in the *Drosophila* circadian system. *Mol. Cell Biol., 21*, 7287–7294.

Lin, J. M., Kilman, V. L., Keegan, K., Paddock, B., Emery-Le, M., et al. (2002). A role for casein kinase 2alpha in the *Drosophila* circadian clock. *Nature, 420*, 816–820.

Lin, J. M., Schroeder, A., & Allada, R. (2005). *In vivo* circadian function of casein kinase 2 phosphorylation sites in *Drosophila* PERIOD. *J. Neurosci., 25*, 11175–11183.

Lin, Y., Han, M., Shimada, B., Wang, L., Gibler, T. M., et al. (2002). Influence of the period-dependent circadian clock on diurnal, circadian, aperiodic gene expression in *Drosophila melanogaster. Proc. Natl. Acad. Sci. USA, 99*, 9562–9567.

Lin, Y., Stormo, G. D., & Taghert, P. H. (2004). The neuropeptide pigment-dispersing factor coordinates pacemaker interactions in the *Drosophila* circadian system. *J. Neurosci., 24*, 7951–7957.

Lindauer, M. (1954). Dauertaenze im Bienenstock und ihre Beziehung zur Sonnenbahn. *Natwissen, 41*, 506–507.

Lindauer, M. (1960). Time-compensated sun orientation in bees. *Cold Spring Harb. Symp. Quant. Biol., 25*, 371–377.

Low, K. H., Lim, C., Ko, H. W., & Edery, I. (2008). Natural variation in the splice site strength of a clock gene and species-specific thermal adaptation. *Neuron, 60*, 1054–1067.

Lowrey, P. L., Shimomura, K., Antoch, M. P., Yamazaki, S., Zemenides, P. D., et al. (2000). Positional syntenic cloning and functional characterization of the mammalian circadian mutation tau. *Science, 288*, 483–492.

Maier, B., Wendt, S., Vanselow, J. T., Wallach, T., Reischl, S., et al. (2009). A large-scale functional RNAi screen reveals a role for CK2 in the mammalian circadian clock. *Genes. Dev., 23*, 708–718.

Majercak, J., Sidote, D., Hardin, P. E., & Edery, I. (1999). How a circadian clock adapts to seasonal decreases in temperature and day length. *Neuron, 24*, 219–230.

Majercak, J., Chen, W. F., & Edery, I. (2004). Splicing of the period gene 3′-terminal intron is regulated by light, circadian clock factors, and phospholipase C. *Mol. Cell Biol., 24*, 3359–3372.

Martinek, S., Inonog, S., Manoukian, A. S., & Young, M. W. (2001). A role for the segment polarity gene shaggy/GSK-3 in the *Drosophila* circadian clock. *Cell, 105*, 769–779.

Matsumoto, A., Ukai-Tadenuma, M., Yamada, R. G., Houl, J., Uno, K. D., et al. (2007). A functional genomics strategy reveals clockwork orange as a transcriptional regulator in the *Drosophila* circadian clock. *Genes. Dev., 21*, 1687–1700.

Mazzoni, E. O., Desplan, C., & Blau, J. (2005). Circadian pacemaker neurons transmit and modulate visual information to control a rapid behavioral response. *Neuron, 45*, 293–300.

McDonald, M. J., & Rosbash, M. (2001). Microarray analysis and organization of circadian gene expression in *Drosophila*. *Cell, 107*, 567–578.

McDonald, M. J., Rosbash, M., & Emery, P. (2001). Wild-type circadian rhythmicity is dependent on closely spaced E boxes in the *Drosophila* timeless promoter. *Mol. Cell Biol., 21*, 1207–1217.

McFarlane, R. J., Mian, S., & Dalgaard, J. Z. (2010). The many facets of the Tim-Tipin protein families' roles in chromosome biology. *Cell Cycle, 9*, 700–705.

Meissner, R. A., Kilman, V. L., Lin, J. M., & Allada, R. (2008). TIMELESS is an important mediator of CK2 effects on circadian clock function *in vivo. J. Neurosci., 28*, 9732–9740.

Menet, J. S., Abruzzi, K. C., Desrochers, J., Rodriguez, J., & Rosbash, M. (2010). Dynamic PER repression mechanisms in the *Drosophila* circadian clock: From on-DNA to off-DNA. *Genes Dev., 24*, 358–367.

Merlin, C., Lucas, P., Rochat, D., Francois, M. C., Maibeche-Coisne, M., & Jacquin-Joly, E. (2007). An antennal circadian clock and circadian rhythms in peripheral pheromone reception in the moth *Spodoptera littoralis. J. Biol. Rhythms, 22*, 502–514.

Merlin, C., Gegear, R. J., & Reppert, S. M. (2009). Antennal circadian clocks coordinate sun compass orientation in migratory monarch butterflies. *Science, 325*, 1700–1704.

Mertens, I., Vandingenen, A., Johnson, E. C., Shafer, O. T., Li, W., et al. (2005). PDF receptor signaling in *Drosophila* contributes to both circadian and geotactic behaviors. *Neuron, 48*, 213–219.

Meyer, P., Saez, L., & Young, M. W. (2006). PER-TIM interactions in living *Drosophila* cells: An interval timer for the circadian clock. *Science, 311*, 226–229.

Minis, D. H., & Pittendrigh, C. S. (1968). Circadian oscillation controlling hatching: Its ontogeny during embryogenesis of a moth. *Science, 159*, 534–536.

Miyasako, Y., Umezaki, Y., & Tomioka, K. (2007). Separate sets of cerebral clock neurons are responsible for light and temperature entrainment of *Drosophila* circadian locomotor rhythms. *J. Biol. Rhythms, 22*, 115–126.

Moore, D., Siegfried, D., Wilson, R., & Rankin, M. A. (1989). The influence of time of day on the foraging behavior of the honeybee, *Apis mellifera. J. Biol. Rhythms, 4*, 305–325.

Moriyama, Y., Sakamoto, T., Karpova, S. G., Matsumoto, A., Noji, S., & Tomioka, K. (2008). RNA interference of the clock gene period disrupts circadian rhythms in the cricket *Gryllus bimaculatus. J. Biol. Rhythms, 23*, 308–318.

Moriyama, Y., Sakamoto, T., Matsumoto, A., Noji, S., & Tomioka, K. (2009). Functional analysis of the circadian clock gene period by RNA interference in nymphal crickets *Gryllus bimaculatus. J. Insect Physiol., 55*, 183–187.

Mouritsen, H., & Frost, B. J. (2002). Virtual migration in tethered flying monarch butterflies reveals their orientation mechanisms. *Proc. Natl. Acad. Sci. USA, 99*, 10162–10166.

Murad, A., Emery-Le, M., & Emery, P. (2007). A subset of dorsal neurons modulates circadian behavior and light responses in *Drosophila. Neuron, 53*, 689–701.

Myers, M. P., Wager-Smith, K., Wesley, C. S., Young, M. W., & Sehgal, A. (1995). Positional cloning and sequence analysis of the *Drosophila* clock gene timeless. *Science, 270*, 805–808.

Myers, M. P., Wager-Smith, K., Rothenfluh-Hilfiker, A., & Young, M. W. (1996). Light-induced degradation of TIMELESS and entrainment of the *Drosophila* circadian clock. *Science, 271*, 1736–1740.

Myers, E. M., Yu, J., & Sehgal, A. (2003). Circadian control of eclosion: Interaction between a central and peripheral clock in *Drosophila melanogaster. Curr. Biol., 13*, 526–533.

Nagoshi, E., Saini, C., Bauer, C., Laroche, T., Naef, F., & Schibler, U. (2004). Circadian gene expression in individual fibroblasts: Cell-autonomous and self-sustained oscillators pass time to daughter cells. *Cell, 119*, 693–705.

Nagoshi, E., Sugino, K., Kula, E., Okazaki, E., Tachibana, T., et al. (2009). Dissecting differential gene expression within the circadian neuronal circuit of *Drosophila. Nat. Neurosci., 13*, 60–68.

Naidoo, N., Song, W., Hunter-Ensor, M., & Sehgal, A. (1999). A role for the proteasome in the light response of the timeless clock protein. *Science, 285*, 1737–1741.

Nakajima, M., Imai, K., Ito, H., Nishiwaki, T., Murayama, Y., et al. (2005). Reconstitution of circadian oscillation of cyanobacterial KaiC phosphorylation *in vitro. Science, 308*, 414–415.

Nawathean, P., & Rosbash, M. (2004). The doubletime and CKII kinases collaborate to potentiate *Drosophila* PER transcriptional repressor activity. *Mol. Cell, 13*, 213–223.

Nawathean, P., Stoleru, D., & Rosbash, M. (2007). A small conserved domain of *Drosophila* PERIOD is important for circadian phosphorylation, nuclear localization, and transcriptional repressor activity. *Mol. Cell Biol., 27*, 5002–5013.

Newby, L. M., & Jackson, F. R. (1991). *Drosophila* ebony mutants have altered circadian activity rhythms but normal eclosion rhythms. *J. Neurogenet, 7*, 85–101.

Nitabach, M. N., & Taghert, P. H. (2008). Organization of the *Drosophila* circadian control circuit. *Curr. Biol., 18*, R84–93.

Oishi, K., Shiota, M., Sakamoto, K., Kasamatsu, M., & Ishida, N. (2004). Feeding is not a more potent Zeitgeber than the light–dark cycle in *Drosophila*. *Neuroreport, 15*, 739–743.

Ousley, A., Zafarullah, K., Chen, Y., Emerson, M., Hickman, L., & Sehgal, A. (1998). Conserved regions of the timeless (tim) clock gene in *Drosophila* analyzed through phylogenetic and functional studies. *Genetics, 148*, 815–825.

Ozturk, N., Song, S. H., Ozgur, S., Selby, C. P., Morrison, L., et al. (2007). Structure and function of animal cryptochromes. *Cold Spring Harb. Symp. Quant. Biol., 72*, 119–131.

Ozturk, N., Song, S. H., Selby, C. P., & Sancar, A. (2008). Animal type 1 cryptochromes. Analysis of the redox state of the flavin cofactor by site-directed mutagenesis. *J. Biol. Chem., 283*, 3256–3263.

Page, T. L. (1982). Transplantation of the cockroach circadian pacemaker. *Science, 216*, 73–75.

Panda, S., Provencio, I., Tu, D. C., Pires, S. S., Rollag, M. D., et al. (2003). Melanopsin is required for non-image-forming photic responses in blind mice. *Science, 301*, 525–527.

Pandian, R. S. (1994). Circadian rhythm in the biting behaviour of a mosquito *Armigeres subalbatus* (Coquillett). *Indian J. Exp. Biol., 32*, 256–260.

Parisky, K. M., Agosto, J., Pulver, S. R., Shang, Y., Kuklin, E., et al. (2008). PDF cells are a GABA-responsive wake-promoting component of the *Drosophila* sleep circuit. *Neuron, 60*, 672–682.

Park, J. H., Helfrich-Forster, C., Lee, G., Liu, L., Rosbash, M., & Hall, J. C. (2000). Differential regulation of circadian pacemaker output by separate clock genes in *Drosophila*. *Proc. Natl. Acad. Sci. USA, 97*, 3608–3613.

Pelc, D., & Steel, C. G. (1997). Rhythmic steroidogenesis by the prothoracic glands of the insect *Rhodnius prolixus* in the absence of rhythmic neuropeptide input: Implications for the role of prothoracicotropic hormone. *Gen. Comp. Endocrinol., 108*, 358–365.

Peng, Y., Stoleru, D., Levine, J. D., Hall, J. C., & Rosbash, M. (2003). *Drosophila* free-running rhythms require intercellular communication. *PLOS Biol., 1*, E13.

Peschel, N., Veleri, S., & Stanewsky, R. (2006). Veela defines a molecular link between Cryptochrome and Timeless in the light-input pathway to *Drosophila*'s circadian clock. *Proc. Natl. Acad. Sci. USA, 103*, 17313–17318.

Peschel, N., Chen, K. F., Szabo, G., & Stanewsky, R. (2009). Light-dependent interactions between the *Drosophila* circadian clock factors cryptochrome, jetlag, and timeless. *Curr. Biol., 19*, 241–247.

Petri, B., & Stengl, M. (1997). Pigment-dispersing hormone shifts the phase of the circadian pacemaker of the cockroach *Leucophaea maderae*. *J. Neurosci., 17*, 4087–4093.

Phillips, J. B., & Sayeed, O. (1993). Wavelength-dependent effects of light on magnetic compass orientation in *Drosophila melanogaster*. *J. Comp. Physiol. A, 172*, 303–308.

Picot, M., Cusumano, P., Klarsfeld, A., Ueda, R., & Rouyer, F. (2007). Light activates output from evening neurons and inhibits output from morning neurons in the *Drosophila* circadian clock. *PLoS Biol., 5*, e315.

Picot, M., Klarsfeld, A., Chelot, E., Malpel, S., & Rouyer, F. (2009). A role for blind DN2 clock neurons in temperature entrainment of the *Drosophila* larval brain. *J. Neurosci., 29*, 8312–8320.

Pittendrigh, C. S. (1954). On temperature independence in the clock-system controlling emergence time in *Drosophila*. *Proc. Natl. Acad. Sci. USA, 40*, 1018–1029.

Pittendrigh, C. S. (1960). Circadian rhythms and the circadian organization of living systems. *Cold Spring Harbor Symp. Quant. Biol., 25*, 159–182.

Pittendrigh, C. S. (1967). Circadian systems. I. The driving oscillation and its assay in *Drosophila* pseudoobscura. *Proc. Natl. Acad. Sci. USA, 58*, 1762–1767.

Pittendrigh, C. S. (1993). Temporal organization: Reflections of a Darwinian clock-watcher. *Annu. Rev. Physiol., 55*, 17–54.

Pittendrigh, C. S., Kyner, W. T., & Takamura, T. (1991). The amplitude of circadian oscillations: Temperature-dependence, latitudinal clines, and the photoperiodic time measurement. *J. Biol. Rhythms, 6*, 299–313.

Plautz, J. D., Kaneko, M., Hall, J. C., & Kay, S. A. (1997). Independent photoreceptive circadian clocks throughout *Drosophila*. *Science, 278*, 1632–1635.

Preuss, F., Fan, J. Y., Kalive, M., Bao, S., Schuenemann, E., et al. (2004). *Drosophila* doubletime mutations which either shorten or lengthen the period of circadian rhythms decrease the protein kinase activity of casein kinase I. *Mol. Cell Biol., 24*, 886–898.

Price, J. L., Blau, J., Rothenfluh-Hilfiker, A., Abodeely, M., Kloss, B., & Young, M. W. (1998). double-time is a novel *Drosophila* clock gene that regulates PERIOD protein accumulation. *Cell, 94*, 83–95.

Pszczolkowski, M. A., & Dobrowolski, M. (1999). Circadian dynamics of locomotor activity and deltamethrin susceptibility in the pine weevil, *Hylobius abietis*. *Phytoparasitica, 27*, 19–25.

Reddy, P., Zehring, W. A., Wheeler, D. A., Pirrotta, V., Hadfield, C., et al. (1984). Molecular analysis of the period locus in *Drosophila melanogaster* and identification of a transcript involved in biological rhythms. *Cell, 38*, 701–710.

Reischig, T., & Stengl, M. (2003). Ectopic transplantation of the accessory medulla restores circadian locomotor rhythms in arrhythmic cockroaches (*Leucophaea maderae*). *J. Exp. Biol., 206*, 1877–1886.

Reischl, S., Vanselow, K., Westermark, P. O., Thierfelder, N., Maier, B., et al. (2007). Beta-TrCP1-mediated degradation of PERIOD2 is essential for circadian dynamics. *J. Biol. Rhythms, 22*, 375–386.

Remy, S., Tesson, L., Menoret, S., Usal, C., Scharenberg, A. M., & Anegon, I. (2010). Zinc-finger nucleases: A powerful tool for genetic engineering of animals. *Transgenic Res., 19*, 363–371.

Renn, S. C. P., Park, J. H., Rosbash, M., Hall, J. C., & Taghert, P. H. (1999). A pdf neuropeptide gene mutation and ablation of PDF neurons each cause severe abnormalities of behavioral circadian rhythms in *Drosophila*. *Cell, 99*, 791–802.

Reppert, S. M., Gegear, R. J., & Merlin, C. (2010). Navigational mechanisms of migrating monarch butterflies. *Trends Neurosci., 33*, 399–406.

Richier, B., Michard-Vanhee, C., Lamouroux, A., Papin, C., & Rouyer, F. (2008). The clockwork orange *Drosophila* protein functions as both an activator and a repressor of clock gene expression. *J. Biol. Rhythms, 23*, 103–116.

Richter, K. (2001). Daily changes in neuroendocrine control of moulting hormone secretion in the prothoracic gland of the cockroach *Periplaneta americana* (L.). *J. Insect Physiol.*, *47*, 333–338.

Rieger, D., Stanewsky, R., & Helfrich-Forster, C. (2003). Cryptochrome, compound eyes, Hofbauer-Buchner eyelets, and ocelli play different roles in the entrainment and masking pathway of the locomotor activity rhythm in the fruit fly *Drosophila melanogaster*. *J. Biol. Rhythms*, *18*, 377–391.

Rieger, D., Wulbeck, C., Rouyer, F., & Helfrich-Forster, C. (2009). Period gene expression in four neurons is sufficient for rhythmic activity of *Drosophila melanogaster* under dim light conditions. *J. Biol. Rhythms*, *24*, 271–282.

Ritz, T., Dommer, D. H., & Phillips, J. B. (2002). Shedding light on vertebrate magnetoreception. *Neuron*, *34*, 503–506.

Rorth, P., Szabo, K., Bailey, A., Laverty, T., Rehm, J., et al. (1998). Systematic gain-of-function genetics in *Drosophila*. *Development*, *125*, 1049–1057.

Rosato, E., Trevisan, A., Sandrelli, F., Zordan, M., Kyriacou, C. P., & Costa, R. (1997). Conceptual translation of timeless reveals alternative initiating methionines in *Drosophila*. *Nucleic Acids Res.*, *25*, 455–458.

Rosato, E., Codd, V., Mazzotta, G., Piccin, A., Zordan, M., et al. (2001). Light-dependent interaction between *Drosophila* CRY and the clock protein PER mediated by the carboxy terminus of CRY. *Curr. Biol.*, *11*, 909–917.

Rosbash, M. (2009). The implications of multiple circadian clock origins. *PLoS Biol.*, *7*, e62.

Rothenfluh, A., Young, M. W., & Saez, L. (2000). A TIMELESS-independent function for PERIOD proteins in the *Drosophila* clock. *Neuron*, *26*, 505–515.

Rubin, E. B., Shemesh, Y., Cohen, M., Elgavish, S., Robertson, H. M., & Bloch, G. (2006). Molecular and phylogenetic analyses reveal mammalian-like clockwork in the honey bee (*Apis mellifera*) and shed new light on the molecular evolution of the circadian clock. *Genome Res.*, *16*, 1352–1365.

Rutila, J. E., Suri, V., Le, M., So, W. V., Rosbash, M., & Hall, J. C. (1998). CYCLE is a second bHLH-PAS protein essential for circadian transcription of *Drosophila* period and timeless. *Cell*, *93*, 805–814.

Rymer, J., Bauernfeind, A. L., Brown, S., & Page, T. L. (2007). Circadian rhythms in the mating behavior of the cockroach, *Leucophaea maderae*. *J. Biol. Rhythms*, *22*, 43–57.

Saez, L., & Young, M. W. (1996). Regulation of nuclear entry of the *Drosophila* clock proteins PERIOD and TIMELESS. *Neuron*, *17*, 911–920.

Sakai, T., & Ishida, N. (2001a). Circadian rhythms of female mating activity governed by clock genes in *Drosophila*. *Proc. Natl. Acad. Sci. USA*, *98*, 9221–9225.

Sakai, T., & Ishida, N. (2001b). Time, love and species. *Neuro. Endocrinol. Lett.*, *22*, 222–228.

Sakamoto, T., Uryu, O., & Tomioka, K. (2009). The clock gene period plays an essential role in photoperiodic control of nymphal development in the cricket *Modicogryllus siamensis*. *J. Biol. Rhythms*, *24*, 379–390.

Sancar, A. (2003). Structure and function of DNA photolyase and cryptochrome blue-light photoreceptors. *Chem. Rev.*, *103*, 2203–2237.

Sancar, A. (2008). Structure and function of photolyase and *in vivo* enzymology: 50th anniversary. *J. Biol. Chem.*, *283*, 32153–32157.

Sandrelli, F., Tauber, E., Pegoraro, M., Mazzotta, G., Cisotto, P., et al. (2007). A molecular basis for natural selection at the timeless locus in *Drosophila melanogaster*. *Science, 316*, 1898–1900.

Sandrelli, F., Costa, R., Kyriacou, C. P., & Rosato, E. (2008). Comparative analysis of circadian clock genes in insects. *Insect Mol. Biol.*, *17*, 447–463.

Sathyanarayanan, S., Zheng, X., Xiao, R., & Sehgal, A. (2004). Posttranslational regulation of *Drosophila* PERIOD protein by protein phosphatase 2A. *Cell, 116*, 603–615.

Sathyanarayanan, S., Zheng, X., Kumar, S., Chen, C. H., Chen, D., et al. (2008). Identification of novel genes involved in light-dependent CRY degradation through a genome-wide RNAi screen. *Genes. Dev.*, *22*, 1522–1533.

Sauman, I., & Reppert, S. M. (1996). Period protein is necessary for circadian control of egg hatching behavior in the silmoth *Antheraea pernyi*. *Neuron, 17*, 901–909.

Saunders, D. S. (1990). The circadian basis of ovarian diapause regulation in *Drosophila melanogaster*: Is the period gene causally involved in photoperiodic time measurement? *J. Biol. Rhythms*, *5*, 315–331.

Saunders, D. S., & Thomson, E. J. (1977). "Strong" phase response curve for the circadian rhythm of locomotor activity in a cockroach (*Nauphoeta cinerea*). *Nature, 270*, 241–243.

Saunders, D. S., Henrich, V. C., & Gilbert, L. I. (1989). Induction of diapause in *Drosophila melanogaster*: Photoperiodic regulation and the impact of arrhythmic clock mutations on time measurement. *Proc. Natl. Acad. Sci. USA, 86*, 3748–3752.

Sehadova, H., Glaser, F. T., Gentile, C., Simoni, A., Giesecke, A., et al. (2009). Temperature entrainment of *Drosophila*'s circadian clock involves the gene nocte and signaling from peripheral sensory tissues to the brain. *Neuron, 64*, 251–266.

Sehgal, A., Price, J. L., Man, B., & Young, M. W. (1994). Loss of circadian behavioral rhythms and per RNA oscillations in the *Drosophila* mutant timeless. *Science, 263*, 1603–1606.

Sehgal, A., Rothenfluh-Hilfiker, A., Hunter-Ensor, M., Chen, Y., Myers, M., & Young, M. W. (1995). Circadian oscillations and autoregulation of timeless RNA. *Science, 270*, 808–810.

Shafer, O. T., Rosbash, M., & Truman, J. W. (2002). Sequential nuclear accumulation of the clock proteins period and timeless in the pacemaker neurons of *Drosophila melanogaster*. *J. Neurosci.*, *22*, 5946–5954.

Shafer, O. T., Helfrich-Forster, C., Renn, S. C., & Taghert, P. H. (2006). Reevaluation of *Drosophila melanogaster*'s neuronal circadian pacemakers reveals new neuronal classes. *J. Comp. Neurol.*, *498*, 180–193.

Shafer, O. T., Kim, D. J., Dunbar-Yaffe, R., Nikolaev, V. O., Lohse, M. J., & Taghert, P. H. (2008). Widespread receptivity to neuropeptide PDF throughout the neuronal circadian clock network of *Drosophila* revealed by real-time cyclic AMP imaging. *Neuron, 58*, 223–237.

Shang, Y., Griffith, L. C., & Rosbash, M. (2008). Light-arousal and circadian photoreception circuits intersect at the large PDF cells of the *Drosophila* brain. *Proc. Natl. Acad. Sci. USA, 105*, 19587–19594.

Sheeba, V., Sharma, V. K., Gu, H., Chou, Y. T., O'Dowd, D. K., & Holmes, T. C. (2008a). Pigment dispersing factor-dependent and -independent circadian locomotor behavioral rhythms. *J. Neurosci.*, *28*, 217–227.

Sheeba, V., Fogle, K. J., Kaneko, M., Rashid, S., Chou, Y. T., et al. (2008b). Large ventral lateral neurons modulate arousal and sleep in *Drosophila. Curr. Biol.*, *18*, 1537–1545.

Sheeba, V., Kaneko, M., Sharma, V. K., & Holmes, T. C. (2008c). The *Drosophila* circadian pacemaker circuit: Pas de deux or tarantella? *Crit. Rev. Biochem. Mol. Biol.*, *43*, 37–61.

Sheeba, V., Fogle, K. J., & Holmes, T. C. (2010). Persistence of morning anticipation behavior and high amplitude morning startle response following functional loss of small ventral lateral neurons in *Drosophila. PLoS One*, *5*, e11628.

Shirasu-Hiza, M. M., Dionne, M. S., Pham, L. N., Ayres, J. S., & Schneider, D. S. (2007). Interactions between circadian rhythm and immunity in *Drosophila melanogaster. Curr. Biol.*, *17*, R353–355.

Silvegren, G., Lofstedt, C., & Qi Rosen, W. (2005). Circadian mating activity and effect of pheromone pre-exposure on pheromone response rhythms in the moth *Spodoptera littoralis. J. Insect Physiol.*, *51*, 277–286.

Siwicki, K. K., Eastman, C., Petersen, G., Rosbash, M., & Hall, J. C. (1988). Antibodies to the period gene product of *Drosophila* reveal diverse tissue distribution and rhythmic changes in the visual system. *Neuron*, *1*, 141–150.

Smith, E. M., Lin, J. M., Meissner, R. A., & Allada, R. (2008). Dominant-negative CK2alpha induces potent effects on circadian rhythmicity. *PLoS Genet.*, *4*, e12.

So, W. V., & Rosbash, M. (1997). Post-transcriptional regulation contributes to *Drosophila* clock gene mRNA cycling. *EMBO J.*, *16*, 7146–7155.

Sogawa, K., Nakano, R., Kobayashi, A., Kikuchi, Y., Ohe, N., et al. (1995). Possible function of Ah receptor nuclear translocator (Arnt) homodimer in transcriptional regulation. *Proc. Natl. Acad. Sci. USA*, *92*, 1936–1940.

Sothern, R. B., Hermida, R. C., Nelson, R., Mojon, A., & Koukkari, W. L. (1998). Reanalysis of filter-feeding behavior of caddis fly (Brachycentrus) larvae reveals masking and circadian rhythmicity. *Chronobiol. Intl.*, *15*, 595–606.

Standfast, H. A. (1967). Biting times of nine species of New Guinea Cullicidae (Diptera). *J. Med. Entomol.*, *4*, 192–196.

Stanewsky, R., Jamison, C. F., Plautz, J. D., Kay, S. A., & Hall, J. C. (1997). Multiple circadian-regulated elements contribute to cycling period gene expression in *Drosophila. EMBO J.*, *16*, 5006–5018.

Stanewsky, R., Kaneko, M., Emery, P., Beretta, M., Wager-Smith, K., et al. (1998). The cryb mutation identifies cryptochrome as a circadian photoreceptor in *Drosophila. Cell*, *95*, 681–692.

Stanewsky, R., Lynch, K. S., Brandes, C., & Hall, J. C. (2002). Mapping of elements involved in regulating normal temporal period and timeless RNA expression patterns in *Drosophila melanogaster. J. Biol. Rhythms*, *17*, 293–306.

Steel, C. G., & Vafopoulou, X. (2006). Circadian orchestration of developmental hormones in the insect, *Rhodnius prolixus. Comp. Biochem. Physiol. A Mol. Integr. Physiol.*, *144*, 351–364.

Stelzer, J. R., Stanewsky, R., & Chittka, L. (2010). Circadian foraging rhythms of bumblebees monitored by radio-frequency identification. *J. Biol. Rhythms*, *25*, 257–267.

Stengl, M., & Homberg, U. (1994). Pigment-dispersing hormone-immunoreactive neurons in the cockroach *Leucophaea maderae* share properties with circadian pacemaker neurons. *J. Comp. Physiol. A*, *175*, 203–213.

Stoleru, D., Peng, Y., Agusto, J., & Rosbash, M. (2004). Coupled oscillators control morning and evening locomotor behaviour of *Drosophila. Nature*, *431*, 862–868.

Stoleru, D., Peng, Y., Nawathean, P., & Rosbash, M. (2005). A resetting signal between *Drosophila* pacemakers synchronizes morning and evening activity. *Nature*, *438*, 238–242.

Stoleru, D., Nawathean, P., Fernandez Mde, L., Menet, J. S., Ceriani, M. F., & Rosbash, M. (2007). The *Drosophila* circadian network is a seasonal timer. *Cell*, *129*, 207–219.

Suh, J., & Jackson, F. R. (2007). *Drosophila* ebony activity is required in glia for the circadian regulation of locomotor activity. *Neuron*, *55*, 435–447.

Sullivan, W. N., Cawley, B., Hayes, D. K., Rosenthal, J., & Halberg, F. (1970). Circadian rhythm in susceptibility of house flies and Madeira cockroaches to pyrethrum. *J. Econ. Entomol.*, *63*, 159–163.

Sun, W. C., Jeong, E. H., Jeong, H. J., Ko, H. W., Edery, I., & Kim, E. Y. (2010). Two distinct modes of PERIOD recruitment onto dCLOCK reveal a novel role for TIMELESS in circadian transcription. *J. Neurosci.*, *30*, 14458–14469.

Suri, V., Qian, Z., Hall, J. C., & Rosbash, M. (1998). Evidence that the TIM light response is relevant to light-induced phase shifts in *Drosophila melanogaster. Neuron*, *21*, 225–234.

Suri, V., Lanjuin, A., & Rosbash, M. (1999). TIMELESS-dependent positive and negative autoregulation in the *Drosophila* circadian clock. *EMBO J.*, *18*, 675–686.

Suri, V., Hall, J. C., Rosbash, M. (2000). Two novel doubletime mutants alter circadian properties and eliminate the delay between RNA and protein in *Drosophila. J. Neurosci.*, *20*, 7547–7555

Swanson, H. I., Chan, W. K., & Bradfield, C. A. (1995). DNA binding specificities and pairing rules of the Ah receptor, ARNT, and SIM proteins. *J. Biol. Chem.*, *270*, 26292–26302.

Takasu, Y., Kobayashi, I., Beumer, K., Uchino, K., Sezutsu, H., et al. (2010). Targeted mutagenesis in the silkworm *Bombyx mori* using zinc finger nuclease mRNA injection. *Insect Biochem. Mol. Biol.*, *40*, 759–765.

Tang, C. H., Hinteregger, E., Shang, Y., & Rosbash, M. (2010). Light-mediated TIM degradation within *Drosophila* pacemaker neurons (s-LNvs) is neither necessary nor sufficient for delay zone phase shifts. *Neuron*, *66*, 378–385.

Tanoue, S., Krishnan, P., Chatterjee, A., & Hardin, P. E. (2008). G-protein Receptor Kinase 2 is required for rhythmic olfactory responses in *Drosophila. Curr. Biol.*, 787–794.

Tauber, E., Zordan, M., Sandrelli, F., Pegoraro, M., Osterwalder, N., et al. (2007). Natural selection favors a newly derived timeless allele in *Drosophila melanogaster. Science*, *316*, 1895–1898.

Tischkau, S. A., & Gillette, M. U. (2005). Oligodeoxynucleotide methods for analyzing the circadian clock in the suprachiasmatic nucleus. *Methods Enzymol.*, *393*, 593–610.

Tomioka, K., & Matsumoto, A. (2010). A comparative view of insect circadian clock systems. *Cell Mol. Life Sci.*, *67*, 1397–1406.

Truman, J. W. (1972). Physiology of insect rhythms: II. The silkmoth brain as the location of the biological clock controlling eclosion. *J. Comp. Physiol.*, *99*, 99–114.

Truman, J. W. (1974). Physiology of insect rhythms: IV. Role of the brain in the regulation of the flight rhythm of the giant silkmoths. *J. Comp. Physiol.*, *95*, 281–296.

Tu, D. C., Batten, M. L., Palczewski, K., & Van Gelder, R. N. (2004). Nonvisual photoreception in the chick iris. *Science*, *306*, 129–131.

Ueda, H. R., Matsumoto, A., Kawamura, M., Iino, M., Tanimura, T., & Hashimoto, S. (2002). Genome-wide transcriptional orchestration of circadian rhythms in *Drosophila*. *J. Biol. Chem.*, *277*, 14048–14052.

Unsal-Kacmaz, K., Mullen, T. E., Kaufmann, W. K., & Sancar, A. (2005). Coupling of human circadian and cell cycles by the timeless protein. *Mol. Cell Biol.*, *25*, 3109–3116.

Uryu, O., & Tomioka, K. (2010). Circadian oscillations outside the optic lobe in the cricket *Gryllus bimaculatus*. *J. Insect Physiol.*, *56*, 1284–1290.

Vafopoulou, X., & Steel, C. G. (1991). Circadian regulation of synthesis of ecdysteroids by prothoracic glands of the insect *Rhodnius prolixus*: Evidence of a dual oscillator system. *Gen. Comp. Endocrinol.*, *83*, 27–34.

Vafopoulou, X., & Steel, C. G. (1996). Circadian regulation of a daily rhythm of release of prothoracicotropic hormone from the brain retrocerebral complex of *Rhodnius prolixus* (hemiptera) during larval-adult development. *Gen. Comp. Endocrinol.*, *102*, 123–129.

van der Horst, G. T., Muijtjens, M., Kobayashi, K., Takano, R., Kanno, S., et al. (1999). Mammalian Cry1 and Cry2 are essential for maintenance of circadian rhythms. *Nature*, *398*, 627–630.

VanVickle-Chavez, S. J., & Van Gelder, R. N. (2007). Action spectrum of *Drosophila* cryptochrome. *J. Biol. Chem.*, *282*, 10561–10566.

Veleri, S., Brandes, C., Helfrich-Forster, C., Hall, J. C., & Stanewsky, R. (2003). A self-sustaining, light-entrainable circadian oscillator in the *Drosophila* brain. *Curr. Biol.*, *13*, 1758–1767.

Veleri, S., Rieger, D., Helfrich-Forster, C., & Stanewsky, R. (2007). Hofbauer-Buchner eyelet affects circadian photosensitivity and coordinates TIM and PER expression in *Drosophila* clock neurons. *J. Biol. Rhythms*, *22*, 29–42.

Vosshall, L. B., Price, J. L., Sehgal, A., Saez, L., & Young, M. W. (1994). Specific block in nuclear localization of period protein by a second clock mutation, timeless. *Science*, *263*, 1606–1609.

Wehner, R., & Labhart, T. (1970). Perception of the geomagnetic field in the fly *Drosophila melanogaster*. *Experientia*, *26*, 967–968.

Wheeler, D. A., Hamblen-Coyle, M. J., Dushay, M. S., & Hall, J. C. (1993). Behavior in light–dark cycles of *Drosophila* mutants that are arrhythmic, blind, or both. *J. Biol. Rhythms*, *8*, 67–94.

Wijnen, H., & Young, M. W. (2006). Interplay of circadian clocks and metabolic rhythms. *Annu. Rev. Genet.*, *40*, 409–448.

Xu, K., Zheng, X., & Sehgal, A. (2008). Regulation of feeding and metabolism by neuronal and peripheral clocks in *Drosophila*. *Cell Metab.*, *8*, 289–300.

Xu, Y., Padiath, Q. S., Shapiro, R. E., Jones, C. R., Wu, S. C., et al. (2005). Functional consequences of a CKIdelta mutation causing familial advanced sleep phase syndrome. *Nature*, *434*, 640–644.

Yang, H. Q., Wu, Y., Tang, R. H., Liu, D., Liu, Y., & Cashmore, A. R. (2000). The C termini of *Arabidopsis* cryptochromes mediate a constitutive light response. *Cell*, *103*, 815–827.

Yang, M., Lee, J. E., Padgett, R. W., & Edery, I. (2008). Circadian regulation of a limited set of conserved microRNAs in *Drosophila*. *BMC Genomics*, *9*, 83.

Yang, Z., & Sehgal, A. (2001). Role of molecular oscillations in generating behavioral rhythms in *Drosophila*. *Neuron*, *29*, 453–467.

Yang, Z., Emerson, M., Su, H. S., & Sehgal, A. (1998). Response of the timeless protein to light correlates with behavioral entrainment and suggests a nonvisual pathway for circadian photoreception. *Neuron*, *21*, 215–223.

Yildiz, O., Doi, M., Yujnovsky, I., Cardone, L., Berndt, A., et al. (2005). Crystal structure and interactions of the PAS repeat region of the *Drosophila* clock protein PERIOD. *Mol. Cell*, *17*, 69–82.

Yoshii, T., Heshiki, Y., Ibuki-Ishibashi, T., Matsumoto, A., Tanimura, T., & Tomioka, K. (2005). Temperature cycles drive *Drosophila* circadian oscillation in constant light that otherwise induces behavioural arrhythmicity. *Eur. J. Neurosci.*, *22*, 1176–1184.

Yoshii, T., Todo, T., Wulbeck, C., Stanewsky, R., & Helfrich-Forster, C. (2008). Cryptochrome is present in the compound eyes and a subset of *Drosophila*'s clock neurons. *J. Comp. Neurol.*, *508*, 952–966.

Yoshii, T., Ahmad, M., & Helfrich-Forster, C. (2009a). Cryptochrome mediates light-dependent magnetosensitivity of *Drosophila*'s circadian clock. *PLoS Biol.*, *7*, e1000086.

Yoshii, T., Vanin, S., Costa, R., & Helfrich-Forster, C. (2009b). Synergic entrainment of *Drosophila*'s circadian clock by light and temperature. *J. Biol. Rhythms*, *24*, 452–464.

Yoshii, T., Wulbeck, C., Sehadova, H., Veleri, S., Bichler, D., et al. (2009c). The neuropeptide pigment-dispersing factor adjusts period and phase of *Drosophila*'s clock. *J. Neurosci.*, *29*, 2597–2610.

Yu, W., Zheng, H., Houl, J. H., Dauwalder, B., & Hardin, P. E. (2006). PER-dependent rhythms in CLK phosphorylation and E-box binding regulate circadian transcription. *Genes. Dev.*, *20*, 723–733.

Yu, W., Zheng, H., Price, J. L., & Hardin, P. E. (2009). DOUBLETIME plays a noncatalytic role to mediate CLOCK phosphorylation and repress CLOCK-dependent transcription within the *Drosophila* circadian clock. *Mol. Cell Biol.*, *29*, 1452–1458.

Yuan, Q., Lin, F., Zheng, X., & Sehgal, A. (2005). Serotonin modulates circadian entrainment in *Drosophila*. *Neuron*, *47*, 115–127.

Yuan, Q., Metterville, D., Briscoe, A. D., & Reppert, S. M. (2007). Insect cryptochromes: Gene duplication and loss define diverse ways to construct insect circadian clocks. *Mol. Biol. Evol.*, *24*, 948–955.

Zehring, W. A., Wheeler, D. A., Reddy, P., Konopka, R. J., Kyriacou, et al. (1984). *P*-element transformation with period locus DNA restores rhythmicity to mutant, arrhythmic *Drosophila melanogaster*. *Cell*, *39*, 369–376.

Zeng, H., Hardin, P. E., & Rosbash, M. (1994). Constitutive overexpression of the *Drosophila* period protein inhibits period mRNA cycling. *EMBO J.*, *13*, 3590–3598.

Zeng, H., Qian, Z., Myers, M. P., & Rosbash, M. (1996). A light-entrainment mechanism for the *Drosophila* circadian clock. *Nature*, *380*, 129–135.

Zerr, D. M., Hall, J. C., Rosbash, M., & Siwicki, K. K. (1990). Circadian fluctuations of period protein immunoreactivity in the CNS and the visual system of *Drosophila*. *J. Neurosci.*, *10*, 2749–2762.

Zhang, L., Lear, B. C., Seluzicki, A., & Allada, R. (2009). The CRYPTOCHROME photoreceptor gates PDF neuropeptide signaling to set circadian network hierarchy in *Drosophila*. *Curr. Biol.*, *19*, 2050–2055.

Zhang, L., Chung, B. Y., Lear, B. C., Kilman, V. L., Liu, Y., et al. (2010a). DN1(p) circadian neurons coordinate acute light and PDF inputs to produce robust daily behavior in *Drosophila*. *Curr. Biol.*, *20*, 591–599.

Zhang, Y., Liu, Y., Bilodeau-Wentworth, D., Hardin, P. E., & Emery, P. (2010b). Light and temperature control the contribution of specific DN1 neurons to *Drosophila* circadian behavior. *Curr. Biol.*, *20*, 600–605.

Zheng, X., Koh, K., Sowcik, M., Smith, C. J., Chen, D., et al. (2009). An isoform-specific mutant reveals a role of PDP1 epsilon in the circadian oscillator. *J. Neurosci.*, *29*, 10920–10927.

Zhu, H., Yuan, Q., Briscoe, A. D., Froy, O., Casselman, A., & Reppert, S. M. (2005). The two CRYs of the butterfly. *Curr. Biol.*, *15*, R953–954.

Zhu, H., Sauman, I., Yuan, Q., Casselman, A., Emery-Le, M., et al. (2008). Cryptochromes define a novel circadian clock mechanism in monarch butterflies that may underlie sun compass navigation. *PLoS Biol.*, *6*, e4.

Zimmerman, W. F., & Goldsmith, T. H. (1971). Photosensitivity of the circadian rhythm and of visual receptors in carotenoid-depleted *Drosophila*. *Science*, *171*, 1167–1168.

Zwiebel, L. J., Hardin, P. E., Liu, X., Hall, J. C., & Rosbash, M. (1991). A post-transcriptional mechanism contributes to circadian cycling of per B-galactosidase fusion protein. *Proc. Natl. Acad. Sci. USA*, *88*, 3882–3886.

Index

Note: Page numbers suffixed by *t* and *f* refer to tables and figures respectively.

Printed and bound by CPI Group (UK) Ltd, Croydon, CR0 4YY

21/10/2024

01777131-0001